THE PERIODIC TABLE OF ELEMENTS (Long Period Form)

There are seven periods (horizontal rows) and eighteen groups (vertical columns) of elements. The number of electrons in filled shells is shown in the column at the extreme left; the remaining electrons for each element are shown immediately below the symbol for each element. Atomic numbers are enclosed by brackets. Atomic weights are shown above the symbols. Atomic weight values in parentheses are those of the isotopes of longest half-life for certain radioactive elements whose atomic weights cannot be quoted precisely without knowledge of origin of the element. Atomic weights are based on carbon-12. (From *Pure and Applied Chemistry*, **51**(2), 405 (1979). By permission of *International Union of Pure and Applied Chemistry*.)

METALS

NONMETALS

TRANSITION METALS

PERIODS	IA	IIA	IIIB	IVB	VB	VIB	VIIB	VIIIB	VIIIB	VIIIB	IB	IIB	IIIA	IVA	VA	VIA	VIIA	VIIIA
1 0	1.0079 H[1] 1																1.0079 H[1] 1	4.00260 He[2] 2
2 2	6.941 Li[3] 1	9.01218 Be[4] 2											10.81 B[5] 3	12.011 C[6] 4	14.0067 N[7] 5	15.9994 O[8] 6	18.99840 F[9] 7	20.179 Ne[10] 8
3 2, 8	22.98977 Na[11] 1	24.305 Mg[12] 2											26.98154 Al[13] 3	28.0855 Si[14] 4	30.97376 P[15] 5	32.06 S[16] 6	35.453 Cl[17] 7	39.948 Ar[18] 8
4 2, 8	39.0983 K[19] 8, 1	40.08 Ca[20] 8, 2	44.9559 Sc[21] 9, 2	47.90 Ti[22] 10, 2	50.9415 V[23] 11, 2	51.996 Cr[24] 13, 1	54.9380 Mn[25] 13, 2	55.847 Fe[26] 14, 2	58.9332 Co[27] 15, 2	58.70 Ni[28] 16, 2	63.546 Cu[29] 18, 1	65.38 Zn[30] 18, 2	59.72 Ga[31] 18, 3	72.59 Ge[32] 18, 4	74.9216 As[33] 18, 5	78.96 Se[34] 18, 6	79.904 Br[35] 18, 7	83.80 Kr[36] 18, 8
5 2, 8, 18	85.4678 Rb[37] 8, 1	87.62 Sr[38] 8, 2	88.9059 Y[39] 9, 2	91.22 Zr[40] 10, 2	92.9064 Nb[41] 12, 1	95.94 Mo[42] 13, 1	[98] Tc[43] 14, 1	101.07 Ru[44] 15, 1	102.9055 Rh[45] 16, 1	106.4 Pd[46] 18	107.868 Ag[47] 18, 1	112.41 Cd[48] 18, 2	114.82 In[49] 18, 3	118.69 Sn[50] 18, 4	121.75 Sb[51] 18, 5	127.60 Te[52] 18, 6	126.9045 I[53] 18, 7	131.30 Xe[54] 18, 8
6 2, 8, 18	132.9054 Cs[55] 8, 1	137.33 Ba[56] 18, 8, 2	[57-71] *	178.49 Hf[72] 32, 10, 2	180.9479 Ta[73] 32, 11, 2	183.85 W[74] 32, 12, 2	186.207 Re[75] 32, 13, 2	190.2 Os[76] 32, 14, 2	192.22 Ir[77] 32, 15, 2	195.09 Pt[78] 32, 17, 1	196.9665 Au[79] 32, 18, 1	200.59 Hg[80] 32, 18, 2	204.37 Tl[81] 32, 18, 3	207.2 Pb[82] 32, 18, 4	208.9804 Bi[83] 32, 18, 5	[209] Po[84] 32, 18, 6	[210] At[85] 32, 18, 7	[222] Rn[86] 32, 18, 8
7 2,8,18,32	[223] Fr[87] 18, 8, 1	226.0254 Ra[88] 18, 8, 2	[89-103] †	[261] Rf[104] 32, 10, 2	[262] Ha[105] 32, 11, 2	[263] [106] 32, 12, 2												

*LANTHANIDE SERIES	138.9055 La[57] 18, 9, 2	140.12 Ce[58] 20, 8, 2	140.9077 Pr[59] 21, 8, 2	144.24 Nd[60] 22, 8, 2	[145] Pm[61] 23, 8, 2	150.4 Sm[62] 24, 8, 2	151.96 Eu[63] 25, 8, 2	157.25 Gd[64] 25, 9, 2	158.9254 Tb[65] 27, 8, 2	162.50 Dy[66] 28, 8, 2	164.9304 Ho[67] 29, 8, 2	167.26 Er[68] 30, 8, 2	168.9342 Tm[69] 31, 8, 2	173.04 Yb[70] 32, 8, 2	174.967 Lu[71] 32, 9, 2
†ACTINIDE SERIES	227.0278 Ac[89] 18, 9, 2	232.0381 Th[90] 18, 10, 2	231.0359 Pa[91] 20, 9, 2	238.029 U[92] 21, 9, 2	237.0482 Np[93] 23, 8, 2	238.029 Pu[94] 24, 8, 2	[243] Am[95] 25, 8, 2	[247] Cm[96] 25, 9, 2	[247] Bk[97] 26, 9, 2	[251] Cf[98] 28, 8, 2	[252] Es[99] 29, 8, 2	[257] Fm[100] 30, 8, 2	[258] Md[101] 31, 8, 2	[259] No[102] 32, 8, 2	[260] Lr[103] 32, 9, 2

GENERAL CHEMISTRY

$SO_2(g) + \frac{1}{2}O_2(g) \rightleftharpoons SO_3(g)$ # 13

$$K_c = \frac{[SO_3]}{[SO_2] \times [O_2]^{\frac{1}{2}}} = 25$$

$[SO_2]$ $\frac{2 \text{ moles}}{5 \text{ l}} = .4 \text{ mole } l$

$[SO_3]$ $\frac{2.0 \text{ mole}}{5 \text{ l}} = .4 \text{ mole/l}$

$[O_2] = 0$ $K_c =$

$$\frac{[.40]}{(.48)(0)^{\frac{1}{2}}} = \infty$$

6th edition

GENERAL CHEMISTRY

WILLIAM H. NEBERGALL
Late of Indiana University

HENRY F. HOLTZCLAW, JR.
University of Nebraska—Lincoln

WILLIAM R. ROBINSON
Purdue University

D.C. HEATH AND COMPANY
Lexington, Massachusetts   Toronto

The cover picture shows a portion of the crystal structure of potassium strontium tetrametaphosphate, $K_2Sr(PO_3)_4$. The linked tetrahedra represent tetrametaphosphate anions, $P_4O_{12}{}^{4-}$; the blue spheres, Sr^{2+} ions; and the red spheres, K^+ ions located above the anions. Illustrated by Leonard Preston.

Preface

The Sixth Edition of *General Chemistry* represents an extensive revision to improve further a textbook that has been widely accepted in each of its previous editions. Theory and principles have been increased, but the text retains its excellent balance between theory and reaction chemistry. Particular attention has been paid to emphasizing the role theory plays in explaining the chemical behavior of the elements and how the behavior of the elements follows from the principles developed. The authors have added two completely new chapters and have rewritten much of the material in many chapters. All subject matter is presented at a sound pedagogical level, with close attention given to achieving a clear presentation that students can understand.

General Chemistry, Sixth Edition, as in previous editions, presents the study of the metals organized according to the groups of the Periodic Table. Its companion edition, *College Chemistry,* Sixth Edition, is identical in level and in most of the content but differs in organization, with discussion of the metals presented according to the qualitative analysis scheme and including a section of complete qualitative analysis laboratory procedures. The first 32 chapters are the same in both texts.

Of special note among the many new features are the two new chapters. One of these new chapters, "Applications of Chemical Stoichiometry" (Chapter 3), expands on the material of Chapter 2 by treating more advanced concepts of stoichiometry. The other new chapter, "The Relationship of the Periodic Classification to Chemical Behavior" (Chapter 8), describes the relationship of the location of an element in the Periodic Table to its chemical behavior. This new chapter, expanding on an earlier discussion of the Periodic Table in Chapter 4, identifies classes of chemical compounds (salts, electrolytes and nonelectrolytes, acids, bases), classes of chemical reactions (addition, decomposition, acid-base, oxidation-reduction), and metallic and nonmetallic behavior. The chapter also discusses periodic variation of oxidation state. This new chapter attempts to give the student a basic understanding of the types of reactions commonly observed in the laboratory, and to help the student begin to make some elementary predictions of the kinds of reactions an element or compound is likely to undergo.

Other special features of the Sixth Edition include the following:

1. Significant figures, dimensional analysis, and unit conversion factors are introduced in Chapter 1 when conversion between systems of units is discussed. This prepares the student for the consistent use of dimensional analysis and significant figures throughout the remainder of the text. Dimensional analysis is used extensively in problem-solving in all chapters.

2. Both metric and SI units are introduced and used in appropriate sections since both systems are currently in use in the scientific literature. Later in the text the emphasis tends toward greater use of SI units.

3. Elementary chemical stoichiometry based on mass and concentration, including a discussion of molarity, is introduced in Chapter 2 in order to prepare the student for quantitative laboratory work. A more extensive discussion of concentration measurements is provided later, in Chapter 13. All stoichiometry calculations in Chapter 2 and in the new Chapter 3 are presented using dimensional analysis.

4. The chapter on the structure of the atom (Chapter 4) has been extensively rewritten. Material on the Periodic Law is introduced here, and the discussion on periodicity of physical properties of the elements has been shifted into this chapter for an earlier presentation.

5. Chapters 5 and 6 on chemical bonding continue the successful treatment of earlier editions, with an added emphasis in Chapter 5 on directions for writing Lewis structures and in Chapter 6 on treating molecular orbitals for diatomic systems with two different elements. At the option of the instructor, Chapter 6 ("Molecular Orbitals") can be postponed to a later time in the course.

6. In Chapter 7, "Molecular Structure," we have chosen to use the student's new knowledge of Lewis structures to predict molecular geometries instead of introducing the set of rules used in the Fifth Edition. This reinforces the concept of Lewis structures and illustrates one use of such structures.

7. The treatment of oxygen and hydrogen has been combined into one chapter (Chapter 9), instead of two as in previous editions. This is the first chapter on truly descriptive chemistry in the text, and it follows the new Chapter 8 on relationships of the periodic classification to chemical behavior. The chapter on oxygen and hydrogen is written in a way that reinforces and illustrates the concepts introduced in the first eight chapters.

8. Chapter 10 ("The Gaseous State and the Kinetic-Molecular Theory") proceeds from experimental observations (Boyle's Law and Charles' Law) through the

Ideal Gas Law to the molecular behavior of gases and the kinetic-molecular theory. The Ideal Gas Law is introduced earlier than in previous editions, and greater use is made of it in gas-law problems.

9. A section on the structures of crystalline solids has been added to the chapter on the liquid state and the solid state (Chapter 11). Crystal structures can be omitted or postponed at the option of the instructor.

10. Chapter 12 ("Water and the Environment; Hydrogen Peroxide") has been reordered to present the structure, physical properties, and chemical properties of water first, followed by sections on natural waters, water pollution, purification, and softening.

11. The two chapters on solutions have been combined into a single chapter (Chapter 13) in the Sixth Edition. The section on colloids is at the end of the chapter.

12. The discussion of equivalent weights of acids and bases has been shifted into the chapter on acids and bases (Chapter 14). A new discussion on the strengths of acids and bases has been added. It should be noted that a preliminary discussion of some of the concepts of acids and bases is presented earlier in the new Chapter 8.

13. The chapter on chemical kinetics and equilibrium (Chapter 15) has been kept as a single chapter, but the kinetics sections have been rewritten to emphasize the molecular phenomena that give rise to reaction rates and rate laws.

14. The material on equilibria of weak electrolytes and solubility product (Chapters 16 and 17) has been retained with minor changes.

15. The chapter on thermodynamics (Chapter 18) has been significantly rewritten to emphasize the application of thermodynamics to chemical systems. Thus the student is extensively exposed to the idea that thermodynamics is useful in describing how chemicals behave. The chapter first treats thermochemistry, then entropy and free energy.

16. The discussion of oxidation-reduction reactions has been shifted to the chapter on electrochemistry (Chapter 20).

17. An expanded section on structure and isomerism of organic compounds has been added to the chapter on carbon compounds (Chapter 25). In addition, more interconversions between the various organic functional groups have been included. The mechanisms by which organic reactions occur have been described as a guide to correlation of various types of organic reactions.

18. The biochemistry chapter (Chapter 26) has been completely rewritten and updated.

19. The material on nuclear chemistry (Chapter 29) has been significantly revised and updated with new material on nuclear power reactors, breeder reactors, and fusion reactors.

20. Increased emphasis on the occurrence and use of coordination compounds has been added to an updated chapter on coordination chemistry (Chapter 31). The chapter now includes a discussion of the colors and magnetic moments of complex ions in terms of elementary crystal field theory.

21. Descriptive chapters have been updated, with emphasis on the relationship of the behavior of the elements to their locations in the Periodic Table.

22. New material on natural waters and water pollution, air pollution, fuel cells, energy, nuclear reactors, and some problems associated with the storage of nuclear waste has been introduced where appropriate throughout the book.

23. All artwork has been redrawn, and many new figures have been added.

24. Problems have been written so that the first problems in a given set provide straightforward no-frill problems for drill on the concepts presented in the chapter. The problems become more involved and challenging later in each problem set with increased emphasis on relevant topics, or on the application of a concept to determine the composition, properties, or behavior of a chemical system. Over 600 new problems have been written; these, along with more than 1000 questions, provide complete sets of end-of-chapter exercises.

25. References to pertinent articles in the literature have been extensively updated at the ends of chapters.

26. The index, as in previous editions, is unusually complete and hence truly useful, with over 6000 separate entries.

27. The authors have meticulously examined the present edition word-by-word and have rewritten many phrases and sentences for added clarity in the new edition.

The extensive supplemental materials especially designed for use with the text are: (1) A problems manual, *Problems and Solutions for General and College Chemistry, Sixth Editions,* by F. Keith Ault, John H. Meiser (both of Ball State University), Henry F. Holtzclaw, Jr. (University of Nebraska-Lincoln), and William R. Robinson (Purdue University); the manual includes the worked-out solutions to approximately one-third of the text problems, those marked $\boxed{S}$, at the ends of chapters. (2) A study guide to assist the student, written by Norman E. Griswold of Nebraska Wesleyan University. (3) An instructor's guide by the same author. (4) A laboratory manual, *Basic Laboratory Studies in College Chemistry,* Sixth Edition, by Grace Hered, William H. Nebergall, and William Hered; this manual complements the text but is also suitable for use with most other beginning chemistry texts.

The authors wish to express their special gratitude to George Bodner (Purdue University), who prepared the biochemistry chapter, and to those who read all or parts of the manuscript and made many valuable suggestions: In addition to Ault, Meiser, Griswold, and Hered, they include David Brooks and Sheldon Schuster of the University of Nebraska-Lincoln; Jack Dalton, Boise State University; James Erman, Northern Illinois University; Philip A. Kinsly, University of Evansville; Ronald Marks, Indiana University, Pennsylvania; and R. Thomas Myers, Kent State University.

We are also indebted to many present users of the text for a large number of good ideas based upon their own teaching experience. The authors also express their appreciation to Paul P. Bryant and to the editorial staff at D.C. Heath and Company for their gracious help and willing cooperation. The several revisions of the manuscript were typed by Roberta Molander and Martha Moll. Charles W. McLaughlin checked independently the answers to problems at the ends of chapters and the worked-out examples of problems within the chapters.

Finally, the authors especially wish to pay a warm tribute to our colleague, William H. Nebergall, whose unexpected death came shortly after the initial planning for the Sixth Edition and his participation in selecting and welcoming William R. Robinson of Purdue University as a new co-author for our texts. Bill Nebergall was a fine teacher, author, and person, an esteemed colleague, and a good friend.

HENRY F. HOLTZCLAW, JR.
WILLIAM R. ROBINSON

Contents

24 Phosphorus and Its Compounds 639

25 Carbon and Its Compounds 653

GENERAL
CHEMISTRY

1

Some Fundamental Concepts

Energy, pollution, and *conservation* are words we hear frequently these days, and science, including chemistry, is significantly involved in both the problems and the solutions implied by these terms. Within the last two decades, science and technology have acquired an ambivalent reputation in that some of their many valuable contributions to our lives have had undesirable side effects. For example, although energy is said to be the key to our existence, we deplore the destruction of natural beauty caused by strip mining for coal; the pollution of oceans, lakes, and beaches by oil spills; the possibility of excess radioactive contamination from accidents in nuclear plants; and the alarmingly rapid depletion of natural sources of energy. The fertilizers and insecticides that made possible great increases in the production of food have polluted our air and water. The automobile and the jet airplane revolutionized travel and the movement of goods, but they also have contaminated much of the air we breathe. We worry justifiably about the toxicity of some of the very medicines that have saved countless lives.

We have sometimes been slow to recognize and almost always slow to act on the vexing and often unexpected problems that result from the interaction and interdependence of various aspects of our society. Still, most of the problems created by science in the development of new products and processes can also be solved by science through cooperative effort with others in many different occupations.

Today, chemists as members of the scientific community face great challenges. They can legitimately take well-deserved credit for many scientific and technological advances, but they must also share in the responsibility for solving the problems created by some of these advances. The responsibility is not the chemists' alone, however. It is the *joint* responsibility of the physicists, the sociologists, the biologists, the theologians, the political scientists, the humanists, *and* the chemists. Together we must work to solve the problems of a world that sometimes seems to be trying to destroy itself in the process of its own technological progress. No single discipline offers the total perspective or the expertise to accomplish this task. Each discipline has specialized knowledge which, together with that of other disciplines, can contribute to bringing about a better life for all peoples. We must draw upon all of our knowledge if we are to achieve that goal and still preserve ethical and moral standards.

There is therefore a great need for chemists not only to know their field extremely well but also to have the necessary understanding and perspective in other fields to work effectively for the wise use of their discoveries. Equally important is the need for persons in other fields to know enough chemistry and other sciences to allow them to apply their specialized knowledge to problems involving science. The question as to why a person should today study chemistry or any other science, regardless of the intended main field of endeavor, and as to why a scientist should study other fields in addition to science, has never had a more demanding or obvious answer.

Introduction

1.1 Chemistry

In our study of chemistry we shall be concerned with the composition and structure of matter. Such forms of matter as wood and glass, water and gasoline, salt and sugar, coal and granite, and iron and gold differ strikingly from each other in many ways. These differences result from differences in the composition and structure of the various substances. This is not the whole story of chemistry, however, for matter can and does undergo changes. Our existence, in fact, depends upon changes that occur in matter. The changes that take place when food is digested and assimilated by the body are critical to the life process. The changes occurring during the combustion in a gasoline engine cause the formation of pollutant gases that can be extremely harmful to our health. Detergents in sewage-waste can produce changes in the composition of rivers that may ultimately cause huge fish-kills, thereby affecting our food supply.

As we proceed in our study of chemistry, we shall examine some of the changes in the composition and structure of matter, the causes that produce these changes, the changes in energy that accompany them, and the principles and laws involved in these changes. In brief, then, the science of **chemistry is the study of the composition, structure, and properties of matter, and the changes that matter undergoes.**

The amount of chemical knowledge has become so vast during the last century that chemists usually specialize in one of several principal branches. **Analytical chemistry** is concerned with the identification, separation, and quantitative determination of the composition of different substances. **Physical chemistry** is primarily concerned with the structure of matter, energy changes, and the laws, principles, and theories that explain the transformations of one form of matter into another. **Organic chemistry** is the branch dealing with the compounds of carbon. **Inorganic chemistry** is concerned with the chemistry of elements other than carbon and their compounds. **Biochemistry** is the chemistry of the substances comprising living organisms. A course in **general chemistry** is a survey of all of the branches of chemistry and introduces the student to the entire field of the science.

It should be mentioned that the boundaries between the branches of chemistry are arbitrarily defined. Much of the work in chemistry cuts across these defined boundaries.

1.2   Matter and Energy

Matter, by definition, is anything that occupies space and has mass. All the objects in the universe, since they occupy space and have mass, are composed of matter. The property of occupying space is often easily perceived by our senses of sight and touch. The **mass** of an object pertains to the quantity of matter that the object contains. The force required to give an object a given acceleration, or the resistance of the object to being moved (inertia), is a measure of its mass.

Energy can be defined as the capacity for doing work, where **work** is simply the process of moving matter against an opposing force. We recognize heat, electricity, and light as forms of energy when we see that heat generates the steam that drives a steam turbine, electricity causes electric motors to turn, and the energy of light is consumed in the manufacture of food by green plants.

Scientists recognize two kinds of energy: potential energy and kinetic energy. A piece of matter is said to possess **potential energy** by virtue of its position, condition, or composition. Water *at the top* of a waterfall possesses potential energy because of its position; if it falls in a hydroelectric plant it does work that leads to the production of electricity. A compressed spring, because of its compression, can do work such as making a clock run. Natural gas is recognized as having potential energy, because it will burn, producing another kind of energy, heat, that can do work, for example, in the boiler of a steam turbine. When a body is in motion, it also has energy—the capacity for doing work. The energy that a body possesses because of its motion is called **kinetic energy.** As water falls from the top of a waterfall, its potential energy becomes kinetic energy.

1.3   Law of Conservation of Matter and Energy

When a piece of copper metal is heated in air, it unites with oxygen in the air. If the product (copper oxide) is collected and weighed, it is found to have a greater mass than the original piece of metal. If, however, the mass of the oxygen of the air that combines with the metal is taken into consideration, it can be shown that the mass of the product (copper oxide) is, within the limits of accuracy of any weighing instrument, equal to the sum of the masses of the copper and oxygen that combine. This behavior of matter is in accord with what is called the **Law of Conservation of Matter: During an ordinary chemical change, there is no detectable increase or decrease in the quantity of matter.**

Conversions of one type of matter into another (chemical changes) are always accompanied by the conversion of one form of energy into another. Usually heat is evolved or absorbed, but sometimes the conversion involves light or electrical energy instead of, or in addition to, heat. Many transformations of energy, of course, do not involve chemical changes. Electrical energy can be changed into either mechanical, light, heat, or potential energy without chemical changes. Mechanical energy is converted into electrical energy in a generator. Potential and kinetic energy can be converted into one another. Many other conversions are possible, but all of the energy involved in any change always appears in some form after the change is completed. This fact is expressed in the **Law of Conservation of Energy: Energy cannot be created or destroyed, although it can be changed in form.**

In very precise terms, it is necessary to regard matter and energy not as distinct entities, but as two forms of a single entity. In the production of nuclear energy, matter is actually converted into energy. The reverse conversion of energy into matter can also be demonstrated. Actually, this interconversion of matter and energy also takes place in ordinary chemical reactions, but the total mass of the products differs so slightly from the mass of the reactants that it is impossible to measure the difference experimentally. Thus, for practical purposes we say that the law of conservation of matter holds in ordinary chemical reactions. However, to be more precise, we may combine the separate laws of conservation of matter and energy into one statement: **The total quantity of matter and energy available in the universe is fixed.**

1.4   States of Matter

Matter can exist in three different states, designated as solid, liquid, and gas (Fig. 1-1). Each state can be distinguished by certain characteristics.

Figure 1-1 The three states of matter as illustrated by water.

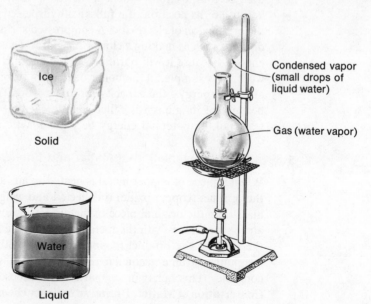

Ice

Solid

Water

Liquid

Condensed vapor (small drops of liquid water)

Gas (water vapor)

A **solid** is rigid, possesses a definite shape, and has a volume that is very nearly independent of changes in temperature and pressure.

A **liquid** flows and thus takes the shape of its container, except that it assumes a horizontal surface. Liquids, like solids, are only slightly compressible and so for practical purposes have definite volumes.

A **gas** takes both the shape and volume of its container. Gases are readily compressible and capable of infinite expansion.

1.5   Chemical and Physical Properties

The characteristics that enable us to distinguish one substance from another are known as **properties.** Those characteristics involved in a transformation of one substance into another are known as **chemical properties.** Thus a chemical property of wood is exhibited when wood burns, for the constituents of the wood combine with oxygen to form substances different from the wood or the oxygen. A chemical property of iron is also exhibited when iron combines with oxygen and water to form the reddish-brown iron oxide which we know as iron rust.

In addition to chemical properties, every substance also possesses definite properties that can be used to distinguish it from other substances. These properties, which can be measured without changing the substance into a different substance, are called **physical properties.** Some familiar physical properties are color, hardness, crystalline form, physical state, melting temperature, boiling temperature, density, electrical and thermal conductivity, and specific heat.

A **physical change** is a change in a substance that occurs without the formation of a different substance. The melting of ice, the freezing of water, the conversion of water to steam, the condensation of steam to water, and the magnetization of iron are all examples of physical change. In each of these there is a change in properties but there is no alteration of the chemical composition of the substances involved. Water in either the solid, liquid, or gaseous state has the same chemical composition. Iron is still the same substance whether magnetized or not.

Changes in physical conditions such as temperature or pressure may modify the physical properties of a substance. For example, a substance in the gaseous state has certain physical properties such as density, specific heat, and thermal conductivity. By decreasing the temperature of the substance or compressing it to a smaller volume, we can change it from a gas to a liquid, in which condition it has an entirely different density, specific heat, and thermal conductivity.

A **chemical change,** on the other hand, always produces one or more different substances than those that existed before the change occurred. When charcoal, a solid black substance composed mostly of carbon, burns in air, an invisible gas consisting of both carbon and oxygen (carbon dioxide) is formed. When milk sours, the sugar in the milk is converted into lactic acid, and the composition of the properties of the acid differ from those of the sugar. Iron rust, formed by the corrosion of iron metal, contains oxygen as well as iron, and it is therefore a different substance with different properties than those of iron metal. All such changes are chemical changes.

1.6   Substances

A pure substance is defined as any matter, all specimens of which have identical composition as well as identical chemical and physical properties. Pure water is an example of a substance. All samples of pure water, regardless of their source, have exactly the same composition, 2.0158 parts by weight of hydrogen to 15.9994 parts of oxygen, and are identical in melting point, boiling point, and all other properties. Pure iron, pure aluminum, pure carbon, pure sugar, pure oxygen, and pure carbon dioxide are representative substances. Two substances can be distinguished from each other by a study of their characteristic properties. For example, sugar and salt may be distinguished by taste, iron and gold by their colors, and silver and mercury by their physical states at normal temperatures.

1.7   Mixtures

A mixture is composed of two or more substances each of which retains its chemical identity and specific properties. The composition of a mixture can be varied continuously. Granite is a mixture of quartz, feldspar, and mica; a solution of sugar and water is a mixture; and air is a mixture of nitrogen, oxygen, carbon dioxide, water vapor, and other gases. Milk, butter, cement, flour, and gasoline are other examples of mixtures.

Because each component of a mixture possesses and retains its own set of characteristic properties, the various components can be separated by physical methods. The components of black gunpowder (a heterogeneous mixture of carbon, sulfur, and potassium nitrate) are readily detected by examining it under a microscope. Treatment of a sample of gunpowder with water causes the potassium nitrate to dissolve, leaving the sulfur and charcoal as solid particles; subsequent treatment of the residue with carbon disulfide causes the sulfur to dissolve, leaving only the carbon in the form of solid particles. The potassium nitrate and sulfur may be reclaimed as crystalline particles by evaporating their respective solutions to dryness. An intimate mixture of iron filings and sulfur may be separated either by dissolving the sulfur in carbon disulfide, leaving the iron, or by removing the iron with a magnet, leaving the sulfur.

1.8   Elements

Chemists divide substances into two classes: elements and compounds. **Elements are pure substances which cannot be decomposed by a chemical change.** Familiar examples are iron, silver, gold, aluminum, sulfur, oxygen, and carbon.

One hundred and six elements are known at the present time; a list of these is printed on the inside front cover of this book. Eighty-eight elements occur naturally on the earth and the other eighteen have been synthesized. Eleven of the eighty-eight elements make up about 99% of the earth's crust and the atmosphere (Table 1-1). Oxygen constitutes nearly one-half and silicon about one-fourth of the total quantity of the elements in the atmosphere and the earth's crust together. Only about one-fourth of the elements ever occur on the earth in the free state; the others are found only in chemical combination with other elements.

TABLE 1-1 Percentages of Elements in the Atmosphere and the Earth's Crust by Mass			
Oxygen	49.20%	Chlorine	0.19%
Silicon	25.67	Phosphorus	0.11
Aluminum	7.50	Manganese	0.09
Iron	4.71	Carbon	0.08
Calcium	3.39	Sulfur	0.06
Sodium	2.63	Barium	0.04
Potassium	2.40	Nitrogen	0.03
Magnesium	1.93	Fluorine	0.03
Hydrogen	0.87	Strontium	0.02
Titanium	0.58	All others	0.47

1.9   Compounds

Compounds are substances that are composed of two or more different elements and that can be decomposed by chemical changes. Elements in combination are different from elements in the free, or uncombined, state. However, the term *element* is used to designate an elemental substance whether free or in combination. For example, white crystalline sugar is a compound consisting of the element carbon, which is usually a black solid when free, and the two elements hydrogen and oxygen, which are colorless gases when uncombined. If heated sufficiently in the absence of air, sugar decomposes to carbon and water. Water, a compound, can be decomposed by an electric current into its two constituent elements, hydrogen and oxygen. Sodium chloride, a compound that is the principal constituent of table salt, can be broken down by an electric current into sodium and chlorine.

Whereas there are only 106 known elements, there are hundreds of thousands of chemical compounds resulting from different combinations of these elements. Each compound possesses definite chemical and physical properties by which chemists can distinguish it from all other compounds.

1.10   Molecules

Water has a definite composition and a set of chemical and physical properties that enable us to recognize it as a distinct substance. One might ask the question, "To what extent can a drop of water be subdivided and still be water?" The limit to which such a subdivision could be carried is a particle called a **molecule** of water. Subdivision of a molecule of water would result in the formation of the gases hydrogen and oxygen, each with a set of properties quite different from water and from each other. **The smallest particle of an element or compound that can have a stable, independent existence is called a molecule** (Fig. 1-2). Molecules are too small to be seen even with very powerful optical microscopes. Only relatively large molecules can be seen with an electron microscope. However, by means of new electron microscope techniques reported late in 1978, individual platinum molecules were seen for the first time. An idea of the minute size of molecules can be

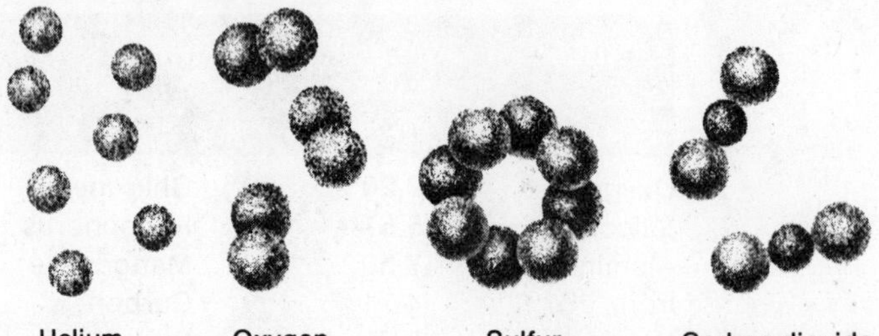

Figure 1-2 Representations of molecules for the elements helium, oxygen, and sulfur and for the compound carbon dioxide.

Helium   Oxygen   Sulfur   Carbon dioxide

appreciated from the fact that if a drop of water were to be magnified to the size of the earth, its constituent molecules would appear to be about the size of baseballs. One hundred million molecules of water laid side by side would make a row about 1 inch long.

1.11   Atoms

We define an atom as the smallest particle of an element that can enter into a chemical combination. For example, the particles of carbon and oxygen that combine *to form one molecule* of carbon dioxide are atoms.

An atom of an element may or may not be capable of independent existence. In some cases, an atom of an element and a molecule of it are identical. When the molecule of an element contains only one atom, the molecule is said to be monatomic. Examples of elements that are composed of monatomic molecules are helium, neon, and xenon. On the other hand, the elements hydrogen, oxygen, nitrogen, fluorine, chlorine, bromine, and iodine consist of diatomic molecules (two atoms per molecule). Molecules of phosphorus and sulfur normally contain four and eight atoms, respectively. It follows, then, that the term *molecule* applies to small particles of either elements or compounds, whereas the term *atom* applies only to the smallest particle of an element that retains the identity of the element. We shall see in Chapter 4 that even atoms are not the fundamental units of matter. Atoms are made up of still smaller particles, comprising a very complex system.

The word *atom* comes from the Greek word *atomos,* which means indivisible. The early Greek philosophers were the originators of the concept of atoms, and they taught that matter, since it is composed of atoms, is therefore finitely divisible. Although generally credited as the father of the atomic theory, John Dalton (1766–1844), an English chemist and physicist, revived the old Greek atomic hypothesis and put it on a quantitative basis. Like the Greeks, Dalton made no distinction between atoms and molecules as we do today. He applied the word *atom* to particles of both elements and compounds.

1.12   The Scientific Method

Your study of chemistry will be concerned with the hypotheses, theories, and laws that give this science its foundation and framework, a framework into which bits of information fit to make an integrated area of knowledge.

This framework has developed from a logical pursuit of the answers to the many questions that can be subjected to experimental inquiry and investigation. That logical approach is called the **scientific method.** The first step in applying the scientific method to the solution of a problem involves the carrying out of carefully planned experiments to gather facts that give information about all phases of the problem. The second step consists of an attempt to formulate a simple general statement, known as a **hypothesis,** which correlates these facts. The hypothesis is then tested by further experiments, and if it is capable of explaining the larger body of data it is dignified by the name of **theory.** Theories serve as guides for still further work by serving as the bases for predicting new information or the direction in which additional information must be sought. Finally, a theory may become so well established that it can be accepted as a general truth, which then is often referred to as a **law.** Theories and laws help to simplify the study of chemistry by systematizing the ordering of the vast body of chemical knowledge.

The theories and laws of chemistry are, in general, less precise than those of physics. This lack of perfection results from the complexity of the systems that chemists study and serves to stimulate interest in chemistry. It also points up the fact that there are still great opportunities for discovery in chemistry.

Measurements in Chemistry

1.13   Units of Measurement

The hypotheses, theories, and laws that describe the behavior of matter and energy are usually based on quantitative measurements. When properly reported, these measurements convey three ideas: the size or magnitude of the property being measured (a number), an indication of the possible error in the measurement, and the units of the measurement. Indicating the possible error in a measurement will be discussed in a subsequent section.

Units provide a standard of comparison for a measurement. A well-known sandwich is prepared from a quarter-pound of hamburger meat (0.250 pound). The mass of the meat has a magnitude of 0.250 times the mass of the arbitrary standard of 1 pound. If the unit of reference is changed, the property being measured (in this case the mass) does not change. The mass of hamburger meat noted above can be reported as either a quarter of a pound, 4.00 ounces, or 0.000125 ton (a 0.000125 tonner?), but the actual amount of meat is the same no matter what the unit of measurement.

The results of scientific measurements are usually reported in the metric system originally devised in France in 1799. The original metric units for length, mass, and time were the meter, the gram, and the second. Various combinations of these units are still in use. One such system that has been used frequently by chemists is called the centimeter-gram-second, or cgs, system. More recently, scientists have begun to use an updated version of the metric system of units in which the units for length, mass, and time are the meter, the kilogram, and the second. This system of units is known as the **International System of Units (SI units).** SI units were adopted in 1960 by the Bureau International des Poids et Mesures, the international organization that defines metric units. These units have been in use by the U.S. National Bureau of Standards since 1964.

The SI system uses the following seven base units, from which other units of weights and measures can be derived.

Physical Property	Name of Unit	Symbol
Length	Meter	m
Mass	Kilogram	kg
Time	Second	s
Electric current	Ampere	A
Thermodynamic temperature	Kelvin	K
Luminous intensity	Candela	cd
Quantity of substance	Mole	mol

Both in the SI and in other metric systems, measurements may be reported as fractions or multiples of 10 times a base unit by using the appropriate prefix with the name of the base unit. For example, a distance can be reported in units of meters, kilometers (10^3 meters), millimeters (10^{-3} meters), or picometers (10^{-12} meters). The prefixes and their symbols denoting the powers to which 10 is raised are given in the following table.

Factor	Prefix	Symbol	Factor	Prefix	Symbol
10^{-12}	pico	p	10	deka	da
10^{-9}	nano	n	10^2	hecto	h
10^{-6}	micro	μ	10^3	kilo	k
10^{-3}	milli	m	10^6	mega	M
10^{-2}	centi	c	10^9	giga	G
10^{-1}	deci	d	10^{12}	tera	T

Since both SI and cgs units are currently in use by chemists, it is necessary to be familiar with both systems. Consequently, both systems of units will be used in the text.

1.14   Conversion of Units; Dimensional Analysis

In addition to the two systems of units in use within the scientific community, many other units are employed throughout other disciplines. For example, units of mass range from grams and kilograms in science, to pounds and tons in engineering, to grains and drams in medicine. By using a technique referred to as dimensional analysis, conversions between these various units are not difficult.

If units are treated like algebraic quantities, they can be carried through calculations and multiplied together, divided into each other, or canceled. The final units resulting from such a treatment will provide valuable assistance in determining whether or not a calculation has been set up correctly. Consider the conversion of a length from meters to yards. As given in Appendix C (a table of conversion factors), 1 meter = 1.094 yards. This may be read as indicating that there is exactly 1 meter per 1.094 yards or, alternatively, 1.094 yards per 1 meter.

Very often a slash (/) or a division line is used instead of the word *per*. This gives the following unit conversion factors:

$$1 \text{ meter}/1.094 \text{ yards}, \quad \text{or} \quad \frac{1 \text{ meter}}{1.094 \text{ yards}}$$

$$1.094 \text{ yards}/1 \text{ meter}, \quad \text{or} \quad \frac{1.094 \text{ yards}}{1 \text{ meter}}$$

A **unit conversion factor** is used to convert a quantity in one system of units to the corresponding quantity in another system of units. For example, the second factor above can be used to convert a quantity with meters as the standard of comparison (unit) to the corresponding quantity with yards as the standard of comparison by multiplying the quantity in meters by the conversion factor.

EXAMPLE   **How long in yards is a run of 100.0 meters?**

From the relationship 1 meter = 1.094 yards, we can get the unit conversion factor 1.094 yards/1 meter. Multiplication gives

$$100.0 \text{ meters} \times \frac{1.094 \text{ yards}}{1 \text{ meter}} = 109.4 \text{ yards}$$

Note that the unit of meters in 100.0 meters and in the conversion factor cancel, as indicated by the slash cancel lines in color. Because the answer is in the correct unit, it is highly likely that the conversion has been set up correctly. If an incorrect unit conversion factor had been used, an incorrect unit would have been obtained. Consider the following conversion:

$$100.0 \text{ meters} \times \frac{1 \text{ meter}}{1.094 \text{ yards}} = 91.41 \frac{\text{meters}^2}{\text{yard}}$$

The final unit, meters²/yard, indicates that the conversion does not convert meters into yards since the units do not cancel to give yards.

◆ ◆ ◆ ◆ ◆

1.15   Uncertainty in Measurements and Significant Figures

Suppose you weighed an object on an inexpensive balance and found, as best you could determine, that its mass is closer to 13.4 grams than to either 13.3 or 13.5 grams. You would report the mass as 13.4 grams. The uncertainty in this measurement is almost 0.1 gram or 1 part in 134 (about 0.75%), since the true mass of the object could lie anywhere between 13.35 and 13.45 grams. If the same object were weighed on an analytical balance in your laboratory, your best efforts might show that its mass is 13.384 grams; that is, the mass is closer to 13.384 grams than to either 13.383 or 13.385 grams. The uncertainty in this case would be about 0.001 gram (1 part in 13,384 or about 0.0075%) since the true mass could lie anywhere between 13.3835 and 13.3845 grams. Any experimental measurement, no matter how carefully carried out, may contain this type of uncertainty. **All**

measurements may be taken to have an uncertainty of at least one unit in the last digit of the measured quantity.

The mass 13.4 grams has an uncertainty of 0.1 gram; the mass 13.384, an uncertainty of 0.001 gram; a volume of 25 milliliters, an uncertainty of 1 milliliter; and a length of 0.001378 meter, an uncertainty of 0.000001 meter. All of the measured digits in the determination are called **significant figures.**

Results calculated from measurements are as uncertain as the measurement itself. Suppose a sample weighs 78.7 grams, and exactly one-third of it is iron. In order to calculate how much iron is in the sample, we need only multiply 78.7 grams by one-third. On an electronic calculator, this comes out as 26.233333 grams. This is not a correct answer, however. Since the mass of the sample is uncertain to 1 part in 787, or about 0.1%, the mass of iron in the sample is subject to the same uncertainty. To report the mass of iron as 26.233333 grams indicates an uncertainty of 0.000001 gram, or about 0.000004% (1 part in 26,233,333). The mass of iron should be reported as 26.2 grams (three significant figures). In calculations involving experimental quantities, the significant figures in these quantities must be taken into account in order not to overestimate or underestimate the uncertainty in the calculated result.

The following rules may be used to determine the number of significant figures in a measured quantity:

1. *All digits that are not zero are significant.* The number 78.7 contains three significant figures; 13,384, five significant figures.
2. *Zeros between nonzero digits are significant.* 103 contains three significant figures; 1029.03, six significant figures.
3. *Zeros to the left of the first nonzero digit are not significant;* they simply indicate where the decimal point should be. The number 0.0038 contains two significant figures; 0.02081, four significant figures.
4. *Zeros to the right of a decimal point are significant if they are at the end of the number* (or, of course, if they are followed by nonzero digits): The number 0.0120 contains three significant figures; 820.0, four significant figures; 90.00370, seven significant figures.
5. *When a number ends in zeros that are not to the right of a decimal point, the trailing zeros may or may not be significant.* A mass of 1300 grams may indicate a mass closer to 1300 grams than to 1200 or 1400 grams. In this case the zeros serve only to locate the decimal point and are not significant. If the mass is closer to 1300 grams than to 1290 or 1310 grams, then the zero to the right of the 3 is significant. If the mass is closer to 1300 than to 1299 or 1301, both zeros are significant. For convenience in this text, we will assume that when a number ends in zeros that are not to the right of a decimal point, these trailing zeros are not significant unless otherwise specified. The ambiguity can be avoided by using exponential notation: 1.3×10^3 (two significant figures), 1.30×10^3 (three significant figures), 1.300×10^3 (four significant figures).

The correct use of significant figures in a calculation carries the uncertainty of measured quantities into the calculated results. The following rules govern the number of significant figures in the answer:

1. *When adding or subtracting, the result should be reported with the same number of decimal places as that of the number with the least number of decimal places.*

EXAMPLE 1   **a. Add 4.383 g and 0.0023 g.**

b. Subtract 421 g from 486.39 g.

$$
\begin{array}{ll}
\text{a.} \quad 4.383 \ \text{g} & \text{b.} \quad 486.39 \ \text{g} \\
\phantom{\text{a.} \quad} \underline{0.0023 \ \text{g}} & \phantom{\text{b.} \quad} \underline{-421 \ \text{g}} \\
\phantom{\text{a.} \quad} 4.385 \ \text{g} & \phantom{\text{b.} \quad 48} 65 \ \text{g}
\end{array}
$$

◆ ◆ ◆ ◆ ◆

2. *When multiplying or dividing, the product or quotient should contain no more digits than the least number of significant figures in the numbers involved in the computation.*

EXAMPLE 2   **Multiply 0.6238 by 6.6.**

$$0.6238 \times 6.6 = 4.1$$

◆ ◆ ◆ ◆ ◆

In rounding off numbers, simply drop numbers less than 5; 8.72 rounds off to 8.7. If the number to be dropped is 6 or more, increase the preceding number by 1; 3.8479 rounds off to 3.85. If the number to be dropped is 5, a common practice is to increase the preceding number by 1 if it is odd and to leave it unchanged if it is even. Dropping a 5 always leaves an even number. Thus 3.425 rounds off to 3.42 while 7.535 rounds off to 7.54.

Note that conversion factors within the same system of units are exact; that is, they are good to any number of significant figures. One mile is exactly 5280 feet, one kilometer is exactly 1000 meters. Conversions from one system of units to another may not be exact. The conversion 1 kilogram = 2.2 pounds is to be taken as having two significant figures unless you know it to be exact. Since this conversion was determined experimentally, answers involving it should be given to two significant figures.

1.16   Length and Volume

The standard unit of length in both metric and SI units is the **meter** (m). In 1899, the meter was defined as the length of the platinum-iridium alloy bar kept at Sèvres, France, a town near Paris. In 1960, the meter was redefined as the length equal to exactly 1,650,763.73 wavelengths, in vacuum, of the orange line in the emission spectrum of ^{86}Kr, a particular isotope of krypton (spectra and isotopes will be discussed in subsequent sections). Within the error of the measurement, the length of the meter has not changed; only the standard used to define a meter has been changed.

In more familiar units of reference, a meter is about 3 inches longer than a yard; to four significant figures, exactly 1 meter is approximately 1.094 yards.

Volumes are defined in terms of the standard length. Common metric volumes are the **liter** (ℓ) and **milliliter** (ml). A liter is exactly one one-thousandth (10^{-3}) times the volume of a cube that is exactly 1 meter on each edge, or the volume of a cube with an edge length of exactly 10 cm (1 decimeter, dm). Thus a liter is 1 dm^3 or 1000 cm^3. In common terms, a liter is about 1.06 quarts. A milliliter

$(10^{-3}\ \ell)$ is the volume of a cube (Fig. 1-3) with an edge length of exactly 1 cm and thus is equal in volume to 1 cubic centimer (cm³).

EXAMPLE   **What is the volume in liters of 1.000 ounce (oz), given that 1 ℓ = 1.06 qt and that 32 oz = 1 qt?**

From the two relationships it is possible to determine the exact unit conversion factor 1 qt/32 oz and the inexact unit conversion factor 1 ℓ/1.06 qt. First convert ounces to quarts.

$$1.000\ \text{oz} \times \frac{1\ \text{qt}}{32\ \text{oz}} = 0.03125\ \text{qt}$$

Then convert quarts to liters.

$$0.03125\ \text{qt} \times \frac{1\ \ell}{1.06\ \text{qt}} = 2.95 \times 10^{-2}\ \ell$$

Thus

$$1.000\ \text{oz} = 2.95 \times 10^{-2}\ \ell.$$

Note that these two steps could have been combined into one operation as follows:

$$1.000\ \text{oz} \times \frac{1\ \text{qt}}{32\ \text{oz}} \times \frac{1\ \ell}{1.06\ \text{qt}} = 2.95 \times 10^{-2}\ \ell$$

Note also that the number of significant figures in the result is determined by the inexact experimental relationship that exactly 1 liter is equal to 1.06 quarts.

◆ ◆ ◆ ◆ ◆

Dime

1 cubic centimeter
= 1 milliliter

Figure 1-3 Comparison between a cubic centimeter block and a dime. One cubic centimeter is equal to one milliliter. One milliliter of water at 4°C weighs one gram.

1.17   Mass and Weight

We saw in Section 1.2 that the **mass** of a body is the quantity of matter that it contains. The mass of a body of matter is an invariable quantity. On the other hand, the **weight** of a body is the force that gravity exerts on the body and is variable, since the attraction is dependent upon the distance of the body from the earth's center (or on another planet, from that planet's center). If this book were taken up in an airplane it would weigh less than it does at sea level. If it were to be taken far out into space its weight would become negligible. It is well known that astronauts experience weightlessness while in outer space. However, neither the mass of the book nor that of the astronauts changes as the distance from the center of the earth changes. Scientists measure quantities of matter in terms of mass rather than weight because the mass of a body remains constant whereas the weight of a body is an "accident of its environment." It should be noted, however, that in common terminology the term "weight" is often used loosely for the mass of a substance.

The mass of an object is found by comparing it to another object of known mass. The instrument used to make this comparison is called a **balance.** One type

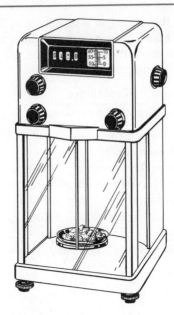

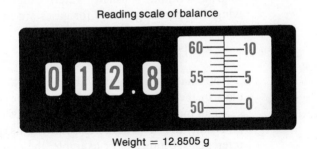

Reading scale of balance

Weight = 12.8505 g

Figure 1-4 A modern balance with scale for direct reading.

of balance is shown in Fig. 1-4. With this type of balance the object whose mass is to be determined is placed on the pan of the balance. "Weights" of known mass within the balance are adjusted by means of knobs, to balance the object on the pan. "Weighing" an object on a balance on the earth's surface makes use of the fact that the gravitational attraction for objects of equal mass is the same; i.e., their weights are equal.

The standard object to which all SI and other metric units of mass are referred is a cylinder of platinum-iridium alloy which also is kept at Sèvres. The unit of mass, the **kilogram,** is defined as the mass of this cylinder. The gram (10^{-3} kg) is equal to exactly one one-thousandth of the mass of the standard kilogram and is very nearly equal to the weight of 1 cubic centimeter of water at 4° Celsius, the temperature at which water has its maximum density. One kilogram is about 2.2 pounds. An American one-cent coin has a mass of about 3.1 grams.

1.18   Density and Specific Gravity

One of the physical properties of a solid, liquid, or gas is its density. **Density is defined as mass per unit volume.** This may be expressed mathematically as

$$\text{Density} = \frac{\text{mass}}{\text{volume}} \qquad \text{or} \qquad D = \frac{M}{V}$$

Substances can usually be distinguished by measuring their densities because it is rare that any two substances have identical densities. One cubic centimeter of mercury (a liquid) at 25°C has a mass of 13.5939 grams. We say, therefore, that mercury has a density of 13.5939 g/cm³ (or 13.5939 g/ml because the milliliter and cubic centimeter are the same size). The density of water at 25°C is 0.99707 g/cm³ and at 4°C it is 1.000 g/cm³. At 0°C and 1 atmosphere of pressure, 1 liter of hydrogen has a mass of 0.08987 gram; its density is 0.08987 g/ℓ. The density of air is 1.2929 g/ℓ at 0°C and 1 atmosphere of pressure.

The term **specific gravity** denotes the ratio of the density of a substance to the density of an equal volume of a reference substance. The reference substance for solids and liquids is usually water. Common reference substances used in specifying the specific gravities for gases are air and hydrogen.

$$\text{Specific gravity of substance} = \frac{\text{density of substance}}{\text{density of reference substance}}$$

Note that specific gravity, being the ratio of two densities, has no units. The units in the numerator and denominator cancel each other.

When measured in the metric system of units, the density of any substance has practically the same numerical value as its specific gravity referred to water as the reference substance. For example, the density of glycerin is 1.26 g/cm³ whereas its specific gravity referred to water as the reference substance is 1.26 (no units).

EXAMPLE 1   **Calculate the density and specific gravity of a body that has a mass of 321 g and a volume of 45.0 cm³ at 25°C.**

Step 1.

$$\text{Density} = \frac{\text{mass}}{\text{volume}} = \frac{321 \text{ g}}{45.0 \text{ cm}^3} = 7.13 \text{ g/cm}^3$$

Step 2.

$$\text{Specific gravity} = \frac{\text{density of sample}}{\text{density of water}}$$

The density of water at 25°C is 0.99707 g/cm³.

$$\text{Specific gravity} = \frac{7.13 \text{ g/cm}^3}{0.99707 \text{ g/cm}^3} = 7.15$$

◆ ◆ ◆ ◆ ◆

EXAMPLE 2   **The density of a solution that contains 38.0% by mass of a substance called sulfuric acid (H_2SO_4) is 1.30 g/ml. How many grams of sulfuric acid are contained in 400.0 ml of this solution?**

One milliliter of the solution of sulfuric acid has a mass of 1.30 g. The mass of 400.0 ml of acid may be calculated as follows:

$$\text{Mass} = \text{volume} \times \text{density} = 400.0 \text{ ml} \times 1.30 \text{ g/ml} = 5.20 \times 10^2 \text{ g}$$

Because 38.0% by mass of the solution is pure H_2SO_4 (38.0 g H_2SO_4 per exactly 100 g solution), the number of grams of H_2SO_4 in 400.0 ml of the solution of the acid is

$$5.20 \times 10^2 \text{ g solution} \times \frac{38.0 \text{ g } H_2SO_4}{100 \text{ g solution}} = 198 \text{ g } H_2SO_4$$

◆ ◆ ◆ ◆ ◆

EXAMPLE 3   Calculate the mass of 50.0 ml of kerosene, specific gravity 0.82.

Because kerosene has a specific gravity of 0.82, the mass of 1 ml of kerosene is 0.82 times that of 1 ml of water or 0.82 × 1.0 g/ml = 0.82 g/ml. Then 50.0 ml has a mass of 50.0 ml × 0.82 g/ml = 41 g.

◆ ◆ ◆ ◆ ◆

1.19   Temperature and Its Measurement

The word **temperature** refers to the "hotness" or "coldness" of a body of matter. In order to measure a change in temperature, some physical property of a substance that varies with temperature must be used. Practically all substances expand with an increase in temperature and contract when the temperature decreases. It is this property that is the basis for the common thermometer. The mercury or alcohol in the tube of a common glass thermometer rises when the temperature increases because the volume of the mercury or alcohol expands more than that of the glass container.

1.20   Standard Reference Temperatures

In order that we may agree on a set of temperature values, it is necessary to have **standard reference temperatures** that can be readily determined. Two such temperatures are the freezing and boiling points of water at an atmospheric pressure of 760 torricelli (torr), or one atmosphere. On the Celsius scale the freezing point of water at 760 torr is taken as 0° and the boiling point as 100°. The heights of the mercury column at these two reference temperatures determine the 0° and 100° points on a thermometer. The space between these two points is divided into 100 equal intervals, or degrees. On the Fahrenheit scale the freezing point of water is taken as 32° and the boiling point as 212°. The space between these two points on a thermometer is divided into 180 equal parts, or degrees. Thus, a degree Fahrenheit is exactly 100/180, or 5/9, of a degree Celsius (Fig. 1-5). The relationships are shown by the equations

$$\frac{F - 32}{180} = \frac{C}{100}, \qquad C = \frac{5}{9}(F - 32), \qquad F = \frac{9}{5}C + 32$$

The readings below 0° on either scale are treated as negative.

Another temperature scale, used mainly by scientists, is called the **Kelvin** scale, named after Lord Kelvin, a British physicist. On the Kelvin scale, the zero point on the scale is −273.15°C. The size of the degree on the Kelvin scale is the same as that on the Celsius scale. The freezing temperature of water on the Kelvin scale is 273.15 K (0°C), and the boiling temperature is 373.15 K (100°C). A temperature on the Celsius scale is converted to the Kelvin scale by adding 273.15 to the Celsius reading; to convert from Kelvin to Celsius, 273.15 is subtracted from the Kelvin reading. (Note that temperatures on the Kelvin scale are, by convention, reported without the degree sign.)

$$K = C + 273.15, \qquad C = K - 273.15$$

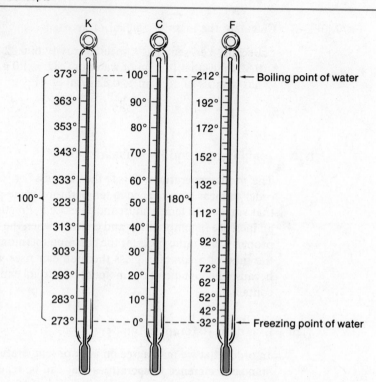

Figure 1-5 The relationships between the Kelvin, Celsius, and Fahrenheit temperature scales.

The Kelvin (K) is the basic unit of thermodynamic temperature recommended in the International System of Units (SI). The Kelvin temperature scale is discussed in greater detail in Section 10.5.

Figure 1-5 shows the relationships between the three temperature scales. Temperatures in this book are in Celsius unless otherwise specified. The following are examples of conversions between the three scales.

EXAMPLE 1   **The boiling temperature of a compound called ethyl alcohol is 78.5°C at 1 atmosphere of pressure. What is its boiling point on the Fahrenheit scale and on the Kelvin scale?**

$$F = \tfrac{9}{5}C + 32 = (\tfrac{9}{5} \times 78.5) + 32 = 141 + 32 = 173°F$$
$$K = C + 273.15 = 78.5 + 273.15 = 351.6 \text{ K}$$

◆ ◆ ◆ ◆ ◆

EXAMPLE 2   **Convert 50°F to the Celsius scale and the Kelvin scale.**

$$C = \tfrac{5}{9}(F - 32) = \tfrac{5}{9}(50 - 32) = \tfrac{5}{9} \times 18 = 10°C$$
$$K = C + 273.15 = 10 + 273.15 = 283 \text{ K}$$

◆ ◆ ◆ ◆ ◆

1.21   The Measurement of Heat

Chemical reactions are accompanied by either the evolution or the absorption of energy, usually in the form of heat. A **joule** (J), the SI unit of heat and other forms of energy, is equal to the kinetic energy ($\frac{1}{2}Mv^2$) of an object with mass M of 2 kilograms moving with a velocity v of exactly 1 meter per second. The **kilojoule** (kJ) is equal to 10^3 joules. The older metric unit for heat, as for other forms of energy, is the **calorie** (cal). For many years the calorie was defined as the amount of heat, or other energy, necessary to raise the temperature of exactly 1 gram of water from 14.5° to 15.5°C. With the adoption of SI units, the calorie was redefined as exactly 4.184 joules.

The **heat capacity** of a body of matter is the quantity of heat required to increase its temperature by 1 degree Celsius. The greater the mass of a substance the greater its heat capacity. The heat capacity of 1 gram of water is very nearly 4.184 joules (or 1 calorie) per gram. The amount of heat necessary to raise the temperature of 1 gram of water 1 degree Celsius is not quite the same at all temperatures, but for our purposes it is sufficiently accurate to assume that it is the same.

The specific heat of a substance is a physical property that may be used in describing the substance. The **specific heat** of a substance is the quantity of heat required to raise the temperature of 1 gram of the substance 1 degree Celsius. Every substance has its own specific heat. Metals have much lower specific heats than does water. Only a very few substances have higher specific heats than water. The specific heat of copper is 0.38 joule per gram per degree (0.38 J/g °C), that of aluminum is 0.88 joule per gram per degree, and that of zinc is 0.39 joule per gram per degree. The following problems are based upon the measurement of heat.

EXAMPLE 1   **Calculate the quantity of heat in joules required to raise the temperature of 100 g of water from 11°C to 41°C.**

Heat required = mass × specific heat × temperature change

$$= 100\,g \times 4.184\,\frac{J}{g\,°C} \times (41 - 11)\,°C = 1 \times 10^4\,J$$

◆ ◆ ◆ ◆ ◆

EXAMPLE 2   **Calculate the quantity of heat in calories required to raise the temperature of 3.00×10^2 g of mercury from 15°C to 95°C. The specific heat of mercury is 0.033 calorie per gram per degree.**

Heat required = mass × specific heat × temperature change

$$= 3.00 \times 10^2\,g \times 0.033\,\frac{cal}{g\,°C} \times (95 - 15)\,°C = 790\,cal$$

Note that the calculated answer is 792 calories. However, only two significant figures are justified, and hence the answer is properly expressed in two significant figures as 790 calories. It could better be expressed exponentially, as 7.9×10^2 calories.

◆ ◆ ◆ ◆ ◆

EXAMPLE 3   Calculate the temperature rise that results when 1625 J are supplied to 75 g of ethyl alcohol, the specific heat of which is 2.4 J/g °C.

$$\text{Heat required} = \text{mass} \times \text{specific heat} \times \text{temperature change}$$

$$\text{Temperature change} = \frac{\text{heat required}}{\text{mass} \times \text{specific heat}}$$

$$= \frac{1625 \text{ J}}{75 \text{ g} \times 2.4 \dfrac{\text{J}}{\text{g °C}}} = 9.0 \text{ °C}$$

• • • • •

Questions

1. With what is the science of chemistry concerned?
2. List the five principal branches of chemistry and describe the area covered by each branch.
3. What is the meaning of the following statement? Matter and energy are not distinct entities but are different forms of a single entity.
4. Describe a chemical change that illustrates the Law of Conservation of Matter.
5. State the Law of Conservation of Matter, the Law of Conservation of Energy, and the combined Law of Conservation of Matter and Energy.
6. Why is a consideration of energy, which is not matter, so important in chemistry, the study of matter?
7. Explain how you would distinguish between an element, a compound, and a mixture.
8. Classify each of the following as an element, compound, or mixture: crude oil, gasoline, iron, a diamond, table salt, air, hydrogen, coal, ethyl alcohol, sugar, concrete, blood, oxygen.
9. Classify each of the following as a physical or chemical change: condensation of steam, emission of light by a flashbulb, emission of light by an electric light bulb, souring of milk, dissolving sugar in water, melting of gold, "turning" of leaves in color, explosion of a firecracker, magnetizing a screwdriver, melting of ice.
10. Define the terms "atom" and "molecule."
11. Explain why an atom and a molecule of helium are identical while an atom and a molecule of sulfur are not. Figure 1-2 may prove helpful.
12. Name three different compounds and list the elements of which each is composed.
13. How do molecules of elements and compounds differ?
14. What properties distinguish each of the three states of matter?
15. Compare the three physical states of matter with regard to volume, shape, and compressibility.
16. Classify each of the following as possessing potential energy, kinetic energy, or both: a coasting automobile, a book on a bookshelf, a stretched rubber band, a pendulum at the end of its swing, a pendulum at the middle of its swing, gasoline, falling water, a battery, natural gas.
17. Name some properties that are useful in identifying specimens of pure substances.
18. Trace the progression of a concept from hypothesis to theory to law.
19. Distinguish between the terms "mass" and "weight."
20. Why is it unnecessary to consider the distance from the earth's center when the mass of an object is determined by weighing it on an analytical balance as described in Section 1.17?

21. The addition of two numbers, each of which contains three significant figures, can give an answer that may have three or four significant figures. Explain how both cases can arise.
22. Much of Chapter 1 deals with measurement. Why is measurement so important in chemistry?
23. A burning match and a bonfire may have the same temperature yet you would not sit around a burning match on a fall evening in order to stay warm. Why not?
24. Explain how you could determine if the outside temperature were warmer or cooler than 0°C without using a thermometer.
25. The density of mercury is 13.6 g/cm^3, whereas its specific gravity is simply 13.6. Why is specific gravity a dimensionless number?
26. How do the densities of most substances change as a result of an increase in temperature?
27. Explain the difference between "heat capacity" and "specific heat" of a substance.

Problems

NOTE: In all problem sets at the ends of chapters the symbol [S] before the number of a problem indicates that the solution to that problem is worked out in the manual prepared by Ault, Meiser, Holtzclaw, and Robinson titled *Problems and Solutions for General and College Chemistry, 6th editions, by Nebergall, Holtzclaw, and Robinson.*

You should make a special effort to develop good habits with respect to expressing answers for problems to the correct number of significant figures. To this end, we have paid careful attention to significant figures in the answers provided for the problems in all problem sets. See Section 1.15 for a discussion of significant figures.

[S]1. Express 6.28 meters in centimeters, millimeters, kilometers, and angstrom units. (See Appendix C.)
Ans. 628 cm; 6.28 × 10^3 mm; 6.28 × 10^{-3} km; 6.28 × 10^{10} Å

2. What is the height in centimeters of a person who is 5 feet, 8 inches tall? (See Appendix C.)
Ans. 1.7 × 10^2 cm

3. Calculate the length of a football field (100.0 yards) in meters and kilometers. (See Appendix C.)
Ans. 91.41 m; 9.141 × 10^{-2} km

4. The diameter of a red blood cell is about 3 × 10^{-4} inches. What is the diameter in centimeters? (See Appendix C.)
Ans. 8 × 10^{-4} cm

5. The distance between the centers of the two oxygen atoms in an oxygen molecule is 1.21 Å. What is this distance in centimeters and in inches? (See Appendix C.)
Ans. 1.21 × 10^{-8} cm; 4.76 × 10^{-9} in.

[S]6. Express 1.39 quarts in liters, milliliters, and cubic centimeters. (See Appendix C.)
Ans. 1.31 ℓ; 1.31 × 10^3 ml; 1.31 × 10^3 cm^3

7. Calculate the number of milliliters in exactly 1 pt. (See Appendix C.)
Ans. 473.2 ml

8. A barrel of oil is exactly 42 gallons. How many liters of oil are in a barrel? (See Appendix C.)
Ans. 159.0 ℓ

9. Solutions in chemical laboratories are commonly prepared in 250.0-ml flasks. What is the volume of such a flask in fluid ounces? in pints? (See Appendix C.)
Ans. 8.454 oz; 0.5284 pt

10. Laboratory acids commonly come in 5-pint bottles. How many liters of acid are contained in 5.0 pt? (See Appendix C.)
Ans. 2.4 ℓ

[S]11. Express 90.8 pounds in kilograms, grams, milligrams, and metric tons. (See Appendix C.)
Ans. 41.2 kg; 4.12 × 10^4 g; 4.12 × 10^7 mg; 4.12 × 10^{-2} metric tons

12. What is the mass in kilograms and in grams of a 3.3-oz object? (See Appendix C.)
Ans. 9.4 × 10^{-2} kg; 94 g

13. What is the mass in ounces of 250.0 g of bromine? (See Appendix C.)
Ans. 8.818 oz

14. A 5-pint bottle of sulfuric acid contains 9.0 lb of sulfuric acid. What is that mass of sulfuric acid in kilograms? (See Appendix C.)
Ans. 4.1 kg

15. The mass of one oxygen atom is 2.66 × 10^{-23} g. What is this mass in ounces? (See Appendix C.)
Ans. 9.38 × 10^{-25} oz

16. If a line of 1.0 × 10^8 water molecules is 1.00 inches long, what is the average diameter, in centimeters,

angstrom units, and nanometers, of a water molecule? (See Appendix C.)

Ans. 2.5 × 10⁻⁸ cm; 2.5 Å; 0.25 nm

Ⓢ17. A student places 28.70 g of iron, 0.3807 oz of aluminum, and 0.00389 lb of copper in a beaker that weighs 138 g. What is the total mass in grams of the beaker and its contents?   *Ans. 179 g*

18. If gasoline sells for 99.9 cents per gallon, what is its cost per liter?   *Ans. 26.4¢/ℓ*

19. The density of liquid bromine is 2.928 g/cm³. What is the mass of 25 ml of bromine? What is the volume of 100.0 g of bromine?

Ans. 73 g; 34.15 cm³

20. Use the proper exponential form to express each of the following:
    (a) 0.00010230   *Ans. 1.0230 × 10⁻⁴*
    (b) 806000   *Ans. 8.06 × 10⁵*
    (c) The number of angstrom units in 1 m
       Ans. 10¹⁰ Å/m
    (d) The number of pounds in exactly one gram
       Ans. 2.2046226 × 10⁻³ lb/g
    (e) The distance from the earth to the moon (238,900 mi) in kilometers
       Ans. 3.844 × 10⁵ km
    (f) The number of feet in 1.000 centimeter
       Ans. 3.281 × 10⁻² ft/cm

21. Calculate the density of aluminum if 27.6 cm³ has a mass of 74.6 g.   *Ans. 2.70 g/cm³*

22. What is the density of lithium if 0.238 g occupies a volume of 0.4457 cm³?   *Ans. 0.534 g/cm³*

23. What is the mass of each of the following?
    Ⓢ(a) 6.00 cm³ of mercury; density = 13.5939 g/cm³
       Ans. 81.6 g
    (b) 25.0 ml of octane; density = 0.702 g/cm³
       Ans. 17.6 g
    (c) 4.00 cm³ of sodium; density = 0.97 g/cm³
       Ans. 3.9 g
    Ⓢ(d) 125 ml of gaseous chlorine; density = 3.16 g/ℓ
       Ans. 0.395 g

24. What is the volume of each of the following?
    (a) 25 g of iodine; density = 4.93 g/cm³
       Ans. 5.1 ml
    (b) 3.28 g of gaseous hydrogen; density = 0.089 g/ℓ
       Ans. 37 ℓ
    (c) 11.3 g of graphite; density = 2.25 g/cm³
       Ans. 5.02 ml

25. Osmium is the densest element known. What is the density of osmium if 2.72 g of osmium has a volume of 0.121 cm³?   *Ans. 22.5 g/cm³*

26. What is the mass of 25.0 ml of ethyl alcohol with a density of 0.789 g/cm³ at 20°?   *Ans. 19.7 g*

Ⓢ27. What is the specific gravity of ethyl alcohol at 20° relative to water with a density of 0.9982 g/cm³? (See Problem 26.)   *Ans. 0.790*

28. What is the specific gravity of nickel if 2.35 cm³ of nickel weighs as much as 20.92 ml of water?

Ans. 8.90

Ⓢ29. Using water as a reference at 15°C (density = 0.9991 g/cm³), what is the specific gravity of acetone at 25° if 25.0 ml of acetone at 25° has the same mass as 22.5 ml of benzene at the same temperature? The density of benzene is 0.8787 g/cm³.   *Ans. 0.793*

30. What is the density of sugar (sucrose) in g/cm³ if 100.0 lb occupies 1.008 ft³?   *Ans. 1.589 g/cm³*

31. The density of copper is 8.94 g/cm³. Which of the following amounts of copper will have the greatest volume: 1.0 lb; 0.50 kg; 0.50 ℓ?   *Ans. 0.50 kg*

Ⓢ32. What mass in kilograms of concentrated hydrochloric acid is contained in a standard 5.0-pt container. The specific gravity of concentrated hydrochloric acid is 1.21.   *Ans. 2.9 kg*

Ⓢ33. Copper melts at 1083°C. What is its melting temperature in °F and K?   *Ans. 1981°F; 1356 K*

34. Acetic acid freezes at 289.8 K. What is its freezing temperature in °F and °C?

Ans. 61.9°F; 16.6°C

35. Anhydrous ammonia boils at −28.1°F. What is its boiling temperature in °C and K?

Ans. −33.4°C; 239.8 K

36. Butane freezes at −138.3°C. What is its freezing temperature in °F and K?

Ans. −216.9°F; 134.8 K

Ⓢ37. How many calories would be required to raise the temperature of 235 ml of water from 26.8°C to 39.9°C? How many joules? Assume that the specific heat of water is 4.184 J/g °C.

Ans. 3.08 × 10³ cal; 1.29 × 10⁴ J

38. If 1025 joules is added to 35.0 g of water at 288 K, what is the resulting temperature? Assume that the specific heat of water is 4.184 J/g °C.

Ans. 295 K; 22°C

39. What is the specific heat of aluminum if 2400 J of heat will increase the temperature of a 1.10-kg block by 2.5 °C? *Ans. 0.87 J/g °C*

[S] 40. If 5 g of copper cools from 35.0 °C to 22.6 °C and loses 23.6 joules, what is the specific heat of copper in J/g °C? *Ans. 0.4 J/g °C*

41. Calculate the specific heat of water in units of kcal/lb °F. The specific heat of water is 1.00 cal/g °C. *Ans. 0.252 kcal/lb °F*

[S] 42. What is the final temperature of the combination when 50.0 g of chromium at 15 °C (specific heat = 0.107 cal/g °C) is added to 25 ml of water at 45 °C? Assume the specific heat of water is 1.00 cal/g °C. *Ans. 40 °C*

43. How many kilograms of water at 95 °C must be added to 1325 g of water at 25 °C so that the resulting combination will have a temperature of 65 °C? Assume the specific heat of water is 4.184 J/g °C. *Ans. 1.8 kg*

References

"Impacts of Technology," R. W. Peterson, *Amer. Scientist,* **67,** 28 (1979).

"Creativity, Discovery and Science," R. A. Brown, *J. Chem. Educ.,* **54,** 720 (1977).

"Nuclear Wastes and Public Acceptance," R. P. Hammond, *Amer. Scientist,* **67,** 146 (1979).

"No Risk Is the Highest Risk of All," A. Wildavsky, *Amer. Scientist,* **67,** 32 (1979).

"Freshman-Level Chemistry Shapes the Nuclear Power Industry," R. C. Plumb, *J. Chem. Educ.,* **52,** 523 (1975).

"Recombinant DNA: Scientific and Social Perspectives," V. Vandegriff, *J. Chem. Educ.,* **56,** 77 (1979).

"Analytical Chemistry," W. Worthy, *Chem. and Eng. News,* July 31, 1978, p. 29.

"Biochemistry," W. Worthy, *Chem. and Eng. News,* April 17, 1978, p. 16.

"Inorganic Chemistry," R. Rawls, *Chem. and Eng. News,* March 6, 1978, p. 19.

"Organic Chemistry," S. Stinson, *Chem. and Eng. News,* August 28, 1978, p. 26.

"Physical Chemistry," M. Waldrop, *Chem. and Eng. News,* June 5, 1978, p. 20.

"Polymer Chemistry," S. Stinson, *Chem. and Eng. News,* January 8, 1979, p. 21.

"Dreams, Daydreams, and Discovery," R. A. Brown and R. G. Luckcock, *J. Chem. Educ.,* **55,** 694 (1978).

"A Historical Note on the Conservation of Mass," R. D. Whitaker, *J. Chem. Educ.,* **52,** 658 (1975).

"Scientific Methods in Art and Archeology," A. E. Werner, *Chem. in Britain,* **6,** 55 (1970).

"Description of Fahrenheit's Thermometer" (provided by D. B. Murphy from *Encyclopedia Britannica,* first American edition, Thomas Dobson, Philadelphia, 1798, Vol. XVIII, pp. 496–7), *J. Chem. Educ.,* **46,** 192 (1969).

"SI Units" (International System of Units), G. Socrates, *J. Chem. Educ.,* **46,** 710 (1969).

"Physical versus Chemical Change," W. J. Gensler, *J. Chem. Educ.,* **47,** 154 (1970).

"Differentiating Physical and Chemical Changes," L. E. Strong, *J. Chem. Educ.,* **47,** 689 (1970).

"Policy for NBS Usage of SI Units," National Bureau of Standards, *J. Chem. Educ.,* **48,** 569 (1971).

"Chemistry is Real and Relevant," E. Wolthuis, *J. Chem. Educ.,* **50,** 423 (1973).

"The Origin of Chemical Elements, 1," J. Selbin, *J. Chem. Educ.,* **50,** 306 (1973).

"The Origin of Chemical Elements, 2," J. Selbin, *J. Chem. Educ.,* **50,** 380 (1973).

"The Chemical Elements of Life," E. Frieden, *Sci. American,* July 1972, p. 52.

"Chemistry and Science: The Next Hundred Years," A. L. Hammond, *Science,* **185,** 847 (1974).

"Precision Balances—A Chapter in their Development," J. T. Stock, *Chem. in Britain,* **7,** 385 (1971).

"Dalton Revisited" (Staff), *Chem. in Britain,* **9,** 228 (1973).

"The Mole and Avogadro's Number," R. M. Hawthorne, Jr., *J. Chem. Educ.,* **50,** 282 (1973).

"Time, Frequency, and Physical Measurement," H. Hellwig, K. M. Everson, and D. J. Wineland, *Phys. Today,* December 1978, p. 23.

"James Curtis Booth and his Balance," J. T. Stock, *J. Chem. Educ.,* **53,** 497 (1976).

"Milli Meter," J. Wotiz and E. Siegel, *J. Chem. Educ.,* **53,** 291 (1976).

"A m^3 is Bigger than a Breadbox," J. L. Lambert, *J. Chem. Educ.,* **55,** 638 (1978).

"SI Units? A Camel is a Camel," A. W. Adamson, *J. Chem. Educ.,* **55,** 634 (1978).

"Timetable Outlined for New Temperature Scale" (Staff) *Chem. and Eng. News,* August 7, 1978, p. 21.

2

Symbols, Formulas, and Equations; Elementary Stoichiometry

When speaking and writing about matter and the changes that it undergoes, chemists use symbols, formulas, and equations to indicate the elements present, the amounts of each element, and how the combinations of elements vary during a chemical change. It is essential, therefore, that you become skilled in the use of the symbolism of the field in order to study chemistry effectively.

2.1 Symbols

As a matter of convenience in indicating elements the chemist usually uses abbreviations, since they are more quickly written than names. These abbreviations are called **symbols.** No symbol contains more than two letters, and the first letter is always capitalized. For hydrogen, the symbol is H; for oxygen, O; and for carbon, C. Examples of two-letter symbols are Ca for calcium, Co for cobalt, Cr for chromium, Cl for chlorine, and Ni for nickel. Note that the second letter of a two-letter symbol is never capitalized; Co is the symbol for cobalt; CO is the notation for a molecule of carbon monoxide. Some symbols are abbreviations of the Latin names of the elements, as Fe for iron (Latin, *ferrum*), Na for sodium (Latin, *natrium*), and Cu for copper (Latin, *cuprum*). Tungsten has the symbol W from the German word *wolfram*. The symbols for all known elements are given in the table of atomic weights on the inside front cover of this book.

2.2   Formulas

A **formula** is a single symbol or a group of symbols that represents the composition of a substance. The symbols in a formula identify the elements present in the substance. Thus NaCl is the formula for sodium chloride (common table salt) and so identifies the elements sodium and chlorine as the constituents of salt. Subscripts are used in formulas to indicate the relative numbers of atoms of each type in the compound, but only if more than one atom of a given element is present. For example, the formula for water, H_2O, indicates that each molecule contains two atoms of hydrogen and one atom of oxygen, or that there are two atoms of hydrogen per one atom of oxygen (2 atoms H/1 atom O). The formula of sodium chloride, NaCl, indicates the presence of equal numbers of atoms of the elements sodium and chlorine (1 atom Na/1 atom Cl). Note that in the formulas for water and sodium chloride the subscript 1 is omitted, as is the usual practice. The formula for aluminum sulfate, $Al_2(SO_4)_3$, specifies two atoms of aluminum, Al, for each three sulfate, SO_4, groups. Each sulfate group contains one atom of sulfur and four atoms of oxygen. Hence, the formula shows a total of two atoms of aluminum, three atoms of sulfur, and twelve atoms of oxygen. The formula H_2SO_4 indicates the presence of atoms in the exact ratio of two atoms of hydrogen, one atom of sulfur, and four atoms of oxygen. The ratios obtained from such formulas are exact to any number of significant figures.

A **molecular formula** gives the actual numbers of atoms of each element in a molecule. An **empirical formula** (sometimes called a **simplest formula**) gives the simplest whole-number ratio of atoms. For example, the molecular formula of water, H_2O, indicates the presence of exactly two atoms of hydrogen and one atom of oxygen in one molecule. Since the simplest ratio of hydrogen atoms to oxygen atoms in water is 2 to 1, the empirical formula is also H_2O. The molecular formula of benzene, C_6H_6, identifies the benzene molecule as being composed of exactly six carbon atoms and six hydrogen atoms. The simplest ratio of carbon atoms to hydrogen atoms in benzene is one carbon atom per one hydrogen atom. Thus the empirical formula of benzene is CH.

Many compounds do not contain discrete molecules but instead are composed of particles called **ions.** Ions are atoms or groups of atoms that are electrically charged. (See Section 5.2.) Thus the composition of compounds such as NaCl, $Al_2(SO_4)_3$, KOH, and $CuSO_4$ cannot be identified in terms of the number of atoms of each type in a molecule, but only in terms of a *formula unit,* inasmuch as such compounds do not contain physically distinct and electrically neutral molecules. The formula units of such compounds are empirical formulas because they give only the simplest whole-number ratios of elements in the compound. The formula for copper sulfate, $CuSO_4$, indicates that there is exactly one atom of copper for each atom of sulfur and each four atoms of oxygen present in a sample of the compound, but it is an empirical formula since copper sulfate does not contain distinct molecules composed of a copper atom, a sulfur atom, and four oxygen atoms. The total numbers of atoms of each element in any sample of copper sulfate will be proportional to these numbers.

The molecular formula of the most common form of sulfur is S_8, showing that each molecule of this element consists of eight atoms (see Fig. 1-2 in Chapter 1). The molecular formulas for elemental hydrogen and oxygen (diatomic molecules) are H_2 and O_2, while the formula for neon (a monatomic molecule) is Ne. Note that H_2 and 2 H do not mean the same thing. H_2 represents a molecule of hydrogen consisting of two atoms of the element, chemically combined. The expression 2 H, on the other hand, indicates that the two hydrogen atoms are not in combination as a unit but are separate particles.

EXAMPLE 1   **A molecule of the sugar glucose contains 6 carbon atoms, 12 hydrogen atoms, and 6 oxygen atoms. What are the empirical and molecular formulas of glucose?**

The simplest whole-number ratio of C to H to O atoms in glucose is 1 C per 2 H per 1 O, so the *empirical* formula is CH_2O. The *molecular* formula is $C_6H_{12}O_6$ since one molecule actually contains 6 C, 12 H, and 6 O atoms.

◆ ◆ ◆ ◆ ◆

EXAMPLE 2   **How many oxygen atoms are found in a sample that contains 3.18×10^{13} H_2SO_4 molecules?**

The formula H_2SO_4 indicates that there are *exactly* 4 atoms of oxygen per H_2SO_4 molecule (4 atoms O/1 molecule H_2SO_4). This ratio can be used as a unit conversion factor to find the answer.

$$3.18 \times 10^{13} \text{ molecules } H_2SO_4 \times \frac{4 \text{ atoms O}}{1 \text{ molecule } H_2SO_4} = 1.27 \times 10^{14} \text{ atoms O}$$

◆ ◆ ◆ ◆ ◆

2.3   Atomic Weights and Moles of Atoms

Although a chemical formula tells a chemist *how many* atoms of each type must be assembled to make a particular compound, it is not necessary actually to count the number of atoms involved in the chemical reaction used to produce the compound. It is sufficient to get the correct ratios of the numbers of atoms. Making calcium chloride, $CaCl_2$, simply requires that there be twice as many chlorine atoms as calcium atoms available. The correct ratio of Cl to Ca can be obtained by weighing. Before we consider how to weigh out relative numbers of atoms, let us examine the relative masses of atoms.

The mass of an atom is very small. Rather than describe such a small mass in grams, chemists use another arbitrary unit called either the **atomic mass unit (amu)** or the **dalton** (after John Dalton). By international agreement, an atomic mass unit (or dalton) is taken to be exactly $\frac{1}{12}$ of the mass of a ^{12}C atom, a particular carbon isotope. **Isotopes** are atoms that exhibit identical chemical properties but that differ in mass (see Section 4.8). On the average, the mass of a hydrogen atom is 1.0079 amu (or daltons) on this arbitrary scale. The average mass of a carbon atom is 12.011 amu, that of fluorine is 18.998 amu, and that of phosphorus, 30.974 amu. Average masses are given here because many elements consist of mixtures of atoms of differing mass (different isotopes). For example, 98.89% of naturally occurring carbon atoms are ^{12}C atoms with a mass of exactly 12 amu (by definition); the remainder have different masses, with 1.11% possessing a mass of

13.00335 amu, and less than $10^{-8}\%$ having a mass of 14.0032 amu. The average mass of all of the carbon atoms in the naturally occurring mixure is 12.011 amu.

The term **atomic weight** is customarily used to identify a number without units that is numerically equal to the average mass of an atom in atomic mass units. Thus the atomic weight of hydrogen is 1.0079, that of fluorine is 18.998, that of carbon is 12.011, etc.

Let us now look at the relationship between masses of atoms and numbers of atoms. A mass of 30.974 amu of phosphorus contains one atom of phosphorus; a mass of 18.998 amu of fluorine contains one atom of fluorine. A mass of 123.90 amu of phosphorus contains four phosphorus atoms. The same number of fluorine atoms (four) could be obtained by weighing out 75.992 amu (4×18.998) of fluorine. A mass of 7.7435×10^{19} amu of phosphorus and a mass of 4.7495×10^{19} amu of fluorine also contain equal numbers of atoms, 2.5000×10^{18} atoms. As long as the ratio of the mass of a sample of phosphorus to that of a sample of fluorine is the same as the ratio of their atomic weights (30.974 to 18.998), the same number of atoms will be contained in the samples. This is true of any pair of elements. **If the masses of samples of each of two elements have the same ratio as the ratio of the atomic weights of the elements, the samples will contain identical numbers of atoms.**

Balances in chemical laboratories read in units of grams, not atomic mass units. But this is of little consequence in weighing out relative numbers of atoms. Just as 123.90 amu of phosphorus and 75.992 amu of fluorine contain the same number of atoms, 123.90 grams of phosphorus and 75.992 grams of fluorine also contain identical numbers of atoms, because the ratio of the masses is still 30.974 to 18.998. A convenient mass of an element for laboratory use is a mass in grams equal to the atomic weight of an element. This quantity of an element is referred to as a **mole of atoms** of the element. A mole of atoms is sometimes called a **gram-atomic weight** or **gram-atom.** The quantity 30.974 g of phosphorus contains one mole (abbreviated mol) of phosphorus atoms (one gram-atom or gram-atomic weight of phosphorus). Similarly, 18.998 g of fluorine contains a mole of fluorine atoms, and 12.011 g of carbon contains a mole of carbon atoms. Since the mass of a mole of carbon atoms (12.011 g) and of a mole of fluorine atoms (18.998 g) have the same ratio as the atomic weights of carbon and fluorine, a mole of carbon atoms and a mole of fluorine atoms both contain the same number of atoms. **A mole of atoms of any element contains the same number of atoms as a mole of atoms of any other element.**

From the definition of a mole, it is easy to see how to "count out" relative numbers of atoms. To get the same number of zinc atoms as of oxygen atoms found in 15.999 g of oxygen, we need only recognize that 15.999 g of oxygen is one mole of oxygen atoms and then weigh out one mole of zinc atoms. One mole of zinc atoms is 65.38 grams of zinc, inasmuch as the atomic weight of zinc is 65.38. Fractional amounts of moles may be handled equally well.

EXAMPLE 1   **How many moles of nitrogen atoms are contained in 9.34 g of nitrogen?**

From the atomic weights given on the inside front cover, the atomic weight of nitrogen is found to be 14.0067; hence, 1 mol of nitrogen atoms consists of 14.0067 g, or there is 1 mol of nitrogen atoms per 14.0067 g (1 mol N atoms/

14.0067 g). The quantity of nitrogen in 9.34 g can be converted to moles of N atoms as follows:

$$9.34 \text{ g} \times \frac{1 \text{ mol N atoms}}{14.0067 \text{ g}} = 0.667 \text{ mol N atoms}$$

Notice that the units cancel, indicating that the correct conversion factor has been used, and notice also that three significant figures are justified by the mass of nitrogen (9.34 g).

◆ ◆ ◆ ◆ ◆

EXAMPLE 2   **How much sodium contains the same number of atoms as 18.29 g of chlorine?**

From the table of atomic weights, we find that 1 mol of Na atoms = 22.99 g (22.99 g/1 mol Na atoms) and 1 mol of Cl atoms = 35.45 g (1 mol Cl atoms/ 35.45 g). First find the number of moles of chlorine atoms present.

$$18.29 \text{ g} \times \frac{1 \text{ mol Cl atoms}}{35.45 \text{ g}} = 0.5159 \text{ mol Cl atoms}$$

A quantity of 0.5159 mol of Cl atoms contains the same number of atoms as 0.5159 mol of Na atoms. The next step, then, is to find the mass of 0.5159 mol of Na atoms.

$$0.5159 \text{ mol Na atoms} \times \frac{22.99 \text{ g}}{1 \text{ mol Na atoms}} = 11.86 \text{ g}$$

◆ ◆ ◆ ◆ ◆

2.4   Avogadro's Number

It can be shown experimentally that one amu (or one dalton) is equal to 1.6606×10^{-24} g (the unit conversion factor, therefore, is 1.6606×10^{-24} g/1 amu). To four significant figures, the mass of one atom of oxygen in grams would be

$$16.00 \text{ amu} \times \frac{1.6606 \times 10^{-24} \text{ g}}{1 \text{ amu}} = 2.657 \times 10^{-23} \text{ g}$$

From the mass of one atom of oxygen, we can calculate the number of oxygen atoms in 16.00 g of oxygen (one mole of oxygen atoms) as

$$16.00 \text{ g} \times \frac{1 \text{ atom}}{2.657 \times 10^{-23} \text{ g}} = 6.022 \times 10^{23} \text{ atoms}$$

The number is called **Avogadro's number** in honor of the Italian professor of physics Amedeo Avogadro (1776–1856). **Avogadro's number represents the number of atoms in one mole of atoms of any element;** for example, Avogadro's number of atoms (6.022×10^{23}) is contained in 15.999 g of oxygen, in 18.998 g of fluorine, in 1.0079 g of hydrogen, and in 208.980 g of bismuth, respectively. All samples containing one mole of atoms, whatever the masses happen to be, contain 6.022×10^{23} atoms.

A mole of atoms can now be defined in two ways. **A mole of atoms is the mass in grams of an element that corresponds to the atomic weight in atomic mass units, and it also is the mass in grams that contains 6.022×10^{23} atoms.**

EXAMPLE   How many chlorine atoms are contained in 18.29 g of chlorine?

In Example 2, Section 2.3, 18.29 g of chlorine was shown to contain 0.5159 mol of chlorine atoms. Since 1 mol = 6.022×10^{23} atoms, the conversion factor is 6.022×10^{23} atoms/1 mol.

$$0.5159 \text{ mol} \times \frac{6.022 \times 10^{23} \text{ atoms}}{1 \text{ mol}} = 3.107 \times 10^{23} \text{ atoms}$$

◆ ◆ ◆ ◆ ◆

2.5   Molecular Weights, Empirical Weights, and Moles of Molecules

The **molecular weight** of a molecule is the sum of the atomic weights of all the atoms in the molecule. The molecular weight of chloroform, $CHCl_3$, equals the sum of the atomic weights of one carbon atom, one hydrogen atom, and three chlorine atoms, or $12.011 + 1.0079 + (3 \times 35.453) = 119.378$. The molecular weight of vitamin C, $C_6H_8O_6$, is 176.126 [$(6 \times 12.011) + (8 \times 1.0079) + (6 \times 15.9994)$]. For most of our work it will be permissible to round off the atomic weights of the elements to one digit after the decimal point; e.g., 12.011 to 12.0 for carbon, 1.0079 to 1.0 for hydrogen, and 35.453 to 35.5 for chlorine. However, we should be alert for those occasions when more than one decimal place is required by the number of significant figures in other data.

As we have pointed out earlier, ionic compounds such as NaCl, $Al_2(SO_4)_3$, KOH, and $CuSO_4$ do not contain discrete molecules and cannot be characterized properly by a molecular weight. But for each we can calculate the **formula weight,** or **empirical weight,** which is the sum of the atomic weights of the atoms found in one unit as indicated by the empirical formula. The formula weight of $Al_2(SO_4)_3$ is $(2 \times 27.0) + (3 \times 32.1) + (12 \times 16.0) = 342.3$. Many chemists refer loosely to the formula weight of an ionic compound as its molecular weight, but there is a difference that you should recognize.

A **mole of molecules** is contained in the mass of a molecular compound, in grams, which is equal to the molecular weight. This is the **gram-molecular weight** of the substance. Thus 119.378 grams of $CHCl_3$ (chloroform) contains a mole of $CHCl_3$ molecules and is the gram-molecular weight of chloroform; a mole of vitamin C molecules is contained in 176.126 grams of vitamin C. A mole of molecules also contains Avogadro's number of molecules; hence, 119.378 g of $CHCl_3$ and 176.126 g of $C_6H_8O_6$ each contains 6.022×10^{23} molecules. It is also possible to speak of a mole of an ionic compound such as NaCl or KOH. A mole of each of these substances (or any other ionic compound) contains 6.022×10^{23} of the units described by the empirical formula. A mole of KOH would, therefore, contain Avogadro's number of KOH units, a unit being composed of one potassium atom, one oxygen atom, and one hydrogen atom. The mass of a mole of an ionic compound in grams is equal numerically to the formula weight and is referred to as the **gram-formula weight.**

We have seen previously that a chemical formula indicates the number of atoms in one molecular or empirical unit of that compound. **A chemical formula also indicates the number of moles of atoms in one mole of the compound.** Ethanol (C_2H_5OH, often called ethyl alcohol) contains two carbon atoms in one molecule. It also contains two moles of carbon atoms in one mole of ethanol molecules ($2 \times 6.022 \times 10^{23}$ C atoms in 6.022×10^{23} C_2H_5OH molecules).

The following examples illustrate the use of the concepts of chemical formulas, atomic weights, molecular weights, and moles of atoms or molecules.

EXAMPLE 1   **How many moles of sulfur atoms are in 80.3 g of sulfur?**

The atomic weight of sulfur is 32.1, so 1 mol S atoms = 32.1 g (1 mol S atoms/32.1 g).

$$80.3 \text{ g} \times \frac{1 \text{ mol S atoms}}{32.1 \text{ g}} = 2.50 \text{ mol S atoms}$$

◆ ◆ ◆ ◆ ◆

EXAMPLE 2   **How many moles of sulfur molecules are in 80.3 g of sulfur if the molecular formula is S_8?**

The molecular weight of S_8 is $8 \times 32.1 = 256.8$; hence, 1 mol of S_8 = 256.8 g (1 mol S_8/256.8 g).

$$80.3 \text{ g} \times \frac{1 \text{ mol } S_8}{256.8 \text{ g}} = 0.313 \text{ mol } S_8$$

◆ ◆ ◆ ◆ ◆

EXAMPLE 3   **How many grams of ethanol, C_2H_5OH, is contained in 0.923 mol of that substance?**

The molecular weight of C_2H_5OH is calculated as follows:

$$2 \text{ C} = 2 \times 12.0 = 24.0$$
$$6 \text{ H} = 6 \times 1.0 = 6.0$$
$$1 \text{ O} = 1 \times 16.0 = 16.0$$

Molecular weight of C_2H_5OH = 46.0

Hence 1 mol C_2H_5OH = 46.0 g (46.0 g C_2H_5OH/1 mol C_2H_5OH).

$$0.923 \text{ mol } C_2H_5OH \times \frac{46.0 \text{ g } C_2H_5OH}{1 \text{ mol } C_2H_5OH} = 42.5 \text{ g } C_2H_5OH$$

◆ ◆ ◆ ◆ ◆

EXAMPLE 4   **How many moles of hydrogen atoms are contained in 1.38 g of ethanol?**

First, calculate the number of moles of C_2H_5OH; then use the information from the chemical formula, which indicates exactly 6 mol of H atoms per 1 mol of C_2H_5OH.

$$1.38 \text{ g } C_2H_5OH \times \frac{1 \text{ mol } C_2H_5OH}{46.0 \text{ g } C_2H_5OH} = 0.0300 \text{ mol } C_2H_5OH$$

$$0.0300 \text{ mol } C_2H_5OH \times \frac{6 \text{ mol H atoms}}{1 \text{ mol } C_2H_5OH} = 0.180 \text{ mol H atoms}$$

◆ ◆ ◆ ◆ ◆

2.6   Per Cent Composition from Formulas

All samples of pure water, H_2O, no matter what their source may be, contain the same elements, hydrogen and oxygen, united in the same proportion by mass, 11 parts of hydrogen to 89 parts of oxygen. A similar statement may be made about any pure compound. This is one of the fundamental concepts of chemistry and is known as the **Law of Definite Proportions,** or the **Law of Definite Composition.** This law may be stated as follows: **Different samples of a pure compound always contain the same elements in the same proportions by mass.** This law, among others, convinced Dalton of the atomic nature of matter and led him to outline his atomic theory.

Since all samples of a pure compound always contain the same relative amounts of constituents by mass, the per cent by mass of each element in a compound can be used to identify the compound. The per cent by mass may be calculated from the formula of the compound. This per cent is the fraction of the total mass of a sample of the compound which is attributable to the element, multiplied by 100. For example, suppose we calculate the per cent of hydrogen and that of oxygen in the compound water, H_2O. A convenient quantity to use is one mole of water.

$$\text{Per cent hydrogen by mass in } H_2O = \frac{2 \times \text{atomic weight of H}}{\text{molecular weight of } H_2O} \times 100$$

$$= \frac{2 \times 1.01}{(2 \times 1.01) + (1 \times 16.0)} \times 100 = 11.2\%$$

$$\text{Per cent oxygen by mass in } H_2O = \frac{1 \times \text{atomic weight of O}}{\text{molecular weight of } H_2O}$$

$$= \frac{1 \times 16.0}{(2 \times 1.01) + (1 \times 16.0)} \times 100 = 88.8\%$$

2.7   Derivation of Formulas

Since an empirical formula shows the relative numbers of moles of atoms in a compound, in order to write the empirical formula of a compound we must know the relative numbers of moles of atoms in a sample of it.

EXAMPLE 1   **A sample of a compound contains 0.36 mol of H and 0.090 mol of C. What is the empirical formula of the compound?**

For every 0.090 mol of C in the compound there is 0.36 mol of H; thus the formula might be written $C_{0.090}H_{0.36}$. However, since chemical formulas are customarily written in terms of whole-number ratios, this formula must be reduced to whole numbers. The ratio of C to H is 1 to 4, so the *empirical* formula of the compound must be CH_4. The simplest way to reduce two noninteger numbers to integers is to divide by the smaller noninteger. For this example,

$$0.36/0.090 = 4$$
$$0.090/0.090 = 1$$

◆ ◆ ◆ ◆ ◆

It is rare to find information about the composition of a compound given in moles. Most often, either the masses of the elements in a sample of the compound or the per cent by mass of the compound is known. If the masses of the elements are given, convert these to moles and reduce the mole ratio to simplest whole numbers in order to find the empirical formula. If the per cent by mass is given, find the mass of each element present in a specific mass of sample (100 g is a convenient mass), convert to moles, and reduce the mole ratios to simplest whole numbers.

EXAMPLE 2   **A sample of a gaseous compound is found to consist of 27.29% carbon and 72.71% oxygen. What is the empirical formula of the compound?**

Unless otherwise specified, per cent composition is understood to indicate per cent by mass. The mass of an element in a 100.0-g sample of a compound is equal in grams to the per cent of that element in the sample; hence, 100.0 g of this sample would contain 27.29 g of carbon and 72.71 g of oxygen. [100.0 g sample × (27.29 g C/100 g sample) = 27.29 g C; 100.0 g sample × (72.71 g O/100 g sample) = 72.71 g O.] The relative number of moles of carbon and oxygen atoms in the compound can be obtained by conversion of grams to moles as shown in the following table.

Element	Mass of Element in 100.0 g of Sample	Relative Number of Moles		Divide by the Smaller Number	Smallest Integral Number of Moles
Carbon	27.29 g	$27.29 \text{ g C} \times \dfrac{1 \text{ mol C}}{12.01 \text{ g C}}$	$= 2.27$ mol	$\dfrac{2.27}{2.27} = 1.00$	1
Oxygen	72.71 g	$72.71 \text{ g O} \times \dfrac{1 \text{ mol O}}{16.00 \text{ g O}}$	$= 4.54$ mol	$\dfrac{4.54}{2.27} = 2.00$	2

For every 2.27 mol of carbon there are 4.54 mol of oxygen. Reducing to smallest whole numbers makes it evident that the number of moles of oxygen is twice the number of moles of carbon in the compound. Hence, the number of oxygen *atoms* is twice the number of carbon *atoms*. The simplest formula for the compound must therefore be CO_2.

The empirical formula CO_2 indicates a formula weight of 44. It can be shown by experiment that the compound actually has a molecular weight of 44. Thus in this case the empirical formula and the molecular formula are one and the same. If the molecular weight for the compound were 88, then the molecular formula would indicate twice as many atoms as the simplest formula, i.e., C_2O_4 rather than CO_2. The correct formula is CO_2.

◆ ◆ ◆ ◆ ◆

EXAMPLE 3   **A sample of an oxide of iron contains 34.97 g of iron and 15.03 g of oxygen. What is the empirical formula of this iron oxide?**

The steps in the solution of this problem are outlined in the following table.

Element	Mass of Element	Relative Number of Moles	Divide by the Smaller Number	Smallest Integral Number of Moles
Iron	34.97 g	$34.97 \text{ g Fe} \times \dfrac{1 \text{ mol Fe}}{55.85 \text{ g Fe}}$ $= 0.6261 \text{ mol}$	$\dfrac{0.6261}{0.6261} = 1.00$	$2 \times 1.0 = 2$
Oxygen	15.03 g	$15.03 \text{ g O} \times \dfrac{1 \text{ mol O}}{16.00 \text{ g O}}$ $= 0.9394 \text{ mol}$	$\dfrac{0.9394}{0.6261} = 1.50$	$2 \times 1.5 = 3$

Unlike the previous example, division by the smaller number of moles does not give two integers, so a third step is necessary. This step involves multiplying both 1.0 and 1.5 by the smallest whole number that will give whole numbers for the relative numbers of moles, and hence atoms, of each element. Thus, $2 \times 1.0 = 2$ and $2 \times 1.5 = 3$. Hence the simplest whole-number mole ratio (and also whole-number atom ratio) is 2 to 3, and the empirical formula is Fe_2O_3.

◆ ◆ ◆ ◆

EXAMPLE 4   **Derive the formula of a hydrated salt, $Na_2SO_4 \cdot xH_2O$ where x indicates the number of moles of water in one mole of the hydrated salt. When a 1.500-g sample of the salt was heated to drive off the water, a residue of Na_2SO_4 weighing 0.795 g remained.**

The mass of H_2O in the 1.500-g sample is $1.500 \text{ g} - 0.795 \text{ g} = 0.705 \text{ g}$. The relative number of moles of Na_2SO_4 and H_2O in the salt may be calculated by dividing the mass of each of these compounds by the appropriate molecular weight, and then continuing as indicated in the following table.

Compound	Mass	Relative Number of Moles	Divide by the Smaller Number	Smallest Integral Number of Moles
Na_2SO_4	0.795 g	$0.795 \text{ g Na}_2\text{SO}_4 \times \dfrac{1 \text{ mol Na}_2\text{SO}_4}{142 \text{ g Na}_2\text{SO}_4}$ $= 0.0056 \text{ mol}$	$\dfrac{0.0056}{0.0056} = 1.0$	1
H_2O	0.705 g	$0.705 \text{ g H}_2\text{O} \times \dfrac{1 \text{ mol H}_2\text{O}}{18.0 \text{ g H}_2\text{O}}$ $= 0.0392 \text{ mol}$	$\dfrac{0.0392}{0.0056} = 7.0$	7

For each mole of Na_2SO_4 in the salt there are 7 moles of H_2O. Hence, the simplest formula of the hydrated salt is $Na_2SO_4 \cdot 7H_2O$.

◆ ◆ ◆ ◆ ◆

2.8 Chemical Equations

Atoms are the fundamental particles of the elements that enter into chemical changes. Substances that take part in chemical changes are made up of atoms in the form of molecules or ions. **Chemical changes** involve the regrouping of atoms, molecules, or ions to form other substances. To describe a chemical change, the chemist uses a shorthand type of expression called a **chemical equation.** Symbols and formulas are used to indicate the composition of each substance involved in the change. The writer of an equation must know what substances react and what substances are formed, as well as the correct formulas for these substances.

A statement of the decomposition of water by electrolysis, "When water is decomposed by an electric current, hydrogen and oxygen are formed," can be expressed as

$$H_2O \longrightarrow H_2 + O_2 \tag{1}$$

The formula for the reactant is written to the left of the arrow and the formulas for the products are written to the right. The arrow is read as either "gives," "produces," "yields," or "forms." The $+$ on the right side of the expression is read as "and"; it does not imply mathematical addition. When a $+$ appears between the formulas for two reactants on the left side of an equation, it implies "reacts with."

As it now stands, expression (1) does not conform with the law of conservation of matter. Two atoms of oxygen in the molecule O_2 cannot be formed from one molecule of water containing but one oxygen atom. To conform with the law, the equation must be **balanced** by introducing the proper number (coefficient) before each formula. Thus two molecules of water will furnish two oxygen atoms for the production of one diatomic oxygen molecule. Two water molecules will then supply four hydrogen atoms for the production of two diatomic molecules of hydrogen. The balanced equation is

$$2H_2O \longrightarrow 2H_2 + O_2 \tag{2}$$

It is important to remember that the subscripts in a formula cannot be changed in order to make an equation balance because compounds have definite atomic compositions.

In addition to identifying reactants and products, chemical equations often indicate other information such as the state of the reactants and products, whether heat is evolved or absorbed during the reaction, the temperature of the reaction, or other special conditions. For example, the formula for a gaseous reactant or product is followed by (g); a liquid by (l); a solid by (s); and a reactant or product that is dissolved in water by (aq). The reaction of solid sodium, Na, with liquid water to give hydrogen gas, H_2, and a solution of sodium hydroxide, NaOH, in water would be indicated by

$$2Na(s) + 2H_2O(l) \longrightarrow H_2(g) + 2NaOH(aq)$$

Sometimes it is useful to identify the amount of heat absorbed or evolved during a reaction. A reaction that proceeds with absorption of heat is called an **endothermic reaction;** a reaction that proceeds with evolution of heat, an **exothermic reaction.**

Sometimes the amount of heat is written as if it were a reactant or product, as in the following equations.

$$H_2(g) + Cl_2(g) \longrightarrow 2HCl(g) + 184,610 \text{ joules}$$

$$131,300 \text{ joules} + H_2O(g) + C(s) \longrightarrow CO(g) + H_2(g)$$

However, it is more conventional to write this information to the right of the equation using the symbol ΔH to indicate a heat change. A positive value of ΔH indicates that heat is absorbed (an endothermic reaction) while a negative value of ΔH indicates that heat is produced (an exothermic reaction).

$$H_2(g) + Cl_2(g) \longrightarrow 2HCl(g) \qquad \Delta H = -184,610 \text{ J}$$
$$\text{or } \Delta H = -44,124 \text{ cal}$$

$$H_2O(g) + C(s) \longrightarrow CO(g) + H_2(g) \qquad \Delta H = 131,300 \text{ J}$$
$$\text{or } \Delta H = 31,380 \text{ cal}$$

Recall that calories (cal) and joules (J) are simply different units for amounts of heat (Section 1.21).

Special conditions under which a reaction proceeds are written above or below the arrow. Such conditions may include the temperature, the presence of a catalyst, or any other special circumstances that characterize the reaction. The electrolysis of water described above is sometimes written

$$2H_2O(l) \xrightarrow{\text{Elect.}} 2H_2(g) + O_2(g)$$

where the "Elect." over the arrow indicates that the reaction occurs by means of an electric current. A reaction carried out by heating may be indicated with a triangle over the arrow.

$$CaCO_3(s) \xrightarrow{\triangle} CaO(s) + CO_2(g)$$

2.9   Mole Relationships Based on Equations

As discussed above, the chemical reaction

$$2H_2O \longrightarrow 2H_2 + O_2 \tag{1}$$

indicates that exactly two molecules of H_2O are decomposed to give two molecules of H_2 and one molecule of O_2. But that which can be said of the relative proportions of molecules, atoms, or ions can also be said of the relative proportions of *moles* of molecules, atoms, or ions. Thus the equation also indicates that two moles of H_2O molecules ($2 \times 6.022 \times 10^{23}$ H_2O molecules, are decomposed to give two moles of H_2 molecules ($2 \times 6.022 \times 10^{23}$ H_2 molecules) and one mole of O_2 molecules (6.022×10^{23} O_2 molecules). **The coefficients in a chemical equation indicate the relative numbers of moles of atoms or molecules that react or are formed.** Thus, Equation (1) indicates that two moles of H_2 are formed for each two moles of H_2O decomposed (2 mol H_2/2 mol H_2O), that one mole of O_2 is formed for each two moles of H_2O decomposed (1 mol O_2/2 mol H_2O), and that one mole of O_2 forms for each two moles of H_2 formed (1 mol O_2/2 mol H_2). These unit conversion factors are exact to any number of significant figures.

It is therefore quite straightforward to determine, from the information provided by a chemical equation, the numbers of atoms and molecules, or the numbers of moles of these atoms and molecules, which react or are produced in a chemical process.

EXAMPLE 1   **How many moles of Al_2I_6 are produced by the reaction of 4.0 mol of Al according to the following equation?**

$$2Al(s) + 3I_2(s) \longrightarrow Al_2I_6(s)$$

The balanced equation indicates that exactly one mole of Al_2I_6 is produced per two moles of Al consumed (1 mol Al_2I_6/2 mol Al). This relationship can be used to convert the number of moles of reacting Al to the number of moles of Al_2I_6 produced.

$$4.0 \text{ mol Al} \times \frac{1 \text{ mol } Al_2I_6}{2 \text{ mol Al}} = 2.0 \text{ mol } Al_2I_6$$

The use of the correct factor can be checked by verifying that the unwanted units cancel. Note that the significant figures are determined by the amount of Al since conversion factors determined from equations are exact.

◆ ◆ ◆ ◆ ◆

EXAMPLE 2   **How many moles of I_2 are required to react exactly with 0.429 mol of Al?**

The balanced equation indicates that exactly 3 mol of I_2 react with 2 mol of Al. The conversion factor is 3 mol I_2/2 mol Al.

$$0.429 \text{ mol Al} \times \frac{3 \text{ mol } I_2}{2 \text{ mol Al}} = 0.644 \text{ mol } I_2$$

◆ ◆ ◆ ◆ ◆

2.10   Calculations Based on Equations

There was little real progress in the development of chemistry before the French chemist Lavoisier (1743–1794) put this science on a quantitative basis. His simple observation that mass was gained when metals were burned (oxidized) in air had a profound impact upon the chemical theories of his time and provided the basis for the later work of John Dalton and his contemporaries.

In this section, we shall become familiar with a method of chemical calculation that is based on chemical equations. All such calculations proceed from the facts that (1) equations give the ratios of moles of reactants and products involved in chemical reactions, and (2) the masses of the reactants and products can be determined from the number of moles involved, using the atomic, molecular, or formula weights of the substances participating in the reaction.

EXAMPLE 1   Calculate the moles of oxygen produced during the thermal decomposition of 100.0 g of potassium chlorate to form potassium chloride and oxygen according to the following reaction:

$$2KClO_3(s) \xrightarrow{\triangle} 2KCl(s) + 3O_2(g)$$

If one knows the number of moles of $KClO_3$ that react, the number of moles of O_2 produced can be determined since the equation indicates that exactly 3 mol of O_2 are produced for each 2 mol of $KClO_3$ that react (3 mol O_2/2 mol $KClO_3$; Section 2.9). The number of moles of $KClO_3$ used can be found from the known mass of $KClO_3$ (100.0 g), the formula weight of $KClO_3$ [39.1 + 35.5 + (3 × 16.0) = 122.6], and the relationship 1 mol $KClO_3$ = 122.6 g $KClO_3$ (Section 2.5).

$$100.0 \text{ g } KClO_3 \times \frac{1 \text{ mol } KClO_3}{122.6 \text{ g } KClO_3} = 0.8157 \text{ mol } KClO_3$$

$$0.8157 \text{ mol } KClO_3 \times \frac{3 \text{ mol } O_2}{2 \text{ mol } KClO_3} = 1.224 \text{ mol } O_2$$

Thus 100.0 g of $KClO_3$ will produce 1.224 mol of O_2. In this case, the chain of calculations involves the following unit conversions:

$$\boxed{\text{Mass of } KClO_3} \longrightarrow \boxed{\text{Moles of } KClO_3} \longrightarrow \boxed{\text{Moles of } O_2}$$

Note how dimensional analysis with the resulting cancellation of units is helpful in checking to see if the correct conversion factors have been used.

It is not necessary to write each step of the calculation separately. Although the logic of the problem is perhaps best approached in two steps, the calculation could have been written in one step as follows:

$$100.0 \text{ g } KClO_3 \times \frac{1 \text{ mol } KClO_3}{122.6 \text{ g } KClO_3} \times \frac{3 \text{ mol } O_2}{2 \text{ mol } KClO_3} = 1.223 \text{ mol } O_2$$

(The discrepancy in the last significant figure, 1.224 and 1.223, results because of rounding twice in the first case and rounding only once in the second case. This slight discrepancy could have been avoided by carrying one more figure in the first step of the two-step solution, 0.81566, and rounding the final answer to the justified number of four significant figures, 1.223.)

◆ ◆ ◆ ◆ ◆

EXAMPLE 2   Calculate the mass of oxygen required for the combustion of 10.0 g of ethane, C_2H_6.

$$2C_2H_6(g) + 7O_2(g) \longrightarrow 4CO_2(g) + 6H_2O(l)$$

The chain of calculations requires the following conversions:

$$\boxed{\text{Mass of } C_2H_6} \longrightarrow \boxed{\text{Moles of } C_2H_6} \longrightarrow \boxed{\text{Moles of } O_2} \longrightarrow \boxed{\text{Mass of } O_2}$$

From the molecular weight of C_2H_6, 1 mol C_2H_6 = 30.0 g C_2H_6. Thus

$$10.0 \text{ g } C_2H_6 \times \frac{1 \text{ mol } C_2H_6}{30.0 \text{ g } C_2H_6} = 0.3333 \text{ mol } C_2H_6$$

From the balanced chemical equation, 2 mol of C_6H_6 react for each 7 mol of O_2, so

$$0.3333 \text{ mol } C_2H_6 \times \frac{7 \text{ mol } O_2}{2 \text{ mol } C_2H_6} = 1.166 \text{ mol } O_2$$

From the molecular weight of O_2, 1 mol O_2 = 32.0 g O_2. Hence

$$1.166 \text{ mol } O_2 \times \frac{32.0 \text{ g } O_2}{1 \text{ mol } O_2} = 37.3 \text{ g } O_2$$

In this example, one more figure than justified by the data was carried through the problem and the final answer was rounded off to three significant figures. Although the logic of the problem involves three steps, the calculation could have been handled in one step as follows:

$$10.0 \text{ g } C_2H_6 \times \frac{1 \text{ mol } C_2H_6}{30.0 \text{ g } C_2H_6} \times \frac{1 \text{ mol } O_2}{2 \text{ mol } C_2H_6} \times \frac{32.0 \text{ g } O_2}{1 \text{ mol } O_2} = 37.3 \text{ g } O_2$$

◆ ◆ ◆ ◆ ◆

EXAMPLE 3   **What mass of sodium hydroxide, NaOH, would be required to produce 16 g of magnesium hydroxide, $Mg(OH)_2$, by the reaction of magnesium chloride, $MgCl_2$, with NaOH?**

$$MgCl_2(aq) + 2NaOH(aq) \longrightarrow Mg(OH)_2(s) + 2NaCl(aq)$$

In order to determine the mass of NaOH required, we need the following steps:

$$\boxed{\begin{array}{c} \text{Mass of} \\ Mg(OH)_2 \end{array}} \longrightarrow \boxed{\begin{array}{c} \text{Moles of} \\ Mg(OH)_2 \end{array}} \longrightarrow \boxed{\begin{array}{c} \text{Moles of} \\ \text{NaOH} \end{array}} \longrightarrow \boxed{\begin{array}{c} \text{Mass of} \\ \text{NaOH} \end{array}}$$

From the formula weight of $Mg(OH)_2$, 58.3 g $Mg(OH)_2$ = 1 mol $Mg(OH)_2$. Therefore

$$16 \text{ g } Mg(OH)_2 \times \frac{1 \text{ mol } Mg(OH)_2}{58.3 \text{ g } Mg(OH)_2} = 0.274 \text{ mol } Mg(OH)_2$$

From the chemical equation, two moles of NaOH react to give one mole of $Mg(OH)_2$ [2 mol NaOH/1 mol $Mg(OH)_2$]. Hence

$$0.274 \text{ mol } Mg(OH)_2 \times \frac{2 \text{ mol NaOH}}{1 \text{ mol } Mg(OH)_2} = 0.548 \text{ mol NaOH}$$

From the formula weight of NaOH, 40.0 g NaOH = 1 mol NaOH. Thus

$$0.548 \text{ mol NaOH} \times \frac{40.0 \text{ g NaOH}}{1 \text{ mol NaOH}} = 22 \text{ g NaOH}$$

Two significant figures are required by the data [16 g $Mg(OH)_2$ used initially].

The calculation also could have been handled in one step, as demonstrated in previous examples.

◆ ◆ ◆ ◆ ◆

2.11 Solutions

In the preceding section we used chemical equations to relate quantities of reactants and products that were measured by weighing. In the following two sections we will describe how chemical equations can be used to relate quantities of materials present in solution.

When crystals of sugar are stirred with a sufficient quantity of water, the sugar disappears and a clear mixture of sugar in water is formed. The sugar is said to have dissolved (or gone into solution) in the water. This **solution** consists of two components, the **solute** (the dissolved sugar) and the **solvent** (the water). In this solution the molecules of sugar are uniformly distributed among the molecules of water; i.e., all portions of the solution have the same composition. The solution is said to be a *homogeneous* mixture of sugar molecules and water molecules.

An aqueous solution of sugar that contains only a small amount of sugar (the solute) in comparison with the amount of water is said to be **dilute;** the addition of more sugar makes the solution more **concentrated. A solution, then, can be defined as a homogeneous mixture in which the composition of the individual components can be varied within limits.** These limits depend on the nature of the substance dissolved (the solute) and the medium in which the solute is dissolved (the solvent). The solute or the solvent can be either a gas, a liquid, or a solid. In this chapter, we will limit our attention to a gaseous, liquid, or solid solute dissolved in a liquid solvent. Solutions will be discussed in greater detail in Chapter 13.

The **concentration** of a solution is an expression of the relative amounts of solute and solvent present in the solution. One useful way of expressing the concentration of solute in a solution is to use molarity. By definition, **the molarity, *M*, of a solution is the number of moles of solute per exactly one liter (or dm³) of solution.** The molarity of a solution is found by dividing the number of moles of solute in a given volume of solution by the volume of the solution in liters.

$$\text{Molarity} = \frac{\text{moles of solute}}{\text{liters of solution}} \qquad (1)$$

EXAMPLE 1   **What is the concentration of a hydrochloric acid solution that contains 18.23 g of HCl (0.5000 mol) per 0.250 ℓ of solution?**

$$\text{Molarity} = \frac{\text{moles of HCl}}{\text{liters of solution}}$$

$$= \frac{0.5000 \text{ mol}}{0.250 \text{ ℓ}}$$

$$= 2.00 \ M, \text{ or } 2.00 \text{ mol/ℓ, or } 2.00 \text{ mol/dm}^3$$

For calculations, it would be convenient to write the molarity as the following unit conversion factor:

$$\frac{2.00 \text{ moles HCl}}{1 \text{ liter}}$$

Although this example was presented with the volume in liters, many experimental measurements are in milliliters and must be converted to liters before being used to find molarities.

◆ ◆ ◆ ◆ ◆

EXAMPLE 2   **How many moles of sulfuric acid, H_2SO_4, are contained in 0.80 ℓ of a solution of 0.050 M sulfuric acid?**

Rearrangement of Equation (1) gives

$$\text{Molarity} \times \text{liters of solution} = \text{moles of solute}$$

$$\frac{0.050 \text{ mol } H_2SO_4}{1 \ell} \times 0.80 \ell = 0.040 \text{ mol } H_2SO_4$$

◆ ◆ ◆ ◆ ◆

EXAMPLE 3   **What mass of iodine is required to make 5.00×10^2 ml of a 0.250 M solution of I_2 with chloroform, $CHCl_3$, as the solvent?**

The moles of I_2 necessary to make a 0.250 M solution can be found by rearranging Equation (1). The weight of I_2 required can then be calculated. The steps required are

$$\boxed{\text{Volume of solution}} \longrightarrow \boxed{\text{Moles of } I_2} \longrightarrow \boxed{\text{Mass of } I_2}$$

Note that the volume is given in milliliters and must be converted to liters.

$$5.00 \times 10^2 \text{ ml} \times \frac{1 \ell}{1000 \text{ ml}} = 0.500 \ell$$

$$\frac{0.250 \text{ mol } I_2}{1 \ell} \times 0.500 \ell = 0.125 \text{ mol } I_2$$

$$0.125 \text{ mol } I_2 \times \frac{253.8 \text{ g } I_2}{1 \text{ mol } I_2} = 31.7 \text{ g } I_2$$

◆ ◆ ◆ ◆ ◆

2.12   Calculations Based on Concentration

Just as a chemical equation can be used to relate the masses of the reactants and the products of a chemical reaction, it can be used also to relate concentrations or volumes of solutions of reactants and products. An equation gives the ratios of moles of reactants and products entering into a chemical reaction, while Equation (1) in Section 2.11 relates the concentration, volume, and number of moles of solute in a solution. A number of ways in which this information can be combined are illustrated in the following examples.

EXAMPLE 1   The first step in the production of metallic uranium involves the reaction of U_3O_8, "yellow cake," with a solution of nitric acid, HNO_3, in water.

$$U_3O_8(s) + 8HNO_3(aq) \longrightarrow 3UO_2(NO_3)_2(aq) + 2NO_2(g) + 4H_2O(l)$$

What mass of "yellow cake" is required to react with the nitric acid present in 2.50 ℓ of a 3.00 M solution?

If we know the number of moles of HNO_3 used in this reaction, the number of moles of U_3O_8 required can be determined, since the equation indicates that exactly one mole of U_3O_8 reacts per eight moles of HNO_3. The mass of U_3O_8 required can be determined from the moles of U_3O_8 that react. The number of moles of HNO_3 available can be determined from Equation (1) in Section 2.11. The chain of calculations would be

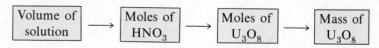

Rearrangement of Equation (1), Section 2.11, gives molarity × liters of solution = moles of HNO_3.

$$\frac{3.00 \text{ mol } HNO_3}{1 \text{ } \ell} \times 2.50 \text{ } \ell = 7.500 \text{ mol } HNO_3$$

The number of moles of U_3O_8 is determined from the conversion factor indicated by the equation, i.e., 1 mol U_3O_8 per 8 mol HNO_3.

$$7.500 \text{ mol } HNO_3 \times \frac{1 \text{ mol } U_3O_8}{8 \text{ mol } HNO_3} = 0.9375 \text{ mol } U_3O_8$$

Finally, the mass of 0.9375 mol of U_3O_8 is determined using the formula weight of U_3O_8 in the conversion factor 842.1 g U_3O_8 per 1 mol U_3O_8.

$$0.9375 \text{ mol } U_3O_8 \times \frac{842.1 \text{ g } U_3O_8}{1 \text{ mol } U_3O_8} = 789 \text{ g } U_3O_8$$

◆ ◆ ◆ ◆ ◆

EXAMPLE 2   What is the molar concentration of sodium hydroxide, NaOH, in a solution if 25.81 ml of the solution reacts exactly with 0.1184 g of oxalic acid, $H_2C_2O_4$, according to the following equation?

$$2NaOH(aq) + H_2C_2O_4(s) \longrightarrow Na_2C_2O_4(aq) + 2H_2O(l)$$

To determine the concentration of sodium hydroxide solution proceed as follows: Convert the mass of $H_2C_2O_4$ to moles of $H_2C_2O_4$ (1 mol $H_2C_2O_4$/90.03 g $H_2C_2O_4$); convert the moles of $H_2C_2O_4$ to the equivalent amount of NaOH

(2 mol NaOH/1 mol $H_2C_2O_4$); and using the number of moles of NaOH and the volume of the solution, determine the concentration of the solution (M = mol NaOH/ℓ of solution). The following conversions are required.

$$\boxed{\begin{array}{c}\text{Mass of}\\ H_2C_2O_4\end{array}} \longrightarrow \boxed{\begin{array}{c}\text{Moles of}\\ H_2C_2O_4\end{array}} \longrightarrow \boxed{\begin{array}{c}\text{Moles of}\\ \text{NaOH}\end{array}} \longrightarrow \boxed{\begin{array}{c}\text{Concentration}\\ \text{of solution}\end{array}}$$

$$0.1184 \text{ g } H_2C_2O_4 \times \frac{1 \text{ mol } H_2C_2O_4}{90.03 \text{ g } H_2C_2O_4} = 1.315 \times 10^{-3} \text{ mol } H_2C_2O_4$$

$$1.315 \times 10^{-3} \text{ mol } H_2C_2O_4 \times \frac{2 \text{ mol NaOH}}{1 \text{ mol } H_2C_2O_4} = 2.630 \times 10^{-3} \text{ mol NaOH}$$

Now use Equation (1), Section 2.11, but note that the volume is given in milliliters and must be converted to liters.

$$25.81 \text{ ml} \times \frac{1 \text{ } \ell}{1000 \text{ ml}} = 0.02581 \text{ } \ell$$

$$M = \frac{\text{moles of NaOH}}{\text{liters of solution}} = \frac{2.630 \times 10^{-3} \text{ mol}}{0.02581 \text{ } \ell}$$

$$= 0.1019 \text{ mol}/\ell$$

◆ ◆ ◆ ◆ ◆

EXAMPLE 3   **What volume of a 0.2089 M solution of KI will contain enough KI to react exactly with the $Cu(NO_3)_2$ in 43.88 ml of a 0.3842 M $Cu(NO_3)_2$ solution according to the following equation?**

$$2Cu(NO_3)_2(aq) + 4KI(aq) \longrightarrow 2CuI(s) + I_2(s) + 4KNO_3(aq)$$

In order to determine the necessary volume of the KI solution, we need to know the moles of KI required to react. This can be determined from the moles of $Cu(NO_3)_2$ present in 43.88 ml of the $Cu(NO_3)_2$ solution using the conversion factor obtained from the equation. The calculation involves the following steps:

$$\boxed{\begin{array}{c}\text{Volume of}\\ Cu(NO_3)_2 \text{ soln}\end{array}} \longrightarrow \boxed{\begin{array}{c}\text{Moles of}\\ Cu(NO_3)_2\end{array}} \longrightarrow \boxed{\begin{array}{c}\text{Moles of}\\ \text{KI}\end{array}} \longrightarrow \boxed{\begin{array}{c}\text{Volume of}\\ \text{KI soln}\end{array}}$$

Note that the volumes must be expressed in liters.
From Equation (1), Section 2.11,

$$\frac{0.3842 \text{ mol } Cu(NO_3)_2}{1 \text{ } \ell} \times 0.04388 \text{ } \ell = 1.686 \times 10^{-2} \text{ mol } Cu(NO_3)_2$$

$$1.686 \times 10^{-2} \text{ mol } Cu(NO_3)_2 \times \frac{4 \text{ mol KI}}{2 \text{ mol } Cu(NO_3)_2} = 3.372 \times 10^{-2} \text{ mol KI}$$

The volume of KI solution required may be obtained by rearranging Equation (1), Section 2.11, to solve for volume.

$$\text{Liters of solution} = \text{moles of KI} \times \frac{1}{\text{molarity}}$$

Since a 0.2089 M solution contains 0.2089 mole KI per exactly 1 liter (0.2089 mol KI/1 ℓ)

$$\frac{1}{\text{molarity}} = \frac{1\,\ell}{0.2089\text{ mol KI}}$$

Thus

$$3.372 \times 10^{-2}\text{ mol KI} \times \frac{1\,\ell}{0.2089\text{ mol KI}} = 0.1614\,\ell$$

The reaction requires 0.1614 ℓ or 161.4 ml of KI solution.

◆ ◆ ◆ ◆ ◆

Questions

1. What are the names of the elements represented by the following symbols? H, C, N, O, Ne, Na, Mg, Al, Si, K, Ca, Fe, Cu, Zn, Ag, I, Pt, Au, Pb, U

2. Define the following terms: amu, molecular formula, empirical formula, mole of lithium atoms, Avogadro's number, a mole of sulfate ions.

3. Under what circumstances are the molecular and empirical formulas of a compound identical?

4. Write the chemical formulas for molecules of the following: (a) water (b) sulfur (c) sulfuric acid (d) ethanol (e) ethane

5. The symbol for the element sulfur differs from the formula for a molecule of sulfur. Why?

6. Write out in words the meaning of the following chemical equations:
   (a) $CH_4(g) + 2O_2(g) \longrightarrow CO_2(g) + 2H_2O(g)$
   (b) $2Al(s) + 3H_2SO_4(aq) \longrightarrow 3H_2(g) + Al_2(SO_4)_3(aq)$
   (c) $Ca(s) + Cl_2(g) \longrightarrow CaCl_2(s)$
   (d) $2Na(s) + 2H_2O(l) \longrightarrow 2NaOH(aq) + H_2(g)$
   (e) $SO_3(g) + H_2O(l) \longrightarrow H_2SO_4(l)$

7. Balance the following equations:
   (a) $Al_2O_3 + Cl_2 + C \longrightarrow AlCl_3 + CO$
   (b) $CaCO_3 \longrightarrow CaO + CO_2$
   (c) $NH_4NO_3 \longrightarrow N_2O + H_2O$
   (d) $H_3PO_3 \longrightarrow H_3PO_4 + PH_3$
   (e) $N_2 + H_2 \longrightarrow NH_3$
   (f) $Cu(NO_3)_2 \longrightarrow CuO + NO_2 + O_2$
   (g) $MgO + Si \longrightarrow Mg + SiO_2$
   (h) $HI + Cl_2 \longrightarrow HCl + I_2$
   (i) $(NH_4)_2Cr_2O_7 \longrightarrow Cr_2O_3 + H_2O + N_2$

8. Balance the following equations:
   (a) $PCl_3 + Cl_2 \longrightarrow PCl_5$
   (b) $TiCl_4 + H_2O \longrightarrow TiO_2 + HCl$
   (c) $Fe + H_2O \longrightarrow Fe_3O_4 + H_2$
   (d) $Sc_2O_3 + SO_3 \longrightarrow Sc_2(SO_4)_3$
   (e) $Al + H_2SO_4 \longrightarrow Al_2(SO_4)_3 + H_2$
   (f) $KClO_3 \longrightarrow KCl + KClO_4$
   (g) $Ca_3(PO_4)_2 + C \xrightarrow{\Delta} Ca_3P_2 + CO$
   (h) $Ca_3(PO_4)_2 + H_3PO_4 \longrightarrow Ca(H_2PO_4)_2$

9. Balance the following equations:
   (a) $PtCl_4 \longrightarrow Pt + Cl_2$
   (b) $H_2 + Br_2 \longrightarrow HBr$
   (c) $Sb + O_2 \longrightarrow Sb_4O_6$
   (d) $P_4 + O_2 \longrightarrow P_4O_{10}$
   (e) $Pb + H_2O + O_2 \longrightarrow Pb(OH)_2$
   (f) $Ag + H_2S + O_2 \longrightarrow Ag_2S + H_2O$
   (g) $Cu + HNO_3 \longrightarrow Cu(NO_3)_2 + H_2O + NO$
   (h) $PCl_5 + H_2O \longrightarrow POCl_3 + HCl$

10. Write balanced chemical equations that represent the reactions described below:
    (a) Gaseous hydrogen (H_2) and gaseous oxygen (O_2) combine to form liquid water (H_2O).
    (b) Solid carbon burns in gaseous oxygen (O_2) to produce gaseous carbon monoxide (CO).
    (c) Solid carbon burns in gaseous oxygen to produce gaseous carbon dioxide (CO_2). Compare

with (b). Would the use of excess oxygen favor the production of CO or CO_2? Would an excess of carbon favor the production of CO or CO_2?

(d) Solid magnesium carbonate ($MgCO_3$) is heated to drive off gaseous carbon dioxide (CO_2), leaving behind a solid residue of magnesium oxide (MgO).

(e) Water vapor reacts with sodium metal to produce gaseous hydrogen (H_2) and solid sodium hydroxide (NaOH).

(f) Propane gas (C_3H_8) burns in oxygen (O_2) to produce gaseous carbon dioxide (CO_2) and liquid water (H_2O).

(g) Water and hot carbon react to form gaseous hydrogen (H_2) and gaseous carbon monoxide (CO).

11. Extract the maximum amount of qualitative and quantitative information from the chemical formula for sulfuric acid (H_2SO_4).

12. Does one mole of water molecules contain the same number of hydrogen atoms as the number of iron atoms in one mole of iron atoms? Show why.

13. If the mass of a ^{12}C atom were redefined to be exactly 18 "nmu" (new mass units), how would the values of the masses of the other atoms change when converted to nmu? Give several examples.

14. Explain why the following equation does not conform to the Law of Conservation of Matter and change the equation so that it does:

$$C_{12}H_{22}O_{11} + O_2 \longrightarrow 12CO_2 + 11H_2O + heat$$

15. If 28.1 g of Si reacts with N_2 giving 0.333 mole of Si_3N_4 according to the equation

$$3Si + 2N_2 \longrightarrow Si_3N_4$$

which of the following can be determined using only the information given in this problem: (a) the moles of Si reacting, (b) the moles of N_2 reacting, (c) the atomic weight of N, (d) the atomic weight of Si, (e) the mass of Si_3N_4 produced?

Problems

NOTE: Atomic weights used to determine the answers to these problems have been rounded off to one decimal place unless the number of significant figures in other data requires more significant figures.

1. Calculate the molecular weight of each of the following compounds:
   - ⓢ(a) Hydrogen chloride, HCl   *Ans. 36.4*
   - (b) Methane, CH_4 *Ans. 16.0*
   - (c) Dinitrogen pentaoxide, N_2O_5 *Ans. 108.0*
   - ⓢ(d) Sulfuric acid, H_2SO_4 *Ans. 98.1*
   - ⓢ(e) Tetraethyl lead, $Pb(C_2H_5)_4$ *Ans. 323.2*
   - (f) Glucose, $C_6H_{12}O_6$ *Ans. 180.0*
   - (g) Aspirin, $C_6H_4(CO_2H)(CO_2CH_3)$ *Ans. 180.0*

2. Determine the mass in grams of a mole of each of the following compounds:
   - (a) Hydrogen peroxide, H_2O_2 *Ans. 34.0 g*
   - (b) Phosphoric acid, H_3PO_4 *Ans. 98.0 g*
   - (c) Sodium cyanide, NaCN   *Ans. 49.0 g*
   - (d) Hydrated magnesium nitrate, $Mg(NO_3)_2 \cdot 7H_2O$ *Ans. 274.3 g*
   - (e) Acetic acid, CH_3CO_2H *Ans. 60.0 g*
   - (f) Styrene, $C_6H_5CHCH_2$ *Ans. 104.0 g*
   - (g) Propane, $CH_3CH_2CH_3$ *Ans. 44.0 g*

3. The formula weight of carnallite, $KMgCl_3 \cdot 6H_2O$, is 277.85. Calculate the atomic weight of K if H = 1.01, Mg = 24.31, Cl = 35.45, and O = 16.00.   *Ans. 39.07*

ⓢ4. A sample of FeO contains 0.777 g of Fe and 0.223 g of O. The atomic weight of O is 16.00. Determine the atomic weight of Fe.   *Ans. 55.7*

5. Determine the number of
   - ⓢ(a) moles of ammonia, NH_3, in 8.5 g of NH_3.   *Ans. 0.50 mol*
   - (b) moles of sodium hydroxide, NaOH, in 378 g of NaOH.   *Ans. 9.45 mol*
   - (c) moles of boric acid, $B(OH)_3$, in 37.0 g of $B(OH)_3$.   *Ans. 0.599 mol*
   - ⓢ(d) moles of caffeine, $C_8H_{10}N_4O_2$, in 0.381 g of caffeine.   *Ans. 1.96×10^{-3} mol*
   - (e) moles of ethylene glycol, $C_2H_4(OH)_2$, in 254 g of $C_2H_4(OH)_2$.   *Ans. 4.10 mol*
   - (f) moles of nitrogen molecules, N_2, in 28 g of nitrogen.   *Ans. 1.0 mol*

(g) moles of phosphorus molecules, P_4, in 0.119 g of phosphorus. *Ans.* 9.60×10^{-4} *mol*

6. What is the mass of each of the following in grams?

[S](a) 0.178 mol of potassium bromide, KBr *Ans. 21.2 g*

(b) 2.35 mol of calcium carbonate, $CaCO_3$ *Ans. 235 g*

(c) 0.81 mol of tin(II) fluoride, SnF_2 *Ans.* 1.3×10^2 *g*

(d) 1.07 mol of acetylene, C_2H_2 *Ans. 27.8 g*

[S](e) 2.60×10^{-4} mol of the amino acid glycine, $CH_2(NH_2)CO_2H$ *Ans.* 1.95×10^{-2} *g*

(f) 38.4 mol of aluminum sulfate, $Al_2(SO_4)_3$ *Ans.* 1.31×10^4 *g*

(g) 0.6612 mol of nitrogen atoms *Ans. 9.261 g*

(h) 0.661 mol of nitrogen molecules, N_2 *Ans. 18.5 g*

[S](i) 1.378 mol of oxygen molecules, O_2 *Ans. 44.10 g*

(j) 1.378 mol of ozone, O_3 *Ans. 66.14 g*

7. How many moles of water are contained in a 22.6-gram sample of the mineral tschermigite, $(NH_4)Al(SO_4)_2 \cdot 12H_2O$? *Ans. 0.598 mol*

[S]8. Determine the moles of nickel phosphate, $Ni_3(PO_4)_2$, the moles of nickel atoms, the moles of oxygen atoms, the number of nickel atoms, and the number of oxygen atoms in 9.37 g of $Ni_3(PO_4)_2$. *Ans.* 2.56×10^{-2}; 7.68×10^{-2}; *0.205;* 4.62×10^{22}; 1.23×10^{23}

9. Determine the number of

(a) moles of sulfur dioxide in 36 g of sulfur dioxide, SO_2. *Ans. 0.56 mol*

(b) grams of magnesium in 2.3 mol of magnesium atoms. *Ans. 56 g*

(c) moles of fluorine atoms in 0.532 mol of F_2. *Ans. 1.06 mol*

(d) moles of chlorine atoms in 17.8 g of Cl_2. *Ans. 0.502 mol*

(e) moles of selenium atoms in 118.53 g of selenium. *Ans. 1.501 mol*

10. Show by calculations which of the following contains the greatest mass of aluminum: one mole of aluminum atoms; one mole of aluminum phosphate, $AlPO_4$; 1.5 mol of aluminum chloride, Al_2Cl_6; 0.75 mol of aluminum sulfate, $Al_2(SO_4)_3$. *Ans. 1.5 mol* Al_2Cl_6

11. Show by calculations which of the following contains the greatest number of oxygen atoms: 34 g of hydrogen peroxide, H_2O_2; 34 g of calcium carbonate, $CaCO_3$; or 34 g of water. *Ans. 34 g of* H_2O_2

12. Show by calculations which of the following contains the largest mass of carbon atoms: 0.10 mol of glucose, $C_6H_{12}O_6$; 3.0 g of ethane, C_2H_6; or a 1.0-g diamond (diamond is pure carbon). *Ans. 0.10 mol* $C_6H_{12}O_6$

13. Show by calculations

(a) which has the greatest mass in one mole of atoms: nitrogen, oxygen, phosphorus, or argon. *Ans. Ar*

(b) which has the greatest mass in one mole of molecules: N_2, O_2, P_4, or Ar. *Ans.* P_4

14. Show by calculations which of the following contains the greatest total number of atoms: 10.0 g of potassium metal; 1.02×10^{23} molecules of nitrogen, N_2; 0.10 mol of chloroform, $CHCl_3$; or 5 g of carbon dioxide, CO_2. *Ans. 0.10 mol* $CHCl_3$

[S]15. Calculate the number of moles of ethanol, CH_3CH_2OH, in 7.55 kg of ethanol. *Ans. 164 mol*

16. Calculate the molar concentration of each of the following solutions:

[S](a) 98.1 g of sulfuric acid, H_2SO_4, in 1.00 ℓ of solution *Ans. 1.00 M*

(b) 180 g of glucose, $C_6H_{12}O_6$, in 1.000 ℓ of solution *Ans. 1.0 M*

[S](c) 24.5 g of sodium cyanide, NaCN, in 2.000 ℓ of solution *Ans. 0.250 M*

(d) 90.0 g of acetic acid, CH_3CO_2H, in 0.750 ℓ of solution *Ans. 2.00 M*

[S](e) 2.12 g of potassium bromide, KBr, in 458 ml of solution *Ans.* 3.89×10^{-2} *M*

(f) 0.1374 g of copper sulfate, $CuSO_4$, in 13 ml of solution *Ans.* 6.6×10^{-2} *M*

17. Determine the moles of solute present in each of the following solutions:

(a) 0.450 ℓ of 1.00 *M* NaCl solution *Ans. 0.450 mol NaCl*

[S](b) 2.0 ℓ of 0.480 *M* $MgCl_2$ solution *Ans. 0.96 mol* $MgCl_2$

(c) 455 ml of 3.75 *M* HCl solution *Ans. 1.71 mol HCl*

[S](d) 2.50 ml of 0.1812 $M\ KMnO_4$ solution
Ans. 4.53×10^{-4} *mol* $KMnO_4$

(e) 13.98 ℓ of 3.189×10^{-4} $M\ HClO_4$ solution
Ans. 4.458×10^{-3} *mol* $HClO_4$

(f) 25.38 ml of 9.721×10^{-2} M KOH solution
Ans. 2.467×10^{-3} *mol KOH*

18. Determine the mass of each of the following solutes required to make the indicated amount of solution:

[S](a) 1.00 ℓ of 1.00 $M\ LiNO_3$ solution
Ans. 68.9 *g* $LiNO_3$

(b) 1.00 ℓ of 0.525 $M\ CaCl_2$ solution
Ans. 58.3 *g* $CaCl_2$

(c) 4.2 ℓ of 2.45 $M\ C_2H_5OH$ solution
Ans. 4.7×10^2 *g* C_2H_5OH

[S](d) 275 ml of 0.5151 $M\ KClO_4$ solution
Ans. 19.6 *g* $KClO_4$

(e) 25 ml of 0.1881 $M\ H_2O_2$ solution
Ans. 0.16 *g* H_2O_2

(f) 1856 ml of 0.1475 $M\ H_3PO_4$ solution
Ans. 26.82 *g* H_3PO_4

19. Calculate the average mass in grams of one atom of uranium from its atomic weight and Avogadro's number. *Ans.* 3.953×10^{-22} *g*

20. Calculate the per cent composition of each of the following to three significant figures.

[S](a) Lead sulfide, PbS *Ans.* 86.6% *Pb*; 13.4% *S*

(b) Propylene, C_3H_6 *Ans.* 85.7% *C*; 14.3% *H*

[S](c) Iron(III) nitrate, $Fe(NO_3)_3$
Ans. 23.1% *Fe*; 17.4% *N*; 59.6% *O*

(d) Lithium cobalt(II) chloride, Li_2CoCl_4
Ans. 6.47% *Li*; 27.4% *C*; 66.1% *C*

(e) Carnotite, $K_2(UO_2)_2(VO_4)_2 \cdot 12H_2O$
Ans. 7.35% *K*; 44.7% *U*; 36.1% *O*; 9.57% *V*; 2.26% *H*

21. Calculate the per cent composition of each of the following to three significant figures:

(a) Acetic acid, CH_3CO_2H
Ans. 40.0% *C*; 53.3% *O*; 6.67% *H*

(b) Vitamin C, ascorbic acid, $C_6H_8O_6$
Ans. 40.9% *C*; 54.5% *O*; 4.54% *H*

(c) Trinitrotoluene, TNT, $C_6H_2(CH_3)(NO_2)_3$
Ans. 37.0% *C*; 2.2% *H*; 42.3% *O*; 18.5% *N*

(d) Codeine, $C_{18}H_{21}NO_3$
Ans. 72.2% *C*; 7.02% *H*; 4.68% *N*; 16.0% *O*

(e) Aspirin, $C_6H_4(CO_2H)(CO_2CH_3)$
Ans. 60.0% *C*; 4.44% *H*; 35.6% *O*

22. Determine each of the following to three significant figures:

(a) The per cent of water in plaster of paris, $CaSO_4 \cdot \frac{1}{2}H_2O$ *Ans.* 6.20%

(b) The per cent of ammonia in $[Co(NH_3)_6]Cl_3$
Ans. 38.2%

(c) The per cent of P_2O_5 in $(NH_4)_2HPO_4$
Ans. 53.8%

23. From the per cent composition data given, work out the empirical formulas of the following compounds:

[S](a) Potassium bromide: K, 32.85%; Br, 67.15%
Ans. KBr

(b) Nitrogen dioxide: N, 30.4%; O, 69.6%
Ans. NO_2

(c) Carbon disulfide: C, 15.8%; S, 84.2%
Ans. CS_2

[S](d) Aluminum oxide: Al, 52.9%; O, 47.1%
Ans. Al_2O_3

(e) Formaldehyde: C, 40.0%; H, 6.7%; O, 53.3%
Ans. CH_2O

(f) Cryolite: Na, 32.79%; Al, 13.02%; F, 54.19%
Ans. Na_3AlF_6

(g) Magnesium diphosphate: Mg, 21.84%; P, 27.83%; O, 50.33% *Ans.* $Mg_2P_2O_7$

[S]24. (a) What is the empirical formula of hydrazine, which contains 87.5% N and 12.5% H?
Ans. NH_2

[S](b) The molecular weight of hydrazine is 32. What is its molecular formula? *Ans.* N_2H_4

[S]25. A 4.00-g sample of an oxide of cobalt contains 2.937 g of cobalt. What is the empirical formula of this oxide? *Ans.* Co_3O_4

26. Vinyl chloride is a compound containing carbon, hydrogen, and chlorine. What is its simplest formula if it contains 38.4% carbon and 4.80% hydrogen? *Ans.* C_2H_3Cl

27. Dichloroethane, a compound containing carbon, hydrogen, and chlorine, is often used for dry-cleaning. The molecular weight of dichloroethane is 99. If a sample of dichloroethane is found to contain 24.3% carbon and 71.6% chlorine, what are its empirical and molecular formulas?
Ans. CH_2Cl; $C_2H_4Cl_2$

28. (a) What is the empirical formula of a compound that contains 85.6% carbon and 14.4% hydrogen? *Ans.* CH_2

(b) The molecular weight of the compound is 84. What is its molecular formula? *Ans. C_6H_{12}*

29. A 0.709-g sample of iron reacts with chlorine to give 1.610 g of a compound containing iron and chlorine. Identify the product and write the balanced equation for the reaction.

$$Ans.\ Fe + Cl_2 \longrightarrow FeCl_2$$

30. The 1.610-g sample of iron(II) chloride described in Problem 29 reacts with another 0.450 g of chlorine to give another iron chloride. Show by calculation the empirical formula of this compound.

Ans. $FeCl_3$

31. A 161-g sample of chromium(III) oxide contains 110 g of chromium. What is the empirical formula of this oxide? *Ans. Cr_2O_3*

s 32. The mineral gypsum contains 20.91% water and 79.09% $CaSO_4$. What is the empirical formula of gypsum? *Ans. $CaSO_4 \cdot 2H_2O$*

33. A sample containing potassium, chlorine, and oxygen weighs 3.22 g. As the sample is heated, it decomposes with evolution of oxygen. The residue consists of 1.96 g of KCl. What is the empirical formula of the sample? *Ans. $KClO_3$*

34. Nicotine contains 74.9% carbon, 8.7% hydrogen, and 17.3% nitrogen. It is known that this compound contains two nitrogen atoms per molecule. What are the empirical and molecular formulas of nicotine? *Ans. C_5H_7N; $C_{10}H_{14}N_2$*

s 35. The reaction of chromium metal with H_3PO_4 produces hydrogen, H_2, and a green solid containing 35.38% Cr, 21.07% P, and 43.55% O. Identify the product and write the balanced equation for the reaction.

$$Ans.\ 2Cr + 2H_3PO_4 \longrightarrow 3H_2 + 2CrPO_4$$

36. The reaction of Li_2CO_3 with Co_2O_3 at 1200° produces CO_2 and a black compound containing 7.09% Li, 60.2% Co, and 32.7% O. Identify the product and write the balanced equation for the reaction.

$$Ans.\ Li_2CO_3 + Co_2O_3 \longrightarrow 2LiCoO_2 + CO_2$$

37. The reaction of $Mg(OH)_2$ with SiO_2 at 1200° yields water, H_2O, and a white material containing 34.55% Mg, 19.96% Si, and 45.48% O. Identify the product and write the balanced equation for the reaction.

$$Ans.\ 2Mg(OH)_2 + SiO_2 \longrightarrow Mg_2SiO_4 + 2H_2O$$

38. A 0.00138-mol sample of atoms of an element weighs 0.164 g. What is the atomic weight of the element? Which element is it?

Ans. At. wt. 119; Sn

39. Fungal laccase, a blue protein found in wood-rotting fungi, is approximately 0.39% copper. If a laccase molecule contains four copper atoms, what is its approximate molecular weight? *Ans. 65,000*

s 40. What mass of calcium chloride, $CaCl_2$, contains 17.8 g of chlorine? *Ans. 27.9 g*

41. What mass of hydrogen is contained in 0.971 g of ammonium nitrate, NH_4NO_3?

Ans. 2.43×10^{-2} g

42. A 4.13-mmol sample of a compound of boron and hydrogen weighs 0.114 g. What is the molecular weight of the compound? *Ans. 27.6*

s 43. How many moles of phosphorus(V) sulfide are produced when 0.50 mol of S_8 reacts according to the following equation?

$$4P_4 + 5S_8 \longrightarrow 4P_4S_{10}$$

Ans. 0.40 mol

44. How many moles of N_2 are produced by the decomposition of 0.219 mol of NaN_3 according to the following equation?

$$3NaN_3 \xrightarrow{\triangle} Na_3N + 4N_2$$

Ans. 0.292 mol

s 45. How many moles of HF are produced by the reaction of 1.5×10^{23} H_2 molecules in the following reaction?

$$H_2 + F_2 \longrightarrow 2HF$$

Ans. 0.50 mol

46. How many molecules of I_2 are required to react with 0.360 mol of $Na_2S_2O_3$ in the following reaction?

$$2Na_2S_2O_3(aq) + I_2(s) \longrightarrow$$
$$Na_2S_4O_6(aq) + 2NaI(aq)$$

Ans. 1.08×10^{23}

47. How many moles of NaN_3 are required to produce 25.0 g of N_2 by the reaction shown in Problem 44?

Ans. 0.670 mol

48. How many moles of P_4 are required to react with 16 g of sulfur by the reaction shown in Problem 43?

Ans. 5.0×10^{-2} mol

s 49. How many grams of O_2 will be produced by the

thermal decomposition of 7.79 mol of potassium nitrate, KNO_3?

$$2KNO_3 \xrightarrow{\Delta} 2KNO_2 + O_2$$

Ans. 125 g

50. What mass of calcium carbonate is required to produce 71 kg of quicklime, CaO?

$$CaCO_3(s) \xrightarrow{\Delta} CaO(s) + CO_2(g)$$

Ans. 1.3 × 10² kg

51. Calculate the mass of hydrogen formed when 25 g of the active metal aluminum reacts with excess HCl by the following equation:

$$2Al + 6HCl \longrightarrow Al_2Cl_6 + 3H_2$$

Ans. 2.8 g

s 52. What mass of ammonia, NH_3, is required to react with 1.33 kg of H_3PO_4 according to the following equation?

$$2NH_3 + H_3PO_4 \longrightarrow (NH_4)_2HPO_4$$

Ans. 0.461 kg

53. Propane, C_3H_8, burns in air, reacting with oxygen of the air to give carbon dioxide, CO_2, and water. Write the balanced equation for the reaction and calculate the mass of CO_2 produced when 42.0 g of propane is burned.

Ans. $C_3H_8(g) + 5O_2(g) \longrightarrow 3CO_2(g) + 4H_2O(g)$;
126 g CO_2

s 54. How many moles of nitric acid, HNO_3, are required to react with the calcium hydroxide, $Ca(OH)_2$, in 250 ml of a 0.1 M $Ca(OH)_2$ solution?

$$2HNO_3(aq) + Ca(OH)_2(aq) \longrightarrow$$
$$Ca(NO_3)_2(aq) + 2H_2O(l)$$

Ans. 5 × 10⁻² mol

55. How many moles of chlorine dioxide, ClO_2, are required to react with the sodium hydroxide present in 2.34 l of a 1.38 M NaOH solution?

$$2ClO_2(g) + 2NaOH(aq) \longrightarrow$$
$$NaClO_2(aq) + NaClO_3(aq) + H_2O(l)$$

Ans. 3.23 mol

s 56. What mass of silver nitrate, $AgNO_3$, is required to react exactly with the calcium chloride in 25.00 ml of a 0.5285 M solution of $CaCl_2$?

$$2AgNO_3(aq) + CaCl_2(aq) \longrightarrow$$
$$2AgCl(s) + Ca(NO_3)_2(aq)$$

Ans. 4.489 g

57. What mass of the active metal magnesium is required to react exactly with the hydrochloric acid in a 125.0 ml sample of a 0.2110 M solution of HCl?

$$Mg(s) + 2HCl(aq) \longrightarrow MgCl_2(aq) + H_2(g)$$

Ans. 0.3205 g

58. What mass of CO_2 is produced by the reaction of 8.797 l of a 2.7 M solution of H_2SO_4 with an excess of $CaCO_3$?

$$CaCO_3(s) + 2H_2SO_4(aq) \longrightarrow$$
$$Ca(HSO_4)_2(aq) + CO_2(g) + H_2O(l)$$

Ans. 5.2 × 10² g

s 59. An excess of barium chloride reacted with 50.0 ml of a dilute solution of sulfuric acid, producing 0.482 g of $BaSO_4$. What is the molarity of the sulfuric acid solution?

$$BaCl_2(aq) + H_2SO_4(aq) \longrightarrow$$
$$BaSO_4(s) + 2HCl(aq)$$

Ans. 0.0413 M

60. What is the concentration of barium hydroxide in a solution formed by the reaction of 0.1700 g of barium with enough water to give 100.0 ml of solution?

$$Ba(s) + 2H_2O(l) \longrightarrow Ba(OH)_2(aq) + H_2(g)$$

Ans. 1.238 × 10⁻² M

s 61. The nitric acid in 125 ml of a 0.600 M solution reacts exactly with the potassium hydroxide in 47.0 ml of KOH solution. What is the molar concentration of KOH?

$$KOH(aq) + HNO_3(aq) \longrightarrow KNO_3(aq) + H_2O(l)$$

Ans. 1.60 M

62. The lead nitrate, $Pb(NO_3)_2$, in 25.49 ml of a 0.1338 M solution reacts with all the aluminum sulfate, $Al_2(SO_4)_3$, in 25.00 ml of solution. What is the molar concentration of the $Al_2(SO_4)_3$ in the original $Al_2(SO_4)_3$ solution?

$$3Pb(NO_3)_2(aq) + Al_2(SO_4)_3(aq) \longrightarrow$$
$$3PbSO_4(s) + 2Al(NO_3)_3(aq)$$

Ans. 4.547 × 10⁻² M

References

"The Poisonous Nature of Carbon Dioxide (128 Years Ago)," F. L. Pilar, *J. Chem. Educ.*, **52,** 791 (1975).

"The Changing Role of the Symbol in the Evolution of Chemical Notation," D. H. Rouvray, *Endeavour,* **1,** 23 (1977).

"The Chemical Formula: Part 1, Development," D. Kolb, *J. Chem. Educ.*, **55,** 44 (1978).

"The Chemical Formula: Part II, Determination," D. Kolb, *J. Chem. Educ.*, **55,** 109 (1978).

"Avogadro's Number: Early Values of Loschmidt and Others," R. M. Hawthorne, Jr., *J. Chem. Educ.*, **47,** 751 (1970).

"History of the Chemical Sign Language," R. Winderlich (transl. by R. E. Oesper), *J. Chem. Educ.*, **30,** 58 (1953).

The Mole Concept in Chemistry, W. F. Kieffer, Reinhold Publishing Corp., New York, 1962 (paperback).

"Size and Shape of a Molecule," M. J. Demchik and V. C. Demchik, *J. Chem. Educ.*, **48,** 770 (1971).

"Avogadro's Hypothesis and the Duhemian Pitfall," R. L. Causey, *J. Chem. Educ.*, **48,** 365 (1971).

"Atomic Weights of the Elements; Changes in Atomic Weight Values," (Report of the Inorganic Chemistry Division Commission on Atomic Weights, International Union of Pure and Applied Chemistry), *Pure and Applied Chemistry,* **51** (2), 405 (1979).

"Metrology: A More Accurate Value for Avogadro's Number," A. L. Robinson, *Science,* **185,** 1037 (1974).

"The Mole," D. Kolb, *J. Chem. Educ.*, **55,** 728 (1978).

"A Famous Hypothesis," T. I. Williams, *Endeavour,* January, 1976, p. 2.

3

Applications of Chemical Stoichiometry

Chemistry is a quantitative science. Both the study and use of the principles of chemistry involve a great deal of calculation. However, the purpose of these calculations is not merely to get a number, but to provide a quantitative prediction or description of how materials behave.

3.1 Chemical Calculations

Although the categories overlap extensively, it is helpful to separate chemical calculations into general categories according to the objectives of the calculations.

One category involves the calculation of quantities of reactants and products of a chemical reaction based upon the balanced equation representing the reaction. These are calculations of the type introduced in Chapter 2. Determination of the masses of reactants required for a reaction, choosing the amount of reactant necessary to give a desired amount of product, calculation of the concentration of a solution, determination of per cent composition as a check for purity, and identification of the empirical and molecular formula of an unknown product all fall into this category. Calculations such as these are generally referred to as **chemical stoichiometry**. In this chapter, we will be primarily concerned with a detailed treatment of this category.

A second category of calculation includes those calculations used to determine the utility of a theory. Most theories of chemical behavior contain quantitative concepts that can be used to predict the behavior of materials from the assumptions used to derive the theory. Agreement of the calculated properties with those determined experimentally provides strong support for the validity of the theory. Although we will not attempt to prove any new theories, examples showing why chemists accept various current theories or models of chemical behavior will be found in several chapters. We shall see in Chapter 4, for example, that the model proposed to explain the structure and behavior of atoms leads to equations that can be used to calculate the energies of light emitted by atoms under certain conditions. The model is regarded as a valid description of the structure of atoms, in part because the values calculated agree with the results of experiment. Similar quantitative considerations have been applied to the kinetic-molecular model of the behavior of gases (Chapter 10). Since the equations derived from the assumptions of the theory can be used to predict correctly the quantitative behavior of gases, the kinetic-molecular model is taken as a satisfactory description of the nature of a gas.

A third category of calculation is that of converting experimental measurements into information about a chemical system or determining properties of a chemical system under a given set of conditions. For example, measurements of the density of a gas under specific conditions can be used to determine its molecular weight. The dimensions of molecules can be calculated from the relative intensities of the x rays that they scatter. The rate at which glucose is metabolized by the body can be determined from the amount of copper sulfate that reacts with the glucose remaining in the blood at measured times after a patient eats a known (measured) amount of glucose. The energy available from a lead-acid battery (an electrochemical cell) may be calculated from the density of the acid in the battery. All of these calculations convert experimental data into information about the chemical or physical properties of the systems involved. Examples of calculations to determine the properties of a chemical system include evaluation of the pressure of a known amount of a particular gas at a given temperature, the amount of heat produced by combustion of methane (natural gas), the freezing point of a mixture of the antifreeze ethylene glycol and water, the amount of ammonia produced from hydrogen and nitrogen under given conditions, and the amount of heat necessary to break a chemical bond. Calculations of these types will be discussed in future chapters.

3.2   Chemical Calculations Involving Stoichiometry

Stoichiometry involves the calculation of both material balances and energy balances in a chemical system. In this chapter, we will limit the discussion to material balances; the subject of energy balances will be discussed in future chapters.

Calculations of material balances in a chemical system make possible the determination of the quantity of a reactant or of a product from a known (measured) quantity of one of the other reactants or products in a chemical reaction. As we saw in Chapter 2, such calculations depend on the Law of Conservation of Mass and make use of the balanced equation that describes the reaction. In this present chapter, stoichiometry calculations will be discussed in more detail, beginning with an additional example of the most fundamental sort and then progressing to a variety of examples of calculations that involve several kinds of additional information about the reaction.

Most stoichiometry problems involve the basic transformations or conversions illustrated below. The arrows indicate that one quantity is converted to the next.

$$\boxed{\text{Quantity A}} \longrightarrow \boxed{\text{Moles A}} \longrightarrow \boxed{\text{Moles B}} \longrightarrow \boxed{\text{Quantity B}}$$

These conversions, which were introduced in Sections 2.5, and 2.9–2.12, involve three steps, each of which may require one or more unit conversion factors: (1) conversion of a measured quantity of a material (for example, a quantity in grams) into the number of moles of that material, (2) determination of how many moles of a second material are equivalent to the number of moles of the first, and (3) conversion of the number of moles of the second material into a quantity of the second material, commonly in terms of the mass of the material or the volume of a solution of known concentration.

EXAMPLE   **What volume of a 0.750 M solution of hydrochloric acid can be prepared from the HCl produced by the reaction of 25.0 g of NaCl with an excess of sulfuric acid?**

$$NaCl(s) + H_2SO_4(l) \longrightarrow HCl(g) + NaHSO_4(s)$$

In order to determine the volume of the 0.750 M solution, we need the number of moles of HCl(g) that are available to make the solution. This requires the following steps, each of which requires a single unit conversion factor:

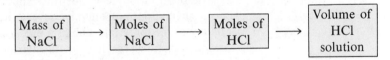

Step 1. Convert the quantity of NaCl in grams to moles of NaCl. As described in Section 2.5, this conversion requires calculating the formula weight of NaCl, 58.44; thus 1 mol NaCl = 58.44 g NaCl. The conversion is then

$$25.0 \text{ g NaCl} \times \frac{1 \text{ mol NaCl}}{58.44 \text{ g NaCl}} = 0.428 \text{ mol NaCl}$$

Step 2. Convert moles of NaCl to moles of HCl. The chemical equation indicates that one mole of HCl is produced per mole of NaCl that reacts. The conversion factor is 1 mol HCl/1 mol NaCl (Section 2.9).

$$0.428 \text{ mol NaCl} \times \frac{1 \text{ mol HCl}}{1 \text{ mol NaCl}} = 0.428 \text{ mol HCl}$$

Step 3. Convert moles of HCl to volume of a 0.750 M solution of HCl. As shown in Section 2.11,

$$\text{Molarity } (M) = \frac{\text{moles of solute}}{\text{liters of solution}}$$

This can be rearranged to solve for the volume of solution.

$$\text{Liters of solution} = \text{moles of solute} \times \frac{1}{\text{molarity}}$$

Since a 0.750 M solution contains 0.750 mole HCl per exactly one liter (0.750 mol HCl/1 ℓ), the reciprocal of the molarity ($1/M$) would be expressed as

$$\frac{1}{\text{molarity}} = \frac{1\ \ell}{0.750\ \text{mol HCl}}$$

Thus

$$0.428\ \text{mol HCl} \times \frac{1\ \ell}{0.750\ \text{mol HCl}} = 0.571\ \ell$$

• • • • •

This example and the examples that follow illustrate a basic concept of stoichiometry. **The conversion between the amounts of two different species in a chemical system is based upon the relative numbers of atoms or molecules of the two species; that is, it is based on the relative number of moles (units of 6.022 × 10²³ atoms or molecules) of the two species.** This follows from the fact that in a reaction a given number of molecules or atoms of one type is rearranged into or combined with some number of molecules or atoms of another type. The relative number of atoms or molecules (or the relative numbers of moles) of the various species participating in the reaction is provided by the chemical equation that describes the system (Section 2.9).

Although Example 1 illustrates the basic flow of any stoichiometry problem, such calculations often involve several additional steps. These may result from the nature of the material (such as a mixture rather than a pure compound) used in a reaction or may result from the various ways of measuring out a quantity of material (measuring a volume rather than a mass, for example). In such cases, the additional steps usually involve relating moles to measured quantities.

3.3   Common Operations in Stoichiometry Calculations

Conversion of a measured quantity of a substance to moles of the substance, or the reverse conversion, may involve one or more of the unit conversion factors discussed below. Careful attention to units and their cancellation will be a great help in correctly setting up the calculations.

▶ **1. INTERCONVERSION BETWEEN MASS OF A SUBSTANCE AND MOLES.** The number of moles in a given mass of a pure substance or the mass of a pure substance corresponding to a given number of moles can be determined as described in Section 2.5 and illustrated in Example 1 of Section 3.2.

▶ **2. INTERCONVERSION BETWEEN MASS OF A MIXTURE AND MASS OF A COMPONENT.** The per cent by mass of a component, A, in a sample of a mixture or a compound is given by the expression

$$\text{Per cent A} = \frac{\text{mass A}}{\text{mass sample}} \times 100$$

where 100 is an exact multiplier. It is often convenient to use the fraction of component A per 100 g of sample (g A/100 g sample) when using per cent by mass in problems. The mass of a component per 100 g of a sample is numerically equal to the per cent of component by mass. For example, a sample of coal that contains 0.23% sulfur by mass contains 0.23 g of sulfur per 100 g of coal.

The mass of A or the mass of the sample can be determined from the following expressions:

$$\text{Mass A} = \frac{\text{g A}}{100 \text{ g sample}} \times \text{mass sample}$$

$$\text{Mass sample} = \frac{100 \text{ g sample}}{\text{g A}} \times \text{mass A}$$

One way of expressing the amount of solute in a solution is to report the per cent composition of the solution by mass. Since a solution is a homogeneous mixture (Section 2.11), this conversion can be used both with solid mixtures and with solutions.

▶ **3. INTERCONVERSION BETWEEN VOLUME AND MASS USING DENSITY.** In Section 1.18 density was defined as the mass of a sample divided by its volume (density = mass/volume). The equation may be rearranged to find the mass of a given volume of a substance of known density (mass = density × volume), or to give the volume of a given mass of a substance of known density (volume = mass/density).

EXAMPLE   **Calculate the mass of hydrogen chloride, HCl, in 100.0 ml of hydrochloric acid, a solution prepared by dissolving HCl in water. The solution has a density of 1.19 g/cm³ and contains 37.23% HCl by mass.**

Two steps are required to solve this problem: a conversion from volume and density to mass of a mixture (the solution); then a conversion from the mass of a mixture to the mass of a pure substance.

$$\boxed{\begin{array}{c}\text{Volume of}\\\text{solution}\end{array}} \longrightarrow \boxed{\begin{array}{c}\text{Mass of}\\\text{solution}\end{array}} \longrightarrow \boxed{\begin{array}{c}\text{Mass of}\\\text{HCl}\end{array}}$$

Step 1.   (Paragraph 3, this section) Since 1 ml is the same as 1 cm³, we will express the volume in cm³.

$$100.0 \text{ cm}^3 \text{ solution} \times \frac{1.19 \text{ g solution}}{1 \text{ cm}^3 \text{ solution}} = 119 \text{ g solution}$$

Step 2.   (Paragraph 2, this section) A 37.23% HCl solution by mass contains 37.23 g HCl per 100 g solution.

$$119 \text{ g solution} \times \frac{37.23 \text{ g HCl}}{100 \text{ g solution}} = 44.3 \text{ g HCl}$$

◆ ◆ ◆ ◆ ◆

▶ **4. INTERCONVERSION BETWEEN VOLUME OF SOLUTION AND MOLES OF SOLUTE USING MOLARITY.** The number of moles of solute in a given volume of a solution of known concentration, the concentration of a solution, and the volume of a solution containing a given number of moles of solute can each be determined as described in Sections 2.11 and 2.12.

3.4   Applications of Chemical Stoichiometry

Many of the stoichiometry problems that a student sees at the beginning of a chemistry course appear formidable. However, the majority of these problems involve nothing more complicated than the unit conversions noted above. It is true that there may be several of these conversions in a given problem and getting them in the correct order requires a little practice, but if you first plan and write down the individual steps, the difficulties are reduced considerably.

EXAMPLE 1   How much of the white pigment rutile, TiO_2, can be prepared from 379 g of an ore containing ilmenite, $FeTiO_3$, if the ore is 88.3% ilmenite by mass? TiO_2 is prepared by the reaction

$$2FeTiO_3 + 4HCl + Cl_2 \longrightarrow 2FeCl_3 + 2TiO_2 + 2H_2O$$

This calculation involves the following unit conversions:

$$\boxed{\begin{array}{c}\text{Mass of}\\\text{ore}\end{array}} \xrightarrow{2} \boxed{\begin{array}{c}\text{Mass of}\\FeTiO_3\end{array}} \xrightarrow{1} \boxed{\begin{array}{c}\text{Moles of}\\FeTiO_3\end{array}} \longrightarrow \boxed{\begin{array}{c}\text{Moles of}\\TiO_2\end{array}} \xrightarrow{1} \boxed{\begin{array}{c}\text{Mass of}\\TiO_2\end{array}}$$

(The numbers above the arrows indicate the paragraphs in Section 3.3 that relate to these conversions. The calculations for each step follow.)

$$379 \text{ g ore} \times \frac{88.3 \text{ g } FeTiO_3}{100 \text{ g ore}} = 335 \text{ g } FeTiO_3$$

$$335 \text{ g } FeTiO_3 \times \frac{1 \text{ mol } FeTiO_3}{151.7 \text{ g } FeTiO_3} = 2.21 \text{ mol } FeTiO_3$$

$$2.21 \text{ mol } FeTiO_3 \times \frac{2 \text{ mol } TiO_2}{2 \text{ mol } FeTiO_3} = 2.21 \text{ mol } TiO_2$$

$$2.21 \text{ mol } TiO_2 \times \frac{79.9 \text{ g } TiO_2}{1 \text{ mol } TiO_2} = 176 \text{ g } TiO_2$$

Thus 176 g of TiO_2 can be prepared from the 379-g sample of ore.

◆ ◆ ◆ ◆ ◆

EXAMPLE 2   The reaction of calcium hydroxide, $Ca(OH)_2$, with acetic acid, CH_3CO_2H, proceeds according to the equation

$$Ca(OH)_2 + 2CH_3CO_2H \longrightarrow Ca(CH_3CO_2)_2 + 2H_2O$$

How many grams of a sample that contains 75.0% $Ca(OH)_2$ by mass is required to react with the acetic acid, CH_3CO_2H, in 25.0 ml of a solution with a density of 1.065 g/ml and which contains 58.0% acetic acid by mass?

The following conversions are used in the calculation:

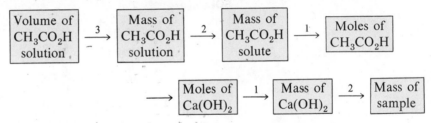

(The numbers above the arrows indicate the paragraphs in Section 3.3 that relate to these conversions. The calculations for each step follow.)

$$25.0 \text{ ml solution} \times \frac{1.065 \text{ g solution}}{1 \text{ ml solution}} = 25.62 \text{ g solution}$$

$$26.62 \text{ g solution} \times \frac{58.0 \text{ g CH}_3\text{CO}_2\text{H}}{100 \text{ g solution}} = 15.44 \text{ g CH}_3\text{CO}_2\text{H}$$

$$15.44 \text{ g CH}_3\text{CO}_2\text{H} \times \frac{1 \text{ mol CH}_3\text{CO}_2\text{H}}{60.0 \text{ g CH}_3\text{CO}_2\text{H}} = 0.2574 \text{ mol CH}_3\text{CO}_2\text{H}$$

$$0.2574 \text{ mol CH}_3\text{CO}_2\text{H} \times \frac{1 \text{ mol Ca(OH)}_2}{2 \text{ mol CH}_3\text{CO}_2\text{H}} = 0.1287 \text{ mol Ca(OH)}_2$$

$$0.1287 \text{ mol Ca(OH)}_2 \times \frac{74.1 \text{ g Ca(OH)}_2}{1 \text{ mol Ca(OH)}_2} = 9.536 \text{ g Ca(OH)}_2$$

$$9.536 \text{ g Ca(OH)}_2 \times \frac{100 \text{ g sample}}{75.0 \text{ g Ca(OH)}_2} = 12.7 \text{ g sample}$$

◆ ◆ ◆ ◆ ◆

EXAMPLE 3   **What is the molar concentration of acetic acid in the solution described in Example 2?**

This problem can be solved by the following conversions.

| Volume of CH$_3$CO$_2$H solution | →3 | Mass of CH$_3$CO$_2$H solution | →2 | Mass of CH$_3$CO$_2$H solute | →1 | Moles of CH$_3$CO$_2$H | →4 | Molarity of CH$_3$CO$_2$H solution |

(The numbers above the arrows indicate the paragraphs in Section 3.3 that relate to these conversions.) Steps 1 through 3 have been worked in Example 2, which shows that the sample contains 0.2574 mol of CH$_3$CO$_2$H.

$$\text{Molarity} = \frac{\text{moles of CH}_3\text{CO}_2\text{H}}{\text{liters of solution}}$$

$$M = \frac{0.2574 \text{ mol}}{0.0250 \text{ } \ell} = 10.3 \text{ mol}/\ell$$

Note that molarity is expressed in moles per liter; hence, the volume of the solution must be expressed in liters.

◆ ◆ ◆ ◆ ◆

EXAMPLE 4   **Sulfuric acid, H_2SO_4, may be prepared from FeS_2 by the following sequence of reactions:**

$$4FeS_2(s) + 11O_2(g) \longrightarrow 2Fe_2O_3(s) + 8SO_2(g) \tag{1}$$

$$2SO_2(g) + O_2(g) \longrightarrow 2SO_3(g) \tag{2}$$

$$SO_3(g) + H_2O(l) \longrightarrow H_2SO_4(l) \tag{3}$$

What mass of FeS_2 is required to prepare 1.00 ℓ of H_2SO_4 (density = 1.85 g/ml)?

The problem is solved by converting the volume of H_2SO_4 into the mass of FeS_2 necessary to produce it. The following conversions are required:

$$\boxed{\text{Volume of } H_2SO_4(l)} \xrightarrow{3} \boxed{\text{Mass of } H_2SO_4} \xrightarrow{1} \boxed{\text{Moles of } H_2SO_4} \longrightarrow \boxed{\text{Moles of } SO_3}$$

$$\longrightarrow \boxed{\text{Moles of } SO_2} \longrightarrow \boxed{\text{Moles of } FeS_2} \xrightarrow{1} \boxed{\text{Mass of } FeS_2}$$

(The numbers above the arrows indicate the paragraphs in Section 3.3 that relate to these steps.)

$$1.00 \, \ell \; H_2SO_4 \times \frac{1000 \text{ ml}}{1 \, \ell} \times \frac{1.85 \text{ g}}{1 \text{ ml}} = 1850 \text{ g } H_2SO_4$$

$$1850 \text{ g } H_2SO_4 \times \frac{1 \text{ mol } H_2SO_4}{98.0 \text{ g } H_2SO_4} = 18.88 \text{ mol } H_2SO_4$$

From Equation (3):

$$18.88 \text{ mol } H_2SO_4 \times \frac{1 \text{ mol } SO_3}{1 \text{ mol } H_2SO_4} = 18.88 \text{ mol } SO_3$$

From Equation (2):

$$18.88 \text{ mol } SO_3 \times \frac{2 \text{ mol } SO_2}{2 \text{ mol } SO_3} = 18.88 \text{ mol } SO_2$$

From Equation (1):

$$18.88 \text{ mol } SO_2 \times \frac{4 \text{ mol } FeS_2}{8 \text{ mol } SO_2} = 9.440 \text{ mol } FeS_2$$

$$9.440 \text{ mol } FeS_2 \times \frac{120.0 \text{ g } FeS_2}{1 \text{ mol } FeS_2} = 1.13 \times 10^3 \text{ g } FeS_2$$

◆ ◆ ◆ ◆ ◆

3.5   Theoretical Yield, Actual Yield, and Per Cent Yield

In the preceding sections, we have seen how to calculate the mass of product produced by a chemical reaction. These calculations are based on the assumptions that the reaction is the only one involved, that all of the product can be collected, and that all of the reactant is converted into product. The calculated amount of product based on these assumptions is called the **theoretical yield.** These conditions are generally not met in the laboratory, and the actual mass of product isolated from a reaction usually is less than the theoretical yield. The mass of product actually isolated is called the **actual yield** and is used in calculating the **per cent yield** of a reaction from the relationship

$$\text{Per cent yield} = \frac{\text{actual yield}}{\text{theoretical yield}} \times 100$$

EXAMPLE   A volume of 2.27 ml of $SiCl_4$ (density = 1.483 g/ml) reacts with $H_2S(g)$ giving $HSSiCl_3$ according to the reaction

$$SiCl_4(l) + H_2S(g) \longrightarrow HSSiCl_3(l) + HCl(g) \qquad (1)$$

If the HCl produced reacts with 8.267×10^{-3} mol of NaOH according to the equation

$$HCl + NaOH \longrightarrow NaCl + H_2O \qquad (2)$$

what is the per cent yield of $HSSiCl_3$?

To calculate per cent yield requires two values, (1) the theoretical yield of $HSSiCl_3$ and (2) the actual yield.

Theoretical Yield. The theoretical yield can be calculated from the following steps:

$$\boxed{\begin{array}{c}\text{Volume of}\\ SiCl_4\end{array}} \longrightarrow \boxed{\begin{array}{c}\text{Mass of}\\ SiCl_4\end{array}} \longrightarrow \boxed{\begin{array}{c}\text{Moles of}\\ SiCl_4\end{array}} \longrightarrow \boxed{\begin{array}{c}\text{Moles of}\\ HSSiCl_3\end{array}} \longrightarrow \boxed{\begin{array}{c}\text{Mass of}\\ HSSiCl_3\end{array}}$$

$$2.27 \text{ ml } SiCl_4 \times \frac{1.483 \text{ g } SiCl_4}{1 \text{ ml } SiCl_4} = 3.366 \text{ g } SiCl_4$$

$$3.366 \text{ g } SiCl_4 \times \frac{1 \text{ mol } SiCl_4}{169.9 \text{ g } SiCl_4} = 1.982 \times 10^{-2} \text{ mol } SiCl_4$$

$$1.982 \times 10^{-2} \text{ mol } SiCl_4 \times \frac{1 \text{ mol } HSSiCl_3}{1 \text{ mol } SiCl_4} = 1.982 \times 10^{-2} \text{ mol } HSSiCl_3$$

$$1.982 \times 10^{-2} \text{ mol } HSSiCl_3 \times \frac{167.5 \text{ g } HSSiCl_3}{1 \text{ mol } HSSiCl_3} = 3.32 \text{ g } HSSiCl_3$$

Thus the theoretical yield is 3.32 g.

Actual Yield. The actual yield of $HSSiCl_3$ may be calculated if it is assumed that the actual yield of NaCl from the reaction of HCl with NaOH is 100%. As will be seen in Chapter 14, this is indeed the case. From the moles of NaOH reacting in Equation (2), we can get the moles of HCl produced by Equation (1). From the moles of HCl produced, we can get the moles of $HSSiCl_3$ actually produced and, from this, the mass of $HSSiCl_3$ produced. The steps are

$$\boxed{\begin{array}{c}\text{Moles of} \\ \text{NaOH}\end{array}} \longrightarrow \boxed{\begin{array}{c}\text{Moles of} \\ \text{HCl}\end{array}} \longrightarrow \boxed{\begin{array}{c}\text{Moles of} \\ \text{HSSiCl}_3\end{array}} \longrightarrow \boxed{\begin{array}{c}\text{Mass of} \\ \text{HSSiCl}_3\end{array}}$$

From Equation (2):

$$8.267 \times 10^{-3} \text{ mol NaOH} \times \frac{1 \text{ mol HCl}}{1 \text{ mol NaOH}} = 8.267 \times 10^{-3} \text{ mol HCl}$$

From Equation (1):

$$8.267 \times 10^{-3} \text{ mol HCl} \times \frac{1 \text{ mol HSSiCl}_3}{1 \text{ mol HCl}} = 8.267 \times 10^{-3} \text{ mol HSSiCl}_3$$

$$8.267 \times 10^{-3} \text{ mol HSSiCl}_3 \times \frac{167.5 \text{ g HSSiCl}_3}{1 \text{ mol HSSiCl}_3} = 1.38 \text{ g HSSiCl}_3$$

Thus the actual yield of $HSSiCl_3$ is 1.38 g.

Per Cent Yield.

$$\text{Per cent yield HSSiCl}_3 = \frac{\text{actual yield HSSiCl}_3}{\text{theoretical yield HSSiCl}_3} \times 100$$

$$= \frac{1.38 \text{ g}}{3.32 \text{ g}} \times 100 = 41.6\%$$

◆ ◆ ◆ ◆ ◆

3.6   Limiting Reagents

The equation $4Na + O_2 \longrightarrow 2Na_2O$ indicates that exactly four moles of Na react per mole of O_2. If the mole ratio of reactants in this reaction is different from 4 to 1 then one of the reactants is present in excess and not all of it will be consumed. If, for example, we have five moles of Na and one mole of O_2, one mole of Na must remain unreacted at the completion of the reaction inasmuch as 1 mole of O_2 can react with only 4 moles of Na. The reactant that will be completely consumed is called the **limiting reagent.** This reactant limits the amount of product that can be formed. Other reactants are said to be present in excess.

To determine which reagent is the limiting reagent, calculate, in turn, the amount of product expected on the basis of the quantity of each reactant. The reactant that gives the smallest theoretical yield is the one that limits the amount of product actually formed.

EXAMPLE   A mixture of 5.0 g $H_2(g)$ and 10.0 g $O_2(g)$ is ignited. Water forms according to the following equation:

$$2H_2(g) + O_2(g) \longrightarrow 2H_2O(g)$$

Which reactant is limiting? How much water will be produced by the reaction?

First, assume that all the H_2 will react and calculate the theoretical yield of water, using the following steps:

$$\boxed{\begin{array}{c}\text{Mass of} \\ \text{H}_2\end{array}} \longrightarrow \boxed{\begin{array}{c}\text{Moles of} \\ \text{H}_2\end{array}} \longrightarrow \boxed{\begin{array}{c}\text{Moles of} \\ \text{H}_2\text{O}\end{array}} \longrightarrow \boxed{\begin{array}{c}\text{Mass of} \\ \text{H}_2\text{O}\end{array}}$$

$$5.0 \text{ g } H_2 \times \frac{1 \text{ mol } H_2}{2.0 \text{ g } H_2} = 2.5 \text{ mol } H_2$$

$$2.5 \text{ mol } H_2 \times \frac{2 \text{ mol } H_2O}{2 \text{ mol } H_2} = 2.5 \text{ mol } H_2O$$

$$2.5 \text{ mol } H_2O \times \frac{18 \text{ g } H_2O}{1 \text{ mol } H_2O} = 45 \text{ g } H_2O$$

Now, assume that all of the O_2 will react and calculate the theoretical yield of water, using the following steps:

Mass of O_2	→	Moles of O_2	→	Moles of H_2O	→	Mass of H_2O

$$10.0 \text{ g } O_2 \times \frac{1 \text{ mol } O_2}{32.0 \text{ g } O_2} = 0.3125 \text{ mol } O_2$$

$$0.3125 \text{ mol } O_2 \times \frac{2 \text{ mol } H_2O}{1 \text{ mol } O_2} = 0.625 \text{ mol } H_2O$$

At this point it should be apparent that O_2 is the limiting reactant since 10.0 g O_2 gives 0.625 mole of H_2O whereas 5.0 g H_2 would give 2.5 mole of H_2O if the hydrogen were all converted to water. The following expression gives the mass of water formed from 10.0 g (0.625 mol) of H_2O:

$$0.625 \text{ mol } H_2O \times \frac{18.0 \text{ g } H_2O}{1 \text{ mol } H_2O} = 11.2 \text{ g } H_2O$$

◆ ◆ ◆ ◆ ◆

It should be obvious from the discussions in this chapter that chemistry is a quantitative science. However, it should be emphasized that the objectives of the study and applications of chemistry are not merely the calculation of a great many different kinds of numbers. Rather, the principal objectives are to use the calculations as tools in the use of matter and in understanding the behavior of matter, as will be described in the remainder of this text.

Questions

1. What are the general categories of chemical calculations presented in this chapter?
2. How may calculations be useful in understanding the behavior of matter?
3. Define the following terms: stoichiometry, molarity, per cent yield, limiting reagent.
4. Begin with the expression for each of the following conversions, and derive the expression for the reverse conversion:

(a) Moles A to grams A
(b) Volume of a liquid of known density to mass
(c) Moles of solute to molar concentration (given volume of solution)
(d) Per cent A and mass of A to mass of sample containing A

5. Why is the actual yield of a reaction not always equal to the theoretical yield?
6. Derive a one-step conversion factor for the following:

(a) Moles A to mass of a mixture containing a known percentage of A

(b) Volume of pure A and its density to volume of a mixture of known density containing a known percentage of A

(c) Volume of a solution of known density with a known molarity of A to percentage A in the solution

Problems

[s]1. How many moles of calcium sulfate, $CaSO_4$, will be required to prepare 50.0 g of plaster of paris? Plaster of paris contains 93.8% $CaSO_4$ by mass.

Ans. 0.344 mol

2. How many grams of an insect repellent containing 14% N, N-dimethyl-metatoluamide, $C_9H_{13}N$, can be prepared from 1.5 mol of $C_9H_{13}N$?

Ans. 1.4×10^3 g

3. What is the volume of 4.88×10^{-2} mol of copper (density = 8.94 g/cm³)?   *Ans. 0.347 cm³*

[s]4. What volume of a solution of 37.0% HCl in water with a density of 1.181 g/ml (concentrated hydrochloric acid) can be prepared from 2.744 mol of HCl?

Ans. 229 ml

5. A 5.0-ml sample of human plasma with a density of 1.0299 g/ml was found to contain 0.478 g of dissolved protein and 0.077 g of dissolved nonprotein material. What are the per cents of protein, of nonprotein materials, and of all dissolved solids in the plasma?   *Ans. 9.3%; 1.5%; 10.8%*

6. What is the molarity of CO_2 in a solution formed by dissolving 10.0 g of CO_2 in enough water to make 1500 ml of solution?   *Ans. 0.15 M*

[s]7. How many grams of CaO are required for reaction with the HCl in 275 ml of a 0.523 M HCl solution? The equation for the reaction is

$$CaO + 2HCl \longrightarrow CaCl_2 + H_2O$$

Ans. 4.03 g

8. Limestone is almost pure calcium carbonate, $CaCO_3$. How much quicklime, CaO, can be prepared by roasting (heating) a metric ton, 1.000×10^3 kg, of limestone that contains 94.6% $CaCO_3$. $CaCO_3$ decomposes into CaO and CO_2 during the roasting process.

Ans. 5.30×10^2 kg

[s]9. Elemental phosphorus, P_4, is prepared by the following reaction:

$$2Ca_3(PO_4)_2 + 6SiO_2 + 10C \longrightarrow$$
$$6CaSiO_3 + 10CO + P_4$$

How much phosphorus can be prepared from a 374-g sample of an ore that is 75.9% $Ca_3(PO_4)_2$?

Ans. 56.7 g

10. Butane, C_4H_{10}, reacts with O_2 to form $CO_2(g)$ and $H_2O(g)$. What volume of oxygen (density = 1.43 g/ℓ) is required to combine with (burn) 37.0 g of C_4H_{10}?   *Ans. 92.8 ℓ*

[s]11. Calculate the mass of sodium nitrate required to produce 5.00 ℓ of O_2 (density = 1.43 g/ℓ) according to the reaction

$$2NaNO_3 \xrightarrow{\triangle} 2NaNO_2 + O_2$$

if the per cent yield of the reaction is 78.4%.

Ans. 48.4 g

12. A volume of 10.0 ml of phosphorus trichloride, PCl_3 (density = 1.574 g/cm³), reacts with water according to the equation

$$PCl_3 + 3H_2O \longrightarrow P(OH)_3 + 3HCl$$

What mass of HCl is produced and what mass of water is required for the reaction?

Ans. 12.5 g HCl; 6.19 g H_2O

[s]13. Benzene, $C_6H_6(l)$, combines with oxygen (burns) giving $CO_2(g)$ and $H_2O(g)$. How many liters of $CO_2(g)$, with a density of 1.428 g/ℓ, will be produced by burning 1.00 ℓ of benzene (density 0.88 g/ml)?   *Ans. 2.1×10^3 ℓ*

[s]14. A sample of a compound containing oxygen, boron, and fluorine decomposed, giving 152 ml of BF_3 (density = 2.99 g/ℓ), 150 ml of O_2 (density = 1.43 g/ℓ), and 76.2 ml of F_2 (density = 1.69 g/ℓ). What is the empirical formula of the compound?

Ans. O_2BF_4

15. Stibine, a compound of antimony and hydrogen, decomposes when heated. A sample of stibine decomposed to 0.136 g of Sb and 37.6 ml of H_2 (density = 0.090 g/ℓ). What is the empirical formula of stibine?   *Ans. SbH_3*

16. A compound found in cast iron is cementite, which contains iron and carbon. A 3.113-g sample of ce-

mentite was roasted (heated) in air producing 4.153 g of Fe_2O_3. The CO_2 gas, another product of the reaction, escaped. What is the empirical formula of cementite? *Ans. Fe_3C*

[s]17. Reaction of rhenium metal with Re_2O_7 gives a solid of metallic appearance which conducts electricity almost as well as copper. A 0.788-g sample of this material, which contains only rhenium and oxygen, was oxidized in an acidic solution of hydrogen peroxide. Addition of an excess of KOH gave 0.973 g of $KReO_4$. What is the equation for the reaction of Re with Re_2O_7? *Ans. $Re + 3Re_2O_7 \longrightarrow 7ReO_3$*

[s]18. Bronzes used in bearings are often alloys (solid solutions) of copper and aluminum. A 2.053-g sample of one such bronze was analyzed for its copper content by the sequence of reactions given below. The aluminum also dissolves, but we do not need to consider it because aluminum sulfate, $Al_2(SO_4)_3$, does not react with KI.

$$Cu + 2H_2SO_4 \longrightarrow CuSO_4 + 2H_2O + SO_2$$
$$2CuSO_4 + 4KI \longrightarrow 2CuI + I_2 + 2K_2SO_4$$
$$I_2 + 2Na_2S_2O_3 \longrightarrow Na_2S_4O_6 + 2NaI$$

If 35.11 ml of 0.8375 M $Na_2S_2O_3$ is required to react with the I_2 formed, what is the per cent Cu in the sample? *Ans. 91.02%*

19. A mixture of morphine ($C_{17}H_{19}NO_3$) and an inert solid is analyzed by combustion with O_2. The *unbalanced* equation for the reaction of morphine with O_2 is

$$C_{17}H_{19}NO_3 + O_2 \longrightarrow CO_2 + H_2O + NO_2$$

The inert solid does not react with O_2. If 4.000 g of the mixture yields 8.72 g of CO_2, calculate the per cent morphine by mass in the mixture.

Ans. 83.1%

[s]20. What mass of a sample that is 98.0% sulfur would be required in the production of 75.0 kg of H_2SO_4 by the following reaction sequence?

$$S_8 + 8O_2 \longrightarrow 8SO_2$$
$$2SO_2 + O_2 \longrightarrow 2SO_3$$
$$SO_3 + H_2O \longrightarrow H_2SO_4 \quad \textit{Ans. 25.0 kg}$$

21. For many years, the standard medical laboratory technique for determination of the concentration of calcium (as Ca^{2+}) in blood serum used the following reaction sequence involving ammonium oxalate, $(NH_4)_2C_2O_4$, and potassium permanganate, $KMnO_4$:

$$Ca^{2+}(aq) + (NH_4)_2C_2O_4(s) \longrightarrow$$
$$CaC_2O_4(s) + 2NH_4^+(aq)$$
$$CaC_2O_4(s) + H_2SO_4(aq) \longrightarrow$$
$$H_2C_2O_4(aq) + CaSO_4(aq)$$
$$2KMnO_4(aq) + 5H_2C_2O_4(aq) + 3H_2SO_4(aq) \longrightarrow$$
$$K_2SO_4(aq) + 2MnSO_4(aq) + 10CO_2(g) + 8H_2O(l)$$

What is the molar concentration of calcium in a 2.0-ml serum sample if 0.68 ml of 2.44×10^{-3} M $KMnO_4$ is required for the final reaction? How many milligrams of calcium is contained in 1.00×10^2 ml of the serum?

Ans. 0.0021 M; 8.3 mg

[s]22. Zinc reacts with nitric acid according to the equation

$$Zn + 4HNO_3 \longrightarrow Zn(NO_3)_2 + 2H_2O + 2NO_2$$

If 1.05 g of Zn reacts with 2.50 ml of nitric acid (density = 1.503 g/cm^3), which reagent is the limiting reagent? What mass of $Zn(NO_3)_2$ is produced?

Ans. HNO_3; 2.82 g

23. Uranium may be isolated from the mineral pitchblende, which contains U_3O_8. The pitchblende in a 4.835-g sample of an ore containing uranium was subjected to the following sequence of reactions:

$$2U_3O_8 + O_2 + 12HNO_3 \longrightarrow$$
$$6(UO_2)(NO_3)_2 + 6H_2O$$
$$(UO_2)(NO_3)_2 + 4H_2O + H_3PO_4 \longrightarrow$$
$$(UO_2)HPO_4 \cdot 4H_2O + 2HNO_3$$
$$2(UO_2)HPO_4 \cdot 4H_2O \xrightarrow{\triangle} (UO_2)_2P_2O_7 + 9H_2O$$

What is the percentage of U_3O_8 in the ore by mass if 1.432 g of $(UO_2)_2P_2O_7$ is isolated? *Ans. 23.29%*

24. Calculate the molarity of each of the following solutions:

(a) 13.0 g of KOH in 5.0 l of solution
Ans. 4.6×10^{-2} M

(b) 20.5 g of H_2SO_4 in 1.0 l of solution
Ans. 0.21 M

[s](c) 50.0 mg of HNO_3 in 15.1 ml of solution
Ans. 5.26×10^{-2} M

[s](d) 7.0 millimoles (mmol) of I_2 in 100 ml of solution
Ans. 7.0×10^{-2} M

(e) 18.0 g of HCl in 75 ml of solution *Ans. 6.6 M*

25. What is the molar concentration of H_2O in pure water (density = 1.00 g/ml)? *Ans. 55.5 M*

[S] 26. What mass of solid NaOH of 97.0% purity would be required to prepare 250.0 ml of 0.350 *M* NaOH solution? *Ans. 3.61 g*

27. A 35.86-ml volume of a solution of CsOH reacts exactly with 21.34 ml of a 0.0542 *M* H_2SO_4 solution ($2CsOH + H_2SO_4 \longrightarrow Cs_2SO_4 + 2H_2O$). What is the concentration of the CsOH solution? *Ans. 0.0645 M*

28. What volume of 0.600 *M* HCl would be required to react completely with 2.50 g of sodium hydrogen carbonate, $NaHCO_3$?

$$NaHCO_3 + HCl \longrightarrow NaCl + H_2O + CO_2$$

Ans. 49.6 ml

[S] 29. Crystalline potassium hydrogen phthalate, $KHC_8H_4O_4$, is often used as a "standard" acid because it is easy to purify and to weigh. If 1.5428 g of this salt reacts with a solution of $Ca(OH)_2$, the reaction is complete when 42.37 ml of the solution has been added. What is the concentration of the $Ca(OH)_2$ solution?

$$2KHC_8H_4O_4 + Ca(OH)_2 \longrightarrow CaK_2(C_8H_4O_4)_2 + 2H_2O$$

Ans. 8.914 × 10⁻² M

30. In a common laboratory determination of the concentration of free chloride ion, Cl^-, in blood, the cells and dissolved protein in a blood sample are removed and $\frac{1}{10}$ (0.100) of the remaining serum is allowed to react with a $Hg(NO_3)_2$ solution. The reaction is stopped when all of the chloride ion has reacted according to the equation

$$2Cl^- + Hg(NO_3)_2 \longrightarrow HgCl_2 + 2NO_3^-$$

What is the concentration of Cl^- in a 1.50-ml blood sample that requires 1.62 ml of 0.00500 *M* $Hg(NO_3)_2$ solution for the reaction? *Ans. 0.108 M*

[S] 31. How many milliliters of concentrated sulfuric acid (a solution of H_2SO_4 with a density of 1.841 g/cm³ and which contains 98.0% H_2SO_4 by mass) is required to produce 1.000 ℓ of a 0.500 *M* H_2SO_4 solution? *Ans. 27.2 ml*

32. A 25-ml volume of a 0.100 *M* solution of $H_2Cr_2O_7$ is added to 75 ml of a 0.100 *M* solution of H_2FeCl_4. The two react according to the following equation:

$$H_2Cr_2O_7 + 6H_2FeCl_4 + 5H_2O \longrightarrow 2[Cr(H_2O)_6]Cl_3 + 6FeCl_3$$

Which reagent is the limiting reagent? What mass of $[Cr(H_2O)_6]Cl_3$ is produced? *Ans. H_2FeCl_4; 0.66 g*

33. Hydrogen peroxide may be prepared by the following reactions:

$$2NH_4HSO_4 \xrightarrow{\text{Elect.}} H_2 + (NH_4)_2S_2O_8$$
$$(NH_4)_2S_2O_8 + 2H_2O \longrightarrow 2NH_4HSO_4 + H_2O_2$$

What mass of NH_4HSO_4 is initially required to prepare 1.00 mol of H_2O_2? What mass of H_2O is required? *Ans. 2.30 × 10² g; 36.0 g*

34. Potassium perchlorate, $KClO_4$, may be prepared from KOH and Cl_2 by the following series of reactions:

$$2KOH + Cl_2 \longrightarrow KCl + KClO + H_2O$$
$$3KClO \longrightarrow 2KCl + KClO_3$$
$$4KClO_3 \longrightarrow 3KClO_4 + KCl$$

What mass of $KClO_4$ can be prepared from 25.0 g of KOH? *Ans. 7.72 g*

[S] 35. If 1.00 g of sodium carbonate, Na_2CO_3, reacts with sulfuric acid according to the following equation and produces 1.301 g of sodium sulfate, Na_2SO_4, what is the per cent yield?

$$Na_2CO_3 + H_2SO_4 \longrightarrow Na_2SO_4 + CO_2 + H_2O$$

Ans. 97.1%

36. When 1.11 g of vanadium is dissolved in HCl, 0.066 g of H_2 is produced. Calculate the value of x for the compound VCl_x produced according to the following equation:

$$V + xHCl \longrightarrow VCl_x + \frac{x}{2}H_2$$

Ans. x = 3; VCl_3

37. A 25-ml volume of a solution of triethylaluminum, $Al(C_2H_5)_3$, a catalyst used in the production of polyethylene, was allowed to react with 25.0 ml of a 0.103 *M* HCl solution.

$$Al(C_2H_5)_3 + 3HCl \longrightarrow AlCl_3 + 3C_2H_6$$

The $Al(C_2H_5)_3$ was the limiting reagent; hence, some HCl remained unreacted. The unreacted HCl was allowed to react with a 0.142 *M* solution of NH_3.

$$HCl + NH_3 \longrightarrow NH_4Cl$$

The reaction was complete when 16.75 ml of the NH_3 solution had been added. What was the concentration of $Al(C_2H_5)_3$ in the catalyst solution? *Ans. 2.6 × 10⁻³ M*

4

Structure of
the Atom and
the Periodic Law

A series of discoveries beginning in the latter part of the nineteenth century has modified the Daltonian concept of the atom. Atoms are no longer considered to be simple, compact bodies as supposed by Dalton; they are complex systems composed of a number of smaller particles. The only particles of the atoms we shall consider here, however, are electrons, protons, and neutrons.

Our present knowledge of the structure of atoms has made it possible to systematize the facts of chemistry in such a way that the subject is easier to understand and remember. Chemical reactions that occur between atoms, and the forces that hold atoms together in molecules, can be explained in terms of atomic structure.

The Structure of the Atom

4.1 Electrons

Much of the information concerning the structure of atoms resulted from experiments involving the passage of electricity through different gases at very low pressures. The apparatus used for experiments of this type, called a discharge tube, or **Crookes tube,** was developed by Sir William Crookes (1832–1919). Crookes took a gas-filled glass tube with electrodes sealed into both ends [Fig. 4-1(a)] and

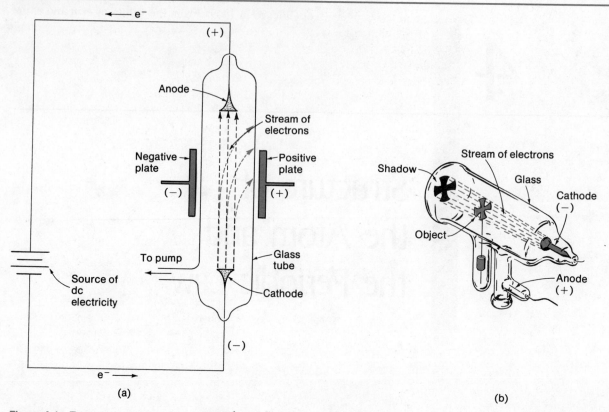

(a)

(b)

Figure 4-1 Two types of discharge tubes. (a) In the absence of a charge on the plates, electrons flow from the cathode to the anode (black path). When the plates are charged, the electron stream curves toward the positive plate (colored path). (b) A shadow being cast by an object in a stream of electrons.

passed an electric current through it. As gas was pumped out of the tube, a pressure was reached at which the remaining gas glowed. Continued pumping resulted in a pressure at which the gas ceased to glow. However, close observation revealed the glass at one end of the tube to be glowing. Crookes suggested that this glow was produced by negative particles, which he called **cathode rays,** passing from the negative electrode **(cathode)** and moving toward the positive electrode **(anode).** Those particles that did not strike the anode hit the glass and caused it to glow.

Cathode rays were shown to have the following characteristics: (1) They travel from the cathode to the anode in straight lines, as indicated by the fact that an object placed in their path casts a sharp shadow on the end of the tube [Fig. 4-1(b)]. (2) They cause a piece of metal foil to become hot after striking it for awhile. (3) They are deflected by magnetic and electrical fields [Fig. 4-1(a)]. The direction of the deflection is the same as that shown by negatively charged particles passing through such fields (a charged object will attract an object of opposite charge, whereas objects having charges of the same sign repel each other). These characteristics of cathode rays are best explained by assuming that they are streams of small negatively charged particles. These particles are called **electrons.**

In 1897, the English physicist Sir J. J. Thomson determined the ratio of the charge to mass, e/m, of electrons, and found e/m to be identical for all cathode-ray particles, irrespective of the kind of gas in the tube or the metal the electrodes are made of. The charge on an electron was measured by the American physicist R. A.

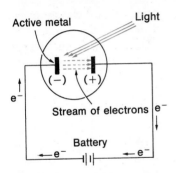

Figure 4-2 The photoelectric cell in an electric circuit. Light striking the active metal electrode causes electrons to be emitted and a current to flow.

Millikan at the University of Chicago in 1909. From the values of e/m and e, the mass of the electron was calculated to be 1/1837 of that of the hydrogen atom. Thus the mass of the electron is 0.00055 amu.

Quantitative evaluation of their charge and mass suggests that all electrons are identical. This observation that all electrons are identical, irrespective of their source or method of liberation from matter, and that they can be liberated from any kind of atom, proves quite conclusively that they are parts of all types of matter.

Electrons are emitted by certain metals when the metals are heated to high temperatures. This process is called **thermal emission** and is the source of electrons in vacuum tubes and cathode-ray tubes such as television picture tubes. Electrons are also emitted from very active metals such as cesium, sodium, and potassium when they are exposed to light (Fig. 4-2). This type of emission is known as the **photoelectric effect** and is the basis of the "electric eye," or photoelectric cell.

4.2 Protons

In 1886, Eugen Goldstein, a German physicist, using a discharge tube with a perforated cathode, observed that another kind of "ray" was being emitted from the anode and passing through the holes in the cathode. Wilhelm Wien, in 1898, showed these rays to be positively charged particles; they are today called **positive ions.** The ratio of the charge on one of the particles to its mass was found to be much smaller than that for the electron and to vary with the kind of gas in the tube. Since the ratio varied, it follows that either the charge varied or the mass varied or both. Actually, it was found that both the charge and the mass varied. Measurements showed the charge on each particle to be either a positive unit charge or some whole-number multiple of this unit. The unit positive charge was shown to be equal to that on the electron but opposite in sign. The mass of the positive particle was found to be less when hydrogen was used in the discharge tube than when any other gas was employed. From the values of e and e/m for the positive particles when hydrogen is used, the value of m can be calculated; the mass is 1.0073 amu. This particle is called the **proton** and is one of the basic units of structure of all atoms.

The formation of positively charged particles from neutral atoms or molecules of whatever gas is used in a discharge tube is caused by the loss of electrons by these neutral atoms or molecules when they are struck by high-speed cathode rays. These charged atoms are known as **ions.** From the fact that any neutral atom can be made to form positive ions by the loss of one or more electrons, it follows that every atom must contain one or more positive units. The simplest atom is that of hydrogen. It contains one proton (hydrogen ion) and one electron.

4.3 Neutrons

In 1932, the English physicist James Chadwick observed the third basic unit of the atom. He noted that when atoms of beryllium (and other elements) are bombarded with high-velocity helium atoms with all electrons removed (α particles), uncharged particles are emitted. Chadwick called these neutral particles **neutrons;** they were determined to have a mass of 1.0087 amu. Under certain conditions a neutron may disintegrate and form a proton and an electron. This may be the origin of the electrons composing the β rays, which are emitted by radioactive elements.

4.4 Summary of Three Basic Particles in an Atom

In the preceding sections we have seen that three basic particles are present in an atom—electrons, protons, and neutrons (the ordinary hydrogen atom, which contains one proton, one electron, and no neutrons is the only exception).

Particle	Mass, amu	Charge
Electron	0.00055	−1
Proton	1.0073	+1
Neutron	1.0087	0

4.5 Radioactivity and Atomic Structure

The first conclusive evidence that atoms are complex rather than "indivisible," as stated in the original atomic theory of matter, came with the discovery of radioactivity by Antoine Becquerel, a French physicist, in 1896. **Radioactivity** is the term applied to the spontaneous decomposition of atoms of certain elements, such as radium and uranium, into other elements with the simultaneous production of so-called "rays" of particles. There are three types of these rays, and they can be studied and characterized by placing samples of radioactive material in a narrow hole bored in a block of lead and allowing the emitted rays to pass through a strong electric field (Fig. 4-3). One type of ray is observed to curve toward the negative part of the electric field and so must consist of positively charged particles. These are called **alpha (α) particles.** Experiment has shown that alpha particles have a mass of approximately 4 amu and a charge of +2. They are helium ions, i.e., helium atoms that have lost two electrons each. A second type of ray is deflected toward the positive part of the field, with a deflection much greater than that of the alpha particles. These facts indicate that these particles, called **beta (β)**

Figure 4-3 The effect of an electric field on rays from the radioactive substance radium. Beta (β) rays are electrons. Alpha (α) rays are nuclei of helium atoms. Gamma (γ) rays are similar to x rays.

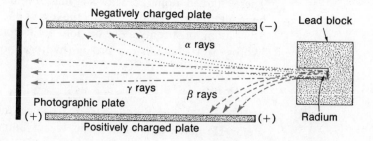

particles, are negatively charged and have much less mass than alpha particles. Beta particles have been shown to be high-speed electrons with velocities approaching that of the speed of light (about 186,000 mi/sec, or 3.00×10^8 m/sec). The third type of ray is not deflected at all when it passes through the electric field, indicating that the particles are neither negatively nor positively charged. These rays are similar to x rays but have higher energies than x rays (and higher energies than α particles and β particles as well). These high-energy particles are called **gamma (γ) rays.** Several other types of rays (radiation) are also known and will be discussed in Chapter 29.

Convincing proof that atoms contain protons was obtained in England by Ernest Rutherford in 1919. Using high-velocity α particles from radium as projectiles, he bombarded atoms such as nitrogen and aluminum and found that protons were ejected from the atoms as a result of these collisions. These experiments indicated quite definitely that the proton is a unit of the atom.

4.6   The Nuclear Atom

Although the experiments described above indicate that atoms can be described as assemblies of electrons, protons, and neutrons, they give no clue as to how these particles are arranged. The first insight into the architecture of atoms resulted from α-particle scattering experiments by Ernest Rutherford.

When Rutherford projected a beam of α particles from a radioactive source upon very thin gold foil, he found that most of the particles passed through the solid foil without deflection; only a few of them were diverted from their paths. From the results of a series of such experiments, Rutherford concluded that (1) the volume occupied by an atom must be largely empty space, as indicated by the fact that most of the α particles pass through the foil undeflected, and (2) there must be located within each atom a heavy, positively charged body (the nucleus). This follows because an abrupt change in path (as noted for a few α particles) of a relatively heavy and positively charged α particle can result only from its hitting or from its close approach to another particle (the nucleus) with a highly concentrated, positive charge. The atom, then, was presumed to consist of a very small, positively charged **nucleus,** in which most of the mass of the atom is concentrated, surrounded by a number of electrons necessary to produce an electrically neutral atom. This is Rutherford's nuclear theory of the atom, proposed in 1911. We still use this model today.

Rutherford determined the diameter of the nucleus to be about $\frac{1}{10,000}$ of the diameter of the atom (the diameter of an atom is of the order of 10^{-8} cm). Thus a nucleus is almost unbelievably small. If the nucleus of an atom were as large as the period at the end of this sentence, the atom would have a radius of about 50 yards.

From this same series of experiments, Rutherford found that the number of the positive charges on the nucleus (and thus the number of protons in the nucleus) is, for many of the lighter elements, approximately one-half the atomic weight of the element. The number of positive charges on the nucleus of an atom of a given element is equal to the **atomic number** of that element.

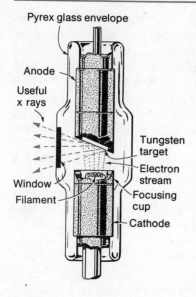

Pyrex glass envelope

Anode

Useful
x rays

Tungsten
target

Electron
stream

Window

Focusing
cup

Filament

Cathode

Figure 4-4 In the x-ray tube, electrons striking a target such as tungsten cause the production of x rays. In Moseley's experiment, various elements were used as targets.

Figure 4-5 Sunlight passing through a prism is separated into its component colors by refraction.

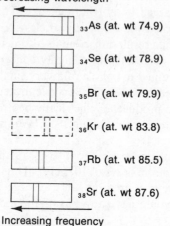

Decreasing wavelength

$_{33}$As (at. wt 74.9)

$_{34}$Se (at. wt 78.9)

$_{35}$Br (at. wt 79.9)

$_{36}$Kr (at. wt 83.8)

$_{37}$Rb (at. wt 85.5)

$_{38}$Sr (at. wt 87.6)

Increasing frequency

Figure 4-6 Diagram representing several x-ray spectra as found by Moseley. Note the greater shift in wavelength between Br and Rb. The element having atomic number 36, krypton, is a gas and could not be used in an x-ray tube.

4.7   Moseley's Determination of Atomic Numbers

In 1914, another English physicist, Henry Moseley, worked out a method for determining the number of positive charges on the nucleus of the atom for all elements. This method involved the use of x rays. A modern x-ray tube (Fig. 4-4) is a modified cathode-ray tube in which electrons are obtained by thermal emission from a filament heated by an electric current. When a solid target is placed in the path of the speeding electrons, very penetrating rays are produced. These penetrating rays are called **x rays.** In order to study the x rays produced when different elements (or their compounds) were made the targets in the x-ray tube, Moseley used certain salt crystals as diffraction gratings to spread the x rays into a **spectrum.** This effect is similar to that produced when a beam of sunlight is spread into its component colors (a spectrum) by passing it through a glass prism (Fig. 4-5). When the spectrum of x rays was recorded on a photographic film, it appeared as lines characteristic of the material of which the target was made. Moseley ar-

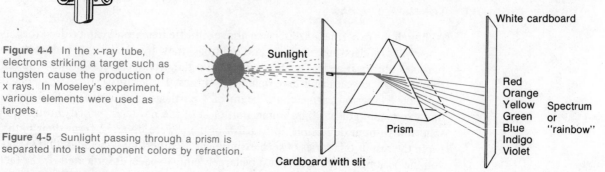

White cardboard

Sunlight

Red
Orange
Yellow   Spectrum
Green   or
Blue   "rainbow"
Indigo
Violet

Prism

Cardboard with slit

ranged the x-ray spectra of different elements in order of their increasing atomic weights and found that (with three exceptions) the heavier the element, the shorter the wavelengths of the principal lines making up the spectrum. He also found that the wavelength of an element differed from that of the element next to it by the same amount (Fig. 4-6). This result suggested that nuclei of atoms of elements adjacent in the series differ from each other by one proton. It then became possible to arrange all the elements in order of increasing positive charge on the nuclei of their atoms, beginning with hydrogen as 1. **The atomic number of an element, then, is the number of positive charges in the nucleus of each of its atoms.** Because an atom is electrically neutral, it follows that the atomic number represents the number of electrons outside its nucleus, as well as the number of protons in its nucleus. The atomic numbers for the elements used in Fig. 4-6 are 33 (As), 34 (Se), 35 (Br), 37 (Rb), and 38 (Sr). The atomic number for krypton is 36.

4.8   Isotopes

The nuclei of atoms contain both protons and neutrons except for the nucleus of the ordinary hydrogen atom, which only contains a proton. The particles inside the nucleus are often referred to as **nucleons.** Because each proton and each neutron has a mass of approximately 1 amu, the atomic weight is approximately equal to the sum of the number of protons and neutrons in the nucleus, or, in other words, the number of nucleons (neglecting the small mass of the electron). The number of nucleons in an atom is called the **mass number** of the atom.

Since each proton and each neutron contributes approximately one unit of atomic mass to an atom, we expect to find the mass of an atom to be approximately a whole number. The fact that many atomic weights (the average mass of a great many atoms of an element) are far from being whole numbers is a consequence of the fact that most elements exist as mixtures of two or more kinds of atoms of different atomic masses but of similar chemical properties. For example, chlorine, with an atomic weight of 35.453 exists as two kinds of chlorine atoms. These have masses very close to the whole numbers 35 and 37. Both kinds of chlorine atoms have an atomic number of 17, which means that they each have 17 protons in the nucleus. The difference must lie, then, in the number of neutrons in the nuclei of the two types of atoms; chlorine-35 has 18 neutrons and chlorine-37 has 20 neutrons. **Atoms of the same atomic number and different masses are called isotopes.** It is important to remember that the only difference in composition between isotopes of the same element is in the number of neutrons in the nucleus. The atomic weight of an element is, therefore, an average of the weights of the isotopes of the element in the proportions in which they normally occur in nature. The atomic weight of 35.453 for chlorine indicates that the atoms of weight 35 are almost three times as abundant as the atoms of weight 37.

A symbolism has been devised to distinguish between isotopes of an element. The two isotopes of chlorine are designated by the symbols $^{35}_{17}\text{Cl}$ and $^{37}_{17}\text{Cl}$. Note that the mass number is made the superscript and the atomic number the subscript. The composition of each of the atomic nuclei of the elements that occur naturally on the earth with atomic numbers 1 through 10 is given in Table 4-1.

4.9 The Bohr Model of the Atom

As described by Rutherford, the nuclear model of the atom with its very small, massive nucleus surrounded by a diffuse sea of lightweight electrons accounted nicely for the properties of the atom as revealed by the discharge-tube experiments of Crookes and others, by the α-particle scattering experiments, and by the varying weights of an element due to the presence of isotopes. However, one basic problem remained. According to the known physical principles of the time (about 1911), such an atom should not be stable. Since there is an attractive force between the oppositely charged nucleus and electrons, the electrons would be expected to fall into the nucleus. On the other hand, if it is assumed that the electrons move around the nucleus in circular orbits, centrifugal force acting on the electrons should counterbalance the force of attraction, and the electrons would stay in their orbits. However, according to classical physics, a system consisting of an electron moving with respect to another charged particle would radiate energy continuously; and, since this means it loses energy, it should move in smaller and smaller orbits and finally fall into the nucleus.

In 1913, Niels Bohr, a Danish scientist, proposed a solution to the problem by suggesting a new theory for the behavior of matter. He suggested that the energy of an electron in an atom cannot vary continuously, but is restricted to a limited

TABLE 4-1   Nuclear Compositions of Atoms of the Very Light Elements

	Symbol	Atomic Number	Number of Protons	Number of Neutrons	Mass, amu	% Natural Abundance
Hydrogen	$^1_1$H	1	1	0	1.0078	99.985
	$^2_1$D	1	1	1	2.0141	0.015
Helium	$^3_2$He	2	2	1	3.0160	0.00013
	$^4_2$He	2	2	2	4.0026	100
Lithium	$^6_3$Li	3	3	3	6.0151	7.42
	$^7_3$Li	3	3	4	7.0160	92.58
Beryllium	$^9_4$Be	4	4	5	9.0122	100
Boron	$^{10}_5$B	5	5	5	10.0129	19.6
	$^{11}_5$B	5	5	6	11.0093	80.4
Carbon	$^{12}_6$C	6	6	6	12.0000	98.89
	$^{13}_6$C	6	6	7	13.0033	1.11
	$^{14}_6$C	6	6	8	14.0032	$<10^{-8}$
Nitrogen	$^{14}_7$N	7	7	7	14.0031	99.63
	$^{15}_7$N	7	7	8	15.0001	0.37
Oxygen	$^{16}_8$O	8	8	8	15.9949	99.759
	$^{17}_8$O	8	8	9	16.9991	0.037
	$^{18}_8$O	8	8	10	17.9992	0.204
Fluorine	$^{19}_9$F	9	9	10	18.9984	100
Neon	$^{20}_{10}$Ne	10	10	10	19.9924	90.92
	$^{21}_{10}$Ne	10	10	11	20.9940	0.257
	$^{22}_{10}$Ne	10	10	12	21.9914	8.82

number of discrete or individual values. The energy of the electron is said to be **quantized.** The success of this proposal in explaining the spectra of the hydrogen atom (Section 4.10) and of hydrogenlike ions containing only one electron moving about a nucleus led to the general acceptance of the proposal.

The Bohr model of the hydrogen atom assumed that the electron moves about the nucleus in a circular orbit (sometimes called an **energy level**) and that the centrifugal force due to this motion counterbalances the electrostatic attraction between the nucleus and the electron. The energy of the electron was thought by Bohr to be related to its angular momentum, that is, the product of its mass, its velocity, and the radius of its orbit.

$$\text{Angular momentum} = \text{mass} \times \text{velocity} \times \text{radius of orbit}$$

The new concept introduced by Bohr was that angular momentum values for electrons in an atom are quantized; that is, they are restricted to a limited number of discrete values that are integral multiples (n) of a constant known as Planck's constant (h, which is 6.626×10^{-34} J sec) divided by 2π.

$$\text{Angular momentum} = n\left(\frac{h}{2\pi}\right), \qquad \text{where } n = 1, 2, 3, 4 \ldots$$

Using this assumption about restrictions on the angular momentum of an electron, the energy of an electron in a hydrogen atom or hydrogenlike ion was shown by Bohr to be given by the expression

$$E = -kZ^2/n^2$$

where k is a constant and Z is the atomic number (the number of units of positive charge on the nucleus). For a hydrogen atom, $Z = 1$ and $k = 2.179 \times 10^{-18}$ J. Other units of energy commonly used include electron volts (eV; $1\,\text{eV} = 1.602 \times 10^{-19}$ J) and ergs ($1\,\text{erg} = 10^{-7}$ J). The radius of the orbit could also be related to the value of n.

$$\text{Radius of orbit} = n^2a_0/Z$$

where a_0 is the size of the orbit in the hydrogen atom for which $n = 1$ (0.529 Å, where $1\,\text{Å} = 10^{-10}$ m or 100 pm). According to this model, the closest an electron can get to the nucleus in a Bohr atom is in an orbit with $n = 1$ and with a radius of 0.529 Å; thus, an electron cannot fall into the nucleus.

A tabulation of some of the energies and radii calculated from the Bohr model for an electron in a hydrogen atom is given in Table 4-2. A representation of a hydrogen atom is shown in Fig. 4-7.

Note that as n increases the difference in energy between the levels decreases, while the distance of the electron from the nucleus becomes increasingly greater. When the electron is an infinite distance from the nucleus, its energy is zero.

Thus the Bohr model of the hydrogen atom postulates a single electron that moves in circular orbits about the nucleus. The electron usually moves in the energy level for which $n = 1$, the energy level in which it has the lowest energy. The atom is said to be in its **ground state.** If the atom picks up energy from an outside source it changes to an **excited state,** and the electron moves to one of the higher energy orbits.

The integer n is called a **quantum number,** or, more specifically, the **principal quantum number.** (We shall see in Sec. 4.12 that there are also other kinds of quantum numbers.) The properties of an electron in an atom are often specified by giving its principal quantum number. If the principal quantum number of the electron in a hydrogen atom is known, we can easily evaluate the angular momentum of the electron, its energy, or the size of the orbit that it occupies. Thus an electron with the quantum number $n = 1$ resides in the first orbit with an energy of -13.395 eV and an orbital radius of 0.529 Å. An older nomenclature is sometimes

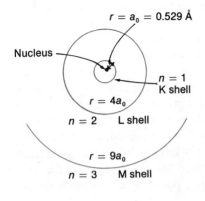

$r = a_0 = 0.529$ Å

Nucleus

$n = 1$ K shell

$r = 4a_0$

$n = 2$ L shell

$r = 9a_0$

$n = 3$ M shell

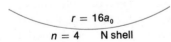

$r = 16a_0$

$n = 4$ N shell

$r = 25a_0$

$n = 5$ O shell

$r = 36a_0$

$n = 6$ P shell

Figure 4-7 A sketch of the circular orbits of the Bohr model of the hydrogen atom drawn to scale. If the nucleus were drawn to scale it would be invisible.

TABLE 4-2	Calculated Values for a Hydrogen Atom		
n Shell	Energy, eV	Distance from Nucleus, Å	Distance from Nucleus, pm
1 K	−13.595	0.529	52.9
2 L	− 3.399	2.116	211.6
3 M	− 1.511	4.761	476.1
4 N	− 0.850	8.464	846.4
5 O	− 0.544	13.225	1322.5
∞ —	0	∞	∞

used as well. As noted in Table 4-2, the $n = 1$ orbit is sometimes called the K shell, the $n = 2$ orbit the L shell, the $n = 3$ orbit the M shell, and the $n = 4$ orbit the N shell, etc., so we can speak of a K electron, an L electron, and so forth.

4.10   Atomic Spectra and Atomic Structure

The Bohr model of the hydrogen atom would be only an interesting intellectual curiosity if it could not be checked against experimental data. The fact is, it explains the spectrum of hydrogen atoms very well, and this observation tends to substantiate the new assumption in the model. Before we look at the agreement of the model with experiment, however, a brief introduction to spectra is required.

Sir Isaac Newton showed that sunlight is a mixture of different kinds of light. He separated sunlight into its component colors (its spectrum) by allowing it to pass through a refracting glass prism. Sunlight contains all wavelengths of visible light and hence gives the **continuous spectrum** (Fig. 4-5) so familiar in the rainbow. Sunlight also contains ultraviolet (very short wavelengths) and infrared light (very long wavelengths), both of which may be detected and recorded with instruments or photographically but which are beyond the range of the human eye. Incandescent solids, liquids, and gases under great pressure give continuous spectra. When an electric current is passed through a gas contained in a vacuum tube at low pressure, the gas gives off light, which when passed through a prism shows a spectrum made up of a number of bright lines (**a bright line spectrum**). These lines can be recorded photographically and the wavelength of the light producing each line can be calculated from the positions of the lines on the photograph.

Light is one form of the radiation known as **electromagnetic radiation** (heat, radio waves, ultraviolet light, x rays, and γ rays are other examples). All of these forms of electromagnetic radiation exhibit properties that are associated with wavelike behavior. They all can be characterized by a wavelength, λ, and a frequency, ν, which are inversely proportional ($\nu \propto 1/\lambda$). The product $\lambda\nu$ is equal to the speed, c, with which all forms of electromagnetic radiation move (2.998×10^8 m/sec in a vacuum). Thus the different forms of electromagnetic radiation are characterized by different wavelengths and frequencies although they all move with the same speed in a vacuum. Electromagnetic radiation also has properties associated with particles called **quanta.** The energy of a quantum of light (or a quantum of any other form of electromagnetic radiation) with wavelength λ and frequency ν may be determined from the expression

$$E = h\nu = \frac{hc}{\lambda}$$

where h is Planck's constant (6.626×10^{-34} J sec) and c is the velocity of light.

If a sealed tube containing hydrogen at low pressure is subjected to an electric discharge, light of a blue-pink hue is produced. Passage of this light through a prism produces a line spectrum, indicating that the light is composed of light of several different wavelengths. After careful measurement of the frequencies of the lines in the visible spectrum, J. R. Rydberg found that these frequencies could

Figure 4-8 Relative energies of some of the circular orbits in the Bohr model with electronic transitions which give rise to atomic spectra. Note the decreasing spacing between levels as n increases.

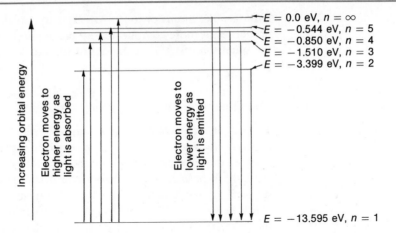

be related by a general equation. One form of this equation gives the energies of the lines in joules as

$$E = h\nu = 2.179 \times 10^{-18} \left(\frac{1}{n_1{}^2} - \frac{1}{n_2{}^2} \right) \text{ J}$$

where n_1 and n_2 are integers with n_1 being smaller than n_2. This equation is empirical; that is, it is derived from observation rather than theory.

Bohr suggested that a hydrogen atom radiates energy as light only when the electron suddenly changes from its present orbit to one that has a lower energy (Fig. 4-8). If a hydrogen atom is excited with an electric discharge, the electron gains energy and moves to an orbit of higher energy (and higher n). As the atom relaxes, the electron moves from this higher energy level to one in which its energy is lower and emits a quantum of light with an energy equal to the difference in energy between the two orbits. According to Bohr's theory (Section 4.9), the electron in a hydrogen atom can have only those *discrete* energies, E_n, permitted by the equation

$$E_n = \frac{-kZ^2}{n^2} = \frac{-2.179 \times 10^{-18}}{n^2} \text{ J}$$

When an electron falls from an outer orbit characterized by n_2 to an orbit of lower energy characterized by n_1 (n_1 less than n_2) the difference in energy will be given by the difference in energy of the two orbits, $E_{n_1} - E_{n_2}$. This is the energy, E, emitted as a quantum of light, or

$$E = E_{n_1} - E_{n_2} = 2.179 \times 10^{-18} \left(\frac{1}{n_1{}^2} - \frac{1}{n_2{}^2} \right) \text{ J}$$

This theoretical expression is exactly the expression determined from the experimental observations of Rydberg. The agreement of the two equations provides powerful and indispensable evidence for the validity of the Bohr concept of atomic structure involving discrete electronic energy levels.

It should be emphasized that, although the idea of discrete energy levels applies to all atoms, many other aspects of the Bohr model explain only the spectra of hydrogenlike atoms or ions. Attempts to explain the spectra and other physical properties of more complicated atoms require the use of a more complex model, the quantum mechanical model of the atom.

4.11   The Quantum Mechanical Model of the Atom

The simple Bohr equation given above is derived for hydrogenlike systems with only one electron. If more than one electron is present in an atom, repulsive forces between the electrons complicate the calculations to the extent that predictions are virtually impossible. The theory can account for the spectra of only the very simplest atomic systems, those that contain only one electron.

It is now known that the laws of ordinary mechanics do not adequately explain the properties of such small particles as electrons. Bohr's theory of the electron moving in a circular or elliptical orbit about a nucleus has been largely supplanted by the more mathematical theories of modern quantum mechanics. Unfortunately, the quantum mechanical concept of the atom does not lend itself readily to a mechanical model that can be visualized.

For our purposes, it is enough to consider that the quantum mechanical treatment indicates that the hydrogen atom consists of one proton (as a nucleus) and one electron moving about the proton. Larger atoms have several electrons moving about a positive nucleus, which consists of several protons and neutrons. Although the electrons are known to move about the nucleus, the exact path they take within a particular energy level cannot be determined. The German physicist Werner Heisenberg expressed this in a form which has come to be known as the **Heisenberg Uncertainty Principle: It is impossible to determine accurately both the momentum and the position of an electron simultaneously.** (The momentum is the mass of the particle multiplied by its velocity.) The Heisenberg principle maintains that the more accurately we measure the momentum of a moving electron, the less accurately we can determine its position (and conversely). An experimental procedure that is designed to measure the position of an electron exactly will alter the momentum of the electron, and an experiment that seeks to measure accurately the momentum of an electron unavoidably changes its position. If measurements are obtained for both position and momentum at the same time, the values are inexact for one or the other or both. Therefore, since the exact position or path of an electron cannot be determined, the best that can be done is to speak of **the probability of finding an electron at a given location within the atom.**

In spite of the limitations described by the uncertainty principle, the behavior of electrons can be determined in a useful way using the mathematical tools of quantum mechanics. Both the energy of an electron in an atom and the region of space in which it may be found can be determined.

The results of a quantum mechanical treatment of the problem developed by another German physicist, Erwin Schrödinger, show that the electron may be visualized as being in rapid motion within a relatively large region around the nucleus, but spending most of its time in certain high-probability regions. For example, in a hydrogen atom, the single electron effectively occupies all the space within about 1 Å (1×10^{-8} cm) of the nucleus, with a greatest probability of being

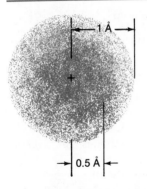

at a distance of 0.529 Å, the radius of the orbit with $n = 1$ in the Bohr model. This gives the hydrogen atom a spherical shape. Some chemists prefer to consider the electron in terms of a cloud of negative charge (**electron cloud**), with the cloud being dense in regions of high electron probability and more diffuse in regions of low probability.

Each electron in an atom can be described using quantum mechanical techniques by a mathematical expression called a **wave function,** which is given the symbol ψ. The square of the wave function, ψ^2, is a measure of the probability of finding the electron at a given point at a distance r from the nucleus; and $4\pi r^2 \psi^2$ **(the radial probability density)** is a measure of the probability of finding the electron within the volume of a thin spherical shell (somewhat like a layer of an onion) of radius r and thickness dr (where dr is a very small fraction of r).

Figure 4-9 illustrates the electron density distribution of the electron cloud nearest the positive nucleus. Most of the time, but not always, the electron will be located inside the spherical outline. Putting it another way, the probability of finding the electron inside the spherical boundary is high. For hydrogen, the probability of finding the electron in a thin shell very close to the nucleus is practically zero, but increases rapidly just beyond the nucleus and becomes highest in a thin shell at a distance of 0.529 Å from the nucleus. This distance may be considered as the radius of the electron shell. The probability then decreases rapidly as the distance of the thin shell from the nucleus increases and becomes exceedingly small at a distance of 1 Å. Figure 4-10 is a plot of the radial probability density, $4\pi r^2 \psi^2$, as a function of distance, r, from the nucleus for the $n = 1$, $n = 2$, and $n = 3$ energy levels.

Figure 4-9 Electron cloud, or electron density distribution, for an electron of lowest possible energy in a hydrogen atom. The electron occupies the 1s orbital.

Figure 4-10 The probability of finding the electron at a given distance r from the nucleus for the 1s, 2p, and 3d subshells. Note the overlap of subshells.

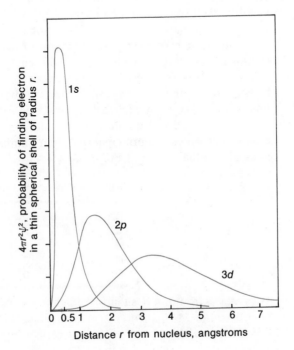

$4\pi r^2 \psi^2$, probability of finding electron in a thin spherical shell of radius r.

1s

2p

3d

Distance r from nucleus, angstroms

4.12   Results of the Quantum Mechanical Model of the Atom

The general results of the quantum mechanical description of the atom are described in the following paragraphs.

1. Electrons are located in energy levels, or shells, which are characterized by **n**, **the principal quantum number,** which may take on any integral value: $n = 1, 2, 3, 4, 5, \ldots$. The shells may be identified by the principal quantum number or by the letters K, L, M, N, O, $\ldots$, which correspond to $n = 1, 2, 3, 4, 5, \ldots$, respectively. As n increases the electrons are farther from the nucleus; that is, the average location of each electron is farther from the nucleus (Fig. 4-10).

2. There are a limited number of discrete values of energy that the electron may possess. The mathematics is far more complicated than the simple $-k/n^2$ of the Bohr model, but the energy is still quantized. The energy of an electron can be determined from the wave function that describes the behavior of the electron.

3. Electrons do not move in circular paths. In fact, we cannot say anything about the exact paths in which the electrons move. All we can identify is the region, or volume, of space (the charge cloud) around a nucleus where an electron with a given energy is most likely to be found and the region where it is not likely to be found.

4. Depending upon the particular shell, electrons may be found within a shell in regions of space with different shapes. In those atoms with more than one electron, electrons in each of these regions will have slightly different energies. Each of these different regions may be identified by a second quantum number, ℓ, called the **azimuthal, or subsidiary, quantum number.** An electron with $\ell = 0$ (called an s electron) may most probably be found within a sphere with the nucleus at its center [Fig. 4-11(a)]. A p electron is one with $\ell = 1$ and is most likely to be found in a dumbbell-shaped region of space [Fig. 4-11(b)]; a d electron with $\ell = 2$, in a volume of space usually drawn with four lobes [Fig. 4-11(c)]; and an f electron with $\ell = 3$, in a rather complex-looking volume of space usually shown with eight lobes [Fig. 4-11(d)]. Within a given shell, the energies of electrons in these regions increase in the order s electrons $< p$ electrons $< d$ electrons $< f$ electrons. Although higher ℓ values than 3 are possible, electrons with ℓ greater than 3 are not found unless they have been excited by absorption of energy by an atom.

The mathematics of quantum mechanics limits the values that ℓ can have: an electron for which the principal quantum number has a value of n may have an integral ℓ value ranging from 0 to $(n - 1)$.

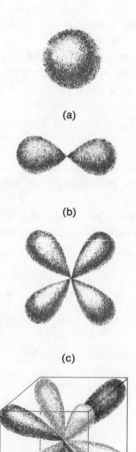

(a)

(b)

(c)

(d)

Figure 4-11 Shapes of the electron clouds for (a) s subshells, (b) p subshells (c) d subshells, (d) f subshells. To improve perspective, the f electron cloud is shown within a cube with alternate lobes shown in color.

Shell	n Value	Possible ℓ Values
K	1	0
L	2	0, 1
M	3	0, 1, 2
N	4	0, 1, 2, 3
O	5	0, 1, 2, 3, 4

Thus an electron in a K shell for which $n = 1$ must have an ℓ value of zero. An electron in an M shell ($n = 3$) can be either an s electron ($\ell = 0$), a p electron ($\ell = 1$), or a d electron ($\ell = 2$), but cannot have an ℓ value higher than 2.

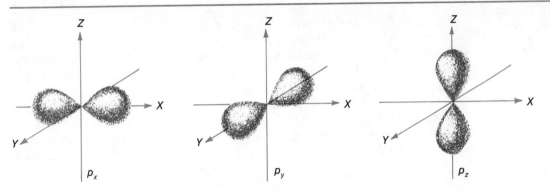

Figure 4-12 The different orientations of the three equivalent atomic *p* orbitals. Each of these orientations can be characterized by a different value of *m*.

5. For ℓ values larger than zero there are several $(2\ell + 1)$ regions of space that have the same shape and the same energy but differ in their orientation. Each of these regions is called an **orbital**, or an **atomic orbital.** Each orientation is characterized by a third quantum number *m*, the **magnetic quantum number,** which may have any integral value from $-\ell$ through zero to $+\ell$. A sphere may have only one orientation in space, and thus there is only one type of spherical or *s* orbital. This orbital is characterized by an ℓ value of zero; hence *m* can only be zero. There are three possible values of *m* $(-1, 0, +1)$ for an orbital with $\ell = 1$ (a *p* orbital), and thus three possible orientations. Figure 4-12 shows how the dumbbell that characterizes a *p* orbital can be oriented in three ways along the *x*, *y*, and *z* axes of an *x-y-z* coordinate system. The three *p* orbitals are designated, as in the figure, p_x, p_y, and p_z to indicate their directional character. Figure 4-13 shows the

Figure 4-13 The different orientations of the five equivalent atomic *d* orbitals. The drawing is oriented so that the lobes of the orbitals designated d_{z^2} and $d_{x^2-y^2}$ lie along the axes, whereas the lobes of those labeled d_{xz}, d_{yz}, and d_{xy} lie in between the axes.

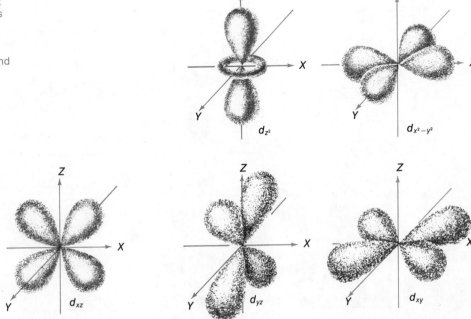

five orientations of the d orbitals ($\ell = 2$; $m = -2, -1, 0, +1, +2$). The seven orientations of the f orbitals ($\ell = 3$; $m = -3, -2, -1, 0, +1, +2, +3$) are shown in Fig. 4-14. There is no general agreement as how best to represent the seven f orbitals. The figure shows one way with each orbital in a cube to help you see the three-dimensional aspect of this representation.

In a free atom, p orbitals in the same shell are referred to as degenerate; that is, all three p orbitals have the same energy. Likewise, the set of five d orbitals in a shell is degenerate, as is a set of seven f orbitals. Each set of degenerate orbitals, that is, orbitals with the same ℓ value in a given shell, is referred to as a **subshell.** Thus the N shell ($n = 4$) contains an s subshell consisting of one s orbital ($\ell = 0$; $m = 0$); a p subshell consisting of three p orbitals ($\ell = 1$; $m = -1, 0,$ or $+1$), a d subshell consisting of five d orbitals ($\ell = 2$; $m = -2, -1, 0, +1,$ or $+2$) and an f subshell of the seven f orbitals ($\ell = 3$; $m = -3, -2, -1, 0, +1, +2,$ or $+3$).

6. The maximum number of electrons that may be found in a shell with a principal quantum number of n is $2n^2$. Each atomic orbital can hold a maximum of two electrons. The number of atomic orbitals in a shell, therefore, is given by n^2, as shown in Table 4-3.

7. The electrons within a particular orbital also have a fourth quantum number, the **spin quantum number, s.** Electrons behave as if they were spinning about their own axis. The spin quantum number specifies the direction of spin of the electron about *its own axis* as the electron moves around the nucleus. The spin can be either counterclockwise or clockwise and is designated arbitrarily by either $s = +\frac{1}{2}$ or $s = -\frac{1}{2}$.

Figure 4-14 The different orientations of the seven equivalent atomic f orbitals. To improve perspective, each of the seven orbitals is shown within a cube with one lobe of each pair shown in color.

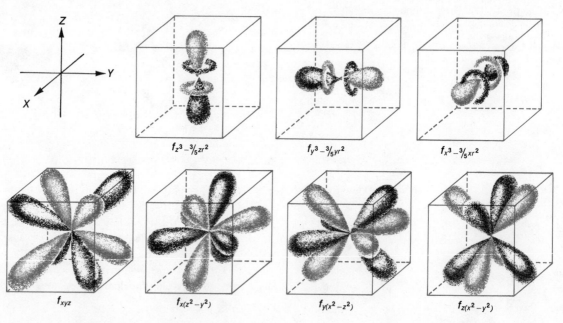

$f_{z^3 - \frac{3}{5}zr^2}$ $f_{y^3 - \frac{3}{5}yr^2}$ $f_{x^3 - \frac{3}{5}xr^2}$

f_{xyz} $f_{x(z^2-y^2)}$ $f_{y(x^2-z^2)}$ $f_{z(x^2-y^2)}$

4.13 Summary of Electronic Quantum Numbers

In general, the properties of the electrons about a nucleus can be described in terms of four quantum numbers.

▶ **1. PRINCIPAL QUANTUM NUMBER, *n*.** The principal quantum number for an electron designates, in general, the energy of the shell as well as the volume of space in which the electron moves. An increase in the value of *n* indicates an increase in the energy associated with the electron energy level, or major shell, and an increase in the *average* distance of the electron from the nucleus (Fig. 4-10). Values of *n* are whole numbers which can vary from 1 to ∞.

Shell:	K	L	M	N	O	P	Q
n:	1	2	3	4	5	6	7

▶ **2. AZIMUTHAL, OR SUBSIDIARY, QUANTUM NUMBER, *ℓ*.** The subsidiary quantum number designates the shape of the region, or subshell, that the electron occupies. Values of *ℓ* are whole numbers which can vary from 0 to $(n - 1)$. Thus, if $n = 1$, *ℓ* can be only 0; if $n = 2$, *ℓ* can be either 0 or 1; etc.

Electron subshells and orbitals are often designated by the letters *s, p, d, f*, as indicated earlier. These letters correspond to the values of *ℓ*, and continue alphabetically following *f*.

	ℓ:	1	2	3	4	5	6	7
Electron Designation:		*s*	*p*	*d*	*f*	*g*	*h*	*i*

In an atom with more than one electron the energies of the subshells within a shell increase with increasing values of *ℓ*.

▶ **3. MAGNETIC QUANTUM NUMBER, *m*.** The magnetic quantum number designates in a general way the orientation of the charge cloud in space. Values of *m* are whole numbers and can vary from $-ℓ$ through zero to $+ℓ$.

▶ **4. SPIN QUANTUM NUMBER, *s*.** The spin quantum number specifies the direction of spin of the electron about *its own axis*. The spin can be either counterclockwise or clockwise and is designated arbitrarily by either $+\frac{1}{2}$ or $-\frac{1}{2}$. For every possible combination of *n*, *ℓ*, and *m*, there can be two electrons differing only in the direction of spin about their own axes. The two electrons that can occupy any one given orbital differ only in spin and have identical *n*, *ℓ*, and *m* values.

The notation commonly used in describing electronic structures of atoms consists of a number in front of the subshell letter to designate the number of the principal, or major, shell and a superscript to designate the number of electrons in that particular subshell. For example, the notation $2p^4$ (read two-*p*-four) indicates four electrons in the *p* subshell of the second principal shell from the nucleus; the notation $3d^8$ (three-*d*-eight) indicates eight electrons in the *d* subshell of the third principal shell.

TABLE 4-3   Summary of Permissible Values For Each Quantum Number

Shell	n	ℓ	m	s	Number of Electrons in Subshell	Total Electrons in Shell
K	1	0 (s)	0	$+\frac{1}{2}$	$\left.\begin{array}{c}1\\1\end{array}\right\}2$	2
	1	0	0	$-\frac{1}{2}$		
L	2	0 (s)	0	$+\frac{1}{2}, -\frac{1}{2}$	2	
	2	1 (p)	−1	$+\frac{1}{2}, -\frac{1}{2}$	$\left.\begin{array}{c}2\\2\\2\end{array}\right\}6$	8
	2	1	0	$+\frac{1}{2}, -\frac{1}{2}$		
	2	1	+1	$+\frac{1}{2}, -\frac{1}{2}$		
M	3	0 (s)	0	$+\frac{1}{2}, -\frac{1}{2}$	2	
		1 (p)	−1, 0, +1	$\pm\frac{1}{2}$ for each value of m	6	18
		2 (d)	−2, −1, 0, +1, +2	$\pm\frac{1}{2}$ for each value of m	10	
N	4	0 (s)	0	$\pm\frac{1}{2}$ for each value of m	2	
		1 (p)	−1, 0, +1	$\pm\frac{1}{2}$ for each value of m	6	
		2 (d)	−2, −1, 0, +1, +2	$\pm\frac{1}{2}$ for each value of m	10	32
		3 (f)	−3, −2, −1, 0, +1, +2, +3	$\pm\frac{1}{2}$ for each value of m	14	
O	5	0 (s)	0	$\pm\frac{1}{2}$ for each value of m	2	
		1 (p)	−1, 0, +1	$\pm\frac{1}{2}$ for each value of m	6	
		2 (d)	−2, −1, 0, +1, +2	$\pm\frac{1}{2}$ for each value of m	10	50[a]
		3 (f)	−3, −2, −1, 0, +1, +2, +3	$\pm\frac{1}{2}$ for each value of m	14	
		4 (g)	−4, −3, −2, −1, 0, +1, +2, +3, +4	$\pm\frac{1}{2}$ for each value of m	18[a]	

[a] The total number of 50 electrons for the O shell, including the 18 electrons for the g subshell, is the number of electrons theoretically possible. No element presently known contains more than 32 electrons in the O shell.

4.14   An Energy-Level Diagram

Although it is not possible to depict the electron energy levels exactly, a diagram representing roughly the relative energy values of all electrons in atoms is given in Fig. 4-15. The energy of an electron in an orbital is indicated by the vertical coordinate in the diagram, the orbital with electrons of lowest energy and greatest stability being the 1s orbital at the bottom of the diagram. Levels of nearly the same energy are enclosed by vertical braces at the right of the diagram. In general, electrons may be expected to occupy the orbitals in order as they appear in the diagram, starting at the bottom and working up, each set of orbitals being filled before electrons enter the next set immediately above. This scheme holds strictly true only for the elements of low atomic numbers; as the atomic number increases, the relative energies of the levels change somewhat, but not all change to the same extent. This means that the elements of higher atomic number have electron arrangements slightly different from that depicted in the energy-level diagram. It should also be noted that the actual energies of the electrons in these orbitals vary from atom to atom. For example, the energy of an electron in a 1s orbital becomes lower and lower as the atomic numbers of the atoms increase.

The device shown in Fig. 4-16 is quite useful in recalling the order of occupancy of energy levels in an atom. In general, the energy increases in the order indicated by the connecting lines.

4.15   Atomic Structures

The distribution of electrons in the atomic orbitals of an atom can be predicted using the following guidelines.

1. Electrons in a stable atom occupy the lowest possible energy levels, or orbitals. The first electron that is placed in a set of atomic orbitals will go into the 1s orbital. When the 1s orbital is filled, the next electron will go into the 2s orbital, and so on up through the 2p orbital, the 3s orbital, etc. The order of orbital energies is given in Figs. 4-15 and 4-16.

2. The maximum number of electrons in an orbital is limited to two. This is expressed by a statement known as the **Pauli Exclusion Principle: No two electrons in the same atom can have the same set of four quantum numbers.** Thus, a 1s orbital is filled when it contains two electrons; one with $n = 1$, $l = 0$, $m = 0$, and $s = +\frac{1}{2}$ and the other with $n = 1$, $l = 0$, $m = 0$, and $s = -\frac{1}{2}$. Since no other combination of quantum numbers is possible for an electron in a 1s orbital, a third electron would be forced to occupy the 2s orbital. The 2s orbital, in turn, is filled when it contains two electrons, so that the fifth electron is forced into one of the three 2p orbitals, and so forth.

3. Subshells containing more than one orbital are filled as described by **Hund's Rule: Each orbital in a subshell is singly occupied (filled) with one electron before any one orbital is doubly occupied, and all electrons in partly occupied orbitals**

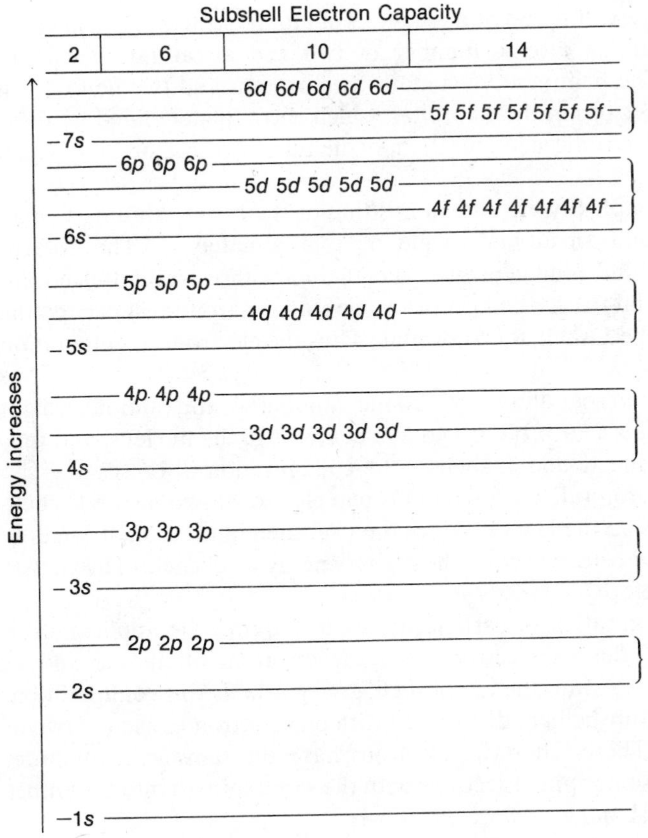

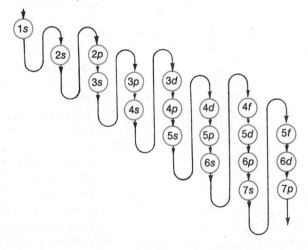

Figure 4-15 Generalized energy-level diagram for atomic orbitals.

Figure 4-16 Order of occupancy of atomic orbitals.

within a subshell have their spins aligned; that is, their spin quantum numbers are the same. Hund's Rule indicates that an electronic structure of the type p^2 would have one electron in each of two different p orbitals. Both electrons would have the same spin and thus the same spin quantum number. An electronic structure with a d^6 arrangement would have two electrons in one d orbital with different spins and one electron in each of the remaining four d orbitals, each electron with the same spin.

4.16 The Aufbau Process

In arriving at the electronic structure of atoms of the various elements, it is convenient to consider the subshells that electrons would enter if these atoms were built up in order of increasing atomic number, beginning with hydrogen, by adding one proton to the nucleus and one electron to the atom at a time. Such a process is referred to as the **Aufbau Process** from the German word *aufbau* (to build up). Each additional electron added will occupy the available subshell of lowest energy according to the rules given in Section 4.15. Electrons enter higher energy subshells only after lower energy subshells have been filled to capacity.

As mentioned previously, a single hydrogen atom consists of one proton and one electron. The single electron is found in the $1s$ orbital around the proton. The electronic configuration of a hydrogen atom is customarily represented as $1s^1$ (read one-s-one).

After hydrogen, the next simplest atom in terms of atomic structure is that of the noble gas helium, with an atomic number of two and an atomic weight of approximately four. The helium atom contains two protons and two neutrons in its nucleus and two electrons in the $1s$ orbital, which is completely filled by these two electrons. The electronic configuration of helium is, therefore, $1s^2$ (read one-s-two).

The next atom in complexity after helium is that of the metal lithium, with an atomic number of 3 and an atomic weight of approximately 7. The nucleus contains three protons and four neutrons; and there are three orbital electrons. Two of these electrons fill the K shell, so the remaining electron must occupy the lowest lying orbital in the L shell, the $2s$ orbital. Thus the electronic configuration of lithium is $1s^2 2s^1$.

An atom of the metal beryllium, of atomic number 4 and atomic weight approximately 9, contains four protons and five neutrons in the nucleus, and four electrons, two each in the K and L shells. The configuration is $1s^2 2s^2$.

A neutral atom of boron (at. no. 5) contains five electrons: two in the K shell, which is filled, and three in the L shell. Since the s subshell in the L shell can only contain two electrons, one must occupy the higher energy p subshell. The configuration of boron, therefore, is $1s^2 2s^2 2p^1$.

To describe the configuration of carbon (at. no. 6) requires the application of Hund's Rule. Four of the six electrons in a carbon atom fill the $1s$ and $2s$ subshells. The other two go into the $2p$ subshell, giving a $1s^2 2s^2 2p^2$ configuration. The electrons in the $2p$ subshell are distributed with one electron in each of two of the three $2p$ orbitals. These latter two electrons have the same spin quantum number and hence the same spin. Electrons with the same spin quantum number are said to have **parallel spins.**

Nitrogen (at. no. 7) has a $1s^2 2s^2 2p^3$ configuration with one electron in each of the three $2p$ orbitals, in accordance with Hund's Rule. These three electrons have parallel spins. Oxygen atoms (at. no. 8) have the configuration $1s^2 2s^2 2p^4$ with a pair of electrons in one of the $2p$ orbitals and a single electron in each of the other two. Fluorine atoms (at. no. 9) have the configuration $1s^2 2s^2 2p^5$ with only one $2p$ orbital containing an unpaired electron. All of the electrons in neon (at. no. 10) are paired, with a configuration $1s^2 2s^2 2p^6$. Both the K and L shells of neon are filled (two electrons in the K shell and eight in the L shell).

The sodium atom (at. no. 11) has one more electron than the neon atom. This electron must go into the M shell in the lowest lying available subshell, the $3s$ orbital. The sodium atom resembles the lithium atom in that it has one electron in an s orbital in its outermost shell.

The magnesium atom (at. no. 12) has two electrons in its outer shell (the M shell) and so is analogous to beryllium, which also has two electrons in its outer shell (the L shell). Aluminum (at. no. 13), with three electrons in the outer shell, silicon with four, phosphorus with five, sulfur with six, chlorine with seven, and argon with eight, correspond respectively to boron, carbon, nitrogen, oxygen, fluorine, and neon. Any shell that is an outermost shell (other than the K shell, which is completely filled with two electrons) will contain a maximum of only eight electrons. Hence, argon has a complete *outer shell* electron configuration with eight electrons in the M shell. The M shell, when it becomes an inner shell in larger atoms, can hold up to eighteen electrons.

Potassium and calcium (at. nos. 19 and 20) have one and two electrons, respectively, in the N shell. Hence, potassium corresponds to lithium and sodium in outer configuration, whereas calcium corresponds to beryllium and magnesium.

Beginning with scandium (at. no. 21), additional electrons are added successively to the $3d$ subshell after two electrons have already occupied the $4s$ subshell. After the $3d$ subshell is filled to its capacity with ten electrons, the $4p$ subshell fills. The electronic configurations of the elements potassium through gallium are given in Table 4-4.

TABLE 4-4

Atomic Number	Element	Electronic Configuration of Atoms
19	Potassium	$1s^2\ 2s^2\ 2p^6\ 3s^2\ 3p^6\ 3d^0\quad 4s^1$
20	Calcium	$3d^0\quad 4s^2$
21	Scandium	$3d^1\quad 4s^2$
22	Titanium	$3d^2\quad 4s^2$
23	Vanadium	$3d^3\quad 4s^2$
24	Chromium	$3d^5\quad 4s^1$
25	Manganese	$3d^5\quad 4s^2$
26	Iron	$3d^6\quad 4s^2$
27	Cobalt	$3d^7\quad 4s^2$
28	Nickel	$3d^8\quad 4s^2$
29	Copper	$3d^{10}\ 4s^1$
30	Zinc	$3d^{10}\ 4s^2$
31	Gallium	$3d^{10}\ 4s^2\ 4p^1$

TABLE 4-5 Electron Distribution, in Terms of Subshells, for the Known Elements

		K	L		M			N				O				P			Q
		1s	2s	2p	3s	3p	3d	4s	4p	4d	4f	5s	5p	5d	5f	6s	6p	6d	7s
H	1	1																	
He	2	2																	
Li	3	2	1																
Be	4	2	2																
B	5	2	2	1															
C	6	2	2	2															
N	7	2	2	3															
O	8	2	2	4															
F	9	2	2	5															
Ne	10	2	2	6															
Na	11	2	2	6	1														
Mg	12	2	2	6	2														
Al	13	2	2	6	2	1													
Si	14	2	2	6	2	2													
P	15	2	2	6	2	3													
S	16	2	2	6	2	4													
Cl	17	2	2	6	2	5													
Ar	18	2	2	6	2	6													
K	19	2	2	6	2	6		1											
Ca	20	2	2	6	2	6		2											
Sc	21	2	2	6	2	6	1	2											
Ti	22	2	2	6	2	6	2	2											
V	23	2	2	6	2	6	3	2											
Cr	24	2	2	6	2	6	5	1											
Mn	25	2	2	6	2	6	5	2											
Fe	26	2	2	6	2	6	6	2											
Co	27	2	2	6	2	6	7	2											
Ni	28	2	2	6	2	6	8	2											
Cu	29	2	2	6	2	6	10	1											
Zn	30	2	2	6	2	6	10	2											
Ga	31	2	2	6	2	6	10	2	1										
Ge	32	2	2	6	2	6	10	2	2										
As	33	2	2	6	2	6	10	2	3										
Se	34	2	2	6	2	6	10	2	4										
Br	35	2	2	6	2	6	10	2	5										
Kr	36	2	2	6	2	6	10	2	6										
Rb	37	2	2	6	2	6	10	2	6			1							
Sr	38	2	2	6	2	6	10	2	6			2							
Y	39	2	2	6	2	6	10	2	6	1		2							
Zr	40	2	2	6	2	6	10	2	6	2		2							
Nb	41	2	2	6	2	6	10	2	6	4		1							
Mo	42	2	2	6	2	6	10	2	6	5		1							
Tc	43	2	2	6	2	6	10	2	6	6		1							
Ru	44	2	2	6	2	6	10	2	6	7		1							
Rh	45	2	2	6	2	6	10	2	6	8		1							
Pd	46	2	2	6	2	6	10	2	6	10									
Ag	47	2	2	6	2	6	10	2	6	10		1							
Cd	48	2	2	6	2	6	10	2	6	10		2							
In	49	2	2	6	2	6	10	2	6	10		2	1						
Sn	50	2	2	6	2	6	10	2	6	10		2	2						
Sb	51	2	2	6	2	6	10	2	6	10		2	3						
Te	52	2	2	6	2	6	10	2	6	10		2	4						
I	53	2	2	6	2	6	10	2	6	10		2	5						
Xe	54	2	2	6	2	6	10	2	6	10		2	6						
		2	8		18			18				8							

TABLE 4-5   (Continued)

		K	L	M	N				O				P			Q
					4s	4p	4d	4f	5s	5p	5d	5f	6s	6p	6d	7s
Cs	55	2	8	18	2	6	10		2	6			1			
Ba	56	2	8	18	2	6	10		2	6			2			
La	57	2	8	18	2	6	10		2	6	1		2			
Ce	58	2	8	18	2	6	10	2	2	6			2			
Pr	59	2	8	18	2	6	10	3	2	6			2			
Nd	60	2	8	18	2	6	10	4	2	6			2			
Pm	61	2	8	18	2	6	10	5	2	6			2			
Sm	62	2	8	18	2	6	10	6	2	6			2			
Eu	63	2	8	18	2	6	10	7	2	6			2			
Gd	64	2	8	18	2	6	10	7	2	6	1		2			
Tb	65	2	8	18	2	6	10	9	2	6			2			
Dy	66	2	8	18	2	6	10	10	2	6			2			
Ho	67	2	8	18	2	6	10	11	2	6			2			
Er	68	2	8	18	2	6	10	12	2	6			2			
Tm	69	2	8	18	2	6	10	13	2	6			2			
Yb	70	2	8	18	2	6	10	14	2	6			2			
Lu	71	2	8	18	2	6	10	14	2	6	1		2			
Hf	72	2	8	18	2	6	10	14	2	6	2		2			
Ta	73	2	8	18	2	6	10	14	2	6	3		2			
W	74	2	8	18	2	6	10	14	2	6	4		2			
Re	75	2	8	18	2	6	10	14	2	6	5		2			
Os	76	2	8	18	2	6	10	14	2	6	6		2			
Ir	77	2	8	18	2	6	10	14	2	6	7		2			
Pt	78	2	8	18	2	6	10	14	2	6	9		1			
Au	79	2	8	18	2	6	10	14	2	6	10		1			
Hg	80	2	8	18	2	6	10	14	2	6	10		2			
Tl	81	2	8	18	2	6	10	14	2	6	10		2	1		
Pb	82	2	8	18	2	6	10	14	2	6	10		2	2		
Bi	83	2	8	18	2	6	10	14	2	6	10		2	3		
Po	84	2	8	18	2	6	10	14	2	6	10		2	4		
At	85	2	8	18	2	6	10	14	2	6	10		2	5		
Rn	86	2	8	18	2	6	10	14	2	6	10		2	6		
Fr	87	2	8	18	2	6	10	14	2	6	10		2	6		1
Ra	88	2	8	18	2	6	10	14	2	6	10		2	6		2
Ac	89	2	8	18	2	6	10	14	2	6	10		2	6	1	2
Th	90	2	8	18	2	6	10	14	2	6	10		2	6	2	2
Pa	91	2	8	18	2	6	10	14	2	6	10	2	2	6	1	2
U	92	2	8	18	2	6	10	14	2	6	10	3	2	6	1	2
Np	93	2	8	18	2	6	10	14	2	6	10	4	2	6	1	2
Pu	94	2	8	18	2	6	10	14	2	6	10	6	2	6		2
Am	95	2	8	18	2	6	10	14	2	6	10	7	2	6		2
Cm	96	2	8	18	2	6	10	14	2	6	10	7	2	6	1	2
Bk	97	2	8	18	2	6	10	14	2	6	10	8	2	6	1	2
Cf	98	2	8	18	2	6	10	14	2	6	10	10	2	6		2
Es	99	2	8	18	2	6	10	14	2	6	10	11	2	6		2
Fm	100	2	8	18	2	6	10	14	2	6	10	12	2	6		2
Md	101	2	8	18	2	6	10	14	2	6	10	13	2	6		2
No	102	2	8	18	2	6	10	14	2	6	10	14	2	6		2
Lr	103	2	8	18	2	6	10	14	2	6	10	14	2	6	1	2
Rf	104	2	8	18	2	6	10	14	2	6	10	14	2	6	2	2
Ha	105	2	8	18	2	6	10	14	2	6	10	14	2	6	3	2
—	106	2	8	18	2	6	10	14	2	6	10	14	2	6	4	2
		2	8	18	32				32				12			2

Students are sometimes troubled by exceptions to the expected order for the filling of orbitals shown in Fig. 4-16. For instance, the electronic configurations of chromium and copper (Table 4-4) and of lanthanum (at. no. 57, Table 4-5) do not conform with the behavior expected from Fig. 4-16. It is helpful to notice that these exceptions involve subshells with very similar energies. In some atoms, certain combinations of repulsions between electrons or particular electronic configurations lead to minor exceptions in the expected order of filling of these subshells as the energies due to these effects become greater than the small differences in energy between the subshells.

It has been shown by quantum mechanics that half-filled and completely filled subshells represent conditions of particular stability. The stability of filled or half-filled subshells is such that an electron shifts from the $4s$ into the $3d$ orbitals in order to gain the extra stability of a half-filled $3d$ subshell in chromium and a filled $3d$ subshell in copper.

One would logically expect the electronic configuration of lanthanum (at. no. 57) to be $1s^2$, $2s^2$, $2p^6$, $3s^2$, $3p^6$, $3d^{10}$, $4s^2$, $4p^6$, $4d^{10}$, $4f^1$, $5s^2$, $5p^6$, $6s^2$. Instead, as shown in Table 4-5, it has no $4f$ electron but has one $5d$ electron: $1s^2$, $2s^2$, $2p^6$, $3s^2$, $3p^6$, $3d^{10}$, $4s^2$, $4p^6$, $4d^{10}$, $5s^2$, $5p^6$, $5d^1$, $6s^2$.

This minor exception in the case of lanthanum reflects the fact that the $4f$ and $5d$ subshells have very similar energies (see Fig. 4-15) and that electrons change from one sublevel to the other easily. A close inspection of Table 4-5 will disclose a considerable number of such "exceptions." You should realize that these exceptions result from the similar energies of various shells, and remember a few of the specific instances where they occur. Beyond this, a complete understanding of the general principles of electronic configuration and an ability to write the generalized structure for each element, based upon the expected order of addition of electrons, is sufficient for most purposes.

Table 4-5 lists the lowest energy, or ground state, electronic structures of atoms for each of the known elements. Note that for 3 series of elements, scandium (Sc) through copper (Cu), yttrium (Y) through silver (Ag), and lutetium (Lu) through gold (Au), a total of 10 d electrons are successively added to the principal shell next to the outer shell to bring the shell from 8 to 18 electrons. For 2 series, lanthanum (La) through lutetium (Lu) and actinium (Ac) through lawrencium (Lr), 14 f electrons are successively added to the third shell from the outside to bring that shell from 18 to 32 electrons.

The Periodic Law

4.17   The Periodic Table

The known elements vary greatly in their physical and chemical properties and in the nature of the compounds that they form. The study of the individual properties and the many compounds of each of the elements would prove to be extremely laborious and time consuming, but though every element is different from every other element, similarities make possible groupings that simplify such a study. The most widely used grouping is that called the Periodic Table.

It became evident early in the development of chemistry that certain elements known at that time could be grouped together by reason of their similar properties. The members of one such grouping are lithium, sodium, and potassium (a portion of the group called the **alkali metals**). These elements all look like metals, all conduct electricity well, all react with chlorine, forming white water-soluble compounds with one chlorine atom per metal atom, and all react with water, giving hydrogen gas and the metal hydroxides lithium hydroxide (LiOH), sodium hydroxide (NaOH), and potassium hydroxide (KOH), respectively. A second grouping, calcium, strontium, and barium (a portion of the group called **alkaline earth metals**), also look like metals and conduct electricity well. They react with chlorine, giving the white solids, calcium chloride (CaCl$_2$), strontium chloride (SrCl$_2$), and barium chloride (BaCl$_2$), and they give the hydroxides calcium hydroxide [Ca(OH)$_2$], strontium hydroxide [Sr(OH)$_2$], and barium hydroxide [Ba(OH)$_2$] when they react with water. In each case, there are either two chlorine atoms or two hydroxyl (OH) groups per metal atom. Fluorine, chlorine, bromine, and iodine (in the group called the **halogens**) also exhibit similar properties. They do not conduct electricity, and are nonmetallic. They form hydrogen fluoride (HF), hydrogen chloride (HCl), hydrogen bromide (HBr), and hydrogen iodide (HI) when they react with hydrogen, and form sodium fluoride (NaF), sodium chloride (NaCl), sodium bromide (NaBr), and sodium iodide (NaI) when they react with sodium.

The German chemist Johann Dobereiner, in 1829, was the first to suggest the existence of a relationship between the atomic weights and the properties of the elements. He showed that the atomic weight of strontium lies about midway between the atomic weights of calcium and barium, and that the properties of strontium are intermediate between those of calcium and barium. He later recognized the existence of other **triads** of similar elements. Among these are the triads chlorine, bromine, and iodine; and lithium, sodium, and potassium. By 1854, other chemists had shown that oxygen, sulfur, selenium, and tellurium could be classed as a **family** of elements, and that nitrogen, phosphorus, arsenic, antimony, and bismuth make up another family.

In 1865, John Newlands in England recognized a correlation between the magnitude of the atomic weights and the properties of the elements. By 1870, Dimitri Mendeleev in Russia and Lothar Meyer in Germany, working independently and apparently unaware of the work of Newlands, outlined the nature of this relationship between properties and atomic weights in some detail. Their conclusions led to the statement that the properties of the elements are periodic functions of their atomic weights. The idea is illustrated in the following two series of elements. Note that the elements are arranged in order of increasing atomic weights from left to right.

Li	Be	B	C	N	O	F	Ne
6.941	9.01218	10.81	12.011	14.0067	15.9994	18.998403	20.179
Na	Mg	Al	Si	P	S	Cl	Ar
22.98977	24.305	26.98154	28.0855	30.97376	32.06	35.453	39.948

These two rows of elements comprise what are called the second and third **periods** of the modern Periodic Table (Table 4-6 and the front endleaf of this book). Elements in each horizontal row differ decidedly in both physical and chemical properties from one another. There is, however, an interesting and useful gradation in properties of the elements in a given row as the atomic weight increases. Furthermore, when the elements are arranged in order of increasing atomic weight, elements in the same vertical columns have similar properties. For example, lithium (Li) is very similar physically and chemically to sodium (Na), and beryllium (Be) is very similar to magnesium (Mg).

If you examine the arrangement of the Periodic Table on the front endleaf of this text and in Table 4-6 you will note that argon (at. wt 39.948) precedes potassium (at. wt 39.098), cobalt (at. wt 58.9332) precedes nickel (at. wt 58.70), tellurium (at. wt 127.60) precedes iodine (at. wt 126.9045), thorium (at. wt 232.0381) precedes protactinium (at. wt 231.0359), and uranium (at. wt 238.029) precedes neptunium (at. wt 237.0482). These reverses, on the basis of atomic weights, reflect the work of Moseley (Section 4.7) who showed that atomic numbers rather than atomic weights are fundamental in determining the chemical properties of the elements. The modern statement of the **Periodic Law** is: **The properties of the elements are periodic functions of their atomic numbers.**

The modern Periodic Table as presented on the front endleaf of this text, and in Table 4-6 for convenient reference and study, consists of 7 horizontal rows (often referred to as **periods**) of elements and 32 vertical columns (referred to as **families** or **groups**) of elements. Note that in order to fit the table onto a single page some of the groups are written below the table at the bottom of the page. The elements labeled "lanthanide series" and "actinide series" fit between elements number 56 (barium, Ba) and 72 (hafnium, Hf), and between elements number 88 (radium, Ra) and 104 (rutherfordium, Rf), respectively.

When arranged in order of increasing atomic numbers the elements with similar chemical properties recur at definite intervals, i.e., periodically. In regard to electronic structure, this suggests a periodicity in the number of electrons in the outer shells of the atoms of the elements. The electrons in the outermost shell or shells of an atom (referred to as **valence electrons**) are largely responsible for its chemical behavior. If elements having the same number of valence electrons are grouped together, the elements falling within each group have similar chemical properties. Table 4-7 reviews the electron grouping in the shells for the elements of atomic numbers 1 to 18, inclusive. Note the periodicity in regard to the number of valence electrons (shown in boldface).

It is seen that elements in any one vertical column (known as a **group,** or **family**) have the same number of electrons in their valence shells. We can now understand the similarity in chemical properties among elements of the same group, because the number of electrons in the valence shell of an atom is very important in determining its chemical properties.

We shall consider the overall arrangement of elements in the Periodic Table. In Section 4.18, we will describe the relationship of this arrangement to the electronic structures of the elements. As this discussion proceeds, you will find it exceedingly helpful to refer to the Periodic Table as each idea is developed.

The **first short period** of the table contains the element hydrogen and the noble gas helium.

TABLE 4-6 The Periodic Table of Elements (Long Period Form)

There are seven periods (horizontal rows) and eighteen groups (vertical columns) of elements. The number of electrons in filled shells is shown in the column at the extreme left; the remaining electrons for each element are shown immediately below the symbol for each element. Atomic numbers are enclosed by brackets. Atomic weights are shown above the symbols. Atomic weight values in parentheses are those of the isotopes of longest half-life for certain radioactive elements whose atomic weights cannot be quoted precisely without knowledge of origin of the element. Atomic weights are based on carbon-12. (*From Pure and Applied Chemistry,* **51**(2), 405 (1979). By permission of International Union of Pure and Applied Chemistry.)

METALS · TRANSITION METALS · NONMETALS

PERIODS	IA	IIA	IIIB	IVB	VB	VIB	VIIB	VIIIB	VIIIB	VIIIB	IB	IIB	IIIA	IVA	VA	VIA	VIIA	VIIIA
1 — 0	1.0079 H[1] 1																	4.00260 He[2] 2
2 — 2	6.941 Li[3] 1	9.01218 Be[4] 2											10.81 B[5] 3	12.011 C[6] 4	14.0067 N[7] 5	15.9994 O[8] 6	18.998403 F[9] 7	20.179 Ne[10] 8
3 — 2,8	22.98977 Na[11] 1	24.305 Mg[12] 2											26.98154 Al[13] 3	28.0855 Si[14] 4	30.97376 P[15] 5	32.06 S[16] 6	35.453 Cl[17] 7	39.948 Ar[18] 8
4 — 2,8	39.0983 K[19] 8,1	40.08 Ca[20] 8,2	44.9559 Sc[21] 9,2	47.90 Ti[22] 10,2	50.9415 V[23] 11,2	51.996 Cr[24] 13,1	54.9380 Mn[25] 13,2	55.847 Fe[26] 14,2	58.9332 Co[27] 15,2	58.70 Ni[28] 16,2	63.546 Cu[29] 18,1	65.38 Zn[30] 18,2	69.72 Ga[31] 18,3	72.59 Ge[32] 18,4	74.9216 As[33] 18,5	78.96 Se[34] 18,6	79.904 Br[35] 18,7	83.80 Kr[36] 18,8
5 — 2,8,18	85.4678 Rb[37] 8,1	87.62 Sr[38] 8,2	88.9059 Y[39] 9,2	91.22 Zr[40] 10,2	92.9064 Nb[41] 12,1	95.94 Mo[42] 13,1	[98] Tc[43] 14,1	101.07 Ru[44] 15,1	102.9055 Rh[45] 16,1	106.4 Pd[46] 18	107.868 Ag[47] 18,1	112.41 Cd[48] 18,2	114.82 In[49] 18,3	118.69 Sn[50] 18,4	121.75 Sb[51] 18,5	127.60 Te[52] 18,6	126.9045 I[53] 18,7	131.30 Xe[54] 18,8
6 — 2,8,18	132.9054 Cs[55] 18,8,1	137.33 Ba[56] 18,8,2	[57-71] *	178.49 Hf[72] 32,10,2	180.9479 Ta[73] 32,11,2	183.85 W[74] 32,12,2	186.207 Re[75] 32,13,2	190.2 Os[76] 32,14,2	192.22 Ir[77] 32,15,2	195.09 Pt[78] 32,17,1	196.9665 Au[79] 32,18,1	200.59 Hg[80] 32,18,2	204.37 Tl[81] 32,18,3	207.2 Pb[82] 32,18,4	208.9804 Bi[83] 32,18,5	[209] Po[84] 32,18,6	[210] At[85] 32,18,7	[222] Rn[86] 32,18,8
7 — 2,8,18,32	[223] Fr[87] 18,8,1	226.0254 Ra[88] 18,8,2	[89-103] †	[261] Rf[104] 32,10,2	[262] Ha[105] 32,11,2	[263] [106] 32,12,2												

* LANTHANIDE SERIES

138.9055 La[57] 18,9,2	140.12 Ce[58] 20,8,2	140.9077 Pr[59] 21,8,2	144.24 Nd[60] 22,8,2	[145] Pm[61] 23,8,2	150.4 Sm[62] 24,8,2	151.96 Eu[63] 25,8,2	157.25 Gd[64] 25,9,2	158.9254 Tb[65] 27,8,2	162.50 Dy[66] 28,8,2	164.9304 Ho[67] 29,8,2	167.26 Er[68] 30,8,2	168.9342 Tm[69] 31,8,2	173.04 Yb[70] 32,8,2	174.967 Lu[71] 32,9,2

† ACTINIDE SERIES

227.0278 Ac[89] 18,9,2	232.0381 Th[90] 18,10,2	231.0359 Pa[91] 20,9,2	238.029 U[92] 21,9,2	237.0482 Np[93] 23,8,2	244 Pu[94] 24,8,2	[243] Am[95] 25,8,2	[247] Cm[96] 25,9,2	[247] Bk[97] 26,9,2	[251] Cf[98] 28,8,2	[252] Es[99] 29,8,2	[257] Fm[100] 30,8,2	[258] Md[101] 31,8,2	[259] No[102] 32,8,2	[260] Lr[103] 32,9,2

TABLE 4-7 Periodicity of Valence Electrons

Element	H							He
Electrons in shell	**1**							**2**
Element	Li	Be	B	C	N	O	F	Ne
Electrons in shells	2,**1**	2,**2**	2,**3**	2,**4**	2,**5**	2,**6**	2,**7**	2,**8**
Element	Na	Mg	Al	Si	P	S	Cl	Ar
Electrons in shells	2,8,**1**	2,8,**2**	2,8,**3**	2,8,**4**	2,8,**5**	2,8,**6**	2,8,**7**	2,8,**8**

The **second short period** contains eight elements beginning with lithium and ending with the noble gas neon.

The **third short period** also contains eight elements beginning with sodium and ending with the noble gas argon.

The **fourth period** is the first of two long periods each of which contains 18 elements. This period includes the elements from potassium through the noble gas krypton. Within this period are the elements from scandium (at. no. 21) through copper (at. no. 29), which are known as the **first transition series.**

The **fifth period,** beginning with rubidium and ending with the noble gas xenon, also contains 18 elements; within the period are the elements yttrium (at. no. 39) through silver (at. no. 47), which comprise the second transition series.

The **sixth period,** beginning with cesium and ending with the noble gas radon, contains 32 elements. The third transition series, made up of lanthanum (at. no. 57) and the elements hafnium (at. no. 72) through gold (at. no. 79) occurs in the sixth period. Notice that the third transition series is split, however; between lanthanum and hafnium is a series of 14 elements, cerium (at. no. 58) through lutetium (at. no. 71). These 14 elements constitute the **first inner transition series** and are referred to as the **lanthanide series,** or the **rare earth** elements. Inasmuch as lanthanum has an outer electron configuration similar to those of the elements cerium through lutetium, it behaves very much like these elements. Hence, lanthanum is often included in the lanthanide series even though in terms of electron structure it is more properly considered as the first element of the third transition series.

The **seventh period** is incomplete. It extends from francium through element number 106 and includes a portion of the fourth transition series (actinium, plus rutherfordium through the last element to be reported, element 106), analogous to the sixth period. The elements between actinium (at. no. 89) and rutherfordium (at. no. 104), namely the elements thorium (at. no. 90) through lawrencium (at. no. 103), make up a second inner transition series referred to as the **actinide elements.** Actinium is sometimes included with the actinide elements because of its similarity in properties to those elements. It is logical to assume that, although element 106 is the last element to be reported, elements 104, 105, and 106 and any elements that are later discovered immediately beyond them (elements 107 through 111) will constitute a fourth transition series.

Many of the groups (or families) in the Periodic Table are identified by Roman numerals and letters at the top of the vertical columns. Some of these groups also have common names. The members of Group IA, other than hydrogen, are called the **alkali metals.** The elements of Group IIA are called **alkaline earth**

metals. The elements of Group VIA are **chalcogens;** the elements of Group VIIA, the **halogens;** and the elements of Group 0, the **noble gases.** Within a group, or family, there are striking similarities in the chemical behavior of the elements.

4.18 Electronic Structure and the Periodic Law

During the following discussion of the relationship of electronic structures of the atoms to their positions in the Periodic Table, it will be helpful to refer to Fig. 4-17. This figure shows in Periodic Table form the electronic configuration of the last subshell to be filled if the atoms were built up in order of increasing atomic number (the Aufbau Principle).

When arranged in a Periodic Table, elements with similar chemical properties occur in the same regions of the table. As may be seen in Fig. 4-17, elements in the same group, or family, of the table have similar electronic structures. For example, lithium, sodium, and potassium each has only one electron in its outermost shell. Calcium, strontium, and barium each has two. Chlorine, bromine, and iodine each has seven electrons in its outermost shell. Valence electrons are involved in chemical changes. We expect those elements with similar numbers of

Figure 4-17 The order of occupancy of atomic orbitals in the Periodic Table. The configuration of the last subshell to be occupied as the atoms are built up by the Aufbau technique is shown.

s

H	He
$1s^1$	$1s^2$

s | *d* | *p*

1	H $1s^1$																	
2	Li $2s^1$	Be $2s^2$											B $2p^1$	C $2p^2$	N $2p^3$	O $2p^4$	F $2p^5$	Ne $2p^6$
3	Na $3s^1$	Mg $3s^2$											Al $3p^1$	Si $3p^2$	P $3p^3$	S $3p^4$	Cl $3p^5$	Ar $3p^6$
4	K $4s^1$	Ca $4s^2$	Sc $3d^1$	Ti $3d^2$	V $3d^3$	Cr $3d^5$	Mn $3d^5$	Fe $3d^6$	Co $3d^7$	Ni $3d^8$	Cu $3d^{10}$	Zn $3d^{10}$	Ga $4p^1$	Ge $4p^2$	As $4p^3$	Se $4p^4$	Br $4p^5$	Kr $4p^6$
5	Rb $5s^1$	Sr $5s^2$	Y $4d^1$	Zr $4d^2$	Nb $4d^4$	Mo $4d^5$	Tc $4d^6$	Ru $4d^7$	Rh $4d^8$	Pd $4d^{10}$	Ag $4d^{10}$	Cd $4d^{10}$	In $5p^1$	Sn $5p^2$	Sb $5p^3$	Te $5p^4$	I $5p^5$	Xe $5p^6$
6	Cs $6s^1$	Ba $6s^2$	*	Hf $5d^2$	Ta $5d^3$	W $5d^4$	Re $5d^5$	Os $5d^6$	Ir $5d^7$	Pt $5d^9$	Au $5d^{10}$	Hg $5d^{10}$	Tl $6p^1$	Pb $6p^2$	Bi $6p^3$	Po $6p^4$	At $6p^5$	Rn $6p^6$
7	Fr $7s^1$	Ra $7s^2$	†	104 $6d^2$	105 $6d^3$	106 $6d^4$												

f

*	La $5d^1$	Ce $4f^2$	Pr $4f^3$	Nd $4f^4$	Pm $4f^5$	Sm $4f^6$	Eu $4f^7$	Gd $4f^7$	Tb $4f^9$	Dy $4f^{10}$	Ho $4f^{11}$	Er $4f^{12}$	Tm $4f^{13}$	Yb $4f^{14}$	Lu $4f^{14}$
†	Ac $6d^1$	Th $6d^2$	Pa $5f^2$	U $5f^3$	Np $5f^4$	Pu $5f^6$	Am $5f^7$	Cm $5f^7$	Bk $5f^8$	Cf $5f^{10}$	Es $5f^{11}$	Fm $5f^{12}$	Md $5f^{13}$	No $5f^{14}$	Lr $5f^{14}$

valence electrons to exhibit similar chemical behavior since **it is the loss, gain, or sharing of valence electrons that determines how elements react.** Thus, the Periodic Table is simply an arrangement of atoms that puts elements with the same number of valence electrons in the same group.

It is convenient to classify the elements in the Periodic Table into four categories according to their atomic structures.

1. *Noble gases.* Elements in which the outer shell is complete. The noble gases are the members of Group 0. The valence electrons of the noble gases are found only in the outermost shell.
2. *Representative elements.* Elements in which the last electron added enters the outermost shell but in which the outermost shell is incomplete. The outermost shell of these elements is the valence shell. The representative elements are those in Groups IA, IIA, IIB, IIIA, IVA, VA, VIA, and VIIA of the Periodic Table.
3. *Transition elements.* Elements in which the second shell counting from the outside is building from 8 to 18 electrons. The outermost s subshell and the d subshell of the second shell from the outside contain the valence electrons in these elements. Thus the $(n - 1)d$ and ns subshells are regarded as the valence shells in the transition elements. The four transition series are

    First transition series: Scandium (Sc) through copper (Cu) ($3d$ subshell filling).
    Second transition series: Yttrium (Y) through silver (Ag) ($4d$ subshell filling).
    Third transition series: Lanthanum (La); and hafnium (Hf) through gold (Au) ($5d$ subshell filling).
    Fourth transition series (incomplete): Actinium (Ac); and rutherfordium (Rf), hahnium (Ha) and element number 106 ($6d$ subshell filling).

4. *Inner transition elements.* Elements in which the third shell counting from the outside is building from 18 to 32 electrons. The valence shells of the inner transition elements consist of the $(n - 2)f$, $(n - 1)d$, and ns subshells in these elements. The two inner transition series are

    First inner transition series: Cerium (Ce) through lutetium (Lu) ($4f$ subshell filling).
    Second inner transition series: Thorium (Th) through lawrencium (Lr) ($5f$ subshell filling).

    (Lanthanum and actinium, because of their similarities to the other members of the series, are sometimes included as the first elements of the first and second inner transition series, respectively.)

4.19   Variation of Properties Within Periods and Groups

As has already been pointed out, the properties of the elements are determined largely by their atomic structures. Elements within a group have identical numbers and generally identical distributions of electrons in their valence shells. Thus we expect them to exhibit very similar chemical behavior. We might expect a smoothly varying change in chemical behavior across a period, because each element differs from the preceding element by only one electron. However, sometimes the similarities or differences within a group or within a series are not as clear-cut as we might expect. This is a reflection of the fact that the loss, gain, or sharing of valence electrons depends upon several factors, including (1) the number of valence electrons, (2) the magnitude of the nuclear charge and the total number of electrons surrounding the nucleus, (3) the number of filled shells lying between the nucleus and the valence shell, and (4) the distances of the electrons in the various shells from each other and from the nucleus.

Examples of these effects on the periodic variation of some physical properties will be considered in the following sections.

▶**1. VARIATION IN COVALENT RADII.** (See table inside back cover.) There are several ways to define the radii of atoms and thus to determine their relative sizes. We will use the concept of the **covalent radius.** The covalent radius of an atom is taken as one-half the distance between the nuclei of two of these atoms when joined by a single covalent bond. The concept of a covalent bond will be discussed in Chapter 5; for now, we will simply use the covalent radius as one measure of the size of an atom. In general, from left to right across the periods of the Periodic Table, each element has a smaller covalent radius than the one preceding it (Table 4-8). This change in size can be attributed to increasing nuclear charge across the periods, with the added electrons going into partially occupied shells. Each element (except the first element, hydrogen) in the Periodic Table has a nuclear charge one higher than the preceding element. Each element also has one more electron than the preceding element. Within a period, however, the number of shells is constant. In general, within a given period, since the additional electrons are in the same shell, the larger nuclear charge will result in a larger force of electrostatic attraction between the nucleus and the electrons. This in theory causes the decrease in covalent radii across the period.

Proceeding down the groups of the Periodic Table, succeeding elements have increasing covalent radii as a result of larger numbers of electron shells (Table 4-9), a factor that more than offsets the effect of the increased nuclear charge. The nuclear charge of each succeeding element increases down a group. This might cause us to expect the electrons to be held more tightly and to be pulled closer to the nucleus. However, the total number of shells also increases down a group, and shells with larger principal quantum numbers have larger radii. The larger size of the additional shell coupled with repulsions between the increasing number of

TABLE 4-8

Atom	Covalent Radius, Å	Nuclear Charge	Electronic Configuration
Na	1.86	$+11$	$1s^2 2s^2 2p^6 3s^1$
Mg	1.60	$+12$	$3s^2$
Al	1.43	$+13$	$3s^2 3p^1$
Si	1.17	$+14$	$3s^2 3p^2$
P	1.10	$+15$	$3s^2 3p^3$
S	1.04	$+16$	$3s^2 3p^4$
Cl	0.99	$+17$	$3s^2 3p^5$

TABLE 4-9

Atom	Covalent Radius, Å	Nuclear Charge	Electronic Configuration Number of Electrons in Each Shell
F	0.64	$+ 9$	2, 7
Cl	0.99	$+17$	2, 8, 7
Br	1.14	$+35$	2, 8, 18, 7
I	1.33	$+53$	2, 8, 18, 18, 7
At	1.4	$+85$	2, 8, 18, 32, 18, 7

electrons overcomes the increased nuclear attraction so that the atoms increase in size down a group.

Values of covalent radii are based on interatomic distances between atoms held closely together by strong chemical bonds. Noble gases, however, do not exhibit this behavior. Even in a solid noble gas such as frozen argon, only a weak force of attraction holds the atoms together. This force, known as the **London force,** is the electrostatic attraction of the positive nucleus of one atom for the negative electron cloud of a neighboring atom (see Section 11.4). Thus covalent radii are not available for the noble gases.

▶ **2. VARIATION IN IONIC RADII.** A positive ion forms when one or more electrons is removed from an atom. Usually the representative elements form positive ions by loss of all of their valence electrons. As shown in the table inside the back cover, the radius of a positive ion is less than the covalent radius of its parent atom. The loss of all the electrons from the outermost shell as the positive ion is formed results in the smaller radius. Thus the covalent radius of the sodium atom ($1s^22s^22p^63s^1$) is 1.86 Å, whereas the ionic radius of the sodium ion ($1s^22s^22p^6$) is 0.95 Å. Not only does the outer electron shell of the sodium atom disappear as the sodium ion is formed, but also the radii of the two remaining electron shells decrease because of an increase in *effective* nuclear charge as the valence electron is removed, giving rise to a greater average attraction of the nucleus per remaining electron. Down the groups of the Periodic Table, positive ions of succeeding elements have larger radii corresponding to greater numbers of shells.

A simple negative ion is formed by the addition of one or more electrons to the valence shell of an atom. This results in a greater force of repulsion among the electrons and also a decrease in the *effective* nuclear charge per electron. Both effects operate in the same direction to cause the radius of a negative ion to be greater than that of the parent atom. For example, the chlorine atom ($1s^22s^22p^63s^23p^5$) has a covalent radius of 0.99 Å, whereas the ionic radius of the chloride ion, Cl^- ($1s^22s^22p^63s^23p^6$), is 1.81 Å. For succeeding elements down the groups, negative ions have more electron shells, greater nuclear charge, and larger radii. These effects are apparent in the table of covalent and ionic radii presented inside the back cover.

Ions and atoms that have the same electron configuration, such as those in the series N^{3-}, O^{2-}, F^-, Ne, Na^+, Mg^{2+}, and Al^{3+} and those in the series P^{3-}, S^{2-}, Cl^-, Ar, K^+, Ca^{2+}, and Sc^{3+}, are termed **isoelectronic.** The greater the nuclear charge, the smaller the ionic radius in a series of isoelectronic ions. This trend is illustrated in Table 4-10 for these series of isoelectronic ions and atoms.

TABLE 4-10

Species:	N^{3-}	O^{2-}	F^-	Ne	Na^+	Mg^{2+}	Al^{3+}
Radius, Å:	1.71	1.40	1.36	1.12	0.95	0.65	0.50
Electronic configuration:	$1s^22s^22p^6$						

Species:	P^{3-}	S^{2-}	Cl^-	Ar	K^+	Ca^{2+}	Sc^{3+}
Radius, Å:	2.12	1.84	1.81	1.54	1.33	0.99	0.81
Electronic configuration:	$1s^22s^22p^63s^23p^6$						

As we shall see later, many of the properties of ions can best be explained in terms of their sizes and charges.

▶ **3. VARIATION IN IONIZATION POTENTIALS.** The amount of energy required to remove *the most loosely bound* electron from an atom in an endothermic process ($\Delta H > 0$) is called its **ionization potential** or, more precisely, its **first ionization potential.** This change may be represented by

$$X + Energy \longrightarrow X^+ + e^-$$

In general, first ionization potentials increase from left to right across a period (Table 4-11). This may be attributed to the fact that the electrons lost come from the same shell while the nuclear charge increases.

Figure 4-18 shows the relationships between first ionization potentials and atomic numbers of several elements. The values of the first ionization potentials are provided in Table 4-12. Note that the ionization potential of boron is less than that of beryllium. This is explained in terms of the relative attraction by the positive nucleus of an atom for electrons of different subshells. On the average, an s electron will be attracted to the nucleus more tightly than a p electron of the same

TABLE 4-11								
Element	Li	Be	B	C	N	O	F	Ne
Ionization potential, eV	5.39	9.32	8.30	11.26	14.54	13.61	17.42	21.60
Nuclear charge	+3	+4	+5	+6	+7	+8	+9	+10
Valence shell configuration	$2s^1$	$2s^2$	$2s^22p^1$	$2s^22p^2$	$2s^22p^3$	$2s^22p^4$	$2s^22p^5$	$2s^22p^6$

Figure 4-18 A graphic illustration of the periodic relationships between first ionization potentials and the atomic numbers for some of the elements.

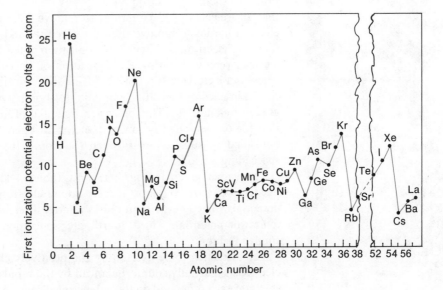

TABLE 4-12 First Ionization Potentials of Some of the Elements (Potentials are given in electron volts per atom; 1 eV per atom corresponds to 96.49 kJ per mole of atoms.)

1 H 13.6																	2 He 24.6
3 Li 5.4	4 Be 9.3											5 B 8.3	6 C 11.3	7 N 14.5	8 O 13.6	9 F 17.4	10 Ne 21.6
11 Na 5.1	12 Mg 7.6											13 Al 6.0	14 Si 8.1	15 P 11.0	16 S 10.4	17 Cl 13.0	18 Ar 15.8
19 K 4.3	20 Ca 6.1	21 Sc 6.5	22 Ti 6.8	23 V 6.7	24 Cr 6.8	25 Mn 7.4	26 Fe 7.9	27 Co 7.9	28 Ni 7.6	29 Cu 7.7	30 Zn 9.4	31 Ga 6.0	32 Ge 8.1	33 As 10	34 Se 9.8	35 Br 11.8	36 Kr 14.0
37 Rb 4.2	38 Sr 5.7	39 Y 6.4	40 Zr 6.8	41 Nb 6.9	42 Mo 7.1	43 Tc 7.3	44 Ru 7.4	45 Rh 7.5	46 Pd 8.3	47 Ag 7.6	48 Cd 9.0	49 In 5.8	50 Sn 7.3	51 Sb 8.6	52 Te 9.0	53 I 10.5	54 Xe 12.1
55 Cs 3.9	56 Ba 5.2	[57-71] *	72 Hf 5.7	73 Ta 7.9	74 W 8.0	75 Re 7.9	76 Os 8.7	77 Ir 9.2	78 Pt 9.0	79 Au 9.2	80 Hg 10.4	81 Tl 6.1	82 Pb 7.4	83 Bi 8	84 Po 8.4	85 At ...	86 Rn 10.7
87 Fr ...	88 Ra 5.3	[89-103] †	104 Rf ...	105 Ha ...	106 —												

	57 La 5.6	58 Ce 6.9	59 Pr 5.8	60 Nd 6.3	61 Pm ...	62 Sm 5.6	63 Eu 5.7	64 Gd 6.2	65 Tb 6.7	66 Dy 6.8	67 Ho ...	68 Er ...	69 Tm ...	70 Yb 6.2	71 Lu 5.0
*LANTHANIDE SERIES															
†ACTINIDE SERIES	89 Ac 6.9	90 Th ...	91 Pa ...	92 U 4	93 Np ...	94 Pu ...	95 Am ...	96 Cm ...	97 Bk ...	98 Cf ...	99 Es ...	100 Fm ...	101 Md ...	102 No ...	103 Lr ...

principal shell, a *p* electron more closely than a *d* electron, and so on. This means that an *s* electron will be harder to remove from an atom than a *p* electron, a *p* electron harder to remove than a *d* electron, and a *d* electron harder to remove than an *f* electron, and hence the ionization potentials decrease in this order. The electron removed during the ionization of beryllium ($1s^2, 2s^2$) is an *s* electron whereas a *p* electron is removed during the ionization of boron ($1s^2, 2s^2, 2p^1$); this results in a lower ionization potential for boron even though its nuclear charge is greater by one unit. The ionization potential of nitrogen is abnormally high and that for oxygen slightly lower than for nitrogen because of the stability of the half-filled $2p$ subshell in nitrogen (Section 4-16). Analogous changes occur in succeeding periods.

The values given in Fig. 4-18 and Tables 4-11 and 4-12 are for the energy needed to remove the most loosely bound electron and are called **first ionization potentials.** The energy required to remove a second electron is called the **second ionization potential.** Third, fourth, etc., ionization potentials can be defined in a similar fashion.

The attractive force exerted by the positively charged nucleus on the valence electrons is partially counterbalanced by the repulsive forces of electrons in inner shells on each other and on the valence electrons. An electron being removed from an atom is thus shielded from the nucleus by these inner shells. This shielding, and the increasing distance of the outer electron from the nucleus, is the apparent explanation of the fact that down the groups succeeding elements generally have smaller ionization potentials. The elements in a group as a rule have the same outer electronic configuration and the same number of valence electrons.

▶ **4. VARIATION IN ELECTRON AFFINITIES.** Just as the ionization potential is a measure of the energy required to remove an electron from an atom to form a positive ion, another quantity, the **electron affinity,** is a measure of the energy involved when an extra electron is added to an atom to form a negative ion. The change is expressed by the equation

$$X + e^- \longrightarrow X^- + \text{energy}$$

A positive electron affinity indicates that energy is produced (*exothermic process,* $\Delta H < 0$) when an electron is placed on an atom. A negative electron affinity indicates that energy must be added (*endothermic process*) to force the electron onto the atom. Elements lying to the left in a period have little tendency to form negative ions by gaining extra electrons; thus their electron affinities tend to be small or negative (Table 4-13). Elements to the right in a period tend to have larger electron affinities.

Although succeeding elements across the periods of the Periodic Table generally have progressively greater electron affinities, exceptions are found with the elements of Group IIA, Group VA, and Group 0. These are the groups with filled *ns* subshells, half-filled *np* subshells, and all subshells filled, respectively, cases in which completely filled or half-filled subshells represent relatively stable configurations.

TABLE 4-13	Electron Affinities of Some Elements. (Electron affinities in electron volts per atom; 1 eV per atom = 96.49 kJ per mole.)						
IA							0
H 0.75	IIA	IIIA	IVA	VA	VIA	VIIA	He −0.2[a]
Li 0.62	Be −2.5[a]	B 0.24	C 1.27	N 0.0	O 1.46	F 3.34	Ne −0.3
Na 0.55	Mg −2.4[a]	Al 0.46	Si 1.24	P 0.77	S 2.08	Cl 3.61	Ar −0.36[a]
K 0.50	Ca −1.6[a]	Ga 0.4[a]	Ge 1.20	As 0.80	Se 2.02	Br 3.36	Kr −0.4[a]
Rb 0.48	Sr −1.7[a]	In 0.4[a]	Sn 1.25	Sb 1.05	Te 1.97	I 3.06	Xe −0.4[a]
Cs 0.47	Ba −0.5[a]	Tl 0.5	Pb 1.05	Bi 1.05	Po 1.8[a]	At 2.8[a]	Rn −0.4[a]

[a] Calculated value.

Questions

1. Describe the production of cathode rays in a discharge tube and describe the evidence that these rays are negatively charged particles.
2. What is the evidence that all atoms contain electrons?
3. Describe the experiments and interpret the results of the experiments that show that the electrons in an atom are not located in its nucleus.
4. Compare the properties of cathode rays and β particles.
5. How are positive rays and α rays similar? How do they differ?
6. What is the experimental evidence that electrons are small, negatively charged particles?
7. Define the following: atomic number, proton, neutron, nuclear charge, isotope, nucleon.
8. In what way do the isotopes of a given element differ?
9. In what way is the lightest hydrogen nucleus unique?
10. Identify the following elements: $^{13}_{6}X$, $^{39}_{19}X$, $^{56}_{26}X$, $^{79}_{35}X$, $^{196}_{78}X$.
11. Why does an element that consists of a single kind of atom (no isotopes) have an atomic weight very close to an integral value?
12. What is the relationship between the number of electrons in a neutral atom and its atomic number?
$\boxed{s}$ 13. What is the general relationship between the atomic weight and atomic number of an element? Why are there occasional exceptions to this relationship?
14. Explain why atomic weights of elements often differ markedly from whole numbers, although the nucleons have masses that are very nearly 1 on the atomic weight scale.
15. Determine the number of protons and neutrons in the nuclei of each of the following elements: $^{9}_{4}Be$, $^{32}_{16}S$, $^{112}_{48}Cd$, $^{157}_{64}Gd$, $^{210}_{82}Pb$, $^{235}_{92}U$.
16. Describe the Bohr model of the hydrogen atom.
17. How does the Bohr model of an atom differ from Rutherford's model of an atom?
18. What is the importance of Rydberg's spectral measurements to the Bohr model of the atom?
19. What is meant when it is said that the possible energies of an electron in an atom are quantized?

20. (a) Are the charges on the nuclei of atoms quantized?
    (b) Are the masses of individual atoms quantized?
    (c) Must the atomic weight of an element that is a mixture of isotopes be quantized?
21. Using the Bohr model, what is the expression for the energy of the electron in the ion Be^{3+}? What is the expression for the radius of the $n = 1$ circular orbit?
22. How does the Bohr model of the atom differ from the quantum mechanical model? How are the two similar?
23. Distinguish between electron shells, subshells, and orbitals.
24. Discuss the following: electron or charge cloud, probable electron density, atomic orbital, radial electron density.
25. What can be determined about the location of an electron using the quantum mechanical model of the atom?
26. Identify the four quantum numbers used in the quantum mechanical model of the atom and indicate what each describes.
27. (a) State Hund's Rule.
    (b) Indicate those groups in the representative elements where Hund's Rule must be used to determine the correct electronic structures.
28. Sketch the shapes of an s and a p orbital and the two possible shapes of a d orbital.
29. Consider the atomic orbitals, (a), (b), and (c) shown below in outline.

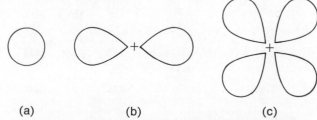

(a)   (b)   (c)

$\boxed{s}$ (a) What is the maximum number of electrons that can be contained in atomic orbital (c)?
$\boxed{s}$ (b) How many orbitals with the same value of l as orbital (a) can be found in the shell with $n = 4$? How many as orbital (b)? How many as orbital (c)?

S(c) What is the smallest n value possible for an electron in an orbital of type (c)? of type (b)? of type (a)?

S(d) What are the l values that characterize each of these three orbitals?

S(e) Arrange these orbitals in order of increasing energy in the M shell. Is this order different in other shells?

30. Identify the subshell in which electrons with the following quantum numbers are found:
(a) $n = 3, l = 0$ S(d) $n = 7, l = 4$
S(b) $n = 3, l = 2$ S(e) $n = 4, l = 3$
S(c) $n = 2, l = 1$

31. What type of orbital is occupied by an electron with the quantum numbers $n = 5, l = 2$? How many equivalent (degenerate) orbitals of this type are in an atom?

S 32. Write the quantum numbers for each electron found in a nitrogen atom. For example, the quantum numbers for one of the 2s electrons will be $n = 2$, $l = 0, m = 0, s = +\frac{1}{2}$.

33. The quantum numbers that describe the electron in the lowest energy level of a hydrogen atom (the ground state) are $n = 1, l = 0, m = 0, s = +\frac{1}{2}$. Excitation of the electron can promote it to energy levels described by other sets of quantum numbers. Which of the following sets of quantum numbers *cannot* exist in a hydrogen atom (or any other atom)?
S(a) $n = 2, l = 1, m = -1, s = +\frac{1}{2}$
S(b) $n = 2, l = 0, m = 1, s = +\frac{1}{2}$
S(c) $n = 7, l = 3, m = 0, s = -\frac{1}{2}$
S(d) $n = 3, l = 3, m = -2, s = -\frac{1}{2}$
S(e) $n = 4, l = 2, m = 3, s = +\frac{1}{2}$
S(f) $n = 3, l = 2, m = -1, s = +\frac{3}{2}$

34. Using subshell notation ($1s^2, 2s^2, 2p^6$, etc.) predict the electron distribution for each of the following atoms and classify each as a representative element, noble gas, transition element, or inner transition element: $^{19}_{9}F$, $^{24}_{12}Mg$, $^{31}_{15}P$, $^{40}_{18}Ar$, $^{51}_{23}V$, $^{96}_{42}Mo$, $^{99}_{43}Tc$, $^{208}_{82}Pb$, $^{144}_{60}Nd$, $^{222}_{86}Rn$.

35. Using subshell notation ($1s^2, 2s^2, 2p^6$, etc.) predict the electronic structures for each of the following ions: Na^+, Si^{2+}, P^{3-}, Cl^-, Ti^{4+}, Tl^+, La^{3+}.

36. The 4s orbitals fill before the 3d orbitals when building up electronic structures by the Aufbau Process. However, electrons are lost from the 4s orbital before they are lost from the 3d orbitals when transition elements are ionized. The ionization of copper ($1s^2 2s^2 2p^6 3s^2 3p^6 4s^1 3d^{10}$) gives Cu^+ ($1s^2 2s^2 2p^6 3s^2 3p^6 3d^{10}$), for example. The electronic structures of a number of transition metal ions are given below; identify the transition metals.
(a) M^{3+}, $1s^2 2s^2 2p^6 3s^2 3p^6$
(b) M^{2+}, $1s^2 2s^2 2p^6 3s^2 3p^6 3d^7$
(c) M^{3+}, $1s^2 2s^2 2p^6 3s^2 3p^6 3d^6$
(d) M^{2+}, $1s^2 2s^2 2p^6 3s^2 3p^6 3d^2$
(e) M^{2+}, $1s^2 2s^2 2p^6 3s^2 3p^6 3d^4$

37. Using subshell notation ($1s^2, 2s^2, 2p^6$, etc.), predict the electronic structures of the following ions: Ti^{3+}, Ni^{2+}, Cr^{3+}, Ru^{3+}, Ag^+. Read the introduction to Problem 36 before attempting this problem.

38. In what way are the ions H^+ and He^{2+} similar? How do they differ from Li^+ and Ne^{2+}?

39. A number of apparent exceptions occur in the expected order of filling of electronic orbitals. Explain why this is so.

40. Define noble gases, representative elements, inner transition elements, and transition elements in terms of s, p, d, and f subshells.

41. State the Periodic Law. Why does the Periodic Law correlate with electronic structure?

42. Define each of the following terms as applied to the Periodic Table: period, group, short period, long period, transition series, inner transition series, noble gas, representative element.

43. In terms of electronic structure, explain why there are 2 elements in the first period of the Periodic Table, 8 elements in the second and third periods, 18 elements in the fourth and fifth periods, and 32 elements in the sixth period.

44. Find the places in the Periodic Table where the positions of the elements are not in keeping with their atomic weights. Account for the positions of these elements.

45. Identify the groups that have the following electronic structures in their valence shells (n represents the principal quantum number).
S(a) ns^2
S(b) $ns^2 np^2$
S(c) $ns^2 np^6$
S(d) $ns^2 (n-1)d^2$
S(e) $ns^2 (n-1)d^5$

[S]46. Why is the radius of a positive ion smaller than the radius of its parent atom?

47. Why do negative ions have larger radii than their parent atoms?

48. Explain the increasing radius of the ions Li^+, Na^+, K^+.

49. Account for the decrease in radius of the isoelectronic ions Na^+, Mg^{2+}, Al^{3+}.

50. Which of the following would have the higher ionization potential: (a) K or Ca? (b) Mg or Ba? (c) Be or B? (d) S or Cl? (e) P or O? Explain each.

51. *Without* looking at the table of sizes of atoms and ions inside the back cover, and basing your answers on locations of the atoms and ions in the Periodic Table, arrange each of the atoms or ions in the following groups in order of increasing size:
(a) Ba, Be, Ca, Mg, Sr
(b) Al, Cl, Mg, Na, P, S, Si
(c) Br^-, Ca^{2+}, K^+, Kr, Se^{2-}
(d) As^{3-}, Be^{2+}, F^-, Na^+, S^{2-}
(e) Ce, Eu, La, Nd, Pm, Pr

52. Which of the following sets of quantum numbers describes the most easily removed electron in an unexcited aluminum atom? Which of the electrons described is most difficult to remove?
[S](a) $n = 1,\ l = 0,\ m = 0,\ s = -\frac{1}{2}$
[S](b) $n = 2,\ l = 1,\ m = 0,\ s = -\frac{1}{2}$
[S](c) $n = 3,\ l = 0,\ m = 0,\ s = \frac{1}{2}$
[S](d) $n = 3,\ l = 1,\ m = 1,\ s = -\frac{1}{2}$
[S](e) $n = 4,\ l = 1,\ m = 1,\ s = \frac{1}{2}$

53. $N^+(g)$ can be produced from $N(g)$ by removing one electron from any of the occupied orbitals of the nitrogen atom. Several of these processes are
(a) $1s^22s^22p^3 \longrightarrow 1s^12s^22p^3$
(b) $1s^22s^22p^3 \longrightarrow 1s^22s^12p^3$
(c) $1s^22s^22p^3 \longrightarrow 1s^22s^22p^2$
The first ionization potential of N is the energy of which of these processes? Which process will require the most energy?

54. A gaseous ion, X^+, in its most stable state has three unpaired electrons. If X is a representative element, to which group of the Periodic Table does it belong?

55. The gaseous ion X^{2+} has no unpaired electrons in its most stable state. If X is a representative element it could be a member of one of two possible groups. Which ones are they?

56. The ion X^- has three unpaired electrons. If X is a representative element, to which group does it belong?

Problems

[S]1. From the per cent abundances and masses given in Table 4-1, calculate the atomic weight of naturally occurring boron.
Ans. 10.81

[S]2. Light that looks green has a frequency of 5×10^{14} per second. What is the wavelength of green light in centimeters and in angstroms? What is the energy of a quantum of green light in joules?
Ans. 6×10^{-5} *cm;* 6×10^3 Å; 3×10^{-19} *J*

[S]3. Does a quantum of the green light described in Problem 2 have enough energy to excite the electron in a hydrogen atom from the K shell to the L shell?
Ans. No, 1.634×10^{-18} *J is required.*

4. Calculate the amount of energy in joules and electron volts required (absorbed) to pull an electron in the hydrogen atom from the energy level closest to the nucleus ($n_1 = 1$) to the third energy level from the nucleus ($n_2 = 3$).
Ans. 12.08 eV; 1.937×10^{-18} *J*

[S]5. Consider a collection of hydrogen atoms with electrons randomly distributed in either the $n = 1, 2, 3, 4$, or 5 shells. How many different wavelengths of light will be emitted by these atoms as the electrons fall into the lower energy states? *Ans. 10*

6. Calculate the lowest and highest energies, in electron volts, of the light produced by the transitions described in Problem 5.
Ans. From $n = 5$: *0.306 eV; 13.06 eV*

7. Calculate the frequencies and wavelengths of the light produced by the transitions described in Problem 6.
Ans. $\nu = 7.40 \times 10^{13}$; 3.157×10^{15} *per second*

8. Calculate the energy required to ionize a hydrogen atom. *Ans. 13.60 eV; 2.179×10^{-18} J*
9. Calculate the energy in joules and in kilocalories required to ionize a mole of hydrogen atoms. *Ans. 1312 kJ/mol; 313.6 kcal/mol*
10. Calculate the diameter of a He^+ ion using the Bohr model of the electronic structure of this ion. *Ans. 0.264 Å*
11. Using the Bohr model, calculate the energy of the light in electron volts emitted by a transition from the L to the K shell in Li^{2+}. *Ans. 91.81 eV*
12. Recently, H atoms with electrons in very high energy levels have been isolated. How large (in centimeters) is a H atom with an electron characterized by an n of 110. *Ans. 6.40×10^{-5} cm*

[S] 13. X rays are produced when the electron stream in an x-ray tube knocks an electron out of a low-lying shell of an atom in the target, and an electron in a higher shell falls into the lower lying shell. The x ray is the energy given off as the electron jumps into the lower shell. The most intense x rays produced by an x-ray tube with a copper target have wavelengths of 1.542 Å and 1.392 Å. These x rays are produced by an electron from the L or M shell falling into the K shell of a copper atom. Calculate the energy separation of the K, L, and M shells in copper. *Ans. K, L: 8.908×10^3 eV; L, M: 8.68×10^2 eV; K, M: 8.040×10^3 eV*

References

"Atoms in Living Color" (Staff), *Chemistry,* January 1979, p. 2.

"What is Smaller Than . . . ," J. O'Reilly, *Chemistry,* January 1979, p. 4.

"The Structure of the Proton and the Neutron," H. W. Kendall and W. Panofsky, *Sci. American,* June 1971, p. 60.

"Atomic Number before Moseley," J. W. van Spronsen, *J. Chem. Educ.*, **56,** 106 (1979).

"Chemical Aspects of Bohr's 1913 Theory," H. Kragh, *J. Chem. Educ.*, **54,** 208 (1977).

"The Flash of Genius: The Bohr Atomic Model: Niels Bohr," A. B. Garrett, *J. Chem. Educ.*, **39,** 534 (1962).

"The Flash of Genius: The Neutron Identified: Sir James Chadwick," A. B. Garrett, *J. Chem. Educ.*, **39,** 638 (1962).

"The Background of Dalton's Atomic Theory," M. B. Hall, *Chem. in Britain,* **2,** 341 (1966).

"On the Discovery of the Electron," B. A. Morrow, *J. Chem. Educ.*, **46,** 584 (1969).

"Robert A. Millikan," D. J. Kevles, *Sci. American,* January 1979, p. 142.

"Before Neutrons," R. A. Heller, *J. Chem. Educ.*, **53,** 714 (1976).

"What Is the Electron, Really?" J. J. Morwick, *J. Chem. Educ.*, **55,** 662 (1978).

"Moseley and the Numbering of the Elements," W. A. Smeaton, *Chem. in Britain,* **1,** 353 (1965).

"Demonstration of the Uncertainty Principle," W. Laurita, *J. Chem. Educ.*, **45,** 461 (1968).

"Chemical Dynamics, Accurate Quantum Mechanical Calculations at Last," A. L. Robinson, *Science,* **191,** 275 (1976).

"Particles, Waves, and the Interpretation of Quantum Mechanics," N. D. Christoudouleas, *J. Chem. Educ.*, **52,** 573 (1975).

"Lasers, The Light Fantastic" (Staff), *J. Chem. Educ.*, **55,** 529 (1978).

"High Rydberg Atoms: Newcomers to the Atomic Physics Scene," R. F. Stebbings, *Science,* **193,** 537 (1976).

"The Spectrum of Atomic Hydrogen," T. W. Hänsch, A. L. Schawlow, and G. W. Series, *Sci. American,* March 1979, p. 94.

"The Absorption Spectra of Alkali Metal Vapors," R. A. Ashby, *J. Chem. Educ.*, **55,** 500 (1978).

"The Zeeman Effect: A Unique Approach to Atomic Absorption," T. H. Maugh, *Science,* **198,** 39 (1977).

"Principles, Methodologies, and Applications of Atomic Fluorescence Spectrometry," J. D. Winfordner, *J. Chem. Educ.*, **55,** 72 (1978).

"Quantum Mechanics You Can See" (Staff), *Science News,* **106,** 68 (1974).

"A Pattern of Chemistry. Hundred Years of the Periodic Table," F. Greenaway, *Chem. in Britain,* **5,** 97 (1969).

"On the Shapes of *f* Orbitals," E. A. Ogryzlo, *J. Chem. Educ.*, **42,** 150 (1965).

"Lord Ernest Rutherford," R. H. Cragg, *Chem. in Britain,* **7,** 518 (1971).

"Structure of the Proton," R. F. Feynman, *Science,* **183,** 601 (1974).

"The Experimental Values of Electron Affinities: Their Selection and Periodic Behavior," E. C. M. Chen and W. E. Wentworth, *J. Chem. Educ.*, **52,** 486 (1975).

"Doubly Charged Negative Atomic Ions of Hydrogen," M. Anbar and R. Schnitzer, *Science,* **191,** 463 (1976).

"Anions of the Alkali Metals," J. L. Dye, *Sci. American,* July 1977, p. 92.

"Laser Detection of Single Atoms," M. H. Nayfeh, *Amer. Scientist,* **67,** 204 (1979).

"What Happened to Alabamine, Virginium and Illinium?" R. F. Trimble, *J. Chem. Educ.*, **52,** 585 (1975).

"Origin of the Elements" (Staff), *J. Chem. Educ.*, **54,** 702 (1977).

"The Physics and Chemistry of a Candle Flame," J. Walker, *Sci. American,* April 1978, p. 154.

5

Chemical Bonding, Part I-General Concepts

The establishment of chemical bonds between the atoms of elements results in the union of elements to form compounds. When the atoms separate, the bonds are destroyed, and the compound no longer exists. Before the discovery of the electrical nature of the atom the character of the forces holding atoms together was a mysterious one. Now we know that these forces are electrical in nature and that the chemical reactions that occur between atoms involve changes in their electronic structures.

5.1 Valence Electronic Symbols and Formulas

The electrons involved in bond formation between atoms are those in the outermost shell (sometimes in the next to the outermost shell) of the neutral atom; these electrons are called **valence electrons.**

The changes in electronic structure that take place during chemical bonding can be expressed by a system of notation in which the symbol of an atom represents all of the atom except the valence electrons; the valence electrons are indicated by small dots (·), crosses (×), or circles (○), written around the symbol for the atom. Such notations are referred to as **valence electronic symbols.** Valence electronic symbols can be used to show the distribution of valence electrons in compounds. Such formulas are referred to as **valence electronic formulas,** or **Lewis formulas.** For clarity, the different symbols for electrons (· , × , ○) may be used as a convenience to distinguish their sources; it must be remembered, however, that all electrons are identical, regardless of their origin.

The valence electronic symbols for the atoms of the second period are:

$$\text{Li}\cdot \quad \text{Be}\colon \quad \dot{\text{B}}\colon \quad \cdot\dot{\text{C}}\colon \quad \cdot\dot{\text{N}}\colon \quad \cdot\ddot{\text{O}}\colon \quad \cdot\ddot{\text{F}}\colon \quad \colon\ddot{\text{Ne}}\colon$$

When two electrons are indicated adjacent to each other, as in Be:, the two electrons are paired with opposite spins in the same orbital.

5.2   Ionic Bonds; Chemical Bonding by Electron Transfer

When an element that loses electrons easily reacts with an element that gains electrons easily, one or more electrons are completely transferred from one element to the other, and ions are produced. The compound formed by this transfer is stabilized by the strong electrostatic forces between the ions of opposite charge present in it. The ions in the compound are said to be held together by **ionic bonds.** The term **electrovalence** has also been used to designate this type of bonding.

An atom that has lost one or more valence electrons possesses a positive charge and is called a **positive ion,** or **cation.** For example, the sodium atom loses its one valence electron and acquires a positive charge of 1 when it enters into chemical combination with an atom of an element such as chlorine. The calcium atom loses its two valence electrons and assumes a positive charge of 2.

$$\text{Na}\cdot \longrightarrow \text{Na}^+ + e^- \qquad \text{Ca}\colon \longrightarrow \text{Ca}^{2+} + 2e^-$$

| Sodium atom | Sodium cation | Calcium atom | Calcium cation |

The smaller the number of valence electrons in an atom the greater the tendency for the atom to lose electrons and thus to form positive ions during reactions with atoms of other elements. Atoms that form cations easily are characterized by relatively low ionization potentials because the ionization potential is simply a quantitative measure of how easily an atom loses an electron. As noted in Section 4.19, atoms with low ionization potentials tend to be those lying to the left in a period or those lying toward the bottom of a group of the Periodic Table. These elements are identified as metals in the table on the inside front cover.

The atoms that pick up electrons readily have relatively high electron affinities and lie to the upper right in the Periodic Table (Section 4.19). These elements, known as nonmetals, lack only a few electrons of having a filled valence shell, and can pick up the electrons lost by metals. A neutral atom, as it fills its valence shells by gaining one or more electrons, becomes a **negative ion,** or **anion.** Nonmetals, such as F, Cl, Br, I, O, and S, form negative ions.

$$\cdot\ddot{\text{Cl}}\colon + e^- \longrightarrow \colon\ddot{\text{Cl}}\colon^- \qquad \cdot\ddot{\text{S}}\colon + 2e^- \longrightarrow \colon\ddot{\text{S}}\colon^{2-}$$

| Chlorine atom | Chloride anion | Sulfur atom | Sulfide anion |

Thus, when metals react with nonmetals, electrons are transferred from the metals to the nonmetals and **ionic compounds** (compounds that contain ions) are formed. The use of valence electronic symbols and Lewis formulas to show the transfer of electrons during formation of some ionic compounds is illustrated as follows:

Metal	Nonmetal	Ionic Compound	
Na ·	+ ×C̈l ⁝	⟶ Na⁺ [⁝C̈l ⁝]⁻	(1)
Sodium atom	Chlorine atom	Sodium chloride (sodium ion and chloride ion)	
Mg :	+ ×Ö ⁝	⟶ Mg²⁺ [×Ö ⁝]²⁻	(2)
Magnesium atom	Oxygen atom	Magnesium oxide (magnesium ion and oxide ion)	
Ca :	+ 2 ×F̈ ⁝	⟶ [⁝F̈ ⁝]⁻ Ca²⁺ [⁝F̈ ⁝]⁻	(3)
Calcium atom	Fluorine atoms	Calcium fluoride (calcium ion and two fluoride ions)	

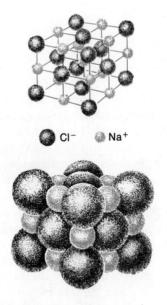

Cl⁻   Na⁺

Figure 5-1 The arrangement of sodium and chloride ions in a crystal of sodium chloride (common salt). The smaller spheres represent sodium ions and the larger ones chloride ions in this drawing. The upper structure is an "expanded" view to show the geometry more clearly.

The actual reactions of sodium, magnesium, and calcium with chlorine, oxygen, and fluorine are not exactly as written since the metal atoms do not occur in solids as isolated metal atoms. Furthermore, the nonmetals in these reactions exist as the diatomic molecules Cl_2, O_2, and F_2. However, the transfer of electrons from the metals to the nonmetals is an important feature of the reaction, and the products have the electronic structures shown.

Equations (1)–(3) show clearly that the ions in an ionic compound have electronic structures quite different from their parent atoms. The sodium ion, for example, has the structure $1s^2 2s^2 2p^6$, whereas the sodium atom has a structure with one more electron, $1s^2 2s^2 2p^6 3s^1$. A chlorine atom changes from the electronic structure $1s^2 2s^2 2p^6 3s^2 3p^5$ to that for the chloride ion, $1s^2 2s^2 2p^6 3s^2 3p^6$, as the atom picks up an electron to form the anion.

These electronic differences result in distinctly different physical and chemical properties for the ion than for the atom of an element. Individual sodium *atoms* combine to give sodium metal, a soft, silvery white metal that burns vigorously in air and reacts rapidly with water. Chlorine *atoms* combine to give chlorine gas, Cl_2, a greenish-yellow gas extremely corrosive to most metals and very poisonous to animals and plants. Sodium chloride, common table salt, is formed in a vigorous reaction between sodium and chlorine. This compound, which contains *ions* of sodium and chlorine, exists as a crystalline solid with the ions exhibiting properties entirely different from those of sodium and chlorine as elements. Chlorine is poisonous, but sodium chloride is essential to life; sodium atoms react vigorously with water, but sodium chloride is stable in water.

A crystal of sodium chloride consists of a regular geometrical arrangement of sodium ions and chloride ions (Fig. 5-1). This arrangement results from each anion and each cation pulling the maximum number of oppositely charged ions around itself. Each sodium ion is surrounded by six chloride ions and each chloride ion is surrounded by six neighboring sodium ions. The force that holds the ions together in the crystal is the electrostatic attraction between ions of opposite charge. This electrostatic force of attraction is very strong in ionic compounds. For example, it requires 776 kilojoules of energy to separate the Na^+ and Cl^- ions in a mole of NaCl.

Any given ion in a crystal of sodium chloride exerts a similar force on all of its six immediate neighbors of opposite charge, and it is therefore impossible to identify any one sodium ion and chloride ion as constituting a molecule of sodium chloride. In truly ionic compounds, no molecules are present and a crystal of an ionic compound is an aggregation of ions. The formula of an ionic compound represents the relative number of ions necessary to give an algebraic balance of ionic charges. Since a crystal of sodium chloride is electrically neutral, it must contain the same numbers of Na^+ and Cl^- ions, and its formula is NaCl. A crystal of sodium oxide contains twice as many Na^+ as O^{2-} ions. Its empirical formula is Na_2O. It follows that the term "molecular weight" has no significance in connection with ionic substances. The formula NaCl represents one formula weight of sodium chloride; it cannot be said to represent a molecular weight, for there are no molecules of sodium chloride. The International System of Units (SI) recommends that the term **molar mass** be used instead of either "molecular weight" or "formula weight" for all compounds.

5.3   The Electronic Structures of Ions

With a few exceptions found among the heavy atoms at the bottom of Groups IIB, IIIA, IVA, and VA, those *representative elements* (Section 4.18) that form cations do so by the loss of all of the electrons in their valence shell. The positive ions thus produced (with the exceptions of those atoms noted) will have a **noble gas electronic configuration** (the same electronic structure as a noble gas), in which the outermost shell in the ion has the electronic structure ns^2np^6 (or $1s^2$ for cations of the second period), or they will have a **pseudo noble gas electronic configuration,** in which the outermost shell has the structure $ns^2np^6nd^{10}$. An example of the latter is the zinc ion (Zn^{2+}), with electronic structure $1s^22s^22p^63s^23p^63d^{10}$, which forms from a zinc atom, with electronic structure $1s^22s^22p^63s^23p^63d^{10}4s^2$. This tendency to lose all valence electrons during the formation of a positive ion of a representative metal simplifies remembering the charges on the various positive ions. The charge is equal to the number of the group in which the ion is found inasmuch as the group number is the same as the number of electrons in the outermost shell for the representative elements. Li^+, Na^+, K^+, Rb^+, Cs^+, and Fr^+ are members of Group IA; Be^{2+}, Mg^{2+}, Ca^{2+}, Ba^{2+}, and Ra^{2+} are members of Group IIA; Zn^{2+}, Cd^{2+}, and Hg^{2+} are members of Group IIB; etc. The exceptions are Hg, Tl, Sn, Pb, and Bi. In addition to the expected formation of Hg^{2+}, Tl^{3+} Sn^{4+}, Pb^{4+}, and Bi^{5+}, these elements can also form Hg_2^{2+}, Tl^+, Sn^{2+}, Pb^{2+}, and Bi^{3+}. It should be pointed out that highly charged cations such as Pb^{4+} and Bi^{5+} are not formed except under quite high energy conditions.

Transition and inner transition metal elements exhibit behavior different from that of the representative elements. Most transition metal cations have $+2$ or $+3$ charges resulting from loss of their outermost s electron (or electrons) sometimes followed by loss of one or two d electrons from the next-to-outer shell. For example, copper ($1s^2 2s^2 2p^6 3s^2 3p^6 3d^{10} 4s^1$) forms the ions Cu^+ ($1s^2 2s^2 2p^6 3s^2$ $3p^6 3d^{10}$) and Cu^{2+} ($1s^2 2s^2 2p^6 3s^2 3p^6 3d^9$). Although the d orbitals of the transition elements are the last to fill when building up electronic structures by the Aufbau Principle, the outermost s electrons are the first to be lost when these atoms ionize.

The formation of monatomic anions by representative elements always involves formation of negative ions with noble gas electronic structures. Thus negative ions such as N^{3-}, S^{2-}, I^-, and O^{2-} form when the neutral atom picks up enough electrons to fill its valence shell completely. With this in mind, it is simple to determine the charge on any monatomic negative ion; it is equal to the number of electrons necessary to fill the valence orbitals of the parent atom. Phosphorus, for example, has the electronic structure $1s^2 2s^2 2p^6 3s^2 3p^3$. The phosphorus anion, P^{3-}, has the noble gas configuration $1s^2 2s^2 2p^6 3s^2 3p^6$, and since three additional electrons are required to fill the valence shell of the atom, the ion has a charge of -3.

5.4   Covalent Bonds; Chemical Bonding by Electron Sharing

In Section 5.2 we considered chemical compounds made up of ions and held together by strong electrostatic forces. There are many compounds, however, that do not consist of ions. These nonionic compounds consist of atoms bonded tightly together in the form of molecules. The bonds holding the atoms together form when pairs of electrons are shared between atoms. Such bonds are called **covalent bonds.** Electron pairs are shared, and a covalent bond forms, between two atoms when both atoms in the bond have about the same tendency to give up electrons to form positive ions or to pick up electrons to form negative ions.

The simplest substance in which atoms are covalently bonded is the hydrogen molecule, H_2. A hydrogen atom has one electron in its $1s$ shell. The two electrons (an electron pair) from the two hydrogen atoms in a hydrogen molecule are shared by the two nuclei (Fig. 5-2), and the shared electrons spend much of their time in the region between the two nuclei. It is the electrostatic attraction that each of the two positively charged nuclei has for the two negatively charged electrons that holds the molecule together. The bond resulting from this attraction is very strong, as evidenced by the large amount of energy required to break the covalent bonds in a mole of hydrogen molecules—436 kilojoules per mole of bonds. Conversely, this same quantity of energy is evolved when a mole of molecular hydrogen is formed from hydrogen atoms.

It is evident that the bonding in the hydrogen molecule cannot be the result of electron transfer, as in ionic compounds, because the two hydrogen atoms have identical ionization potentials and identical electron affinities. Thus no ions can form when two atoms unite by the sharing of a pair of electrons; the product of the union is a molecule. We saw in the preceding section that there is a strong tendency for certain metals and the nonmetals to gain stability by assuming the electronic arrangements of noble gases through the transfer of electrons. This same tendency is operative when atoms unite to form covalent molecules by electron-pair sharing. If the two electrons of the covalent bond are considered to

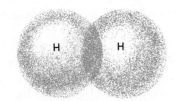

$$H\cdot \; + \; \times H \qquad H\overset{\times}{\cdot}H$$
Hydrogen atoms $\longrightarrow$ Hydrogen molecule

Figure 5-2 Combination of two hydrogen atoms to form a hydrogen molecule, H_2, by covalent bonding.

Figure 5-3 (a) Separate atomic orbitals of hydrogen. (b) Atomic orbitals of hydrogen approach each other closely enough to begin to act upon each other. (c) Distribution of charge within the hydrogen molecules. The electron pair occupies the whole molecule, spending an equal amount of time near each nucleus.

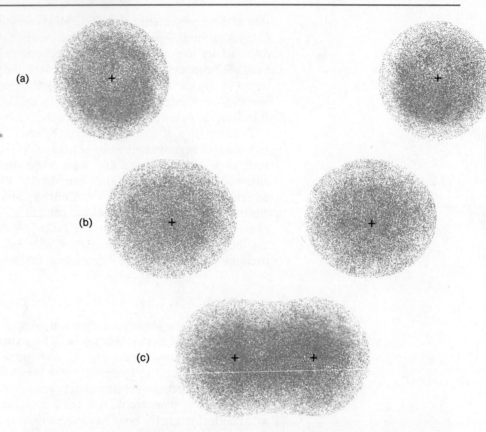

be on each hydrogen atom, each atom has the electronic arrangement of the stable helium atom. The $1s$ orbital of each hydrogen atom in the H_2 molecule is, in effect, occupied by both electrons of the shared pair. The electron pair occupies the whole molecule, spending an equal amount of time near each nucleus. Figure 5-3 illustrates the distribution of charge in the H_2 molecule. The shading represents the intensity of negative charge, that is, the relative probability of finding the electron pair at a given location. In Chapter 6 we shall discover that when two hydrogen atoms combine to form a hydrogen molecule the electron pair goes into a molecular orbital, which is formed from the $1s$ orbitals of the two hydrogen atoms (see Section 6.3).

The bonding in a molecule of chlorine, Cl_2, furnishes a second example of covalent bonding. Each atom of chlorine has seven electrons in its outer shell and differs from the noble gas argon in its electronic configuration by one electron. The sharing of one pair of electrons between two atoms in a molecule of chlorine gives each atom the same stable electronic structure as an atom of argon.

$$:\!\overset{..}{\underset{..}{Cl}}\cdot \;+\; \overset{\times\times}{\underset{\times\times}{\times Cl}}\overset{}{\underset{}{\times}} \longrightarrow \;:\!\overset{..}{\underset{..}{Cl}}\times\overset{\times\times}{\underset{\times\times}{Cl}}\times$$

Chlorine   Chlorine
atoms   molecule

The Lewis formula for Cl_2 indicates that each Cl atom contains three pairs of electrons that are not used in bonding (referred to as **lone pairs**) and one pair of electrons (written between the atoms) that is shared between them. A dash is sometimes used to indicate a shared pair of electrons.

$$H—H \qquad \overset{..}{\underset{..}{:Cl}}—\overset{xx}{\underset{xx}{Cl}}\overset{x}{}$$

A single shared pair of electrons is referred to as a **single bond.**

The bonding in the molecules of the other elements in the same family with chlorine, F_2, Br_2, I_2, and At_2, is like that in the chlorine molecule (one single bond between atoms and three lone pairs per atom), but with the formation of electronic structures like those of the noble gases Ne, Kr, Xe, and Rn, respectively.

The number of covalent bonds (shared electron pairs) that an atom will form often can be predicted by counting the number of electrons that the atom needs to fill its valence shell. Each atom of Group IVA has four electrons in its valence shell and can accept four more electrons to achieve a noble gas configuration as does carbon in CCl_4 (carbon tetrachloride), or silicon in SiH_4 (silane).

Note that the Lewis formula of a molecule in general does not indicate its three-dimensional structure. As will be discussed in Chapter 7, carbon tetrachloride and silane are not flat molecules.

Elements of Group VA, such as nitrogen, need only three electrons to achieve a noble gas configuration. These three electrons can be gained by formation of three covalent bonds, as in NH_3 (ammonia). Oxygen and the other atoms in Group VIA need only two electrons to fill the valence shell; thus they form two single covalent bonds. The elements in Group VIIA such as fluorine need only form one bond in order to fill the valence shell.

A pair of atoms is not limited to sharing a single pair of electrons. Two or three pairs of electrons may be shared between pairs of atoms in order to give the appropriate noble gas structure. For example, two pairs of electrons are shared between C and O in CH_2O (formaldehyde) and between the two C atoms in C_2H_4 (ethylene), giving rise to a **double bond.**

When three electron pairs are shared as in CO (carbon monoxide), or in a nitrogen molecule, a **triple bond** is formed.

$$:C:::O:\quad \text{or} \quad :C\!\equiv\!O:\qquad :N:::N:\quad \text{or}\quad :N\!\equiv\!N:$$

Under normal conditions, the ability to share two or three pairs of electrons and thereby to form double or triple bonds is limited almost exclusively to bonds between carbon, nitrogen, and oxygen atoms. For example, the element nitrogen forms the N_2 molecule, which contains a triple bond, whereas phosphorus forms the P_4 molecule, containing only single bonds. Phosphorus, sulfur, and selenium will sometimes form double bonds with carbon, nitrogen, and oxygen.

The molecules of CO and N_2 are said to be **isoelectronic;** they have the same arrangement of lone pairs and bonding pairs of electrons. Other examples of isoelectronic systems are HCl and OH$^-$; and NH_3 and H_3O^+.

$$H\!-\!\ddot{\underset{..}{Cl}}: \qquad H\!-\!\ddot{\underset{..}{O}}:^- \qquad \overset{\displaystyle H}{\underset{\displaystyle H}{:N\!-\!H}} \qquad \overset{\displaystyle H}{\underset{\displaystyle H}{:O\!-\!H^+}}$$

Note that OH$^-$ and H_3O^+ are ions that both possess several atoms. Such ions are referred to as **polyatomic ions** (Section 5.13). There are many such polyatomic ions, and each is held together by covalent bonds. The compounds containing these polyatomic ions are stabilized both by covalent bonds and by ionic bonds. For example, potassium nitrate, KNO_3, contains the K$^+$ cation and the polyatomic NO_3^- anion. It thus exhibits ionic bonding resulting from the electrostatic attraction between K$^+$ and NO_3^-, and it also contains covalent bonds between nitrogen and oxygen in NO_3^-.

5.5   Covalently Bonded Atoms Without a Noble Gas Structure

Some stable compounds exist in which some atoms of the molecule do not have the noble gas arrangement. For example, boron, with three valence electrons, shares electron pairs with three chlorine atoms in the stable molecule BCl_3.

$$\dot{B}: + 3\cdot\ddot{Cl}: \longrightarrow \; :\overset{}{Cl}\!\overset{120°}{\underset{}{-}}\!B\!\!\begin{array}{c}\nearrow \ddot{Cl}\cdot \\ \searrow \\ \cdot\ddot{Cl}\cdot\end{array}$$

This union of atoms gives each chlorine atom an argon structure; however, boron with six electrons in its outer shell does not have the electron arrangement of a noble gas.

Although the elements in the second row of the Periodic Table never form compounds with more than eight electrons in their valence shells, larger atoms in which the outermost electron shell is the M shell or higher can participate in covalent bonding in which more than four pairs of electrons are shared with other atoms. For example, the phosphorus atom in PF_5 shares five pairs of electrons, or ten electrons in all, whereas atoms of the noble gases contain eight electrons in their outer shells. The outermost, or M, shell of phosphorus has a theoretical maximum capacity of eighteen electrons, or nine electron pairs, but no examples of phosphorus compounds are known in which this condition exists. Sulfur shares six electron pairs (twelve electrons) in the SF_6 molecule; and iodine shares seven electron pairs in IF_7. There are eight shared pairs of electrons in OsF_8.

The three lone pairs on each fluorine atom on these Lewis structures have been omitted for clarity.

In some cases, the number of electrons in the outer shell exceeds eight, even though some of the pairs on the central atom are not shared. This is the case with IF_5 and XeF_4.

As in the preceding paragraph, the lone pairs on each fluorine atom have been omitted.

The formation of covalent compounds such as BCl_3, PCl_5, SF_6, and IF_7 conforms to a rule that supplements the noble gas structure, or octet rule; namely, electrons tend to occur in pairs in molecular structures. In other words, we may state that *the atoms in most covalent molecules appear to have reached a stable condition by sharing pairs of electrons with each other.* Thus the boron atom, which can form only three bonds by sharing its electrons because it has only three valence electrons, attains a condition of stability by forming three electron-pair bonds in the molecule BCl_3.

5.6   Electronegativity of the Elements

When a covalent bond is formed between identical atoms, the electrons in the bond are shared equally between the atoms. However, when a bond is formed between unlike atoms, the electron pair need not necessarily be shared equally. The electrons may spend more time closer to one atom than to the other. For example, one hydrogen atom combines covalently with one chlorine atom in the formation of a molecule of hydrogen chloride.

Although the electrons of the pair are shared between the hydrogen atom and chlorine atom, the electrons are not shared equally, as they are in H_2 and Cl_2. A chlorine atom attracts electrons more strongly than a hydrogen atom does, causing the electrons of the shared pair to be more closely associated with the chlorine nucleus. This results in the development of a small positive charge (often referred to as a partial positive charge) on the hydrogen atom and a partial negative charge on the chlorine atom. This does not imply, however, that the hydrogen atom has lost its electron; it means that the electrons of the pair spend more time *on the average* in the vicinity of the chlorine nucleus than they do near the hydrogen nucleus. Another way of stating this is to say that the electron density, or the density of the electron cloud, is greater around the chlorine nucleus than around the hydrogen nucleus.

The extent of attraction that an atom has for a shared pair of electrons is known as its electronegativity. **Electronegativity** is a measure of the attraction of an atom for electrons *in a covalent bond*. The values of the electronegativities of many of the elements are given in Table 5-1. These values are based upon an arbitrary scale, meaning that we cannot say, for example, that fluorine (4.1) is twice as electronegative as boron (2.0). The electronegativity values are not a measure of absolute electronegativity, but they do provide a measure of differences in electronegativity. For example, the difference in electronegativity between boron (2.0) and nitrogen (3.0) is the same as that between carbon (2.5) and oxygen (3.5). Electronegativity values for a given element vary a little from compound to compound. Thus the values in Table 5-1 are average values.

TABLE 5-1 Approximate Electronegativity Values of Some of the Elements, According to the Periodic Table Arrangement

METALS · NONMETALS

H 2.1																H 2.1	He ...
Li 1.0	Be 1.5			TRANSITION METALS								B 2.0	C 2.5	N 3.0	O 3.5	F 4.1	Ne ...
Na 1.0	Mg 1.2											Al 1.5	Si 1.7	P 2.1	S 2.4	Cl 2.8	Ar ...
K 0.9	Ca 1.0	Sc 1.2	Ti 1.3	V 1.4	Cr 1.6	Mn 1.6	Fe 1.6	Co 1.7	Ni 1.8	Cu 1.8	Zn 1.7	Ga 1.8	Ge 2.0	As 2.2	Se 2.5	Br 2.7	Kr ...
Rb 0.9	Sr 1.0	Y 1.1	Zr 1.2	Nb 1.2	Mo 1.3	Tc 1.4	Ru 1.4	Rh 1.4	Pd 1.4	Ag 1.4	Cd 1.5	In 1.5	Sn 1.7	Sb 1.8	Te 2.0	I 2.2	Xe ...
Cs 0.9	Ba 1.0	La-Lu 1.1-1.2	Hf 1.2	Ta 1.3	W 1.4	Re 1.5	Os 1.5	Ir 1.6	Pt 1.4	Au 1.4	Hg 1.4	Tl 1.4	Pb 1.6	Bi 1.7	Po 1.8	At 2.0	Rn ...
Fr 0.9	Ra 1.0	Ac-Lr 1.1-															

In general, for the representative elements, electronegativity increases across a period from left to right and up a group. Thus the nonmetals, which lie toward the right in the Periodic Table, tend to have higher electronegativities than the metals, although there are some irregularities in the center of the table. This is another example of the periodic variation in properties of the elements.

Fluorine, the most chemically active nonmetal, has the highest electronegativity (4.1), and cesium, the most chemically active metal (with the possible exception of francium), has the lowest electronegativity (0.9). Because the metals have relatively low electronegativities and tend to assume positive charges in compounds, they are often spoken of as being **electropositive;** conversely, nonmetals are said to be **electronegative.**

5.7   Writing Lewis Formulas

Many Lewis formulas for compounds of the representative elements can be written by inspection. If we consider a molecule or ion as being formed from individual atoms, when writing these formulas, unpaired electrons on the reacting atoms are paired up giving single or multiple bonds.

$$\text{H} \cdot + \cdot \ddot{\underset{..}{\text{Br}}} : \longrightarrow \text{H} : \ddot{\underset{..}{\text{Br}}} :$$

$$2\text{H} \cdot + \cdot \ddot{\underset{.}{\text{S}}} : \longrightarrow \text{H} : \ddot{\underset{\underset{\displaystyle \text{H}}{|}}{\text{S}}} :$$

$$: \dot{\text{N}} \cdot + \cdot \dot{\text{N}} : \longrightarrow : \text{N} : : : \text{N} :$$

Sometimes it is helpful to follow the general procedure outlined below. Let us determine the Lewis formulas of $PO_2F_2^-$ and CO as examples of the use of this procedure.

Step 1.   Draw a skeleton structure of the molecule or ion showing the arrangement of atoms and connect each atom with a single (one electron pair) bond.

$$\begin{array}{c} \text{F} \\ | \\ \text{F} - \text{P} - \text{O}^- \qquad \text{C} - \text{O} \\ | \\ \text{O} \end{array}$$

Sometimes only one arrangement of atoms will be possible. In other cases, experimental evidence must be used to decide among several alternative possibilities. As a guideline, we can use the general rule that the less electronegative element is often found as the central atom (as in $PO_2F_2^-$, SF_6, ClO_4^-, and PCl_5), except that hydrogen does not serve as a central atom.

Step 2.   Determine the total number of valence electrons in the molecule or ion. For a molecule, this is equal to the sum of the number of valence electrons on each atom. For a positive ion, it is equal to the sum of the number of valence electrons minus the positive charge on the ion. For a negative ion, it is equal to the sum of the valence electrons plus the number of negative charges on the ion.

$PO_2F_2^-$:   Number of valence electrons = 5 (P atom) + 6 (O atom)
                    + 6 (O atom) + 7 (F atom) + 7 (F atom) + 1 (negative charge) = 32

CO: Number of valence electrons = 4 (C atom) + 6 (O atom) = 10

Step 3. Deduct the two valence electrons that are used in each of the bonds written in Step 1. Distribute the remaining electrons as unshared electron pairs, so that each atom (except a hydrogen atom) has eight electrons if possible. If there are too few electrons to give each atom eight electrons, convert single bonds to multiple bonds. Note that the covalent compounds of beryllium, magnesium, and the elements of Group IIIA (B, Al, Ga, In, Tl) only form single bonds. These elements may have fewer than eight electrons in their valence shells when they function as central atoms. For example, see the Lewis formula for BCl_3 in Section 5.5.

In the ion $PO_2F_2^-$, the total of 32 valence electrons can be distributed as four electron pairs (eight electrons) in bonds and three lone pairs around each of the fluorine and oxygen atoms. The ten valence electrons in CO cannot be distributed as one bonding pair and four lone pairs with eight electrons about each atom.

Filled valence shells about the carbon and oxygen atoms can only be obtained if each atom has three shared pairs in a triple bond and one lone pair.

Step 4. In those molecules in which there are too many electrons to have only eight electrons around each atom, the central atom may have more than eight electrons in its valence shell (see, for example, the Lewis formulas of PF_5, SF_6, IF_5, and IF_7 in Section 5.5). However, note that the outer atoms normally contain eight electrons in their valence shells. A Lewis formula with more than eight electrons for an atom of the first or second period is almost certainly incorrect and should be reexamined.

5.8   Resonance

Sometimes two or more equivalent Lewis formulas with the same arrangement of atoms but with differing arrangements of electrons can be drawn. Such is the case with sulfur dioxide, SO_2, and with the nitrite ion, NO_2^-.

Each of the structures for SO_2, and each of the structures for NO_2^-, have their atoms in identical positions, but differ in their arrangement of electrons.

A double bond between two atoms is shorter in length than a single bond between the same two atoms. However, experiment shows that both sulfur-oxygen bonds in SO_2 have the same length and are identical in all other properties. Likewise both nitrogen-oxygen bonds in NO_2^- are identical. Since it is not possible to write a single Lewis formula for these species in which both bonds are equivalent, we handle the problem with the concept of **resonance.** If two or more Lewis formulas can be written for a molecule or ion, the actual distribution of electrons in the molecule or ion *is an average* of that shown by the various formulas. The distribution of electrons in the sulfur-oxygen bonds in SO_2 and in the nitrogen-oxygen bonds in NO_2^- is an average of that shown in the two Lewis formulas for each species; that is, an average of a double bond and a single bond. The individual Lewis formulas are referred to as **resonance formulas.** The actual electronic structure of the molecule (the average of the resonance formulas) is called a **resonance hybrid** of the individual resonance formulas. A double-headed arrow is used between Lewis formulas to indicate that they are resonance formulas and that *the true distribution of electrons is an average of the individual resonance formulas.*

The distribution of electrons in N_2O, nitrous oxide, may be represented by two resonance formulas as follows:

$$:\ddot{N}=N=\ddot{O}: \quad \longleftrightarrow \quad :N\equiv N-\ddot{\ddot{O}}:$$

From these two resonance formulas, the distribution of electrons in the bond between the two nitrogen atoms may be seen to be greater than that found in an ordinary double bond but less than that in an ordinary triple bond. The distribution of electrons in the bond between nitrogen and oxygen lies between that in a single bond and in a double bond.

At no time does a molecule described as a resonance hybrid possess an electronic structure described by a single resonance formula. The actual electronic structure is always an average of that shown by all resonance formulas. A homely analogy of resonance is the mule, which is the hybrid offspring of a jackass and a mare. Just as the characteristics of a mule are fixed, so the properties of a resonance hybrid are fixed, with no oscillation between the contributing electronic structures. The mule is not a jackass part of the time and a mare part of the time. It is always a mule. Correspondingly, a molecule with an electronic structure described by a resonance hybrid does not exhibit one electronic structure part of the time and another electronic structure the remainder of the time. Instead, the material is always in the form of the intermediate resonance hybrid, which cannot be written with a single Lewis structure.

5.9   Polar Covalent Bonds and Polar Molecules

A bond between a pair of atoms is said to be polar if its center of positive charge does not coincide with its center of negative charge. When a covalent bond is formed between atoms of different electronegativities, the pair of electrons will be more closely associated with the more electronegative atom, and the resulting covalent bond will be somewhat polar. As mentioned previously, the chlorine atom in the hydrogen chloride molecule attracts the pair of electrons of the covalent bond more strongly than does the hydrogen atom (Section 5.6). The hydrogen-chlorine bond is polar, the chlorine atom becoming somewhat negative and the hydrogen atom becoming somewhat positive as the bond is formed.

The greater the difference between the electronegativities of the two atoms involved in a bond, the greater the polarity of the bond. Thus, the polarity of the bond in the hydrogen halides increases in the order HI < HBr < HCl < HF, corresponding to an increase in electronegativity of the halogens: I (2.2), Br (2.7), Cl (2.8), and F (4.1). If the difference in electronegativity between the two atoms is sufficiently large, the electron furnished by the atom of lower electronegativity will be transferred completely to the more electronegative atom, and ionic bonding, rather than covalent bonding, will result. Such is the case with LiF, NaCl, K_2O, Li_3N, and CsBr, where the electronegativity difference is greater than about 2. The other extreme may be achieved when identical atoms share a pair of electrons as in H—H, where the bond is covalent with no polarity. Bonds between atoms with small differences in electronegativities are primarily covalent in character. This is the case with CO_2, CCl_4, I_2O_5, NI_3, ICl, NO, and BN. It follows that bonds that are primarily ionic are formed when elements at the extreme left of the Periodic Table react with elements at the extreme right of the table, whereas bonds that are primarily covalent are formed when elements close together in the table react with one another.

There is no sharp dividing line between compounds in which the bonding is covalent and those in which the bonding is ionic. In the intermediate cases the molecules will have bonds that possess some of the nature of both covalent and ionic bonds and are often referred to as **covalent bonds with partial ionic character,** or **polar covalent bonds.**

If the centers of positive and negative charge in a molecule do not coincide, the molecule may be polar. Such a molecule is said to possess a **dipole.** Molecules with dipoles contain polar covalent bonds. For example, since the centers of positive and negative charges do not coincide in a molecule of hydrogen chloride or of one of the other hydrogen halides, the molecule is electrically unsymmetrical. Because of the separation of centers of charge, polar molecules tend to turn when placed in an electric field, with the positive end of the molecule oriented toward the negative plate and the negative end toward the positive plate (Fig. 5-4).

It is possible to have bonds that are of the polar covalent type but where the molecule as a whole is nonpolar. If a molecule contains several polar covalent bonds directed in such a way as to give a symmetrical molecule, then the molecule is nonpolar. This is illustrated by $HgCl_2$, in which each of the covalent bonds is polar while the molecule as a whole is nonpolar. The centers of positive and negative charge within each molecule are identical. Each chlorine is negative with respect to positive mercury, and each mercury-chlorine bond has some polar character. However, these bond polarities counterbalance each other because the bonds are directed in such a manner as to give an electrically symmetrical molecule.

$$Cl^- —^+Hg^+ —^-Cl$$

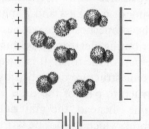

Polar molecules randomly oriented in the absence of an electric field

Positive plate   Negative plate

Polar molecules tending to line up in an electric field

Figure 5-4 Polar molecules, such as hydrogen chloride, tend to line up when in an electric field, with the positive ends oriented toward the negative plate and the negative ends toward the positive plate.

5.10   Coordinate Covalence

We have noted that covalent bonding involves the sharing of electron pairs between atoms, with both atoms involved in the bond furnishing equal numbers of electrons. When only one of the two atoms involved in the linkage furnishes both electrons of the electron-pair bond, the bonding is called **coordinate covalence.** An example of coordinate covalence is provided by the ammonium ion, NH_4^+. The bonds in the ammonia molecule itself are of the covalent type. The unshared pair of electrons originally belonging to the nitrogen atom is available for use in bond formation as indicated by the readiness with which ammonia will combine with a hydrogen ion to form the ammonium ion.

$$
\begin{array}{c}
H \\
\overset{\cdot\times}{H \times N \overset{\times}{\underset{\cdot\times}{}}} \\
H
\end{array}
+ H^+ \longrightarrow
\left[
\begin{array}{c}
H \\
\overset{\cdot\times}{H \times N \overset{\times}{}H} \\
\overset{\cdot\times}{H}
\end{array}
\right]^+
$$

Ammonia   Ammonium
molecule, NH_3 ion, NH_4^+

Because NH_3 is a neutral molecule, the union with a hydrogen ion (proton) gives a unit positive charge to the resulting ammonium ion. In a similar fashion, water molecules combine with hydrogen ions to form hydronium ions.

$$
\begin{array}{c}
\overset{\times\times}{H \times O \overset{\times}{\underset{\cdot\times}{}}} \\
H
\end{array}
+ H^+ \longrightarrow
\left[
\begin{array}{c}
\overset{\times\times}{H \times O \overset{\times}{}H} \\
\overset{\cdot\times}{H}
\end{array}
\right]^+
$$

Water   Hydronium
molecule   ion, H_3O^+

Even though one electron pair on the H_3O^+ is still available for bonding, a second hydrogen ion is not likely to be attracted to the hydronium ion due to the electrostatic repulsion between these two positive ions.

The formation of a coordinate covalent bond is possible only between an atom or ion with an unshared pair of electrons in its valence shell and an atom or ion that needs a pair of electrons to acquire a more stable electronic configuration. The chief difference between the coordinate covalent bond and the normal covalent bond is in the mode of formation. Once established, they are indistinguishable since electrons are indistinguishable regardless of their source. There is no difference between any of the N—H bonds in NH_4^+ or any of the O—H bonds in H_3O^+.

In our discussion of bonding, we have thus far considered the electrons as being located within atomic orbitals of the atoms undergoing bonding. Actually, the valence electrons move about over the molecule as a whole in what are called **molecular orbitals** instead of being restricted to the atomic orbitals of specific individual atoms. In the next chapter we shall discuss how atomic orbitals combine to form molecular orbitals and also point out some of the ways in which this concept is especially useful in explaining the properties of molecules.

5.11   Oxidation Numbers

Oxidation numbers, sometimes referred to as oxidation states, are a useful book-keeping concept and are of value in writing chemical formulas, naming compounds, and keeping track of the redistribution of electrons during chemical reactions. The **oxidation number,** or **oxidation state,** can be determined by application of the following rules:

1. The oxidation number of any element in its elemental form is zero no matter what the complexity of the molecule in which the element is found. The atoms of Na, N_2, P_4, or S_8 all have oxidation numbers of zero.
2. The oxidation number of a monatomic ion is equal to the charge on the ion. The ions Li^+, Co^{2+}, Cl^-, and N^{3-} have oxidation numbers $+1$, $+2$, -1, and -3, respectively.
3. The oxidation number of fluorine in a compound is always -1.
4. The oxidation numbers of the elements of Group IA (except hydrogen) are always $+1$ in compounds.
5. The oxidation numbers of the elements of Group IIA are always $+2$ in compounds.
6. The elements of Group VIIA have oxidation numbers of -1 in compounds in which they are combined with less electronegative elements. Chlorine has an oxidation number of -1 in NaCl, CCl_4, PCl_3, and HCl, for example.
7. Oxygen usually has an oxidation state of -2 with three exceptions:
   (a) In compounds with fluorine it has a positive oxidation number (fluorine is more electronegative than oxygen). An example is oxygen difluoride, OF_2.
   (b) In peroxides (compounds which contain an O—O single bond) it has an oxidation number of -1. An example is hydrogen peroxide, H_2O_2.
   (c) In superoxides (compounds containing O_2^-) it has an oxidation number of $-\frac{1}{2}$. An example is potassium superoxide, KO_2.
8. Hydrogen has an oxidation number of -1 in compounds with less electronegative elements; an oxidation number of $+1$ with more electronegative elements.
9. The sum of the oxidation numbers of all of the atoms in a compound is zero. The sum of the oxidation numbers of all atoms in an ion is equal to the charge on the ion.

If the oxidation numbers are known for all but one kind of atom in a compound, the remaining oxidation number can be calculated. In Na_2SO_4, for example, the oxidation number for sulfur can be calculated from the known oxidation numbers for sodium and oxygen. The two sodium atoms, each with an oxidation number of $+1$ (Group IA, rule 4), total $+2$; the four oxygen atoms, each with an oxidation number of -2 (rule 7) total -8. For the sum of the oxidation numbers to be zero sulfur must have an oxidation number of $+6$. For Na_2SO_3 a similar calculation shows the oxidation number of sulfur in that compound to be $+4$. In H_2S, the oxidation number of sulfur is -2.

It should be emphasized that, although the concept of oxidation numbers is of great convenience in writing formulas and in balancing oxidation-reduction equations, the concept is quite arbitrary. This is especially apparent in a compound in which the calculation of oxidation number results in a fractional value.

For example, with Fe_3O_4 the four oxygen atoms, each -2, would contribute a total of -8. For the unit to be neutral, the three iron atoms must together contribute a total of $+8$ units of charge, or $+\frac{8}{3}$ or $+2\frac{2}{3}$ each. The oxidation number always refers to one atom of an element and hence is $+2\frac{2}{3}$ for iron in this example. It can be noted for this particular case that if one atom of iron is considered to be $+2$ (as in FeO) and two atoms of iron are considered to be $+3$ each (as in Fe_2O_3) the *average weighted* oxidation state for the iron atoms is $2\frac{2}{3}$. Fe_3O_4 can thus be considered as if it were a composite of one unit of FeO combined with one unit of Fe_2O_3. However, analogous reasoning cannot be applied to all molecules possessing an atom with a fractional oxidation number.

Many elements exhibit more than one oxidation number in their various compounds. For example, iron has an oxidation number of $+2$ in $FeCl_2$ and an oxidation number of $+3$ in $FeCl_3$. Tin exhibits oxidation numbers of $+2$ and $+4$ in $SnCl_2$ and $SnCl_4$, respectively. The oxidation number of chlorine in each of the above examples is -1. Chlorine, however, exhibits an oxidation number of $+1$ in NaClO, $+3$ in $NaClO_2$, $+5$ in $NaClO_3$, and $+7$ in $NaClO_4$.

The distribution of electrons in the molecule is a more fundamental property of a molecule than oxidation number, but in many cases a close relationship exists between electron distribution and oxidation number. We should always think of the calculation of oxidation numbers as a useful but quite arbitrary concept.

5.12   Application of Oxidation Numbers to Writing Formulas

We can write the formulas of a great many compounds by knowing the oxidation numbers of the constituent elements of each compound. The principal oxidation number of an element is, in general, predictable from the position of the element in the Periodic Table and from a knowledge of the electronic structure of the atom. The writing of formulas by using oxidation numbers is possible because the algebraic sum of the units of positive and negative oxidation number for any molecule must be equal to zero.

EXAMPLE 1   **Write the formula for magnesium chloride using the oxidation numbers of its constituent elements.**

Magnesium is a member of Group IIA and so will have an oxidation number of $+2$. Chlorine, a member of Group VIIA, is combined with a less electronegative element and so has an oxidation number of -1.

The formula of magnesium chloride cannot be MgCl, because $+2$ and -1 do not add up to 0. For the total of the oxidation numbers to be zero for the compound, the ions must be in a ratio of one magnesium ion to two chloride ions, or $MgCl_2$.

◆ ◆ ◆ ◆ ◆

EXAMPLE 2   **Write the formula for aluminum oxide.**

Aluminum, the less electronegative element, is a member of Group IIIA and will lose three electrons, giving Al^{3+} with an oxidation number of $+3$. Oxygen, a

This guy is positively the most boring teacher I've have [handwritten note]

member of Group VIA, needs two electrons to fill its valence shell, giving the ion O^{2-} with an oxidation number of -2. Because $+3$ plus -2 does not give 0, AlO is not the correct formula for aluminum oxide. By inspection it is readily seen that two atoms of aluminum would give a total of six units of positive oxidation number, that three atoms of oxygen would give six units of negative oxidation number, and that the algebraic sum of the oxidation numbers would be zero. The correct simplest formula for aluminum oxide is, therefore, Al_2O_3.

◆ ◆ ◆ ◆ ◆

5.13   Polyatomic Ions

As discussed in Section 5.4, many ions, referred to as polyatomic ions, contain more than one atom. Several examples, selected from many such ions, are given in Table 5-2.

TABLE 5-2   Some Common Polyatomic Ions

Ammonium	NH_4^+	Carbonate	CO_3^{2-}	Phosphate	PO_4^{3-}
Acetate	$CH_3CO_2^-$	Sulfate	SO_4^{2-}	Diphosphate	$P_2O_7^{4-}$
Nitrate	NO_3^-	Sulfite	SO_3^{2-}	Arsenate	AsO_4^{3-}
Nitrite	NO_2^-	Thiosulfate	$S_2O_3^{2-}$	Arsenite	AsO_3^{3-}
Hydroxide	OH^-	Peroxide	O_2^{2-}		
Hypochlorite	ClO^-	Chromate	CrO_4^{2-}		
Chlorite	ClO_2^-	Dichromate	$Cr_2O_7^{2-}$		
Chlorate	ClO_3^-	Silicate	SiO_3^{2-}		
Perchlorate	ClO_4^-				
Permanganate	MnO_4^-				

For ions containing more than one atom, the algebraic sum of the positive and negative oxidation numbers of the constituent atoms must equal the charge on the ion. Hence, for the OH^- ion, the -2 oxidation number of oxygen and the $+1$ oxidation number of hydrogen add to give the -1 charge for the ion.

It is customary in writing formulas of compounds that include more than one unit of a given polyatomic ion to enclose the formula of the ion in parentheses and to indicate with a subscript the number of such ions in the compound. Examples are $(NH_4)_2CO_3$ and $Al_2(SO_4)_3$. In $(NH_4)_2CO_3$, two ammonium ions, each with a $+1$ ionic charge, are necessary to balance the -2 ionic charge of the carbonate ion. In $Al_2(SO_4)_3$, two aluminum ions, each with a charge of $+3$, and three sulfate ions, each with a charge of -2, are required to balance the charges. It should be noted that the sum of the total positive and negative oxidation numbers for the various atoms, as well as the sum of the total charges on the ions, equals zero for each compound.

The following examples illustrate the process of writing formulas for compounds containing polyatomic ions.

EXAMPLE 1   Write the formula for iron perchlorate, given that the oxidation number for iron in the compound is $+3$.

The oxidation number of a monatomic ion is equal to the charge on the ion. Thus the iron ion is Fe^{3+}. Consideration of the charge of $+3$ for the iron ion and the charge of -1 for the perchlorate ion (see Table 5-2) shows the formula $Fe(ClO_4)$ to be incorrect. By using three perchlorate ions and one iron ion, the sum of the positive and negative charges becomes zero; the correct formula is $Fe(ClO_4)_3$.

◆ ◆ ◆ ◆ ◆

EXAMPLE 2   **Write the formula for calcium phosphate.**

Because the charge for the calcium ion is $+2$ (Ca is a member of Group IIA) and the charge on the phosphate ion is -3, the formula cannot be $Ca(PO_4)$, because the algebraic sum of the charges on the ions must be zero for the compound. By using three calcium ions and two phosphate ions the algebraic sum becomes zero, $3(+2) + 2(-3) = 0$. Thus $Ca_3(PO_4)_2$ is the correct formula for calcium phosphate.

◆ ◆ ◆ ◆ ◆

5.14   The Names of Compounds

▶**1. BINARY COMPOUNDS.**   Binary compounds contain two different elements. The name of a binary compound consists of the name of the more electropositive element followed by the name of the more electronegative element with its ending replaced by the suffix "ide."   Some examples are

NaCl, sodium chloride	Na_2O, sodium oxide
KBr, potassium bromide	CdS, cadmium sulfide
CaI_2, calcium iodide	Mg_3N_2, magnesium nitride
AgF, silver fluoride	Ca_3P_2, calcium phosphide
HCl, hydrogen chloride	Al_4C_3, aluminum carbide
HBr, hydrogen bromide	LiH, lithium hydride
H_2S, hydrogen sulfide	Mg_2Si, magnesium silicide

A few polyatomic ions have special names and are treated as if they were single atoms in naming their compounds; thus NaOH is called sodium hydroxide; HCN, hydrogen cyanide; and NH_4Cl, ammonium chloride.

If a binary hydrogen compound is an acid when it is dissolved in water, the prefix "hydro" is used, and the suffix "ic" replaces the suffix "ide" when we are referring to the solution.

HCl(aq), hydrochloric acid	H_2S(aq), hydrosulfuric acid
HBr(aq), hydrobromic acid	HCN(aq), hydrocyanic acid

When an element of variable oxidation number unites with another element to form more than one compound, the compounds may be distinguished from each other by the Greek prefixes *mono-* (meaning one), *di-* (two), *tri-* (three), *tetra-* (four), *penta-* (five), *hexa-* (six), *hepta-* (seven), and *octa-* (eight). The prefixes precede the name of the constituent to which they refer.

CO, carbon monoxide

PbO, lead monoxide

CO_2, carbon dioxide

PbO_2, lead dioxide

NO_2, nitrogen dioxide

SO_2, sulfur dioxide

N_2O_4, dinitrogen tetraoxide

SO_3, sulfur trioxide

N_2O_5, dinitrogen pentaoxide

BCl_3, boron trichloride

A second method of naming different binary compounds containing the same elements involves the use of Roman numerals placed in parentheses to indicate the oxidation number of the more electropositive element, and following the names of the elements to which they refer. This method of naming binary compounds is usually applied to those in which the electropositive element is a metal, but it is occasionally applied to other compounds as well.

$FeCl_2$, iron(II) chloride

SO_2, sulfur(IV) oxide

$FeCl_3$, iron(III) chloride

SO_3, sulfur(VI) oxide

Hg_2O, mercury(I) oxide

NO, nitrogen(II) oxide

HgO, mercury(II) oxide

NO_2, nitrogen(IV) oxide

Although the system of nomenclature used in this book is for the most part a system formulated by a committee of the International Union of Pure and Applied Chemistry, it is essential that you also become familiar with an older system, because you will still encounter it frequently elsewhere.

According to the older system, when two elements form more than one compound with each other, and when both elements are nonmetals, the distinction is made by indicating only the number of atoms of the more electronegative element by Greek prefixes. NO_2, N_2O_3, N_2O_4, and N_2O_5 are nitrogen dioxide, trioxide, tetraoxide, and pentaoxide, respectively. When the more electropositive element is a metal the lower oxidation number of the metal is indicated by using the suffix -ous on the name of the metal. The higher oxidation number is designated by the suffix -ic. Thus $FeCl_2$ is ferrous chloride, and $FeCl_3$ is ferric chloride; Hg_2O is mercurous oxide, and HgO is mercuric oxide.

▶ **2. TERNARY COMPOUNDS.** Ternary compounds are those containing three different elements. It has already been noted that a few ternary compounds, such as NH_4Cl, KOH, and HCN, are named as if they were binary compounds. Chlorine, nitrogen, sulfur, phosphorus, and several other elements each form oxyacids (ternary compounds with hydrogen and oxygen) that usually differ from each other in their oxygen content. Normally, the most common acid of a series bears the name of the acid-forming element ending with the suffix -ic. This may be noted in the names chloric acid ($HClO_3$), sulfuric acid (H_2SO_4), nitric acid (HNO_3), and phosphoric acid (H_3PO_4). If the "central" element (Cl, S, etc.) of an acid has a higher oxidation number than the -ic acid the suffix -ic is retained and the prefix per- is added. The name perchloric acid for $HClO_4$ illustrates this rule. If the "central" element has a lower oxidation number than the -ic acid, the suffix -ic is replaced with the suffix -ous. Examples are chlorous acid ($HClO_2$), sulfurous acid (H_2SO_3), nitrous acid (HNO_2), and phosphorous acid (H_3PO_3). If the same central element in two acids should have lower oxidation numbers than the -ic acid for the central element, the lower of the two is named by adding the prefix hypo-

TABLE 5-3 Names of Oxyacids of Chlorine and the Corresponding Sodium Salts

Acids	Salts
HClO, hypochlorous acid	NaClO, sodium hypochlorite
$HClO_2$, chlorous acid	$NaClO_2$, sodium chlorite
$HClO_3$, chloric acid	$NaClO_3$, sodium chlorate
$HClO_4$, perchloric acid	$NaClO_4$, sodium perchlorate

and retaining the ending *-ous*. Thus HClO is hypochlorous acid and H_3PO_2 is hypophosphorous acid.

Metal salts of the oxyacids (compounds in which a metal replaces the hydrogen of the acid) are named by identifying the metal and then the negative acid ion. The ending *-ic* of the oxyacid name is changed to *-ate* and the ending *-ous* of the acid is changed to *-ite* for the salt. The salts of perchloric acid are perchlorates, those of sulfuric acid are sulfates, those of nitrous acid are nitrites, and those of hypophosphorous acid are hypophosphites. This system of naming applies to all inorganic oxyacids and their salts. The names of the oxyacids of chlorine and their corresponding sodium salts are given in Table 5-3.

The system of nomenclature for a class of compounds known as **coordination compounds,** or **complexes,** will be described in Chapter 31.

Questions

[S] 1. Why does a cation have a positive charge?

[S] 2. Why does an anion have a negative charge?

3. In what portion of the Periodic Table are those elements that tend to form positive ions located? What classes of elements form positive ions?

4. In what portion of the Periodic Table are those elements that tend to form negative ions located? What class of elements forms negative ions?

5. From the following list of atoms select the ones you would expect to form positive ions and the ones you would expect to form negative ions: Al, Ba, Cl, Co, K, La, N, Ni, O, P, Se.

[S] 6. Predict the charge on the monatomic ions formed from the following elements in ionic compounds containing these elements: Ag, Al, Ba, Br, K, Zn, N, O, P, Se.

7. Why is the number of valence electrons in an atom of an element related to its tendency to gain or lose electrons during formation of an ionic compound?

8. Why should a sample of calcium chloride, $CaCl_2$, contain two chloride ions for each calcium ion?

9. Why is it incorrect to consider sodium chloride as a molecular compound?

10. What are the characteristics of two atoms that will form a covalent bond?

11. Define the term "covalence." Give examples of covalent bonding between like atoms and between unlike atoms.

[S] 12. How do single, double, and triple bonds differ? How are they similar?

13. When is formation of a covalent bond likely? When is a compound unlikely to contain a covalent bond?

14. What is meant by the electronegativity of an element?

15. How is the polarity of a covalent bond related to the electronegativity difference of the atoms in the bond?

16. How is the polarity of a covalent bond related to the polarity of a molecule containing the bond?

[S] 17. Many molecules that contain polar bonds are nonpolar. How can this happen?

18. Give examples of polar covalent molecules and nonpolar covalent molecules that contain polar covalent bonds.

19. Which of the following molecules or ions contain polar bonds? F_2, P_4, SO_2, NO_3^-, N_2O, H_2S, NH_4^+, HCCH

20. Which of the following molecules or ions are polar? NO_2^- (a bent ion); CN^-; NNN^- (a linear ion); HCCH (linear); H_2S (a bent molecule); Br_2; BrCl

21. From the location in the Periodic Table of the elements in the following compounds, predict which compounds are ionic and which are covalent.

$$CaF_2 \quad SiCl_4 \quad COCl_2 \quad Cs_2S$$
$$CO_2 \quad C_2H_4 \quad ClF_3 \quad MnO$$
$$CuS \quad SF_6 \quad MgI_2 \quad Al_2O_3$$

22. Do the suffixes -ous and -ic correspond to any particular oxidation state? How are they used?

23. Name the following compounds: CaF_2, CO_2, CuS, $SiCl_4$, SF_6, ClF_3, MgI_2, Cs_2S, MnO, AgCl, HF, ZnO, Tl_2O, Tl_2O_3, $Fe(NO_3)_2$, $Ca(OH)_2$, NH_4CN, Na_2SO_4, Na_2SO_3, $AgNO_2$, $Cu(NO_3)_2$.

24. Determine the oxidation number of N in each of the following compounds: Na_3N, NH_4Cl, N_2O, N_2H_4, KNO_2, $Ca(NO_3)_2$.

25. Determine the oxidation number of Mn in each of the following compounds: $LiMnO_2$, K_2MnCl_4, $ZnMn_2O_4$, Mn_2O_7, $K_5Mn(CN)_6$ (CN is an ion with a charge of -1).

26. With the aid of oxidation numbers, write chemical formulas for the following compounds: sodium phosphide, potassium oxide, magnesium hydroxide, hydrogen chloride, tin(II) nitrate, lead(IV) acetate, ammonium sulfate, aluminum carbonate, manganese(III) fluoride, silicon hydride, strontium phosphate, lithium perchlorate, nitrogen dioxide, osmium tetraoxide.

27. Write Lewis formulas for the following: BrCl, H_2S, ClO_4^-, NO^+, PCl_3.

28. Write Lewis formulas for the compounds given in Question 21. (Omit MnO.)

29. Write Lewis formulas for the compounds given in Question 23. (Omit those compounds that contain transition metals.)

30. How does the electronic structure of an isolated nitrogen atom differ from that of a nitrogen atom in a molecule of nitrogen, N_2?

31. Complete and balance the following equations:
    (a) $Na + I_2 \longrightarrow$
    (b) $Cs + S_8 \longrightarrow$
    (c) $Mg + P_4 \longrightarrow$
    (d) $H^+ + NH_3 \longrightarrow$
    (e) $Al + O_2 \longrightarrow$
    (f) $Zn + Cl_2 \longrightarrow$

32. Draw the resonance structures of the following: nitric acid, $HONO_2$; selenium dioxide, OSeO; the nitrate ion, NO_3^-; the acetate ion, $HCCO_2^-$.

33. X may indicate a different representative element in each of the following Lewis formulas. To which group does X belong in each case?

(a) :F—X—F: (with F above and below)

(b) :X—Br:

(c) :Cl—X—Cl: (with Cl above and below)

(d) :O—X—O: (with O above and below, top O has charge −)

(e) :X with three F attached

(f) X=O with Cl attached

(g) :I—X—I:

(h) $X^{3+} \left(:Cl:^- \right)_3$

(i) $X^{2+} \left(:Cl:^- \right)_2$

(j) :X—X—X:$^-$

34. The number of covalent bonds between two atoms is called bond order. Bond order may be an integer as with :N≡N: (bond order = 3) or a fraction as in SO_2 (bond order = $1\frac{1}{2}$ due to the resonance O=S—O ⟷ O—S=O). Arrange the molecules or ions in each of the following groups in increasing bond order of the bond indicated. Lone pairs have been omitted for clarity.

(a) C—O bond order: C≡O; O=C=O; CH_3—O—O—CH_3; HCO_2^-

$$\left(\begin{array}{ccc} & \overset{\displaystyle O}{\underset{\displaystyle O^-}{\parallel}} & \\ H\!-\!C & \longleftrightarrow & H\!-\!C \underset{\displaystyle O}{\overset{\displaystyle O^-}{}} \end{array} \right).$$

(b) N—N bond order: CH_3—N=N—CH_3; N_2; H_2N—NH_2; N_2O (N=N=O ⟷ N≡N—O).

(c) C—N bond order: H_3C—NH_2; CH_3CONH^-

$$\left(\begin{array}{ccc} CH_3\!-\!C\underset{\displaystyle NH^-}{\overset{\displaystyle O}{}} & \longleftrightarrow & CH_3\!-\!C\underset{\displaystyle NH}{\overset{\displaystyle O^-}{}} \end{array} \right);$$

H_2C=NH; H_3C—C≡N

(d) C—C bond order: H_3C—CH_3; HC≡CH; H_2C=CH_2;

C_6H_6

35. Arrange the following ions or molecules in increasing order of their C—O bond order (see Question 34): CO; CO_3^{2-} (C is the central atom since it is less electronegative); H_2CO (C is the central atom); CH_3OH (with a CO bond, three CH bonds, and one OH bond.

36. Why is the bond order (Question 34) of a bond between H and another atom never greater than 1?

37. As the bond order (Question 34) between two atoms of the same type increases, the distance between the two atoms, the bond distance, decreases. How would you expect the bond distance in NO_2^- to compare to that in NO_3^-?

38. Separate the following species into three different groups of isoelectronic molecules and ions: NCN^{2-}, CH_3^-, NNO, H_3O^+, NO_2^- (ONO⁻), O_3 (OOO), NH_3, CO_2 (OCO), ClNO.

References

"Early Views on Forces Between Atoms," L. Holliday, *Sci. American,* May 1970, p. 116.

"Principles of Chemical Bonding," R. T. Sanderson, *J. Chem. Educ.,* **38,** 382 (1961).

"A Faithful Couple: The Electron Pair," L. Salem, *J. Chem. Educ.,* **55,** 344 (1978).

"The Three-Electron Bond," N. C. Baird, *J. Chem. Educ.,* **54,** 291 (1977).

"An Alternative Procedure to Writing Lewis Structures," K. Imkampe, *J. Chem. Educ.,* **52,** 429 (1975).

"Alkali Metal Anions, An Unusual Oxidation State," J. L. Dye, *J. Chem. Educ.,* **54,** 332 (1977).

"The Chemical Bond and the Geochemical Distribution of the Elements," L. H. Ahrens, *Chem. in Britain,* **2,** 14 (1966).

"Hydrogen-Like Wave Functions," B. Perlmutter-Hayman, *J. Chem. Educ.,* **46,** 428 (1969).

"Where Does Resonance Energy Come From?" D. J. Sardella, *J. Chem. Educ.,* **54,** 217 (1977).

"The Chemical Origins of Color," M. V. Orna, *J. Chem. Educ.,* **55,** 478 (1978).

"Chemistry and the Spinning Electron" (Staff), *New Scientist,* **60** (867), 128 (1973).

"Strengths of Chemical Bonds," J. D. Christian, *J. Chem. Educ.,* **50,** 176 (1973).

"Group Electronegativity by NMR," J. C. Greever, *J. Chem. Educ.,* **55,** 538 (1978).

"The First Successful Explanation of some Intramolecular Rearrangements," M. D. Saltzman, *J. Chem. Educ.,* **54,** 25 (1977).

"The Appreciation of Molecular Transformation in Organic Chemistry," S. Ranganathan and D. Ranganathan, *J. Chem. Educ.,* **52,** 424 (1975).

"Non-Covalent Interactions: Key to Biological Flexibility and Specificity," E. Frieden, *J. Chem. Educ.,* **52,** 754 (1975).

6

Chemical Bonding, Part 2-Molecular Orbitals

Before we consider molecular orbital theory, we should remind ourselves about the basic forces in a chemical bond. We know that a hydrogen atom consists of a nucleus (a proton) and an electron that moves somewhere within a sphere centered about the nucleus. This atomic structure results from, and is stable because of, the electrostatic attraction of the positively charged nucleus for the negatively charged electron. Now, suppose we let two hydrogen atoms approach each other to form a hydrogen molecule (H_2). When the two atoms are close together, two new electrostatic attractions are added: the attractions between the nucleus on each atom and the electron on the other atom. In addition, two repulsions are added: the electrostatic repulsion between the two electrons and the electrostatic repulsion between the two nuclei. Because the electrons in a hydrogen molecule spend most of their time between the two nuclei, the new attractions are stronger than the repulsions, and a stable molecule results. On the other hand, if the electrons did not spend most of their time between the two nuclei, the two new attractions would be relatively weak since the distance between the nucleus on one atom and the electron on the other would be relatively large. Consequently, the molecule would be destabilized by the repulsions of the nuclei. Thus a covalent bond results only when a pair of electrons are situated (shared) between two nuclei.

In Chapter 5 we considered bonding from the viewpoint that the electrons participating in bonding are principally located in the atomic orbitals of the bonded atoms. The distribution of valence electrons in a covalent molecule was treated simply in terms of lone pairs of electrons and shared pairs of electrons. Except for noting that shared pairs of electrons spend much of their time between the atoms they bond, little was said about the distribution of the electrons in a molecule, and nothing was said about the relative energies of electrons in molecules. Molecular Orbital Theory provides a model for describing the distribution of electrons throughout the molecule and the energies of those electrons.

6.1   Molecular Orbital Theory

The Molecular Orbital Theory treats the distribution of electrons in molecules in much the same way as the distribution of electrons in atomic orbitals is treated for atoms. Using the techniques of quantum mechanics, the behavior of an electron in a molecule is described by a wave function, ψ, which may be used to determine the energy of the electron and the shape of the charge cloud that describes the regions of the space within which the electron moves around a molecule. As in an atom, the region of space in which the electron is likely to be found is called an orbital. However, since an electron in a molecule might be found around any of the nuclei of the atoms in the molecule, the orbital extends over all of the atoms in the molecule and is thus called a **molecular orbital.** Like an atomic orbital, a molecular orbital is full when it contains two electrons of opposite spin.

Several different types of molecular orbitals are found in molecules. Since these orbitals vary with the shape and composition of the molecule, let us first consider only molecules composed of two identical atoms (H_2 or Cl_2, for example). Such molecules are referred to as **homonuclear diatomic molecules.** The types of molecular orbitals commonly observed in these diatomic molecules are shown in Fig. 6-1. In three of these orbitals, called σ_s, σ_p, and π_p orbitals (referred to in words as sigma-s, sigma-p, and pi-p), the electrons are more or less between the nuclei. Adding electrons to these orbitals stabilizes the molecule so the orbitals are called **bonding orbitals.** Electrons in the orbitals called $\sigma_s{}^*$, $\sigma_p{}^*$, and $\pi_p{}^*$ (referred to in words as sigma-s-star, sigma-p-star, and pi-p-star) are located well away from the region between the two nuclei. Electrons in these latter orbitals destabilize the molecule, and hence the orbitals are called **antibonding orbitals.**

The exact wave functions that describe the behavior of an electron in molecular

Figure 6-1 Molecular orbitals found in a diatomic molecule containing identical atoms.

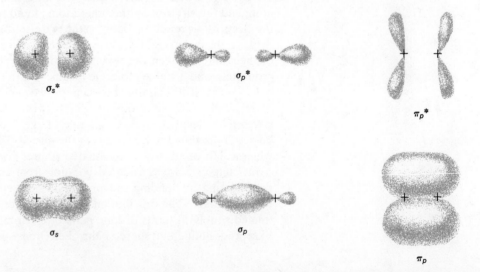

orbitals such as those shown in Fig. 6-1 are very difficult to determine. Thus it is common to use approximate wave functions for the molecular orbitals. One approximation involves using the sum of the wave functions of the atoms in the molecule as a wave function to describe the behavior of the electron in the molecule. The wave functions of any number of atomic orbitals may be added together to give the wave function of a molecular orbital. At this stage, however, we will use only one atomic orbital on each of the two atoms of a diatomic molecule to approximate the molecular orbitals in those molecules.

The basic principles of quantum mechanics state that if the sum of two atomic orbitals is a molecular orbital, then the difference is also a molecular orbital. When a molecular orbital is approximated by the addition of two atomic orbitals, the resulting molecular orbital wave function has a higher value between the two nuclei, indicating that electrons in such an orbital would spend a large amount of time between the nuclei (Fig. 6-2, bottom). This is a **bonding** situation. If the wave function for a molecular orbital is approximated by subtracting the two atomic orbitals, the resulting molecular orbital wave function will have a lower value between the nuclei, indicating that the electrons spend little time between the nuclei (Fig. 6-2, top). Such a situation is unstable and is an **antibonding** situation.

In a diatomic molecule, two molecular orbitals result from the combination of each two atomic orbitals. The two molecular orbitals can be thought of intuitively as being described by the addition or by the subtraction, respectively, of the parts of the two atomic orbitals that overlap. The molecular orbital represented by addition is a bonding orbital; the increase of electron density between the nuclei of the two atoms increases the total electron density between the nuclei. The second molecular orbital is described by subtraction of the overlapping parts of the atomic orbitals. The resulting lowering of electron density between the nuclei results in an antibonding orbital.

Of the two molecular orbitals formed by the combination of two atomic orbitals, the bonding orbital is of lower energy than the antibonding orbital. Electrons tend, therefore, to fill the lower energy bonding molecular orbital in preference to the higher energy antibonding molecular orbital, just as electrons tend to fill atomic orbitals of lower energy before filling those of higher energy.

Figure 6-2 illustrates how a σ_s orbital is represented by addition of two s orbitals

Figure 6-2 A representation of the formation of two molecular orbitals by the combination of two atomic s orbitals.

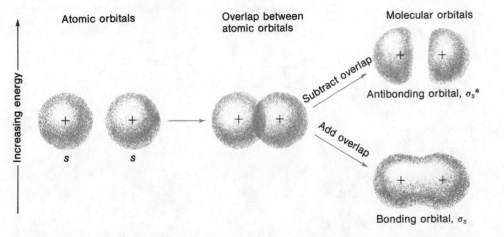

on adjacent atoms and how a σ_s^* orbital is represented by subtraction of two s orbitals. The letter subscript is used to designate the type of atomic orbitals added or subtracted to give the molecular orbital. Adding the overlap produces a bonding orbital of lower energy than that of either of the atomic orbitals that combine to produce it. The antibonding orbital produced by subtraction of overlap is of higher energy than either of the two atomic orbitals. As might be expected, electrons favor the lower energy bonding orbital. This is because a given electron is attracted by both nuclei in the bonding orbital, but largely by only one nucleus either in the antibonding orbital or in the two atomic orbitals. Notice in Fig. 6-2 that an electron in an antibonding orbital has a low probability of being found in the space between the nuclei.

Four different kinds of molecular orbitals can be represented as combinations of atomic p orbitals: σ_p, σ_p^*, π_p, and π_p^*. Two atomic p orbitals, each of which possesses two lobes (Section 4.12), can be combined in either of two ways—end-to-end or side-by-side—to produce molecular orbitals (Fig. 6-3). When two p atomic orbitals of two atoms, are combined end-to-end (bottom part of Fig. 6-3), the resulting two molecular orbitals are σ orbitals (as is the case with atomic s

Figure 6-3 A representation of the formation of sigma (σ) and pi (π) molecular orbitals by the combination of two p orbitals.

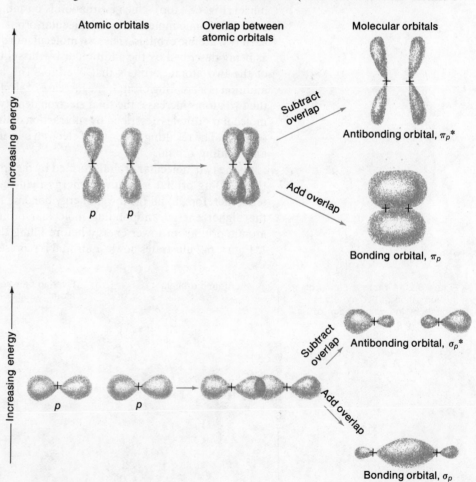

Atomic orbitals

Overlap between atomic orbitals

Molecular orbitals

Increasing energy

p p

Subtract overlap

Antibonding orbital, π_p^*

Add overlap

Bonding orbital, π_p

Increasing energy

p p

Subtract overlap   Antibonding orbital, σ_p^*

Add overlap

Bonding orbital, σ_p

orbitals). However, when two p orbitals are combined side-by-side (top part of Fig. 6-3), the resulting molecular orbitals are π orbitals.

Each atom contains three p atomic orbitals—p_x, p_y, and p_z (see Section 4.12). One of these (e.g., p_x) combines end-to-end with a corresponding p_x atomic orbital of another atom to form two σ molecular orbitals, σ_{p_x} and $\sigma_{p_x}{}^*$. The other two p atomic orbitals, p_y and p_z, each combine side-by-side with a corresponding p atomic orbital of another atom, giving rise to a set of two π bonding molecular orbitals and a set of two π^* antibonding molecular orbitals. The two π bonding orbitals are oriented at right angles to each other as are also the two π^* antibonding orbitals. It is as if one molecular orbital of a set (either π or π^*) were oriented along the y axis of a set of coordinates and the other orbital of the set along the z axis. The notations π_{p_y} and π_{p_z} (commonly referred to as "pi-p-y" and "pi-p-z") are thus applied to the bonding orbitals, and $\pi_{p_y}{}^*$ and $\pi_{p_z}{}^*$ ("pi-p-y star" and "pi-p-z star") to the antibonding orbitals. Except for their orientation, the π_{p_y} and π_{p_z} orbitals are identical and have the same energy (that is, they are **degenerate**). The $\pi_{p_y}{}^*$ and $\pi_{p_z}{}^*$ antibonding orbitals likewise are the same except for their orientation, and they too are degenerate.

Thus a total of six molecular orbitals results from the combination of the six atomic p orbitals in two atoms. If the σ_p and $\sigma_p{}^*$ orbitals are thought of as being oriented along the x axis, and the two sets of π_p and $\pi_p{}^*$ orbitals as being oriented along the y axis and z axis, respectively, the six molecular orbitals are σ_{p_x} and $\sigma_{p_x}{}^*$, π_{p_y} and $\pi_{p_y}{}^*$, and π_{p_z} and $\pi_{p_z}{}^*$.

6.2   Molecular Orbital Energy Diagrams

The relative energy levels of the lower energy atomic and molecular orbitals of a homonuclear diatomic molecule are typically as shown in Fig. 6-4. Each colored disk represents one atomic or molecular orbital that can hold one or two electrons but no more than two. Bonding and antibonding orbitals are joined by dashed lines to the atomic orbitals that combine to form them. Inasmuch as there are three $2p$ orbitals on each atom, a total of six molecular orbitals can be constructed from the $2p$ orbitals.

The distribution of electrons in the molecular orbitals of a homonuclear diatomic molecule can be predicted by filling these molecular orbitals in much the same way that atomic orbitals are filled following the Aufbau Principle (Section 4.16). As with atomic electronic configurations, the number of electrons in each orbital is indicated with a superscript. Thus a molecule containing seven electrons would have the configuration $(\sigma_{1s})^2(\sigma_{1s}{}^*)^2(\sigma_{2s})^2(\sigma_{2s}{}^*)^1$.

It should be mentioned here that in some cases the energy levels of the σ_p and the two π_p bonding orbitals are reversed, with the π_{p_y} and π_{p_z} bonding orbitals being slightly lower in energy than the σ_{p_x} orbital. There is a logical reason for this. We have assumed up to now in our discussion that s orbitals interact only with s orbitals to form σ_s molecular orbitals and that p orbitals interact only with p orbitals to form σ_p or π_p molecular orbitals. The main interactions do indeed occur between orbitals identical in energy. It is, however, also possible for an s orbital to interact with a p orbital if the s and p orbitals are similar in energy.

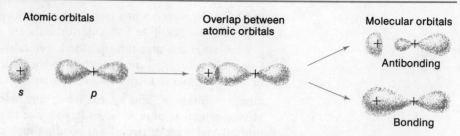

Atomic orbitals

Overlap between
atomic orbitals

Molecular orbitals

s   *p*

Antibonding

Bonding

This results in the σ_s molecular orbital no longer having pure *s* character and the σ_p molecular orbital no longer having pure *p* character. The mixing of *s* and *p* character shifts the energies of the molecular orbitals and in some cases even changes the relative positions of the energy levels of the orbitals. Hence the π_{p_y} and π_{p_z} molecular orbitals, normally at a higher energy than the σ_{p_x} orbital, sometimes are slightly lower than the σ_{p_x} orbital. Such energy shifts will be the

Figure 6-4 Molecular orbital energy diagram. A colored disk represents an atomic or molecular orbital that can hold one or two electrons.

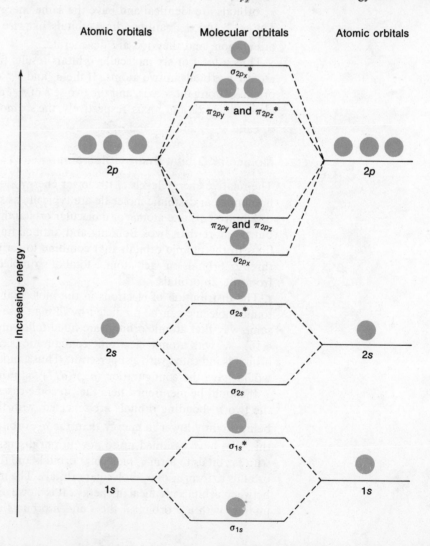

Atomic orbitals   Molecular orbitals   Atomic orbitals

$\sigma_{2p_x}^*$

$\pi_{2p_y}^*$ and $\pi_{2p_z}^*$

2p   2p

π_{2p_y} and π_{2p_z}

σ_{2p_x}

σ_{2s}^*

2s   2s

σ_{2s}

σ_{1s}^*

1s   1s

σ_{1s}

Increasing energy

greatest when the energy difference between the s and p orbitals is low so that they may interact easily.

6.3   The Hydrogen Molecule, H_2

The hydrogen molecule (H_2) is made from two hydrogen atoms, each of which possesses one electron in a $1s$ atomic orbital. When the atomic orbitals for the two hydrogen atoms combine, the two electrons seek the molecular orbital of lowest energy, which is the σ_{1s} bonding orbital. Each molecular orbital can hold two electrons. Hence both electrons are in the σ_{1s} bonding orbital for the hydrogen molecule, and the electronic configuration is $(\sigma_{1s})^2$. This can be represented by a molecular orbital energy diagram (Fig. 6-5) in which each electron within an orbital is indicated by an arrow, ⬆. Two electrons of opposite spin within an orbital are designated ⬍.

Figure 6-5   Molecular orbital energy diagram for the hydrogen molecule.

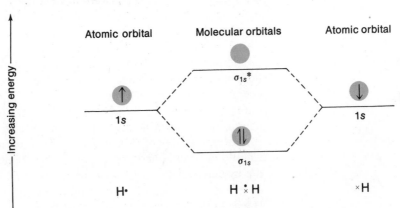

The σ_{1s} orbital, in which both electrons of the hydrogen molecule are most likely to be found, is lower in energy than either of the two $1s$ orbitals. Hence the hydrogen molecule, H_2, is readily formed from two hydrogen atoms.

In general, the difference in energy of the atomic orbitals and the bonding molecular orbital will depend upon the extent of overlap of the two atomic orbitals. The greater the amount of overlap the greater the energy difference will be, and the stronger the resulting bond will be.

6.4   Two Diatomic Helium Species: He_2^+, He_2

What happens when three electrons are present in the two combining atoms or ions? Because the σ_{1s} bonding orbital can hold only two electrons, one of the three electrons in such a case goes to the σ_{1s}^* antibonding orbital even though it is higher in energy than either the bonding molecular orbital or the two atomic orbitals. This use of the antibonding orbital is illustrated by the combination of a helium atom (He) and a helium ion (He^+) to form the dihelium ion (He_2^+).

The helium atom, as we have seen, has two electrons, both of which are in the $1s$ orbital (Table 4-5). The helium ion (He^+) is a helium atom after it has lost one of its two $1s$ electrons. The charge of $+1$ occurs as a result of the loss of the negatively charged electron. When the helium atom and the helium ion combine,

Figure 6-6 Molecular orbital energy diagram for the He_2^+ ion.

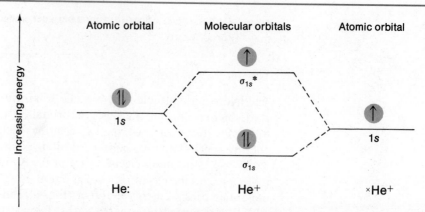

two of the electrons go to the lower energy σ_{1s} bonding orbital. The third electron must go into the molecular orbital next lowest in terms of energy, the σ_{1s}^* anti-bonding orbital; hence, He_2^+ has the configuration $(\sigma_{1s})^2(\sigma_{1s}^*)^1$. The locations of the electrons are illustrated in the molecular orbital energy-level diagram of Fig. 6-6.

The net energy for the dihelium ion is slightly lower than the energy of the helium atom or of the helium ion because two electrons are present in the lower energy molecular orbital and only one in the higher energy molecular orbital. The lower net energy for the dihelium ion indicates an appreciable tendency for a helium atom and a helium ion to combine to form a dihelium ion.

By a similar line of reasoning, the dihelium molecule (He_2) with four electrons would not be expected to form by the combination of two helium atoms because the two electrons in the lower energy bonding orbital would be balanced by two electrons in the higher energy antibonding orbital [configuration $(\sigma_{1s})^2(\sigma_{1s}^*)^2$]. Hence the net energy change would be zero, indicating no appreciable tendency for helium atoms to combine to form the diatomic molecule. Indeed, as predicted, the element helium exists as discrete atoms, He, rather than as diatomic molecules, He_2.

It is instructive to apply the principles we have just discussed to the possible formation of diatomic molecules of the second period of elements (from lithium to neon).

6.5   The Lithium Molecule, Li_2

The combination of two lithium atoms to form a lithium molecule, Li_2, is analogous to the formation of H_2, except that the $2s$ atomic orbital is principally involved instead of the $1s$ orbital. Each of the two lithium atoms, with an electronic configuration $1s^2 2s^1$, has one valence electron. Hence two valence electrons are available to go into the σ_{2s} bonding molecular orbital. The lower lying $1s$ electrons in the $1s$ orbital of each atom are not appreciably involved in the bonding.

Since both valence electrons would be in the lower energy σ_{2s} bonding orbital (Fig. 6-7), it should be possible to form Li_2. The molecule is, in fact, present in appreciable concentration in lithium vapor at temperatures near the boiling point of the element.

In general, electrons in inner shells of atoms are not involved in forming

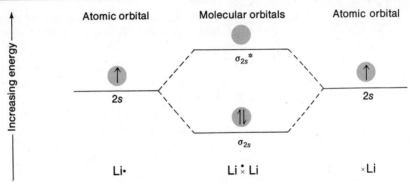

Figure 6-7 Molecular orbital energy diagram for Li_2. Lower lying electrons in the $1s$ orbital of each atom are not shown.

molecular orbitals. Because of the relatively high effective nuclear charge experienced by the electrons in inner shells, the inner shells have small radii. Consequently, inner shells on adjacent atoms do not overlap. In order to indicate that the inner shells of the atoms do not form molecular orbitals, when writing the electronic configuration of diatomic molecules the filled inner shells are indicated with the letter by which they are identified. The configuration for Li_2 would be $KK(\sigma_{2s})^2$, where the two K's indicate that the $1s$ orbitals on each atom are filled but do not enter into the bonding.

6.6   The Instability of the Beryllium Molecule, Be_2

The diatomic molecule of beryllium, Be_2, in contrast to the diatomic lithium molecule, would not be expected to be particularly stable, inasmuch as two of the four valence electrons would go to the σ_{2s} bonding orbital and the other two would be forced to go to the σ_{2s}^* antibonding orbital (Fig. 6-8). The configuration would be $KK(\sigma_{2s})^2(\sigma_{2s}^*)^2$. The net energy change, just as is the case with diatomic helium, is zero, indicating no tendency for beryllium atoms to combine to form the diatomic beryllium molecule. This is in accord with the experimental finding that no stable Be_2 molecule is known.

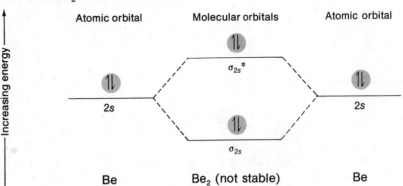

Figure 6-8 Molecular orbital energy diagram for Be_2 (not stable). Lower lying electrons in the $1s$ orbital of each atom are not shown.

6.7   The Boron Molecule, B_2

The element boron has the electronic structure $1s^2 2s^2 2p^1$. Hence, beginning with boron in this period, the p orbitals make an important contribution to the bonding. We have noted that combinations of p atomic orbitals give rise to both σ and π molecular orbitals and that the two π_p bonding orbitals are of equal energy (degenerate). **Whenever there are two or more molecular orbitals at the same energy level, electrons fill the orbitals of that type singly before any pairing of electrons takes place within these orbitals.** Recall that an exactly analogous situation pertains to atomic orbitals (see Section 4.15).

We have noted earlier that usually the σ_p energy level is slightly lower than the energy level of the two π_p orbitals, but that sometimes the reverse is true. Experimental magnetic data for the B_2 molecule, which is known to exist, tell us that the molecule contains two unpaired electrons. This is an indication that *in this particular case* the two π_p orbitals are lower in energy than the σ_p orbital and are filled first, with one electron going into each π_p orbital singly. If the σ_p orbital were lower in energy than the two π_p orbitals, the two electrons would be expected to pair within the one orbital. Figure 6-9 is the molecular orbital energy diagram for B_2 and shows the two unpaired electrons. The configuration of B_2 is $KK(\sigma_{2s})^2(\sigma_{2s}{}^*)^2(\pi_{2p_y}, \pi_{2p_z})^2$.

The four electrons in the σ_{2s} and $\sigma_{2s}{}^*$ orbitals balance each other in terms of energy and do not make a significant contribution to the stability of the B_2 molecule. However, the fact that the two π_p electrons are in bonding orbitals

Figure 6-9 Molecular orbital energy diagram for B_2. Lower lying electrons in the 1s orbital of each atom are not shown.

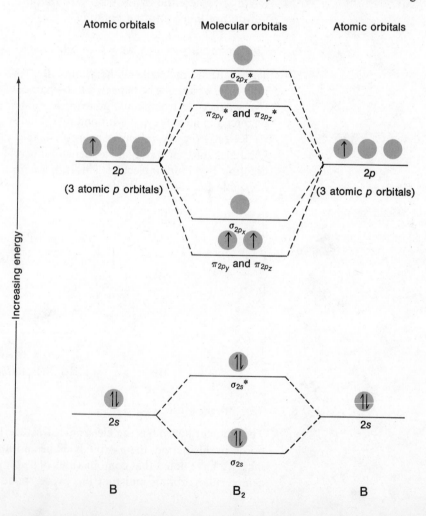

indicates that it should be possible to form the diboron molecule by combination of boron atoms. This is verified by the fact that the molecule is known to exist.

6.8 The Carbon Molecule, C_2

For the formation of the carbon molecule, C_2, the situation is similar to that of B_2. As with B_2, the π_{2p} orbitals of C_2 are at a lower energy level than the σ_{2p} orbital. The two additional electrons (one from each atom) will fill the π_{2p_y} and π_{2p_z} molecular orbitals, giving the electronic structure $KK(\sigma_{2s})^2(\sigma_{2s}{}^*)^2(\pi_{2p_y}, \pi_{2p_z})^4$.

6.9 The Nitrogen Molecule, N_2

Each nitrogen atom has three $2p$ electrons. The nitrogen molecule, N_2, is well known and very stable. Experiment shows that all electrons are paired in the molecular orbitals and that the π_{2p} orbitals are lower in energy than the σ_{2p} orbital. The molecular orbital energy-level diagram is given in Fig. 6-10; the configuration is $KK(\sigma_{2s})^2(\sigma_{2s}{}^*)^2(\pi_{2p_y}, \pi_{2p_z})^4(\sigma_{2p_x})^2$. The fact that the molecule is stable is in accord with the assumption that the six electrons arising from the $2p$ atomic orbitals are all in bonding molecular orbitals.

Figure 6-10 Molecular orbital energy diagram for N_2. Lower lying electrons in the $1s$ orbital of each atom are not shown.

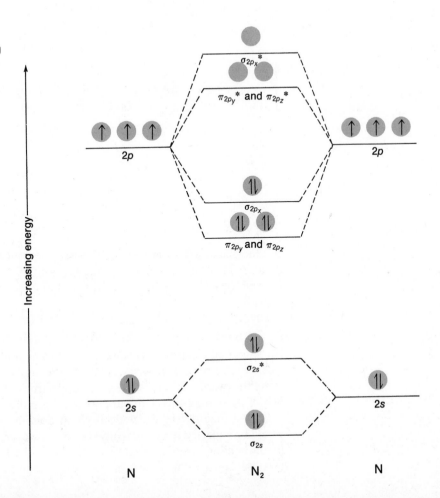

Figure 6-11 Molecular orbital energy diagram for O_2. Lower lying electrons in the 1s orbital of each atom are not shown.

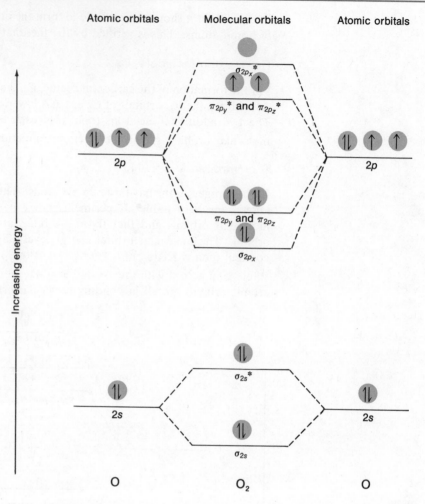

6.10   The Oxygen Molecule, O_2

The molecular orbital energy-level diagram for the oxygen molecule, O_2, is shown in Fig. 6-11. The order of the σ_{2p} and π_{2p} orbitals changes from that for B_2, C_2, and N_2, and the σ_{2p} orbital in oxygen is believed to be at a lower energy level than the π_{2p} orbitals.

The electronic configuration for O_2 $[KK(\sigma_{2s})^2(\sigma_{2s}^*)^2(\sigma_{2p_x})^2(\pi_{2p_y}, \pi_{2p_z})^4(\pi_{2p_y}^*, \pi_{2p_z}^*)^2]$ is in accord with the known experimental fact that the oxygen molecule has two unpaired electrons. This has proved to be difficult to explain on the basis of valence electronic formulas, but with the molecular orbital theory it is quite straightforward. In fact, the unpaired electrons of the oxygen molecule provide one of the strong pieces of support for the molecular orbital theory.

With a majority of the p electrons in bonding orbitals, the oxygen molecule would be expected to be relatively stable. However, the bond, with two p electrons in antibonding orbitals, might be expected to be less strong than in the nitrogen molecule, in which no p electrons are in antibonding orbitals, and this is the case as is verified by the experimental observation that much less energy is required to break the oxygen molecule bond than the nitrogen molecule bond (see Sec. 6.14).

6.11   The Fluorine Molecule, F$_2$

The fluorine molecule has no unpaired electrons (Fig. 6-12). With six of the p electrons in bonding orbitals and four in antibonding orbitals [electronic structure

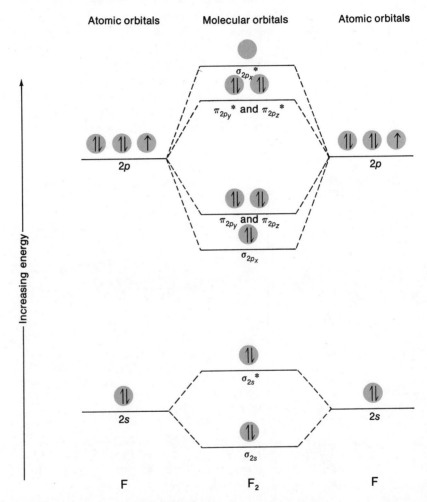

Figure 6-12 Molecular orbital energy diagram for F$_2$. Lower lying electrons in the 1s orbital of each atom are not shown.

$KK(\sigma_{2s})^2(\sigma_{2s}{}^*)^2(\sigma_{2p_x})^2(\pi_{2p_y}, \pi_{2p_z})^4(\pi_{2p_y}{}^*, \pi_{2p_z}{}^*)^4]$, fluorine atoms should combine to form the fluorine molecule but with a somewhat weaker bond than in the oxygen molecule. Such is the case. In fact, the F$_2$ bond is one of the weakest of the covalent bonds.

6.12   The Instability of the Neon Molecule, Ne$_2$

The Ne$_2$ molecule has no appreciable tendency to form because it would have the same number of electrons in antibonding orbitals as in bonding orbitals. As with helium, the atoms remain as discrete atoms. Were the molecule to form, it would have the configuration $KK(\sigma_{2s})^2(\sigma_{2s}^*)^2(\sigma_{2p_x})^2(\pi_{2p_y}, \pi_{2p_z})^4(\pi_{2p_y}^*, \pi_{2p_z}^*)^4(\sigma_{2p_x}^*)^2$.

6.13   Diatomic Systems with Two Different Elements

Thus far we have considered only molecular orbitals in molecules or ions containing two identical atoms. It is of interest to look briefly at some of the factors involved in an extension of the Molecular Orbital Theory to diatomic molecules or ions with bonds between atoms of two different elements (**heteronuclear diatomic species).**

The treatment of molecular orbitals of heteronuclear diatomic species is not fundamentally different from that of the homonuclear diatomic molecules described in the previous sections. Electrons still move over the whole molecule in molecular orbitals that can be approximated by combinations of atomic orbitals from each atom. However, for heteronuclear molecules the equivalent atomic orbitals for each contributing atom do not have the same energy. Orbitals for the more electronegative element are at a lower energy level. As may be seen in Fig. 6-13, which shows a molecular orbital energy-level diagram for nitric oxide, NO,

Figure 6-13 Molecular orbital energy diagram for NO. Lower lying electrons in the 1s orbital of each atom are not shown.

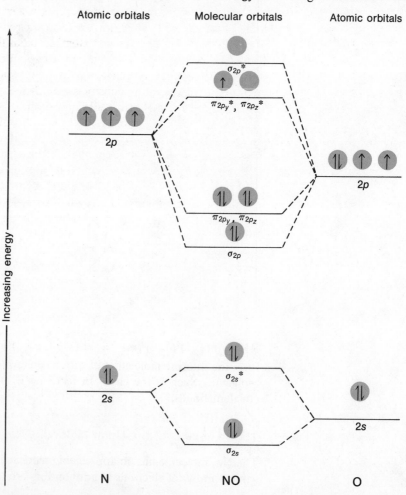

Figure 6-14 Sigma (σ) molecular orbitals found in N_2 and NO.

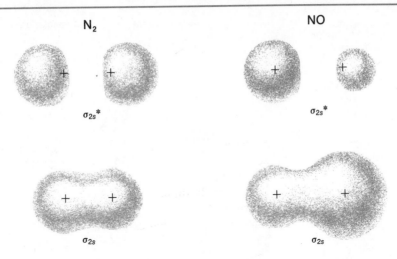

the energies of the atomic orbitals of oxygen are lower than those of nitrogen due to the greater charge of the oxygen nucleus (see Section 4.19). Consequently, some molecular orbitals are closer in energy to the energy of the oxygen atomic orbitals than to the energy of nitrogen atomic orbitals, whereas other molecular orbitals are closer in energy to the nitrogen atomic orbitals than to the oxygen atomic orbitals.

Unlike the molecular orbitals in a diatomic molecule composed of identical atoms, the molecular orbitals in a heteronuclear diatomic molecule are not symmetric. The electron density is greater nearer the atom whose atomic orbital is closest to the energy of the molecular orbital. Figure 6-14 illustrates the distribution of electron density in the σ_{2s} and σ_{2s}^* orbitals of N_2 and of NO. In the N_2 molecule, the σ_{2s} and σ_{2s}^* orbitals may be described by combination of two $2s$ orbitals that have identical energies, so the σ_{2s} and σ_{2s}^* orbitals in N_2 are symmetrical. In NO the σ_{2s} and σ_{2s}^* orbitals are represented by combination of the $2s$ orbital of oxygen with the higher energy $2s$ orbital of nitrogen, and the σ_{2s} and σ_{2s}^* orbitals are not symmetrical. The σ_{2s}^* orbital is closer in energy to the nitrogen $2s$ atomic orbital (Fig. 6-13); hence the antibonding σ_{2s}^* molecular orbital has the greatest electron density near the nitrogen atom. The bonding σ_{2s} orbital, on the other hand, has the greatest electron density near the oxygen atom because its energy is closer to that of the oxygen $2s$ atomic orbital. The electron pair in the σ_{2s} molecular orbital spends more time, on the average, closer to the oxygen atom than to the nitrogen atom. The electron pair in the σ_{2s}^* molecular orbital spends more time closer to the nitrogen atom.

In spite of the asymmetry of the orbitals in heteronuclear diatomic molecules, the filling of molecular orbitals in them is similar to that in homonuclear diatomic molecules. Each molecular orbital still can hold only two electrons, and the lowest energy orbitals are filled first. The configuration of NO is $KK(\sigma_{2s})^2(\sigma_{2s}^*)^2$ $(\sigma_{2p_x})^2(\pi_{2p_y}, \pi_{2p_z})^4(\pi_{2p_y}^*, \pi_{2p_z}^*)^1$. Since there are more electrons in bonding molecular orbitals with greater electron density near the oxygen atom than in antibonding orbitals with greater electron density near the nitrogen atom, the bond is polar (Section 5.9).

Molecular orbital energy-level diagrams similar to the one for NO may be used to describe CO, NO^+, CN^-, and NO^-.

Figure 6-15 Molecular orbital energy diagram for HF. The filled $1s$ orbital of F is not shown.

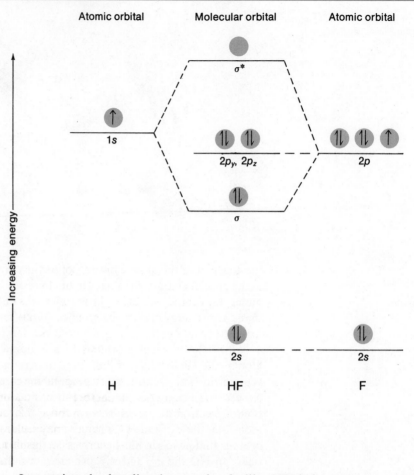

In treating the bonding in a molecule like HF, it is necessary to introduce another idea: only atomic orbitals that are similar in energy can be combined to give molecular orbitals; if the energies of two atomic orbitals are very different, molecular orbitals cannot be formed from them. In HF, the $1s$ orbital of hydrogen is about the same energy as each of the $2p$ orbitals of fluorine (Fig. 6-15). Thus the molecular orbitals that form may be represented as combinations of the $1s$ hydrogen orbital and the $2p_x$ fluorine orbital (Fig. 6-16). The energy difference, however, between the hydrogen $1s$ orbital and the fluorine $2s$ orbital is too great for these atomic orbitals to combine.

When an s orbital combines with a p orbital, it combines in the head-to-tail fashion shown in Fig. 6-16. In HF, this gives the bonding and antibonding orbitals shown in Fig. 6-15. The $2s$, $2p_y$, and $2p_z$ orbitals of the fluorine atom in the HF molecule are shown at the same energy as for the original fluorine atomic orbitals in the middle of the energy-level diagram of HF (Fig. 6-15). To a first approximation, the energies of the $2s$, $2p_y$, and $2p_z$ orbitals in the molecule are not different from their energies in a free fluorine atom since they do not combine with orbitals on the hydrogen atom. As may be seen in the HF molecular orbital energy diagram (Fig. 6-15), one bonding and one antibonding orbital are formed. The

Figure 6-16 Sigma molecular orbitals found in HF.

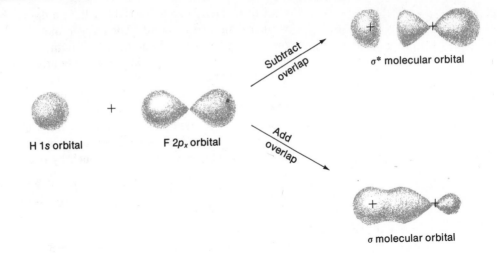

H 1s orbital   +   F 2p_x orbital

Subtract overlap → σ* molecular orbital

Add overlap → σ molecular orbital

eight valence electrons are located in the lowest lying orbitals in the molecule—the 2s, 2p_y, and 2p_z atomic orbitals on fluorine and the σ bonding orbital. The molecule is stable because the two electrons in the σ orbital have a lower energy than they would have in the isolated atoms.

6.14   Bond Order

When using Lewis formulas to describe the distribution of electrons in molecules **the order of a bond (bond order) between two atoms can be defined as the number of bonding pairs of electrons between the atoms** (see Problem 34, Chapter 5). Bond order is defined differently when the molecular orbital description of the distribution of electrons is used; however, the resulting bond order is the same.

From our study of nitrogen, oxygen, and fluorine molecules we have learned that the bond between the atoms is weaker in O_2 than in N_2 and weaker in F_2 than in O_2 (Sections 6.9, 6.10, and 6.11). This order of bond energies was predicted because of an observed increase in the ratio of the number of antibonding orbital electrons to that of the bonding orbital electrons, in the order N_2, O_2, and F_2. Values for the bond energies of the three molecules support the predictions.

The following data show the energy necessary to break the bonds in these atoms.

Molecule	Bond Energy
N_2	955 kJ (226 kcal) per mole
O_2	498 kJ (119 kcal) per mole
F_2	159 kJ (38 kcal) per mole

Nitrogen, with three pairs of bonding electrons in σ_p and π_p molecular orbitals, has a triple bond (see Fig. 6-10 and Section 5.4) and is said to have a *bond order of three*. In oxygen, the two additional electrons are in the π_p* antibonding orbitals (Fig. 6-11), thereby weakening the bond. The effect of six bonding electrons minus the partially offsetting effect of two antibonding electrons results in a net effect approximately equivalent to four bonding electrons (or two pairs of bonding electrons). Hence oxygen is said to have a double bond, which is on the average weaker than a triple bond. The oxygen molecule, therefore, has a *bond order of two*. Fluorine, with six bonding and four antibonding electrons in the *p* levels (Fig. 6-12) has a net effect equivalent to two bonding electrons (or one pair of bonding electrons). The fluorine molecule, therefore, has a single bond and a *bond order of one*. A single bond is, in general, weaker than either a double bond or a triple bond.

By way of summary, **the bond order for a given bond may be determined by dividing the number of bonding electrons in the outer shell minus the number of antibonding electrons in the outer shell by two.**

$$\text{Bond order} = \frac{\text{number of bonding electrons minus number of antibonding electrons}}{2}$$

$$\text{Bond order in } N_2 = \frac{6-0}{2} = 3$$

$$\text{Bond order in } O_2 = \frac{6-2}{2} = 2$$

$$\text{Bond order in } F_2 = \frac{6-4}{2} = 1$$

These bond orders are the same as calculated from Lewis formulas for N≡N, O=O, and F—F.

Questions

1. In Section 4.15, Hund's Rule was discussed in connection with atomic orbitals. Formulate a similar rule for the filling of molecular orbitals and give specific examples of its application.
2. How do the following differ and how are they similar? molecular orbitals and atomic orbitals; bonding orbitals and antibonding orbitals; σ orbitals and π orbitals.
3. Describe the similarities and differences of σ orbitals

formed from two *s* atomic orbitals, from two *p* atomic orbitals, and from an *s* and a *p* atomic orbital.
4. Under what conditions does an *s* atomic orbital on one atom combine with a *p* atomic orbital on a second atom rather than with the *s* atomic orbital on the second atom?
5. Draw diagrams showing the molecular orbitals that can be formed by combining two *s* orbitals, by

combining two p orbitals, and by combining an s and a p orbital.

6. Using molecular orbital energy-level diagrams and the electron occupancy in the molecular orbitals, compare H_2, He_2^+, and He_2 with respect to stability.

7. Determine the bond orders of the homonuclear diatomic molecules formed by the first 10 elements of the Periodic Table.

8. Consider the diatomic *ions* X_2^{2+}, where X is any one of the first 10 elements of the Periodic Table. Write the molecular orbital electronic configurations of these ions. Identify those ions which might have a bond order greater than zero. What are the bond orders?

9. Predict whether each of the following atoms, ions, or molecules should be paramagnetic or diamagnetic. (A paramagnetic species contains unpaired electrons; a diamagnetic species has all of its electrons paired.)

(a) He	(b) B	(c) B_2
(d) O	(e) O_2	(f) F
(g) F_2	(h) Co	(i) CO
(j) NO	(k) H_2^+	(l) F_2^+

10. How does the molecular orbital energy-level diagram for a diatomic molecule involving atoms of two different elements differ from the diagram for a diatomic molecule made up of two atoms of the same element?

11. Draw a molecular orbital energy-level diagram for P_2 using only the orbitals in the valence shell of phosphorus that are occupied in the free atom. What is the configuration of P_2?

12. Draw a molecular orbital energy-level diagram for the ion Si_2^+ using only the orbitals in the valence shell of silicon that are occupied in the free atom. What is the electronic configuration of Si_2^+?

13. Arrange the following in increasing order of bond energy: F_2, F_2^+, F_2^-.

14. What does the presence of a polar bond in a molecule imply about the shapes of the molecular orbitals in the molecule?

15. Identify the homonuclear diatomic molecules or ions that have the following electronic configurations:

(a) X_2^+: $KK(\sigma_{2s})^1$

(b) X_2: $KK(\sigma_{2s})^2(\sigma_{2s}{}^*)^2(\pi_{2p_y}, \pi_{2p_z})^4$

(c) X_2^-: $KK(\sigma_{2s})^2(\sigma_{2s}{}^*)^2(\pi_{2p_y}, \pi_{2p_z})^3$

(d) X_2^+: $KK(\sigma_{2s})^2(\sigma_{2s}{}^*)^2(\sigma_{2p_x})^2(\pi_{2p_y}, \pi_{2p_z})^4(\pi_{2p_y}{}^*, \pi_{2p_z}{}^*)^3$

16. Draw the molecular orbital energy-level diagrams for CO and CN^-. Show that CO and CN^- are isoelectronic and have the same bond order. Compare these molecular orbital energy-level diagrams to that for NO.

17. The Lewis formula of the nitric oxide anion ($N{=}O^-$) indicates a bond order of two for this ion. Beginning with the energy-level diagram for NO, work out the bond order for NO^- using the molecular orbital model.

18. What is the bond order of NO^+? Is the bond in NO^+ stronger or weaker than that in NO?

19. Use the molecular orbital model and determine the bond orders of O_2^+, O_2, O_2^-, and O_2^{2-}. Arrange these species in order of increasing bond energy.

20. Sketch the molecular orbital energy-level diagram of OH^-.

21. Consider the molecular orbitals represented by the following outlines (the symbol + represents a nucleus):

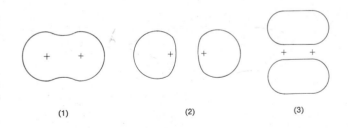

(1)   (2)   (3)

(a) What is the maximum number of electrons that can be placed in molecular orbital (1)? In orbital (2)? in orbital (3)?

(b) How many orbitals of type (1) are found in the valence shell of F_2? How many orbitals of type (3)?

(c) What homonuclear diatomic molecule formed by an element in the third period of the Periodic Table has its two highest energy electrons in an orbital of type (1)?

(d) What homonuclear diatomic molecule formed by the elements in the third period of the Periodic Table has its four highest energy electrons in orbitals like type (3)?

References

"A Simple, Quantitative Molecular Orbital Theory," W. F. Cooper, G. A. Clark, and C. R. Hare, *J. Chem. Educ.*, **48**, 247 (1971).

"Explanation of the Permutation of the σ_p and π Levels in Homonuclear Diatomic Molecules," O. Henri-Rousseau and B. Boulil, *J. Chem. Educ.*, **55**, 571 (1978).

"Back-of-the-Envelope Molecular Orbital 'Calculations'," R. B. Davidson, *J. Chem. Educ.*, **54**, 531 (1977).

"An Alternative Procedure to Writing Lewis Structures," K. Imkampe, *J. Chem. Educ.*, **52**, 429 (1975).

"Chemical Bonds," N. N. Greenwood, *Educ. in Chemistry*, **4**, 164 (1967).

"The Three-Electron Bond," N. C. Baird, *J. Chem. Educ.*, **54**, 291 (1977).

"Molecular Orbital Symmetry Rules," R. G. Pearson, *Chem. and Eng. News*, September 28, 1970, p. 66.

"Polar Bonds," L. Melander, *J. Chem. Educ.*, **49**, 687 (1972).

"Where Does Resonance Energy Come From?" D. J. Sardella, *J. Chem. Educ.*, **54**, 217 (1977).

"The Chemical Origins of Color," M. V. Orna, *J. Chem. Educ.*, **55**, 478 (1978).

"Electronic Structure of Organic Conductors and Semiconductors," Z. G. Soos, *J. Chem. Educ.*, **55**, 546 (1978).

"The Appreciation of Molecular Transformation in Organic Chemistry," S. Ranganathan and D. Ranganathan, *J. Chem. Educ.*, **52**, 424 (1975).

7

Molecular Structure and Hybridization

For the most part, the chemist is interested in the properties of molecules or ions that contain two or more atoms. This interest includes information on the arrangement of the atoms when they are bonded together, that is, the **molecular structure.** The chemical, physical, and biological properties that depend on the structures of molecules often are related in a very sensitive manner to the exact spatial distribution of the constituent atoms. Many chemical reactions are best interpreted in terms of the detailed structures of the molecules or ions of reactants and products. In this chapter we will consider the relationship of molecular structure to the distribution of the valence electrons in a covalent molecule or ion.

Valence Shell Electron-Pair Repulsion Theory

7.1 Prediction of Molecular Structures

By **molecular structure** we mean the geometrical arrangement of the atoms of a molecule or ion in three dimensions. The molecular structure is normally described in terms of bond distances and bond angles. A **bond distance** is the distance between the nuclei of two bonded atoms along a straight line joining the nuclei. Bond distances are often expressed in nanometers (1 nm = 10^{-9} m) or in angstrom units (1 Å = 10^{-10} m). A **bond angle** is the angle between the two lines,

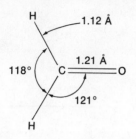

Figure 7-1 Bond distances and angles in formaldehyde, H_2CO. Note that $1 Å = 10^{-10}$ m = 0.1 nm.

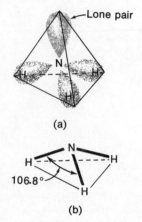

(a)

(b)

Figure 7-2 (a) Regions occupied by the lone pair (shown in color) and bonds (black) in NH_3 and (b) the resulting trigonal pyramidal molecular structure.

defining the bond distances, which meet at a given atom. The bond distances and angles in formaldehyde, H_2CO, are illustrated in Fig. 7-1.

Approximate bond angles in a molecule can be predicted using the **Valence Shell Electron-Pair Repulsion Theory.** According to this theory, the electron pairs associated with the valence shell of the central atom of a molecule are present either as bonding pairs, located primarily between the bonded atoms, or as lone pairs that occupy a restricted region of space shaped rather like that occupied by the bonding pairs. The electrostatic repulsion of the negatively charged electrons in bonds and those in the regions occupied by the lone pairs is reduced to a minimum when the various regions of high electron density assume positions as far from each other as possible. For example, in a molecule with only two bonds to the central atom, the bonds are as far apart as possible when they are on opposite sides of the central atom and the bond angle is 180°. Another example is provided in Fig. 7-2, which shows the regions of space occupied by the lone pair and bonding pairs in the ammonia molecule. The repulsions between the four regions of electron density in NH_3 are minimized if they are directed toward the corners of a tetrahedron, as shown in Fig. 7-2(a). These and the other geometries expected from reducing the repulsions between regions of high electron density (bonding pairs and/or lone pairs) to a minimum are illustrated in Table 7-1. Linear, trigonal planar, tetrahedral, trigonal bipyramidal, and octahedral arrangements are found for two, three, four, five, and six regions of high electron density, respectively.

If the regions of high electron density are not identical, the angles in their arrangement may differ by several degrees from the ideal values given in Table 7-1. Nevertheless, these structures will usually still be good approximations for the distribution of electron density around the central atom. In NH_3, for example, the central nitrogen atom is surrounded by three single N—H bonds and one lone pair of electrons. Since the regions of high electron density are not alike (three are bonds and one is a lone pair), the bond angle is 106.8°, as shown in Fig. 7-2(b), instead of 109.5°, as shown in Table 7-1 for symmetrical tetrahedral structures. In formaldehyde, H_2CO, the regions of high electron density consist of two single bonds and one double bond, and the bond angles differ from each other by 1 or 2 degrees (Fig. 7-1).

Although the arrangement of lone pairs and bonding pairs in a molecule may correspond to one of the arrangements in Table 7-1, the molecular structure may appear different. The presence of a lone pair affects the structure of a molecule, but the lone pair is invisible to the experimental techniques used to determine the structure of the molecule. Consequently, the molecular structure is described as if the lone pair were not there. The molecule NH_3, for example, has a tetrahedral arrangement of regions of electron density, Fig. 7-2(a). However, one of these regions is a lone pair and is not observed when the molecular structure is determined experimentally. The molecular structure (the arrangement of atoms only) is a trigonal pyramid, Fig. 7-2(b), with a nitrogen at the apex and three hydrogen atoms forming the base. Table 7-2 illustrates the molecular structures that are observed with various combinations of lone pairs and bonding pairs.

In several of the examples in Table 7-2, a different arrangement of the location of lone pairs and bonds would give a different molecular structure. As shown in Fig. 7-3, there are three different arrangements possible for the three bonds and two lone pairs in ClF_3. Figure 7-3(a) illustrates the T-shaped molecule actually

TABLE 7-1 Arrangement of Lone Pairs and Bonds as a Result of Electron-Pair Repulsions

Two regions of high electron density (bonding pairs and/or lone pairs)	180° M	Linear, 180° angle
Three regions of high electron density	120° M	Trigonal planar, all angles 120°
Four regions of high electron density	109.5° M	Tetrahedral, all angles 109.5°
Five regions of high electron density	90° M 120°	Trigonal bipyramidal, angles of 90° or 120° (an attached atom may be equatorial, in the plane of the triangle, or axial, above or below the plane of the triangle)
Six regions of high electron density	90° M	Octahedral, all angles 90°

Figure 7-3 Possible arrangements of two lone pairs (shown in color) and three bonds (black) in ClF_3, and also the resulting molecular structures showing only the positions of the atoms without the lone pairs. (a) shows the stable T-shaped arrangement.

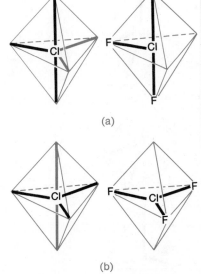

(a)

(b)

(c)

found experimentally. However, a trigonal planar arrangement, Fig. 7-3(b), and a trigonal pyramidal arrangement, Fig. 7-3(c), can be written. The stable structure actually observed is the one that puts the lone pairs as far apart as possible and thus minimizes electron-pair repulsions. Lone pairs appear to require more room than bonding pairs. In a trigonal bipyramidal arrangement, as found in ClF_3, the positions that minimize repulsions are those in the triangular plane of the molecule because the electrons in these positions are 120° apart.

The repulsions of the various types can be ordered as follows:

1. Lone pair–lone pair repulsions; these have the greatest repulsion.
2. Lone pair–bond pair repulsions; intermediate in repulsive force.
3. Bond pair–bond pair repulsions; smallest in repulsive force.

TABLE 7-2 Molecular Structures Based on the Valence Shell Electron-Pair Repulsion Theory

Example	Regions of High Electron Density			Arrangement of Regions of High Electron Density	Molecular Structure (Chemical Bonds are Indicated in Black, Lone Pairs in Color)
	Lone Pairs	Bonds	Total		
BF_3, CO_3^{2-}	0	3	3	Trigonal planar	Trigonal planar
ClNO, NO_2^-	1	2	3	Trigonal planar	Angular (120°)
CH_4, NH_4^+	0	4	4	Tetrahedral	Tetrahedral
NH_3, $SnCl_3^-$, PCl_3	1	3	4	Tetrahedral	Trigonal pyramidal
H_2O, NH_2^-	2	2	4	Tetrahedral	Angular (109.5°)
PF_5, $SnCl_5^-$	0	5	5	Trigonal bipyramidal	Trigonal bipyramidal

TABLE 7-2 (Continued)

Example	Regions of High Electron Density			Arrangement of Regions of High Electron Density	Molecular Structure (Chemical Bonds are Indicated in Black, Lone Pairs in Color)	
	Lone Pairs	Bonds	Total			
SF_4, ClF_4^+	1	4	5	Trigonal bipyramidal	Seesaw	
ClF_3, ICl_3	2	3	5	Trigonal bipyramidal	T-shaped	
I_3^-, ClF_2^-	3	2	5	Trigonal bipyramidal	Linear	
SF_6, PCl_6^-, IF_6^+	0	6	6	Octahedral	Octahedral	

TABLE 7-2   (Continued)

Example	Regions of High Electron Density			Arrangement of Regions of High Electron Density	Molecular Structure (Chemical Bonds are Indicated in Black, Lone Pairs in Color)
	Lone Pairs	Bonds	Total		
IF_5, ClF_5	1	5	6	Octahedral	Square pyramidal
XeF_4, ICl_4^-	2	4	6	Octahedral	Square planar
XeO_3	3	3	6	Octahedral	Trigonal pyramidal

When several possible structures can be written, the one that minimizes as many of the strong repulsions as possible will be the most stable.

7.2   Rules for Predicting Molecular Structures

To use the Valence Shell Electron-Pair Repulsion Theory to predict molecular structures apply the following procedure:

1. Write the Lewis structure of the molecule as described in Section 5.7.
2. From the Lewis structure determine the number of regions of high electron density (lone pairs or chemical bonds) around the central atom. A single, double, or triple bond is counted as one region of high electron density. An unpaired electron is counted the same as a lone pair.
3. Identify the most stable arrangement of the regions of high electron density (linear, trigonal planar, tetrahedral, trigonal bipyramidal, or octahedral) as indicated in Table 7-1.
4. If more than one arrangement of lone pairs and chemical bonds is possible, select the one that will minimize lone-pair repulsions. In trigonal bipyramidal arrangements, the repulsion is minimized when every lone pair is in the plane of the triangle. In an octahedral arrangement with two lone pairs, repulsion is minimized when the lone pairs are on opposite sides of the central atom. With three lone pairs in an octahedral arrangement of lone pairs and bonding pairs, the lone pairs are observed to be adjacent (see the figure for XeO_3 in Table 7-2).
5. Identify the arrangement of atoms (the molecular structure) from their locations at the ends of chemical bonds (Table 7-2).

EXAMPLE 1   **Predict the molecular structure of $BeCl_2$.**

The Lewis structure of $BeCl_2$ is

$$: \ddot{Cl} - Be - \ddot{Cl} :$$

The central beryllium atom has no lone pairs but is bonded to two atoms. There are therefore two regions of high electron density about the beryllium atom. Table 7-1 indicates that two regions of high electron density arrange themselves on opposite sides of the central atom so the bond angle will be 180°. $BeCl_2$ will be a linear molecule.

◆ ◆ ◆ ◆ ◆

EXAMPLE 2   **Predict the structure of BCl_3.**

The Lewis structure of BCl_3 is

$$: \ddot{Cl} :$$
$$| $$
$$: \ddot{Cl} - B - \ddot{Cl} :$$

BCl_3 contains three bonds, and there are no lone pairs on boron. Table 7-1 indicates that the arrangement of three regions of electron density will be trigonal planar. The bonds in BCl_3 will lie in a plane with 120° angles between them, a trigonal planar arrangement (Fig. 7-4).

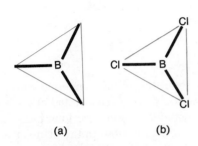

(a)   (b)

Figure 7-4 (a) Trigonal planar arrangement of three bonds in BCl_3 and (b) the resulting trigonal planar molecular structure.

◆ ◆ ◆ ◆ ◆

EXAMPLE 3   **Predict the molecular structure of CH_4.**

The Lewis structure of CH_4 is

$$
\begin{array}{c}
H \\
| \\
H-C-H \\
| \\
H
\end{array}
$$

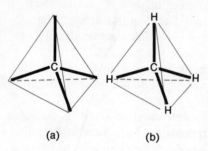

CH_4 contains four bonds from the carbon atom to hydrogen atoms. Table 7-1 indicates that four regions of electron density will arrange themselves so that they point at the corners of a tetrahedron with the central atom in the middle. The hydrogen atoms located at the ends of the bonds are located at the corners of a tetrahedron. This molecular structure (Fig. 7-5) is tetrahedral.

(a)   **(b)**

Figure 7-5 (a) Tetrahedral arrangement of four bonds in CH_4 and (b) the resulting tetrahedral molecular structure.

◆ ◆ ◆ ◆ ◆

EXAMPLE 4   **Predict the structure of H_2O.**

The Lewis structure of H_2O

$$
\begin{array}{c}
H \\
| \\
:\ddot{O}-H
\end{array}
$$

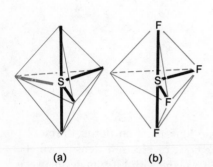

indicates that there are four regions of high electron density around the oxygen atom—two lone pairs and two chemical bonds. These four regions are arranged in a tetrahedral fashion (Fig. 7-6) as indicated in Table 7-1. However, the arrangement of the atoms themselves (the molecular structure) in H_2O is angular with a bond angle of approximately $109.5°$ (Table 7-2).

(a)   **(b)**

Figure 7-6 (a) Tetrahedral arrangement of two lone pairs (shown in color) and two bonds (black) in H_2O and (b) the resulting bent molecular structure.

◆ ◆ ◆ ◆ ◆

EXAMPLE 5   **Predict the molecular structure of SF_4.**

The Lewis structure of SF_4

indicates five regions of high electron density about the sulfur—one lone pair and four chemical bonds. These regions of electron density are directed at the corners of a trigonal bipyramid (Table 7-1). In order to minimize lone pair-bond pair repulsions the lone pair occupies one of the locations in the plane of the triangle. The molecular structure (Fig. 7-7) will be that of a seesaw (Table 7-2).

(a)   **(b)**

Figure 7-7 (a) Trigonal bipyramidal arrangement of one lone pair (shown in color) and four bonds (black) in SF_4 and (b) the resulting seesaw-shaped molecular structure.

◆ ◆ ◆ ◆ ◆

EXAMPLE 6   **Predict the molecular structure of XeF$_4$.**

The Lewis structure of XeF$_4$

indicates six regions of high electron density around xenon; two lone pairs and four bonds. These six regions adopt an octahedral arrangement. The two possible arrangements of lone pairs and bonds are illustrated in Figs. 7-8(a) and 7-8(b). Since to minimize repulsions the lone pairs are on opposite sides of the central atom, structure 7-8(a) is the more stable of the two possible structures for XeF$_4$. The five atoms are all in the same plane and form what is usually referred to as a "square planar" configuration.

Figure 7-8 (a),(b) Possible octahedral arrangements of two lone pairs (shown in color) and four bonds (black) in XeF$_4$ and (c) the stable square planar molecular structure.

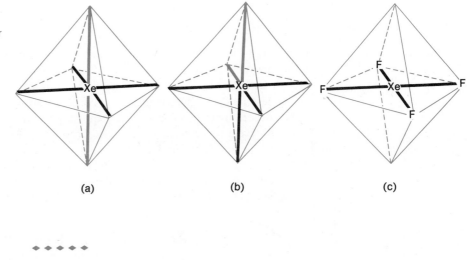

(a)   (b)   (c)

(a)

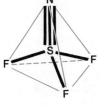

(b)

Figure 7-9 (a) Tetrahedral arrangement of four bonds (three single bonds and one triple bond) around S in SF$_3$N and (b) the resulting tetrahedral molecular structure.

EXAMPLE 7   **Predict the molecular structure of SF$_3$N.**

The Lewis structure of SF$_3$N

indicates four regions of high electron density (one triple bond and three single bonds) around the central sulfur atom, the atom of least electronegativity. These four regions will be distributed in a tetrahedral arrangement giving a tetrahedral molecular structure as shown in Fig. 7-9.

EXAMPLE 8   **Predict the molecular structure of ClO_2F.**

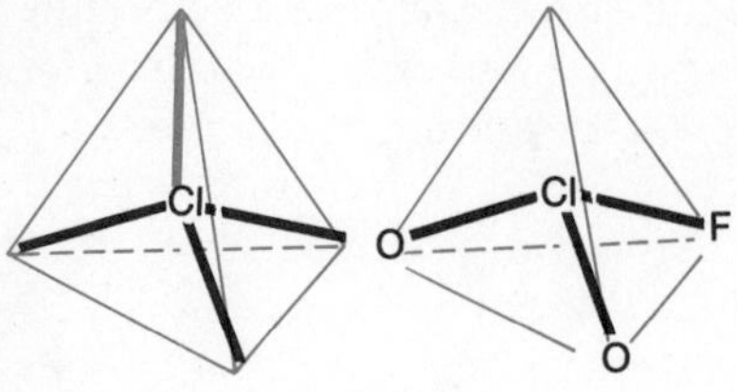

The least electronegative element, chlorine, is the central element as shown in the Lewis structure.

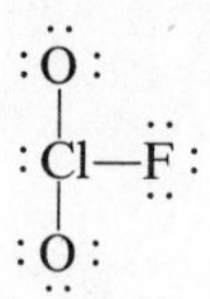

$$:\ddot{O}:$$
$$:\ddot{Cl}\!-\!\ddot{F}:$$
$$:\ddot{O}:$$

The four regions of high electron density are arranged tetrahedrally. The oxygen and fluorine atoms are arranged to give a trigonal pyramidal structure, as shown in Fig. 7-10.

◆ ◆ ◆ ◆ ◆

(a)   (b)

Figure 7-10 (a) Tetrahedral arrangement of one lone pair (shown in color) and three bonds in ClO_2F and (b) the resulting trigonal pyramidal molecular structure.

Hybridization of Atomic Orbitals

Many of the molecules discussed in previous sections contain identical bonds between a central atom and two to six other atoms. The formation of these identical bonds from the various atomic orbitals on the central atom can be described by hybridization, a concept introduced by Linus Pauling in 1935. According to this concept, nonequivalent atomic orbitals can hybridize (combine) to give a set of equivalent hybrid orbitals whose orientation reflects the geometry of the molecule or ion. Although the theory uses atomic orbitals rather than molecular orbitals, it is nevertheless especially useful in explaining why the bonds in many molecules are equivalent and how the atomic orbitals on a central atom interact with the orbitals on the other atoms in a molecule.

7.3   Methane, CH_4, and Ethane, C_2H_6 (Tetrahedral Hybridization)

A molecule of methane, CH_4, consists of a carbon atom surrounded by four hydrogen atoms at the corners of a tetrahedron (Fig. 7-5). The structure of ethane (Fig. 7-13) is similar to that of methane. Each carbon atom in ethane, C_2H_6, has four neighbors arranged at the corners of a tetrahedron: three hydrogen atoms and one carbon atom. The bonding in CH_4 and C_2H_6 is very similar. We will consider how a carbon atom can form four single bonds and why these bonds are directed at the corners of a tetrahedron. Each atomic orbital can accommodate two electrons of opposing spin. As we learned earlier (Section 4.15), the electrons enter each orbital of a given type singly before any pairing of electrons occurs within those orbitals (Hund's Rule). The carbon atom therefore has only two unpaired electrons in its normal state (ground state) with an electronic configuration $1s^2 2s^2 2p^1 2p^1$. With this structure, one would expect carbon to form only two covalent bonds, but it rarely does. The divalent compounds of carbon are not nearly so prevalent as are the tetravalent compounds containing four covalent bonds. We might ask, then, how the carbon atom attains four unpaired electrons with which to form four covalent bonds. The answer is, by promoting (by excitation) one of the electrons in the 2s orbital to the third 2p orbital with the resulting configuration $1s^2 2s^1 2p^1 2p^1 2p^1$. It should be noted that the energy required for promotion is more than offset by the energy given off during bond formation. Therefore the energy state after electron promotion *and* bond formation is lower than before these steps; the decrease in energy corresponds to a more stable state.

The electron distribution for the carbon atom in the ground state and after promotion of the electrons can be shown as follows by representing each orbital by a circle and each electron by an arrow; two arrows pointing in opposite directions within an orbital designate two electrons of opposing spin.

C atom (ground state)

1s   2s   2p

The three *p*-orbitals are oriented at right angles (see Section 4.12); the *s* orbital is undirected because of its spherical shape.

In the formation of the methane molecule, each of the four hydrogen atoms covalently bonded to the carbon provides one electron (indicated by a colored arrow) to each one of the four orbitals that contain an unpaired electron.

CH$_4$ molecule

1s   2s   2p

sp^3 **hybridization**

Tetrahedral structure

Experimental evidence shows that the four bonds of methane are equal in length and in strength and that they are pointed toward the corners of a regular tetrahedron. According to quantum mechanical calculations, the one 2s orbital and the three 2p orbitals of the carbon atom **hybridize** (combine) to form four new hybridized orbitals, which are exactly equivalent to one another and are so arranged in space that their lobes point to the corners of a regular tetrahedron (Fig. 7-11). They are referred to as *sp*^{3} hybridized orbitals, to designate the hybridization of one *s* orbital and three *p* orbitals; *sp*^{3} hybridization always results in a tetrahedral orientation of four orbitals.

The 1s orbital of each of the four hydrogen atoms of the methane molecule can be thought of as overlapping with each one of the four *sp*^{3} orbitals of a carbon atom to form an *sp*^{3}-*s* sigma bond. (Notice the similarity between this and the overlap to produce molecular orbitals, Section 6.1). This results in the formation

(a)

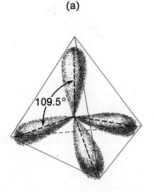

109.5°

(b)

Figure 7-11 (a) An individual hybridized *sp*^{3} orbital. (b) The four tetrahedral hybridized *sp*^{3} orbitals of the carbon atom. For clarity, only the large lobe of each hybrid orbital is shown.

Figure 7-12 The methane molecule. (a) Diagram showing the overlap of the four tetrahedral hybridized sp^3 orbitals with four s orbitals of four hydrogen atoms to produce the methane molecule. (b) The overall outline of the four bonding orbitals in methane.

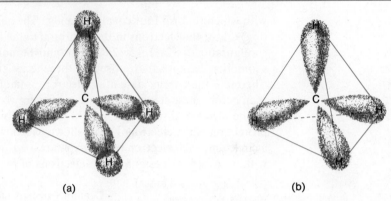

(a)   (b)

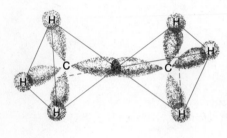

(a)

(b)

Figure 7-13 The ethane molecule. (a) The overlap diagram for the ethane molecule. (b) The overall outline of the seven bonding orbitals in ethane.

of four very strong covalent bonds between one carbon atom and four hydrogen atoms to produce the methane molecule, CH_4 (Fig. 7-12).

In ethane, C_2H_6, an sp^3 orbital of one carbon atom overlaps end-to-end with an sp^3 orbital of a second carbon atom to form a sigma bond. Each of the other three sp^3 orbitals of each carbon atom overlaps with an s orbital of a hydrogen atom to form additional sigma bonds. The structure and overall outline of the bonding orbitals of ethane are shown in Fig. 7-13. Ethane is made up of two tetrahedra with one corner in common.

Hybridization can occur *only* when the orbitals involved have very similar energies. It is possible to hybridize $2s$ with $2p$ orbitals, or $3s$ with $3p$ orbitals, for example, but not $2s$ with $3p$ (see Chapter 4, Fig. 4-15).

7.4   Beryllium Chloride, $BeCl_2$ (Digonal Hybridization)

The electronic configuration of Be is $1s^2 2s^2$, and it appears that in the normal, or ground, state the element should not form covalent bonds at all. However, in an excited state an electron in the $2s$ orbital can be promoted to a $2p$ orbital so that the configuration is $1s^2 2s^1 2p^1$, which means that two unpaired electrons are available for forming covalent bonds with atoms that can share electrons.

In the gaseous state, $BeCl_2$ is a linear molecule, Cl—Be—Cl; all three atoms lie in a straight line and, inasmuch as the two Be—Cl bonds have the same length and strength they are equivalent. It is assumed that the $2s$ and $2p$ atomic orbitals hybridize to form two sp hybrid orbitals. These sp orbitals overlap with p orbitals from the chlorine atoms to form sigma bonds. Because there are only two covalent bonds to the beryllium atom, and no lone pairs, the maximum angle separating the axes of the two sp hybrid atomic orbitals (regions of high electron density) should be $180°$ (Table 7-1), which is the observed value. Because of the spatial orienta-

Figure 7-14 Hybridization of an *s* orbital and a *p* orbital (of the same atom) to produce two *sp* hybrid orbitals.

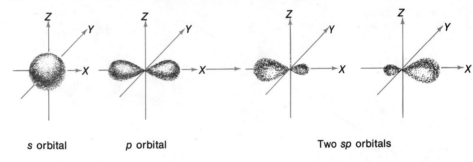

s orbital   *p* orbital   Two *sp* orbitals

tion, the *sp* hybrid atomic orbitals are also referred to as **digonal hybrid orbitals** and always correspond to a linear (straight-line) structure. (Fig. 7-14).

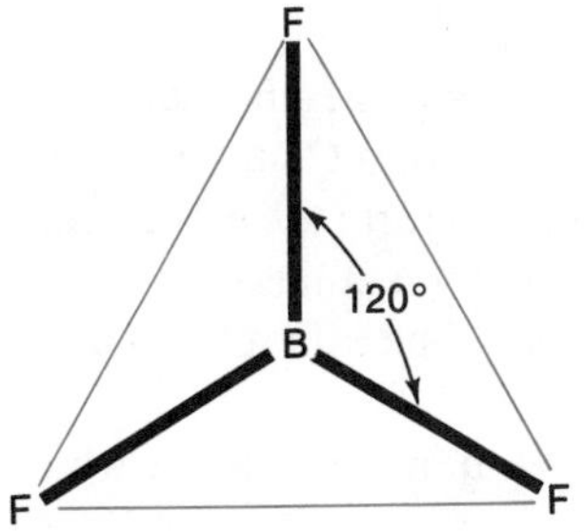

7.5   Boron Trifluoride, BF_3 (Trigonal Hybridization)

Experimental evidence shows that there are three equivalent B—F bonds in boron trifluoride, BF_3. All four atoms of this molecule lie in the same plane, with the boron atom in the center and the three fluorine atoms at the corners of an equilateral triangle. The F—B—F bond angle, therefore, is 120° (see Fig. 7-15). A similar structure is observed for BCl_3 (see Fig. 7-4).

The electronic structure of boron in the ground state is $1s^2 2s^2 2p^1$. During chemical reaction with fluorine (or chlorine), a $2s$ electron of boron is postulated as being promoted to a $2p$ orbital giving the structure $1s^2 2s^1 2p^1 2p^1$. The $2s$ and two of the $2p$ orbitals of boron hybridize to form three sp^2 hybrid orbitals. The sp^2 hybrid orbitals always have a trigonal planar orientation (see Fig. 7-16).

Figure 7-15 The trigonal planar structure of BF_3.

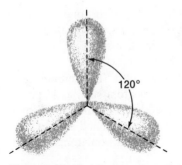

Figure 7-16 The shape and spatial orientation of trigonal planar, sp^2, hybrid orbitals.

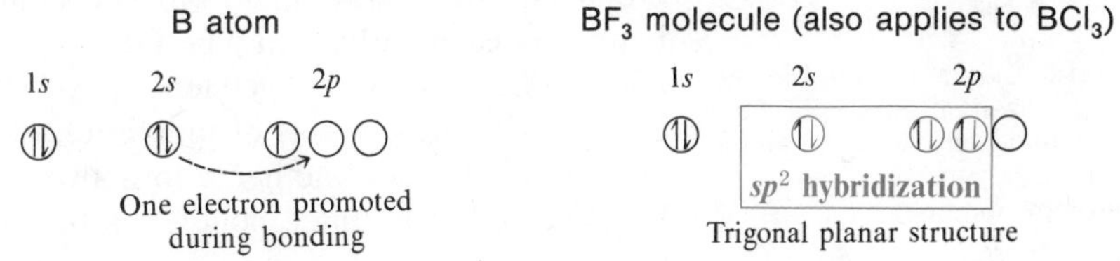

B atom

$1s$	$2s$	$2p$

One electron promoted during bonding

BF_3 molecule (also applies to BCl_3)

$1s$	$2s$	$2p$

sp^2 hybridization

Trigonal planar structure

7.6   Ethylene, $H_2C=CH_2$ (Trigonal Hybridization)

As we discovered from reading Section 7.3, tetrahedral hybridization of carbon in methane and ethane is due to the mixing of a $2s$ orbital with three $2p$ orbitals of carbon. In ethylene a different situation arises, inasmuch as only two $2p$ orbitals are involved in the mixing with the $2s$ orbital, giving sp^2 hybridization as in BF_3 (see Section 7.5). This leaves one $2p$ orbital unhybridized. The three sp^2 orbitals of carbon are pictured in Fig. 7-17. As in BF_3, the three sp^2 orbitals (of carbon in ethylene) lie in the same plane and are arranged so that the angles between them are 120°. The unhybridized p orbital (shown in color in Fig. 7-17) is perpendicular to the sp^2 plane.

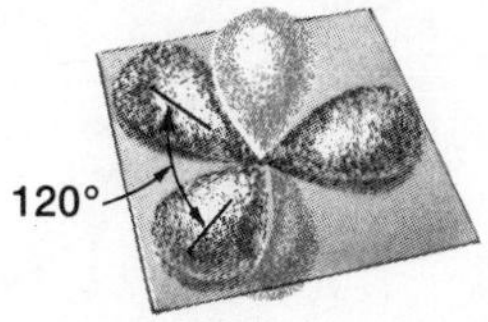

When we bring the two sp^2 hybridized carbon atoms together in the formation of ethylene, it should be noted first (see Fig. 7-18) that the overlap of lobes of the two sp^2 orbitals end-to-end forms a strong σ bond. Second, it should be noted that

Figure 7-17 Diagram illustrating three trigonal sp^2 hybridized orbitals of the carbon atom, which lie in the same plane, and the one unhybridized *p* orbital (shown in color), which is perpendicular to the plane.

Figure 7-18 The ethylene molecule. (a) The overlap diagram of two sp^2 hybridized carbon atoms and four s orbitals from four hydrogen atoms. There are four σ C—H bonds, one σ C—C bond, and one π C—C bond (making the net carbon-carbon bond a double bond). The dashed lines, each connecting two lobes, indicate the side-by-side overlap of the two unhybridized p orbitals. The hybridized sp^2 orbitals are shown in grey, and the unhybridized p orbitals are shown in color. The hybridized sp^2 orbitals lie in a plane with the unhybridized p orbitals extending above and below the plane and perpendicular to it. (b) The overall outline of the bonding molecular orbitals in ethylene. The two portions of the π bonding orbital (shown in color), resulting from the side-by-side overlap of the unhybridized p orbitals, are above and below the plane.

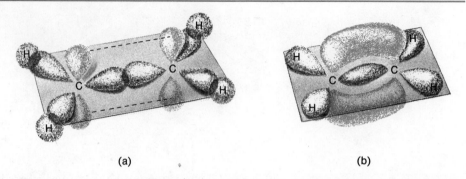

(a)   (b)

the unhybridized p orbitals are in a position to interact with one another, side-by-side. This overlap results in a π bond of the type discussed in Chapter 6. The overlap to form the π bond is not as efficient as the overlap to form the σ bond and the π bond is therefore weaker than the σ bond. The two carbon atoms of ethylene are thus bound together by two kinds of bonds—one σ and one π bond, a **double bond.**

Note that in an ethylene molecule the four hydrogen atoms and two carbon atoms are all in the same place.

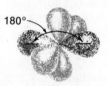

Figure 7-19 Diagram of the two linear, hybridized sp orbitals of the carbon atom, which lie in a straight line, and the two unhybridized p orbitals (shown in color).

7.7   Acetylene, H—C≡C—H (Digonal Hybridization)

We can now go one step further. As we have seen for $BeCl_2$, if just one 2p orbital hybridizes with a 2s orbital we obtain two hybrid orbitals, designated as sp orbitals. This arrangement leaves two 2p orbitals unhybridized (Fig. 7-19).

When an sp hybridized orbital on each of two carbon atoms combine, the two sp orbitals overlap end-to-end to form a sigma bond (Fig. 7-20). The remaining sp hybrid on each carbon may be used to bond to another atom such as H giving linear molecules such as acetylene, H—C≡C—H. In addition to this, as indicated in Fig. 7-20, the two sets of unhybridized p orbitals are positioned so that they overlap side-by-side and hence form two π bonds. The two carbon atoms are thus bound together by three bonds, a **triple bond.**

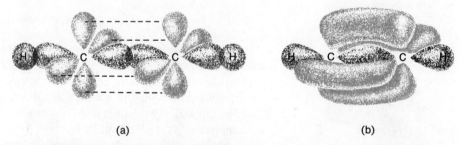

(a)   (b)

Figure 7-20 The acetylene molecule. (a) The overlap diagram of two sp hybridized carbon atoms and two s orbitals from two hydrogen atoms. There are two σ C—H bonds, one σ C—C bond, and two π C—C bonds (making the net carbon-carbon bond a triple bond). The dashed lines, each connecting two lobes, indicate the side-by-side overlap of the four unhybridized p orbitals. The hybridized sp orbitals are shown in grey, and the unhybridized p orbitals are shown in color. (b) The overall outline of the bonding orbitals in acetylene. The π bonding orbitals (in color) are positioned with one above and below the line of the σ bonds and the other behind and in front of the line of the σ bonds.

7.8   Phosphorus Pentachloride, PCl$_5$ (Trigonal Bipyramidal Hybridization)

In a molecule of phosphorus pentachloride, PCl$_5$, there are five P—Cl bonds and thus five pairs of valence electrons about the central phosphorus atom. To accommodate five pairs of electrons, either lone or bonding, an atom must have five atomic orbitals available. The outermost s and p orbitals of a phosphorus atom total only four. More than four hybrid orbitals can be made available by hybridization of these four atomic orbitals with one or more d orbitals. The electronic structure of phosphorus in the ground state is $1s^2 2s^2 2p^6 3s^2 3p^1 3p^1 3p^1 3d^0$. A $3s$ electron is postulated as being promoted to the $3d$ orbital giving the structure $1s^2 2s^2 2p^6 3s^1 3p^1 3p^1 3p^1 3d^1$, thereby making five orbitals available for electron-pair sharing with five chlorine atoms. This is called sp^3d hybridization.

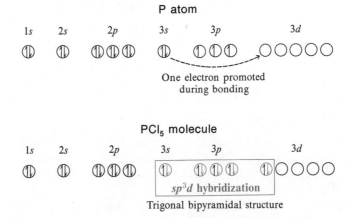

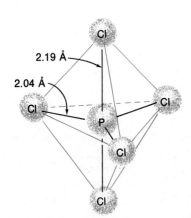

Figure 7-21 Structure of phosphorus pentachloride, PCl$_5$; sp^3d hybridization.

It is known that the PCl$_5$ molecule is a trigonal bipyramid (Fig. 7-21). Three of the chlorine atoms (equatorial) lie at the corners of an equilateral triangle, whereas the other two chlorine atoms (axial) lie above and below the center of the triangle. It is interesting to note that the two axial bonds are not equivalent geometrically to the equatorial bonds and also are longer than the equatorial bonds. In this respect the sp^3d hybrid orbitals differ from the hybrid orbitals discussed in Sections 7.3 through 7.7; sp, sp^2, and sp^3 orbital hybridization results in geometrically equivalent hybrid orbitals. The sp^3d (or dsp^3) hybridization always results in a trigonal bipyramidal set of hybrid orbitals in which the two axial hybrid orbitals differ from the equatorial orbitals.

7.9   Sulfur Hexafluoride, SF$_6$ (Octahedral Hybridization)

A molecule of sulfur hexafluoride has six fluorine atoms surrounding a single sulfur atom. To accommodate six pairs of electrons, in bonding with six fluorine atoms, the sulfur atom must provide six bonding orbitals. The electronic structure of sulfur in the ground state is $1s^2 2s^2 2p^6 3s^2 3p^2 3p^1 3p^1$. Promotion of an s and a p electron to d orbitals gives the structure $1s^2 2s^2 2p^6 3s^1 3p^1 3p^1 3p^1 3d^1 3d^1$. The one $3s$,

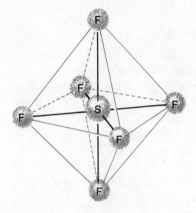

three $3p$, and two $3d$ orbitals hybridize to form six hybridized sp^3d^2 orbitals. Each orbital points to the corner of an octahedron (Fig. 7-22), and hence the sp^3d^2 (or d^2sp^3) hybridization is called octahedral. The angle between any two bonds is $90°$.

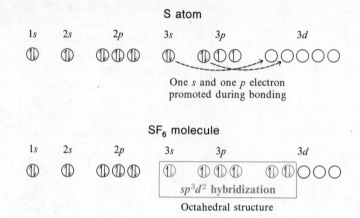

Figure 7-22 The octahedral configuration of SF_6; sp^3d^2 hybridization.

Questions

1. Predict the structure of each of the following molecules or ions: SiH_4, CI_4, TeF_6, NH_2Cl, SbF_5, O_3, CO_3^{2-}, H_3O^+, $ClNO$, PO_4^{3-}, $AlCl_3$, SiF_6^{2-}, SO_3^{2-}, TeF_4, SeO_2, BrF_3, ClO_3^-, ICl_4^-, XeO_2F_2, SO_2, ICl_2^-, $XeOF_4$, BeF_4^{2-}, SCl_3^+, CH_2Cl_2. [For Xe compounds, Xe is the central atom. In all other cases the less electronegative atom (except H) is also the central atom.]

2. Predict the geometry around the indicated atom or atoms:
   (a) Nitrogen in nitric acid
   (b) The N atom and two C atoms in glycine, the simplest amino acid. The *framework* of glycine is

$$
\begin{array}{ccc}
& H & O \\
& | & \parallel \\
H\!-\!\!&C\!-\!C \\
& | & \\
& NH_2 & OH
\end{array}
$$

   (c) Oxygen atoms in hydrogen peroxide, H_2O_2
   (d) Carbon atoms in the benzene ring (see Question 34, Chapter 5)
   (e) Boron atoms in B_2Cl_4 (B_2Cl_4 contains a B—B bond.)

3. What is a hybrid orbital? Illustrate with the hybrid orbitals that may be formed by phosphorus.

4. Show the distribution of valence electrons in the orbitals of the central atom in each of the following molecules or ions just prior to bonding to the other atoms in the compound, and show the distribution after bonding takes place: $GeCl_4$, $AlCl_3$, BrF_3, BrF_5, IF_6^+ (consider this ion as if formed from I^+ and 6 F atoms)

5. Write Lewis formulas for the molecules or ions listed below. Utilizing both the Valence Shell Electron-Pair Repulsion Theory and the concept of hybridization and remembering that single bonds are σ bonds, double bonds are a σ plus a π bond, and triple bonds are a σ plus two π bonds, indicate the type of hybridization (sp,sp^2,sp^3) for each central atom in the molecules or ions listed: BH_4^-, $AlCl_3$, CN^-, CN_2^{2-}, $ClNO$, BO_3^{3-}, PO_4^{3-}, $SnCl_3^-$, SeO_2, N_2O_4 (contains a N—N bond), $POCl_3$, HCN, $COCl_2$

6. XF_3 has a trigonal pyramidal molecular structure. To which main group of the Periodic Table does X belong?

7. In the compound H—X≡X—H, X exhibits sp^2 hybridization. To which main group of the Periodic Table does X belong?

8. $B(OH)_3$, boric acid, reacts with water according to the following equation: $B(OH)_3 + H_2O \longrightarrow H^+(aq) + B(OH)_4^-(aq)$. How does the hybridization of the boron atom change during this reaction?

9. Elemental sulfur consists of covalently bonded, puckered rings of eight sulfur atoms, S_8. The S—S—S angles in the ring are $107.9°$. The production of sulfuric acid, $SO_2(OH)_2$, from sulfur involves oxidation of sulfur to SO_2, then to SO_3, and finally reaction with water to give sulfuric acid. Trace the changes in hybridization of the sulfur atom during this sequence of reactions.

10. Although the location of the lone pairs in the hybrid orbitals cannot be observed, the oxygen atom in water is believed to exhibit sp^3 hybridization. Explain.

11. The structure of graphite consists of layers of carbon atoms. Each carbon atom has three other carbon atoms as its near neighbors. These are arranged in a trigonal planar arrangement (Fig. 25-2). The structure of diamond consists of a three-dimensional network with each carbon located in the center of a tetrahedron of carbon atoms.

(a) How does the hybridization of the carbon atoms in graphite differ from that in diamond?

(b) Which of these two forms of elemental carbon contains only single bonds?

References

"Hybrid Orbitals in Molecular Orbital Theory," I. Cohen and J. D. Bene, *J. Chem. Educ.*, **46,** 487 (1969).

"Simplified Molecular Orbital Approach to Inorganic Stereochemistry," R. M. Gavin, *J. Chem. Educ.*, **46,** 413 (1969).

"Size and Shape of a Molecule," M. J. Demchik and V. C. Demchik, *J. Chem. Educ.*, **48,** 770 (1971).

"Accurate Molecular Geometry," D. H. Whiffen, *Chem. in Britain,* **7,** 57 (1971).

"The Electron-Pair Repulsion Model for Molecular Geometry," R. J. Gillespie, *J. Chem. Educ.*, **47,** 18 (1970).

"Dissociation Energies of π Bonds in Alkenes," S. I. Miller, *J. Chem. Educ.*, **55,** 778 (1978).

"Some Structural Principles for Introductory Chemistry," A. F. Wells, *J. Chem. Educ.*, **54,** 273 (1977).

"Carbenes," M. Jones, Jr., *Sci. American,* February 1976, p. 101.

"Chirality in Chemistry," V. Prelog, *Science,* **193,** 17 (1976).

"Unusual Structures Predicted for Carbon Compounds," T. H. Maugh, *Science,* **194,** 413 (1976).

8

The Relationship of the Periodic Classification to Chemical Behavior

As mentioned previously (Section 4.16), we now know of 106 elements that have either been discovered in nature or made artificially. It was pointed out that though each element is different from every other element, similarities in electronic structures make possible arrangements such as the Periodic Table that aid greatly in correlating the study of the properties and compounds of the elements. In Chapter 4 we considered the periodic variation of physical properties of the elements. In this chapter we will examine the periodic behavior of their chemical properties.

As an example of the utility of the Periodic Table, consider the remarkable use that Mendeleev made of his periodic table. Mendeleev's table included only 62 elements, the number known at that time. However, he predicted the existence and properties of 6 additional elements corresponding to 6 vacant places in his table—places that were left vacant so that known elements of similar chemical properties would fall in the same group. Mendeleev predicted that elements would subsequently be discovered whose chemical properties would fit these gaps. This is indeed the case. The elements are scandium (Sc), gallium (Ga), germanium (Ge), technetium (Tc), rhenium (Re), and polonium (Po), and they all have properties similar to those predicted by Mendeleev.

A comparison of the properties predicted by Mendeleev for the substance that we know as germanium, but which he called eka-silicon, and those determined experimentally for the element after it was actually discovered is given in Table 8-1.

It is unlikely that you will have the opportunity to determine the properties of a new element, but the same principles that you would use to predict these

Predicted Properties for Ek, Eka-silicon (1871)	Observed Properties of Ge, Germanium (discovered in 1886)
Atomic weight, 72	Atomic weight, 72.59
Specific gravity, 5.5	Specific gravity, 5.36
A gray-colored metal	A grayish white metal
Oxidation number of 4 with oxygen	Oxidation number of 4 with oxygen
EkO_2, a white solid of sp gr 4.7 and high mp	GeO_2, a white solid of sp gr 4.70, mp 1100°
$EkCl_4$, a volatile liquid, bp below 100°, sp gr 1.9	$GeCl_4$, a volatile liquid, bp 83°, sp gr 1.88
Ek will be acted upon by acids only slightly, and not at all by alkalies like NaOH	Ge does not react with HCl or NaOH, but dissolves in concentrated HNO_3

TABLE 8-1

NaCl(*s*)

(a)

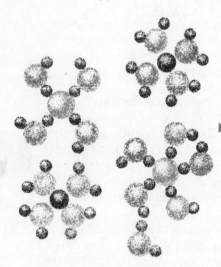

Na⁺(*aq*) + Cl⁻(*aq*)

(b)

Figure 8-1 (a) The distribution of ions in solid sodium chloride. Black spheres represent Na⁺; colored spheres represent Cl⁻. Each ion is in contact with six ions of opposite charge. (b) The distribution of ions in an aqueous solution of sodium chloride. Each ion is free to move independently because it is surrounded by water molecules and separated from ions of opposite charge.

properties can be applied to the correlation and recall of the behavior of the elements. The Periodic Table provides a powerful framework for organizing the chemical behavior of the elements.

8.1   Classification of Chemical Compounds

Before we begin the examination of the properties of the elements, let us consider several of the types of compounds that elements form.

▶**1. SALTS.** A **salt** is an ionic compound composed of positive ions (cations) and negative ions (anions)—see Section 5.2. The cations may be monatomic metal ions, such as K^+ and Co^{2+}, or polyatomic positive ions, such as NH_4^+ and PCl_4^+. The anions may be monatomic nonmetal ions, such as Cl^- and S^{2-}, or polyatomic negative ions, such as NO_3^- and SO_4^{2-} (with the exception of OH^- or O^{2-}). As discussed in part 2 of this section, ionic compounds that contain hydroxide ions or oxide ions are referred to as bases rather than as salts.

Ionic compounds are held together in the solid state by ionic bonds, strong electrostatic attractions between oppositely charged ions that are in contact with each other. When soluble salts dissolve in water, the ions separate (Fig. 8-1). Compounds that dissolve and exist in solution as ions are called **electrolytes.** Chemical equations may be used to indicate that a compound behaves as an electrolyte. The process for the dissolution (dissolving) of the salt sodium chloride in water, for example, may be represented by the following equation, in which *aq* indicates that the compound is present in aqueous solution:

$$NaCl(s) \xrightarrow{\text{H}_2\text{O}(l)} NaCl(aq) \tag{1}$$

However, in order to emphasize the fact that sodium chloride exists as ions in solution and that the ions are separated, the equation is usually written

$$NaCl(s) \xrightarrow{\text{H}_2\text{O}(l)} Na^+(aq) + Cl^-(aq) \tag{2}$$

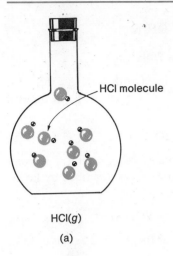

HCl molecule

HCl(g)

(a)

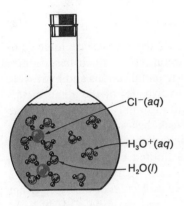

Cl⁻(aq)

H₃O⁺(aq)

H₂O(l)

$H_3O^+(aq)$ + $Cl^-(aq)$

(b)

Figure 8-2 (a) Molecules of hydrogen chloride in the gas phase. Each chlorine atom (colored sphere) is bonded to a hydrogen atom (black sphere) by a covalent bond. (b) A solution of hydrogen chloride in water (hydrochloric acid). The hydrogen ions are bonded to water molecules by a coordinate covalent bond giving a solution of H_3O^+ and Cl^- ions.

▶ **2. ACIDS AND BASES.** Acids and bases were first identified as specific types of compounds because of their behavior in aqueous solutions (solutions in water). Other definitions have since been developed that emphasize other aspects of the behavior of acids and bases. We shall examine these various definitions in Chapter 14. For now, however, let us concentrate on two particularly useful descriptions of these compounds.

a. Protonic Acids and Hydroxide Bases. A compound that increases the concentration of hydronium ions (H_3O^+) in aqueous solution is an **acid.** A hydronium ion is a hydrated proton (hydrated hydrogen ion), but it is sometimes written H^+ or $H^+(aq)$. It is important to remember, however, that H^+ and $H^+(aq)$ are abbreviated symbols and that a hydrogen ion is always associated with at least one water molecule in an aqueous solution. Hydrogen chloride, HCl, and sulfuric acid, H_2SO_4, are examples of acids; since they contain hydrogen that can be released as protons they are called **protonic acids.** These acids increase the hydronium ion concentration of water by the following reactions when they dissolve (Fig. 8-2):

$$HCl(g) + H_2O(l) \longrightarrow H_3O^+(aq) + Cl^-(aq) \qquad (3)$$

$$H_2SO_4(l) + H_2O(l) \longrightarrow H_3O^+(aq) + HSO_4^-(aq) \qquad (4)$$

Some compounds contain no hydrogen but increase the hydronium ion concentration in solution by reacting with water to give protonic acids; these protonic acids then behave as electrolytes, providing hydrogen ions to the solution. Sulfur trioxide is an example of such a compound. One mole of SO_3 reacts with one mole of water to give a mole of sulfuric acid, H_2SO_4.

$$SO_3(g) + H_2O(l) \longrightarrow H_2SO_4(l) \qquad (5)$$

If excess water is present, the hydronium ion concentration is increased by the reaction of H_2SO_4 with water, as indicated in Equation (4). The net equation can be written as the sum of Equation (5) plus Equation (4).

$$SO_3(g) + 2H_2O(l) \longrightarrow H_3O^+(aq) + HSO_4^-(aq) \qquad (6)$$

When a covalent molecule separates into ions it is said to **ionize.** Hydrogen chloride dissolves in water and ionizes into hydronium ions and chloride ions; similarly, nitric acid ionizes into hydronium ions and nitrate ions. A few protonic acids, such as hydrogen chloride, sulfuric acid, nitric acid (HNO_3), and perchloric acid ($HClO_4$), ionize completely. Such acids are called **strong acids.** Most acids, however, are **weak acids;** only a small fraction of the molecules ionize when dissolved in water. Examples of weak acids include hydrogen fluoride (HF), acetic acid (CH_3CO_2H), boric acid (H_3BO_3), and hydrogen cyanide (HCN). Although the hydronium ion concentration in an aqueous solution of a weak acid is greater than that of pure water, the hydronium ion concentration is only a small fraction of that which would result if the weak acid were completely ionized.

A compound that increases the concentration of hydroxide ion (OH^-) in solution is a **base.** Sodium hydroxide, NaOH, and calcium hydroxide, $Ca(OH)_2$, are examples of bases; since they contain hydroxide ions, they are sometimes called **hydroxide bases.** They increase the hydroxide ion concentration by the following reactions when they dissolve in water:

$$NaOH(s) \xrightarrow{H_2O(l)} Na^+(aq) + OH^-(aq) \tag{7}$$

$$Ca(OH)_2(s) \xrightarrow{H_2O(l)} Ca^{2+}(aq) + 2OH^-(aq) \tag{8}$$

Some compounds containing a metal and oxygen react with water to form hydroxides. For example, one mole of sodium oxide, Na_2O, reacts with one mole of water to give two moles of sodium hydroxide.

$$Na_2O(s) + H_2O(l) \longrightarrow 2NaOH(s) \tag{9}$$

If more than one mole of water is present, so that the excess can act as a solvent, the sodium hydroxide dissolves with the formation of ions as described by Equation (7). The net equation is

$$Na_2O(s) + H_2O(l) \longrightarrow 2Na^+(aq) + 2OH^-(aq) \tag{10}$$

Ammonia, NH_3, is a base because it reacts with water to a limited extent to form ammonium ions, NH_4^+, and hydroxide ions.

$$NH_3(g) + H_2O(l) \rightleftharpoons NH_4^+(aq) + OH^-(aq) \tag{11}$$

Many hydroxide bases resemble salts inasmuch as these bases are ionic compounds containing a cation and the hydroxide anion, OH^-. Such bases ionize completely in water to give solutions that contain the metal cations and hydroxide anions; they are classified as **strong bases.** The hydroxides of the alkali metals (Group IA) and of calcium, strontium, barium, and radium (Group IIA, alkaline earth metals) are examples of strong bases. **Weak bases** are bases, such as beryllium hydroxide, $Be(OH)_2$, which ionize only slightly in water, or bases such as ammonia, NH_3, which react with water only to a limited extent (Equation 11). Solutions of weak bases contain only a small amount of hydroxide ion and a relatively large amount of undissociated or unreacted molecules of the base. Other weak bases include aluminum hydroxide, $Al(OH)_3$, and pyridine, C_5H_5N.

b. Lewis Acids and Lewis Bases. During formation of a coordinate covalent bond (Section 5.10), one of the two atoms involved in forming the bond provides both electrons of the electron pair that bonds the atoms together. These atoms are, respectively, a base and an acid according to the **Lewis Theory** of acids and bases. A **Lewis base** is a molecule or ion that provides an electron pair. A **Lewis acid** is a molecule or ion that accepts the pair of electrons to form the bond. In the reaction of an oxide ion (O^{2-}) with sulfur trioxide (SO_3) to form a sulfate ion, the oxide ion behaves as a Lewis base in that it provides a pair of electrons to form the sulfur-oxygen bond; sulfur trioxide acts as a Lewis acid in that it accepts the pair of electrons from the Lewis base.

$$\tag{12}$$

Lewis base   Lewis acid

Other examples of Lewis bases and Lewis acids are provided in the following equations:

$$:C\equiv O: \; + \; \overset{\displaystyle H}{\underset{\displaystyle H}{B}}\!-\!H \; \longrightarrow \; H\!-\!\overset{\displaystyle H}{\underset{\displaystyle H}{B}}\!-\!C\equiv O: \tag{13}$$

Lewis   Lewis
base   acid

$$2:\ddot{F}:^- \; + \; \overset{\displaystyle F}{\underset{\displaystyle F}{F\!-\!Si\!-\!F}} \; \longrightarrow \; \overset{\displaystyle F\;F}{\underset{\displaystyle F\;\;F}{F\!-\!Si\!-\!F}}\;^{2-} \tag{14}$$

Lewis   Lewis
base   acid

The three lone pairs of electrons on each fluorine atom bonded to silicon in Equation (14) have been omitted for clarity.

The relationship between acids and bases defined in terms of hydrogen and hydroxide ions, Lewis acids and bases, and still other definitions of acids and bases will be explored more fully in Chapter 14.

▶ **3. ELECTROLYTES AND NONELECTROLYTES.** As we have seen, **electrolytes** may be either ionic compounds (such as sodium hydroxide and potassium nitrate) that dissolve in water, giving solutions of ions, or they may be covalent compounds (such as sulfuric acid and ammonia) that react with water to form ions in solution. **Nonelectrolytes,** however, are compounds that do not ionize when they dissolve in water. Nonelectrolytes are limited to covalent compounds. Many compounds of carbon such as methane, CH_4, benzene, C_6H_6, ethanol, C_2H_5OH, ether, $(C_2H_5)_2O$, and formaldehyde, CH_2O, are nonelectrolytes. A few inorganic compounds such as nitrous oxide, N_2O, phosphine, PH_3, and nitrogen(III) chloride, NCl_3, are nonelectrolytes.

8.2 Classification of Chemical Reactions

The chemical behavior of the elements is described by the way in which they and their compounds react. Several types of reactions are discussed below.

▶ **1. ADDITION, OR COMBINATION, REACTIONS.** In such reactions two or more substances add together, or combine, to give another substance.

$$Mg + Br_2 \longrightarrow MgBr_2 \tag{1}$$

$$K_2S + 2O_2 \overset{\triangle}{\longrightarrow} K_2SO_4 \tag{2}$$

$$Na_2O + SiO_2 \overset{\triangle}{\longrightarrow} Na_2SiO_3 \tag{3}$$

$$2Ni + 4Al + S_8 \overset{\triangle}{\longrightarrow} 2NiAl_2S_4 \tag{4}$$

▶ **2. DECOMPOSITION REACTIONS.** One compound breaks down (decomposes) into two or more substances.

$$2HgO \xrightarrow{\triangle} 2Hg + O_2 \tag{5}$$

$$CaCO_3 \xrightarrow{\triangle} CaO + CO_2 \tag{6}$$

$$2Cu(NO_3)_2 \xrightarrow{\triangle} 2CuO + 4NO_2 + O_2 \tag{7}$$

$$2NaHSO_4 \xrightarrow{\triangle} Na_2SO_4 + SO_3 + H_2O \tag{8}$$

▶ **3. METATHETICAL REACTIONS.** Two compounds exchange parts.

$$CaCl_2(aq) + 2AgNO_3(aq) \longrightarrow 2AgCl(s) + Ca(NO_3)_2(aq) \tag{9}$$

$$Al_2(SO_4)_3(aq) + 6KOH(aq) \longrightarrow 2Al(OH)_3(s) + 3K_2SO_4(aq) \tag{10}$$

Metathetical reactions usually, though not always, involve ionic compounds that exchange ions in forming new compounds. Equation (11) illustrates a reaction involving two covalent compounds, PCl_3 and PF_3, and two ionic compounds, $AgCl$ and AgF.

$$PCl_3(l) + 3AgF(s) \longrightarrow PF_3(g) + 3AgCl(s) \tag{11}$$

▶ **4. OXIDATION-REDUCTION REACTIONS.** Many reactions of the types described above can also be classified as oxidation-reduction reactions. When an atom, molecule, or ion loses electrons, it is **oxidized.** The oxidation number (Section 5.11) of at least one atom increases as an atom, molecule, or ion loses electrons and is thereby oxidized. When an atom, molecule, or ion gains electrons, it is **reduced.** At least one atom undergoes a decrease in its oxidation number as an atom, molecule, or ion gains electrons and is thereby reduced.

Oxidation and reduction always occur simultaneously, for if one atom gains electrons and is thereby reduced, a second atom must be present to provide the electrons. Because the second atom loses electrons, it is oxidized. Reactions involving oxidation and reduction are referred to as **oxidation-reduction,** or **redox,** reactions. An example is the reaction between sodium and chlorine (also an addition reaction), in which sodium is oxidized by chlorine and the chlorine is correspondingly reduced, as described by the following equation:

$$\overset{0}{2Na} + \overset{0}{Cl_2} \longrightarrow \overset{+1\ -1}{2NaCl} \tag{12}$$

As sodium is oxidized, the oxidation number of each sodium atom is increased from 0 to $+1$, as indicated by the superscripts. As chlorine is reduced, its oxidation number decreases from 0 to -1. Other examples of oxidation-reduction reactions are given below. In each case, the species that is oxidized is written first.

$$\overset{0}{Zn} + \overset{+1}{H_2SO_4} \longrightarrow \overset{+2}{ZnSO_4} + \overset{0}{H_2} \tag{13}$$

$$\overset{+2}{SnCl_2} + \overset{+4}{PbCl_4} \longrightarrow \overset{+4}{SnCl_4} + \overset{+2}{PbCl_2} \tag{14}$$

$$\overset{+2}{2CO} + \overset{+2}{2NO} \longrightarrow \overset{+4}{2CO_2} + \overset{0}{N_2} \tag{15}$$

The species that accepts electrons and thereby causes a second species to be oxidized in an oxidation-reduction reaction is called the **oxidizing agent.** In Equations (12)–(15), chlorine, Cl_2, sulfuric acid, H_2SO_4, lead(IV) chloride, $PbCl_4$, and nitrogen(II) oxide, NO, respectively, are the oxidizing agents. Note that a species that acts as an oxidizing agent is itself reduced. The species that provides electrons and thereby causes a second species to be reduced is called a **reducing agent.** The reducing agents in Equations (12)–(15) are, respectively, sodium, zinc, tin(II) chloride, $SnCl_2$, and carbon monoxide, CO. The reducing agent is itself oxidized during the course of an oxidation-reduction reaction.

As mentioned before, Equation (12) may be classified as an additon reaction as well as an oxidation-reduction reaction. Other reactions that fall into both categories include those indicated by Equations (1), (2), and (4) in this section. Decomposition reactions may also be redox reactions. Equations (5) and (7) in this section illustrate such reactions.

▶ **5. ACID-BASE REACTIONS.** An **acid-base** reaction involving a protonic acid may be described as a reaction in which a hydrogen ion is transferred from an acid to a base. Thus the following reactions may be classified as acid-base reactions:

$$H_2SO_4(l) + Ca(OH)_2(s) \longrightarrow CaSO_4(s) + 2H_2O(l) \qquad (16)$$

$$HCl(aq) + NaOH(aq) \longrightarrow NaCl(aq) + H_2O(l) \qquad (17)$$

$$H_2SO_4(l) + 2NH_3(g) \longrightarrow (NH_4^+)_2SO_4^{2-}(s) \qquad (18)$$

$$HBr(g) + C_5H_5N(l) \longrightarrow (C_5H_5NH^+)Br^-(s) \qquad (19)$$

$$HCl(g) + H_2O(l) \longrightarrow H_3O^+(aq) + Cl^-(aq) \qquad (20)$$

In all of these reactions, the first reactant is the acid and the second, the base. Note that acid-base reactions such as those illustrated by Equations (16), (18), and (19) need not occur in water.

Lewis acid-base reactions are reactions that proceed with the formation of a coordinate covalent bond (Sections 5.10 and 8.1). Since a proton adds to a base by sharing a pair of electrons donated by the base, the reactions illustrated by Equations (16)–(20) involve Lewis acid-base reactions as well as proton-transfer acid-base reactions. The following reactions, which do not involve the transfer of hydrogen at all, are also Lewis acid-base reactions since a coordinate covalent bond is formed:

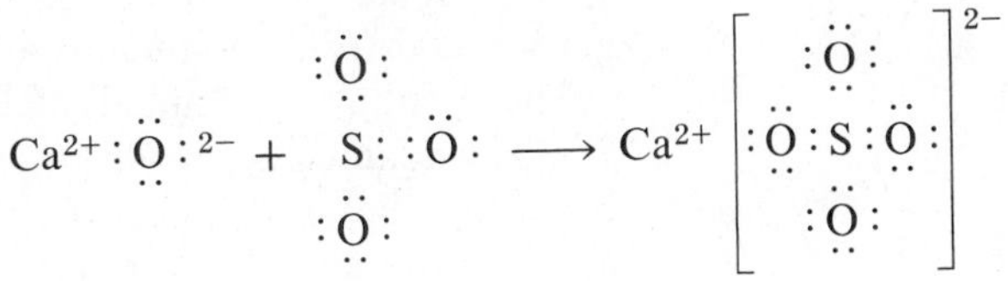

$$Ca^{2+} : \overset{..}{\underset{..}{O}} :^{2-} + \quad \overset{:\overset{..}{O}:}{\underset{:\overset{..}{O}:}{S : : \overset{..}{O}:}} \longrightarrow Ca^{2+} \left[\overset{:\overset{..}{O}:}{\underset{:\overset{..}{O}:}{: \overset{..}{O} : S : \overset{..}{O} :}} \right]^{2-} \qquad (21)$$

$$K^+ : \overset{..}{\underset{..}{Cl}} :^- + \quad \overset{Cl}{\underset{Cl}{|}} \overset{..}{\underset{|}{I}} - Cl \longrightarrow K^+ \left[\overset{Cl}{\underset{Cl}{Cl - I - Cl}} \right]^- \qquad (22)$$

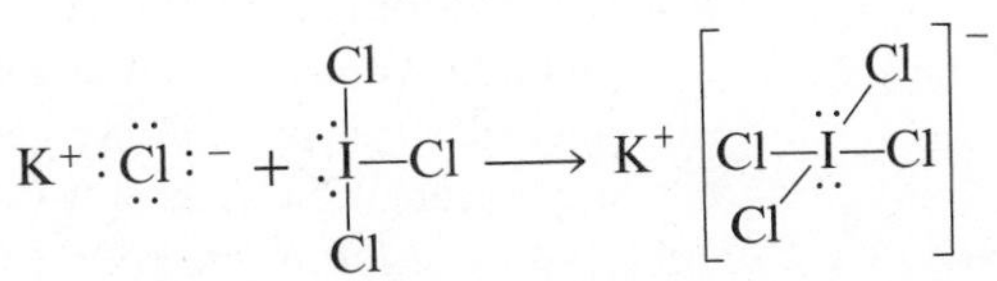

The three lone pairs of electrons on each chlorine atom bonded to iodine have been omitted for clarity in Equation (22). Equations (12)–(14) in Section 8.1 illustrate other Lewis acid-base reactions.

8.3   Metals and Nonmetals

The elements may be subdivided into two broad groups, metals and nonmetals, by the types of compounds they form and the reactions that these compounds undergo. One of the first attempts by chemists to classify the elements was based upon the observation that the oxides of some of the elements react with water to form bases, whereas other oxides react with water to form acids. For example, sodium oxide and calcium oxide react with water to form the bases sodium hydroxide and calcium hydroxide, respectively. Sulfur dioxide and carbon dioxide react with water to form sulfurous acid and carbonic acid, respectively. It soon became apparent that other types of chemical behavior could also be used to classify elements. For instance, the hydroxides of some elements are bases while the hydroxides of others are acids; $Ca(OH)_2$ is a base, but ClOH is an acid. The chlorides of some elements are ionic (NaCl is an example) while the chlorides of others (such as NCl_3) are covalent. Some elements are reducing agents (Li, Na, Mg, and Al, for example); others are oxidizing agents (F_2, Cl_2, O_2, S_8, and I_2, for example). Each type of chemical behavior divides the elements into one of the two broad groups, metals or nonmetals. With only a few exceptions, the same elements are classified as metals or as nonmetals by different kinds of chemical classifications.

Elements that exhibit many or all of the following properties are classified as **metals.** Metals are reducing agents (many reduce the hydrogen in an acid to hydrogen gas, leaving a salt of the acid, Section 9.15); generally, their oxides and hydroxides are bases; their hydrides (compound of the elements with hydrogen) either are inert in water or reduce water, giving hydrogen and the hydroxide of the metal; and often their oxides, hydrides, and halides are ionic. Metals tend to have low electronegativities (Section 5.6) and low ionization energies (Section 4.19), to exhibit a metallic luster, and to conduct heat and electricity well. Nonmetals tend to exhibit the opposite properties. In general, **nonmetals** (except the noble gases) are good oxidizing agents; their oxides and hydroxides are acids; their hydrides are acids; their oxides, hydrides, and halides are covalent; and they tend to have high electronegativities, to show no metallic luster, and to conduct heat and electricity poorly. Some of the differences in chemical and physical properties that distinguish metals from nonmetals are summarized in Table 8-2.

The distribution of metals and nonmetals in the Periodic Table is shown in Fig. 8-3. The nonmetals lie to the right and above the broad line running from between beryllium (Be) and boron (B) at the upper left to between polonium (Po) and astatine (At) at the lower right. The metals lie to the left and below this line. About three-fourths of the elements exhibit metallic behavior.

In Section 4.18, the elements were classified into four categories according to their atomic structures: noble gases, representative elements, transition elements, and inner transition elements. Nonmetals are either representative elements or noble gases. Metals are found in each category except the noble gases. In the

TABLE 8-2   Comparison of Metallic and Nonmetallic Elements

Metals	Nonmetals
Chemical Behavior of the Elements	
1. Are reducing agents	1. Are oxidizing agents (except noble gases)
2. Form oxides that may react with water giving hydroxides	2. Form oxides that may react with water giving acids
3. Form basic hydroxides	3. Form acidic hydroxides
4. React with O, F, H and other non-metals, giving ionic compounds	4. React with O, F, H and other non-metals, giving covalent compounds
5. React with other metals, giving metallic compounds	5. React with metals, giving ionic compounds
6. Exhibit lower electronegativities	6. Exhibit higher electronegativities
7. Have one to five electrons in outermost shell; usually not more than three	7. Usually have four to eight electrons in outermost shell
8. Exhibit low ionization potentials; readily form cations by loss of electrons	8. Exhibit high electron affinities; readily form anions by accepting electrons to fill the outermost shell (except noble gases)
Physical Behavior of the Elements	
1. Are good conductors of heat and electricity	1. Are poor conductors
2. Are malleable and ductile in solid state	2. Are brittle and nonductile in solid state
3. Show metallic luster	3. Show no metallic luster
4. Are opaque	4. May be transparent or translucent
5. Have high density	5. Have low density
6. Are solids (except mercury)	6. Are gases, liquids, or solids
7. Have crystal structure in which each atom is surrounded by eight to twelve near neighbors; metallic bonds exist between atoms	7. Form molecules that consist of atoms covalently bonded; the noble gases are monatomic

remainder of this chapter we will concentrate on the chemistry of the representative, or main group, elements.

As we shall see in the next section, there is a gradual change from metallic behavior of the elements at the left end of a period to nonmetallic behavior of the elements at the right end. Since the change between metallic and nonmetallic behavior is not abrupt, those elements (the elements shaded grey in Fig. 8-3) lying near the dividing line between metals and nonmetals cannot neatly be classified either as metals or nonmetals. These elements, boron (B), silicon (Si), germanium

1 H																	2 He
3 Li	4 Be											5 B	6 C	7 N	8 O	9 F	10 Ne
11 Na	12 Mg											13 Al	14 Si	15 P	16 S	17 Cl	18 Ar
19 K	20 Ca	21 Sc	22 Ti	23 V	24 Cr	25 Mn	26 Fe	27 Co	28 Ni	29 Cu	30 Zn	31 Ga	32 Ge	33 As	34 Se	35 Br	36 Kr
37 Rb	38 Sr	39 Y	40 Zr	41 Nb	42 Mo	43 Tc	44 Ru	45 Rh	46 Pd	47 Ag	48 Cd	49 In	50 Sn	51 Sb	52 Te	53 I	54 Xe
55 Cs	56 Ba	[57-71] *	72 Hf	73 Ta	74 W	75 Re	76 Os	77 Ir	78 Pt	79 Au	80 Hg	81 Tl	82 Pb	83 Bi	84 Po	85 At	86 Rn
87 Fr	88 Ra	[89-103] †	104 Rf	105 Ha	106												

*LANTHANIDE SERIES	57 La	58 Ce	59 Pr	60 Nd	61 Pm	62 Sm	63 Eu	64 Gd	65 Tb	66 Dy	67 Ho	68 Er	69 Tm	70 Yb	71 Lu
†ACTINIDE SERIES	89 Ac	90 Th	91 Pa	92 U	93 Np	94 Pu	95 Am	96 Cm	97 Bk	98 Cf	99 Es	100 Fm	101 Md	102 No	103 Lr

Figure 8-3 Metals, metalloids, and nonmetals in the Periodic Table. The nonmetals are shaded in color, the metalloids in gray. The metals are not shaded.

(Ge), arsenic (As), antimony (Sb), polonium (Po), and astatine (At), are called semimetallic elements or **metalloids.** They exhibit **amphoteric** behavior; that is, they may behave either as metals or nonmetals depending on the conditions under which they react. Boron, for example, has three electrons in its outermost shell (a metallic characteristic); it reacts with fluorine giving the covalent molecule boron(III) fluoride, BF_3 (a nonmetallic property); its hydride reacts with water to give hydrogen gas (a metallic property); but the hydroxide that forms is boric acid, $B(OH)_3$, an acidic hydroxide (a nonmetallic property).

8.4   Variation in Metallic and Nonmetallic Behavior of the Representative Elements

Elements that lose electrons easily exhibit metallic behavior. As discussed in Section 4.19, the elements with only one or two electrons in the valence shell or the larger elements lying at the bottom of the groups of the Periodic Table generally lose electrons most readily. Atoms of nonmetals readily fill their valence shells either by sharing electrons with other nonmetals or by using electrons transferred from metal atoms. These are the atoms that lie in the upper right-hand portion of the table (Section 4.19). Sodium, the active metal at the left end of the third period, enters into chemical combination by the loss of its single valence electron. An atom of chlorine, a typical nonmetal at the right end of the third period, combines by adding one electron to an outer shell of seven electrons, either by gaining an electron from a metal or by sharing an electron with a nonmetal.

Proceeding across the third period from sodium to chlorine, we encounter five other elements. In general, across the period, valence electrons are lost with increasing difficulty. Thus the chemical behavior becomes decreasingly metallic and increasingly nonmetallic as we proceed from left to right. The changeover from metallic to nonmetallic behavior is gradual, but generally is considered to appear between aluminum and silicon in this period. Both aluminum and silicon, however, may exhibit some metallic and some nonmetallic properties under the appropriate conditions. As examples of gradual changeover from metallic to nonmetallic behavior upon proceeding from left to right across a period of the representative elements, let us examine the variation in some of the chemical properties of the elements of the third period. Sodium, magnesium, and aluminum are shiny metals that conduct heat and electricity well. Silicon is an element with a luster characteristic of a metal, but it is a semiconductor (a poor conductor of electricity). Phosphorus, sulfur, and chlorine are dull in appearance and are nonconducting elements. Sodium oxide, Na_2O, reacts with water, giving the strong base sodium hydroxide, NaOH. Magnesium oxide, MgO, reacts slowly with water, giving the less strongly basic magnesium hydroxide, $Mg(OH)_2$. Aluminum oxide, Al_2O_3, does not react with water, but the reaction of a solution of an aluminum salt such as aluminum nitrate, $Al(NO_3)_3$, with a solution of a base produces very weakly basic aluminum hydroxide, $Al(OH)_3$. Likewise, silicon dioxide, SiO_2, will not react with water, but covalent silicon compounds such as silicon tetrachloride, $SiCl_4$, react with water, giving gelatinous precipitates of silicic acid, $Si(OH)_4$, a very weak acid. The covalent oxides of phosphorus, sulfur, and chlorine (P_4O_{10}, SO_3, and Cl_2O_7) react with water, giving phosphoric acid (H_3PO_4), sulfuric acid (H_2SO_4), and perchloric acid ($HClO_4$), respectively. Of these acids, phosphoric acid is the weakest and perchloric acid, the strongest. *Increasing metallic character from right to left across a period of the representative elements and, conversely, increasing nonmetallic character from left to right across a period of the representative elements are observed in all periods.*

Metallic character also increases vertically within a group (family) of representative elements as the ionization energy of the elements decreases. Cesium, near the bottom of Group IA, is a more active metal than lithium, the element at the top of the group. Cesium is also a better reducing agent than lithium because it loses its outer shell electron more easily (outer shell farther from the nucleus). Nonmetallic character increases as we go up a group. For example, elemental fluorine, at the top of Group VIIA, is a stronger oxidizing agent than is elemental iodine, near the bottom of the group. It is interesting to note that iodine has sufficient metallic character to possess a metallic luster. Likewise, elements in Groups IIA, IIIA, IVA, VA, and VIA change from nonmetallic behavior at the top of the group to metallic behavior at the bottom.

The extent of metallic behavior or of nonmetallic behavior of an element varies with its oxidation number. The metallic behavior of an element decreases and its nonmetallic behavior increases as the oxidation number of the element in its compounds increases. For instance, in hypochlorous acid, HOCl, chlorine has an oxidation number of $+1$; in perchloric acid, $HClO_4$ ($HOClO_3$), the oxidation number of chlorine is $+7$. Hypochlorous acid is a weak acid (a characteristic of a compound of a poor nonmetal) whereas perchloric acid is a very strong acid (a characteristic of a compound of a good nonmetal). Thallium(I) chloride, TlCl, is an ionic compound (a characteristic of a chloride of a metal) whereas thallium(III) chloride, $TlCl_3$, is covalent (a characteristic of a chloride of a nonmetal). Likewise the chlorides of tin(IV) and lead(IV), $SnCl_4$ and $PbCl_4$, are covalent tetrahedral molecules whereas the chlorides of tin(II) and lead(II), $SnCl_2$ and $PbCl_2$, are solids with much more ionic character in their bonds. Thus, in addition to the position of an element in the Periodic Table, we must consider its oxidation number when predicting metallic or nonmetallic behavior of its compounds.

8.5   Periodic Variation of Oxidation Number

The oxidation numbers commonly observed for representative elements in their compounds are tabulated in Table 8-3. Although this array of numbers may appear formidable at first, there are a number of regularities related to the position of the elements in the Periodic Table that simplify recall of common oxidation numbers. You will find it helpful to refer to Table 8-3 as you study the regularities outlined below.

1. The maximum positive oxidation number found in any group of representative elements is equal to the group number. Thus the maximum possible positive oxidation number increases from $+1$ for the alkali metals (Group IA) to $+7$ for all of the halogens, except fluorine (Group VIIA), which is so electronegative that it in fact exhibits no oxidation number other than -1 in its compounds. With the exception of thallium at the bottom of Group IIIA and mercury at the bottom of Group IIB, the maximum positive oxidation number is the only common oxidation number displayed by the elements of Groups IA, IIA, IIB, and IIIA.

2. Metallic representative elements commonly exhibit only positive oxidation numbers.

3. The most negative oxidation number found in any group of representative elements is equal to the group number minus 8. Thus for the elements in Group VA the most negative possible oxidation number is $5 - 8 = -3$. This corresponds to the number of electrons that would have to be gained to provide an outer shell of eight electrons for the element.

4. Negative oxidation numbers are commonly limited to nonmetals and metalloids and are observed only when these elements are combined with a less electronegative element. Inasmuch as the nonmetal fluorine is the most electronegative element known, it never exhibits a positive oxidation number.

5. Elements commonly exhibit positive oxidation numbers only when combined with more electronegative elements. Oxygen can only exhibit a positive oxidation number in the few compounds it forms with the more electronegative element fluorine.

6. With the exception of carbon, nitrogen, oxygen, and mercury, each representative element that exhibits multiple oxidation numbers in its compounds commonly has oxidation numbers that are either all even or all odd. Carbon commonly exhibits oxidation numbers from $+4$ to -4; nitrogen, from $+5$ to

TABLE 8-3   Commonly Observed Oxidation Numbers of the Representative Elements in Their Compounds

IA	IIA		IIB	IIIA	IVA	VA	VIA	VIIA	0
H +1 −1								H +1 −1	He
Li +1	Be +2			B +3	C +4 to −4	N +5 to −3	O −1 −2	F −1	Ne
Na +1	Mg +2			Al +3	Si +4 −4	P +5 +3 −3	S +6 +4 −2	Cl +7 +5 +3 +1 −1	Ar
K +1	Ca +2		Zn +2	Ga +3	Ge +4 −4	As +5 +3 −3	Se +6 +4 −2	Br +7 +5 +3 +1 −1	Kr +4 +2
Rb +1	Sr +2		Cd +2	In +3	Sn +4 +2	Sb +5 +3 −3	Te +6 +4 −2	I +7 +5 +3 +1 −1	Xe +8 +6 +4 +2
Cs +1	Ba +2		Hg +2 +1	Tl +3 +1	Pb +4 +2	Bi +5 +3	Po +2	At −1	Rn
Fr +1	Ra +2								

−3; and oxygen, −2 or −1. Mercury forms a diatomic ion, Hg_2^{2+}, in which the oxidation number of the individual atoms is +1, as well as the monatomic mercury(II) ion, Hg^{2+}.

Before we proceed, a word of caution about the word "common" in "common oxidation states" may be helpful. The common oxidation states of an element are the oxidation states observed in a majority of its compounds. In some cases, as with the elements in Groups IA and IIA, this may be the vast majority or all of the known compounds of the element. Of all the thousands of known sodium compounds, for example, there is only one very reactive compound in which sodium has been shown to exhibit an oxidation state of −1. For other elements, the minority may be relatively large. Boron exhibits an oxidation state of $+3$ in the majority of its common compounds; however, there are a series of compounds in which it has an oxidation state of $+2$ and another series in which it exhibits nonintegral oxidation states. The large majority of compounds that you study in this text will involve elements in their common oxidation states. The ability to recall these common oxidation states will simplify your study considerably.

8.6   Prediction of Reaction Products

Many features determine whether or not a chemical reaction will occur and what the products will be. These features depend greatly upon the kinds of reactants and the conditions under which the reaction is run. Some chemists spend years developing a "feel" for the types of reactants and conditions that are likely to give a desired product. However, even they must resort to the ultimate test of their ideas; they try a proposed reaction in the laboratory to see if it works. The design and testing of reactions in order to prepare specific types of compounds or to improve the ease, yield, or simplicity with which known compounds are prepared constitutes the field known as **chemical synthesis.**

In theory it should be possible to predict accurately the products of any chemical reaction using ideas of chemical bonding, kinetics, thermodynamics, solution behavior, and electrochemistry. In practice, however, many chemical systems are simply too complicated to be treated precisely by these tools; nevertheless, as we shall see in subsequent chapters, the tools do provide valuable insights into the behavior of chemical systems.

In this section we shall examine some of the general guidelines that can be helpful in answering the question, What are the likely products, if any, which will result from the reaction of two or more substances? These guidelines are based upon the metallic or nonmetallic behavior of the representative elements involved and the common oxidation states that representative elements are likely to exhibit. We shall consider only those reactions between pure substances, in water solution, that occur at room temperature or when the pure substances are heated. Obviously, different conditions may result in different products, but these guidelines will serve as a foundation for our considerations of chemical reactions to be discussed in subsequent chapters. The following example will illustrate the approach:

EXAMPLE 1   **Predict the product of the reaction that occurs when elemental gallium (Ga) and sulfur (S_8) are warmed together.**

First consider what we know about gallium and sulfur from their respective positions in the Periodic Table. Gallium is located in Group IIIA and thus should exhibit metallic behavior. As a metal it would be expected to form compounds in which it exhibits a positive oxidation number. Since it is a member of Group IIIA, the expected oxidation number is $+3$. Sulfur is a member of Group VIA and thus should exhibit nonmetallic behavior. The common oxidation numbers of sulfur are $+6$ (its maximum positive oxidation number, equal to the group number), $+4$, and -2 (its most negative oxidation number, equal to the group number minus 8). However, since sulfur in this instance will be combined with the less electronegative gallium, it would be expected to exhibit the negative oxidation number of -2. As a metal, gallium can act as a reducing agent; as a nonmetal, sulfur can act as an oxidizing agent. Thus we can expect an oxidation-reduction reaction between gallium and sulfur, giving a product containing gallium with an oxidation number of $+3$ and sulfur with an oxidation number of -2. This leads us to formulate the product (Section 5.12) as Ga_2S_3.

The predicted chemical reaction therefore would be

$$16Ga + 3S_8 \xrightarrow{\triangle} 8Ga_2S_3$$

◆ ◆ ◆ ◆ ◆

As a general approach to predicting the products of a reaction, use the following steps:

1. Identify each element in the reactants as a metal or a nonmetal from its position in the Periodic Table. Determine the oxidation number of each element in the reactants (as described in Section 5.11) as well as the oxidation numbers each commonly exhibits in its compounds.

2. Consider the possibility of an oxidation-reduction reaction (Sec. 8.2, part 4). The following general considerations are helpful: (a) An elemental metal will reduce an elemental nonmetal as shown in the example above and in Equation (1) below. (b) A less electronegative metal (a more active metal) in its elemental state will reduce the ion of a more electronegative metal [Equation (2)]. (c) A more electronegative nonmetal in its elemental form will oxidize a less electronegative nonmetal [Equations (3) and (4)].

$$Mg + Cl_2 \longrightarrow MgCl_2 \tag{1}$$

$$Mg + SnCl_2 \longrightarrow MgCl_2 + Sn \tag{2}$$

$$P_4 + 3O_2 \longrightarrow P_4O_6 \tag{3}$$

$$P_4O_6 + 2O_2 \longrightarrow P_4O_{10} \tag{4}$$

The metals of Group IA and calcium, strontium, and barium of Group IIA are sufficiently active that they will reduce the hydrogen in water and acids to hydrogen gas.

$$Ba + 2H_2O \longrightarrow Ba(OH)_2 + H_2 \tag{5}$$

$$Ba + 2HCl \longrightarrow BaCl_2 + H_2 \tag{6}$$

The other representative metals, with the exception of bismuth and lead (the least metallic representative metals), will only reduce the hydrogen in acids.

$$4Al + 6H_2SO_4 \longrightarrow 2Al_2(SO_4)_3 + 6H_2 \tag{7}$$

3. Consider the possibility of an acid-base reaction (Section 8.2, part 5). Remember that the oxides and hydroxides of metals are generally basic and that the oxides and hydroxides of nonmetals are generally acidic. Reactions such as

(8)–(10) that involve either a protonic acid, a hydroxide base, or both, produce a salt and water.

$$2HI + Sr(OH)_2 \longrightarrow SrI_2 + 2H_2O \tag{8}$$

$$2HNO_3 + K_2O \longrightarrow 2KNO_3 + H_2O \tag{9}$$

$$SO_3 + Ca(OH)_2 \longrightarrow CaSO_4 + H_2O \tag{10}$$

Reactions between Lewis acids and Lewis bases that proceed without the transfer of a proton will only produce a salt.

$$K_2O + SO_3 \longrightarrow K_2SO_4 \tag{11}$$

$$LiF + BF_3 \longrightarrow LiBF_4 \tag{12}$$

4. Consider the possibility of a metathetical reaction (Section 8.2, part 3). Metathetical reactions between ionic compounds generally occur when one product is a solid, a gas, a weak electrolyte, or a nonelectrolyte. For examples, consider the following metathetical reactions:

$$Ba^{2+}(aq) + [2Cl^-(aq)] + [2K^+(aq)] + SO_4^{2-}(aq) \longrightarrow$$
$$BaSO_4(s) + [2K^+(aq)] + [2Cl^-(aq)] \tag{13}$$

$$KCl(s) + H_2SO_4(l) \longrightarrow KHSO_4(s) + HCl(g) \tag{14}$$

$$CH_3CO_2^-(aq) + [Na^+(aq)] + H^+(aq) + [Cl^-(aq)] \longrightarrow$$
$$CH_3CO_2H(aq) + [Na^+(aq)] + [Cl^-(aq)] \tag{15}$$

The ions enclosed in brackets are present, but are unchanged during the course of the reaction. Such ions, sometimes referred to as "spectator" ions, are often not included when writing the equation, but are included here to show the full composition of each compound participating in the reaction.

The acetic acid, CH_3CO_2H, produced in reaction (15) is a weak electrolyte. Most of the molecules of acetic acid in solution are not ionized.

5. Consider whether or not the products of the reaction may react with each other or with the solvent (if any).

Some specific examples of the application of these ideas are illustrated in the following examples:

EXAMPLE 2   **Write the balanced chemical equation for the reaction of elemental calcium with iodine.**

Calcium is a member of Group IIA and is thus an active metal. Iodine is a member of Group VIIA and is thus a nonmetal. Active metals react with nonmetals in an oxidation-reduction reaction. Iodine will be reduced to its only available negative oxidation number, -1, as a member of Group VIIA, while calcium will be oxidized to the only oxidation number that it exhibits in compounds, $+2$. With these oxidation numbers, the product must be CaI_2. The equation will be

$$Ca + I_2 \longrightarrow CaI_2$$

◆ ◆ ◆ ◆ ◆

EXAMPLE 3   Write a balanced chemical equation for the reaction which occurs when tin(II) oxide is added to a solution of perchloric acid, $HClO_4$.

Tin is a member of Group IVA. Inasmuch as it is located at the bottom of the group, it exhibits metallic properties. The oxides of metals are basic; hence we can expect an acid-base reaction between SnO and $HClO_4$ to produce a salt and water.

$$SnO + 2HClO_4 \longrightarrow Sn(ClO_4)_2 + H_2O$$

◆ ◆ ◆ ◆ ◆

EXAMPLE 4   What are the two possible products of the reaction of phosphorus, P_4, with chlorine, Cl_2?

Both phosphorus (Group VA) and chlorine (Group VIIA) are nonmetals. Phosphorus exhibits common oxidation numbers of +5, +3, and −3; chlorine, +7, +5, +3, +1, and −1. Chlorine lies to the right of phosphorus in the third period, so it is more electronegative than phosphorus; thus we recognize that it is able to oxidize phosphorus. Because phosphorus is oxidized, it will exhibit a positive oxidation number—probably +3 or +5 since these are the common positive oxidation numbers exhibited by a member of Group VA. Chlorine will exhibit a negative oxidation number because it is combined with a less electronegative element. The only negative oxidation number formed by chlorine is −1. Thus P_4 and Cl_2 can combine to give two compounds; in one, phosphorus has an oxidation number of +3 and chlorine has an oxidation number of −1. In the other, phosphorus has an oxidation number of +5 and chlorine has an oxidation number of −1. Thus the two compounds are PCl_3 and PCl_5, respectively.

Note that either phosphorus(III) chloride or phosphorus(V) chloride can be formed by the appropriate choice of reaction conditions. One mole of P_4 will react with 6 moles of Cl_2 to give PCl_3. If the amount of Cl_2 is increased to 10 moles, then PCl_5 is produced. Hence an excess of Cl_2 favors the formation of PCl_5; conversely, an excess of P_4 favors the formation of PCl_3.

$$P_4 + 6Cl_2 \longrightarrow 4PCl_3$$
$$P_4 + 10Cl_2 \longrightarrow 4PCl_5$$

◆ ◆ ◆ ◆ ◆

EXAMPLE 5   Predict the likely products for the reaction of sodium with sulfur trioxide.

Sodium (Group IA) is an active metal. Sulfur and oxygen are both nonmetals located in Group VIA. Elemental sodium has an oxidation number of zero. It can be oxidized only to an oxidation number of +1. Sulfur has an oxidation number of +6 in sulfur trioxide, SO_3; oxygen, an oxidation number of −2.

The only reaction possible with sodium in this system is for sodium to function as a reducing agent. The only element that can be reduced is sulfur. Oxygen already exists with its minimum (most negative) oxidation number of −2. Thus we can expect the reaction of Na with SO_3 to give products in which the sulfur is reduced to an oxidation number of $+4, 0$, or -2. With an oxidation number of $+4$, sulfur must still be combined with oxygen, the only element in the system that would exhibit a negative oxidation number. The reaction in this case would be

$$2Na + SO_3 \longrightarrow Na_2SO_3 \quad \text{(ratio of moles of Na to moles of } SO_3, \text{ 2 to 1)}$$

With additional sodium the reaction could give sulfur with an oxidation number of zero; that is, elemental sulfur, S_8.

$$48Na + 8SO_3 \longrightarrow 24Na_2O + S_8 \quad \text{(ratio of Na to } SO_3, \text{ 6 to 1)}$$

The sodium in this reaction must be combined with oxygen as sodium oxide since sodium is oxidized to an oxidation number of $+1$ in the reaction. Finally, even more sodium could reduce the sulfur in SO_3 to an oxidation number of -2.

$$8Na + SO_3 \longrightarrow Na_2S + 3Na_2O \quad \text{(ratio of Na to } SO_3, \text{ 8 to 1)}$$

The reaction that actually occurs will depend on the relative amounts of sodium and sulfur trioxide reacting, but in each case sodium functions as a reducing agent and is itself oxidized.

◆ ◆ ◆ ◆ ◆

Questions

1. In what ways does a salt differ from an acid? from a base? In what way are salts and hydroxide bases similar?
2. Identify the cations and anions in the following compounds: KI, MgF_2, K_2SO_4, NH_4Cl, LiOH, Al_2O_3, $Sr(HSO_4)_2$, $KICl_4$, $[N(CH_3)_4]I$.
3. When the following compounds dissolve in water, do they give solutions of acids, bases, or salts? H_2S, $In(NO_3)_3$, $Ba(OH)_2$, SO_3, Li_2O, NH_4NO_3, Cl_2O_7, I_2O_5, CaO, KIO_4, H_3PO_3, NH_3
4. Hydrogen chloride is a covalent molecule, yet it dissolves in water and behaves as an electrolyte. Explain.
5. How do addition reactions that are also oxidation-reduction reactions differ from addition reactions that are acid-base reactions? Write balanced chemical equations illustrating each type.
6. Classify the following reactions as oxidation-reduction or acid-base reactions:

(a) $Li_3N + 3HCl \longrightarrow 3LiCl + NH_3$
(b) $2Li + 2HCl \longrightarrow 2LiCl + H_2$
(c) $Mg + SO_3 \longrightarrow MgSO_3$
(d) $MgO + SO_2 \longrightarrow MgSO_3$
(e) $KI + 2Cl_2 \longrightarrow KICl_4$
(f) $KCl + ICl_3 \longrightarrow KICl_4$

7. Identify the atoms that are oxidized and reduced, the change in oxidation number for each, and the oxidizing and reducing agents in each of the following equations:

(a) $Mg + NiCl_2 \longrightarrow MgCl_2 + Ni$
(b) $PCl_3 + Cl_2 \longrightarrow PCl_5$
(c) $C_2H_4 + 3O_2 \longrightarrow 2CO_2 + 2H_2O$
(d) $Zn + H_2SO_4 \longrightarrow ZnSO_4 + H_2$
(e) $2K_2S_2O_3 + I_2 \longrightarrow K_2S_4O_6 + 2KI$
(f) $3Cu + 8HNO_3 \longrightarrow$
$$3Cu(NO_3)_2 + 2NO + 4H_2O$$

8. From the positions of their component elements in the Periodic Table, predict which of the following compounds will be (a) more basic, TlOH or BrOH; (b) more acidic, KOH or ClOH; (c) more acidic, CO_2 or SnO_2; (d) more basic, $Ba(OH)_2$ or $Be(OH)_2$; (e) an oxidizing agent, Mg or Cl_2; (f) a better reducing agent, Al or P_4.

9. Which of the following elements will exhibit a positive oxidation number when combined with phosphorus? Al, Br, Ca, Cl, F, O, Si

10. With which elements will sulfur form compounds in which sulfur has a positive oxidation number?

11. Write the formula of a binary compound (a compound containing only two elements) for each of the following elements so that the element has the indicated oxidation number: Cs, +1; Ga, +3; P, −3; As, +3; Se, −2; I, +1; Cl, +7; C, +2.

12. From their positions in the Periodic Table and without reference to Table 8-3, predict the common oxidation numbers of the following elements in their compounds: Al, Br, Ca, Cs, Sb, Se, Si, Sn.

13. Without reference to Table 8-3, identify those representative metals that may display two or more common oxidation states in their compounds.

14. Which of the following oxides or hydroxides are likely to show amphoteric behavior? $Al(OH)_3$, $B(OH)_3$, Cl_2O_7, RaO, Sb_2O_5, $Si(OH)_4$, $Sr(OH)_2$, XeO_3

15. Write balanced chemical equations for the preparation of each of the following compounds (a) by a metathetical reaction, (b) by an oxidation-reduction reaction, and (c) by an acid-base reaction: KCl, $SrSO_4$, Cs_2SiF_6, NaH_2PO_4 ($H_2PO_4^-$ is the anion resulting from removal of one proton from phosphoric acid, H_3PO_4).

16. Complete and balance the following oxidation-reduction equations. In some cases, there may be more than one correct answer depending upon the amounts of reactants used.
   (a) $K + S_8 \longrightarrow$ /8ᵢ)
   (b) $Li + Cl_2 \longrightarrow$
   (c) $Mg + N_2 \longrightarrow$
   (d) $Rb + P_4 \longrightarrow$
   (e) $Sn + F_2 \longrightarrow$
   (f) $S_8 + F_2 \longrightarrow$
   (g) $P_4 + O_2 \longrightarrow$
   (h) $SO_2 + O_2 \longrightarrow$

   (i) $Na(s) + HCl(g) \longrightarrow$
   (j) $Mg(s) + H_2SO_4(aq) \longrightarrow$
   (k) $Ga(s) + HClO_4(aq) \longrightarrow$
   (l) $Cs(s) + H_2O(l) \longrightarrow$

17. Complete and balance the following acid-base reactions. If the reactions are run in water as a solvent, write the reactants and products as solvated ions. In some cases, there may be more than one correct answer depending upon the amounts of reactants used.

   (a) $Ca(OH)_2(s) + HBr(g) \longrightarrow$
   (b) $CaO(s) + SO_2(g) \longrightarrow$
   (c) $Ba(OH)_2(aq) + HCl(aq) \longrightarrow$
   (d) $Al(OH)_3(s) + H_2SO_4(aq) \longrightarrow$
   (e) $Cs_2O(s) + H_2O(l) \longrightarrow$
   (f) $SrO(s) + HNO_3(aq) \longrightarrow$
   (g) $SO_3(g) + H_2O(l) \longrightarrow$
   (h) $NH_3(g) + H_2SO_4(aq) \longrightarrow$
   (i) $Li_2O(s) + N_2O_5(g) \longrightarrow$
   (j) $CaF_2(s) + BF_3(g) \longrightarrow$

18. Complete and balance the following equations. If the reactions are run in water as a solvent, write the reactants and products as solvated ions. In some cases, there may be more than one correct answer depending upon the amounts of reactants used.
   (a) $HCl(g) + In_2O_3(s) \longrightarrow$
   (b) $Mg(OH)_2 + HNO_3 \xrightarrow{H_2O}$
   (c) $Ca + P_4 \xrightarrow{\triangle}$
   (d) $Cs + S_8 \longrightarrow$
   (e) $CsOH + Al(OH)_3 \xrightarrow{H_2O}$
   (f) $Mg + SnSO_4 \xrightarrow{H_2O}$
   (g) $Si + F_2 \longrightarrow$
   (h) $Ba(OH)_2 + B(OH)_3 \xrightarrow{H_2O}$
   (i) $Ca(CH_3CO_2)_2 + H_2SO_4 \xrightarrow{H_2O} 2H4C_2O_2 + CaSO4$
   (j) $Li_2O + SO_3 \longrightarrow$
   (k) $H_2 + S_8 \longrightarrow$
   (l) $Sb + F_2 \longrightarrow$
   (m) $NH_3(g) + H_2S(g) \longrightarrow$
   (n) $KF + BF_3 \longrightarrow$
   (o) $As + S_8 \longrightarrow$
   (p) $Li + SeO_3 \longrightarrow$
   (q) $Cl_2 + I_2 \longrightarrow$
   (r) $NaF + HNO_3 \xrightarrow{H_2O}$

References

"Experiments in Alchemy: Part I, Ancient Arts," A. T. Schwartz and G. B. Kauffman, *J. Chem. Educ.,* **53,** 136 (1976).

"Experiments in Alchemy: Part II, Medieval Discoveries and Transmutations," A. T. Schwartz and G. B. Kauffman, *J. Chem. Educ.,* **53,** 235 (1976).

"The Chemical Equation: Part I, Simple Reactions," D. Kolb, *J. Chem. Educ.,* **55,** 184 (1978).

"Chemical Equations: Part II, Oxidation-Reduction Equations," D. Kolb, *J. Chem. Educ.,* **55,** 326 (1978).

"Oxidation of Organic Compounds," R. O. C. Norman, *Educ. in Chem.,* January 1979, p. 12.

"Reduction of Organic Compounds, R. O. C. Norman, *Educ. in Chem.,* September 1978, p. 154.

"A Pattern of Chemistry. Hundred Years of Periodic Table," F. Greenaway, *Chem. in Britain,* **5,** 97 (1969).

"Mendeleev's Other Prediction," H. Foldwhite, *J. Chem. Educ.,* **56,** 35 (1979).

"Redox Revisited," K. L. Lockwood, *J. Chem. Educ.,* **38,** 326 (1961).

"The Stoichiometry of an Oxidation-Reduction Reaction," W. C. Child, Jr. and R. W. Ramette, *J. Chem. Educ.,* **44,** 109 (1967).

"Anions of the Alkali Metals," J. L. Dye, *Sci. American,* July 1977, p. 92.

"Ailing Copper Industry Seeks New Technology" (Staff), *Chem. and Eng. News,* March 13, 1978, p. 30.

9

Oxygen (O₂), Ozone (O₃), and Hydrogen (H₂)

Oxygen (O$_2$),
Ozone (O$_3$), and
Hydrogen (H$_2$)

Oxygen

A study of the chemical elements appropriately begins with oxygen because it is the most abundant of all the elements in the earth's crust and atmosphere (see Table 1-1). It is essential to the processes of respiration in most plants and animals and the combustion of fuels and other substances. The discovery of oxygen marked the beginning of modern chemistry.

9.1 History and Occurrence

Credit for the discovery of oxygen is usually given to Joseph Priestley, an English clergyman and scientist, who prepared oxygen in 1774 by focusing the sun's rays on mercury(II) oxide by means of a lens. He tested the gaseous product with a burning candle and noted that the candle burned more brightly than in ordinary air. Shortly thereafter, Antoine-Laurent Lavoisier correctly interpreted the role played by oxygen in the process of combustion.

Oxygen is the most abundant and widely distributed of the terrestrial elements. It forms about 23% of the air as the free element, 89% of water in which it is combined with hydrogen, and 50% of the earth's crust by mass. About 90% of the volume of the earth's crust is occupied by oxygen, combined with other elements, principally silicon. In combination with carbon, hydrogen, and nitrogen, oxygen constitutes a large part of the weight of the bodies of plants and animals.

9.2   Preparation of Oxygen

Oxygen may be prepared from air or from certain oxygen-containing compounds. Because of the abundance and cheapness of air and water and the simplicity of the processes involving their use, approximately 97% of commercial oxygen is produced from air and 3% by the electrolysis of water. Total production of oxygen in 1978 amounted to 18 million tons, third in production for all chemicals.

▶ **1. BY FRACTIONAL EVAPORATION OF LIQUID AIR.** Commercial quantities of oxygen are obtained from air by cooling and compressing air until it liquefies and then evaporating off the lower-boiling nitrogen and some other elements that are a part of air (see Section 22.1). Oxygen is often stored and shipped in Dewar flasks (see Fig. 22-2) as a liquid (boiling point −183°). These flasks are constructed so as to be self-refrigerating by the evaporation of some of the oxygen. Much commercial oxygen, however, is stored and shipped as a compressed gas in steel cylinders. Liquid oxygen is one of the propellant components for today's rockets.

▶ **2. BY ELECTROLYSIS OF WATER.** Pure water is a very poor conductor of electricity; but when a small amount of a strong acid, a strong base, or a salt (electrolytes, see Section 8.1) is dissolved in water, the resulting solution readily conducts an electric current. When a direct current of electricity is passed through a solution of an electrolyte, the ions of the electrolyte are the agencies that carry the current; the ions migrate toward the two electrodes, the positive ions **(cations)** moving toward the negative electrode **(cathode)** and the negative ions **(anions)** moving toward the positive electrode **(anode).** The process is called **electrolysis** (see Chapter 20). When an electric current is passed through water containing a small amount of an electrolyte, such as H_2SO_4, $NaOH$, or Na_2SO_4, bubbles of hydrogen are formed at the cathode and oxygen is evolved at the anode (Fig. 9-1). The volume of hydrogen produced is twice that of the oxygen. The net reaction can be summarized by the equation

$$2H_2O(l) + \text{electrical energy} \longrightarrow 2H_2(g) + O_2(g)$$

Although the use of electrolysis for the industrial production of oxygen is limited by the high cost of electricity, electrolytic oxygen is very pure and is used where high purity is required.

▶ **3. BY DECOMPOSITION OF METAL OXIDES AND PEROXIDES.** A few metal oxides are not thermally stable and lose oxygen when heated. For example, when red mercury(II) oxide, HgO, is heated (Fig. 9-2), metallic mercury and oxygen are formed.

$$2HgO \xrightarrow{\triangle} 2Hg + O_2$$

This is the method Priestley and Lavoisier used to produce oxygen. It is important only for its historic interest because it is too expensive to be used commercially. Similar decomposition reactions that give the metallic element and oxygen occur when silver oxide (Ag₂O), gold(I) oxide (Au₂O), gold(III) oxide (Au₂O₃), and the platinum(II) and platinum(IV) oxides (PtO and PtO₂) are heated. As we shall see

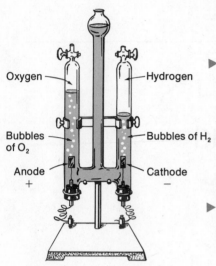

Oxygen ── ── Hydrogen

Bubbles ── ── Bubbles of H₂
of O₂

Anode ── ── Cathode
+   −

Figure 9-1 Laboratory apparatus for the electrolysis of water.

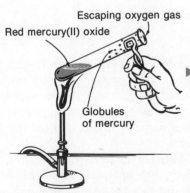

Escaping oxygen gas

Red mercury(II) oxide

Globules of mercury

Figure 9-2 When mercury(II) oxide is strongly heated it decomposes, yielding gaseous oxygen and liquid mercury.

in Section 9.21, these are oxides of metals located near the bottom of the electromotive series. Metal oxides that contain the metal in a relatively unstable high oxidation state (oxidation number) undergo thermal decomposition in which only part of the oxygen is liberated. Such reactions produce oxygen and an oxide of the metal in which the metal has a lower oxidation state than in the original oxide.

$$2PbO_2 \xrightarrow{\Delta} 2PbO + O_2$$

Lead(IV) oxide   Lead(II) oxide

$$4CrO_3 \xrightarrow{\Delta} 2Cr_2O_3 + 3O_2$$

Chromium(VI) oxide   Chromium(III) oxide

Ionic peroxides (compounds containing O_2^{2-} ions with an O—O single bond), such as barium peroxide (BaO_2) and cesium peroxide (Cs_2O_2), decompose to the corresponding oxides with evolution of oxygen.

$$2Ba^{2+}\left(:\ddot{O}:\ddot{O}:^{2-}\right) \xrightarrow{\Delta} 2Ba^{2+}\left(:\ddot{O}:^{2-}\right) + O_2$$

$$2(Cs^+)_2\left(:\ddot{O}:\ddot{O}:^{2-}\right) \xrightarrow{\Delta} 2(Cs^+)_2\left(:\ddot{O}:^{2-}\right) + O_2$$

▶ **4. BY HEATING CERTAIN SALTS THAT CONTAIN OXYGEN.** The nitrate salts of certain metals yield oxygen when heated.

$$2NaNO_3 \xrightarrow{\Delta} 2NaNO_2 + O_2$$

Sodium nitrate   Sodium nitrite

$$2Cu(NO_3)_2 \xrightarrow{\Delta} 2CuO + 4NO_2 + O_2$$

Copper(II) nitrate   Copper(II) oxide   Nitrogen dioxide

Oxygen is often prepared in the laboratory on a small scale (Fig. 9-3) by heating potassium chlorate to about 50° above its melting point of 368.4°. It should be

Figure 9-3 In the laboratory, oxygen is usually prepared as shown. The manganese(IV) oxide catalyst and potassium chloride remain after the potassium chlorate has decomposed.

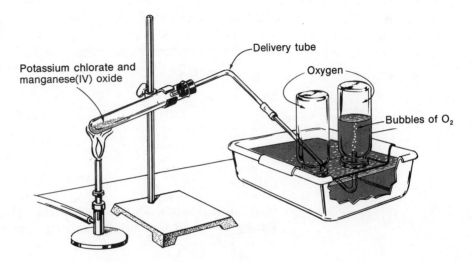

Potassium chlorate and manganese(IV) oxide

Delivery tube

Oxygen

Bubbles of O_2

noted that when sodium nitrate is heated only part of the oxygen is lost, but when potassium chlorate is heated all of the oxygen escapes. If manganese(IV) oxide is mixed with the chlorate, the latter decomposes quite rapidly at about 270°, nearly 100° below its melting point.

$$2KClO_3 \xrightarrow[MnO_2]{\Delta} 2KCl + 3O_2$$

Potassium   Potassium
chlorate   chloride

The manganese(IV) oxide may be reclaimed chemically unchanged after the reaction is completed; it has served to "catalyze" the reaction by causing it to take place more rapidly at a lower temperature. Many other metal oxides, such as Fe_2O_3 and Cr_2O_3, also serve as catalysts for the decomposition of potassium chlorate. **A catalyst is a substance that changes the rate of a reaction without undergoing a permanent chemical change itself.** Catalysts are usually specific in their action; a substance that will catalyze one reaction is often without effect upon another. Catalysis is discussed in more detail in Section 15.15.

The preparation of oxygen by heating potassium chlorate can be very dangerous when done carelessly. Explosions can occur when combustible materials such as carbon, sulfur, rubber, or a glowing wood splint come in contact with fused (melted) potassium chlorate. The danger involved in this experiment cannot be overemphasized. Proceed with caution when preparing oxygen by this method.

▶**5. BY THE ACTION OF WATER ON SODIUM PEROXIDE.** A convenient but expensive method for the preparation of oxygen involves the action of water upon sodium peroxide. Sodium peroxide is a white solid formed by burning sodium in a high pressure of oxygen.

$$2Na_2O_2(s) + 2H_2O(l) \longrightarrow 4Na^+(aq) + 4OH^-(aq) + O_2(g)$$

The fact that the sodium hydroxide is present in the water solution as ions rather than molecules is indicated by designating it as $Na^+(aq) + OH^-(aq)$. Evaporation of the solution to dryness gives the ionic compound sodium hydroxide (NaOH).

▶**6. BY PHOTOSYNTHESIS.** The oxygen in the atmosphere is continually replenished through the action of algae and higher plants by a process called **photosynthesis**. The products of photosynthesis may vary, but the process can be generalized as the conversion of carbon dioxide and water to glucose (a sugar) and oxygen by chlorophyll using the energy of light.

$$6CO_2 + 6H_2O \xrightarrow[\text{light}]{\text{Chlorophyll}} C_6H_{12}O_6 + 6O_2$$

Carbon   Water   Glucose   Oxygen
dioxide

9.3   Physical Properties

Oxygen is a colorless, odorless, and tasteless gas at ordinary temperatures. It is slightly more dense than air; one liter of oxygen measured at 0° and a pressure of

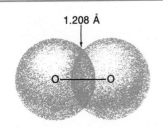

1.208 Å

Figure 9-4 A representation of the oxygen molecule, O_2. Each sphere represents an oxygen atom.

1 atmosphere weighs 1.429 g, whereas a liter of air under the same conditions weighs 1.292 g. Oxygen dissolves in water without reaction. Although it is only slightly soluble (49 ml of gas dissolve in one liter of water at 0°), its solubility is very important to marine life and in the decomposition of sewage recklessly dumped into streams and lakes. Oxygen is pale blue in the liquid state and boils at −183° at atmospheric pressure. Solid oxygen, also pale blue, melts at −218.4°. The oxygen molecule is diatomic (O_2) (Fig. 9-4) and it is **paramagnetic,** i.e., attracted by a magnetic field, particularly when in the solid or liquid state. The paramagnetism is an indication that the diatomic molecule contains one or more unpaired electrons (see Section 6.10).

Recalling that the oxygen atom possesses six valence electrons, we might expect the sharing of two pairs of electrons between the two atoms in the diatomic molecule with all electrons paired. The degree of paramagnetic character of the diatomic oxygen molecule, however, indicates *two unpaired electrons* in the structure, and, as we have seen in Section 6.10, the oxygen molecule does have two unpaired electrons, one each in the two π_p^* antibonding molecular orbitals.

9.4   Chemical Properties

Oxygen is a very electronegative nonmetal; only fluorine is more electronegative. It forms ionic compounds with metals and covalent compounds with other nonmetals. The electronic configuration of an oxygen atom is $1s^2 2s^2 2p^4$, with two of the p orbitals singly occupied with one electron each. Thus oxygen can complete the octet of electrons in its valence shell by picking up two electrons to form the oxide ion, O^{2-}, or by the formation of two single covalent bonds as in water, H_2O, or by the formation of a double bond as in formaldehyde, $H_2C{=}O$. Oxygen also forms a triple bond in carbon monoxide, $C{\equiv}O$.

Oxygen forms compounds with all of the elements except He, Ne, Ar, and Kr. In compounds, oxygen exhibits an oxidation state of −2 except in compounds with the most electronegative element, fluorine, and in compounds containing O—O bonds. The product formed when an element combines with oxygen is called an oxide if the oxygen exhibits an oxidation number of −2. It is called a peroxide if the oxygen has an oxidation state of −1 by virtue of being in a unit with an O—O single bond. Superoxide ions, O_2^-, contain oxygen in a formal oxidation state of $-\frac{1}{2}$.

▶ **1. ACTION WITH ELEMENTS.**   Oxygen reacts directly at room temperature or at elevated temperatures with all of the other elements except the noble gases, the halogens, and a few second and third row transition metals of low reactivity (those below copper in the electromotive series, Section 9.21) whose oxides decompose upon heating. The most active metals (Cs, Rb, K) react with oxygen giving superoxides (RbO_2, for example). Their neighbors in the Periodic Table (Na, Ca, Sr, Ba) give peroxides when they react with oxygen. Less active metals and the nonmetals give oxides.

$$Rb + O_2 \longrightarrow RbO_2$$

Rubidium
superoxide

$$Ba + O_2 \longrightarrow BaO_2$$

Barium
peroxide

$$4Li + O_2 \longrightarrow 2Li_2O$$

Lithium oxide

$$4Ga + 3O_2 \xrightarrow{\triangle} 2Ga_2O_3$$

Gallium oxide

$$C + O_2 \xrightarrow{\triangle} CO_2$$

Carbon dioxide

$$P_4 + 5O_2 \longrightarrow P_4O_{10}$$

Phosphorus(V)
oxide

Oxides of the halogens, of some of the noble gases, and of the metals at the bottom of the electromotive series can be prepared, but not by the direct action of the element with oxygen.

As we saw in Chapter 8, the formulas of the binary oxides, peroxides, and superoxides generally can be determined from the Periodic Table position of the element that combines with oxygen and the oxidation state of the oxygen. Thus the reaction of lithium, a Group IA element, with oxygen at ordinary pressures gives Li$_2$O (oxidation state of Li = +1, oxidation state of O = −2); gallium, a Group IIIA element gives Ga$_2$O$_3$ (oxidation state of Ga = +3, oxidation state of O = −2); and phosphorus, a Group VA element, gives P$_4$O$_6$ or P$_4$O$_{10}$ (oxidation state of P = +3 or +5, oxidation state of O = −2), depending on the amount of oxygen available. As we also saw in Chapter 8, the oxides of the metals are basic, whereas the oxides of the nonmetals are acidic.

▶ **2. ACTION WITH COMPOUNDS.** If a compound is composed of elements that when free will combine with oxygen, the compound may be expected to react with oxygen to form oxides of the constituent elements, *provided* that at least one of the atoms in the compound is not in its maximum oxidation state. For example, hydrogen sulfide, H$_2$S, contains sulfur in an oxidation state of −2. Since the sulfur is not in its maximum oxidation state and sulfur will react with oxygen in its free state, H$_2$S will react with oxygen, giving water and sulfur dioxide.

$$2H_2S + 3O_2 \longrightarrow 2H_2O + 2SO_2$$

Additional examples of compounds that react with oxygen and the equations for the reactions follow:

$$CS_2 + 3O_2 \longrightarrow CO_2 + 2SO_2$$

Carbon
disulfide

$$C_{12}H_{22}O_{11} + 12O_2 \longrightarrow 12CO_2 + 11H_2O$$
Sugar

$$CH_4 + 2O_2 \longrightarrow CO_2 + 2H_2O$$
Methane

$$2ZnS + 3O_2 \longrightarrow 2ZnO + 2SO_2$$
Zinc sulfide

Those oxides that contain an element in a lower oxidation state than is usually formed when the element combines with an excess of oxygen will also react with additional oxygen. Examples are

$$2CO + O_2 \longrightarrow 2CO_2$$
Carbon
monoxide

$$P_4O_6 + 2O_2 \longrightarrow P_4O_{10}$$
Phosphorus(III)
oxide

Other oxides, however, such as carbon dioxide (CO_2), silicon dioxide (SiO_2), sulfur trioxide (SO_3), and magnesium oxide (MgO), do not react with oxygen, because in these compounds the maximum oxidation state of the element combined with oxygen has been reached.

An important point that can be made now is that an equation usually represents only the *main* reaction that takes place under a given set of conditions. There may be by-products, but these are not included in writing the equation for the principal reaction. In the equation for burning sulfur in oxygen, for example, we write $S + O_2 \longrightarrow SO_2$; however, when the reaction takes place, small amounts of sulfur trioxide are produced. When hydrogen sulfide is burned, the chief products are water and sulfur dioxide, but we may also get small amounts of elemental sulfur or of sulfur trioxide. In the burning of carbon disulfide, possible products, in addition to carbon dioxide and sulfur dioxide, are elemental carbon, carbon monoxide, elemental sulfur, and sulfur trioxide.

9.5   Combustion

The term **combustion** is applied to chemical reactions that are accompanied by the evolution of both light and heat. Common examples of combustion are the burning of wood, coal, or magnesium in air. Combustion, however, is not restricted to reactions involving oxygen. For example, hydrogen will burn in an atmosphere of chlorine, hydrogen chloride being formed ($H_2 + Cl_2 \longrightarrow 2HCl$).

Before combustion can take place, the substances involved must be heated to the **kindling temperature,** i.e., the temperature at which the burning is sufficiently rapid to proceed without further addition of heat from the outside. The kindling temperature for many substances, especially solids, is not definite but depends upon the extent of subdivision and other factors. For example, iron in the form of fine wire will burn readily when heated in a flame and then placed in pure oxygen. However, a rod of iron a few millimeters in diameter will not ignite under the same conditions.

Spontaneous combustion may occur when the heat evolved during a reaction is not carried away from the system and thus accumulates to raise the temperature of the reacting substances to the kindling temperature. For example, linseed oil unites with oxygen of the air at ordinary temperatures **(slow oxidation)** in an exothermic reaction. Rags soaked with linseed oil and stored in locations where there is insufficient circulation of air to carry off the heat produced by slow oxidation may ignite spontaneously; i.e., the heat of reaction accumulates and raises the temperature of the system above the kindling point. Many costly and disastrous fires have been started by spontaneous combustion of such materials as coal in large piles, uncured hay stored in unventilated barns, and waste rags containing paints or drying oils.

Whenever combustion takes place extremely rapidly, the heat of reaction is liberated almost instantly and usually a large increase in gaseous volume results. An explosion results if the gases are confined so that they cannot escape readily. The mixture of gasoline and air in the cylinder of an automobile engine explodes when ignited by a spark. The rapid burning of gunpowder confined in a small space results in an explosion. Disastrous explosions sometimes occur in flour mills, grain elevators, and coal mines when dry finely divided particles are ignited. Any process, whether a chemical reaction or simply the overheating of a steam boiler, that leads to a sudden, large increase in gas pressure in a confined space can create an explosion.

9.6   Heat of Reaction

The combustion of magnesium, hydrogen, carbon, sulfur, methane, carbon disulfide, and sugar are reactions that are accompanied by the production of heat and are referred to as exothermic reactions (Section 2.8). Once started, exothermic reactions usually proceed without addition of energy from the outside. For example, carbon will continue to burn in oxygen with the evolution of heat until either the supply of carbon or oxygen is exhausted.

Reactions accompanied by the absorption of heat are called endothermic reactions. Reactions of this kind require a continuous supply of energy from the outside to keep them going. For example, the decomposition of mercury(II) oxide into mercury and oxygen will continue only as long as the compound is heated.

The quantity of energy liberated or absorbed (usually in the form of heat) during a chemical change is referred to as the **heat of reaction.** As noted in Section 2.8, the heat of reaction is indicated as a ΔH value following the equation for the reaction. A negative value of ΔH indicates an exothermic reaction; a positive value of ΔH, an endothermic reaction. The formation of *one mole* of water from hydrogen and oxygen is accompanied by the production of 285.8 kJ of heat (68.3 kcal). The equation is

$$H_2(g) + \tfrac{1}{2}O_2(g) \longrightarrow H_2O(l) \qquad \Delta H = -285.8 \text{ kJ/mol}$$

The negative sign of ΔH indicates that the reaction is exothermic. The energy involved in the formation of one mole of a compound from the constituent elements is known as the **heat of formation** of the compound. The heat of formation of water is -285.8 kJ/mol, that of CO_2 is -393 kJ/mol, and that of MgO is -601.83 kJ/mol. The same amount of energy is required to decompose a mole of a compound into its constituent elements as is involved in the formation of a mole of the compound. This follows from the Law of Conservation of Energy (Section 1.3), a very important concept in chemistry. Thus a quantity of electrical energy equivalent to 285.8 kJ is needed to decompose one mole of water in an endothermic process (positive ΔH).

$$H_2O(l) \longrightarrow H_2(g) + \tfrac{1}{2}O_2(g) \qquad \Delta H = +285.8 \text{ kJ/mol}$$

Compounds formed from their elements by highly exothermic reactions, like H_2O and CO_2, are stable toward heat; they are said to be **thermally stable.** Usually a very high temperature is required to decompose them into their constituent elements. On the other hand, compounds resulting from endothermic reactions, such as hydrogen peroxide (H_2O_2), are thermally unstable; i.e., only moderate heating is necessary to break the bonds holding the atoms together, and the molecule may decompose even at room temperature.

Additional points concerning the energy relationships in chemical reactions will be brought out in the discussion of enthalpy, free energy, and entropy in Chapter 18 (Chemical Thermodynamics).

9.7   Rate of Reaction

Oxygen combines slowly with finely divided bituminous, or soft, coal at ordinary temperatures. Such a reaction is called **slow oxidation.** When ignited, a piece of coal may burn quietly with a flame. Finely powdered coal and oxygen unite explosively when ignited by a spark. It is evident, then, that the speed of a given reaction may vary greatly. The rate at which substances are used up or are formed during a chemical change is known as the **rate of reaction.** Among the factors that influence the rate of a reaction are the temperature, the concentration of the reactants, the presence of a catalyst (see Chapter 15), and the area of contact between the reactants.

▶ **1. TEMPERATURE.**   Heat affects the rate of chemical reactions. Almost all reactions are speeded up when the temperature is increased, although for very fast reactions, such as $Ag^+(aq) + Cl^-(aq) \longrightarrow AgCl(s)$, the effect is negligible. There is a rough rule that for reactions that do not appear to be instantaneous, a 10 degree rise in temperature will double the rate of reaction.

▶ **2. CONCENTRATION.**   The fact that a heated iron wire will burn in pure oxygen but not in air, which is only 21% oxygen by volume, shows the effect of concentration on the rate of oxidation of iron. The rate of a reaction increases as the concentration of the reactants is increased. The reaction of zinc with hydrochloric acid to produce zinc chloride and hydrogen proceeds more rapidly if the concentration of hydrochloric acid is increased.

▶**3. CATALYSTS.** The effect of manganese(IV) oxide as a catalyst upon the thermal decomposition of potassium chlorate was mentioned in Section 9.2. The success of many industrial processes is due to the use of catalysts for increasing the rate of reactions that are normally too slow to be commercially feasible.

▶**4. STATE OF SUBDIVISION.** Reactions between substances take place only when they are in contact, and the more extensive the contact the more rapid the reaction may become. As the state of subdivision of a solid substance increases, its surface area increases so that more of the substance is in contact with a reactant. Wood shavings burn much more rapidly than a massive wood chunk. Finely divided zinc will react more rapidly with hydrochloric acid than large pieces of zinc.

9.8   The Importance of Oxygen to Human Life

In human beings and other animals, the energy required for maintenance of normal body temperature is derived from the slow oxidation of materials in the body. Oxygen passes from the lungs into the blood, where it combines with hemoglobin, producing oxyhemoglobin. In this form, oxygen is carried by the blood to the various tissues of the body, where it is released and consumed in reactions with oxidizable materials. Products are mainly carbon dioxide and water. The blood, which at that stage contains hemoglobin, carbon dioxide, and some oxyhemoglobin, returns through the veins to the lungs. There it gives up carbon dioxide and collects another supply of oxygen. Slow oxidation in the cells of the body involves a series of complicated reactions catalyzed by a group of substances called **enzymes** (Chapter 26). The digestion and assimilation of food regenerates the materials consumed by oxidation in the body, with the same amount of energy being liberated as if the food had been burned by rapid oxidation outside the body.

9.9   Uses of Oxygen

The many applications of oxygen make use of the oxygen in the air, oxygen-enriched air, or pure oxygen. Animals use the oxygen in the air in the metabolic process known as respiration. Oxygen in the air is essential in combustion processes such as the burning of fuels for the production of heat; it is also essential in the decay of organic matter. Oxygen-enriched air is used in medical practice when the blood receives an inadequate supply of oxygen because of such things as shock, pneumonia, tuberculosis, and heart ailments.

Approximately 70% of all oxygen produced commercially is used in the removal of carbon from iron in the production of steel. Large quantities of pure oxygen are also consumed in the cutting and welding of metals with oxyhydrogen and oxyacetylene torches. Worldwide, more than three-fourths of the pure oxygen produced is used in making other substances.

Liquid oxygen is commonly used in rocket engines of spacecraft as an oxidizing agent. Additional quantities are carried along to provide oxygen for life support in space.

Ozone

9.10   Allotropy

When dry oxygen is passed between the two electrically charged plates of an apparatus called an **ozonizer** (Fig. 9-5), a decrease in volume of the gas occurs and a substance possessing a distinctive odor is formed. This substance is known as **ozone;** its molecules are each composed of three oxygen atoms; i.e., it is triatomic, O_3.

Ozone and oxygen are referred to as **allotropes.** The term **allotropy** is used to designate the existence of an element in two or more forms in the same physical state. Ordinary oxygen, O_2, and ozone, O_3, are different gaseous forms of the same element.

Figure 9-5 Laboratory apparatus for preparing ozone.

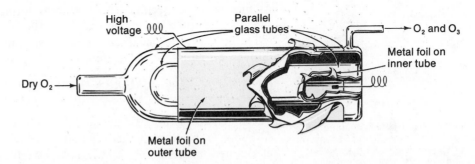

9.11   Preparation of Ozone

The formation of ozone from oxygen is an endothermic reaction in which the energy is furnished in the form of an electrical discharge, heat, or ultraviolet light.

$$3O_2 \xrightarrow[\text{discharge}]{\text{Electric}} 2O_3 \qquad \Delta H = 285 \text{ kJ}$$

Ozone is prepared commercially by an electrical ozonizer that is more complicated in design and more efficient than the laboratory apparatus. An alternative method of commercial production of ozone is by electrolysis of very cold, concentrated sulfuric acid. Ozone is also formed in small quantities by several methods not suitable for commercial production; for example, it is produced by the slow oxidation of phosphorus, by a jet of burning hydrogen, by lightning, and by most electrical discharges in air. Significant amounts of ozone are formed in the upper atmosphere by the action of ultraviolet light from the sun upon the oxygen there. Ozone introduced in the lower atmosphere has been found to react with engine exhaust products to form substances that are important factors contributing to smog (see Section 22.4). A great deal of the deterioration of automobile tires is due to contact with ozone in the air. Considerable discussion has arisen recently in connection with the fear that exhaust gases from supersonic airliners or the residual chlorofluorocarbon propellants from aerosol cans may react and catalyze the decomposition of ozone in the upper atmosphere, causing a decrease in the concentration to such an extent that the ozone layer would no longer constitute a protective layer against the harmful rays of the sun. It is not yet known whether this would happen, but if it does it could have a drastic effect on the processes involved in life on the earth.

9.12 Properties of Ozone

Ozone is a pale blue gas with a pungent (sharp, irritating) odor. The liquid is deep blue. Ozone produces headache and is poisonous. Ozone is 1.5 times as dense as molecular oxygen, O_2. Because energy is absorbed when ozone is formed from oxygen, it it not surprising that ozone is more active chemically than oxygen and that it decomposes readily into oxygen by the exothermic reaction

$$2O_3 \longrightarrow 3O_2 \qquad \Delta H = -285 \text{ kJ}$$

As mentioned in Section 9.6, substances formed by endothermic reactions in general tend to be less stable than those formed by exothermic reactions.

The presence of ozone in gas mixtures can readily be detected by passing the gas through a solution of potassium iodide (KI) containing some starch.

$$[2K^+(aq)] + 2I^-(aq) + O_3(g) + H_2O(l) \longrightarrow$$
$$[2K^+(aq)] + 2OH^-(aq) + I_2 + O_2(g)$$

The O_3 oxidizes I^- to I_2, elemental iodine, which imparts a blue color to the starch. Although O_2 reacts with almost all other elements, in most cases it will do so only at elevated temperatures. Ozone is a powerful oxidizer and forms oxides with many elements under conditions where O_2 will not react.

The molecular structure of ozone is given in Fig. 9-6. The ozone molecule is angular as shown, and the two bonds are indistinguishable. The electronic structure of ozone can be described by the following resonance forms (Section 5.8):

$$\ddot{O}\!\!:\!\!\ddot{O}\!\!:\!\!\ddot{O}\!\!: \longleftrightarrow :\!\!\ddot{O}\!\!:\!\!\ddot{O}\!\!:\!\!\ddot{O}\!\!:$$

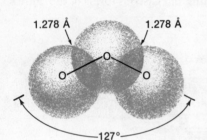

1.278 Å   1.278 Å

127°

Figure 9-6 Representation of an ozone molecule. The arrangement of spheres indicates the angular nature (127° angle) of the molecule.

9.13 Uses of Ozone

The uses of ozone depend on its readiness to react with other substances. Its use as a bleaching agent for oils, waxes, fabrics, and starch involves reactions that oxidize the colored compounds in these substances to colorless compounds. It is sometimes used instead of chlorine in the purification of water.

Hydrogen

Early in the sixteenth century the Swiss-German physician Paracelsus noted that a flammable gas was formed by the reaction of sulfuric acid with iron. However, it was not until 1766 that Cavendish, an Englishman, recognized this gas as a distinct substance and prepared it by the rather novel method of passing steam through a red-hot gun barrel and also by the action of various acids on certain metals.

Lavoisier, a French chemist, named the gas **hydrogen,** meaning "water producer," because water was formed when the gas burned in air.

The hydrogen atom, which consists of one proton and one electron, has the simplest structure of any of the atoms. Because hydrogen has one valence electron in an *s* orbital it is usually placed in Group IA of the Periodic Table with the alkali metals. However, a hydrogen atom can also pick up an electron to give H^-, so sometimes it is also placed with the halogens in Group VIIA.

9.14   Occurrence

Because it is a chemically active element, hydrogen is found in the free state on the earth in only negligible quantities. It occurs free in very small amounts in the atmosphere, in the gases of active volcanoes, in natural gas, in the gases present in some coal mines, and trapped in meteorites. The sun and other stars appear to be composed largely of hydrogen, as does the gas found in interstellar space. It is estimated that 90% of the atoms in the universe are hydrogen atoms.

More compounds containing hydrogen are known than those of any other element. In the combined form hydrogen comprises nearly 11% of the weight of water, its most abundant compound. Hydrogen is found combined with carbon and oxygen in the tissues of all plants and animals. It is an important part of petroleum, many minerals, cellulose and starch, sugar, fats, oils, alcohols, acids and bases, and thousands of other substances.

9.15   Preparation of Hydrogen

Hydrogen is more expensive than oxygen since useful amounts of free hydrogen are not found in nature. Hydrogen must be obtained from compounds by breaking chemical bonds, and this requires much more energy than simply condensing an element found free in the air. There are many ways of liberating hydrogen from its compounds, particularly from hydrocarbons, acids, bases, and water. Where small quantities of hydrogen are required, as in the laboratory, it is usually generated from acids. When commercial quantities of hydrogen are needed, water or petroleum is generally used as the raw material.

▶**1. FROM NATURALLY OCCURRING HYDROCARBONS.** Hydrogen is produced commercially in large quantities from the hydrocarbons in oil and natural gas. **Hydrocarbons** are compounds such as methane, CH_4, and octane, C_8H_{18}, which contain only carbon and hydrogen. When a mixture of methane (the principal component of natural gas) and steam is heated to a high temperature in the presence of catalysts, a gaseous mixture of carbon monoxide, carbon dioxide, and hydrogen is produced.

$$CH_4 + H_2O \xrightarrow[\text{Catalyst}]{\Delta} CO + 3H_2$$

$$CO + H_2O \xrightarrow[\text{Catalyst}]{\Delta} CO_2 + H_2$$

These are typical reactions and other hydrocarbons may be substituted for methane.

Hydrogen may also be obtained from catalyzed thermal decomposition of hydrocarbons. Such reactions are called **cracking reactions.** Methane is decomposed into carbon and hydrogen when heated in the presence of suitable catalysts.

$$CH_4(g) \xrightarrow[\text{Catalyst}]{\Delta} C + 2H_2(g)$$

Other cracking reactions, like those used in the refining of oil, may produce hydrogen as a by-product. For example,

$$2C_4H_{10}(g) \xrightarrow[\text{Catalyst}]{\Delta} C_8H_{18} + H_2(g)$$

▶ **2. BY ELECTROLYSIS.** The decomposition of water into its constituent elements by means of a direct current of electricity was described in Section 9.2. Hydrogen is liberated at the cathode when water containing a small amount of an electrolyte is electrolyzed.

$$2H_2O(l) + \text{electrical energy} \longrightarrow 2H_2(g) + O_2(g) \tag{1}$$

If water is replaced by a concentrated solution of sodium chloride (NaCl) in the electrolysis, hydrogen is formed but, instead of oxygen, sodium hydroxide and chlorine are formed.

$$[2Na^+(aq)] + 2Cl^-(aq) + 2H_2O + \text{electrical energy} \longrightarrow$$
$$[2Na^+(aq)] + 2OH^-(aq) + Cl_2(g) + H_2(g) \tag{2}$$

This is a commercial method for producing NaOH (caustic soda) and chlorine; hydrogen is a by-product (Section 20.4). If the salt solution used is not fairly concentrated, oxygen is liberated at the anode by the reaction shown in Equation (1). This fact points up an important aspect of the process of electrolysis: The concentration of the electrolyte is an important determining factor in product formation.

▶ **3. BY THE ACTION OF CERTAIN METALS ON ACIDS.** Hydrogen is conveniently produced in the laboratory by the reaction of an active metal with an acid. The metal reduces the hydrogen in the acid to H_2. Although any metal above hydrogen in the electromotive series (Section 9.21) could be used, zinc and iron are the metals frequently used with dilute solutions of either hydrochloric acid or sulfuric acid.

$$Zn(s) + 2H^+(aq) + [2Cl^-(aq)] \longrightarrow Zn^{2+}(aq) + [2Cl^-(aq)] + H_2(g) \tag{3}$$
$$\text{Zinc chloride}$$

$$Zn(s) + 2H^+(aq) + [SO_4^{2-}(aq)] \longrightarrow Zn^{2+}(aq) + [SO_4^{2-}(aq)] + H_2(g) \tag{4}$$
$$\text{Zinc sulfate}$$

$$Fe(s) + 2H^+(aq) + [2Cl^-(aq)] \longrightarrow Fe^{2+}(aq) + [2Cl^-(aq)] + H_2(g) \tag{5}$$
$$\text{Iron(II) chloride}$$

$$2Al(s) + 6H^+(aq) + [3SO_4^{2-}(aq)] \longrightarrow 2Al^{3+}(aq) + [3SO_4^{2-}(aq)] + 3H_2(g) \tag{6}$$
$$\text{Aluminum sulfate}$$

$H^+(aq)$, $Cl^-(aq)$, and $SO_4^{2-}(aq)$ are used in these equations to indicate that HCl

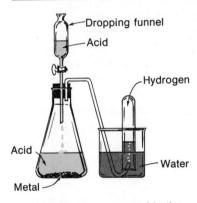

Figure 9-7 Apparatus used in the laboratory preparation of hydrogen.

and H_2SO_4 undergo ionization in water giving $H^+(aq)$ and the corresponding anions. In a **net ionic equation,** an equation that shows only those species that change during the reaction, the ions enclosed in brackets would not be shown. For example,

$$2Al(s) + 6H^+(aq) \longrightarrow 2Al^{3+}(aq) + 3H_2(g)$$

In the laboratory preparation of hydrogen, an apparatus like that shown in Fig. 9-7 is used. The acid solution is added to the metal, and the hydrogen is collected by the downward displacement of water. It is the hydrogen ion of the dilute aqueous acids that reacts with the active metals. The negative ion does not enter the reaction. However, on evaporation of the solution to dryness after the reaction is complete, a crystalline ionic salt composed of the metal cation and the negative ion of the acid is obtained. In the reactions described by Equations (3)–(6), the salts are zinc chloride, zinc sulfate, iron(II) chloride, and aluminum sulfate.

▶ **4. BY THE ACTION OF CERTAIN METALS AND NONMETALS ON WATER.** The very active metals of Groups IA and IIA, such as sodium, potassium, and calcium, rapidly displace hydrogen from water at room temperature, producing a solution of the corresponding base.

$$2Na + 2H_2O \longrightarrow 2Na^+(aq) + 2OH^-(aq) + H_2(g)$$
$$2K + 2H_2O \longrightarrow 2K^+(aq) + 2OH^-(aq) + H_2(g)$$
$$Ca + 2H_2O \longrightarrow Ca^{2+}(aq) + 2OH^-(aq) + H_2(g)$$

The reaction of a small piece of sodium or potassium with water produces sufficient heat to ignite the metal as well as the hydrogen as it is produced. It is dangerous to bring large pieces of either metal into contact with water because explosions may result. An alloy of lead and sodium, or one of mercury and sodium, either of which reacts with water less vigorously than does pure sodium, is sometimes used for the preparation of hydrogen in small quantities.

Other less active metals will displace hydrogen from water at higher temperatures. Magnesium reacts only slowly with boiling water, but when steam is passed over magnesium or red-hot iron, hydrogen is liberated rapidly (Fig. 9-8).

Figure 9-8 Some metals, such as magnesium and iron, reduce the hydrogen in steam, $H_2O(g)$, forming H_2 and metallic oxides. The apparatus for carrying out the reaction of iron with steam to produce hydrogen and iron(II,III) oxide, Fe_3O_4, is illustrated.

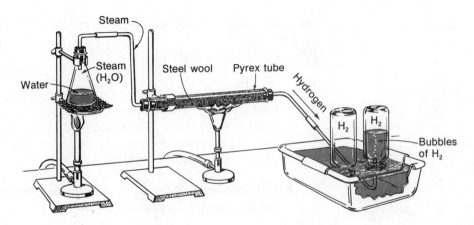

$$Mg + H_2O \xrightarrow{\Delta} H_2 + MgO$$

$$3Fe + 4H_2O \xrightarrow{\Delta} 4H_2 + Fe_3O_4$$

Note that all of the hydrogen of the water molecule is replaced by the metals in the reactions described by the above equations, whereas the active metals sodium and potassium replace only half of it.

Some of the hydrogen that is used in industry is produced by the reaction of iron with steam. After a mass of iron has been largely converted to iron(II,III) oxide (Fe$_3$O$_4$) by this reaction, the iron is regenerated by passing carbon monoxide, CO, over the heated oxide.

$$Fe_3O_4 + 4CO \xrightarrow{\Delta} 3Fe + 4CO_2$$

By the alternate use of steam and carbon monoxide the iron can be used over and over again.

Carbon, a nonmetal, when white hot (1500–1600°C) will react with steam in an endothermic reaction, producing a mixture of carbon monoxide and hydrogen. The mixture is commonly known as **water gas.**

$$C(s) + H_2O(g) \xrightarrow{\Delta} CO(g) + H_2(g)$$

Because both carbon monoxide and hydrogen will burn in oxygen or in air and produce much heat, water gas is a valuable industrial fuel. When water gas is mixed with steam and passed over a catalyst such as iron oxide or thorium oxide at a fairly high temperature (500°C), carbon dioxide and additional hydrogen are produced.

$$CO + H_2O + [H_2] \xrightarrow{\text{Catalyst}} CO_2 + H_2 + [H_2]$$

▶ **5. BY THE ACTION OF CERTAIN ELEMENTS ON HYDROXIDE BASES.** Certain elements, such as aluminum, zinc, and silicon, react with sodium hydroxide, or other strong bases, in concentrated aqueous solution with the liberation of hydrogen.

$$2Al(s) + [2Na^+(aq)] + 2OH^-(aq) + 6H_2O(l) \longrightarrow [2Na^+(aq)] + 2Al(OH)_4^-(aq) + 3H_2(g)$$
$$\text{Sodium aluminate}$$

$$Zn(s) + [2Na^+(aq)] + 2OH^-(aq) + 2H_2O(l) \longrightarrow [2Na^+(aq)] + Zn(OH)_4^{2-}(aq) + H_2(g)$$
$$\text{Sodium zincate}$$

$$Si(s) + 2Na^+(aq) + 2OH^-(aq) + H_2O(l) \longrightarrow Na_2SiO_3(s) + 2H_2(g)$$
$$\text{Sodium silicate}$$

Aluminum and zinc alone do not displace hydrogen from water, because the oxide that forms produces a film on the metal surface preventing further contact of the metal with the water. Acids and strong bases react with aluminum and zinc to produce hydrogen, because they dissolve the oxide film and allow the metal to come in direct contact with the acid or base.

▶ **6. BY THE ACTION OF ACTIVE METAL HYDRIDES ON WATER.** In Chapter 8 we saw that the hydrides (which contain H$^-$ anions) of the active metals react with water liberating hydrogen and forming basic solutions because H^- is a very strong base.

$$NaH(s) + H_2O(l) \longrightarrow Na^+(aq) + OH^-(aq) + H_2(g)$$

$$CaH_2(s) + 2H_2O(l) \longrightarrow Ca^{2+}(aq) + 2OH^-(aq) + 2H_2(g)$$

Metal hydrides are expensive but convenient sources of very pure hydrogen, especially where space and weight are important factors. Examples are for the inflation of life jackets, life rafts, military balloons, and weather balloons. Calcium hydride is so much used for these and other purposes that it is sold in commercial quantities.

9.16   Physical Properties of Hydrogen

At ordinary temperatures, hydrogen is a colorless, odorless, and tasteless gas consisting of diatomic molecules, H_2. (See Fig. 5-2.) The bond between the two hydrogen atoms involves a pair of electrons in the bonding sigma molecular orbital (σ_s) as described in Section 6.3.

With a density of 0.08987 g/liter at 0° and 760 torr, hydrogen is the lightest known substance. Because of its low density it can be collected by the downward displacement of air. The fact that hydrogen is so much lighter than air (which has a density of 1.293 g/liter) makes it useful as the lifting agent in balloons.

If sufficiently cooled and compressed, hydrogen changes to a liquid that boils at atmospheric pressure at $-252.7°$. The low temperature of liquid hydrogen makes it useful in cooling other materials to low temperatures. It may be used to transform all other gases into solids with the single exception of helium. Hydrogen freezes to a transparent solid at $-259.14°$.

Hydrogen diffuses (Section 10.14) faster than any other gas because its molecules are lighter and move faster than those of any other gas.

9.17   Chemical Properties of Hydrogen

At ordinary temperatures hydrogen is relatively inactive chemically, but when heated it enters into many chemical reactions.

▶**1. REACTION WITH OXYGEN.** When a mixture of hydrogen and oxygen is ignited a vigorous reaction occurs, and water is formed in an exothermic reaction (negative ΔH).

$$2H_2 + O_2 \longrightarrow 2H_2O \qquad \Delta H = -571.5 \text{ kJ}$$
$$= -136.6 \text{ kcal}$$

Because of the violence of the reaction, often resulting in explosion, great caution should be used in handling hydrogen (or any other combustible gas). Hydrogen will burn without explosion under some conditions. The fact that water is formed when hydrogen is burned in either air or oxygen can be demonstrated by allowing a jet of hydrogen to burn inside an inverted, cold, dry beaker (Fig. 9-9).

The very high heat of combustion of hydrogen in pure oxygen makes it possible

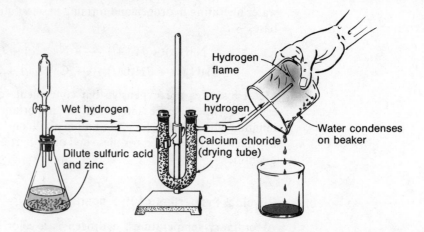

Figure 9-9 When hydrogen burns in air, water vapor is formed. The large beaker here serves as a condenser.

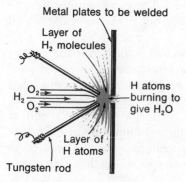

Figure 9-10 Atomic hydrogen torch. An electric arc between the tungsten electrodes converts H_2 molecules to H atoms.

to achieve temperatures up to 2800° with the oxyhydrogen blowtorch. The hot flame of this torch can be used in "cutting" thick sheets of many metals.

When hydrogen gas, consisting of diatomic molecules, is passed through an electric arc, energy is absorbed by the hydrogen, causing the diatomic molecules to be broken up into hydrogen atoms in an endothermic reaction.

$$H_2 \xrightarrow[\text{arc}]{\text{Electric}} 2H \qquad \Delta H = 435 \text{ kJ}$$
$$= 104 \text{ kcal}$$

The single atoms of hydrogen readily recombine upon collision with one another and release the heat that was absorbed when the molecule was broken up in the electric arc.

$$2H \longrightarrow H_2 \qquad \Delta H = -435 \text{ kJ}$$
$$= -104 \text{ kcal}$$

The energy released in this reaction, added to that from the burning of the hydrogen in pure oxygen, is utilized in the atomic hydrogen torch (Fig. 9-10). This torch is useful in cutting and welding metals which melt at temperatures up to 5000°.

▶ **2. REACTION WITH NONMETALS OTHER THAN OXYGEN.** In addition to reacting with oxygen, hydrogen reacts directly with the halogens, sulfur, and nitrogen. The reactions, except for that with fluorine, are very slow at room temperature, but the rates increase upon heating. Under some conditions hydrogen reacts explosively with fluorine and chlorine.

$$H_2 + F_2 \longrightarrow 2HF \qquad \text{(hydrogen fluoride)}$$
$$H_2 + Cl_2 \longrightarrow 2HCl \qquad \text{(hydrogen chloride)}$$
$$H_2 + Br_2 \longrightarrow 2HBr \qquad \text{(hydrogen bromide)}$$
$$H_2 + I_2 \longrightarrow 2HI \qquad \text{(hydrogen iodide)}$$
$$8H_2 + S_8 \longrightarrow 8H_2S \qquad \text{(hydrogen sulfide)}$$
$$3H_2 + N_2 \longrightarrow 2NH_3 \qquad \text{(ammonia)}$$

As described in Section 8.3, the resulting hydrides are acidic. Ammonia, however, is a very weak acid, and it can also function as a base. These hydrides (as well as hydrides of elements that do not react *directly* with hydrogen) also can be formed by acid-base reactions.

$$KF + H_2SO_4 \longrightarrow KHSO_4 + HF \qquad \text{(hydrogen fluoride)}$$
$$Li_3N + 3H_2O \longrightarrow 3LiOH + NH_3 \qquad \text{(ammonia)}$$
$$Na_3P + 3H_2O \longrightarrow 3NaOH + PH_3 \qquad \text{(phosphine)}$$

With the exception of the halide ions, all monatomic negative ions such as the nitride ion, N^{3-}, and phosphide ion, P^{3-}, are very strong bases and can pick up hydrogen ions from very weak acids such as water.

▶ **3. REACTION WITH ACTIVE METALS.** When heated, hydrogen reacts with the elements of Group IA and the more active elements in Group IIA, Ca, Sr, and Ba. The compounds formed are crystalline ionic hydrides that contain hydrogen as the anion H^-.

$$2Li + H_2 \longrightarrow 2LiH \qquad \text{(lithium hydride)}$$
$$Ca + H_2 \longrightarrow CaH_2 \qquad \text{(calcium hydride)}$$

The reactions are exothermic and the resulting hydrides are relatively stable toward heat but react vigorously with water (see Section 9.15, part 6).

▶ **4. INTERACTION WITH OTHER METALS.** Hydrogen is **absorbed** (dissolved) at elevated temperatures by many transition metals, lanthanide metals, and actinide metals. The materials thus produced are called metallic hydrides because they exhibit many metallic properties. For example, they are electrical conductors. In most cases these materials can be described as solutions of hydrogen atoms in the metal since hydrogen atoms are located in the spaces between the metal atoms. Since these hydrides behave like solutions, they have no fixed atomic composition, and there is a range of solubility of hydrogen in the metal. In palladium hydride, for example, the H to Pd ratio can vary from 0.4 to 0.7. Because of this solubility, hydrogen readily passes through the heated walls of a palladium container.

Hydrogen is also **adsorbed** by certain metals, the process being most pronounced with gold, platinum, and tungsten. **Adsorption** is the adhesion of molecules of a gas, liquid, or dissolved substance to the surface of a solid. The quantity of hydrogen adsorbed by a given mass of metal depends upon the condition of the metal, such as extent of subdivision, and the temperature and pressure under which the adsorption takes place. Adsorbed hydrogen is very active chemically as indicated by its rapid union with oxygen when in this state. Ordinary hydrogen and oxygen do not combine noticeably unless the mixture is ignited. There is evidence that adsorbed hydrogen is activated because metals such as platinum are good catalysts for many chemical reactions involving hydrogen as a reactant.

▶ **5. REACTION WITH COMPOUNDS.** Hydrogen reacts with heated oxides of many metals with the formation of the free metal and water. For example, when hydrogen is passed over heated copper(II) oxide, copper and water are formed.

$$CuO + H_2 \longrightarrow Cu + H_2O$$

This is an example of an oxidation-reduction reaction (Section 8.2). Here, hydrogen reduces copper and is itself oxidized. The hydrogen molecule acts as the reducing agent, each atom losing an electron, and therefore undergoes oxidation. The copper in the oxide gains electrons from the hydrogen and hence causes the hydrogen to be oxidized; the copper(II) oxide thus is the oxidizing agent. As pointed out in Section 8.2, the processes of reduction and oxidation always occur simultaneously.

Hydrogen may also react with certain metal oxides, reducing them to lower oxides.

$$MnO_2 + H_2 \longrightarrow MnO + H_2O$$

9.18   Reversible Reactions

When hydrogen is passed over hot Fe$_3$O$_4$, the iron is reduced, and iron and water (steam) are formed.

$$Fe_3O_4 + 4H_2 \longrightarrow 3Fe + 4H_2O \tag{1}$$

This reaction is exactly the reverse of the one described in Section 9.15, i.e., the production of hydrogen by passing steam over heated iron.

$$3Fe + 4H_2O \longrightarrow Fe_3O_4 + 4H_2 \tag{2}$$

When these reactions are utilized in the laboratory, a stream of hydrogen gas is passed over a heated solid in a tube. In Reaction (1), hydrogen is in excess and the steam is swept out of the reaction tube by the current of hydrogen. Thus, Reaction (1) can go to completion. In Reaction (2) the hydrogen is swept out of the tube by the steam and thus this reaction also can go to completion. If a mixture of iron and steam is heated in a closed tube, neither the steam nor hydrogen that forms can escape. At first, Reaction (2) will be the only one to take place because only iron and steam are present, but as this reaction proceeds iron and steam are used up while Fe$_3$O$_4$ and hydrogen are formed. It follows that the rate of Reaction (2) decreases while that of Reaction (1) increases, and after a time the two reaction rates will become equal. The two reactions will reach a state of **chemical equilibrium.** The quantity of each of the four substances in the closed tube remains unchanged at equilibrium because each is constantly being formed at the same rate as it is being consumed. Reactions of this type are called **reversible reactions,** and it is customary to represent the reversibility by means of double arrows:

$$3Fe + 4H_2O \rightleftharpoons Fe_3O_4 + 4H_2$$

A more complete account of the very important phenomenon of chemical equilibrium is given in Chapter 15.

9.19   Uses of Hydrogen

Although hydrogen was once extensively used as the lifting agent in lighter-than-air craft such as blimps and dirigibles, it has largely been replaced by helium gas. Because of its inertness, helium is much safer than the highly combustible hydrogen (Fig. 9-11).

Figure 9-11 Explosion of hydrogen on the airship *Hindenberg* while landing at Lakehurst, New Jersey, May 6, 1937. Following the ignition of hydrogen in one compartment, ignition of the other two compartments took place within seconds, resulting in the complete destruction of the giant airliner. This picture was taken only 22 seconds after the first explosion. (*World Wide Photos*)

Two-thirds of the world's hydrogen production is used in the manufacture of ammonia, which is used primarily as a fertilizer. Ammonia is manufactured by the Haber process (Section 23.6):

$$3H_2(g) + N_2(g) \underset{500°C}{\overset{\text{High pressure}}{\rightleftharpoons}} 2NH_3(g)$$

Hydrogen is extensively used in a process called **hydrogenation,** in which vegetable oils are changed from liquids to solids. When vegetable oils are treated with hydrogen under pressure and in the presence of nickel as a catalyst, the hydrogen combines chemically with the oil, producing solid fats used in cooking. Crisco is an example of a hydrogenated oil.

An increasingly important use of hydrogen is in the catalytic conversion of coal dust to liquid hydrocarbons, which supplement petroleum as a source of liquid fuel. Methyl alcohol is produced synthetically by the catalyzed reaction of hydrogen with carbon monoxide:

$$2H_2 + CO \xrightarrow{\text{Catalyst}} CH_3OH$$

Water gas, a mixture of hydrogen with carbon monoxide, is an important industrial fuel. Hydrogen is used in the reduction of certain metal oxides to obtain the free metal. The use of hydrogen in nuclear fusion reactions is discussed in Chapter 29 (Nuclear Chemistry).

9.20   Isotopes of Hydrogen

Using an instrument known as the mass spectrograph (see Chapter 32) it is possible to show that hydrogen is composed of three isotopes. They are ordinary hydrogen, or **protium,** $_1^1H$; heavy hydrogen, or **deuterium,** $_1^2H$; and **tritium,** $_1^3H$. In a sample of hydrogen, there is only one atom of deuterium for every 5000 atoms of ordinary hydrogen and one atom of tritium for every 10^7 atoms of ordinary hydrogen. The chemical properties of the different isotopic forms of hydrogen are essentially the same because they have identical electronic structures. The differences in their atomic masses give rise to differences in physical properties, however. Deuterium and tritium have lower vapor pressures than does ordinary hydrogen. Consequently, when liquid hydrogen evaporates, the heavier isotopes are concentrated in the last portions.

Deuterium is of interest in connection with heavy water in atomic reactors (Section 12.6) and tritium with nuclear fusion reactions (Section 29.11).

9.21   The Activity, or Electromotive, Series

We noted in Section 9.15 that certain metals, such as sodium and potassium, react readily with cold water, displacing hydrogen and forming metal hydroxides. It was also pointed out that certain other metals, such as magnesium and iron, react with water only when they are heated. Sodium and potassium react much more vigorously with acids than do magnesium and iron. From experimental observations of this sort it is possible to arrange the metals in order of their chemical activities and thereby to establish an **activity, or electromotive, series.** A brief form of the electromotive series containing only the common elements is given in Table 9-1. Note that elements with the more metallic behavior are found at the top of the electromotive series (Groups IA and IIA) and, in general, as the metallic character of the elements decrease they appear lower in the series.

Potassium is the most reactive of the common metals and so heads the series. Each succeeding metal in the series is less reactive, and gold, the least reactive of all, is found at the bottom of the series. In principle, any metal in the series will displace any other element below it in the series from dilute aqueous solutions of its soluble compounds, if simple ions are involved. That is, any metal in its elemental form will reduce the ion, in water, of any metal below it in the series. Any metal above hydrogen in the series will liberate hydrogen from aqueous acids (solutions of acids in water); those from cadmium on up will liberate hydrogen from hot water or steam; and those from sodium on up will liberate hydrogen even from cold water. As discussed in Section 9.15, for example, Zn, Fe, and Al reduce the hydrogen in acids. The metals below hydrogen do not displace hydrogen from water or from aqueous acids. Another example is the displacement of copper by

TABLE 9-1   Activity, or Electromotive, Series of Common Metals

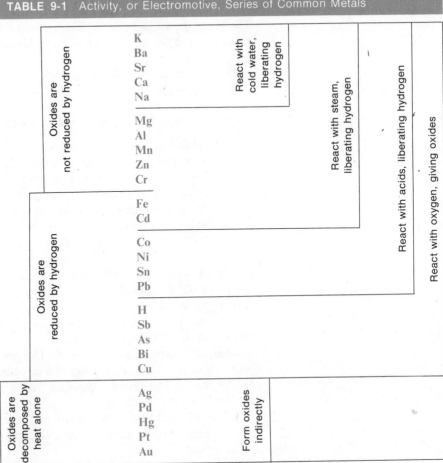

	Metal		
Oxides are not reduced by hydrogen	K, Ba, Sr, Ca, Na	React with cold water, liberating hydrogen	
	Mg, Al, Mn, Zn, Cr	React with steam, liberating hydrogen	React with acids, liberating hydrogen
Oxides are reduced by hydrogen	Fe, Cd		
	Co, Ni, Sn, Pb		React with oxygen, giving oxides
	H, Sb, As, Bi, Cu		
Oxides are decomposed by heat alone	Ag, Pd, Hg, Pt, Au	Form oxides indirectly	

iron, which is above copper in the series, from aqueous solutions of copper(II) salts (Fig. 9-12).

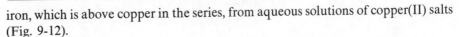

$$Fe(s) + Cu^{2+}(aq) \longrightarrow Cu(s) + Fe^{2+}(aq)$$

In a similar manner, aqueous silver ions are reduced by metallic copper and aqueous gold ions are reduced by silver.

The reactivity of the metals toward oxygen, sulfur, and the halogens is less for each succeeding metal as they become less metallic in progressing down the series. The heat of formation and stability of the compounds formed decreases in a similar manner. It is evident then that the activity series is very useful because it indicates the possibility of a reaction of a given metal with water, acids, salts of other metals, oxygen, sulfur, and the halogens (fluorine, chlorine, bromine, and iodine); in addition, it provides some indication of the stability of the compounds formed. It should be noted, however, that the order in which the elements are placed in the series depends somewhat on the conditions under which the activity is observed. A series determined from observation of the activity of the metals

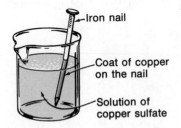

Iron nail

Coat of copper on the nail

Solution of copper sulfate

Figure 9-12 When an iron nail is immersed in a solution of copper sulfate, copper (a less active metal than iron) is deposited as the iron reduces the Cu^{2+} ion to metallic copper and is itself oxidized to Fe^{2+}.

with respect to their ions in water solution will be slightly different from that showing the order of activity at high temperatures and in the absence of a solvent. The series shown in Table 9-1 is not applicable to reactions in solvents other than water (e.g., in fused salts), or when stable complex ions are formed, except in a general way.

The activity series of metals is discussed in more detail in Chapter 20 (Electrochemistry).

9.22   Hydrogen as a Secondary Energy Source

Hydrogen, like electricity, is a secondary source of energy, because it must be produced by using energy from a primary source such as coal, nuclear fusion, or the sun. When mass-produced, hydrogen may in a few years cost no more than increasingly scarce and costly gasoline and natural gas. Unlike electricity, hydrogen can be stored conveniently and then burned to produce energy where it is needed. It is pollution-free when burned because the only product is water. When our supply of natural gas is depleted, hydrogen conceivably can be used just as natural gas is, if safety problems can be solved. It can be transported in the nationwide network of underground pipelines that already exists for the transport of gases such as natural gas. These lines are now linked to over 60% of all U.S. homes and industries.

As an internal-combustion fuel, hydrogen is in many ways superior to gasoline and diesel oil. Engines using hydrogen start more easily, particularly in cold weather, and experience less wear. The greatest advantage is that almost no pollution is generated. A number of hydrogen-powered research cars are already on the roads. The greatest problem is the carrying of the fuel in a safe and practical manner. Because liquid hydrogen is 40% lighter than conventional jet fuel, it particularly benefits aircraft performance. It is estimated that a hydrogen-powered aircraft can be designed to carry twice the current ratio of payload to plane weight.

In recent years hydrogen has been used as a fuel in the production of electricity in fuel cells (Section 20.19), a process much more efficient than that involving generators powered by steam.

Questions

1. Both hydrogen and oxygen are colorless, odorless, and tasteless gases. How may we distinguish between them using physical properties? using chemical properties?

2. Both H$_2$ and O$_2$ are diatomic molecules. In what ways do the chemical bonds in the molecules differ and how are they similar?

3. Write the equation that describes the preparation of hydrogen when steam is passed through a red-hot gun barrel.

4. Write the reaction describing the preparation of oxygen by focusing the sun's rays on a sample of mercury(II) oxide by means of a lens.

5. Why is it easier and cheaper to prepare commercial quantities of liquid oxygen than to prepare commercial quantities of liquid hydrogen?

6. Which of the various laboratory techniques used in the preparation of oxygen should be used if a sample of oxygen free of traces of water (dry oxygen) were required for an experiment? What laboratory preparations of hydrogen would give dry hydrogen?
7. Write balanced equations for the preparation of oxygen from $Cu(NO_3)_2$, $NaNO_3$, H_2O, and BaO_2.
8. Write balanced equations showing the release of oxygen gas upon heating the following oxides: Au_2O, Ag_2O, PtO_2, PbO_2, CrO_3.
9. Write balanced equations for the preparation of hydrogen from $C + H_2O$; H_2SO_4; H_2O + electrical energy; NaH.
10. Write balanced equations for the reaction, if any, of each of the following metals with water, perchloric acid ($HClO_4$ in water), and hydrobromic acid (HBr in water), respectively: K, Mg, Ga, Bi, Fe, Pt. Name the compounds formed.
11. (a) Describe the oxygen-oxygen bonding in the peroxide ion in terms of the Lewis structure of O_2^{2-}. (b) Using the molecular orbital model describe the bonding in O_2, O_2^-, and O_2^{2-}. Arrange these species in decreasing order of bond energy.
12. Write balanced equations for the reaction of an excess of oxygen with each of the following (keep in mind that oxygen tends to give the highest oxidation state when it combines in excess with an element): Ca, Cs, As, Ge, Na_2SO_3, AlP, C_2H_6, CO.
13. Write balanced equations for the reaction of an excess of hydrogen with each of the following: PtO, Na, Fe_2O_3, Cl_2, O_3.
14. Which of the compounds listed at the end of this question will react with water to give an acid solution? Which will give a basic solution? Write a balanced equation for the reaction of each with water. NaH, H_2S, HCl, SO_3, CaO, Na_2O_2, P_4O_{10}
15. Define the term "combustion." Give examples of combustion in an O_2 atmosphere and in a Cl_2 atmosphere.
16. Which of the following materials will burn in O_2? $SiH_4(g)$, SiO_2, CO, CO_2, Mg, CaO. Why won't some of these materials burn in oxygen?
17. Under what conditions may spontaneous combustion occur?
18. Why will magnesium ribbon burn more rapidly in pure oxygen than in air?
19. Why will wood shavings ignite and burn more rapidly than a log of the same kind of wood?

20. The heat of reaction for the combustion of carbon to produce carbon dioxide has a value that is the same as the heat of formation of carbon dioxide. The heat of reaction for the combustion of hydrogen to produce water is twice that of the heat of formation of water. Explain why.
21. Show that the heat of formation of ozone is 143 kJ and that of H(g) (atomic hydrogen) is 218 kJ. Are these endothermic or exothermic reactions?
22. Is the heat of formation of MgO positive or negative?
23. Pure $KClO_3$ must be heated to about 400°C before it will decompose to O_2 and KCl. When a small amount of MnO_2 is added, $KClO_3$ decomposes smoothly at about 270°C. Why does the presence of MnO_2 reduce the temperature at which O_2 may be readily prepared?
24. Steam reacts with iron to give Fe_3O_4 and hydrogen. The Fe_3O_4 may be reduced to iron by reaction with CO giving CO_2. The overall change in this sequence is the conversion of carbon monoxide and steam to hydrogen and carbon dioxide. Write a single balanced equation for the overall change. Is iron a catalyst in this system?
25. When heated, sodium reacts with H_2 to form NaH. NaH reacts with H_2O to give NaOH and H_2. Can H_2 be regarded as a catalyst in the conversion of Na to NaOH?
26. What is the function of a catalyst in a chemical reaction? Under what conditions is it useful to use a catalyst in a chemical reaction?
27. With the aid of the activity (electromotive) series (Table 9-1), predict whether or not the following reactions will take place:
(a) $Mg + Co^{2+} \longrightarrow Mg^{2+} + Co$
(b) $Al_2O_3 + 3H_2 \longrightarrow 2Al + 3H_2O$
(c) $2Au + Fe^{2+} \longrightarrow 2Au^+ + Fe$
(d) $Ni + 2H^+ \longrightarrow Ni^{2+} + H_2$
(e) $PtO_2 \xrightarrow{\Delta} Pt + O_2$
(f) $K_2O + H_2 \longrightarrow 2K + H_2O$
(g) $3Cd + 2Bi^{3+} \longrightarrow 2Bi + 3Cd^{2+}$
(h) $3Pd + 2Au^{3+} \longrightarrow 2Au + 3Pd^{2+}$
28. Lithium and beryllium do not appear in the electromotive series given in Table 9-1. From the position of these elements in the Periodic Table, locate their approximate position in the electromotive series and write equations for their reactions, if any, with water, steam, oxygen, and acids. Will the oxides of

Li and Be be reduced by hydrogen? Are the oxides reduced by heat alone?

29. Write the balanced equation or equations necessary to carry out the following transformations. (H_2O, H_2, and/or O_2 may be used as needed.)
   (a) $KAl(OH)_4$ from Al and K
   (b) $Zn_3(PO_4)_2$ from Zn and P
   (c) NaCl from Na_2O_2 and Cl_2
   (d) $ZnSO_4$ from Zn and H_2S
   (e) Fe from Fe_3O_4

30. Under what chemical conditions may the activity series be relied upon to predict chemical behavior?

31. What is meant by a "reversible reaction"? What is meant by the term "chemical equilibrium"?

32. Explain: "Metal hydrides are convenient and portable sources of hydrogen."

33. The reaction of NaH with H_2O may be characterized as an acid-base reaction; identify the acid and the base in the reactants. The reaction is also an oxidation-reduction reaction. Identify the oxidizing agent, the reducing agent, and the oxidation number changes that occur in the reaction.

34. One of the elements in the activity series (Table 9-1) will not produce hydrogen when it reacts with steam, but it will form hydrogen when it reacts with hydrochloric acid. Reaction of this element with a solution of $Ni(NO_3)_2$ produces Ni(s). What is this element?

35. Zinc and mercury are both members of Group IIB, but ZnS and HgS behave differently when roasted in air. Write the equations for the reactions of ZnS and HgS with O_2 when roasted.

36. Increases in the price of petroleum result in increases in the price of fertilizers. Why?

37. The bond length in the O_2 molecule is 1.208 Å while that in the O_3 molecule is 1.278 Å. Why does ozone have a longer bond?

Problems

1. Assume that when O_2 dissolves in water the volume of the solution is the same as the volume of water from which the solution was made. What is the molar concentration of O_2 in a solution at 0° which is saturated with O_2; that is, one that contains the maximum possible amount of dissolved O_2?
   Ans. 2.2×10^{-3} M

2. Hydrogen is not very soluble. At 25°C, a saturated solution of H_2 in water is 7.86×10^{-5} M. How many milliters of H_2 at 25° (density, 0.0823 g/ℓ) are dissolved in a liter of solution?   *Ans. 1.91 ml*

3. What mass and volume of $O_2(g)$ (density 1.429 g/ℓ) is required to prepare 41.3 g of MgO?
   Ans. 16.4 g; 11.5 ℓ

4. What mass of oxygen can be obtained by the thermal decomposition of 48.19 g of $Cu(NO_3)_2$? What is the volume of O_2 produced if the density of O_2 is 1.429 g/ℓ?   *Ans. 4.111 g; 2.878 ℓ*

5. Elements generally exhibit their highest oxidation states in compounds containing oxygen or fluorine. What is the oxidation number of osmium in 0.3789 g of an osmium oxide prepared by the reaction of 0.2827 g of Os with O_2? Write the equation for the reaction that gives this compound.
   Ans. +8; Os + 2O_2 $\longrightarrow$ OsO_4

6. What mass of hydrogen would result from the reaction of 27.2 g of BaH_2 with water?
   Ans. 0.787 g

7. What mass of LiH is required to provide enough hydrogen by reaction with water to fill a balloon at 0° and 760 mm pressure with a volume of 3.50 ℓ?
   Ans. 1.24 g

8. What is the concentration of the barium hydroxide solution formed when 1.38 g of barium reacts with 483 ml of water. Assume that the volume of the solution is the same as that of the water used in its preparation.   *Ans. 2.08×10^{-2} M*

9. What is the concentration of hydrogen ion in a solution that contains 31.9 g of H_2SO_4 per liter of solution. Assume that the H_2SO_4 completely dissociates to hydrogen ion and sulfate ion.
   Ans. 0.650 M

10. What mass of copper can be recovered from a solution of copper(II) sulfate by addition of 2.11 kg of aluminum?   *Ans. 7.45 kg*

11. How many grams of hydrogen and of zinc acetate,

$Zn(CH_3CO_2)_2$, can be prepared by the reaction of 22 g of zinc with acetic acid, CH_3CO_2H?

Ans. 0.68 g H_2; 62 g $Zn(CH_3CO_2)_2$

12. The combustion of 9.180 g of sodium to sodium oxide is exothermic; 100.74 kJ of heat is evolved. What is the heat of formation of sodium oxide?

Ans. 504.6 kJ/mol

13. The heat of formation of $CS_2(l)$ is +21.4 kcal/mol. How much heat is required for the reaction of 10.00 g of S_8 with carbon according to the following equation?

$$4C(s) + S_8(s) \longrightarrow 4CS_2(l)$$

Is the reaction exothermic or endothermic?

Ans. 3.34 kcal; endothermic

14. Deuterium, ^2_1H, usually indicated as D, may be separated from ordinary hydrogen by repeated distillation of water. Ultimately, pure D_2O can be isolated. What is the molecular weight of D_2O to three significant figures. (The atomic weight of D may be found in Table 4-1.)   *Ans. 20.0*

15. What mass of PD_3 may be prepared by the reaction of 0.498 g of Na_3P with D_2O? (See problem 14.)

Ans. 0.184 g

16. The equation for the combustion of methane in oxygen is

$$CH_4(g) + 2O_2(g) \longrightarrow CO_2(g) + 2H_2O(l)$$
$$\Delta H = -212.7 \text{ kcal}$$

ΔH for this reaction is sometimes called the heat of combustion of methane. The combustion of hydrogen proceeds by the following equation:

$$2H_2(g) + O_2(g) \longrightarrow 2H_2O(l) \quad \Delta H = -571.5 \text{ kJ}$$

Which will produce more heat, the combustion of one cubic foot of methane (density 0.7143 g/ℓ) or the combustion of one cubic foot of hydrogen (density 0.08987 g/ℓ)? How much more?

Ans. methane; 761.2 kJ, or 181.9 kcal, more

References

"Oxygen" (Staff), *Chem. and Eng. News,* June 26, 1978, p. 11.

"The Oxygen Cycle," P. Cloud and A. Gibor, *Sci. American,* September 1970, p. 110.

"Oxygen in Steelmaking," J. K. Stone, *Sci. American,* April 1968, p. 24.

"Recent Improvements in Explaining the Periodicity of Oxygen Chemistry," R. T. Sanderson, *J. Chem. Educ.,* **46,** 635 (1969).

"Lavoisier," D. I. Duveen, *Sci. American,* May 1956, p. 84.

"Combustion and Flame," R. C. Anderson, *J. Chem. Educ.,* **44,** 248 (1967).

"Why Does Methane Burn?" R. T. Sanderson, *J. Chem. Educ.,* **45,** 423 (1968).

"Medieval Uses of Air," L. White, Jr., *Sci. American,* August 1970, p. 92.

"Life Support Systems for Manned Space Flights," W. H. Bowman and R. M. Lawrence, *J. Chem. Educ.,* **48,** 260 (1971).

"The Cabin Atmosphere in Manned Space Vehicles," W. H. Bowman and R. M. Lawrence, *J. Chem. Educ.,* **48,** 152 (1971).

"High Temperature Chemistry of Simple Metallic Oxides," C. B. Alcock, *Chem. in Britain,* **5,** 216 (1969).

"Catalysis," V. Haensel and R. L. Burwell, Jr., *Sci. American,* December 1971, p. 46.

"A Simple Demonstration of O_2 Paramagnetism," G. H. Saban and T. F. Moran, *J. Chem. Educ.,* **50,** 217 (1973).

"Ozone: Properties, Toxicity, and Applications," F. Leh, *J. Chem. Educ.,* **50,** 404 (1973).

"Ozone Formation Related to Power Plant Emissions," D. F. Miller *et al., Science,* **202,** 1186 (1978).

"The Ozone Layer: The Threat from Aerosol Cans is Real," T. H. Maugh, *Science,* **194,** 170 (1976).

"Priestley," M. Wilson, *Sci. American,* October 1954, p. 68.

"Joseph Priestley and the Discovery of Oxygen," S. F. Mason, *Chem. in Britain,* **10,** 286 (1974).

"Cornelis Drebbel and Oxygen," J. W. van Spronsen, *J. Chem. Educ.,* **54,** 157 (1977).

"Scheele's Priority for the Discovery of Oxygen," H. Cassebaum and J. A. Schufle, *J. Chem. Educ.,* **52,** 442 (1975).

"Oxygen: Boon and Bane," I. Fridovich, *Amer. Scientist,* **63,** 54 (1975).

"Fluorinated Peroxides," J. M. Shreeve, *Endeavour,* May 1979, p. 79.

"Hydrogen Staging Comeback," W. J. Storch, *Chem. and Eng. News,* May 15, 1978, p. 8.

"Significance of Hydrogen Isotopes," H. C. Urey (Willard Gibbs Medal Address), *Ind. Eng. Chem.,* **26,** 803 (1934).

"The Flash of Genius: Deuterium: Harold C. Urey," A. B. Garrett, *J. Chem. Educ.,* **39,** 583 (1962).

"Principles of Hydrogen Chemistry," R. T. Sanderson, *J. Chem. Educ.,* **41,** 331 (1964).

"Hydrogen and Oxygen from Water," E. A. Fletcher and R. L. Moon, *Science,* **197,** 1050 (1977).

"Thermochemical Production of Hydrogen from Water," C. E. Banberger, J. Brannstein, and D. M. Richardson, *J. Chem. Educ.,* **55,** 561 (1978).

"Hydrogen Advocates Focus on Practical Goals" (Staff), *Chem. and Eng. News,* August 14, 1978, p. 28.

"The Hydrogen Economy and the Chemist," C. Marchetti, *Chem. in Britain,* **13,** 219 (1977).

"Raising the Titanic by Electrolysis," L. C. Plumb, *J. Chem. Educ.,* **50,** 61 (1973).

"The Hydrogen Economy," D. P. Gregory, *Sci. American,* January 1973, p. 13.

"The Stability of the Hydrogen Atom," F. Rioux, *J. Chem. Educ.,* **50,** 550 (1973).

"Interstellar Hydrogen in Galaxies," M. S. Roberts, *Science,* **183,** 371 (1974).

"Deuterium in the Universe," J. M. Pasachoff and W. A. Fowler, *Sci. American,* May 1974, p. 108.

"Atomic Hydrogen on Mars," J. S. Levine *et al., Science,* **200,** 1047 (1978).

"Hydride Reductions: A 40-Year Revolution in Organic Chemistry," H. C. Brown, *Chem. and Eng. News,* March 5, 1979, p. 24.

10

The Gaseous State and the Kinetic-Molecular Theory

Those elements and compounds that are gases at room temperature have played an important part in the history of the development of chemistry. The identification of oxygen as a component of air was crucial to the development of the atomic theory. Water was shown to be a compound rather than an element when it was prepared by the reaction of the gases oxygen and hydrogen. Atomic weights were determined in part from a study of the densities of gases. Conclusions regarding chemical stoichiometry and the molecular nature of matter followed from observations of the volumes of gases that combine in chemical reactions. The variations in the pressure and volume of a gas with temperature led to the discovery of the concept of absolute zero and the development of the kinetic-molecular theory of gaseous behavior. Studies of gases also led to the first quantitative models for the description of the behavior of matter.

In this chapter we shall consider the equations that relate temperature, pressure, volume, and mass of a gas present in a container. These equations are the tools used to describe the quantitative behavior of gases. We will see how to use these tools to convert physical measurements to moles of gas present, to the molecular weight of a gas, or to the quantities of gases involved in chemical changes (stoichiometry calculations of the third type, Section 3.1). We will examine the development of the theoretical kinetic-molecular model of gases and compare the experimental behavior of gases with the predictions of the theoretical model. This will help us to determine if the assumptions used in the theory to describe the nature of molecules of gases lead to a useful model for the behavior of gases.

The Physical Behavior of Gases

10.1   Behavior of Matter in the Gaseous State

We noted in Section 1.4 that matter exists in three different physical states: **solid, liquid,** and **gas.** Water in the solid state is known as ice, in the liquid state as water, and in the gaseous state as water vapor. When water in the form of ice takes up enough heat energy, it changes to the liquid state; and when liquid water is heated at 100° at sea level, it changes to the gaseous state. These changes are reversed when heat energy is removed. Such changes in physical state are possible for a great many but not all substances. For example, we have observed in our study of the preparation of oxygen that both solid mercury(II) oxide and solid copper(II) nitrate decompose upon heating. Most substances that can exist in the gaseous state, however, can be liquefied and solidified under suitable conditions of temperature and pressure.

The volume of a gas, unlike that of a solid or a liquid, can be decreased greatly by increasing the pressure upon the gas. This property is known as **compressibility.** Increasing the amount of weight on a piston, such as that shown in Fig. 10-1, causes the volume of the gas confined in the cylinder to decrease until the pressure the gas exerts is sufficient to support the piston and the greater weight.

If a gas in a cylinder with a piston is heated, the pressure of the gas increases. By raising the piston we can increase the volume and keep the pressure constant, thus allowing the gas to expand. **Expansibility** is characteristic of all gases. If the piston is kept from moving by adding more weight to it, the volume remains constant. In this case the pressure of the gas increases with a rise in temperature at constant volume.

A small quantity of a pungent gas released into a room in which the air is still can, in time, be detected by its odor in other parts of the room. When a sample of gas is introduced into an evacuated container, it almost instantly distributes itself and fills the container completely. These are examples of the **diffusion** of gases. Diffusion results because the molecules of a gas are in constant high-speed motion within a container. Two gases introduced into the same container quickly mix by diffusion. Each gas is said to permeate the other. Thus, we say that gases possess the properties of **diffusibility** (or **diffusion**) and **permeability.**

Figure 10-1 These figures illustrate what happens when a fixed quantity of gas at constant temperature is confined in a cylinder and subjected to more and more pressure by placing heavier and heavier weights on a movable, gas-tight piston. The number of molecules represented is but a minute fraction of the actual number in such a volume. It should be noted also that the molecules, shown lined up in the diagram for clarity, are actually randomly distributed in the gaseous state.

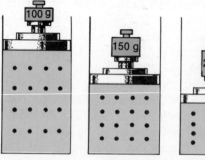
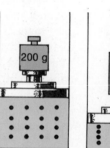

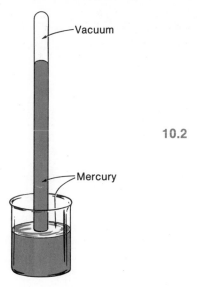

Figure 10-2 A mercury barometer.

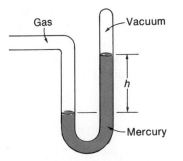

Figure 10-3 A simple two-arm mercury barometer. Such barometers are sometimes called manometers.

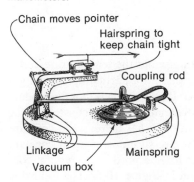

Figure 10-4 Diagram of an aneroid barometer. Changes in air pressure cause changes in thickness of the vacuum box to which the coupling rod is attached. The pointer reacts to changes in position of the coupling rod.

The different gases of the atmosphere differ in density, but diffusion prevents a separation of the gases. Mixtures of gases are homogeneous; they do not separate into layers because of differences in densities.

10.2 Measurement of Gas Pressures

The pressure exerted by the atmosphere (a mixture of gases) may be measured by a simple mercury **barometer** (Fig. 10-2). Such an instrument can be made by filling a glass tube, closed at one end and about 80 cm long, with mercury and inverting it in a dish of mercury. The mercury falls in the tube to the level where the pressure exerted by the air upon the surface of the mercury in the dish is just sufficient to support the mercury in the tube. The pressure of the air is thus measured in terms of the height of the mercury column, which is the vertical distance between the surface of the mercury in the tube and that in the open vessel. (Pressure means the force exerted upon a unit area of surface.) Because the pressure of the atmosphere is proportional to the height of the mercury column, pressure is sometimes expressed in terms of millimeters of mercury (mm Hg, or simply mm). A pressure of 1 mm Hg is generally referred to as a **torricelli** (torr).

A mercury barometer may be constructed with two arms, one closed and one open to the atmosphere or connected to a container filled with gas (Fig. 10-3). In this type of barometer the pressure exerted by the atmosphere or a gas in a container is equal to that of a column of mercury whose height is the distance between the mercury levels in the two arms of the tube (h in the diagram).

The pressure of the atmosphere varies with the distance above sea level and with climatic changes. At a higher elevation, the air is less dense, and thus the pressure is less. The atmospheric pressure at 20,000 feet is only one-half of that at sea level because about half of the entire atmosphere is below this elevation. Portable **aneroid barometers** (Fig. 10-4) are made with scales graduated in feet for measuring elevations. Mountain climbers and air pilots use such barometers to determine height above sea level.

The average pressure of the atmosphere at sea level at a latitude of 45° will support a column of mercury 760 mm in height. The average sea-level pressure at the latitude of 45° is thus 760 mm Hg, which is defined as **one atmosphere** (atm).

The pressure unit recommended for use with the International System of Units (SI) is the **pascal** (Pa), based on the newton unit for force. [A **newton** (N) is the force which, when applied for one second, will give a one-kilogram mass a speed of one meter per second. Hence, a newton is expressed in the units kg m/sec², or kg m sec⁻².] Since pressure is force per unit area, the pressure can be expressed in terms of newtons per square meter, which are called pascals. Thus,

$$\text{Pa} = \text{N/m}^2 = \text{kg m sec}^{-2}/\text{m}^2 = \text{kg sec}^{-2}/\text{m}$$
$$= \text{kg/sec}^2\,\text{m}$$

The conversions between the various units we have considered in this section for expressing pressure are

$$1 \text{ atmosphere} = 760 \text{ mm Hg} = 760 \text{ torr} = 101,325 \text{ Pa}$$

EXAMPLE   A typical barometric pressure in a Midwestern city 1000 feet above sea level is 740 torr. Express this pressure in atmospheres, mm Hg, and pascals.

$$740 \text{ torr} \times \frac{1 \text{ atm}}{760 \text{ torr}} = 0.974 \text{ atm}$$

$$740 \text{ torr} \times \frac{1 \text{ mm Hg}}{1 \text{ torr}} = 740 \text{ mm Hg}$$

$$740 \text{ torr} \times \frac{101,325 \text{ Pa}}{760 \text{ torr}} = 9.87 \times 10^4 \text{ Pa}$$

◆ ◆ ◆ ◆ ◆

10.3   Relation of Volume to Pressure at Constant Temperature; Boyle's Law

It has been observed from experiment that, at constant temperature, the volume of a given mass of gas is reduced to half when the pressure on the gas is doubled. Conversely, the volume is doubled by a decrease in pressure to one-half. Observations of this sort were summarized by the English physicist and chemist Robert Boyle in 1660 in a statement now known as **Boyle's Law: The volume of a given mass of gas held at constant temperature is inversely proportional to the pressure under which it is measured** or

$$V \propto \frac{1}{P}$$

where V is the volume of the gas, P, the pressure, and $\propto$ means "is proportional to." The proportionality may be changed to an equality by including a constant, k, referred to as a proportionality constant.

$$V = \text{constant} \times \frac{1}{P}, \quad \text{or} \quad V = k \times \frac{1}{P}$$

Hence $PV = \text{constant},$ or $PV = k$ (1)

The value of the constant depends on the mass of gas and the temperature, but it does not vary if only the pressure or volume is varied for a particular mass of gas at constant (specific) temperature. The meaning of Equation (1) is that **the product of the pressure of a given mass of gas times its volume is always constant if the temperature does not change.** Thus as the pressure of a gas increases at constant temperature the volume must decrease, or as the pressure decreases at constant temperature the volume must increase, in order for the product to remain constant. A graph showing this relationship is shown in Figure 10-5. The following examples illustrate the application of Boyle's Law.

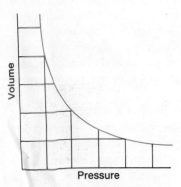

Figure 10-5 A graphical illustration of Boyle's Law. Note how the volume decreases as the pressure increases and conversely, increases as the pressure decreases in an inversely proportional relationship.

EXAMPLE 1   A sample of methane has a volume of 202.4 ml when measured at a temperature of 25°C and a pressure of 1.01×10^5 Pa. What volume will it occupy at 25°C and 5.05×10^4 Pa?

From Boyle's Law we know that PV = constant. The value of the constant may be determined from the volume (202.4 ml) at a pressure of 1.01×10^5 Pa.

$$PV = k$$

$$1.01 \times 10^5 \text{ Pa} \times 202.4 \text{ ml} = 2.044 \times 10^7 \text{ Pa ml} = k$$

Using the value of the constant, the volume at 50,500 Pa can be determined.

$$5.05 \times 10^4 \text{ Pa} \times V = k$$

Thus
$$50,500 \text{ Pa} \times V = 2.044 \times 10^7 \text{ Pa ml}$$

or
$$V = \frac{2.044 \times 10^7 \text{ Pa ml}}{5.05 \times 10^4 \text{ Pa}} = 405 \text{ ml}$$

There is a less formal way of solving this problem. From Boyle's Law, Equation (1), we know that a reduction of the pressure from 101,000 Pa to 50,500 Pa at constant temperature will result in an increase in the volume. The new volume can be determined by multiplying the original volume by the ratio of the two pressures. The ratio must have the larger pressure in the numerator and the smaller pressure in the denominator so that when the original volume is multiplied by this ratio the calculation will show a change to a larger volume in accord with our prediction.

$$202.4 \text{ ml} \times \frac{1.01 \times 10^5 \text{ Pa}}{5.05 \times 10^4 \text{ Pa}} = 405 \text{ ml}$$

Thus the new volume is two times the original volume as the result of decreasing the pressure to one-half its initial value. Note that because the unit Pa appears in both numerator and denominator of the expression, it cancels out and leaves only ml as the unit of measurement in the answer to the problem.

◆ ◆ ◆ ◆ ◆

EXAMPLE 2

What pressure must be applied to a sample of Freon gas having a volume of 325 ℓ at 20°C and 723 torr to permit the expansion of the gas to a volume of 975 ℓ at 20°C?

First, determine the constant in Boyle's Law, PV = constant, at the first pressure and volume.

$$723 \text{ torr} \times 325 \text{ ℓ} = 234,975 \text{ ℓ torr} = k$$

At the second pressure and volume,

$$PV = 234,975 \text{ ℓ torr}$$

$$P \times 975 \text{ ℓ} = 234,975 \text{ ℓ torr}$$

$$P = \frac{234,975 \text{ ℓ torr}}{975 \text{ ℓ}} = 241 \text{ torr}$$

There is also a less formal way of solving this problem. A decrease in pressure is required to permit the volume of a gas to increase at constant temperature. The new pressure will be determined by the ratio of the two volumes. By making the smaller volume the numerator and the larger volume the denominator of the ratio, and multiplying the original pressure by this ratio, the new (lower) pressure may be found.

$$723 \text{ torr} \times \frac{325 \ \cancel{l}}{975 \ \cancel{l}} = 241 \text{ torr}$$

The new pressure must be one-third the initial pressure to permit the volume to increase threefold.

◆ ◆ ◆ ◆ ◆

10.4   Relation of Volume to Temperature at Constant Pressure; Charles' Law

Studies of the effect of temperature upon the volume of confined gases at constant pressure by the French physicist S. A. C. Charles in 1787 led to the generalization known as **Charles' Law,** which we can state as follows: **The volume of a given mass of gas is directly proportional to its temperature on the Kelvin scale when the pressure is held constant** or

$$V \propto T$$

If a proportionality constant that depends upon the mass of gas and its pressure is used, we get the equation

$$V = \text{constant} \times T, \quad \text{or} \quad V = k \times T$$

Hence
$$\frac{V}{T} = \text{constant}, \quad \text{or} \quad \frac{V}{T} = k \tag{1}$$

This is a different constant than that in Boyle's Law. It does not vary if the mass of the gas does not change. Equation (1) means that as the Kelvin temperature of a given mass of gas changes with no change in pressure, its volume also changes so that the ratio V/T remains the same. A decrease in T results in a decrease in V, and an increase in T results in an increase in V. The relationship of volume and temperature is shown in Figs. 10-6 and 10-8.

Note that Charles' Law applies to the volume of a gas and **its Kelvin temperature.** In Section 1.20 we saw that the relationship between the Kelvin and Celsius temperature scales is

$$K = {}^\circ C + 273.15$$

The following examples illustrate the application of Charles' Law.

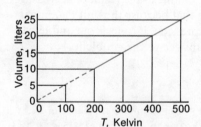

Figure 10-6 A graphical illustration of Charles' Law. Note how the volume increases with increasing Kelvin temperature at constant pressure in a directly proportional relationship.

EXAMPLE 1   A sample of CO_2 occupies 300 ml at 10°C and 750 torr. What volume will the gas have at 30°C and 750 torr? (Assume all figures are significant.)

The Celsius temperatures are first converted to the Kelvin scale, 10°C + 273.15 = 283 K, and 30°C + 273.15 = 303 K. (Recall that temperatures on the Kelvin scale are by convention reported without the degree sign.) Using the Kelvin temperature (283 K) and the corresponding volume (300 ml), evaluate the constant in the Charles' Law equation.

$$\frac{V}{T} = k$$

$$\frac{300 \text{ ml}}{283 \text{ K}} = 1.06 \frac{\text{ml}}{\text{K}} = k$$

Now solve for V at the second temperature.

$$\frac{V}{303 \text{ K}} = 1.06 \frac{\text{ml}}{\text{K}}$$

$$V = 303 \text{ K} \times 1.06 \frac{\text{ml}}{\text{K}} = 321 \text{ ml}$$

This problem can also be solved in a less formal way. From Charles' Law, we know that when the temperature of a gas is raised at constant pressure, the gas expands. The final volume may be found by multiplying the initial volume by that ratio of the Kelvin temperatures that gives the larger volume (high temperature on top in the ratio).

$$300 \text{ ml} \times \frac{303 \text{ K}}{283 \text{ K}} = 321 \text{ ml}$$

◆ ◆ ◆ ◆ ◆

EXAMPLE 2   **A sample of oxygen occupies 0.110 ℓ at 27°C and 98,000 Pa pressure. What temperature will the gas have when its volume is changed to 0.0900 ℓ at 98,000 Pa?**

Since the initial temperature of 27°C is 300 K, the constant for this sample in the Charles' Law equation is

$$\frac{V}{T} = \frac{0.110 \text{ } \ell}{300 \text{ K}} = 3.667 \times 10^{-4} \frac{\ell}{\text{K}}$$

The new temperature is then determined.

$$\frac{V}{T} = k$$

$$T = \frac{V}{k} = \frac{0.0900 \text{ } \ell}{3.667 \times 10^{-4} \frac{\ell}{\text{K}}} = 245 \text{ K}$$

By the less formal approach, we know the temperature must decrease to cause a reduction in volume at constant pressure. Thus the ratio of volumes must be less than 1; i.e., the larger volume must be in the denominator:

$$300 \text{ K} \times \frac{0.0900 \text{ } \ell}{0.110 \text{ } \ell} = 245 \text{ K}$$

Subtracting 273.15 from 245 K, we find that the Celsius temperature is $-28°$.

◆ ◆ ◆ ◆ ◆

10.5   The Kelvin Temperature Scale

In Section 1.20 we discussed the Kelvin temperature scale and its relationship to the Celsius temperature scale. Now, with some knowledge of the effect of temperature on gases, we are ready to understand more about the basis for the Kelvin scale.

As we have just learned, many experiments have shown that if any gas is heated, the gas must expand in order for its pressure to remain constant. Likewise, cooling any gas at constant pressure results in a decrease in its volume. It has been found that when 546 ml of gas at 0°C is warmed to 1°C, its volume increases by 2 ml to 548 ml; at 20°C, its volume increases to 586 ml; at 273° its volume increases to 746 ml (double that at 0°), and so on, provided that in each case the pressure remains constant (Fig. 10-7). Note that the volume of the gas at 0° increases $\frac{1}{273}$ of its volume for each increase of 1° on the Celsius scale. The volume of a gas decreases in the same proportion when the temperature falls. If the temperature of 273 ml of a gas could be lowered from 0° to −273°, then the gas should have no volume at −273°, because its volume should decrease at the rate of $\frac{1}{273}$ of its volume at 0° for each degree of fall in temperature. Before the temperature of −273° is reached, however, all gases become liquids or solids, to which this rate of change in volume does not apply. A plot of data obtained from several such experiments is shown in Fig. 10-8, where the volume of the gas at constant pressure is plotted against the temperature in degrees Celsius. Each line represents a measurement with a given amount of a particular gas. The straight lines reflect the direct proportionality between the volume and temperature of these samples, as described by Charles' Law. The data stop at those temperatures at which the gases condense. If the straight lines are extended back (extrapolated) to the temperature at which the gases would have zero volume if they did not condense, the lines all intersect the temperature axis at −273.15°C. Regardless of the gas, these measurements yield the same temperature at which the gas would have a volume of

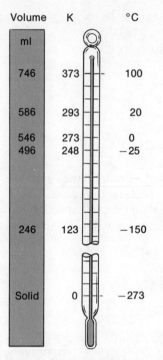

Figure 10-7 The volume of a gas at constant pressure is directly proportional to its Kelvin temperature.

Figure 10-8 Charles' Law behavior for several gases, each gas at constant pressure. All gases extrapolate to zero volume at −273.15°C. The dashed lines represent the extrapolated portion of each line for temperatures below the boiling point.

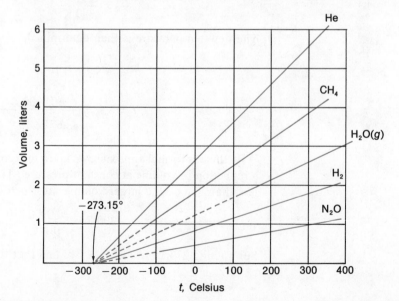

zero if it did not condense. Below this temperature, gases would theoretically have a negative volume. Since negative volumes are impossible, $-273.15\,°C$ must be the lowest temperature possible. This temperature, called **absolute zero,** is taken as the zero point of the Kelvin temperature scale. The freezing point of water $(0.00\,°C)$ is therefore 273.15 Kelvin, and the boiling point $(100.00\,°C)$ is 373.15 Kelvin (Fig. 10-7).

In order to lower the temperature of a substance we must remove heat. Absolute zero (0 K) is the temperature reached when all of the heat energy has been removed from a substance. Obviously, it is not possible to cool a substance any further after all the heat energy has been removed from it.

10.6   Reactions Involving Gases; Gay-Lussac's Law

Gases combine, or react, in definite and simple proportions by volume. For example, it has been determined by experiment that one volume of nitrogen will combine with three volumes of hydrogen to give two volumes of ammonia gas, provided the volumes of the reactants and product are measured under the same conditions of temperature and pressure.

$$N_2(g) \ + \ 3H_2(g) \ \longrightarrow \ 2NH_3(g)$$

1 volume   3 volumes   2 volumes

The term "volume" is used here in a general sense, and if the volume of nitrogen is measured in liters, the volumes of hydrogen and ammonia must also be measured in liters. Experimental observations on volumes of combining gases were summarized by **Joseph Louis Gay-Lussac** (1788–1850) in the **Law of Combining Volumes: The volumes of gases involved in a reaction, at constant temperature and pressure, can be expressed as a ratio of small whole numbers.** It is important to remember that this law applies only to substances in the gaseous state and measured at the same temperature and pressure. The volumes of any solids or liquids involved in the reactions are not considered. Additional examples illustrating Gay-Lussac's Law are as follows:

1. Two volumes of hydrogen and one volume of oxygen react to give two volumes of steam.

$$2H_2(g) \ + \ O_2(g) \ \longrightarrow \ 2H_2O(g)$$

2 volumes   1 volume   2 volumes

2. One volume of hydrogen combines with one volume of chlorine to form two volumes of hydrogen chloride.

$$H_2(g) \ + \ Cl_2(g) \ \longrightarrow \ 2HCl(g)$$

1 volume   1 volume   2 volumes

3. Carbon (a solid) reacts with one volume of oxygen to give one volume of carbon dioxide.

$$C(s) \ + \ O_2(g) \ \longrightarrow \ CO_2(g)$$

A solid   1 volume   1 volume

4. Four volumes of steam react with iron (a solid) to yield four volumes of hydrogen and Fe_3O_4 (a solid).

$$4H_2O(g) + 3Fe(s) \longrightarrow 4H_2(g) + Fe_3O_4(s)$$
$$\text{4 volumes} \quad \text{A solid} \quad \text{4 volumes} \quad \text{A solid}$$

The simplest ratio, of course, for the gaseous volumes of water and hydrogen in this case is 1 to 1.

Gay-Lussac's Law can be used to determine the volumes of gases involved in a reaction.

EXAMPLE   **Calculate the number of liters of hydrogen that will combine with 20 ℓ of nitrogen to form ammonia, the gaseous volumes being measured at the same temperature and pressure.**

First, write the balanced equation for the reaction.

$$N_2 + 3H_2 \longrightarrow 2NH_3$$

From the equation we see that one volume of nitrogen will combine with three volumes of hydrogen, so 20 ℓ of nitrogen will combine with 60 ℓ of hydrogen:

$$20 \; \ell \; N_2 \times \frac{3 \text{ volumes } H_2}{1 \text{ volume } N_2} = 60 \; \ell \; H_2$$

◆ ◆ ◆ ◆ ◆

10.7   An Explanation of Gay-Lussac's Law; Avogadro's Law

The law of combining volumes of gases can be satisfactorily explained in terms of the molecular nature of gases if we assume that **equal volumes of all gases, measured under the same conditions of temperature and pressure, contain the same number of molecules.** The Italian physicist Amadeo Avogadro advanced this hypothesis in 1811 to account for the behavior of gases. His hypothesis, which has since been experimentally proven, is now accepted as fact and is known as **Avogadro's Law.**

Avogadro's Law may be illustrated by the reaction of hydrogen and oxygen at 200°C to give gaseous water, with the volumes of the reactants and product measured at the same temperature and pressure.

$$2H_2(g) + O_2(g) \longrightarrow 2H_2O(g)$$
$$\text{2 volumes} \quad \text{1 volume} \quad \text{2 volumes}$$

According to Avogadro's Law, equal volumes of gaseous H_2, O_2, and H_2O contain the same number of molecules. During reaction, then, two molecules of water are formed from two molecules of hydrogen and one molecule of oxygen. This relationship is illustrated in Fig. 10-9, in which each volume is shown with four molecules.

10.8   The Ideal Gas Law Equation

Boyle's Law, Charles' Law, and Avogadro's Law are special cases that can be covered by one equation that relates pressure, volume, temperature, and number of moles of an ideal gas (a gas that follows these gas laws perfectly). This equation

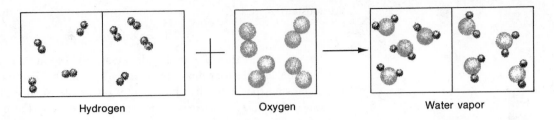

Hydrogen   Oxygen   Water vapor

Figure 10-9 Two volumes of hydrogen combine with one volume of oxygen to yield two volumes of water vapor.

is referred to as the **General Gas Law Equation,** or the **Ideal Gas Law Equation.** Although it is derived (Section 10.18) for an ideal gas, under normal conditions of temperature and pressure, it also applies very well to real gases, because real gases behave very nearly like ideal gases under these conditions. In this text, it will generally be assumed that all gases exhibit ideal behavior. Although we will not attempt to describe deviations from this behavior in a quantitative fashion, we shall consider these deviations qualitatively in Section 10.19.

The Ideal Gas Law Equation is

$$PV = nRT$$

where P is the pressure of a gas; V, its volume; n, the number of moles of the gas; T, the temperature of the gas on the Kelvin temperature scale; and R is a constant often referred to as the **Universal Gas Constant.** The following paragraphs show how the Ideal Gas Law Equation reduces to either Boyle's Law, Charles' Law, or Avogadro's Law under the appropriate conditions.

Boyle's Law (Section 10.3) states that for a given amount of an ideal gas, the product of its volume, V, and its pressure, P, is a constant at constant temperature. Since n, R, and T do not change under the conditions for which Boyle's Law holds, their product is a constant. So the Ideal Gas Law Equation, $PV = nRT$, becomes

$$PV = \text{constant}$$

which is the mathematical expression for Boyle's Law.

Charles' Law (Section 10.4) states that the volume, V, of n moles of an ideal gas divided by the temperature, T, on the Kelvin scale is equal to a constant at constant pressure. If we rearrange the Ideal Gas Law Equation so that the terms which do not vary (n, R, and P) are on the right side and V and T are on the left, we obtain the Charles' Law equation.

$$\frac{V}{T} = \frac{nR}{P} = \text{constant}$$

According to Avogadro's Law (Section 10.7) the volume of a gas is proportional to the number of molecules, and hence the number of moles, n, of the gas at constant pressure, P, and constant temperature, T. This may be written as

$$V = \text{constant} \times n$$

The Ideal Gas Law Equation may be rearranged to the same expression of Avogadro's Law by placing all the constant quantities (R, T, and P) together:

$$V = \frac{RT}{P} \times n$$

At constant P and T,

$$V = \text{constant} \times n$$

The numerical value of R in the Ideal Gas Law Equation can be determined by substituting actual experimental values for P, V, n, and T into the equation. It has been shown experimentally that exactly one mole of any ideal gas occupies a volume of 22.413 ℓ at 273.15 K and a pressure of exactly 1 atm. Substituting these values in $PV = nRT$ gives

$$(1\ \text{atm})(22.413\ \ell) = (1\ \text{mol})(R)(273.15\ \text{K})$$

$$R = \frac{(22.413\ \ell)(1\ \text{atm})}{(1\ \text{mol})(273.15\ \text{K})}$$

$$R = 0.08205\ \frac{\ell\ \text{atm}}{\text{mol K}}$$

The numerical value for R in other units for P, V, n, and T will be different. For example, if we take the pressure in pascals instead of atmospheres (1 atm = 101,325 Pa), the value of R is calculated as

$$R = \frac{(22.413\ \ell)(101,325\ \text{Pa})}{(1\ \text{mol})(273.15\ \text{K})}$$

$$R = 8314\ \frac{\ell\ \text{Pa}}{\text{mol K}}$$

It is absolutely essential to remember that the units of pressures, volumes, temperatures, and amounts of gas used in the Ideal Gas Law Equation must always be the same as the units used in R.

The following problems illustrate some of the uses of the Ideal Gas Law Equation.

EXAMPLE 1   Calculate the volume occupied by **1.00 mol of an ideal gas at exactly 0°C and at 1.00 atm of pressure. Use the value of R with the units ℓ atm/mol K.**

Since the Ideal Gas Law Equation requires temperatures to be on the Kelvin scale, convert the temperature from °C to K.

$$T = 0.00°C + 273.15 = 273.15\ \text{K}.$$

Now rearrange $PV = nRT$ to solve for V.

$$V = \frac{nRT}{P}$$

Substitution gives

$$V = \frac{(1.00\ \text{mol})\left(0.08205\ \frac{\ell\ \text{atm}}{\text{mol K}}\right)(273.15\ \text{K})}{(1.00\ \text{atm})} = 22.4\ \ell$$

Note that the units cancel to liters, an appropriate unit for volume.

◆ ◆ ◆ ◆ ◆

22.4 liters

32 g of O_2

22.4 liters

2 g of H_2

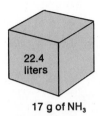

22.4 liters

17 g of NH_3

Figure 10-10 A mole of any gas occupies a volume of approximately 22.4 liters at 0°C and 1 atmosphere pressure.

The volume of one mole of any gas at 0°C and 1 atm pressure is 22.4 ℓ and is referred to as the **gram-molecular volume** (see Fig. 10-10).

EXAMPLE 2   Calculate the volume occupied by 15 g of N_2 at 15.0°C and 735 torr. Use the value of R with the units ℓ atm/mol K.

As shown in Example 1, the volume of an ideal gas may be obtained from the expression

$$V = \frac{nRT}{P}$$

The units of R (liters, atmospheres, moles, and Kelvin degrees) require that the pressure be expressed in atmospheres, the amount of gas in moles, and the temperature in Kelvin degrees. Once these conversions are completed, the values can be substituted into the equation and the volume calculated.

$$P = 735 \text{ torr} \times \frac{1 \text{ atm}}{760 \text{ torr}} = 0.967 \text{ atm}$$

$$n = 15 \text{ g } N_2 \times \frac{1 \text{ mol } N_2}{28 \text{ g } N_2} = 0.54 \text{ mol}$$

$$K = 15°C + 273.15 = 288 \text{ K}$$

$$V = \frac{nRT}{P} = \frac{(0.54 \text{ mol})\left(0.08205 \frac{\ell \text{ atm}}{\text{mol K}}\right)(288 \text{ K})}{(0.967 \text{ atm})}$$

$$V = 13 \ \ell \text{ (to two significant figures as justified by the data)}$$

Note that the units cancel out to liters, an appropriate unit for volume. Cancellation of units is a very simple way of checking to see that the units of R are consistent with those of P, V, n, and T.

◆ ◆ ◆ ◆ ◆

EXAMPLE 3   While resting, the average human male consumes 200 ml of O_2 per hour at 25°C and 1.0 atm for each kilogram of weight. How many moles of O_2 are consumed by a 70-kg man in 1 hr while resting?

For the purpose of this problem let us assume the oxygen consumption per kilogram per hour is good to two significant figures. The volume of O_2 consumed by a resting 70-kg male is then 14,000 ml/hr, or 14 ℓ/hr. Since the pressure and volume are given in atmospheres and liters, take $R = 0.08205$ ℓ atm/mol K. The temperature is $(273.15 + 25) = 298$ K. The Ideal Gas Law Equation can be rearranged to give n.

$$n = \frac{PV}{RT} = \frac{(1.0 \text{ atm})(14 \ \ell)}{\left(0.08205 \frac{\ell \text{ atm}}{\text{mol K}}\right)(298 \text{ K})} = 0.57 \text{ mol } O_2$$

◆ ◆ ◆ ◆ ◆

EXAMPLE 4   **What is the pressure in pascals in a 35.0-ℓ balloon at 25°C which was filled with dry hydrogen produced by the reaction of 38.9 g of NaH with water?**

To use the Ideal Gas Law Equation, $PV = nRT$, for determining P requires that we know V, n, R, and T. V and T are given, R must be 8314 ℓ Pa/mol K in order for the pressure units to be pascals, and n can be calculated from the weight of NaH and the equation for the reaction.

$$NaH + H_2O \longrightarrow NaOH + H_2$$

$$38.9 \text{ g NaH} \times \frac{1 \text{ mol NaH}}{23.99 \text{ g NaH}} = 1.62 \text{ mol NaH}$$

$$1.62 \text{ mol NaH} \times \frac{1 \text{ mol H}_2}{1 \text{ mol NaH}} = 1.62 \text{ mol H}_2$$

so $n = 1.62 \text{ mol}$

Now convert T to the Kelvin scale.

$$T = 25°C + 273.15 = 298 \text{ K}$$

Using the Ideal Gas Law Equation gives

$$P = \frac{nRT}{V} = \frac{(1.62 \text{ mol}) \left(8314 \frac{\ell \text{ Pa}}{\text{mol K}}\right)(298 \text{ K})}{35.0 \ell} = 1.15 \times 10^5 \text{ Pa}$$

The pressure in the balloon is thus about 14% greater than atmospheric pressure (101,325 Pa).

◆ ◆ ◆ ◆ ◆

EXAMPLE 5   **What volume of hydrogen at 27°C and 723 torr may be prepared by the reaction of 8.88 g of gallium with an excess of hydrochloric acid?**

$$2Ga(s) + 6HCl(aq) \longrightarrow 2GaCl_3(aq) + 3H_2(g)$$

This problem requires the following steps:

$$\boxed{\text{Mass of Ga}} \longrightarrow \boxed{\text{Moles of Ga}} \longrightarrow \boxed{\text{Moles of H}_2} \longrightarrow \boxed{\text{Volume of H}_2}$$

$$8.88 \text{ g Ga} \times \frac{1 \text{ mol Ga}}{69.72 \text{ g Ga}} = 0.1274 \text{ mol Ga}$$

$$0.1274 \text{ mol Ga} \times \frac{3 \text{ mol H}_2}{2 \text{ mol Ga}} = 0.1910 \text{ mol H}_2$$

Now use the Ideal Gas Law to get the volume of H_2.

$$V = \frac{nRT}{P} = \frac{(0.1910 \text{ mol H}_2) \left(0.08205 \frac{\ell \text{ atm}}{\text{mol K}}\right)(300 \text{ K})}{(0.951 \text{ atm})} = 4.94 \ell \text{ H}_2$$

◆ ◆ ◆ ◆ ◆

10.9   Standard Conditions of Temperature and Pressure

From the foregoing discussion of the variation of the volume of a given mass of gas with changes in pressure and temperature, it should be clear that the volume and the density (mass per unit volume) of a gas vary with these conditions. Thus, to be able to compare different gases with regard to their densities or to fix a definite density for any gas, it is desirable to adopt a set of **standard conditions of temperature and pressure** (STP) for all measurements of gases. Accordingly, 0°C temperature and 1 atm (273.15 K and 101,325 Pa) are universally used as standard conditions for measurements of gases.

10.10   Correction of the Volume of a Gas to Standard Conditions

EXAMPLE   Let us suppose that a sample of ammonia is found to occupy 250 ml under laboratory conditions of 27°C and 740 torr. Correct the volume to standard conditions of 0°C and 760 torr. (Assume all figures in the data are significant.)

1. *Using the Ideal Gas Law Equation.* Put all constant terms in the equation on the right-hand side and the variable terms on the left. In this example the moles of ammonia and R do not vary, so

$$\frac{PV}{T} = nR = \text{constant}$$

Now evaluate the constant term, nR, at the initial conditions (250 ml, 300 K, and 740 torr).

$$\frac{(740 \text{ torr})(250 \text{ ml})}{(300 \text{ K})} = nR = 616.7 \frac{\text{ml torr}}{\text{K}}$$

Now calculate the volume at standard conditions (0°C = 273.15 K, 760 torr = 1 atm). Since the amount of ammonia does not change, the equation

$$\frac{PV}{T} = \text{constant}$$

still holds. Rearranging the equation gives

$$V = \frac{\text{constant} \times T}{P}$$

$$V = \frac{\left(616.7 \frac{\text{ml torr}}{\text{K}}\right)(273.15 \text{ K})}{(760 \text{ torr})}$$

$$V = 222 \text{ ml}$$

Notice that the units cancel leaving ml, the correct units for volume.

2. *Using a combination of Boyle's and Charles' Laws.* First convert the Celsius temperatures to the Kelvin scale: $27°C = (27 + 273.15) \text{ K} = 300 \text{ K}$, and $0°C = (0 + 273.15) \text{ K} = 273.15 \text{ K}$. A decrease in temperature from 300 K

to 273.15 K alone will cause a decrease in the volume of ammonia. Therefore we multiply the original volume by a fraction made up of the two temperatures and having a value less than unity to correct for the temperature change:

$$250 \text{ ml} \times \frac{273.15 \text{ K}}{300 \text{ K}}$$

The increase in pressure from 740 torr to 760 torr alone will also decrease the volume, and this factor may be included in the same expression with the temperature factor to obtain the corrected volume:

$$250 \text{ ml} \times \frac{273.15 \text{ K}}{300 \text{ K}} \times \frac{740 \text{ torr}}{760 \text{ torr}} = 222 \text{ ml}$$

Note that these procedures apply to correction to any second set of temperature and pressure values and are not restricted to correction to standard conditions.

◆ ◆ ◆ ◆ ◆

10.11   Densities of Gases

The density of a gas usually is expressed as mass of the gas per liter of gas. Since molecules of different gases have different masses, and equal volumes of different gases under the same conditions of temperature and pressure contain the same number of molecules, it follows that the densities of different gases will not be the same. For example, the density of oxygen at STP is 1.429 g/ℓ, whereas that of hydrogen at STP is 0.08987 g/ℓ. Note that it is necessary to specify both the temperature and pressure of the gas when reporting a density, since the number of moles of gas (and thus the mass of the gas) in a liter changes with both temperature and pressure. Gas densities are commonly reported at STP.

The density of a gas that exhibits ideal behavior can be calculated in several ways. Two of these are illustrated in the following examples.

EXAMPLE 1   **The density of oxygen at STP is 1.429 g/ℓ. Calculate the density of carbon dioxide at STP.**

Since equal volumes of these gases contain the same number of molecules, the ratio of the densities of the two gases is the same as the ratio of their molecular weights.

$$\frac{\text{Density of CO}_2}{\text{density of O}_2} = \frac{\text{mol. wt CO}_2}{\text{mol. wt O}_2}$$

$$\text{Density of CO}_2 = \text{density of O}_2 \times \frac{\text{mol. wt CO}_2}{\text{mol. wt O}_2}$$

$$\text{Density of CO}_2 = 1.429 \text{ g/}\ell \times \frac{44.01}{32.00}$$

$$\text{Density of CO}_2 = 1.965 \text{ g/}\ell$$

◆ ◆ ◆ ◆ ◆

EXAMPLE 2   Calculate the density of ethane, C_2H_6, at a pressure of 183,400 Pa and at a temperature of 25°C.

The density of C_2H_6 under these conditions is the mass of 1 ℓ of C_2H_6. The moles of C_2H_6 in exactly 1 ℓ can be determined using the Ideal Gas Law Equation and then can be converted to the mass of C_2H_6.

$$n = \frac{PV}{RT}$$

$$n = \frac{(183{,}400 \text{ Pa})(1 \ell)}{\left(8314 \frac{\ell \text{ Pa}}{\text{mol K}}\right)(298 \text{ K})} = 0.0740 \text{ mol}$$

$$\text{g of } C_2H_6 = 0.0740 \text{ mol } C_2H_6 \times \frac{30.1 \text{ g } C_2H_6}{1 \text{ mol } C_2H_6} = 2.23 \text{ g } C_2H_6$$

$$\text{Density} = \frac{2.23 \text{ g}}{1.00 \ell} = 2.23 \text{ g}/\ell$$

Note that since the pressure of C_2H_6 was expressed in pascals, the value of R with pascals in the units was used.

◆ ◆ ◆ ◆ ◆

10.12   Determination of Molecular Weights of Gases or Volatile Compounds

The molecular weight of a gas can be found readily by determining experimentally the mass of a known volume of the gas under known conditions of temperature and pressure. The number of moles of gas is determined by using the Ideal Gas Law Equation and the molecular weight determined from the number of moles of gas and its mass (Section 2.5).

EXAMPLE 1   Cyclopropane is a commonly used anesthetic. If a 2.00-ℓ flask contains 3.11 g of cyclopropane gas at 684 torr and 23°C, what is the molecular weight of cyclopropane?

As we have noted previously, in order to use the Ideal Gas Law Equation the temperature must be expressed on the Kelvin scale, and the P and V units must correspond to the units of R. Using R as 0.08205 ℓ atm/mol K requires that P be expressed in atmospheres.

$$P = 684 \text{ torr} \times \frac{1 \text{ atm}}{760 \text{ torr}} = 0.900 \text{ atm}$$

$$T = 23°C + 273.15 = 296 \text{ K}$$

Now calculate the number of moles of cyclopropane using the Ideal Gas Law Equation.

$$n = \frac{PV}{RT} = \frac{(0.900 \text{ atm})(2.00 \ell)}{\left(0.08205 \frac{\ell \text{ atm}}{\text{mol K}}\right)(296 \text{ K})} = 0.0741 \text{ mol}$$

Use the mass of cyclopropane and the number of moles to get the molecular weight.

$$\text{G-mol. wt} = \frac{3.11 \text{ g}}{0.0741 \text{ mol}} = 42.0 \text{ g/mol}$$

Thus the molecular weight is 42.0. (Cyclopropane is C_3H_6.)

◆ ◆ ◆ ◆ ◆

EXAMPLE 2   A sample of phosphorus vapor weighing 3.243×10^{-2} g at 550°C exerted a pressure of 31,890 Pa in a 56.0-ml bulb. What is the molecular weight and molecular formula of phosphorus vapor?

Since the pressure is given in pascals we will take R as 8314 l Pa/mol K. This will require that the units of volume be in liters and that the temperature be expressed on the Kelvin scale.

$$T = 550 + 273.15 = 823 \text{ K}$$

$$V = 56.0 \text{ ml} \times \frac{1 \text{ } l}{1000 \text{ ml}} = 0.0560 \text{ } l$$

$$n = \frac{PV}{RT} = \frac{(31,890 \text{ Pa})(0.0560 \text{ } l)}{(8314 \text{ } l \text{ Pa/mol K})(823 \text{ K})}$$

$$n = 2.61 \times 10^{-4} \text{ mol}$$

Using the mass of phosphorus vapor,

$$\text{G-mol. wt} = \frac{3.243 \times 10^{-2} \text{ g}}{2.61 \times 10^{-4} \text{ mol}} = 124 \text{ g/mol}$$

Hence the molecular weight is 124.

Since the molecular weight of phosphorus vapor is 124 and the atomic weight of a single phosphorus atom is 31, there must be 124/31 or 4 phosphorus atoms in a molecule of phosphorus vapor. The molecular formula is therefore P_4. The P_4 molecule is a tetrahedron (Fig. 10-11).

◆ ◆ ◆ ◆ ◆

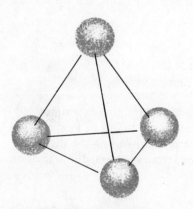

Figure 10-11 The molecular structure of gaseous phosphorus, P_4.

10.13   The Pressure of a Mixture of Gases; Dalton's Law

Suppose, for example, that we have gas A at a pressure of 100 torr in a 1-l container, and gas B also at a pressure of 100 torr in a second 1-l container. After transferring gas B to the 1-l container containing gas A, the temperature being held constant, the total pressure of the mixture is found to be 200 torr. It is logical to assume that each gas is exerting the same pressure as before. In the absence of chemical interaction between the components of a mixture of gases, the individual gases do not affect the pressures of one another. The pressure exerted by each gas in a mixture is called the **partial pressure** of that gas, and the total pressure of the mixture of gases is the sum of the partial pressures of all the gases present in the mixture. This law is known as **Dalton's Law of Partial Pressures** and may be stated as follows: **The total pressure of a mixture of ideal gases is equal to the sum of the partial pressures of the component gases.**

$$P_T = P_A + P_B + P_C + \cdots$$

In the equation, P_T is the total pressure of a mixture of gases. P_A is the pressure of gas A; P_B, the pressure of gas B; etc.

EXAMPLE 1   **What is the total pressure in atmospheres in a 10.0-ℓ vessel containing 2.50×10^{-3} mol of H_2, 1.00×10^{-3} mol of He, and 3.00×10^{-4} mol of Ne at 35°C?**

The pressure of each of these gases individually in a 10.0-ℓ container can be determined from the Ideal Gas Law Equation.

$$P_{H_2} = \frac{nRT}{V} = \frac{(2.50 \times 10^{-3}\,\text{mol})\left(0.08205\,\frac{\ell\,\text{atm}}{\text{mol K}}\right)(308\,\text{K})}{(10.0\,\ell)} = 6.32 \times 10^{-3}\,\text{atm}$$

$$P_{He} = \frac{(1.00 \times 10^{-3}\,\text{mol})\left(0.08205\,\frac{\ell\,\text{atm}}{\text{mol K}}\right)(308\,\text{K})}{10.0\,\ell} = 2.53 \times 10^{-3}\,\text{atm}$$

$$P_{Ne} = \frac{(3.00 \times 10^{-4}\,\text{mol})\left(0.08205\,\frac{\ell\,\text{atm}}{\text{mol K}}\right)(308\,\text{K})}{10.0\,\ell} = 7.58 \times 10^{-4}\,\text{atm}$$

The total pressure in the 10.0-ℓ vessel is the sum of these individual pressures since the gases do not react with each other.

$$P_T = P_{H_2} + P_{He} + P_{Ne}$$
$$P_T = (0.00632 + 0.00253 + 0.00076)\,\text{atm} = 9.61 \times 10^{-3}\,\text{atm}$$

◆ ◆ ◆ ◆ ◆

A convenient method of measuring the pressure of a given volume of a gas or a mixture of gases that do not react with water is to collect the gas over water and make its pressure equal to the existing air pressure (which can be measured conveniently by a barometer). This is easily accomplished by adjusting the water level (Fig. 10-12) so that it is the same both inside and outside the container. The gas is then at the existing atmospheric pressure, which can be read on a laboratory barometer.

There is, however, another factor to consider when determining pressure by this method. Above a sample of water there is always a little gaseous water (water vapor). The pressure of the water vapor above a sample of water in a closed container like that in Fig. 10-12 depends upon the temperature, as shown in Fig. 10-13 and Table 10-1.

When a gas is collected over water, it will soon become saturated with water vapor. The total pressure of the mixture then will be equal to the sum of the partial pressure of the gas and that of the water vapor, according to Dalton's Law.

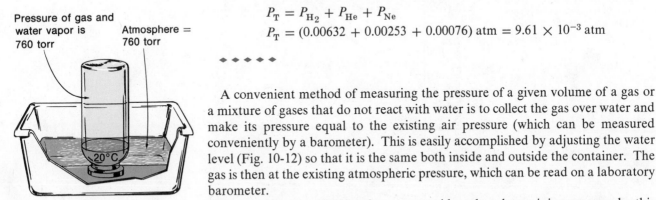

Pressure of gas and water vapor is 760 torr

Atmosphere = 760 torr

20°C

Figure 10-12 Method by which the pressure on a confined mixture of gases can be made equal to the atmospheric pressure. The level of water inside and outside the vessel is made the same. In the diagram, a typical atmospheric pressure of 760 torr is indicated. The actual atmospheric pressure depends upon temperature and altitude.

TABLE 10-1   Vapor Pressure of Ice and Water at Various Temperatures

Temperature, °C	Pressure, torr	Temperature, °C	Pressure, torr	Temperature, °C	Pressure, torr
−10	1.95	18	15.5	80	355.1
−5	3.0	19	16.5	90	525.8
−2	3.9	20	17.5	95	633.9
−1	4.2	21	18.7	96	657.6
0	4.6	22	19.8	97	682.1
1	4.9	23	21.1	98	707.3
2	5.3	24	22.4	99	733.2
3	5.7	25	23.8	99.1	735.9
4	6.1	26	25.2	99.2	738.5
5	6.5	27	26.7	99.3	741.2
6	7.0	28	28.3	99.4	743.9
7	7.5	29	30.0	99.5	746.5
8	8.0	30	31.8	99.6	749.2
9	8.6	31	33.7	99.7	751.9
10	9.2	32	35.7	99.8	754.6
11	9.8	33	37.7	99.9	757.3
12	10.5	34	39.9	100.0	760.0
13	11.2	35	42.2	100.1	762.7
14	12.0	40	55.3	100.2	765.4
15	12.8	50	92.5	100.3	768.2
16	13.6	60	149.4	100.5	773.7
17	14.5	70	233.7	101.0	787.6

Figure 10-13 A graph illustrating change of water vapor pressure with change in temperature.

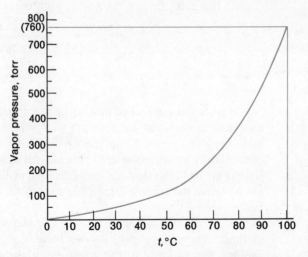

The pressure of the pure gas is therefore equal to the total pressure minus the water vapor pressure.

EXAMPLE 2   Suppose that 0.200 ℓ of argon was collected over water at a temperature of 26°C and a pressure of 750 torr in a system like that shown in Fig. 10-12. What is the partial pressure of argon?

According to Dalton's Law of Partial Pressures, the total pressure in the bottle (750 torr) is the sum of the pressure of argon and the pressure of gaseous water.

$$P_T = P_{Ar} + P_{H_2O}$$

Rearranging this equation to solve for the pressure of argon gives

$$P_{Ar} = P_T - P_{H_2O}$$

The pressure of water vapor above a sample of liquid water at 26° is given as 25.2 torr in Table 10-1. So,

$$P_{Ar} = 750 \text{ torr} - 25.2 \text{ torr}$$
$$P_{Ar} = 725 \text{ torr}$$

◆ ◆ ◆ ◆ ◆

EXAMPLE 3   **A 1.34-g sample of calcium carbide, CaC_2, was allowed to react with water. The acetylene produced by the reaction**

$$CaC_2(s) + 2H_2O(l) \longrightarrow C_2H_2(g) + Ca(OH)_2(s)$$

was collected over water in a system like that shown in Fig. 10-12. If 471 ml of acetylene gas was collected at a temperature of 23°C and a pressure of 743 torr, what was the per cent yield of the reaction?

To determine per cent yield (Section 3.5), we need the actual yield and the theoretical yield.

Theoretical yield.

$$1.34 \text{ g CaC}_2 \times \frac{1 \text{ mol CaC}_2}{64.1 \text{ g CaC}_2} = 2.09 \times 10^{-2} \text{ mol CaC}_2$$

$$2.09 \times 10^{-2} \text{ mol CaC}_2 \times \frac{1 \text{ mol C}_2\text{H}_2}{1 \text{ mol CaC}_2} = 2.09 \times 10^{-2} \text{ mol C}_2\text{H}_2$$

Actual Yield. To determine the actual yield, we need to get the moles of C_2H_2 produced. This is available from the Ideal Gas Law Equation, $PV = nRT$, and the data in the problem. However, the pressure in the equation must be the pressure of C_2H_2, not the pressure of a mixture of $H_2O(g)$ and $C_2H_2(g)$. So we use Dalton's Law of Partial Pressures to get the pressure of C_2H_2. The vapor pressure of water at 23°C is 21.1 torr (Table 10-1). So,

$$P_{C_2H_2} = P_T - P_{H_2O}$$
$$P_{C_2H_2} = 743 \text{ torr} - 21.1 \text{ torr} = 722 \text{ torr}$$

Using the Ideal Gas Law Equation with P in atmospheres, T on the Kelvin scale, and V in liters, gives n, the moles of C_2H_2 produced:

$$n_{C_2H_2} = \frac{PV}{RT} = \frac{(0.950 \text{ atm})(0.471 \text{ l})}{\left(0.08205 \dfrac{\text{l atm}}{\text{mol K}}\right)(296 \text{ K})} = 1.84 \times 10^{-2} \text{ mol}$$

Now calculate the per cent yield.

$$\text{Per cent yield} = \frac{\text{actual yield}}{\text{theoretical yield}} \times 100$$

$$\text{Per cent yield} = \frac{1.84 \times 10^{-2}\ \text{mol}}{2.09 \times 10^{-2}\ \text{mol}} \times 100 = 88.0\%$$

◆ ◆ ◆ ◆ ◆

10.14   Diffusion of Gases; Graham's Law

When a sample of gas is set free in one part of a closed container, it very quickly diffuses throughout the container (see Fig. 10-14). If a mixture of gases is placed in a container the walls of which are porous to gases, diffusion of the gases through the porous walls will take place. The lighter gases will diffuse through the small openings of the porous walls more rapidly than the heavier ones. In 1832, Thomas Graham studied the rates of diffusion of different gases and showed that **the rates of diffusion of gases are inversely proportional to the square roots of their densities (or molecular weights).**

$$\frac{\text{Rate of diffusion of gas A}}{\text{rate of diffusion of gas B}} = \frac{\sqrt{\text{density B}}}{\sqrt{\text{density A}}} = \frac{\sqrt{\text{mol. wt B}}}{\sqrt{\text{mol. wt A}}}$$

EXAMPLE **Calculate the ratio of the rate of diffusion of hydrogen to the rate of diffusion of oxygen.**

Using densities. The density of hydrogen is 0.08987 g/ℓ; that of oxygen is 1.429 g/ℓ.

$$\frac{\text{Rate of diffusion of hydrogen}}{\text{rate of diffusion of oxygen}} = \frac{\sqrt{1.429\ \text{g/liter}}}{\sqrt{0.08987\ \text{g/liter}}} = \frac{1.20}{0.30} = \frac{4}{1}$$

Using molecular weights.

$$\frac{\text{Rate of diffusion of hydrogen}}{\text{rate of diffusion of oxygen}} = \frac{\sqrt{32}}{\sqrt{2}} = \frac{\sqrt{2}\sqrt{16}}{\sqrt{2}} = \frac{\sqrt{16}}{1} = \frac{4}{1}$$

This means that hydrogen diffuses four times as rapidly as oxygen. The rates of diffusion must depend upon the speeds of the molecules; the rate of diffusion is greater for molecules of smaller mass and higher speeds than for molecules of larger mass and lower speeds. Also note that the density of oxygen is almost exactly 16 times that of hydrogen.

◆ ◆ ◆ ◆ ◆

Practical application of differences in diffusion rates of gaseous substances is made in the separation of light isotopes from heavy ones. Fractional diffusion through a porous barrier from a region of higher pressure to one of lower pressure is used for the large-scale separation of gaseous $^{235}_{92}UF_6$ from $^{238}_{92}UF_6$ at the Oak Ridge atomic energy installation in Tennessee. It is said that for separation to be complete, a given volume of UF_6 must be diffused some 2 million times.

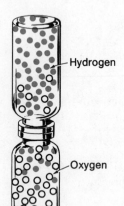

Figure 10-14 Despite the difference in densities of these two gases, their molecules intermingle; that is, the gases mix by diffusing together. The rapid motion of the molecules and the relatively large spaces between them explain why diffusion occurs. Note that H_2, the lighter of the two gases, diffuses faster than O_2.

Hydrogen

Oxygen

The Molecular Behavior of Gases

10.15   The Kinetic-Molecular Theory

The idea that the properties of gases such as compressibility, expansibility, and diffusibility could be accounted for by considering the gas molecules to be in continuous motion occurred to several people in the past (Bernoulli in 1738, Poule in 1851, and Kronig in 1856). During the latter half of the nineteenth century, Clausius, Maxwell, Boltzmann, and others developed this hypothesis into the detailed **kinetic-molecular theory** of gases.

The assumptions about the nature of gas molecules used to develop the theory are summarized below. These are the properties attributed to the molecules of an **ideal gas.**

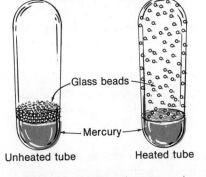

Pyrex tube
(partially evacuated)

Glass beads

Mercury

Unheated tube   Heated tube

Figure 10-15 When the mercury in the tube is heated sufficiently to provide an appreciable vapor pressure, the glass beads move toward the top of the tube in violent random motion and their motion provides visual evidence of the upward motion of gaseous mercury molecules.

1. Gases are composed of separate particles called molecules. The volume actually occupied by the individual molecules of a gas under ordinary conditions is insignificant compared with the total volume of the gas. Thus the molecules of a gas are relatively far apart and they have no significant attraction for one another except when near the temperature at which they become liquids.
2. The molecules of a gas are in continuous, completely random motion in all directions (Fig. 10-15) with varying speeds. They move in straight lines between collisions and behave as perfectly elastic bodies when they collide with the walls of the container or with each other.
3. Although the molecules in a sample of a gas move with different speeds, we can calculate their *average* speed, *u*, and their *average* kinetic energy. The kinetic-molecular theory postulates that the average kinetic energy of the molecules (energy due to their motion) of different gases is the same at the same temperature, regardless of differences in mass. The kinetic energy of gaseous molecules increases with a rise in temperature and decreases as the temperature falls. The average kinetic energy in joules of any moving body is equal to $\frac{1}{2}mu^2$, where *m* is its mass in kilograms and *u* is its average speed in meters per second. It follows that molecules of small mass, like hydrogen, must move at higher speeds than molecules of larger mass, like oxygen, since both have the same average kinetic energy at the same temperature.

As seen in the next section, the kinetic-molecular theory explains the gas laws in a qualitative way. The true test of the theory, however, is its ability to describe the behavior of a gas quantitatively. As we shall see in Section 10.18, the Ideal Gas Law Equation can be derived from the assumptions of the theory, and this has led chemists to believe that the assumptions of the theory do represent the true properties of gas molecules.

10.16   Relation of the Behavior of Gases to the Kinetic-Molecular Theory

The gas laws may be readily explained in light of the kinetic theory.

▶**1. BOYLE'S LAW.** The pressure exerted by a gas upon the walls of its container is caused by the bombardment of the walls by rapidly moving molecules of the gas and varies directly with the number of molecules, confined within a given volume, hitting the walls per unit time. Reducing the volume of a given mass of gas to one-half will double the number of molecules per unit volume. It follows that the number of impacts per unit time upon the same area of wall surface will also be doubled. The doubling of the number of impacts per unit area results in twice as much pressure. These results are in accordance with Boyle's law relating volume and pressure.

▶ **2. CHARLES' LAW.** It is a common observation that a rise in temperature increases the pressure of a gas at constant volume. The increase in pressure of a gas with increase in temperature reflects the increase in average kinetic energy of the molecules as the temperatue is raised. An increase in the average speed of the molecules results in more frequent and harder impacts upon the walls of the container, i.e., greater pressure.

If the pressure remains constant with increasing temperature, the volume must increase so that each molecule on the average travels farther before hitting the wall. The smaller number of molecules striking the wall at a given time exactly offsets the greater force with which each molecule hits, making it possible for the pressure to remain constant. This may be visualized by picturing a cylinder fitted with a piston. As the temperature increases, the molecules hit the piston harder. The piston must move out, thereby increasing the volume, if the pressure is to remain the same as before.

A study of the effect of temperature upon kinetic-molecular motion leads to the belief that heat is a manifestation of the total kinetic energy that molecules possess by virtue of their random motion. The temperature of a gas is a measure of the average kinetic energy of the molecules.

▶ **3. DALTON'S LAW.** In a mixture of gases the molecules of one component will bombard the walls of the container just as frequently in the presence of other kinds of molecules as in their absence. Thus the total pressure of a mixture of gases will be the sum of the partial pressures of the individual gases.

▶ **4. GRAHAM'S LAW.** The fact that the molecules of a gas are in rapid motion and that the free space between the molecules is very great explains the phenomenon of diffusion. At the same temperature, molecules of different gases have the same average kinetic energy. This means that molecules of small mass move with higher speeds than those of large mass. Consequently, diffusion rates of gases are related inversely to their molecular weights and densities. This can be shown mathematically by the following derivation.

For two different gases, at the same temperature and pressure,

$$\text{Average kinetic energy for first gas} = \tfrac{1}{2}m_1u_1^2$$

$$\text{Average kinetic energy for second gas} = \tfrac{1}{2}m_2u_2^2$$

where m_1 and m_2 are the masses of individual molecules of the two gases, and u_1 and u_2 are their *average* speeds.

The kinetic energies of the two gases are equal according to the kinetic-molecular theory. Hence

$$\tfrac{1}{2}m_1u_1^2 = \tfrac{1}{2}m_2u_2^2$$

Dividing both sides of the equation by $\tfrac{1}{2}$, we obtain

$$m_1u_1^2 = m_2u_2^2$$

On rearranging, we have

$$\frac{u_1^2}{u_2^2} = \frac{m_2}{m_1}$$

Extracting the square roots gives

$$\frac{u_1}{u_2} = \sqrt{\frac{m_2}{m_1}}$$

We may assume that the rate at which a gas will diffuse, R, is proportional to the average speed of the gas molecules. The molecular weights, M_1 and M_2, of the two gases are proportional to the masses of the individual molecules. Thus

$$\frac{R_1}{R_2} = \sqrt{\frac{M_2}{M_1}}$$

This equation states that the ratio of the rates of diffusion of two gases, held at the same temperature and pressure, equals the inverse ratio of the square roots of the molecular weights of the two gases.

As shown in Section 10.11, the ratio of densities (d_1 and d_2) of two gases is the same as the ratio of their molecular weights. Since $d_2/d_1 = M_2/M_1$,

$$\frac{R_1}{R_2} = \sqrt{\frac{d_2}{d_1}}$$

Therefore Graham's Law can equally well be expressed either in terms of the square roots of the molecular weights or the square roots of the densities of the gases.

Hydrogen is the lightest of all the gases and therefore diffuses the most rapidly. The average speed of its molecules at room temperature is about one mile per second and that of oxygen molecules is about one-fourth mile per second. However, diffusion rates are much lower than would be expected from such fast-moving molecules due to collisions between molecules. There are about 11 billion collisions per molecule per second in hydrogen gas at standard conditions. Hydrogen molecules travel about 1.7×10^{-5} cm between collisions and, on the average, collide about 60,000 times in traveling one linear centimeter.

10.17   The Distribution of Molecular Velocities

On the basis of the kinetic-molecular theory of gases, we must assume that the individual molecules of a particular gas travel at different speeds. Because of the enormous number of collisions, the speeds of the molecules will vary from practically zero to nearly the speed of light. Because of the collisions of molecules and the consequent exchanges of energy, the speed of a given molecule changes with each collision. However, because a large number of molecules is involved, the

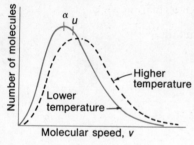

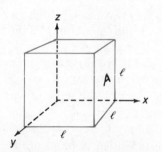

Figure 10-16 Distributions of molecular speed versus number of molecules of a gas at different temperatures.

distribution of molecular speeds of the total number of molecules is constant. The nature of the distribution of molecular speeds is described by the **Maxwell-Boltzmann Distribution Law.**

The distribution of molecular speeds is shown in Fig. 10-16. The vertical axis represents the number of molecules and the horizontal axis represents molecular speed, v. Alpha (α) is the most probable speed and u the average speed of molecules. The graph indicates that very few molecules move at very low or very high speeds. The number of molecules with intermediate speeds increases rapidly up to a maximum and then drops off rapidly.

At the higher temperature the curve is flattened and shifted to higher speeds. Thus at a higher temperature more molecules have higher speeds and fewer molecules have the lower speeds. This is in accord with expectations based upon kinetic-molecular theory.

10.18   Ideal Gas Law Equation Derived from the Kinetic-Molecular Theory

The mathematical expression of the Ideal Gas Law Equation, $PV = nRT$, may be derived from the kinetic-molecular theory of gases as follows: Consider **N** molecules, each having a mass m, confined within a cubical container as shown in Fig. 10-17, an edge of which has a length of l cm. Although the molecules are moving in all possible directions, we may assume that one-third of the molecules (**N/3**) are moving in the direction of the X axis, one-third in the Y direction, and one-third in the Z direction. This assumption is valid because the motion of the molecules is entirely random and no particular direction is preferred.

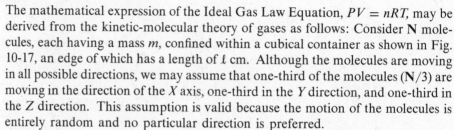

Consider the collisions of molecules with wall A in Fig. 10-17. A molecule, on the average, travels $2l$ cm between two consecutive collisions with A as it moves back and forth across the container. If the average speed of the molecule is u cm per second, it will collide $u/2l$ times per second with wall A. Since the collisions are perfectly elastic, the molecule will rebound with a speed of $-u$, having lost no kinetic energy as a result of the collision. Because momentum is defined as the product of mass and speed, the average momentum before a collision is mu and after a collision is $-mu$. Thus the average change in momentum per molecule per collision is $2mu$. There are $u/2l$ collisions per second, making the average change in momentum per molecule per second $2mu \times u/2l = mu^2/l$. The total change in momentum per second for the **N/3** molecules that can collide with wall A is **N**/3 $\times mu^2/l$; this represents the average force on A, because force may be defined as the time rate of change of momentum. Because pressure is defined as force per unit area, and the area of A is l^2, the pressure P on A is

$$P = \frac{\text{force}}{\text{area}} = \frac{1}{l^2} \times \frac{\mathbf{N}}{3} \times \frac{mu^2}{l} = \frac{\mathbf{N}}{3} \times \frac{mu^2}{l^3}$$

However, l^3 is equal to the volume of the cubical container, V; thus we have

$$P = \frac{\mathbf{N}}{3} \times \frac{mu^2}{V}$$

or

$$PV = \frac{1}{3}\mathbf{N}mu^2$$

Figure 10-17 Drawing illustrating the container used in the derivation of the Ideal Gas Law Equation from the kinetic-molecular theory.

This is the fundamental equation of the kinetic-molecular theory of gases. It must be remembered that the equation is only as exact as the gas laws themselves. It holds for real gases, however, in very good approximation at ordinary pressures.

The equation above may be written $PV = (2N/3)(mu^2/2)$. The kinetic energy, KE, of a body in motion is equal to one-half the product of its mass times the square of its velocity. Thus $KE = mu^2/2$, and

$$PV = \left(\frac{2N}{3}\right)\left(\frac{mu^2}{2}\right) = \frac{2}{3}N(KE)$$

The average kinetic energy of the molecules, KE, is directly proportional to the Kelvin temperature, T.

$$KE = kT$$

The number of molecules, **N,** is proportional to the number of moles of molecules, n.

$$N = k'n$$

Substituting the expressions for KE and N in the equation for PV gives

$$PV = \frac{2}{3}(k'n)(kT) = n(\frac{2}{3}k'k)T$$

The expression $\frac{2}{3}k'k$ is a constant (being made up entirely of constants), which we designate R and refer to as the **universal gas constant.**

Therefore

$$PV = nRT$$

Thus from the assumptions of the kinetic-molecular theory (Section 10.15) we can derive the Ideal Gas Law Equation. Since this equation describes the behavior of gases very well, the assumptions used in its derivation are thereby given very strong support.

10.19   Deviations from the Gas Laws

When gases at ordinary temperatures and pressures are compressed, the volume is reduced by crowding the molecules closer together. This reduction in volume is really a reduction in the amount of empty space between the molecules. At high pressures, the molecules are crowded so closely together that the volume they occupy is a relatively large fraction of the total volume, including the empty space between molecules. Because the volume of the molecules themselves cannot be compressed, only a fraction of the entire volume is affected by a further increase in pressure. Thus at very high pressures the whole volume is not inversely proportional to the pressure as predicted by Boyle's Law. It might be noted, nevertheless, that even at moderately high pressures, most of the volume is not occupied by molecules. The molecules in a real gas at relatively low pressures have practically no attraction for one another because they are far apart. Thus they behave almost like ideal gases. However, as the molecules are crowded closer together at high pressures, the force of attraction between the molecules increases. This attraction has the same effect as an increase in external pressure. Consequently, when external pressure is applied to a volume of gas, especially at low temperatures, there is a slightly greater decrease in volume than would be achieved by the pressure alone. This slightly greater decrease in volume caused by intermolecular attraction is more pronounced at low temperatures because the molecules move more slowly and have less tendency to fly apart after collisions with one another.

In 1879, van der Waals expressed the deviations of real gases from the ideal gas laws quantitatively with the following equation, which bears his name:

$$(P + n^2a/V^2)(V - nb) = nRT$$

The constant a represents the attraction between molecules, and van der Waals assumed that this force varies inversely as the square of the total volume of the gas. (See Section 11.4 for a discussion of the nature of the van der Waals force.) Since this force augments the pressure and thus tends to make the volume smaller, it is added to the term P.

The term b in the equation represents the volume of the molecules themselves and is subtracted from V, the total volume of the gas. When V is large, both nb and n^2a/V^2 become negligible, and the van der Waals equation reduces to the simple gas equation, $PV = nRT$.

At low pressures the correction for intermolecular attraction, a, is more important than the correction for molecular volume, b. At high pressures and small volumes the correction for the volume of the molecules becomes important because the molecules themselves are relatively incompressible and constitute an appreciable fraction of the total volume. At some intermediate pressure the two corrections cancel one another, and the gas appears to follow the relation $PV = nRT$ over a small range of pressures.

Strictly speaking, then, the gas laws apply exactly only to gases whose molecules do not attract each other and which occupy no appreciable part of the whole volume. Because there are no gases that have these properties, we can speak of such hypothetical gases only as **ideal**, or **perfect, gases**. Under ordinary conditions, however, the deviations from the gas laws are so slight that they may be neglected.

Questions

1. Using Boyle's Law, set up a mathematical equation relating the initial volume, V_1, and final volume, V_2, of a given quantity of an ideal gas at constant temperature as the pressure changes from the initial pressure, P_1, to the final pressure, P_2.

2. Using Charles' Law, set up a mathematical equation relating the initial volume, V_1, and temperature, T_1, of a given quantity of an ideal gas at constant pressure to some other volume, V_2, and temperature, T_2.

3. State the common dimensional units used to express pressure.

4. How does the definition of 1 atmosphere differ from the definition of 1 pascal?

5. One liter of methane, CH_4, at STP contains more hydrogen than one liter of pure H_2 at STP. Verify or disprove this statement by calculation.

6. Show that the ratio of the rate of diffusion of gas 1 to the rate of diffusion of gas 2, R_1/R_2, is the same at $0°C$ and $100°C$.

7. For one mole of H_2 showing ideal gas behavior, draw labeled graphs of
   (a) the variation of P with V at $T = 273$ K.
   (b) the variation of V with T at $P = 1.00$ atm.
   (c) the variation of P with T at $V = 22.4$ ℓ.
   (d) the variation of the average velocity of the gas molecules with T.

8. State the postulates of the kinetic-molecular theory.
9. Show how Boyle's Law, Charles' Law, and Dalton's Law follow from the assumptions of the kinetic-molecular theory.
10. Using the kinetic-molecular theory, explain why gases exhibit the following properties: compressibility, expansibility, diffusibility, homogeneity.
11. Describe what happens to the average kinetic energy of ideal gas molecules when the conditions are changed as follows:
    (a) The pressure of the gas is increased by decreasing the volume at constant temperature.
    (b) The pressure of the gas is increased by increasing the temperature at constant volume.
    (c) The average velocity of the molecules is increased by a factor of 2.
12. Can the speed of a given molecule in a gas double at constant temperature? Explain your answer.
13. What is the ratio of the kinetic energy of a helium atom and a hydrogen molecule in a mixture of the two gases?
14. Describe the factors responsible for the deviation of the behavior of real gases from that of an ideal gas.
15. For which of the following sets of conditions will a real gas behave most like an ideal gas and for which conditions would a real gas be expected to deviate from ideal behavior? Explain your answer for each.
    (a) High pressure, small volume
    (b) High temperature, low pressure
    (c) Low temperature, high pressure
16. How could you show by experiment that the molecular formula of benzene is C_6H_6 and not CH?
17. State Gay-Lussac's Law. Does this law pertain to substances in the liquid and solid states as well as in the gaeous state? Explain.
18. Apply Gay-Lussac's law to the following reactions, assuming that conditions are such that all reactants and products are gases at the same temperature and pressure.
    (a) $C_2H_4 + H_2 \longrightarrow C_2H_6$
    (b) $2CO + O_2 \longrightarrow 2CO_2$
    (c) $N_2 + 3H_2 \longrightarrow 2NH_3$
19. A certain balanced chemical equation describes a reaction of gaseous reactants giving gaseous products. What information given by the equation can be used in determining volumetric relations between the reactants and products?
20. Graphs showing the behavior of several different gases are given below. Which of these gases exhibit behavior significantly different than that expected for ideal gases?

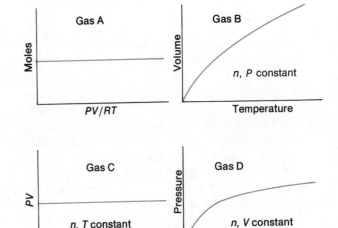

Problems

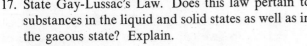

[S] 1. Weather reports in the United States often give barometric pressures in inches of mercury rather than mm Hg or torr; 29.92 in. = 760 mm. Convert a pressure of 29.20 inches of mercury to torr, atm, and Pa.
Ans. 742 torr; 0.976 atm; 9.90 × 10⁴ Pa

2. The volume of a sample of ethane is 3.24 ℓ at 477 torr and 27°C.
    (a) What volume will the ethane occupy at 87°C and 477 torr? *Ans. 3.89 ℓ*
    [S] (b) What volume will it occupy at 27°C if the pressure is 831 torr? *Ans. 1.86 ℓ*

(c) At what temperature will the volume be 5.00 l if the pressure is 477 torr?

Ans. 463 K or 190°C

(d) At what pressure will the volume be 5.00 l at 27°C?   *Ans. 309 torr*

3. The volume of a sample of carbon monoxide is 405 ml at 10.0 atm and 467 K.

(a) What volume will it occupy at 4.29 atm and 467 K?   *Ans. 946 ml*

s(b) What volume will it occupy at 1873 torr and 467 K?   *Ans. 1.64 × 10³ ml*

(c) What volume will it occupy at 238,000 Pa and 467 K?   *Ans. 1.72 × 10³ ml*

s4. A 2.50-l volume of hydrogen measured at the normal boiling point of nitrogen, −210.0°C, is warmed to the normal boiling point of water, 100°C. Calculate the new volume of the gas, assuming ideal behavior and no change in pressure.

Ans. 14.8 l

5. A sample of oxygen occupies 38.9 l at STP.

s(a) What volume will it occupy at 456 K and 1 atm?   *Ans. 65.0 l*

s(b) What volume will it occupy at 0°C and 917 torr?   *Ans. 32.2 l*

(c) What volume will it occupy at 55°C and 725 torr?   *Ans. 49.0 l*

s(d) The density of oxygen is 1.429 g/l at STP. What is the mass of oxygen in the sample?

Ans. 55.6 g

s6. A sample of a gas occupies 60.0 ml at −10°C and 720 mm. What pressure will the gas exert in a 101-ml sealed bulb at 25°C?   *Ans. 485 mm*

s7. What volume of O_2 at STP is required to oxidize 14.0 l of CO at STP to CO_2? What volume of CO_2 is produced at STP?   *Ans. 7.00 l O_2; 14.0 l CO_2*

8. Calculate the volume of oxygen required to burn 7.00 l of propane gas (C_3H_8) to produce carbon dioxide and water, if the volumes of both the propane and oxygen are measured under the same conditions.

Ans. 35.0 l

s9. A 2.50-l sample of a colorless gas at STP decomposed to give 2.50 l of N_2 and 1.25 l of O_2 at STP. What is the colorless gas?   *Ans. N_2O*

s10. How many moles of hydrogen sulfide, H_2S, are contained in a 327.3-ml bulb at 48.1°C if the pres-

sure is 1.493 × 10⁵ Pa?   *Ans. 1.830 × 10⁻² mol*

s11. What is the volume of a bulb which contains 8.17 g of neon at 13°C with a pressure of 8.73 atm?

Ans. 1.09 l

12. What is the temperature of a 0.274-g sample of methane, CH_4, confined in a 300.0-ml sample bulb at a pressure of 198,700 Pa?

Ans. 419 K or 146°C

13. A gas occupies 275 ml at 0°C and 305 torr. What final temperature would be required to increase the pressure to 380 torr, if the volume is held constant?

Ans. 340 K or 67°C

s14. What volume of oxygen at 15°C and 745 torr is produced by the decomposition of 23.6 g of HgO to Hg and O_2?   *Ans. 1.31 l*

15. The density of a certain phosphorus fluoride gas is 3.93 g/l at STP.

s(a) Calculate the molecular weight of this phosphorus fluoride, given that the density of oxygen is 1.429 g/l at STP.   *Ans. 88.0*

s(b) Calculate the molecular weight of this phosphorus fluoride using the Ideal Gas Law Equation.

Ans. 88.0

s(c) What is the formula of this phosphorus fluoride?

Ans. PF_3

s16. Calculate the density of $C_6H_6(g)$ at 100°C and a pressure of 1.012 × 10⁴ Pa.   *Ans. 0.255 g/l*

17. At what temperature is the density of H_2 equal to 0.0800 g/l at 1.00 atm pressure?

Ans. 307 K or 34°C

18. A 5.73-l flask at 25°C contains 0.0388 mol of N_2, 0.147 mol of CO, and 0.0803 mol of H_2. What is the pressure in the flask in atm, torr, and Pa?

Ans. 1.14 atm; 863 torr; 1.15 × 10⁵ Pa

s19. What mass of $KClO_3$ must be decomposed to KCl and O_2 to give 638 ml of O_2 at a temperature of 18°C and a pressure of 752 torr?   *Ans. 2.16 g*

s20. Calculate the relative rates of diffusion of the gases CO and CO_2.

Ans. CO diffuses 1.25 times faster than CO_2

21. Show by calculation which of the following gases diffuses more slowly than oxygen: F_2, Ne, N_2O, C_2H_2, NO, Cl_2, H_2S.   *Ans. F_2, Cl_2, N_2O, H_2S*

22. Calculate the relative rate of diffusion of HF as compared to HCl, of SO_2 as compared to SO_3, and of O_2 as compared to O_3.   *Ans. 1.35; 1.12; 1.22*

23. The average velocity of H_2 molecules at 25°C is about 1.6 km/sec. What is the average speed of a CH_4 molecule at 25°C?   *Ans. 0.57 km/sec*

24. A sample of oxygen collected over water at a temperature of 29.0°C and a pressure of 764 torr has a volume of 0.560 ℓ. What volume would the dry oxygen have under the same conditions of temperature and pressure?   *Ans. 0.538 ℓ*

25. The volume of a sample of a gas collected over water at 32.0°C and 752 torr is 627 ml. What will the volume of the gas be when dried and measured at STP?   *Ans. 529 ml*

26. Calculate the density of the gas methane, CH_4, at STP.   *Ans. 0.7156 g/ℓ*

27. Methanol (sometimes called wood alcohol), CH_3OH, is produced industrially by the following reaction.

$$CO(g) + 2H_2(g) \xrightarrow[\text{300°C, 300 atm}]{\substack{\text{Copper} \\ \text{catalyst}}} CH_3OH(g)$$

Assuming that the gases behave as ideal gases, what is the ratio of the total volume of the reactants to the final volume?   *Ans. 3 to 1*

28. What volume of oxygen at 423.0 K and a pressure of 127,400 Pa will be produced by the decomposition of 129.7 g of BaO_2 to BaO and O_2?   *Ans. 10.57 ℓ*

29. What volume of SO_2 at 343°C and 1.21 atm is produced by the combustion of 1.83 kg of sulfur?   *Ans. 2.38×10^3 ℓ*

30. (a) What volume of hydrogen at 195 atm and 35°C is required to reduce 1.00 metric ton (1000 kg = 1 metric ton) of Fe_3O_4 to Fe (Section 9.18)?   *Ans. 2.24×10^3 ℓ*

    (b) If the reduction is run at 500°C and 1.0 atm, what volume of gaseous water is produced?   *Ans. 1.1×10^6 ℓ*

31. Assume that 1.00 lb of Dry Ice (solid CO_2) is placed in an evacuated 50.0-ℓ closed tank. What is the pressure in the tank in atmospheres at a temperature of 45°C after all the $CO_2(s)$ has been converted to gas?   *Ans. 5.38 atm*

32. A 265-ml sample of gaseous nitric oxide, NO, was collected over mercury at 19°C and 714.0 torr.

    (a) What is the volume of the nitric oxide at STP?   *Ans. 233 ml*

    (b) If the same sample of nitric oxide were collected over water at 19°C and 714.0 torr, what would the volume be?   *Ans. 271 ml*

33. (a) What mass of nitrosyl chloride, ClNO, occupies a volume of 0.250 ℓ at a temperature of 325 K and a pressure of 113,000 Pa?   *Ans. 0.684 g*

    (b) What is the density of ClNO under these conditions?   *Ans. 2.74 g/ℓ*

34. Calculate the densities of each of the following gases at STP and at 30.0° and 725 torr: F_2, N_2, CF_2Cl_2, SF_6.

Ans.	STP	30.0° and 725 torr
F_2	1.70 g/ℓ	1.46 g/ℓ
N_2	1.25 g/ℓ	1.07 g/ℓ
CF_2Cl_2	5.39 g/ℓ	4.64 g/ℓ
SF_6	6.52 g/ℓ	5.60 g/ℓ

35. (a) What is the total volume of $CO_2(g)$ and $H_2O(g)$ at 600°C and 735 torr produced by the combustion of 1.00 ℓ of $C_3H_8(g)$ measured at STP?   *Ans. 23.1 ℓ*

    (b) What is the partial pressure of CO_2 in the product gases?   *Ans. 315 torr*

36. Calculate the volume of the following quantities of gas at STP.

    (a) 6.72 g of C_2H_2 *Ans. 5.78 ℓ*
    (b) 0.588 g of NH_3 *Ans. 0.774 ℓ*
    (c) 1.47 kg of C_2H_6 *Ans. 1.10×10^3 ℓ*
    (d) 0.720 mol of BF_3 *Ans. 16.1 ℓ*
    (e) 13.5 mol of SO_3 *Ans. 303 ℓ*
    (f) 0.027 mol of PH_3 *Ans. 0.61 ℓ*

37. How many grams of gas are present in each of the following?

    (a) 0.100 ℓ of NO at 703 torr and 62°C.   *Ans. 0.101 g*

    (b) 8.75 ℓ of CH_4 at 278,300 Pa and 843 K.   *Ans. 5.57 g*

    (c) 73.3 ml of I_2 at 0.462 atm and 125°C.   *Ans. 0.263 g*

    (d) 4341 ℓ of BF_3 at 2.22 atm and 788 K.   *Ans. 10.1 kg*

    (e) 221 ml of Ne at 0.23 torr and −54°C.   *Ans. 7.5×10^{-5} g*

$\boxed{s}$ 38. (a) What is the concentration of the atmosphere in molecules per milliliter at STP?

Ans. 2.69 × 10¹⁹ molecules/ml

(b) At a height of 150 km (about 94 miles) the atmospheric pressure is about 3.0×10^{-6} torr with a temperature of 420 K. What is the concentration of the atmosphere in molecules per milliliter at 150 km?

Ans. 6.9 × 10¹⁰ molecules/ml

39. A mixture of 0.200 g of hydrogen, 1.00 g of nitrogen, and 0.820 g of argon is stored in a closed container at STP. What is the volume of the container assuming that the gases exhibit ideal behavior?

Ans. 3.50 ℓ

40. A gas of unknown identity diffuses at the rate of 83.3 ml per second in a diffusion apparatus in which a second gas, whose molecular weight is 44.0, diffuses at the rate of 102 ml per second. Calculate the molecular weight of the first gas.   *Ans. 66.0*

$\boxed{s}$ 41. (a) When two cotton plugs, one moistened with ammonia and the other with hydrochloric acid, are simultaneously inserted into opposite ends of a glass tube 87.0 cm long a white ring of NH_4Cl forms where gaseous NH_3 and gaseous HCl first come into contact $[NH_3(g) + HCl(g) \longrightarrow NH_4Cl(s)]$. At what distance from the ammonia-moistened plug does this occur?   *Ans. 51.7 cm*

(b) In an experiment, a student is trying to identify an amine that is one of the three possible compounds, CH_3NH_2, $(CH_3)_2NH$, or $(CH_3)_3N$. In this case, a white ring due to the amine hydrochloride forms at a distance of 41.2 cm from the amine-moistened plug. What is the true formula for the amine?   *Ans. (CH₃)₂NH*

42. What mass of Na_3P is necessary to produce 4.50 ℓ of $PH_3(g)$ at STP?

$$Na_3P + 3H_2O \longrightarrow PH_3 + 3NaOH$$

Ans. 20.1 g

43. What mass of H_2SO_4 is required to produce 4.76 ℓ of HCl(g) at 28°C and 101,000 Pa?

$$NaCl + H_2SO_4 \longrightarrow NaHSO_4 + HCl$$

Ans. 18.8 g

44. What volume of hydrogen at 680 torr and 43°C can be prepared by the reaction of 13.1 g of aluminum with sulfuric acid?

$$2Al + 3H_2SO_4 \longrightarrow Al_2(SO_4)_3 + 3H_2$$

Ans. 21.1 ℓ

45. Chlorine may be prepared by the reaction

$$2KMnO_4 + 16HCl \longrightarrow 2KCl + 2MnCl_2 + 5Cl_2 + 8H_2O$$

How many grams of $KMnO_4$ are required to produce 2.80 ℓ of chlorine at 100,800 Pa and 21°C?

Ans. 7.30 g

$\boxed{s}$ 46. What is the molecular weight of methylamine if 0.157 g of methylamine occupies 125 ml with a pressure of 99,500 Pa at 22°C?   *Ans. 31.0*

$\boxed{s}$ 47. The density of a gaseous compound containing 45.4% bromine and 54.5% fluorine is 6.13 g/ℓ at 75°C and 1.00 atm. What is the molecular formula of the compound?   *Ans. BrF₅*

48. Ethanol, C_2H_5OH, is produced industrially from ethylene, C_2H_4, by the following sequence of reactions:

$$3C_2H_4 + 2H_2SO_4 \longrightarrow C_2H_5HSO_4 + (C_2H_5)_2SO_4$$
$$C_2H_5HSO_4 + (C_2H_5)_2SO_4 + 3H_2O \longrightarrow 3C_2H_5OH + 2H_2SO_4$$

What volume of ethylene at STP is required to produce 1.000 metric ton (1000 kg) of ethanol if the overall yield of ethanol is 90.1%?   *Ans. 5.40 × 10⁵ ℓ*

$\boxed{s}$ 49. Thin films of amorphous silicon for electronic applications are prepared by decomposing silane gas, SiH_4, on a hot surface at low pressures.

$$SiH_4(g) \xrightarrow{\triangle} Si(s) + 2H_2(g)$$

What volume of SiH_4 at 130 Pa and 800 K is required to produce a 10.0 cm by 10.0 cm film which is 200 Å thick (1 Å = 10^{-8} cm)? The density of amorphous silicon is 1.9 g/cm³.   *Ans. 0.69 ℓ*

50. A 0.0124-g sample of a compound of carbon and hydrogen gives 0.0390 g of CO_2 and 0.0159 g of H_2O when burned in an oxygen atmosphere. If the volume of a 0.125-g gaseous sample of the compound at 735 torr and 45°C is 60.2 ml, what is the molecular formula of the compound?   *Ans. C₄H₈*

51. If the oxygen consumed by a resting human male

(Section 10.8) is used to produce energy by the oxidation of glucose,

$$C_6H_{12}O_6 + 6O_2 \longrightarrow 6CO_2 + 6H_2O$$

what is the mass of glucose required per hour for a resting 70-kg male? *Ans. 17 g*

52. The anesthetic, cyclopropane, is a gas containing only carbon and hydrogen. If 0.45 ℓ of cyclopropane at 120°C and 0.72 atm reacts with O_2 to give 1.35 ℓ of CO_2 and 1.35 ℓ of $H_2O(g)$ at the same temperature and pressure, determine the molecular formula of cyclopropane. *Ans.* C_3H_6

53. One molecule of hemoglobin will combine with four molecules of oxygen. If 1.0 g of hemoglobin combines with 1.53 ml of oxygen at body temperature (37°C) and a pressure of 743 torr, what is the molecular weight of hemoglobin? *Ans. 68,000*

54. In a laboratory determination, a 0.1009-g sample of a compound containing boron and chlorine gave 0.3544 g of silver chloride upon reaction with silver nitrate. A 0.06237-g sample of the compound exerted a pressure of 6520 Pa at 27°C in a volume of 147 ml. What is the molecular formula of the compound? *Ans.* B_2Cl_4

[S] 55. Ethanol, C_2H_5OH, is often produced by the fermentation of sugars. For example, the preparation of ethanol from the sugar, glucose, is represented by the equation

$$C_6H_{12}O_6(aq) \xrightarrow{\text{Yeast}} 2C_2H_5OH(aq) + 2CO_2(g).$$

What volume of CO_2 at STP is produced by the fermentation of 125 g of glucose if the fermentation has a yield of 97.5%? *Ans. 30.3 ℓ*

56. A sample prepared by pumping 10.0 ℓ of air at STP and 1.0 ℓ of CO at STP into a 2.0-ℓ closed flask was heated until all the CO had been converted to CO_2. Assuming that air is 20% oxygen and 80% nitrogen, calculate the partial pressures of N_2, O_2, and CO_2 in the flask at 0°C after the reaction has taken place. *Ans.* N_2, *4.0 atm;* O_2, *0.75 atm;* CO_2, *0.5 atm.*

[S] 57. One method of analysis of amino acids is the van Slyke method. The characteristic amino groups ($-NH_2$) in protein material are allowed to react with nitrous acid (HNO_2) to form N_2 gas. From the volume of the gas, the amount of amino acid can be determined. A 0.0604-g sample of a biological material containing glycine, $CH_2(NH_2)COOH$, was analyzed by the van Slyke method giving 3.70 ml of N_2 collected over water at a pressure of 735 torr and 29°C. What was the percentage of glycine in the sample?

$$CH_2(NH_2)COOH + HNO_2 \longrightarrow CH_2(OH)COOH + H_2O + N_2.$$

Ans. 17.2%

58. Natural gas often contains hydrogen sulfide, H_2S, a gas that itself is a pollutant and that produces another pollutant, sulfur dioxide, upon combustion. Hydrogen sulfide is removed from raw natural gas by the reaction

$$HOC_2H_4NH_2 + H_2S \longrightarrow (HOC_2H_4NH_3)HS$$

How much ethanolamine, $HOC_2H_4NH_2$, is required to remove the H_2S from 1.00×10^3 ft^3 (1 ft^3 = 28.3 ℓ) of natural gas at STP if the partial pressure of H_2S is 233 Pa? *Ans. 177 g*

59. One step in the production of sulfuric acid (Section 21.14) involves oxidation of $SO_2(g)$ to $SO_3(g)$ using air at 400°C with vanadium(V) oxide as a catalyst. Assuming that air is 21% oxygen by volume, what volume of air at 19°C and 754 torr is required to produce enough SO_3 to give 1.00 metric ton (1000 kg = 1 metric ton) of sulfuric acid, H_2SO_4? *Ans.* 5.9×10^5 *ℓ*

[S] 60. A sample of a compound of xenon and fluorine was confined in a bulb with a pressure of 24 torr. Hydrogen was added to the bulb until the pressure was 96 torr. Passage of an electric spark through the mixture produced Xe and HF. After the HF was removed by reaction with solid KOH, the final pressure of xenon and unreacted hydrogen in the bulb was 48 torr. What is the empirical formula of the xenon fluoride in the original sample? *Note:* Xenon fluorides contain only one xenon atom per molecule. *Ans.* XeF_4

61. A 0.250-ℓ bottle contains 75.0 ml of a 3.0% by mass solution of hydrogen peroxide. How much will the pressure (in torricelli) in the bottle increase if the H_2O_2 decomposes to $H_2O(l)$ and $O_2(g)$ at 34°C?

Assume the density of the solution is 1.00 g/cm^3, that the solubility of the oxygen can be neglected, and that the volume of the liquid does not change during the decomposition.

Ans. 3.6×10^3 torr, or 4.8 atm

62. The pressure in a sample of hydrogen collected above water in a 425-ml bottle at 35°C is 763 torr. What is the sample volume when the temperature falls to 23°C, assuming the sample pressure does not change?

Ans. 397 ml

References

"Gay-Lussac after 200 Years," H. Goldwhite, *J. Chem. Educ.,* **55,** 366 (1978).

"Gay-Lussac (1778–1850): A View of Chemistry, Industry, and Society in Post-Revolutionary France," M. Crosland, *Endeavour,* **2,** 52 (1978).

"Some Early Thermometers," E. H. Brown, *J. Chem. Educ.,* **11,** 448 (1934).

"Thomas Graham's Study of the Diffusion of Gases," A. Ruckstuhl, *J. Chem. Educ.,* **28,** 594 (1951).

"The Discovery of Boyle's Law, 1661–62," R. G. Neville, *J. Chem. Educ.,* **39,** 356 (1962).

"Robert Boyle," M. B. Hall, *Sci. American,* August 1967, p. 97.

"Graham's Laws of Diffusion and Effusion," E. A. Mason and B. Kronstadt, *J. Chem. Educ.,* **44,** 740 (1967).

"The Range of Validity of Graham's Law," A. D. Kirk, *J. Chem. Educ.,* **44,** 745 (1967).

"Atmospheric Pressure from Barometer Readings," G. F. Kinney, *J. Chem. Educ.,* **46,** 321 (1969).

"Graham's Laws: Simple Demonstrations of Gases in Motion—Part I, Theory," E. A. Mason and R. B. Evans III, *J.* *Chem. Educ.,* **46,** 358 (1969). "—Part II, Experiments," R. B. Evans III, D. L. Love, and E. A. Mason, **46,** 423 (1969).

"The Critical Temperature: A Necessary Consequence of Gas Non-Ideality," F. L. Pilar, *J. Chem. Educ.,* **44,** 284 (1967).

"Weak Intermolecular Interactions," J. E. House, Jr., *Chemistry,* **45 (4),** 13 (1972).

"Determination of the Molar Volume of a Gas at Standard Temperature and Pressure," L. M. Zaborowski, *J. Chem. Educ.,* **49,** 361 (1972).

"Scuba Diving and the Gas Laws," E. D. Cooke and C. Baranowski, *J. Chem. Educ.,* **50,** 425 (1973).

"The Mole and Avogadro's Number," R. M. Hawthorne, Jr., *J. Chem. Educ.,* **50,** 282 (1973).

"The Chemistry of Planetary Atmospheres," W. T. Huntress, Jr., *J. Chem. Educ.,* **53,** 204 (1976).

"Structural Theories of Fluids," C. A. Croxton, *Endeavour,* May 1975, p. 79.

"Atmospheric Carbon Dioxide and Carbon Reservoir Changes," M. Stuiver, *Science,* **199,** 253 (1978).

11

The Liquid and Solid States

The Liquid State

11.1 The Kinetic-Molecular Theory and the Liquid State

We learned in Section 10.15 that the molecules of a substance in the gaseous state are in constant and very rapid motion and that the space between the molecules is large compared to the sizes of the molecules themselves. As the molecules of a gas are brought closer together by increased pressure, the average distance between the molecules is decreased and **intermolecular attractive forces** become stronger. As a gas is cooled, the average speed of the molecules decreases and their tendency to move apart after collision decreases. If the pressure is sufficiently high and the temperature sufficiently low, the intermolecular attraction overcomes the tendency of the molecules to fly apart, and the gas changes to the liquid state. Although molecules in the liquid state are pulled close together, they still retain a limited amount of motion as reflected in the capacity of liquids to flow, to take the shape of a container, to diffuse, and to evaporate.

The molecules in a liquid are held in close contact by their mutual attractive forces. An increase in pressure on a liquid can only reduce the distance between the molecules slightly, so the volume of a liquid decreases very little with increased pressure; liquids are relatively incompressible compared to gases. The molecules in a liquid are able to move past one another in random fashion (diffusion) but, because of the much more limited freedom of molecular motion existing in liquids, they diffuse much more slowly than do gases.

249

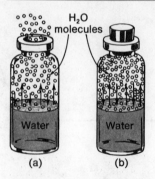

Figure 11-1 (a) Molecules of water evaporate from the liquid, escape from the open bottle, and the water level is lowered. (b) Molecules of water do not escape from the closed bottle and a dynamic equilibrium between vapor and liquid occurs.

11.2   Evaporation of Liquids

Almost everyone knows that the amount of water present in an open vessel decreases upon standing. We say that **evaporation** has occurred. Evaporation may be explained in terms of the motion of molecules. Just as in a gas, at any given temperature the molecules of a liquid move—some slowly, many at intermediate rates, and some very rapidly. A rapidly moving molecule near the surface of the liquid may possess sufficient kinetic energy to overcome the attraction of its neighbors and escape—i.e., evaporate—to the space above the liquid [Fig. 11-1(a)]. More fast-moving molecules will leave the liquid phase and appear in the gaseous phase above the liquid as the process of evaporation continues. When the space above the liquid is confined [Fig. 11-1(b)], molecules cannot escape into the open but strike the walls of the container, rebound, and eventually strike the surface of the liquid and move once again into the liquid state. The return of the molecules from the vapor state to the liquid state is known as **condensation.** As evaporation proceeds, the number of molecules in the vapor state increases, and in turn the rate of condensation increases. The rate of condensation soon becomes equal to the rate of evaporation, whereupon the vapor in the closed container is in **equilibrium** with the liquid. At equilibrium, neither the amount of the liquid nor the amount of the vapor in the closed container changes with time. Although the amounts of vapor and liquid do not change at equilibrium, some molecules from the vapor condense at the same time that an equal number of molecules from the liquid evaporate. Since both processes are in full operation but counterbalance each other, the equilibrium is called a **dynamic equilibrium.**

$$\text{Liquid} \underset{\text{Condensation}}{\overset{\text{Evaporation}}{\rightleftharpoons}} \text{Vapor}$$

At equilibrium, the space above the liquid is saturated with respect to molecules of the vapor. The pressure exerted by the vapor in equilibrium with its liquid, at a given temperature, is called the **vapor pressure** of the liquid. The area of the surface of the liquid in contact with the vapor and the size of the vessel have no effect upon the vapor pressure. Vapor pressures are commonly measured by introducing a liquid into a closed container and measuring the increase in pressure due to the vapor in equilibrium with the liquid by means of a manometer (Fig. 10-3). Some vapor pressure data for water, alcohol, and ether are given in Table 11-1 and Fig. 11-2.

11.3   Boiling Points of Liquids

It will be seen by studying Table 11-1 and Fig. 11-2 that, at a given temperature, the vapor pressure of ether is greater than that of alcohol, and that of alcohol greater than that of water. These differences in vapor pressure are related to the intermolecular attractive forces in each liquid, which for the three substances are least for ether and greatest for water. Thus the vapor pressure of a liquid is dependent upon the particular kind of molecule composing the liquid.

Data in Table 11-1 and Fig. 11-2 show that the vapor pressures of the three liquids increase as the temperature is raised. Indeed this is true for all liquids and

TABLE 11-1	Vapor Pressures (torr) of Some Common Substances at Various Temperatures					
	0°C	20°C	40°C	60°C	80°C	100°C
Water	4.6	17.5	55.3	149.4	355.1	760.0
Ethyl Alcohol	12.2	43.9	135.3	352.7	812.6	1693.3
Ethyl Ether	185.3	442.2	921.1	1730.0	2993.6	4859.4

Figure 11-2 The vapor pressures of three common substances at various temperatures. The intersection of a curve with the dashed line at 760 torr pressure, referred to the temperature axis, indicates the normal boiling point of that substance. (The normal boiling point of ethyl ether is 34.6°; of ethyl alcohol, 78.4°; of water, 100°.)

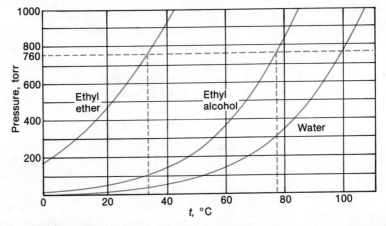

Figure 11-3 In a closed system, water does not boil if an equilibrium exists between water vapor pressure and hot liquid water. Cooling the water vapor, however, lowers its pressure and boiling occurs.

is due to an increase in the rate of molecular motion that accompanies an increase in temperature. This results in the escape of more molecules from the surface of the liquid per unit of time and a greater speed for each molecule that escapes. Each result contributes toward a higher vapor pressure.

When bubbles of vapor form within a liquid and rise to the surface where they burst and release the vapor, the liquid is said to boil. A liquid exposed to the air will boil when its vapor pressure becomes equal to the pressure of the atmosphere. The **normal boiling point** of a liquid is that temperature at which its vapor pressure becomes exactly equal to the standard atmospheric pressure of 760 torr, or 101,325 Pa (Fig. 11-2). A liquid may boil at temperatures higher than normal under external pressures greater than 1 atm; alternatively, the boiling point of a liquid may be lowered below normal by decreasing the pressure on the surface of the liquid below 1 atm (Fig. 11-3). Thus, at high altitudes where the atmospheric pressure is less than 760 torr, water boils at temperatures below its normal boiling point of 100°C. Food cooked in boiling water cooks more slowly at high altitudes because the temperature of boiling water is lower than it would be nearer sea level. The temperature of boiling water in pressure cookers is higher than normal due to higher equilibrium vapor pressures, thus making it possible to cook foods faster than in open vessels.

11.4   Intermolecular Forces and the Boiling Point

We saw in the preceding section that differences in vapor pressures of liquids are related to differences in the forces of attraction between molecules of the liquids.

Figure 11-4 A schematic representation of nearly instantaneous fluctuating dipoles that give rise to London forces. The large circle represents the position of the electron cloud about a nucleus (the small circle) before the dipole forms. The colored area indicates the shift of the electron cloud after the dipole has formed.

All molecules exert an attraction for each other. What is the nature of these attractive forces, which are collectively called **van der Waals forces,** and why do they differ in magnitude for molecules of different substances?

It was noted in Section 5.9 that molecules whose centers of positive and negative electric charge do not coincide are polar. They possess a **dipole moment.** The HCl molecule, because chlorine has a greater electronegativity than hydrogen, has a partial negative charge on the chlorine atom and a partial positive charge on the hydrogen atom; the HCl molecule has a permanent dipole moment. The electrostatic attraction of the positive end of one HCl molecule for the negative end of another constitutes an attractive force that causes HCl to have a higher boiling point than nonpolar molecules of similar molecular weight. This **dipole-dipole attraction** is one type of van der Waals attraction.

Additional components of the van der Waals attractions between molecules, which occur for nonpolar molecules as well as polar molecules, are called **dispersion forces** or **London forces** (after the person who first explained them). Because of the constant motion of the electrons in an atom, an atom that is on the average nonpolar becomes periodically polar. At this moment a temporary dipole exists in the atom. A second atom, in turn, is distorted at the same moment with its nucleus attracted toward the negative end of the first atom (Fig. 11-4) and its negative end opposite from the negative end of the first atom. A mutual attraction occurs between the two rapidly fluctuating dipoles. If the two atoms are located in different molecules, the two molecules experience a mutual attraction.

London forces are weak and are significant only when molecules are very close together, i.e., very nearly in contact with each other. The London forces are opposed by (1) the repulsive force of the electron clouds of the adjacent molecules and (2) the repulsion of the nuclei of neighboring atoms for one another. However, the attractive forces are somewhat stronger than the repulsive forces.

The magnitude of the London forces increases with an increase in the number of electrons per molecule, and therefore with the molecular weight. This increase in intermolecular attraction with molecular weight is reflected in the rise in boiling point in series of related substances such as He, Ne, Ar, Kr, Xe, Rn and such as H_2, F_2, Cl_2, Br_2, I_2 (see Table 11-2).

The London forces acting between molecules are of decreasing effectiveness with an increase in temperature, inasmuch as the additional thermal energy increases the kinetic energy of the molecules and overcomes the attraction. Another way of

TABLE 11-2 Molecular Weights and Boiling Points						
Substance	He	Ne	Ar	Kr	Xe	Rn
Molecular weight	4.0	20.18	39.94	83.7	131.1	222
Boiling point, °C	−268.9	−245.9	−185.7	−152.9	−107.1	−61.8
Substance	H_2	F_2	Cl_2	Br_2	I_2	
Molecular weight	2.016	38.0	70.91	159.8	253.8	
Boiling point, °C	−252.7	−187	−34.6	58.78	184.4	

putting it is that the kinetic energy at the higher temperature is greater than the energy associated with the London forces.

It is the London forces that cause substances such as the noble gases and the halogens to condense to liquids and to freeze into solids when the temperature is lowered sufficiently.

In general, liquids composed of discrete molecules that have no permanent dipole moments have low boiling points relative to their molecular weights, because only the weak London forces must be overcome during vaporization. In such molecules the centers of positive and negative electric charge coincide. Examples are the molecules of the noble gases and the halogens, and other symmetrical molecules such as CH_4, SiH_4, CF_4, SiF_4, SF_6, and UF_6. Molecular substances such as H_2O, HF, and C_2H_5OH (ethyl alcohol), which have permanent dipole moments, have rather high boiling points relative to their molecular weights (see Section 19.10, part 1).

11.5   Heat of Vaporization

In Section 11.2 it was noted that evaporation involves the escape of the molecules of high kinetic energy (faster moving molecules) from the surface of a liquid. The loss of such molecules through evaporation results in a lower average kinetic energy for those remaining behind and, consequently, a lowering of the temperature of the liquid. The cooling effect due to evaporation of water is very evident to a swimmer emerging from the water. In this case, the skin is supplying heat energy to the water molecules, which causes the skin to feel cool.

In order that a liquid may evaporate at a constant temperature, heat must be supplied in sufficient quantities to offset the cooling effect brought about by the escape of the molecules possessing high kinetic energies. The heat energy that must be supplied to evaporate a unit mass of liquid at a constant temperature is known as the **heat of vaporization.** The heat of vaporization of water, for example, is 2258 joules per gram, or 40.67 kilojoules per mole. That of ammonia is 1368 joules per gram or 23.31 kilojoules per mole.

$$H_2O(l) \longrightarrow H_2O(g) \qquad \Delta H = 40.67 \text{ kJ/mol}$$
$$= 9.72 \text{ kcal/mol}$$

$$NH_3(l) \longrightarrow NH_3(g) \qquad \Delta H = 23.31 \text{ kJ/mol}$$
$$= 5.57 \text{ kcal/mol}$$

The quantity of heat evolved during the condensation of a liquid is the same as that absorbed during evaporation.

$$H_2O(g) \longrightarrow H_2O(l) \qquad \Delta H = -40.67 \text{ kJ/mol}$$
$$NH_3(g) \longrightarrow NH_3(l) \qquad \Delta H = -23.31 \text{ kJ/mol}$$

The operation of mechanical refrigerators is based on the loss of heat energy to evaporating refrigerants. Heat is taken from within a refrigerator as a circulating refrigerant (usually ammonia, NH_3, or CCl_2F_2, one of the Freons) absorbs the energy needed to change it to a gas. In the gaseous state, the refrigerant is then circulated through a compressor outside the refrigerator and again liquefied by combined cooling and compression. To be an effective refrigerant, a substance must be readily convertible from the gaseous to the liquid state at the working temperature and have a high heat of vaporization.

11.6   Critical Temperature and Pressure

We noted in Section 11.1 that compressing a gas and lowering its temperature favor the transition from the gaseous to the liquid state. In some cases it is possible to liquefy a gas at normal temperatures by simply compressing it. However, for each substance there exists a temperature, called the **critical temperature,** above which it cannot be liquefied no matter how much pressure is applied. The pressure required to liquefy a gas at its critical temperature is called the **critical pressure.** The critical temperatures and critical pressures of some common substances are given in Table 11-3.

TABLE 11-3 Critical Temperatures and Pressures of Some Common Substances		
	Critical Temperature, K	Critical Pressure, atm
Hydrogen	33.24	12.8
Nitrogen	126.0	33.5
Oxygen	154.3	49.7
Carbon dioxide	304.2	73.0
Ammonia	405.5	111.5
Water	647.1	217.7
Sulfur dioxide	430.3	77.7

Above the critical temperature of a substance the average kinetic energy of the gaseous molecules is sufficient to overcome their mutually attractive forces, and the molecules will not cling together closely enough to form a liquid no matter how great the pressure. If the temperature is decreased, the average kinetic energy of the molecules is decreased. At the critical temperature the intermolecular forces are sufficiently large, relative to the average kinetic energy, to liquefy the gas provided the substance is under a pressure equal to or greater than its critical pressure. The pressure aids the intermolecular forces in bringing the molecules sufficiently close together to make liquefaction possible. Below the critical temperature, the pressure required for liquefaction decreases with decreasing temperature until it reaches 1 atm at the normal boiling temperature. Substances possessing strong intermolecular forces, such as water and ammonia, have high critical temperatures; on the other hand, substances with weak intermolecular attraction, such as hydrogen and nitrogen, have low critical temperatures.

11.7   Distillation

Liquids may contain dissolved materials that make them unsuitable for a particular purpose. For example, water containing dissolved mineral matter should not be used in storage batteries (Section 20.17) because it shortens the life of the battery. Water and other liquids may be purified by a process known as **distillation.** By heating impure water in a distilling flask (Fig. 11-5), the liquid is converted to vapor, which passes over into the condenser. The vapor is condensed to the liquid in the water-cooled condenser, and the liquid flows into the receiving vessel. The dissolved mineral matter, such as calcium sulfate or magnesium chloride, is not volatile at the boiling point of water and remains in the distillation flask. Distillation makes use of the facts that the addition of heat to a liquid speeds up the rate of evaporation, an endothermic change, and that cooling a vapor favors condensation, an exothermic change. The separation of volatile substances by distillation is described more fully in Section 13.5.

Figure 11-5 Laboratory distillation apparatus. When impure water is distilled with this apparatus, nonvolatile substances remain in the distilling flask. The water is vaporized, condensed, and finally collected as distillate in the receiving flask.

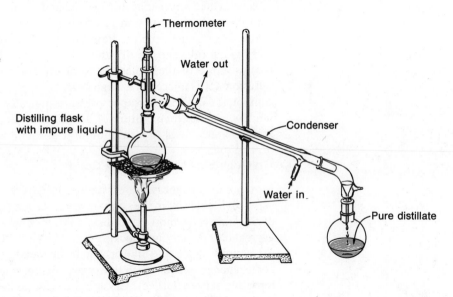

Thermometer

Water out

Distilling flask
with impure liquid

Condenser

Water in

Pure distillate

11.8   Surface Tension

The molecules within the bulk of a liquid are attracted equally in all directions by neighboring molecules; the resultant force on any one molecule within the liquid is therefore zero. However, the molecules on the surface of a liquid are attracted only inward and sideways. This unbalanced molecular attraction pulls some of the surface molecules into the bulk of the liquid, and a condition of equilibrium is reached only when as many molecules as possible have been pulled into the liquid. At this point the minimum number of molecules possible are on the surface and the surface area is reduced to a minimum. The surface of a liquid, therefore, behaves as if it were under a strain, or tension. This contracting force is called **surface tension.** A small drop of liquid tends to assume a spherical shape because in a sphere the ratio of surface area to volume is at a minimum. We may define surface tension as the force that causes the surface of a liquid to contract. A liquid surface acts as if it were a stretched membrane. A steel needle carefully placed on water will float. Some insects, even though they are heavier than water, can move on its surface, being supported by the surface tension. One of the forces causing water to rise in capillary tubes (tubes with a very small bore) is its surface tension. Water is brought from the soil up through the roots and into the portion of a plant above the soil by this capillary action.

The Solid State

11.9   The Kinetic-Molecular Theory and the Solid State

When enough heat is removed from water to cause its temperature to be lowered to 0° and then more heat is removed, the molecular motion decreases and the molecules take up comparatively fixed and ordered positions relative to each other, with only vibratory motion remaining. When this happens, the water is said to freeze, i.e., change from the liquid to the solid state. Ice crystals, which are hard and rigid, form. The rigidity and small compressibility of crystalline solids reflect the resistance of the molecules to change of position. However, the facts that diffusion takes place to a slight extent in the solid state and that crystalline compounds show some vapor pressure indicates that the molecules are not motionless. As the temperature of a solid is lowered, the motion of the molecules gradually decreases to a minimum at absolute zero ($-273°C$).

11.10   Crystalline and Amorphous Solids

Most solid substances are crystalline in nature. Although some solids, such as sugar and table salt in the form in which we are familiar with them, are composed of single crystals, most crystalline solids with which we come in daily contact are aggregates of many interlocking small crystals. Common examples of the latter are chunks of ice and objects made of metals. **A crystal may be defined as a homogeneous solid in which the atoms, ions, or molecules are arranged in a definite repeating pattern (lattice).** Crystals are three-dimensional solids bounded by plane surfaces. The angles at which the surfaces of the crystal intersect are always the same for a given substance, and are characteristic of that substance. Figure 11-6 illustrates the appearance of single crystals of sodium chloride, copper metal, and calcite ($CaCO_3$).

The temperature of many liquids may be lowered below their freezing points before crystallization begins. A liquid existing at a temperature below its freezing point is said to be **undercooled,** or **supercooled.** An undercooled liquid is in a metastable condition; i.e., it is not at equilibrium with its solid. Mechanical agitation, such as vigorous stirring, or the introduction of a "seed" crystal of the substance often induces crystallization by providing an ordered structure to which the slow-moving molecules can become attached.

Liquid materials such as fused glass, melted butter, or molten asphalt, which contain large cumbersome particles that cannot move readily into the positions of a regular crystal lattice, often show great tendencies to undercool. As the temperature is lowered, the presence of the large and irregular structural units composing

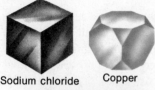

Sodium chloride   Copper

Calcite

Figure 11-6 The appearance of single crystals of sodium chloride, copper metal, and calcite ($CaCO_3$).

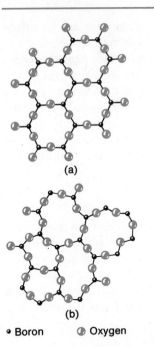

Figure 11-7 (a) A two-dimensional illustration of the ordered arrangement of atoms in a crystal of boric oxide. (b) An illustration showing the disorder in vitreous (amorphous) boric oxide.

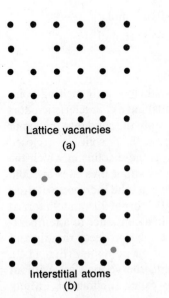

Figure 11-8 Representations of two types of crystal defects.

the material causes the undercooled liquid to become less and less mobile, and finally to become rigid. Such materials are often spoken of as **amorphous solids,** or **glasses.** True solids are crystalline in structure with a definite internal ordered arrangement of their building units; they have sharp melting points and resist change of shape under pressure. Amorphous solids, on the other hand, lack an ordered internal structure (Fig. 11-7). They do not melt at a definite temperature, but gradually soften and become less viscous when heated. They are really supercooled liquids; the term "amorphous solid," though often used, is not strictly correct.

11.11   Crystal Defects

Crystals are actually never perfect; several types of defects occur. It is quite common for some positions that should contain atoms or ions to be vacant [Fig. 11-8(a)]. Less commonly, some atoms or ions in a crystal may be located in positions, called **interstitial sites,** which are located between the regular positions for atoms [Fig. 11-8(b)]. Certain distortions occur in some impure crystals, as for example when the impurity cations or anions are too large to fit into the regular positions without distortion of the structure. Minute amounts of impurities are sometimes deliberately introduced into a crystal to cause imperfections in the structure in order that changes in the electrical conductivity of the crystal (Section 20.18) or in other physical properties will result. This type of imperfection gives rise to practical applications such as the manufacture and use of semiconductors for electrical circuits.

11.12   Melting of Solids

When a crystalline solid is heated sufficiently, the vibrational energy of some of the molecules becomes great enough to overcome the intermolecular forces holding the molecules in their fixed positions in the crystal, and the solid begins to melt (fuse). If heating is continued, all of the solid will pass into the liquid state even though the temperature does not rise. If, however, the heating is stopped and no heat is withdrawn, the solid and liquid phases will remain in equilibrium, the rate of melting being just balanced by the rate of freezing. The changes will continue, but the quantities of solid and liquid will remain constant. The temperature at which the solid and liquid phases of a given substance are in equilibrium is known as the **melting point** of the solid, or the **freezing point** of the liquid.

The temperature at which a solid melts is determined by the strength of the forces of attraction between the units present in the crystal. Crystals consisting of small symmetrical molecules, such as H_2, N_2, O_2, and F_2, have low melting points because the intermolecular forces are of the weak London type. Crystals consisting of larger nonpolar molecules melt at higher temperatures because the number of electrons within the molecules, and thus the London forces, increase. Crystalline solids built up of unsymmetrical molecules with permanent dipole moments melt at still higher temperatures; examples are ice and sugar. Diamond is an atomic crystal in which the small carbon atoms are held together in the crystal by strong covalent bonds (Section 25.1 and Fig. 25-1); the melting point of diamond is very high. The atoms in the crystals of metals are strongly bonded together by metallic bonds (Section 30.4), which are modified covalent bonds. In general the metals have high melting points. The electrostatic forces of attraction between the ions in ionic solids are quite strong; thus ionic crystals have high melting points.

Crystalline solids such as NaCl and $NaClO_3$ melt sharply when heated because the forces holding the ions together are all of the same strength, and thus all the ions break apart at once. The gradual softening of glasses, as opposed to the sharp melting of crystalline solids, results from the structural nonequivalence of the atoms. When a glass is heated, the weakest bonds break first; as the temperature is further increased, the stronger bonds are broken. This causes a gradual decrease in the size and a corresponding increase in the mobility of the structural units as the temperature is raised.

11.13   Heat of Fusion

When heat is applied to a crystalline solid, the temperature rises until the melting point is reached. The temperature then remains constant as additional heat is applied, until all of the solid has been changed to liquid, after which the temperature rises again. The quantity of heat that must be supplied to change a unit mass of a substance from the solid to the liquid state at constant temperature is known as the **heat of fusion** of the substance. The heat of fusion of ice is approximately 333.6 joules per gram, or 6.01 kJ per mole.

$$H_2O(s) \longrightarrow H_2O(l) \qquad \Delta H = 6.01 \text{ kJ/mol}$$
$$= 1.44 \text{ kcal/mol}$$

In this case the heat of fusion represents the difference between the heat content of water, in which the molecules have considerable freedom of motion, and that of ice, in which the molecules vibrate about fixed positions in the crystal. The quantity of heat liberated during crystallization (freezing) is exactly the same as that absorbed during fusion.

$$H_2O(l) \longrightarrow H_2O(s) \qquad \Delta H = -6.01 \text{ kJ/mol}$$
$$= -1.44 \text{ kcal/mol}$$

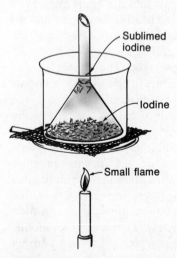

Sublimed iodine

Iodine

Small flame

Figure 11-9 A simple method of demonstrating the sublimation of iodine.

11.14   Vapor Pressure of Solids

The fact that snow and ice evaporate at temperatures below their melting point and that certain solids such as naphthalene (moth balls) have characteristic odors is evidence that molecules of some solids pass directly from the solid into the vapor state. The pressure of such a vapor in equilibrium with its solid in a closed container is called the **vapor pressure** of the solid. It is those solids in which the intermolecular forces are weak that exhibit measurable vapor pressures at room temperature. As one might predict, the vapor pressure of a solid increases with rise in temperature. The vapor pressure of solid iodine is 0.2 torr at 20° and 90 torr at 114°, its melting point. If iodine crystals are heated in a container to a temperature just below their melting point, evaporation of the solid proceeds rapidly and the vapor may condense to crystals in a cooler part of the container (Fig. 11-9). The combined process of a solid passing directly into the vapor state without melting and the recondensing into the solid state is called **sublimation.** Many substances, such as iodine, may be purified by sublimation if the impurities have low vapor pressures.

The Structures of Crystalline Solids

11.15   The Structures of Metals

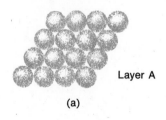

Layer A

(a)

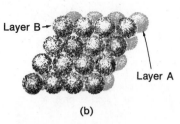

Layer B →

Layer A

(b)

Figure 11-10 (a) A portion of a layer of closest packed spheres in the same plane. Each sphere touches six others. (b) Spheres in two closest packed layers. Each sphere in layer B touches three spheres in layer A (the colored layer). In an actual crystal, many more than two planes would exist. Each sphere contacts six spheres in its own layer, three spheres in the layer below, and three spheres in the layer above, making a total contact for each sphere of twelve other spheres.

A pure metal is a crystalline solid in which the metal atoms are packed closely together in a repeating array, or pattern. In most metals, the atoms pack together as if they were spheres. If spheres of equal size are packed together as closely as possible in a plane, they arrange themselves as shown in Fig. 11-10(a), with each sphere in contact with six others. This arrangement, called **closest packing,** can extend indefinitely in a single layer. Crystals of many metals can be described as stacks of closest packed layers.

Two types of stacking of closest packed planes are observed in simple metallic structures. In both types, a second layer (B) is placed on the first layer (A) so that each sphere in the second layer is in contact with three spheres in the first layer, as shown in Fig. 11-10(b). A third layer can be positioned in one of two ways. In one type of stacking, each sphere in the third layer lies directly above a sphere in the first layer [Fig. 11-11(a)]. The spheres in the third layer are also of type A. The stacking continues with alternate close-packed layers of type B and type A; ABABAB···. This arrangement is called **hexagonal closest packing.** Metals that crystallize with this structure are said to have a **hexagonally closest packed structure.** Examples include Be, Cd, Co, Li, Mg, Na, and Zn. Those elements or compounds that crystallize with the same structure are said to be **isomorphous.**

In the second type of stacking, the third layer is positioned so that its spheres are

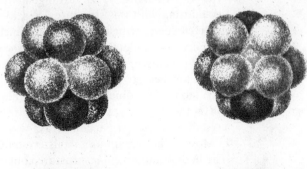

Figure 11-11 A portion of two types of crystal structures in which spheres are packed as compactly as possible. The lower diagrams show the structures expanded for clarification. Note that the first and third layers have identical orientations in (a). The first and third layers have different orientations in (b). In both structures, each sphere is surrounded by 12 others in an infinite extension of the structure and is said to have a coordination number of 12.

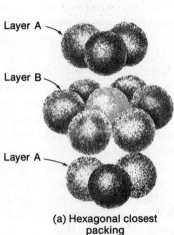

Layer A

Layer B

Layer A

(a) Hexagonal closest packing

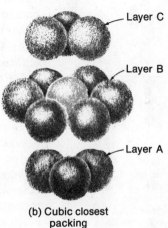

Layer C

Layer B

Layer A

(b) Cubic closest packing

Figure 11-12 (a) A portion of a plane of spheres found in a body-centered cubic structure. Note that the spheres do not touch. (b) Spheres in two layers of a body-centered cubic structure. Each sphere in the one layer touches four spheres in the adjacent layer but none in its own layer.

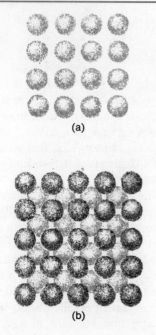

(a)

(b)

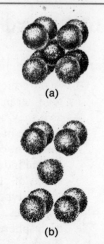

(a)

(b)

Figure 11-13 (a) A portion of a body-centered cubic structure, showing parts of three layers. (b) An expanded view of a body-centered cubic structure. In this structure, each sphere in an infinite extension of the structure touches four spheres in the layer above it and four spheres in the layer below and is said to have a coordination number of 8.

(a)

(b)

Figure 11-14 (a) A portion of a plane of spheres found in a simple cubic structure. Note that the spheres are in contact. (b) A portion of a simple cubic structure, showing two of the planes.

not directly above those in either layer A or layer B [Fig. 11-11(b)]. This layer is of type C. The stacking continues with alternating layers of type A, type B, and type C; ABCABC···, an arrangement called **cubic closest packing**. Metals crystallizing with this structure are said to have a **cubic closest packed,** or **face-centered cubic, structure.** Examples include Ag, Al, Ca, Cu, Ni, Pb, and Pt.

In crystals of metals with either of the closest packed arrangements, hexagonal closest packing or cubic closest packing, each atom has 12 equidistant neighbors that it touches, six in its own plane and three in each adjacent plane. This gives each atom a coordination number of 12. The **coordination number** of an atom or ion is the number of neighbors nearest to it. About two-thirds of all metals crystallize in closest packed arrays with coordination numbers of 12.

Most of the remaining metals crystallize with a **body-centered cubic** structure. The body-centered cubic structure contains planes of spheres that are not closest packed. Each sphere in the plane is surrounded by four neighbors [Fig. 11-12(a)] rather than six as in closest packed planes [Fig. 11-10(a)]. Note that the spheres in this plane in Fig. 11-12(a) *do not touch.* The structure contains alternating layers of these planes. The second layer is stacked on top of the first so that a sphere in the second layer touches four spheres in the first layer [Fig. 11-12(b)]. The third layer is positioned directly above the atoms in the first layer (Fig. 11-13); the fourth layer, above the second; etc. Any atom in this structure touches four atoms in the layer above it and four atoms in the layer below it. An atom in a body-centered cubic structure thus has a coordination number of 8. Isomorphous metals with a body-centered cubic structure include Ba, Cr, Mo, W, and Fe at room temperature.

Polonium crystallizes in an array rare for metals, but which is well known with other substances. This structure, the **simple cubic structure,** is also not closest packed. It contains planes of spheres, with each sphere in a plane touching its four near neighbors in that same plane [Fig. 11-14(a)]. The planes are stacked directly above each other, so that an atom in the second layer touches only one atom in the first layer [Fig. 11-14(b)]. The coordination number of a polonium atom in a simple cubic array is 6; an atom touches four other atoms in its own layer, one atom in the layer above, and one atom in the layer below.

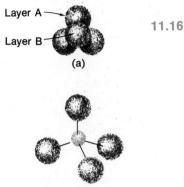

Figure 11-15 (a) Spheres in two adjacent closest packed layers that form a tetrahedral hole. (b) A cation (smaller sphere) located in a tetrahedral hole surrounded by four anions (larger spheres) from a different perspective. The structure has been expanded to show the geometrical relationships.

11.16   Ionic Crystal Structures

Ionic crystals consist of two or more different kinds of ions, usually with the cations and anions having different sizes. Consequently, the packing of these ions into a crystal structure is more complex than that in a metal in which all of the atoms are the same size and kind.

Since most monatomic ions behave as charged spheres, their attraction for ions of opposite charge is the same in every direction. Thus the most stable structures result when ions of one charge are surrounded by as many ions as possible of the other charge. Furthermore, the most stable arrangements of ions appear to be those in which the cations and anions are in contact with each other. The structures are determined by two factors: the relative sizes of the ions and the relative numbers of positive and negative ions required to maintain electrical neutrality in the crystal as a whole.

In simple ionic structures the anions, which are usually larger than the cations, commonly are arranged in a closest packed array; the smaller cations fill in the spaces that remain between the anions. Such spaces remaining between the spheres in a closest packed array are called **holes,** or **interstices.** Figures 11-15 and 11-16 illustrate the two types of holes that are commonly present and that can then be occupied by other atoms or ions. The smaller of these holes is found between three spheres in one plane and one sphere in the adjacent plane [Fig. 11-15(a)]. The four spheres that bound this hole are arranged at the corners of a tetrahedron [Fig. 11-15(b)]; the hole is called a **tetrahedral hole.** The larger type of hole is found at the center of six spheres (three in one layer and three in an adjacent layer) located at the corners of an octahedron [Fig. 11-16]. Such a hole is called an **octahedral hole.**

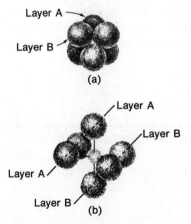

Figure 11-16 (a) Spheres in two adjacent closest packed layers that form an octahedral hole. (b) A cation (smaller sphere) located in an octahedral hole surrounded by six anions (larger spheres) from a different perspective. The structure has been expanded to show the geometrical relationships.

Depending upon the relative sizes of the cations and anions, the cations of an ionic compound occupy tetrahedral or octahedral holes. As will be discussed in Section 11.17, relatively small cations occupy tetrahedral holes while larger cations occupy octahedral holes. If a cation is too large to fit into an octahedral hole, the packing of the anions may change to give a more open structure such as a simple cubic array (Fig. 11-14) wherein the larger cation can occupy the larger cubic hole made possible by the more open spacing (Fig. 11-17).

In either a hexagonal closest packed or a cubic closest packed array of anions, for each anion in the array there are two tetrahedral holes that can accommodate cations. The isomorphous compounds Li_2O, Na_2O, Li_2S, Na_2S, and Li_2Se, among others, crystallize with a cubic closest packed array of anions. The relatively small cations occupy tetrahedral holes in the array. Since the ratio of tetrahedral holes to anions is 2 to 1, all of these holes must be filled by cations inasmuch as the cation to anion ratio in these salts is 2 to 1. A compound such as Li_2O, which crystallizes with a closest packed array of anions and with cations in the tetrahedral holes, can have a maximum cation-to-anion ratio of 2 to 1; all of the tetrahedral

Figure 11-17 A cation in the cubic hole in a simple cubic array of anions.

holes are filled at this ratio. Compounds with a cation-to-anion ratio of less than 2 to 1 also can crystallize in a closest packed anion array with cations in tetrahedral holes, if the ionic sizes are correct. In these compounds, however, some of the tetrahedral holes remain vacant. Examples of such systems include ZnS and BeO (one-half of the tetrahedral holes in the close-packed array of anions are occupied by smaller Zn^{2+} or Be^{2+} cations, respectively), SiO_2 (one-fourth of the tetrahedral holes in the close-packed array of the larger oxygen atoms are occupied by smaller silicon atoms), and $TiCl_4$ (one-eighth of the tetrahedral holes in the chloride array are occupied by titanium).

The ratio of octahedral holes to anions in either a hexagonal or a cubic closest packed structure is 1 to 1. Thus isomorphous compounds such as NiO, MnS, NaCl, KH, and AgBr, which crystallize in a closest packed anion array with cations in octahedral holes, can have a maximum cation-to-anion ratio of 1 to 1. Cation-to-anion ratios of less than 1 to 1 are observed when some of the octahedral holes remain empty. Examples include Ga_2O_3 (two-thirds of the octahedral holes in the closest packed array of oxide ions are occupied by Ga^{3+}), CdI_2 and TiO_2 (one-half of the octahedral holes in the anion arrays are occupied by Cd^{2+} or Ti^{4+}, respectively), and $FeCl_3$ (one-third of the octahedral holes are occupied).

In some compounds both tetrahedral and octahedral holes are occupied. Spinel, $MgAl_2O_4$, for example, crystallizes as a cubic closest packed array of oxide anions with Mg^{2+} in one-eighth of the tetrahedral holes and Al^{3+} in one-half of the octahedral holes.

In a simple cubic array of anions (Fig. 11-17), there is one cubic hole that can be occupied by a cation for each anion in the array. In CsCl and other compounds with the CsCl structure, all of the cubic holes are occupied. One-half of the cubic holes are occupied in SrH_2, UO_2, $SrCl_2$, and CaF_2.

The reason that so many different types of ionic compounds crystallize with the same structures (that is, are isomorphous) is simply a reflection of the fact that in many different compounds the features that determine structure (the relative sizes of the ions and the stoichiometry) are similar. More about this will be found in the next section.

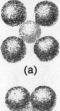

(a)

(b)

(c)

Figure 11-18 Packing of anions (black spheres) around cations of varying size (colored spheres). Decreasing size of cations is illustrated successively in (a), (b), and (c).

11.17   The Radius Ratio Rule for Ionic Compounds

The structure assumed by an ionic compound, as previously described, is largely the result of stoichiometry and of simple geometric and electrostatic relationships that depend on the relative sizes of the cation and anion. A relatively large cation can be in contact with a large number of negative ions and occupy a cubic or an octahedral hole, whereas a relatively small cation can only contact a few negative ions in a tetrahedral hole.

Consider the situation where a cation, M^+, has a coordination number of 6. As shown in Fig. 11-18(a), the M^+ ion is in contact with four X^- ions in a plane. In addition, although they are not shown in the figure, there is an X^- ion above the M^+ ion and in contact with it, and another in contact below it. Note that the M^+ ion is large enough to expand the array of X^- ions so the X^- ions are not in contact with each other. As long as the expansion is not great enough to get still another anion in contact with the cation, this is a stable situation; the cation-anion contacts

TABLE 11-4 Limiting Values for the Radius Ratio for Ionic Compounds (r^+ is radius of cation; r^- is radius of anion)

Coordination Number	Type of Hole Occupied	Approximate Limiting Values of r^+/r^-
8	Cubic	Above 0.732
6	Octahedral	0.414 to 0.732
4	Tetrahedral	0.225 to 0.414

are maintained. Figure 11-18(b) illustrates what happens when the size of the M^+ ion is decreased somewhat. Now the X^- ions are in contact with each other, as are the M^+ and X^- ions. If the size of M^+ is decreased still more, it becomes impossible to get a structure with six-coordination. The negative ions are in contact [Fig. 11-18(c)], but there is no contact between the M^+ and X^- ions; this is an unstable structure. In this case, a more stable structure would be formed, with only four anions about the cation. The limiting condition for the formation of a six-coordinate structure is illustrated in Fig. 11-18(b) and is defined by the relative sizes of the ions where the radius ratio (the radius of the positive ion, r^+, divided by the radius of the negative ion, r^-) is equal to 0.414.

There is a minimum radius ratio (r^+/r^-) for each coordination number below which the structure with that coordination number is generally not stable. The approximate limiting values for the radius ratios for ionic compounds are given in Table 11-4.

Some examples of salts in which the cation occupies a cubic hole in a simple cubic anion array and their actual radius ratios are CsCl (0.93); CsBr (0.87); TlCl (0.83); SrH_2 (0.82); BaF_2 (0.99). Salts with cations in octahedral holes include NaCl (0.52); NaI (0.44); KCl (0.73); AgCl (0.70); MnO (0.51); CaS (0.54); $CdCl_2$ (0.54); and NiF_2 (0.51). Salts with cations in tetrahedral holes include ZnS (0.40); ZnSe (0.37); and BeO (0.22). It would appear, then, that the structure of an ionic crystal depends primarily upon the relative numbers and sizes of its ions. Many ionic compounds with quite different ions but with the same numbers of cations and anions and the same radius ratios crystallize in the same structure.

The radius ratio rule applies strictly only to ionic crystals. In compounds in which the covalent character of the bonds is pronounced, the rule for predicting structures does not hold.

11.18   Crystal Systems

Since the atoms in a crystal are arranged in a definite repeating pattern, we can start at one point in the crystal and by traveling in specific directions come to other points with environments identical to that of the starting point. The collection of all of the points within the crystal that have identical environments is called a **space lattice**. A simple three-dimensional cubic space lattice is shown in Fig. 11-19. The atoms around these points have been omitted so you can see the lattice. Figure 11-20 illustrates a portion of the structure of sodium chloride. This structure will be described more fully later in this section. For the present, let us

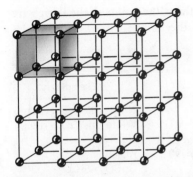

Figure 11-19 A portion of a simple cubic space lattice. One unit cell is shaded in color.

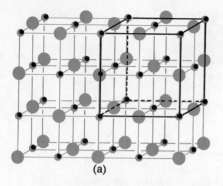

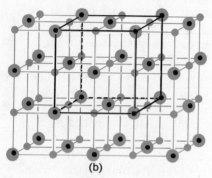

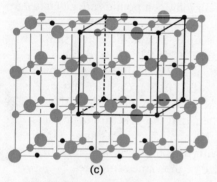

Figure 11-20 A portion of the structure of NaCl illustrating three ways to position a space lattice in the structure. Small spheres represent Na^+; large spheres, Cl^-. (a) Lattice points in the center of the Na^+ ions. (b) Lattice points in the center of the Cl^- ions. (c) Lattice points with a chloride ion to the left and a sodium ion to the right. Black lines connect points that define a unit cell in each lattice. The unit cell of NaCl is usually described as indicated in (a) or (b). Note that this structure has been expanded for clarity; the ions touch in the actual structure.

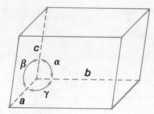

Figure 11-21 A unit cell.

consider only the space lattice associated with the structure. There are an infinite number of ways to construct a space lattice in the sodium chloride structure. The points of one possible space lattice are located at the centers of the sodium ions, as shown in Fig. 11-20(a). Alternatively, points with identical environments could be located at the centers of chloride ions [Fig. 11-20(b)] or between sodium and chloride ions [Fig. 11-20(c)]. In each case the resulting space lattice is the same; only the locations of the points of the lattice differ. The unit cell of sodium chloride is most commonly described as indicated in (a) or (b) of Fig. 11-20.

That part of the space lattice that will generate the entire lattice if it is repeated in three dimensions is called a **unit cell.** The cube outlined in Fig. 11-19 is the unit cell for the simple cubic space lattice. The cubes in Figs. 11-20(a), (b), and (c) illustrate three ways in which unit cells may be selected for the sodium chloride structure. The structure of a crystal is specified by describing the size and shape of the unit cell and indicating the arrangement of its contents because the crystal can be built up by repeating the unit cell in three dimensions.

Thus far we have considered only unit cells shaped like a cube, but there are others. In general, a unit cell is a parallelepiped for which the size and shape are defined by lengths of three axes (a, b, c) and the angles (α, β, γ) between the axes (Fig. 11-21). The axes are defined as lengths between points in the space lattice. Consequently, *unit cell axes join points with identical environments.* Unit cells must be one of seven shapes. The unit cells of the seven crystal systems are indicated in Table 11-5. Variations in the number and location of lattice points in these seven unit cells give rise to fourteen types of space lattices. In some of these, points with identical surroundings (lattice points) are found only at the corners of the unit cell. In others, the points of the lattice occur at the corners as well as in the centers of some or all of the faces of the unit cell. In still others, the points of the lattice are found at the corners as well as in the center of the unit cell.

TABLE 11-5   Unit Cells of the Seven Crystal Systems

System			Examples
Cubic	$a = b = c$	$\alpha = \beta = \gamma = 90°$	NaCl, CsCl, ZnS, CaF$_2$
Tetragonal	$a = b \neq c$	$\alpha = \beta = \gamma = 90°$	White tin, TiO$_2$ (rutile)
Orthorhombic	$a \neq b \neq c$	$\alpha = \beta = \gamma = 90°$	HgCl$_2$
Monoclinic	$a \neq b \neq c$	$\alpha = \gamma = 90°$; $\beta \neq 90°$	KClO$_3$
Triclinic	$a \neq b \neq c$	$\alpha \neq \beta \neq \gamma \neq 90°$	CuSO$_4 \cdot$ 5H$_2$O
Hexagonal	$a = b \neq c$	$\alpha = \beta = 90°$; $\gamma = 120°$	ZnS, SiO$_2$
Rhombohedral	$a = b = c$	$\alpha = \beta = \gamma \neq 90°$	CaCO$_3$, Al$_2$O$_3$

Students sometimes have difficulty in reconciling the number of points or atoms stated as the number per unit cell with the number shown in a diagram of the unit cell or in a figure such as Fig. 11-22. It should be noted that some of the points or atoms shown in a diagram of the unit cell are actually shared by other unit cells and therefore do not lie in their entirety within one unit cell. It is helpful to keep in mind the following rules:

1. A point or atom that lies completely within the unit cell belongs to that unit cell only.
2. A point or atom lying on a face of a unit cell belongs equally to two unit cells and therefore counts as one-half for a particular unit cell.
3. A point or atom lying on an edge is shared equally by four unit cells and thus counts as one-fourth for a particular unit cell.
4. A point or atom lying at a corner is shared by eight unit cells, and counts as one-eighth for each particular unit cell.

Now let us look more closely at the contents of some cubic unit cells. The lattice points associated with the space lattice found in each of the three cubic unit cells are indicated in Fig. 11-22. There is one lattice point $(8 \times \frac{1}{8})$ associated with the unit cell of the simple cubic lattice [Fig. 11-22(a)]. Since a unit cell containing one lattice point is called a **primitive cell,** the simple cubic lattice is sometimes called a **primitive cubic lattice.** The second unit cell [Fig. 11-22(b)] has two lattice points, one at the corners $(8 \times \frac{1}{8})$ and one in the center (1×1) of the cube. This is called a **body-centered cubic** cell. Such a cell has points with identical surroundings at the corners and at its center. The third cell [Fig. 11-22(c)] has four lattice points (points with identical environments), one at the corners $(8 \times \frac{1}{8})$ and three for the six face centers $(6 \times \frac{1}{2})$. The cell is called **face-centered cubic.**

The structure of polonium, Section 11.15, may be described as consisting of a simple cubic space lattice with an atom located at each of the lattice points [Fig. 11-22(a)]. The length of the unit cell edge is 3.36 Å. Since the polonium atoms touch along the edges of the cell, the nearest distance between the centers of polonium atoms is 3.36 Å, and the radius of each polonium atom is thus 1.68 Å. It should be noted that the distance between two atoms or ions is measured between the nuclei of the two atoms or ions.

The structure of a metal with a body-centered cubic structure may be described as consisting of a space lattice with a body-centered cubic unit cell [Fig. 11-22(b)]. One metal atom is located on each lattice point, so there are two identical metal atoms (atoms with identical environments) in the unit cell. The atoms touch along the diagonal of the cubic unit cell.

A metal with a cubic closest packed structure may be described as consisting of a space lattice with a face-centered cubic unit cell [Fig. 11-22(c)]. One metal atom is located on each lattice point, so there are four equivalent metal atoms in each unit cell. The structure shown in Fig. 11-22(c) is the same as that shown in Fig. 11-11(b); the perspective is different, however. Note that the atoms touch along the diagonals of the faces of the cell.

Ionic compounds also can crystallize with cubic unit cells; as indicated in Table 11-5, CsCl, NaCl, and ZnS (zinc blende) crystallize with cubic space lattices. Another form of ZnS, wurtzite, crystallizes with a hexagonal space lattice. The

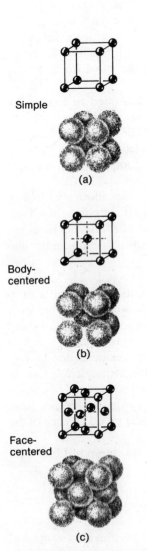

Simple

(a)

Body-
centered

(b)

Face-
centered

(c)

Figure 11-22 Cubic unit cells showing the locations of lattice points in the upper figures and the locations of metal atoms in the lower figures.

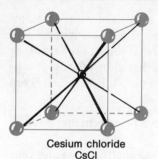

Cesium chloride
CsCl

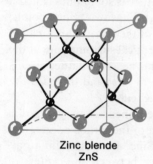

Sodium chloride
NaCl

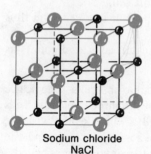

Zinc blende
ZnS

Figure 11-23 The unit cells of some ionic compounds of the general formula MX. The black spheres represent positive ions (cations) and colored spheres represent negative ions (anions). These structures have been expanded to show the geometrical relationships. In the crystal, the cations and anions touch.

assumption of two or more crystal structures by the same substance is called **polymorphism.**

The structure of CsCl is simple cubic. The contents of the unit cell are shown in Fig. 11-23. Chloride ions are located on the lattice points at the corners of the cell, and the cesium ion is located at the center of the cell. The cesium and chloride ions touch along the diagonal of the cubic cell. There is no lattice point in the center of the cell because a cesium ion is not identical to a chloride ion. There are one cesium ion and one chloride ion per unit cell, giving the 1 to 1 stoichiometry required by the formula for cesium chloride.

The unit cell contents of sodium chloride are also illustrated in Fig. 11-23. Chloride ions are located on the lattice points of a face-centered cubic unit cell. Sodium ions are located in the octahedral holes in the middle of the cell edges and in the center of the cell. The sodium and chloride ions touch each other. A count of the number of atoms in the unit cell shows four sodium ions and four chloride ions, giving a 1 to 1 stoichiometry as required by the formula, NaCl.

The contents of the unit cell of zinc blende, ZnS, also are illustrated in Fig. 11-23. This cubic structure contains sulfide ions on the lattice points of a face-centered cubic lattice. Zinc ions are located in alternate tetrahedral holes (one-half of the tetrahedral holes). There are four zinc ions and four sulfide ions in the unit cell, making the unit cell neutral in net charge.

The unit cell contents of calcium fluoride are indicated in Fig. 11-24. This cubic structure has equivalent calcium ions located on the lattice points of a face-centered cubic lattice. All of the tetrahedral sites in the face-centered cubic array of calcium ions are occupied by fluoride ions. There are four calcium ions and eight fluoride ions in a unit cell, giving a calcium to fluorine ratio of 1 to 2 as required by the formula of calcium fluoride, CaF_2. Close examination of Fig. 11-24 will reveal the simple cubic array of fluoride ions with calcium ions in one-half of the cubic holes. The structure cannot be described in terms of a space lattice of points on the fluoride ions inasmuch as all fluoride ions do not have identical environments. The orientation of the four calcium ions about each of the fluoride ions is not always the same.

The contents of the tetragonal unit cell of rutile, TiO_2, are shown in Fig. 11-25.

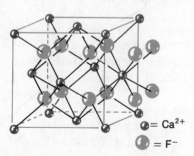

= Ca^{2+}

= F^-

Figure 11-24 The unit cell of CaF_2. The black spheres represent calcium ions, Ca^{2+}, and the colored spheres, fluoride ions, F^-. Note the face-centered cubic array of Ca^{2+} and the simple cubic array of F^-. This structure has been expanded to show the geometrical relationships. In the crystal, the cations and anions touch.

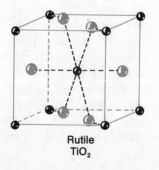

Rutile
TiO_2

Figure 11-25 The unit cell of rutile, TiO_2. The black spheres represent the titanium ions, Ti^{4+}, and the colored spheres the oxygen ions, O^{2-}. This structure has been expanded to show the geometrical relationships. In the crystal, the cations and anions touch.

11.19   Calculation of Ionic Radii

If the edge length of the cubic cell is known, ionic radii for the ions in the crystal lattice can be calculated, making certain assumptions.

EXAMPLE   **a. The unit cell cube edge length for LiCl (NaCl-like structure, face-centered cubic) is 5.14 Å. Assuming anion-anion contact, calculate the ionic radius for chloride ion.**

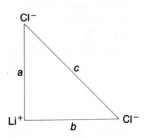

By looking at the diagram for the NaCl structure in Fig. 11-23, we can visualize for a section of the unit cell of LiCl a right triangle involving two chloride ions and one lithium ion. The lithium ion is so small that all ions in the structure touch as in Fig. 11-18b.

Because a (the distance between the center of a chloride ion and the center of a lithium ion) is one-half the edge length of the cubic unit cell,

$$a = \frac{5.14 \text{ Å}}{2}$$

Similarly, b is the distance between the center of a chloride ion and the center of a lithium ion and hence also is one-half the edge length of the cubic unit cell.

$$b = \frac{5.14 \text{ Å}}{2}$$

By the Pythagorean Theorem, the length c (which is the distance between the centers of two chloride ions) can be calculated.

$$c^2 = a^2 + b^2$$

$$c^2 = \left(\frac{5.14}{2}\right)^2 + \left(\frac{5.14}{2}\right)^2 = 13.21$$

$$c = \sqrt{13.21} = 3.63 \text{ Å}$$

Since the anions are assumed to touch each other, c is twice the radius of one chloride ion. Hence, the radius of the chloride ion is $\frac{1}{2}c$.

$$r_{Cl^-} = \tfrac{1}{2}c = \tfrac{1}{2}(3.63 \text{ Å}) = 1.81 \text{ Å}$$

Note that this agrees with the value given on the inside back cover for the radius of the Cl^- ion.

◆ ◆ ◆ ◆ ◆

b. The unit cell cube edge length for potassium chloride (NaCl structure, face-centered cubic) is 6.28 Å. Assuming anion-cation contact, calculate the ionic radius for K^+.

Inspection of the diagram in Fig. 11-23 shows that the distance between the center of a potassium ion and the center of a chloride ion is one-half the cubic unit cell edge length for the KCl unit cell, or $\frac{1}{2}(6.28) = 3.14$ Å.

Assuming anion-cation contact, 3.14 Å is the sum of the radii for K^+ and Cl^-.

$$r_{K^+} + r_{Cl^-} = 3.14 \text{ Å}$$

In part (a) of this example, r_{Cl^-} was calculated as 1.81 Å. Therefore

$$r_{K^+} = 3.14 - 1.81 = 1.33 \text{ Å}$$

Note that the actual radius for K^+, given on the inside back cover, is also 1.33 Å. Also note that the chloride ions do not touch in solid potassium chloride.

◆ ◆ ◆ ◆ ◆

It is important to realize that values for ionic radii calculated from crystal unit cell edge lengths depend also on numerous other assumptions, such as a perfect spherical shape for ions, which are approximations at best. Hence such calculated values are themselves approximate, and comparisons cannot be pushed too far. Nevertheless, this is one of the methods for calculating ionic radii from experimental measurements such as x-ray crystallographic determinations, and the method has proved to be very useful.

11.20   The Lattice Energies of Ionic Crystals

The **lattice energy** of an ionic compound may be defined as the energy required to separate the ions in a mole of the compound by infinite distances. Alternatively, it is also the energy released when a mole of a compound is formed by bringing together from infinite distances the necessary number of positive and negative ions. Lattice energies can be calculated from basic principles or they can be measured experimentally.

The calculation of lattice energies is based primarily on the work of Born, Lande, and Mayer. They derived the following equation to express the lattice energy, U, of an ionic crystal:

$$U = C\frac{Z^+ Z^-}{R_0}$$

where C is a constant that depends upon the type of crystal structure and the electronic structures of the ions, Z^+ and Z^- are the charges on the ions, and R_0 is the interionic distance (the sum of the radii of the positive and negative ions).

By inspection of the lattice energy equation, we may readily deduce that for a given sructure the principal factors determining the lattice energy are Z^+, Z^-, and R_0. The lattice energy increases rapidly with the charges on the ions. Keeping all other parameters constant, doubling the charges on both the cation and anion quadruples the lattice energy.

The lattice energy of LiF is 1023 kJ per mole, while that for MgO is 3900 kJ per mole (R_0 is nearly the same for the two compounds). The lattice energy also increases rapidly with decreasing interionic distance in the lattice, which may result from decreasing either the cation or anion radius, or both. Examples of crystals that exhibit a large difference in interionic distances are lithium fluoride and rubidium chloride.

	Interionic Distance, R_0	Lattice Energy, U
LiF	2.008 Å	1023 kJ/mol
RbCl	3.28 Å	680 kJ/mol

The large lattice energies of ionic compounds result from strong electrostatic forces between the ions in the crystal. These forces must be overcome in order to vaporize the crystal to separated gaseous ions, to melt the crystal, or to vaporize the liquid or solid to gaseous ion pairs. In the case of either fusion or vaporization, however, the energy requirement is less than the total lattice energy because of the interaction, respectively, of the neighboring ions in the liquid phase or the two ions of a pair in the gas phase. However, the strong interionic forces in ionic compounds are responsible for relatively large heats of fusion, sublimation, and vaporization and for the high melting and boiling temperatures as compared to those of most covalent compounds. These same strong forces are responsible for the fact that ionic crystals are generally hard, dense, rigid, and nonvolatile. We should note, however, that while ionic compounds are characterized by such properties, certain covalent structures, such as diamond and silicon carbide, also exhibit them (see Sections 25.1 and 27.6). In such cases, the very strong forces in the crystal are due to strong covalent bonds operative in three dimensions throughout the crystal rather than the weaker London forces usually found between molecules in covalent compounds.

11.21   The Born-Haber Cycle

The direct determination of the lattice energy of an ionic crystal has been made for only a few compounds. In most cases it is not possible to measure the lattice energy directly; however, a cyclic process has been developed independently by Born and Haber; it relates the lattice energy to other thermochemical quantities and makes its calculation possible. The **Born-Haber cycle** is a thermodynamic cycle that involves the heat of formation of the compound (Section 9.6), the ionization potential of the metal (Section 4.19, part 3), the electron affinity of the nonmetal (Section 4.19, part 4), the heats of fusion and vaporization of the metal, the heat of dissociation of the nonmetal, and the lattice energy of the compound.

The Born-Haber cycle for sodium chloride may be expressed diagrammatically as follows:

$$NaCl(s) \xleftarrow{\;\;-U\;\;} Na^+(g) + Cl^-(g)$$
$$\uparrow Q \qquad\qquad\qquad \uparrow I \qquad \uparrow -E$$
$$Na(s) + \tfrac{1}{2}Cl_2(g) \xrightarrow{\;+S+\tfrac{1}{2}D\;} Na(g) \;\; + Cl(g)$$

The symbols shown for each step will be defined later in this section. This diagram indicates hypothetical steps in the formation of sodium chloride from one mole of sodium metal and one-half mole of chlorine gas. We can assume that first the sodium metal is vaporized and the diatomic chlorine is dissociated. Then the sodium atoms are ionized, and the electrons thus obtained are transferred to the chlorine atoms to form chloride ions. Hence sodium ions and chloride ions are the species present in the gaseous phase. These ions are allowed to come together to form solid sodium chloride. The energy change in this step is the negative of the lattice energy, which is the amount of energy required to produce one mole of gaseous sodium ions and one mole of gaseous chloride ions from one mole of solid sodium chloride. The total energy evolved in the preceding hypothetical preparation of sodium chloride is equal to the experimentally determined heat of formation, Q, of the compound from its elements.

The calculation of the lattice energy of sodium chloride may be carried out using experimental data as follows.

1. Heat of formation of sodium chloride:

$$Na(s) + \tfrac{1}{2}Cl_2(g) \longrightarrow NaCl(s) \qquad \Delta H = Q = -411 \text{ kJ}$$

2. Heats of fusion and vaporization of sodium metal:

$$Na(s) \longrightarrow Na(g) \qquad \Delta H = S = 109 \text{ kJ}$$

3. Dissociation energy of molecular chlorine:

$$\tfrac{1}{2}Cl_2(g) \longrightarrow Cl(g) \qquad \Delta H = \tfrac{1}{2}D = 121 \text{ kJ}$$

4. Ionization energy of sodium atom:

$$Na(g) \longrightarrow Na^+(g) + e^- \qquad \Delta H = I = 496 \text{ kJ}$$

5. Electron affinity of chlorine atom:

$$Cl(g) + e^- \longrightarrow Cl^-(g) \qquad \Delta H = -E = -368 \text{ kJ}$$

6. Lattice energy of sodium chloride:

$$Na^+(g) + Cl^-(g) \longrightarrow Na^+Cl^-(s) \qquad \Delta H = -U = ? \text{ (to be calculated)}$$

The total energy of formation of the crystal from its elements is given by the equation

$$Q = S + \tfrac{1}{2}D + I - E - U$$

The value of the heat of formation, Q, is accurately known for many substances. If the other thermochemical values are available, we can solve for the lattice energy (U) by rearranging the preceding equation as follows:

$$U = -Q + S + \tfrac{1}{2}D + I - E$$

For sodium chloride, using the above data, the lattice energy is

$$U = (411 + 109 + 121 + 496 - 368) \text{ kJ} = 769 \text{ kJ}$$

The Born-Haber cycle may be used to calculate any one of the quantities in the equation for lattice energy, provided all of the others are known. Usually, Q, S, I, and D are known. The direct measurement of electron affinities is rather difficult, and only for the halogens have really accurate values of E been determined. For halogen compounds, then, the Born-Haber cycle can be used to calculate lattice energies, the values of which are found to agree well with those obtained by other means. The electron affinity of oxygen, an important quantity, cannot be measured directly. This value has been obtained by applying the Born-Haber cycle to the formation of various ionic oxides using the values of U calculated by the method of Born, Lande, and Mayer.

11.22   X-Ray Diffraction and Crystals

Much information concerning the arrangements of atoms, molecules, and ions within crystals has been obtained from measurements of the transmission and scattering **(diffraction)** of x rays by crystals. X rays are electromagnetic radiations of very short wavelengths (Section 4.7) of the same order of magnitude as the distance between neighboring atoms in crystals (around 2 Å).

William Bragg, an English physicist, showed that the diffraction of x rays by the atoms or ions in a crystal can be interpreted simply by considering the crystal as a reflection grating. The reflection of x rays from the face of a crystal may be compared with the reflection of a beam of light from a stack of thin glass plates of equal thickness. It is known that monochromatic (single-wavelength) light is reflected from a stack of glass plates only at definite angles, the values of which depend upon the wavelength of the incident light and upon the thickness of the plates. For crystals, the planes of atoms or ions within the crystal correspond to the plates of glass. A simple mathematical relation exists between the wavelength, λ, of the x rays, the distance between the atomic planes in the crystal, and the angle of reflection.

As is shown in Fig. 11-26, when a beam of monochromatic x rays impinges at an angle θ upon two planes of a crystal it is reflected, the rays from the lower plane having traveled farther than those from the upper plane by a distance equal to $BC + CD$. If this distance is equal to 1, 2, or 3 or any other integral number of wavelengths ($n\lambda = BC + CD$), the two beams emerging from the crystal will be in phase and will reinforce each other. From the right triangle ABC of Fig. 11-26,

$$\sin BAC = \frac{BC}{AC}$$

Rearranging gives

$$BC = AC \sin BAC$$

The distance AC is the distance between the crystal planes. We shall call this distance d. It can be shown that angle BAC is equal to θ, which can be called the angle of grazing incidence.

Then,

$$BC = d \sin \theta$$

Figure 11-26   Diffraction of a monochromatic beam of x rays by two planes of a crystal.

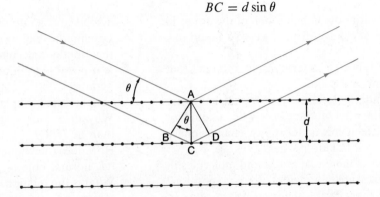

But the difference in length of path of the two beams is $BC + CD$. Because $BC = CD$, the total difference in length of path is $2BC$. X rays will be reflected, provided that, as indicated previously,

$$n\lambda = BC + CD = 2BC$$

Hence, $$n\lambda = 2d \sin \theta \quad \text{(the } \textbf{Bragg equation)}$$

If the x rays strike the crystal at angles other than θ, the extra distance $BC + CD$ will not be an integral number of wavelengths, and there will be interference of the rays reflected from the different crystal planes rather than reinforcement. This will result in weakening the intensity of the reflected rays.

The reflection corresponding to $n = 1$ is called the first-order reflection, that corresponding to $n = 2$ is the second-order reflection, and so forth. Each successive order gives a larger angle and weaker intensity.

The maximum intensities are found by rotating the crystal so as to change the angle θ at which the x-ray beam strikes the surface. If the wavelength of the incident radiation is known, it becomes possible to determine the spacing, d, of the planes of atoms or ions within the crystal. On the other hand, if the x rays are incident upon the face of a crystal for which d is known, the value of λ for the radiation may be determined by measuring the angles at which the reflections of maximum intensity occur.

Questions

1. Describe the variation in the motion of the atoms in a substance as it changes from a gas to a liquid to a solid.
2. Describe the types of "intermolecular forces" that can be present in a solid.
3. The types of intermolecular forces in a substance are identical whether the substance is a solid, a liquid, or a gas. Why then does a substance change phase from a solid to a liquid or a gas?
4. What evidence is there that atoms and molecules in solid substances are not rigidly fixed in position?
5. What feature characterizes the equilibrium between a solid and its liquid as being dynamic?
6. Explain the cooling effect of evaporation in terms of the kinetic-molecular theory.
7. Is it possible to liquefy sulfur dioxide at room temperature? Is it possible to liquefy oxygen at room temperature? Explain your answers.
8. Why does iodine sublime more rapidly at 110°C than at 25°C?
9. How does the "boiling point" of a liquid differ from its "normal boiling point"?
10. What is the relationship between the intermolecular forces in liquids and their boiling points?

11. Why do the boiling points of the halogens increase as their atomic weights increase?

12. The melting points of NaCl, KCl, and RbCl decrease with increasing atomic number of the alkali metal. Suggest an explanation.

13. Account for the fact that steam produces much more severe burns than does the same mass of boiling water.

14. By referring to Fig. 11-2, determine the boiling point of ethyl ether at 500 torr, the boiling point of ethyl alcohol at 0.5 atm, and the boiling point of water at 55,000 Pa. At what temperature will the vapor pressure of ethyl alcohol be equal to the vapor pressure of ethyl ether at 20°?

15. A closed piston like that shown in Fig. 10-1 is filled with liquid ether so that there is no space for any vapor. Describe what will happen if the plunger is withdrawn somewhat, forming a volume that can be occupied by vapor. If the system is insulated so that no heat can enter or leave, will the temperature of the liquid increase, decrease, or remain constant? Explain.

16. What are the origins of the van der Waals forces of attraction?

17. Describe how the water in a solution of sodium chloride can be separated from the sodium chloride by distillation.

18. If a mixture of ethyl ether and ethyl alcohol is distilled, is the resulting distillate pure?

19. Silane (SiH_4), phosphine (PH_3), and hydrogen sulfide (H_2S) melt at $-185°C$, $-133°C$, and $-85°C$, respectively. What does this suggest about the polar character and intermolecular attractions in the three compounds?

20. How do crystalline solids and "amorphous solids" (supercooled liquids) differ with regard to internal structure?

21. Contrast crystalline and "amorphous" solids (supercooled liquids) with regard to melting points.

22. How do we know that certain solids such as ice and naphthalene have vapor pressures?

23. What types of liquid materials tend to undercool readily and form glasses?

24. Copper crystallizes with four equivalent metal atoms in a cubic unit cell. Describe the crystal structure of copper.

25. The carbon atoms in the unit cell of diamond occupy the same positions as both the zinc and sulfur atoms in cubic zinc sulfide. How many carbon atoms are found in the unit cell of diamond? The bonds between the carbon atoms are covalent. What is the hybridization of a carbon atom?

26. In terms of its internal structure, explain the high melting point (and hardness) of diamond. The structure of diamond is described in Question 25.

27. Silicon carbide can be described as a cubic closest packed array of silicon atoms, with carbon atoms in one-half of the tetrahedral holes. What is the empirical formula of silicon carbide? What is the coordination number of carbon in silicon carbide?

28. Anatase is a mineral that contains titanium and oxygen. The structure of anatase consists of a closest packed array of oxygen atoms with titanium in one-half of the octahedral holes. What is the formula of anatase? What is the oxidation number of titanium?

29. Why are compounds that are isomorphous with CsCl generally not observed when the radius ratio for the compound is less than 0.73?

30. The free space in a lattice of spheres may be found by subtracting the volume of the spheres in a unit cell from the volume of the cell. Calculate the per cent of free space in each of the three cubic lattices if all atoms in each lattice are of equal radius and touch their nearest neighbors. Which of these lattice structures represents the most efficient packing of spheres into a volume?

31. The chemically similar alkali metal chlorides NaCl and CsCl have different crystal structures, while chemically different NaCl and MnS have the same structure. Explain.

32. Divide the following compounds according to the type of crystal structure they form, assuming that they can be treated as ionic compounds: NaCl, CsCl, ZnS, RbH, MgO, ZnTe, CaTe, TlBr, CsI, AlP, BeO.

33. What changes in a given type of crystal structure result in an increase in the lattice energy of the structure?

34. Compare the features that determine the geometry of the nearest neighbors around the nonmetal in a nonmetal halide (Chapter 7) with those that determine the geometry of the nearest neighbors about a metal in an ionic metal halide.

Problems

NOTE: In Problems 1–11, the following data for water will be useful.

Heat of vaporization of water (100°C)	2258 J/g
Heat of vaporization of water (25°C)	2436 J/g
Heat of fusion of water (0°C)	333.6 J/g
Specific heat of water (solid)	2.04 J/g K
Specific heat of water (liquid)	4.18 J/g K
Specific heat of water (gas)	2.00 J/g K

s1. In hot, dry climates water is cooled by allowing some of it to evaporate slowly. How much water must evaporate to cool 1.00 kg from 37°C to 21°C? Assume that the heat of vaporization of water is constant between 21° and 37° and is equal to the value at 25°.　*Ans. 27 g*

2. How much heat is required (a) to change 3.00 mol of water at 100°C to steam at 100°C? (b) to change 0.2500 mol of ice at 0°C to water at 0°C?
　Ans. 122 kJ; 1.502 kJ

s3. How much heat is produced when 75 g of steam at 135°C is converted to ice at −40°C?
　Ans. 240 kJ

4. How much heat is required to convert 35.0 g of ice at −8.0°C to steam at 105°C?　*Ans. 106 kJ*

s5. The specific heat of copper is 0.0931 cal/(g K). What weight of steam at 100°C must be converted to water at 100°C to raise the temperature of a 1.00×10^2 g copper block from 20°C to 100°C?
　Ans. 1.38 g

6. How much more heat is contained in 5.0×10^2 kg of steam at 100°C than in the same amount of ice at −10°C?　*Ans. 1.5×10^6 kJ*

7. How much water at 12°C could be converted to ice at 0°C by the cooling action of vaporizing 25 g of liquid ammonia? The heat of vaporization of liquid ammonia is 1368 J/g.　*Ans. 89 g*

s8. During the fermentation step in the production of beer, 4.5×10^6 kcal of heat are evolved per 1000 gal of beer produced. How many liters of cooling water are required to maintain the optimum fermentation temperature of 58°F for 1000 gal of beer? How many gallons? The cooling water enters with a temperature of 5°C and is discharged with a temperature of 13°C.　*Ans. 6×10^5 ℓ; 2×10^5 gal*

s9. A river is 30 ft wide and has an average depth of 5 ft and a current of 2 mi/hr. A power plant dissipates

2.1×10^5 kJ of waste heat into the river every second. What is the temperature difference between the water upstream and downstream from the plant?　*Ans. 4°C*

10. If the heat dissipated by the power plant in Problem 9 is to be removed by evaporating water at 25°C, how many gallons of water per day would be needed?　*Ans. 2.0×10^6 gal/day*

11. If 135 g of steam at 100°C and 475 g of ice at 0°C are combined, what is the temperature of the resultant water, assuming that no heat is lost?
　Ans. 79.5°C

s12. Silver crystallizes in a face-centered cubic unit cell with a silver atom on each lattice point. (a) If the unit cell dimension is 4.0862 Å, what is the atomic radius of silver? (b) Calculate the density of silver.
　Ans. (a) 1.4447Å; (b) 10.50 g/cm³

s13. One form of tungsten crystallizes in a body-centered cubic unit cell with one tungsten atom on each lattice point. If the unit cell edge is 3.165 Å, what is the atomic radius of tungsten in this structure?
　Ans. 1.370 Å

s14. The atomic radius of lead is 1.75 Å. If lead crystallizes in a face-centered cubic unit cell with the nearest neighbors in contact, what is the length of the unit cell edge?　*Ans. 4.95 A*

15. TlI crystallizes with the same structure as CsCl. The unit cell dimension of TlI is 4.20 Å. Assuming cation-anion contact, calculate the ionic radius of Tl+. (The ionic radius of I− may be found on the inside of the back cover.)　*Ans. 1.48 Å*

s16. LiH crystallizes with the same crystal structure as NaCl. The cubic unit cell dimension of LiH is 4.08 Å. Assuming anion-anion contact, calculate the ionic radius of H−.　*Ans. 1.44 Å*

17. The unit cell edge length of CaF_2 is 5.46295 Å. The density of CaF_2 is 3.1805 g/cm³. From these data and the atomic weights of calcium and fluorine, calculate Avogadro's Number.
　Ans. 6.023×10^{23}

18. A cubic unit cell contains rhenium ions at the corners and oxide ions at the center of each edge. What is the empirical formula of the oxide?
　Ans. ReO_3

19. A cubic unit cell contains oxide ions in the center of each face, a titanium ion in the center of the cell, and calcium ions at the corners of the cell. What is

the empirical formula of this oxide? What is the oxidation number of the titanium ion?

Ans. CaTiO₃; +4

20. A compound containing zinc, aluminum, and sulfur crystallizes with a closest packed sulfide array. Zinc ions are found in one-eighth of the tetrahedral holes and aluminum ions in one-half of the octahedral holes. What is the empirical formula of the compound? *Ans. ZnAl₂S₄*

21. The following compounds crystallize with either a NaCl, CsCl, ZnS, CaF₂, or TiO₂-type structure. From their radius ratios, predict the structure formed by each.

s (a) CoF₂   *Ans. TiO₂ str.*
s (b) ZnTe   *Ans. ZnS str.*
s (c) BaF₂   *Ans. CaF₂ str.*
s (d) KBr   *Ans. NaCl str.*
s (e) AlP   *Ans. ZnS str.*
s (f) CaS   *Ans. NaCl str.*
(g) SrF₂   *Ans. CaF₂ str.*
(h) NiO   *Ans. NaCl str.*

(i) CsBr   *Ans. CsCl str.*
(j) ZnF₂   *Ans. TiO₂ str.*

s 22. What x-ray wavelength would be diffracted at a first-order θ angle of 10.40° by planes with a spacing of 2.00 Å? *Ans. 0.722 Å*

23. What is the spacing between planes which diffract 2.28 Å x rays at an angle, θ, of 21°31′ (first order)? *Ans. 3.11 Å*

24. Gold crystallizes in a face-centered cubic unit cell. The first-order x-ray reflection ($n = 1$) for the planes of atoms which make up the tops and bottoms of the unit cells is at $\theta = 10.89°$. The wavelength of the x rays is 1.54 Å. What is the density of the metallic gold? *Ans. 19.3 g/cm³*

s 25. When an electron in an excited Mo atom falls from the L to the K shell, an x ray is emitted. These x rays are diffracted at an angle of 7.75° by planes with a separation of 2.64 Å. What is the difference in energy between the K and the L shell in molybedenum? *Ans. 2.79 × 10⁻⁸ erg or 1.74 × 10⁴ eV*

References

"Ion-Ion, Ion-Dipole, and Dipole-Dipole Interactions," C. H. Yoder, *J. Chem. Educ.,* **54,** 402 (1977).

"Non-Crystalline Materials," E. A. Davis, *Endeavour,* **1,** 103 (1977).

"Thermotropic Liquid Crystals," G. R. Van Hecke, *J. Chem. Educ.,* **53,** 161 (1976).

"Ionic Hydration Enthalpies," D. W. Smith, *J. Chem. Educ.,* **54,** 540 (1977).

"An Introduction to Principles of the Solid State," P. F. Weller, *J. Chem. Educ.,* **47,** 501 (1970).

"Chemical Aspects of Dislocations in Solids," J. M. Thomas, *Chem. in Britain,* **6,** 60 (1970).

"The Undercooling of Liquids," D. Turnbull, *Sci. American,* January 1965, p. 38.

"The Solid State," Sir Neville Mott, *Sci. American,* September 1967, p. 80.

"Solid Phase Synthesis," D. C. Neckers, *J. Chem. Educ.,* **52,** 695 (1975).

"Other Views of Unit Cells," L. Suchow, *J. Chem. Educ.,* **53,** 226 (1976).

"The Influence of Lattice Defects on the Properties of Solids," M. I. Pope, *Educ. in Chemistry,* **2,** 127 (1965).

"The Nature of Glasses," R. J. Charles, *Sci. American,* September 1967, p. 127.

"X-Ray Crystallography," Sir Lawrence Bragg, *Sci. American,* July 1968, p. 58.

"Ionic Conduction and Diffusion in Solids," J. Kummer and M. E. Milberg, *Chem. and Eng. News,* May 12, 1969, p. 90.

"Crystal Chemistry of d^0 Cations," M. R. Truter, *Chem. in Britain,* **7,** 203 (1971).

"Accurate Molecular Geometry," D. H. Whiffen, *Chem. in Britain,* **7,** 57 (1971).

"Opals," P. J. Darragh, A. J. Gaskin, and J. V. Sanders, *Sci. American,* April 1976, p. 84.

"Ambidextrous Crystals," R. E. Newnham and L. E. Cross, *Endeavour,* January 1974, p. 8.

"Perovskite Oxides, Materials Science in Catalysis," R. J. H. Voorhoeve *et al., Science,* **195,** 827 (1977).

"Crystal Chemistry of Si—O Bonds at High Pressure: Implications for the Earth's Mantle Mineralogy," R. M. Hazen and L. W. Finger, *Science,* **201,** 1122 (1978).

"Species in Layers, Cavities, and Channels (or Trapped Species)," J. E. P. Davies, *J. Chem. Educ.,* **54,** 536 (1977).

"Simple Calculation of the Lattice Energy of Lithium Hydride," F. Rioux, *J. Chem. Educ.,* **54,** 555 (1977).

"The Tensile Strength of Liquids," R. E. Apfel, *Sci. American,* December 1972, p. 58.

"Snow Crystals," C. Knight and N. Knight, *Sci. American,* January 1973, p. 100.

"Close-Packing of Atoms," H. C. Benedict, *J. Chem. Educ.,* **50,** 419 (1973).

"Isoelectronic Principle," R. R. Heald, *Chemistry,* **46** (6), 10 (1973).

"The Chemistry of Liquid Metals," C. C. Addison, *Chem. in Britain,* **10,** 331 (1974).

"X-Ray Crystallography: A Refinement of Technique," T. H. Maugh II, *Science,* **186,** 913 (1974).

"X-Ray Diffraction: Solving the Structure," J. J. Guy, *Educ. in Chem.,* July 1978, p. 126.

"More U.S. Chemists Using Neutron Diffraction" (Staff), *Chem. and Eng. News,* January 29, 1979, p. 26.

"Heterogeneous Catalysis, Some Recent Developments," J. H. Sinfelt, *Science,* **195,** 641 (1977).

"Surface Science," G. A. Somorjai, *Science,* **201,** 489 (1978).

12

Water and the Environment; Hydrogen Peroxide

Water

Water covers nearly three-fourths of the earth's surface; it is present in the atmosphere and the earth's crust and composes a large part of all living plant and animal matter. Of all the hundreds of thousands of chemical substances, none is more important to plant and animal life than water.

12.1 The Structures of Gaseous, Liquid, and Solid Water

Until the latter part of the eighteenth century, water was thought to be an element. In 1781, Henry Cavendish showed that water is formed when hydrogen burns in air, and, a few years later, Antoine-Laurent Lavoisier determined the composition of water by mass. The formula for water is H_2O, and the relative masses of hydrogen and oxygen in the compound have been determined with great accuracy to be $1.0079:8.000$.

Water has an abnormally low vapor pressure, a high boiling point, a high heat of vaporization, and remarkable solvent properties when compared to molecules of similar formula such as H_2S, H_2Se, and H_2Te. These unusual properties are all closely related to the structure of the water molecule. The oxygen atom has the electronic structure $1s^2 2s^2 2p^4$, with two unpaired electrons in its valence shell (Section 4.16). Each of these unpaired electrons pair with the $1s$ valence electron in each of the two hydrogen atoms as indicated by the Lewis electronic structure.

(a)

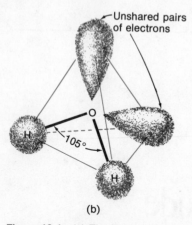

Unshared pairs
of electrons

105°

(b)

Figure 12-1 (a) The Lewis valence electron formula and (b) the bent structure of the water molecule. Note that the Lewis formula shows only the distribution of electrons and does not represent the three-dimensional structure of the molecule.

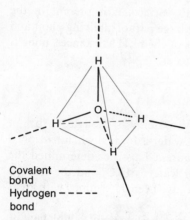

Covalent ———
bond
Hydrogen - - - - -
bond

Figure 12-2 The tetrahedral arrangement of four hydrogen atoms about an oxygen atom in ice. Two hydrogen atoms participate in a covalent bond and two in a hydrogen bond to the oxygen. Each bond leading from a hydrogen atom is to an oxygen of another water molecule.

The water molecule contains two bonding electron pairs and two lone pairs, as shown in Fig. 12-1(a). As discussed in Chapter 7, these four regions of high electron density are arranged tetrahedrally about the oxygen atom, giving the water molecule a bent structure [Fig. 12-1(b)]. The reduction of the H—O—H angle from 109.5° (the bond angle in a regular tetrahedral molecule) to 105° can be attributed to the balance between the repulsions of the lone pairs and bond pairs. In spite of the repulsions between them, the electron pairs in the hydrogen-oxygen bonds are pushed closer together by the stronger lone pair–bond pair repulsions (Section 7.1). Since the distortion from a regular tetrahedral arrangement is small, the hybridization of the oxygen atom can be described as essentially sp^3 (Section 7.3). Two of the four hybrid orbitals form σ bonds with the hydrogen atoms. The two remaining hybrid orbitals contain lone pairs.

The atoms in a molecule of water are held together by covalent bonds that are distinctly polar in character, because oxygen is more electronegative (3.5) than hydrogen (2.2). (See Table 5-1.) Thus the oxygen atom possesses a partial negative charge and each hydrogen atom a partial positive charge. If the molecule were linear (H—O—H), the two bond polarities would cancel each other, and the molecule would be nonpolar. However, since the two bonds make an angle of 105° with each other, the molecule is highly polar. In solid or liquid water, the water molecules are associated closely together. This association occurs because the molecules are sufficiently close together and moving slowly enough for the partial positive charges on the hydrogen atoms of one molecule to be weakly attracted electrostatically to the partial negative charges on the oxygen atoms of two neighboring molecules. In the gaseous state the larger distances between molecules and their larger kinetic energy (due to their faster motion) make the attractions negligible; the molecules of water in the gaseous state therefore exist largely as discrete molecules.

The attraction between molecules in the solid and liquid states of water forms what is known as a **hydrogen bond** (see Section 19.10, part 1). Hydrogen bonds are not strong; the strength of a hydrogen bond between water molecules is about 5% that of a covalent hydrogen-oxygen bond. The hydrogen bond between water molecules nevertheless is among the strongest of *intermolecular* forces of attraction (attractive forces between molecules) and is an important contributing factor in the distinctive physical properties of both liquid and solid water.

X-ray diffraction studies have shown that the water molecules in ice (solid water) are so arranged that each oxygen atom has four hydrogen atoms as close neighbors in an approximately tetrahedral arrangement; two are attached by electron pair bonds and two by hydrogen bonds (Fig. 12-2). Each hydrogen atom participates in a covalent bond to one oxygen atom and in a hydrogen bond to another. Such an arrangement leads to an open structure, i.e., one with relatively large holes in it. This makes ice (Fig. 12-3) a substance with a relatively low density. When ice at 0° is heated, some of the hydrogen bonds are broken, the open structure is destroyed, and the water molecules pack more closely together in the liquid. This accounts for the unusual observation that liquid water is more dense than the solid (ice). It is rare for a solid substance to be less dense than the liquid.

Although the structure of liquid water is still not completely understood, it appears that clusters of varying numbers of hydrogen-bonded water molecules

Figure 12-3 The arrangement of water molecules in ice, showing each oxygen atom with four hydrogen atoms as close neighbors, two attached by electron-pair bonds and two by hydrogen bonds. The oxygen atoms shown on the corners of a prism are also on the corners of adjoining prisms, as this is only a portion of a continuous array.

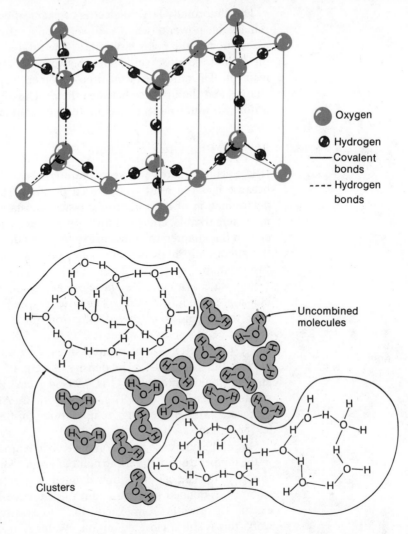

Figure 12-4 The structure of liquid water showing hydrogen-bonded clusters and uncombined molecules.

exist throughout the liquid. Probably some uncombined water molecules are present as well (Fig. 12-4). One theory of the structure of liquid water suggests a dynamic equilibrium between the clusters and uncombined molecules, with the clusters continually varying in size and shape as they pick up uncombined molecules and lose other molecules that become uncombined. As the temperature is raised, more of the hydrogen bonds are disrupted, and the larger clusters are broken down into smaller ones. In the vapor state water is composed almost entirely of single H_2O molecules. Association of the polar water molecules is responsible for the abnormally low vapor pressure, high boiling point, high heat of fusion, and high heat of vaporization of water relative to the same properties of nonassociated compounds of similar formulas such as H_2S, H_2Se, and H_2Te (see Fig. 19-6). These abnormal physical properties of water are due to the fact that energy is required to break the hydrogen bonds holding together clusters of associated molecules.

The abnormally large dielectric constant of water, which is responsible for the remarkable power of water to dissolve ionic substances (Chapter 13), is due to the association of the polar molecules. The product of the charge on either end of a dipole and the effective distance between the centers of the charges is the **dipole moment.** Thus the size of the dipole moment depends upon the magnitude of the charges and the distance between them. The dipole moment of a complex made up of two water molecules is more than double that of a single molecule.

12.2   Physical Properties of Water

It is essential that we have a thorough knowledge of the properties of water because it is the most widely used of all substances and is a standard for the determination of a number of physical constants and units. In addition, it has many remarkable, almost unique, properties. These properties, which are considered in this chapter, include the expansion of the liquid on cooling below 4°C and the abnormally high melting point, boiling point, heat of fusion, and heat of vaporization for a molecule of such small molecular weight.

Pure water is an odorless, tasteless, and colorless liquid. Natural waters appear bluish-green in large bodies. The taste of drinking water is due to dissolved gases from the air and to dissolved salts from the earth.

The freezing point of water is 0°C (273.15 K), and its boiling point at 1 atm is 100°C (373.15 K). The vapor pressures of ice and water at various temperatures are listed in Table 10-1. The density of ice at 0° is 0.9168 g/cm^3. The density of air-free liquid water at 0°C is 0.99984 g/cm^3. As it warms, water reaches its maximum density of 0.99997 g/cm^3 at 3.98°C. Above this temperature its density decreases with increasing temperature to 0.99704 g/cm^3 at 25°C and to 0.95837 g/cm^3 at 100°C.

One of the reference points on the Kelvin temperature scale, 273.16 K, is defined as the temperature of the triple point of water. The triple point is discussed later in this section. The calorie is now defined as exactly 4.184 J. Earlier, however, the calorie was defined as the amount of heat necessary to raise the temperature of exactly 1 g of water from exactly 14.5° to exactly 15.5°C; except for very precise work, this is still a close definition. Water is also the reference substance in the measurement of the specific gravity of liquids and solids (Section 1.18). The heat of fusion (Section 11.13) of water is 333.6 J/g (79.71 cal/g) at 0°, and its heat of vaporization (Section 11.5) is 2257.5 J/g (539.55 cal/g) at 100°.

A diagram relating the pressures and temperatures at which the gaseous, liquid, and solid states (sometimes referred to as phases) can exist is called a **phase diagram.** Such a diagram is constructed by plotting the points corresponding to pressures and temperatures at which changes of state take place. A part of the phase diagram for water is given in Fig. 12-5.

As heat is added to ice at 760 torr, the ice changes at 0°C from a solid to a liquid; hence, these values locate one point on the line (in color) that separates the regions for solid and liquid in the diagram. At the same pressure, a further increase of temperature causes a change of state from liquid to gas at 100°, these values corresponding to a point on the line (also in color) that separates the liquid and gas regions in the diagram. At a lower pressure, 600 torr, the change from

Figure 12-5   Phase diagram for water.

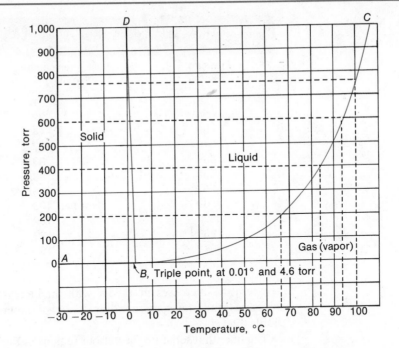

liquid to gas takes place at 93.5°; these two values then locate another point. Other points are determined similarly in order to produce the complete diagram.

For each value of pressure and temperature falling within the portion of the diagram labeled "Solid," water exists only in the solid form (ice). The boundary of this region, the line *AB*, gives the vapor pressure of ice; at any point along this line ice and water vapor are in equilibrium. The boundary *BD* gives the temperature at which ice and liquid water are at equilibrium as the pressure changes. Similarly for values of pressure and temperature within the portion labeled "Liquid," water exists only as a liquid. The boundary *BC* gives the vapor pressure of liquid water as the temperature varies; this boundary indicates the temperature and pressure conditions under which liquid water and water vapor are in equilibrium. For any combination of pressure and temperature in the region labeled "Gas," water exists only in the gaseous state. For example, at −5° and 760 torr the point on the diagram is in the region indicating that only solid water is stable. Moving across the diagram horizontally at a constant pressure of 760 torr, we see that as the temperature increases water becomes a liquid at 0°, which is its *normal melting temperature.* At this temperature and pressure, both solid and liquid water can exist in equilibrium. Water is stable only as a liquid at a pressure of 760 torr for temperatures between 0° and 100°. It changes to a gas at 100°, its *normal boiling temperature,* at which liquid water at 760 torr and water vapor at 760 torr can exist at equilibrium. At temperatures above 100° at 760 torr, water is stable only as a gas.

The almost vertical colored line in Fig. 12-5 shows us that as the pressure increases, the melting temperature remains almost constant, actually decreasing very slightly. On the other hand, the diagram indicates clearly that the boiling

Pressure, torr	Boiling Temperature of Water, °C
TABLE 12-1 Variation of the Boiling Temperature of Water with Pressure	
5	1
20	22
80	44
310	77
596	93
600	93.5
702	98
760	(Normal) 100
822	102
1960	129
5042	163

temperature increases markedly with an increase in pressure. Representative pressures and the corresponding boiling temperatures for water are given in Table 12-1.

The lines that separate the various regions in Fig. 12-5 represent points at which an equilibrium exists between two states. Combinations of pressure and temperature falling on the line between the solid and liquid regions, for example, represent conditions under which the solid and liquid are in equilibrium with each other.

Note that at just one point on the diagram the three lines intersect. This point is referred to as the **triple point.** At the pressure and temperature corresponding to the triple point [4.6 torr and 0.01°C (273.16 K), respectively], all three states are in equilibrium with each other. Reference to the diagram shows that at temperatures and pressures below those at the triple point, water changes directly from solid to gas without going through the liquid state (sublimation—see Section 11.14) when the pressure on it is reduced below the triple point pressure. Frost forms under these conditions as water vapor condenses to ice.

Phase diagrams can also be constructed for compounds other than water, and such diagrams prove to be very useful to scientists working with the various physical states of these materials.

12.3   Chemical Properties of Water

Water is a medium in which many chemical reactions take place. The majority of the chemical changes that we study in a course in general chemistry are reactions taking place in aqueous solutions.

▶ **1. THE THERMAL STABILITY OF WATER.** Water is the product of a highly exothermic reaction, the burning of hydrogen in oxygen; i.e., its heat of formation is high and it is accordingly a very stable compound. It decomposes to the extent of only 11.1% at the very high temperature of 2727°, the reaction being reversible. As the temperature of the system is lowered, the constituent elements recombine. The reversibility of the reaction can be represented by double arrows.

$$2H_2O \rightleftharpoons 2H_2 + O_2 \qquad \Delta H = 571.5 \text{ kJ}$$

Thermal stability is not to be confused with nonreactivity. Although water is stable toward decomposition by heat, it is a reactive substance and enters readily into a great number and variety of reactions, even at ordinary temperatures.

▶**2. IONIZATION OF WATER.** As described in Chapter 8, water can accept protons from an acid to form the hydronium ion.

$$HCl(g) + H_2O(l) \longrightarrow H_3O^+(aq) + Cl^-(aq)$$

Water can also act as an acid and provide protons, to a limited extent, to a base such as ammonia.

$$NH_3(g) + H_2O(l) \rightleftharpoons NH_4^+(aq) + OH^-(aq)$$

Thus water can both accept protons and, to a limited extent, provide protons. In pure water, there is a very limited transfer of protons from one water molecule to another, giving a very low concentration of hydronium ion and hydroxide ion.

$$2H_2O(l) \rightleftharpoons H_3O^+(aq) + OH^-(aq)$$

Even in completely pure water, there are very small, *but equal,* amounts of H_3O^+ and OH^-.

The addition of an acid to water causes an increase in the hydronium ion concentration and a corresponding decrease in the hydroxide ion concentration, but not to zero. With the addition of a base to water, the hydroxide ion concentration becomes large and the hydronium ion concentration very small. Many of the chemical properties of water depend upon the hydronium ion and hydroxide ion content. A more complete account of the ionization of water is given in Chapter 16.

▶**3. ACTION OF WATER WITH CERTAIN ELEMENTS TO GIVE HYDROGEN.** The reactivity of various metals with water was mentioned in connection with the preparation of hydrogen (Chapter 9). Only those metals that lie above cobalt in the activity series (Section 9.21) displace hydrogen from water. The more active metals react with cold water, but the less active ones must be heated to high temperatures in steam.

$$2Na + 2H_2O \text{ (cold)} \longrightarrow 2Na^+ + 2OH^- + H_2$$
$$3Fe + 4H_2O \text{ (steam)} \longrightarrow Fe_3O_4 + 4H_2$$

Magnesium reacts with hot water to give the hydroxide and hydrogen.

$$Mg + 2H_2O \text{ (hot)} \longrightarrow Mg(OH)_2 + H_2$$

When magnesium burns in steam, the oxide rather than the hydroxide results.

$$Mg + H_2O \text{ (steam)} \longrightarrow MgO + H_2$$

White-hot carbon also reacts with steam and produces a mixture of carbon monoxide and hydrogen called **water gas.**

$$C(s) + H_2O(g) \longrightarrow CO(g) + H_2(g)$$

Because coal can be used as the source of carbon, water gas is used extensively as a fuel gas.

▶ **4. REACTION OF WATER WITH OXIDES.** The oxide ion, O^{2-}, is a very strong base that will remove a hydrogen ion from the weakly acidic water molecule.

$$O^{2-} + H_2O \longrightarrow 2OH^-$$

Since oxides of the metals contain oxide ions (Section 8.3), those metal oxides that are soluble react with water to give hydroxides. The oxides of the alkali metals (Li, Na, K, Rb, and Cs) react readily with one mole of water per mole of oxide in exothermic reactions to form hydroxides.

$$Na_2O(s) + H_2O(l) \longrightarrow 2Na^+(aq) + 2OH^-(aq)$$

The alkali metal hydroxides thus produced are too stable to be decomposed by heating, even at their boiling points, which are high. The oxides of the alkaline earth metals (Be, Mg, Ca, Sr, Ba, and Ra) react less readily with water than oxides of the alkali metals and give hydroxides that may be decomposed into metal oxides and water by heating. Beryllium and magnesium oxides, which are highly insoluble in water, react with water only slowly and incompletely. Metal oxides that react with water to form bases are sometimes referred to as **basic anhydrides.**

The oxides of the nonmetals are covalent molecules that do not contain oxide ions. Most of these molecules react with water to form acids. Hence, they are sometimes referred to as **acidic anhydrides.** Some examples are shown by the following equations:

$$SO_2 + H_2O \longrightarrow \quad H_2SO_3$$
Sulfurous acid

$$CO_2 + H_2O \longrightarrow \quad H_2CO_3$$
Carbonic acid

$$P_4O_{10} + 6H_2O \longrightarrow \quad 4H_3PO_4$$
Phosphoric acid

▶ **5. HYDROLYSIS.** **Hydrolysis** is a general term for any reaction in which the water molecule is split and a hydrogen atom, an oxygen atom, and/or a hydroxide group from the water is incorporated into the products that result from the reaction. Hydrolysis is generally observed in certain compounds with ionic or polar bonds. The positive end of the bond is attracted to the oxygen at the negative end of a dipolar water molecule, and the negative end of the bond is attracted to a hydrogen at the positive end of a water molecule. Compounds that undergo hydrolysis include almost all polar nonmetal halides (except those of carbon) and salts of weak acids and weak bases. The hydrolysis of nonmetal halides is illustrated by the following equations:

$$SiBr_4(l) + 4H_2O(l) \longrightarrow Si(OH)_4(s) + 4HBr(aq)$$
$$PCl_5(l) + 4H_2O(l) \longrightarrow (HO)_3PO(aq) + 5HCl(aq)$$

One product of the second reaction, phosphoric acid, is more commonly written as H_3PO_4.

Examples of the hydrolysis of salts include the following:

$$Li_3P(s) + 3H_2O(l) \longrightarrow 3Li^+(aq) + 3OH^-(aq) + PH_3(g)$$

$$MgS(s) + 2H_2O(l) \longrightarrow Mg(OH)_2(s) + H_2S(g)$$

Reactions of this type will be considered more fully in Chapter 16.

Chlorine and bromine also undergo hydrolysis to a limited extent. The equation for the reaction of chlorine with water at room temperature is

$$Cl_2(g) + 2H_2O(l) \rightleftharpoons H_3O^+(aq) + Cl^-(aq) + \underset{\text{Hypochlorous acid}}{HOCl(aq)}$$

$$\underset{\text{Hydrochloric acid}}{}$$

Hydrochloric acid, a strong acid, is completely ionized in dilute aqueous solution. On the other hand, hypochlorous acid, a weak acid, is only slightly ionized, most of it being in the form of molecules in aqueous solution.

Bromine reacts similarly; but fluorine, with its unusually strong oxidizing power, reacts with water in a complex way, producing ozone, O_3.

▶ **6. HYDRATION.** The combination of water with other substances without splitting of the water molecule is called **hydration.** Hydrates, the products of such a reaction, are discussed in the next section.

12.4 Hydrates

When aqueous solutions of many of the soluble salts are evaporated, the salt separates as crystals that contain the salt and water combined in definite proportions. Such compounds are known as **hydrates,** and the water is called **water of hydration.** The formation of hydrates is not limited to salts but is also common with acids and bases and even some free elements. Furthermore, it is not confined to crystals. Familiar examples of hydrates are blue vitriol, $CuSO_4 \cdot 5H_2O$; Epsom salts, $MgSO_4 \cdot 7H_2O$; alum, $KAl(SO_4)_2 \cdot 12H_2O$; Glauber's salt, $Na_2SO_4 \cdot 10H_2O$; $H_2SO_4 \cdot H_2O$; $Ba(OH)_2 \cdot 8H_2O$; $BaCl_2 \cdot 6H_2O$; and $Cl_2 \cdot 8H_2O$.

When blue crystals of copper(II) sulfate 5-hydrate, $CuSO_4 \cdot 5H_2O$, are heated, water is given off, the crystalline structure characteristic of the hydrated salt breaks down, and the white anhydrous salt, $CuSO_4$, remains.

$$CuSO_4 \cdot 5H_2O \rightleftharpoons CuSO_4 + 5H_2O$$

The preceding reaction is reversible in that the addition of water to the anhydrous salt produces the original hydrated salt.

X-ray diffraction has shown that in $CuSO_4 \cdot 5H_2O$ the copper(II) ion, Cu^{2+}, is combined with four water molecules, and the sulfate ion (SO_4^{2-}) with one. We may formulate the hydrate, then, as $[Cu(H_2O)_4][SO_4(H_2O)]$. All ions are undoubtedly hydrated in aqueous solution, but in many cases the water of hydration is lost when the compound separates from solution.

Water may exist in a crystalline hydrate in four different forms.

▶ **1. WATER OF COORDINATION.** Water of this type is joined to the metal ion by coordinate covalent bonds. A crystal of $BeSO_4 \cdot 4H_2O$ has as its building units

$[Be(H_2O)_4]^{2+}$ and SO_4^{2-}, and one of $NiSO_4 \cdot 6H_2O$ contains $[Ni(H_2O)_6]^{2+}$ and SO_4^{2-} as structural units. Water of coordination is an essential part of the crystal structure; the crystal structure collapses and a new crystal structure forms when even part of the water is driven off.

▶ **2. ANION WATER.** This type of water is attached to the anion by hydrogen bonding, the attraction of the partially positive hydrogen atom for the negatively charged anion (see Sections 12.1 and 19.10). In a crystal of $CuSO_4 \cdot 5H_2O$, the four water molecules that are attached to the copper ion are water of coordination; the fifth molecule of water, which is attached to the sulfate ion, is anion water.

▶ **3. LATTICE WATER.** In some hydrates, water molecules occupy definite positions in the crystal lattice but are not attached directly to either a cation or an anion. In $KAl(SO_4)_2 \cdot 12H_2O$, six water molecules are attached to the aluminum ion by coordinate covalent bonding; the other six are lattice water, arranged in definite positions around the potassium ion but at distances too great to allow direct attachment. Lattice water is often hydrogen-bonded to other water molecules in the crystal.

▶ **4. ZEOLITIC WATER.** Such water molecules occupy relatively open spaces in the crystal structure. Zeolites, for example, contain three-dimensional networks of SiO_4 tetrahedra, which lose or take up water in the holes without any apparent change in crystal structure.

12.5   Deliquescence and Efflorescence

Many hydrates exhibit measurable vapor pressures, as can be demonstrated by placing a crystal of a hydrate in the vacuum above the mercury in a barometer tube. As the water vapor from the hydrate exerts pressure on the mercury, the mercury level is depressed to an extent dependent upon the magnitude of the **vapor pressure of the hydrate.** The vapor pressure of a hydrate increases with a rise in temperature and decreases as the temperature is lowered.

When a hydrate has a vapor pressure higher than the partial pressure of water in the atmosphere, the hydrate will lose part or all of its water of hydration if exposed to the air. Hydrates that lose water of hydration when exposed to the air are said to **effloresce,** and **efflorescence** is said to have taken place. $Na_2SO_4 \cdot 10H_2O$, which has a vapor pressure of 14 torr, effloresces rapidly when exposed to the air if the partial pressure of the water vapor in the air is less than 14 torr. $CuSO_4 \cdot 5H_2O$, with a vapor pressure of 7.8 torr, is stable in air as long as the partial pressure of the water in the air is greater than 7.8 torr. It follows that some hydrates are stable when the humidity is high but effloresce when it is low. A salt such as copper(II) sulfate may form more than one hydrate, each of which possesses its own definite vapor pressure at a given temperature. The following hydrates of copper(II) sulfate are known: $CuSO_4 \cdot 5H_2O$; $CuSO_4 \cdot 3H_2O$; and $CuSO_4 \cdot H_2O$.

When certain hydrated salts of low vapor pressure, such as $CaCl_2 \cdot H_2O$, are exposed to moist air, they form higher hydrates. Such salts may be used in the removal of moisture from air or other gases. A substance that can remove moisture from the air is said to be **hygroscopic.** Concentrated sulfuric acid, a liquid, and phosphorus(V) oxide, a solid, are hygroscopic and are used as drying agents, or **desiccants.** Certain water-soluble hygroscopic solids remove sufficient water from the air to dissolve completely in this water and form solutions. Such substances are said to be **deliquescent,** and the process is termed **deliquescence.** Very soluble salts, such as $CaCl_2 \cdot 6H_2O$, are often extremely deliquescent.

12.6 Heavy Water

Water composed of deuterium, ^2_1H, and oxygen is known as **heavy water,** or deuterium oxide, D_2O. Deuterium oxide is present in ordinary water in the ratio of one part D_2O to 6900 parts H_2O. Heavy water was first obtained by the electrolysis of ordinary water. Deuterium oxide molecules migrate to the cathode more slowly than do ordinary water molecules during electrolysis; therefore the heavy water tends to remain behind while the ordinary water is decomposed. Heavy water is now produced in the atomic energy program by an exchange equilibrium process, which is less costly and more efficient than the electrolytic method. In the exchange process, advantage is taken of the fact that when liquid water (which contains small quantities of D_2O) and gaseous hydrogen sulfide are mixed, the deuterium atoms exchange between sulfur and oxygen, preferentially combining with oxygen as D_2O at low temperatures and with sulfur as D_2S at elevated temperatures.

$$H_2S + D_2O \underset{\text{Cold}}{\overset{\text{Hot}}{\rightleftharpoons}} D_2S + H_2O$$

If steam and hydrogen sulfide are passed through each other in opposite directions at an elevated temperature, the hydrogen sulfide becomes saturated with water and through exchange is enriched with respect to D_2S. After the enriched hydrogen sulfide is drawn off, the reverse exchange can be accomplished at lower temperatures, producing water that is enriched with respect to D_2O and which thereby contains a larger proportion of D_2O than before. Many repetitions of the process yield a D_2O concentrate of about 15%, which is then enriched to 99.8% D_2O by fractional distillation and electrolysis. To produce one ton of D_2O, a plant must process 45,000 tons of water and must cycle 150,000 tons of H_2S. Several hundred tons of heavy water are now produced per year.

Heavy water resembles ordinary water in appearance but differs from it slightly in other physical properties (Table 12-2).

TABLE 12-2 Some Physical Properties of Ordinary Water and Heavy Water

Property	Ordinary Water	Heavy Water
Density, 20°	0.997	1.108
Boiling point, °C	100.00	101.41
Melting point, °C	0.00	3.79
Heat of vaporization, kJ/mol	40.7	41.61
Heat of fusion, kJ/mol	6.01	6.3

Heavy water is one of the "moderators" used in nuclear reactors to reduce the speed of neutrons to that required for the fission process to take place (see Section 29.12).

Because heavy water is about 10% heavier than ordinary water, it is possible to detect the presence of as little as 1 part of it in 100,000 parts of an aqueous solution. For this reason, heavy water and deuterium serve as valuable tracers in the study of both chemical and physiological changes. By replacing ordinary hydrogen with deuterium in the molecules of food, the processes of digestion and metabolism in the body can be studied.

Salts have slightly lower solubilities in D_2O than in H_2O, and the rates of reactions of D_2O are somewhat lower than those of similar reactions involving H_2O.

12.7   Water, an Important Natural Resource

The yearly use of water in the United States averages about 600,000 gallons per person. It is estimated that in the year 2000, the world's consumption of fresh water will be 9700 km^3 (2300 cubic miles) per year. Although there is enough water in the oceans to cover the entire earth to a depth of 2 miles, for the most part human needs generally can be satisfied only by fresh water. Unfortunately, however, the supply of fresh water is rapidly becoming critically low in many parts of the United States as the population of the country increases.

About a half-gallon of water per day is required to satisfy the biological needs of a human being. Nearly 150 gallons per day per person are used in maintaining cleanliness, in the cooking of food, and in heating and air-conditioning homes. These quantities are dwarfed by the 750 gallons of water used per person per day by industries in the United States. Five gallons of water are used in the production of 1 gallon of milk, 10 gallons for 1 of gasoline, 80 gallons for 1 kilowatt of electricity, 65,000 gallons for 1 ton of steel, and 3,000,000 gallons for 1 ton of nylon. More water is used for irrigation in the United States than for any other one purpose; the average consumption is over 750 gallons per person per day for this purpose.

Fresh water, unlike most mineral resources, is renewable in that it is a part of a gigantic cycle. Water evaporates from the earth into the atmosphere, the energy required being supplied by radiation from the sun; it is transported in the atmosphere by the winds; and it finally falls back to the earth in some form of precipitation. A large part of the water that precipitates is lost to our daily use, however, through evaporation and transpiration by plants.

At the present time, the oceans are becoming increasingly important as potential sources of fresh water because of recent advances in devising economical and practical desalting processes. The United States government has participated in establishing several desalting plants in various parts of the country.

The city of Key West, Florida, now obtains from a large sea water desalting plant a significant part of the water supply for the city—over 2,700,000 gallons of pure drinking water per day.

12.8   Natural Waters

All natural waters, even when not polluted by people, are impure since they contain many dissolved substances. Rain water is relatively pure, the chief impurities being dust and dissolved gases. After rain falls for awhile, the air will have been washed free of dust and bacteria, and any rain that falls thereafter is quite free of such impurities. Sea water contains about 3.6% of dissolved solids, principally sodium chloride. Perhaps all naturally occurring elements exist in the sea. Seventy-two elements are definitely known to exist in sea water, ranging in concentration from chlorine as Cl^- (19,000 mg/ℓ), sodium as Na^+ (10,600 mg/ℓ), and magnesium as Mg^{2+} (1300 mg/ℓ), to radon (9×10^{-5} mg/ℓ).

The impurities in the water on the earth's surface depend upon the nature of the soil and rocks that the water has passed over or through, and, increasingly, impurities added by people. The natural impurities may include dissolved gases such as the components of air (for example, oxygen, nitrogen, and carbon dioxide), ammonia, and hydrogen sulfide; dissolved salts, particularly the chlorides, sulfates, and the hydrogen carbonates of sodium, potassium, calcium, magnesium, aluminum, and iron; dissovled organic substances from the decay of plant and animal matter; and suspended solids such as sand, clay, silt, organic material, and microorganisms.

During the normal life cycle of a body of fresh water such as a lake, organic sediment slowly accumulates from algae, bacteria, aquatic plants, and animal by-products. A variety of bacteria and other organisms that constitute the lowest level of the food chain in the lake feed on this sediment and, in the presence of oxygen, decompose the sediment to various molecules and ions including carbon dioxide, water, nitrate ion, sulfate ion, and phosphates. This decomposition of organic matter by bacteria in the presence of oxygen is called **aerobic decomposition.** If the supply of oxygen is reduced or if the amount of organic sediment increases to the point where aerobic decomposition cannot keep up with it, the aerobic bacteria die and other bacteria that live in the absence of oxygen thrive. They too decompose the sediment, but the products are much less pleasant. Decomposition of organic matter in the absence of oxygen is called **anerobic decomposition.** Anerobic decomposition is accompanied by the smell of rotten eggs due to hydrogen sulfide (which in fact is the substance responsible for the unpleasant odor of rotten eggs); bubbles of methane are usually visible, and the water is black and often filled with slime.

The process by which a lake grows rich in nutrients (phosphates and nitrates, primarily) and gradually fills with organic sediment and aquatic plants is called **eutrophication.** As a lake ages naturally, it becomes shallower as it fills with undecomposed sediment, and plant life abounds on the banks and within the lake itself. Slowly, the lake is transformed into a marsh and ultimately into dry land by the processes accompanying eutrophication.

12.9   Water Pollution

In addition to the natural contaminants in water, many pollutants are being added to our water supplies by people. For too long, neither the magnitude nor even the existence of the problems created by these pollutants was realized. Unfortunately, very little is known of the long-term effects on human health of the many, largely unidentified chemical compounds that enter water supply sources. A whole new field of chemistry is developing rapidly whereby pollutants in our *overall* environment, not just in our water supplies, are studied in connection with both their short- and long-range effects on human health and with the efficient removal of the pollutants.

Several important water pollutants are now causing concern, and their effects are worthy of our attention.

▶ **1. PHOSPHATES AND OTHER NUTRIENTS.** Phosphates have been used widely as water softeners in laundry detergents (see Section 12.12) and hence have been released in large quantities in municipal wastes into rivers and streams. Furthermore, some fertilizers contain phosphate [for example, a mixture of $Ca(H_2PO_4)_2$ and $CaSO_4 \cdot 2H_2O$ called "superphosphate of lime"—see Section 24.10]. A number of phosphate compounds also are present in certain rocks. Thus additional phosphates reach streams through natural soil runoff. The quantity of phosphates reaching streams from natural sources alone may, in the opinion of some, be great enough to make phosphates from detergents an insignificant factor in comparison. A high concentration of phosphates in streams and lakes presents a problem because such substances are significant nutrients for algae, bacteria, and aquatic plants, leading to eutrophication (Section 12.8).

Other nutrients for microorganisms and aquatic plants include nitrogen (in the form of ammonia, ammonium salts such as ammonium sulfate, and nitrate salts such as calcium nitrate) and potassium (in potassium salts such as potassium nitrate and potassium sulfate). These nutrients also have been released in municipal wastes, in the runoff from fertilizer application, and in the wastes from cattle.

The growth of algae and other aquatic plants is usually controlled by the amount of nutrients available. Excess nutrients lead to rapid plant growth with a consequent rapid increase in the rate of eutrophication of natural waters. In the worst cases, the deposition of organic sediment can become so rapid that anerobic bacteria replace the aerobic bacteria and the lake is no longer able to support life.

Development of effective long-term controls of eutrophication will require a better understanding of the balance and form of significant nutrients in streams and lakes, the natural plant and animal populations in bodies of water, and how these are affected by externally applied factors such as the added nutrients. In the short range, substitutes for phosphates in detergents are now in use, and methods of removing phosphates and controlling their undesirable effects in municipal wastes are increasingly being used.

▶ **2. SEWAGE WASTE MATERIALS.** In addition to nutrients, municipal sewage contains microorganisms as well as the by-products from many industrial processes that are discharged into sewers. In some communities, "raw sewage" is dumped directly into lakes that serve as public water supply sources, or into rivers that are then the sources of public water supply a short distance downstream. Other contamination of natural waters may result from breakdowns in sewage treatment plants or from flooding of such plants during periods of heavy rainfall. Chlorination of municipal water supplies (Section 12.10) has largely eliminated dangers from bacteria and viruses in drinking water. However, many bodies of natural water are still contaminated with microorganisms, and the chemical by-products present may pose hazards in domestic water supplies.

The presence of organic matter, nutrients, and microorganisms in the output of sewage treatment plants is measured by three tests: *coliform count, algal count,* and *biological oxygen demand* (BOD). The coliform count describes the number of *E. coli* (the characteristic bacteria in animal wastes) present. The algal count is a biological test for microorganisms other than bacteria and viruses which may be present. The BOD measures the volume of oxygen gas taken up by a given amount of water in five days at 20°C. Since the oxygen is consumed by microorganisms that decompose organic materials in the water, the BOD is a measure of the total quantity of microorganisms and the nutrients available to the microorganisms. "Pure" water is considered to be that with a BOD of 1 part per million (equivalent to 1 mg of O_2 per liter of water) or less. Acceptable effluent from industrial or sewage treatment plants has a BOD of 20 parts per million or less. Raw sewage exhibits a BOD of 100–400 parts per million.

The coliform count, the algal count, and the BOD do not determine the presence of small amounts of organic compounds, often resulting from industrial wastes, in water. Specialized tests for compounds such as benzene, acetone, toluene, chloroform, and carbon tetrachloride have revealed small amounts (1–40 parts per billion) of these materials in the drinking water of cities in industrial river basins. The normal water purification procedures do not remove these compounds. In fact, the presence of trihalomethanes such as chloroform ($CHCl_3$) in domestic water supplies may result from the reaction of chlorine (used in water purification processes to kill microorganisms) with naturally occurring organic compounds such as humic acid, which results from the decay of plant matter in the soil. Some of these organic pollutants can be removed by filtering domestic water through activated charcoal.

▶ **3. TOXIC POLLUTANTS.** Many waters contain organic compounds that were used to kill insects, weeds, and fungi. These materials, through soil runoff, reach streams either directly or through city storm sewers. The chief hazard of pesticides in the environment is their concentration in the food chain of fish and wildlife, which may result in death of these species or a decrease in their ability to reproduce. Some of these compounds such as *Aldrin, Dieldrin,* and *Kepone,* have proved so toxic that their manufacture and use has been prohibited. The use of others such as *DDT* has been severely restricted.

The very poisonous compound methylmercury is produced in inland waters by the action of microorganisms in the bottom mud with mercury, which may be present as a result of prior discharge of waste mercury and mercury compounds into sewers or natural waters. The discharge of mercury and mercury compounds is now prohibited, but much remains from that previously dumped. Metallic mercury usually attacks the liver and kidneys, but methylmercury can be even more dangerous, because it attacks the central nervous system. Furthermore, methylmercury is retained in the human body for long periods so that, even with small rates of intake, toxic amounts can be accumulated.

Microorganisms in the intestines of some animals, including chickens, can methylate mercury. It is not known whether such a conversion can take place in fish and mammals without the microorganisms, but it is known that fish take in methylmercury both through their food and through their gills to the extent that the concentration in their flesh is several thousand times that in the surrounding water. Hence, the possibility of mercury poisoning in humans from eating fish (or chicken and other animals) is greater than might be expected from the actual concentrations of the element in water, air, or soil.

Considerable attention has recently been devoted to pollution by cadmium, an element in the same periodic family as mercury. The discharge of cadmium as well as other toxic pollutants such as cyanide, benzidine, and polybromobiphenyls (PBB), has also been prohibited, but it may be some time before these materials are absent from natural waters.

▶ **4. THERMAL POLLUTION.** Industrial discharges that markedly raise the temperature of a body of water may result in harm to fish and in increased growth of algae and other microorganisms. In 1970, the Department of the Interior stated a new policy for Lake Michigan: "The minimum possible waste heat shall be added to the waters of Lake Michigan. In no event will heat discharges be permitted to exceed 1°F rise over ambient at the point of discharge." This caused great consternation in Chicago, for the implications of such a standard are far-reaching. If this standard were strictly and immediately enforced under present operating circumstances, virtually all electric power plants, both nuclear and conventional, would have to be shut down. A reasonable amount of time, however, is normally allowed for the industries, acting in good faith, to make the modifications necessary for compliance.

Interestingly, it is at least conceivable that thermal pollution, under certain special circumstances, could be beneficial. For example, the Ralston Purina Company and the Florida Power Company, which has replaced a conventional power plant with a nuclear power plant, are cooperating in a study to determine whether hot water discharges from power plants can be put to profitable use. They envision the possibility that selected saltwater fish and shellfish might thrive in the heated water, thereby improving the future of the food supply from the sea. Nevertheless, even a few degrees rise in temperature of the water through thermal pollution can decrease the solubility of oxygen in the water to a level too low for many forms of marine life to survive. Achieving a proper balance is an exceedingly delicate problem.

12.10   Purification of Water

Water is purified for city water supplies by first allowing it to stand in large reservoirs where most of the mud, clay, and silt settle out, a process called **sedimentation,** and then filtering it through beds of sand and gravel. Often, prior to filtration, lime and aluminum sulfate are added to the water in the settling reservoirs or filters. These chemicals react to form aluminum hydroxide, a gelatinous precipitate, which settles slowly carrying with it much of the suspended matter, including most of the bacteria. The equation for the precipitation of aluminum hydroxide (an example of a metathesis reaction, Section 8.2) is

$$2Al^{3+}(aq) + [3SO_4{}^{2-}(aq)] + [3Ca^{2+}(aq)] + 6OH^-(aq) \longrightarrow$$
$$2Al(OH)_3(s) + [3Ca^{2+}(aq)] + [3SO_4{}^{2-}(aq)]$$

It is common practice to kill the bacteria that remain in the water after filtration by adding chlorine.

Various synthetic materials such as cationic polyamines (substances which contain several $-NH_2$ groups) may displace aluminum sulfate as water clarifiers in some installations in the near future. They possess the advantage of being easier to remove from the purified water than aluminum hydroxide.

Relatively pure water for laboratory use is commonly prepared by distillation of tap water (see Section 11.7). Because the basic constituents of glass slowly dissolve in water, glass equipment is not satisfactory for the preparation and storage of very pure water. Distillation apparatus made of fused silica or pure tin is often used in making pure water. Gases from the air, particularly carbon dioxide, often contaminate distilled water and must be removed for some laboratory operations.

12.11   Hard Water

Water containing dissolved calcium, magnesium, and iron salts is known as **hard water.** The negative ions present in hard water are usually chloride, sulfate, and hydrogen carbonate. Hardness in water is objectionable for two reasons: (1) The calcium, magnesium, and iron ions form insoluble soaps by reaction with soluble soaps such as sodium stearate, $C_{17}H_{35}COONa$.

$$2C_{17}H_{35}COO^-(aq) + [2Na^+(aq)] + Ca^{2+}(aq) \longrightarrow$$
$$Ca(C_{17}H_{35}COO)_2(s) + [2Na^+(aq)]$$

Insoluble soaps have no cleansing power, and due to their sticky nature adhere to fabrics, giving them a dingy appearance. Insoluble soaps also make up the "ring" in the bathtub. Soap in excess of that needed to precipitate the calcium and magnesium in hard water must be added in order to obtain cleansing action. (2) Hard water is responsible for the formation of a tightly adherent scale in boilers. At high temperatures, much of the mineral matter dissolved in hard water is precipitated as scale [insoluble substances such as calcium carbonate, magnesium carbonate, iron(II) carbonate, and calcium sulfate]. Much of the scale is calcium sulfate, which is *less* soluble in hot water than in cold and precipitates partly because of that fact. The scale is a poor conductor of heat and thus causes a waste of fuel. Furthermore, boiler explosions are often due to the presence of scale. Since the scale is not a good heat conductor, the metal must be very hot (often red hot) in order to bring the water to the desired temperature. If the scale cracks, the water seeps through and comes in contact with the hot metal.

$$4H_2O + 3Fe \text{ (hot)} \longrightarrow Fe_3O_4 + 4H_2$$

As it is formed, the hydrogen gas breaks the scale loose and more water gets in, producing still more hydrogen, thereby setting the stage for a violent explosion.

It is important that the substances responsible for hardness in water be removed before the water is used for washing or in boilers. The removal of the metallic ions responsible for the hardness in hard water is known as **water-softening.**

12.12   Water-Softening

When hydrogen carbonate ions are present in water, boiling the water drives off carbon dioxide, and the hardness is removed as the metal carbonates then formed precipitate.

$$Ca^{2+} + 2HCO_3^- \xrightarrow{\triangle} CaCO_3(s) + CO_2 + H_2O$$

The carbonates of calcium and magnesium thus precipitated form the deposits found in teakettles and boilers. Such water is said to possess **carbonate,** or **temporary, hardness**—temporary, because most of the hardness can be removed by boiling the water. The hydrogen carbonate ion may also be converted to the carbonate ion by the addition of a basic substance. Commercially, calcium hydroxide is added in the exact quantity needed to react with the hydrogen carbonate ions.

$$Ca^{2+} + 2OH^- + Ca^{2+} + 2HCO_3^- \longrightarrow 2CaCO_3(s) + 2H_2O$$

Similar equations may be written where the positive ion in the hard water is either Mg^{2+} or Fe^{2+}. The precipitated carbonates are readily removed from the water by filtration. An aqueous solution of ammonia is a basic solution that can be used in the home to remove temporary hardness in water.

$$Ca^{2+} + 2HCO_3^- + 2NH_3 \longrightarrow CaCO_3(s) + 2NH_4^+ + CO_3^{2-}$$

Water that contains chloride or sulfate ions in addition to the calcium, magnesium, or iron ions is said to possess **noncarbonate hardness.** Chloride and sulfate ions are not removed by boiling as is the hydrogen carbonate ion. Noncarbonate hardness can be eliminated by the addition of washing soda (sodium carbonate) to the water.

$$Ca^{2+} + [SO_4^{2-}] + [2Na^+] + CO_3^{2-} \longrightarrow CaCO_3(s) + [2Na^+] + [SO_4^{2-}]$$
$$Ca^{2+} + [2Cl^-] + [2Na^+] + CO_3^{2-} \longrightarrow CaCO_3(s) + [2Na^+] + [2Cl^-]$$

The sodium sulfate and sodium chloride produced during the reaction do not interfere with the cleansing action of soap, nor do they form boiler scale.

Crude sodium hydroxide (caustic soda) is often used in large installations to remove both carbonate and noncarbonate hardness.

$$Ca^{2+} + 2HCO_3^- + [2Na^+] + 2OH^- \longrightarrow CaCO_3(s) + [2Na^+] + CO_3^{2-} + 2H_2O$$

The Na_2CO_3 produced in the removal of carbonate hardness then reacts to remove any remaining noncarbonate hardness.

$$Ca^{2+} + [SO_4^{2-}] + [2Na^+] + CO_3^{2-} \longrightarrow CaCO_3(s) + [2Na^+] + [SO_4^{2-}]$$

Trisodium phosphate and borax also act as water softeners by forming insoluble calcium and magnesium phosphates and borates. Polymetaphosphates, such as $(NaPO_3)_x$, form soluble complexes with Ca^{2+} and Mg^{2+} and thus prevent them from reacting with soap. However, the use of phosphates in detergents is declining because of undesirable nutrient effects on algae and other plant life in rivers and streams (Section 12.9).

Ion exchange provides another useful and highly important method of softening

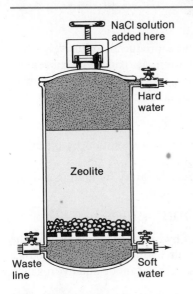

Figure 12-6 Water-softening by the ion exchange method. When regeneration is necessary, the soft-water outlet valve is closed and a concentrated NaCl solution run through the Ca-zeolite and out the waste line until regeneration of the Na-zeolite is complete.

water. Naturally occurring insoluble minerals, the sodium aluminosilicates known as **zeolites,** are used as ion-exchangers. When water containing calcium and magnesium ions filters slowly through thick layers of coarse granules of the zeolites, the sodium in the zeolite is replaced by calcium or magnesium, and the water is softened.

$$2NaAlSi_2O_6 + Ca^{2+} \longrightarrow Ca(AlSi_2O_6)_2 + 2Na^+$$

Sodium aluminosilicate   Calcium aluminosilicate
(insoluble)   (insoluble)

The calcium aluminosilicate can be reconverted into the sodium compound by treating the zeolite with a concentrated solution of sodium chloride.

$$Ca(AlSi_2O_6)_2 + 2Na^+ \longrightarrow 2NaAlSi_2O_6 + Ca^{2+}$$

Note that this reaction is the reverse of the preceding one; the reversibility can be represented by double arrows. The direction the reaction takes is controlled by the excess of either Ca^{2+} or Na^+ ions.

$$2Na\ zeolite + Ca^{2+} \rightleftharpoons Ca(zeolite)_2 + 2Na^+$$

After the sodium zeolite is regenerated, it is ready for use again. The ion exchange method of water-softening is effective for the removal of both carbonate and noncarbonate hardness (Fig. 12-6).

Synthetic resin ion-exchangers have also been developed to soften hard water. These exchangers remove both cations (positive ions) and anions (negative ions) from hard water; that is, they demineralize the water completely. Examples of these insoluble resins are Amberlite and Zeo-Carb. Metal ions in the hard water displace hydrogen ions from one type of resin.

$$2RCOOH + Ca^{2+} + [SO_4^{2-}] \longrightarrow (RCOO)_2Ca + 2H^+ + [SO_4^{2-}]$$

Another type of insoluble resin then removes the ions of the acid from the water.

$$2RNH_2 + 2H^+ + [SO_4^{2-}] \longrightarrow 2RNH_3^+ + [SO_4^{2-}]$$

Hence, if hard water (or a salt solution) is passed first through an exchanger that replaces all metal ions with H^+ and then through a resin that combines with the resulting acids, it can be completely demineralized. The result is essentially the same as that achieved through distillation of water. In recent years the use of ion-exchangers in the treatment of water for industrial and domestic purposes has grown tremendously.

Hydrogen Peroxide

Hydrogen forms another compound with oxygen, called **hydrogen peroxide,** H_2O_2, in which there is twice as much oxygen for the same weight of hydrogen as there is in water. Hydrogen peroxide was first prepared in 1818 by Thénard, who obtained it by treating barium peroxide with hydrochloric acid. Very small quantities of hydrogen peroxide are present in dew, rain, and snow, probably as a result of the action of ultraviolet light upon oxygen mixed with water vapor.

12.13   Preparation

In the nineteenth century, hydrogen peroxide was produced exclusively on a commercial scale by the reaction of an acid upon barium peroxide, which is readily formed by heating barium oxide in air ($2BaO + O_2 \longrightarrow 2BaO_2$). For the reaction of sulfuric acid with barium peroxide, the equation is

$$BaO_2(s) + 2H^+(aq) + SO_4^{2-}(aq) \longrightarrow BaSO_4(s) + H_2O_2(aq)$$

Barium sulfate is insoluble and can be separated from the solution by filtration. The hydrogen peroxide thus produced is quite dilute.

Hydrogen peroxide also can be prepared in the laboratory by adding sodium peroxide to cold water or cold dilute hydrochloric acid.

$$Na_2O_2 + 2H_2O \longrightarrow 2Na^+ + 2OH^- + H_2O_2$$

Electrochemical processes for the production of hydrogen peroxide involve the electrolysis of solutions of sulfuric acid or ammonium hydrogen sulfate to produce peroxydisulfuric acid, $H_2S_2O_8$, or its ammonium salt, at the anode and hydrogen at the cathode.

$$2H_2SO_4(aq) + \text{electrical energy} \longrightarrow H_2S_2O_8(aq) + H_2(g)$$

$$[2NH_4^+(aq)] + 2HSO_4^-(aq) + \text{electrical energy} \longrightarrow [2NH_4^+(aq)] + S_2O_8^{2-}(aq) + H_2(g)$$

The peroxydisulfuric acid is subsequently hydrolyzed (allowed to react with water) to produce hydrogen peroxide.

$$H_2S_2O_8 + 2H_2O \longrightarrow 2H_2SO_4 + H_2O_2$$

The sulfuric acid formed is recycled; i.e., it is used over again in the production of more hydrogen peroxide.

For many years, the electrochemical process accounted for 80 to 90% of U.S. production of hydrogen peroxide. However, a different industrial method has gained importance rapidly in recent years and now accounts for most of the hydrogen peroxide produced. In this process, an organic compound referred to as anthraquinone is reduced by hydrogen in a benzene solution, using palladium as a catalyst, to a compound known as dihydroanthraquinone. When the dihydroanthraquinone is allowed to react with oxygen, the original anthraquinone is obtained along with hydrogen peroxide. The anthraquinone is then used over again in the production of more hydrogen peroxide.

$$\underset{\substack{\text{2-Ethyl} \\ \text{anthraquinone}}}{C_{16}H_{12}O_2} + H_2 \xrightarrow{\text{Pd catalyst}} \underset{\substack{\text{2-Ethyl} \\ \text{dihydroanthraquinone}}}{C_{16}H_{12}(OH)_2}$$

$$C_{16}H_{12}(OH)_2 + O_2 \longrightarrow C_{16}H_{12}O_2 + H_2O_2$$

When the hydrogen peroxide reaches a concentration, in the benzene solution, of about 0.15 mole per liter, it is extracted with water to produce an 18% water solution of hydrogen peroxide, which may be further concentrated by multiple vacuum distillation. Hydrogen peroxide decomposes when heated, necessitating vacuum distillation, which takes place at relatively low temperatures (lower boiling points at reduced pressures—see Section 11.3).

When very pure, H_2O_2 is quite stable. Solutions of 90% concentration and higher are produced and used routinely as a propellant in rockets, as a chemical reagent, and in a number of other applications. However, when impurities are present, the concentrated solutions are dangerously explosive. For storage purposes, a **stabilizer,** such as a small amount of acid or the organic compound acetanilide, is added.

12.14   Properties

Pure hydrogen peroxide is a colorless, syrupy liquid with a sharp odor and an astringent taste. It has a density of 1.438 g/cm^3 at 20°C, boils at 62.8° at 21 torr, and freezes at −1.70°. The relatively low vapor pressure and high boiling point are caused by hydrogen bonding between hydrogen peroxide molecules. It is miscible (mutually soluble) with water, alcohol, and ether in all proportions. As stated in the preceding section, hydrogen peroxide is thermally unstable and decomposes spontaneously with the evolution of much heat.

$$2H_2O_2(l) \longrightarrow 2H_2O(l) + O_2(g) \qquad \Delta H = -196 \text{ kJ}$$

Its instability may be explained by the fact that its formation is a highly endothermic process.

Hydrogen peroxide is an active oxidizing agent. It oxidizes sulfurous acid (H_2SO_3) to sulfuric acid (H_2SO_4) and sulfides to sulfates.

$$H_2SO_3 + H_2O_2 \longrightarrow H_2SO_4 + H_2O$$
$$PbS + 4H_2O_2 \longrightarrow PbSO_4 + 4H_2O$$

When used as an oxidizing agent, it has an advantage over other oxidizing agents, because the only by-product of the oxidizing reactions is water.

Hydrogen peroxide not only acts as an oxidizing agent with strong reducing agents, but also as a reducing agent with strong oxidizing agents. For example, it will reduce silver oxide to the metal.

$$Ag_2O + H_2O_2 \longrightarrow 2Ag + H_2O + O_2$$

Hydrogen peroxide reduces many other substances with the evolution of oxygen. These substances include the oxides of mercury, gold, and platinum, and the strong oxidizing agents manganese dioxide and potassium permanganate. The reaction of hydrogen peroxide with the permanganate ion in acid solution proceeds according to the equation

$$6H^+ + 2MnO_4^- + 5H_2O_2 \longrightarrow 5O_2 + 2Mn^{2+} + 8H_2O$$

Hydrogen peroxide is a weak acid; in water it ionizes to a very slight extent in two steps.

$$H_2O_2(aq) \rightleftharpoons H^+(aq) + \underset{\text{Hydroperoxide ion}}{HO_2^-(aq)}$$

$$HO_2^-(aq) \rightleftharpoons H^+(aq) + \underset{\text{Peroxide ion}}{O_2^{2-}(aq)}$$

Its acidic character is demonstrated by the equation

$$Ba^{2+}(aq) + 2OH^-(aq) + H_2O_2(l) \longrightarrow BaO_2(s) + 2H_2O(l)$$

12.15   The Structure of Peroxides

The valence electronic formula for hydrogen peroxide and a diagram of its molecular structure are given in Fig. 12-7. Its structure can be likened to that of a sheet of paper folded to an angle of 94°, with the oxygen atoms located in the fold and one hydrogen atom located in each plane.

Note in the valence electronic formula in Fig. 12-7 that two atoms of oxygen are connected to each other through an electron-pair bond. The valence electronic formulas of the covalent peroxides peroxymonosulfuric acid (H_2SO_5), and peroxydisulfuric acid ($H_2S_2O_8$) and the ionic peroxide barium peroxide (BaO_2) illustrate that other peroxides also contain oxygen-oxygen bonds. Note that the charge on the peroxide ion, containing two oxygen atoms, is -2.

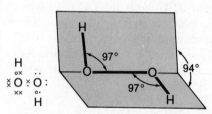

Figure 12-7   The electronic and molecular structures of hydrogen peroxide, H_2O_2.

Peroxymonosulfuric acid   Peroxydisulfuric acid   Barium peroxide

The usual charge for a *single* oxygen ion in an oxide is -2, as indicated in the valence electronic structures for barium oxide (BaO) and manganese dioxide (MnO_2).

Barium oxide   Manganese dioxide

Manganese dioxide, unlike barium peroxide, does not yield hydrogen peroxide when treated with acids. This fact indicates that MnO_2 is a dioxide rather than a peroxide; i.e., the two oxygen atoms are not joined together but are present as oxide ions. There is no oxygen-oxygen bond in MnO_2 as there is in BaO_2.

12.16   Uses of Hydrogen Peroxide

Three per cent solutions of hydrogen peroxide by mass are used in the home as a mild antiseptic, a deodorizer, and a germicide; a somewhat more concentrated solution is a bleach for hair. Its effectiveness as a germicide is questionable, however. A 30% solution of hydrogen peroxide is commonly used in the laboratory as an oxidizing agent. In commerce, the principal consumption of hydrogen peroxide is as an oxidizing agent, particularly as a bleaching agent. Hydrogen peroxide is often used to bleach substances of animal origin such as wool and hair, which are harmed by most other bleaching agents. At present, most cotton cloth is bleached with hydrogen peroxide because of the innocuous character of the decomposition products and the permanency of the whiteness produced by peroxide bleaching.

Ninety per cent hydrogen peroxide is used as an oxidizing agent in certain high explosives and in rockets. It is an ideal liquid oxidant because it has a high energy release on reaction, it is relatively stable to shock, and it is noncorrosive, nontoxic, and has a high boiling point and low freezing point.

Concentrated hydrogen peroxide may be used as a monopropellant, since it is decomposed under pressure yielding a gaseous mixture of oxygen and superheated steam.

Questions

1. Account for the polar nature of the water molecule and of the hydrogen peroxide molecule in terms of their structures.
2. Under what conditions does water exist as single H_2O molecules?
3. Account for the abnormally high boiling point of water.
4. Describe the hydrogen bond.
5. Why does the H—O—H angle in water deviate from the angle expected for a tetrahedral molecule?
6. Using information about the structure of water molecules and the interaction of water molecules with one another, discuss the fact that the heat of fusion of water is only 334 J/g whereas the heat of vaporization of water is 2258 J/g.
7. The density of water increases when it melts; the density of sodium chloride decreases. Explain.
8. The hydrogen bonds in liquid hydrogen fluoride (HF) are stronger than those in liquid water, yet the heat of vaporization of liquid hydrogen fluoride is less than that of water. Explain.
9. Using the phase diagram for water (Fig. 12-5), determine the following:
   (a) At 400 torr, what is the approximate temperature necessary to convert water from a solid to a liquid? from a liquid to a gas?
   (b) At 20°C, what is the approximate pressure at which water changes from a gas to a liquid? What pressure is necessary at 66°C?
10. Using the phase diagram for water (Fig. 12-5), answer the following questions:
    (a) Holding the temperature constant, how may water be changed from gas to liquid and from liquid to solid?
    (b) At a temperature of −10°C, can water exist as a gas? as a liquid? as a solid? Explain.
    (c) At each of the following temperatures can each of the changes listed in part (a) be accomplished: −5°C; +20°C? Explain.
11. Explain what is meant by the term "triple point" of a substance.
12. Determine from the phase diagram for water (Fig. 12-5) what physical state water would have at
    (a) 600 torr pressure and 40°C.
    (b) 400 torr and −25°C.
    (c) 400 torr and 50°C.
    (d) 200 torr and 90°C.
13. (a) Can water be converted from a solid to a liquid at 3 torr? from a gas to a liquid? from a solid to a gas? from a gas to a solid?
    (b) If the pressure is 10 torr, can the changes in state listed in part (a) be accomplished?
    (c) Explain both (a) and (b) in terms of the phase diagram for water.
14. What is heavy water and how is it prepared?
15. Explain the difference in vapor density of H_2O and D_2O at 1 atm and 115°C.
16. Define and illustrate the terms "acidic anhydride" and "basic anhydride."
17. Complete and balance the following equations. (Note that, depending upon the stoichiometry, different products may result from reaction of the same reagents.)
    (a) $K + H_2O \longrightarrow$
    (b) $SO_3 + H_2O \longrightarrow$
    (c) $SiCl_4 + H_2O \longrightarrow$
    (d) $Na_3P + H_2O \longrightarrow$
    (e) $BaS + H_2O \longrightarrow$
    (f) $BaS + H_2O_2 \longrightarrow$
    (g) $SO_2 + H_2O_2 \longrightarrow$
    (h) $Al_2(SO_4)_3(aq) + Ca(OH)_2(aq) \longrightarrow$
18. Explain how P_4O_{10} acts as a desiccant and converts $CuSO_4 \cdot 5H_2O$ to $CuSO_4$.
19. Distinguish between the terms "hygroscopic" and "deliquescent."
20. Relate vapor pressure of hydrates to temperature.
21. Define the following terms: BOD, eutrophication, anerobic decomposition, hard water, temporary hardness, zeolites.
22. What is the effect of excess nutrients upon natural waters?
23. What are some of the substances of primary concern in water pollution?
24. What is "thermal pollution" of water? How does it arise? Is it necessarily undesirable?
25. Can the purification of water lead to its pollution? Can the pollution of water produce beneficial results?
26. What metal ions are commonly responsible for the hardness of water?

27. What naturally occurring impurities are typically present in rain water? in sea water? in distilled water?
28. Water possessing "temporary hardness" may be softened by boiling. Explain by use of an equation.
29. How may the noncarbonate hardness of water be removed? Use an equation to illustrate your answer.
30. Distinguish between "softening" and "demineralizing" water. Does demineralization soften the water?
31. Explain how aluminum sulfate and calcium hydroxide are used in the purification of water.
32. The heat of fusion of H_2O is 334 J/g, whereas heats of fusion of H_2S and H_2Te are 70 J/g and 32 J/g, respectively. (a) Explain why the heats of fusion would normally be expected to decrease in that order. (b) Explain why, even taking into account the expected trend, the heat of fusion for water is abnormally high (nearly five times as high as that for H_2S).
33. How is hydrogen peroxide prepared in the laboratory? How is it prepared commercially? Write balanced equations.
34. What is the advantage of using hydrogen peroxide as an oxidizing agent over other such agents?
35. Why is hydrogen peroxide purified by distillation under reduced pressure?
36. Distinguish between dioxides and peroxides in terms of electronic structures.
37. Under what conditions does hydrogen peroxide act as a reducing agent? as an oxidizing agent? Illustrate with balanced equations.

Problems

1. The density of water vapor at 100°C is 0.58 g/ℓ. What is the density of D_2O vapor at 100°C?
   Ans. 0.64 g/ℓ
2. What mass of steam at 100°C must be converted to water at 100°C in order to convert 1.00 kg of $D_2O(l)$ at 101.4°C to $D_2O(g)$ at 101.4°C?   *Ans. 0.920 kg*
3. How many grams of the metal hydroxide can be produced when excess water is added to each of the following?
   (a) 3.00 g of Na_2O *Ans. 3.87 g*
   (b) 15.0 mole of SrO *Ans. 1.82×10^3 g*
   (c) 15.0 g of K_2O *Ans. 17.9 g*
   (d) 2.25 mole of CaO *Ans. 167 g*
4. What mass of a solution of hydrogen peroxide (25.0% by mass) is required to convert 1.50 mol of sulfurous acid to sulfuric acid if the yield of the reaction is 100%?   *Ans. 204 g*
5. What mass of hydrogen is produced by the reaction of 29.6 g of hot magnesium with steam to produce magnesium oxide and hydrogen?   *Ans. 2.45 g*
6. What is the per cent of water by mass in each of the following hydrates?
   (a) $CuSO_4 \cdot 5H_2O$ *Ans. 36.1%*
   (b) $MgSO_4 \cdot 7H_2O$ *Ans. 51.2%*
   (c) $Na_2SO_4 \cdot 10H_2O$ *Ans. 55.9%*
   (d) $Mg(ClO_4)_2 \cdot 6H_2O$ *Ans. 32.6%*
7. How many liters of dry hydrogen at 25.0°C and 760 torr are produced by the reaction of 23.0 g of sodium with water?   *Ans. 12.2 ℓ*
8. What volume of dry oxygen is produced when 68.00 g of hydrogen peroxide decomposes to give oxygen and water at 25°C and 740.0 torr?
   Ans. 25.1 ℓ
9. What mass of sulfuric acid can be prepared by the reaction of 75.00 ℓ of sulfur trioxide at 25.0°C and 750.0 torr with enough water to produce pure sulfuric acid?   *Ans. 296.7 g*
10. Calculate the number of moles of hydrogen peroxide that can be prepared from 2.5×10^2 g of barium peroxide and an excess of sulfuric acid, assuming the yield of the reaction is 92%.   *Ans. 1.4 moles*
11. (a) To break each hydrogen bond in ice requires 3.5×10^{-20} J. The measured heat of fusion of ice is 334 J/g. Essentially all of the energy involved in the heat of fusion goes to break hydrogen bonds. What per cent of the hydrogen bonds are broken when ice is converted to liquid water?   *Ans. 14%*
    (b) How many additional calories would be required, per gram, to break the remaining hydrogen bonds?   *Ans. 2.0 kJ, or 4.8×10^2 cal*
12. Calculate the molar concentration of chlorine (as Cl^-), magnesium (as Mg^{2+}), and radon in sea water.
    Ans. 0.54 M; 0.053 M; 4×10^{-10} M

References

"The Unique Chemical," F. Franks, *Chem. in Britain,* **12,** 278 (1976).

"Water Structure and Ion Binding: A Role in Cell Physiology?" G. B. Kolata, *Science,* **192,** 1221 (1976).

"Vapor Pressure of Water at its Triple Point: A Highly Accurate Value," L. A. Guildner, D. P. Johnson, and F. E. Jones, *Science,* **191,** 1261 (1976).

"Water Pollution," R. J. Bazell, *Science,* **171,** 266 (1971).

"Mercury in the Environment: Natural and Human Factors," A. L. Hammond, *Science,* **171,** 788 (1971).

"The Biology of Heavy Water," J. J. Katz, *Sci. American,* July 1960, p. 106.

"Chemistry in Natural Water Systems," L. Coyne, *J. Chem. Educ.,* **52,** 796 (1975).

"Thermal Pollution and Aquatic Life," J. R. Clark, *Sci. American,* March 1969, p. 18.

"The Hydrogen Bond," J. N. Murrell, *Chem. in Britain,* **5,** 107 (1969).

"Chemical Reactions and the Composition of Sea Water," K. E. Chave, *J. Chem. Educ.,* **48,** 148 (1971).

"Pollution Problems, Resources, Policy, and the Scientist," A. W. Eipper, *Science,* **169,** 11 (1970).

"Mercury in the Environment," L. J. Goldwater, *Sci. American,* May 1971, p. 15.

"The Water Cycle," H. L. Penman, *Sci. American,* September 1970, p. 98.

"Ion Exchange," T. V. Arden, *Chem. in Britain,* **12,** 285 (1976).

"The Calefaction of a River" (Thermal pollution), D. Merriman, *Sci. American,* May 1970, p. 42.

"Mechanisms Controlling World Water Chemistry," R. J. Gibbs, *Science,* **170,** 1088 (1970).

"Pollution of the Seas," H. A. Cole, *Chem. in Britain,* **7,** 232 (1971).

"The Responsibilities of Industry" (Regarding pollution), E. Challis, *Chem. in Britain,* **7,** 235 (1971).

"The Control of the Water Cycle," J. Peixoto and M. A. Kettani, *Sci. American,* April 1973, p. 46.

"Power, Fresh Water, and Food from Cold, Deep Sea Water," D. F. Othmer and O. A. Roels, *Science,* **182,** 121 (1973).

"The Disposal of Waste in the Ocean," W. Bascom, *Sci. American,* August 1974, p. 16.

"Salt Solution" (Ocean as a source of drugs, food, salt, and fresh water), K. Jacques, *Chem. in Britain,* **11,** 12 (1975).

"The Environment Today," R. E. Train, *Chem. and Eng. News,* May 29, 1978, p. 27.

"Dioxins, Dioxins Everywhere" (Staff), *Chemistry,* January 1979, p. 3.

"Chemical Waste Disposal a Costly Problem" (Staff), *Chem. and Eng. News,* March 12, 1979, p. 11.

"Humates and Other Natural Organic Substances in the Aquatic Environment," C. Steelink, *J. Chem. Educ.,* **54,** 599 (1977).

13

Solutions; Colloids

13.1 The Nature of Solutions

In Section 2.11 we noted that when crystals of sugar are stirred with a sufficient quantity of water, the sugar disappears and a solution of sugar in water is formed. The sugar is said to have **dissolved** (or gone into solution) in the water. The solution consists of two components, the **solute** (the substance that dissolves—in this case, sugar) and the **solvent** (the substance in which a solute dissolves—in this case, water). In a solution of sugar in water, the molecules of sugar are uniformly distributed among the molecules of water, i.e., the solution is a **homogeneous mixture** of solute and solvent molecules. The molecules of sugar diffuse continuously through the water, and although each of them is heavier than a single molecule of water, the sugar does not settle out on standing.

When sodium chloride, an ionic substance, dissolves in water, sodium ions and chloride ions become uniformly distributed throughout the water. This solution is a homogeneous mixture of water molecules, sodium ions, and chloride ions. As for a molecular solute (for example, sugar), there is diffusion of the solute particles (ions in the case of NaCl and other ionic substances) through the water; no settling of the solute particles takes place upon standing.

A solution that contains only a small amount of solute in comparison with the amount of solvent is said to be **dilute;** the addition of more solute increases its concentration, and the solution becomes more **concentrated.** A solution is said to be **saturated** when the concentration is such that the dissolved solute exists in equilibrium with excess undissolved solute. The **solubility** of a given solute is

303

defined as that quantity of the solute that will dissolve in a specified quantity of solvent to produce a saturated solution. Since we can have solutions of different concentrations, it is evident that the composition of a solution may be varied between certain limits. Thus solutions are not compounds, because the latter always contain the same elements in the same proportions by mass.

The solubility of a pure solid substance in a pure solvent at a given temperature is quantitatively a definite physical property of the substance. The solubility of sodium chloride is 35.8 g per 100 g of water at 20°; that of sodium fluoride is 4.2 g; and that of silver bromide is 0.00002 g. A general idea of relative solubilities is conveyed by use of the terms quite soluble, moderately soluble, slightly soluble, and insoluble. Strictly speaking, no substance is absolutely insoluble, although for all practical purposes many substances appear to be so.

Because the physical state of both the solute and solvent may be gas, liquid, or solid, many kinds of solutions are possible. Almost any gas, liquid, or solid can act as a solvent for other gases, liquids, and solids. Many alloys are solid solutions of one metal dissolved in another: Nickel coins contain nickel dissolved in copper. Air is a homogeneous mixture of gases and thus is a gaseous solution. Oxygen (a gas), alcohol (a liquid), and sugar (a solid) will each dissolve in water (a liquid) to form liquid solutions.

Liquid solutions exhibit many of the general properties characteristic of pure liquids. The most common solutions with which we work in chemistry are those in which the solvent is a liquid. The use of water as a solvent is so general that the word "solution" has come to imply a water solution unless some other solvent is designated.

All solutions are characterized by (1) homogeneity, (2) absence of settling, (3) the molecular or ionic state of subdivision of the components, and (4) a composition that can be varied continuously within limits.

13.2   Importance of Solutions

Many reactions take place at an appreciable rate only when the reactants are in solution. For example, when dry powdered barium chloride and sodium sulfate are mixed at ordinary temperatures, there is no perceptible reaction. However, when aqueous solutions of these compounds are mixed, a reaction takes place rapidly, as indicated by the immediate formation of a precipitate of barium sulfate.

$$Ba^{2+} + [2Cl^-] + [2Na^+] + SO_4{}^{2-} \longrightarrow BaSO_4(s) + [2Na^+] + [2Cl^-]$$

When solutions of these two ionic compounds are brought together, their ions mix freely by diffusion, and when the barium ions collide with sulfate ions, they form slightly soluble barium sulfate. When reactants are in the solid state, reaction is possible only between the molecules (or ions) in the surface layers of the particles, because collision between the molecules (or ions) is necessary before reaction can take place. In solids, the movement of the particles is greatly restricted, and the chance of collision between molecules or ions in adjacent solid particles is very slight. Consequently, reactions between most solids are slow.

Solutions are useful in the separation of substances of wide differences in solubility in a given solvent. Thus barium sulfate, which is only slightly soluble in water, is readily separated from the quite soluble sodium chloride by filtration.

The sodium and chloride ions pass through the pores of filter paper with the water, but the crystals of barium sulfate are retained by the paper. **Gravimetric analysis,** a part of the field of analytical chemistry, is based upon the separation of compounds that differ greatly in their solubilities.

Solutions are extremely important in life processes. For example, carbon dioxide and nutrients are carried throughout the body in solution in the blood. Digestion is a process in which food is converted into a form that the body can use, with the useful nutrients going into solution in order that they can pass through the walls of the digestive tract into the blood.

The action of natural waters in dissolving substances from air and earth is one of the important processes involved in the conversion of rocks to soil, in altering the fertility of the soil, and in changing the form of the earth's surface. The deposits of many minerals in the earth's crust are the result of reactions that have taken place in solution followed by the evaporation of the solvent.

Solutions of Gases in Liquids

13.3   Conditions Affecting the Dissolution of Gases in Liquids

The extent to which a gas dissolves in a liquid depends on the following factors: (1) the nature of the gas and the solvent, (2) the pressure, and (3) the temperature.

The solubility of a gas in a given liquid is considered to be a specific property of the gas, because its solubility differs from that of other gases in the same liquid. For example, at 0°C and 1 atm, 1 liter of water dissolves 0.0489 liter of oxygen, or 1.713 liters of carbon dioxide, or 79.789 liters of sulfur dioxide, or 1176.0 liters of ammonia.

The solubility of a gas in a liquid can be increased by increasing the pressure of the gas. This relationship is quantitatively expressed by **Henry's Law: The mass of a gas that dissolves in a definite volume of liquid is directly proportional to the pressure of the gas.** This means that if 1 gram of a gas dissolves in 1 liter of water at 1 atmosphere of pressure, 5 grams will dissolve at 5 atmospheres of pressure. The effect of pressure does not follow Henry's Law when a chemical reaction takes place between the gas and the solvent. Thus the solubility of ammonia in water does not increase as rapidly with increasing pressure as predicted by the law, because ammonia, being a base, reacts to some extent with water to form ammonium ions and hydroxide ions.

$$NH_3 + H_2O \rightleftharpoons NH_4^+ + OH^-$$

The effect of increased pressure upon the solubility of a gas in a liquid is illustrated by the behavior of carbonated beverages. Carbon dioxide is forced under pressure into the beverage, and the bottle is tightly capped to maintain the pressure and prevent escape of carbon dioxide. When the cap is removed, the pressure is decreased and some of the gas escapes. The escape of bubbles of the gas from the liquid is known as **effervescence.**

Finally, the solubility of gases in liquids decreases with an increase in temperature. For example, carbon dioxide at 760 torr dissolves in 1 liter of water to the extent of 1.713 liters at 0°, 1.194 liters at 10°, 0.878 liter at 20°, and 0.359 liter at 60°. This relationship is not one of inverse proportion, however, and the solubility of a gas in a liquid at a given temperature must be determined experimentally. This decrease in solubility of gases with increasing temperature is a very important factor in thermal pollution, particularly in connection with the quantity of dissolved oxygen in water (see Section 12.9, part 4). Most gases can be expelled from solvents by boiling their solutions. Thus the gases oxygen, nitrogen, carbon dioxide, and sulfur dioxide can be removed from water by boiling it for a few minutes. The fact that a gas is expelled from a solution by boiling is not due just to the rise in temperature of the solution. When the solution is boiled in an open container, part of the vapor of the solvent escapes, carrying with it some of the solute gas that was dissolved. This lowers the pressure of that gas above the solution, and, in accord with Henry's Law, more gas escapes.

Solutions of Liquids in Liquids

13.4   Miscibility of Liquids

Miscibility is a term used for liquids to indicate the extent of solubility of one liquid in another. Some liquids will mix with water in all proportions. Such liquids are usually polar substances (see Section 5.9) with polar character similar to that of water. For such polar liquids, the negative ends of the polar solute molecules attract the positive ends of the polar solvent molecules, and vice versa, with about the same degree of attraction that like molecules of either substance have for each other. Hence the two kinds of molecules mix easily in all proportions. Liquids that mix with water in all proportions are said to be completely **miscible** with water. Ethyl alcohol, sulfuric acid, and glycerin are examples of such liquids.

Some nonpolar covalent liquids, such as gasoline, carbon disulfide, and carbon tetrachloride, do not have a *net* charge separation in the molecule as a whole, and hence there is no effective attraction between the molecules of such liquids and the polar water molecules. Thus the only appreciable attractions in a mixture of a nonpolar covalent liquid and water are between the water molecules, and so they effectively "squeeze out" the molecules of the nonpolar liquid, making these liquids very nearly insoluble in water. Such liquids are said to be **immiscible** with water (Fig. 13-1). Other liquids, such as ether and bromine, are slightly soluble in

Figure 13-1 Immiscible, partially miscible, and miscible liquids.

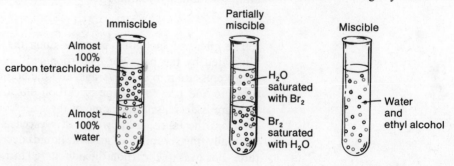

water and are said to be partially miscible. Two layers are formed when two immiscible liquids are in contact with each other. In cases of partial miscibility, each layer is a solution of one liquid in the other (Fig. 13-1).

13.5   Fractional Distillation

It will be recalled from Chapter 11 that a pure liquid exhibits a definite vapor pressure at a given temperature and that the normal boiling point of a pure liquid is that temperature at which the vapor pressure of the liquid is 760 torr. The total pressure exerted by the vapor of a solution made by dissolving one liquid in another liquid depends upon the concentration of its components. That solution vapor pressure may be (1) greater than that of either component taken separately, (2) less than that of either component, or (3) between those of the two liquids concerned. The boiling point of the solution is the temperature at which the total vapor pressure of the mixture is equal to the atmospheric pressure. Usually the boiling point of a mixture of two liquids lies between the boiling points of the two components.

When a solution of two liquids is boiled, the vapor produced at the boiling point is richer in the lower-boiling component (the component with the higher vapor pressure) than was the original mixture. When this vapor is condensed (Fig. 11-5), the resulting distillate is therefore richer in the liquid of the lower-boiling component than was the original mixture. This means that the composition of the boiling mixture is constantly changing, that the amount of the lower-boiling component of the mixture is constantly decreasing, and that the boiling point rises as distillation is continued. The vapor (and distillate) contains more and more of the higher-boiling component and less and less of the lower-boiling component as distillation proceeds. By changing the receiver at intervals, successive fractions, each increasingly richer in the less volatile (higher boiling) component, are obtained. If this process (**fractional distillation**) is repeated several times, relatively pure samples of the two liquids may be obtained.

Fractionating columns have been devised in which a single operation achieves separation of liquids that would require a great number of simple fractional distillations of the type just described. Crude oil, a complex mixture of hydrocarbons, is separated into its components by fractional distillation on an enormous industrial scale (Fig. 13-2).

13.6   Constant Boiling Solutions

When a dilute aqueous solution of nitric acid is distilled, the first fraction of distillate that is formed consists mostly of water because water has a higher vapor pressure than nitric acid under distillation conditions. As distillation is continued, the solution remaining in the distilling flask becomes richer in nitric acid. After a time a concentration of 68% HNO_3 is reached, and thereafter the solution boils at a constant temperature of 120.5°C (at 760 torr). At this temperature the vapor pressures of the nitric acid and water in the solution are equal, the solution and the vapor have the same composition, and the solution completely distills without any further change in composition.

Figure 13-2 Fractional distillation of crude oil. Oil heated to about 425°C in the furnace vaporizes when it enters the tower at the right. The vapors rise through bubble caps in a series of trays in the tower. As the vapors gradually cool, fractions of higher, then of lower, boiling points condense to liquids and are drawn off. The fraction of highest boiling point is drawn off at the bottom as a residue. It is heavy fuel oil. In modern refineries these fractions, which still consist of mixtures of hydrocarbons, are further processed.

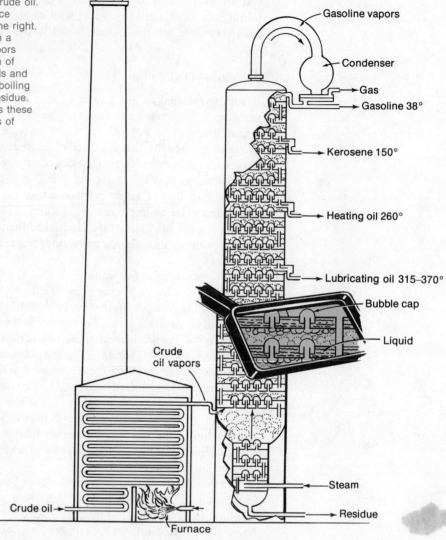

If a nitric acid solution more concentrated than 68% is distilled, the vapor first formed contains a large amount of HNO_3. The solution that remains in the distilling flask contains a greater percentage of water than at first, and the concentration of the nitric acid in the distilling flask decreases as the distillation is continued. Finally, a concentration of 68% HNO_3 is reached and the solution again boils at the constant temperature of 120.5° (at 760 torr). The 68% solution of nitric acid is referred to as a **constant boiling solution.** At this specific concentration, the solution has a lower vapor pressure, and thus a higher boiling point, than at any other possible concentration. Solutions that distill without change in composition or temperature are also called **azeotropic mixtures.**

Other common and important substances that form azeotropic mixtures with water are HCl (20.24%, 110°C at 760 torr) and H_2SO_4 (98.3%, 338°C at 760 torr).

Solutions of Solids in Liquids

13.7   The Effect of Temperature on Solubility

When a solid dissolves in a liquid, a change in the physical state of the solid analogous to melting takes place. Energy is absorbed in overcoming the forces that hold the molecules, atoms, or ions in their lattice positions in the crystal. This is an endothermic change that accompanies the dissolution of all crystalline solids in liquids. The physical process of solution is often accompanied by a second change, that of a chemical reaction between the solute molecules or ions and the solvent. This second change is commonly exothermic in character. If the heat evolved in the chemical change is greater than that absorbed in the physical change, the heat of solution is negative; i.e., the net process is exothermic. In other cases, the heat of solution is positive; i.e., the net process is endothermic (see Section 9.6). Furthermore, heat is either absorbed or evolved during the process of crystallization of a solid from solution, a process opposite to that of a solid going into solution.

The dependence of solubility upon temperature for a number of inorganic substances in water is shown graphically by the solubility curves in Fig. 13-3.

Figure 13-3 Graph showing the effect of temperature on the solubility of several inorganic substances.

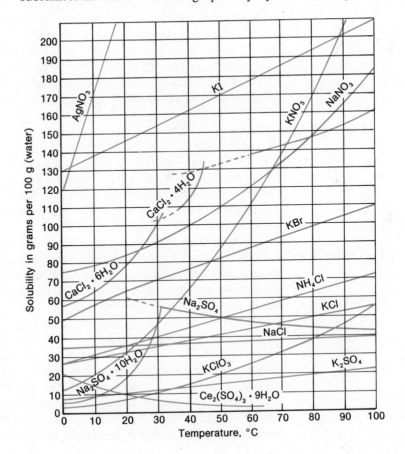

Generally, the solubility of a solid increases with increasing temperature, although there are important exceptions (calcium sulfate, Section 12.11, for example). Note that the solubility of sodium chloride increases very slightly with a rise in temperature, whereas that of potassium nitrate increases greatly. A sharp break in a solubility curve indicates the formation of a compound whose solubility is different from that of the substance from which it was formed. For example, when $Na_2SO_4 \cdot 10H_2O$ (Glauber's salt) is heated to 32.4°, it loses its water of hydration and forms the anhydrous salt, Na_2SO_4. The curve up to 32.4°, the **transition point,** represents the effect of a rise in temperature upon the solubility of $Na_2SO_4 \cdot 10H_2O$ and the curve for temperatures above this point represents the effect of a rise in temperature upon the solubility of Na_2SO_4.

13.8   Saturated and Unsaturated Solutions

It was noted in Section 13.1 that sugar dissolves in water at a given temperature until no more will dissolve, and the solution is said to be **saturated.** When more sugar is added to a saturated solution, excess sugar falls to the bottom of the container, and it appears that the dissolving process stops. Actually, molecules of sugar continue to leave the solid under the solvent action of the water and go into solution. However, as the molecules of sugar in solution move about by diffusion, some of them collide with the solid on the bottom and take up positions in the crystal lattice. This process of crystallization is the opposite of dissolution. If enough molecules return to the solid state from the saturated solution, the process of crystallization just counterbalances that of dissolution, and we have a state of dynamic equilibrium. The same dynamic equilibrium occurs with saturated solutions of partially miscible liquid systems such as bromine and water. At equilibrium, as bromine molecules leave the bromine layer and enter the water layer, the same number of bromine molecules leave the water layer and return to the bromine layer. A **saturated solution** is a solution in which the dissolved solute is in equilibrium with the undissolved solute. If a crystal of imperfect shape is placed in a saturated solution of the crystalline material, it will slowly "mend" its shape by loss of particles to the solution and gain of particles from the solution. This is experimental evidence that a saturated solution in contact with its solid solute is a system in dynamic equilibrium.

An **unsaturated solution** contains less solute dissolved in a given quantity of solvent than would be the case for a saturated solution. Excess solute will not exist in an unsaturated solution. Instead, the solute dissolves, and the concentration of the solution increases until either the solution becomes saturated or all the solute has gone into solution.

13.9   Supersaturated Solutions

When saturated solutions of solid solutes are prepared at elevated temperatures and then permitted to cool, the excess solute usually separates from the solution by crystallizing. However, if a saturated solution is prepared at an elevated temperature and excess solute removed, crystallization often does not take place if the solution is allowed to cool undisturbed. The solution contains more of the solute than it does when it is in equilibrium with the undissolved state and is called a **supersaturated solution.** Such solutions are **metastable systems** (unstable, but not so unstable that they cannot exist). Agitation of the solution or the addition of a "seed" crystal of the solute may start crystallization of the excess solute; after crystallization, a saturated solution in equilibrium with the crystals of solute remains.

Gases also form supersaturated solutions. If a solution of a gas in a liquid is prepared either at low temperature or under pressure (or both), as the solution warms or as the gas pressure is reduced, the solution may become supersaturated. For example, a bottle of a carbonated beverage may not liberate the excess carbon dioxide when opened (thereby reducing the carbon dioxide pressure above the solution), but if the beverage is shaken or stirred, it does.

13.10   Rate of Solution

Rate of solution pertains to the amount of solute entering the solution per unit of time. It depends upon several factors. First, more soluble substances dissolve more rapidly than less soluble ones. Second, because the dissolution of a solid solute can take place only at the surface of the particles, the more finely divided the solute is (the greater the surface area per unit mass of the substance) the greater will be its rate of solution. Third, the rate at which the dissolved molecules diffuse away from the solid solute is relatively slow, and so the solution in the immediate vicinity of the solid phase approaches the saturation concentration. Stirring or shaking the mixture brings unsaturated solution in contact with the solute and thus increases the rate of solution. Fourth, heating results in convection currents, which produce the same effects as agitation of the mixture, and heating also usually increases the solubility of the solid; if the solute and solvent react during the solution process, heating causes the reaction to go faster and hence increases the rate of solution.

Solutions of Electrolytes

13.11   Electrolyte Solutions

Substances that give solutions that conduct an electric current are known as **electrolytes.** Molten substances that conduct an electric current are also electrolytes. The process of passing an electric current through a solution or through a molten substance so that a chemical reaction results is called **electrolysis.** Most electrolytes are acids, bases, or salts. Aqueous solutions of some substances, such as sugar and alcohol, do not conduct an electric current. Substances that form nonconducting solutions are called **nonelectrolytes.**

The classification of substances as electrolytes and nonelectrolytes may be carried out experimentally by setting up a simple conductivity apparatus as shown in Fig. 13-4. The terminals of a storage battery or a 110-volt circuit are connected through an electric lamp to two electrodes in a beaker. When the beaker is filled with pure water, not enough current flows through the circuit to cause the electric

Figure 13-4 Apparatus for demonstrating the conductivity of solutions.

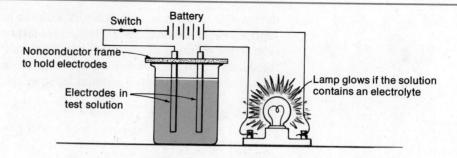

lamp to emit light. When an aqueous solution of a nonelectrolyte such as sugar, alcohol, or glycerin is tested, the lamp again does not emit light. If, however, an electrolyte such as hydrochloric acid, sodium hydroxide, or sodium chloride is dissolved in the water, the lamp glows brightly.

If the beaker in Fig. 13-4 is filled with a 0.1 M solution of hydrochloric acid, the lamp in the circuit will glow brightly, showing that the solution is a good conductor of electricity. The same is true of 0.1 M solutions of nitric and sulfuric acids, as well as bases such as potassium, sodium, and barium hydroxides. Most salts behave in a similar fashion. Substances whose aqueous solutions are good conductors of electricity are known as **strong electrolytes.** When a 0.1 M solution of acetic acid or ammonia is placed in the beaker, it is found that the lamp glows much less brightly than it does with acids like hydrochloric acid or bases like sodium hydroxide. Substances whose aqueous solutions are poor conductors of electricity are called **weak electrolytes.**

Svante Arrhenius, a Swedish chemist, first successfully explained electrolytic conduction. His theory also explained why acids, bases, and salts, when dissolved in water, affect certain physical properties of water more than do nonelectrolytes. Although Arrhenius' theory has since been modified in the light of our present knowledge of atomic structure and chemical bonding, the modern theory of electrolytes embodies most of the principal postulates of Arrhenius' theory.

According to the modern theory, a water solution of an electrolyte contains positive and negative ions that can move independently. As previously defined, an ion is an atom or group of atoms carrying an electric charge. When a solution conducts an electric current, the positive ions (the cations) migrate toward the negative electrode while the negative ions (the anions) migrate toward the positive electrode. The movement of the ions toward the electrodes of opposite charge during electrolysis accounts for electrolytic conduction. The total number of positive charges on the cations in solution is just equal to the total number of negative charges on the anions. The solutions of electrolytes are therefore electrically neutral. A solution of a nonelectrolyte contains *molecules* of the nonelectrolyte rather than ions.

According to the theory of electrolytes, if a solution is a good conductor of electricity the solute consists entirely or principally of ions, and if a solution is a poor conductor the solute consists principally of molecules. Thus salts such as KBr, $NaNO_3$, and $CoSO_4$, strong acids such as HCl, H_2SO_4, and HNO_3, and strong bases such as $LiOH$, KOH, and $Ba(OH)_2$, all of which dissolve in water, forming solutions composed principally of ions (Section 8.1), behave as strong

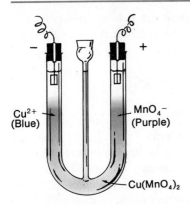

Figure 13-5 When an electric current flows in the system shown, blue Cu^{2+} ions and purple MnO_4^- ions travel toward different electrodes, visual evidence that the current is carried by the ions.

Cu²⁺ (Blue)

MnO₄⁻ (Purple)

Cu(MnO₄)₂

electrolytes. Weak acids such as HF, HCN, and acetic acid, and weak bases such as NH_3 and pyridine, which exist in solution in large part as unreacted molecules and only in small part as ions, behave as weak electrolytes. The partial ionization of acetic acid, a weak acid and a weak electrolyte, is expressed by the equation

$$CH_3CO_2H + H_2O \rightleftharpoons H_3O^+ + CH_3CO_2^-$$

The double arrow indicates that an equilibrium exists between the ions and the un-ionized molecules of acetic acid. A 0.1 M solution of acetic acid is only 1.34% ionized at 25°C. This means that 98.66% of the acid is in the molecular form. As the solution is diluted, the percentage of ionization increases.

Pure water is an extremely poor conductor of electricity, indicating very slight ionization (actually about 0.00001%). As indicated in Section 12.3, water ionizes when one molecule of water gives up a proton to another molecule of water, yielding hydronium and hydroxide ions.

$$H_2O + H_2O \rightleftharpoons H_3O^+ + OH^-$$

The migration of ions during electrolytic conduction can be demonstrated in a striking manner. A U-tube (Fig. 13-5) is partially filled with a solution of potassium nitrate, which is colorless and which has been acidified with a few drops of sulfuric acid. Then a solution of copper(II) permanganate is carefully introduced into the bottom of the U-tube without mixing of the two solutions. When the electrical circuit is closed, the blue hydrated copper(II) ions $[Cu(H_2O)_4^{2+}]$ begin to migrate through the colorless potassium nitrate toward the negative electrode, and the purple permanganate ions (MnO_4^-) migrate toward the positive electrode. This demonstration offers visual evidence that the electric current is carried by ions, and that ions of opposite charge exist independently in the solution.

13.12   Dissolution of Molecular Electrolytes

We learned in Chapter 5 that a hydrogen atom combines with a chlorine atom with the formation of a molecule containing a covalent bond.

$$H \cdot + \times \overset{\times\times}{\underset{\times\times}{Cl}} \times \longrightarrow H \times \overset{\times\times}{\underset{\times\times}{Cl}} \times$$

However, the electron pair involved in the covalent bond is not shared equally between the hydrogen and chlorine atoms. The chlorine atom attracts the electrons more strongly than does the hydrogen due to the greater electronegativity of the chlorine. The hydrogen chloride bond is therefore somewhat polar, with the hydrogen end positive and the chlorine end negative. Since dry hydrogen chloride in the liquid state does not conduct an electric current, we conclude that no ions are present. But an aqueous solution conducts electricity and possesses acidic properties, because molecules of hydrogen chloride react with water to produce hydronium ions and chloride ions in the solution. Since no ionization of HCl occurs in a nonpolar solvent such as benzene, the polar water molecules evidently play an important role in bringing about ionization.

The ionization of hydrogen chloride in water may be represented by the equation

$$H_2O(l) + HCl(g) \longrightarrow H_3O^+(aq) + Cl^-(aq)$$

A proton (hydrogen ion) is shifted from a polar hydrogen chloride molecule to a lone pair of electrons on the water molecule, forming a hydronium ion (H_3O^+) and a chloride ion (Cl^-). As shown by the Lewis formulas,

$$H:\overset{..}{\underset{..}{O}}: + H \overset{\times\times}{\underset{\times\times}{\times Cl}} \overset{\times}{\times} \longrightarrow H:\overset{..}{\underset{H}{O}}:H^+ + \overset{\times\times}{\underset{\times\times}{\times Cl}} \overset{\times-}{\times}$$

the pair of electrons that bonded the hydrogen and chlorine together in the HCl molecule remains with the chlorine, making it a chloride ion. Either of the two unshared pairs of electrons on the water molecule can be shared with the proton, and the proton may pass quite readily from one molecule of water to another. Both the hydronium ions and chloride ions resulting from the ionization of hydrogen chloride in water are undoubtedly hydrated in solution, as are all other ions. All molecular acids are polar substances and ionize in solution in the same manner as described for hydrogen chloride.

Many substances composed of molecules dissolve in water as hydrated molecules rather than as hydrated ions. In many instances, however, a fraction of the hydrated molecules undergo ionization. For example, cyanic acid, a weak acid that consists of polar molecules, dissolves extensively in water as hydrated molecules. Ionization of these dissolved molecules takes place to a slight degree under ordinary conditions.

$$H_2O(l) + HOCN(aq) \rightleftharpoons H_3O^+(aq) + OCN^-(aq)$$

$$H:\overset{..}{\underset{H}{O}}: + H:\overset{..}{\underset{..}{O}}:C:::N: \rightleftharpoons H:\overset{..}{\underset{H}{O}}:H^+ + :\overset{..}{\underset{..}{O}}:C:::N:^-$$

Because the cyanate ion, OCN^-, holds a proton (hydrogen ion) more strongly than does a water molecule, the reaction proceeds only to the extent of 5.6% in a 0.1 M solution of HOCN at 25°C. Acids such as acetic acid (CH_3CO_2H), nitrous acid (HNO_2), and hydrocyanic acid (HCN) are also soluble in water, but at any one time only a small fraction of their hydrated polar molecules undergoes ionization. Hence they too are classified as weak acids.

Weak bases also dissolve to give solutions of hydrated molecules, some of which undergo ionization. For example, a solution of ammonia in water consists primarily of hydrated molecules, $NH_3(aq)$, with small amounts of solvated ammonium ions, $NH_4^+(aq)$, and hydroxide ions, $OH^-(aq)$, which result from the following reaction:

$$NH_3(aq) + H_2O(l) \rightleftharpoons NH_4^+(aq) + OH^-(aq)$$

$$H:\overset{H}{\underset{H}{N}}: + H:\overset{..}{\underset{..}{O}}: \rightleftharpoons H:\overset{H}{\underset{H}{N}}:H^+ + :\overset{..}{\underset{..}{O}}:H^-$$

Certain other inorganic compounds readily dissolve in water by forming hydrated molecules. Among these compounds are the halides and cyanides of mercury, cadmium, and zinc. Molecules of these compounds have no dipole moments even though the bonds are of the polar covalent type. Their centers of positive and negative electric charge coincide because of a high degree of molecular symmetry, as in the linear (straight line) molecule Cl—Hg—Cl. There are, however, both a net positive electric charge on the mercury atom and net negative charges on the chlorine atoms of $HgCl_2$. Water dipoles are attracted to these points of electric charge concentration and the $HgCl_2$ molecule becomes hydrated. A small fraction of the bonds between mercury and chlorine breaks, with the formation of a few hydrated mercury(II) ions and hydrated chloride ions. Compounds of this type also are classed as weak electrolytes.

13.13   Dissolution of Ionic Compounds

The process of dissolution of ionic compounds in water is essentially one of separation of the ions that are associated in the solid. Water reduces the strong electrostatic forces between the ions and allows their separation. Let us consider the dissolution of potassium chloride in water. The hydrogen (positive) side of a polar water molecule is strongly attracted to a negative chloride ion at the surface of the solid, and the oxygen (negative) side of a water molecule is strongly attracted to a positive potassium ion at the surface of the solid. We may picture the water molecules as surrounding individual K^+ and Cl^- ions at the surface of the crystal and penetrating between them, thereby reducing the strong interionic forces of attraction that bind them together in the crystal and thus permitting them to move off into the water as hydrated ions (Fig. 13-6). Several water molecules become associated with each ion in solution as a result of the electrostatic attraction between the charged ions and the dipole of the water. Such an attraction is called an **ion-dipole attraction.** The increase in the distance between oppositely charged ions due to the layer of water molecules around each ion reduces the electrostatic attraction between the ions. In addition, water is a good insulator (it has a high dielectric constant), which further reduces the electrostatic attraction between the ions. This reduced electrostatic attraction permits the independent motion of each hydrated ion in a dilute solution.

Figure 13-6 The dissolution of potassium chloride and the solvation of ions in water. Water molecules in front of and behind the ions are not shown. The positive spheres represent potassium ions; the negative spheres, chloride ions.

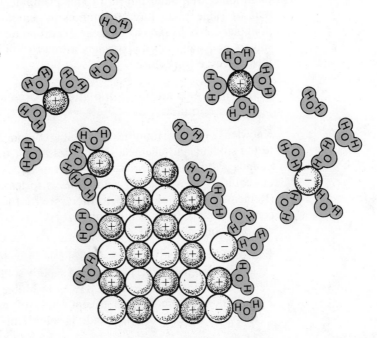

TABLE 13-1   The Solubility of Sodium Chloride and the Dielectric Constant of the Solvent

Solvent	Solubility of NaCl (grams per 100 g of solvent, 25°)	Dielectric Constant of the Solvent
Water, H_2O	36.12	80.0
Methyl alcohol, CH_3OH	1.3	33.1
Carbon tetrachloride, CCl_4	0.00	2.2

The solubility of an ionic compound in water is determined in large part by (1) the magnitude of the crystal forces and (2) the energy of hydration of the ions. A soluble salt is one for which the attraction of the ions for water molecules is greater than the attraction of the oppositely charged ions for each other, or, putting it another way, one for which the energy of hydration is greater than the crystal lattice energy. On the other hand, a slightly soluble salt is characterized by strong crystal forces and slight tendency of the ions to hydrate.

Now is a good time to point out that many other solvents, in addition to water, solvate ions. In this chapter, we shall be concerned mainly with solvation by water (hydration). When we discuss acids and bases in Chapter 14 we shall consider solvation by some other solvents such as liquid ammonia and liquid sulfur dioxide (at low temperatures) to produce solvated ions such as NH_4^+ (the ammonium ion), $[Ag(NH_3)_2]^+$, and $SO^{2+} \cdot SO_2$. Note that *solvation* is the general term; *hydration* is the specific term referring to solvation by water.

Ionic compounds usually dissolve only in polar solvents in which the polar solvent molecules can solvate and insulate the ions. In general, the higher the dielectric constant (greater polar character) of the solvent, the greater the solubility of an ionic compound put in it. This phenomenon is strikingly illustrated in the data of Table 13-1. Ionic substances in general do not dissolve appreciably in nonpolar solvents, such as benzene or carbon tetrachloride, because the nonpolar solvent molecules are not strongly attracted to ions, and nonpolar solvents have low dielectric constants.

13.14   Generalizations on the Solubilities of Common Metal Compounds

Knowledge of the solubilities of metallic compounds is very important and useful to the student and chemist. Memorization of the solubility of individual compounds is difficult, laborious, and unnecessary. A simple and worthwhile method of acquiring knowledge of solubilities is to learn the following generalizations. (Remember that these generalizations are for the simple compounds of the more common metals, there being many exceptions when the less common metals and complex compounds are considered.)

1. Most nitrates and acetates are soluble in water; silver acetate, chromium(II) acetate, and mercury(I) acetate are slightly soluble; bismuth acetate hydrolyzes to bismuth oxyacetate, $BiOC_2H_3O_2$, insoluble in water.
2. All chlorides are soluble except those of mercury(I), silver, lead(II), and copper(I) ions; lead(II) chloride is soluble in hot water.

3. All sulfates, except those of strontium, barium, and lead(II) are soluble; calcium sulfate and silver sulfate are slightly soluble.
4. Carbonates, phosphates, borates, arsenates, and arsenites—except those of ammonium and the alkali metals—are insoluble.
5. The sulfides of ammonium and the alkali metals are soluble and other sulfides are insoluble; the alkaline earth metal sulfides are hydrolyzed in water.
6. The hydroxides of the alkali metals, barium, and strontium are soluble and other hydroxides are insoluble; calcium hydroxide is slightly soluble.

Expressing Concentration

13.15   Per Cent Composition of Solutions

One method of expressing concentration in physical units involves giving the mass of solute in a given mass of solvent; for example, 1 g of NaCl in 100 g of water. A related system, per cent composition by mass, has been described in Chapter 3 (Section 3.3). For example, a 10% NaCl solution by mass may contain 10 g of NaCl in 100 g of solution (90 g of water and 10 g of NaCl). Or it may contain any other ratio of grams of NaCl to grams of solution for which the mass of NaCl is 10% of the total mass of the solution; for example, 20 g of NaCl in 200 g of solution, 15 g of NaCl in 150 g of solution, 7.4 g of NaCl in 74 g of solution.

$$\% \text{ solute} = \frac{\text{mass solute}}{\text{mass solution}} \times 100$$

When using per cent by mass in problems it is often convenient to substitute grams of solute per 100 g of solution (g solute/100 g solution) for per cent, since the grams of solute in 100 g of solution is numerically equal to the per cent of solute by mass.

EXAMPLE 1   Calculate the mass of NaCl solution that will contain 75.0 g of NaCl if the solution is 15.0% NaCl by mass.

A solution that is 15.0% NaCl by mass contains 15.0 g of NaCl per 100 g of solution (15.0 g NaCl/100 g solution). Thus we have

$$\text{Mass solution} = 75.0 \text{ g NaCl} \times \frac{100 \text{ g solution}}{15.0 \text{ g NaCl}}$$

$$= 5.00 \times 10^2 \text{ g solution}$$

Hence, 500 g of 15.0% NaCl solution by mass will contain 75.0 g of NaCl (and 425 g of water).

◆ ◆ ◆ ◆ ◆

In order to calculate the mass of solute in a given *volume* of solution from the per cent composition by mass, it is necessary to know the specific gravity (or density) of the solution.

EXAMPLE 2   Calculate the mass of hydrogen chloride (HCl) in 115 ml of concentrated hydro-chloric acid of specific gravity 1.19 and containing 37.23% HCl by mass.

This problem requires the following steps.

$$\boxed{\begin{array}{c}\text{Volume}\\\text{of}\\\text{solution}\end{array}} \longrightarrow \boxed{\begin{array}{c}\text{Mass}\\\text{of}\\\text{solution}\end{array}} \longrightarrow \boxed{\begin{array}{c}\text{Mass}\\\text{of}\\\text{HCl}\end{array}}$$

From the specific gravity, we know that the mass of 1 ml of concentrated hydrochloric acid, a concentrated solution of HCl in water, is 1.19 g.

$$\frac{1.19 \text{ g solution}}{1 \text{ ml}} \times 115 \text{ ml} = 136.8 \text{ g solution}$$

The solution contains 37.23% HCl by mass (37.23 g HCl/100.0 g solution)

$$136.8 \text{ g solution} \times \frac{37.23 \text{ g HCl}}{100.0 \text{ g solution}} = 50.9 \text{ g HCl}$$

◆ ◆ ◆ ◆ ◆

EXAMPLE 3   What volume of concentrated hydrochloric acid of specific gravity 1.19 and containing 37.23% HCl by mass contains 125 g of HCl?

This problem requires the following steps.

$$\boxed{\begin{array}{c}\text{Mass}\\\text{of}\\\text{HCl}\end{array}} \longrightarrow \boxed{\begin{array}{c}\text{Mass}\\\text{of}\\\text{solution}\end{array}} \longrightarrow \boxed{\begin{array}{c}\text{Volume}\\\text{of}\\\text{solution}\end{array}}$$

Since the solution contains 37.23% HCl, there are 37.23 g HCl/100.0 g solution.

$$125 \text{ g HCl} \times \frac{100.0 \text{ g solution}}{37.23 \text{ g HCl}} = 335.8 \text{ g solution}$$

The mass of 1 ml of solution is 1.19 g (1 ml/1.19 g).

$$335.8 \text{ g solution} \times \frac{1 \text{ ml solution}}{1.19 \text{ g solution}} = 282 \text{ ml solution}$$

Thus 282 ml of the concentrated hydrochloric acid contains 125 g of HCl.

◆ ◆ ◆ ◆ ◆

13.16   Molar Solutions

In Section 2.11, concentration was described in terms of moles of solute using units of molarity. The **molarity,** M, of a solution is the number of moles of solute in exactly one liter of solution. Molarity may be calculated using the following expression:

$$M = \frac{\text{moles of solute}}{\text{liters of solution}}$$

Figure 13-7 One method of preparation of a 1.00 M solution of NaHCO$_3$. 84.0 g of NaHCO$_3$ (1 mole) is added to a flask that is calibrated to hold 1.000 ℓ (a) and enough water is added to make 1.000 ℓ of solution (b).

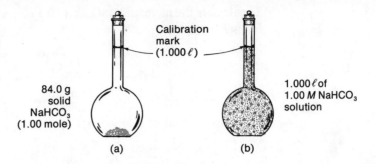

Calibration mark (1.000 ℓ)

84.0 g solid NaHCO$_3$ (1.00 mole)

1.000 ℓ of 1.00 M NaHCO$_3$ solution

(a)   (b)

A solution containing one mole of the solute in one liter of solution is called a **one-molar** solution. Note that a liter of solution rather than a liter of solvent is specified in this definition. Because one gram-formula weight of different substances contains the same number of molecules, it follows that equal volumes of one-molar solutions will contain the same number of molecules of the solute. It should be obvious that by using the molar method of expressing concentration of solutions, it is easy to select a desired number of moles, molecules, or ions of the solute by measuring out the appropriate volume of solution. For example, if 1 mole of sodium hydroxide is needed for a given reaction, we can use 40 g of solid sodium hydroxide (1 mole), or 1 liter of a 1-molar (1 M) solution or 2 liters of a 0.5 M solution of the base. If the 2 liters of 0.5 M sodium hydroxide solution were to be used in a reaction with 0.4 M hydrochloric acid, 2.5 liters of the acid would be needed, because 2.5 liters of 0.4 M acid will furnish 1 mole of the solute HCl.

$$\text{NaOH} \quad + \quad \text{HCl} \quad \longrightarrow \text{NaCl} + \text{H}_2\text{O}$$

1 mole   1 mole

$$\left(2\ \ell \times \frac{0.5\ \text{mol}}{1.0\ \ell} = 1\ \text{mol} \right) \quad \left(2.5\ \ell \times \frac{0.4\ \text{mol}}{1.0\ \ell} = 1\ \text{mol} \right)$$

One method of preparing a 1 M solution of sodium bicarbonate (NaHCO$_3$) is to weigh out accurately 84.0 g (1 mole) of pure sodium bicarbonate and dissolve it in sufficient water to form 1 liter of solution (Fig. 13-7). A 1 M solution of hydrochloric acid contains 36.5 g of the acid per liter of solution, and a 2 M solution of hydrochloric acid contains 73.0 g (2 moles) of HCl per liter. A 0.1 M solution of sodium bicarbonate contains 8.4 g of NaHCO$_3$ per liter.

Examples of the use of molar concentrations in stoichiometry calculations are presented in Sections 2.11 and 2.12. However, it may prove helpful to work through an additional example.

EXAMPLE   **Calculate the molarity of a concentrated sulfuric acid solution of specific gravity 1.84 containing 98.3% H$_2$SO$_4$ by mass.**

If we know the number of moles of H$_2$SO$_4$ in a given volume of solution, we can calculate the molarity by using the equation

$$M = \frac{\text{moles of solute}}{\text{liters of solution}}$$

To calculate the moles of H_2SO_4 in 1.00 ℓ of a concentrated solution requires the following steps.

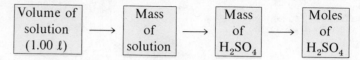

The mass of 1 ml of solution is 1.84 g.

$$\frac{1.84 \text{ g solution}}{1.00 \text{ ml solution}} \times 1000 \text{ ml solution} = 1.84 \times 10^3 \text{ g solution}$$

There are 98.3 g H_2SO_4 per 100 g of solution since the solution is 98.3% H_2SO_4 by mass.

$$1.84 \times 10^3 \text{ g solution} \times \frac{98.3 \text{ g } H_2SO_4}{100 \text{ g solution}} = 1.81 \times 10^3 \text{ g } H_2SO_4$$

$$1.81 \times 10^3 \text{ g } H_2SO_4 \times \frac{1 \text{ mol } H_2SO_4}{98.0 \text{ g } H_2SO_4} = 18.5 \text{ mol } H_2SO_4$$

$$\frac{18.5 \text{ mol } H_2SO_4}{1.00 \text{ } \ell} = 18.5 \text{ } M$$

The calculation can all be carried out in one step as follows:

$$\frac{1.84 \text{ g solution}}{1 \text{ ml solution}} \times 1000 \text{ ml solution} \times \frac{98.3 \text{ g } H_2SO_4}{100 \text{ g solution}} \times \frac{1 \text{ mol } H_2SO_4}{98.0 \text{ g } H_2SO_4}$$

$$= 18.5 \text{ mol } H_2SO_4$$

◆ ◆ ◆ ◆ ◆

A knowledge of the molar concentration of a solution is obviously useful because it permits us to measure out a definite number of moles of solute by volume. However, solutions change in volume as the temperature changes, and thus molarity changes with temperature. Solutions of a given molarity must be used at or near the temperature at which they were prepared in order for the value of the molarity to be meaningful.

A concentration unit called normality should be mentioned here. It is related to molarity, but differs from it in one important way. Normality is often used to describe the concentration of acids and bases or oxidizing and reducing agents. Since the normality of a solute is dependent on the reaction that the solute undergoes, this unit will be described in the chapters dealing with acid-base reactions (Section 14.12) and with electrochemical reactions (Section 20.7).

13.17   Molal Solutions

The concept of molality, m, is quite different from that of molarity, M, and is useful because it does not change its value as the temperature changes. The **molality, m,** of a solution is the number of moles of solute in exactly 1 kg of solvent. Note that a *kilogram of solvent* rather than a liter of solution is specified, and that this constitutes the difference between molality and molarity.

$$m = \frac{\text{moles of solute}}{\text{kilograms of solvent}}$$

A solution that contains one mole of solute in 1 kg of solvent is a **one-molal** solution.

EXAMPLE 1   **What is the molality of a solution that contains 0.850 g of ammonia (NH_3) in 125 g of water?**

Once the mass of NH_3 has been converted to moles of NH_3, the molality may be determined by dividing by the mass of the solvent, water, in kilograms.

$$0.850 \text{ g NH}_3 \times \frac{1 \text{ mol NH}_3}{17.0 \text{ g NH}_3} = 5.00 \times 10^{-2} \text{ mol NH}_3$$

$$m = \frac{\text{mol NH}_3}{\text{kg H}_2\text{O}} = \frac{5.00 \times 10^{-2} \text{ mol NH}_3}{0.125 \text{ kg H}_2\text{O}} = 0.400 \, m$$

◆ ◆ ◆ ◆ ◆

EXAMPLE 2   **Calculate the molality of an aqueous NaCl solution if 0.250 kg of the solution contains 40.0 g of sodium chloride.**

The mass of water is $0.250 - 0.0400 = 0.210$ kg. The molality is determined by converting the mass of NaCl to moles of NaCl and dividing by the mass of the solvent, water, in kilograms.

$$40.0 \text{ g NaCl} \times \frac{1 \text{ mol NaCl}}{58.5 \text{ g NaCl}} = 0.684 \text{ mol NaCl}$$

$$m = \frac{0.684 \text{ mol NaCl}}{0.210 \text{ kg H}_2\text{O}} = 3.26 \, m$$

◆ ◆ ◆ ◆ ◆

13.18   Mole Fraction

The **mole fraction, X,** of each component in a solution is the number of moles of the component divided by the total number of moles of all components present. The mole fraction of substance A in a solution (or other mixture) of substances A, B, C, ... is expressed as follows:

$$\text{Mole fraction of A} = X_A = \frac{\text{moles A}}{\text{moles A} + \text{moles B} + \text{moles C} + \cdots}$$

The mole fractions of all components of a system added together always equal 1. Like molality, the mole fraction of a component of a solution does not change with temperature.

EXAMPLE 1   Calculate the mole fraction of each component in a solution of 42.0 g CH_3OH, 35 g C_2H_5OH, and 50.0 g C_3H_7OH.

$$42.0 \text{ g } CH_3OH \times \frac{1 \text{ mol } CH_3OH}{32.0 \text{ g } CH_3OH} = 1.31 \text{ mol } CH_3OH$$

$$35 \text{ g } C_2H_5OH \times \frac{1 \text{ mol } C_2H_5OH}{46.0 \text{ g } C_2H_5OH} = 0.76 \text{ mol } C_2H_5OH$$

$$50.0 \text{ g } C_3H_7OH \times \frac{1 \text{ mol } C_3H_7OH}{60.0 \text{ g } C_3H_7OH} = 0.833 \text{ mol } C_3H_7OH$$

$$\text{Mole fraction of } CH_3OH = X_{CH_3OH} = \frac{1.31}{1.31 + 0.76 + 0.833} = \frac{1.31}{2.90} = 0.452$$

$$X_{C_2H_5OH} = \frac{0.76}{1.31 + 0.76 + 0.833} = \frac{0.76}{2.90} = 0.26$$

$$X_{C_3H_7OH} = \frac{0.833}{2.90} = 0.287$$

Note that the sum of the mole fractions, $0.452 + 0.26 + 0.287$, is 1.00.

◆ ◆ ◆ ◆ ◆

EXAMPLE 2   Calculate the mole fraction of solute and solvent for a 3.0-molal solution of sodium chloride.

A 3.0 m solution of sodium chloride contains 3.0 mol of NaCl dissolved in exactly 1 kg of water.

$$1 \text{ kg } H_2O \times \frac{1000 \text{ g } H_2O}{1 \text{ kg } H_2O} = 1000 \text{ g } H_2O$$

$$1000 \text{ g } H_2O \times \frac{1 \text{ mol } H_2O}{18.0 \text{ g } H_2O} = 55.5 \text{ mol } H_2O$$

$$X_{NaCl} = \frac{3.0}{3.0 + 55.5} = 0.051$$

$$X_{H_2O} = \frac{55.5}{3.0 + 55.5} = 0.949$$

Note that the sum of the two mole fractions, $0.051 + 0.949$, is 1.000.

◆ ◆ ◆ ◆ ◆

13.19   Applications of Concentration Calculations

When a solution is diluted, the volume is increased by adding more solvent and the concentration is decreased, but the total amount of solute is constant.

EXAMPLE 1   If 0.750 ℓ of 5.00 M silver nitrate is diluted to a volume of 1.80 ℓ by adding water, what is the molarity of the resulting diluted silver nitrate solution?

Since the number of moles of silver nitrate does not change on dilution, the problem can be solved by the following steps:

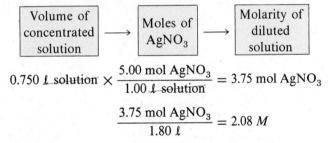

$$0.750 \; \ell \; \text{solution} \times \frac{5.00 \; \text{mol AgNO}_3}{1.00 \; \ell \; \text{solution}} = 3.75 \; \text{mol AgNO}_3$$

$$\frac{3.75 \; \text{mol AgNO}_3}{1.80 \; \ell} = 2.08 \; M$$

Note that the solution was diluted from 5.00 M to 2.08 M.

◆ ◆ ◆ ◆ ◆

EXAMPLE 2   **What volume of water in milliters would be required to dilute 11 ml of 0.45 M acid solution to a concentration of 0.12 M?**

Again the number of moles of solute does not change. The following steps are necessary to solve this problem.

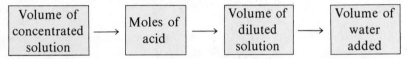

Convert volume to liters:

$$11 \; \text{ml} \times \frac{1 \; \ell}{1000 \; \text{ml}} = 1.1 \times 10^{-2} \; \ell$$

$$1.1 \times 10^{-2} \; \ell \; \text{solution} \times \frac{0.45 \; \text{mol acid}}{1.00 \; \ell \; \text{solution}} = 4.95 \times 10^{-3} \; \text{mol acid}$$

Rearrangement of the expression for molarity (molarity = moles/liters), gives the expression

$$\text{Liters} = \frac{\text{moles}}{\text{molarity}}$$

$$\text{Liters of dilute solution} = \frac{4.95 \times 10^{-3} \; \text{mol acid}}{\dfrac{0.12 \; \text{mol acid}}{1.00 \; \ell \; \text{solution}}}$$

$$= 4.12 \times 10^{-2} \; \ell \; \text{solution}$$

Convert to milliters:

$$4.12 \times 10^{-2} \; \ell \times \frac{1000 \; \text{ml}}{1.00 \; \ell} = 41.2 \; \text{ml solution}$$

The volume of water added in the dilution is equal to the final volume of the solution minus the original volume.

$$41.2 \; \text{ml} - 11 \; \text{ml} = 30 \; \text{ml}$$

◆ ◆ ◆ ◆ ◆

EXAMPLE 3   A sulfuric acid solution containing 571.6 g of H_2SO_4 per liter of solution at 20°C has a density of 1.3294 g/ml. Calculate (a) the molarity, (b) the molality, (c) the per cent by mass, and (d) the mole fractions for the solution.

a. $571.6 \text{ g } H_2SO_4 \times \dfrac{1 \text{ mol } H_2SO_4}{98.08 \text{ g } H_2SO_4} = 5.828 \text{ mol } H_2SO_4$

$$M = \dfrac{5.828 \text{ mol } H_2SO_4}{1.000 \text{ } \ell} = 5.828 \text{ } M$$

b. Since we know the moles of H_2SO_4 in 1 ℓ of solution, to get the molality we need the mass of water in 1 ℓ of solution. To find the molality we then need the following steps.

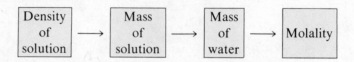

The mass of 1 ℓ of solution is given by rearranging the expression for density, $D = M/V$.

$$\text{Density} \times \text{volume} = \text{mass}$$

$$\dfrac{1.3294 \text{ g}}{1 \text{ ml}} \times 1000 \text{ ml} = 1329.4 \text{ g}$$

Thus 1 ℓ of solution weighs 1329.4 g and contains 571.6 g of H_2SO_4. The mass of solvent is therefore

$$1329.4 \text{ g} - 571.6 \text{ g} = 757.8 \text{ g} \qquad \text{(or 0.7578 kg)}$$

$$\dfrac{5.828 \text{ mol } H_2SO_4}{0.7578 \text{ kg } H_2O} = 7.691 \text{ } m$$

c. The solution contains 571.6 g of H_2SO_4 in 1329.4 g of solution.

$$\dfrac{571.6 \text{ g } H_2SO_4}{1329.4 \text{ g solution}} \times 100 = 43.00\% \text{ } H_2SO_4 \text{ by mass}$$

d. The number of moles of water present in 1 ℓ of the solution is given by

$$757.8 \text{ g } H_2O \times \dfrac{1 \text{ mol}}{18.02 \text{ g } H_2O} = 42.05 \text{ mol } H_2O$$

$$X_{H_2SO_4} = \dfrac{\text{mol } H_2SO_4}{\text{mol } H_2SO_4 + \text{mol } H_2O} = \dfrac{5.828}{5.828 + 42.05} = \dfrac{5.828}{47.88} = 0.1217$$

$$X_{H_2O} = \dfrac{42.05}{47.88} = 0.8783$$

Note that the sum of the mole fractions, $0.1217 + 0.8783$, is 1.0000.

★ ★ ★ ★ ★

Effect of Solutes on Properties of the Solvent

13.20   Lowering the Vapor Pressure of the Solvent

It has been found by experiment that when a nonvolatile substance (or one with such a low vapor pressure that we can disregard it) is dissolved in a liquid the vapor pressure of the liquid is lowered. Solid solutes exhibit only negligible vapor pressures, so they may be considered to be nonvolatile. Thus, for example, the vapor pressure of an aqueous sugar solution at 20°C is less than that of pure water at 20°C. The vapor pressure of a liquid is determined by the frequency of escape of molecules from the surface of the liquid. The presence of sugar molecules in the solution lowers the frequency of escape of water molecules from the surface of the liquid. This is the reason the vapor pressure of the solution is less than that of pure water. The kind or size of solute molecule has little to do in determining the extent of this effect. For a dilute solution, the decrease in vapor pressure is proportional to the ratio of the number of molecules of solute to the total number of solute and solvent molecules. The greater the number of molecules of solute, the lower the vapor pressure exerted by the solution. These considerations lead to **Raoult's Law,** which is stated as follows: **The vapor pressure of the solvent in a dilute solution, P_{solv}, is equal to the mole fraction of the solvent, X_{solv}, times the vapor pressure of the pure solvent, P^0_{solv}.**

$$P_{solv} = X_{solv} P^0_{solv} \qquad (1)$$

The *decrease* in vapor pressure, ΔP, of the solution, compared to that of the pure solvent, is equal to the mole fraction of solute, X_{solute}, times the vapor pressure of the pure solvent, P^0_{solv}.

$$\Delta P = X_{solute} P^0_{solv} \qquad (2)$$

EXAMPLE   Calculate the vapor pressure of a solution of **92.1 g of glycerin ($C_3H_8O_3$) in 184 g of ethanol (ethyl alcohol, C_2H_5OH) at 40°C. The vapor pressure of pure ethanol at 40° is 135.3 torr. Glycerin is essentially nonvolatile at this temperature.**

First we need to determine the mole fraction of ethanol (the solvent) in the solution.

$$92.1 \text{ g } C_3H_8O_3 \times \frac{1 \text{ mol } C_3H_8O_3}{92.1 \text{ g } C_3H_8O_3} = 1.00 \text{ mol } C_3H_8O_3$$

$$184 \text{ g } C_2H_5OH \times \frac{1 \text{ mol } C_2H_5OH}{46.0 \text{ g } C_2H_5OH} = 4.00 \text{ mol } C_2H_5OH$$

$$X_{C_2H_5OH} = \frac{4.00}{1.00 + 4.00} = 0.800$$

Now calculate the vapor pressure of ethanol:

$$P_{C_2H_5OH} = X_{C_2H_5OH} P^0_{C_2H_5OH}$$

$$P_{C_2H_5OH} = 0.800 \times 135.5 \text{ torr} = 108 \text{ torr}$$

Note that the vapor pressure of the solvent (ethanol) has been lowered from 135.3 torr to 108 torr by adding the glycerin. This change of 27 torr can also be calculated from Equation (2).

$$\Delta P = X_{\text{solute}} P^0_{\text{solvent}} = 0.200 \times 135.3 \text{ torr} = 27 \text{ torr}$$

◆ ◆ ◆ ◆ ◆

13.21   Elevation of the Boiling Point of the Solvent

The boiling point of a liquid is the temperature at which the vapor pressure of the liquid is equal to the external pressure upon its surface (Section 11.3). Because the addition of a solute lowers the vapor pressure of a liquid, a higher temperature is required to bring the vapor pressure of the solution in an open vessel up to the atmospheric pressure and make the solution boil. According to **Raoult's Law, the lowering of the vapor pressure of the solvent in a dilute solution is directly proportional to the mole fraction of the solute.** It follows, then, that the elevation of the boiling point of the solvent is also proportional to the mole fraction of the solute. For substances that do not dissociate into ions in solution and which either are nonvolatile or have a very low vapor pressure, the elevation of the boiling point of the solvent is the same when solutions of the same mole fraction are considered. For example, 0.100 mole of sucrose ($C_{12}H_{22}O_{11}$) and 0.100 mole of glucose ($C_6H_{12}O_6$), each dissolved in 1 kg of water, have the same mole fraction and form solutions that have the same boiling point, 100.0512°C at 760 torr. The elevation of the boiling point of the water, then, is $100.0512° - 100° = 0.0512°$. On the basis of one mole of solute, we can say that one mole of any nonelectrolyte dissolved in 1 kg of water raises the boiling point by 0.512°C. In general, the difference between the boiling point of a very dilute solution of a nonelectrolyte and the boiling point of the pure solvent is directly proportional to the *molal* concentration of the solute. Thus, 0.200 mole of a nonelectrolyte dissolved in 1 kg of water (a 0.200 molal solution) will increase the boiling point by $0.200 \times 0.512° = 0.102°$; one-half mole of a nonelectrolyte in 1 kg of water (a 0.5 molal solution) will increase the boiling point by $0.5 \times 0.512 = 0.256°$.

The change in boiling point of a dilute solution, ΔT, from that of the pure solvent is given by the expression

$$\Delta T = K_b m \tag{1}$$

where m is the molal concentration of the solute in the solvent and K_b is the increase in boiling point for a one-molal solution. K_b is referred to as the **molal boiling-point elevation constant.** Values of K_b for several solvents are listed in Table 13-2. Note that the value of K_b depends upon the solvent.

It should be emphasized that the extent to which the vapor pressure of a solvent is lowered and the boiling point is elevated depends upon the number of solute particles present in a given amount of solvent and not upon the mass or size of the particles. Properties of solutions that depend upon the number and not the kind of particles concerned are spoken of as **colligative** properties.

It is not surprising to find that a mole of sodium chloride, which exists as two ions in the solution, causes nearly twice as great a rise in boiling point as does a

TABLE 13-2 Boiling Points, Freezing Points, and Molal Boiling- and Freezing-Point Constants for Several Solvents

Solvent	Boiling Point, °C (760 torr)	K_b	Freezing Point, °C	K_f
Water	100.0	0.512	0	1.86
Acetic acid	118.1	3.07	16.6	3.9
Benzene	80.1	2.53	5.48	5.12
Chloroform	61.26	3.63	−63.5	4.68
Nitrobenzene	210.9	5.24	5.67	8.1

mole of a nonelectrolyte. One mole of table sugar consists of 6.022×10^{23} particles (as molecules), whereas one mole of sodium chloride consists of $2 \times 6.022 \times 10^{23}$ particles (as ions). Hence, calcium chloride ($CaCl_2$), which consists of three ions, causes nearly three times as great a rise in boiling point as does sugar. Why the elevation is not exactly twice (for NaCl) and three times (for $CaCl_2$) that of the molecular boiling point elevation is explained in Section 13.27.

EXAMPLE 1   **How much does the boiling point of water change when 1.00 g of glycerin, $C_3H_8O_3$, is dissolved in 47.8 g of water?**

Since the change in boiling point is proportional to the molal concentration of the glycerin, we first calculate the molal concentration of the glycerin.

$$1.00 \text{ g } C_3H_8O_3 \times \frac{1 \text{ mol } C_3H_8O_3}{92.1 \text{ g } C_3H_8O_3} = 1.08 \times 10^{-2} \text{ mol } C_3H_8O_3$$

$$\frac{1.08 \times 10^{-2} \text{ mol } C_3H_8O_3}{0.0478 \text{ kg } H_2O} = 0.226 \, m$$

The change in boiling point is equal to the molal concentration of the glycerin multiplied by K_b (the change in boiling point produced by a one-molal concentration of glycerin, 0.512° for water).

$$\Delta T = 0.226 \times 0.512 = 0.116°$$

This is simply the application of Equation (1) when $m = 0.226$ and $K_b = 0.512$.

◆ ◆ ◆ ◆ ◆

EXAMPLE 2   **What is the boiling point of a solution of 92.1 g of iodine in 800.0 g of chloroform, $CHCl_3$, assuming that the iodine is nonvolatile?**

First calculate the molality of iodine, I_2, in the solution since the change in boiling point of a dilute solution is proportional to m.

$$92.1 \text{ g } I_2 \times \frac{1 \text{ mol } I_2}{253.8 \text{ g } I_2} = 0.3629 \text{ mol } I_2$$

$$\frac{0.3629 \text{ mol } I_2}{0.8000 \text{ kg } CHCl_3} = 0.454 \, m$$

From the values of molal boiling-point elevation constants in Table 13-2, a one-molal solution of a nonelectrolyte in chloroform will display an increase of 3.63° in its boiling point. Thus, for a 0.454 m solution,

$$\Delta T = 0.454 \times 3.63 = 1.65°$$

Since the boiling point of chloroform (61.26°) is raised by 1.65°, the boiling point of the solution will be 61.26° + 1.65° = 62.91°.

◆ ◆ ◆ ◆ ◆

13.22   Depression of the Freezing Point of the Solvent

It is a common observation that solutions freeze at lower temperatures than do pure liquids. We use aqueous solutions of various antifreezes such as ethylene glycol in place of pure water in automobile radiators because such solutions freeze at lower temperatures. Sea water, with its large salt content, freezes at a lower temperature than fresh water. The depression of the freezing point of a solvent by an added solute is a reflection of the vapor pressure lowering caused by the solute. In Section 11.12 we learned that the freezing point of a pure liquid is the temperature at which the liquid is in equilibrium with its solid. A pure liquid and its solid have the same vapor pressure at the freezing point. For example, pure water and ice have the same vapor pressure at 0°C. However, the water *in a solution* has a lower vapor pressure than ice at this temperature. Consequently, if ice and an aqueous solution at 0°C are placed in contact, the ice melts. As the temperature is lowered below 0°C the vapor pressure of the ice decreases more rapidly than does that of the water in the solution. At a temperature somewhat below 0°C, the ice and the water have the same vapor pressure, and this is the temperature at which the solution and ice are in equilibrium; it is the freezing point of the solution. This property of solutes gives rise to the practice of using such substances as sodium chloride and calcium chloride to melt ice on streets and highways.

It has been found that one mole of such nonelectrolytes as sucrose, glycerin, ethylene glycol, and alcohol, when dissolved in 1 kg of water, gives solutions that freeze at −1.86°. In general, the difference, ΔT, between the freezing point of a pure solvent and the freezing point of a solution of a nonelectrolyte dissolved in that solvent is directly proportional to the *molal* concentration of the solute.

$$\Delta T = K_f m$$

The constant K_f, the **molal freezing-point depression constant,** is the change in freezing point for a one-molal solution and is different for each solvent. Values of K_f for several solvents are listed in Table 13-2.

A gram-formula weight of sodium chloride in 1 kg of water will show nearly twice the freezing-point depression characteristic of molecular compounds. Each ion individually produces about the same effect as a molecule upon the freezing point of a solution. In Section 13.27 we shall consider why the lowering produced by sodium chloride is not exactly twice that produced by a similar amount of a nonelectrolyte.

EXAMPLE   **What is the freezing point of the solution of I_2 in $CHCl_3$ described in Example 2, Section 13.21?**

The molal concentration of I_2 was shown to be 0.454 m. The value of the molal freezing-point depression constant, K_f, in Table 13-2 indicates that a one-molal solution in $CHCl_3$ reduces the freezing point by 4.68° below that of −63.5° for pure $CHCl_3$. Thus, for a 0.454 m solution

$$\Delta T = 0.454 \times 4.68 = 2.12°$$

Thus the freezing point of the solution will be 2.12° lower than that of pure $CHCl_3$, or,

$$\text{Freezing point of solution} = -63.5° - 2.12° = -65.6°$$

◆ ◆ ◆ ◆ ◆

13.23   Phase Diagram for an Aqueous Solution of a Nonelectrolyte

In Chapter 12, we discussed the phase diagram for pure water (see Section 12.2 and Fig. 12-5). A phase diagram is given in Fig. 13-8 for an aqueous solution of a nonelectrolyte solute, such as sucrose ($C_{12}H_{22}O_{11}$). The phase diagram for pure water is reproduced for comparison (broken line). The lower freezing points for the solution are shown by the solid line separating solid and liquid states being displaced to the left of the broken one (lower temperatures). Correspondingly, the higher boiling points for the solution are shown by the displacement of the solid line separating the liquid and gas states to the right of the broken one (higher

Figure 13-8 Phase diagram for a one-molal aqueous solution of a nonelectrolyte (solid lines) compared to that for pure water (broken lines).

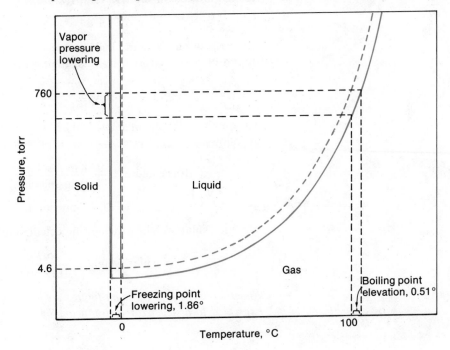

temperatures). The decrease in vapor pressure for the solution, at any given temperature, is indicated by the vertical distance between the broken line and the solid line. The normal freezing-point depression and the normal boiling-point elevation may be seen on the diagram as the horizontal distance between the broken line and the solid line near 0°C and near 100°C, respectively, at a pressure of 760 torr.

13.24 Determination of Molecular Weights of Substances in Solution

Molecular weights of nonelectrolytes can be determined by measuring the effect they have upon the freezing point or boiling point of a solvent. Since the change in freezing or boiling point of a solution is directly proportional to the molal concentration of solute present, the molecular weight of the solute can be calculated if the mass of solute and solvent in a solution is known.

EXAMPLE 1 **A solution of 35.7 g of an organic nonelectrolyte in 220.0 g of chloroform has a boiling point of 64.5°C. What is the molecular weight of this organic compound?**

This problem requires the following steps:

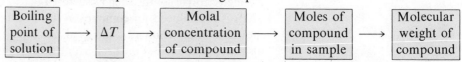

From the boiling point of pure chloroform (61.26°, Table 13-2) we can calculate ΔT, the increase in boiling temperature.

$$\Delta T = 64.5° - 61.26° = 3.2°$$

From K_b for chloroform (3.63°, Table 13-2) we can calculate the molal concentration of the electrolyte by rearranging the equation $\Delta T = K_b m$:

$$m = \frac{\Delta T}{K_b} = \frac{3.2}{3.63} = 0.88 \, m$$

The number of moles of solute in 0.2200 kg (220.0 g) of solvent is then calculated as follows:

$$\text{Moles solute} = \frac{0.88 \text{ mol solute}}{1.00 \text{ kg solvent}} \times 0.2200 \text{ kg solvent}$$

$$= 0.19 \text{ mol solute}$$

The molecular weight can then be calculated:

$$\text{Molecular weight} = \frac{35.7 \text{ g}}{0.19 \text{ mol}} = 180, \text{ or } 1.8 \times 10^2$$

Note that only two significant figures are justified, despite the fact that all data are expressed to at least three significant figures. Why is this?

◆ ◆ ◆ ◆ ◆

EXAMPLE 2   If 4.00 g of a particular nonelectrolye is dissolved in 55.0 g of benzene, the resulting solution freezes at 2.32°. Calculate the molecular weight of the nonelectrolyte.

The steps for this problem are

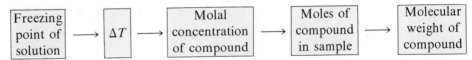

According to Table 13-2, K_f for benzene is 5.12° and the freezing point of benzene is 5.48°. Since the freezing point of the solution described is 2.32°, the 4.00 g of solute has lowered the freezing point of 55.0 g of benzene from 5.48° to 2.32°, so $\Delta T = 3.16°$.

The molality of the solution is

$$m = \frac{\Delta T}{K_f} = \frac{3.16}{5.12} = 0.617\ m$$

The number of moles of solute in 0.0550 kg (55.0 g) of solvent is

$$\text{Moles solute} = \frac{0.617\ \text{mol}}{1.000\ \text{kg solvent}} \times 0.0550\ \text{kg solvent}$$

$$= 0.0339\ \text{mol}$$

The molecular weight can then be calculated:

$$\text{Molecular weight} = \frac{4.00\ \text{g}}{0.0339\ \text{mol}} = 118$$

◆ ◆ ◆ ◆ ◆

13.25   Osmosis and Osmotic Pressure of Solutions

When a solution and its pure solvent are separated by a semipermeable membrane (one through which the solvent but not the solute can pass), the pure solvent will diffuse through the membrane and dilute the solution. This process is known as **osmosis.** Actually, the solvent diffuses through the membrane in both directions simultaneously; however, the rate of diffusion is greater from the pure solvent to the solution than in the opposite direction so there is an increase in the number of solvent molecules in the solution.

A force sufficient to prevent the osmosis of solvent molecules into a solution produces a pressure referred to as **osmotic pressure,** which can be measured in the following way: A piece of cellophane is securely fastened over the end of a thistle tube to serve as a semipermeable membrane, and the bowl of the thistle tube is filled with a concentrated sugar solution. When the thistle tube is inverted in a beaker of water (Fig. 13-9), water will diffuse through the membrane into the solution, and a slow rise of liquid in the tube will be observed. If the membrane is sufficiently strong to withstand the pressure and if the stem of the thistle tube is long enough, the liquid will rise until its hydrostatic pressure (weight of the column of water in the tube) is equal to the osmotic pressure. The height to which the

Figure 13-9 Apparatus for
demonstrating osmosis. Drawing at
the right gives some of the details
of the process.

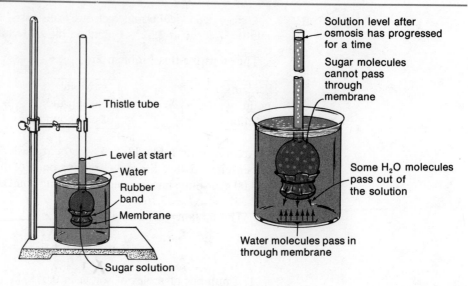

liquid rises is proportional to the concentration of the solution; in other words, the osmotic pressure is proportional to the number of solute particles in a definite volume of liquid. When gram-molecular weights of different nonelectrolytes are dissolved in 1 kg of water, their solutions exert equal osmotic pressures, approximately 22.4 atm.

13.26   The Effect of Electrolytes on the Colligative Properties of Solutions

We have seen that the effect of nonelectrolytes on the colligative properties of a solvent (vapor pressure, boiling point, freezing point, and osmotic pressure) is dependent only upon the number, and not on the kind, of particles dissolved. For example, one mole of any nonelectrolyte in solution in 1 kg of water (a one-molal solution) produces the same lowering of the freezing point, $1.86°C$, as one mole of any other nonelectrolyte because one mole of any nonelectrolyte contains the same number of molecules (6.022×10^{23}). However, the molal lowering of the freezing point produced by electrolytes is much greater than for nonelectrolytes since an electrolyte ionizes when it dissolves. The water in a solution which contains one mole of sodium chloride dissolved in 1 kg of water freezes at $-3.37°$. This lowering of the freezing point of water is $3.37 \div 1.86 = 1.81$ times the molal depression for a nonelectrolyte. This illustrates the fact that one mole of an electrolyte produces more than 6.022×10^{23} solute particles. All acids, bases, and salts behave in this fashion when placed in solution.

13.27   Ion Activities

In 1923, Peter J. W. Debye and Erich Hückel accepted the idea that strong electrolytes completely ionize in aqueous solution, and they proposed a theory to explain the *apparent* incomplete ionization of strong electrolytes. If a strong

Figure 13-10 Diagrammatic representation of the various species thought to be present in a water solution of potassium chloride.

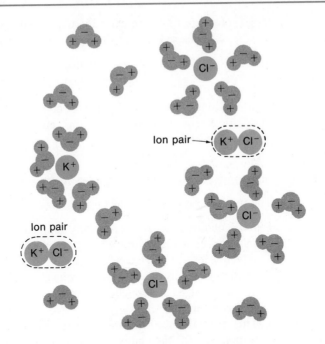

electrolyte, such as sodium chloride, is completely dissociated in aqueous solution, it should lower the freezing point of water and raise the boiling point twice as much as an equal molal concentration of a nonelectrolyte, since it gives two ions per mole. However, a mole of sodium chloride lowers the freezing point of water only 1.81 times as much as a mole of a nonelectrolyte does, instead of twice as much. A similar discrepancy occurs in the boiling-point elevation. Debye and Hückel accounted for these discrepancies between calculated and observed values for the colligative properties of solutions of electrolytes by the following theory.

Although the forces of interionic attraction in aqueous solution are very greatly reduced by hydration of the ions and the insulating action of the polar solvent, they are not completely nullified. The residual interionic forces of attraction prevent the ions from behaving as totally independent particles insofar as the colligative properties of the solution or their function as carriers of the electric current are concerned (Fig. 13-10). In some cases, a positive and negative ion may actually touch, giving a solvated unit called an **ion pair.** Thus the **activities** of the ions as independent units, or their "effective concentrations," are less than indicated by their actual concentrations. The more dilute the solution the greater the separation of oppositely charged ions and the less the residual interionic attraction. Thus in extremely dilute solutions the effective concentrations of the ions (activities) are essentially equal to the actual concentrations.

The factor by which the actual ion concentration must be multiplied to obtain the activity of the ion is called the **activity coefficient.**

$$\text{Activity} = f \times \text{concentration}$$

The activity coefficient, f, approaches unity with increasing dilution. For a 0.100

molal solution of sodium chloride, the activity coefficient has a value of 0.778 at 25.0°C, and the activity of the ions is 0.0778 molal.

$$a = f \times \text{concentration}$$
$$a = 0.778 \times 0.100 = 0.0778 \text{ molal}$$

In solutions of weak electrolytes the concentrations of the ions are small, and the interionic forces are so slight that the activities *of the ions* are essentially equal to their concentrations.

13.28   Summary of the Modern Theory of Electrolytes

The modern concepts of electrolytes may be summarized as follows:

1. Ionic compounds are completely ionized in the solid state and in solution.
2. In aqueous solution the ions of an electrolyte are hydrated (in many other solvents, ions of an electrolyte are similarly *solvated*) and the electrostatic force of attraction between ions of opposite charge is so reduced that the charged ions move about as independent particles except for a slight residual attraction between them.
3. The residual force of attraction between ions of opposite charge in solution is responsible for the "activity," or "effective concentration," being less than the "actual concentration" of the ions. The residual force of attraction between the ions decreases upon dilution, because the ions are on the average further apart.
4. Acids ionize by reacting with water to form hydronium ions and acid anions of the acids.
5. Weak electrolytes react with water to only a very limited extent and are thus only slightly ionized.

Colloid Chemistry

Solutions have been characterized by their homogeneity, absence of settling, and the molecular or ionic state of subdivision of their components. We shall now consider dispersions of particles somewhat larger than ordinary molecules and ions, yet not large enough to be seen under an ordinary microscope.

13.29   Colloidal Matter

Although the mixture obtained when powdered starch is heated with water is not homogeneous, the particles of insoluble starch do not settle out but remain in suspension indefinitely. Such a system is called a **colloidal dispersion;** the finely divided starch is called the **dispersed phase,** and the water is called the **dispersion medium.**

The term **colloid**—from the Greek word *kolla,* meaning glue, and *eidos,* meaning like—was first used in 1861 by Thomas Graham to classify substances such as starch and gelatin, which usually exist in an amorphous condition. It is now known that the properties characterized as colloidal are not peculiar to a special

TABLE 13-3 Colloidal Systems

Dispersed Phase	Dispersion Medium	Examples	Common Name
Solid	Gas	Smoke, dust	Solid aerosol
Solid	Liquid	Starch suspension, some inks, paints, milk of magnesia	Sol
Solid	Solid	Colored gems, some alloys	Solid sol
Liquid	Gas	Clouds, fogs, mists, sprays	Liquid aerosol
Liquid	Liquid	Milk, mayonnaise, butter	Emulsion
Liquid	Solid	Jellies, gels, opal (SiO_2 and H_2O), pearl ($CaCO_3$ and H_2O)	Solid emulsion
Gas	Liquid	Foams, whipped cream, beaten egg whites	Foam
Gas	Solid	Pumice, floating soaps	

class of substances, but that any substance may be obtained in the colloidal form if suitable means are employed. Colloidal particles are usually aggregates of hundreds, or even thousands, of molecules. However, some colloidal particles such as viruses and polymer molecules consist of single, well-defined molecules. The viruses are giant molecules with molecular weights ranging from several hundred thousand up to billions. Tobacco mosaic virus has a molecular weight of about 40,000,000. Protein and synthetic polymer molecules may have weights in the range of a few thousand to many million.

Colloidal matter is not limited to dispersions of solid particles in liquid media; a gas or a solid may also be the dispersion medium, and the dispersed phase may be either a gas, a liquid, or a solid. A classification of colloidal systems is given in Table 13-3. A gas dispersed in another gas is not a colloidal system because the particles are of molecular dimensions. The most important colloidal systems are those involving a solid dispersed in a liquid, and such systems are called **sols.**

13.30   Preparation of Colloidal Systems

The preparation of a colloidal system is accomplished by producing particles of colloidal dimensons and distributing these particles through the dispersion medium. Particles of colloidal size are formed: (1) by **dispersion methods,** i.e., the subdivision of larger particles or masses, or (2) by **condensation methods,** i.e., growth from smaller units, such as molecules or ions.

1. Dispersion can be accomplished by grinding large particles in special mills designed for this purpose, called colloid mills. Paint pigments are produced by this method. Colloidal systems are sometimes formed by grinding a solid substance in a liquid.

A few solid substances, when brought into contact with water, disperse spontaneously to form colloidal systems. Gelatin, glue, and starch behave in this manner, and are said to undergo **peptization.** The particles are already of colloidal size; the water simply disperses them. Some atomizers produce colloidal dispersions by a spraying process. Powdered milk of colloidal particle size is produced by dehydrating milk spray.

An **emulsion** may be prepared by shaking together two immiscible liquids. Agitation breaks one liquid into droplets of colloidal dimensions, which then disperse throughout the mass of the other liquid. The droplets of the dispersed phase, however, tend to coalesce, forming drops too large to remain in the dispersed condition, and separation of the liquids into two layers follows. Therefore, emulsions must usually be stabilized by the addition of **emulsifying agents.** These emulsifying agents may either decrease the surface tension of the two liquids and thereby reduce the tendency of the tiny droplets to coalesce and form drops, or they may form protecting layers, or films, around the droplets, preventing the formation of large drops. The addition of a little soap will stabilize an emulsion of kerosene in water.

Milk is an emulsion of droplets of butterfat in water, with casein acting as the emulsifying agent. Mayonnaise is an emulsion of olive oil in vinegar, with egg yolk serving as the emulsifying agent. Oil spills in the ocean are particularly difficult to clean up because of the formation of an emulsion between the oil and the water.

2. Condensation methods involve the formation of colloidal particles by causing smaller particles to aggregate. If the particles grow beyond the colloidal range, no colloidal system will result and larger aggregates **(precipitates)** are formed.

Condensation methods usually employ chemical reactions. A dark red colloidal suspension of iron(III) hydroxide may be prepared by mixing a concentrated solution of iron(III) chloride with hot water.

$$Fe^{3+} + [3Cl^-] + 6H_2O \longrightarrow Fe(OH)_3 + 3H_3O^+ + [3Cl^-]$$

A colloidal suspension of arsenic(III) sulfide is produced by the reaction of hydrogen sulfide with arsenic(III) oxide dissolved in water.

$$As_2O_3 + 3H_2S \longrightarrow As_2S_3 + 3H_2O$$

The formation of a colloidal gold sol is accomplished by the reduction of a very dilute solution of gold chloride by such reducing agents as formaldehyde, tin(II) chloride, or iron(II) sulfate ($Au^{3+} + 3e^- \longrightarrow Au$). Some gold sols prepared by Faraday in 1857 are still perfectly clear after 123 years.

The formation of synthetic polymers and protein colloids from smaller molecules is described in Chapters 25 and 26.

13.31   Detergents and Their Cleansing Action

Although the term detergent means cleansing agent and includes soap, it is now commonly used to refer to soap substitutes. Soaps are made by boiling either fats or oils with a strong base such as sodium hydroxide. [Pioneer women made soap by boiling fat with a strong basic solution made by leaching potassium carbonate (K_2CO_3) from wood ashes with hot water.] When animal fat is treated with sodium hydroxide, glycerol and sodium salts of the fatty acids (palmitic, oleic, and stearic) are formed. The sodium salt of stearic acid, sodium stearate, has the formula $C_{17}H_{35}CO_2Na$.

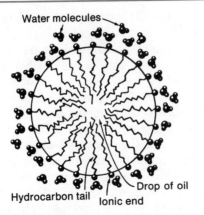

Figure 13-11 Diagrammatic cross section of an emulsified drop of oil in water, with soap or detergent as the emulsifier. The negative ions of the emulsifier are oriented at the interface between the oil particle and water. The nonpolar hydrocarbon end of the ion is in oil, and the ionic end ($-CO_2^-$ for soap, $-SO_3^-$ for detergent) is in water.

Water molecules

Hydrocarbon tail   Ionic end   Drop of oil

Many detergents, or soap substitutes, are in common use. Each of these compounds contains a hydrocarbon chain or ring structure group which is nonpolar, such as $C_{12}H_{25}-$, or $C_{12}H_{25} \cdot C_6H_4-$; and an ionic group, such as a sulfate, $-OSO_3^-$, or a sulfonate, $-SO_3^-$. Whereas soaps form insoluble calcium and magnesium compounds in hard water, these detergents form water-soluble products—a definite advantage for detergents. Numerous detergents with familiar commercial names are available in stores throughout many countries of the world.

The cleansing action of soaps and detergents can be explained in terms of the structures of the molecules involved. The molecules of both soaps and synthetic detergents consist of long hydrocarbon chains attached to an ionic group.

Hydrocarbon chain   Ionic end

$$CH_3-CH_2-CH_2-CH_2-CH_2-CH_2-CH_2-CH_2-CH_2-CH_2-CH_2-CH_2-CH_2-CH_2-CH_2-CH_2-CO_2^- \; Na^+$$

Sodium stearate (soap)

$$CH_3-CH_2-CH_2-CH_2-CH_2-CH_2-CH_2-CH_2-CH_2-CH_2-CH_2-CH_2-OSO_3^- \; Na^+$$

Sodium lauryl sulfate (detergent)

The hydrocarbon end of such a molecule is attracted by the dirt, oil, or grease particles and the ionic end is attracted by the water (Fig. 13-11). The result is an orientation of the molecules of the cleansing agent at the interface between the dirt particles and the water in such a way that the **interfacial tension** is lowered. This enables the dirt particles to become suspended in the solution as colloidal particles, in which form they are readily washed away. Substances that lower the surface tension of liquids are sometimes called **wetting agents.**

13.32   The Brownian Movement

When a beam of intense light is passed through a colloidal system in a darkened space and viewed with a powerful microscope against a black background and at right angles to the beam of light, the colloidal particles are observed as tiny bright flashes of light. The particles are seen to be in a state of irregular rapid, dancing

Figure 13-12 A diagram showing how six particles in suspension might move due to unequal bombardment by molecules of the dispersion medium.

motion (Fig. 13-12). This motion is called the **Brownian movement** after the botanist Robert Brown, who first observed it in 1828. He could not explain it, but we now know that the colloidal particles are small enough that bombardment by molecules of the dispersion medium gives them an irregular motion. Any particle in suspension will be bombarded on all sides by the moving molecules of the dispersion medium. For particles larger than colloidal size the bombardment will not impart motion because bombardment on one side of the particle is likely to be counterbalanced by an equal force on the opposite side. For particles of usual colloidal size, however, the probability of equal and simultaneous bombardments on opposite sides of the colloidal size particle is slight. Hence the irregular motion results. The Brownian movement explains the absence of settling of the dispersed particles in colloidal systems, even though these particles may be more dense than the medium in which they are dispersed. The kinetic-molecular theory of matter received one of its earliest confirmations as a result of studies of the Brownian movement.

13.33   Electrical Properties of Colloidal Particles

One of the most important properties of dispersed colloidal particles is that they are usually electrically charged. When an iron(III) hydroxide sol is placed in an electrolytic cell (Fig. 13-13), the dispersed particles move to the negative electrode. Because opposite charges of electricity attract, this is good evidence that the iron(III) hydroxide particles are positively charged. At the cathode the particles lose their charge and the colloidal particles coagulate as a precipitate. All particles in any one colloidal system have the same kind of charge with respect to sign. The similar charges on the particles help keep them dispersed, because like charges repel each other. Most hydroxides of metals have positive charges, while most sulfides of metals and the metals themselves form negatively charged colloidal dispersions.

The charges on colloidal particles result from the adsorption of ions that exist in the dispersion medium. If a colloidal particle preferentially adsorbs positive ions, it acquires a positive charge; if it preferentially adsorbs negative ions, it acquires a negative charge. Thus iron(III) hydroxide particles become positively charged because of a preferential adsorption of iron(III) ions (Fe^{3+}) when $FeCl_3$ hydrolyzes in hot water. Arsenic(III) sulfide (As_2S_3) particles preferentially adsorb sulfide ions (S^{2-}), resulting from the ionization of H_2S, and become negatively charged.

Colloidal clay or silt particles in a river are negatively charged by the adsorption of negative hydroxide ions. When the river water reaches the salty ocean water, the negatively charged particles are neutralized by the positive sodium and magnesium ions of the sea water, and the clay and silt particles precipitate. Enough precipitate accumulates during thousands of years to form deltas at the mouths of rivers.

The carbon and dust particles in the smoke from furnace fires are often colloidally dispersed and electrically charged. A process for precipitating smoke particles was developed by Cottrell, an American chemist, to lessen the smoke nuisance in industrial centers. In this process the charged particles are attracted to highly charged electrodes, where they are neutralized and deposited as dust (Fig.

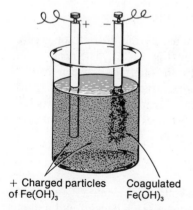

+ Charged particles   Coagulated
of Fe(OH)₃   Fe(OH)₃

Figure 13-13 Colloidal iron(III) hydroxide is coagulated in an electrolytic cell.

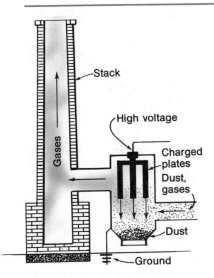

Figure 13-14   The principle of a Cottrell precipitator. Positively and negatively charged particles in smoke are precipitated as they pass over the electrically charged plates.

13-14). The process is also important in the recovery of many valuable products that would otherwise escape from the flues of smelters, furnaces, and kilns.

13.34   The Adsorption Phenomenon

One of the properties of the surface of any substance is its specific ability to hold other substances to itself. This phenomenon is called **adsorption.** The atoms, molecules, or ions that compose a surface differ from those in the interior of the substance in that they are surrounded only on one side by like particles that possess equal and similar cohesive forces. The particles within the body, however, are surrounded on *all* sides by particles having equal attractive forces. The surface particles, therefore, have some attractive forces that are not satisfied. These unsatisfied forces are largely responsible for the phenomenon of adsorption.

When matter is subdivided to the extent that its particles become colloidal in dimensions, there is a tremendous increase in the surface area exposed to the surrounding medium. A cube 1 cm along each edge has a surface area of 6 cm². If this cube is subdivided until the particles are within the colloidal range of 10 mμ on an edge, the total surface area becomes 6,000,000 cm², or approximately the surface area of a square 80 feet on each side. Because colloidal particles have a very large surface area for a given volume of material, they are good adsorbers. In fact, colloidal chemistry is largely the chemistry of surface effects. Hence the principles of colloid chemistry apply to films and to filaments, both of which have a very large surface area for a given volume of material. Dispersed solids such as charcoal adsorb vast quantities of gas. As pointed out in Section 13.33, colloidal dispersions of arsenic(III) sulfide and iron(III) hydroxide are stabilized by adsorption of ions from solution. Furthermore, emulsions are stabilized by adsorption of an emulsifying agent upon the surface of the droplets of the dispersed phase (Section 13.30).

13.35   Gels

Under certain conditions, the dispersed phase in colloidal systems coagulates in a manner such that the whole mass, including the liquid, sets to an extremely viscous gelatinous body known as a **gel.** For example, a hot aqueous "solution" of gelatin sets to a gel upon cooling. Because the formation of a gel is accompanied by the taking up of water or some other solvent, the gel is said to be hydrated or solvated. Apparently the fibers of the dispersed substance form a complex three-dimensional network, the interstices being filled with the liquid medium or a dilute solution of the dispersed phase.

A carbohydrate known as **pectin,** from fruit juices, is a gel-forming substance important in jelly making. Silica gel, a colloidal dispersion of hydrated silicon dioxide, is formed when dilute hydrochloric acid is added to a dilute solution of sodium silicate. Canned heat is a gel made by mixing alcohol and a saturated aqueous solution of calcium acetate. The wall of the living cell is colloidal in character, and within the cell there is a jellylike dispersion. In fact, all living tissue is colloidal, and the various life processes—nutrition, digestion, secretion—are largely those concerned with the chemistry of the colloidal state.

Questions

1. How do solutions differ from compounds? from ordinary mixtures?
2. Identify the various types of solutions that can form with solid, liquid, and gaseous solvents.
3. Show how solutions of (a) carbon dioxide in air, (b) ethanol (ethyl alcohol, C_2H_5OH) in water, and (c) potassium chloride in water exhibit the principal characteristics of solutions.
4. Why are the majority of chemical reactions carried out in solutions?
5. Explain the following terms as applied to solutions: solute, solvent, weak electrolyte, strong electrolyte, nonelectrolyte, saturated, supersaturated, and unsaturated.
6. Describe the factors that affect the solubility of gases in liquids.
7. For which of the following gases is solubility in water not directly proportional to the pressure of the gas: HCl, H_2, SO_2, NH_3, CH_4, Ne? Explain.
8. Describe the effect of (a) the nature of the gas and its solvent, (b) the pressure on the gas, (c) the temperature, and (d) the reaction of the gas with the solvent, upon the solubility of a gas.
9. In order to prepare supersaturated solutions of most solids, saturated solutions are cooled. Supersaturated solutions of gases are prepared by warming saturated solutions. Explain the reasons for the differences in the two procedures.
10. What is a constant boiling solution? Describe two ways to prepare a constant boiling solution of hydrochloric acid.
11. Why is it not possible to prepare a constant boiling solution of sodium chloride?
12. Compare the processes that occur when glycerin, hydrogen chloride, ammonia, and sodium hydroxide dissolve in water. Write equations and prepare sketches showing the form in which each of these compounds is present in its respective solution.
13. Solid water (ice) is soluble in ethanol (C_2H_5OH). Compare the process that occurs when ice dissolves in ethanol with that which occurs when sodium chloride dissolves in water. (See Section 12.1 for a description of the structure of ice.)
14. A one-molal solution of HCl in benzene has a freezing point of 0.4°C. Is HCl an electrolyte in benzene? Explain.
15. Why is a solution of nitric acid (HNO_3) in water a good conductor of an electric current when both pure nitric acid and water are not?
16. Why are solid ionic compounds nonconductors, whereas ionic compounds that are fused (melted) are good conductors?
17. Why are nonpolar solvents such as benzene and carbon tetrachloride generally poor solvents for ionic compounds?
18. What are colligative properties? Give examples.
19. State and explain Raoult's Law.
20. The triple point of air-free water is defined as 273.16 K. Why is it important to have the water air-free?
21. Distinguish between one-molar and one-molal solutions.
22. Why will a mole of sodium chloride depress the freezing point of 1 kg of water almost twice as much as a mole of glycerin?
23. Why will ice melt when placed in an aqueous solution that is kept at 0°C?
24. Demonstrate, or take issue with, the following statement: "In all cases, the molality of a solution has a value that is larger than the molarity of the same solution."
25. Suppose you are presented with a clear solution and told what the solute and solvent are but not told whether the solution is unsaturated, saturated, or supersaturated. How could you determine which of these three conditions exist?
26. What is meant by mole fraction?
27. (a) Such substances as sodium chloride (NaCl), calcium chloride ($CaCl_2$), and the nonelectrolyte urea (NH_2CONH_2) are frequently used to melt ice on streets and highways. Rank the three from high to low in terms of their effectiveness per pound in lowering the freezing point. Explain your ranking.
    (b) Would these substances be as effective in melting ice on a day when the maximum temperature is −20°F? Explain why or why not.
28. Explain what is meant by the term "osmotic pressure." How is the osmotic pressure of a solution related to its concentration?
29. The cell walls of red and white blood cells are semipermeable membranes. The concentration of solute

particles in the blood is about 0.6 M. What will happen to blood cells that are placed in pure water? in a 1 M sodium chloride solution?

30. The salt lithium nitrate, $LiNO_3$, does not give twice the molal depression of the freezing point of water that nonelectrolytes do. Explain.

31. The approximate radius of a water molecule, assuming a spherical shape, is 1.40 Å. (1 Å = 10^{-8} cm.) Assume that water molecules cluster around metal ions in solution so that the water molecules essentially touch both the metal ion and each other. On this basis, and assuming that 4, 6, 8, and 12 are the only possible coordination numbers, what is the maximum number of water molecules that can hydrate each of the following ions: Mg^{2+} (radius 0.65 Å); Al^{3+} (0.50 Å); Rb^+ (1.48 Å); and Sr^{2+} (1.13 Å)?

32. Distinguish between dispersion methods and condensation methods for preparing colloidal systems.

33. How do colloidal "solutions" differ from true solutions with regard to dispersed particle size and homogeneity?

34. Identify the dispersed phase and the dispersion medium in each of the following colloidal systems: starch dispersion, smoke, fog, pearl, whipped cream, floating soap, jelly, milk, and ruby.

35. Explain the cleansing action of soap.

36. How can it be demonstrated that colloidal particles are electrically charged?

37. What is the structure of a gel?

38. Explain the phenomenon of Brownian movement. What is its relationship to the stability of the dispersed particles of a colloidal system?

Problems

1. Calculate the number of moles and the mass of solute in each of the following solutions.
   (a) 1.20 l of 1.30 M $HClO_4$.
     Ans. 1.56 mol; 157 g
   (b) 0.500 l of 0.0125 M $C_{12}H_{22}O_{11}$.
     Ans. 6.25×10^{-3} mol; 2.14 g
   (c) 15.0 ml of 0.1600 M $MgCl_2$.
     Ans. 2.40×10^{-3} mol; 0.229 g
   (d) 40.0 ml of 0.0700 M $Ca(OH)_2$.
     Ans. 2.80×10^{-3} mol; 0.207 g
   (e) 4.25 l of 0.75 M $Co(NO_3)_2 \cdot 6H_2O$.
     Ans. 3.2 mol; 0.93 kg

2. Calculate the molarity of each of the following solutions:
   (a) 13.0 g of potassium hydroxide, KOH, in 5.0 l of solution. *Ans. 0.046 M*
   (b) 20.0 g of sulfuric acid, H_2SO_4, in 1.0 l of solution. *Ans. 0.20 M*
   (c) 0.0500 g of nitric acid, HNO_3, in 10.0 ml of solution. *Ans. 0.0794 M*
   (d) 7.0×10^{-3} mol of iodine, I_2, in 1.00×10^2 ml of solution. *Ans. 0.070 M*
   (e) 18.00 g of hydrogen chloride, HCl, in 75 ml of solution. *Ans. 6.6 M*

3. What mass of solute is present in each of the following solutions? The per cent concentration is by mass.

   (a) 1.34 l of 19.00% acetic acid, CH_3CO_2H (density 1.0267 g/cm³). *Ans. 261 g*
   (b) 0.250 l of 13.5% ammonium chloride, NH_4Cl (specific gravity 1.040). *Ans. 35.1 g*
   (c) 3.748 ml of 4.5% ammonium sulfate, $(NH_4)_2SO_4$ (density 1.027 g/cm³). *Ans. 0.17 g*
   (d) 50.0 ml of 8.75% lactose, $C_{12}H_{22}O_{11}$ (specific gravity 1.0355). *Ans. 4.53 g*
   (e) 353 ml of 40.0% sodium hydroxide, NaOH (density 1.432 g/cm³). *Ans. 202 g*

4. Calculate the molality of each of the following solutions:
   (a) 97.58 g of lead acetate, $Pb(CH_3CO_2)_2$ in 1.000 kg of water. *Ans. 0.3000 m*
   (b) 46.85 g of codeine, $C_{18}H_{21}NO_3$, in 125.5 g of ethanol, C_2H_5OH. *Ans. 1.247 m*
   (c) 0.372 g of histamine, C_5H_9N, in 125 g of chloroform, $CHCl_3$. *Ans. 3.58×10^{-2} m*
   (d) 1.65 g of silicon tetraiodide, SiI_4, in 7.5 g of carbon disulfide, CS_2.
     Ans. 0.41 m

5. Calculate the mole fraction of solute and solvent in each of the solutions in Problem 4.
   Ans. (a) $Pb(CH_3CO_2)_2$, 0.005375; H_2O, 0.9946
   (b) $C_{18}H_{21}NO_3$, 0.05433; C_2H_5OH, 0.9457
   (c) C_5H_9N, 0.00426; $CHCl_3$, 0.996
   (d) SiI_4, 0.0303; CS_2, 0.97

[S] 6. What mass of sulfuric acid (95% by mass) is needed to prepare 200.0 g of a 20.0% solution of the acid by mass? *Ans. 42 g*

7. What mass of a 4.00% NaOH solution by mass contains 15.0 g of NaOH? *Ans. 375 g*

8. What mass of HCl is contained in 45.0 ml of an HCl solution with a specific gravity of 1.19 and containing 37.21% HCl by mass? *Ans. 19.9 g*

9. What mass of solid NaOH (97.0% NaOH by mass) is required to prepare 1.00 ℓ of a 10.0% solution of NaOH by mass? The density of the 10.0% solution is 1.109 g/cm³. *Ans. 114 g*

[S] 10. The hardness of water (hardness count) is usually expressed as parts per million (by mass) of $CaCO_3$, which is equivalent to milligrams of $CaCO_3$ per liter of water. What is the molar concentration of Ca^{2+} ions in a water sample with a hardness count of 175? *Ans.* 1.75×10^{-3} *M*

11. The Safe Drinking Water Act of 1974 sets the maximum amount of cadmium in drinking water at 0.01 mg per liter. What is the maximum permissible molar concentration of cadmium in drinking water? *Ans.* 9×10^{-8} *M*

12. The concentration of glucose, $C_6H_{12}O_6$, in normal spinal fluid is 75 mg per 100 g. What is the molal concentration? *Ans.* 4.2×10^{-3} *m*

13. Hydrogen gas dissolves in the metal palladium with hydrogen atoms going into the holes between metal atoms. Determine the molarity, molality, and per cent by mass of hydrogen atoms in a solution (specific gravity 10.8) of 0.94 g of hydrogen gas in 215 g of palladium metal. *Ans. 47 M; 4.3 m; 0.44%*

[S] 14. How many liters of $NH_3(g)$ at 25°C and 1.46 atm are required to prepare 3.00 ℓ of a 2.50 *M* solution of NH_3? *Ans. 126 ℓ*

15. What volume of 0.20 *M* K_2SO_4 solution would contain 57 g of K_2SO_4? *Ans. 1.6 ℓ*

16. A 0.553 *M* solution of $NaHCO_3$ has a density of 1.032 g/cm³. What is the molality of the solution? *Ans. 0.561 m*

[S] 17. Calculate the volume of sulfuric acid solution (specific gravity 1.070, and containing 10.00% H_2SO_4 by mass) that would contain 18.50 g of pure H_2SO_4 at a temperature of 25°. *Ans. 172.9 ml*

18. A 13.0% solution of K_2CO_3 by mass has a density of 1.09 g/cm³. Calculate the molarity of the solution. *Ans. 1.03 M*

19. Concentrated hydrochloric acid is 37.0% HCl by mass and has a specific gravity of 1.19 at 25°C. What is the mole fraction of HCl and of H_2O in the solution? *Ans. HCl, 0.225;* H_2O*, 0.775*

20. The density of a 2.50% by mass solution of aqueous ammonia is 0.994 g/ml. What volume of a concentrated solution of ammonia (28.0% by mass; specific gravity 0.900 at 25°C) would be required in the preparation of 1.50 ℓ of this solution? *Ans. 0.148 ℓ*

21. A 0.200-ℓ volume of gaseous ammonia measured at 30.0° and 764 torr was absorbed in 0.100 ℓ of water. How many ml of 0.0100 *M* hydrochloric acid would be required in the neutralization of this aqueous ammonia? What is the molarity of the aqueous ammonia solution? (Assume no change in volume when the gaseous ammonia is added to the water.) *Ans. 808 ml; 0.0808 M*

22. Calculate the per cent by mass and the molarity in terms of $CuSO_4$ for a solution prepared by dissolving 11.5 g of $CuSO_4 \cdot 5H_2O$ in 0.100 kg of water. *Remember to consider the water released from the hydrate.* *Ans. 6.59%; 0.442 molal*

[S] 23. It is desired to produce 1.000 ℓ of 0.050 *M* nitric acid by diluting 10.00 *M* nitric acid. Calculate the volume of the concentrated acid and the volume of water required in the dilution. *Ans. 5.0 ml of* HNO_3*; 995 ml of* H_2O

24. What is the molarity of a solution that is prepared by dissolving 10.0 g of P_4O_{10} in sufficient water to make 0.500 ℓ of solution? Assume that the product of the reaction of P_4O_{10} with water is H_3PO_4 (orthophosphoric acid). Would it make a difference in the answer if the product of the reaction were HPO_3 (metaphosphoric acid)? Explain. *Ans. 0.282 M*

25. A solution of sodium carbonate having a volume of 0.400 ℓ was prepared from 4.032 g of $Na_2CO_3 \cdot 10H_2O$. Calculate the molarity of this solution. *Ans. 0.0352 M*

26. A solution of $Ba(OH)_2$ has a molarity of 0.1055. What volume of 0.211 *M* nitric acid would be required in the neutralization of 15.5 ml of the $Ba(OH)_2$ solution? *Ans. 15.5 ml*

[S] 27. The sulfate in 50.0 ml of dilute sulfuric acid was precipitated using an excess of barium chloride. The mass of $BaSO_4$ formed was 0.482 g. Calculate the molarity of the sulfuric acid solution. *Ans. 0.0413 M*

28. Equal volumes of 0.50 M Ca(OH)$_2$ and 0.40 M HCl are mixed. Calculate the molarity of each ion present in the final solution.

Ans. 0.30 M OH$^-$; 0.25 M Ca^{2+}; 0.20 M Cl$^-$

29. What volume of 0.600 M HCl would be required to react completely with 2.50 g of sodium hydrogen carbonate?

(NaHCO$_3$ + HCl $\longrightarrow$ NaCl + CO$_2$ + H$_2$O)

Ans. 49.6 ml

30. Calculate the volume of 0.050 M hydrochloric acid necessary to precipitate the silver contained in 12.0 ml of 0.050 M AgNO$_3$. [Ag$^+$ + Cl$^-$ $\longrightarrow$ AgCl(s)]

Ans. 12.0 ml

31. What volume of a 0.33 M solution of hydrobromic acid would be required to neutralize completely 1.00 ℓ of 0.15 M magnesium hydroxide?

Ans. 0.91 ℓ

s 32. What volume of 0.100 M HCl would be required to precipitate the silver in 0.634 g of 98.0% purity AgNO$_3$?

Ans. 36.6 ml

33. To 10.0 ml of a 0.100 M K$_2$Cr$_2$O$_7$ solution was added 10.0 ml of a 0.100 M Pb(NO$_3$)$_2$ solution. What mass of PbCr$_2$O$_7$ forms?

Ans. 0.423 g

s 34. A gaseous solution was found to contain 15% H$_2$, 10.0% CO, and 75% CO$_2$ by mass. What is the mole fraction of each component?

Ans. H$_2$, 0.78; CO, 0.037; CO$_2$, 0.18

35. Calculate the mole fraction of solute and solvent for a 1.6 m solution of calcium nitrate.

Ans. 0.028; 0.97

36. Calculate the mole fraction of methanol (CH$_3$OH), ethanol (C$_2$H$_5$OH), and water in a solution that is 40% methanol, 40% ethanol, and 20% water, by mass. (Assume the data are good to two significant figures.)

Ans. 0.39; 0.27; 0.34

37. Concentrated hydrochloric acid is 37.0% HCl by mass and has a specific gravity of 1.19. Calculate (a) the molarity, (b) the molality, and (c) the mole fraction of HCl and of H$_2$O.

Ans. (a) 12.1 M; (b) 16.1 m; (c) 0.225 mole fraction HCl, 0.775 mole fraction H$_2$O

38. Calculate (a) the per cent composition and (b) the molality of an aqueous solution of NaNO$_3$, if the mole fraction of NaNO$_3$ is 0.20.

Ans. (a) 54% NaNO$_3$, 46% H$_2$O; (b) 14 m

s 39. A 1.80-g sample of an acid, H$_2$X, required 14.00 ml of KOH solution for neutralization of all the hydro-

gen ion. Exactly 14.2 ml of this same KOH solution was found to neutralize 10.0 ml of 0.750 M H$_2$SO$_4$. Calculate the molecular weight of H$_2$X.

Ans. 243

s 40. A solution of 5.00 g of an organic compound in 25.00 g of carbon tetrachloride (bp 76.8°C; K_b 5.02) boils at 81.5°C at 760 torr. What is the molecular weight of the compound?

Ans. 2.1 × 10^2

s 41. A solution contains 5.00 g of urea, CO(NH$_2$)$_2$, per 0.100 kg of water. If the vapor pressure of pure water at 25° is 23.7 torr, what is the vapor pressure of the solution?

Ans. 23.4 torr

42. Twelve grams of a nonelectrolyte is dissolved in 80.0 g of water. The solution freezes at −1.94°C. Calculate the molecular weight of the substance.

Ans. 144

43. A sample of an organic compound (nonelectrolyte) weighing 1.350 g lowered the freezing point of 10.0 g of benzene by 3.66°C. Calculate the molecular weight of the organic compound.

Ans. 189

s 44. Calculate the boiling-point elevation of 0.100 kg of water containing 0.010 mol of NaCl, 0.020 mol of Na$_2$SO$_4$, and 0.030 mol of MgCl$_2$, assuming complete dissociation of these electrolytes.

Ans. 0.87°C

45. What would be the approximate freezing point of a 0.27 m aqueous solution of sodium bromide? Assume the activity coefficient is 1.00.

Ans. −1.0°C

s 46. How would you prepare a 3.08 m aqueous solution of glycerin (C$_3$H$_8$O$_3$)? What would be the freezing point of this solution?

Ans. Dissolve 284 g of glycerin in 1.00 kg of water; −5.73°C

47. If 26.4 g of the nonelectrolyte C$_6$H$_4$Br$_2$ is dissolved in 0.250 kg of benzene, what is (a) the freezing point of the solution and (b) the boiling point of the solution at 760 torr?

Ans. (a) 3.19°C; (b) 81.2°C

s 48. A sample of sulfur weighing 0.210 g was dissolved in 17.8 g of carbon disulfide, CS$_2$ (K_b 2.34). If the boiling-point elevation was 0.107°, what is the formula of a sulfur molecule in carbon disulfide?

Ans. S$_8$

49. What would be the boiling point, at 760 torr, of a solution containing 140.0 g of sucrose (C$_{12}$H$_{22}$O$_{11}$) in 400.0 g of water?

Ans. 100.524°C

50. A sample of $HgCl_2$ weighing 9.41 g is dissolved in 32.75 g of ethanol, C_2H_5OH (K_b 1.20). The boiling point elevation of the solution is 1.27°C. Is $HgCl_2$ an electrolyte in ethanol? Show your calculation. *Ans. No*

51. A salt is known to be an alkali metal fluoride. A quick approximate freezing point determination indicates that 4 g of the salt dissolved in 100 g of water produces a solution which freezes at about −1.4°C. What is the formula of the salt? Show your calculation. *Ans. RbF*

Ⓢ52. A solution of 0.045 g of an unknown organic compound in 0.550 g of camphor melts at 158.4°C. The melting point of pure camphor is 178.4°C. K_f for camphor is 37.7. The solute contains 93.46% C and 6.54% H by mass. What is the molecular formula of the solute? Show your calculation. *Ans. $C_{12}H_{10}$*

Ⓢ53. The activity of the ions in a 0.20 m solution of $LiNO_3$ at 25.0°C is 0.15 m. Calculate the activity coefficient for the ions. *Ans. 0.75*

54. The activity coefficient for the ions in 0.0050 m CsCl is 0.921 at 25.0°C. Calculate the activities of these ions. *Ans. 0.0046 m*

Ⓢ55. A solution is made by dissolving 0.0745 g of potassium chloride in 100.0 g of water. The solution is observed to freeze at −0.0358°C. If one assumes that the activity coefficient of the potassium ion is equal to that of the chloride ion, what is the activity coefficient of the potassium ion under these conditions? *Ans. 0.960*

References

"Ideal Solutions," W. A. Oates, *J. Chem. Educ.,* **46**, 501 (1969).

"Superfluid Helium-3," N. D. Mermin and D. M. Lee, *Sci. American,* December 1976, p. 56.

"Interfacial Spreading," J. Ahmad, *J. Chem. Educ.,* **52**, 534 (1975).

"Relationship between Rate of Evaporation and Vapor Pressure of Binary Systems," J. S. Shapiro, E. C. Watton, and J. M. Kilford, *J. Chem. Educ.,* **52**, 439 (1975).

"Drops of Liquid Can Be Made to Float on the Liquid. What Makes Them Do So?" J. Walker, *Sci. American,* June 1978, p. 151.

"Colligative Properties of a Solution," H. T. Hammel, *Science,* **192**, 748 (1976).

"Liquid-Liquid Extraction," P. Joseph-Nathan, *J. Chem. Educ.,* **44**, 176 (1967).

"Demonstrating Osmotic and Hydrostatic Pressures in Blood Capillaries," J. W. Ledbetter, Jr., and H. D. Jones, *J. Chem. Educ.,* **44**, 362 (1967).

"Reactions in Solutions Under Pressure," W. J. le Noble, *J. Chem. Educ.,* **44**, 729 (1967).

"Chemiluminescent Reactions in Solution," J. W. Haas, Jr., *J. Chem. Educ.,* **44**, 396 (1967).

"The Osmotic Pump," O. Levenspiel and N. de Nevers, *Science,* **183**, 157 (1974).

"The History of Colloid Science," E. A. Haùser, *J. Chem. Educ.,* **32**, 2 (1955).

"Syndets and Surfactants," F. D. Snell and C. T. Snell, *J. Chem. Educ.,* **35**, 271 (1958).

"Poliomyelitis Virus," F. L. Schaffer, *J. Chem. Educ.,* **36**, 469 (1959).

"How Giant Molecules Are Measured," Peter Debye, *Sci. American,* September 1957, p. 90.

"Brownian Motion and Potential Theory," R. Hersh and R. J. Griego, *Sci. American,* March 1969, p. 67.

"Physical Adsorption—A Tool in the Study of the Frontiers of Matter," A. W. Adamson, *J. Chem. Educ.,* **44**, 710 (1967).

"Monomolecular Layers and Light," K. H. Drexhage, *Sci. American,* March 1970, p. 108.

"Brownian Movement and Molecular Reality Prior to 1900," M. Kerker, *J. Chem. Educ.,* **51**, 764 (1974).

"Electrostatic Effects in Proteins," M. F. Perutz, *Science,* **201**, 1187 (1978).

"Synthetic-Membrane Technology," H. P. Gregor and C. D. Gregor, *Sci. American,* July 1978, p. 112.

"Serious Fun with Polyox, Silly Putty, Slime, and other Non-Newtonian Fluids," J. Walker, *Sci. American,* November 1978, p. 186.

"Effect of Ionic Strength on Equilibrium Constants," M. D. Seymour and Q. Fernando, *J. Chem. Educ.,* **54**, 225 (1977).

"Household soaps and Detergents" (Staff), *J. Chem. Educ.,* **55**, 598 (1978).

"Colloids: New Life from Old Roots," J. A. Kitchener, *Chem. in Britain,* **13**, 105 (1977).

14

Acids, Bases, and Salts

Functional definitions of acids and bases were introduced in Section 8.1. Protonic acids were defined in a preliminary fashion as compounds that increase the concentration of hydrogen ion in solution, whereas hydroxide bases were defined as compounds that increase the hydroxide ion concentration. In this chapter we will explore the concepts of acids and bases more fully.

14.1 History of the Acid-Base Concept

Historically, the concepts of "acid" and "base" have been defined in a number of ways. The first significant characterization of acids and bases were made by Boyle in 1680. He noted that acids dissolve many substances, that they change the color of certain natural dyes (for example, litmus from blue to red), and that they lose these characteristic properties after coming in contact with alkalies. In the eighteenth century it was recognized that acids have a sour taste, that they react with limestone with the liberation of a gaseous substance (CO_2), and that neutral substances result from their interaction with alkalies. Lavoisier, in 1787, proposed that acids are binary compounds of oxygen, and he considered oxygen to be responsible for the acidic properties of this class of substances. The essentiality of oxygen was disproved by Davy in 1811 when he showed that hydrochloric acid contains no oxygen. Davy contributed greatly to the development of the acid-base concept by concluding that hydrogen, rather than oxygen, is the essential constituent of acids. In 1814, Gay-Lussac concluded that acids are substances that can neutralize alkalies and that these two classes of substances can be defined only in terms of each other. The ideas of Davy and Gay-Lussac provide the foundation for our modern concepts of acids and bases in water solution.

The significance of hydrogen was reemphasized in 1884 by Arrhenius when he defined an acid as a compound that dissolves in water to yield hydrogen ions, and a base as a compound that dissolves in water to yield hydroxide ions. The close relationship of acids and bases was pointed out independently in 1923 by Johannes Brönsted, a Danish chemist, and Thomas Lowry, an English chemist. The Brönsted-Lowry model for acid-base behavior defines acids as hydrogen ion donors and bases as hydrogen ion acceptors. An even broader view of the relationship between acids and bases was developed by the American chemist G. N. Lewis. The Lewis theory defines acids as electron-pair acceptors and bases as electron-pair donors.

In this chapter we shall consider first some of the properties of acids that contain hydrogen (protonic acids), hydroxide bases, and salts in water solution. Then we shall proceed to a consideration of some of the very useful more generalized concepts of acid-base behavior.

Protonic Acids, Hydroxide Bases, and Salts in Aqueous Solution

14.2   Properties of Protonic Acids in Aqueous Solution

A **protonic acid** (Arrhenius acid) is an acid which contains one or more hydrogen atoms that ionize to give hydronium ion, H_3O^+, when the acid is dissolved in water. Examples are hydrogen chloride (HCl), perchloric acid ($HClO_4$), nitric acid (HNO_3), and sulfuric acid (H_2SO_4). The properties that are common to protonic acids in aqueous solution are those of the hydronium ion. Although the hydronium ion is actually formed, the symbol H^+ is often used instead of H_3O^+ for the sake of convenience. It is important to remember, however, that H^+ is an abbreviated symbol, and that a hydrogen ion is always hydrated in water solution. Even the H_3O^+ symbol is in a strict sense an abbreviation. For example, there is good evidence for the existence of $H_9O_4^+$, corresponding to $H_3O(H_2O)_3^+$

Since aqueous solutions of protonic acids contain hydronium ions, these solutions all exhibit the following properties:

1. They have a sour taste.
2. They change the color of certain indicators. For example, they change litmus from blue to red and phenolphthalein from red to colorless.
3. They react with metals above hydrogen in the electromotive series (Section 9.21) and liberate hydrogen.

$$2H_3O^+(aq) + Zn(s) \longrightarrow Zn^{2+} + H_2(g) + 2H_2O(l)$$

4. They react with many metal oxides and hydroxides, which are bases, forming salts and water.

$$2H_3O^+ + [2Cl^-] + FeO \longrightarrow Fe^{2+} + [2Cl^-] + 3H_2O$$

$$2H_3O^+ + [2Cl^-] + Fe(OH)_2 \longrightarrow Fe^{2+} + [2Cl^-] + 4H_2O$$

5. They react with salts of either weaker or more volatile acids, such as carbonates or sulfides, to give a new salt and a new acid.

$$2H_3O^+ + [2Cl^-] + CaCO_3 \longrightarrow H_2CO_3 + Ca^{2+} + [2Cl^-] + 2H_2O$$

$$\qquad\qquad\qquad\qquad\qquad \lfloor\!\!\!\rightarrow H_2O + CO_2(g)$$

$$2H_3O^+ + [2Cl^-] + FeS \longrightarrow H_2S(g) + Fe^{2+} + [2Cl^-] + 2H_2O$$

6. Their aqueous solutions conduct an electric current because they contain ions; they are electrolytes.

14.3   Formation of Protonic Acids

Protonic acids may be prepared by one or more of the following methods:

1. *By the direct union of the elements.* Since binary compounds of hydrogen with the more electronegative nonmetals are acids (Section 8.3), the direct reaction of hydrogen with such nonmetals as F_2, Cl_2, Br_2, and S_8 will give an acid. For example,

$$H_2 + Br_2 \longrightarrow 2HBr$$

$$8H_2 + S_8 \overset{\triangle}{\rightleftharpoons} 8H_2S$$

2. *By the reaction of water with an oxide of a nonmetal.* Most oxides of nonmetals are acidic (Section 8.3). The action of water on such a nonmetal oxide forms an acid.

$$CO_2 + H_2O \longrightarrow H_2CO_3$$

$$SO_3 + H_2O \longrightarrow H_2SO_4$$

$$P_4O_{10} + 6H_2O \longrightarrow 4H_3PO_4$$

3. *By the metathetical reaction of a salt of a volatile acid with a nonvolatile or slightly volatile acid.*

$$NaF(s) + H_2SO_4(l) \longrightarrow NaHSO_4(s) + HF(g)$$

$$KNO_3(s) + H_3PO_4(l) \overset{\triangle}{\longrightarrow} KH_2PO_4(s) + HNO_3(g)$$

4. *By the metathetical reaction of a salt with another acid to produce a second acid and an insoluble precipitate.*

$$H_3O^+ + Cl^- + Ag^+ + [NO_3^-] \longrightarrow AgCl(s) + H_3O^+ + [NO_3^-]$$

$$2H_3O^+ + SO_4^{2-} + Ba^{2+} + [2ClO_3^-] \longrightarrow BaSO_4(s) + 2H_3O^+ + [2ClO_3^-]$$

5. *By hydrolysis of a compound of nonmetals containing polar bonds.*

$$PBr_3 + 3H_2O \longrightarrow H_3PO_3 + 3HBr$$
$$PCl_5 + 4H_2O \longrightarrow H_3PO_4 + 5HCl$$
$$SiI_4 + 4H_2O \longrightarrow Si(OH)_4 + 4HI$$

6. *By oxidation or reduction of an element or compound.*

$$H_2S + I_2 \longrightarrow 2HI + S$$
$$2HNO_3 + 2SO_2 + H_2O \longrightarrow 2H_2SO_4 + NO(g) + NO_2(g)$$

14.4   Polyprotic Acids

Acids can be classified in terms of the number of protons per molecule that can be given up in a reaction. Acids, such as HCl, HNO_3, and HCN, that contain one ionizable hydrogen atom in one molecule of acid are called **monoprotic acids.** Their reaction with water is given by

$$HCl + H_2O \longrightarrow H_3O^+ + Cl^-$$
$$HNO_3 + H_2O \longrightarrow H_3O^+ + NO_3^-$$
$$HCN + H_2O \rightleftharpoons H_3O^+ + CN^-$$

Acetic acid, CH_3CO_2H, is monoprotic because only one of the four hydrogen atoms in a single molecule is given up as a proton in reacting with bases.

$$H-\underset{\underset{H}{|}}{\overset{\overset{H}{|}}{C}}-CO_2H + H_2O \rightleftharpoons H_3O^+ + H-\underset{\underset{H}{|}}{\overset{\overset{H}{|}}{C}}-CO_2^-$$

Diprotic acids contain two ionizable hydrogen atoms in one molecule of the acid; ionization of such acids occurs in two stages. The primary ionization always takes place to a greater extent than the secondary. For example, carbonic acid ionizes as follows:

$$H_2CO_3 + H_2O \rightleftharpoons H_3O^+ + HCO_3^- \qquad \text{(primary ionization)}$$
$$HCO_3^- + H_2O \rightleftharpoons H_3O^+ + CO_3^{2-} \qquad \text{(secondary ionization)}$$

Triprotic acids, such as phosphoric acid, ionize in three steps:

$$H_3PO_4 + H_2O \rightleftharpoons H_3O^+ + H_2PO_4^- \qquad \text{(primary ionization)}$$
$$H_2PO_4^- + H_2O \rightleftharpoons H_3O^+ + HPO_4^{2-} \qquad \text{(secondary ionization)}$$
$$HPO_4^{2-} + H_2O \rightleftharpoons H_3O^+ + PO_4^{3-} \qquad \text{(tertiary ionization)}$$

Often the terms monobasic, dibasic, and tribasic are used instead of monoprotic, diprotic, and triprotic.

14.5   Properties of Hydroxide Bases in Aqueous Solution

A hydroxide base is an ionic compound that contains hydroxide ion and dissolves in water to give a solution containing hydroxide ions. In common usage, the term "base" often refers to a hydroxide base, although as we shall see later, many other anions may properly be classified as bases.

The hydroxides of the alkali metals (Group IA: Li, Na, K, Rb, Cs) are called alkali bases, and those of the alkaline earth metals (Group IIA: Be, Mg, Ca, Sr, Ba) are known as alkaline earth bases. The hydroxides of Group IA, strontium hydroxide, and barium hydroxide are strong bases because they give a high concentration of hydroxide ions in aqueous solution. The hydroxides of the alkali metals are very soluble in water, those of the alkaline earth metals strontium and barium are moderately soluble, and those of all of the other metals are essentially insoluble.

The properties common to hydroxide bases in aqueous solution are due to the presence of the hydroxide ion:

1. They have a bitter taste.
2. They change the colors of certain indicators. For example, they change litmus from red to blue, and phenolphthalein from colorless to red.
3. They neutralize aqueous acids, forming water and a solution of a salt.

$$[M^+] + OH^- + H_3O^+ + [A^-] \longrightarrow [M^+] + [A^-] + 2H_2O$$

4. They give aqueous solutions that conduct an electric current; they are electrolytes.

14.6   Formation of Hydroxide Bases

Hydroxide bases may be prepared by the following methods:

1. *By the reaction of active metals with water.* The active metals that give strong hydroxide bases lie above magnesium in the electromotive series (Section 9.21) and react directly with a stoichiometric amount of water to give bases.

$$2K + 2H_2O \longrightarrow 2KOH + H_2$$
$$Ca + 2H_2O \longrightarrow Ca(OH)_2 + H_2$$

If an excess of water is present, solutions of the bases are formed.

2. *By the reaction of oxides of active metals with water.* The oxides of active metals react with water to give hydroxide bases (Section 8.3).

$$Li_2O + H_2O \longrightarrow 2LiOH$$
$$SrO + H_2O \longrightarrow Sr(OH)_2$$

If an excess of water is present, solutions of the bases are formed.

3. *By the metathetical reaction of a salt with a base to give a solution of a second base and an insoluble precipitate.*

$$[Ca^{2+}] + SO_4{}^{2-} + Ba^{2+} + [2OH^-] \longrightarrow BaSO_4(s) + [Ca^{2+}] + [2OH^-]$$

A hydroxide base

4. *By electrolysis of certain salt solutions.*

$$[2Na^+] + 2Cl^- + 2H_2O \xrightarrow{\text{Electrolysis}} [2Na^+] + 2OH^- + H_2(g) + Cl_2(g)$$

A hydroxide base

14.7   Acid-Base Neutralization

The term **neutralization** is commonly used to describe the reaction of a solution of a protonic acid with a stoichiometric amount of a solution of a hydroxide base.

When aqueous solutions of hydrochloric acid and sodium hydroxide are mixed in the proper proportion, a reaction takes place during which the acidic and basic properties disappear. The hydronium ion, which is responsible for the acidic properties, has reacted with the hydroxide ion, which is responsible for the basic properties, producing water. The sodium and chloride ions have undergone no chemical change and appear in the form of crystalline sodium chloride upon evaporation of the solution. Sodium chloride is an example of the class of compounds called **salts.**

$$H_3O^+ + [Cl^-] + [Na^+] + OH^- \longrightarrow 2H_2O + [Na^+] + [Cl^-]$$

A salt

Because the only change that takes place is the reaction of the hydronium and hydroxide ions, the neutralization may be represented simply as

$$H_3O^+ + OH^- \longrightarrow 2H_2O$$

This equation may be used to represent the reaction of any completely ionized hydroxide base with any completely ionized protonic acid. The fact that the amount of heat evolved per mole of hydronium ion or of hydroxide ion that reacts is the same for *any* completely ionized acid and metal hydroxide is evidence that acid-base neutralizations involve only the hydronium ion from the acid and the hydroxide ion from the base; neutralizations are independent of the acid anion and the base cation.

$$H_3O^+ + [Cl^-] + [Na^+] + OH^- \longrightarrow 2H_2O + [Na^+] + [Cl^-] \qquad \Delta H = -55.8\ \text{kJ}$$
$$H_3O^+ + [NO_3^-] + [Na^+] + OH^- \longrightarrow 2H_2O + [Na^+] + [NO_3^-] \qquad \Delta H = -55.8\ \text{kJ}$$
$$H_3O^+ + [Cl^-] + [K^+] + OH^- \longrightarrow 2H_2O + [K^+] + [Cl^-] \qquad \Delta H = -55.8\ \text{kJ}$$
$$2H_3O^+ + [2Cl^-] + [Ca^{2+}] + 2OH^- \longrightarrow 4H_2O + [Ca^{2+}] + [2Cl^-] \qquad \Delta H = -111.7\ \text{kJ}$$

In each of the preceding equations, the metal ion and the acid anion appear on both sides of the equation, showing that they do not enter the reaction. Accordingly, these ions may be omitted, and the same exothermic equation may be written for all the reactions.

$$H_3O^+ + OH^- \longrightarrow 2H_2O \qquad \Delta H = -55.8\ \text{kJ}$$

In the event that either the acid or base is not completely ionized, the heat of neutralization will be less than 55.8 kJ. For example, a solution of the very weak acid hydrogen cyanide consists of HCN molecules with very little H_3O^+ and CN^- present. The neutralization of such a solution with sodium hydroxide effectively involves the reaction

$$HCN(aq) + OH^- \rightleftharpoons H_2O + CN^-$$

rather than the reaction between hydronium ions and hydroxide ions. In addition, the reaction does not proceed to completion; some hydrogen cyanide molecules remain in solution after a stoichiometric amount of sodium hydroxide has been added. For these reasons, the heat of neutralization is only 12.2 kJ in this particular case.

$$HCN + [Na^+] + OH^- \rightleftharpoons H_2O + [Na^+] + CN^- \qquad \Delta H = -12.2 \text{ kJ}$$

If the salt formed is insoluble the heat evolved is *greater* than 55.8 kJ per mole of hydronium ion or hydroxide ion reacting because of the energy evolved in the process of crystallization.

$$2H_3O^+ + SO_4^{2-} + Ba^{2+} + 2OH^- \rightleftharpoons 4H_2O + BaSO_4(s) \qquad \Delta H = -131 \text{ kJ}$$

In this reaction the two moles of hydronium ion reacted account for 111.7 kJ. The remainder of the 131 kJ is the energy evolved in the crystallization of barium sulfate.

14.8   Salts

Salts are ionic compounds (Section 8.1). Among the positive ions most frequently found in salts are the simple metal ions such as Na^+, K^+, Ca^{2+}, and Ba^{2+}; solvated metal ions such as $[Al(H_2O)_6]^{3+}$ and $[Ag(NH_3)_2]^+$ and solvated hydrogen ions such as NH_4^+ and $(CH_3)_3NH^+$ also are found. The simple negative ions of salts include F^-, Cl^-, Br^-, I^-, S^{2-}, P^{3-}, N^{3-}, C_2^{2-}, and H^-. Many salts contain polyatomic negative ions such as NO_3^-, SO_4^{2-}, PO_4^{3-}, and CN^-. Salts should be regarded as compounds made up of positive and negative ions that need not be formed during acid-base reactions. For example, the salt NaCl may be formed by the direct union of sodium metal with elemental chlorine. At this point, it must be evident that a more correct formula for sodium chloride is Na^+Cl^-. This is true also for other salts. However, it is not generally customary to use the charge signs in such formulas except in special circumstances where their use might be especially helpful.

Beginners in the study of chemistry quite often get the erroneous impression that all compounds composed of a metal and a nonmetal are salts. However, many such compounds have building units that are not ions but instead may be atoms or molecules; examples are $AlCl_3$, $SnCl_4$, $PbCl_4$, and $GeCl_4$. The bonding is either covalent or polar covalent in such compounds, and the absence of any ions is indicated by the fact that these compounds are nonconductors in the liquid state. In contrast to anhydrous $AlCl_3$ (a covalent compound), the hexahydrate $Al(H_2O)_6Cl_3$ is a salt composed of $[Al(H_2O)_6]^{3+}$ and Cl^- ions.

Salts vary greatly in all of their properties except their ionic character. Salts may taste salty, sour, bitter, astringent, or sweet, or be tasteless. Solutions of salts may be acidic, or basic, or neutral to acid-base indicators. Fused salts and aqueous solutions of salts conduct an electric current. The reactions of salts are numerous and varied.

14.9   Normal Salts, Hydrogen Salts, Hydroxy- and Oxysalts

A salt that contains neither a replaceable hydrogen nor a hydroxide group is called a **normal salt;** examples are NaCl, K_2SO_4, and $Ca_3(PO_4)_2$.

When only a part of the acidic hydrogen of an acid has been replaced by a metal, the compound is known as a **hydrogen salt;** examples are sodium hydrogen carbonate, $NaHCO_3$ (generally called sodium bicarbonate, or baking soda), and sodium dihydrogen phosphate, NaH_2PO_4.

When only a part of the hydroxide groups of an ionic metal hydroxide have been replaced, the compound is known as a **hydroxysalt;** examples are barium hydroxychloride, $Ba(OH)Cl$, and bismuth dihydroxychloride, $Bi(OH)_2Cl$. Some hydroxysalts readily lose the elements of one or more molecules of water and form **oxysalts.** Thus bismuth dihydroxychloride breaks down to give bismuth oxychloride, $BiOCl$, sometimes called bismuthyl chloride.

$$Bi(OH)_2Cl \longrightarrow BiOCl + H_2O$$

14.10   Preparation of Salts

Salts may be prepared by one or more of the following methods:

1. *By the direct union of the elements.*

$$2Na + Cl_2 \longrightarrow 2NaCl$$
$$8Fe + S_8 \longrightarrow 8FeS$$

2. *By the reaction of acids with metals, hydroxides of metals, or oxides of metals.*

$$Zn + 2H^+ + [SO_4^{2-}] \longrightarrow Zn^{2+} + [SO_4^{2-}] + H_2(g)$$
$$Fe(OH)_3 + 3H^+ + [3Cl^-] \longrightarrow Fe^{3+} + [3Cl^-] + 3H_2O$$
$$CuO + 2H^+ + [SO_4^{2-}] \longrightarrow Cu^{2+} + [SO_4^{2-}] + H_2O$$

The solid salts, often in the hydrated form, may be obtained by evaporating the solutions to dryness.

3. *By the reaction of metal oxides with nonmetal oxides.*

$$BaO + SO_3 \longrightarrow BaSO_4$$
$$CaO + CO_2 \longrightarrow CaCO_3$$

4. *By the reaction of acids with salts.*

$$BaCO_3 + 2H^+ + [2Cl^-] \longrightarrow Ba^{2+} + [2Cl^-] + H_2O + CO_2(g)$$
$$Ba^{2+} + [2Cl^-] + [2H^+] + SO_4^{2-} \longrightarrow BaSO_4(s) + [2H^+] + [2Cl^-]$$

5. *By the metathetical reaction of salts with other salts.*

$$Ag^+ + [NO_3^-] + [Na^+] + Cl^- \longrightarrow AgCl(s) + [Na^+] + [NO_3^-]$$
$$Zn^{2+} + [2Cl^-] + [2Na^+] + S^{2-} \longrightarrow ZnS(s) + [2Na^+] + [2Cl^-]$$

14.11   Quantitative Reactions of Acids and Bases; Titration

The use of chemical reactions to determine the concentrations of solutions (to **standardize** solutions) and the determination of the amount of a substance present in a sample using a solution of known concentration (a **standard solution**) make up the branch of quantitative analysis called **volumetric analysis.** A solution of reactant is added to a solution of a sample, drop by drop from a buret (Fig. 14-1), in a process called **titration.** An added indicator, which changes color when the

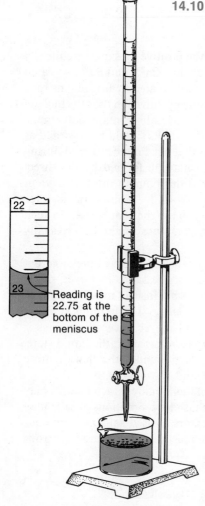

Reading is 22.75 at the bottom of the meniscus

Figure 14-1 A buret is used for accurately measuring the volume of a solution. Accurate reading (see insert) is necessary.

reaction is completed, is used to determine the **end point** of the titration, i.e., the point when the exact amount of reactant necessary to react completely with the sample has been added. Since the end point of a titration involving a neutralization reaction can be detected readily, solutions of acids and bases are often used in volumetric analyses.

EXAMPLE 1   A liter of approximately 0.2 M HCl is prepared by diluting 17 ml of concentrated HCl solution to about 1 liter. The solution is standardized by titrating a 0.5015-g sample of pure, dry Na_2CO_3. What is the exact concentration of the HCl solution if the titration requires 48.47 ml of acid to reach the end point?

$$Na_2CO_3 + 2HCl \longrightarrow 2NaCl + H_2O + CO_2$$

At the end point, the amount of HCl in 48.47 ml of the solution has reacted exactly with the 0.5015 g of Na_2CO_3. The balanced equation tells us that two moles of HCl have been added for each mole of Na_2CO_3 present. The concentration of the solution can be determined by the following steps:

$$\boxed{\text{Mass of } Na_2CO_3} \longrightarrow \boxed{\text{Moles of } Na_2CO_3} \longrightarrow \boxed{\text{Moles of } HCl} \longrightarrow \boxed{\text{Molarity of solution}}$$

$$0.5015 \text{ g } Na_2CO_3 \times \frac{1 \text{ mol } Na_2CO_3}{105.99 \text{ g } Na_2CO_3} = 4.7316 \times 10^{-3} \text{ mol } Na_2CO_3$$

$$4.7316 \times 10^{-3} \text{ mol } Na_2CO_3 \times \frac{2 \text{ mol } HCl}{1 \text{ mol } Na_2CO_3} = 9.4632 \times 10^{-3} \text{ mol } HCl$$

$$\frac{9.4632 \times 10^{-3} \text{ mol } HCl}{0.04847 \text{ } l} = 0.1952 \text{ } M$$

The exact concentration of the HCl solution is 0.1952 M.

◆ ◆ ◆ ◆ ◆

EXAMPLE 2   Titration of a 40.00-ml sample of a solution of H_3PO_4 requires 35.00 ml of 0.1500 M KOH to reach the end point. Determine the molar concentration of H_3PO_4 if the reaction is

$$2KOH + H_3PO_4 \longrightarrow K_2HPO_4 + 2H_2O$$

The following steps are required for this problem:

$$\boxed{\text{Volume of } KOH} \longrightarrow \boxed{\text{Moles of } KOH} \longrightarrow \boxed{\text{Moles of } H_3PO_4} \longrightarrow \boxed{\text{Molarity of } H_3PO_4}$$

$$0.03500 \text{ } l \text{ } KOH \times \frac{0.1500 \text{ mol } KOH}{1 \text{ } l \text{ } KOH} = 5.250 \times 10^{-3} \text{ mol } KOH$$

$$5.250 \times 10^{-3} \text{ mol } KOH \times \frac{1 \text{ mol } H_3PO_4}{2 \text{ mol } KOH} = 2.625 \times 10^{-3} \text{ mol } H_3PO_4$$

$$\frac{2.625 \times 10^{-3} \text{ mol } H_3PO_4}{0.04000 \text{ } l} = 0.06562 \text{ } M$$

◆ ◆ ◆ ◆ ◆

14.12   Equivalent Weights of Acids and Bases

The **gram-equivalent weight of a protonic acid** is defined as the mass of the acid in grams that will provide one mole of protons (H^+) in a reaction. The **gram-equivalent weight of a base** is defined as the mass of the base in grams that will provide one mole of hydroxide ions in a reaction or react with one mole of protons. A gram-equivalent weight is often called an **equivalent** (equiv). In an acid-base reaction, one equivalent of an acid will react with one equivalent of a base.

In an acid-base reaction, protons (H^+) are transferred from the acid to the base, and the number of transferred protons is used in the calculation of the gram-equivalent weights of the acid and the base. When hydrochloric acid reacts with sodium hydroxide, a proton is transferred to the hydroxide ion.

$$HCl + NaOH \longrightarrow NaCl + H_2O$$

When one mole of hydrochloric acid reacts with one mole of sodium hydroxide, one mole of protons (6.022×10^{23} protons) is transferred to one mole of hydroxide ions (6.022×10^{23} hydroxide ions). It follows, then, that one gram-equivalent weight of hydrochloric acid is the same as one mole (36.5 g) of HCl (1 equiv HCl/1 mol HCl), and one gram-equivalent weight of sodium hydroxide is the same as one mole (40.0 g) of NaOH (1 equiv NaOH/1 mol NaOH). However, one mole of sulfuric acid reacts with *two* moles of sodium hydroxide, and *two* moles of protons are transferred from the acid to the base.

$$H_2SO_4 + 2NaOH \longrightarrow Na_2SO_4 + H_2O$$

Hence one gram-equivalent weight of sulfuric acid is equal to one-half mole of the acid (2 equiv H_2SO_4/1 mol H_2SO_4).

It should be emphasized that the number of protons *actually transferred* in an acid-base reaction determines the equivalent weight. For example, phosphoric acid may react with potassium hydroxide in one of three ways:

$$H_3PO_4 + KOH \longrightarrow KH_2PO_4 + H_2O \tag{1}$$

$$H_3PO_4 + 2KOH \longrightarrow K_2HPO_4 + 2H_2O \tag{2}$$

$$H_3PO_4 + 3KOH \longrightarrow K_3PO_4 + 3H_2O \tag{3}$$

The numbers of moles of protons transferred, per mole of H_3PO_4, in the three reactions are one, two, and three moles, respectively. Thus the equivalent weight of H_3PO_4 in the three cases is equal to one mole, one-half mole, and one-third mole, respectively. *It must be emphasized that the gram-equivalent weight of an acid or a base must be deduced from the reaction, not merely from the formula of the substance.*

EXAMPLE 1   **Calculate the equivalent weight of H_3PO_4 in Equations (1), (2), and (3).**

The molecular weight of H_3PO_4 is 97.99. In Equation (1), one mole of H_3PO_4 provides one mole of H^+ (1 equiv H_3PO_4/1 mol H_3PO_4).

$$\frac{97.99 \text{ g } H_3PO_4}{1 \text{ mol } H_3PO_4} \times \frac{1 \text{ mol } H_3PO_4}{1 \text{ equiv } H_3PO_4} = \frac{97.99 \text{ g } H_3PO_4}{1 \text{ equiv } H_3PO_4}$$

Thus the equivalent weight of H_3PO_4 in Equation (1) is 97.99.

In Equation (2), one mole of H_3PO_4 provides two moles of H^+ (2 equiv H_3PO_4/1 mol H_3PO_4).

$$\frac{97.99 \text{ g } H_3PO_4}{1 \text{ mol } H_3PO_4} \times \frac{1 \text{ mol } H_3PO_4}{2 \text{ equiv } H_3PO_4} = \frac{48.99 \text{ g } H_3PO_4}{1 \text{ equiv } H_3PO_4}$$

The equivalent weight of H_3PO_4 in Equation (2), therefore, is 48.99.

In Equation (3), one mole of H_3PO_4 provides three moles of H^+ (3 equiv H_3PO_4/1 mol H_3PO_4).

$$\frac{97.99 \text{ g } H_3PO_4}{1 \text{ mol } H_3PO_4} \times \frac{1 \text{ mol } H_3PO_4}{3 \text{ equiv } H_3PO_4} = \frac{32.66 \text{ g } H_3PO_4}{1 \text{ equiv } H_3PO_4}$$

Hence the equivalent weight of H_3PO_4 in Equation (3) is 32.66.

◆ ◆ ◆ ◆ ◆

The **normality, N,** of a solution is the number of equivalents of solute per liter of solution. (Compare this with the definition of molarity, Section 13.16.)

$$N = \frac{\text{equivalents of solute}}{\text{liters of solution}} = \frac{\text{equiv}}{\ell}$$

Normality may also be determined from the molarity of the solution, provided the reaction is known.

$$N = \frac{\text{moles of solute}}{\text{liters of solution}} \times \frac{\text{equivalents of solute}}{\text{moles of solute}} = \frac{\text{equiv}}{\ell}$$

A solution that contains one gram-equivalent weight (one equivalent) of solute in one liter of solution is called a **one-normal** (1 N) solution. Provided that both hydrogens react, a one-normal solution of sulfuric acid contains one gram-equivalent weight (98 g/2 = 49 g) of H_2SO_4 per liter; a 2 N solution of H_2SO_4 contains 98 g; and a 0.01 N solution contains 0.49 g. Note that the 1 N solution of sulfuric acid contains the same amount of sulfuric acid as a 0.5 M solution of sulfuric acid; each has 49 g of solute per liter of solution.

It should be evident that a 1 M solution of hydrochloric acid is also 1 N, because a gram-equivalent weight of HCl is the same as the gram-molecular weight. However, a 1 M solution of sulfuric acid is $2N$, because one mole of H_2SO_4 is equal to two equivalents of H_2SO_4, if both hydrogens react.

EXAMPLE 2   **Calculate the normality of a solution of HCl containing 3.65 g of HCl per 0.500 ℓ of solution.**

The equivalent weight of HCl is 36.5.

$$3.65 \text{ g HCl} \times \frac{1 \text{ equiv HCl}}{36.5 \text{ g HCl}} = 0.100 \text{ equiv HCl}$$

$$\frac{0.100 \text{ equiv HCl}}{0.500 \ell} = 0.200 \text{ } N$$

◆ ◆ ◆ ◆ ◆

EXAMPLE 3   How many grams of H_2SO_4 are contained in 1.2 ℓ of 0.50 N H_2SO_4 solution that reacts according to the reaction

$$H_2SO_4 + Ba(OH)_2 \longrightarrow BaSO_4 + 2H_2O$$

This problem may be solved by the following steps.

$$\boxed{\text{Volume of solution}} \longrightarrow \boxed{\text{Equivalents of } H_2SO_4} \longrightarrow \boxed{\text{Moles of } H_2SO_4} \longrightarrow \boxed{\text{Mass of } H_2SO_4}$$

A 0.50 N solution contains 0.50 equivalents per liter

$$1.2\,\ell \times \frac{0.50 \text{ equiv } H_2SO_4}{1.0\,\ell} = 0.60 \text{ equiv } H_2SO_4$$

Since each mole of H_2SO_4 provides two protons in the acid-base reaction, one mole of H_2SO_4 is equal to two equivalents of H_2SO_4.

$$0.60 \text{ equiv } H_2SO_4 \times \frac{1 \text{ mol } H_2SO_4}{2 \text{ equiv } H_2SO_4} = 0.30 \text{ mol } H_2SO_4$$

$$0.30 \text{ mol } H_2SO_4 \times \frac{98.0 \text{ g } H_2SO_4}{1 \text{ mol } H_2SO_4} = 29 \text{ g } H_2SO_4$$

◆ ◆ ◆ ◆ ◆

EXAMPLE 4   What is the normality of a solution of $Ba(OH)_2$ that contains 17.14 mg of $Ba(OH)_2$ per 0.1000 ℓ of solution?

$$Ba(OH)_2 + 2HCl \longrightarrow BaCl_2 + 2H_2O$$

Since both of the hydroxide ions react, one mole of $Ba(OH)_2$ provides two moles of OH^-. Hence there is one mole of $Ba(OH)_2$ per 2 equivalents of $Ba(OH)_2$. This problem requires the following steps.

$$\boxed{\text{Mass of } Ba(OH)_2} \longrightarrow \boxed{\text{Moles of } Ba(OH)_2} \longrightarrow \boxed{\text{Equivalents of } Ba(OH)_2} \longrightarrow \boxed{\text{Normality of solution}}$$

$$0.01714 \text{ g } Ba(OH)_2 \times \frac{1 \text{ mol } Ba(OH)_2}{171.34 \text{ g } Ba(OH)_2} = 1.000 \times 10^{-4} \text{ mol } Ba(OH)_2$$

$$1.000 \times 10^{-4} \text{ mol } Ba(OH)_2 \times \frac{2 \text{ equiv } Ba(OH)_2}{1 \text{ mol } Ba(OH)_2} = 2.000 \times 10^{-4} \text{ equiv } Ba(OH)_2$$

$$\frac{2.000 \times 10^{-4} \text{ equiv } Ba(OH)_2}{0.1000 \text{ } \ell} = 2.000 \times 10^{-3} \text{ } N$$

◆ ◆ ◆ ◆ ◆

The concepts of "equivalents" and "normality" also are used with other types of reactions such as oxidation-reduction reactions, which will be discussed in Chapter 20.

The Brönsted-Lowry Concept of Acids and Bases

14.13   The Protonic Concept of Acids and Bases

As a result of the work of several chemists, especially Brönsted and Lowry, the classical acid-base concept has been extended to a more general one of acids and bases. From the Brönsted-Lowry point of view, **an acid is any species (molecule or ion) that can give up a proton to another species. A base is any species that can combine with a proton.** Stated simply, an acid is a proton donor, and a base is a proton acceptor. An acid-base reaction is the transfer of a proton from a proton donor (acid) to a proton acceptor (base). The Brönsted-Lowry view of acids and bases hardly changes the number of molecules or ions that are regarded as acids. However, it does include a great many more bases. For instance, all negative ions can combine, in theory, with a hydrogen ion, so all negative ions must be considered as bases in this concept.

In accordance with the definition that acids are proton donors, there may be molecular acids, such as HCl, HNO_3, H_2SO_4, CH_3CO_2H, HCN, H_2S, H_2CO_3, HF, and H_2O; anion acids, such as HSO_4^-, HCO_3^-, $H_2PO_4^-$, HPO_4^{2-}, and HS^-; and cation acids, such as H_3O^+, NH_4^+, $[Cu(H_2O)_4]^{2+}$, and $[Fe(H_2O)_6]^{3+}$. These acids can lose a proton to form what is called the **conjugate base** of the acid. The conjugate base is a base in that it can pick up a proton to re-form the acid.

Acid		Proton	Conjugate base
HCl	$\longrightarrow$	H^+	$+$ Cl^-
HCN	$\longrightarrow$	H^+	$+$ CN^-
H_2O	$\longrightarrow$	H^+	$+$ OH^-
HSO_4^-	$\longrightarrow$	H^+	$+$ SO_4^{2-}
NH_4^+	$\longrightarrow$	H^+	$+$ NH_3
$[Fe(H_2O)_6]^{3+}$	$\longrightarrow$	H^+	$+$ $[Fe(H_2O)_5(OH)]^{2+}$

As in the case of acids, there are molecular bases, such as H_2O, NH_3, and CH_3NH_2; anion bases, such as OH^-, HS^-, S^{2-}, HCO_3^-, CO_3^{2-}, HSO_4^-, SO_4^{2-}, HPO_4^{2-}, Cl^-, F^-, NO_3^-, and PO_4^{3-}; and cation bases, such as $[Fe(H_2O)_5OH]^{2+}$ and $[Cu(H_2O)_3OH]^+$. The most familiar base is the hydroxide ion, OH^-, which accepts protons from acids, forming water. A base adds a proton to form what is called the **conjugate acid** of the base. The conjugate acid is an acid in that it can give up the proton to re-form the base.

Base		Proton		Conjugate acid
H_2O	$+$	H^+	$\longrightarrow$	H_3O^+
CH_3NH_2	$+$	H^+	$\longrightarrow$	$CH_3NH_3^+$
OH^-	$+$	H^+	$\longrightarrow$	H_2O
S^{2-}	$+$	H^+	$\longrightarrow$	HS^-
CO_3^{2-}	$+$	H^+	$\longrightarrow$	HCO_3^-
F^-	$+$	H^+	$\longrightarrow$	HF
$[Cu(H_2O)_3OH]^+$	$+$	H^+	$\longrightarrow$	$[Cu(H_2O)_4]^{2+}$

In order for an acid to act as a proton donor, a proton acceptor (base) must be present to receive the proton. An acid does not form its conjugate base unless a second base is present to accept the proton. When the second base accepts the proton, it forms its conjugate acid, a second acid. When hydrogen chloride (HCl) reacts with anhydrous ammonia (NH_3) forming ammonium ions (NH_4^+) and chloride ions (Cl^-), hydrogen chloride ($acid_1$) gives up a proton forming chloride ion, its conjugate base ($base_1$); ammonia acts as a proton acceptor and therefore is a base ($base_2$). The proton combines with ammonia to give its conjugate acid, the ammonium ion ($acid_2$). The equations for this and other acid-base reactions are given below.

$Acid_1$		$Base_2$		$Acid_2$		$Base_1$
HCl	+	NH_3	$\rightleftharpoons$	NH_4^+	+	Cl^-
HNO_3	+	F^-	$\rightleftharpoons$	HF	+	NO_3^-
HSO_4^-	+	CO_3^{2-}	$\rightleftharpoons$	HCO_3^-	+	SO_4^{2-}
NH_4^+	+	S^{2-}	$\rightleftharpoons$	HS^-	+	NH_3

In all of these acid-base reactions, the forward reaction is the transfer of a proton from $acid_1$ to $base_2$. The reverse reaction is the transfer of a proton from $acid_2$ to $base_1$. In effect, $base_1$ and $base_2$ are in competition for the proton, and the species present in the greatest concentration will depend on whether $acid_1$ or $acid_2$ binds the proton more strongly. In all of the examples given above, $base_2$ is the stronger base. As a result, at equilibrium the system will consist primarily of a mixture of $acid_2$ and $base_1$.

When an acid such as HCl dissolves in water, the proton acceptor is a water molecule. Thus water functions as a base in this reaction.

$$HCl + H_2O \rightleftharpoons H_3O^+ + Cl^-$$
$$acid_1 \quad base_2 \quad\quad acid_2 \quad base_1$$

The hydronium ion is the conjugate acid of the base, water, in this reaction.

When a base such as ammonia dissolves in water, water functions as a proton donor and hence as an acid.

$$H_2O + NH_3 \rightleftharpoons NH_4^+ + OH^-$$
$$acid_1 \quad base_2 \quad\quad acid_2 \quad base_1$$

The hydroxide ion is the conjugate base of the acid, water, in this reaction. Under the appropriate conditions, then, water can function either as an acid or as a base. (But note that in the reaction with ammonia the hydroxide ion is a stronger base than ammonia so that this reaction gives only a small amount of NH_4^+ and OH^-; the majority of ammonia does not react.)

14.14   Amphiprotic Species

From the examples of acids and bases given in the preceding section, it is evident that certain molecules and ions may be classed as both acids and bases. A species that may either gain or lose a proton is said to be **amphiprotic.** For example, water may lose a proton to a base, such as NH_3, or gain a proton from an acid, such as HCl (Section 14.13), and so is classified as amphiprotic.

The proton-containing negative ions listed in Section 14.13 are also amphiprotic, as is readily seen from the following equations:

Acid$_1$		Base$_2$		Acid$_2$		Base$_1$
HS^-	+	OH^-	$\rightleftharpoons$	H_2O	+	S^{2-}
HBr	+	HS^-	$\rightleftharpoons$	H_2S	+	Br^-
HCO_3^-	+	CN^-	$\rightleftharpoons$	HCN	+	CO_3^{2-}
H_3O^+	+	HCO_3^-	$\rightleftharpoons$	H_2CO_3	+	H_2O

The hydroxides of certain metals, especially those near the boundary between metals and nonmetals, are amphiprotic and so react either as acids or bases.

Acid$_1$		Base$_2$		Acid$_2$		Base$_1$
$[Al(H_2O)_3(OH)_3]$	+	OH^-	$\rightleftharpoons$	H_2O	+	$[Al(H_2O)_2(OH)_4]^-$
H_3O^+	+	$[Al(H_2O)_3(OH)_3]$	$\rightleftharpoons$	$[Al(H_2O)_4(OH)_2]^+$	+	H_2O

In the first reaction, one of the water molecules of the hydrated aluminum hydroxide loses a proton to the hydroxide ion. In the second reaction, the aluminum hydroxide receives a proton from the hydronium ion. In both cases the insoluble hydrated aluminum hydroxide is dissolved. More will be said about this kind of reaction in the discussion on hydrolysis in Chapter 16.

14.15   The Strengths of Acids and Bases

In the fundamental acid-base relationship, $HA + H_2O \rightleftharpoons H_3O^+ + A^-$, the anion A^- is the conjugate base of the acid HA. The strength of the acid is a measure of its tendency to liberate protons. This depends upon the ability of its conjugate base to combine with protons; the stronger the acid the more protons it donates and the more A^- it forms. If A^- is a strong base and accepts protons readily, there will be relatively little A^- and H_3O^+ in solution and the acid, HA, is classified as weak. If A^- is a poor proton acceptor (a weak base), the solution will contain primarily A^- and H_3O^+ and the acid is classified as strong. Strong acids form weak conjugate bases, and weak acids form strong conjugate bases.

The greater the extent of its ionization in water the stronger the acid. Thus, from the data given below, the order of acid strength is $HCl > HNO_2 > CH_3CO_2H$.

$$HCl + H_2O \longrightarrow H_3O^+ + Cl^- \qquad \text{(complete in dil soln)}$$
$$HNO_2 + H_2O \rightleftharpoons H_3O^+ + NO_2^- \qquad \text{(6.5\% in 0.1 } M \text{ soln)}$$
$$CH_3CO_2H + H_2O \rightleftharpoons H_3O^+ + CH_3CO_2^- \qquad \text{(1.3\% in 0.1 } M \text{ soln)}$$

It follows that strong bases are characterized by high proton affinities. The nitrite ion is a good base, as indicated by the weakness of its conjugate acid, HNO_2; i.e., the nitrite ion holds the proton firmly in the nitrous acid molecule. On the other hand, the chloride ion is such a weak base that HCl readily loses its proton to the water molecule, which is a stronger base than the chloride ion. Thus, the ionization of HCl is extensive, and the acid is strong. Other weak bases are Br^-, I^-,

TABLE 14-1   The Relative Strengths of Conjugate Acid-Base Pairs

	Acid			Base
Tetrafluoroboric acid	HBF_4	BF_4^-	Tetrafluoroborate ion	
Perchloric acid	$HClO_4$	ClO_4^-	Perchlorate ion	
Sulfuric acid	H_2SO_4	HSO_4^-	Hydrogen sulfate ion	
Hydrogen iodide	HI	I^-	Iodide ion	
Hydrogen bromide	HBr	Br^-	Bromide ion	
Hydrogen chloride	HCl	Cl^-	Chloride ion	
Nitric acid	HNO_3	NO_3^-	Nitrate ion	
Hydronium ion	H_3O^+	H_2O	Water	
Trichloroacetic acid	CCl_3CO_2H	$CCl_3CO_2^-$	Trichloroacetate ion	
Hydrogen sulfate ion	HSO_4^-	SO_4^{2-}	Sulfate ion	
Phosphoric acid	H_3PO_4	$H_2PO_4^-$	Dihydrogen phosphate ion	
Nitrous acid	HNO_2	NO_2^-	Nitrite ion	
Hydrogen fluoride	HF	F^-	Fluoride ion	
Formic acid	HCO_2H	HCO_2^-	Formate ion	
Acetic acid	CH_3CO_2H	$CH_3CO_2^-$	Acetate ion	
Carbonic acid	H_2CO_3	HCO_3^-	Hydrogen carbonate ion	
Hydrogen sulfide	H_2S	HS^-	Hydrogen sulfide ion	
Ammonium ion	NH_4^+	NH_3	Ammonia	
Hydrogen cyanide	HCN	CN^-	Cyanide ion	
Hydrogen carbonate ion	HCO_3^-	CO_3^{2-}	Carbonate ion	
Hydrogen sulfide ion	HS^-	S^{2-}	Sulfide ion	
Water	H_2O	OH^-	Hydroxide ion	
Ammonia	NH_3	NH_2^-	Amide ion	
Hydrogen	H_2	H^-	Hydride ion	
Methane	CH_4	CH_3^-	Methide ion	

Increasing Acid Strength ↑ (left margin)   *Increasing Base Strength* ↓ (right margin)

NO_3^-, ClO_4^-, and HSO_4^-; thus, HBr, HI, HNO_3, $HClO_4$, and H_2SO_4 are strong acids; i.e., they are almost 100% ionized in dilute aqueous solution. The acetate ion, $CH_3CO_2^-$, is a stronger base than is the water molecule; it is therefore the conjugate base of a weak acid. Acetic acid is ionized only 1.3% in a 0.1 M solution. Other examples of weak acids are H_2O, HS^-, HCO_3^-, HCN, H_2S, and H_2SO_3; thus OH^-, S^{2-}, CO_3^{2-}, CN^-, HS^-, and HSO_3^- are strong to very strong bases.

According to the Brönsted-Lowry concept, then, the strongest acids have the weakest conjugate bases and the strongest bases have the weakest conjugate acids. Table 14-1 illustrates the relationships between conjugate acid-base pairs. An acid will donate a proton to any base lying below it in the table. Thus hydrogen fluoride, HF, will react with hydrogen sulfide ion, HS^-, to form fluoride ion, F^-, and hydrogen sulfide, H_2S.

An **acid-base neutralization reaction** occurs when equivalent quantities of an acid and a base are mixed. This statement does not imply, however, that the resulting solution is always neutral and does not contain either excess hydronium ions or excess hydroxide ions. The nature of the particular acid and base involved in the reaction determines whether or not the resulting aqueous solution is acidic or basic. The following equations illustrate some reactions that take place when equal numbers of equivalents of acids and bases are brought together in aqueous solution:

1. *A strong acid plus a strong base gives a neutral solution.*

$$H_3O^+ + [Cl^-] + [Na^+] + OH^- \longrightarrow 2H_2O + [Na^+] + [Cl^-]$$

Both acid and base are completely ionized, the reaction goes to completion, and only un-ionized water and sodium ions and chloride ions remain as the final products of the reaction. The solution is neutral.

2. *A strong acid plus a weak base gives a weakly acidic solution.*

$$H_3O^+ + [Cl^-] + NH_3(aq) \rightleftharpoons NH_4^+ + [Cl^-] + H_2O$$

Because ammonia is a moderately weak base, its conjugate acid (the ammonium ion) is a weak acid. A small fraction of the ammonium ions give up protons to water molecules, giving an acidic solution.

3. *A weak acid plus a strong base gives a weakly basic solution.*

$$CH_3CO_2H + [Na^+] + OH^- \rightleftharpoons H_2O + [Na^+] + CH_3CO_2^-$$

Because acetic acid is a moderately weak acid, its conjugate base, the acetate ion, is a moderate base. A small fraction of the acetate ions pick up protons from water, producing OH^- ions and giving a basic solution.

4. *A weak acid plus a weak base.*

$$CH_3CO_2H + NH_3(aq) \rightleftharpoons NH_4^+ + CH_3CO_2^-$$

This is the most complex of the four combinations and will be discussed in detail in Chapter 16, but it is appropriate to note here that if the weak acid has the same strength as an acid that the weak base has as a base, the solution will be neutral. Ammonia, for example, takes up about the same amount of protons in the formation of ammonium ion as acetic acid gives up and, although the reaction does not go to completion, the solution is neutral.

If the weak acid and the weak base are *not* of equal strengths, the solution will either be acidic or basic, depending upon the relative acid-base strengths of the two substances.

14.16   The Relative Strengths of Strong Acids and Bases

The strongest acids, such as HCl, HBr, and HI, appear to have about the same strength in water solution. The water molecule is such a strong base compared to the conjugate bases Cl^-, Br^-, and I^- of the strong acids HCl, HBr, and HI, that ionization is nearly complete for all these acids in aqueous solutions. Hence in water these acids are all strong and appear to have equal strengths. When HCl, HBr, and HI are studied in solvents less strongly basic than water, however, they are observed to differ markedly in tendency to give up a proton to the solvent. In methyl alcohol, a weaker base than water, the extent of ionization increases in the order $HCl < HBr < HI$. Because water tends to level off any differences in strength among strong acids, the effect is known as the **leveling effect** of water.

To a somewhat lesser extent, water exerts a leveling effect upon the base strength of very strong bases. For example, the oxide ion, O^{2-}, and the amide ion, NH_2^-, are such strong bases that reaction with water by the removal of protons from the water is complete.

$$O^{2-} + H_2O \longrightarrow OH^- + OH^-$$

$$NH_2^- + H_2O \longrightarrow NH_3 + OH^-$$

Thus O^{2-} and NH_2^- appear to have the same base strength, as indicated by their complete reaction with water. The relative strengths of weaker bases may be measured by the extent to which they yield hydroxide ions when they react with water. This point is illustrated by the following examples for 0.1 M solutions:

$$CH_3CO_2^- + H_2O \rightleftharpoons CH_3CO_2H + OH^- \qquad \text{(0.0075\% ionized)}$$

$$NH_3 + H_2O \rightleftharpoons NH_4^+ + OH^- \qquad \text{(1.34\% ionized)}$$

$$CO_3^{2-} + H_2O \rightleftharpoons HCO_3^- + OH^- \qquad \text{(3.9\% ionized)}$$

In binary compounds with hydrogen, the tendency to lose a proton, and thus the acid strength, increases as the H—A bond strength decreases down a group of nonmetals in the Periodic Table. Thus for Group VIIA the order of increasing acidity is $HF < HCl < HBr < HI$, in the absence of any leveling effect due to solvent. Likewise, in Group VIA the order of increasing acid strengths is $H_2O < H_2S < H_2Se < H_2Te$.

Across a row in the Periodic Table, the acid strength of the binary hydrogen compounds increases with increasing electronegativity of the nonmetal atom as the polarity of the H—A bond increases. Thus the order of increasing acidity (for removal of one proton) across the second row is $CH_4 < NH_3 < H_2O < HF$ and across the third row is $SiH_4 < PH_3 < H_2S < HCl$.

Compounds containing oxygen and hydroxide groups can be acidic, basic, or amphoteric depending upon the position in the Periodic Table (and thus the electronegativity) of the element, E, to which the oxygen and hydroxide groups are bonded. These compounds have the general formula $O_nE(OH)_m$ and include sulfuric acid, $O_2S(OH)_2$, sulfurous acid, $OS(OH)_2$, nitric acid, O_2NOH, perchloric acid, O_3ClOH, and aluminum hydroxide, $Al(OH)_3$, for example. If the central atom, E, has a low electronegativity, its attraction for electrons is low, little tendency exists for it to form a strong covalent bond with the oxygen atom, and the bond between the element and oxygen is weaker than that between oxygen and hydrogen. Hence, bond a is ionic, hydroxyl ions are released to the solution, and the material behaves as a base. Large atomic size, small nuclear charge, and low oxidation number are factors that operate in the direction of low electronegativity for E and are characteristic of the more metallic elements.

$$\overset{a \qquad b}{E \; \vdots \ddot{O} \vdots \; H}$$

If, on the other hand, the element E has a relatively high electronegativity, it attracts the electrons that are available for sharing between it and the oxygen atom rather tightly, giving rise to a relatively strong bond between the element E and the oxygen atom. The oxygen-hydrogen bond is thereby weakened through the displacement of electrons toward E, bond *b* is polar and readily releases hydrogen ions to the solution. The material behaves as an acid. Small atomic size, large nuclear charge, and high oxidation number operate in the direction of high electronegativity and are characteristic of the more nonmetallic elements. As the electronegativity of E increases the O—H bond becomes weaker, and the acid strength increases.

Increasing the oxidation number of a given element in a class of acids also increases the acidity due to weakening of the O—H bond. Sulfuric acid, $\overset{+6}{O_2S(OH)_2}$, is more acidic than sulfurous acid, $\overset{+4}{OS(OH)_2}$; likewise nitric acid, $\overset{+5}{O_2NOH}$, is more acidic than nitrous acid, $\overset{+3}{ONOH}$. In each pair, the oxidation number of the central atom (indicated by the superscript) is larger for the stronger acid.

The hydroxides of elements of intermediate electronegativity with relatively high oxidation numbers (for example, elements near the heavy diagonal line separating the metals from the nonmetals in the Periodic Table) are usually amphoteric. This means that the hydroxides act as acids toward strong bases and as bases toward strong acids.

$$Al(OH)_3(s) + OH^- \longrightarrow [Al(OH)_4]^-$$
$$Al(OH)_3(s) + 3H_3O^+ \longrightarrow Al^{3+}(aq) + 6H_2O$$

The amphoterism of aluminum hydroxide is reflected in its solubility in both strong acids and strong bases. In strong bases, the relatively insoluble $Al(OH)_3$ is converted to the soluble $[Al(OH)_4]^-$ ion by reaction with hydroxide ion; in strong acids, the $Al(OH)_3$ is converted to the soluble Al^{3+} ion by reaction with hydronium ion.

The different direction in the trend of acid strengths for the binary hydrogen halide acids arises from the fact that the halogen is attached directly to a hydrogen in the binary acid (instead of to an oxygen, which is in turn attached to a hydrogen, as in the oxyacids). In either kind of compound, as we go from the larger halogen to the smaller (i.e., from I to Br to Cl), a progressive shrinking of the electron shells of the halogen takes place, including a pulling in of the adjacent atom, whether the adjacent atom be hydrogen (as in the hydrogen halides) or oxygen (as in the oxyacids). As the adjacent atom is pulled in it is brought into a region of higher electron density. Intuitively, this can be understood by visualizing the valence electrons as residing within a smaller volume for the smaller halogen, giving rise to a higher density of electrons. If the adjacent atom being pulled in is a hydrogen atom (as in the case with the hydrogen halides), the hydrogen is held more tightly by the higher electron density in the smaller halogen and is pulled loose with greater difficulty. Conversely, when attached to a larger halogen the hydrogen is in a region of lower electron density (more diffuse region of electrons), is attracted less tightly and is pulled away more easily.

The Lewis Concept of Acids and Bases

14.17   Definitions and Examples

In 1923, G. N. Lewis proposed a generalized acid-base theory in which acids and bases are not restricted to proton donors and proton acceptors, respectively. According to the **Lewis concept, an acid is any species (molecule or ion) that can accept a pair of electrons; a base is any species (molecule or ion) that can donate a pair of electrons.** An acid-base reaction then takes place with the donation of a pair of electrons by the base to the acid, with the formation of an acid-base adduct, a compound that contains a coordinate covalent bond between the Lewis acid and the Lewis base.

The following equations show the general application of the acid-base theory, as proposed by Lewis:

$$2H \overset{H}{\underset{H}{\ce{:N:}}} + Ag^+ \longrightarrow \left[H \overset{H}{\underset{H}{\ce{:N:}}} Ag \overset{H}{\underset{H}{\ce{:N:}}} H \right]^+$$

Base   Acid   Acid-base adduct

$$\left[\ce{:F:} \right]^- + \overset{\ce{:F:}}{\underset{\ce{:F:}}{\ce{:B:F:}}} \longrightarrow \left[\overset{\ce{:F:}}{\underset{\ce{:F:}}{\ce{:F:B:F:}}} \right]^-$$

Base   Acid   Acid-base adduct

$$Ca^{2+} \left[\ce{:O:} \right]^{2-} + \overset{\ce{:O:}}{\underset{\ce{:O:}}{\ce{:S:O:}}} \longrightarrow Ca^{2+} \left[\overset{\ce{:O:}}{\underset{\ce{:O:}}{\ce{:O:S:O:}}} \right]^{2-}$$

Base   Acid   Acid-base adduct

Many Lewis acid-base reactions are displacement reactions in which one Lewis base displaces another Lewis base from an acid-base adduct, or in which one Lewis acid displaces another Lewis acid.

$$\left[H \overset{H}{\underset{H}{\ce{:N:}}} Ag \overset{H}{\underset{H}{\ce{:N:}}} H \right]^+ + 2\, \ce{:C≡N:^-} \longrightarrow \left[\ce{:N≡C:Ag:C≡N:} \right]^- + 2\, \overset{H}{\underset{H}{\ce{:N:}}} H$$

Acid-base adduct   Base   New adduct   New base

$$Ca^{2+} \begin{bmatrix} :\ddot{O}: \\ :C::\ddot{O}: \\ :\ddot{O}: \end{bmatrix}^{2-} + SO_3 \longrightarrow Ca^{2+} \begin{bmatrix} :\ddot{O}: \\ :\ddot{O}:S:\ddot{O}: \\ :\ddot{O}: \end{bmatrix}^{2-} + CO_2$$

Acid-base adduct   Acid   New adduct   New acid

$$H:\ddot{Cl}: + :\ddot{O}:H \longrightarrow H:\ddot{O}:H^+ + :\ddot{Cl}:^- $$
$$\qquad\qquad H \qquad\qquad\quad H$$

Acid-base   Base   New adduct   New base
adduct

The last displacement reaction given above illustrates how the reaction of a protonic acid with a base fits into the Lewis acid-base concept. This concept also explains why oxides of active metals are basic oxides; they contain the oxide ion, which is a strong base. It also explains why oxides of active nonmetals are acidic oxides; the oxide ion can bring in a pair of electrons to share with the central nonmetal atom.

The Solvent System Concept of Acids and Bases

14.18   Definitions and Examples

Work in the early twentieth century showed that reactions analogous to known acid-base reactions in water can also occur in nonaqueous solutions. Such reactions can be classified as acid-base reactions according to the Brönsted-Lowry concept if the solvent molecules (liquid ammonia, for example) can release or accept protons. However, a considerable number of solvents, such as liquid phosgene ($COCl_2$) and liquid sulfur dioxide (SO_2), cannot release or accept protons.

To take into account the existence of acids and bases in a variety of solvents, including nonprotonic ones, the Solvent System concept of acids and bases was developed. It is based upon the idea of parent solvents from which acids and bases can be derived. The solvents themselves are postulated as undergoing slight auto-ionization, resulting in ions that exist in the pure solvent, as shown for typical solvents in Table 14-2.

In the Solvent System concept, **an acid is any material that gives the cation characteristic of the solvent.** The cation may originate either by dissociation of the solute in the solvent or by reaction of the solute with the solvent. **A base is any material that gives the anion characteristic of the solvent. Neutralization thus becomes the reaction of the solvent cation (the acid) with the solvent anion (the base) to provide the solvent.**

TABLE 14-2 Auto-ionization in Various Solvents

Examples of Protonic Solvents	Solvated	Unsolvated
Water	$H_2O + H_2O \rightleftharpoons H_3O^+ + OH^-$	$H^+ + OH^-$
Liquid ammonia	$NH_3 + NH_3 \rightleftharpoons NH_4^+ + NH_2^-$	$H^+ + NH_2^-$
Glacial acetic acid	$CH_3CO_2H + CH_3CO_2H \rightleftharpoons CH_3C(OH)_2^+ + CH_3CO_2^-$	$H^+ + CH_3CO_2^-$

Examples of Nonprotonic Solvents	Solvated	Unsolvated
Liquid phosgene	$COCl_2 + COCl_2 \rightleftharpoons (COCl \cdot COCl_2)^+ + Cl^-$	$COCl^+ + Cl^-$
Liquid sulfur dioxide	$SO_2 + 2SO_2 \rightleftharpoons (SO \cdot SO_2)^{2+} + SO_3^{2-}$	$SO^{2+} + SO_3^{2-}$

TABLE 14-3 Acids and Bases in Various Solvents

Solvent	Acid Ion	Base Ion	Typical Neutralization Reaction Acid + Base $\rightleftharpoons$ Salt + Solvent
Protonic			
Water	H^+ or H_3O^+	OH^-	$HNO_3 + NaOH \rightleftharpoons NaNO_3 + H_2O$
Liquid ammonia	H^+ or NH_4^+	NH_2^-	$NH_4NO_3 + NaNH_2 \rightleftharpoons NaNO_3 + 2NH_3$
Glacial acetic acid	H^+ or $CH_3C(OH)_2^+$	$CH_3CO_2^-$	$HClO_4 + CH_3CO_2Na \rightleftharpoons NaClO_4 + CH_3CO_2H$
Nonprotonic			
Liquid phosgene	$COCl^+$	Cl^-	$[COCl][AlCl_4] + KCl \rightleftharpoons K[AlCl_4] + COCl_2$
Liquid sulfur dioxide	SO^{2+}	SO_3^{2-}	$SOCl_2 + Cs_2SO_3 \rightleftharpoons 2CsCl + 2SO_2$
			Thionyl chloride

Several equations illustrating the Solvent System concept of acids and bases are given in Table 14-3.

Many reactions in solvents other than water are quite analogous to acid-base reactions in water. The following equations illustrate this by showing the amphoteric nature of $Al(OH)_3$ in water (Section 14.16) and the similar amphoteric nature of $Al_2(SO_3)_3$ in liquid sulfur dioxide.

▶ **1. ALUMINUM HYDROXIDE IN WATER.** Aluminum ion and hydroxyl ion (the base) react to form a precipitate of aluminum hydroxide, which is insoluble in water.

$$Al^{3+} + 3OH^- \longrightarrow Al(OH)_3(s)$$

Aluminum hydroxide can be dissolved by adding acid (H_3O^+) again to produce aluminum ions.

$$Al(OH)_3(s) + 3H_3O^+ \longrightarrow Al^{3+} + 6H_2O$$

Aluminum hydroxide also can be dissolved by adding an excess of the base, thereby producing the soluble negative ion $[Al(OH)_4]^-$.

$$Al(OH)_3(s) + OH^- \longrightarrow [Al(OH)_4]^-$$

The aluminum hydroxide can be precipitated again along with production of water (the solvent) by treating the $[Al(OH)_4]^-$ ion with H_3O^+ (the acid).

$$[Al(OH)_4]^- + H_3O^+ \longrightarrow Al(OH)_3(s) + 2H_2O$$

Sometimes the formulas are written to show that the substances are solvated; e.g., $[Al(H_2O)_3(OH)_3]$, $[Al(H_2O)_2(OH)_4]^-$. (See Section 14.14.)

▶ **2. ALUMINUM SULFITE IN LIQUID SULFUR DIOXIDE.** Aluminum ion and sulfite ion (the base in liquid sulfur dioxide) react to form a precipitate of aluminum sulfite, $Al_2(SO_3)_3$.

$$2Al^{3+} + 3SO_3^{2-} \longrightarrow Al_2(SO_3)_3(s)$$

Aluminum sulfite can be dissolved by the addition of the acid SO^{2+} (available from $SOCl_2$) to form aluminum ions.

$$Al_2(SO_3)_3(s) + 3SO^{2+} \longrightarrow 2Al^{3+} + 6SO_2$$

Aluminum sulfite can also be dissolved by the addition of excess base, thereby producing the soluble negative ion $[Al(SO_3)_3]^{3-}$.

$$Al_2(SO_3)_3(s) + 3SO_3^{2-} \longrightarrow 2[Al(SO_3)_3]^{3-}$$

The aluminum sulfite can be precipitated again along with the production of sulfur dioxide (the solvent) by treating the $[Al(SO_3)_3]^{3-}$ ion with SO^{2+} (the acid).

$$2[Al(SO_3)_3]^{3-} + 3SO^{2+} \longrightarrow Al_2(SO_3)_3(s) + 6SO_2$$

Many such analogies occur in a variety of solvents. In Section 23.9, we shall consider a number of reactions in liquid ammonia.

Questions

1. Define the terms acid and base according to the classical, or Arrhenius, concept; the Brönsted-Lowry protonic concept; the Lewis concept; and the Solvent System concept.
2. Define the term *salt* and give examples of normal salts, hydrogen salts, hydroxy salts, and oxysalts.
3. Write equations to illustrate three typical and characteristic reactions of aqueous acids.
4. Contrast the action of water upon oxides of metals with that of water upon oxides of nonmetals.
5. Give four methods of preparing hydroxide bases; of preparing protonic acids.
6. Describe the chemical reaction, using equations where appropriate, that you would expect from the combination of a solution of HBr with each of the following:
   (a) Magnesium
   (b) Calcium oxide
   (c) A potassium hydroxide solution
   (d) A blue-colored litmus solution
   (e) Solid aluminum hydroxide
   (f) Solid aluminum sulfide (Al_2S_3)
   (g) Ammonia gas
7. Write equations for four methods of preparing the salt calcium fluoride, CaF_2.
8. What is the conjugate acid for each of the following

bases: OH^-, F^-, H_2O, HSO_4^-, HCO_3^-, NH_3, NH_2^-, H^-, N^{3-}? Is each a strong or weak acid?

9. What is the conjugate base for each of the following acids: OH^-, H_2O, HCO_3^-, HBr, HSO_4^-, NH_3, HS^-, PH_3, H_2O_2? It is stated that each substance listed is an acid. Is this correct? Explain your answer for each substance listed.

10. What is the conjugate base of each of the following acids: NH_4^+, HCl, HNO_3, $HClO_4$, CH_3CO_2H, and HCN? Is each a strong or weak base?

11. Write equations for the reaction of each of the following with water: HCl, HNO_3, NH_3, NH_2^-, $HClO_4$, F^-, and NH_4^+. What is the role played by water in each of these acid-base reactions?

12. Show by suitable equations that the following species are Brønsted bases: OH^-, NH_3, Cl^-, NO_3^-, and H_2O.

13. Define acid-base "neutralization." Does your definition imply that the resulting solution is always neutral? Explain.

14. What is meant by titration? end point?

15. Predict which acid in each of the following pairs is the stronger: (a) H_2O or H_2Te, (b) NH_3 or PH_3, (c) $B(OH)_3$ or $Al(OH)_3$, (d) HSO_4^- or $HSeO_4^-$, (e) HSO_3^- or HSO_4^-, (f) H_2S or PH_3. Explain your reasoning for each.

16. What are amphiprotic substances? Illustrate with suitable equations.

17. State which of the following species are amphiprotic and write chemical equations illustrating the amphiprotic character of the species: H_2O, $H_2PO_4^-$, S^{2-}, CH_4, HSO_4^-, H_2CO_3.

18. Relate the extent of ionization of an acid in aqueous solution to the strength of the acid.

19. Write the equation for the essential reaction between aqueous solutions of acids and bases.

20. Write equations to show the stepwise ionization of the following polyprotic acids: H_2S, H_2CO_3, and H_3PO_4.

21. Write an equation to illustrate an acid-base reaction by the Brønsted-Lowry concept; the Lewis concept; the Solvent System concept.

22. What is meant by the leveling effect of water on strong acids and strong bases?

23. How can the relative strengths of strong acids be measured?

24. State and explain the order of increasing acidity for each of the following series of compounds:
(a) $HOCl$, $HOBr$, HOI
(b) HCl, HBr, HI
(c) $HOCl$, $HOClO$, $HOClO_2$, $HOClO_3$
(d) HF, H_2O, NH_3, CH_4
(e) $Mg(OH)_2$, $Si(OH)_4$, $ClO_3(OH)$

25. Using Lewis valence electronic configurations, write balanced equations for the following reactions:
(a) $HF + NH_3 \longrightarrow$
(b) $H_3O^+ + CN^- \longrightarrow$
(c) $Na_2O + SO_3 \longrightarrow$
(d) $Li_3N + H_2O \longrightarrow$
(e) $NH_3 + CH_3^- \longrightarrow$

26. Predict the products of the following reactions and write balanced equations for each. There may be more than one reasonable equation depending on the stoichiometry assumed for the reactants.
(a) $Fe_2O_3(s) + H_2SO_4(l) \longrightarrow$
(b) $Li_2O(s) + SiO_2(s) \xrightarrow{\triangle}$
(c) $HCN(g) + Na \longrightarrow$
(d) $NaHCO_3(s) + NaNH_2(s) \longrightarrow$
(e) $Li_3N(s) + NH_3(l) \longrightarrow$
(f) $BaO(s) + Cl_2O_7(l) \longrightarrow$
(g) $NaF(s) + H_3PO_4(l) \longrightarrow$
(h) $MgH_2(s) + H_2S(g) \longrightarrow$
(i) $NaCH_3(s) + NH_3(g) \longrightarrow$
(j) $KHCO_3(s) + KHS(s) \longrightarrow$
(k) $H_2SO_4(l) + NaCH_3CO_2(s) \longrightarrow$

27. Trichloroacetic acid, CCl_3CO_2H, is amphiprotic.

$$CCl_3CO_2H + B \longrightarrow CCl_3CO_2^- + BH$$
$$HA + CCl_3CO_2H \longrightarrow CCl_3CO_2H_2^+ + A^-$$

Write equations for the reaction of pure $CCl_3CO_2H(l)$ with $H_2O(l)$, $HClO_4(l)$, $HBr(g)$, $NH_3(g)$, and $CH_3CO_2H(l)$. (See Table 14.1.)

28. A solution of $Al_2(SO_4)_3$ in water is acidic. Write the equations that lead to formation of H_3O^+ when $Al_2(SO_4)_3$ is dissolved in water.

29. Addition of CO_2 to an aqueous solution of $Na[Al(OH)_4]$ causes the precipitation of $Al(OH)_3$. Write balanced equations for the processes that occur.

30. Write the Lewis valence electron formulas for sulfuric acid, nitric acid, nitrous acid, perchloric acid, carbonic acid, and the conjugate bases of these acids.

Problems

1. Calculate the equivalent weight of each of the reactants in the following equations:

    (a) $Ba(OH)_2 + 2HClO_4 \longrightarrow Ba(ClO_4)_2 + 2H_2O$
    Ans. Ba(OH)$_2$, 85.7; HClO$_4$, 100.5

    (b) $NaOH + NaH_2PO_4 \longrightarrow Na_2HPO_4 + H_2O$
    Ans. NaOH, 40.0; NaH$_2$PO$_4$, 120.0

    (c) $Ni(OH)_2 + NaH_2PO_4 \longrightarrow NaNiPO_4 + 2H_2O$
    Ans. Ni(OH)$_2$, 46.4; NaH$_2$PO$_4$, 60.0

    (d) $2LiOH + H_2SO_4 \longrightarrow Li_2SO_4 + 2H_2O$
    Ans. LiOH, 23.9; H$_2$SO$_4$, 49.0

    (e) $KOH + H_2SO_4 \longrightarrow KHSO_4 + H_2O$
    Ans. KOH, 56.1; H$_2$SO$_4$, 98.1

    (f) $H_2S + Li_3N \longrightarrow Li_2S + NH_3$ (not balanced)
    Ans. H$_2$S, 17.0; Li$_3$N, 11.6

2. Calculate the normality of each of the following solutions.

    (a) 5.0 equivalents of LiOH in 2.5 ℓ of solution
    Ans. 2.0 N

    (b) 0.50 equivalents of H_2SO_4 in 0.67 ℓ of solution
    Ans. 0.75 N

    (c) 0.0015 equivalents of $Ca(OH)_2$ in 100.0 ml of solution
    Ans. 0.015 N

    (d) 0.0450 equivalents of $HClO_4$ in 75.0 ml of solution
    Ans. 0.600 N

3. A 0.144 N solution of KOH is titrated with H_2SO_4 to give K_2SO_4. If a 47.0-ml sample of the KOH solution is used, what volume of 0.144 M H_2SO_4 is required to reach the end point? What volume of 0.144 N H_2SO_4? (Assume both protons react.)
    Ans. 23.5 ml of 0.144 M H$_2$SO$_4$; 47.0 ml of 0.144 N H$_2$SO$_4$

4. What volume of 0.600 N H_2SO_4 would be required to titrate a solution containing 2.50 g of sodium hydrogen carbonate? $NaHCO_3 + H_2SO_4 \longrightarrow NaHSO_4 + CO_2 + H_2O$
    Ans. 49.6 ml

5. A particular standard acid solution is 0.800 N. What volume of 0.300 N base would be required to neutralize 47.0 ml of the acid?
    Ans. 125 ml

6. A titration of 0.1500 g of an acid requires 47.00 ml of 0.0120 N NaOH to reach the end point. What is the equivalent weight of the acid?
    Ans. 266

7. Titration of 10.0 ml of a 0.444 M HCl solution with LiSH requires 23.2 ml of the LiSH solution to reach the end point. What is the molar concentration of LiSH in the solution?
    Ans. 0.191 M

8. What mass of magnesium will produce the same volume of hydrogen gas as 1.00 g of aluminum, each metal being treated with an excess of dilute sulfuric acid?
    Ans. 1.35 g

9. The reaction of 0.871 g of sodium with an excess of liquid ammonia containing a trace of $FeCl_3$ as a catalyst produced 0.473 ℓ of pure H_2 measured at 25°C and 745 torr. What is the equation for the reaction of sodium with liquid ammonia? Show your calculations.
    Ans. 2Na + 2NH$_3$ $\longrightarrow$ 2NaNH$_2$ + H$_2$

10. The reaction of WCl_6 with Al at about 400° gives black crystals of a compound containing only tungsten and chlorine. A sample of this compound, when reduced with hydrogen, gives 0.2232 g of tungsten metal and hydrogen chloride, which is absorbed in water. Titration of the hydrochloric acid thus produced requires 46.2 ml of 0.1051 M NaOH to reach the end point. What is the empirical formula of the black tungsten chloride?
    Ans. WCl$_4$

References

"Acids and Bases," D. Kolb, *J. Chem. Educ.*, **55**, 459 (1978).

"Trends in the Acidities of Some Binary Hydrides in Aqueous Solutions," S. D. Lessley and R. O. Ragsdale, *J. Chem. Educ.*, **53**, 19 (1976).

"The Strength of Hydrohalic Acids," R. T. Meyers, *J. Chem. Educ.*, **53**, 17 (1976).

"The Standard Free Energies of Solution of Anhydrous Salts in Water," D. A. Johnson, *J. Chem. Educ.*, **45**, 236 (1968).

"The Intrinsic Basicity of the Hydroxide Ion," W. L. Jolly, *J. Chem. Educ.*, **44**, 304 (1967).

"The Theory of Acids and Bases," F. M. Hall, *Educ. in Chemistry*, **1**, 91 (1964).

"Lewis Acid-Base Titration in Fused Salts," J. M. Schlegel, *J. Chem. Educ.*, **43**, 362 (1966).

"Hard and Soft Acids and Bases," R. G. Pearson, *Chem. in Britain,* **3,** 103 (1967).

"Donor-Acceptor Interactions," R. J. Drago, *Chem. in Britain,* **3,** 516 (1967).

"Ionizing Solvents," V. Gutmann, *Chem. in Britain,* **7,** 102 (1971).

"Selected Properties of Selected Solvents," G. P. Nilles and R. D. Shuetz, *J. Chem. Educ.,* **50,** 267 (1973).

"Conjugate Acid-Base and Redox Theory," R. A. Pacer, *J. Chem. Educ.,* **50,** 178 (1973).

"A Modern Approach to Acid-Base Chemistry," R. S. Drago, *J. Chem. Educ.,* **51,** 300 (1974).

"The Chemistry of Liquid Ammonia," J. J. Lagowski, *J. Chem. Educ.,* **55,** 752 (1978).

"Equivalents—A Winner or a Dead Horse," F. Brescia, *J. Chem. Educ.,* **53,** 362 (1956).

"The Great Titration Contest" (Staff), *Science,* **203,** 38 (1979).

"The *p*H of Hair Shampoos," J. J. Griffin, R. F. Corcoran, and K. K. Akana, *J. Chem. Educ.,* **54,** 553 (1977).

15

Chemical Kinetics and Chemical Equilibrium

When considering a proposed chemical reaction, we should ask whether the reaction will produce the desired products in useful quantities. We can answer the question quantitatively through the use of equilibrium and thermodynamic considerations, as will be discussed in the latter part of this chapter and in Chapters 16–18. We may also use qualitative considerations, such as the fact that the reaction of a strong acid with a strong base will give a salt, or that an active metal (a reducing agent) generally reacts with a nonmetal (an oxidizing agent).

The question of whether the reaction will proceed sufficiently rapidly for the process to be useful is equally important. A reaction that takes 50 years to produce a product is about as useless to a chemist as one that will never give a product.

The first part of this chapter deals with the rate at which a chemical reaction proceeds to give products **(chemical kinetics).** We will examine the factors influencing the rate of chemical reactions, the mechanisms by which reactions proceed, and the quantitative techniques used to determine and to describe the rate at which reactions occur. The second part of the chapter will consider chemical equilibrium and will describe how equilibrium is related to kinetics.

The general principle of equilibrium may be stated as: **A condition of equilibrium is reached in a system when two opposing changes occur simultaneously at the same rate.** Although we have used this general principle of equilibrium often in preceding chapters, most of the examples have been systems involving physical rather than chemical changes. For example, we have defined the vapor pressure of a liquid (Section 11.2) as the pressure exerted by a vapor in equilibrium with its liquid at a given temperature. The two opposing changes in this case are evaporation and condensation. We did encounter an example of a chemical equilibrium in Section 9.18; this equilibrium involved the reduction of steam by hot iron.

$$3Fe + 4H_2O \rightleftharpoons Fe_3O_4 + 4H_2$$

At equilibrium, the reaction between Fe and H_2O (the forward reaction) and the reaction between Fe_3O_4 and H_2 (the reverse reaction) occur at the same rate.

Chemical Kinetics

15.1   The Rate of Reaction

All rates involve a division by some unit of time. For example, the rate of production of a well can be measured in gallons per minute. The rate of use of coal by an electric generating plant can be measured in tons per day. **The rate of a chemical reaction can be defined as the decrease in concentration of a reactant or the increase in concentration of a product in a unit of time.** When pure hydrogen peroxide is dissolved in water, it slowly decomposes according to the following equation:

$$2H_2O_2(aq) \longrightarrow 2H_2O(l) + O_2(g)$$

The change in concentration with time at 40°C of a solution that is initially one-molar in hydrogen peroxide is shown in Fig. 15-1. From these data, which must be determined experimentally, we can determine the rate, or speed, at which the hydrogen peroxide decomposes.

$$\text{Rate} = -\frac{\text{change in concentration of reactant}}{\text{time interval}}$$

$$= -\frac{[H_2O_2]_{t_2} - [H_2O_2]_{t_1}}{t_2 - t_1} = -\frac{\Delta[H_2O_2]}{\Delta t}$$

In these equations, brackets are used to indicate molar concentrations and delta (Δ) to indicate "the change in." Thus $[H_2O_2]_{t_1}$ represents the molar concentration of hydrogen peroxide at some time t_1; $[H_2O_2]_{t_2}$, the molar concentration of hydrogen peroxide at a later time t_2; and $\Delta[H_2O_2]$, the change in molar concen-

Figure 15-1 The decomposition of H_2O_2 ($2H_2O_2 \longrightarrow 2H_2O + O_2$) at 40°C. The intensity of the color represents the concentration of H_2O_2 at the indicated times after the reaction begins. The color fades to indicate that the concentration of H_2O_2 decreases with time.

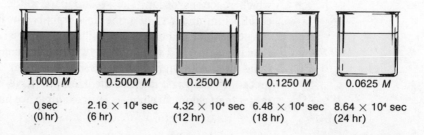

1.0000 M	0.5000 M	0.2500 M	0.1250 M	0.0625 M
0 sec (0 hr)	2.16 × 10⁴ sec (6 hr)	4.32 × 10⁴ sec (12 hr)	6.48 × 10⁴ sec (18 hr)	8.64 × 10⁴ sec (24 hr)

TABLE 15-1 The Variation in the Rate of Decomposition of H_2O_2 at 40°

Time, sec	$[H_2O_2]$, mol ℓ^{-1}	$\Delta[H_2O_2]$, mol ℓ^{-1}	Δt, sec	Rate, mol ℓ^{-1} sec^{-1}
0	1.0000			
2.16×10^4	0.5000	-0.5000	2.16×10^4	2.31×10^{-5}
4.32×10^4	0.2500	-0.2500	2.16×10^4	1.16×10^{-5}
6.48×10^4	0.1250	-0.1250	2.16×10^4	0.579×10^{-5}
8.64×10^4	0.0625	-0.0625	2.16×10^4	0.289×10^{-5}

tration of hydrogen peroxide during the time interval Δt (i.e., $t_2 - t_1$). Since the concentration of hydrogen peroxide decreases during the reaction, $\Delta[H_2O_2]$ is negative. For the period of the reaction from zero to 21,600 sec, $[H_2O_2]_{t_1} = 1.0000$ mol ℓ^{-1} ($t_1 = 0$ sec), $[H_2O_2]_{t_2} = 0.5000$ mol ℓ^{-1} ($t_2 = 21{,}600$ sec), $\Delta t = 21{,}600$ sec, and $\Delta[H_2O_2] = 0.5000 - 1.0000 = -0.5000$ mol ℓ^{-1}. Hence,

$$\text{Rate} = -\left(\frac{-0.5000 \text{ mol } \ell^{-1}}{21{,}600 \text{ sec}}\right) = 2.31 \times 10^{-5} \text{ mol } \ell^{-1} \text{ sec}^{-1}$$

Note that the unit mol/ℓ (mole per liter, or molarity) has been written as mol ℓ^{-1} inasmuch as $1/X = X^{-1}$, according to the mathematical behavior of exponents. The unit sec^{-1} (meaning "per second") is similarly equivalent to 1/sec.

The rates of almost all chemical reactions are not constant, but change continuously with time until equilibrium is established. The rates usually decrease as the reactants are consumed. At some time the rate of a reaction becomes zero; that is, the concentrations of reactants and products no longer change with time, and the reaction is at equilibrium. The variation in the rate of decomposition of hydrogen peroxide in water is illustrated in Table 15-1. Each average rate is calculated for the time intervals shown in Fig. 15-1.

Since the rate of a chemical reaction changes with time, the reaction rates presented in Table 15-1 are average reaction rates for the time intervals indicated. The rate of decomposition of hydrogen peroxide is faster at 21,600 sec than it is at 43,200 sec; it is the average rate between 21,600 and 43,200 sec, which is equal to 1.16×10^{-5} mol ℓ^{-1} sec^{-1}. If we calculate the rate for the time interval from 0 to 86,400 sec, we find

$$\text{Rate} = -\left(\frac{0.0625 - 1.0000}{86{,}400 - 0}\right) = -\left(\frac{-0.9375 \text{ mol } \ell^{-1}}{86{,}400 \text{ sec}}\right) = 1.08 \times 10^{-5} \text{ mol } \ell^{-1} \text{ sec}^{-1}$$

This value is the average of the rates for the four time intervals in Table 15-1.

We can obtain the rate of the reaction at any *instant* of time from a graph of the concentration of hydrogen peroxide versus time. The rate at any particular time is given by the slope of a straight line segment tangent to the curve at that time. Such a graph is shown in Fig. 15-2, with a tangent drawn at $t = 40{,}000$ sec. The slope of this line ($\Delta[H_2O_2]/\Delta t$) is -8.90×10^{-6}. Thus the rate of decomposition at 40,000 sec is 8.90×10^{-6} mol ℓ^{-1} sec^{-1}. The initial rate of the reaction is equal numerically (but opposite in sign) to the slope of the tangent at zero time.

The rates of chemical reactions vary greatly. The initial rate for the decompo-

Figure 15-2 A graph of concentration versus time for the 1.0000 M solution of H_2O_2 described in Fig. 15-1. The rate at any instant of time is equal to the slope of a line tangent to this curve at that time. A tangent is shown for $t = 40,000$ sec.

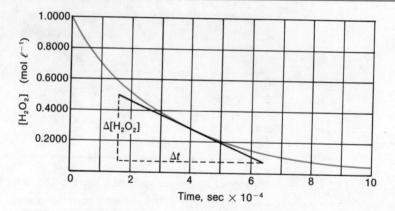

sition of the $1 M$ solution of hydrogen peroxide at 40° is 3.2×10^{-5} mol ℓ^{-1} sec^{-1}, and the reaction takes many hours to go to completion. The initial rate of formation of water when 1 M hydrochloric acid and sodium hydroxide react at 25° is 1.4×10^{11} mol ℓ^{-1} sec^{-1}. This acid-base reaction occurs almost instantaneously, as rapidly as the two solutions can be mixed. The initial rate for the reaction of hydrochloric acid with sodium hydroxide is 10^{17} times faster than that for the decomposition of hydrogen peroxide. In the following sections we will examine some of the factors that influence the rates of chemical reactions.

15.2 Rate of Reaction and the Nature of the Reacting Substances

Similar reactions have different reaction rates under the same conditions if different reacting substances are involved. Calcium and sodium both react with water at ordinary temperatures, calcium at a moderate rate and sodium so rapidly that the reaction is of almost explosive violence. Sodium, an active metal, reacts much more rapidly with chlorine than does iron, a moderately active metal. We noted in the preceding section that the reaction of an acid with a base, a process involving the simple combination of two ions of opposite charge ($H_3O^+ + OH^- \longrightarrow 2H_2O$), is much faster than the decomposition of hydrogen peroxide, a process involving the rearrangement of molecules ($2H_2O_2 \longrightarrow 2H_2O + O_2$). It is evident that the rate of a reaction is strongly influenced by the nature of the substances participating in the reaction.

15.3 Rate of Reaction and the State of Subdivision

Except for substances in the gaseous state or in solution, reactions occur at the boundary, or interface, between two phases. The rate of a reaction between two phases depends to a great extent on the area of surface contact between phases. Reactions involving solids occur on the surfaces of the solids. Finely divided solids, because of the greater surface area available, react more rapidly than massive specimens of the same substances. For example, large pieces of coal burn slowly, smaller pieces more rapidly, and powdered coal may burn at an explosive rate.

Figure 15-3 The effect of temperature upon reaction rate, shown graphically. The rate of reaction is often approximately doubled for each 10° rise in temperature.

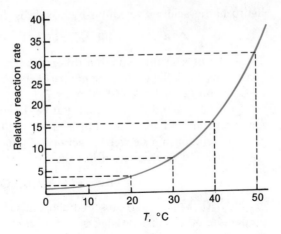

15.4   Rate of Reaction and Temperature

It is a common observation that chemical reactions are accelerated by increases in temperature. The initial rate of decomposition of a $1\,M$ hydrogen peroxide solution is 3.2×10^{-5} mol ℓ^{-1} sec^{-1} at 40° and is slightly more than twice as fast, 7.2×10^{-5} mol ℓ^{-1} sec^{-1}, at 50°. The oxidation of either iron or coal is very slow at ordinary temperatures but proceeds rapidly at high temperatures. Students use a burner or a hot plate in the laboratory to increase the speed of reactions that may proceed slowly or even at an imperceptible rate at ordinary temperatures. It is a familiar fact that foods cook faster at higher temperatures than at lower ones. In many cases the rate of a reaction in a homogeneous system is *approximately* doubled by an increase in temperature of only 10° (Fig. 15-3). This rule, however, is a rough approximation and applies only to those reactions that do not appear to be instantaneous.

15.5   Rate of Reaction and Concentration; Rate Laws

At a fixed temperature and in the absence of a catalyst (catalysts will be discussed in greater detail in Section 15.15), the rate of a given reaction in solution is largely dependent on the concentrations of the reacting substances. Many familiar facts might be cited to illustrate this principle. For example, the gas methane burns much more rapidly in pure oxygen than in air, in which only about 20% of the molecules are oxygen.

The rate of decomposition of hydrogen peroxide (Table 15-1) decreases with decreasing concentration. The rate at any instant of time is directly proportional to the concentration of hydrogen peroxide at that time.

$$\text{Rate} = k[\text{H}_2\text{O}_2] \tag{1}$$

The proportionality constant, k, is called the **rate constant.** The brackets are used to indicate the molar concentration of the reactant. The rate constant is independent of the concentration of the reactant, but does vary with temperature. At 40°, k is 3.2×10^{-5} sec^{-1}; at 50°, 7.2×10^{-5} sec^{-1}. The rate for the reaction of hydrochloric acid with sodium hydroxide is described by a similar expression:

$$\text{Rate} = k[\text{H}_3\text{O}^+][\text{OH}^-] \tag{2}$$

The value of k in this equation differs from that in Equation (1) since the rate constant is dependent upon the nature of the reacting substances. The value of k in Equation (2) is 1.4×10^{11} ℓ mol sec^{-1}.

Equations (1) and (2) are quantitative descriptions based upon experimental measurements and are referred to as **rate equations,** or **rate laws.** These equations give the relationship between reaction rate and concentration of reactant.

In general, a rate equation has the form

$$\text{Rate} = k[\text{A}]^m[\text{B}]^n[\text{C}]^p \cdots$$

in which [A], [B], [C] $\cdots$ represent molar concentrations of reactants (or sometimes products or other substances), k is the rate constant for the particular reaction, and the exponents m, n, p $\cdots$ are usually positive integers (although fractions and negative numbers sometimes appear). *Both k and the exponents m, n, p $\cdots$ must be determined experimentally from the variation of the rate of a reaction as the concentrations of the reactants are varied.*

EXAMPLE   One of the reactions occurring in the ozone layer in the upper atmosphere is the combination of NO with O_3.

$$\text{NO} + \text{O}_3 \longrightarrow \text{NO}_2 + \text{O}_2$$

This reaction has been studied in the laboratory and the following rate data obtained at 25°:

[NO], mol ℓ^{-1}	[O_3], mol ℓ^{-1}	$\dfrac{\Delta[\text{NO}_2]}{\Delta t}$, mol ℓ^{-1} sec^{-1}
1.00×10^{-6}	3.00×10^{-6}	0.660×10^{-4}
1.00×10^{-6}	6.00×10^{-6}	1.32×10^{-4}
1.00×10^{-6}	9.00×10^{-6}	1.98×10^{-4}
2.00×10^{-6}	9.00×10^{-6}	3.96×10^{-4}
3.00×10^{-6}	9.00×10^{-6}	5.94×10^{-4}

Determine the rate equation for the reaction at 25°.

The rate equation will have the form

$$\text{Rate} = k[\text{NO}]^n[\text{O}_3]^m$$

The values of n and m cannot be determined from the chemical equation, but must be evaluated from the experimental data. In the first three entries in the data table, [NO] is held constant and [O_3] is allowed to vary. The reaction rate changes in direct proportion to the change in [O_3]. When [O_3] doubles, the rate doubles; when the [O_3] increases by a factor of 3, the rate increases by a factor of 3. Thus the rate is directly proportional to [O_3], or

$$\text{Rate} = k'[\text{O}_3]$$

Hence m in the rate equation must be equal to 1.

In the last three entries in the table [NO] varies as [O_3] is held constant. When [NO] doubles the rate doubles, and when [NO] triples the rate also triples. Thus the rate is also directly proportional to [NO] and n in the rate equation is also equal to 1.

The equation is thus

$$\text{Rate} = k[\text{NO}][\text{O}_3]$$

◆ ◆ ◆ ◆ ◆

It is not unusual to find that a rate equation does not reflect the overall stoichiometry of a chemical reaction. In some cases one or more of the reactants may not even appear in the rate equation.

15.6   Order of Reaction

The order of a reaction is determined from the exponents n, m $\cdots$ found in the rate equation for the reaction. The order can be expressed either in terms of the **order with respect to each specific reactant** or in terms of the **overall order of the reaction.** The order with respect to one of the reactants is equal to the power to which the concentration of that particular reactant is raised.

Consider a reaction for which the rate equation is

$$\text{Rate} = k[\text{A}]^m[\text{B}]^n$$

If the exponent m is 1, the reaction is said to be **first order** with respect to A. If $m = 2$, the reaction is said to be **second order** with respect to A. If $n = 2$, the reaction is referred to as second order with respect to B. If a reaction is **zero order** with respect to a reactant, the rate of the reaction does not change as the concentration of the reactant changes. If this reaction were zero order in B, n would be zero. Note that [B]^0 is equal to 1 whatever the value of [B].

The overall order of a reaction is the sum of the orders with respect to each reactant.

If $m = 1$ and $n = 1$, the reaction is first order in A and first order in B. The overall order of the reaction is given by $m + n$; therefore the overall order of the reaction in this particular case is second order.

EXAMPLE 1   **Experiment shows that the reaction of NO_2 with CO**

$$\text{NO}_2 + \text{CO} \longrightarrow \text{NO} + \text{CO}_2$$

is second order in NO_2 and zero order in CO at 200°. What is the rate equation for the reaction?

The rate equation will have the form Rate $= k[\text{NO}_2]^n[\text{CO}]^m$. The reaction is second order in NO_2; thus $n = 2$. The reaction is zero order in CO; thus $m = 0$. The rate equation is

$$\text{Rate} = k[\text{NO}_2]^2[\text{CO}]^0 = k[\text{NO}_2]^2$$

Recall that a number raised to the zero power is equal to 1, so [CO]$^0 = 1$.

◆ ◆ ◆ ◆ ◆

EXAMPLE 2   **What is the order with respect to each reactant and the overall order of the reaction $NO + O_3 \longrightarrow NO_2 + O_2$ (Example, Section 15.5)?**

The rate equation for the reaction was found to be

$$\text{Rate} = k[NO][O_3]$$

Thus the reaction is first order with respect to both NO and O_3 (writing [NO] is the same as writing $[NO]^1$). The overall reaction is second order since $n + m = 2$.

◆ ◆ ◆ ◆ ◆

15.7   Half-Life of a First-Order Reaction

The equation for a first-order reaction relating the rate constant k to the initial concentration $[A_0]$ and to the concentration $[A]$ present after any given time t can be derived and is

$$\log \frac{[A_0]}{[A]} = \frac{kt}{2.303} \tag{1}$$

An example illustrating the use of the equation follows:

EXAMPLE 1   **The rate constant for the first-order decomposition of C_4H_8 (Section 15.11) at 500° is 9.2×10^{-3} sec^{-1}. How long will it take for 90.0% of a 0.100 M sample of C_4H_8 at 500° to decompose, that is, for the concentration of C_4H_8 to decrease to 0.0100 M?**

The initial concentration of C_4H_8, $[A_0]$, is 0.100 mol ℓ^{-1}; [A], the concentration at time t, is 0.0100 mol ℓ^{-1}; and k is 9.2×10^{-3} sec^{-1}.

$$\log \frac{[A_0]}{[A]} = \frac{kt}{2.303}$$

$$t = \log \frac{[A_0]}{[A]} \times \frac{2.303}{k}$$

$$t = \log \left(\frac{0.100 \text{ mol } \ell^{-1}}{0.0100 \text{ mol } \ell^{-1}} \right) \times \frac{2.303}{9.2 \times 10^{-3} \text{ sec}^{-1}} = 2.5 \times 10^2 \text{ sec}$$

◆ ◆ ◆ ◆ ◆

The half-life of a reaction, $t_{1/2}$, is the time required for half the original concentration of the limiting reactant to be consumed as the reaction takes place. In each succeeding half-life period, half the remaining concentration of the reactant will be used up. The decomposition of hydrogen peroxide in water at 40° illustrated in Fig. 15-1 displays the concentration after each of several successive half-life periods, 2.16×10^4 sec. During the first half-life period (from 0 to 2.16×10^4 sec) the concentration decreases from 1.0000 M to 0.5000 M. During the second half-life period (from 2.16×10^4 to 4.32×10^4 sec), it decreases from 0.5000 M to 0.2500 M; during the third half-life period, from 0.2500 M to 0.1250 M. The concentration decreases by one-half during each successive period of 2.16×10^4 sec.

The equation for determining the half-life for a first-order reaction can be derived from Equation (1) as follows.

$$\log \frac{[A_0]}{[A]} = \frac{kt}{2.303}$$

$$t = \log \frac{[A_0]}{[A]} \times \frac{2.303}{k}$$

If the time t is the half-life time, $t_{1/2}$, the concentration of the limiting reactant at the end of this time, $[A]$, is equal to one-half the initial concentration. Hence, at time $t_{1/2}$, $[A] = \frac{1}{2}[A_0]$.

Therefore

$$t_{1/2} = \log \frac{[A_0]}{\frac{1}{2}[A_0]} \times \frac{2.303}{k}$$

$$= \log 2 \times \frac{2.303}{k}$$

$$= 0.301 \times \frac{2.303}{k} = \frac{0.693}{k}$$

Thus

$$t_{1/2} = \frac{0.693}{k} \tag{2}$$

You should note from Equation (2) that for a first-order reaction the half-life is inversely proportional to the rate constant k. Hence, a fast reaction with short half-life has a large k value; a slow reaction with a longer half-life has a smaller k value.

EXAMPLE 2   Calculate the rate constant for the first-order decomposition of H_2O_2 in water at 40° using the data from Fig. 15-1.

The half-life for the decomposition of H_2O_2 is 2.16×10^4 sec.

$$t_{1/2} = \frac{0.693}{k}$$

$$k = \frac{0.693}{t_{1/2}} = \frac{0.693}{2.16 \times 10^4 \text{ sec}} = 3.21 \times 10^{-5} \text{ sec}^{-1}$$

◆ ◆ ◆ ◆ ◆

Similar, somewhat more complex, quantitative relationships can be derived for reactions of higher order than first.

15.8   Collision Theory of the Reaction Rate

Chemists believe that before two atoms, molecules, or ions can react, they must collide with one another. The rate at which they collide is determined by how rapidly they can diffuse together. In only a few reactions does every collision between reactants lead to products. The rate of these reactions is determined only by how rapidly the reactants can diffuse together. The rate of such reactions is said to be **diffusion-controlled.** Diffusion-controlled reactions are very fast; hence they are characterized by large rate constants. For a typical diffusion-controlled second order gas-phase reaction at 25° (such as the reaction of an oxygen atom with a nitrogen molecule, $O + N_2 \longrightarrow NO + N$), the rate constant falls in the range 10^{10}–10^{12} ℓ mol sec^{-1}. The diffusion-controlled reaction between hydronium ions and hydroxide ions in water at 25° ($H_3O^+ + OH^- \longrightarrow 2H_2O$) has a rate constant of 1.4×10^{11} ℓ mol sec^{-1}. The rates of these reactions with one-molar concentrations of reactants are such that initially 10^{10} to 10^{12} moles per liter of reactants would be consumed per second. At these rates over 95% of the reactants would be consumed in 10^{-11} sec.

Almost all reactions, however, occur at a rate much slower than the diffusion-controlled limit. In these reactions only a very small fraction of the collisions that occur between molecules result in reaction. In the majority of collisions, the reactants simply bounce away unchanged. In order for a collision to lead to a reaction, the following conditions must be fulfilled:

1. The molecules frequently must have a particular orientation so that the atoms that are bonded together in the product come in contact during the collision.
2. The collision must occur with enough energy for the electron shells of the reacting atoms to penetrate into each other so that bonding electrons can be rearranged with the formation of new bonds and a new chemical species.

The gas-phase reaction of nitrogen dioxide with carbon monoxide

$$NO_2 + CO \longrightarrow NO + CO_2$$

above 225° illustrates the factors necessary for effective collision. During the course of the reaction, an oxygen atom is transferred from an NO_2 molecule to a CO molecule. There are many orientations of the NO_2 and CO molecules during collision that do not place an oxygen atom of the NO_2 molecule close to the carbon atom of the CO molecule. Three of these are indicated in Fig. 15-4. These collisions will not be effective in producing a chemical reaction. Only a collision in which an oxygen atom strikes the carbon atom [Fig. 15-4(d)] may be effective in producing a reaction.

Even if the orientation is correct, a collision may still not lead to reaction. As the oxygen atom of an NO_2 molecule approaches the carbon atom of a CO molecule, the electrons in the two molecules begin to repel each other. Unless the molecules possess a kinetic energy greater than a certain minimum, the two molecules bounce

Figure 15-4 Some possible collisions between NO_2 and CO molecules. Only in (d) are the molecules correctly oriented for transfer of an oxygen atom from NO_2 to CO to give NO and CO_2.

away from each other before they can get close enough to react. If the molecules are moving fast enough that their kinetic energy exceeds the necessary minimum, then the repulsion of the electron clouds is not strong enough to keep them apart, and the molecules can get sufficiently close together for a C—O bond to begin to form as the N—O bond begins to break. Using dots to represent partially formed or broken bonds, we can write the resulting species as follows:

$$NO_2 + CO \longrightarrow \left[\substack{O \\ \diagdown} N\cdots O \cdots C\equiv O \right]$$

The species $O\!=\!N\cdots O\cdots C\equiv O$, which is formed from the NO_2 and CO molecules and which contains the partially formed $C\cdots O$ and partially broken $N\cdots O$ bonds, is called the **activated complex,** or **transition state.** An activated complex is a combination of reacting molecules intermediate between reactants and products, in which some bonds have weakened and new bonds have begun to form. Ordinarily, an activated complex cannot be isolated. It breaks down to give either reactants or products, depending upon the conditions under which the reaction takes place.

$$NO_2 + CO \longrightarrow \left[\substack{O \\ \diagdown} N\cdots O \cdots C\equiv O \right] \substack{\nearrow NO_2 + CO \\ \text{or} \\ \searrow NO + CO_2}$$

The decrease in reaction rate that accompanies a decrease in concentration of reacting substances is readily explained in terms of the collision theory. With decreased concentration of any or all of the reacting substances, the chances for collision between molecules are decreased due to the presence of a smaller number of molecules per unit volume. Fewer collisions per unit time means a slower reaction rate. The concentration of a substance, when expressed in moles per liter, provides a direct measure of the number of molecules per unit space and is therefore appropriately used in rate equations (Sections 15.5 and 15.6).

15.9   Activation Energy and the Arrhenius Law

The minimum energy necessary to form an activated complex during a collision between reactants is called the **activation energy,** E_a (Fig. 15-5). In a slow reaction, the activation energy is much larger than the average energy content of the molecules. The energies of a large fraction of the molecules in a system are close to the average value. However, a few of the molecules, the fast-moving ones, have relatively high energies. The collisions between the fast-moving molecules are most apt to result in reactions. In very fast reactions the fraction of molecules in the system possessing the necessary energy of activation is large, and most collisions between the molecules result in reaction.

The energy relationships for the general reaction of a molecule of A with a molecule of B to form molecules of C and D are shown in Fig. 15-6.

$$A + B \longrightarrow C + D$$

The figure shows that after the activation energy, E_a, is exceeded, and as C and D

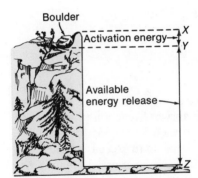

Figure 15-5 Illustration, by analogy, of activation energy in relation to available energy release. The boulder, potentially, can release the amount of energy created by its falling the distance from height Y to height Z. However, an amount of energy must be put into the system (the activation energy) sufficient to lift the boulder over the barrier through the distance Y to X before the boulder can fall to Z. The activation energy is released as the boulder falls through the distance X to Y, and the additional energy is then released in the continued fall from Y to Z. The net energy released is that provided by the fall from Y to Z.

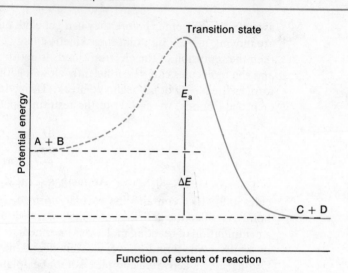

Figure 15-6 Potential energy relationships for the reaction A + B $\rightleftharpoons$ C + D. The energy represented by the broken curve is that for the system with a molecule of A and a molecule of B present; the energy represented by the solid curve is that for the system with a molecule of C and a molecule of D present. The activation energy for the forward reaction is represented by E_a, the activation energy for the reverse reaction by $(E_a + \Delta E)$. The species present at the height of the peak energy corresponds to the transition state.

begin to form, the system loses energy, falling to a point where it has a lower total energy than the initial mixture of A and B. The forward reaction (that between molecules A and B to form molecules C and D) therefore tends to take place readily if sufficient energy is available in any one collision to exceed the activation energy, E_a. In Fig. 15-6, ΔE represents the difference in energy between the two molecules of reactants (A and B) and the two molecules of products (C and D). The sum of $E_a + \Delta E$ represents the activation energy for the reverse reaction, C + D $\longrightarrow$ A + B.

For a given reaction, the rate constant is related to the activation energy by a relationship known as the **Arrhenius Law.**

$$k = A \times 10^{-E_a/2.303RT}$$

In this equation R is a constant with the value 8.314 J mol^{-1} K^{-1}, T is temperature in Kelvin degrees, E_a the activation energy in joules per mole, and A is a constant, called the **frequency factor,** which is related to the frequency of collisions and the orientation of the reacting molecules. In effect, A indicates the number of collisions with the correct orientation for reaction and the remainder of the equation, $10^{-E_a/2.303RT}$, gives the fraction of the collisions in which the energy of the molecules is greater than E_a, the activation energy for the reaction.

The Arrhenius Law describes quantitatively much of what we have discussed about reaction rates. Reactions with similar frequency factors but with different activation energies will be characterized by different rate constants because the rate constant for a particular reaction is proportional to $10^{-E_a/2.303\,RT}$. For two reactions at the same temperature, the reaction with the larger activation energy, E_a, will have the lower rate constant and the slower rate. The larger value of E_a decreases the magnitude of the term $10^{-E_a/2.303\,RT}$, reflecting the smaller fraction of molecules with sufficient energy to react. Alternatively, the reaction with the lower E_a will have a larger fraction of molecules with the necessary energy to react. This will be reflected as a larger value of $10^{-E_a/2.303\,RT}$, a higher rate constant, and a faster rate for the reaction. An increase in temperature has the same effect as a decrease in E_a. A larger fraction of molecules has the necessary energy to react as indicated by an increase in the value of $10^{-E_a/2.303\,RT}$. A rate constant is also directly proportional to the frequency factor, A. Hence a change in conditions or reactants that increases the fraction of collisions in which the orientation of the molecules is satisfactory for reaction results in an increase in A and, consequently, an increase in k.

The increase in k with increasing T is consistent with the observation that the rates of almost all chemical reactions increase with temperature (Section 15.4). The sensitivity of the rate of a reaction to changes in temperature depends on the value of the activation energy for the reaction. If the value of E_a is high, the reaction is more sensitive to temperature changes. The generalization that a change of temperature of 10° doubles the rate of a reaction applies to reactions with an activation energy of about 50 kJ mol^{-1}.

In order to determine E_a for a reaction it is necessary to measure k at different temperatures and evaluate E_a from the Arrhenius Law. The Arrhenius Law may be rewritten

$$\log k = \log A - \frac{E_a}{2.303\,RT}$$

A plot of $\log k$ against $1/T$ gives a straight line whose slope is $-E_a/2.303\,R$, from which E_a may be determined.

EXAMPLE   **The variation of the rate constant with temperature for the decomposition of HI(g) to $H_2(g)$ and $I_2(g)$ is given below. What is the activation energy for the reaction?**

T, K	$1/T$, K^{-1}	k, ℓ mol^{-1} sec^{-1}	$\log k$
555°	1.80×10^{-3}	3.52×10^{-7}	-6.453
575°	1.74×10^{-3}	1.22×10^{-6}	-5.913
645°	1.55×10^{-3}	8.59×10^{-5}	-4.066
700°	1.43×10^{-3}	1.16×10^{-3}	-2.936
781°	1.28×10^{-3}	3.95×10^{-2}	-1.403

A graph of $\log k$ vs $1/T$ is given in Fig. 15-7. The slope of this line is given by the following expression:

$$\text{Slope} = \frac{\Delta(\log k)}{\Delta(1/T)} = \frac{(-6.453) - (-1.403)}{(1.80 \times 10^{-3}) - (1.28 \times 10^{-3})} = \frac{-5.050}{0.52 \times 10^{-3}} = -9.71 \times 10^3 \text{ K}$$

$$\text{Slope} = -\frac{E_a}{2.303 \, R}$$

Thus

$$-E_a = \text{slope} \times 2.303 \, R = -9.71 \times 10^3 \, K \times 2.303 \times 8.314 \text{ J mol}^{-1} K^{-1}$$
$$E_a = 186 \text{ kJ mol}^{-1}$$

Figure 15-7 A graph of the linear relationship of $\log k$ with $1/T$ for the reaction $2HI \longrightarrow H_2 + I_2$ according to the Arrhenius Law.

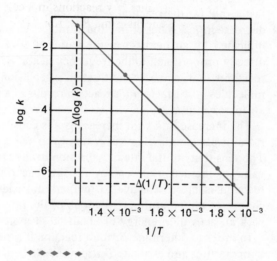

15.10   Elementary Reactions

A balanced equation for a chemical reaction indicates the materials that are reacting and the materials that are produced, but it provides no information about the process by which the reaction actually takes place. The process, or pathway, by which a reaction occurs is called the **reaction mechanism,** or the **reaction path.**

A reaction often occurs in steps. The decomposition of ozone is believed to occur by a mechanism involving two steps.

$$O_3 \longrightarrow O_2 + O \tag{1}$$
$$O + O_3 \longrightarrow 2O_2 \tag{2}$$

The two steps add up to the overall reaction for the decomposition ($2O_3 \longrightarrow 3O_2$). Note that the O atom produced in the first step is a reactant in the second step; although it is produced in the first step, it is used up in the second step and thus does not appear in the final products of the overall reaction. Species that are produced in one step and consumed in a subsequent step are called **intermediates.**

Each of the steps in a reaction mechanism is called an **elementary reaction.** Elementary reactions occur exactly as written and cannot be broken down into simpler steps, whereas an overall reaction can often be broken down into a number of steps. Thus, although the overall reaction indicates that two molecules of ozone react to give three molecules of oxygen, the reaction path does not involve the collision of two ozone molecules. Instead, a molecule of ozone decomposes to an oxygen molecule and an intermediate oxygen atom, then the oxygen atom reacts with a second ozone molecule to give two oxygen molecules. These two elementary reactions occur exactly as written in Equations (1) and (2).

15.11   Unimolecular Reactions

An elementary reaction is **unimolecular** if the rearrangement of a single molecule or ion produces a molecule of product or products. A unimolecular reaction may be one of several elementary reactions in a complex mechanism or it may be the only reaction in a mechanism. Reaction (1) in Section 15.10 ($O_3 \longrightarrow O_2 + O$) illustrates a unimolecular elementary reaction occurring in a complex reaction mechanism. The gas-phase decompositions of dinitrogen pentaoxide (N_2O_5) and cyclobutane (C_4H_8) occur with a one-step unimolecular mechanism.

All that is required for each of these three reactions to occur is the separation of parts of single reactant molecules into products. The latter two reactions also illustrate the fact that, upon occasion, an overall reaction also may be an elementary reaction.

During chemical reactions chemical bonds do not simply "fall apart." The decomposition of C_4H_8, for instance, requires the input of 261 kJ of energy per mole of C_4H_8 to distort the molecules into activated complexes that can decompose into products.

Thus the activation energy for this reaction is 261 kJ per mole of C_4H_8 reacting. In a sample of C_4H_8, very few molecules contain enough energy to form an activated complex. Those that do, react by the process indicated above. Other molecules can pick up additional energy by collision with rapidly moving molecules. The kinetic energy of the rapidly moving molecules is converted into energy of activation in the other molecules. If the energy these other molecules pick up is sufficient to form the activated complex, then they too can undergo reaction. In effect, an energetic collision will "knock" the reacting molecule into the geometry of the activated complex. However, you may recall from Section 10.17 and Fig. 10-16 that only a small fraction of gas molecules are traveling at high speeds with large kinetic energies. Hence at any one time only a few additional molecules can pick up enough energy from collisions to climb the activation energy barrier and react.

If we double the concentration of C_4H_8 molecules in a sample with a given volume, there will be twice as many molecules per liter. Although the fraction of molecules with enough energy to react will be the same, the total number that can react will be twice as great. Consequently, the change in the amount of C_4H_8 per liter and thus the reaction rate ($-\Delta[C_4H_8]/\Delta t$) will be twice as great. The reaction rate is directly proportional to the concentration of the C_4H_8.

$$\text{Rate} = k\,[C_4H_8]$$

This relationship applies to any *unimolecular elementary reaction;* the reaction rate is directly proportional to the concentration of the reactant, and the reaction exhibits first-order behavior. The proportionality constant is the rate constant for the particular unimolecular reaction.

15.12   Bimolecular Reactions

A second commonly observed type of elementary reaction is a reaction that involves the collision *and combination* of two reactants to give an activated complex. Such a process is called a **bimolecular** reaction. Reaction (2) in Section 15.10 ($O + O_3 \longrightarrow 2O_2$) is an example of a bimolecular elementary reaction that occurs in a complex reaction mechanism. The reaction of NO_2 with CO (Section 15.8) and the decomposition of 2HI molecules to give H_2 and I_2 (Fig. 15-8) are examples of reactions that occur with a single bimolecular elementary reaction in the mechanism.

For the bimolecular elementary reaction, $A + B \longrightarrow$ products, the rate equation is first order in A and first order in B.

$$\text{Rate} = k\,[A][B]$$

In the reaction between the two reactants A and B, doubling the concentration of either A or B doubles the number of total molecular collisions and the number of effective collisions between molecules A and B. Thus if the concentration of A is doubled (and that of B is left the same), the rate of reaction is doubled. Doubling the concentration of both A and B quadruples the number of collisions and the

Figure 15-8 Probable mechanism for the dissociation of two HI molecules to produce one molecule of H_2 and one molecule of I_2.

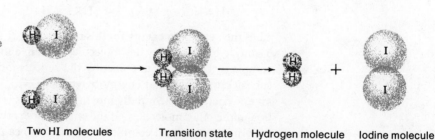

Two HI molecules   Transition state   Hydrogen molecule   Iodine molecule

Figure 15-9 A schematic representation of the effect of concentration on the number of possible collisions and hence the rate of reaction.

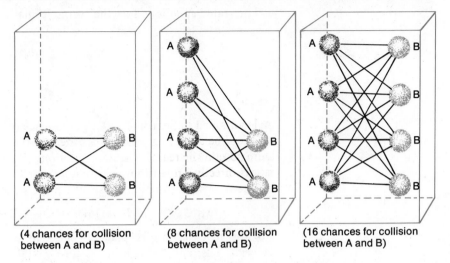

(4 chances for collision between A and B)   (8 chances for collision between A and B)   (16 chances for collision between A and B)

rate of reaction (Fig. 15-9). If the initial concentrations of both A and B are tripled, then the reaction proceeds nine times as fast. For the bimolecular elementary reaction A + A $\longrightarrow$ products, the rate equation is second order in A.

$$\text{Rate} = k\,[\text{A}][\text{A}] = k\,[\text{A}]^2$$

15.13   Termolecular Reactions

An elementary termolecular reaction involves the simultaneous collision of any combination of three atoms, molecules, or ions. Termolecular elementary reactions are rare because the probability of three particles colliding at one time is less than a thousandth that of two particles colliding. There are, however, a few established termolecular elementary reactions. The reaction of NO with O_2 ($2NO + O_2 \longrightarrow 2NO_2$), the reaction of NO with Cl_2 ($2NO + Cl_2 \longrightarrow 2NOCl$), and the reaction of H_2 with I_2 ($H_2 + I_2 \longrightarrow 2HI$) all involve termolecular steps.

The reaction of H_2 with I_2 looks as if it might involve a bimolecular reaction of H_2 with I_2. In fact, the reaction actually involves a hydrogen molecule and two iodine atoms (Fig. 15-10). The iodine atoms are produced by the dissociation of an iodine molecule ($I_2 \longrightarrow 2I$). Two iodine atoms combine with one hydrogen molecule to give a transition state which then splits to give two hydrogen iodide molecules.

Figure 15-10 The termolecular step in the mechanism for the reaction of H_2 and I_2 to produce two HI molecules. The first step (not shown) involves the dissociation of an iodine molecule into two iodine atoms ($I_2 \longrightarrow 2I$). In the termolecular step (shown here) two iodine atoms and one hydrogen molecule combine to produce two HI molecules.

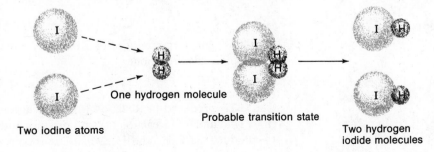

One hydrogen molecule

Probable transition state

Two iodine atoms

Two hydrogen iodide molecules

The rate equation for the termolecular elementary reaction $A + B + C \longrightarrow$ products is

$$\text{Rate} = k\,[A][B][C]$$

15.14   Reaction Mechanisms

As mentioned in Section 15.10, the sequence of elementary reactions (steps) that converts reactants into products is called the reaction mechanism, or reaction path, for a reaction. Since elementary reactions involving three or more reactants are rare, it is reasonable to suppose that a complex reaction such as that represented by the following equation must take place in several steps:

$$2MnO_4^- + 10Cl^- + 16H_3O^+ \longrightarrow 2Mn^{2+} + 5Cl_2 + 24H_2O$$

For this reaction to take place in one step, 2 permanganate ions (MnO_4^-), 10 chloride ions, and 16 hydronium ions would have to collide with one another simultaneously in an effective collision. This, of course, is very unlikely.

Even an apparently simple reaction such as the formation of ethylene (C_2H_4) from ethane (C_2H_6) may proceed by a complex mechanism. Ethylene is one of the most important reagents used by the chemical industry for manufacture of polyethylene, polyester fiber, synthetic rubber, and some detergents. Ethylene may be prepared by the pyrolysis (decomposition by heating) of ethane, a component of natural gas, at temperatures of 500°–800°. In the early stages of the pyrolysis, the reaction may be represented by the following equation:

$$C_2H_6 \longrightarrow C_2H_4 + H_2$$

The actual mechanism is much more complex than suggested by this equation and involves the following elementary reactions:

a. *Initiation.* The reaction begins with a unimolecular reaction when a C_2H_6 molecule splits at its weakest point, the C—C bond.

$$C_2H_6 \xrightarrow{\triangle} 2CH_3 \tag{1}$$

The CH_3 group abstracts a H atom from another C_2H_6 molecule in a bimolecular reaction.

$$CH_3 + C_2H_6 \longrightarrow CH_4 + C_2H_5 \tag{2}$$

b. *Propagation.* The C_2H_5 group undergoes a unimolecular dissociation.

$$C_2H_5 \longrightarrow C_2H_4 + H \tag{3}$$

The hydrogen atom produced by Reaction (3) reacts with another C_2H_6 molecule in a bimolecular reaction.

$$H + C_2H_6 \longrightarrow C_2H_5 + H_2 \tag{4}$$

The C_2H_5 produced in Reaction (4) reacts according to Equation (3) producing more H to undergo Reaction (4) and produce additional C_2H_5. Reactions (3) and (4) produce the C_2H_4 and H_2, which are the principal products of the pyrolysis.

Normally, each initiation [Equations (1) and (2)] is followed by about 100 cycles of Equations (3) and (4) before termination.

   c. *Termination.* The chain of Reactions (3) and (4) may be stopped by one of the following bimolecular reactions:

$$2C_2H_5 \longrightarrow C_4H_{10}$$
$$2C_2H_5 \longrightarrow C_2H_4 + C_2H_6$$

   Note that the C_2H_5 produced in Reaction (4) can react according to Reaction (3) to produce more H for Reaction (4). This provides the possibility of initiating once again the series of elementary reactions [Equations (3) and (4)] that can repeat themselves over and over. Such a mechanism, involving repeating reactions, is known as a **chain mechanism.**

   Some of the elementary reactions in a reaction path are slow relative to others. The slowest reaction step is the one that determines the maximum rate of a reaction because a stepwise reaction can proceed no faster than its slowest step. The slowest step, therefore, is referred to as the **rate-determining step** of the reaction.

   Whereas we can write the rate equation for each elementary reaction in a reaction mechanism (Sections 15.11–15.13), *we cannot ordinarily write a correct rate equation or establish the order for an overall reaction that involves several steps simply by inspection of the overall balanced equation.* It is necessary in such cases to determine the overall rate equation from experimental data. An excellent example is the reaction considered in Sections 15.6 and 15.8 for the overall reaction of NO_2 and CO:

$$NO_2 + CO \longrightarrow CO_2 + NO$$

For temperatures above 225° the reaction proceeds by a single one-step bimolecular elementary reaction (Section 15.8). For temperatures above 225°C, therefore, the reaction is first order with respect to NO_2 and first order with respect to CO. The rate equation is

$$\text{Rate} = k\,[NO_2][CO]$$

The overall order of the reaction is second order. It is interesting to note, however, that the reaction *at temperatures below 225°* proceeds by two elementary reactions, the first of which is slow and is therefore the rate-determining step on which the rate equation must be based.

$$NO_2 + NO_2 \longrightarrow NO_3 + NO \quad \text{(slow)}$$
$$NO_3 + CO \longrightarrow NO_2 + CO_2 \quad \text{(fast)}$$

(A summation of the two equations represents the net overall reaction.) Hence, for lower temperatures, the rate equation, based upon the first step as the rate-determining step, is

$$\text{Rate} = k\,[NO_2]^2$$

Note that the rate equation for the lower temperatures does not even include the concentration of CO. At temperatures below 225°C, therefore, the reaction is second order with respect to NO_2 and zero order with respect to CO; the overall order of the reaction still happens to be second order.

In general, when the rate-determining step is the first step in a reaction mechanism, the rate equation for the overall mechanism is identical to that for the rate-determining elementary reaction. However, when the rate-determining step occurs later in the mechanism, the rate equation may be complex. The oxidation of iodide ion by hydrogen peroxide illustrates this point.

$$H_2O_2 + 3I^- + 2H^+ \longrightarrow 2H_2O + I_3^-$$

In a solution that has a high concentration of acid, one reaction pathway is found to have the following rate equation:

$$\text{Rate} = k\,[H_2O_2][I^-][H^+]$$

This rate equation is consistent with several mechanisms, three of which follow:

Mechanism A

$H_2O_2 + H^+ + I^- \longrightarrow H_2O + HOI$	(slow)
$HOI + H^+ + I^- \longrightarrow H_2O + I_2$	(fast)
$I_2 + I^- \longrightarrow I_3^-$	(fast)

Mechanism B

$H^+ + I^- \longrightarrow HI$	(fast)
$H_2O_2 + HI \longrightarrow H_2O + HOI$	(slow)
$HOI + H^+ + I^- \longrightarrow H_2O + I_2$	(fast)
$I_2 + I^- \longrightarrow I_3^-$	(fast)

Mechanism C

$H^+ + H_2O_2 \longrightarrow H_3O_2^+$	(fast)
$H_3O_2^+ + I^- \longrightarrow H_2O + HOI$	(slow)
$HOI + H^+ + I^- \longrightarrow H_2O + I_2$	(fast)
$I_2 + I^- \longrightarrow I_3^-$	(fast)

In Mechanism A the slow step is the first elementary reaction, so the rate equation for the overall process is equal to the rate equation for this rate-determining step. In Mechanisms B and C the rate-determining step is the second elementary reaction in the mechanism, and the rate equation is not simply the rate equation for the rate-determining step. Mechanisms A, B, and C result in the same rate equation; hence it is not possible to distinguish between these mechanisms on the basis of the rate law alone. Additional experimental information is required to distinguish which path actually leads to product. Moreover, the rate equation provides no information about what happens after the rate-determining step. The following steps must be worked out from other chemical knowledge or from other measurements.

The determination of the mechanism of a reaction is important in selecting reaction conditions that provide a good yield of the desired product. A knowledge of the mechanism of a reaction sometimes makes it possible for the chemist to devise a suitable procedure for the preparation of a compound previously unknown. The study of reaction mechanisms is a very interesting but complex subject. Rather few reaction mechanisms have been completely characterized, relative to the number of chemical reactions that are known. This kind of study, referred to as a study of the **kinetics of reactions,** is a very active current research area.

15.15   Catalysts

The rate of many reactions can be accelerated by the presence of small amounts of substances that are not themselves permanently used up by the reaction. Such substances, called **catalysts** (Section 9.2), may be divided into two general classes: (1) **homogeneous catalysts,** which are present in the same phase as the reactants, and (2) **heterogeneous catalysts,** which are present in a different phase than the reactants.

A homogeneous catalyst interacts with a reactant, forming an intermediate substance which then decomposes, or reacts with more reactant, to regenerate the original catalyst and product. The ozone in the stratosphere that protects the earth from ultraviolet radiation is formed by the following process:

$$O_2(g) \quad \xrightarrow{h\nu} \quad 2O(g) \tag{1}$$

$$O(g) + O_2(g) \longrightarrow O_3(g) \tag{2}$$

As shown in Section 15.10, ozone decomposes ($2O_3 \longrightarrow 3O_2$) in the stratosphere by the mechanism

$$O_3(g) \longrightarrow O_2(g) + O(g) \tag{3}$$

$$O(g) + O_3(g) \longrightarrow 2O_2(g) \tag{4}$$

The rate of decomposition of ozone is also influenced by the presence of nitric oxide, NO, inasmuch as NO acts as a catalyst in the decomposition.

$$NO(g) + O_3(g) \longrightarrow NO_2(g) + O_2(g) \tag{5}$$

$$O_3(g) \longrightarrow O_2(g) + O(g) \tag{6}$$

$$NO_2(g) + O(g) \longrightarrow NO(g) + O_2(g) \tag{7}$$

The overall chemical change for Equations (5)–(7) also is $2O_3 \longrightarrow 3O_2$. Since the nitric oxide is not permanently used up in these reactions, it functions as a catalyst. The rate of decomposition of ozone is greater in the presence of nitric oxide due to the catalytic activity of NO. Under normal conditions, the rate at which ozone is formed in the atmosphere by Reactions (1) and (2) is equal to the rate at which it decomposes by Reactions (3) and (4) and (5)–(7)—a state of equilibrium exists; hence the total concentration of ozone does not change significantly over long periods. However, if additional nitric oxide were introduced into the stratosphere, perhaps by atmospheric nuclear explosions or from the exhaust of high-flying supersonic aircraft, the rate of decomposition would be increased and the total amount of ozone present would be reduced. Certain compounds that contain chlorine also catalyze the decomposition of ozone.

Many reactions of organic compounds are catalyzed by enzymes, which are complex substances produced by living organisms. A solution of pure sugar in pure water will not react with oxygen, even over a period of years. However, the enzyme zymase, which is produced by yeast cells, catalyzes the reaction between sugar and oxygen to form alcohol in a period of hours. The energy used by the body is produced, in part, by the oxidation of sugar by oxygen in a multistep mechanism that is catalyzed by a variety of enzymes. Many of the other chemical reactions that take place in living organisms also are catalyzed by enzymes. It has been estimated that there may be as many as 30,000 different enzymes in the human body, each of which is a protein constructed in such a way as to make it effective as a catalyst for a specific chemical reaction useful to the body. The subject of enzymes and their kinetic activity will be examined in greater detail in Chapter 26.

In contrast to the mode of action of homogeneous catalysts, heterogeneous catalysts act by furnishing a surface at which a reaction can occur. The characteristic feature of gas- and liquid-phase reactions catalyzed by heterogeneous catalysts is the fact that the reactions occur on the surface of the catalyst rather than within the gas or liquid phase. For this reason heterogeneous catalysts are times called contact catalysts. Heterogeneous catalysis is characterized by at least four steps: (1) adsorption of the reactant onto the surface of the catalyst, (2) activation of the adsorbed reactant, (3) reaction of the adsorbed reactant, and (4) diffusion of the product from the surface into the gas or liquid phase (desorption). Any one of these steps may be slow and thus be the rate-determining step, but in general the overall rate of the reaction is faster than that between the reactants in the gas or liquid phases. The steps that are believed to occur in the reaction of compounds containing a carbon-carbon double bond with hydrogen on a nickel catalyst are illustrated in Fig. 15-11. This is the catalyst used in the hydrogenation of edible fats and oils containing several carbon-carbon double bonds (polyunsaturated fats and oils) to edible fats and oils containing only single carbon-carbon bonds (saturated fats and oils).

Other significant industrial processes involving the use of contact catalysts include the preparation of sulfuric acid from sulfur and oxygen using vanadium(V) oxide (Section 21.14); the preparation of ammonia from hydrogen and nitrogen using a mixture of iron oxide and potassium aluminate (Section 23.6); the oxidation of ammonia to nitric acid using a mixture of platinum and rhodium (Section 23.15); and the synthesis of methanol, CH_3OH, from carbon monoxide and hydrogen using silver or copper (Section 25.18).

Both homogeneous and heterogeneous catalysts function by providing a reaction path with a lower activation energy than in the absence of the catalyst (Fig. 15-12). This lower activation energy results in an increase in rate (Section 15.9). Note that a catalyst decreases the activation energy for both the forward and the reverse reactions and hence *accelerates both the forward and the reverse reactions.*

Some substances, called **inhibitors,** decrease the rate of a chemical reaction. In many cases these are substances that react with and "poison" some catalyst in the system and thereby prevent its action. For example, the presence of certain organic substances, in very small amounts, greatly decreases the rate of deterioration of rubber. These organic substances are complexing agents (see Chapter 31)

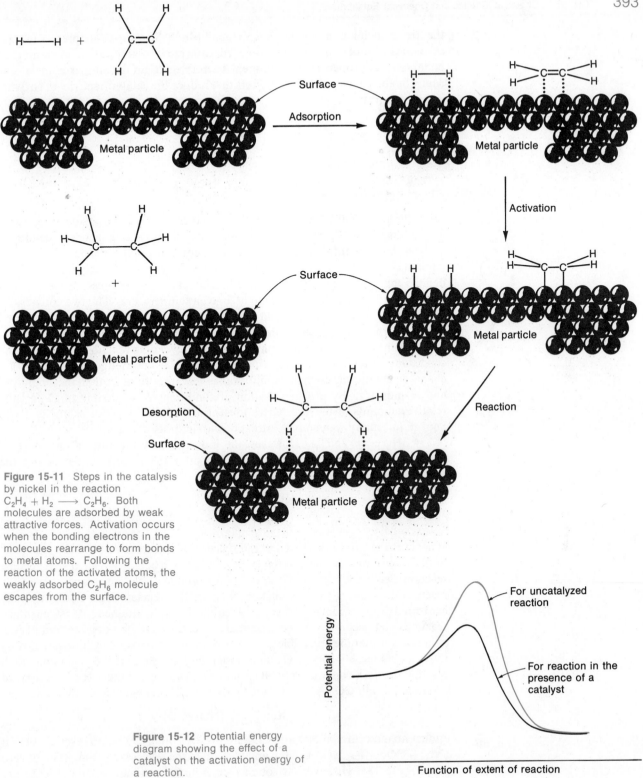

Figure 15-11 Steps in the catalysis by nickel in the reaction $C_2H_4 + H_2 \longrightarrow C_2H_6$. Both molecules are adsorbed by weak attractive forces. Activation occurs when the bonding electrons in the molecules rearrange to form bonds to metal atoms. Following the reaction of the activated atoms, the weakly adsorbed C_2H_6 molecule escapes from the surface.

Figure 15-12 Potential energy diagram showing the effect of a catalyst on the activation energy of a reaction.

that tie up such ions as Cu^{2+}, Fe^{3+}, V^{3+}, and Mn^{2+}. These ions are present in trace amounts in rubber and, if not complexed, catalyze its deterioration. A complexed metal ion has properties very different from those of an uncomplexed metal ion. Many biological poisons are inhibitors that reduce the catalytic activity of enzymes in an organism.

Chemical Equilibrium

15.16   The State of Equilibrium

A condition of equilibrium is reached in a system when the conversion of reactants into products and the conversion of products back into reactants occur simultaneously at the same rate. As a consequence, at equilibrium there is no change in the amounts of reactants and products present in a system. For example, we have defined the vapor pressure of a liquid (Section 11.2) as the pressure exerted by a vapor in equilibrium with its liquid at a given temperature. The two conversions in this case are evaporation and condensation, and at equilibrium there is no change in the quantity of liquid or vapor.

The point at which equilibrium occurs in a reversible reaction varies with the conditions under which the reaction takes place. A knowledge of the effect of a change in conditions upon an equilibrium is very important, both in the laboratory and in industry. By proper selection of conditions it is possible to control the relative amounts of the substances present at equilibrium. For example, it was known for many years that nitrogen and hydrogen react to form ammonia ($N_2 + 3H_2 \rightleftharpoons 2NH_3$). However, the industrial manufacture of ammonia by means of this reaction was achieved only after the factors that influence the equilibrium were understood (Section 23.6).

15.17   Law of Mass Action

Whenever the products of a chemical reaction can react to re-form the reactants, the two reactions occur simultaneously. Such a reaction is called a reversible reaction (Section 9.18). As long as one of the products does not escape, a reversible reaction does not go to completion; that is, the reactants are not completely converted into products. A state of equilibrium is attained. Most chemical reactions that occur in a closed system are reversible and do not go to completion.

Let us consider the reversible oxidation of carbon monoxide by nitrogen dioxide ($NO_2 + CO \rightleftharpoons NO + CO_2$) in a closed flask at 300°. At the beginning of the reaction the flask contains only NO_2 and CO. Above 225° the rate of the forward reaction, $NO_2 + CO \longrightarrow NO + CO_2$ (Section 15.14), is given by the expression

$$\text{Rate}_1 = k_1[NO_2][CO]$$

At first no molecules of NO and CO_2 are present, there can be no reverse reaction ($NO + CO_2 \longrightarrow NO_2 + CO$), and the rate of the reverse reaction, Rate_2, is zero (see Fig. 15-13). However, as soon as some of the products (NO and CO_2) are

Figure 15-13 Rates of reaction of forward and reverse reactions for $NO_2 + CO \rightleftharpoons NO + CO_2$, assuming only NO_2 and CO are present initially.

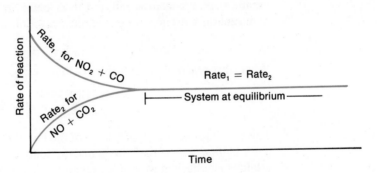

formed they begin to react. The rate of this reaction above $225°$ is given by

$$\text{Rate}_2 = k_2[NO][CO_2]$$

Because the initial molar concentrations of NO and CO_2 are small, the rate of the reverse reaction, Rate_2, will be relatively slow. But as the reaction between NO_2 and CO proceeds, the concentrations of NO and CO_2 will increase and, likewise, the rate of the reverse reaction will increase. Meanwhile, the concentrations of NO_2 and CO are becoming less and less, so that the rate of the forward reaction, Rate_1, falls off. Consequently, the two reaction rates approach each other and finally become equal—a condition of **dynamic equilibrium,** which means that the opposing reactions are in full operation but at the same rate. Thus at equilibrium $\text{Rate}_2 = \text{Rate}_1$, and we may write

$$k_2[NO][CO_2] = k_1[NO_2][CO]$$

or, by rearranging,

$$\frac{[NO][CO_2]}{[NO_2][CO]} = \frac{k_1}{k_2}$$

Because k_1 and k_2 are constants, the ratio k_1/k_2 is also constant, and the expression may be written

$$\frac{[NO][CO_2]}{[NO_2][CO]} = K$$

K is called the **equilibrium constant** for the reaction. Just as the rate constants k_1 and k_2 are specific for each reaction at a definite temperature, K is likewise a constant specific to this system in equilibrium at a given temperature. The values for the molar concentrations used to evaluate K from the above mathematical expression must always be the concentrations present after the reaction has reached a state of equilibrium.

It is important to note that at equilibrium the rates of reaction, Rate_1 and Rate_2, are equal, but the molar concentrations of the reactants and products in the equilibrium mixture are usually not equal. It is true, however, that the individual concentrations of each reactant and product *at equilibrium* remain constant because the rate at which any one reactant is being used up in one reaction is equal to the rate at which it is being formed by the opposite reaction. As pointed out before, such a system is referred to as being in a state of dynamic equilibrium. A general equation for a chemical reaction may be written

$$m\text{A} + n\text{B} + \cdots \rightleftharpoons x\text{C} + y\text{D} + \cdots$$

When this reaction has reached equilibrium

$$\frac{[\text{C}]^x[\text{D}]^y \cdots}{[\text{A}]^m[\text{B}]^n \cdots} = K$$

This is a mathematical expression of the **law of chemical equilibrium,** or **the law of mass action,** which may be stated as follows: **When a reversible reaction has attained equilibrium at a given temperature, the product of the molar concentrations of the substances to the right of the arrow in the equation divided by the product of the molar concentrations of the substances to the left, each concentration raised to the power equal to the number of molecules of each substance appearing in the equation, is a constant.**

To be precise, the law of mass action should be expressed in terms of the activities of the reactants and products (Section 13.27) rather than their concentrations. However, as we have seen, the activity of a dilute solute is closely approximated by its molar concentration, so that concentrations are commonly used in defining equilibrium constants involving dissolved species. The activity of a gas is approximated by its pressure (in atmospheres). Since the molar concentration of a gas is directly proportional to its pressure at constant temperature, either the pressure of a gas or its molar concentration can be used in the expression for an equilibrium constant. The activity of a pure solid or of a solvent is defined as 1.

The equilibrium constant expression is valid even if the reactions take place in steps, with rates involving powers of the reactant concentrations differing from the coefficients appearing in the balanced equation. Below 225°, the reaction of NO_2 with CO occurs by a two-step mechanism (Section 15.14). The rate equation for the forward reaction below 225° is

$$\text{Rate}_1 = k_1[\text{NO}_2]^2$$

The rate equation for the reverse reaction below 225° is more complicated than most we have seen in this chapter.

$$\text{Rate}_2 = k_2[\text{NO}_2][\text{CO}_2][\text{NO}][\text{CO}]^{-1}$$

Thus at equilibrium $\text{Rate}_2 = \text{Rate}_1$ and

$$k_2[\text{NO}_2][\text{CO}_2][\text{NO}][\text{CO}]^{-1} = k_1[\text{NO}_2]^2$$

or, by rearranging

$$\frac{[\text{NO}_2][\text{CO}_2][\text{NO}][\text{CO}]^{-1}}{[\text{NO}_2]^2} = \frac{[\text{CO}_2][\text{NO}]}{[\text{NO}_2][\text{CO}]} = \frac{k_1}{k_2} = K$$

This equation is identical to that obtained from the law of mass action and to that for the equilibrium constant above 225°.

The meaning of the mathematical expression of the law of mass action is that, regardless of how the individual concentrations might be varied, thereby upsetting the equilibrium temporarily, the composition of the system with regard to its various components will *at a constant temperature* always adjust itself to a new condition of equilibrium for which the quotient $[C]^x[D]^y \cdots /[A]^m[B]^n \cdots$ will again have the value K. When a mixture of A, B, C, and D is prepared in such proportions that the quotient $[C]^x[D]^y \cdots /[A]^m[B]^n \cdots$ is not equal to K, then the system is not in a condition of equilibrium. However, its composition will change in such a direction that equilibrium will be established. If the ratio is less than K, the rate of reaction of A with B is greater than that of C with D, so that A and B will be used up faster than they are formed until the ratio becomes equal to K; conversely, if the ratio is greater than K, the rate of reaction of C and D will be greater than that of A with B until the ratio becomes equal to K.

The value of an equilibrium constant is a measure of the completeness of a reversible reaction. A large value for K indicates that equilibrium is attained only after the reactants A and B have been largely converted into the products C and D. When K is very small—much less than unity—equilibrium is attained when only a small proportion of A and B has been converted to C and D.

The following examples illustrate the form of the equilibrium constant expression for two other reactions:

$$3O_2 \rightleftharpoons 2O_3 \qquad K = \frac{[O_3]^2}{[O_2]^3}$$

$$3H_2 + N_2 \rightleftharpoons 2NH_3 \qquad K = \frac{[NH_3]^2}{[H_2]^3[N_2]}$$

The use of square brackets indicates that molar concentrations (mol ℓ^{-1}) are used in these equations.

15.18   Determination of Equilibrium Constants

Although the values of equilibrium constants can be determined from the forward and reverse rate constants for a reaction, they are more commonly determined by measuring, *at equilibrium,* the concentrations of reactants and products of a reaction. As one example of the determination of an equilibrium constant, let us consider the reversible reaction $H_2(g) + I_2(g) \rightleftharpoons 2HI(g)$. [As mentioned in Section 15.17, the law of mass action applies, and the equilibrium constant expression is valid, even if the reaction takes place in steps as is the case with this particular reaction (Section 15.13).]

In order to determine the equilibrium constant for the reaction of hydrogen with gaseous iodine, an equimolar mixture of hydrogen and iodine was heated at 400° until no further change in the concentration of H_2, I_2, or HI was observed. At this point it was assumed that equilibrium had been reached, and it was found by analysis that $[H_2] = 0.221\ M$, $[I_2] = 0.221\ M$, and $[HI] = 1.563\ M$. Substituting these values in the expression for the equilibrium constant for this system, we find

$$K = \frac{[HI]^2}{[H_2][I_2]} = \frac{(1.563\ \text{mol}\ \ell^{-1})^2}{(0.221\ \text{mol}\ \ell^{-1})(0.221\ \text{mol}\ \ell^{-1})} = 50.0 \qquad \text{(units cancel out)}$$

If we start with pure HI at any molar concentration, or with any mixture of H_2 and I_2, or with any mixture of H_2, I_2, and HI and hold the temperature of the system at 400° until equilibrium is established, the molar concentrations of the three substances will have changed so that the quotient $[HI]^2/[H_2][I_2]$ is equal to 50.0.

As the following example illustrates, it is not always necessary to measure the concentration of each species present at equilibrium. The equation for the chemical reaction can be used to relate their concentrations.

EXAMPLE   **Iodine molecules react reversibly with iodide ions to produce triiodide ions, $I_2(aq) + I^-(aq) \rightleftharpoons I_3^-(aq)$. If a liter of solution prepared from 0.1000 mol of I_2 and 0.1000 mol of the strong electrolyte KI contains 9.33×10^{-2} mol of iodine at equilibrium, what is the equilibrium constant for the reaction?**

From the law of mass action we know the equilibrium constant is given by the equation

$$K = \frac{[I_3^-]}{[I_2][I^-]}$$

In order to determine the value of K we need to know the values of $[I_3^-]$, $[I_2]$, and $[I^-]$ at equilibrium. The data tell us that at equilibrium $[I_2] = 9.33 \times 10^{-2}$ M. The other two concentrations must be determined from the other data given in the problem.

At the start of the reaction (when the solution was prepared) the concentration of I_2, $[I_2]_i$, was 0.1000 M; however, it decreased to 9.33×10^{-2} M at equilibrium. The change in concentration of I_2 ($[I_2]_i - [I_2]$) was 0.1000 $M - 9.33 \times 10^{-2}$ $M = 6.7 \times 10^{-3}$ M. Thus 6.7×10^{-3} mol of I_2 per liter of solution was consumed as the reaction proceeded to equilibrium. From the chemical equation we know that one mole of I_3^- is formed for each mole of I_2 that reacts. Thus the concentration of I_3^- at equilibrium, $[I_3^-]$, must be

$$\frac{6.7 \times 10^{-3} \text{ mol } I_2}{1\,\ell} \times \frac{1 \text{ mol } I_3^-}{1 \text{ mol } I_2} = \frac{6.7 \times 10^{-3} \text{ mol } I_3^-}{1\,\ell} = 6.7 \times 10^{-3}\ M$$

KI is a strong electrolyte; it dissociates completely into K^+ and I^- ions. Thus at the start of the reaction the concentration of I^-, $[I^-]_i$, was 0.1000 M. (From the formula of KI we know that there is one mole of I^- per mole of KI.) The concentration of I^- at equilibrium, $[I^-]$, is that amount that did not react to form I_3^-. Since the chemical equation tells us that one mole of I^- reacts per mole of I_3^- formed, the amount of I^- that reacts is equal to the amount of I_3^- formed, 6.7×10^{-3} M. Therefore

$$[I^-] = [I^-]_i - [I_3^-] = 0.1000\ M - (6.7 \times 10^{-3}\ M) = 9.33 \times 10^{-2}\ M$$

Now we need only substitute these equilibrium concentrations into the expression for the equilibrium constant and evaluate it.

$$K = \frac{[I_3^-]}{[I_2][I^-]} = \frac{(6.7 \times 10^{-3} \text{ mol } \ell^{-1})}{(9.33 \times 10^{-2} \text{ mol } \ell^{-1})(9.33 \times 10^{-2} \text{ mol } \ell^{-1})} = 0.770 \text{ mol}^{-1}\ \ell$$

◆ ◆ ◆ ◆ ◆

Units for equilibrium constants (mol^{-1} ℓ in the example) differ depending on the form of the particular mathematical expression for the equilibrium constant. In some cases they may even cancel out, as in the HI example near the beginning of this section. It is common practice not to include units with the value of K, and this practice will be followed henceforth in this book.

15.19   Effect of Change of Concentration on Equilibrium

A chemical system may be shifted out of equilibrium by increasing the rate of the forward or the reverse reaction. Consider the reversible reaction

$$A + B \rightleftharpoons C + D$$

When the system is in equilibrium and an additional quantity of A is added, the rate of the forward reaction is increased because the concentration of the reacting molecules is increased (Fig. 15-14). This means that the rate of the forward reaction will for a time be greater than that of the reverse reaction; the system, then, is temporarily out of equilibrium. However, as the concentrations of C and D increase, the rate of the reverse reaction increases, whereas the decrease in the concentrations of A and B causes the rate of the forward reaction to decrease. The rates of the two reactions thereby will become equal again, and a second state of equilibrium attained in which the molar concentrations of A, B, C, and D have changed; however, the ratio [C][D] to [A][B] is again equal to the original value of K. The equilibrium is said to have been shifted to the right. In the new state of equilibrium, the substances C and D are present in greater concentration than originally, B is present in smaller concentration than originally, and A is present in greater concentration than before the addition of excess A. By increasing the concentration of B, the equilibrium can be shifted to the right in similar fashion. Increasing the concentration of either C or D, or both, will have the effect of shifting the equilibrium to the left.

The equilibrium may also be shifted to the right by the removal of either C or D, or both. This removal of products causes a reduction in the reaction rate of the reaction to the left, and thus the reaction to the left proceeds more slowly than the reaction to the right until the reaction rates again become equal as a new condition of equilibrium is attained.

Figure 15-14 The effect of adding reactant A on the rates of the forward and reverse reactions for $A + B \rightleftharpoons C + D$ when the system is initially at equilibrium.

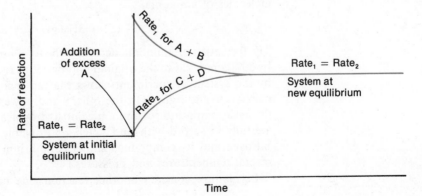

As an example of the effect of a change in concentration upon a system in equilibrium, let us consider the equilibrium $H_2 + I_2 \rightleftharpoons 2HI$, for which K was found to be 50.0 at 400° with $[H_2] = [I_2] = 0.221 \, M$ and $[HI] = 1.563 \, M$ (Section 15.18). If enough H_2 is introduced quickly into the system to double its concentration, the rate of reaction of H_2 with I_2 to form HI increases. When equilibrium is again reached, $[H_2] = 0.374 \, M$, $[I_2] = 0.153 \, M$, and $[HI] = 1.692 \, M$. If these new values are substituted in the expression for the equilibrium constant for this system, we find

$$\frac{[HI]^2}{[H_2][I_2]} = \frac{(1.692)^2}{(0.374)(0.153)} = 50.0 = K$$

Hence, by doubling the concentration of H_2, we have caused the formation of more HI, used up about one-third of the I_2 present at the first equilibrium and used up some, but not all, of the excess H_2 added.

The effect of a change in concentration upon a system in equilibrium is an important application of the **principle of Le Châtelier,** which may be stated as follows: **If a stress (such as a change in concentration, pressure, or temperature) is applied to a system in equilibrium, the equilibrium shifts in a way that tends to undo the effect of the stress.**

15.20   Effect of Change in Pressure on Equilibrium

Changes in pressure measurably affect systems in equilibrium only when gases are involved and, in such cases, only when the chemical reaction involves a change in the total number of molecules in the system. This follows from Avogadro's Law, which states that equal volumes of gases at constant temperature and pressure contain equal numbers of molecules (see Section 10.7). As the pressure on a gaseous system is increased, the gases composing the system undergo compression, and their molar concentrations and thus the total number of molecules per unit volume increase. In accord with the principle of Le Châtelier, the chemical reaction that reduces the total number of molecules per unit of volume will be the one favored by an increase in pressure.

Consider the effect of an increase in pressure upon the system in which one molecule of nitrogen and three molecules of hydrogen interact to form two molecules of ammonia.

$$N_2(g) + 3H_2(g) \rightleftharpoons 2NH_3(g)$$

The formation of ammonia decreases the total number of molecules of the system by 50% compared to the original number, thus reducing the total pressure exerted by the system. Experiment shows that an increase in pressure does drive the reaction to the right. On the other hand, lowering the pressure on the system favors decomposition of ammonia into hydrogen and nitrogen. These observations are fully in accord with Le Châtelier's Principle. (Table 23-2 in the chapter on nitrogen and its compounds shows equilibrium concentrations of ammonia at several temperatures and pressures.)

Let us now consider another reaction, one in which a molecule of nitrogen interacts with a molecule of oxygen with the formation of two molecules of nitric oxide.

$$N_2(g) + O_2(g) \rightleftharpoons 2NO(g)$$

Because there is no change in the total number of molecules in the system during reaction, a change in pressure does not favor either formation or decomposition of nitric oxide at temperatures high enough for all the materials to be present in the gaseous state.

Whenever gases are involved in a system in equilibrium, the pressure of each gas can be substituted for its concentration in the expression for the equilibrium constant because the concentration of a gas at constant temperature varies directly with the pressure. Thus for the system

$$N_2(g) + 3H_2(g) \rightleftharpoons 2NH_3(g)$$

we may write

$$\frac{p_{NH_3}^2}{p_{N_2}p_{H_2}^3} = K_p$$

It is important to note that when partial pressures of gases are substituted for their concentrations the equilibrium constant, K_p, will still be a constant, but it will usually have a different numerical value and different units from those when concentrations are used. Thus an equilibrium constant using gas pressures will be indicated with a subscript p in this text.

The units for K_p for the reaction to produce ammonia, if the three partial pressures are expressed in atmospheres, would be

$$\frac{atm^2}{atm \times atm^3} = \frac{1}{atm^2} = atm^{-2}$$

Although equilibrium constants have units (unless they happen to cancel out as they do in the reaction $N_2 + O_2 \rightleftharpoons 2NO$), the values for equilibrium constants are often given without the units.

EXAMPLE 1   A vessel contains gaseous CO, CO_2, H_2, and H_2O in equilibrium at 980°. The pressures of these gases are CO, 0.150 atm; CO_2, 0.200 atm; H_2, 0.090 atm; H_2O, 0.200 atm. Hydrogen is pumped into the vessel and the equilibrium pressure of CO changes to 0.230 atm. Calculate the partial pressures of the other substances at the new equilibrium assuming that no change in temperature occurs.

$$CO_2(g) + H_2(g) \rightleftharpoons CO(g) + H_2O(g)$$

Before we can calculate pressures after the addition of hydrogen, we need the equilibrium constant. This can be determined from the equilibrium pressures of the reactants and products before the additional hydrogen is added.

$$K_p = \frac{p_{CO}p_{H_2O}}{p_{CO_2}p_{H_2}} = \frac{(0.150\ atm)(0.200\ atm)}{(0.200\ atm)(0.090\ atm)} = 1.67$$

(Units cancel out, and the equilibrium constant in this particular case has no units.)

At the new equilibrium $p_{CO} = 0.230$ atm. Therefore p_{CO} has increased in proportion to the increased concentration, from 0.150 atm to 0.230 atm, an increase of 0.080 atm.

The balanced equation tells us that for each increase of one mole of CO, the H_2O concentration must also increase by one mole; hence, the pressure increase for H_2O from its initial pressure of 0.200 atm will be the same (0.080 atm) as for CO. Thus p_{H_2O} at the new equilibrium is 0.200 atm + 0.080 atm = 0.280 atm. The balanced equation also tells us that for every mole of CO produced, one mole of CO_2 must be consumed, and hence the pressure of CO_2 must decrease from its initial pressure of 0.200 atm by the same amount as the increase of pressure for CO (0.080 atm). Thus p_{CO_2} at the new equilibrium is 0.200 atm − 0.080 atm = 0.120 atm.

Now the partial pressure of hydrogen at the new equilibrium can be calculated by substituting p_{CO}, p_{H_2O}, and p_{CO_2} for the new equilibrium into the expression for K_p and solving for p_{H_2}:

$$K_p = \frac{(0.230 \text{ atm})(0.280 \text{ atm})}{(0.120 \text{ atm})p_{H_2}} = 1.67$$

$$p_{H_2} = \frac{(0.230 \text{ atm})(0.280 \text{ atm})}{(0.120 \text{ atm})(1.67)} = 0.321 \text{ atm}$$

Therefore the partial pressures for CO, H_2O, CO_2, and H_2 at the new equilibrium are 0.230 atm, 0.280 atm, 0.120 atm, and 0.321 atm, respectively. Note that sufficient hydrogen had to be pumped into the vessel to increase the pressure of the hydrogen from 0.090 atm to 0.321 atm.

◆ ◆ ◆ ◆ ◆

EXAMPLE 2   In a 3.0-ℓ vessel, the following equilibrium partial pressures are measured: N_2, 0.380 atm; H_2, 0.400 atm; NH_3, 2.000 atm. Hydrogen is removed from the vessel until the partial pressure of nitrogen, at equilibrium, is equal to 0.450 atm. Calculate the partial pressures of the other substances under the new conditions.

$$N_2(g) + 3H_2(g) \rightleftharpoons 2NH_3(g)$$

$$K_p = \frac{p_{NH_3}^2}{p_{N_2}p_{H_2}^3}$$

$$K_p = \frac{(2.000 \text{ atm})^2}{(0.380 \text{ atm})(0.400 \text{ atm})^3} = 1.645 \times 10^2 \text{ atm}^{-2}$$

The removal of H_2 shifts the equilibrium to the left, producing additional N_2 and therefore increasing the pressure of N_2 in proportion to the increased molar concentration. At the new equilibrium $p_{N_2} = 0.450$ atm. Hence the pressure of N_2 has increased by 0.070 atm.

The balanced equation tells us that two moles of NH_3 must be used up for each mole of N_2 that is produced. The pressure of NH_3 must decrease in proportion to the decrease in molar concentration. The pressure of NH_3 must therefore decrease by twice the amount that the pressure of N_2 increases; thus the pressure of NH_3 must decrease by 0.140 atm. At the new equilibrium $p_{NH_3} = 2.000 - 0.140 = 1.860$ atm.

To calculate p_{H_2} at the new equilibrium, substitute p_{N_2} and p_{NH_3} in the expression for K_p:

$$K_p = \frac{(1.860 \text{ atm})^2}{(0.450 \text{ atm})p_{H_2}^3} = 1.645 \times 10^2 \text{ atm}^{-2}$$

$$p_{H_2}^3 = \frac{(1.860 \text{ atm})^2}{(0.450 \text{ atm})(1.645 \times 10^2 \text{ atm}^{-2})}$$

$$= 4.674 \times 10^{-2} \text{ atm}^3 = 46.74 \times 10^{-3} \text{ atm}^3$$

$$p_{H_2} = \sqrt[3]{46.74 \times 10^{-3} \text{ atm}^3} = 0.360 \text{ atm}$$

Hence at the new equilibrium the partial pressures of NH_3 and H_2 are 1.860 atm and 0.360 atm, respectively. Note that the quantity of hydrogen that was removed is the amount that would reduce the pressure of the hydrogen from 0.400 to 0.360 atm.

◆ ◆ ◆ ◆ ◆

15.21   Effect of Change in Temperature on Equilibrium

All chemical changes involve either the evolution of energy or the absorption of energy. In every system in equilibrium, an endothermic and an exothermic reaction are taking place simultaneously. The endothermic reaction ($\Delta H > 0$) is favored by an increase in temperature, i.e., an increase in energy, and the exothermic reaction ($\Delta H < 0$) is favored by a decrease in temperature.

The effect of temperature changes upon systems in equilibrium may be summarized by the statement: **When the temperature of a system in equilibrium is raised, the equilibrium is displaced in such a way that heat is absorbed.** This generalization is known as **van't Hoff's Law,** which is a special case of Le Châtelier's Principle. It is important to remember that the effect of changing the temperature is a change in the value of the equilibrium constant.

In the reaction between gaseous hydrogen and gaseous iodine, heat is evolved.

$$H_2(g) + I_2(g) \rightleftharpoons 2HI(g) \qquad \Delta H = -9.4 \text{ kJ}$$

Lowering the temperature of the system favors the formation of hydrogen iodide, whereas raising the temperature favors the decomposition of hydrogen iodide. Thus raising the temperature of the system will decrease the value of the equilibrium constant, since the molar concentration of HI at the new equilibrium will be decreased while the molar concentrations of H_2 and I_2 will be increased. The value of the equilibrium constant

$$\frac{[HI]^2}{[H_2][I_2]} = K$$

decreases from 67.5 at 357° to 50.0 at 400°.

The equation for the formation of ammonia from hydrogen and nitrogen is

$$N_2 + 3H_2 \rightleftharpoons 2NH_3 \qquad \Delta H = -92.2 \text{ kJ}$$

Since the reaction is exothermic, the equilibrium can be shifted to the right to favor the formation of more ammonia by lowering the temperature (see Table 23-2). However, it must be remembered that, because of the large decrease of reaction rates with decreasing temperature, equilibrium is attained more slowly at the lower temperature. In the commercial production of ammonia from nitrogen and hydrogen, it is not feasible to use temperatures much lower than 500°, because at such temperatures, even in the presence of a catalyst, the reaction proceeds too slowly to be practical.

15.22   Effect of a Catalyst on Equilibrium

Iron powder is used as a catalyst in the production of ammonia from nitrogen and hydrogen to increase the rate of reaction of these two elements.

$$N_2 + 3H_2 \xrightarrow[\text{Fe}]{\triangle} 2NH_3$$

However, this same catalyst serves equally well to increase the rate of the reverse reaction, that is, the decomposition of ammonia into its constituent elements.

$$2NH_3 \xrightarrow[\text{Fe}]{\triangle} N_2 + 3H_2$$

Thus the net effect of iron in the reversible reaction

$$N_2 + 3H_2 \rightleftharpoons 2NH_3$$

is to cause equilibrium to be reached more rapidly. *Catalysts have no effect on the value of the equilibrium constant.* They merely increase the rate of both the forward and the reverse reactions to the same extent so that equilibrium is reached more rapidly when a catalyst is present.

15.23   Homogeneous and Heterogeneous Equilibria

A **homogeneous system,** or a **homogeneous phase,** is a portion of matter in which any sample of the matter has the same composition. A phase may be a solid, a liquid, or a gas. Furthermore, any phase may be composed of an element, a compound, or a homogeneous mixture (e.g., a solution of two or more compounds). **A homogeneous equilibrium** is an equilibrium within a single homogeneous phase. Most of the equilibria that have been considered in this chapter are homogeneous equilibria involving reversible changes in only one phase, the gas phase.

A **heterogeneous equilibrium** is an equilibrium between two or more different phases. Liquid water in equilibrium with ice, liquid water in equilibrium with water vapor, and a solid in contact with its saturated solution are examples of heterogeneous equilibria. Each of these equilibria involves some kind of boundary surface between two phases.

Suppose we consider the example of a heterogeneous equilibrium provided by the decomposition of calcium carbonate into calcium oxide and carbon dioxide according to the equation

$$CaCO_3(s) \rightleftharpoons CaO(s) + CO_2(g)$$

Let us write the expression for the equilibrium constant for this reversible reaction.

$$\frac{[CaO][CO_2]}{[CaCO_3]} = K$$

The concentration of a solid substance is proportional to its density and hence remains constant as long as the temperature is not changed. It follows then that the concentrations of whatever solid substances are involved in the equilibrium, since they are constant, may be included in the value of the equilibrium constant, and need not appear at all in the expression for the equilibrium constant. We may write

$$[CO_2] = \frac{[CaCO_3]}{[CaO]} \times K$$

Now, we have on the right-hand side of the equation only constant terms. These can be combined into a single constant, K.

$$[CO_2] = K$$

The expression for the equilibrium constant in terms of pressure is written

$$p_{CO_2} = K_p$$

This equation means that at a given temperature there is only one pressure at which carbon dioxide gas can be in equilibrium with the two solids $CaCO_3$ and CaO. Suppose a sample of $CaCO_3$ is placed in a cylinder with a movable piston, as shown in Fig. 15-15, and that the temperature of the system is raised to 900°. Under these conditions calcium carbonate will continue to decompose into calcium oxide and carbon dioxide until there is sufficient carbon dioxide present in the system to exert a pressure equal to K_p (actually 1.04 atm), if the piston is held stationary.

$$p_{CO_2} = K_p = 1.04 \text{ atm} \quad \text{(at 900°)}$$

At this pressure, equilibrium will have been attained, and $CaCO_3$ will decompose into CaO and CO_2 at the same rate that CaO and CO_2 react to produce $CaCO_3$. If we increase the pressure by pushing the piston down and thus compressing the gas, more carbon dioxide will combine with calcium oxide to form calcium carbonate, and the pressure will drop to 1.04 atm again. If, on the other hand, we decrease the pressure on the system by raising the piston, just enough calcium carbonate will decompose to bring the pressure exerted by the carbon dioxide back to its original value of 1.04 atm.

In the commercial production of quicklime, CaO, from limestone the carbon dioxide is continuously removed as fast as it is formed by means of a stream of air; this causes the reaction to proceed essentially only to the right. Since the pressure of carbon dioxide never reaches the equilibrium pressure, equilibrium is never established.

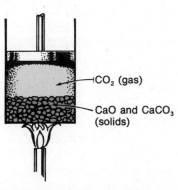

CO_2 (gas)

CaO and CaCO_3 (solids)

Figure 15-15 The thermal decomposition of calcium carbonate in a closed system is an example of a heterogeneous equilibrium.

Tabulated values for equilibrium constants of reactions involving pure gases in equilibrium with pure liquids and/or pure solids are the K_p values based on pressures.

Questions

1. Explain how each of the factors that determine the rate of a reaction is responsible for changing the rate.
2. Account for the increase in reaction rate brought about by catalysts.
3. One liter of a one-molar solution of H_2O_2 slowly decomposes into H_2O and O_2. If 0.50 mol of H_2O_2 decomposes during the first six hours of the reaction, explain why only 0.25 mol decomposes during the next six-hour period.
4. How do the rate of a reaction and its rate constant differ?
5. Doubling the concentration of a reactant increases the rate of a reaction four times. What is the order of the reaction with respect to that reactant?
6. Can we predict the effect of doubling the concentration of A on the rate of the overall reaction $A + B \longrightarrow C$? Can we predict the effect if the reaction is known to be an elementary reaction?
7. What is the effect on the rate of many reactions caused by an increase in temperature of 10°? Explain this effect in terms of the collision theory of reaction rate.
8. What is an activated complex?
9. What is the activation energy of a reaction?
10. Account for the relationship between the rate of a reaction and its activation energy.
11. Define the following: unimolecular reaction, bimolecular reaction, elementary reaction, overall reaction.
12. What is the rate equation for the termolecular reaction $A + 2B \longrightarrow$ products?
13. Why are elementary reactions involving three or more reactants very uncommon?
14. Describe how a homogeneous catalyst and a heterogeneous catalyst function.
15. Using the law of mass action, derive the mathematical expression of the law of chemical equilibrium for the following reversible reactions:

$$N_2(g) + 3H_2(g) \rightleftharpoons 2NH_3(g)$$
$$CH_4(g) + Cl_2(g) \rightleftharpoons CH_3Cl(g) + HCl(g)$$
$$N_2(g) + O_2(g) \rightleftharpoons 2NO(g)$$
$$2SO_2(g) + O_2(g) \rightleftharpoons 2SO_3(g)$$
$$4NH_3(g) + 5O_2(g) \rightleftharpoons 4NO(g) + 6H_2O(g)$$
$$N_2O_4(g) \rightleftharpoons 2NO_2(g)$$
$$CO_2(g) + H_2(g) \rightleftharpoons CO(g) + H_2O(g)$$
$$NH_4Cl(s) \rightleftharpoons NH_3(g) + HCl(g)$$
$$BaSO_3(s) \rightleftharpoons BaO(s) + SO_2(g)$$
$$2Pb(NO_3)_2(s) \rightleftharpoons 2PbO(s) + 4NO_2(g) + O_2(g)$$
$$2H_2(g) + O_2(g) \rightleftharpoons 2H_2O(l)$$
$$S_8(g) \rightleftharpoons 8S(g)$$

16. State Le Châtelier's principle as applied to chemical equilibria.
17. How will an increase in temperature affect the following equilibria? an increase in pressure?
    (a) $N_2(g) + 3H_2(g) \rightleftharpoons 2NH_3(g)$ $\Delta H = -92.2$ kJ
    (b) $H_2O(l) \rightleftharpoons H_2O(g)$ $\Delta H = 41$ kJ
    (c) $N_2(g) + O_2(g) \rightleftharpoons 2NO(g)$ $\Delta H = 181$ kJ
    (d) $3O_2(g) \rightleftharpoons 2O_3(g)$ $\Delta H = 285$ kJ
    (e) $CaCO_3(s) \rightleftharpoons CaO(s) + CO_2(g)$ $\Delta H = 176$ kJ
18. For each of the following reactions between gases at equilibrium, determine the effect upon the equilibrium concentrations of the products when the temperature is decreased and when the external pressure on the system is decreased.
    (a) $2H_2O(g) \rightleftharpoons 2H_2(g) + O_2(g)$ $\Delta H = 484$ kJ
    (b) $N_2(g) + O_2(g) \rightleftharpoons 2NO(g)$ $\Delta H = 181$ kJ
    (c) $N_2(g) + 3H_2(g) \rightleftharpoons 2NH_3(g)$ $\Delta H = -92.2$
    (d) $2O_3(g) \rightleftharpoons 3O_2(g)$ $\Delta H = -285$ kJ
    (e) $H_2(g) + F_2(g) \rightleftharpoons 2HF(g)$ $\Delta H = 541$ kJ
19. Suggest four ways in which the equilibrium concentration of ammonia can be increased for the reaction referred to in Question 18(c).
20. For the reaction in Question 18(b), the equilibrium constant at 2000°C is 6.2×10^{-4}. Consider each of

the two situations below and decide whether a reaction will occur and, if so, in which direction it will proceed predominantly.

(a) A 5.0-ℓ box contains 0.26 mole N_2, 0.0062 mole O_2, and 0.0010 mole NO at 2000°C.

(b) A 2.5-ℓ box contains 0.26 mole N_2, 0.0062 mole O_2, and 0.0010 mole NO at 2500°C.

21. Under what conditions do changes in pressure affect systems in equilibrium?

22. In general, the equilibrium constant for a reaction in the gaseous phase has different units and a different numerical value when pressures rather than concentrations are used in the equilibrium constant expression. Show that such is not the case for the decomposition of HI into H_2 and I_2; show that the numerical value of the equilibrium constant would be independent of the units in which concentration is expressed for this particular reaction.

23. The reaction between hydrogen and sulfur, $S_8(g)$, to produce H_2S is exothermic. How should the pressure and temperature be adjusted in order to improve the equilibrium yield of H_2S, assuming conditions such that all reactants and products are in the gaseous state? Speculate what these conditions will do to the rate of attainment of equilibrium.

24. Under what conditions will the reversible decomposition of $CaCO_3[CaCO_3(s) \rightleftharpoons CaO(s) + CO_2(g)]$ in a closed container proceed to completion so that no $CaCO_3$ remains?

25. Nitric oxide gas (NO) reacts with chlorine gas according to the equation

$$2NO + Cl_2 \longrightarrow 2NOCl$$

The following initial rates of reaction have been observed for the reagent concentrations listed:

NO (mol ℓ^{-1})	Cl_2 (mol ℓ^{-1})	Rate (mol ℓ^{-1} hr^{-1})
0.50	0.50	1.14
1.00	0.50	4.56
1.00	1.00	9.12

What is the mathematical equation (rate equation) describing the rate dependence on the concentrations of NO and Cl_2?

Problems

1. The rate constant at 45° for the decomposition of dinitrogen pentaoxide, N_2O_5, dissolved in chloroform, $CHCl_3$, is 6.2×10^{-4} min^{-1} ($2N_2O_5 \longrightarrow 4NO_2 + O_2$). The decomposition is first order in N_2O_5.

   [S](a) What is the rate of the reaction when $[N_2O_5] =$ 0.40 M?   *Ans. 2.5×10^{-4} mol ℓ^{-1} min^{-1}*

   [S](b) What is the concentration of N_2O_5 remaining at the end of one hour if the initial concentration of N_2O_5 was 0.40 M?   *Ans. 0.38 M*

2. One of the reactions involved in the formation of photochemical smogs is $O_3(g) + NO(g) \longrightarrow O_2(g) + NO_2(g)$. The rate constant for this reaction is 1.2×10^7 ℓ mol^{-1} sec^{-1}. The reaction is first order in O_3 and first order in NO. Calculate the rate of formation of NO_2 in air in which the O_3 concentration is 2×10^{-8} M and the NO concentration 7×10^{-8} M.   *Ans. 2×10^{-8} mol ℓ^{-1} sec^{-1}*

[S]3. Most of the 15.7 billion pounds of HNO_3 produced in the United States during 1977 was prepared by the following sequence of reactions:

$$4NH_3(g) + 5O_2(g) \longrightarrow 4NO(g) + 6H_2O(g) \quad (1)$$
$$2NO(g) + O_2(g) \longrightarrow 2NO_2(g) \quad\quad\quad (2)$$
$$3NO_2(g) + H_2O(l) \longrightarrow 2HNO_3(aq) + NO(g) \,(3)$$

The first reaction is run by burning ammonia in air over a platinum catalyst. This reaction is fast. Reaction (3) is also fast. The second reaction limits the rate at which nitric acid can be prepared from ammonia. If Reaction (2) is second order in NO and first order in O_2, what is the rate of formation of NO_2 when the oxygen concentration is 0.50 M and the nitric oxide concentration is 0.75 M? The rate constant for the reaction is 5.8×10^{-6} ℓ^2 mol^{-2} sec^{-1}.   *Ans. 1.6×10^{-6} mol ℓ^{-1} sec^{-1}*

4. Nitrosyl chloride, NOCl, decomposes to NO and Cl_2 [$2NOCl(g) \longrightarrow 2NO(g) + Cl_2(g)$]. Determine the rate equation and the rate constant for this reaction from the following data:

[NOCl] (M)	0.20	0.40	0.60
Rate (mol ℓ^{-1} sec^{-1})	1.60×10^{-9}	6.40×10^{-9}	1.44×10^{-8}

Ans. Rate = k[NOCl]²;
k = 4.0 × 10⁻⁸ ℓ mol⁻¹ sec⁻¹

$\boxed{s}$5. Hydrogen reacts with nitric oxide to form nitrous oxide, laughing gas, according to the equation $H_2(g) + NO(g) \longrightarrow N_2O(g) + H_2O(g)$. Determine the rate equation and the rate constant for the reaction from the following data:

[NO] (M)	0.30	0.60	0.60
[H₂] (M)	0.35	0.35	0.70
Rate (mol ℓ^{-1} sec^{-1})	2.835×10^{-3}	1.134×10^{-2}	2.268×10^{-2}

Ans. Rate = k[NO]²[H₂];
k = 9.0 × 10⁻² ℓ² mol⁻² sec⁻¹

6. The following data have been determined for the reaction

$$I^- + OCl^- \longrightarrow IO^- + Cl^-$$

[I⁻] (M)	0.10	0.20	0.30
[OCl⁻] (M)	0.050	0.050	0.010
Rate (mol ℓ^{-1} sec^{-1})	3.05×10^{-4}	6.10×10^{-4}	1.83×10^{-4}

Determine the rate equation and rate constant for this reaction. *Ans. Rate = k[I⁻][OCl⁻];*
k = 6.1 × 10⁻² ℓ mol⁻¹ sec⁻¹

$\boxed{s}$7. The rate constant at 325° for the reaction $C_4H_8 \longrightarrow 2C_2H_4$ (Section 15.11) is 6.1×10^{-8} sec^{-1}, and the activation energy is 261 kJ per mole of C_4H_8. Determine the frequency factor for the reaction. *Ans. 3.8 × 10¹⁵ sec⁻¹*

8. The rate constant for the decomposition of acetaldehyde, CH_3CHO, to methane, CH_4, and carbon monoxide, CO, in the gas phase is 1.1×10^{-2} ℓ mol^{-1} sec^{-1} at 703 K and 4.95 ℓ mol^{-1} sec^{-1} at 865 K. Determine the activation energy for this decomposition. *Ans. 190 kJ*

$\boxed{s}$9. The hydrolysis of the sugar sucrose to the sugars glucose and fructose $(C_{12}H_{22}O_{11} + H_2O \longrightarrow C_6H_{12}O_6 + C_6H_{12}O_6)$ follows a first-order rate equation

$$Rate = k[C_{12}H_{22}O_{11}]$$

(The products of the reaction have the same molecular formulas but differ in the arrangement of the

atoms in their molecules.) In neutral solution, $k = 2.1 \times 10^{-11}$ sec^{-1} at 27° and 8.5×10^{-11} sec^{-1} at 37°. Determine the activation energy, the frequency factor, and the rate constant for this equation at 47°. *Ans. Eₐ = 108 kJ;*
A = 1.3 × 10⁸ sec⁻¹; k = 3.1 × 10⁻¹⁰ sec⁻¹
(An alternative and equally acceptable method of working the problem produces the answers
Eₐ = 109 kJ; A = 2.0 × 10⁸ sec⁻¹;
k = 3.2 × 10⁻¹⁰ sec⁻¹.)

$\boxed{s}$10. If the rate of a reaction doubles for every 10° rise in temperature, how much faster would the reaction proceed at 55° than at 25°? at 100° than at 25°? *Ans. 8 times faster; 180 times faster*

11. In an experiment, a sample of $NaClO_3$ was 90% decomposed in 48 minutes. Approximately how long would it have taken had the sample been heated 40° higher? *Ans. 3.0 min*

12. Determine the rate constant for the decomposition of H_2O_2 shown in Fig. 15-1 from the data given in the figure. *Ans. 3.21 × 10⁻⁵ sec⁻¹*

$\boxed{s}$13. The decomposition of SO_2Cl_2 to SO_2 and Cl_2 is a first-order reaction with the rate constant $k = 2.2 \times 10^{-5}$ sec^{-1} at 320°. Determine the half-life of this reaction. At 320°, how much $SO_2Cl_2(g)$ would remain in a 1.00-ℓ flask 90.0 min after the introduction of 0.0238 mol of SO_2Cl_2 into the flask? Assume the rate of the reverse reaction is so slow that it can be ignored.

Ans. t₁/₂ = 3.2 × 10⁴ sec; 2.11 × 10⁻² mol

14. Assume that the rate equations given below apply to each reaction at equilibrium. Determine the rate equation of each of the reverse reactions.

$\boxed{s}$(a) $NO(g) + O_3(g) \longrightarrow NO_2(g) + O_2(g)$
   Rate₁ = k₁[NO][O₃]
Ans. Rate₂ = k₂[NO₂][O₂]

(b) $2NO(g) + O_2(g) \longrightarrow 2NO_2(g)$
   Rate₁ = k₁[NO]²[O₂]
Ans. Rate₂ = k₂[NO₂]²

(c) $2NO_2(g) + F_2(g) \longrightarrow 2NO_2F(g)$
   Rate₁ = k₁[NO₂][F₂]
Ans. Rate₂ = k₂[NO₂F]²/[NO₂]

$\boxed{s}$(d) $2NO(g) + 2H_2(g) \longrightarrow N_2(g) + 2H_2O(g)$
   Rate₁ = k₁[NO][H₂]
Ans. Rate₂ = k₂[N₂][H₂O]²/[NO][H₂]

$\boxed{s}$15. The rate of the reaction $H_2(g) + I_2(g) \longrightarrow 2HI(g)$

at 25° is given by Rate = 1.7×10^{-18} [H_2][I_2]. The rate of decomposition of gaseous HI to $H_2(g)$ and $I_2(g)$ at 25° is given by Rate = 2.4×10^{-21} [HI]^2. What is the equilibrium constant for the formation of gaseous HI from the gaseous elements at 25°?

Ans. 7.1×10^2

S 16. A sample of $NH_3(g)$ was formed from $H_2(g)$ and $N_2(g)$ at 500°. If the equilibrium mixture was found to contain 1.35 mol H_2 per liter, 1.15 mol N_2 per liter and 4.12×10^{-1} mol NH_3 per liter, what is the value of the equilibrium constant for the formation of NH_3?   *Ans. 6.00×10^{-2} (M^{-2}, or mol^{-2} l^2)*

17. Most of the 69 billion pounds of sulfuric acid produced in the United States during 1978 resulted from the reaction sequence

$$S_8(g) + 8O_2(g) \longrightarrow 8SO_2(g) \qquad (1)$$

$$2SO_2(g) + O_2(g) \xrightarrow{V_2O_5} 2SO_3(g) \qquad (2)$$

$$SO_3(g) + H_2O(l) \longrightarrow H_2SO_4(l) \qquad (3)$$

V_2O_5 is required as a catalyst in Reaction (2) because the oxidation of SO_2 to SO_3 is slow in the absence of a catalyst. Each of these reactions has been studied extensively. Under a specific set of conditions at 500° equilibrium pressures of SO_2, O_2, and SO_3 in Reaction (2) have been determined to be 0.342 atm, 0.173 atm, and 0.988 atm, respectively. What is K_p for Reaction (2)?   *Ans. 48.2 (atm^{-1})*

S 18. A 0.72-mol sample of PCl_5 is put into a 1.00-l vessel and heated. At equilibrium, the vessel contains 0.40 mol of $PCl_3(g)$ as well as Cl_2 and undissociated $PCl_5(g)$. What is the equilibrium constant for the decomposition of PCl_5 to PCl_3 and Cl_2 at this temperature?   *Ans. 0.50 (M)*

19. Ethanol and acetic acid interact to form ethyl acetate, an important industrial solvent, and water, according to the equation

$$C_2H_5OH + CH_3CO_2H \rightleftharpoons$$
$$CH_3COOC_2H_5 + H_2O$$

When one mole each of ethanol (C_2H_5OH) and acetic acid (CH_3CO_2H) are allowed to react at 100° in a sealed tube, equilibrium is established when one-third of a mole of each of the reactants remains. Calculate the equilibrium constant for the reaction. (Note: Water is not a solvent in this reaction.)   *Ans. 4*

S 20. The vapor pressure of water is 0.196 atm at 60°. What is the equilibrium constant for the transformation $H_2O(l) \rightleftharpoons H_2O(g)$?

Ans. $K_p = 0.196$ (atm)

21. A sample of ammonium chloride was heated in a closed container [$NH_4Cl(s) \rightleftharpoons NH_3(g) + HCl(g)$]. At equilibrium, the pressure of $NH_3(g)$ was found to be 1.75 atm. What is the equilibrium constant for the decomposition at this temperature?

Ans. $K_p = 3.06$ (atm^2)

22. The equilibrium constant for the reaction represented by $CO + H_2O \rightleftharpoons CO_2 + H_2$ is 5.0 at a given temperature.

S (a) Upon analysis, an equilibrium mixture of the substances present at the given temperature was found to contain 0.20 mol of CO, 0.30 mol of water vapor, and 0.90 mol of H_2 in a liter. How many moles of CO_2 were there in the equilibrium mixture?   *Ans. 0.33 mol*

S (b) Maintaining the same temperature, additional H_2 was added to the system, and some water vapor removed by drying. A new equilibrium mixture was thereby established that contained 0.40 mol of CO, 0.30 mol of water vapor, and 1.2 mol of H_2 in a liter. How many moles of CO_2 were in the new equilibrium mixture? Compare the value with the quantity in Part (a) and discuss whether the second value is reasonable. Explain how it is possible for the water vapor concentration to be the same in the two equilibrium solutions even though some vapor was removed before the second equilibrium was established.   *Ans. 0.50 mol*

23. Hydrated sodium sulfate, $Na_2SO_4 \cdot 10H_2O$, dehydrates according to the equation $Na_2SO_4 \cdot 10H_2O(s) \rightleftharpoons Na_2SO_4(s) + 10H_2O(g)$ with $K = 4.08 \times 10^{-25}$ at 25°C. What is the pressure of water vapor in equilibrium with a sample of $Na_2SO_4 \cdot 10H_2O$?   *Ans. 3.64×10^{-3} atm*

24. Calculate the equilibrium concentration of NO_2 in 1 l of a solution prepared from 0.129 mol of N_2O_4 with chloroform as the solvent. For the reaction $N_2O_4 \rightleftharpoons 2NO_2$, in chloroform, $K = 1.07 \times 10^{-5}$.   *Ans. 1.17×10^{-3} M*

S 25. The equilibrium constant, K, for the reaction $PCl_5(g) \rightleftharpoons PCl_3(g) + Cl_2(g)$ is 0.0211 at a certain temperature. What are the equilibrium concentrations of PCl_5, PCl_3, and Cl_2 starting with a concentration of PCl_5 of 1.00 *M*?

Ans. PCl_5, *0.865 M;* PCl_3, *0.135 M;* Cl_2, *0.135 M*

26. At 990°C, the equilibrium constant for the reaction $H_2(g) + CO_2(g) \rightleftharpoons H_2O(g) + CO(g)$ is 1.6. Calculate the number of moles of each component in the final equilibrium mixture obtained from adding 1.00 mol of H_2, 2.0 mol of CO_2, 0.75 mol of H_2O, and 1.0 mol of CO to a 5.00-ℓ reactor at 990°C.

Ans. 0.60 mol H_2*; 1.6 mol* CO_2*; 1.2 mol* H_2O*; 1.4 mol CO*

[S] 27. The equilibrium constant, K_p, for the decomposition of nitrosyl bromide [$2NOBr(g) \rightleftharpoons 2NO(g) + Br_2(g)$] is 1.0×10^{-2} atm at 25°. What per cent of NOBr is decomposed at 25° and a total pressure of 0.25 atm?

Ans. 34%

[S] 28. A 1.00-ℓ vessel at 400° contains the following equilibrium concentrations: N_2, 1.00 *M*; H_2, 0.50 *M*; and NH_3, 0.50 *M*. How many moles of hydrogen must be removed from the vessel in order to increase the concentration of nitrogen to 1.2 *M*?

Ans. 0.94 (Note that more hydrogen is removed than was originally present as elemental H_2*. Is this possible?)*

29. (a) Using the law of mass action, write the expression for the equilibrium constant for the reversible reaction

$$N_2 + O_2 \rightleftharpoons 2NO \qquad \Delta H = 181 \text{ kJ}$$

(b) What will happen to the concentration of NO at equilibrium if (1) more O_2 is added? (2) N_2 is removed? (3) the pressure on the system is increased? (4) the temperature of the system is increased?

(c) This reaction produces nitrogen oxide pollutants during the operation of an internal combustion engine. NO forms in the cylinders during combustion and is swept out in the exhaust. The reaction for the decomposition of NO ($2NO \longrightarrow N_2 + O_2$) has an equilibrium constant $K_p = 3.0 \times 10^{31}$ at 25° (the temperature of a warm summer day). At equilibrium, what would be the atmospheric NO pressure ($p_{O_2} =$ 0.2 atm, $p_{N_2} = 0.8$ atm)?

Ans. 7×10^{-17} *atm*

(d) The actual concentration of NO along an expressway during peak traffic is higher than this value. Suggest an explanation.

References

"Rate Constants from Initial Concentration Data," K. J. Hall, T. I. Quickenden, and D. W. Watts, *J. Chem. Educ.*, **53**, 493 (1976).

"Interpretation of the Activation Energy," D. G. Truhlar, *J. Chem. Educ.*, **55**, 308 (1978).

"Reactions of Oriented Molecules," P. R. Brooks, *Science*, **193**, 11 (1976).

"The New Look in Chemical Kinetics," D. Edelson, *J. Chem. Educ.*, **52**, 642 (1975).

"Some Common Oversimplifications in Teaching Chemical Kinetics," R. K. Boyd, *J. Chem. Educ*, **55**, 84 (1978).

"Following Second-Order Kinetics by Simple Laboratory Means," G. Schreiber, *J. Chem. Educ.*, **53**, 664 (1976).

"Use of Log-Log Plots in Determination of Reaction Orders," J. P. Birk, *J. Chem. Educ.*, **53**, 704 (1976).

"Recent Progress in the Study of Fast Chemical Reactions," H. Winkler, *Endeavour,* May 1974, p. 73.

"On Chemical Kinetics," C. J. Swartz, *J. Chem. Educ.*, **46**, 308 (1969).

"The Law of Mass Action," S. Berline and C. Bricker, *J. Chem. Educ.*, **46**, 499 (1969).

"Reaction Kinetics and the Law of Mass Action," P. G. Ashmore, *Educ. in Chemistry,* **2**, 160 (1965).

"Orbital Interactions and Reaction Paths," L. Salem, *Chem. in Britain,* **5**, 449 (1969).

"High Pressure Chemistry," H. S. Turner, *Chem. in Britain,* **4**, 245 (1968).

"Recent Developments in Theoretical Chemical Kinetics," R. A. Marcus, *J. Chem. Educ.*, **45**, 356 (1968).

"Some Aspects of Chemical Dynamics in Solution," J. Halpern, *J. Chem. Educ.*, **45**, 372 (1968).

"From Stoichiometry and Rate Law to Mechanism," J. O. Edwards, E. F. Greene, and J. Ross, *J. Chem. Educ*, **45**, 381 (1968).

"Mechanisms of Oxidation-Reduction Reactions," H. Taube, *J. Chem. Educ.* **45**, 452 (1968).

"The Kinetics and Analysis of Very Fast Chemical Reactions," R. G. W. Norrish, *Chem. in Britain,* **1**, 289 (1965).

"Relaxation Methods in Chemistry," L. Faller, *Sci. American,* May 1969, p. 30.

"Encounters and Slow Reactions," C. H. Langford, *J. Chem. Educ.,* **46,** 557 (1969).

"Lasers in Photochemical Kinetics," L. K. Patterson, *Chem. in Britain,* **6,** 246 (1970).

"Hot Water Freezes Faster than Cold Water. Why Does it Do so?" J. Walker, *Sci. American,* September 1977, p. 246.

"Starvation Kinetics," H. Eyring, *Science,* **199,** 740 (1978).

"An Exact Solution to a Consecutive Reaction Sequence," R. L. Anderson, R. S. Nohr, and L. O. Speer, *J. Chem. Educ.,* **52,** 437 (1975).

"Murray Raney of Chattanooga and Nickel Catalysts," D. S. Tarbell and A. T. Tarbell, *J. Chem. Educ.,* **54,** 26 (1977).

"Gas Phase Chemical Equilibria," H. F. Gibbard and M. R. Emptage, *J. Chem. Educ.,* **53,** 218 (1976).

"Determining the K_p for the Ammonia Synthesis as a Function of Temperature," G. Huybrechts and G. Petre, *J. Chem. Educ.,* **53,** 443 (1976).

"The Entropy Gradient: A Heuristic Approach to Chemical Equilibrium," J. S. Wicken, *J. Chem. Educ.,* **55,** 701 (1978).

"Maximizing Profits in Equilibrium Processes," R. J. Rish, *J. Chem. Educ.,* **52,** 441 (1975).

"Catalysis," V. Haenzel and R. L. Burwell, Jr., *Sci. American,* December 1971, p. 46.

"'Heterogenizing' Homogeneous Catalysts," J. C. Bailar, Jr., *Catalysis Reviews—Sci. and Eng.,* **10** (1), 17 (1974).

16

Ionic Equilibria of Weak Electrolytes

We discovered in Section 13.11 that aqueous solutions of both strong and weak electrolytes contain ions, with the strong electrolytes being completely ionic in solution. The strong electrolytes include strong inorganic acids such as aqueous solutions of HCl, HBr, HI, HNO_3, and $HClO_4$; the hydroxides of the alkali metals and the heavier alkaline earth metals; and most salts. The concentrations of the ions in a solution of a strong electrolyte can be calculated readily from the concentration of the electrolyte since a strong electrolyte is effectively 100% ionic in solution.

In solutions of weak electrolytes, both ions and un-ionized molecules of the electrolyte are present. The weak electrolytes include weak inorganic acids such as H_2SO_3, HCN, HF, and H_2CO_3; most organic acids; aqueous ammonia; most divalent and trivalent hydroxides; and most of the organic bases. The polar covalent halides and cyanides of Hg, Cd, Zn, and of a few other metals are also classed as weak electrolytes. The concentrations of ions in a solution of a weak electrolyte must be calculated from the concentration of the weak electrolyte and the extent to which it ionizes in solution. Such calculations involve the concepts of equilibria discussed in Chapter 15.

Many chemical reactions that occur in solution involve ions. The concentration of ions in these solutions is very important in determining the rates, mechanisms, and yields of the reactions. Thus equilibrium calculations take on great significance because they are used to determine the actual concentrations of ions in solutions of weak electrolytes. The weak acids, perhaps, are the most important of all the weak electrolytes. They are critical not only in many processes and reactions of a purely chemical nature, including those of qualitative analysis, but also in the processes in living systems involving the blood, the tissue and cell materials, and the glandular secretions.

Ionic Equilibria of Weak Acids and Bases

16.1   Ionic Concentrations in Solutions of Strong Electrolytes

Because strong electrolytes are completely ionic in aqueous solution, the ion concentrations may be found directly from the molar concentration of the solution. For example, in a $0.01\ M$ solution of hydrochloric acid ($HCl \longrightarrow H^+ + Cl^-$), both the hydrogen ion concentration, $[H^+]$, and the chloride ion concentration, $[Cl^-]$, are equal to the concentration of hydrogen chloride in the solution. From the chemical equation we can find the conversions $1\ mol\ H^+/1\ mol\ HCl$ and $1\ mol\ Cl^-/1\ mol\ HCl$. Thus the concentration of HCl, $0.01\ M$ ($0.01\ mol\ HCl/1\ \ell$), can be converted to $[H^+]$ and $[Cl^-]$ as follows:

$$[H^+] = \frac{0.01\ \text{mol HCl}}{1\ \ell} \times \frac{1\ \text{mol H}^+}{1\ \text{mol HCl}} = \frac{0.01\ \text{mol H}^+}{1\ \ell} = 0.01\ M$$

$$[Cl^-] = \frac{0.01\ \text{mol HCl}}{1\ \ell} \times \frac{1\ \text{mol Cl}^-}{1\ \text{mol HCl}} = \frac{0.01\ \text{mol Cl}^-}{1\ \ell} = 0.01\ M$$

Notice that the brackets here (and later), as explained earlier in Chapter 15, stand for "molar concentration of"; hence $[H^+]$ reads "molar concentration of hydrogen ion." In a $0.01\ M$ solution of the strong electrolyte potassium sulfate ($K_2SO_4 \longrightarrow 2K^+ + SO_4^{2-}$) the potassium ion concentration, $[K^+]$, is $(0.01\ \text{mol}\ K_2SO_4/1\ \ell) \times (2\ \text{mol}\ K^+/1\ \text{mol}\ K_2SO_4)$, or $0.02\ M$. The sulfate ion concentration, $[SO_4^{2-}]$, is $(0.01\ \text{mol}\ K_2SO_4/1\ \ell) \times (1\ \text{mol}\ SO_4^{2-}/1\ \text{mol}\ K_2SO_4)$, or $0.01\ M$. This follows because each mole of potassium sulfate contains two moles of potassium ions and one mole of sulfate ions. In a $0.001\ M$ calcium chloride solution ($CaCl_2 \longrightarrow Ca^{2+} + 2Cl^-$) the calcium ion concentration is $1 \times 10^{-3}\ M$ and the chloride ion concentration is $2 \times 10^{-3}\ M$. This is due to the fact that crystals (and solutions) of calcium chloride contain twice as many chloride ions as calcium ions.

16.2   Ionic Concentrations in Solutions of Weak Acids

In aqueous solutions of a weak acid such as acetic acid (CH_3CO_2H, often abbreviated as HOAc), only a fraction of the molecules are ionized into hydrogen cations and acid anions. The remaining fraction of the acid is present in solution as the un-ionized, or molecular, form. In order to determine the concentration of un-ionized acetic acid, of hydrogen cations, and of acetate anions, we need to use the equilibrium constant for the ionization of acetic acid. For acetic acid the ionization is

$$HOAc(aq) \rightleftharpoons H^+(aq) + OAc^-(aq)$$

(where OAc$^-$ is the acetate ion, $CH_3CO_2{}^-$) or more accurately

$$HOAc(aq) + H_2O(l) \rightleftharpoons H_3O^+(aq) + OAc^-(aq)$$

This equation shows that the ionization reaction is reversible and at equilibrium. Equilibrium is attained almost instantly when the acetic acid is dissolved in the water. Therefore, in any dilute solution of acetic acid, the law of mass action (Section 15.17) may be applied.

$$\frac{[H_3O^+][OAc^-]}{[HOAc][H_2O]} = K$$

Because we deal only with dilute solutions, and because the number of moles of water consumed in the formation of hydronium ions is negligibly small compared to the total number of moles of water present in dilute solution, the molar concentration of water remains practically constant. Therefore the above equation may be written

$$\frac{[H_3O^+][OAc^-]}{[HOAc]} = K[H_2O] = K_i$$

For this particular kind of equilibrium, the constant K_i is called an **ionization constant**. Concentrations of ions and molecules in the equation for K_i are generally expressed in moles per liter. The data in Table 16-1 show that K_i for acetic acid remains practically the same over a considerable range of concentrations. Note that the extent to which acetic acid ionizes increases with decreasing concentration. For example, 0.1 M acetic acid is 1.34% ionized (98.66% is in the molecular form), while the 0.01 M acid is 4.15% ionized (95.85% is in the molecular form). This results because in more dilute solution the ions are farther apart and tend to combine less readily to form undissociated acetic acid molecules in the equilibrium mixture.

The ionization constants of a number of weak acids are given in Table 16-2, with a more complete list given in Appendix G. For convenience, the equations in Table 16-2 are given in simplified form without showing the hydration of the hydrogen ions. Nevertheless, it should be kept in mind that the hydrogen ion is always in the hydrated form (H_3O^+) in aqueous solution.

The acids in Table 16-2 are listed in order of decreasing strength as indicated by the decreasing size of the ionization constant (compare with Table 14-1 in Section 14.15). As the fraction of the acid in the ionized form increases, the value of the

TABLE 16-1   Ionization Constants for Acetic Acid at Different Concentrations

Molarity	Per Cent Ionized	[H$_3$O$^+$] and [OAc$^-$]	[HOAc]	K_i
0.100	1.34	0.00134	0.09866	1.82×10^{-5}
0.0800	1.50	0.00120	0.07880	1.83×10^{-5}
0.0300	2.45	0.000735	0.02927	1.85×10^{-5}
0.0100	4.15	0.000415	0.009585	1.797×10^{-5}

TABLE 16-2	Ionization Constants of Some Weak Acids	
Ionization reaction		K_i at 25°
$HF \rightleftharpoons H^+ + F^-$		7.2×10^{-4}
$HNO_2 \rightleftharpoons H^+ + NO_2^-$		4.5×10^{-4}
$HNCO \rightleftharpoons H^+ + NCO^-$		3.46×10^{-4}
$HCO_2H \rightleftharpoons H^+ + HCO_2^-$		1.8×10^{-4}
$HOAc \rightleftharpoons H^+ + OAc^-$		1.8×10^{-5}
$HClO \rightleftharpoons H^+ + ClO^-$		3.5×10^{-8}
$HBrO \rightleftharpoons H^+ + BrO^-$		2×10^{-9}
$HCN \rightleftharpoons H^+ + CN^-$		4×10^{-10}

constant increases. Thus HF with its constant of 7.2×10^{-4} is a much stronger acid than HCN, which has a K_i of 4×10^{-10}.

16.3 Problems Involving the Ionization of Acetic Acid

The following exemplify the types of problems that are based upon the partial ionization of weak electrolytes.

EXAMPLE 1   **In 0.100 M solution, acetic acid is 1.34% ionized. Calculate [H$^+$], [OAc$^-$], and [HOAc] in the solution.**

First write the equation for the ionization of acetic acid.

$$HOAc \rightleftharpoons H^+ + OAc^-$$

The equation shows that for each mole of HOAc that ionizes, one mole of H$^+$ and one mole of OAc$^-$ are formed. Since 1.34% of 0.100 mole per liter of acid ionizes, then

$$[H^+] = [OAc^-] = 0.100\ M \times 0.0134 = 0.00134\ M$$

and $[HOAc] = (0.100\ M - 0.00134\ M) = 0.099\ M$

◆ ◆ ◆ ◆ ◆

EXAMPLE 2   **Using the concentrations found in Example 1, calculate the ionization constant of acetic acid.**

Write down the expression for the ionization constant of acetic acid, substitute in it the values found above, and solve for K_i.

$$\frac{[H^+][OAc^-]}{[HOAc]} = \frac{(0.00134) \times (0.00134)}{0.099} = 1.8 \times 10^{-5} = K_i$$

◆ ◆ ◆ ◆ ◆

EXAMPLE 3   **Taking the K_i for acetic acid to be 1.8×10^{-5} at 25°C, calculate the $[H^+]$ and the per cent ionization in a 0.010 M solution of the acid.**

In a 0.010 molar solution of acetic acid the total amount of acetic acid, i.e., the ionized plus the un-ionized, contained in 1 l of solution is 0.010 mol. If we let x be the number of moles of acetic acid that ionize to form hydrogen ion and acetate ion to establish equilibrium, then x is the concentration of hydrogen ion $[H^+]$ and also of acetate ion $[OAc^-]$ at equilibrium. The equilibrium concentration of un-ionized acid $[HOAc]$ is $(0.010 - x)$. Then,

$$\frac{[H^+][OAc^-]}{[HOAc]} = \frac{x^2}{0.010 - x} = 1.8 \times 10^{-5} = K_i$$

The small value of K_i indicates that the ratio $x^2/(0.010 - x)$ is small and that x will be small compared with 0.010 from which it is to be subtracted. Thus $0.010 - x$ is virtually equal to 0.010 and the foregoing expression may be simplified for an approximate solution as follows:

$$\frac{x^2}{0.010} = 1.8 \times 10^{-5}$$

$$x^2 = 1.8 \times 10^{-7} = 18.0 \times 10^{-8}$$

$$x = \sqrt{18.0 \times 10^{-8}} = 4.2 \times 10^{-4} \text{ mol } l^{-1}$$

The concentration of H^+ and OAc^- is thus calculated to be $4.2 \times 10^{-4} M$.

It can readily be seen that we were justified in neglecting x in the expression $(0.010 - x)$ since $(0.010 - 0.00042) = 0.00958$, which is very nearly equal to 0.010. We may use the approximate solution when x is less than about 5% of the total concentration of the acid; if x is greater than 5%, then the more complete quadratic equation should be solved (see Appendix A.4 for the solution of quadratic equations).

Calculation of the per cent of ionization in the 0.010 M solution of acetic acid is made by dividing the concentration of the acid in the ionic form, which is equal to the $[H^+]$, by the total concentration of the acid, and then multiplying by 100. Thus we have

$$\frac{\text{Ionic HOAc}}{\text{total HOAc}} \times 100 = \frac{4.2 \times 10^{-4}}{0.010} \times 100 = 4.2\% \text{ ionized}$$

◆ ◆ ◆ ◆ ◆

Note that the values calculated in Examples 2 and 3 do not exactly agree with the values given in Table 16-1. This simply reflects the use of only two significant figures in the examples.

16.4   The Ionization of Weak Bases

Equilibria involving the ionization of weak bases may be treated mathematically in the same manner as those of weak acids. The ionization constants of several weak bases are given in Table 16-3 and in Appendix H.

TABLE 16-3 Ionization Constants of Some Weak Bases

Base	Ionization	K_i at 25°
NH_3 Ammonia	$+ H_2O \rightleftharpoons NH_4^+ + OH^-$	1.8×10^{-5}
CH_3NH_2 Methylamine	$+ H_2O \rightleftharpoons CH_3NH_3^+ + OH^-$	4.4×10^{-4}
$(CH_3)_2NH$ Dimethylamine	$+ H_2O \rightleftharpoons (CH_3)_2NH_2^+ + OH^-$	7.4×10^{-4}
$(CH_3)_3N$ Trimethylamine	$+ H_2O \rightleftharpoons (CH_3)_3NH^+ + OH^-$	7.4×10^{-5}
$C_6H_5NH_2$ Phenylamine	$+ H_2O \rightleftharpoons C_6H_5NH_3^+ + OH^-$	4.6×10^{-10}

The most common weak base is aqueous ammonia. When ammonia gas is dissolved in water the solution is distinctly basic, and the reaction is given by the equation

$$NH_3(aq) + H_2O(l) \rightleftharpoons NH_4^+(aq) + OH^-(aq)$$

Application of the law of mass action to this system gives the expression

$$\frac{[NH_4^+][OH^-]}{[NH_3][H_2O]} = K$$

Because only a small fraction of the water is consumed in the reaction and because we are working with a dilute solution, the concentration of water is practically constant and the above equation may be written

$$K_i = \frac{[NH_4^+][OH^-]}{[NH_3]} = K[H_2O] = 1.8 \times 10^{-5}$$

Aqueous ammonia as a base has about the same strength that acetic acid has as an acid; the ionization constants for the two substances are the same (to two significant figures).

16.5   Salt Effect on Ionization

The expression for the ionization constant holds for dilute solutions of weak electrolytes when the solutions are pure. However, if the solution is not pure, the extent of ionization of a weak electrolyte may be influenced by the presence of other ions in the solution. For example, the ionization constant of acetic acid is increased from 1.8×10^{-5} to 2.2×10^{-5} when the solution is made 0.1 M with respect to sodium chloride. This means that the addition of a salt has slightly increased the degree of ionization of the weak acid, and this effect is known as the **salt effect.** The effect is due to the ions of a strong electrolyte decreasing the activities of the ions (Section 13.27) in a solution of a weak electrolyte as a result of interionic attraction. The ionization constant is, therefore, not strictly constant, but varies slightly with the ionic strength of the solution. For most purposes, and for ours, the expression for the ionization constant may be used without regard to slight errors that may arise due to the salt effect.

16.6   Common Ion Effect on Degree of Ionization

It has been found by experiment that the acidity of an aqueous solution of acetic acid is decreased by the addition of the strong electrolyte sodium acetate. This can be explained by the law of mass action (Section 15.17). The addition of acetate ions to a solution of acetic acid causes the following equilibrium to shift to the left, with a resulting increase in the concentration of HOAc.

$$HOAc \rightleftharpoons H^+ + OAc$$

This decreases the concentration of H^+ and hence the acidity. The NaOAc has the acetate ion in common with HOAc, so the influence is known as the **common ion effect.** Since the equilibrium was disturbed by the addition of OAc^-, a shift in the equilibrium will occur that will reduce the concentration of OAc^-. This is accomplished by the union of H^+ with OAc^- to form molecular acetic acid. Consequently, the $[H^+]$ of the solution is reduced.

The extent to which the concentration of the hydrogen ion is decreased by the addition of the acetate ion may be calculated from the expression for the ionization constant of acetic acid.

EXAMPLE 1   Calculate the $[H^+]$ in a 0.10 M solution of HOAc that is 0.50 M with respect to NaOAc.

At equilibrium the concentration of the acetate ion is equal to the concentration of NaOAc, which is completely dissociated into Na^+ and OAc^-, plus the concentration of the acetate ion derived from the ionization of acetic acid. Let x equal the concentration of acetate ions derived from the ionization of acetic acid. Then $(0.50\ M + x)$ will equal the total concentration of the acetate ion. The concentration of the hydrogen ions will be equal to x and the concentration of un-ionized acetic acid will be equal to $(0.10\ M - x)$. Substituting in the expression for the ionization constant for acetic acid we have

$$\frac{[H^+][OAc^-]}{[HOAc]} = \frac{x(0.50 + x)}{0.10 - x} = 1.8 \times 10^{-5}$$

Even in the absence of NaOAc the concentration of the acetate ion derived from the ionization of 0.10 M acetic acid is small (0.00134 mol ℓ^{-1}) compared to 0.50 mol ℓ^{-1} derived from NaOAc. Since the degree of ionization is even smaller in the presence of the high concentration of sodium acetate, it follows that the concentration of acetate ion may be taken as equal to the concentration of the sodium acetate, or that x may be neglected in the term $(0.50 + x)$. Likewise, the concentration of the un-ionized acetic acid is very nearly equal to

0.10 M, and x in the term $(0.10 - x)$ may be dropped. Thus in an approximate solution to the problem we have

$$\frac{[H^+][OAc^-]}{[HOAc]} = \frac{(0.50)x}{0.10} = 1.8 \times 10^{-5}$$

$$x = \frac{0.10}{0.50} \times 1.8 \times 10^{-5} = 3.6 \times 10^{-6} \, M = [H^+]$$

Table 16-1 shows that the concentration of the hydrogen ion in 0.10 M HOAc is 0.00134 mol ℓ^{-1}. This is reduced to 0.0000036 mol ℓ^{-1} by the presence of 0.50 M sodium acetate in the 0.10 M acetic acid solution. Note that the simplifying assumption that x is negligibly small compared to either 0.50 or 0.10 was well justified.

◆ ◆ ◆ ◆ ◆

EXAMPLE 2   **A 0.10 M solution of aqueous ammonia, also containing ammonium chloride, has a hydroxide ion concentration of 2.8×10^{-6} M. What is the concentration of the ammonium ion in the solution?**

First write the equation for the ionization of ammonia.

$$NH_3 + H_2O \rightleftharpoons NH_4^+ + OH^-$$

The ammonium ion of the ammonium chloride present in the solution causes the ionization of ammonia to be decreased (the equilibrium to be shifted to the left) because of the common ion effect. Let x be the total ammonium ion concentration at equilibrium; this value will be the sum of the ammonium ion concentration resulting from the slight ionization of ammonia and that from the ammonium chloride. The hydroxide ion concentration is given as 2.8×10^{-6} M, and the concentration of ammonia will be $(0.10 - 2.8 \times 10^{-6})$ M, which is very nearly 0.10 M. Substituting in the expression for the ionization constant of ammonia, we have

$$\frac{[NH_4^+][OH^-]}{[NH_3]} = \frac{x(2.8 \times 10^{-6})}{0.10} = 1.8 \times 10^{-5}$$

Solving for x, we find

$$x = \frac{(0.10)(1.8 \times 10^{-5})}{2.8 \times 10^{-6}} = \frac{1.8}{2.8} = 0.64$$

Thus we find the concentration of the ammonium ion to be 0.64 M at equilibrium. Of this amount, 2.8×10^{-6} mole is formed by the ionization of ammonia and the remainder comes from the added ammonium chloride.

◆ ◆ ◆ ◆ ◆

EXAMPLE 3   **Ten milliliters of 4.0 M acetic acid is added to 20 ml of 1.0 M sodium hydroxide. Calculate the hydrogen ion concentration of the resulting solution assuming the volumes are accurate to two significant figures.**

Before mixing, the 10 ml of 4.0 M HOAc contains

$$10 \text{ ml} \times (1 \text{ } \ell/1000 \text{ ml}) \times (4.0 \text{ mol HOAc}/1 \text{ } \ell) = 4.0 \times 10^{-2} \text{ mol HOAc},$$

and the 20 ml of 1.0 M NaOH contains

$$20 \text{ ml} \times (1 \text{ } \ell/1000 \text{ ml}) \times (1.0 \text{ mol NaOH}/1 \text{ } \ell) = 2.0 \times 10^{-2} \text{ mol NaOH}.$$

The 2.0×10^{-2} mol of NaOH neutralize 2.0×10^{-2} mol of HOAc, producing 2.0×10^{-2} mol of NaOAc and leaving $(4.0 \times 10^{-2}) - (2.0 \times 10^{-2}) = 2.0 \times 10^{-2}$ mol of HOAc not neutralized. After reaction, the 2.0×10^{-2} mol of NaOAc and 2.0×10^{-2} mol of HOAc are contained in $(10 + 20)$ ml = 30 ml of solution, so the concentration of each is 2.0×10^{-2} mol/0.030 ℓ, or 0.67 M.

The sodium acetate in the solution exhibits the common ion effect upon the ionization of the acetic acid. Let x equal the concentration of the hydrogen ion. Then the concentration of the molecular acetic acid will be $(0.67 - x)$ M and that of the acetate ion will be $(0.67 + x)$ M. Neglecting the x's in the terms $(0.67 - x)$ and $(0.67 + x)$ and substituting in the expression for the ionization constant of acetic acid, we have

$$\frac{[H^+][OAc^-]}{[HOAc]} = \frac{x(0.67)}{(0.67)} = 1.8 \times 10^{-5}$$

$$x = \frac{(0.67)}{(0.67)} \times 1.8 \times 10^{-5} = 1.8 \times 10^{-5} \text{ } M$$

The value of x calculated using the approximation that $0.67 + x = 0.67 - x = 0.67$ is much less than 5% of 0.67; hence the simplifying approximation is valid. Thus the concentration of the hydrogen ion in the solution is 1.8×10^{-5} M.

◆ ◆ ◆ ◆ ◆

The common ion effect is of importance in the adjustment of the hydrogen ion concentration for many of the precipitations and separations in the qualitative analysis scheme and in blood and other biological fluids.

16.7   The Ionization of Water

Water is an extremely weak electrolyte that undergoes self-ionization (see Section 12.3).

$$H_2O + H_2O \rightleftharpoons H_3O^+ + OH^-$$

or simply

$$H_2O \rightleftharpoons H^+ + OH^-$$

Application of the law of mass action yields the expression

$$\frac{[H^+][OH^-]}{[H_2O]} = K$$

As long as we work with dilute aqueous solutions, the concentration of water may be considered as constant. We may therefore write

$$[H^+][OH^-] = K[H_2O] = K_w$$

In all problems of acidity and basicity of aqueous solutions of electrolytes, K_w is a constant of great importance. It is called the **ion-product constant of water,** and at 25° it has the value 1.0×10^{-14}. The value of K_w increases rapidly with increasing temperature and at 100° is approximately 1×10^{-12}, thus being 100 times as large at 100° as at 25°. This means simply that the degree of ionization of water and the concentration of the hydrogen and hydroxide ions increase with rising temperature.

The ionization of water yields the same number of hydrogen and hydroxide ions. Therefore, in pure water at 25°, $[H^+] = [OH^-]$, and

$$[H^+]^2 = [OH^-]^2 = 1.0 \times 10^{-14}$$
$$[H^+] = [OH^-] = \sqrt{10^{-14}} = 1.0 \times 10^{-7} \, M$$

All aqueous solutions contain both hydrogen ions and hydroxide ions. If by the addition of hydrogen ions (as an acid) the $[H^+]$ is made larger than 10^{-7}, then the $[OH^-]$ will become smaller than 10^{-7}; in basic solutions the $[OH^-]$ will be greater than 10^{-7} and the $[H^+]$ will be less than 10^{-7}. The product of the $[H^+]$ and $[OH^-]$ at a given temperature always remains constant, a fact that chemists find very useful. For example, they may wish to know the $[H^+]$ of $0.01 \, M$ NaOH at 25°. Since NaOH is a strong electrolyte (NaOH $\longrightarrow$ Na$^+$ + OH$^-$), the concentration of OH$^-$ is equal to the total concentration of NaOH in the solution; $[OH^-] = 0.01 \, M$.

$$[H^+][OH^-] = K_w$$
$$[H^+](10^{-2}) = 1.0 \times 10^{-14}$$
$$[H^+] = \frac{10^{-14}}{10^{-2}} = 1 \times 10^{-12} \, M$$

Or, it may be desirable to calculate the $[OH^-]$ in $0.001 \, M$ HCl.

$$[H^+][OH^-] = (10^{-3})[OH^-] = 1.0 \times 10^{-14}$$
$$[OH^-] = \frac{10^{-14}}{10^{-3}} = 1 \times 10^{-11} \, M$$

16.8   The pH Method of Expressing H$^+$ Concentration

The concentration of the hydrogen ion is a measure of the acidity or basicity of a solution. It has been found convenient to express the concentration of the hydrogen ion in terms of the negative logarithm of the hydrogen ion concentration. This is referred to as the **pH** of the solution. Expressed mathematically, we have

$$pH = -\log[H^+]$$

or
$$pH = \log \frac{1}{[H^+]}$$

The pH value is the negative power to which 10 must be raised to equal the hydrogen ion concentration.

$$[H^+] = 10^{-pH}$$

The following examples will serve to illustrate the calculation of pH values from hydrogen and hydroxide ion concentrations:

EXAMPLE 1   **Calculate the pH of 0.01 M HNO₃. Nitric acid is completely ionized in dilute solutions, so the concentration of the hydrogen ion is 0.01 M, or 10^{-2} M.**

Substituting, we have

$$pH = -\log[H^+] = -\log 10^{-2} = -(-2.0) = 2.0$$

(The use of logarithms is explained in Appendix A.3.)

◆ ◆ ◆ ◆ ◆

EXAMPLE 2   **The concentration of the hydrochloric acid secreted by the stomach after a meal is about 1.2×10^{-3} M. What is the pH of stomach acid?**

The hydrogen ion concentration of stomach acid is the same as the molar concentration of HCl because HCl is a strong electrolyte. Thus we have

$$pH = -\log(1.2 \times 10^{-3}) = -(\log 1.2 + \log 10^{-3})$$
$$= -(0.079 - 3.00) = -(-2.92) = 2.92$$

(On a calculator simply take the logarithm of 1.2×10^{-3} and set the pH equal to the negative of this logarithm.)

Note that 2.92 represents *two* significant figures rather than three. The number to the left of the decimal in the logarithm is the "characteristic," which merely establishes the decimal in the number for which 2.92 is the logarithm. The only *significant figures* in the logarithm are those to the right of the decimal. The number of significant figures in a logarithm should be equal to the number of significant figures in the number from which the logarithm is obtained. There are two significant figures in the hydrogen ion concentration (1.2×10^{-3}) and two in the pH.

◆ ◆ ◆ ◆ ◆

EXAMPLE 3   **Calculate the pH of 0.0001 M NaOH.**

Because pH is defined in terms of hydrogen ion concentration, it is first necessary to find the concentration of this ion by substituting in the ion-product expression for water. The [OH⁻] from the sodium hydroxide is given as 10^{-4} M.

$$[H^+][OH^-] = K_w$$
$$[H^+](10^{-4}) = 10^{-14}$$
$$[H^+] = \frac{10^{-14}}{10^{-4}} = 10^{-10}$$

The hydrogen ion concentration thus found is then converted to pH as follows:

$$pH = -\log[H^+] = -\log 10^{-10} = -(-10.0) = 10.0$$

◆ ◆ ◆ ◆ ◆

EXAMPLE 4   **Water in equilibrium with the air contains $4.4 \times 10^{-5}\%$ by mass of carbon dioxide. The resulting carbonic acid, H_2CO_3, gives to the solution a hydrogen ion concentration about 20 times larger than that of pure water, or 2.0×10^{-6} as compared to 1.0×10^{-7}. Calculate the pH of the solution.**

$$pH = -\log[H^+] = -\log(2.0 \times 10^{-6}) = -(\log 2.0 + \log 10^{-6})$$
$$= -(0.30 - 6.00) = -(-5.70) = 5.70$$

Thus we see that water in contact with air is acidic, rather than neutral, due to dissolved carbon dioxide.

◆ ◆ ◆ ◆ ◆

EXAMPLE 5   **Calculate the pH of $0.100\ M$ HOAc, which is 1.34% ionized.**

The hydrogen ion concentration of this weak acid is $0.100\ M \times 0.0134$ $= 0.00134\ M = 1.34 \times 10^{-3}\ M$ (calculated in Section 16.3).

$$pH = -\log[H^+] = -\log(1.34 \times 10^{-3}) = -(\log 1.34 + \log 10^{-3})$$
$$= -(0.127 - 3.000) = -(-2.873) = 2.873$$

◆ ◆ ◆ ◆ ◆

The following examples illustrate the conversion of pH values to hydrogen ion concentrations.

EXAMPLE 6   **Calculate the hydrogen ion concentration of a solution, the pH of which is 9.0.**

$$pH = -\log[H^+] = 9.0$$
$$\log[H^+] = -9.0$$
$$[H^+] = \text{antilog of } (-9.0) = 1 \times 10^{-9}$$

◆ ◆ ◆ ◆ ◆

EXAMPLE 7   **Calculate the hydrogen ion concentration of blood whose pH is 7.3 (slightly alkaline).**

$$pH = -\log[H^+] = 7.3$$
$$\log[H^+] = -7.3$$

It is readily seen that -7.3 is equal to $(-8 + 0.7)$. Then we may write

$$\log[H^+] = 0.7 - 8$$
$$[H^+] = \text{antilog of } (0.7 - 8)$$
$$= (\text{antilog of } 0.7) \times (\text{antilog of } -8)$$
$$= 5 \times 10^{-8}$$

(On a calculator simply take the antilog of -7.3 or calculate $10^{-7.3}$.)

◆ ◆ ◆ ◆ ◆

TABLE 16-4   Relationships of $[H^+]$, $[OH^-]$, pH, and pOH

$[H^+]$	$[OH^-]$	pH	pOH	Solution
10^1	10^{-15}	-1	15	Strongly acidic
10^0 or 1	10^{-14}	0	14	
10^{-1}	10^{-13}	1	13	
10^{-2}	10^{-12}	2	12	
10^{-3}	10^{-11}	3	11	
10^{-4}	10^{-10}	4	10	
10^{-5}	10^{-9}	5	9	
10^{-6}	10^{-8}	6	8	
10^{-7}	10^{-7}	7	7	Neutral
10^{-8}	10^{-6}	8	6	
10^{-9}	10^{-5}	9	5	
10^{-10}	10^{-4}	10	4	
10^{-11}	10^{-3}	11	3	
10^{-12}	10^{-2}	12	2	
10^{-13}	10^{-1}	13	1	Strongly basic

16.9   The pOH Method of Expressing OH^- Concentration

We may define pOH as the negative logarithm of the hydroxide ion concentration ($pOH = -\log[OH^-]$). Similarly, we may define pK_w as the negative logarithm of the ion-product constant (K_w) for water. Now we can write the ion-product expression for water in terms of pH, pOH, and pK_w.

$$[H^+][OH^-] = K_w$$
$$(-\log[H^+]) + (-\log[OH^-]) = -\log K_w$$
$$pH + pOH = pK_w$$

Because K_w has the value 1.0×10^{-14},

$$pK_w = -\log 1.0 \times 10^{-14} = -(-14.00) = 14.00$$

It follows that

$$pH + pOH = 14.00$$
$$pH = 14.00 - pOH$$
$$pOH = 14.00 - pH$$

The relationships between the hydrogen ion and hydroxide ion concentration, and the pH and pOH of certain aqueous solutions are given in Table 16-4.

16.10   Acid-Base Indicators

Certain organic substances have the property of changing color in dilute solution when the hydrogen ion concentration of the solution attains a definite value. For example, phenolphthalein is a colorless substance in any aqueous solution in which the hydrogen ion concentration is greater than 10^{-9} M, or the pH is less than 9. In solutions for which the hydrogen ion concentration is less than 10^{-9} (pH greater than 9), the phenolphthalein imparts a red or pink color to the solution. Substances like phenolphthalein are called **acid-base indicators,** and they are often used to determine the pH of solutions. Acid-base indicators are either weak organic acids, HIn, or weak organic bases, InOH, in which the symbol "In" represents a complex organic group.

The equilibrium existing in a solution of a certain acid-base indicator called **methyl orange,** which is a weak acid, can be represented by the equation

$$\text{HIn} \rightleftharpoons \text{H}^+ + \text{In}^-$$
$$\text{Red} \qquad\qquad\quad \text{Yellow}$$

The anion of the indicator is yellow, and the un-ionized form is red. An increase in the concentration of the hydrogen ion brought about by the addition of an acid to the solution shifts the equilibrium toward the red form in accordance with the law of mass action. Application of the law of chemical equilibrium to this reversible reaction gives the expression for the ionization constant of this acid.

$$\frac{[\text{H}^+][\text{In}^-]}{[\text{HIn}]} = K_i$$

The color exhibited by the indicator is the visible result of the ratio of the concentrations of the two species HIn and In$^-$. For methyl orange,

$$\frac{[\text{In}^-]}{[\text{HIn}]} = \frac{[\text{substance with yellow color}]}{[\text{substance with red color}]} = \frac{K_i}{[\text{H}^+]}$$

When $[\text{H}^+]$ has the same numerical value as K_i, the ratio of $[\text{In}^-]$ to $[\text{HIn}]$ is equal to 1, meaning that 50% of the indicator is present in the acid (yellow form) and 50% in the alkaline (red form). Under these conditions the solution appears orange in color. When the hydrogen ion concentration has been increased until the pH is 3.1, about 90% of the indicator is present in the red form and 10% is present in the yellow form. The eye cannot detect any change in color with further increase in the hydrogen ion concentration and corresponding increase in the concentration of the red form of the indicator.

Addition of a base to the system reduces the concentration of hydrogen ions ($\text{H}^+ + \text{OH}^- \longrightarrow \text{H}_2\text{O}$), and shifts the equilibrium toward formation of the yellow form. At a pH of 4.4 about 90% of the indicator is in the form of the yellow ion, and a further decrease in the hydrogen ion concentration does not produce a color change easily detectable by the eye. The pH range between 3.1 (red) and 4.4 (yellow-orange) is the **color-change interval** of the indicator methyl orange. This means that the pronounced color change takes place between these two pH values.

A large number of acid-base indicators is known, covering a wide range of pH values. A number of indicators may be used in determining the approximate pH of any solution, and by a process of elimination the pH of a solution can be fixed within rather narrow limits. Table 16-5 lists a series of indicators, together with their colors for the corresponding pH values. The selection of indicators for specific purposes will be discussed in Section 16.12.

TABLE 16-5 Some Acid-Base Indicators

Indicator	Color in the More Acid Range	pH Range	Color in the More Basic Range
Methyl violet	Yellow	0–2	Violet
Thymol blue	Pink	1.2–2.8	Yellow
Brom-phenol blue	Yellow	3.0–4.7	Violet
Methyl orange	Pink	3.1–4.4	Yellow
Brom-cresol green	Yellow	4.0–5.6	Blue
Brom-cresol purple	Yellow	5.2–6.8	Purple
Litmus	Red	4.7–8.2	Blue
Phenolphthalein	Colorless	8.3–10.0	Pink
Thymolphthalein	Colorless	9.3–10.5	Blue
Alizarin yellow G	Colorless	10.1–12.1	Yellow
Trinitrobenzene	Colorless	12.0–14.3	Orange

The measurement and control of the hydrogen ion concentration are important in scientific investigations, in industry, and in agriculture. In analytical chemistry the separation and identification of many of the metallic ions depend upon the pH of the solutions containing these ions.

16.11 Buffer Solutions

Mixtures of weak acids and their salts or mixtures of weak bases and their salts are called **buffer solutions.** They resist a change in hydrogen-ion concentration upon the addition of small amounts of acids or bases. An example of a buffer solution is a mixture of 0.10 M acetic acid and 0.10 M sodium acetate. The pH of the solution may be found as follows:

$$HOAc \rightleftharpoons H^+ + OAc^-$$

$$\frac{[H^+][OAc^-]}{[HOAc]} = K_i$$

$$[H^+] = \frac{[HOAc]}{[OAc^-]} K_i = \frac{0.10}{0.10} \times 1.80 \times 10^{-5} = 1.8 \times 10^{-5}$$

$$pH = -\log[H^+] = -\log(1.8 \times 10^{-5}) = 4.745$$

$$= 4.74 \quad \text{(to justifiable limit in significant figures)}$$

Now add 1.0 ml of 0.10 M sodium hydroxide to 100 ml of this buffer mixture. An equivalent amount of acetic acid is neutralized by the sodium hydroxide, and sodium acetate is formed.

$$HOAc + OH^- \longrightarrow H_2O + OAc^-$$

Before reaction, 100 ml of the buffer mixture contains $100 \text{ ml} \times (1 \text{ l}/1000 \text{ ml}) \times (0.10 \text{ mol HOAc}/1 \text{ l}) = 1.0 \times 10^{-2}$ mol HOAc, and 1.0×10^{-2} mol NaOAc. One milliliter of 0.10 M sodium hydroxide contains $1.0 \text{ ml} \times (1 \text{ l}/1000 \text{ ml}) \times (0.10 \text{ mol NaOH}/1 \text{ l}) = 1.0 \times 10^{-4}$ mol NaOH. The 1.0×10^{-4}

mol of NaOH neutralizes 1.0×10^{-4} mol of HOAc, leaving $(1.0 \times 10^{-2}) - (0.01 \times 10^{-2}) = 0.99 \times 10^{-2}$ mol of HOAc, and producing 1.0×10^{-4} mol of NaOAc; this makes a total of $(1.0 \times 10^{-2}) + (0.01 \times 10^{-2}) = 1.01 \times 10^{-2}$ mol of NaOAc. After reaction, the 0.99×10^{-2} mol of HOAc and 1.01×10^{-2} mol of NaOAc are contained in 101 ml of solution, so the concentrations are 9.9×10^{-3} mol/0.101 $\ell = 0.098$ M HOAc, and 1.01×10^{-2} mol/0.101 $\ell = 0.100$ M NaOAc. The final pH of the solution is then calculated as follows:

$$[H^+] = \frac{[HOAc]}{[OAc^-]} \times K_i = \frac{0.098}{0.100} \times 1.80 \times 10^{-5} = 1.76 \times 10^{-5}$$

$$pH = -\log[H^+] = -\log(1.76 \times 10^{-5}) = 4.754$$

$$= 4.75 \quad \text{(to justifiable limit in significant figures)}$$

Thus the addition of the base barely changes the pH of the solution. The explanation for this buffering action lies in the fact that as the sodium hydroxide is added, the hydroxide ions unite with the few hydrogen ions present, and then more of the acetic acid ionizes, thus restoring the concentration of the hydrogen ions to near its original value. If a small amount of hydrochloric acid were to be added to the acetic acid–sodium acetate buffer solution, most of the hydrogen ions from the hydrochloric acid would unit with acetate ions from the large reserve of these ions, forming un-ionized acetic acid.

$$H^+ + OAc^- \longrightarrow HOAc$$

Thus there would be very little increase in the concentration of the hydrogen ion, and the pH would remain practically unchanged.

Some **buffer pairs** that find extensive aplication are HOAc and OAc^-, NH_3 and NH_4^+, H_2CO_3 and HCO_3^-, $H_2PO_4^-$ and HPO_4^{2-}. Blood is an important example of a buffer solution, with the principal acid and ion responsible for the buffering action being H_2CO_3 and HCO_3^-. When an excess of hydrogen ion enters the blood stream, it is removed principally through the reaction

$$H^+ + HCO_3^- \longrightarrow H_2CO_3$$

and when an excess of the hydroxide ion is present, it is absorbed by the reaction

$$OH^- + H_2CO_3 \longrightarrow H_2O + HCO_3^-$$

The pH of human blood thus remains very nearly 7.35, or slightly alkaline.

16.12   Titration Curves

Plots of pH versus volume of acid or base added in an acid-base titration (see Section 14.11) are useful in that they show graphically the point where equivalent quantities of acid and base are present (the equivalence point) and so aid in the choice of a proper indicator. Such plots are referred to as **titration curves.**

The simplest acid-base neutralization reactions are those for titrations involving a strong acid and a strong base. Let us consider the titration of a 25.0-ml sample of a 0.100 M hydrochloric acid solution with a 0.100 M sodium hydroxide solution.

TABLE 16-6   pH Values (a) in the Titration of a Strong Acid with a Strong Base and (b) in the Titration of a Weak Acid with a Strong Base

Volume of 0.100 M NaOH Added, ml	Moles of NaOH Added	pH Values	
		(a) Titration of 25.00 ml of 0.100 M HCl	(b) Titration of 25.00 ml of 0.100 M HOAc
0.0	0.0	1.00	2.87
5.0	0.00050	1.18	4.14
10.0	0.00100	1.37	4.57
15.0	0.00150	1.60	4.92
20.0	0.00200	1.95	5.35
22.0	0.00220	2.20	5.61
24.0	0.00240	2.69	6.13
24.5	0.00245	3.00	6.44
24.9	0.00249	3.70	7.14
25.0	0.00250	7.00	8.72
25.1	0.00251	10.30	10.30
25.5	0.00255	11.00	11.00
26.0	0.00260	11.29	11.29
28.0	0.00280	11.75	11.75
30.0	0.00300	11.96	11.96
35.0	0.00350	12.22	12.22
40.0	0.00400	12.36	12.36
45.0	0.00450	12.46	12.46
50.0	0.00500	12.52	12.52

(a) Titration of 25.00 ml of 0.100 M HCl (0.00250 mol of HCl) with 0.100 M NaOH
(b) Titration of 25.00 ml of 0.100 M HOAc (0.00250 mol of HOAc) with 0.100 M NaOH

Figure 16-1   Titration curve for the titration of 25.00 ml of 0.100 M HCl (strong acid) with 0.100 M NaOH (strong base). The pH ranges for the color change of phenolphthalein, litmus, and methyl orange are indicated by the shaded areas.

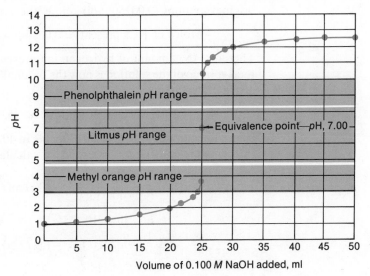

The 25.0-ml solution of 0.100 M HCl contains 0.00250 mol of HCl. Values for pH can be calculated for any given quantity of NaOH solution added. Table 16-6, column (a), provides calculated pH values for a variety of added volumes of NaOH. Figure 16-1 is a plot of the pH values (as the vertical axis) and volume of NaOH added (as the horizontal axis).

It is apparent from Fig. 16-1 that the pH increases very slowly at first, increases very rapidly in the middle portion of the curve, and then increases very slowly again after the middle portion. The point of inflection (the midpoint of the rapid rise in the curve) is the **equivalence point** for the titration and represents the stage of the titration where equivalent quantities of acid and base are present in the titration mixture. For the titration of a strong acid and a strong base, as illustrated in Table 16-6, column (a), and Fig. 16-1, the equivalence point occurs at a pH of 7. The solution at a pH of 7 is neutral, with hydrogen ion concentration equal to hydroxide ion concentration.

The following sample calculations of the pH values in Table 16-6, column (a), are for several points on the titration curve.

EXAMPLE 1   Calculate the pH of the initial solution before any NaOH solution is added.

The initial solution is a 0.100 M HCl solution. Since HCl is a strong acid in water solution, $[H^+] = [HCl] = 0.100\ M$.

$$pH = -\log{[H^+]} = -\log{(0.100)} = 1.000$$

◆ ◆ ◆ ◆ ◆

EXAMPLE 2   Calculate the pH of the solution after 15.0 ml of 0.100 M NaOH solution has been added to the 25 ml of 0.100 M HCl solution.

The 15.0 ml of 0.100 M NaOH solution contains 0.00150 mol of NaOH, which will neutralize 0.00150 mol of HCl. Hence, the remaining HCl present is the original amount (0.00250 mol) minus the amount neutralized (0.00150 mol):

Moles of HCl present = $0.00250 - 0.00150 = 0.00100$ mol

Therefore the amount of hydrogen ion present in solution is 0.00100 mol. The total volume of the solution is now the original volume (25.0 ml) plus the volume of NaOH solution added (15.0 ml):

Total volume of solution = $25.0 + 15.0 = 40.0$ ml

Hence the amount of H^+ is 0.00100 mole in 40.0 ml. The molarity of H^+ (moles H^+ per liter) and the pH can then be calculated:

$$[H^+] = \frac{0.00100\ \text{mol}}{40.0\ \text{ml}} \times 1000\ \text{ml}/1\ \ell = 0.0250\ \text{mol}/\ell$$

$$= 2.50 \times 10^{-2}\ M$$

$$pH = -\log{[H^+]} = -\log{(2.50 \times 10^{-2})} = 1.602$$

◆ ◆ ◆ ◆ ◆

EXAMPLE 3   Calculate the *p*H of the solution when 25.0 ml of 0.100 *M* NaOH has been added.

At this stage, which is the equivalence point of the titration, 0.00250 mol of NaOH has been added to 0.00250 mol of HCl. Hence 0.00250 mol of NaCl and 0.00250 mol of water have been produced, and the NaOH added has exactly neutralized the HCl originally present. A neutral solution has been produced, therefore, in which $[H^+] = [OH^-] = 1.0 \times 10^{-7} M$.

$$pH = -\log(1.0 \times 10^{-7}) = 7.00$$

◆ ◆ ◆ ◆ ◆

EXAMPLE 4   Calculate the *p*H of the solution when 35.0 ml of 0.100 *M* NaOH solution has been added.

At all points beyond the equivalence point, an excess of NaOH is present. When 35.0 ml of NaOH has been added, the excess NaOH is 35.0 − 25.0 ml, or 10.0 ml, and represents an excess of 0.00350 − 0.00250 mol, or a 0.00100 mol excess.

The total volume of the solution is

$$25.0 \text{ ml} + 35.0 \text{ ml} = 60.0 \text{ ml}$$

Hence

$$[OH^-] = \frac{0.00100 \text{ mol}}{60.0 \text{ ml}} \times 1000 \text{ ml}/1 \; \ell$$

$$= 0.0167 \text{ mol}/\ell = 1.67 \times 10^{-2} M$$

$$pOH = -\log(1.67 \times 10^{-2}) = 1.777$$

$$pH = 14.00 - pOH = 14.00 - 1.777 = 12.22$$

◆ ◆ ◆ ◆ ◆

The titration of a weak acid with a strong base or of a weak base with a strong acid is somewhat more complicated than the titrations previously discussed but follows the same basic principles. Let us consider the titration of 25.0 ml of a 0.100 *M* solution of acetic acid with a 0.100 *M* solution of sodium hydroxide and compare the titration curve with that of Fig. 16-1. Table 16-6, column (b), provides calculated *p*H values for several added volumes of NaOH. Figure 16-2 shows the titration curve.

The similarities in the two titration curves are readily apparent, but a close inspection also reveals several important differences. The titration curve for the titration of the weak acid begins at a higher *p*H value (less acidic) and maintains higher *p*H values for corresponding amounts of added NaOH until the equivalence point is reached. This is because the acetic acid in the solution, being a weak acid, releases only a small quantity of hydrogen ions to the solution. The *p*H at the equivalence point is also higher (8.72, rather than 7.00). However, following the equivalence point, the two curves are identical, as shown by the values in Table 16-6, column (b), and by the plot in Fig. 16-2. This is because after the acid is completely neutralized, whether it be a strong acid (like HCl) or a weak acid (like HOAc), it has no further significant effect upon the concentration of the hydroxide

Figure 16-2 Titration curve for the titration of 25.00 ml of 0.100 M HOAc (weak acid) with 0.100 M NaOH (strong base). The pH ranges for the color change of phenolphthalein, litmus, and methyl orange are indicated by the shaded areas.

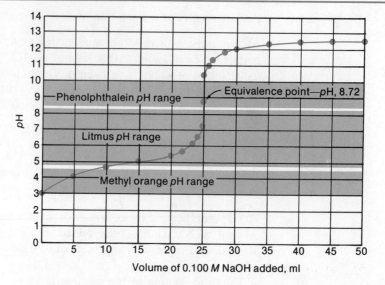

ion in the solution; the pH is dependent upon the excess of hydroxide ion in both cases.

Some sample calculations will help to clarify the points just made. The calculations are very similar to problems for weak acid equilibria discussed earlier in this chapter (see Sections 16.3, 16.6, 16.8, and 16.11).

EXAMPLE 5   **Calculate the pH of the initial solution before any NaOH solution is added.**

The initial solution is a 0.100 M solution of acetic acid. The pH is 2.87. See Section 16.8, Example 5, for the calculation.

◆ ◆ ◆ ◆ ◆

EXAMPLE 6   **Calculate the pH of the solution after 15.0 ml of 0.100 M NaOH solution has been added to the 25 ml of 0.100 M HOAc solution.**

The 15.0 ml of 0.100 M NaOH solution contains 0.00150 mol of NaOH, which will react with 0.00150 mol of HOAc to form 0.00150 mol of NaOAc and 0.00150 mol of water. The amount of undissociated HOAc will thereby decrease from the original 0.00250 mol to 0.00100 mol (0.00250 − 0.00150 = 0.00100). The total volume of solution is 25.0 + 15.0 = 40.0 ml. Hence 0.00150 mol of NaOAc (and hence 0.00150 mol of OAc⁻) and 0.00100 mol of undissociated HOAc are present in 40.0 ml of solution. The molarities of OAc⁻ and HOAc, therefore, are

$$[OAc^-] = \frac{0.00150 \text{ mol}}{40.0 \text{ ml}} \times 1000 \text{ ml}/1 \ \ell = 0.0375 \text{ mol } \ell^{-1}$$

$$= 3.75 \times 10^{-2} \ M$$

$$[HOAc] = \frac{0.00100 \text{ mol}}{40.0 \text{ ml}} \times 1000 \text{ ml}/1 \ \ell = 0.0250 \text{ mol } \ell^{-1}$$

$$= 2.50 \times 10^{-2} \ M$$

Substituting these concentrations in the equilibrium expression for acetic acid:

$$\frac{[H^+][OAc^-]}{[HOAc]} = 1.8 \times 10^{-5}$$

$$\frac{[H^+](3.75 \times 10^{-2})}{2.50 \times 10^{-2}} = 1.8 \times 10^{-5}$$

$$[H^-] = 1.2 \times 10^{-5}\ M$$

$$pH = -\log(1.2 \times 10^{-5}) = 4.92$$

◆ ◆ ◆ ◆ ◆

EXAMPLE 7   **Calculate the pH of the solution when 25.0 ml of 0.100 M NaOH has been added (the equivalence point).**

At this stage in the titration, 0.00250 mol of NaOH has been added to 0.00250 mol of HOAc. Hence all of the HOAc has reacted with NaOH, and in the process 0.00250 mol of NaOAc has been produced. The total volume of the solution is now 50.0 ml. The molarity of OAc$^-$, therefore, is

$$[OAc^-] = \frac{0.00250\ mol}{50.0\ ml} \times 1000\ ml/1\ \ell = 0.0500\ mol\ \ell^{-1} = 5.00 \times 10^{-2}\ M$$

The problem therefore becomes the calculation of the pH of a 0.050 M NaOAc solution. The calculation must take into account the hydrolysis of the OAc$^-$ according to the reaction

$$OAc^- + H_2O \rightleftharpoons HOAc + OH^-$$

The production of hydroxide ions makes the solution basic at the equivalence point. The actual calculation is made in Example 1 in Section 16.17 later in the chapter. The pH is calculated to be 8.72.

◆ ◆ ◆ ◆ ◆

EXAMPLE 8   **Calculate the pH of the solution when 35.0 ml of 0.100 M NaOH solution has been added.**

At all points beyond the equivalence point, the pH is controlled by the excess NaOH present. Hence, the curve is identical to that for the titration of HCl and the calculations of pH are identical. As in the calculation for HCl, the pH is 12.22.

◆ ◆ ◆ ◆ ◆

Titration curves enable us to pick a suitable indicator that will provide a sharp color change at the equivalence point. The pH ranges for the color changes of three indicators listed in Table 16-5 are indicated by shading in Figures 16-1 and 16-2.

Phenolphthalein provides a sharp color change over a small volume of added NaOH and hence would be suitable for titration of either strong acid–strong base or weak acid–strong base. The equivalence point of 25.00 ml of NaOH is located in the steeply rising portion of the titration curve, where the curve passes through the pH range corresponding to the indicator color change.

Litmus would be a suitable indicator for the HCl titration because it would change color, as shown by the titration curve, within less than 0.10 ml addition of NaOH. However, litmus would be a poor choice of indicator for the HOAc titration because, as shown by the plot, the titration curve enters the pH range for color change of litmus when only about 12 ml of NaOH has been added and does not leave the range until 25 ml of NaOH has been added. The end of the color change would therefore be approximately at the equivalence point (a good feature) but the color change would be very gradual, extending over an addition of 13 ml of NaOH, making it almost useless as an indicator of the equivalence point.

Methyl orange could be used for the HCl titration (Fig. 16-1), though its use would have two disadvantages: (1) it would complete its color change slightly before the equivalence point is reached (though very close to it, so this is not particularly serious), and (2) it would change color, as the plot shows, over a range of nearly 0.5 ml of NaOH addition, which is not as sharp a color change as litmus or phenolphthalein would produce in the same system. The titration curve for HOAc shows that methyl orange would be completely useless as an indictor in that system. Its color change would not be sharp, beginning after only about 1 ml of NaOH had been added and being completed when about 8 ml had been added. More serious is the fact that the color change would be completed long before the equivalence point (which occurs when 25.0 ml of NaOH have been added) and hence would provide no evidence of the equivalence point.

You should check Table 16-5 to see whether any of the other indicators listed would be suitable for determining the equivalence point for either of the two titrations we have discussed.

16.13   The pH Meter

The pH of a solution may be measured and the course of a neutralization titration followed by the use of an instrument called a **pH meter.** Each pH determination with the pH meter involves the measurement of the difference in electrical potential between two electrodes. One electrode is called the **reference electrode,** the other the **indicating electrode.** Both are in contact with the solution whose pH is to be determined. Thus the pH meter must consist of at least three parts: a reference electrode, an indicating electrode, and a potential-measuring meter (a voltmeter).

The reference electrode of a pH meter is one whose potential is independent of the hydrogen ion concentration of the solution being measured. The most commonly used reference electrode is the **calomel electrode.** This electrode is composed of metallic mercury and solid mercury(I) chloride (Hg_2Cl_2), called "calomel," in contact with, and in equilibrium with, a solution of potassium chloride (Fig. 16-3). In usual practice, the potassium chloride solution is saturated.

The indicating electrode is an electrode whose potential is dependent upon the concentration of hydrogen ions in the solution being tested. The most widely used indicating electrode is the **glass electrode.** This electrode consists of a glass tube with a thin glass bulb at one end, inside of which is either a buffer solution into

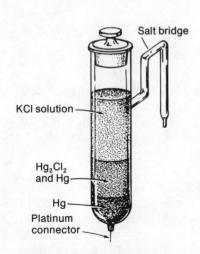

Figure 16-3 A calomel electrode, the most commonly used reference electrode in the pH meter.

which a platinum wire is dipped or a 0.1 M solution of hydrochloric acid in contact with a silver wire coated with a thin layer of silver chloride (Fig. 16-4). An electrical contact is made to the platinum or silver wire for connection to the potential measuring meter. When the glass electrode is in use, the bulb is immersed in the test solution. The glass membrane serves to establish a potential, the size of which is determined by the *relative* hydrogen ion concentrations on the two sides of the membrane. The hydrogen ion concentration inside the electrode is fixed. Before use, each glass electrode must be calibrated empirically by measuring the pH of a standard buffer solution. pH meters are calibrated in pH units rather than volts for direct reading of the pH of the solution being tested.

Figure 16-4 The glass electrode, used extensively as an indicating electrode in the pH meter.

16.14   The Ionization of Weak Diprotic Acids

Polyprotic acids, those from which more than one proton may be removed, ionize in water in successive steps. An example of a **diprotic acid** is hydrogen sulfide in aqueous solution (hydrosulfuric acid). The first step in its ionization yields hydrogen and hydrosulfide ions.

$$H_2S(aq) \rightleftharpoons H^+(aq) + HS^-(aq)$$

The expression for the ionization constant for the primary ionization of H_2S is

$$\frac{[H^+][HS^-]}{[H_2S]} = K_{H_2S} = 1.0 \times 10^{-7}$$

The hydrosulfide ion in turn ionizes and forms hydrogen and sulfide ions.

$$HS^-(aq) \rightleftharpoons H^+(aq) + S^{2-}(aq)$$

The expression for the ionization constant for the hydrosulfide ion is

$$\frac{[H^+][S^{2-}]}{[HS^-]} = K_{HS^-} = 1.3 \times 10^{-13}$$

Note that K_{HS^-} is smaller than K_{H_2S} by almost 10^6 times. This means that very little of the HS^- formed by the ionization of H_2S ionizes to give hydrogen ions and sulfide ions. Thus the concentrations of H^+ and HS^- are practically equal in a pure aqueous solution of H_2S.

EXAMPLE 1   **The concentration of H_2S in a saturated aqueous solution of the gas at 1 atm of pressure and room temperature is approximately 0.1 M. Calculate the $[H^+]$, $[HS^-]$, and $[S^{2-}]$ in the solution.**

If we let x equal $[H^+]$, then $[HS^-]$ will also be equal to x, and $[H_2S]$ will be equal to $0.1 - x$. Since x will be quite small compared to 0.1, it may be dropped from the term $(0.1 - x)$. Solving for x, we have

$$\frac{[H^+][HS^-]}{[H_2S]} = \frac{x^2}{0.1} = 1.0 \times 10^{-7}$$

$$x^2 = 1.0 \times 10^{-8}$$

$$x = 1 \times 10^{-4} \, M = [H^+] = [HS^-]$$

Since, in any solution, all possible equilibria must be satisfied simultaneously, the $[H^+]$ and $[HS^-]$ must be the same for the equilibria involved in the ionization of both HS^- and H_2S. Thus, substituting in the expression for the ionization constant for HS^-, we have

$$\frac{[H^+][S^{2-}]}{[HS^-]} = \frac{(1 \times 10^{-4})}{(1 \times 10^{-4})}[S^{2-}] = K_{HS^-} = 1.3 \times 10^{-13}$$

$$[S^{2-}] = 1.3 \times 10^{-13}\ M$$

(Note that in a pure aqueous solution of H_2S the $[S^{2-}]$ is equal to K_{HS^-}. In fact, for any weak diprotic acid the concentration of the divalent anion is equal to the secondary ionization constant.)

◆ ◆ ◆ ◆ ◆

By multiplying the expressions for the ionization constants K_{H_2S} and K_{HS^-}, we obtain

$$\frac{[H^+][HS^-]}{[H_2S]} \times \frac{[H^+][S^{2-}]}{[HS^-]} = K_{H_2S} \times K_{HS^-}$$

$$\frac{[H^+]^2[S^{2-}]}{[H_2S]} = K_{H_2S}K_{HS^-} = 1.3 \times 10^{-20}$$

This is the equilibrium constant for the reaction $H_2S(aq) \rightleftharpoons 2H^+(aq) + S^{2-}(aq)$.

EXAMPLE 2   Calculate the concentration of sulfide ion in a saturated solution of hydrogen sulfide to which sufficient hydrochloric acid has been added to make the hydrogen ion concentration of the solution 0.10 M. (A saturated H_2S solution is 0.10 M H_2S.)

Substituting in the above equation, we have

$$\frac{[H^+]^2[S^{2-}]}{[H_2S]} = \frac{(0.10)^2[S^{2-}]}{0.10} = 1.3 \times 10^{-20}$$

$$[S^{2-}] = \frac{(1.3 \times 10^{-20})(0.10)}{(0.10)^2} = 1.3 \times 10^{-19}\ M$$

(We may use the same equation in calculating the hydrogen ion concentration necessary to produce a given sulfide ion concentration.)

◆ ◆ ◆ ◆ ◆

16.15   The Ionization of Weak Triprotic Acids

A typical example of a **triprotic acid** is phosphoric acid, which ionizes in water in three steps.

Step 1 $H_3PO_4 \rightleftharpoons H^+ + H_2PO_4^-$

Step 2 $H_2PO_4^- \rightleftharpoons H^+ + HPO_4^{2-}$

Step 3 $HPO_4^{2-} \rightleftharpoons H^+ + PO_4^{3-}$

For each step in the ionization there is a corresponding ionization constant. The expressions for the ionization constants of the three stages in the ionization of phosphoric acid are

$$\frac{[H^+][H_2PO_4^-]}{[H_3PO_4]} = K_{H_3PO_4} = 7.5 \times 10^{-3}$$

$$\frac{[H^+][HPO_4^{2-}]}{[H_2PO_4^-]} = K_{H_2PO_4^-} = 6.3 \times 10^{-8}$$

$$\frac{[H^+][PO_4^{3-}]}{[HPO_4^{2-}]} = K_{HPO_4^{2-}} = 3.6 \times 10^{-13}$$

It will be noted that ionization in the successive steps becomes progressively less extensive. This is a general characteristic of polyprotic acids. Successive ionization constants for many such acids differ by about 10^5–10^6. (See Section 16.14 for the values for H_2S.)

Although this set of three dissociation reactions appears to make equilibrium calculations of equilibrium concentrations very complicated for the phosphoric acid solution, we can be reassured by noting that, as with the hydrogen sulfide calculations (Section 16.14), most equilibrium problems for polyprotic acids can be broken down into a series of calculations similar to those for monoprotic acids.

EXAMPLE   Calculate the concentrations of all species—H_3PO_4, $H_2PO_4^-$, HPO_4^{2-}, PO_4^{3-}, H^+, and OH^-—present at equilibrium in a solution containing a total phosphoric acid concentration of 0.100 M.

$$H_3PO_4 \rightleftharpoons H^+ + H_2PO_4^-$$

Let x = the number of moles of H_3PO_4 that dissociates per liter in establishing the equilibrium. Then x moles of H^+ and x moles of $H_2PO_4^-$ would be produced. A small amount of the $H_2PO_4^-$ produced in the first step will disappear by dissociation into H^+ and HPO_4^- in the second step, and a small amount of additional H^+ will be produced in both the second and third steps. However, because K_2 and K_3 are very small compared to K_1, it is logical to assume that the decrease in $[H_2PO_4^-]$ in the second step and the increase in $[H^+]$ in the second and third steps are negligible compared to the quantity of each ion (x) produced in Step 1.

Hence, at equilibrium,

$$[H_3PO_4] = \text{original amount present minus the amount}$$
$$\text{that dissociates}$$

$$= (0.100 - x) \text{ mol } \ell^{-1}$$

$$[H^+] = x \text{ mol } \ell^{-1}$$

$$[H_2PO_4^-] = x \text{ mol } \ell^{-1}$$

$$K_1 = \frac{[H^+][H_2PO_4^-]}{[H_3PO_4]} = 7.5 \times 10^{-3}$$

$$\frac{(x)(x)}{(0.100 - x)} = \frac{x^2}{(0.100 - x)} = 7.5 \times 10^{-3}$$

With a value of K_1 as large as 7.5×10^{-3}, it is probable that x is too large to ignore compared to 0.100. Hence, the value of x can be calculated by means of the quadratic equation (see Appendix A.4).

$$x^2 = 7.5 \times 10^{-4} - 7.5 \times 10^{-3} x$$

$$x^2 + 7.5 \times 10^{-3} x - 7.5 \times 10^{-4} = 0$$

$$x = \frac{-(7.5 \times 10^{-3}) \pm \sqrt{(7.5 \times 10^{-3})^2 - 4(-7.5 \times 10^{-4})}}{2(1)}$$

$$= 2.4 \times 10^{-2} \, M$$

An alternative, and very useful, method for solving the equation is by **successive approximations.**

First approximation. On a trial basis, we will neglect the x term in the denominator:

$$K_1 = \frac{x^2}{(0.100)} = 7.5 \times 10^{-3}$$

$$x^2 = 7.5 \times 10^{-4}$$

$$x = 2.7 \times 10^{-2} \, M$$

Second approximation. Obviously, x as found above is *not* negligible compared to 0.10 M. A better estimate, using the approximate value of x just obtained, of the equilibrium concentration of phosphoric acid is $(0.100 - x) = (0.100 - 0.027) = 0.073 \, M$:

$$K_1 = \frac{x^2}{(0.073)} = 7.5 \times 10^{-3}$$

$$x^2 = (7.5 \times 10^{-3})(7.3 \times 10^{-2}) = 5.5 \times 10^{-4}$$

$$x = 2.3 \times 10^{-2} \, M$$

Since the values of x obtained from the first and second approximations differ by 15%, a third approximation is desirable.

Third approximation. A still better estimate of the equilibrium phosphoric acid concentration, using x obtained from the second approximation, is $(0.100 - x) = (0.100 - 0.023) = 0.077 \, M$.

$$K_1 = \frac{x^2}{(0.077)} = 7.5 \times 10^{-3}$$

$$x^2 = (7.5 \times 10^{-3})(0.077) = 5.8 \times 10^{-4}$$

$$x = 2.4 \times 10^{-2} \, M$$

(Note that this value of x is the same as that calculated from the quadratic equation.)

The values of x from the last two approximations agree to within approximately 4%. Hence, we can safely conclude that

$$[H^+] = x = 0.024 \, M$$

$$[H_2PO_4^-] = x = 0.024 \, M$$

$$[H_3PO_4] = 0.100 - x = 0.100 - 0.024 = 0.076 \; M$$

$$[OH^-] = \frac{K_w}{[H^+]} = \frac{1.0 \times 10^{-14}}{2.4 \times 10^{-2}} = 4.2 \times 10^{-13} \; M$$

To calculate the concentration of HPO_4^{2-}, we must use the second step of the acid dissociation. Let y be the number of moles of $H_2PO_4^-$ that dissociates in the second step to produce H^+ and HPO_4^{2-}.

	$H_2PO_4^-$	$\rightleftharpoons$	H^+	$+ \; HPO_4^{2-}$
Initial concentrations:	0.024 M		0.024 M	—
Equilibrium concentrations:	(0.024 − y)		(0.024 + y)	y

$$K_2 = \frac{(0.024 + y)(y)}{(0.024 - y)} = 6.3 \times 10^{-8}$$

Since K_2 is so small, we can neglect the additive and subtractive y terms. Then the preceding equation becomes

$$\frac{0.024\,y}{0.024} = y = 6.3 \times 10^{-8} \; M = [HPO_4^-]$$

Note that this provides a check for our earlier assumption that the increase in H^+ and decrease in $H_2PO_4^-$ (now known to be 6.3×10^{-8} mol) is negligible compared to the quantity of each ion produced in the first step (0.024 mol).

Let us now consider the third step of dissociation, to calculate the concentration of PO_4^{3-}. Let z be the number of moles of HPO_4^{2-} that dissociates to produce H^+ and PO_4^{3-}.

	HPO_4^{2-}	$\rightleftharpoons$	H^+	$+ \; PO_4^{3-}$
Initial concentrations:	(6.3×10^{-8} M)		(0.024 M)	—
Equilibrium concentrations:	(6.3×10^{-8} − z)		(0.024 + z)	z

$$K_3 = \frac{(0.024 + z)(z)}{(6.3 \times 10^{-8} - z)} = 3.6 \times 10^{-13}$$

Because K_3 is very small compared to K_2, the extent of the third dissociation is insignificant and we can neglect the additive and subtractive z terms in the preceding equation; thus, we can easily solve the expression for z:

$$\frac{0.024\,z}{6.3 \times 10^{-8}} = 3.6 \times 10^{-13}$$

$$z = [PO_4^{3-}]$$

$$= \frac{(3.6 \times 10^{-13})(6.3 \times 10^{-8})}{(2.4 \times 10^{-2})} = 9.5 \times 10^{-19} \; M$$

Note that the quantity of H^+ added to the solution and the decrease in HPO_4^{2-} in the third step (z) is indeed negligible compared to the amounts previously in the solution.

◆ ◆ ◆ ◆ ◆

Hydrolysis

When a salt is dissolved in water the resulting solution may be either neutral, basic, or acidic, depending upon the nature of the salt. A salt formed from a strong acid and a strong base, such as NaCl, yields an aqueous solution that is practically neutral. A salt of a strong base and a weak acid, such as NaOAc, forms aqueous solutions that are basic. A salt of a weak base and a strong acid, such as NH_4Cl, gives an acidic reaction in water. A salt of a weak acid and a weak base may form either neutral, basic, or acidic solutions, depending upon the relative strengths or solubilities of the acid and the base. Many substances in aqueous solution react with water in such a way as to disturb the normal concentration of the hydrogen and hydroxide ions formed through the auto-ionization of water. A reaction of this type in which water is one of the reactants is called **hydrolysis.** The theory of the hydrolysis of salts of the types described above will be discussed in the sections that follow.

As has been pointed out in earlier chapters, ions known to be hydrated in water solution are often indicated for convenience in abbreviated form without showing the hydration; for example, H^+ for H_3O^+, and Cu^{2+} for $[Cu(H_2O)_4]^{2+}$. However, when water is a participant in the reaction, as in hydrolysis reactions, indication of the hydration of the ions becomes important. In dealing with hydrolysis, therefore, formulas are used that show the extent of hydration when it is important for an understanding of the reaction involved.

16.16   Salt of a Strong Acid and a Strong Base

A salt consisting of a cation from a strong hydroxide base (such as NaOH) and an anion from a strong acid (such as HCl) forms a neutral aqueous solution. This means that the ions of sodium chloride do not react with water in such a way as to alter the normal concentration of the hydrogen and hydroxyl ions formed through the ionization of the water ($2H_2O \rightleftharpoons H_3O^+ + OH^-$). Salts resulting from the reactions of the alkali metals and alkaline earth metals with the hydrohalic acids, nitric acid, sulfuric acid, or perchloric acid are examples of salts that give neutral aqueous solutions; i.e., these salts do not undergo appreciable hydrolysis.

16.17   Salt of a Strong Base and a Weak Acid

A salt formed from a strong base and a weak acid contains the conjugate base of the weak acid. For instance, NaOAc, which contains the acetate ion, is the salt of the strong base sodium hydroxide and the weak acid acetic acid. Such salts dissolve in water to give basic solutions. Sodium acetate dissolves in water with a reaction between water and the acetate ion:

$$OAc^- + H_2O \rightleftharpoons HOAc + OH^-$$

Applying the law of mass action to the reaction of the acetate ion with water, we obtain the expression for the equilibrium constant.

$$\frac{[HOAc][OH^-]}{[OAc^-][H_2O]} = K$$

Because the number of moles of water involved in the reaction is negligibly small compared with the total number of moles of water present in dilute solution, we may assume that the molar concentration of the water remains unchanged during the hydrolysis. Hence the preceding equation can be written

$$\frac{[HOAc][OH^-]}{[OAc^-]} = K[H_2O] = K_h$$

This equilibrium constant is simply the ionization constant (Section 16.4) *for the base OAc⁻*. Since the acid-base reaction occurs during the hydrolysis of a salt, such an equilibrium constant is sometimes called a **hydrolysis constant** and may be designated as K_h. When the value for K_h is small, the degree of hydrolysis is slight; when it is large, hydrolysis is extensive. It should be emphasized that a hydrolysis reaction is simply a Brönsted-Lowry acid-base reaction, which may be treated by the same equilibrium concepts developed earlier in this chapter.

If the value of the ionization constant (hydrolysis constant) for the base OAc⁻ cannot be found in a table, its value may be calculated from the values for the ionization constants of water and of acetic acid. For every aqueous solution

$$[H_3O^+][OH^-] = K_w \quad \text{or} \quad [OH^-] = \frac{K_w}{[H_3O^+]}$$

Substituting $K_w/[H_3O^+]$ for $[OH^-]$ in the expression for the hydrolysis constant, we obtain

$$\frac{[HOAc][OH^-]}{[OAc^-]} = \frac{[HOAc]K_w}{[OAc^-][H_3O^+]} = K_h$$

But the expression $\dfrac{[HOAc]}{[OAc^-][H_3O^+]}$ is the reciprocal of K_i for acetic acid and is equal to $\dfrac{1}{K_i \text{ (for HOAc)}}$. Therefore

$$K_h = \frac{K_w}{K_i} = \frac{1.0 \times 10^{-14}}{1.8 \times 10^{-5}} = 5.6 \times 10^{-10}$$

For any salt of a strong base and a weak acid, the hydrolysis constant (or equivalently, the ionization constant of the conjugate base of the weak acid) is given by the expression $K_h = K_w/K_i$, where K_i is the ionization constant of the weak acid.

Calculations involving hydrolysis are identical to those in other acid-base systems.

EXAMPLE 1   **Calculate the hydroxide ion concentration, the degree of hydrolysis, and the *p*H of a 0.050 *M* solution of sodium acetate.**

The equation for the hydrolysis of sodium acetate is

$$OAc^- + H_2O \rightleftharpoons HOAc + OH^-$$

and the expression for the hydrolysis constant is

$$\frac{[HOAc][OH^-]}{[OAc^-]} = K_h = \frac{K_w}{K_i} = 5.6 \times 10^{-10}$$

If we let x equal the concentration of molecular acetic acid formed by the hydrolysis of the acetate ion, then the concentration of hydroxide ion will also be equal to x, and the concentration of acetate ion at equilibrium will be $0.050 - x$. Substituting in the hydrolysis constant expression above, we have

$$\frac{x^2}{0.050 - x} = 5.6 \times 10^{-10}$$

Since x is very small compared with 0.050, we may drop it from the denominator. Then,

$$x^2 = 0.050 \times 5.6 \times 10^{-10} = 28 \times 10^{-12}$$
$$x = 5.3 \times 10^{-6} = [\text{HOAc}] = [\text{OH}^-]$$

Thus the concentration of hydroxide ion is $5.3 \times 10^{-6}\,M$, and the per cent hydrolysis is equal to

$$\frac{[\text{HOAc}]}{[\text{OAc}^-]} \times 100 = \frac{5.3 \times 10^{-6}}{0.050} \times 100 = 0.011\%$$

We can calculate the $p\text{H}$ of the solution by first finding the hydronium ion concentration

$$[\text{H}_3\text{O}^+] = \frac{K_w}{[\text{OH}^-]} = \frac{1.0 \times 10^{-14}}{5.3 \times 10^{-6}} = 1.9 \times 10^{-9}$$

and then

$$p\text{H} = -\log[\text{H}_3\text{O}^+] = -\log(1.9 \times 10^{-9}) = -(0.28 - 9) = 8.72$$

◆ ◆ ◆ ◆ ◆

EXAMPLE 2   Calculate the ionization constant for sulfide ion ($\text{S}^{2-} + \text{H}_2\text{O} \rightleftharpoons \text{HS}^- + \text{OH}^-$), the sulfide ion concentration, and the degree of hydrolysis in a 0.0010 M solution of sodium sulfide.

The sulfide ion hydrolyzes according to the equation

$$\text{S}^{2-} + \text{H}_2\text{O} \rightleftharpoons \text{HS}^- + \text{OH}^-$$

The hydrosulfide ion thus formed also undergoes hydrolysis, but its extent is so slight compared with that of the sulfide ion that it can be neglected. The expression for the ionization constant (or the hydrolysis constant) of the sulfide ion is

$$\frac{[\text{HS}^-][\text{OH}^-]}{[\text{S}^{2-}]} = K_h = \frac{K_w}{K_i \text{ (for HS}^-)} = \frac{1.0 \times 10^{-14}}{1.3 \times 10^{-13}} = 7.7 \times 10^{-2}$$

The large value for the ionization (or hydrolysis) constant indicates that most of the sulfide ion undergoes hydrolysis. For this reason we shall let x equal the S^{2-} concentration when hydrolytic equilibrium has been established. Then the concentrations of HS^- and OH^-, which are equal to each other, may each be represented by $0.0010 - x$. Substituting these values in the expression, we have

$$\frac{[\text{HS}^-][\text{OH}^-]}{[\text{S}^{2-}]} = \frac{(0.0010 - x)(0.0010 - x)}{x} = 7.7 \times 10^{-2}$$

But since hydrolysis is nearly complete, x is now small as compared to 0.0010 and may be neglected in the numerator. The expression then becomes

$$\frac{(0.0010)^2}{x} = 7.7 \times 10^{-2}$$

$$x = \frac{(0.0010)^2}{7.7 \times 10^{-2}} = 1.3 \times 10^{-5} = [S^{2-}]$$

The degree of hydrolysis may be found by dividing the amount of S^{2-} hydrolyzed ($0.0010\ M - 0.000013\ M = 0.00099\ M$) by the total amount of S^{2-} originally present. Then the per cent hydrolysis is obtained by multiplying by 100.

$$\frac{0.00099}{0.0010} \times 100 = 99\%$$

◆ ◆ ◆ ◆ ◆

16.18   Salt of a Weak Base and a Strong Acid

The salt formed by the reaction of a weak base with a strong acid contains the conjugate acid of the weak base. Ammonium chloride, NH_4Cl, is a salt of a weak base and a strong acid. The ammonium ion, the conjugate acid of ammonia, is acidic and reacts with water according to the equation

$$NH_4^+ + H_2O \rightleftharpoons NH_3 + H_3O^+$$

From the law of mass action we find the following expression for the equilibrium constant for this reaction:

$$\frac{[NH_3][H_3O^+]}{[NH_4^+]} = [H_2O]K = K_h$$

The equilibrium constant, K_h, is simply the ionization constant (Section 16.2) for the acid NH_4^+. Since the acid-base reaction occurs during the hydrolysis of a salt, this equilibrium constant is also called the hydrolysis constant of the ammonium ion.

The value of K_h for NH_4^+ may be determined from K_w and the ionization constant, K_i, for NH_3. The expression for the hydrolysis constant for the hydrolysis of the ammonium ion, based on the equation $NH_4^+ + H_2O \rightleftharpoons NH_3 + H_3O^+$, is written as

$$\frac{[NH_3][H_3O^+]}{[NH_4^+]} = K_h$$

Substituting $K_w/[OH^-]$ for $[H_3O^+]$ in the above equation, we obtain

$$\frac{[NH_3]K_w}{[NH_4^+][OH^-]} = K_h$$

Because the expression $\dfrac{[NH_3]}{[NH_4^+][OH^-]}$ is equal to the reciprocal for K_i for NH_3 and hence equal to $\dfrac{1}{K_i \text{ (for NH}_3)}$, we may write

$$K_h = \frac{K_w}{K_i} = \frac{1.0 \times 10^{-14}}{1.8 \times 10^{-5}} = 5.6 \times 10^{-10}$$

The hydrolysis constant for any salt formed from a weak base and a strong acid (or the ionization constant of the conjugate acid of the weak base) may be calculated from the expression $K_h = K_w/K_i$, where K_i is the ionization constant of the weak base.

16.19   Salt of a Weak Base and a Weak Acid

When ammonium acetate is dissolved in water we have, in solution, ammonium ion (an acid) and acetate ion (a base) both of which undergo acid-base reactions. The equations are

$$NH_4^+ + H_2O \rightleftharpoons NH_3 + H_3O^+$$
$$OAc^- + H_2O \rightleftharpoons HOAc + OH^-$$

The extent to which these reactions take place is approximately the same because ammonium ion and acetate ion have nearly equal ionization constants. Therefore the products are present in nearly equal concentrations at equilibrium. As hydronium and hydroxide ions are produced by these reactions they unite to form water, because the product of their concentrations cannot exceed 1×10^{-14}. As these ions are formed and then immediately removed from solution in equal amounts, the solution remains neutral as hydrolysis proceeds.

The net reaction involved in the hydrolysis may be looked upon as the sum of three equations

$$
\begin{array}{r}
NH_4^+ + H_2O \rightleftharpoons NH_3 + H_3O^+ \\
OAc^- + H_2O \rightleftharpoons HOAc + OH^- \\
H_3O^+ + OH^- \rightleftharpoons 2H_2O \\
\hline
NH_4^+ + OAc^- \rightleftharpoons HOAc + NH_3
\end{array}
$$

Applying the law of chemical equilibrium to the net reaction, we obtain

$$\frac{[NH_3][HOAc]}{[NH_4^+][OAc^-]} = K_h$$

To evaluate the hydrolysis constant in terms of the ion product constant for water and the ionization constants for aqueous ammonia and acetic acid, multiply the numerator and denominator of the above equation by $[H_3O^+][OH^-]$.

$$\frac{[NH_3]}{[NH_4^+][OH^-]} \times \frac{[HOAc]}{[H_3O^+][OAc^-]} \times \frac{[H_3O^+][OH^-]}{1} = K_h$$

Thus it may readily be seen that

$$K_h = \frac{K_w}{K_i \text{ (for NH}_3) \times K_i \text{ (for HOAc)}} = \frac{1.0 \times 10^{-14}}{(1.8 \times 10^{-5})(1.8 \times 10^{-5})} = 3.1 \times 10^{-5}$$

When the hydrolytic products of a salt of a weak base and a weak acid do not have the same ionization constants, the aqueous solution of the salt is either basic or acidic, depending on which electrolyte has the larger ionization constant. For example, ammonium cyanide hydrolyzes to give a basic solution, since K_h for NH_3 is larger than K_i for HCN. This means that the hydrolysis of the cyanide ion, $CN^- + H_2O \rightleftharpoons HCN + OH^-$, is more extensive than that of the ammonium ion, $NH_4^+ + H_2O \rightleftharpoons NH_3 + H_3O^+$, and that an excess of hydroxide ions accumulates in the solution.

EXAMPLE   **Calculate the degree of hydrolysis and the pH of a solution that is 1.0 M in NH_4CN.**

When ammonium cyanide is dissolved in water, both the ammonium ion and the cyanide ion undergo hydrolysis. The equations are

$$NH_4^+ + H_2O \rightleftharpoons NH_3 + H_3O^+$$

$$CN^- + H_2O \rightleftharpoons HCN + OH^-$$

$$H_3O^+ + OH^- \rightleftharpoons 2H_2O$$

The net reaction is given by the equation

$$NH_4^+ + CN^- \rightleftharpoons NH_3 + HCN$$

The expression for the hydrolysis constant may be written

$$\frac{[NH_3][HCN]}{[NH_4^+][CN^-]} = K_h = \frac{K_w}{K_i \text{ (for } NH_3) \times K_i \text{ (for } HCN)}$$

$$= \frac{1.0 \times 10^{-14}}{(1.8 \times 10^{-5})(4 \times 10^{-10})} = 1.4$$

It can be shown that, although the hydrolysis of cyanide ion is somewhat more extensive than that of ammonium ion, the difference in the extent of hydrolysis of the two ions is relatively small compared to the total quantity of each that undergoes hydrolysis. It is justifiable to assume that for every NH_4^+ that hydrolyzes, a CN^- also hydrolyzes, provided the NH_4CN concentration is not extremely low. Let x equal the number of moles of NH_4^+ and the number of moles of CN^- undergoing hydrolysis. At equilibrium then, the $[NH_4^+] = (1.0 - x)$ and the $[NH_3] = [HCN] = x$. Substituting, we have

$$\frac{[NH_3][HCN]}{[NH_4^+][CN^-]} = \frac{x^2}{(1.0 - x)^2} = 1.4$$

Extracting the square root of both sides of the equation, we find

$$\frac{x}{1.0 - x} = 1.18$$

$$x = 1.18 - 1.18x$$

$$2.18x = 1.18$$

$$x = 0.54 = [NH_3] = [HCN]$$

The degree of hydrolysis is the amount of NH_4CN hydrolyzed divided by the total amount of NH_4CN originally present. The per cent hydrolysis is

$$\frac{0.54}{1.0} \times 100 = 54\%$$

At equilibrium, $[NH_4^+] = [CN^-] = (1.0 - x) = (1.0 - 0.54) = 0.46$, and substituting in the expression for the ionization constant for HCN, we find

$$\frac{[H^+][CN^-]}{[HCN]} = \frac{[H^+](0.46)}{0.54} = 4 \times 10^{-10}$$

$$[H^+] = \frac{0.54}{0.46} \times (4 \times 10^{-10}) = 4.7 \times 10^{-10}$$

$$pH = -\log{(4.7 \times 10^{-10})} = 9.3 \qquad \text{(a basic solution)}$$

◆ ◆ ◆ ◆ ◆

16.20   The Ionization of Hydrated Metal Ions

A large number of metal ions behave as acids in solution. For example, the aluminum(III) ion in aluminum nitrate reacts with water to give a hydrated aluminum ion, $[Al(H_2O)_6]^{3+}$, which is an acid.

$$Al(NO_3)_3(s) + 6H_2O(l) \longrightarrow [Al(H_2O)_6]^{3+}(aq) + 3NO_3^-(aq)$$
$$[Al(H_2O)_6]^{3+} + H_2O \rightleftharpoons [Al(OH)(H_2O)_5]^{2+} + H_3O^+$$

Hence an excess of hydronium ions is produced in the solution. Because these reactions involve the reaction of a salt [in this case $Al(NO_3)_3$] with water, they too are sometimes called hydrolysis reactions.

The aluminum ion ionizes in stages just as polyprotic acids ionize in more than one stage, as shown by

$$[Al(H_2O)_6]^{3+} + H_2O \rightleftharpoons [Al(OH)(H_2O)_5]^{2+} + H_3O^+$$
$$[Al(OH)(H_2O)_5]^{2+} + H_2O \rightleftharpoons [Al(OH)_2(H_2O)_4]^+ + H_3O^+$$
$$[Al(OH)_2(H_2O)_4]^+ + H_2O \rightleftharpoons [Al(OH)_3(H_2O)_5] + H_3O^+$$

Just as in the ionization of a polyprotic acid, such as H_3PO_4, the ionization of cations carrying more than one charge is not extensive beyond the first stage. Additional examples of the first stage in the ionization of hydrated metal ions are provided by the equations

$$[Fe(H_2O)_6]^{3+} + H_2O \rightleftharpoons [Fe(OH)(H_2O)_5]^{2+} + H_3O^+$$
$$[Cu(H_2O)_4]^{2+} + H_2O \rightleftharpoons [Cu(OH)(H_2O)_3]^+ + H_3O^+$$
$$[Zn(H_2O)_6]^{2+} + H_2O \rightleftharpoons [Zn(OH)(H_2O)_5]^+ + H_3O^+$$

EXAMPLE   Let us calculate the pH of a **0.10 M** solution of aluminum chloride that dissolves to give the hydrated aluminum ion $[Al(H_2O)_6]^{3+}$ in solution. The ionization constant for the reaction $[Al(H_2O)_6]^{3+} + H_2O \rightleftharpoons [Al(OH)(H_2O)_5]^{2+} + H_3O^+$ is **1.4×10^{-5}**.

The expression for the ionization constant is written as

$$\frac{[Al(OH)(H_2O)_5^+][H_3O^+]}{[Al(H_2O)_6^{3+}]} = K_i = 1.4 \times 10^{-5}$$

Let x equal the concentration of the hydronium ion, which is in turn equal to the concentration of $[Al(OH)(H_2O)_5]^{2+}$. Then $0.10 - x$ will equal the concentration of $[Al(H_2O)_6]^{3+}$. Substituting in the expression for the ionization constant, we have

$$\frac{[Al(OH)(H_2O)_5^{+}][H_3O^{+}]}{[Al(H_2O)_6^{3+}]} = \frac{x^2}{0.10 - x} = 1.4 \times 10^{-5}$$

Because K_h is small, x will also be small, and we may drop the x in the term $0.10 - x$ in an approximate solution of the problem. Thus we obtain

$$\frac{x^2}{0.10} = 1.4 \times 10^{-5}$$

$$x = 1.2 \times 10^{-3} = [H^+]$$

$$pH = -\log(1.2 \times 10^{-3}) = 2.9$$

◆ ◆ ◆ ◆ ◆

Because the constants for the different stages of ionization are not known for most metal ions, we cannot calculate the degree of ionization for many metal ions we know to be ionized in solution. However, if the hydroxide of a metal ion is insoluble in water, we may conclude that the metal ion will hydrolyze extensively to give a very acidic solution. In fact, practically all metal ions other than those of the alkali metals hydrolyze to give acidic solutions. The hydrolysis constants increase as the charge of the metal ion increases or as the size of the metal ion decreases.

Questions

1. Equilibrium calculations are not necessary when dealing with ionic concentrations in solutions of strong electrolytes such as NaOH or HCl. Why not?
2. What ionic and molecular species are present in an aqueous solution of hydrogen fluoride, HF? a solution of sulfuric acid? a solution of SO_2 in water (sulfurous acid)?
3. Classify each of the following compounds as a strong or weak electrolyte: HCl, NaOH, NaCl, HOAc, H_2SO_4, $CaCl_2$, $HgCl_2$, $Ca(OH)_2$, H_3PO_4, NH_3, HCN.
4. How is the extent of ionization of ammonia in a solution changed by addition of HCl? of NH_4Br?
5. Describe the self-ionization of water.
6. Compare the extent of ionization of strong electrolytes and weak electrolytes in water.
7. Define pH both in words and by a mathematical equation.
8. Why does an acid-base indicator change color over a range of pH rather than at a specific pH?
9. Explain why successive ionization constants of polyprotic acids become progressively smaller.
10. Why does the salt of a strong acid and a weak base give an acidic solution?
11. Why does the salt of a weak acid and a strong base give a basic solution?
12. Why does the salt of a strong acid and a strong base give an approximately neutral solution?
13. Does the salt of a weak acid and a weak base give an acidic solution, a basic solution, or a neutral solution? Explain your answer.
14. Why is a hydrolysis reaction simply an acid-base reaction?
15. Arrange the conjugate bases of the acids in Table 16-2 in order of increasing base strength.
16. Arrange the conjugate acids of the bases in Table 16-3 in order of decreasing acid strength.
17. Write equations for the formation of the conjugate

acids or bases (hydrolysis) of the following species: HF, CO_3^{2-}, CN^-, H_3PO_4, NH_4^+, SO_3^{2-}, F^-, HNO_2, H_2CO_3.

18. "Neutralization" is defined in Section 14.7 as the reaction of an acid with a stoichiometric amount of a base. Why don't all neutralization reactions give neutral solutions?

19. Show that the rate of reaction of acetic acid with sodium hydroxide is given by the following expression:

$$\text{Rate} = \frac{kK_i\,[HOAc][OH^-]}{[OAc^-]}$$

20. Using the K_i values in Appendix G and in Section 16.20, place $[Al(H_2O)_6]^{3+}$ in the correct location in Table 14-1, Section 14.15, the table of relative strengths of conjugate acid-base pairs.

Problems

Values of K_i not given may be found in Appendices G and H.

1. Calculate the concentration of each of the ions in the following solutions of strong electrolytes:
   [S] (a) 0.0090 *M* HBr *Ans.* $[H^+]$ *0.0090 M;* $[Br^-]$ *0.0090 M*
   (b) 0.0102 *M* Li_2SO_4 *Ans.* $[Li^+]$ *0.0204 M;* $[SO_4^{2-}]$ *0.0102 M*
   [S] (c) 0.0033 *M* $[Al(H_2O)_6]_2(SO_4)_3$ *Ans.* $[Al(H_2O)_6^{3+}]$ *6.6 × 10⁻³ M;* $[SO_4^{2-}]$ *9.9* $\times 10^{-3}$ *M*

2. Using the ionization constants in Appendix G, calculate the hydrogen ion concentration and the per cent ionization in each of the following solutions:
   [S] (a) 0.0092 *M* HClO *Ans.* 1.8×10^{-5} *M; 0.20%*
   (b) 0.0810 *M* HCN *Ans.* 6×10^{-6} *M;* 7×10^{-3}*%*
   [S] (c) 0.417 *M* HCO_2H *Ans.* 8.7×10^{-3} *M; 2.1%*
   (d) 0.1173 *M* HN_3 *Ans.* 3×10^{-3} *M; 3%*
   (e) 0.0010 *M* HBrO *Ans.* 1×10^{-6} *M; 0.1%*

3. Calculate the hydrogen ion concentration and per cent ionization of the weak acid in each of the following solutions. Note that the ionization constants may be such that the change in electrolyte concentration cannot be neglected and the quadratic equation or successive approximations may be required.
   [S] (a) 0.0184 *M* HCNO ($K_i = 3.46 \times 10^{-4}$) *Ans.* 2.36×10^{-3} *M; 12.8%*
   (b) 0.100 *M* HF ($K_i = 7.2 \times 10^{-4}$) *Ans.* 8.1×10^{-3} *M; 8.1%*
   (c) 0.0655 *M* $[Fe(H_2O)_6]^{3+}$ ($K_i = 4.0 \times 10^{-3}$ for $[Fe(H_2O)_6]^{3+} \rightleftharpoons [Fe(OH)(H_2O)_5]^{2+} + H^+$) *Ans.* 1.4×10^{-2} *M; 22%*
   [S] (d) 0.02173 *M* CH_2ClCO_2H ($K_i = 1.4 \times 10^{-3}$) *Ans.* 4.9×10^{-3} *M; 22%*
   (e) 0.10 *M* HSO_3NH_2 ($K_i = 1.0 \times 10^{-1}$) *Ans. 0.062 M; 62%*

[S] 4. The ionization constant of lactic acid, $CH_3CHOHCO_2H$, is 1.36×10^{-4}. If 20.0 g of lactic acid is used to make a solution with a volume of 1.00 ℓ, what is the concentration of hydrogen ion in the solution? *Ans.* 5.49×10^{-3} *M; note that with three significant figures the answer is different* (5.43×10^{-3} *M*) *if the simplifying assumption is not made.*

5. Calculate the per cent ionization of a 0.50 *M* solution of H_3PO_4. (Consider the second and third ionization steps to be negligible compared to the first.) *Ans. 12%*

6. What is the fluoride ion concentration of 0.750 ℓ of a solution that contains 0.1500 g of HF? *Ans.* 2.3×10^{-3} *M*

7. Calculate the hydroxide ion concentration and the per cent ionization of the weak base in each of the following solutions. (In some cases it may prove necessary to use the quadratic equation or successive approximations.)
   [S] (a) 0.0784 *M* $C_6H_5NH_2$ ($K_i = 4.6 \times 10^{-10}$) *Ans.* 6.0×10^{-6}*M;* 7.6×10^{-3}*%*
   (b) 0.3124 *M* NH_3 ($K_i = 1.8 \times 10^{-5}$) *Ans.* 2.4×10^{-3} *M; 0.77%*
   [S] (c) 4.113×10^{-3} *M* CH_3NH_2 ($K_i = 4.4 \times 10^{-4}$) *Ans.* 1.1×10^{-3}*; 28%*
   (d) 0.222 *M* CN^- ($K_i = 2.5 \times 10^{-5}$) *Ans.* 2.4×10^{-3} *M; 1.1%*
   (e) 0.11 *M* $(CH_3)_2NH$ ($K_i = 7.4 \times 10^{-4}$) *Ans.* 8.7×10^{-3} *M; 7.9%*

(f) 0.300 M N_2H_4 ($K_i = 3 \times 10^{-6}$ for $N_2H_4 + H_2O \rightleftharpoons N_2H_5^+ + OH^-$)
Ans. 9×10^{-4} *M*; 0.3%

8. Calculate the pH and pOH of the following solutions:
[S] (a) 0.100 M HCl
Ans. $pH = 1.000$; $pOH = 13.00$
[S] (b) 1.45 M NaOH
Ans. $pH = 14.16$; $pOH = -0.161$
[S] (c) 0.0071 M $Ba(OH)_2$
Ans. $pH = 12.15$; $pOH = 1.85$
(d) 0.03 M HNO_3 *Ans.* $pH = 1.5$; $pOH = 12.5$

9. Calculate the pH and pOH of the aqueous solutions in problem 2.
Ans. (a) $pH = 4.74$, $pOH = 9.26$; (b) 5.2, 8.8; (c) 2.06, 11.94; (d) 2.5, 11.5; (e) 6.0, 8.0

10. Calculate the pH and pOH of the aqueous solutions in Problem 3.
Ans. (a) $pH = 2.627$, $pOH = 11.37$; (b) 2.09, 11.91; (c) 1.85, 12.15; (d) 2.31, 11.69; (e) 1.21, 12.79

11. Calculate the pH and pOH of the aqueous solutions in Problem 7. *Ans.* (a) $pH = 8.78$, $pOH = 5.22$; (b) 11.38, 2.62; (c) 11.04, 2.96; (d) 11.38, 2.62; (e) 11.94, 2.06; (f) 11.0, 3.0

12. Calculate the ionization constants for each of the following solutes from the per cent ionization and the concentration of the solute:
[S] (a) 0.050 M HClO, 8.4×10^{-2}% ionized
Ans. 3.5×10^{-8}
[S] (b) 0.010 M HNO_2, 19% ionized *Ans.* 4.4×10^{-4}
(c) 1.0 M NH_3, 1.0% ionized *Ans.* 1.0×10^{-4}
(d) 0.10 M HF, 8.1% ionized *Ans.* 7.1×10^{-4}
(e) 0.300 M CH_2ClCO_2H, 6.60% ionized
Ans. 1.40×10^{-3}
(f) 1.0×10^{-4} M H_2O_2, 1.5×10^{-2}% ionized
Ans. 2.2×10^{-12}

[S] 13. What is the concentration of acetic acid in a solution that is 0.30% ionized? *Ans.* 2.0 *M*

14. Calculate the pH, pOH, and total methylamine concentration in a solution of methylamine that is 6.90% ionized. *Ans.* pH 11.77; $pOH = 2.23$; total $CH_3NH_2 = 0.086$ *M*

15. A 0.010 M solution of HF is 23% ionized. Calculate $[H^+]$, $[F^-]$, [HF], and the ionization constant for HF.
Ans. $[H^+] = [F^-] = 2.3 \times 10^{-3}$ *M*; $[HF] = 7.7 \times 10^{-3}$ *M*; $K_i = 6.9 \times 10^{-4}$

16. Calculate the pH of each of the following solutions, which contain two solutes with the concentrations indicated.
[S] (a) 0.50 M HOAc, 0.50 M NaOAc *Ans.* 4.74
(b) 0.100 M HN_3, 0.200 M NaN_3 *Ans.* 4.3
(c) 0.225 M HCNO, 0.315 M NaCNO *Ans.* 3.607
(d) 0.400 M HNO_2, 1.875 M $Ca(NO_2)_2$
Ans. 4.32
[S] (e) 0.125 M NH_3, 1.00 M NaOH *Ans.* 14.00
[S] (f) 0.400 M $NaHSO_4$, 0.400 M Na_2SO_4
Ans. 1.95

[S] 17. What relative number of moles of sodium acetate and acetic acid in water must be used to prepare a buffer with a pH of 5.08? of 4.20?
Ans. 2.2 to 1.0; 0.29 to 1.0

18. Addition of 1.0 ml of 0.10 M NaOH to 0.100 ℓ of the buffer solution discussed in Section 16.11 changed the pH from 4.74 to 4.75. What is the new pH when 1.0 ml of 0.10 M NaOH is added to 0.100 ℓ of an unbuffered solution of hydrochloric acid with a pH of 4.74? *Ans.* 10.99

[S] 19. A buffer solution is made up of equal volumes of 0.100 M acetic acid and 0.500 M sodium acetate. (a) What is the pH of this solution? (b) What is the pH that results from adding 1.00 ml of 0.100 M HCl to 0.200 ℓ of the buffer solution? (Use 1.80×10^{-5} for the ionization constant of acetic acid.)
Ans. 5.444; 5.438

20. Calculate the concentration of each of the species present in each of the following solutions:
(a) 0.050 M H_2S *Ans.* $[H^+] = 7.1 \times 10^{-5}$ *M*; $[OH^-] = 1.4 \times 10^{-10}$ *M*; $[H_2S] = 0.050$ *M*; $[HS^-] = 7.1 \times 10^{-5}$ *M*; $[S^{2-}] = 1.3 \times 10^{-13}$ *M*
(b) 0.010 M $C_6H_4(CO_2H)_2$ (phthalic acid)

$$C_6H_4(CO_2H)_2 \rightleftharpoons C_6H_4(CO_2H)(CO_2^-) + H^+ \quad (K_i = 1.1 \times 10^{-3})$$

$$C_6H_4(CO_2H)(CO_2^-) \rightleftharpoons C_6H_4(CO_2^-)_2 + H^+ \quad (K_i = 3.9 \times 10^{-6})$$

Ans. $[H^+] = 2.8 \times 10^{-3}$ *M*; $[OH^-] = 3.6 \times 10^{-12}$ *M*, $[C_6H_4(CO_2H)_2] = 7.2 \times 10^{-3}$ *M*; $[C_6H_4(CO_2H)(CO_2^-)] = 2.8 \times 10^{-3}$ *M*; $[C_6H_4(CO_2^-)_2] = 3.9 \times 10^{-6}$ *M*

(c) 0.0250 M Na_3PO_4

$$PO_4^{3-} + H_2O \rightleftharpoons HPO_4^{2-} + OH^- \quad (K_i = 8.0 \times 10^{-2})$$

$$HPO_4^{2-} + H_2O \rightleftharpoons H_2PO_4^- + OH^-$$
$$(K_i = 1.6 \times 10^{-7})$$

$$H_2PO_4^- + H_2O \rightleftharpoons H_3PO_4 + OH^-$$
$$(K_i = 1.3 \times 10^{-12})$$

Ans. $[H^+] = 5.0 \times 10^{-13}$ *M;*
$[OH^-] = 2.0 \times 10^{-2}$ *M;*
$[PO_4^{3-}] = 5.0 \times 10^{-3}$ *M;*
$[HPO_4^{2-}] = 2.0 \times 10^{-2}$ *M;*
$[H_2PO_4^-] = 1.6 \times 10^{-7}$ *M;*
$[H_3PO_4] = 1.0 \times 10^{-17}$ *M;* $[Na^+] = 0.0750$ *M*

21. Calculate the *p*H of each of the following solutions:
    (a) 0.4735 *M* NaCN   *Ans. 11.5*
    (b) 0.1050 *M* NH_4NO_3 *Ans. 5.12*
    (c) 0.0270 *M* $Ca(N_3)_2$ *Ans. 8.4*
    (d) 0.333 *M* $[(CH_3)_2NH_2]_2SO_4$ [Note that $(CH_3)_2NH_2^+$ is the conjugate acid of the weak base $(CH_3)_2NH$, just as NH_4^+ is the conjugate acid of the weak base NH_3.]   *Ans. 5.52*
    (e) 0.100 *M* NH_4F *Ans. 6.20*

22. Calculate the ionization constants (hydrolysis constants) for each of the following acids or bases:
    (a) F^- *Ans. 1.4×10^{-11}*
    (b) AsO_4^{3-} *Ans. 1×10^{-1}*
    (c) NO_2^- *Ans. 2.2×10^{-11}*
    (d) NH_4^+ *Ans. 5.6×10^{-10}*
    (e) $(CH_3)_2NH_2^+$ *Ans. 1.4×10^{-11}*
    (f) $HC_2O_4^-$ (as a base)   *Ans. 1.7×10^{-13}*
    (g) S^{2-} *Ans. 7.7×10^{-2}*

23. A 0.0010 *M* solution of KCN is hydrolyzed to the extent of 14.0%. Calculate the value of the ionization constant of HCN.   *Ans. 4.4×10^{-10}*

24. Calculate the *p*H of a 0.470 *M* solution of Li_2CO_3.   *Ans. 11.9*

25. What concentration of calcium acetate, $Ca(OAc)_2$, will give a solution in which the per cent hydrolysis is 0.10%?   *Ans. 2.8×10^{-4} M*

26. White vinegar is a 5.0% by mass solution of acetic acid in water. If the density of white vinegar is 1.007 g/cm^3, what is the *p*H?   *Ans. 2.41*

27. A typical urine sample contains 2.3% by mass of the base urea, $CO(NH_2)_2$

$$CO(NH_2)_2 + H_2O \rightleftharpoons CO(NH_2)(NH_3)^+ + OH^-$$
$$(K_i = 1.5 \times 10^{-14})$$

If the density of the urine sample is 1.06 g/cm^3 and the *p*H of the sample is 6.35, calculate the concentration of $CO(NH_2)_2$ and $CO(NH_2)(NH_3)^+$ in the sample.   *Ans. $[CO(NH_2)_2] = 0.41$ M;* $[CO(NH_2)(NH_3)^+] = 2.7 \times 10^{-7}$ M

28. The solutions of ammonia used for cleaning windows usually contain about 10% NH_3 by mass. If a solution that is exactly 10% by mass has a density of 0.99 g/cm^3, what is the *p*H and hydroxide ion concentration?   *Ans. 12.01; 1.0×10^{-2} M*

29. In many detergents, phosphates have been replaced with silicates as water conditioners. If 125 g of a detergent that contains 8.0% Na_2SiO_3 by weight is used in 4.0 *l* of water, what is the *p*H and hydroxide ion concentration in the wash water?

$$SiO_3^{2-} + H_2O \rightleftharpoons SiO_3H^- + OH^-$$
$$K_i = 1.6 \times 10^{-3}$$

$$SiO_3H^- + H_2O \rightleftharpoons SiO_3H_2 + OH^-$$
$$K_i = 3.1 \times 10^{-5}$$

Ans. pH = 11.70; $[OH^-] = 5.0 \times 10^{-3}$ M

30. In venous blood, the following equilibrium is set up by dissolved carbon dioxide

$$H_2CO_3 \rightleftharpoons H^+ + HCO_3^-$$

If the *p*H of the blood is 7.4, what is the ratio of $[HCO_3^-]$ to $[H_2CO_3]$?   *Ans. 11 to 1*

31. Saccharin, $C_7H_5NSO_3$, is a weak acid ($K_i = 2.1 \times 10^{-12}$). If 0.250 *l* of diet cola with a *p*H of 5.48 was prepared from 2.00×10^{-3} g of sodium saccharide (often referred to as "sodium saccharin" on soft drink bottles), $Na(C_7H_4NSO_3)$, what are the final concentrations of saccharin and "sodium saccharin" in the solution?   *Ans. $[C_7H_5NSO_3] = 3.9 \times 10^{-5}$ M;* $[C_7H_4NSO_3Na] = 2.5 \times 10^{-11}$ M

32. Novocaine, $C_{13}H_{21}O_2N_2Cl$, is the salt of the base procaine with hydrochloric acid. The ionization constant for procaine is 7×10^{-6}. What is the *p*H of a 2.0% solution of novocaine by mass assuming that the density of the solution is 1.0 g/cm^3?   *Ans. 5.0*

33. The ionization constant for water (K_w) is 9.614×10^{-14} at 60°C. Calculate $[H^+]$, $[OH^-]$, *p*H, and *p*OH for pure water at 60°.   *Ans. $[H^+] = [OH^-] = 3.101 \times 10^{-7}$ M;* $pH = pOH = 6.5085$

34. The ionization constant for water, K_w, is 5.474×10^{-14} at 50°C. At 50°, K_i for NH_3 is

1.892×10^{-5}. What is the ionization constant for NH_4^+ at 50°? *Ans.* 2.893×10^{-9}

35. If abominable snowmen were to exist in the world in the same proportion to humans as do hydrogen ions to water molecules in pure water, how many abominable snowmen would there be in the world? Assume a world population of 3.9 billion people. *Ans. 7*

36. Calculate the *p*H of 0.500 ℓ of a 0.0880 *M* solution of $NaHCO_3$ to which has been added 1.10×10^{-2} mole of HCl. *Ans. 6.84*

37. How many moles of sodium acetate must be added to 1 ℓ of a 1.0 *M* acetic acid solution to prepare a buffer solution with a *p*H of 6.0? of 5.08? of 4.20? *Ans. 18 mol; 2.2 mol; 0.29 mol*

[S]38. Calculate the *p*H of a solution resulting from mixing 0.10 ℓ of 0.10 *M* NaOH with 0.40 ℓ of 0.025 *M* HF. *Ans. 7.72*

39. Calculate the *p*H of a solution made by mixing equal volumes of 0.30 *M* NH_3 and 0.030 *M* HNO_3. *Ans. 10.21*

40. Calculate the *p*H of a solution prepared by mixing 10.0 ml of 0.10 *M* LiOH and 20.0 ml of a 0.500 *M* benzoic acid solution. K_i for benzoic acid is 6.7×10^{-5}. *Ans. 3.22*

41. Calculate the *p*H of 0.500 ℓ of a 0.120 *M* solution of HClO to which 0.010 mol of $Ca(OH)_2$ has been added. *Ans. 7.15*

42. How many moles of hydrogen chloride must be added to 1.50 ℓ of a 0.450 *M* solution of CH_3NH_2 to give a *p*H of 10.95? *Ans. 0.22 mol*

43. Given a 0.1 *M* solution of ammonia that is also 0.1 *M* in ammonium chloride (a buffer mixture), and a solution of 1.8×10^{-5} *M* NaOH, calculate (a) the initial *p*H in each solution, (b) the *p*H when a liter of each of the original solutions is treated with 0.03 mol of solid sodium hydroxide, and (c) the *p*H when a liter of each of the original solutions is treated with 0.03 mol of HCl. (Assume no volume change with the addition of the NaOH or the HCl.) Is the buffer solution effective in serving its intended function of holding the [H^+] relatively constant compared to the unbuffered solution? *Ans. (a) 9.3 for each solution; (b) 9.5; 12.5; (c) 9.0; 1.5*

44. A buffer solution is prepared from 5.0 g of NH_4NO_3 and 0.100 ℓ of 0.10 *M* NH_3. What is the *p*H of the buffer? *Ans. 8.46*

45. Calculate the *p*H of 0.750 ℓ of a solution that contains 87.5 g of $Ba(N_3)_2$ and 0.250 mol of HN_3. *Ans. 4.5*

46. What is the *p*H of 1.000 ℓ of a solution of 100.0 g of glutamic acid ($C_5H_9NO_4$, a dibasic acid; $K_1 = 8.5 \times 10^{-5}$, $K_2 = 3.39 \times 10^{-10}$) to which has been added 20.0 g of NaOH during the preparation of monosodium glutamate, the flavoring agent? What is the *p*H when exactly one mole of NaOH per one mole of acid has been added? *Ans. 4.51; 4.82*

47. Calculate the *p*H of a solution made by mixing equal volumes of 0.040 *M* aqueous aniline, $C_6H_5NH_2$, and 0.040 *M* nitric acid. (K_i for $C_6H_5NH_2 = 4.6 \times 10^{-10}$.) *Ans. 3.18*

48. How many moles of CH_3CO_2Na (sodium acetate) must be used to prepare 2.0 ℓ of an aqueous solution with *p*H = 9.40? (K_i for CH_3CO_2H is 1.8×10^{-5}.) *Ans. 2.3 mol*

49. Using the data presented in Table 16-5, arrange the following indicators (which are weak acids) in increasing order of their acid strength: methyl orange, litmus, thymol blue. *Ans. litmus < methyl orange < thymol blue*

50. The indicator dinitrophenol is an acid with a K_i of 1.1×10^{-4}. In a 1.0×10^{-4} *M* solution it is colorless in acid and yellow in base. Calculate the *p*H range over which it converts from 10% ionized (colorless) to 90% ionized (yellow). *Ans. 3.0–4.9*

51. A 0.5000-g sample of an impure monobasic amine is titrated to a good end point with 75.00 ml of 0.100 *M* HCl. The *p*H after 40.00 ml of acid is added is 10.65. The molecular weight of the pure base is known to be 59.1. (a) Calculate the ionization constant of this base and (b) its per cent purity. (c) Which indicator from the following list would be most suitable for this titration? State your reasoning.

(a) Orange IV (red–yellow; *p*H 1.2–2.6)
(b) Methyl orange (red–yellow; *p*H 3.1–4.4)
(c) Methyl red (red–yellow; *p*H 4.8–6.0)
(d) Bromthymol blue (yellow–blue; *p*H 6.0–7.6)
(e) Phenolphthalein (colorless–pink; *p*H 8.3–10.0)
(f) Thymolphthalein (colorless–blue; *p*H 9.3–10.5)

Ans. (a) 5.1×10^{-4}; *(b) 88.7%; (c) methyl red*

[S] 52. Lime juice is among the most acidic of fruit juices with a pH of 1.92. If the acidity is due to citric acid, which we can abbreviate as H_3Cit, what is the ratio of each of the following species to $[Cit^{3-}]$: $[H_3Cit]$; $[H_2Cit^-]$; $[HCit^{2-}]$?

$$H_3Cit \rightleftharpoons H^+ + H_2Cit^- \qquad K_i = 8.4 \times 10^{-4}$$
$$H_2Cit^- \rightleftharpoons H^+ + HCit^{2-} \qquad K_i = 1.8 \times 10^{-5}$$
$$HCit^{2-} \rightleftharpoons H^+ + Cit^{3-} \qquad K_i = 4.0 \times 10^{-6}$$

Ans. 2.9×10^7; 2.0×10^6; 3.0×10^3 *to 1*

53. Sulfuric acid is a strong acid. The ionization, $H_2SO_4 \rightleftharpoons H^+ + HSO_4^-$, is effectively 100% complete. The hydrogen sulfate ion, HSO_4^-, is a weaker acid with $K_i = 1.2 \times 10^{-2}$.
(a) What is the concentration of sulfate ion, SO_4^{2-}, in a 0.102 M solution of $NaHSO_4$?

Ans. 0.029 M

(b) What is the concentration of sulfate ion in a 0.102 M solution of H_2SO_4?

Ans. 9.9×10^{-3} M

54. Sulfuric acid ionizes in two steps:

$$H_2SO_4 \rightleftharpoons H^+ + HSO_4^-$$
$$HSO_4^- \rightleftharpoons H^+ + SO_4^{2-}$$

Since sulfuric acid is a strong acid, the first step is effectively 100% complete ($K_{H_2SO_4}$ is large). The ionization constant for the second step, $K_{HSO_4^-}$, is 1.2×10^{-2}. Check the assumption that the first step goes completely to the right, based upon the approximation that for a polyprotic acid, successive ionization constants often differ by a factor of 1×10^6 and calculate $K_{H_2SO_4}$, (b) the concentration of H_2SO_4, (c) the per cent ionization of H_2SO_4 into H^+ and HSO_4^- in a 0.100 M solution of H_2SO_4.

Ans. (a) 1×10^4; (b) 1×10^{-6} M; (c) 99.999%

References

"The Use of Electron Balance in Ionic Equilibrium Calculations," A.A.S.C. Machado, *J. Chem. Educ.*, **53,** 305 (1976).

"A Criterion for the Simple Approximation in Dissociation Equilibria," A. J. Leffler, *J. Chem. Educ.*, **53,** 460 (1976).

"A Note on the Calculation of Concentrations in the Case of Many Simultaneous Equilibria," T. P. Feenstra, *J. Chem. Educ.*, **56,** 104 (1979).

"Exact Equation for Calculating Titration Curves for Dibasic Salts," G. E. Knudson and D. Nimrod, *J. Chem. Educ.*, **54,** 351 (1977).

"Stepwise Formation Constants of Complex Ions," W. B. Guenther, *J. Chem. Educ.*, **44,** 46 (1967).

"Component Concentrations in Solutions of Weak Acids," D. M. Goldfish, *J. Chem. Educ.*, **47,** 65 (1970).

"The *p*H Concept," D. Kolb, *J. Chem. Educ.*, **56,** 49 (1979).

"On Calculating [H+]," J. D. Burke, *J. Chem. Educ.*, **53,** 79 (1976).

"Hydrolysis of Sodium Carbonate," F. S. Nakayama, *J. Chem. Educ.*, **47,** 67 (1970).

"Errors in Calculating Hydrogen Ion Concentration," J. E. House, Jr. and R. C. Reiter, *J. Chem. Educ.*, **45,** 679 (1968).

"Development of the pH Concept; A Historical Perspective," F. Szabadváry (transl. by Ralph E. Oesper), *J. Chem. Educ.*, **41,** 105 (1964).

"Indicators; A Historical Perspective," F. Szabadváry (transl. by Ralph E. Oesper), *J. Chem. Educ.*, **41,** 285 (1964).

"Equilibrium Constants from Spectrophotometric Data: Principles, Practice, and Programming," R. W. Ramette, *J. Chem. Educ.*, **44,** 647 (1967).

"Rapid Graphical Method for Determining Formation Constants," S. D. Christian, *J. Chem. Educ.*, **45,** 713 (1968).

"Corrections for Simple Equations for Titration Curves of Monoprotic Acids," W. P. Cortelyou, *J. Chem. Educ.*, **45,** 677 (1968).

"Acid-Base Titration and Distribution Curves," J. Waser, *J. Chem. Educ.*, **44,** 274 (1967).

"Dissociation of Weak Acids and Bases at Infinite Dilution," D. I. Stock, *J. Chem. Educ.*, **44,** 764 (1967).

"pK Values for D_2O and H_2O," R. G. Bates, R. A. Robinson, and A. K. Covington, *J. Chem. Educ.*, **44,** 635 (1967).

"Highly Basic Media," J. R. Jones, *Chem. in Britain,* **7,** 336 (1971).

"The Equilibrium between a Solid Solution and an Aqueous Solution of its Ions," A. F. Berndt and R. L. Stearns, *J. Chem. Educ.*, **50,** 415 (1973).

"Effect of Ionic Strength on Equilibrium Constants," M. D. Seymour and Q. Fernando, *J. Chem. Educ.*, **54,** 225 (1977).

17

The Solubility Product Principle

In chemical reactions that involve formation of solids **(precipitation),** the chemist works with slightly soluble substances that are strong electrolytes. Equilibrium systems consisting of slightly soluble strong electrolytes (ionic substances) in contact with their saturated aqueous solutions are particularly important in analytical chemistry. In this chapter we shall be concerned with the theory and application of the precipitation and dissolution of such slightly soluble ionic substances.

The Formation of Precipitates

17.1 The Solubility Product Constant

A slightly soluble electrolyte dissolves in water until a saturated solution of its ions is formed. A **saturated solution** is defined as one that is in equilibrium with undissolved solute. It can be shown by the use of radioactive tracers that at equilibrium the solid (the undissolved solute) continues to dissolve, the process of crystallization occurs, and the opposing processes have equal rates. Let us consider a saturated solution of silver chloride in equilibrium with some crystals of silver chloride. The equilibrium can be expressed by

$$AgCl(s) \underset{\text{Precipitation}}{\overset{\text{Dissolution}}{\rightleftharpoons}} Ag^+(aq) + Cl^-(aq) \quad \text{(saturated solution)}$$

453

Applying the law of mass action to this system, we obtain the mathematical expression

$$\frac{[Ag^+][Cl^-]}{[AgCl]} = K$$

Because the concentration of the solid silver chloride remains constant, we can write

$$[Ag^+][Cl^-] = K\,[AgCl] = K_{sp}$$

The K_{sp} is called the **solubility product constant,** or sometimes simply the **solubility product.** It is important to realize that the product of the concentrations must be equal to the solubility product constant only when a saturated solution is in contact with undissolved solute and when a state of equilibrium exists.

Calcium phosphate dissolves according to the equation

$$Ca_3(PO_4)_2(s) \rightleftharpoons 3Ca^{2+}(aq) + 2PO_4{}^{3-}(aq)$$

The expression for its solubility product constant is

$$[Ca^{2+}]^3[PO_4{}^{3-}]^2 = K_{sp}$$

For a salt like magnesium ammonium phosphate, which dissolves according to the equation

$$MgNH_4PO_4(s) \rightleftharpoons Mg^{2+}(aq) + NH_4{}^+(aq) + PO_4{}^{3-}(aq)$$

we have

$$[Mg^{2+}][NH_4{}^+][PO_4{}^{3-}] = K_{sp}$$

As is the case for all equilibrium constants (Section 15.18), solubility product constants have units. As pointed out earlier, the units are often omitted from equilibrium constant values, and this practice is usually followed in this book.

The solubility product constant is a constant only for solutions of *slightly* soluble electrolytes and is not constant for solutions of moderately or quite soluble salts, such as NaCl and $KClO_3$. Furthermore, even the presence of high concentrations of ions from an added electrolyte not involved in the equilibrium causes an increase in the solubility of a slightly soluble substance—the salt effect (Section 16.5). The salt effect varies with the concentration of the added electrolyte and with the charges on the ions the salt supplies to the solution. Hence the solubility product constant is *not* constant except for saturated solutions in which the total ionic concentration is extremely small.

17.2 Calculation of Solubility Product Constants from Molar Solubilities

When the solubility of a slightly soluble electrolyte is known, its solubility product constant can be calculated. In all calculations involving solubility product constants, the concentrations are expressed in moles per liter. Thus when the concentrations are given in other units, such as grams per liter, they must be converted to moles per liter.

EXAMPLE 1   **Calculate the solubility product constant for silver chromate, Ag_2CrO_4, whose molar solubility is 1.3×10^{-4} mol ℓ^{-1}.**

The concentration of Ag_2CrO_4 in the solution is 1.3×10^{-4} M, and the Ag_2CrO_4 is completely ionized. From the equation for the dissolution of Ag_2CrO_4

$$Ag_2CrO_4(s) \longrightarrow 2Ag^+(aq) + CrO_4^{2-}(aq)$$

we can obtain the conversion factors 2 mol Ag^+/1 mol Ag_2CrO_4 and 1 mol CrO_4^{2-}/1 mol Ag_2CrO_4. The concentration of silver ion will be twice that of the molar solubility of Ag_2CrO_4 and the concentration of CrO_4^{2-} will be equal to the molar solubility of Ag_2CrO_4, or

$$\frac{1.3 \times 10^{-4} \text{ mol } Ag_2CrO_4}{1 \ell} \times \frac{2 \text{ mol } Ag^+}{1 \text{ mol } Ag_2CrO_4} = \frac{2.6 \times 10^{-4} \text{ mol } Ag^+}{1 \ell}$$

$$\frac{1.3 \times 10^{-4} \text{ mol } Ag_2CrO_4}{1 \ell} \times \frac{1 \text{ mol } CrO_4^{2-}}{1 \text{ mol } Ag_2CrO_4} = \frac{1.3 \times 10^{-4} \text{ mol } CrO_4^{2-}}{1 \ell}$$

Thus $[Ag^+] = 2.6 \times 10^{-4}$ M and $[CrO_4^{2-}] = 1.3 \times 10^{-4}$ M. Substituting in the expression for the solubility product,

$$[Ag^+]^2[CrO_4^{2-}] = K_{sp}$$
$$(2.6 \times 10^{-4})^2(1.3 \times 10^{-4}) = 8.8 \times 10^{-12} = K_{sp}$$

Note that $[Ag^+]$ is squared in this expression.

◆ ◆ ◆ ◆ ◆

EXAMPLE 2   **The solubility of $BaSO_4$ in water is 2.42×10^{-3} g ℓ^{-1}. Calculate the solubility product constant for $BaSO_4$.**

First calculate the molar solubility of $BaSO_4$. The formula weight of $BaSO_4$ is 233.4.

$$\frac{2.42 \times 10^{-3} \text{ g } BaSO_4}{1 \ell} \times \frac{1 \text{ mol } BaSO_4}{233.4 \text{ g } BaSO_4} = \frac{1.04 \times 10^{-5} \text{ mol } BaSO_4}{1 \ell} = 1.04 \times 10^{-5} \text{ } M$$

Since the salt ($BaSO_4$) dissolves to give Ba^{2+} and SO_4^{2-} in equal molar amounts (1 mol Ba^{2+}/1 mol $BaSO_4$ and 1 mol SO_4^{2-}/1 mol $BaSO_4$), $[Ba^{2+}] = [SO_4^{2-}] = 1.04 \times 10^{-5}$ M. When these values are substituted in the expression for the solubility product constant of $BaSO_4$, the numerical value of the constant may be obtained.

$$BaSO_4(s) \rightleftharpoons Ba^{2+}(aq) + SO_4^{2-}(aq)$$
$$[Ba^{2+}][SO_4^{2-}] = K_{sp}$$
$$(1.04 \times 10^{-5})(1.04 \times 10^{-5}) = 1.08 \times 10^{-10} = K_{sp}$$

◆ ◆ ◆ ◆ ◆

A table of solubility product constants of some slightly soluble electrolytes at 25°C is given in Appendix E.

17.3   Calculation of the Molar Solubility from the Solubility Product Constant

When the solubility product constant for a slightly soluble electrolyte is known, its solubility in moles per liter can be readily calculated.

EXAMPLE 1   The solubility product constant for silver chloride is 1.8×10^{-10}. Calculate the molar solubility of silver chloride.

Silver chloride dissolves in water to form a saturated solution according to the equation

$$AgCl(s) \rightleftharpoons Ag^+(aq) + Cl^-(aq)$$

Let x be the solubility of AgCl in moles per liter. The molar concentrations of Ag^+ and Cl^- will each also be x since the salt is completely dissociated in aqueous solution. Substituting in the expression for the solubility product constant for AgCl, we have

$$[Ag^+][Cl^-] = K_{sp} = 1.8 \times 10^{-10}$$
$$x^2 = 1.8 \times 10^{-10}$$
$$x = \sqrt{1.8 \times 10^{-10}} = 1.3 \times 10^{-5}$$

Thus the molar solubility of silver chloride is 1.3×10^{-5} M.

◆ ◆ ◆ ◆ ◆

EXAMPLE 2   The solubility product constant of PbI_2 is 8.7×10^{-9}. Calculate the molar solubility of lead(II) iodide.

From the equation for the dissolution of lead(II) iodide, $PbI_2 \longrightarrow Pb^{2+} + 2I^-$, it is apparent that each mole of PbI_2 that dissolves yields one mole of Pb^{2+} and two moles of I^-. If the molar solubility of PbI_2 is x, then

$$[Pb^{2+}] = x \qquad [(x \text{ mol } PbI_2/1 \text{ } \ell) \times (1 \text{ mol } Pb^{2+}/1 \text{ mol } PbI_2) = x \text{ mol } Pb^{2+}/1 \text{ } \ell]$$

and

$$[I^-] = 2x \qquad [(x \text{ mol } PbI_2/1 \text{ } \ell) \times (2 \text{ mol } I^-/1 \text{ mol } PbI_2) = 2x \text{ mol } I^-/1 \text{ } \ell]$$

These values are substituted in the expression for the solubility product constant of PbI_2, and the value of x is calculated.

$$[Pb^{2+}][I^-]^2 = K_{sp} = 8.7 \times 10^{-9}$$
$$x(2x)^2 = 8.7 \times 10^{-9}$$
$$4x^3 = 8.7 \times 10^{-9}$$
$$x = 1.3 \times 10^{-3}$$

Thus 1.3×10^{-3} M is the molar solubility of PbI_2. $[Pb^{2+}] = 1.3 \times 10^{-3}$ M, and $[I^-] = 2.6 \times 10^{-3}$ M.

◆ ◆ ◆ ◆ ◆

17.4   The Precipitation of Slightly Soluble Electrolytes

In order to precipitate a slightly soluble electrolyte, it is necessary to bring together in the solution the ions of that electrolyte in such quantities that the product of the ion concentrations, raised to the appropriate powers, *exceeds the solubility product constant of the compound.* Thus when equal volumes of a 2×10^{-4} M solution of $AgNO_3$ and a 2×10^{-4} M solution of NaCl are mixed, we can show that silver chloride will precipitate. Because the volume of the solution is doubled by mixing equal volumes of the two solutions, each concentration is reduced to half its initial value. Consequently, $[Ag^+]$ and $[Cl^-]$ will each be $(2 \times 10^{-4}/2)M = 1 \times 10^{-4}$ M immediately upon mixing. Substituting in the expression for the solubility product constant of silver chloride,

$$[Ag^+][Cl^-] = K_{sp} = 1.8 \times 10^{-10}$$
$$(1 \times 10^{-4})(1 \times 10^{-4}) = 1 \times 10^{-8} > K_{sp}$$

we see that, momentarily, the product of $[Ag^+]$ and $[Cl^-]$ is greater than ($>$) K_{sp} for AgCl. According to the principle of chemical equilibrium, some of the silver and chloride ions will unite under these conditions with the formation of silver chloride precipitate, and the precipitation will continue until the product of $[Ag^+]$ and $[Cl^-]$ still remaining in the solution attains a value equal to the solubility product constant for silver chloride.

In the above case, the Ag^+ and Cl^- were present in equal concentrations. Since an equal number of Ag^+ and Cl^- combine to form solid AgCl, the saturated solution in contact with the precipitate will contain Ag^+ and Cl^- in equal concentrations, their product being just equal to the solubility product constant of AgCl. If to this saturated solution of silver chloride we add more NaCl, we again shall momentarily have the product of $[Ag^+]$ and $[Cl^-]$ greater than the solubility product constant of AgCl, and then more Ag^+ and Cl^- will unite and form solid AgCl until the product of $[Ag^+]$ and $[Cl^-]$ again attains the value of K_{sp}. In the new equilibrium the $[Ag^+]$ will be less and the $[Cl^-]$ greater than they were in the saturated solution of AgCl in pure water. The greater we make the $[Cl^-]$, the less the $[Ag^+]$ will become. However, we can never reduce $[Ag^+]$ to zero because the product of $[Ag^+]$ and $[Cl^-]$ will always be equal to the solubility product constant. It is interesting to note that adding a *very large* excess of NaCl will cause AgCl to dissolve because of the formation of another species, $AgCl_2^-$. This introduces a new equilibrium reaction to the system.

17.5   Calculation of the Concentration of an Ion Necessary to Form a Precipitate

When the concentration of one of the ions of a slightly soluble electrolyte and the value for the solubility product constant are known, we can calculate the concentration that the other ion must exceed before precipitation can occur.

EXAMPLE   If a solution contains 0.001 mol of $CrO_4{}^{2-}$ per liter, what concentration of Ag^+ ion must be exceeded by adding $AgNO_3$ to the solution before Ag_2CrO_4 will begin to precipitate, neglecting any increase in volume on adding the solid silver nitrate? (K_{sp} is 9×10^{-12}.)

The equilibrium involved in the problem is

$$Ag_2CrO_4(s) \rightleftharpoons 2Ag^+(aq) + CrO_4^{2-}(aq)$$

The expression for the solubility product constant is $[Ag^+]^2[CrO_4^{2-}] = K_{sp}$. Substituting the values for the $[CrO_4^{2-}]$ and K_{sp}, we have (when the solution is saturated)

$$[Ag^+]^2(0.001) = 9 \times 10^{-12}$$

$$[Ag^+]^2 = \frac{9 \times 10^{-12}}{1 \times 10^{-3}} = 9 \times 10^{-9} = 90 \times 10^{-10}$$

$$[Ag^+] = 9.5 \times 10^{-5} \text{ or } 1 \times 10^{-4} \ M \qquad \text{(rounded)}$$

A concentration of Ag^+ greater than $1 \times 10^{-4} \ M$ is necessary to cause precipitation of Ag_2CrO_4 under these conditions.

◆ ◆ ◆ ◆ ◆

17.6   Calculation of the Concentration of an Ion in Solution after Precipitation

It is often advantageous to the chemist to know the concentration of an ion remaining in solution after precipitation. This can also be determined using the solubility product constant.

EXAMPLE   **What is the concentration of Ag^+ left in solution if AgCl is precipitated by adding sufficient hydrochloric acid to a solution of $AgNO_3$ to make the final chloride concentration 0.10 M?**

$$[Ag^+][Cl^-] = K_{sp}$$
$$[Ag^+](0.10) = 1.8 \times 10^{-10}$$
$$[Ag^+] = \frac{1.8 \times 10^{-10}}{1.0 \times 10^{-1}} = 1.8 \times 10^{-9} \ M$$

The concentration of Ag^+ left in solution is 1.8×10^{-9} mol ℓ^{-1}.

◆ ◆ ◆ ◆ ◆

17.7   Supersaturation of Slightly Soluble Electrolytes

When the product of the concentrations of the ions, raised to the proper powers, exceeds the value of the solubility product constant, precipitation usually occurs. Sometimes, however, a supersaturated solution is formed and precipitation does not occur immediately. For example, if magnesium is being precipitated as magnesium ammonium phosphate from a solution of low magnesium ion content, the solution may stand several hours before a visible precipitate appears. Furthermore, even though precipitation may be immediate under ordinary conditions, it does not follow that it is complete. The first crystals to precipitate are often very small and hence more soluble than the larger ones, which form as the solution stands in contact with the precipitate (see Section 17.8). Solutions should be allowed to stand before filtering, to ensure maximum precipitation.

17.8 Solubility and Crystal Size

It has been shown by experiment that small crystals of any substance are more soluble than large ones. A notable example of this is barium sulfate. The solubility of "fine" crystals of barium sulfate has been found to be about twice as great as that of "coarse" crystals of this salt. This results from the fact that ions in the interior of a crystal are bound with greater forces than those of the faces and edges. A greater fraction of the ions in a crystal occupy surface positions when the crystal is small, as opposed to a large crystal, making the tendency for ions to enter solution greater for the small crystal. Because small crystals are more soluble than large crystals, the smallest crystals will in time dissolve and the larger ones will grow still larger. True equilibrium is not reached until large, perfect crystals are formed. The solubility product constants are calculated for solutions in contact with relatively large crystals.

Because very small crystals will pass through filters, or may resist sedimentation by centrifugation, precipitates are often "digested" to increase the size of the crystals. Heat increases the rate at which large crystals will grow from the smaller ones. Allowing the suspended precipitate to stand for a time will also often result in the formation of larger crystals.

17.9 Calculations Involving Both Ionization Constants and Solubility Product Constants

In the theory and practice of precipitation analysis we often deal with systems that involve more than one equilibrium. The following examples illustrate the point.

EXAMPLE 1 **Calculate the concentration of hydrogen ion required to prevent the precipitation of ZnS in a solution that is 0.16 M in ZnCl$_2$ and saturated with H$_2$S (0.10 M H$_2$S).**

The maximum concentration of sulfide ion that can be present without causing ZnS to precipitate may be calculated from the solubility product expression for ZnS.

$$[Zn^{2+}][S^{2-}] = K_{sp} = 1.1 \times 10^{-21}$$

$$[S^{2-}] = \frac{K_{sp}}{[Zn^{2+}]} = \frac{1.1 \times 10^{-21}}{0.16} = 6.9 \times 10^{-21}$$

The sulfide ion concentration varies inversely with the square of the hydrogen ion concentration in a saturated solution of hydrogen sulfide containing a strong acid such as HCl as shown by K_i for H$_2$S.

$$\frac{[H^+]^2[S^{2-}]}{[H_2S]} = K_i = 1.3 \times 10^{-20}$$

For a saturated solution of H_2S,

$$\frac{[H^+]^2[S^{2-}]}{0.10} = K_i = 1.3 \times 10^{-20}$$

$$[H^+]^2[S^{2-}] = 1.3 \times 10^{-21}$$

Substituting in this equation the maximum permissible sulfide ion concentration found above, we have

$$[H^+]^2 = \frac{1.3 \times 10^{-21}}{[S^{2-}]} = \frac{1.3 \times 10^{-21}}{6.9 \times 10^{-21}} = 1.9 \times 10^{-1}$$

$$[H^+] = \sqrt{0.19} = 0.43 \ M$$

Thus it is found that a hydrogen ion concentration of 0.43 M will prevent the precipitation of zinc sulfide in the solution described in the problem.

◆ ◆ ◆ ◆ ◆

EXAMPLE 2   **Calculate the quantity of Cd^{2+} that will remain unprecipitated in a 0.30 M HCl solution that is saturated with H_2S ($[H_2S] = 0.10 \ M$).**

The concentration of $[Cd^{2+}]$ remaining in the solution will be determined by the final sulfide ion concentration, which in turn is determined by the hydrogen ion concentration of the 0.30 M HCl solution. Solving for the sulfide ion concentration, we have

$$\frac{[H^+]^2[S^{2-}]}{[H_2S]} = K_i = \frac{(0.30)^2[S^{2-}]}{(0.10)} = 1.3 \times 10^{-20}$$

$$[S^{2-}] = \frac{(0.10)(1.3 \times 10^{-20})}{(0.30)^2} = \frac{1.3 \times 10^{-21}}{0.090} = 1.4 \times 10^{-20} \ M$$

Substituting the value of the sulfide ion concentration in the expression for the solubility product constant of CdS, we have

$$[Cd^{2+}][S^{2-}] = [Cd^{2+}](1.4 \times 10^{-20}) = K_{sp} = 3.6 \times 10^{-29}$$

$$[Cd^{2+}] = \frac{3.6 \times 10^{-29}}{1.4 \times 10^{-20}} = 2.6 \times 10^{-9} \ M$$

Multiplying the molar concentration of Cd^{2+} by the atomic weight of Cd, we find

$$\frac{2.6 \times 10^{-9} \ \text{mol}}{1 \ l} \times \frac{112.4 \ \text{g}}{1 \ \text{mol}} = 2.9 \times 10^{-7} \ \text{g} \ l^{-1}$$

Hence the precipitation of Cd^{2+} as CdS in a 0.30 M HCl solution saturated with H_2S is nearly complete.

◆ ◆ ◆ ◆ ◆

(Calculations of the type shown in Examples 1 and 2 are important in the separation of the metal ions of Group II from those of Group III in typical qualitative analysis procedures.)

EXAMPLE 3   **Calculate the concentration of ammonium ion, supplied by NH$_4$Cl, required to prevent the precipitation of Mg(OH)$_2$ in a liter of solution containing 0.10 mol of ammonia and 0.10 mol of Mg^{2+}.**

To find the concentration of NH$_4^+$ required to prevent precipitation of Mg(OH)$_2$, we must first find the maximum concentration of OH$^-$ that can be present in solution without precipitating Mg(OH)$_2$. To do so we use the solubility product constant for Mg(OH)$_2$.

$$[\text{Mg}^{2+}][\text{OH}^-]^2 = (0.10)[\text{OH}^-]^2 = K_{\text{Mg(OH)}_2} = 1.5 \times 10^{-11}$$

$$[\text{OH}^-]^2 = \frac{1.5 \times 10^{-11}}{0.10} = 1.5 \times 10^{-10}$$

$$[\text{OH}^-] = 1.2 \times 10^{-5}\,M$$

Thus Mg(OH)$_2$ will not precipitate if the [OH$^-$] does not exceed $1.2 \times 10^{-5}\,M$.

To calculate the [NH$_4^+$] supplied by NH$_4$Cl and needed to repress the ionization of NH$_3$ so that the [OH$^-$] will not exceed 1.2×10^{-5}, we use the expression for the ionization constant of aqueous ammonia.

$$\frac{[\text{NH}_4^+][\text{OH}^-]}{[\text{NH}_3]} = K_{\text{NH}_3} = 1.8 \times 10^{-5}$$

$$[\text{NH}_4^+] = \frac{[\text{NH}_3]}{[\text{OH}^-]} \times K_{\text{NH}_3} = \frac{(0.10)}{(1.2 \times 10^{-5})} \times 1.8 \times 10^{-5} = 0.15\,M$$

The total concentration of ammonium ion that must be present is $0.15\,M$. Because the amount of ammonium ion formed by the ionization of ammonia is very small, it can be neglected.

◆ ◆ ◆ ◆ ◆

17.10   Fractional Precipitation

When two anions form sparingly soluble compounds with the same cation or when two cations form sparingly soluble compounds with the same anion, the less soluble compound will precipitate first on the addition of a precipitant to a solution containing both. An additional quantity of the less soluble compound will precipitate along with the precipitation of the more soluble compound if addition of the precipitant is continued **(coprecipitation).**

Consider the case in which a solution containing both sodium iodide and sodium chloride is treated with silver nitrate. Silver iodide, being less soluble than silver chloride, is precipitated first. Only after most of the iodide is precipitated will the chloride begin to precipitate.

EXAMPLE   A solution contains 0.010 mol of KI and 0.10 mol of KCl per liter. AgNO$_3$ is gradually added to this solution. Which will be precipitated first, AgI or AgCl?

a. Calculate the [Ag$^+$] necessary to start the precipitation of AgI.

$$[Ag^+][I^-] = K_{AgI} = 1.5 \times 10^{-16}$$

$$[Ag^+] = \frac{K_{AgI}}{[I^-]} = \frac{1.5 \times 10^{-16}}{0.010} = 1.5 \times 10^{-14} \, M$$

b. Calculate the [Ag$^+$] necessary to start the precipitation of AgCl.

$$[Ag^+][Cl^-] = K_{AgCl} = 1.8 \times 10^{-10}$$

$$[Ag^+] = \frac{K_{AgCl}}{[Cl^-]} = \frac{1.8 \times 10^{-10}}{0.10} = 1.8 \times 10^{-9} \, M$$

A greater concentration of Ag$^+$ is necessary to cause precipitation of AgCl than for AgI, so AgI will precipitate first.

c. What will be the concentration of I$^-$ in this solution when AgCl begins to precipitate as a result of the continued addition of Ag$^+$? The [Ag$^+$] necessary to initiate the precipitation of AgCl is $1.8 \times 10^{-9} \, M$. For this concentration of Ag$^+$, the I$^-$ concentration will be

$$[I^-] = \frac{K_{AgI}}{[Ag^+]} = \frac{1.5 \times 10^{-16}}{1.8 \times 10^{-9}} = 8.3 \times 10^{-8} \, M$$

d. What fraction of the amount of I$^-$ originally present remains in solution when AgCl begins to precipitate?

$$\frac{[I^-] \text{ when precipitation of AgCl begins}}{[I^-] \text{ originally present}} = \frac{8.3 \times 10^{-8}}{0.010} = 8.3 \times 10^{-6}$$

Only 1 part in about 10^5 of the original I$^-$ remains in solution when AgCl begins to precipitate.

e. What will be the concentration of I$^-$ in the solution after half the Cl$^-$ initially present is precipitated as AgCl? When half the Cl$^-$ initially present has been precipitated, the Cl$^-$ concentration will have been reduced to 0.050 M.

$$[Ag^+] = \frac{K_{AgCl}}{[Cl^-]} = \frac{1.8 \times 10^{-10}}{0.050} = 3.6 \times 10^{-9} \, M$$

$$[I^-] = \frac{K_{AgI}}{[Ag^+]} = \frac{1.5 \times 10^{-16}}{3.6 \times 10^{-9}} = 0.417 \times 10^{-7} = 4.2 \times 10^{-8} \, M$$

Note that half of the I$^-$ remaining in solution when AgCl begins to precipitate is coprecipitated as half of the Cl$^-$ is precipitated.

◆ ◆ ◆ ◆ ◆

The Dissolution of Precipitates

When the molar concentrations of the ions, raised to the appropriate powers, are such that their product is less than ($<$) the solubility product constant, the solid electrolyte either completely dissolves or dissolves until the ion concentration product is equal to the solubility product constant. For example, calcium carbonate dissolves when

$$[Ca^{2+}][CO_3^{2-}] < K_{sp}$$

To achieve this condition it is necessary to make the concentration of at least one of the ions of the electrolyte less than that in a saturated solution. Ions can be removed from saturated solutions of sparingly soluble electrolytes, and hence solid electrolytes dissolved, in the following ways: (1) by the formation of a weak electrolyte; (2) changing an ion to another species; (3) by formation of a complex ion.

17.11   Dissolution of a Precipitate by the Formation of a Weak Electrolyte

Slightly soluble electrolytes derived from weak acids may often be dissolved by strong acids. For example, $CaCO_3$, FeS, and $Ca_3(PO_4)_2$ are dissolved by HCl because the acids that form from the anions (H_2CO_3, H_2S, and $H_2PO_4^-$) are weak acids.

When hydrochloric acid is added to a precipitate of calcium carbonate in equilibrium with a standard solution of its ions, the hydrogen ion from the acid combines with the carbonate ion and forms the hydrogen carbonate ion, a weak electrolyte.

$$H^+(aq) + CO_3^{2-}(aq) \rightleftharpoons HCO_3^-(aq)$$

Then, as the concentration of the hydrogen carbonate ion increases, these ions unite with hydrogen ions from the hydrochloric acid according to the equation

$$H^+(aq) + HCO_3^-(aq) \rightleftharpoons H_2CO_3(aq)$$

Finally, the solution becomes saturated with the slightly ionized and unstable carbonic acid, and carbon dioxide gas is evolved, as shown by the equation

$$H_2CO_3(aq) \rightleftharpoons H_2O(l) + CO_2(g)$$

The preceding reactions cause the carbonate ion concentration to be reduced and maintained at such a low level that the product of the calcium and carbonate ion concentrations is less than the solubility product constant of calcium carbonate.

$$[Ca^{2+}][CO_3^{2-}] < K_{sp}$$

Consequently, the calcium carbonate dissolves. Even an acid as weak as acetic acid gives a concentration of hydrogen ion sufficiently high to bring about the dissolution of calcium carbonate because acetic acid is a stronger acid than either the hydrogen carbonate ion or carbonic acid.

The dissolution of magnesium hydroxide in aqueous ammonium chloride can be brought about by the formation of a weak electrolyte, such as aqueous ammonia.

$$Mg(OH)_2(s) + 2NH_4^+(aq) \rightleftharpoons Mg^{2+}(aq) + 2NH_3(aq) + 2H_2O(aq)$$

When an excess of ammonium ion is added to a suspension of magnesium hydroxide, the following reaction takes place:

$$NH_4^+(aq) + OH^-(aq) \rightleftharpoons NH_3(aq) + H_2O(l)$$

Consequently, the hydroxide ion concentration is lowered to the level such that

$$[Mg^{2+}][OH^-]^2 < K_{sp}$$

and the magnesium hydroxide dissolves.

Slightly soluble lead sulfate dissolves readily in solutions of ammonium acetate when the formation of slightly ionized (but soluble) lead acetate reduces the product of the lead and sulfate ion concentrations below the value of the solubility product constant for lead sulfate.

$$PbSO_4(s) \rightleftharpoons Pb^{2+}(aq) + SO_4^{2-}(aq)$$
$$Pb^{2+}(aq) + 2OAc^-(aq) \rightleftharpoons Pb(OAc)_2(aq)$$
$$[Pb^{2+}][SO_4^{2-}] < K_{sp}$$

The formation of water causes most sparingly soluble metal hydroxides such as $Mg(OH)_2$, $Al(OH)_3$, and $Fe(OH)_3$ to dissolve in solutions of acids.

$$Mg(OH)_2(s) + 2H^+(aq) \longrightarrow Mg^{2+}(aq) + 2H_2O(l)$$
$$[Mg^{2+}][OH^-]^2 < K_{sp}$$
$$Al(OH)_3(s) + 3H^+(aq) \longrightarrow Al^{3+}(aq) + 3H_2O(l)$$
$$[Al^{3+}][OH^-]^3 < K_{sp}$$

17.12   Dissolution of a Precipitate by Changing an Ion to Another Species

Many metal sulfides have solubility product constants sufficiently large that the hydrogen ion provided by strong acids will lower the sulfide ion concentration enough (by forming the weak electrolyte hydrogen sulfide) to dissolve the sulfide. For example, iron(II) sulfide is readily dissolved by hydrochloric acid according to the equation

$$FeS(s) + 2H^+(aq) \rightleftharpoons Fe^{2+}(aq) + H_2S(g)$$
$$[Fe^{2+}][S^{2-}] < K_{sp}$$

However, a number of metal sulfides, such as lead sulfide, furnish such low concentrations of sulfide ion in their saturated solutions that not even the high concentration of hydrogen ion provided by a strong acid is sufficient to exceed the ion product constant of hydrogen sulfide and bring about production of hydrogen sulfide and the subsequent dissolution of the metal sulfide precipitate. For the dissolution of such sulfides, the sulfide ion concentration may be decreased by oxidizing it to elemental sulfur with nitric acid.

$$3S^{2-}(aq) + 2NO_3^-(aq) + 8H^+(aq) \longrightarrow 3S(s) + 2NO(g) + 4H_2O(l)$$

Lead sulfide dissolves in nitric acid, then, because

$$[Pb^{2+}][S^{2-}] < K_{sp}$$

which results from the fact that the sulfide ion is oxidized to sulfur by the nitric acid.

17.13   Dissolution of a Precipitate by the Formation of a Complex Ion

Many slightly soluble electrolytes are dissolved through the formation of complex ions. Several examples which are important to qualitative analysis follow.

$$AgCl(s) + 2NH_3(aq) \rightleftharpoons Ag(NH_3)_2^+(aq) + Cl^-(aq)$$
$$CuCN(s) + CN^-(aq) \rightleftharpoons Cu(CN)_2^-(aq)$$
$$Zn(OH)_2(s) + 2OH^-(aq) \rightleftharpoons Zn(OH)_4^{2-}(aq)$$
$$Sn(OH)_2(s) + OH^-(aq) \rightleftharpoons Sn(OH)_3^-(aq)$$
$$Al(OH)_3(s) + OH^-(aq) \rightleftharpoons Al(OH)_4^-(aq)$$
$$As_2S_3(s) + 3S^{2-}(aq) \rightleftharpoons 2AsS_3^{3-}(aq)$$
$$HgS(s) + S^{2-}(aq) \rightleftharpoons HgS_2^{2-}(aq)$$
$$Sb_2S_3(s) + 3S^{2-}(aq) \rightleftharpoons 2SbS_3^{3-}(aq)$$

The stability of these complex ions is described by an equilibrium constant for the formation of the complex ion from its components in solution. For example, the complex ion $Cu(CN)_2^-$ forms by the reaction

$$Cu^{2+} + 2CN^- \rightleftharpoons Cu(CN)_2^-$$

The equilibrium constant for this reaction is the **stability constant,** or **formation constant, K_f.** The K_f constant is sometimes also referred to as the **association constant.** A table of some formation constants is provided in Appendix F.

$$K_f = \frac{[Cu(CN)_2^-]}{[Cu^{2+}][CN^-]^2} = 1 \times 10^{16}$$

The larger the value of a formation constant, the more stable the complex.

Alternatively, the stability of a complex ion can be described by its **dissociation constant, K_d,** the equilibrium constant for the decomposition of a complex ion into its components in solution. For $Cu(CN)_2^-$ the dissociation is

$$Cu(CN)_2^- \rightleftharpoons Cu^{2+} + 2CN^-$$

and

$$K_d = \frac{[Cu^{2+}][CN^-]^2}{[Cu(CN)_2^-]} = 1 \times 10^{-16}$$

It should be apparent that

$$K_d = \frac{1}{K_f}$$

As an example of dissolution by complex ion formation, let us consider the case of the dissolution of silver chloride in aqueous ammonia. When silver ions and ammonia molecules are brought together in solution the complex diamminesilver ion, $Ag(NH_3)_2^+$, is formed according to the equation

$$Ag^+(aq) + 2NH_3(aq) \rightleftharpoons Ag(NH_3)_2^+(aq)$$

Applying the law of mass action to this system, we obtain

$$\frac{[Ag(NH_3)_2^+]}{[Ag^+][NH_3]^2} = K_f = 1.6 \times 10^7$$

In a saturated solution of silver chloride, the Ag^+ concentration is determined by the value of the solubility product, K_{sp}. This Ag^+ concentration is greater than that which can exist in equilibrium after ammonia is added.

$$\frac{[Ag(NH_3)_2^+]}{[Ag^+][NH_3]^2} < K_f$$

Therefore silver ions and ammonia molecules combine, thus lowering the concentration of silver ions and bringing the product of the silver ion concentration and chloride ion concentration below the solubility product constant for silver chloride.

$$[Ag^+][Cl^-] < K_{sp}$$

Dissolution of some silver chloride then results. If the concentration of ammonia is great enough, equilibrium will be reached only after all the silver chloride has dissolved.

The copper(II) ion forms coordinate bonds with four ammonia molecules to produce the complex $Cu(NH_3)_4^{2+}$. The expression for the formation constant is

$$\frac{[Cu(NH_3)_4^{2+}]}{[Cu^{2+}][NH_3]^4} = K_f$$

When a copper(II) solution is treated with dilute aqueous ammonia, copper(II) hydroxide first precipitates since the hydroxide ion concentration of the dilute ammonia solution is sufficiently large that the solubility product constant of copper(II) hydroxide is exceeded.

$$[Cu^{2+}][OH^-]^2 > K_{sp}$$

As the concentration of ammonia is increased, the value of the formation constant expression drops below the value for the formation constant of the complex ion $Cu(NH_3)_4^{2+}$.

$$\frac{[Cu(NH_3)_4^{2+}]}{[Cu^{2+}][NH_3]^4} < K_f$$

The combination of copper(II) ion with ammonia molecules makes the concentration of Cu^{2+} so small that the condition for the dissolution of copper(II) hydroxide is attained.

$$[Cu^{2+}][OH^-]^2 < K_{sp}$$

17.14   Calculations Involving Complex Ions

EXAMPLE 1   Calculate the concentration of the silver ion in a solution that is $0.10\ M$ with respect to $Ag(NH_3)_2{}^+$.

The formation of the complex ion may be represented by the equilibrium equation

$$Ag^+ + 2NH_3 \rightleftharpoons Ag(NH_3)_2{}^+$$

Let x be the number of moles of $Ag(NH_3)_2{}^+$ that dissociate per liter. Then at equilibrium $[Ag^+]$ will be x, and $[NH_3]$ will be $2x$. From the large value of the formation constant we know that the dissociation must be small, and we can assume that the complex is practically $0.10\ M$ at equilibrium.   Therefore

$$\frac{[Ag(NH_3)_2{}^+]}{[Ag^+][NH_3]^2} = \frac{0.10}{x(2x)^2} = K_f = 1.6 \times 10^7$$

$$0.10 = (4x^3)(1.6 \times 10^7)$$

$$4x^3 = \frac{0.10}{1.6 \times 10^7} = 6.3 \times 10^{-9}$$

$$x^3 = 1.6 \times 10^{-9}$$

$$x = 1.2 \times 10^{-3} = [Ag^+]$$

$$2x = 2.4 \times 10^{-3} = [NH_3]$$

◆ ◆ ◆ ◆ ◆

EXAMPLE 2   Calculate the number of moles of ammonia that must be added to $1\ \ell$ of water just to dissolve 1.0×10^{-3} mol of silver chloride.   (Assume that the volume does not change.)

Writing the equation for the reaction,

$$AgCl(s) + 2NH_3(aq) \rightleftharpoons Ag(NH_3)_2{}^+(aq) + Cl^-(aq)$$

we notice that when 1.0×10^{-3} mol of AgCl is dissolved, 1.0×10^{-3} mol of $Ag(NH_3)_2{}^+$ and 1.0×10^{-3} mol of Cl^- are formed.   Two equilibria are involved when AgCl dissolves in aqueous ammonia.

$$AgCl(s) \rightleftharpoons Ag^+(aq) + Cl^-(aq)$$
$$Ag^+(aq) + 2NH_3(aq) \rightleftharpoons Ag(NH_3)_2{}^+(aq)$$

The $[Ag^+]$ in solution applies to both equilibria when both solid AgCl and $Ag(NH_3)_2{}^+$ ions are present in the system.   Since the $[Cl^-]$ is 1.0×10^{-3}, we can calculate the $[Ag^+]$ using the solubility product expression for AgCl.

$$[Ag^+][Cl^-] = [Ag^+](1.0 \times 10^{-3}) = 1.8 \times 10^{-10}$$

$$[Ag^+] = \frac{1.8 \times 10^{-10}}{1.0 \times 10^{-3}} = 1.8 \times 10^{-7}$$

By substituting in the expression for the formation constant of $Ag(NH_3)_2{}^+$ the values for $[Ag^+]$ and $[Ag(NH_3)_2{}^+]$, we can calculate the $[NH_3]$ required at equilibrium to produce this concentration of Ag^+.

$$\frac{[Ag(NH_3)_2{}^+]}{[Ag^+][NH_3]^2} = \frac{(1.0 \times 10^{-3})}{(1.8 \times 10^{-7})[NH_3]^2} = 1.6 \times 10^7$$

$$[NH_3]^2 = \frac{(1.0 \times 10^{-3})}{(1.8 \times 10^{-7})(1.6 \times 10^7)} = 3.5 \times 10^{-4}$$

$$[NH_3] = 1.9 \times 10^{-2}\ M$$

In this calculation the amount of ammonia consumed in forming the complex ion is $2(1.0 \times 10^{-3})$, or 0.002 mol, which is to be added to the 1.9×10^{-2}, or 0.019, mol of ammonia required at equilibrium. Thus, the total amount of ammonia required for dissolution of 1.0×10^{-3} mol of AgCl in a liter of water is 0.021 mol.

◆ ◆ ◆ ◆ ◆

17.15 Effect of Hydrolysis on Dissolution of Precipitates

Many salts hydrolyze when they are dissolved in water, sometimes with the formation of precipitates, or with the evolution of gases, or with both. For example, when aluminum sulfide, Al_2S_3, is dissolved in water, insoluble aluminum hydroxide is formed and hydrogen sulfide is produced. The aluminum ions and the sulfide ions combine with water according to the equations

$$2Al^{3+} + 12H_2O \longrightarrow 2[Al(H_2O)_6]^{3+}$$
$$2[Al(H_2O)_6]^{3+}(aq) + 6H_2O(l) \rightleftharpoons 2[Al(OH)_3(H_2O)_3](s) + 6H_3O^+(aq)$$
$$3S^{2-}(aq) + 6H_3O^+(aq) \rightleftharpoons 3H_2S(aq) + 6H_2O(l)$$

The net reaction is given by

$$2[Al(H_2O)_6]^{3+}(aq) + 3S^{2-}(aq) \rightleftharpoons 2[Al(OH)_3(H_2O)_3](s) + 3H_2S(aq)$$

If the initial amount of aluminum sulfide is sufficiently high, the solubility of hydrogen sulfide may be exceeded and excess hydrogen sulfide will be evolved as a gas. Many other salts behave similarly toward water.

When a relatively insoluble sulfide such as lead sulfide, PbS, is dissolved in water until a saturated solution is formed, an appreciable amount of the sulfide ion hydrolyzes to form the hydrosulfide ion. In some cases, as we shall see later in Example 2, a significant amount of the hydrosulfide ion hydrolyzes to form hydrogen sulfide.

$$PbS(s) \rightleftharpoons Pb^{2+}(aq) + S^{2-}(aq)$$
$$S^{2-}(aq) + H_2O(l) \rightleftharpoons HS^-(aq) + OH^-(aq)$$
$$HS^-(aq) + H_2O(l) \rightleftharpoons H_2S(aq) + OH^-(aq)$$

For this reason the lead ion concentration is not the same as that of the sulfide ion under these conditions. The concentration of the lead ion is equal to the sum of the concentrations of S^{2-}, HS^-, and H_2S. This means that in calculating the solubility of a relatively insoluble sulfide from the solubility product constant, one must take into account the hydrolysis of the S^{2-} ion.

EXAMPLE 1   **Let us first calculate the solubility of PbS in water, neglecting the hydrolysis of the sulfide ion. (K_{sp} for PbS is 8.4×10^{-28}.)**

Let x equal the molar solubility of lead sulfide in water. Then x will also be equal to the concentration of the lead ion and of the sulfide ion. Substituting in the expression for the solubility product constant, we obtain

$$[Pb^{2+}][S^{2-}] = K_{sp}$$
$$x^2 = 8.4 \times 10^{-28}$$
$$x = 2.9 \times 10^{-14}$$

Thus if the hydrolysis of the sulfide ion were neglected we would find the solubility of PbS in water to be 2.9×10^{-14} mol ℓ^{-1}.

◆ ◆ ◆ ◆ ◆

EXAMPLE 2   **Now let us solve the problem of Example 1 correctly, taking into consideration the hydrolysis of the sulfide ion.**

We shall consider three steps in the dissolution of PbS in water.

$$PbS(s) \rightleftharpoons Pb^{2+} + S^{2-} \qquad K_{sp} = 8.4 \times 10^{-28} \qquad\qquad (1)$$

$$S^{2-} + H_2O \rightleftharpoons HS^- + OH^- \qquad K_{h_1} = \frac{K_w}{K_i \text{ (for } HS^-)} = \frac{1.0 \times 10^{-14}}{1.3 \times 10^{-13}} = 7.7 \times 10^{-2} \qquad (2)$$

$$HS^- + H_2O \rightleftharpoons H_2S + OH^- \qquad K_{h_2} = \frac{K_w}{K_i \text{ (for } H_2S)} = \frac{1.0 \times 10^{-14}}{1.0 \times 10^{-7}} = 1.0 \times 10^{-7} \qquad (3)$$

We might be tempted to say that, since K_{h_2} is quite small compared to K_{h_1}, we could consider only the first stage of the hydrolysis and neglect the second. This is often a valid way to proceed, and indeed we did this in Example 2 of Section 16.17. We found in that example that the sulfide ion in a 0.0010 M sodium sulfide solution was 99% hydrolyzed. Thus the first stage of hydrolysis would produce (0.99×0.0010) mol, or 9.9×10^{-4} mol, of OH^-. Since this quantity of OH^- is large compared to the 1.0×10^{-7} mol furnished by the water, the total concentration of OH^- in solution was 9.9×10^{-4} mol ℓ^{-1} (1.0×10^{-7} is negligible compared to 9.9×10^{-4}, to several significant figures). We can calculate the ratio of $[H_2S]$ to $[HS^-]$, from K_{h_2}, for Example 2 of Section 16.17, then, to establish the relative amounts of HS^- and H_2S produced in the first and second stages of hydrolysis, respectively.

$$\frac{[H_2S][OH^-]}{[HS^-]} = \frac{[H_2S](9.9 \times 10^{-4})}{[HS^-]} = K_{h_2} = 1.0 \times 10^{-7}$$

$$\frac{[H_2S]}{[HS^-]} = 1.0 \times 10^{-4}$$

Hence $[H_2S]$ is only $1/10,000$ $[HS^-]$, indicating that the second stage of the hydrolysis is negligible for Example 2 of Section 16.17, compared to the first stage of the hydrolysis, and could be neglected *in that calculation.*

However, in our present problem, we have a much smaller amount of sulfide ion. Our rough calculation for the solubility of lead sulfide, neglecting hydrolysis, indicated only 2.9×10^{-14} mol of lead sulfide going into solution per liter. The hydroxide ion concentration due to the hydrolysis of the sulfide ion, $S^{2-} + H_2O \rightleftharpoons HS^- + OH^-$, would be of the order of 3×10^{-14} M. However, this hydroxide ion concentration is so small compared with that provided by the water, 1.0×10^{-7} M, that the $[OH^-]$ of the solution will still have a value of 1.0×10^{-7} M. Even some increase in the solubility of PbS and some additional OH^- produced in the second stage of hydrolysis would not likely be enough to add sufficient OH^- to be significant compared to the 1.0×10^{-7} mol ℓ^{-1} furnished by the water. We can now calculate the ratio of $[H_2S]$ to $[HS^-]$ from K_{h_2}, for a solution in which $[OH^-]$ is 1×10^{-7} M (neutral solution).

$$\frac{[H_2S][OH^-]}{[HS^-]} = \frac{[H_2S](1.0 \times 10^{-7})}{[HS^-]} = K_{h_2} = 1.0 \times 10^{-7}$$

$$\frac{[H_2S]}{[HS^-]} = 1$$

Hence, in the case of a neutral solution, $[H_2S] = [HS^-]$, indicating that the second stage *is* important in the calculation and cannot be neglected here. For $[HS^-]$ to equal $[H_2S]$, half of the HS^- ions produced in the first stage of hydrolysis must hydrolyze to produce H_2S in the second stage of hydrolysis.

Now, consider the three equilibria listed earlier for the system. As PbS dissolves, equal quantities of Pb^{2+} and S^{2-} are produced. The S^{2-} produced is essentially all hydrolyzed and hence converted to HS^- in the first stage of hydrolysis. At this point, not yet taking the second stage of hydrolysis into account, $[Pb^{2+}] = [HS^-]$. However, we have established that half of the HS^- then hydrolyzes in the second stage of hydrolysis, thereby reducing the HS^- concentration to half its former value, and thus to half the concentration of Pb^{2+}. Hence, at final equilibrium, $[HS^-] = \frac{1}{2}[Pb^{2+}]$.

Using 1.0×10^{-7} M for the $[OH^-]$, we can calculate the ratio of the concentration of the hydrosulfide ion to that of the sulfide ion, by substituting in K_{h_1} and then solving for $[S^{2-}]$.

$$\frac{[HS^-][OH^-]}{[S^{2-}]} = \frac{[HS^-](1.0 \times 10^{-7})}{[S^{2-}]} = K_{h_1} = 7.7 \times 10^{-2}$$

$$\frac{[HS^-]}{[S^{2-}]} = 7.7 \times 10^5, \text{ or } [S^{2-}] = \frac{[HS^-]}{7.7 \times 10^5}$$

Substituting the value for $[S^{2-}]$ found above in the expression for the solubility product constant for lead sulfide, we have

$$[Pb^{2+}][S^{2-}] = [Pb^{2+}]\frac{[HS^-]}{7.7 \times 10^5} = K_{sp} = 8.4 \times 10^{-28}$$

Because $[HS^-] = \frac{1}{2}[Pb^{2+}]$, we can write

$$[Pb^{2+}]\left(\frac{\frac{1}{2}[Pb^{2+}]}{7.7 \times 10^5}\right) = 8.4 \times 10^{-28}$$

$$\frac{\frac{1}{2}[Pb^{2+}]^2}{7.7 \times 10^5} = 8.4 \times 10^{-28}$$

$$[Pb^{2+}]^2 = 12.94 \times 10^{-22}$$

$$[Pb^{2+}] = 3.6 \times 10^{-11} = \text{molar solubility of PbS}$$

We have found, then, that the solubility of lead sulfide in water is 3.6×10^{-11} mol ℓ^{-1} when hydrolysis of the sulfide ion is considered. This value is over 1000 times greater than that calculated (2.9×10^{-14} mol ℓ^{-1}) when hydrolysis was neglected. If we were to take into account the fact that the lead ion hydrolyzes slightly to give $Pb(OH)^+$ and $Pb(OH)_2$, the solubility of lead sulfide would be found to be somewhat larger yet.

Finally, note that the sulfide hydrolysis would produce 3.6×10^{-11} mol OH^- per liter in the first stage of hydrolysis $+ \frac{1}{2}(3.6 \times 10^{-11})$ additional mole OH^- per liter in the second stage of hydrolysis, for a total of 5.4×10^{-11} mol OH^- per liter. This amount of OH^- is indeed negligible compared to the 1.0×10^{-7} mol of OH^- furnished by the water. Hence, our earlier assumption of $[OH^-] = 1.0 \times 10^{-7}$ M for this particular case was entirely justified.

The relatively insoluble carbonates behave similarly to the sulfides. When a relatively insoluble carbonate such as barium carbonate, $BaCO_3$, is dissolved in water, its solubility is increased due to hydrolysis of the carbonate ion, ($CO_3^{2-} + H_2O \rightleftharpoons HCO_3^- + OH^-$ and $HCO_3^- + H_2O \rightleftharpoons H_2CO_3 + OH^-$). The concentration of the barium ion is equal to the sum of the concentrations of CO_3^{2-}, HCO_3^-, and H_2CO_3.

◆ ◆ ◆ ◆ ◆

Questions

1. A saturated solution of a slightly soluble electrolyte in contact with some of the solid phase of the electrolyte is said to be a system in equilibrium. Explain. Why is such a system called a heterogeneous equilibrium?

2. How does the concentration of Ag^+ and CrO_4^{2-} in a liter of water above 1.0 g of solid Ag_2CrO_4 change when 100 g of solid Ag_2CrO_4 is added to the system? Explain.

3. How does the concentration of Pb^{2+} and S^{2-} change when H_2S is added to a saturated solution of PbS?

4. Under what circumstances, if any, will a sample of solid AgCl completely dissolve in pure water?

5. Refer to Appendix E for solubility product constants for copper salts. Determine which of the copper salts listed is most soluble in moles per liter. Determine which is most soluble in grams per liter.

6. Why is the concentration of a solid constant?

7. If solid silver bromide is in equilibrium with a saturated solution of its ions, Ag^+ and Br^-, will this equilibrium be affected and in what manner if (a) more solid silver bromide is added? (b) silver nitrate is added? (c) sodium bromide is added? (d) the temperature is raised (the solubility increases with temperature)?

8. Write the expression for the solubility product of each of the following slightly soluble electrolytes: $PbCl_2$, $AgBr$, $Ba_3(PO_4)_2$, Ag_2S, $MgNH_4AsO_4$.

9. Explain why the addition of washing soda, Na_2CO_3, is effective in reducing the hardness of water resulting from the presence of dissolved calcium bicarbonate, $Ca(HCO_3)_2$.
10. Explain the formation of the complex ion $Zn(OH)_4^{2-}$ in terms of the Lewis theory of acids and bases.
11. Suggest a reagent that might be used to dissolve each of the following solids and write a balanced chemical equation for each reaction: (a) Ag_2O, (b) $Cu(OH)_2$, (c) CdS, (d) $BaCO_3$, (e) $Pb_3(PO_4)_2$, (f) Hg_2I_2 (HI is a strong acid).

Problems

1. Calculate the solubility product constants of each of the following from the solubility given:
   [S](a) AgBr, 5.7×10^{-7} mol ℓ^{-1} *Ans.* 3.2×10^{-13}
   (b) $CaCO_3$, 6.9×10^{-3} g ℓ^{-1} *Ans.* 4.8×10^{-9}
   [S](c) PbF_2, 2.1×10^{-3} mol ℓ^{-1} *Ans.* 3.7×10^{-8}
   [S](d) Ag_2CrO_4, 4.3×10^{-2} g ℓ^{-1} *Ans.* 8.7×10^{-12}
   (e) CuCl, 4.26×10^{-3} g/100 ml *Ans.* 1.85×10^{-7}
   (f) Ag_2SO_4, 4.47 g ℓ^{-1} *Ans.* 1.18×10^{-5}
2. Calculate the concentrations of ions in a saturated solution of each of the following (see Appendix E for solubility product constants):
   [S](a) AgI *Ans.* $[Ag^+] = [I^-] = 1.2 \times 10^{-8}$ *M*
   [S](b) Ag_2SO_4 *Ans.* $[Ag] = 2.8 \times 10^{-2}$ *M*; $[SO_4^{2-}] = 1.4 \times 10^{-2}$ *M*
   (c) $Mn(OH)_2$ *Ans.* $[Mn^{2+}] = 2.2 \times 10^{-5}$ *M*; $[OH^-] = 4.5 \times 10^{-5}$ *M*
   (d) $PbSO_4$ *Ans.* $[Pb^{2+}] = 1.3 \times 10^{-4}$ *M*; $[SO_4^{2-}] = 1.3 \times 10^{-4}$ *M*
   [S](e) $Sr(OH)_2 \cdot 8H_2O$ *Ans.* $[Sr^{2+}] = 4.3 \times 10^{-2}$ *M*; $[OH^-] = 8.6 \times 10^{-2}$ *M*
[S]3. Calculate the concentration of Sr^{2+} ion when strontium fluoride starts to precipitate from a solution that is 0.0025 *M* in fluoride ion ($K_{sp} = 3.7 \times 10^{-12}$). *Ans.* 5.9×10^{-7} *M*
4. (a) Calculate the silver ion concentration in a saturated aqueous solution of silver bromide for which K_{sp} is 3.3×10^{-3}. *Ans.* 5.7×10^{-7} *M*
   (b) What will be the $[Ag^+]$ when enough KBr has been added to make $[Br^-] = 0.050$ *M*? *Ans.* 6.6×10^{-12} *M*
   (c) What will be the $[Br^-]$ when enough $AgNO_3$ has been added to make $[Ag^+] = 0.020$ *M*? *Ans.* 1.6×10^{-11} *M*
5. Calculate $[Ca^{2+}]$ required to start the precipitation of calcium fluoride from a solution that is 0.0025 *M* in fluoride ions. ($K_{sp} = 3.9 \times 10^{-11}$.) *Ans.* 6.2×10^{-6} *M*
6. Calculate the solubility product constant of each of the following from the molar solubility given: (a) AgI (1.23×10^{-8} mol/ℓ); Ag_2CO_3 (1.27×10^{-4} mol/ℓ). *Ans.* (a) 1.5×10^{-16}; (b) 8.2×10^{-12}
7. The solubility of BaF_2 is 1.292 g/ℓ. Calculate its solubility product. *Ans.* 1.6×10^{-6}
8. Calculate the solubility in moles per liter for each of the following substances from its solubility product constant:
   (a) PbF_2, $K_{sp} = 3.7 \times 10^{-8}$ *Ans.* 2.1×10^{-3} *mol/l*
   (b) $Ag_2(CrO_4)$, $K_{sp} = 9.0 \times 10^{-12}$ *Ans.* 1.3×10^{-4} *mol/l*
   (c) CuCl, $K_{sp} = 1.85 \times 10^{-7}$ *Ans.* 4.3×10^{-4} *mol/l*
   (d) CaC_2O_4, $K_{sp} = 2.27 \times 10^{-9}$ *Ans.* 4.76×10^{-5} *mol/l*
9. In one experiment, a precipitate of $BaSO_4$ was washed with 0.100 ℓ of distilled water; in another experiment, a precipitate of $BaSO_4$ was washed with 0.100 ℓ of 0.010 *M* H_2SO_4. Calculate the quantity of $BaSO_4$ which dissolved in each experiment, assuming that the wash liquid became saturated with $BaSO_4$. (K_{sp} of $BaSO_4 = 1.08 \times 10^{-10}$.) *Ans.* 1.04×10^{-6} *mol*; 1.08×10^{-9} *mol*
10. A 0.800-ℓ volume of a 2×10^{-4} *M* $Ba(NO_3)_2$ solution is added to 0.200 ℓ of 5×10^{-4} *M* Li_2SO_4. Will $BaSO_4$ precipitate? Explain your answer. *Ans. Yes*
11. Calculate the maximum concentration of lead(II) ion in a solution in which the concentration of sulfate ion is 0.0045 *M*. *Ans.* 4.0×10^{-6} *M*

[S]12. A solution of 0.075 M $CoBr_2$ is saturated with H_2S ($[H_2S] = 0.10$ M). What is the minimum pH at which CoS ($K_{sp} = 5.9 \times 10^{-21}$) will precipitate?
Ans. 0.89

13. The K_{sp} for CdS is 3.6×10^{-29}. Calculate the maximum concentration of sulfide ion that can exist in a solution that is 0.0010 M in $CdCl_2$.
Ans. 3.6×10^{-26} M

14. Calculate the solubility of CdS in water. Do not neglect the hydrolysis of sulfide ion.
Ans. 7.4×10^{-12} M

15. Calculate the solubility of $Al(OH)_3$ in water ($K_{sp} = 1.9 \times 10^{-33}$). Do not neglect the hydrolysis of the aluminum ion (Section 16.20).
Ans. 3.8×10^{-5} M

[S]16. What is the concentration of Ca^{2+} and CO_3^{2-} in a saturated solution of calcium carbonate ($K_{sp} = 4.8 \times 10^{-9}$)?
Ans. $[Ca^{2+}] = [CO_3^{2-}] = 6.9 \times 10^{-5}$ M

[S]17. What is the concentration of Ca^{2+} and CO_3^{2-} in a buffer solution with a pH of 4.55, in contact with an excess of $CaCO_3$.
Ans. $[Ca^{2+}] = 0.35$ M; $[CO_3^{2-}] = 1.4 \times 10^{-8}$ M

18. The solubility product of $CaSO_4 \cdot 2H_2O$ is 2.4×10^{-5}. Calculate the mass of this salt that will dissolve in 1.0 ℓ of 0.010 M K_2SO_4.
Ans. 0.41 g

[S]19. Fifty milliliters of 1.8 M NH_3 is mixed with an equal volume of a solution containing 0.95 g of $MgCl_2$. What mass of NH_4Cl must be added to the resulting solution to prevent the precipitation of $Mg(OH)_2$?
Ans. 7.1 g

20. A solution is 0.010 M in both Cu^{2+} and Cd^{2+}. What is the per cent of Cd^{2+} still in the solution when 99.9% of the Cu^{2+} has been precipitated as CuS by addition of sulfide?
Ans. 100%

21. (a) With what volume of water must a precipitate containing $NiCO_3$ be washed to dissolve 0.100 g of this compound? Assume that the wash water becomes saturated with $NiCO_3$. (K_{sp} of $NiCO_3 = 1.36 \times 10^{-7}$.)
Ans. 2.28 ℓ

(b) If the $NiCO_3$ were a contaminant in a sample of $CoCO_3$ ($K_{sp} = 1.0 \times 10^{-12}$), what mass of $CoCO_3$ would have been lost? Keep in mind that both $NiCO_3$ and $CoCO_3$ dissolve in the same solution.
Ans. 7.3×10^{-7} g

[S]22. The calcium ions in human blood serum are necessary for coagulation of the blood. In order to prevent this coagulation when a sample is drawn for laboratory tests, an anticoagulant is added to the sample. Potassium oxalate, $K_2C_2O_4$, can be used as an anticoagulant because, in the proper amounts, it removes the calcium as a precipitate of $CaC_2O_4 \cdot H_2O$. In order to prevent coagulation, it is necessary to remove all but 1.0% of the Ca^{2+} in the serum. If normal blood serum with a buffered pH of 7.40 contains 9.5 mg of Ca^{2+} per 100 ml of serum, what mass of $K_2C_2O_4$ is required to prevent the coagulation of a 10-ml blood sample that is 55% serum by volume? (All volumes are accurate to two significant figures. Note that the volume of fluid in a 10-ml blood sample is 5.5 ml. Assume that the K_{sp} value for CaC_2O_4 is the same in serum as in water.)
Ans. 2.2×10^{-3} g

[S]23. About 50% of all urinary calculi (kidney stones) consist of calcium phosphate, $Ca_3(PO_4)_2$. The normal midrange calcium content excreted in the urine is 0.10 g of Ca^{2+} per day. The normal midrange amount of urine passed may be taken as 1.4 ℓ per day. What is the maximum concentration of phosphate ion possible in urine before a calculus begins to form?
Ans. 4×10^{-9} M

[S]24. If the pH of a normal urine sample is 6.30 and the total phosphate concentration ($[PO_4^{3-}] + [HPO_4^{2-}] + [H_2PO_4^{-}] + [H_3PO_4]$) is 0.020 M, what is the minimum concentration of Ca^{2+} necessary to induce calculus formation? (See Problem 23 for additional information.)
Ans. 3×10^{-3} M

25. Magnesium metal (a component of alloys used in aircraft and a reducing agent used in the production of uranium, titanium, and other active metals) is isolated from seawater by the following sequence of reactions:

$$Mg^{2+}(aq) + Ca(OH)_2(aq) \longrightarrow Mg(OH)_2(s) + Ca^{2+}(aq)$$
$$Mg(OH)_2(s) + 2HCl \longrightarrow MgCl_2(s) + 2H_2O$$
$$MgCl_2(l) \xrightarrow{\text{Electrolysis}} Mg(s) + Cl_2$$

Seawater has a density of 1.026 g/cm³ and contains 1272 parts per million of magnesium as $Mg^{2+}(aq)$ by mass. What mass, in kilograms, of $Ca(OH)_2$ is required to precipitate 99.9% of the magnesium in 1.00×10^3 ℓ of seawater.
Ans. 4.0 kg

26. Calculate the $HgCl_4^{2-}$ concentration in a solution prepared by adding 8.0×10^{-3} mol of NaCl to 0.100 ℓ of a 0.040 M $HgCl_2$ solution. *Ans. 0.040 M*

$\boxed{s}$27. Calculate the cadmium ion concentration in a solution prepared by mixing 0.100 l of 0.0100 M $Cd(NO_3)_2$ solution with 0.150 l of 0.100 M $NH_3(aq)$.
Ans. 2.7 × 10⁻⁴ M

28. Calculate the silver ion concentration, $[Ag^+]$, of a solution prepared by dissolving 1.00 g of $AgNO_3$ and 10.0 g of KCN in sufficient water to make 1.00 l of solution.
Ans. 3 × 10⁻²¹ M

29. Sometimes the equilibrium relationships for complex ions are described in terms of dissociation constants, K_d. For the complex ion AlF_6^{3-} the dissociation reaction is

$$AlF_6^{3-} \rightleftharpoons Al^{3+} + 6F^-$$

and

$$K_d = \frac{[Al^{3+}][F^-]^6}{[AlF_6^{3-}]} = 2 \times 10^{-24}$$

(a) Calculate the value of the formation constant for AlF_6^{3-}.
Ans. 5 × 10²³

(b) From the value of the formation constant for the formation of the complex ion $Co(NH_3)_6^{2+}$, calculate the dissociation constant.
Ans. 1.2 × 10⁻⁵

30. Calculate the concentration of Ni^{2+} in a 1.0 M solution of $[Ni(NH_3)_6](NO_3)_2$.
Ans. 0.014 M

31. Calculate the concentration of Zn^{2+} in a 0.30 M solution of $[Zn(CN)_4]^{2-}$.
Ans. 4 × 10⁻⁵ M

$\boxed{s}$32. (a) What mass of AgCl will dissolve in 1.0 l of 1.0 M NH_3?
Ans. 6.9 g

(b) What mass of AgI will dissolve in 1.0 l of 1.0 M NH_3?
Ans. 1.1 × 10⁻² g

33. Calculate the minimum number of moles of cyanide ion that must be added to 100 ml of solution to dissolve 2 × 10⁻² mol of AgCN.
Ans. 2 × 10⁻² mol

$\boxed{s}$34. Calculate the minimum concentration of ammonia needed in 1.0 l of solution to dissolve 3.0 × 10⁻³ mol of AgBr.
Ans. 1.3 M

35. Calculate the volume of 1.50 M HOAc required to dissolve a precipitate composed of 350 mg each of $CaCO_3$, $SrCO_3$, and $BaCO_3$.
Ans. 10.2 ml

36. In a titration of cyanide ion, 28.72 ml of 0.0100 M $AgNO_3$ is added before precipitation begins. (The reaction of Ag^+ with CN^- goes to completion producing the $[Ag(CN)_2]^-$ complex. Precipitation of solid AgCN takes place when excess Ag^+ is added to the solution, above the amount needed to complete the formation of $[Ag(CN)_2]^-$.) How many grams of NaCN were in the original sample?
Ans. 0.0281 g

37. A 0.010-mol sample of solid AgCN is rendered soluble in 1 l of solution by adding just sufficient excess cyanide ion to form $[Ag(CN)_2]^-$. When all of the solid silver cyanide has just dissolved, the concentration of free cyanide ion is 1.125 × 10⁻⁷ M. Neglecting hydrolysis of the cyanide ion, determine the concentration of free, uncomplexed silver ion in the solution. If more cyanide ion is added (without changing the volume) until the equilibrium concentration of free cyanide ion is 1.0 × 10⁻⁶ M, what will be the equilibrium concentration of free silver ion?
Ans. 8 × 10⁻⁹ M; 1 × 10⁻¹⁰ M

References

"The Relationship between Solubility and Solubility Product Is a Limiting Case," G. P. Haight, *J. Chem. Educ.*, **55**, 452 (1978).

"Determination of the Stability Constants of Nickel(II)-Cysteine," T. L. Rose and R. J. Seyse, *J. Chem. Educ.*, **53**, 728 (1976).

"Ratio Diagrams: A Simple Graphical Representation of Complicated Equilibria," R. de Levie, *J. Chem. Educ.*, **47**, 187 (1970).

"Potentiometric Determination of Solubility Product Constant," S. L. Tackett, *J. Chem. Educ.*, **46**, 857 (1969).

"Are Solubilities and Solubility Products Related?" L. Meites,

J. S. F. Pode, and Henry C. Thomas, *J. Chem. Educ.*, **43**, 667 (1966).

"Student Experiments Involving Unknown Solubility Constants," D. E. Heinz, *J. Chem. Educ.*, **44**, 114 (1967).

"The Stoichiometry of Copper Sulfide Formed in an Introductory Laboratory Exercise," D. Dingledy and W. M. Barnard, *J. Chem. Educ.*, **44**, 242 (1967).

"The Effects of Chloride Ion and Temperature on Lead Chloride Solubility," A. C. West, *J. Chem. Educ.*, **46**, 773 (1969).

"Effect of Ionic Strength on Equilibrium Constants," M. D. Seymour and Q. Fernando, *J. Chem. Educ.*, **54**, 225 (1977).

18

Chemical Thermodynamics

Five important questions often confront a chemist:

1. When two or more substances are put together, will they react?
2. If they do react, what energy changes will be associated with the reaction?
3. If a reaction occurs, what will be the concentrations of the reacting substances and their products at equilibrium?
4. If a reaction occurs, how fast will it go; that is, what will be its rate?
5. What is the mechanism by which the reaction occurs?

Chemical thermodynamics is concerned with the first three of these questions. It has nothing to say about the last two. The field of kinetics (Chapter 15) is concerned with either the rate or the specific manner in which atoms, ions, or molecules come together to form new compounds.

In Section 1.3 it was pointed out that the total quantity of energy available in the universe is constant. This of course does not mean that energy cannot be transformed from one form to another and transferred from one part of the universe to another. There are many examples on both large and small scales, such as the sun, the kitchen stove, and the family automobile, that produce such energy transformations and transfers. **Chemical thermodynamics is that branch of chemistry that studies the energy transformations and transfers that accompany chemical and physical changes.** It is customary to call that part of the universe that we have under study, and with whose properties we are concerned, the **system.** The rest of the universe is defined as the **surroundings. Chemical thermodynamics is then a**

study of the energy transformations that occur in a system and any transfer of energy that may occur between the system and the surroundings.

There are two fundamental laws of nature that are important in thermodynamics and that apply to all systems: (1) systems tend to attain a state of minimum potential energy, and (2) systems tend toward a state of maximum disorder. These two laws can be illustrated in the following way: If you are holding a box and release it, the box falls to the floor. In so doing it goes to a state of lower potential energy. If the box contains a jig-saw puzzle that had been assembled before the box was released, the chances are great that the puzzle will be partly disassembled after the fall, thereby becoming more disordered. The latter fact may not seem very important or fundamental, but most people would agree that no one would try to assemble a jig-saw puzzle by dropping the separated pieces on the floor, whereas we can readily accomplish the reverse process by dropping the assembled puzzle on the floor. A system tends to become less orderly because there are so many more ways to be disorderly than to be orderly; the probability for a system to become more disorderly or more random is greater than for it to become more orderly. We will consider more fully the interplay between these two fundamental laws of nature and their consequences later in this chapter. But before we get into these topics, we will consider **thermochemistry,** the study of the heat effects associated with chemical changes.

Thermochemistry

18.1   Internal Energy, Heat, and Work

In the study of thermodynamics we are interested in *changes* in the internal energy in a system. The **internal energy, E,** is simply the total of all of the possible kinds of energy present in that portion of the universe that we choose to study; that is, the total of heat energy, chemical energy, electrical energy, etc., present in the system.

It is not possible to measure or to calculate the value of E for a system, but it is possible to determine accurately the change in internal energy, ΔE, that accompanies a change in a system. As an example, let us consider a system that consists of a mole of hydrogen molecules. As described in Section 5.4, under certain conditions one mole of hydrogen molecules can be dissociated into two moles of hydrogen atoms by the additon of 436 kJ of energy to the system. Although we know neither the internal energy of the system, E_1, when it exists as a mole of hydrogen molecules nor its energy, E_2, when it exists as two moles of hydrogen atoms, we can find the difference in internal energy, $\Delta E = E_2 - E_1$, between the two states of the system. In this case, the internal energy of two moles of hydrogen atoms is 436 kJ greater than the internal energy of one mole of hydrogen molecules. Note that by common agreement the value of ΔE is designated as positive as energy is transferred from the surroundings to the system. When energy is transferred from the system to the surroundings, ΔE is taken as negative.

Energy is transferred into or out of a system in one or both of two ways: by heat transfer or by work. The transfer of energy as heat occurs as the result of thermal conduction or radiation. The symbol for the amount of heat transferred during a change in the system is q. If heat flows from the surroundings into the system to raise the internal energy of the system (as in the preceding example) the value of q is taken to be positive ($q > 0$). If heat flows from the system to the surroundings, the value of q is negative ($q < 0$). For example, when a system consisting of 18 g of steam, $H_2O(g)$, condenses to liquid water, $H_2O(l)$, at a given temperature, the internal energy of the system decreases as 40.6 kJ of heat flows from the system: $q = -40.6$ kJ.

The other method for changing the internal energy of a system is either (1) to let the system do work on the surroundings, in which case the internal energy of the system is reduced by the amount of work done, w, or (2) to let work be done on the system by the surroundings, in which case the internal energy of the system increases by the amount of work done on it. When energy is transferred from the system to the surroundings as work, and work is done on the surroundings, the value of w is positive ($w > 0$). A system absorbs energy when the surroundings do work on the system. In this case the value of w is negative ($w < 0$). In this chapter we will limit our consideration of work to that of expansion against a constant pressure (so called "expansion work").

It should be apparent from consideration of systems such as the internal combustion engine that gases expanding against a restraining pressure are capable of doing work. This means that if we have a system initially in state E_1 with volume V_1 and which goes to a new energy state E_2 with a larger volume V_2, all at a constant pressure, the system does a certain amount of expansion work on the surroundings. The amount of work done is equal to $P(V_2 - V_1)$, provided the pressure, P, remains constant. The work term $P(V_2 - V_1)$ is an energy term with energy units. This is readily apparent if it is recalled that pressure is defined as force divided by area and therefore has units of force/(length)^2; volume has units of (length)^3. Thus the units for the product are

$$P(V_2 - V_1) = \frac{\text{force}}{(\text{length})^2} \times (\text{length})^3$$

$$= \text{force} \times \text{length}$$

Units of force $\times$ length are units of work, or energy. If the pressure is expressed in pascals (newtons/meter^2) and the volume change in meters^3, the resulting product will have units of joules (newton meters).

It should be noted that a system might contract instead of expand. In this case, work is being done on the system by the surroundings instead of being done on the surroundings by the system. Either way, however, the change in volume is often referred to in thermodynamics as "expansion," in the sense that an increase in volume is a positive expansion, and a decrease in volume is a negative expansion.

18.2   The First Law of Thermodynamics

The **First Law of Thermodynamics** is actually just the Law of Conservation of Energy (see Section 1.3) and can be stated: **The total amount of energy in the universe is constant.** The First Law is often considered in a rather special form. If we have a system in a certain internal energy state, E_1, and we add an amount of heat energy, q, to the system which in turn does some work, w, on the surroundings, the system ends up in a new internal energy state, E_2. The Law of Conservation of Energy requires that for the system

$$E_2 = E_1 + (q - w)$$

This means that the final energy state is equal to the initial energy state plus a quantity composed of the heat added minus the work done. The quantity $q - w$ may be either positive or negative, depending upon the relative values of q and w. Then,

$$E_2 - E_1 = q - w$$

where $E_2 - E_1$ is equal to the change in internal energy for the system, ΔE. The symbol Δ stands for the difference between two quantities—in this case, energies. Therefore,

$$\Delta E = E_2 - E_1 = q - w$$

In other words, we can say that the change in the internal energy of any given system is equal to the heat energy transferred to the system from the surroundings minus the work energy transferred from the system to the surroundings. It is of course true that if ΔE is the energy change for the system and ΔE_{sur} is the energy change for the surroundings, $\Delta E + \Delta E_{sur} = 0$. This emphasizes the fact, expressed by the First Law, that the total amount of energy in the universe is constant. The following example will help clarify the First Law:

EXAMPLE 1   If 600 J of heat energy are added to a system in energy state E_1 and the system does 450 J of work on the surroundings,
a. What is the energy change of the system?

$$\Delta E = q - w = 600 \text{ J} - 450 \text{ J} = 150 \text{ J}$$

b. What is the energy change of the surroundings?

$$\Delta E + \Delta E_{sur} = 0$$
$$150 \text{ J} + \Delta E_{sur} = 0$$
$$\Delta E_{sur} = -150 \text{ J}$$

c. What is the energy of the system in the new energy state, E_2?

$$E_2 - E_1 = \Delta E = 150 \text{ J}$$
$$E_2 = E_1 + 150 \text{ J}$$

◆ ◆ ◆ ◆ ◆

There is a consequence of the First Law that may seem relatively obvious but is very important. If a system is in a state characterized by a certain energy E_1 and the system undergoes a change to a state characterized by E_2, the energy difference is independent of how the system gets from state 1 to state 2.

EXAMPLE 2   A system consisting of 18.015 g of $H_2O(l)$ at 0°C and 1 atm with a density of 0.9998 g/cm³ freezes to $H_2O(s)$ at 0° and 1 atm with a density of 0.9168 g/cm³, and 40,668 J of heat is released to the surroundings. Calculate ΔE for this process.

Since

$$\Delta E = q - w$$

to determine ΔE we need two values: the amount of heat transferred to or from the system and the amount of work done on or by the system. The problem tells us that 40,668 J of heat are transferred from the H_2O to the surroundings, so $q = -40,668$ J. The sign is negative since heat leaves the system.

The amount of work (in joules) involved in this change can be calculated from the expression $w = P(V_2 - V_1)$ if P is expressed in pascals and V_2 and V_1 in cubic meters. From Chapter 10, we know 1 atm = 101,325 Pa. The volumes can be calculated from the densities as follows:

$$V_1 = 18.015 \text{ g} \times \frac{1 \text{ cm}^3}{0.9998 \text{ g}} = 18.02 \text{ cm}^3$$

$$= 18.02 \text{ cm}^3 \times \left(\frac{1 \text{ m}}{100 \text{ cm}}\right)^3 = 1.802 \times 10^{-5} \text{ m}^3$$

$$V_2 = 18.015 \text{ g} \times \frac{1 \text{ cm}^3}{0.9168 \text{ g}} = 19.65 \text{ cm}^3$$

$$= 19.65 \text{ cm}^3 \times \left(\frac{1 \text{ m}}{100 \text{ cm}}\right)^3 = 1.965 \times 10^{-5} \text{ m}^3$$

Now the value of w can be determined:

$$
\begin{aligned}
w &= P(V_2 - V_1) \\
&= 101,325 \text{ Pa} \, (1.965 \times 10^{-5} \text{ m}^3 - 1.802 \times 10^{-5} \text{ m}^3) \\
&= 1.652 \text{ Pa m}^3 = 1.652 \text{ J}
\end{aligned}
$$

As indicated in Section 18.1, units of Pa m^3 are equivalent to joules. Note that the value of w is positive, indicating that work is done by the system as it expands from a volume of 1.802×10^{-5} m^3 to a volume of 1.965×10^{-5} m^3 against a constant pressure of 101,325 Pa (1 atm). Since the amount of expansion in this system is small, the amount of energy transferred as work is also small. In systems with larger changes in volume, the work term is much larger. Now that we know both q ($-40,668$ J) and w (1.652 J), we can calculate ΔE.

$$
\begin{aligned}
\Delta E &= q - w \\
&= (-40,668 \text{ J}) - (1.652 \text{ J}) \\
&= -40,666 \text{ J} = -40.666 \text{ kJ}
\end{aligned}
$$

◆ ◆ ◆ ◆ ◆

18.3   State Functions

The change in internal energy, ΔE, of a mole of water that undergoes the process $H_2O(l) \longrightarrow H_2O(s)$ as in Example 2 in Section 18.2 is simply the difference in internal energy between one mole of $H_2O(l)$ and one mole of $H_2O(s)$. This difference does not depend on how we convert the liquid water to ice. We could convert the liquid directly to the solid at 0° and 1 atm or we could convert the liquid to a mole of steam at 100°C and then condense the steam to ice on a cold metal plate at 0°C and 1 atm. Either way, ΔE for the process would be −40.666 kJ.

When a property of a system (its internal energy, for example) is not dependent on how the system gets from state to state and is only dependent on the state itself, the property is said to be a **state function.** A change in the value of a state function is independent of the path used to carry out the process. The change in the value of the state function is equal to the value of the function at the final state minus its value at the initial state. How the system goes from the initial state to the final state is irrelevant with respect to the final value. To put it another way, state functions have no memory—they forget the path used for the transformation.

An illustration of a state function can be based upon an area consisting of a valley and a high hill. Point A, the starting point, is in the valley while the final point B is at the middle of the hill. The change in potential energy that you would experience in hiking from A to B is independent of the path taken. Your potential energy is a state function.

As opposed to state functions, whose values depend solely upon the initial and final states, there are some functions whose values do depend on the paths followed. The distinction between the two types of functions can be illustrated by the expression in Section 18.2:

$$\Delta E = E_2 - E_1 = q - w$$

The internal energy, E, is a state function; the ΔE of a reaction is the same regardless of how that reaction is carried out. On the other hand, the values of q and w are not constant; they vary with the process used. Although the change in altitude is the same, any walker will tell you that there is a great deal of difference between the amount of effort required to walk directly from a valley to the middle of a hill and that required to walk from a valley to the top of a hill and then to the middle. Analogously, the change in energy, ΔE, associated with the transformation of liquid water to ice is always the same and is a state function. However, q and w associated with the change will depend upon the way in which the transformation is carried out, and thus they are not state functions. Quantities that are state functions are designated by capital letters; those which are not, by small letters.

At this point, we can conclude that a state function is characteristic of a state of matter regardless of how that state is reached.

18.4   Enthalpy Changes; Heat of Reaction

The heat, q, absorbed or released by a system undergoing a chemical change *at constant pressure* is a state function and is called the change in heat content or the **enthalpy change, ΔH,** of the system. Most chemical reactions occur under the essentially constant pressure of the atmosphere; consequently, the enthalpy change of these and other reactions that occur at constant pressure can be determined by subtracting the work term from the value of ΔE. In order to take into account the PV type of work done in many chemical reactions at constant pressure, we add the term $P(V_2 - V_1)$, or $P\,\Delta V$, to the internal energy change, ΔE. The sum $\Delta E + P\,\Delta V$

is a value for the total amount of heat energy that a chemical reaction can provide to the surroundings. Thus $\Delta H = \Delta E + P \Delta V$ for constant pressure processes. Since E, P, and V are state functions, H also is a state function. Note that if only expansion work is involved in going from state 1 to state 2 at constant pressure,

$$w = P \Delta V$$

Since $\Delta E = q - w$, then

$$\Delta E = q - P \Delta V$$

We know that $\Delta H = \Delta E + P \Delta V$; therefore,

$$\Delta H = (q - P \Delta V) + P \Delta V = q$$

At constant volume,

$$P \Delta V = 0 \quad \text{and} \quad \Delta H = \Delta E$$

If ΔH for a chemical reaction conducted at constant pressure (state 1 = reactants, state 2 = products) is negative, the system *evolves* heat energy to the surroundings, and the reaction is said to be **exothermic.** If ΔH for a chemical reaction conducted at constant pressure is positive, heat energy is *absorbed* by the system from the surroundings, and the reaction is said to be **endothermic.**

A number of different enthalpy changes, or changes in heat content, have been mentioned in previous chapters. In Section 9.6 we identified the "heat of reaction" as the quantity of heat liberated or absorbed during a chemical change. "Heat of vaporization" was identified in Section 11.5 as the heat necessary to convert one mole of a liquid to one mole of a gas. The heat of vaporization of ammonia, for example, was identified as the enthalpy change for the process

$$NH_3(l) \longrightarrow NH_3(g) \quad \Delta H = 23.2 \text{ kJ}$$

When one mole of ammonia is converted from a liquid to a gas, 23.2 kJ of heat are absorbed ($\Delta H > 0$) from the surroundings. The heat of fusion of a substance (Section 11.13) is simply the enthalpy change of the process of converting one mole of the solid substance to the liquid state. Although it was not identified as such, the heat of neutralization mentioned in Section 14.7 is the enthalpy change of the reaction

$$H_3O^+(aq) + OH^-(aq) \longrightarrow 2H_2O(l) \quad \Delta H = -55.8 \text{ kJ}$$

Thus, when one mole of solvated hydrogen ions is neutralized by one mole of solvated hydroxide ions, 55.8 kJ of heat is released ($\Delta H < 0$) to the surroundings.

Enthalpy changes for a large number of chemical reactions have been measured and tabulated to simplify the comparison of various reactions. The National Bureau of Standards has been active in the compilation of accurate thermochemical data of this type. To facilitate the tabulation of such data, values have been measured and tabulated for many substances under a specific set of conditions referred to as the **standard state.** The standard state refers to each pure substance as being at 25°C (298.15 K) and at 1 atm pressure. The state of the substance (i.e., gas, liquid, or solid) must be specified. It is customary to use the symbol ΔH°_{298} to indicate the enthalpy changes that have been measured under standard state

TABLE 18-1 Standard Molar Enthalpies of Formation, Standard Molar Free Energies of Formation, and Absolute Standard Molar Entropies (298.15 K, 1 atm). (See also Appendix J for additional values.)

Substance	$\Delta H^{\circ}_{f_{298}},$ kJ mol^{-1}	$\Delta G^{\circ}_{f_{298}},$ kJ mol^{-1}	$S^{\circ}_{298},$ J mol^{-1} K^{-1}
Carbon			
C(s) (graphite)	0	0	5.740
C(g)	716.68	671.29	157.99
CO(g)	−110.5	−137.2	197.56
CO$_2$(g)	−393.5	−394.4	213.6
CH$_4$(g)	−74.81	−50.75	186.15
Chlorine			
Cl$_2$(g)	0	0	222.96
Cl(g)	121.7	105.7	165.09
Copper			
Cu(s)	0	0	33.15
CuS(s)	−53.1	−53.6	66.5
Hydrogen			
H$_2$(g)	0	0	130.57
H(g)	218.0	203.3	114.60
H$_2$O(g)	−241.8	−228.6	188.71
H$_2$O(l)	−285.8	−237.2	69.91
HCl(g)	−92.31	−95.30	186.80
H$_2$S(g)	−20.6	−33.6	205.7
Oxygen			
O$_2$(g)	0	0	205.03
O(g)	249.2	231.8	160.95
Silver			
Ag$_2$O(s)	−31.0	−11.2	121
Ag$_2$S(s)	−32.6	−40.7	144.0

conditions. Probably the most useful tabulation of enthalpy changes is for a particular type of chemical reaction in which *one mole* of a pure substance is formed from the free elements in their most stable states under standard state conditions. This enthalpy change is referred to as the **standard molar enthalpy of formation** of the substance formed and is designated by $\Delta H^{\circ}_{f_{298}}$. In the following discussion, 298.15 K will be implied unless indicated otherwise. Standard molar enthalpies of formation of some common substances are given in Table 18-1. A more extensive compilation is given in Appendix J. Note that the units in these tables are joules or kilojoules. Almost all of the older tables available use kilocalories (1 cal = 4.184 J) as units of energy. If you have occasion to use other tables, be certain to check the units. By making use of these tabulated values and the fact that the enthalpy is a state function, we may obtain the enthalpy changes for a large number of chemical reactions. Several other properties of interest to chemists can be deduced from the enthalpy changes.

EXAMPLE 1   **In Section 9.11, the heat of the reaction $3O_2(g) \longrightarrow 2O_3(g)$ was given as 285 kJ. Assuming that this is the standard enthalpy change, ΔH_{298}° (with both the $O_2(g)$ and $O_3(g)$ in their standard states) what is the standard molar enthalpy of formation, $\Delta H_{f_{298}}^\circ$, of $O_3(g)$?**

$\Delta H_{f_{298}}^\circ$ is the enthalpy change for the formation of one mole of a substance in its standard state from the elements in their standard states. Thus $\Delta H_{f_{298}}^\circ$ for $O_3(g)$ is the enthalpy change for the reaction

$$\tfrac{3}{2}O_2(g) \longrightarrow O_3(g)$$

Since 285 kJ of heat are absorbed when two moles of $O_3(g)$ are formed, to form one mole of $O_3(g)$ would require one-half as much, or 142.5 kJ. Thus the enthalpy change would be 143 kJ and $\Delta H_{f_{298}}^\circ = 143$ kJ mol^{-1}.

◆ ◆ ◆ ◆ ◆

EXAMPLE 2   **What would be the enthalpy change for the reaction of one mole of $H_2(g)$ with one mole of $Cl_2(g)$ to produce two moles of $HCl(g)$ at standard state conditions?**

$$H_2(g) + Cl_2(g) \longrightarrow 2HCl(g) \qquad \Delta H^\circ = \ ?$$

Since $H_2(g)$ and $Cl_2(g)$ are the most stable states of the elements hydrogen and chlorine at standard state conditions and a pure substance is being formed at standard state conditions, ΔH° must be proportional to the standard molar enthalpy of formation of $HCl(g)$. In this example, two moles of $HCl(g)$ are being formed; therefore

$$\Delta H^\circ = 2\,\Delta H_{f_{HCl(g)}}^\circ$$

From Table 18-1, $\Delta H_{f_{HCl(g)}}^\circ = -92.31$ kJ mol^{-1}

Therefore, $\Delta H^\circ = 2 \text{ mol HCl} \times (-92.31 \text{ kJ mol}^{-1}) = -184.6$ kJ

Thus the reaction evolves 184.6 kJ of heat:

$$H_2(g) + Cl_2(g) \longrightarrow 2HCl(g) \qquad \Delta H^\circ = -184.6 \text{ kJ}$$

◆ ◆ ◆ ◆ ◆

18.5   Hess's Law

The use of standard molar enthalpies of formation and other enthalpy changes is possible because of the principles summed up in **Hess's law: For any process that can be looked on as being the sum of several stepwise processes, the enthalpy change for the total process must be equal to the sum of the enthalpy changes for the various steps.** In other words, the heat evolved or absorbed in going from one state to another is the same by whatever route the reactants use to get there. Hess's law tells us that chemical reactions and their enthalpies can be handled algebraically; that is, the enthalpy change for a reaction can be calculated from the enthalpies of a sequence of reactions that can be combined algebraically to give the desired reaction.

Before we apply Hess's law, let us briefly look at two important features of enthalpy changes: (1) ΔH for a reaction in one direction is equal in magnitude and opposite in sign to ΔH for the reaction in the reverse direction, and (2) ΔH is directly proportional to the quantities of reactants or products. The first statement tells us that if we know ΔH for the reaction $H_2O(s) \longrightarrow H_2O(l)$ to be 6.0 kJ, then ΔH for the reverse reaction, $H_2O(l) \longrightarrow H_2O(s)$, is simply the negative of that for the forward reaction, or -6.0 kJ. The second statement indicates that if the heat of fusion of one mole of water is 6.0 kJ

$$H_2O(s) \longrightarrow H_2O(l) \qquad \Delta H = 6.0 \text{ kJ}$$

then the heat of fusion of two moles of water is twice as great.

$$2H_2O(s) \longrightarrow 2H_2O(l) \qquad \Delta H = 12 \text{ kJ}$$

EXAMPLE 1   **Calculate the standard molar enthalpy of formation of $CO_2(g)$ from the following standard enthalpy changes:**

$$2C(s) + O_2(g) \longrightarrow 2CO(g) \qquad \Delta H° = -221.0 \text{ kJ}$$
$$2CO(g) + O_2(g) \longrightarrow 2CO_2(g) \qquad \Delta H° = -566.0 \text{ kJ}$$

The standard molar enthalpy of formation of $CO_2(g)$ is the enthalpy change for the reaction

$$C(s) + O_2(g) \longrightarrow CO_2(g) \qquad \Delta H° = H°_{fCO_2}$$

This equation and the enthalpy change can be obtained by multiplying the first equation in the problem and its ΔH by one-half and adding to this value one-half of the second equation and its enthalpy change.

$$\tfrac{1}{2}[2C(s) + O_2(g) \longrightarrow 2CO(g)] \qquad \Delta H° = \tfrac{1}{2}(-221.0) = -110.5 \text{ kJ}$$
$$\underline{\tfrac{1}{2}[2CO(g) + O_2(g) \longrightarrow 2CO_2(g)] \qquad \Delta H° = \tfrac{1}{2}(-566.0) = -283.0 \text{ kJ}}$$
$$C(s) + O_2(g) \longrightarrow CO_2(g) \qquad \Delta H° = (-110.5) + (-283.0) = -393.5 \text{ kJ}$$

The enthalpy of formation of $CO_2(g)$ at 25° and 1 atm is -393.5 kJ per mole of $CO_2(g)$. Hence $\Delta H°_{fCO_2(g)} = -393.5$ kJ.

◆ ◆ ◆ ◆ ◆

EXAMPLE 2   **What is the $\Delta H°$ for the reaction**

$$CH_4(g) + 2O_2(g) \longrightarrow CO_2(g) + 2H_2O(l)$$

This reaction can be viewed as occurring in several steps:

Step 1. $\qquad CH_4(g) \longrightarrow C(s) + 2H_2(g) \qquad \Delta H°_1 = -\Delta H°_{fCH_4(g)}$

Step 2. $\qquad\quad 2O_2(g) \longrightarrow 2O_2(g) \qquad \Delta H°_2 = 2\,\Delta H°_{fO_2(g)} = 0$

Step 3. $\quad 2H_2(g) + O_2(g) \longrightarrow 2H_2O(l) \qquad \Delta H°_3 = 2\,\Delta H°_{fH_2O(l)}$

Step 4. $\qquad C(s) + O_2(g) \longrightarrow CO_2(g) \qquad \Delta H°_4 = \Delta H°_{fCO_2(g)}$

Adding Steps 1, 2, 3, and 4 gives

$$CH_4(g) + \overset{2}{\cancel{4}O_2(g)} + \cancel{2H_2(g)} + \cancel{C(s)} \longrightarrow \cancel{C(s)} + \cancel{2O_2(g)}$$
$$+ \cancel{2H_2(g)} + 2H_2O(l) + CO_2(g)$$

The net result is

$$CH_4(g) + 2O_2(g) \longrightarrow 2H_2O(l) + CO_2(g) \qquad \Delta H° = ?$$

In either case, i.e., by direct reaction or by the combination of reactions in Steps 1 through 4, we end up with the same result. The enthalpy change is independent of how we get from state 1 to state 2, verifying that H is a state function. Therefore,

$$\Delta H° = \Delta H_1° + \Delta H_2° + \Delta H_3° + \Delta H_4°$$

or

$$\Delta H° = -\Delta H_{f_{CH_4(g)}}° + 2\,\Delta H_{f_{O_2(g)}}° + 2\,\Delta H_{f_{H_2O(l)}}° + \Delta H_{f_{CO_2(g)}}°$$
$$= -1 \text{ mol } \cancel{CH_4(g)} \times (-74.81 \text{ kJ } \cancel{mol^{-1}}) + 2 \text{ mol } O_2$$
$$\times (0 \text{ kJ } \cancel{mol^{-1}}) + 2 \text{ mol } \cancel{H_2O(l)} \times (-285.8 \text{ kJ } \cancel{mol^{-1}})$$
$$+ 1 \text{ mol } \cancel{CO_2(g)} \times (-393.5 \text{ kJ } \cancel{mol^{-1}})$$
$$= -890.3 \text{ kJ}$$

Hence 890.3 kJ of heat are evolved during the combustion of one mole of methane.

$$CH_4(g) + 2O_2(g) \longrightarrow 2H_2O(l) + CO_2(g) \qquad H° = -890.3 \text{ kJ}$$

◆ ◆ ◆ ◆ ◆

A more useful version of Hess's law for the above type of calculation is the following: *For any chemical reaction occurring at standard state conditions, the standard enthalpy change is equal to the sum of the standard molar enthalpies of formation of all the products, each multiplied by the number of moles of the product in the balanced chemical equation, minus the corresponding sum for the reactants.*

EXAMPLE 3   Calculate $\Delta H°$ for the reaction

$$2Ag_2S(s) + 2H_2O(l) \longrightarrow 4Ag(s) + 2H_2S(g) + O_2(g)$$

The standard molar enthalpies of formation, $\Delta H_{f_{298}}°$, of the compounds involved may be found in Table 18-1.

$$\Delta H° = 4\,\Delta H_{f_{Ag(s)}}° + 2\,\Delta H_{f_{H_2S(g)}}° + \Delta H_{f_{O_2(g)}}° - 2\,\Delta H_{f_{Ag_2S(s)}}° - 2\,\Delta H_{f_{H_2O(l)}}°$$

The standard molar enthalpy of formation of an element in its most stable state is zero. Thus

$$\Delta H° = 4 \text{ mol } \cancel{Ag(s)} \times (0 \text{ kJ } \cancel{mol^{-1}}) + 2 \text{ mol } \cancel{H_2S(g)} \times (-20.6 \text{ kJ } \cancel{mol^{-1}})$$
$$+ 1 \text{ mol } \cancel{O_2(g)} \times (0 \text{ kJ } \cancel{mol^{-1}}) - 2 \text{ mol } \cancel{Ag_2S(s)} \times (-32.6 \text{ kJ } \cancel{mol^{-1}})$$
$$- 2 \text{ mol } \cancel{H_2O(l)} \times (-285.8 \text{ kJ } \cancel{mol^{-1}})$$
$$= [-41.2 - (-65.2) - (-571.6)] \text{ kJ} = 595.6 \text{ kJ}$$

Hence the reaction is endothermic.

$$2Ag_2S(s) + 2H_2O(l) \longrightarrow 4Ag(s) + 2H_2S(g) + O_2(g) \qquad \Delta H° = 595.6 \text{ kJ}$$

◆ ◆ ◆ ◆ ◆

EXAMPLE 4   **Calculate $\Delta H°$ for the reaction**

$$2Na(s) + 2H_2O(l) \longrightarrow 2NaOH(s) + H_2(g)$$

The standard molar enthalpies of formation for the reactants and products involved are as follows: $H_2O(l)$, -285.8 kJ mol^{-1}; NaOH(s), -426.8 kJ mol^{-1}; Na and H_2, 0 kJ mol^{-1}.

$$\Delta H° = 2 \text{ mol NaOH}(s) \times (-426.8 \text{ kJ mol}^{-1}) + 1 \text{ mol } H_2(g) \times (0 \text{ kJ mol}^{-1})$$
$$- 2 \text{ mol Na}(s) \times (0 \text{ kJ mol}^{-1}) - 2 \text{ mol } H_2O(l) \times (-285.8 \text{ kJ mol}^{-1})$$
$$= [-853.5 + 571.6] \text{ kJ} = -281.9 \text{ kJ}$$

Hence the reaction evolves heat.

◆ ◆ ◆ ◆ ◆

A positive value of ΔH means that the sum of the enthalpies of the products, taking into account sign as well as numerical value, is greater than the sum of the enthalpies of the reactants; a negative value of ΔH means that the sum of the enthalpies of the products is less than that of the reactants.

In terms of bond strengths within the molecules, a positive value of ΔH indicates that the bonds are stronger in the reactants than in the products, and hence heat must be absorbed for the reaction to take place. A negative value of ΔH, on the other hand, indicates that the bonds are stronger in the products than in the reactants; hence, less energy is required to break the bonds in the reactants than is evolved in the formation of the products. The net effect is evolution of heat.

18.6   Bond Energies

The strength of a chemical bond is measured by the energy required to break the bond, that is, to separate the atoms from their molecular grouping and leave them as distinct isolated gaseous atoms. For a diatomic molecule the bond energy, D, is taken to be equal to the change in standard enthalpy for the reaction

$$XY(g) \longrightarrow X(g) + Y(g) \qquad \Delta H = D$$

When only one bond between atoms X and Y is involved, the bond energy is merely the dissociation energy of the molecule, which may be determined by a variety of spectroscopic or thermochemical methods. For diatomic molecules containing atoms of oxidation number greater than 1 (for example, nitrogen or oxygen), the dissociation energy, and hence the bond energy, may equally well be determined, but it corresponds to the rupture of a multiple bond between the atoms. Bond energies for commonly occurring diatomic molecules range from 946 kJ per mole for N_2 (triple bond) to 150 kJ per mole for I_2 (single bond), and from 569 kJ per mole for HF to 295 kJ per mole for HI.

TABLE 18-2   Some Representative Single Bond Energies (kilojoules per mole of bonds)

H	C	N	O	F	Si	P	S	Cl	Br	I	
436	415	390	464	569	295	320	340	432	370	295	H
	345	290	350	439	290	265	260	330	275	240	C
		160	200	270	—	210	—	200	245(?)	—	N
			140	185	370	350	—	205	—	200	O
				160	540	489	285	255	195(?)	—	F
					175	215	225	359	290	215	Si
						215	230	330	270	215	P
							215	250	215	—	S
								243	220	210	Cl
									190	180	Br
										150	I

Molecules of three or more atoms necessarily have two or more bonds. The heats of formation of such molecules from the isolated gaseous atoms are equal to the sum of all the bond energies; thus, the heat of formation of the S_8 ring from eight sulfur atoms is eight times the energy of formation of a single S—S bond.

Single bond energies for some common bonds appear in Table 18-2. The energies for double or triple bonds between two atoms are generally higher than those for single bonds between the same two atoms, as we would expect; but a double bond is not, in general, quite twice as strong as a comparable single bond, nor a triple bond quite three times as strong as a comparable single bond. Compare the following bond energies (in kilojoules):

C=C, 611 kJ   C=N, 615 kJ   C=O, 741 kJ   N=N, 418 kJ

C≡C, 837 kJ   C≡N, 891 kJ   C≡O, 1070 kJ   N≡N, 946 kJ

Enthalpy changes can be used to obtain chemical bond energies. Returning to the example in Section 18.4 for HCl, we can write two series of steps by which two moles of HCl result from one mole of H_2 and one mole of Cl_2.

$$H_2(g) \quad + \quad Cl_2(g) \quad \xrightarrow{2\,\Delta H^\circ_{f_{HCl(g)}}} \quad 2HCl(g)$$

$$\downarrow 2\,\Delta H^\circ_{f_{H(g)}} \qquad \downarrow 2\,\Delta H^\circ_{f_{Cl(g)}} \quad \Delta H^\circ \nearrow$$

$$2H(g) \quad + \quad 2Cl(g)$$

$$2\,\Delta H^\circ_{f_{HCl(g)}} = 2\,\Delta H^\circ_{f_{H(g)}} + 2\,\Delta H^\circ_{f_{Cl(g)}} + \Delta H^\circ$$

ΔH° is the heat energy evolved in the formation of *two* moles of hydrogen chloride starting from *atomic* hydrogen and *atomic* chlorine at standard state conditions. The corresponding heat energy evolved in the formation of *one* mole of hydrogen chloride from atomic hydrogen and atomic chlorine, therefore, equals $\frac{1}{2}\Delta H^\circ$.

The bond energy of hydrogen chloride is the amount of energy *input* required for the reverse process—that is, to break the bond in HCl to produce atomic hydrogen and atomic chlorine. It is apparent then that the bond energy in the present example is $-\frac{1}{2}\Delta H°$. Calculating the value from the data in Table 18-1, we obtain

$$2\,\Delta H°_{f_{HCl(g)}} = 2\,\Delta H°_{f_{H(g)}} + 2\,\Delta H°_{f_{Cl(g)}} + \Delta H°$$

$$2\text{ mol HCl}(g) \times (-92.31\text{ kJ mol}^{-1}) = 2\text{ mol H}(g) \times (218.0\text{ kJ mol}^{-1})$$
$$+ 2\text{ mol Cl}(g) \times (121.7\text{ kJ mol}^{-1}) + \Delta H°$$
$$\Delta H° = [-184.6 - 436.0 - 243.4]\text{ kJ} = -864.0\text{ kJ}$$
$$\tfrac{1}{2}\Delta H° = -432.0\text{ kJ mol}^{-1}$$

Thus the bond energy for HCl (which is $-\frac{1}{2}\Delta H°$) = $-(-432.0)$ kJ mol^{-1} = 432.0 kJ mol^{-1}.

Note that this is the value, if rounded to three significant figures, which is given in Table 18-2 for the H—Cl bond.

In addition, from the preceding example we can determine the bond energies for H_2 and Cl_2. Referring to Table 18-1, we find that $\Delta H°_{f_{H(g)}} = 218.0$ kJ for one mole of H. Recall that a $\Delta H°_f$ value signifies the enthalpy change for the chemical reaction in which one mole of the pure substance (in this case, atomic hydrogen, H) is formed from the free elements *in their most stable states* under standard state conditions (in this case, the one element, the gaseous diatomic molecule H_2). To form atomic hydrogen, H, from H_2 requires an input of energy to *break* the H—H bond. Note that the $\Delta H°_{f_{H(g)}}$ value of 218.0 kJ is for breaking the bonds in one-half mole of H_2 to produce one mole of H, and no sign change is required in determining the bond energy. To form two moles of H from one mole of H_2 would then require 2(218.0 kJ) = 436.0 kJ. This therefore is the bond energy for $H_2(g)$, and it checks with the rounded-off value listed in Table 18-2. Similarly, $\Delta H°_{f_{Cl(g)}} = 121.7$ kJ for one mole of Cl. To form two moles of Cl from one mole of Cl_2 would require 243.3 kJ. Thus the bond energy for $Cl_2(g)$ is 243.3 kJ. The rounded-off value is listed in Table 18-2.

Entropy and Free Energy

18.7   The Spontaneity of Chemical and Physical Changes

A **spontaneous change,** in a thermodynamic sense, is a change in a system that proceeds without the exertion of any outside influence. Liquid water spontaneously freezes to ice at $-1°$, hydrogen ions spontaneously combine with hydroxide ions in aqueous solution, and carbon spontaneously combines with oxygen to give carbon monoxide or carbon dioxide. Since the definition of a spontaneous change does not include time, changes may be spontaneous although they proceed very slowly. For those who own diamonds (which are essentially pure carbon) it is fortunate that the spontaneous reaction of carbon with oxygen is very slow at room temperature.

Those exothermic changes for which ΔH is large (a large amount of heat is given off) are frequently spontaneous. For the reaction used in Example 3 of Section 18.5 $[2Ag_2S(s) + 2H_2O(l) \longrightarrow 4Ag(s) + 2H_2S(g) + O_2(g)]$, $\Delta H = 595.6$ kJ, and the forward reaction, on the basis of $\Delta H°$, would be expected not to occur spontaneously at 25°C and 1 atm. However, the reverse reaction ($\Delta H = -595.6$ kJ) would be expected to occur spontaneously at 25° and 1 atm. For the reaction of Example 4 of Section 18.5 [$2Na(s) + 2H_2O(l) \longrightarrow 2NaOH(s) + H_2(g)$], $\Delta H = -281.9$ kJ, and the forward reaction would be expected, on the basis of $\Delta H°$ alone, to occur spontaneously at 25° and 1 atm, but the reverse reaction should occur only if external energy is introduced. The predictions for these reactions are in accord with experimental evidence. However, such is not always true, and predictions based upon $\Delta H°$ alone are not always valid.

A reaction that proceeds spontaneously at one temperature and pressure very often does not proceed spontaneously at another temperature and pressure. For example, at 25°C and 1 atm pressure, the reaction $2Ag(s) + \frac{1}{2}O_2(g) \longrightarrow Ag_2O(s)$ proceeds spontaneously. The $\Delta H°$ value is negative, as we would expect for spontaneous reaction. However, although the $\Delta H°$ value for the reaction changes very little with the change in temperature, at temperatures above about 200°C the reaction is not spontaneous. Another example is in the physical change that occurs when a solid crystal melts. At temperatures above the melting temperature the solid melts spontaneously; below the melting temperature it does not. Yet the ΔH value changes very little. The tendency for chemical reactions and physical changes to proceed spontaneously is usually quite dependent upon temperature and pressure changes. Yet temperature and pressure changes often have virtually no effect upon the value of ΔH for a reaction.

It is evident, then, that another factor in addition to the energy change occurring when a reaction takes place must be considered when determining whether or not a given reaction will proceed spontaneously. This factor is the entropy change occurring as the change in the system takes place.

18.8   Entropy and Entropy Changes

Consider two changes that are spontaneous but not exothermic: the melting of ice at room temperature, $H_2O(s) \longrightarrow H_2O(l)$, $\Delta H = 6.0$ kJ, and the decomposition of calcium carbonate at high temperature, $CaCO_3(s) \longrightarrow CaO(s) + CO_2(g)$, $\Delta H = 178$ kJ. Both of these reactions occur with an increase in the disorder of the system.

The molecules of water that make up an ice crystal are held in fixed positions in the crystal in a regular repeating array (Section 12.1). When the ice melts, the water molecules are held much less tightly and are free to move through the liquid as well as to tumble. The molecules in the liquid are much more randomly distributed than in the solid. Thus the amount of order is higher in the solid than in the liquid. As calcium carbonate decomposes, the system changes from an ordered array of calcium ions and carbonate ions in the solid to an ordered array of calcium and oxide ions in solid calcium oxide plus a quite disordered collection of carbon dioxide molecules in the gas phase. The random arrangement (disorder) of the carbon dioxide molecules in the gas phase is even greater than that of an equal number of water molecules in the liquid phase.

The randomness, or the amount of disorder, of a system can be determined quantitatively and is referred to as the entropy, S, of the system. S, like E and H, is a state function. The greater the randomness, or disorder, in a system the higher its entropy. Entropy is one of the most important concepts in science; an entropy increase, corresponding to an increase in disorder, is the major driving force in many chemical and physical processes.

Each substance has an entropy as one of its characteristic properties, just as it has color, hardness, volume, melting point, density, and enthalpy as characteristic properties. The entropy change, ΔS, for a chemical change represents the sum of the entropies of the products of that change minus the sum of the entropies of the reactants. A positive value for ΔS ($\Delta S > 0$) corresponds to an increase in randomness, or disorder; a negative value for ΔS ($\Delta S < 0$) corresponds to a decrease in randomness, or a more ordered structure. The change to a more random and less ordered system when ice melts or when calcium carbonate decomposes corresponds to an increase in entropy ($\Delta S > 0$). Similarly, evaporation is also accompanied by an increase in entropy as the molecules go to the still more disordered and random state characteristic of gases. The amount of change in the entropy value is a measure of the increase in the disorder, or randomness, of a system.

Entropy values for several substances are listed in Table 18-1 and in Appendix J. These values are the actual entropy content of one mole of these substances in their standard states and thus are absolute standard molar entropies. (The superscript in $S°$ indicates the entropy value is for a substance in its standard state.) Note that solids (well ordered) tend to have lower entropies than liquids (less well ordered), and liquids tend to have lower entropies than gases (still less ordered). In the same physical state, substances with simple molecules tend to have lower molar entropies than substances with more complicated molecules. The latter have larger numbers of atoms to move about, can hence create greater randomness or disorder, and have higher entropies. Hard substances, such as diamond, tend to be more ordered and have lower entropies than softer materials, such as graphite or sodium.

The table of absolute standard molar entropies can be used to calculate entropy changes as illustrated below.

EXAMPLE 1   **Determine the entropy change for the change $H_2O(l) \longrightarrow H_2O(g)$ when both $H_2O(l)$ and $H_2O(g)$ are in their standard states.**

The entropy change for this reaction is the sum of the entropies of the products, only $H_2O(g)$ in this case, minus the sum of the entropies of the reactants, $H_2O(l)$ in this case.

$$\Delta S° = S°_{H_2O(g)} - S°_{H_2O(l)}$$
$$= 1 \text{ mol } H_2O(g) \times (188.71 \text{ J mol}^{-1} \text{ K}^{-1}) - 1 \text{ mol } H_2O(l)$$
$$\times (69.91 \text{ J mol}^{-1} \text{ K}^{-1})$$
$$= 118.80 \text{ J K}^{-1}$$

The value for $\Delta S°$ is positive, indicating greater disorder in gaseous H_2O than in liquid H_2O. This is in accord with the fact that the molecules are in much more rapid and random motion in the gaseous state than in the liquid state.

◆ ◆ ◆ ◆ ◆

EXAMPLE 2   **Calculate the entropy change for the following reaction when the reactants and products are in their standard states:**

$$2H_2(g) + O_2(g) \longrightarrow 2H_2O(l)$$

Note that the absolute standard molar entropies given in Table 18-1 are for one mole of substance. The entropy of two moles of hydrogen or two moles of liquid water is twice that of one mole. Thus

$$\Delta S° = 2S°_{H_2O(l)} - 2S°_{H_2} - S°_{O_2}$$
$$= 2 \text{ mol } H_2O(l) \times (69.91 \text{ J mol}^{-1}\text{K}^{-1}) - 2 \text{ mol } H_2(g)$$
$$\times (130.57 \text{ J mol}^{-1}\text{K}^{-1}) - 1 \text{ mol } O_2(g) \times (205.03 \text{ J mol}^{-1}\text{K}^{-1})$$
$$= -326.35 \text{ J K}^{-1}$$

As this reaction proceeds, the entropy *of the system* decreases. As we will see in a subsequent section, the entropy of the surroundings increases.

◆ ◆ ◆ ◆ ◆

18.9   Free Energy Changes

Putting the idea of randomness together with that of enthalpy, we can say that reactions tend to proceed in such a way as to obtain a state of minimum energy (negative ΔH) and maximum disorder (positive ΔS). The difference between the enthalpy change of a reaction, ΔH, and the product of the temperature and the entropy change of a reaction, $T\Delta S$, can be identified with the change in another state function, the **free energy, G.** The relationship between the change in free energy, ΔG, the enthalpy change, and the entropy change for a reaction is given by the expression

$$\Delta G = \Delta H - T\Delta S$$

where T is the temperature of the reaction on the Kelvin scale. The free energy change is the thermodynamic quantity most useful in predicting the spontaneity of a chemical reaction. **Reactions for which the value of ΔG is negative ($\Delta G < 0$) are spontaneous.**

In the free energy equation $\Delta G = \Delta H - T\Delta S$, the term $T\Delta S$ can be thought of in one sense as a correction term for ΔH to determine ΔG. If $T\Delta S$ is small compared to ΔH, then ΔG and ΔH have nearly the same value and serve equally well as a basis for prediction of the tendency of a given reaction to proceed spontaneously. However, when the value of $T\Delta S$ is appreciable compared to that of ΔH, the ΔH value is not an accurate indicator of spontaneity and the ΔG value must be used. The ΔH portion of the free energy equation represents the energy, as heat, involved in a reaction at constant pressure; hence, it corresponds to a difference in energy between the initial and final states in a process. The ΔS portion of the free energy equation (and hence $T\Delta S$), on the other hand, corresponds to the difference in the amount of order of the atoms in the products and reactants.

For processes in which the initial and final states do not differ in enthalpy (ΔH is zero) and ΔS is not zero, the term $T\Delta S$ controls whether the reaction occurs. For processes, on the other hand, where the amount of order of the initial and final states are the same (ΔS, and hence $T\Delta S$, is zero) and ΔH is not zero, the energy term ΔH is considered in predicting whether the reaction will occur. If ΔH and $T\Delta S$ each differ from zero, the algebraic sum of the two (that is, the algebraic sum of the energy and the $T\Delta S$ terms) must be considered in predicting whether the reaction will occur. Since ΔG is the algebraic sum of ΔH and $-T\Delta S$, it is therefore the key quantity in any of the above cases for predicting whether a process will occur spontaneously.

The equation $\Delta G = \Delta H - T\Delta S$ tells us that a reaction that is exothermic (negative ΔH) and produces simpler and more numerous molecules (more disorder, positive ΔS) will proceed spontaneously (both ΔH and ΔS work toward a more negative value for ΔG). A reaction that is endothermic (positive ΔH) and produces a more ordered system (negative ΔS) will not proceed spontaneously (both ΔH and ΔS work toward a more positive value for ΔG).

If the reaction is exothermic (negative ΔH) but produces a more ordered system (negative ΔS), or if the reaction is endothermic (positive ΔH) but produces a less ordered system (positive ΔS), then ΔH and ΔS are working in opposite directions, and the relative sizes of ΔH and $T\Delta S$ will determine the prediction.

It should be noted that T must always be positive. Hence ΔS (which can be either positive or negative) determines the sign of the correction term $T\Delta S$. Also, it should be noted that as T increases, $T\Delta S$ plays an increasingly important role in the value of ΔG.

Again, it should be emphasized that a prediction as to whether a given reaction will proceed spontaneously says *nothing whatever about the rate at which it will occur*. A reaction with a negative ΔG value should proceed spontaneously, but it may do so at a very rapid rate or at an infinitesimally slow rate.

18.10   Determination of Free Energy Changes

Free energy is a state function. Hence values of $\Delta G°$ can be calculated from standard molar free energies of formation, $\Delta G_f°$, in a manner analogous to that used for calculating $\Delta H°$ values from standard molar enthalpies of formation. Several standard molar free energy of formation values are given in Table 18-1 and in Appendix J. The standard molar free energy of formation of any free element is zero for the same reasons as for $\Delta H_f°$. As with ΔH values, many ΔG values have been tabulated, and to facilitate comparison the tabulations are made for standard state conditions using the symbol $\Delta G°$.

EXAMPLE 1   Calculate the free energy change, $\Delta G°$, for the reaction

$$2Ag_2S(s) + 2H_2O(l) \longrightarrow 4Ag(s) + 2H_2S(g) + O_2(g)$$

The standard molar free energies of formation of the compounds involved at 25° and 1 atm are as follows: $Ag_2S(s)$, -40.7 kJ mol^{-1}; $H_2O(l)$, -237.2 kJ mol^{-1}; $H_2S(g)$, -33.6 kJ mol^{-1}; $Ag(s)$ and $O_2(g)$, 0 kJ mol^{-1}.

$$\Delta G^\circ = 4\,\Delta G^\circ_{f_{Ag(s)}} + 2\,\Delta G^\circ_{f_{H_2S(g)}} + \Delta G^\circ_{f_{O_2(g)}} - 2\,\Delta G^\circ_{f_{Ag_2S(s)}} - 2\,\Delta G^\circ_{f_{H_2O(l)}}$$

$$= 4 \text{ mol Ag}(s) \times (0 \text{ kJ mol}^{-1}) + 2 \text{ mol H}_2\text{S}(g) \times (-33.6 \text{ kJ mol}^{-1})$$

$$+ 1 \text{ mol O}_2(g) \times (0 \text{ kJ mol}^{-1}) - 2 \text{ mol Ag}_2\text{S}(s) \times (-40.7 \text{ kJ mol}^{-1})$$

$$- 2 \text{ mol H}_2\text{O}(l) \times (-237.2 \text{ kJ mol}^{-1})$$

$$= 488.6 \text{ kJ}$$

The positive value of ΔG° indicates that the forward reaction should not occur spontaneously, but that the reverse reaction ($\Delta G = -488.6$ kJ) should, at 25°C and 1 atm.

◆ ◆ ◆ ◆ ◆

EXAMPLE 2   **The reaction of calcium oxide with the pollutant, sulfur trioxide, $CaO(s) + SO_3(g) \longrightarrow CaSO_4(s)$, has been proposed as one way of removing SO_3 from the smoke resulting from burning of high-sulfur coal. Using the following ΔH° and S° values, calculate the standard free energy change for the reaction at 298 K: CaO(s): $\Delta H^\circ_f = -635$ kJ mol^{-1}, $S^\circ = 39.7$ J mol^{-1} K^{-1}; $SO_3(g)$: $\Delta H^\circ_f = -395.7$ J mol^{-1}, $S^\circ = 256.6$ J mol^{-1} K^{-1}; $CaSO_4(s)$: $\Delta H^\circ_f = -1432.7$ kJ mol^{-1}, $S^\circ = 106$ J mol^{-1} K^{-1}.**

This calculation can be approached in two ways. ΔH° and ΔS° for the reaction can be calculated from the data using the techniques presented in Sections 18.5 and 18.8, then ΔG° calculated from the expression $\Delta G^\circ = \Delta H^\circ - T\Delta S^\circ$. Alternatively, the standard molar entropy change of formation, $\Delta S^\circ_{f_{298}}$, can be calculated for each compound involved in the reaction. Using $\Delta S^\circ_{f_{298}}$ and ΔH°_f, ΔG°_f for each compound can be determined, and ΔG° for the reaction can be calculated from this data using the technique illustrated in Example 1 of this section. We will use the first approach.

$$\Delta H^\circ = \Delta H^\circ_{f_{CaSO_4(s)}} - \Delta H^\circ_{f_{CaO(s)}} - \Delta H^\circ_{f_{SO_3(g)}}$$

$$= 1 \text{ mol CaSO}_4(s) \times (-1432.7 \text{ kJ mol}^{-1}) - 1 \text{ mol CaO}(s)$$

$$\times (-635 \text{ kJ mol}^{-1}) - 1 \text{ mol SO}_3(g) \times (-395.7 \text{ kJ mol}^{-1})$$

$$= -402 \text{ kJ} = -402,000 \text{ J}$$

$$\Delta S^\circ = S^\circ_{CaSO_4(s)} - S^\circ_{CaO(s)} - S^\circ_{SO_3(g)}$$

$$= 1 \text{ mol CaSO}_4(s) \times (106 \text{ J mol}^{-1}\text{K}^{-1}) - 1 \text{ mol CaO}(s)$$

$$\times (39.7 \text{ J mol}^{-1}\text{K}^{-1}) - 1 \text{ mol SO}_3(g) \times (256.6 \text{ J mol}^{-1}\text{K}^{-1})$$

$$= -190 \text{ J K}^{-1}$$

$$\Delta G^\circ = \Delta H^\circ - T\Delta S^\circ$$

$$= -402,000 \text{ J} - 298 \text{ K} \times (-190 \text{ J K}^{-1})$$

$$= -402,000 \text{ J} + 56,620 \text{ J}$$

$$= -345,000 \text{ J} = -345 \text{ kJ}$$

Note that the units of ΔH° were initially in *kilojoules*, and the units of ΔS° in *joules*. It is necessary to use the same units when ΔH° and ΔS° are used to calculate ΔG°.

◆ ◆ ◆ ◆ ◆

18.11   The Second Law of Thermodynamics

If a process occurs in a closed system, a system in which no heat can get in or out, ΔH must be zero. If the process is spontaneous, the free energy change (ΔG) is negative and, since ΔH is zero, the entropy change (ΔS) must be positive. Thus the entropy increases in a closed system when a spontaneous process occurs. The universe is a closed system; the universe includes everything; nothing can get in or out. Thus **any spontaneous change that occurs in the universe must be accompanied by an increase in the entropy of the universe.** This statement is called the **Second Law of Thermodynamics.**

It is difficult to use the Second Law in its pure form. However, a useful modification can be developed from the free energy equation presented in Section 18.9.

$$\Delta G = \Delta H - T\,\Delta S \qquad \text{(all values pertaining to the system)}$$

Dividing by T gives

$$\frac{\Delta G}{T} = \frac{\Delta H}{T} - \Delta S \qquad \text{(for the system)}$$

It can be shown that

$$\Delta S_{sur} = -\frac{\Delta H_{sys}}{T} \qquad \text{(at constant temperature and pressure)}$$

where ΔS_{sur} is the change in entropy for the surroundings, and ΔH_{sys} is the change in enthalpy for the system.

Thus,

$$\frac{\Delta G_{sys}}{T} = -\Delta S_{sur} - \Delta S_{sys}$$

or

$$-\frac{\Delta G_{sys}}{T} = \Delta S_{sur} + \Delta S_{sys}$$

$$\Delta S_{universe} = \Delta S_{sur} + \Delta S_{sys}$$

Hence,

$$-\frac{\Delta G_{sys}}{T} = \Delta S_{universe}$$

This derived relationship is valid only at constant temperature and pressure, because the relationships from which it is derived are valid only at constant temperature and pressure.

From the derived relationship it can be seen that as the entropy of the universe increases (that is, $\Delta S_{universe}$ is positive), ΔG_{sys} must be negative, corresponding to a spontaneous change. Thus it can be said that **for a spontaneous change at constant temperature and pressure, the entropy of the universe must increase and the free energy change of the system must be negative.**

18.12   Free Energy Changes and Nonstandard States

Many chemical reactions occur under conditions in which the reactants and products are not in their standard states. The free energy change of such reactions,

ΔG, can be determined from the free energy change, $\Delta G°$, of the reaction when both the reactants and the products are in standard states by using the equation

$$\Delta G = \Delta G° + 2.303\ RT \log Q$$

In this equation T is the temperature on the Kelvin scale, 298.15 K, R is a constant with a value of 8.314 J K^{-1} or 1.987 cal K^{-1}, and Q is the **reaction quotient** of the chemical reaction. For a chemical reaction $mA + nB + \cdots \longrightarrow xC + yD + \cdots$,

$$Q = \frac{[C]^x[D]^y \cdots}{[A]^m[B]^n \cdots}$$

Thus the reaction quotient has the same form as the equilibrium constant, K, for the reaction. However, the reaction quotient has no fixed value. The magnitude of Q is determined by the concentrations of reactants and products at whatever stage in the progress of the reaction that we choose to evaluate Q. When A and B are first mixed, no products are present, and $Q = 0$. The value of Q increases as the reaction proceeds. Only when the reaction has reached equilibrium is $Q = K$. As with the equilibrium constant (Section 15.17), the activities of the reactants and products should be used to evaluate Q. However, we will continue to use pressure (in atm) for gases, concentrations (mol/ℓ) for dissolved species, and unity for pure solids and liquids as good approximations of the activities of these species.

EXAMPLE   **The reaction of calcium oxide with sulfur trioxide (Example 2, Section 18.10) rarely occurs under standard state conditions. Calculate ΔG for this reaction at 25°C when the pressure of SO_3 is 0.15 atm.**

   To determine ΔG under nonstandard state conditions at 25°C (298.15 K), use the expression

$$\Delta G = \Delta G° + 2.303\ RT \log Q$$

For the reaction $CaO(s) + SO_3(g) \longrightarrow CaSO_4(s)$

$$Q = \frac{[CaSO_4]}{[CaO][SO_3]}$$

CaO and $CaSO_4$ are solids so their concentrations (activities) are 1. The concentration (activity) of the gas SO_3 is taken as its pressure in atmospheres, 0.15.

$$Q = \frac{1}{(1) \times (0.15)} = 6.67$$

In Example 2, Section 18.10, $\Delta G°$ was shown to be -345 kJ. Thus

$$\Delta G = \Delta G° + 2.303\ RT \log Q$$

$$\Delta G = -345\ \text{kJ} + (2.303)(8.314\ \text{J K}^{-1})\left(\frac{1\ \text{kJ}}{10^3\ \text{J}}\right)(298.15\ \text{K})(\log 6.67)$$

$$= -345\ \text{kJ} + 4.7\ \text{kJ}$$

$$= -340\ \text{kJ}$$

Hence the change of conditions caused a change in ΔG from $-345\,\text{kJ}$ to $-340\,\text{kJ}$; the reaction is also spontaneous under these conditions. Note that the value of R selected has units of $J\,K^{-1}$, and it was necessary to convert the units of R to $kJ\,K^{-1}$.

◆ ◆ ◆ ◆ ◆

18.13   The Relationship Between Free Energy Changes and Equilibrium Constants

If ΔG for any reaction ($A + B \longrightarrow C + D$, for example) is negative, the reaction will occur spontaneously as written, and the quantity of products will increase. If ΔG is positive, the reverse reaction will occur spontaneously; that is, C and D will react to give an increased quantity of A and B. What is the situation at equilibrium which, by definition, is the state when the quantities of reactants and products do not change? At equilibrium, ΔG can be neither positive nor negative; thus, it can only be zero. So at equilibrium we have

$$\Delta G = 0 = \Delta G^\circ + 2.303\,RT \log Q$$

Since the reaction is at equilibrium, the value of the reaction quotient must be equal to that of the equilibrium constant for the reaction, so we have

$$\Delta G = 0 = \Delta G^\circ + 2.303\,RT \log K$$

Hence
$$\Delta G^\circ = -2.303\,RT \log K \tag{1}$$

This derivation shows us that the value of the standard state free energy change, ΔG°, for a reaction can be used to determine the equilibrium constant for that reaction. Values of ΔG° can be determined as shown in Section 18.10 from the data in tables of standard molar free energies of formation such as found in Table 18-1 and Appendix J. Alternatively, ΔG° values can be evaluated from equilibrium constants by means of Equation (1).

EXAMPLE 1   **Calculate the standard state free energy change for the acetic acid ionization, $CH_3CO_2H(aq) \longrightarrow H^+(aq) + CH_3CO_2^-(aq)$ at 25°C. The equilibrium constant for the reaction is 1.8×10^{-5}.**

At equilibrium, $\Delta G^\circ = -2.303\,RT \log K$, where

$$K = \frac{[H^+][CH_3CO_2^-]}{[CH_3CO_2H]} = 1.8 \times 10^{-5}$$

Thus
$$\Delta G^\circ = -(2.303)(8.314\,J\,K^{-1})(298.15\,K)(\log 1.8 \times 10^{-5})$$
$$= -(2.303)(8.314\,J\,K^{-1})(298.15\,K)(-4.74) = 2.7 \times 10^4\,J$$
$$= 27\,kJ$$

Therefore the free energy change for the *complete* transformation of one mole of acetic acid in its standard state in water into one mole of hydrogen ion and one mole of acetate ion, both in their standard states in water, is 27 kJ. Since ΔG° is positive, the reaction is not spontaneous. However, the reverse reaction, $H^+(aq) + CH_3CO_2^-(aq) \longrightarrow CH_3CO_2H(aq)$,   is   spontaneous ($\Delta G° = -27$ kJ mol^{-1}).

◆ ◆ ◆ ◆ ◆

EXAMPLE 2   **Using data in Appendix J, calculate the value of K at 298.15 K for the reaction**

$$H_2(g) + \tfrac{1}{2}O_2(g) \longrightarrow H_2O(g)$$

$$\Delta G° = -2.303\ RT \log K$$

The value of $\Delta G°_{f_{298.15}}$ given in Appendix J for $H_2O(g)$ is -228.59 kJ, which is $-228,590$ J mol^{-1}.   Therefore

$$-228,590\ \text{J} = (-2.303)(8.314\ \text{J K}^{-1})(298.15\ \text{K}) \log K$$

$$\log K = \frac{-228,590\ \text{J}}{(-2.303)(8.314\ \text{J K}^{-1})(298.15\ \text{K})} = 40.067$$

$$K = 1.17 \times 10^{40}$$

◆ ◆ ◆ ◆ ◆

If we combine Equation (1) with the equation relating $\Delta G°$, $\Delta H°$, and $\Delta S°$ we have

$$\Delta G° = \Delta H° - T\Delta S° = -2.303\ RT \log K \qquad (2)$$

Relationship (2) is developed for conditions of constant temperature and pressure. However, $\Delta H°$ and $\Delta S°$ are *approximately* independent of temperature. If it is assumed that they are indeed independent of temperature, we can obtain a simple equation relating the equilibrium constant and temperature.

Consider an equilibrium reaction at two different temperatures, T_1 and T_2, with equilibrium constants K_{T_1} and K_{T_2}.

$$\Delta H° - T_1 \Delta S° = -2.303\ RT_1 \log K_{T_1} \qquad (3)$$

and $$\Delta H° - T_2 \Delta S° = -2.303\ RT_2 \log K_{T_2} \qquad (4)$$

Equations (3) and (4) may be rearranged to

$$\frac{\Delta H°}{T_1} - \Delta S° = -2.303\ R \log K_{T_1} \qquad (5)$$

and $$\frac{\Delta H°}{T_2} - \Delta S° = -2.303\ R \log K_{T_2} \qquad (6)$$

by dividing Equation (3) through by T_1, and Equation (4) through by T_2. Subtracting Equation (6) from Equation (5) gives

$$\frac{\Delta H°}{T_1} - \frac{\Delta H°}{T_2} - \Delta S° - (-\Delta S°) = -2.303\ R \log K_{T_1} - (-2.303\ R \log K_{T_2})$$

or $$\Delta H° \left(\frac{1}{T_1} - \frac{1}{T_2} \right) = 2.303\ R \log \frac{K_{T_2}}{K_{T_1}}$$

or, by multiplying the left side by T_1T_2/T_1T_2, and dividing both sides by 2.303 R,

$$\frac{\Delta H°(T_2 - T_1)}{2.303\ RT_1T_2} = \log \frac{K_{T_2}}{K_{T_1}} \qquad (7)$$

Equations (1) through (7) prove to be quite useful, for if we know $\Delta G°$ for 298 K, we can by means of Equation (1) obtain the equilibrium constant, K, at 298 K. Then, if we know $\Delta H°$ for 298 K (which is essentially independent of temperature), we can obtain K by means of Equation (7) for any other temperature (within limitations of both $\Delta H°$ and $\Delta S°$ being independent of T).

It should be noted that K for temperatures other than 298 K could also be calculated from Equation (1), *provided we have the $\Delta G°$ value for the corresponding temperature.* (Remember that $\Delta G°$ values change significantly with temperature, whereas $\Delta H°$ and $\Delta S°$ values do not.) Values for $\Delta G°$ at temperatures other than 298 K are less frequently available than at 298 K, providing a severe restriction on calculating K values for temperatures other than 298 K by use of $\Delta G°$ alone.

EXAMPLE 3   **For the reaction**

$$CuS(s) + H_2(g) \longrightarrow Cu(s) + H_2S(g)$$

a. Calculate $\Delta G°$ and $\Delta H°$ at 298.15 K and 1 atm pressure.

$$\Delta G° = 1 \text{ mol } Cu(s) \times (0 \text{ kJ mol}^{-1}) + 1 \text{ mol } H_2S(g) \times (-33.6 \text{ kJ mol}^{-1})$$
$$- 1 \text{ mol } CuS(s) \times (-53.6 \text{ kJ mol}^{-1}) - 1 \text{ mol } H_2(g) \times (0 \text{ kJ mol}^{-1})$$
$$= 20.0 \text{ kJ} \quad \text{(reaction is not spontaneous)}$$

$$\Delta H° = 1 \text{ mol } Cu(s) \times (0 \text{ kJ mol}^{-1}) + 1 \text{ mol } H_2S(g) \times (-20.6 \text{ kJ mol}^{-1})$$
$$- 1 \text{ mol } CuS(s) \times (-53.1 \text{ kJ mol}^{-1}) - 1 \text{ mol } H_2(g) \times (0 \text{ kJ mol}^{-1})$$
$$= 32.5 \text{ kJ} \quad \text{(reaction is endothermic)}$$

b. Calculate the value for the equilibrium constant, K, at 298.15 K and 1 atm pressure.

$$K = \frac{p_{H_2S}}{p_{H_2}} \quad \text{(where } p \text{ stands for partial pressure of a gas)}$$

$$\Delta G° = -2.303 \, RT \log K \quad (R = 8.314 \text{ J K}^{-1})$$

Then
$$\log K = \frac{\Delta G°}{-2.303 \, RT \log K}$$

$$= \frac{20.0 \times 10^3 \text{ J}}{-(2.303)(8.314 \text{ J K}^{-1})(298.15 \text{ K})}$$

$$= -3.5038$$

$$K = 3.13 \times 10^{-4} \quad \text{(at 298.15 K)}$$

Note that at 298.15 K the equilibrium constant has a value less than 1, indicating that the elements in the reaction are present in larger quantity as reactants than as products, and that the equilibrium therefore lies far to the left.

c. Estimate the value for K at 798 K and 1 atm pressure.

$$\frac{\Delta H°(T_2 - T_1)}{2.303\, RT_1 T_2} = \log \frac{K_{T_2}}{K_{T_1}}$$

Then
$$\frac{32{,}500\text{ J }(798\text{ K} - 298\text{ K})}{(2.303)(8.314\text{ J K}^{-1})(798\text{ K})(298\text{ K})} = \log \frac{K_{798}}{K_{298}}$$

$$3.5689 = \log \frac{K_{798}}{K_{298}}$$

$$\log K_{798} - \log K_{298} = 3.5689$$

$$\log K_{298} = -3.5038 \quad \text{(calculated in Part b)}$$

Therefore
$$\log K_{798} = 3.5689 + (-3.5038) = 0.0651$$

$$K = 1.16 \quad \text{(at 798 K)}$$

Note that the equilibrium constant at 798 K is greater than 1, indicating the elements are present as products in greater quantity than as reactants and the equilibrium has been displaced to the right.

d. Calculate $\Delta S°$ at 298.15 K and 1 atm pressure.

$$\Delta G° = \Delta H° - T\Delta S°$$

$$\Delta S° = \frac{\Delta H° - \Delta G°}{T} = \frac{32{,}500\text{ J} - 20{,}000\text{ J}}{298.15\text{ K}}$$

$$= 41.9\text{ J K}^{-1}$$

Note that this value for $\Delta S°$ is positive and favors a spontaneous reaction, whereas the positive value of $\Delta H°$ is unfavorable for reaction. This is, therefore, a case where $\Delta S°$ is sufficiently large that $\Delta H°$ is not a valid indicator of spontaneity and $\Delta G°$ must be considered. In particular, as the temperature changes, the reaction will be expected to change with respect to spontaneity.

We have found in part a that $\Delta G° = +20.0$ kJ at 298.15 K, indicating that at this temperature the reaction is not spontaneous. Let us now calculate $\Delta G°$ at a higher temperature assuming that $\Delta H°$ and $\Delta S°$ do not change significantly as the temperature increases.

e. Estimate $\Delta G°$ at 798 K and 1 atm pressure.

$$\Delta G° = \Delta H° - T\Delta S°$$

$$= 32{,}500 - (798\text{ K})(41.9\text{ J K}^{-1})$$

$$= -936\text{ J} = -0.936\text{ kJ}$$

Hence, at the higher temperature $\Delta G°$ is negative, showing that at 798 K the reaction *is* spontaneous. $\Delta H°$, being relatively independent of temperature, still has a value of about 32.5 kJ at 798 K and hence does not indicate the change to spontaneity.

f. Estimate the temperature at which the standard $\Delta G°$ is zero at 1 atm pressure assuming that $\Delta H°$ and $\Delta S°$ do not change significantly as the temperature increases.

$$\Delta G° = \Delta H° - T\Delta S° \qquad (\Delta S° = 41.9 \text{ J K}^{-1}, \text{ as calculated in Part d)}$$

$$0 = 32,500 \text{ J} - T(41.9 \text{ J K}^{-1})$$

$$T = \frac{32,500 \text{ J}}{41.9 \text{ J K}^{-1}} = 776 \text{ K}$$

Hence, $\Delta G° = 0$ at 776 K. At temperatures below 776 K the values of $\Delta G°$ are positive; at temperatures above 776 K they are negative. Therefore, above 776 K the reaction is spontaneous; below 776 K the reaction as written is not spontaneous—indeed, it is spontaneous in the opposite direction.

◆ ◆ ◆ ◆ ◆

18.14   The Third Law of Thermodynamics

In all of the previous discussion we considered changes of state and the changes in the thermodynamic state functions that accompany such changes. So far in this chapter we have not tried to obtain the actual value for any of the state functions E, H, G, or S in a particular state. It is possible to obtain absolute values for the entropy of pure substances at any given temperature. The reason for this is embodied in the **Third Law of Thermodynamics, which states that the entropy of any pure, perfect crystalline substance at absolute zero (0 K) is equal to zero.** This is not true, however, for E, H, or G. The zero value for S means that at absolute zero all molecular motion has stopped and for a pure crystalline substance there is no disorder. For an impure substance, all molecular motion has also stopped, but the impurity can be distributed in different ways, giving rise to disorder.

If we measure the entropy change for the process in which we take a pure crystalline substance from absolute zero to any temperature T, then we have

$$S_T - S_0 = \Delta S = \text{a measured quantity}$$

But since $S_0 = 0$,

$$\Delta S = S_T = \text{a measured quantity}$$

Hence, we obtain *absolute entropies* of the pure substances, in contrast to free energy and enthalpy for which differences between two values can be determined but normally not the absolute values. Absolute entropies allow us to compare the relative amounts of disorder present in different pure substances, and they can be used to determine entropy changes. Caution must be exercised when using absolute entropies, however, since the absolute entropy of pure elemental substances at *standard state conditions* will *not* be equal to zero. Table 18-1 and Appendix J contain absolute standard molar entropies, $S°$, of some common substances at standard state conditions. To illustrate their use consider the following examples:

EXAMPLE 1   **a. Determine the entropy change for the change of liquid water to gaseous water at 298 K and 1 atm pressure.**

$$H_2O(l) \longrightarrow H_2O(g)$$

$S^\circ_{H_2O(l)298}$ = absolute entropy for $H_2O(l)$ at 298 K

$\qquad = S^\circ_{H_2O(l)298} - S^\circ_{H_2O(s)0} = 69.91 - 0 = 69.91$ J mol^{-1} K^{-1}

$S^\circ_{H_2O(g)298}$ = absolute entropy for $H_2O(g)$ at 298 K

$\qquad = S^\circ_{H_2O(g)298} - S^\circ_{H_2O(s)0} = 188.71 - 0 = 188.71$ J mol^{-1} K^{-1}

ΔS° = final state − initial state

$\qquad = S^\circ_{H_2O(g)298} - S^\circ_{H_2O(l)298} = 1 \text{ mol } H_2O(g) \times (188.71 \text{ J mol}^{-1} \text{ K}^{-1})$

$\qquad\qquad - 1 \text{ mol } H_2O(l) \times (69.91 \text{ J mol}^{-1} \text{ K}^{-1})$

$\qquad = 118.80$ J K^{-1}

The value for ΔS° is positive, indicating greater disorder in gaseous H_2O than in liquid H_2O. This is in accord with the fact that the molecules are in much more rapid and random motion in the gaseous state than in the liquid state (see Sections 10.15 and 11.1).

b. Determine the temperature at which liquid water and gaseous water are in equilibrium with each other at 1 atm pressure.

Since the two states are in equilibrium, the free energy change, ΔG°, in going from the liquid to the gaseous water is zero. Assuming as before that ΔH° and ΔS° are independent of temperature, for the reaction $H_2O(l) \longrightarrow H_2O(g)$,

$$\Delta H^\circ = \Delta H^\circ_{f_{H_2O(g)}} - \Delta H^\circ_{f_{H_2O(l)}} = 1 \text{ mol } H_2O(g) \times (-241.8 \text{ kJ mol}^{-1})$$

$$- 1 \text{ mol } H_2O(l) \times (-285.8 \text{ kJ mol}^{-1})$$

$$= 44.0 \text{ kJ} = 44,000 \text{ J}$$

$\Delta S^\circ = 118.80$ J K^{-1} [just calculated in Example 1(a)]

$\Delta G^\circ = \Delta H^\circ - T\Delta S^\circ = 0$ ($\Delta G^\circ = 0$ at equilibrium)

$$T = \frac{\Delta H^\circ}{\Delta S^\circ} = \frac{44,000 \text{ J}}{118.80 \text{ J K}^{-1}}$$

$$= 370 \text{ K} = 97^\circ \text{ C}$$

The correct answer for part b of the preceding example, of course, is 373 K (100°C), the boiling point of water at 1 atm pressure, but the calculation of the value 370 K was based upon the assumptions that ΔH° and ΔS° are independent of temperature. This is a good place to point out that these assumptions are only approximations. They are sufficiently true to be highly useful, as we have seen, in using standard molar enthalpy and free energy values to calculate values at other temperatures; but for really exact calculations, ΔH° and ΔS° values at the temperature in question must be used, as in the following:

c. Recalculate the temperature at which liquid water and gaseous water are in equilibrium with each other at 1 atm pressure, this time using the true tabulated

values of $\Delta H°$ and $\Delta S°$ at the approximate temperature, calculated in part b, of 97°C. ($\Delta H°_{H_2O}$ at 97°C = 40,720 J mol^{-1}; $\Delta S°_{H_2O}$ at 97°C = 109.1 J mol^{-1} K^{-1}.)

Note that the $\Delta H°$ value at 25°C of 44,000 J and the $\Delta S°$ value at 25°C of 118.80 J K^{-1} are close enough, as stated earlier, to the values at 98° for an approximate calculation but are not sufficiently close for an exact calculation. Again using

$$\Delta G° = \Delta H° - T\Delta S° = 0 \quad (\Delta G° = 0 \text{ at equilibrium})$$

$$T = \frac{\Delta H°_{370}}{\Delta S°_{370}} = \frac{40,720 \text{ J}}{109.1 \text{ J K}^{-1}} = 373.2 \text{ K}$$

or $373.2 - 273.2 = 100.0°C$

In this part of the example we obtained the true value of 100°C. If the calculation is repeated with the values of $\Delta H°$ and $\Delta S°$ for 100°C (40,656 J and 108.95 J K^{-1}), a value of $T = 100.0°$ is again obtained indicating that the values for $\Delta H°$ and $\Delta S°$ at the initially calculated approximate temperature are close enough to give the correct temperature of 100.0° (to four significant figures).

◆ ◆ ◆ ◆ ◆

If we were to apply temperatures below 373 K to the relationship $\Delta G° = \Delta H° - T\Delta S°$, we would see that the value for the $T\Delta S°$ becomes smaller, and hence a smaller term is subtracted from $\Delta H°$ to give $\Delta G°$. Thus $\Delta G°$ becomes positive (it was zero for the equilibrium state at 373 K). *Hence, at temperatures below 373 K (100°C) at 1 atm pressure, the spontaneous change is from gaseous H_2O to liquid H_2O.* If we go to higher temperatures than 373 K, $T\Delta S°$ becomes larger and a larger term is subtracted from $\Delta H°$ than at 373 K. Thus $\Delta G°$ is negative, indicating that *at temperatures above 373 K (100°C) at 1 atm pressure the spontaneous change is from liquid H_2O to gaseous H_2O.*

Notice in this example that the heat effect ($\Delta H°$) and entropy effect ($\Delta S°$) work at cross purposes, inasmuch as $\Delta S°$ is positive (maximum disorder, indicating favorable conditions for the change) and $\Delta H°$ is positive (endothermic, indicating unfavorable conditions for the change). We could then suppose that one of these factors (either $\Delta H°$ or $\Delta S°$) is of predominant importance. But we now find ourselves in a quandary, stemming from the fact that the spontaneity changes from unfavorable to favorable at 373 K; however, $\Delta H°$ and $\Delta S°$ are both *relatively* independent of temperature and do not change sufficiently with temperature to reflect this difference in spontaneity that occurs at 373 K. This verifies the earlier statement that $\Delta H°$ alone is not sufficient to predict accurately the spontaneity of a reaction. It is now apparent also that $\Delta S°$ alone is not sufficient to predict spontaneity. However, $\Delta G°$ (which combines $\Delta H°$, $\Delta S°$, and T) is a valid quantity on which to base such predictions. As shown above, $\Delta G°$ is positive at temperatures below 373 K (100°C) and negative above 373 K.

EXAMPLE 2   What is the standard entropy change for the following reaction at 298 K at 1 atm pressure?

$$H_2(g) + \tfrac{1}{2}O_2(g) \longrightarrow H_2O(g)$$

We can consider this reaction as proceeding either through the direct route or through a series of steps: (1) and (2) the $H_2(g)$ and $O_2(g)$ are cooled to $H_2(s)$ and $O_2(s)$ at absolute zero temperature (0 K); (3) the two are allowed to react at absolute zero to form $H_2O(s)$; (4) the $H_2O(s)$ is heated back to $H_2O(g)$ at the original temperature 298 K.

$$H_2(g)_{298} + \tfrac{1}{2}O_2(g)_{298} \xrightarrow{\Delta S°} H_2O(g)_{298}$$

$$(1) \bigg\downarrow \Delta S_1° \qquad (2) \bigg\downarrow \Delta S_2° \qquad (4) \bigg\uparrow \Delta S_4°$$

$$H_2(s)_0 \quad + \tfrac{1}{2}O_2(s)_0 \quad \xrightarrow[(3)]{\Delta S_3°} H_2O(s)_0$$

Since $S°$ is a state function,

$$\Delta S° = \Delta S_1° + \Delta S_2° + \Delta S_3° + \Delta S_4°$$

Since the absolute entropy for H_2 at 0 K is zero.

$$\Delta S_1° = S°_{H_2(s)_0} - S°_{H_2(g)_{298}} = 0 - S°_{H_2(g)_{298}}$$

Therefore, $\Delta S_1° = -S°_{H_2(g)_{298}}$

Similarly, $\Delta S_2° = \tfrac{1}{2}S°_{O_2(s)_0} - \tfrac{1}{2}S°_{O_2(g)_{298}} = 0 - \tfrac{1}{2}S°_{O_2(g)_{298}}$

Therefore, $\Delta S_2° = -\tfrac{1}{2}S°_{O_2(g)_{298}}$

Similarly, $\Delta S_4° = S°_{H_2O(g)_{298}} - S°_{H_2O(s)_0} = S°_{H_2O(g)_{298}} - 0 = S°_{H_2O(g)_{298}}$

Also, $\Delta S_3 = 0$, since the absolute entropies of $H_2(s)$, $O_2(s)$, and $H_2O(s)$ (all pure substances) are zero at absolute zero temperature.

Therefore $\Delta S° = \Delta S_1° + \Delta S_2° + \Delta S_3° + \Delta S_4°$

$$= -S°_{H_2(g)_{298}} + [-\tfrac{1}{2}S°_{O_2(g)_{298}}] + 0 + S°_{H_2O(g)_{298}}$$

$$= -(130.57) - \tfrac{1}{2}(205.03) + 0 + (188.71)$$

$$= -44.37 \text{ J K}^{-1}$$

⁕ ⁕ ⁕ ⁕ ⁕

The relationship of $\Delta G°$ to standard electrode reduction potentials and to equilibrium constants through electrode potentials is discussed in the chapter on Electrochemistry (see Section 20.15).

Questions

1. What are the meanings of the following terms? system, surroundings, state function, enthalpy change, entropy change, internal energy
2. What is the distinction between ΔE and ΔH for systems undergoing a change at constant pressure?
3. What is the distinction between ΔH and ΔG for systems undergoing a change at constant temperature and pressure?

4. What is meant by the term "spontaneous reaction"?
5. What is the connection between entropy and disorder?
6. What is the "standard absolute entropy" of a pure substance?
7. Why is the emphasis in thermodynamics on the change in enthalpy and free energy as a system changes rather than on the values themselves?

8. The enthalpy, H, of a system has been referred to as the "heat content" of a system. Why?
9. What is the difference between ΔH and ΔH°_{298} for a reaction?
10. In which of the following changes at constant pressure is work done by the surroundings on the system? by the system on the surroundings? Is no work done in any of these changes? What is the value of w in each case? Is $w > 0$, $w < 0$ or $w = 0$?
(a) $H_2O(g) \longrightarrow H_2O(l)$
(b) $H_2O(s) \longrightarrow H_2O(g)$
(c) $2Na(s) + Cl_2(g) \longrightarrow 2NaCl(s)$
(d) $H_2(g) + Cl_2(g) \longrightarrow 2HCl(g)$
(e) $3H_2(g) + N_2(g) \longrightarrow 2NH_3(g)$
(f) $Na_2SO_4 \cdot 10H_2O(s) \longrightarrow Na_2SO_4(s) + 10H_2O(g)$
(g) $NO_2(g) + CO(g) \longrightarrow NO(g) + CO_2(g)$
11. Does $\Delta H^\circ_{f_{H_2O(g)}}$ differ from ΔH°_{298} for the reaction $2H_2(g) + O_2(g) \longrightarrow 2H_2O(g)$? If so, how?
12. For the following reactions

$$H_2(g) + Cl_2(g) \longrightarrow 2HCl(g) \quad (1)$$
$$2HCl(g) \longrightarrow H_2(g) + Cl_2(g) \quad (2)$$

show that ΔH°, ΔS°, and ΔG° for Reaction (1) are equal to the negatives of ΔH°, ΔS°, and ΔG°, respectively, for Reaction (2) using the data in Table 18-1.
13. Under what conditions is ΔG equal to ΔG° for the reaction $H_2(g) + O_2(g) \longrightarrow H_2O(l)$?
14. State the First Law of Thermodynamics in two forms.
15. State the Second Law of Thermodynamics in terms of entropy changes for the universe. State the Second Law of Thermodynamics in terms of free energy changes.
16. State the Third Law of Thermodynamics.
17. In each of the following, give the sign of the entropy change for the indicated system going from state 1 to state 2, and explain.

	State 1	State 2
(a)	Egg in shell	Scrambled egg
(b)	Chicken feed and baby chick	Full grown chicken
(c)	$NaCl(s)$ at 298 K	$NaCl(s)$ at 10 K
(d)	$H_2O(g)$ at 273 K and 1 atm	$H_2O(s)$ at 273 K and 1 atm
(e)	1 mole Sn, 1 mole O_2	1 mole SnO_2
(f)	1 mole $CaCO_3$	1 mole CaO, 1 mole CO_2

18. Explain why equilibrium constants change with temperature.
19. No matter what the bond energy, all compounds will decompose if heated to a sufficiently high temperature. Why will the reaction $AB \longrightarrow A + B$, where A and B represent atoms, eventually become spontaneous as the temperature of the system is increased?
20. For the conversion of graphite to diamond, $C(s, \textit{graphite}) \longrightarrow C(s, \textit{diamond})$, $\Delta H = 1.90$ kJ mol^{-1}. Do the heats of combustion of graphite and carbon differ; that is, are the enthalpy changes for the following reactions the same or different?

$$C(s, \textit{graphite}) + O_2(g) \longrightarrow CO_2(g)$$
$$C(s, \textit{diamond}) + O_2(g) \longrightarrow CO_2(g)$$

21. The structure of solid NaCl and a solution of NaCl are illustrated in Figure 8-1. (a) What is the sign of ΔS for the reaction $NaCl(s) \longrightarrow NaCl(aq)$? (b) Sodium chloride spontaneously dissolves in water. What is the sign of ΔG for this reaction? (c) The heat of solution of $NaCl(s)$ is 3.88 kJ mol^{-1}. Is this consistent with a spontaneous reaction? (d) What is the driving force for the reaction?

Problems

1. Calculate the missing value of ΔE, q, or w for a system given the following data:
(a) $q = 570$ J; $w = 300$ J *Ans.* $\Delta E = 270$ *J*
[S] (b) $\Delta E = -7500$ J; $w = -4500$ J *Ans.* $q = -12{,}000$ *J*
(c) $\Delta E = 250$ J; $q = 300$ J *Ans.* $w = 50$ *J*
[S] (d) One kilojoule of heat energy is absorbed by the system and the system does 650 J of work on the surroundings. *Ans. 450 J*

[S] 2. Calculate the work involved compressing a system

consisting of one mole of H_2O as it changes from a gas at 373 K (volume = 30.6 ℓ) to a liquid at 373 K (volume = 18.9 ml) under a constant pressure of 1 atm. Does this work increase or decrease the internal energy of the system?

Ans. 3.10 J; increase

3. A gas in expanding against a constant pressure of 0.50 atm from 10.0 to 16.0 ℓ absorbs 125 J of heat. What is the change in internal energy of the gas?

Ans. −180 J

4. (a) Using the data in Appendix J, calculate the standard enthalpy change for each of the following reactions:

$\boxed{s}$(1) $Fe_2O_3(s) + 13CO(g) \longrightarrow$
$$2Fe(CO)_5(g) + 3CO_2(g)$$
Ans. −387.6 kJ

(2) $2LiOH(s) + CO_2(g) \longrightarrow$
$$Li_2CO_3(s) + H_2O(g)$$
Ans. −89.4 kJ

$\boxed{s}$(3) $CH_4(g) + N_2(g) \longrightarrow HCN(g) + NH_3(g)$
Ans. 164 kJ

(4) $CS_2(g) + 3Cl_2(g) \longrightarrow$
$$CCl_4(g) + S_2Cl_2(g)$$
Ans. −238 kJ

(5) $N_2(g) + O_2(g) \longrightarrow 2NO(g)$
Ans. 180.5 kJ

(b) Which of these reactions are exothermic?

Ans. 1, 2, and 4

$\boxed{s}$5. How many kilojoules of heat energy will be liberated when 49.70 g of manganese are burned to form $Mn_3O_4(s)$ at standard state conditions? ΔH°_{f298} of Mn_3O_4 is equal to -1388 kJ mol^{-1}.

Ans. 418.5 kJ

$\boxed{s}$6. The heat of formation of $OsO_4(s)$, $\Delta H^\circ_{f OsO_4(s)}$, is -391 kJ mol^{-1} at 298 K and the heat of sublimation is 56.4 kJ mol^{-1}. What is ΔH°_{298} for the process $Os(s) + 2O_2(g) \longrightarrow OsO_4(g)$ under standard state conditions? *Ans. −335 kJ*

$\boxed{s}$7. Calculate the standard molar enthalpy of formation of $NO(g)$ from the following data:

$N_2(g) + 2O_2(g) \longrightarrow 2NO_2(g)$ $\quad \Delta H^\circ = 66.4$ kJ

$2NO(g) + O_2(g) \longrightarrow 2NO_2(g)$ $\quad \Delta H^\circ = -114.1$ kJ

Ans. 90.3 kJ mol^{-1}

8. A sample of $WO_2(s)$ with a mass of 0.9745 g was "burned" in oxygen at constant pressure giving $WO_3(s)$ and 1.143 kJ of heat. The enthalpy of formation of $WO_3(s)$ under these conditions is -842.91 kJ mol^{-1}. Calculate the enthalpy of formation of $WO_2(s)$.

Ans. −589.7 kJ mol^{-1}

$\boxed{s}$9. The heat of combustion of a hydrocarbon (a compound of carbon and hydrogen) is the standard state enthalpy change for the reaction of the compound with oxygen to give $CO_2(g)$ and $H_2O(l)$. Determine the heats of combustion of (a) octane [C_8H_{18}, $\Delta H^\circ_f = -208.4$ kJ mol^{-1}], a component of gasoline and (b) methane [$CH_4(g)$, $\Delta H^\circ_f = -74.81$ kJ mol^{-1}], the major component of natural gas. Which has the higher heat content per gram?

Ans. −5512 kJ; −890.3 kJ; CH_4

$\boxed{s}$10. Calculate, using the data in Appendix J, the bond energies of F_2, Cl_2, and FCl. All are gases in their most stable form at standard state conditions.

Ans. F—F = 158.0 kJ per mole of bonds;
Cl—Cl = 243.36 kJ per mole of bonds;
F—Cl = 255.15 kJ per mole of bonds

11. Calculate, using the data in Appendix J, the bond energies of N_2, O_2, and NO. All are gases in their most stable form at standard state conditions.

Ans. N≡N = 945.4 kJ per mole of bonds;
O=O = 498.3 kJ per mole of bonds;
N=O = 632 kJ per mole of bonds

12. Using the data in Appendix J, calculate the Ti—Cl single bond energy in $TiCl_4$. *Ans. 429.9 kJ*

$\boxed{s}$13. (a) Using the bond energies given in Section 18.6, determine the approximate enthalpy change for the formation of ethylene from ethane which is described in Section 15.14; $C_2H_6(g) \longrightarrow C_2H_4(g) + H_2(g)$. (b) Compare this with the standard state enthalpy change.

Ans. 128 kJ; 137.0 kJ

14. The enthalpy of formation of $AsF_5(g)$ has been determined to be -16.46 kJ per gram of arsenic using the reaction $2As(s) + 5F_2(g) \longrightarrow 2AsF_5(g)$. Using this information and the data in Appendix J, calculate the single As—F bond energy in AsF_5.

Ans. 386 kJ

15. (a) Using the data in Appendix J, calculate the standard entropy changes for the reactions given in Problem 4.

Ans. (1) −1124.5; (2) −34.9; (3) 16.4;
(4) −265.5; (5) 24.8 J K^{-1}

(b) For which of the reactions in Problem 4 are the entropy changes favorable for the reaction to proceed spontaneously?

Ans. (3) and (5)

S16. What is the entropy change accompanying the evaporation of one mole of chloroform, $CHCl_3(l) \longrightarrow CHCl_3(g)$, under standard state conditions? *Ans. 94 J K^{-1}*

17. What is the entropy change accompanying the reaction $N_2(g) + 3H_2(g) \longrightarrow 2NH_3(g)$ under standard state conditions? *Ans. −198.6 J K^{-1}*

S18. (a) Using the data in Appendix J, calculate the standard free energy changes for the reactions given in Problem 4.

Ans. (1) −52.4; (2) −78.8; (3) 159; (4) −159; (5) 173.1 kJ

S(b) Which of these reactions are spontaneous? Why? *Ans. (1), (2), and (4)*

19. The standard enthalpies of formation of $NO(g)$, $NO_2(g)$, and $N_2O_3(g)$ are 90.25 kJ mol^{-1}, 33.2 kJ mol^{-1}, and 83.72 kJ mol^{-1}, respectively. Their standard entropies are 210.65, 239.9, and 312.2 J mol^{-1} K^{-1}, respectively.

S(a) Use the data above to calculate the free energy change for the following reaction at 25.0°C.

$$N_2O_3(g) \longrightarrow NO(g) + NO_2(g)$$

Ans. −1.6 kJ

(b) Repeat the above calculation for 0.00°C, 50.0°C, and 100.0°C assuming that the enthalpy and entropy changes do not change with a change in temperature.

Ans. 1.9 kJ; −5.02 kJ; −12.0 kJ

20. Consider the reaction

$$I_2(g) + Cl_2(g) \longrightarrow 2ICl(g)$$

S(a) For this reaction $\Delta H° = −26.9$ kJ, and $\Delta S° = 11.3$ J K^{-1}. Calculate $\Delta G°$ for the reaction. *Ans. −30.3 kJ*

S(b) Calculate the equilibrium constant for this reaction at 25.0°C *Ans. 2.03×10^5*

21. For a certain process at 300 K, $\Delta G = −77.0$ kJ and $\Delta H = −56.9$ kJ. Find the entropy change for this process at this temperature.

Ans. $\Delta S = 67.0$ J K^{-1}

S22. (a) For the vaporization of bromine liquid to bromine gas, calculate the change in enthalpy and the change in entropy at standard state conditions.

Ans. $\Delta H° = 30.91$ kJ mol^{-1}; $\Delta S° = 93.123$ J mol^{-1} K^{-1}

(b) From the calculations in (a) discuss relative disorder in bromine liquid compared to bromine gas. On the basis of the enthalpy change, state what you can about the spontaneity of the vaporization.

S(c) Estimate the value of $\Delta G°$ for the vaporization of bromine from the data in Appendix J.

Ans. 3.14 kJ mol^{-1}

S(d) State what you can about the spontaneity of the process from the value you obtained for $\Delta G°$ in (c).

S(e) Estimate the temperature at which liquid and gaseous Br_2 are in equilibrium with each other at 1 atm (assume $\Delta H°$ and $\Delta S°$ are independent of temperature). *Ans. 331.9 K, or 58.7°C*

(f) From this temperature value [Part (e)] state in which direction the process would be spontaneous at 298 K. at 398 K.

(g) Compare $\Delta H°$, $\Delta S°$, and $\Delta G°$ in terms of their usefulness in predicting spontaneity of the vaporization of Br_2.

23. If the entropy of vaporization of H_2O is equal to 109 J mol^{-1} K^{-1} and the enthalpy of vaporization is 40.62 kJ mol^{-1}, calculate the normal boiling point temperature of water in °C. *Ans. 100°C*

S24. (a) If you wished to decompose $CaCO_3(s)$ into $CaO(s)$ and $CO_2(g)$ at atmospheric pressure, estimate the minimum temperature at which you would conduct the reaction?

Ans. 836°C

(b) Calculate the equilibrium vapor pressure of $CO_2(g)$ above $CaCO_3(s)$ in a closed container at 298 K and 1.00 atm.

Ans. $P_{CO_2} = 1.50 \times 10^{-23}$ atm

S25. If the enthalpy of vaporization of CH_2Cl_2 is 29.0 kJ mol^{-1} at 25.0°C and the entropy of vaporization is 92.5 J mol^{-1} K^{-1}, calculate a value for the normal boiling point temperature of CH_2Cl_2.

Ans. 40.4°C

26. Carbon tetrachloride, an important industrial sol-

vent, is prepared by the chlorination of methane, $CH_4(g) + 4Cl_2(g) \longrightarrow CCl_4(g) + 4HCl(g)$, at 850 K. (a) What is the equilibrium constant for this reaction? (b) Will the reaction vessel need to be heated or cooled to keep the temperature of the reaction constant?

Ans. $K_p = 2.05 \times 10^{23}$; *cooled*, $\Delta H = -397.3$ *kJ*

[S]27. The equilibrium constant, K_p, for the reaction $N_2O_4(g) \rightleftharpoons 2NO_2(g)$ is 0.142 at 298 K. What is $\Delta G°$ for the reaction? *Ans. 4.84 KJ*

28. Calculate $\Delta G°$ at 298 K for the reaction of one mole of $H^+(aq)$ with one mole of $OH^-(aq)$ using the equilibrium constant for the self-ionization of water $H_2O \rightleftharpoons H^+(aq) + OH^-(aq)$, ($K_w = 1.00 \times 10^{-14}$ at 298 K). *Ans.* -79.9 *kJ*

[S]29. Acetic acid, CH_3CO_2H, can form a dimer, $(CH_3CO_2H)_2$, in the gas phase.

$$2CH_3CO_2H(g) \rightleftharpoons (CH_3CO_2H)_2(g)$$

The dimer is held together by two hydrogen bonds

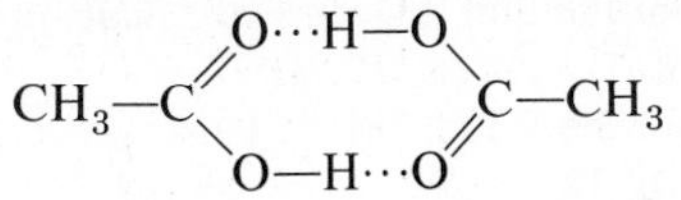

with a total strength of 66.5 kJ per mole of dimer. At 25 °C the equilibrium constant for the dimerization is 1.3×10^3 (pressures in atmospheres). What is ΔS for the reaction at 25°? *Ans.* -163 $J\,K^{-1}$

[S]30. At 1000 K the equilibrium constant for the reaction $Br_2(g) \rightleftharpoons 2Br$ is 2.8×10^4 (pressure in atmospheres). What is $\Delta G°$ for the reaction? Assume that the bond energy of Br_2 does not change between 298 K and 1000 K and calculate $\Delta S°$ for the reaction at 1000 K.

Ans. $\Delta G° = -85$ *kJ*, $\Delta S° = 278$ $J\,K^{-1}$

References

"Thermodynamics and Administration," A. J. White, *Chem. in Britain,* **13,** 150(1977).

"The Chemically Organizing Effects of Entropy Maximization," J. S. Wicken, *J. Chem. Educ.,* **53,** 623 (1976).

"Chemical Equilibrium as a State of Maximal Entropy," L. K. Nash, *J. Chem. Educ.,* **47,** 353 (1970).

"On Squid Axons, Frog Skins, and the Amazing Uses of Thermodynamics," W. H. Cropper, *J. Chem. Educ.,* **48,** 182 (1971),

"The Scope and Limitations of Thermodynamics," K. G. Denbigh, *Chem. in Britain,* **4,** 338 (1968).

"Perpetual Motion Machines" (Ingenious devices, but all doomed by the laws of thermodynamics), S. W. Augrist, *Sci. American,* January 1968, p. 114.

"Bond Energies in the Interpretation of Descriptive Chemistry," R. A. Howald, *J. Chem. Educ.,* **45,** 163 (1968).

"Thermodynamics and Rocket Propulsion," F. H. Verhoek, *J. Chem. Educ.,* **46,** 140 (1969).

"Energy States of Molecules," J. L. Hollenberg, *J. Chem. Educ.,* **47,** 2 (1970).

"Ion Pairs and Complexes: Free Energies, Enthalpies, and Entropies," J. E. Prue, *J. Chem. Educ.,* **46,** 12 (1969).

"Quantities of Work in Thermodynamics Equations," P. G. Wright, *J. Chem. Educ.,* **46,** 380 (1969).

"Thermodynamics of Ionic Solvation and its Significance in Various Systems," C. M. Criss and M. Salmon, *J. Chem. Educ.,* **53,** 763 (1976).

"The Synthesis of Diamond at Low Pressure," B. V. Derjaguin and D. B. Fedoseer, *Sci. American,* November 1975, p. 102.

"Entropy Estimates of Garnets and Other Silicates," S. Cantor, *Science,* **198,** 206 (1977).

"Condensation of Nonequilibrium Phases of Refractory Silicates from the Gas Phase," K. L. Day and B. Donn, *Science,* **202,** 307 (1978).

"Energy Resources Available to the United States, 1985 to 2000," E. T. Hayes, *Science,* **203,** 233 (1979).

"Biomass Potential in 2000 Put at 7 Quads" (Staff), *Chem. and Eng. News,* February 12, 1979, p. 20.

"Energy Review" (Staff), *J. Chem. Educ.,* **55,** 263 (1978).

"Notes on Nutrition" (Staff), *J. Chem. Educ.,* **55,** 113 (1978).

"Energy from Coal" (Staff), *J. Chem. Educ.,* **56,** 186 (1979).

"Energy from the Sun," D. O. Hall, *Educ. in Chem.,* January 1978, p. 8.

"Energy from the Wind," M. A. Patric, *Educ. in Chem.,* January 1978, p. 12.

"Energy from the Tides," M. A. Patric, *Educ. in Chem.,* January 1978, p. 15.

"Energy from the Atom," G. N. Walton, *Educ. in Chem.,* January 1978, p. 18.

"Energy from Fuel Cells," A. C. C. Tseung, *Educ. in Chem.,* January 1978, p. 27.

"Energy from Oil and Gas," C. A. McAuliffe, *Educ. in Chem.,* January 1978, p. 24.

"Energy and Exercise: How Much Work Can a Person Do?" H. A. Bent, *J. Chem. Educ.,* **55,** 456 (1978).

"Energy and Exercise: Caloric Cost of Mass Transport," H. A. Bent, *J. Chem. Educ.,* **55,** 526 (1978).

"Energy and Exercise: Heart Work," H. A. Bent, *J. Chem. Educ.,* **55,** 586 (1978).

"Energy and Exercise: Energy Storage Problems," H. A. Bent, *J. Chem. Educ.,* **55,** 659 (1978).

"Energy and Exercise: Limiting Reagents," H. A. Bent, *J. Chem. Educ.,* **55,** 726 (1978).

19

The Halogens and Their Compounds

The Halogens

The elements of Group VIIA of the Periodic Table are known as the **halogens,** which means *salt formers*. Their binary compounds as a group are called **halides.** Fluorine, chlorine, bromine, and iodine were first isolated between the years 1774 and 1886. Astatine, the fifth halogen, was first prepared artificially in 1940. Salts these elements (excepting astatine) are common in nature.

19.1 Occurrence and Preparation of the Halogens

The halogens never occur free in nature because of their great chemical activity. Chlorine is the most abundant element of the group, and although fluorine, bromine, and iodine are less common, they are reasonably available. The principal occurrences of the halogens are given in Table 19-1.

The best sources of the halogens (except iodine) are the salts in which they exist as anions with an oxidation state of -1. Various methods of oxidizing the halide ions to elemental halogens (zero oxidation state) may be employed. The ease of oxidation of the halide ions to free diatomic halogen molecules increases in the order as follows: $F^- < Cl^- < Br^- < I^- < At^-$. As you would expect, the electron removed during oxidation is closest to the nucleus in F^- and farthest from it in At^-.

509

TABLE 19-1   Occurrences of the Halogens

Fluorine

Fluorite or **fluorspar**, CaF_2
Fluorapatite, $Ca_{10}F_2(PO_4)_6$
Cryolite, Na_3AlF_6
Sea water (small amounts)
Teeth, bones, blood (small amounts)

Chlorine

Sea water (2.8% NaCl; other chlorides 0.8%
Great Salt Lake in Utah (23% NaCl)
Salt beds (NaCl, $MgCl_2$, $CaCl_2$)
Gastric juice (0.2 to 0.4% HCl)

Bromine

Sea water (NaBr, KBr, $MgBr_2$, $CaBr_2$)
Underground brines
Salt deposits

Iodine

Sea water (very small amounts)
$NaIO_3$ in Chilean nitrate deposits
Oil well brines of California
Thyroid gland in human body

▶**1. FLUORINE.** It is difficult to prepare elemental fluorine because it is the most powerful oxidizing agent of the known elements. It is necessary to resort to electrolytic oxidation of a fluoride compound in a nonaqueous electrolyte to oxidize the fluoride ion to fluorine. The electrolyte commonly used is a mixture of three parts potassium hydrogen fluoride (KHF_2) and two parts anhydrous hydrogen fluoride. When electrolysis of the fused electrolyte (melting point 72°C) begins, HF is decomposed to form fluorine gas at the anode and hydrogen at the cathode.

$$2HF + \text{electrical energy} \longrightarrow H_2(g) + F_2(g)$$

The two gases are kept separate by a barrier around the cathode and are drawn from the cell continuously. Anhydrous hydrogen fluoride is added to the cell, either continuously or intermittently, to regenerate the electrolyte. After purification, fluorine gas is compressed in special steel cylinders at 400 pounds pressure.

▶**2. CHLORINE.** The 11.0 million tons of chlorine produced in 1978 made it the seventh highest chemical in amount of commercial production in the United States. The bulk of the commercial chlorine produced in the world is by electrolytic oxidation of the chloride ion in aqueous solutions of sodium chloride. The reactions at the electrodes in the electrolytic cell are given by the equations

At the anode (oxidation)	$2Cl^- \longrightarrow Cl_2(g) + 2e^-$
At the cathode (reduction)	$2H_2O + 2e^- \longrightarrow 2OH^- + H_2(g)$
Net reaction	$2Cl^- + 2H_2O \longrightarrow Cl_2(g) + 2OH^- + H_2(g)$

The sodium hydroxide and hydrogen, which are by-products of the process, must be kept separate from the chlorine so as to prevent them from entering into unwanted secondary reactions.

Chlorine is a by-product in the production of metals such as sodium, calcium, and magnesium by the electrolytic decomposition of their fused chlorides. For example, when fused magnesium chloride is electrolyzed magnesium and chlorine are formed.

$$MgCl_2(l) + \text{electrical energy} \longrightarrow Mg(s) + Cl_2(g)$$

Chlorine is prepared in small quantities in the laboratory by oxidation of the chloride ion in acid solution by either manganese dioxide (MnO_2), potassium permanganate ($KMnO_4$), or sodium dichromate ($Na_2Cr_2O_7$).

$$MnO_2 + 2Cl^- + 4H^+ \longrightarrow Mn^{2+} + Cl_2(g) + 2H_2O$$
$$2MnO_4^- + 10Cl^- + 16H^+ \longrightarrow 2Mn^{2+} + 5Cl_2(g) + 8H_2O$$
$$Cr_2O_7^{2-} + 6Cl^- + 14H^+ \longrightarrow 2Cr^{3+} + 3Cl_2(g) + 7H_2O$$

Sodium chloride and sulfuric acid are usually used as sources of the chloride ion and the hydrogen ion, respectively, for the above reactions.

▶ **3. BROMINE.** The methods for the laboratory scale preparation of bromine are similar to those used for chlorine. In addition, bromine may be prepared by the oxidation of the bromide ion by chlorine.

$$2Br^-(aq) + Cl_2(g) \rightleftharpoons Br_2(l) + 2Cl^-(aq)$$

Chlorine is a stronger oxidizing agent than bromine, and the equilibrium for this reaction lies well to the right. Oxidation of bromide ions with elemental chlorine is used in the production of bromine from sea water. The bromine thus liberated is blown out of the solution by a stream of air. The bromine is then stripped from the air by adsorption in aqueous sodium carbonate, by which a reasonably concentrated solution of sodium bromide, NaBr, and sodium bromate, $NaBrO_3$, is formed.

$$3CO_3^{2-} + 3Br_2 \longrightarrow 5Br^- + BrO_3^- + 3CO_2(g)$$

Acidification of this solution with sulfuric acid liberates the bromine.

$$5Br^- + BrO_3^- + 6H^+ \rightleftharpoons 3Br_2(l) + 3H_2O$$

In this redox reaction, the bromate ion is the oxidizing agent and the bromide ion is the reducing agent. One million pounds of sea water must be processed in order to obtain 70 pounds of elemental bromine.

Bromine can be produced from underground brines (salt solutions) in a manner similar to that described for the production of bromine from sea water.

▶ **4. IODINE.** Elemental chlorine can be used to liberate iodine from iodides. The method is similar to that described for the preparation of bromine.

$$2I^-(aq) + Cl_2(g) \longrightarrow I_2(s) + 2Cl^-(aq)$$

Most of the iodine appears as a solid precipitate and can be separated from the solution by filtration. Care must be taken to avoid the use of an excess of chlorine in this process, otherwise unwanted secondary reactions occur, with the formation of iodine chloride, ICl, and iodic acid, HIO_3. Considerable iodine is obtained from iodides concentrated in kelp and other sea plants, and from oil field brines.

Iodine occurs in the form of sodium iodate ($NaIO_3$) as an impurity in Chile saltpeter ($NaNO_3$) deposits. The iodine is liberated by reducing the iodate with sodium hydrogen sulfite.

$$2IO_3^- + 5HSO_3^- \longrightarrow 3HSO_4^- + 2SO_4^{2-} + H_2O + I_2(s)$$

19.2   General Properties of the Halogens

The properties of the halogens reflect their pronounced active nonmetallic character, which decreases with increasing atomic number. The valence shell of each kind of halogen atom has the configuration ns^2np^5. Thus the halogens can form only one single covalent bond with less electronegative nonmetals. In compounds with metals, halogen atoms tend to pick up one electron each and thereby form stable univalent negative ions. As we would expect for nonmetals, the halogens are oxidizing agents that become progressively weaker as their nonmetallic character decreases down the group; fluorine is the strongest oxidizing agent of the known elements.

The halogens form diatomic molecules in which the atoms are bonded together by a single covalent bond as we would expect from the Lewis structure

$$:\ddot{X}:\ddot{X}: \quad \text{or} \quad :\ddot{X}\!-\!\ddot{X}:$$

where X represents any of the halogen atoms. The molecular orbital model of the bonding in F_2 has been discussed in Section 6.11. The fluorine molecule has the configuration

$$KK(\sigma_{2s})^2(\sigma_{2s}{}^*)^2(\sigma_{2p_x})^2(\pi_{2p_y}, \pi_{2p_z})^4(\pi_{2p_y}{}^*, \pi_{2p_z}{}^*)^4$$

thus having eight of its valence electrons in bonding orbitals and six in antibonding orbitals. The valence electrons in the major energy levels $n = 3, 4, 5,$ and 6 for chlorine, bromine, iodine, and astatine, respectively, have molecular orbital electron distributions analogous to those in the $n = 2$ level for fluorine. Consequently, both the molecular orbital model and the Lewis structures predict an X—X bond with a bond order of one in each halogen molecule.

With one exception, the chemical properties of the halogens differ in degree rather than in kind. For example, each reacts with hydrogen according to the equation $H_2 + X_2 \longrightarrow 2HX$; but the heat (enthalpy) of formation, $\Delta H_{f_{298}}^\circ$, of HX (Table 19-4) changes in the order HF < HCl < HBr < HI, indicating a *corresponding change in the energy evolved* as the hydrogen halides are formed from the free elements. Remember that the enthalpy of formation is negative if heat is liberated (see Sections 9.6 and 18.4). The halogens have such high ionization potentials that the formation of positive ions would be highly unlikely, except possibly for iodine and astatine. On the other hand, positive oxidation states resulting from the sharing of electrons with elements more electronegative in character are quite common for the halogens except fluorine, the most electronegative of all the elements. Oxidation states of $+1, +3, +5,$ and $+7$, and sometimes other oxidation states, are shown when the halogens share electrons with oxygen in their oxides, oxyacids, and oxysalts or with other more electronegative halogens in interhalogen compounds (see Section 19.6).

TABLE 19-2 Physical Properties of the Halogens[a]

	Fluorine	Chlorine	Bromine	Iodine	Astatine
Atomic number	9	17	35	53	85
Atomic weight	18.998	35.453	79.904	126.904	(210)[b]
Electronic structure	2, 7	2, 8, 7	2, 8, 18, 7	2, 8, 18, 18, 7	2, 8, 18, 32, 18, 7
Radius of X^-, Å	1.36	1.81	1.95	2.16	. . .
Covalent bond radius, Å	0.64	0.99	1.14	1.33	1.40
Physical state at 25°	Gas	Gas	Liquid	Solid	Solid
Melting point, °C	−218	−101	−7.3	114	. . .
Boiling point, °C	−188	−34.1	58.78	184	. . .
Density, g/cm³	1.108 (liq.)	1.557 (liq.)	3.119 (liq.)	4.93 (sol.)	. . .
Color	Pale yellow	Greenish-yellow	Reddish-brown	Black (s); violet (g)	. . .
Electronegativity	4.1	2.8	2.7	2.2	. . .

[a] The halogens are represented by the symbol X.
[b] Most stable isotope.

The halogens oxidize a variety of metals, nonmetals, and ions giving ionic or covalent products in which the halogen has an oxidation state of −1. A free halogen will oxidize those halide ions that are formed from less electronegative halogens. Thus fluorine will oxidize chloride, bromide, iodide, and astatide ions, whereas chlorine will oxidize only bromide, iodide, and astatide ions.

19.3 Physical Properties of the Halogens

As the atomic structures of the halogens become more complex with increasing atomic weight, there is a gradation in each physical property (Table 19-2). For example, fluorine is a pale yellow gas of low density; chlorine is a greenish-yellow gas 1.892 times as dense as fluorine gas; bromine is a deep reddish-brown liquid three times as dense as water; iodine is a grayish-black crystalline solid with a metallic appearance; and astatine is a solid with properties that indicate that it is somewhat metallic in character.

Liquid bromine has a high vapor pressure, and the reddish vapor can easily be seen in a bottle partly filled with the liquid. Iodine crystals have a high vapor pressure, and when heated gently these crystals change into a beautiful deep-violet vapor without melting; the vapor condenses readily upon cooling. This sublimation process (Section 11.14) is used in the purification of iodine.

Bromine is slightly soluble in water but is miscible in all proportions with alcohol, ether, chloroform, carbon tetrachloride, and carbon disulfide, forming solutions that vary in color from yellow to reddish-brown, depending upon the concentration.

Iodine dissolves only slightly in water, giving brown solutions. However, it is quite soluble in alcohol, in ether, and in aqueous solutions of iodides, with which it forms brown solutions. The solvents in which iodine combines with the solvent molecules (**solvation**) are the ones that give brown solutions. In chloroform, carbon disulfide, and many hydrocarbons, iodine forms violet solutions. In such solutions, the iodine is present in the molecular state (I_2), and the violet color is like that of iodine in the vapor state, where it is also molecular. Solutions of hydrogen iodide, potassium iodide, or other iodides dissolve iodine very readily. The iodine and iodide ion combine reversibly, forming the complex ion I_3^-.

$$\ddot{\underset{..}{I}}:^- + :\ddot{\underset{..}{I}}:\ddot{\underset{..}{I}}: \rightleftharpoons :\ddot{\underset{..}{I}}:\ddot{I}:\ddot{\underset{..}{I}}:^-$$

The negative iodide ion, when close to a large iodine molecule, whose outer electrons are far from the positive nucleus, disturbs these electronic arrangements enough to form a coordinate-covalent bond. Compounds containing the complexes I_5^-, Br_3^-, Cl_3^-, ICl_2^-, which also contain coordinate-covalent bonds, are known.

19.4   Chemical Properties of the Halogens

▶ **1. FLUORINE.** Fluorine forms binary fluorides with all of the elements except the lighter noble gases (He, Ne, and Ar). Fluorides of the active metals are ionic; nonmetal fluorides, covalent. The extreme reactivity of fluorine gas as an oxidizing agent is demonstrated by the fact that immediately upon contact with many substances, it ignites them. When a stream of the gas flows onto the surface of water it actually causes the water to burn. Wood and asbestos rapidly ignite and burn when held in a stream of fluorine. Heated glass will burn in fluorine, giving off smoke which looks much like that from wood. Most hot metals burn vigorously in fluorine. However, fluorine can be handled at ordinary or moderately elevated temperatures in containers made of copper, iron, magnesium, nickel, or Monel (an alloy of nickel, copper, and a little iron). An adherent film of the metal fluoride protects the metal surfaces from further attack.

Fluorine readily displaces chlorine and other halogens from the solid metal halides. It reacts immediately with water in several simultaneous reactions that involve the formation of O_2, OF_2, H_2O_2, O_3, and HF. Fluorine and hydrogen react explosively, even at temperatures as low as that of liquid air.

Fluorine, as a result of its high electronegativity, reacts with the noble gas xenon to form a variety of compounds such as XeF_2, XeF_4, $XeOF_4$ (Fig. 19-1). Additional compounds of the noble gases are discussed in Chapter 22.

Figure 19-1 Structures of the molecules of XeF_2 (linear), XeF_4 (square planar), and $XeOF_4$ (square pyramidal). Note the close similarity to the isoelectronic species ICl_2^-, ICl_4^-, and IF_5 in Fig. 19-3.

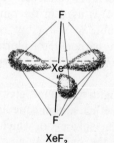

XeF₂

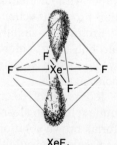

XeF₄

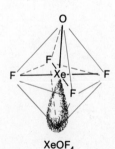

XeOF₄

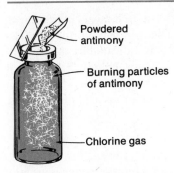

Powdered antimony

Burning particles of antimony

Chlorine gas

Figure 19-2 Powdered antimony burns brightly in chlorine gas, producing antimony(III) chloride.

▶ **2. CHLORINE.** Chlorine is less active as an oxidizing agent than fluorine; this is a reflection of the larger atomic size of the chlorine atom.

a. *Action with hydrogen.* Chlorine may be mixed safely with hydrogen in the dark, as reaction between the two then is imperceptibly slow. However, when the mixture is exposed to light, the reaction is explosive.

$$H_2(g) + Cl_2(g) \longrightarrow 2HCl(g) \qquad \Delta H^\circ = -184.61 \text{ kJ}$$

Chemical reactions of this type, which are caused to proceed more rapidly by the effect of light, are called **photochemical reactions.**

We should note at this point that we have in Chapters prior to Chapter 18 used the general enthalpy symbol, ΔH. Now that we have, in Chapter 18, defined *standard molar enthalpy*, we will henceforth use the symbol ΔH° when a standard molar enthalpy value is provided.

b. *Action with metals.* Chlorine is less active toward metals than fluorine, and higher temperatures are generally required in its oxidation of metals. However, powdered antimony ignites at room temperature in chlorine to form antimony(III) chloride, $SbCl_3$ (Fig. 19-2).

c. *Action with nonmetals.* Chlorine reacts with most nonmetals, forming covalent molecular compounds. For example, it combines with sulfur to give "sulfur monochloride," a liquid used in the vulcanization of rubber.

$$S_8(s) + 4Cl_2(g) \longrightarrow 4S_2Cl_2(l) \qquad \Delta H^\circ = -59.4 \text{ kJ}$$

When the chlorine is in excess, the sulfur is oxidized to the dichloride.

$$S_8(s) + 8Cl_2(g) \longrightarrow 8SCl_2(l) \qquad \Delta H^\circ = -50 \text{ kJ}$$

Chlorine oxidizes phosphorus to the trichloride, PCl_3, when a limited supply of the halogen is employed. When an excess of chlorine is used, the pentachloride, PCl_5, is formed.

$$P_4(\text{excess}) + 6Cl_2 \longrightarrow 4PCl_3$$
$$P_4 + 10Cl_2(\text{excess}) \longrightarrow 4PCl_5$$

d. *Action with compounds.* Chorine generally reacts with hydrocarbons (compounds containing only carbon and hydrogen) by adding to multiple bonds

$$H_2C{=}CH_2 + Cl_2 \longrightarrow CH_2Cl{-}CH_2Cl$$

or by substitution. For example, an exothermic stepwise *substitution* of hydrogen by chlorine takes place when methane, CH_4, reacts with chlorine.

$$CH_4(g) + Cl_2(g) \longrightarrow CH_3Cl(g) + HCl(g) \qquad \Delta H^\circ = -94.14 \text{ kJ}$$
$$CH_3Cl(g) + Cl_2(g) \longrightarrow CH_2Cl_2(l) + HCl(g) \qquad \Delta H^\circ = -132.9 \text{ kJ}$$
$$CH_2Cl_2(l) + Cl_2(g) \longrightarrow CHCl_3(l) + HCl(g) \qquad \Delta H^\circ = -105.3 \text{ kJ}$$
$$CHCl_3(l) + Cl_2(g) \longrightarrow CCl_4(l) + HCl(g) \qquad \Delta H^\circ = -93.26 \text{ kJ}$$

When chlorine is dissolved in water, a **disproportionation,** or **auto-oxidation-reduction, reaction** takes place, and a small amount of the chlorine **disproportionates.**

$$Cl_2 + H_2O \rightleftharpoons H^+ + Cl^- + \underset{\text{Hypochlorous acid}}{HClO}$$

One of the chlorine atoms of Cl_2 is oxidized to the $+1$ state in hypochlorous acid, while the other atom of chlorine is reduced to the -1 state in the chloride ion. The reaction between chlorine and water is incomplete, as is indicated by the equilibrium equation. Thus chlorine water is a mixture of water, chlorine, the ions of hydrochloric acid, and molecules of hypochlorous acid, a weak acid. The solution gives off oxygen in sunlight, as the light-sensitive hypochlorous acid decomposes.

$$2HClO \xrightarrow{\text{Sunlight}} 2H^+ + 2Cl^- + O_2(g)$$

The reaction removes HClO from the equilibrium mixture, causing the reaction to the right to go slowly to completion. For this reason chlorine water is usually stored in dark bottles, which do not permit the passage of light.

▶ **3. BROMINE.**   The chemical properties of bromine are very similar to those of chlorine, as we would expect, but bromine is a weaker oxidizing agent than chlorine. The reactivity of bromine toward hydrogen, the nonmetals, the metals, and methane is less vigorous than that of chlorine. This lower reactivity is reflected in the evolution of less heat in the formation of bromides as compared to chlorides. Bromine also exhibits slight disproportionation in water to form hydrobromic acid and hypobromous acid; the reaction takes place less readily than that between chlorine and water.

$$Br_2 + H_2O \rightleftharpoons H^+ + Br^- + \underset{\substack{\text{Hypobromous} \\ \text{acid}}}{HBrO}$$

Hypobromous acid is also a weak acid and decomposes in sunlight according to the equation

$$2HBrO \longrightarrow 2H^+ + 2Br^- + O_2(g)$$

▶ **4. IODINE.**   Of the four naturally occurring halogens, the iodine atom has the least attraction for an additional electron. This is due to its large radius and great number of electron shells. Thus iodine is the weakest oxidizing agent of the four naturally occurring halogens, and the iodide ion is the most easily oxidized of the four halides.

Iodine combines with the active metals but with the liberation of much less heat than the other halogens; with less active metals, heating is required before reaction occurs. Although iodine does not oxidize the other halide ions (except astatide), it will oxidize certain other nonmetal ions such as sulfide ions, S^{2-}.

$$8S^{2-} + 8I_2 \longrightarrow S_8 + 16\,I^-$$

Iodine shows slight chemical activity with water, compared with the other halogens. A very sensitive qualitative test for elemental iodine depends upon the formation of a deep blue color when even trace quantities react with starch.

▶ **5. ASTATINE.**   Most of our knowledge of the element astatine, the fifth halogen, has come from a study of one of its isotopes (mass number 211). The isotope was first obtained in 1940 by Corson, Mackenzie, and Segre, who prepared it by bombarding bismuth with alpha particles in the 60-inch cyclotron at the University of California, Berkeley.

$$^{209}_{83}Bi + ^{4}_{2}He \longrightarrow ^{211}_{85}At + 2\,^{1}_{0}n$$

Astatine may exist in nature as a short-lived intermediate in a nuclear decay chain, but too little of the element is present from this source at a given time to permit its study. The name **astatine** comes from the Greek word for "unstable."

19.5   Uses of the Halogens

Fluorine gas has been used in the fluorination of organic compounds (replacing hydrogen with fluorine) since the early work of Moissan in 1886. Such reactions are usually difficult to control, however. More commonly now, the organic compounds are dissolved in anhydrous hydrogen fluoride, and the solution is electrolyzed. The **fluorocarbon** compounds thus produced are quite stable and nonflammable. They are used as lubricants, refrigerants, coolants, hydraulic liquids, plastics, and insecticides. Freon-12, which is CCl_2F_2, is widely used as a refrigerant; CCl_3F is used as an insecticide; and Teflon is a polymer composed of $-CF_2CF_2-$ units. One of the most important uses of fluorine gas is in the production of uranium hexafluoride, UF_6, which is used in separating the isotopes of uranium by the gaseous diffusion process in connection with the production of atomic energy. Another use of fluorine is in the making of sulfur(VI) fluoride, SF_6, a stable gas with high dielectric and insulating capacities for high voltage. Fluoride ion is added to many community water supplies to decrease the incidence of dental cavities.

Large quantities of the chlorine produced commercially are used in bleaching wood pulp and cotton cloth. The chlorine reacts with water to form hypochlorous acid, which bleaches by oxidizing colored substances to colorless compounds. Most community water supplies are treated with small amounts of chlorine to kill bacteria. Large quantities of chlorine are used in chlorinating hydrocarbons (replacing hydrogen with chlorine) to produce such compounds as carbon tetrachloride (CCl_4), chloroform ($CHCl_3$), and paradichlorobenzene (Dichlorocide), and in the production of polyvinyl chloride and other polymers.

A principal use of bromine has been in the manufacture of ethylene dibromide, $C_2H_4Br_2$, a constituent of antiknock gasoline along with tetraethyl lead, $Pb(C_2H_5)_4$. When the gaseous fuel in an internal combustion engine explodes with extreme rapidity, "knocking" occurs, with a great loss in efficiency. Tetraethyl lead slows the ignition, with a corresponding increase in the efficiency of the engine. The ethylene dibromide is added to convert the lead from tetraethyl lead to lead bromide, which escapes from the engine as a vapor. This largely prevents the accumulation of lead in the engine, but it increases the lead concentration in the air and is a cause of great concern to many people. "Lead-free" gasolines, containing little or no tetraethyl lead, must be used in automobiles with catalytic converters, introduced in 1975 on many models to clean up the exhaust (see Section 22.5). The tetraethyl lead quickly "poisons" the platinum catalyst in the converters.

Other applications of bromine include the production of certain organic dyes, the preparation of light-sensitive silver bromide for use in making photographic film, and the manufacture of the bromides of sodium and potassium, which are used in medicine as sedatives and soporifics.

Iodine in alcohol solution with potassium iodide **(tincture of iodine)** is used as an antiseptic. In the form of iodide salts, iodine is essential in small amounts in the diet for the proper functioning of the thyroid gland. Iodine deficiency may lead to the development of goiter. Iodized table salt contains about 0.023% potassium iodide. Silver iodide is used in the manufacture of photographic films and in the seeding of clouds to induce rain. Iodoform, CHI_3, is used as an antiseptic in the dressing of wounds.

19.6   Interhalogen Compounds

The compounds formed by the union of two different halogens are called **interhalogen compounds.** Molecules of these compounds consist of an atom of the heavier halogen bonded to an odd number of atoms of the lighter halogen. Examples are

$$Cl_2 + 3F_2 \longrightarrow 2ClF_3$$
$$I_2 + 5F_2 \longrightarrow 2IF_5$$

Formulas for some other interhalogen compounds are given in Table 19-3.

TABLE 19-3 Interhalogen Compounds		
IBr	BrCl	ClF
ICl	BrF	ClF_3
ICl_3	BrF_3	ClF_5
IF_5	BrF_5	
IF_7		

Because the smaller halogen atoms are grouped about the larger, the maximum possible number of smaller atoms per molecule increases as the radius, $r_{larger}/r_{smaller}$, increases (see Section 11.17). Many of these compounds are unstable, and most are extremely reactive chemically. The reactions of the interhalogen compounds are similar to those of the component halogens.

The ionic **polyhalides** of the alkali metals, such as KI_3, $KICl_2$, $KICl_4$, $CsIBr_2$, and $CsClBr_2$, are closely related to the interhalogen compounds.

Typical structures of the polyhalides and interhalogens are shown in Fig. 19-3.

Figure 19-3 Structures of the ions ICl_2^- (linear) and ICl_4^- (square planar), and of the molecules IF_5 (square pyramidal), and IF_7 (pentagonal bipyramidal). (See also Fig. 19-1.)

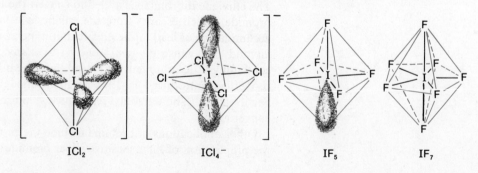

ICl_2^- ICl_4^- IF_5 IF_7

The Hydrogen Halides

The binary compounds that contain only hydrogen and one of the halogens are called **hydrogen halides.** These compounds have the general formula HX, in which X represents the halogen. Figure 19-4 shows the relative sizes of the hydrogen halide molecules.

Figure 19-4 Approximate relative sizes of the hydrogen halide molecules. The bond distance (the distance between the center of the hydrogen atom and the center of the halogen atom) is indicated for each molecule.

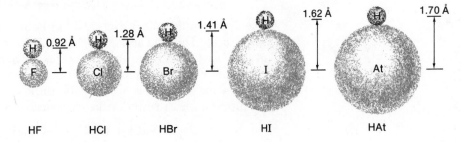

19.7   Preparation of the Hydrogen Halides

Since the hydrogen halides are acids, they can be prepared by the general techniques used to prepare acids (Section 14.3). Because of variations in the properties of the halides, however, all techniques are not equally satisfactory for preparation of all hydrogen halides.

▶**1. BY DIRECT UNION.** One method of preparing the hydrogen halides is by the direct union of their constituent elements according to the equation

$$H_2 + X_2 \longrightarrow 2HX \qquad (X = symbol \ for \ halogens)$$

The difficulty of producing elemental fluorine and its relatively high cost make the direct union method of preparing hydrogen fluoride impractical. On the other hand, both hydrogen and chlorine are readily available from the electrolysis of brine (Section 19.1), and one commercial method of producing hydrogen chloride involves the combustion of hydrogen with chlorine in burners specially designed for this purpose. The reaction of bromine with hydrogen is much less vigorous than that with either chlorine or fluorine. If the reaction between hydrogen and bromine is to proceed appreciably, the mixture must be heated to about 200°C in contact with a catalyst such as platinum or carbon. The reaction is a reversible one that quickly reaches equilibrium, but it may be forced nearly to completion by using an excess of hydrogen. The excess hydrogen is recycled by liquefying the HBr and pumping away the gaseous hydrogen.

The direct union of hydrogen and iodine is unsatisfactory for the preparation of hydrogen iodide, because the reaction is slow and never complete. Heat decomposes hydrogen iodide; this means that the reaction is readily reversible and the equilibrium yield of HI is low. This reaction is not self-sustaining, as are the similar reactions of the other halogens.

▶**2. BY THE ACTION OF CONCENTRATED SULFURIC ACID ON A METALLIC HA-LIDE.** The most convenient laboratory method of producing hydrogen fluoride and hydrogen chloride is the reaction of concentrated sulfuric acid upon a halide of a metal.

$$2MX + H_2SO_4 \longrightarrow M_2SO_4 + 2HX(g)$$

Hydrogen fluoride is usually prepared by heating a mixture of the mineral fluorite, CaF_2, and concentrated sulfuric acid in a lead or platinum retort.

$$CaF_2 + H_2SO_4 \longrightarrow CaSO_4 + 2HF(g)$$

The hydrogen fluoride is evolved as a gas, and if absorbed in water forms hydrofluoric acid. This acid acts upon glass, and for this reason is stored in bottles made of lead, wax, or certain plastics such as polyethylene.

Hydrogen chloride is prepared, both in the laboratory and on a commercial scale, by the action of concentrated sulfuric acid upon sodium chloride, the most plentiful and least costly of the chlorides.

$$NaCl + H_2SO_4 \longrightarrow NaHSO_4 + HCl(g)$$

The reaction goes to completion because hydrogen chloride is insoluble in the reaction mixture. By adding more sodium chloride and heating the mixture to a higher temperature, additional hydrogen chloride is produced.

$$NaCl + NaHSO_4 \longrightarrow Na_2SO_4 + HCl(g)$$

The hydrogen chloride thus produced is usually absorbed in water and marketed as hydrochloric acid.

The reaction of sulfuric acid with salts containing bromide or iodide ions is not a suitable technique for the synthesis of hydrogen bromide or hydrogen iodide. Sulfuric acid is a sufficiently strong oxidizing agent to oxidize the bromide ion and, even more easily, the larger iodide ion to bromine and iodine, respectively. The sulfuric acid is reduced to sulfur dioxide by the bromide ion. Iodide ion reduces sulfuric acid to sulfur dioxide, elemental sulfur, or hydrogen sulfide depending upon the relative amounts of sulfuric acid and iodide ion used in the reaction.

▶ **3. BY THE HYDROLYSIS OF NONMETALLIC HALIDES.** Another method of producing hydrogen halides is by the hydrolysis (reaction with water) of the covalent halides of the nonmetals. Typical of this class of compounds are PCl_3, PBr_3, PI_3, and SCl_4. Halides of this type generally react with water to form two acids—the hydrogen halide and an oxyacid of the other nonmetal; quite often the reaction is a vigorous one.

$$PBr_3 + 3H_2O \longrightarrow H_3PO_3 + 3HBr(g)$$
$$SCl_4 + 3H_2O \longrightarrow H_2SO_3 + 4HCl(g)$$

This method is rarely used for the commercial preparation of hydrogen chloride. However, the preparation of hydrogen bromide and hydrogen iodide by the hydrolysis of phosphorus(III) bromide and phosphorus(III) iodide, respectively, is quite common. The PBr_3 and PI_3 are prepared by the direct union of phosphorus with bromine and iodine, respectively.

▶ **4. BY THE HALOGENATION OF HYDROCARBONS.** When fluorine, chlorine, or bromine (but not iodine) is allowed to react with a saturated or aromatic hydrocarbon, one product of the reaction is the corresponding hydrogen halide. A catalyst is usually required, and the hydrogen halide is often a by-product of a reaction used to produce a desired halogenated hydrocarbon. For example, hydrogen chloride is a by-product of the manufacture of ethyl chloride (C_2H_5Cl) from ethane (C_2H_6) and chlorine.

$$C_2H_6(g) + Cl_2(g) \longrightarrow C_2H_5Cl(l) + HCl(g) \qquad \Delta H° = -144.2 \text{ kJ}$$

About two-thirds of the hydrogen chloride produced in the United States results from the chlorination of hydrocarbons because the demand for chlorinated hydrocarbons is large. The production of chlorinated organic compounds uses about half the chlorine produced each year.

19.8   General Properties of the Hydrogen Halides

HF, HCl, HBr, and HI are colorless gases with sharp, penetrating odors. Their constituent atoms are linked together by polar covalent bonds. The polarity of these molecules is a function of the electronegativity of the halogen atom; it is greatest for HF and least for HI. Some of the more important physical properties of the halides are given in Table 19-4.

All of the hydrogen halides decompose into their constituent elements when heated sufficiently. Of the four hydrogen halides shown in Table 19-4, hydrogen iodide is the most easily decomposed, while hydrogen chloride and hydrogen fluoride show only slight dissociation at 1000°C. The stability of these gases decreases, then, as the formula weight increases.

The hydrogen halides fume in moist air; that is, they combine with the water vapor in the air to give a fog of very small drops of the corresponding acid. All of the hydrogen halides are very soluble in water. With the exception of hydrogen fluoride, the hydrogen halides are strong electrolytes; they ionize completely in dilute aqueous solution. If the concentration of these acids in water is greatly increased, the molecular form (HX) appears in low concentrations and escapes from solution.

The anhydrous hydrogen halides are rather inactive chemically and do not attack dry metals at ordinary temperatures. Neither do they conduct electricity in the liquid state; this fact indicates that they are molecular rather than ionic compounds.

TABLE 19-4   Some Properties of the Hydrogen Halides

	HF	HCl	HBr	HI
Molecular weight	20.006	36.461	80.917	127.91
Melting point, °C	−83.1	−114.2	−86.8	−50.8
Boiling point, °C	19.9	−85.1	−66.7	−35.36
Solubility in water, g/100 g of water	∞ (0°)	82.3 (0°)	221 (0°)	234 (10°)
Enthalpy of formation, $\Delta H°_{f298}$, kJ mol^{-1}	−271.1	−92.307	36	26.5

19.9   The Hydrohalic Acids

Aqueous solutions of the hydrogen halides are called **hydrofluoric acid, hydro-chloric acid, hydrobromic acid,** and **hydriodic acid.**

▶**1. PROPERTIES DEPENDENT ON HYDROGEN ION.** Hydrofluoric acid is a weak acid, but the other hydrohalic acids are very strong. As with other protonic acids, the hydrogen ion of the hydrohalic acids is reduced by metals above hydrogen in the activity series (Section 9.21); halide salts and hydrogen are formed. The general reaction is illustrated with the metal zinc.

$$Zn(s) + 2H^+(aq) + [2X^-(aq)] \longrightarrow Zn^{2+}(aq) + [2X^-(aq)] + H_2(g)$$

Notice that the hydrogen ion acts as an oxidizing agent in the reaction. The hydrohalic acids neutralize aqueous bases, forming salts and water. Their reaction with sodium hydroxide is given by the equation

$$\underset{\text{Base}}{[Na^+(aq)] + OH^-(aq)} + \underset{\text{Acid}}{H^+(aq) + [X^-(aq)]} \longrightarrow \underset{\text{Salt}}{[Na^+(aq)] + [X^-(aq)]} + \underset{\text{Water}}{H_2O(l)}$$

These strong acids also react with other bases.

$$NH_3(aq) + H^+(aq) + [X^-(aq)] \longrightarrow NH_4^+(aq) + [X^-(aq)]$$
$$CH_3CO_2^-(aq) + H^+(aq) + [X^-(aq)] \longrightarrow CH_3CO_2H(aq) + [X^-(aq)]$$

▶**2. PROPERTIES DEPENDENT UPON HALIDE ION.** The halide ion of the hydro-halic acids can serve either as a reducing agent or as a precipitating agent, as illustrated by the following equations:

As reducing agent

$$[2H^+] + 2Br^- + Cl_2 \rightleftharpoons [2H^+] + 2Cl^- + Br_2$$
$$14H^+ + 6Cl^- + Cr_2O_7^{2-} \longrightarrow 2Cr^{3+} + 3Cl_2 + 7H_2O$$

As precipitating agent

$$Ag^+ + [NO_3^-] + [H^+] + Cl^- \longrightarrow AgCl(s) + [H^+] + [NO_3^-]$$

19.10   Unique Properties of Hydrogen Fluoride and Hydrofluoric Acid

Hydrogen fluoride is unique among the hydrogen halides in that its molecules are associated. Furthermore, hydrofluoric acid is weak while the other hydrohalic acids are strong, and it attacks glass whereas the others do not.

▶**1. HYDROGEN BONDING.** Pure hydrogen fluoride differs from the other hydro-gen halides because of the tendency of its molecules to associate through hydrogen bonding. In the solid state hydrogen fluoride exists as zigzag chains of molecules that extend the entire length of the crystal (Fig. 19-5). Liquid hydrogen fluoride likewise exhibits evidence of extensive association of the HF molecules. Vapor density measurements show hydrogen fluoride gas to be a mixture of $(HF)_2$ and $(HF)_3$ at 26°C; only at 88° does the vapor density correspond to the formula HF.

Figure 19-5 Illustration showing the associative nature of HF resulting from hydrogen bonding.

The association of molecules in solid, liquid, and gaseous hydrogen fluoride can be attributed to the large electronegativity difference between fluorine (4.1) and hydrogen (2.2). This difference leads to a highly polar covalent H—F bond (Section 5.9), in which the hydrogen bears a large fractional positive charge and the fluorine, a large fractional negative charge. The electrostatic attraction between a partially positive hydrogen atom in one molecule and a partially negative fluorine atom of another molecule leads to the association of hydrogen fluoride molecules. This attraction is referred to as a hydrogen bond. A similar association between hydrogen and oxygen is found in water (Section 12.1).

Hydrogen bonds can form when hydrogen is bonded to a highly electronegative element such as fluorine, oxygen, nitrogen, or chlorine. Examples include $HF\cdots HF$, $H_2O\cdots HOH$, $H_3N\cdots HNH_2$, $H_3N\cdots HOH$, and $H_2O\cdots HOCH_3$. Although the strengths of hydrogen bonds are only about 5 to 10% of that of ordinary covalent bonds, these intermolecular forces have distinctive effects on physical and chemical properties. The boiling point of hydrogen fluoride is abnormally high compared to that of the other hydrogen halides, and results from the fact that hydrogen bonds must be broken before the hydrogen fluoride vaporizes.

Figure 19-6(a) shows a plot of boiling points and Fig. 19-6(b) heats of vaporization against the number of the period (horizontal row) of the Periodic Table, for

Figure 19-6 Graphic illustrations of (a) the boiling points and (b) the heats of vaporization of several binary hydrides of Groups IVA–VIIIA. Each line connects compounds of the elements in a given group. H_2O, HF, and NH_3 have hydrogen bonding and exhibit abnormally high boiling points and heats of vaporization. CH_4 does not have hydrogen bonding. Plots for four of the noble gases are included to indicate typical behavior due to increasing London forces resulting from increasing mass.

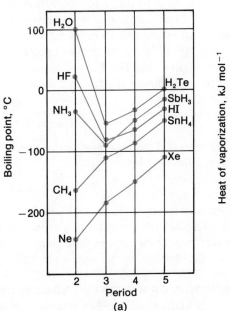

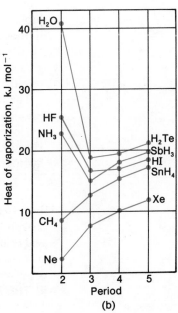

the binary hydrogen compounds of elements in families IVA, VA, VIA, and VIIA. Included for comparison are four of the noble gases, Ne, Ar, Kr, and Xe, which of course do not contain hydrogen. Note that, in general, boiling points and heats of vaporization increase with increasing molecular or atomic weight. However, the compounds H_2O, HF, and NH_3 possess appreciable hydrogen bonding. More energy is required to break the additional hydrogen bonds in order to bring the material to the gaseous (vapor) state. Hence, those substances possessing hydrogen bonds have abnormally high boiling points and heats of vaporization, as shown in the diagrams. Methane (CH_4), however, does not have hydrogen bonding and hence has a lower boiling point and heat of vaporization than the corresponding heavier compounds of other elements in the family. Similarly, the noble gas Ne has lower values than do those noble gases that are heavier than neon.

The tendency for fluorine to form hydrogen bonds with hydrogen makes it possible for many metal fluorides to form stable hydrogen difluoride salts such as $NaHF_2$ and KHF_2, which contain hydrogen bonds in the anions. The hydrogen difluoride ion can be described as a resonance hybrid (Section 5.8).

$$F—H \cdots F^- \longleftrightarrow F^- \cdots H—F$$

Only very weak hydrogen bonds can occur in hydrogen chloride, hydrogen bromide, hydrogen iodide, and hydrogen sulfide, because the electronegativities of chlorine, bromine, iodine, and sulfur are low, and, consequently, the polarities of the bonds are small. The effect of this very weak hydrogen bonding can be seen in the slight deviation in the boiling points and heats of vaporization of HCl and H_2S from the expected behavior (Fig. 19-6).

▶ **2. THE IONIZATION OF HYDROFLUORIC ACID.** Hydrofluoric acid is weak (slightly ionized in water) whereas the other hydrohalic acids are strong (totally ionized in water). In order for ionization to occur, the bond in H—X must be broken, and then hydration of the resultant ions must take place. The H—F bond is about twice as difficult to break as is the H—I bond, and the strengths of the H—Br and H—Cl bonds lie between those of the other two (see Table 18-2). The greater strength of the H—F bond as compared to that of H—I is considered to result from the smaller interatomic distance in HF (HF, 1.0 Å; HI, 1.7 Å).

The ionization of hydrogen fluoride in dilute solution may be written as

$$HF \rightleftharpoons H^+ + F^-$$

However, a second ion is formed in this process due to hydrogen bonding; it is the hydrogen difluoride ion.

$$F^- + HF \rightleftharpoons HF_2^-$$

In very concentrated solutions of hydrofluoric acid, the degree of ionization rises sharply. This behavior is opposite to that observed for all other weak electrolytes and results from an increase in the concentration of the difluoride ion, HF_2^-, which increases the stability of the anion and shifts the equilibrium to the right.

▶ **3. THE ACTION OF HYDROFLUORIC ACID ON GLASS.** The most characteristic behavior of hydrofluoric acid is its action on sand (silicon dioxide) and on glass,

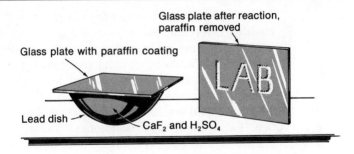

which is a mixture of silicates (mostly calcium silicate). The reactions are

$$SiO_2(s) + 4HF(aq) \longrightarrow SiF_4(g) + 2H_2O(l) \qquad \Delta H° = -191 \text{ kJ}$$

$$CaSiO_3(s) + 6HF(aq) \longrightarrow CaF_2(s) + SiF_4(g) + 3H_2O(l)$$

The silicon escapes in these reactions as a volatile compound, silicon tetrafluoride. When one mole of SiF_4 is formed *from the elements,* the amount of energy released (1510 kJ) is exceedingly large. This large amount of energy provides the driving force for the conversion of silicon-oxygen compounds to silicon tetrafluoride by the action of hydrofluoric acid. For example, in the reaction with SiO_2, 191 kJ per mole of SiO_2 are released, and the escape of the gaseous SiF_4 helps drive the reaction to completion. Glass objects such as electric light bulbs can be frosted, or etched, by hydrofluoric acid. Thermometers, burets, and other glassware may be marked by use of this acid. In order that a restricted area may be etched, the glass is first covered with paraffin, and then, with a sharp-pointed instrument, part of the paraffin is removed from the glass (Fig. 19-7). Extreme care should be taken in handling hydrofluoric acid for it is very corrosive to the skin and causes painful, slow-healing burns. Hydrofluoric acid is also a local anesthetic, and its contact with the skin may not be noticed until much damage has been done.

19.11   The Metal Halides

All metals form halides, but the variation in character among such halides is striking. The halides of the alkali metals and the heavier alkaline earth metals, almost all the metal fluorides, and some halides of the transition metals in their lower oxidation states (e.g., Fe^{2+}, Mn^{2+}, Cr^{2+}) are predominantly ionic, or saltlike, in character. These ionic halides tend to have high melting points (above 400°), high boiling points, and high electrical conductivities in the fused state. Those that dissolve appreciably in water undergo little or no hydrolysis.

On the other hand, the halides of metals with large charge-to-size ratios and/or high oxidation states (more nonmetallic behavior) are likely to be predominantly covalent in character because of the attraction by the metal for the electrons in the outer shell of the halide ion. Aluminum chloride, $AlCl_3$, tin(IV) chloride, $SnCl_4$, titanium(IV) chloride, $TiCl_4$, and uranium(VI) fluoride, UF_6, are typical of this group of halides. These covalent halides are characterized by volatility, by solubility in nonpolar solvents, by lack of electrical conductivity in the fused state, and by extensive hydrolysis in water.

As might be expected, there are metal halides intermediate in character between the ionic and covalent ones. For example, iron(III) chloride, $FeCl_3$, is volatile and readily hydrolyzed, but it is a good conductor of electricity in the fused state.

The anhydrous halides of metals may be prepared by direct union of their constituents. On the other hand, hydrated halo salts are obtained by the action of the hydrohalic acids on metals above hydrogen in the activity series, and by the action of hydrohalic acids on most oxides, hydroxides, and carbonates of these metals. Several slightly soluble metal halides may be prepared by ionic combination. For example, when solutions containing silver nitrate and sodium chloride are mixed, silver chloride precipitates as a curdy white solid.

$$Ag^+ + [NO_3^-] + [Na^+] + Cl^- \longrightarrow AgCl(s) + [Na^+] + [NO_3^-]$$

or simply
$$Ag^+ + Cl^- \longrightarrow AgCl(s)$$

Other slightly soluble halides include $AgBr$, AgI, $PbCl_2$, $PbBr_2$, PbI_2, $CuCl$, CuI, $CuBr$, $TlCl$, Hg_2Cl_2, CaF_2, and BaF_2. It is a striking fact that the solubilities of the fluorides differ markedly from those of the other halides. For example, silver fluoride is readily soluble in water whereas the other silver halides are only slightly soluble. On the other hand, calcium and barium fluorides are insoluble in water while the other halides of these metals are freely soluble.

19.12   Uses of the Hydrohalic Acids and Halides

The largest use for hydrogen fluoride is in the production of fluorocarbons for refrigerants, plastics, and propellants. The second largest use is in the production of cryolite, Na_3AlF_6, which is important in the metallurgy of aluminum. The acid is also used in making UF_6 in the atomic energy program, in the etching of glass, as a catalyst in the petroleum industry, and in the production of other inorganic compounds (such as BF_3) that serve as catalysts in the industrial synthesis of certain organic compounds. The mineral fluorspar, CaF_2, is used as a flux, because it reacts to form easily fusible products with various substances that cannot be melted readily. Sodium fluoride is used as an insecticide, a flux, and a fungicide.

Hydrochloric acid is inexpensive; thus it is the most important acid in industry other than sulfuric acid. It is used in the manufacture of metal chlorides, dyes, glue, glucose, and various other chemicals. A considerable amount is also used in the activation of oil wells and in removing oxide coatings from iron or steel that is to be galvanized, tinned, or enameled.

The amounts of hydrobromic and hydriodic acid used commercially are insignificant compared to the amount of hydrochloric acid used. Hydrobromic acid is used in analytical chemistry and in the synthesis of certain organic compounds.

Oxygen Compounds of the Halogens

Although the halogens tend to gain an eighth electron for their outer shells by taking electrons from many metals, thus becoming negative halide ions, X^-, they also quite frequently complete their outer shells by sharing electrons with other atoms. For example, electrons are shared in X_2, HX, PX_3, and SnX_4. The halogens (except fluorine) also share electrons with oxygen in the formation of oxyacids and oxysalts, in which the halogens assume positive oxidation states (oxygen is more electronegative than all of the halogens except fluorine). The oxygen compounds of the halogens are noted for their strengths as oxidizing agents.

19.13   Binary Halogen-Oxygen Compounds

The halogens do not combine directly with oxygen. However, binary halogen-oxygen compounds can be prepared by indirect methods. Because oxygen is more electronegative than chlorine, bromine, and iodine, it is proper to call their oxygen derivatives oxides. On the other hand, the fluorine compounds with oxygen are called fluorides because fluorine is the more electronegative element. The known binary halogen-oxygen compounds are listed in Table 19-5. These compounds are characteristically reactive and unstable toward heat.

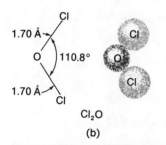

Figure 19-8 Angular structures of the OF_2 and Cl_2O molecules (drawn approximately to scale.) Bond distances are measured from the centers of the atoms.

TABLE 19-5 Binary Halogen-Oxygen Compounds			
Fluorine	Chlorine	Bromine	Iodine
OF_2	Cl_2O	Br_2O	I_2O_4
O_2F_2	ClO_2	Br_3O_8	I_4O_9
	Cl_2O_6	BrO_2	I_2O_5
	Cl_2O_7		

▶ **1. FLUORIDES OF OXYGEN.** Just two oxygen-fluorine compounds are known. They are **oxygen difluoride**, OF_2, a colorless gas, and **oxygen monofluoride**, O_2F_2, a red liquid. Only the difluoride is of sufficient importance for consideration here.

The bonding in OF_2 is covalent and the oxidation states for oxygen and fluorine are $+2$ and -1, respectively. Oxygen difluoride is prepared by the reaction of elemental fluorine with a solution of sodium hydroxide.

$$2F_2 + [2Na^+] + 2OH^- \longrightarrow [2Na^+] + 2F^- + OF_2(g) + H_2O$$

Although oxygen difluoride dissolves appreciably in water, it is not the anhydride of an acid because the resulting solutions are not acidic.

▶ **2. OXIDES OF CHLORINE.** **Chlorine monoxide,** Cl_2O, is a yellowish-red gas that is apt to decompose explosively. This compound is best prepared by passing chlorine over freshly precipitated, dried mercury(II) oxide.

$$2Cl_2 + 2HgO \longrightarrow HgCl_2 \cdot HgO + Cl_2O(g)$$

Chlorine monoxide is an active oxidizing agent and dissolves in water, giving hypochlorous acid, HClO.

The structures of OF_2 and Cl_2O, both angular, are shown in Fig. 19-8.

Chlorine dioxide, ClO_2 (Fig. 19-9), is a yellow gas that is violently explosive when pure. It can be safely handled, however, if diluted with carbon dioxide or air. The most convenient method available for the commercial preparation of

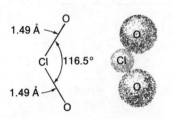

Figure 19-9 Structure of the ClO_2 molecule.

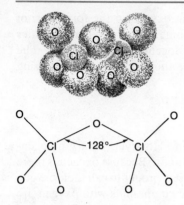

Figure 19-10 The structure of the Cl_2O_7 molecule. Two tetrahedral ClO_4 groups share a common oxygen with a Cl—O—Cl angle of 128°.

chlorine dioxide involves the reaction of sodium chlorite with chlorine diluted with air.

$$2NaClO_2 + Cl_2 \xrightarrow{\text{Air}} 2NaCl + 2ClO_2(g)$$

Chlorine heptaoxide, Cl_2O_7, is a colorless liquid, which is dangerously heat and shock sensitive. It is the acid anhydride of perchloric acid and is obtained by dehydrating perchloric acid with P_4O_{10} and then distilling the product.

$$4HClO_4(l) + P_4O_{10}(s) \longrightarrow 2Cl_2O_7(g) + 4HPO_3(s) \qquad \Delta H° = -376 \text{ kJ}$$

Its structure is shown in Fig. 19-10.

19.14   The Oxyacids of the Halogens and Their Salts

There is no definite evidence for the existence of oxyfluorine acids or oxyfluorine salts. The known oxyacids for three of the other halogens, including their formulas and the oxidation states of the halogen in them, are given in Table 19-6. The nomenclature for these acids and their salts was discussed in Section 5.14.

The oxygen compounds of chlorine, bromine, and iodine differ sufficiently to make generalizations concerning them impossible. However, trends in properties of the oxychlorine acids and salts are well established (Table 19-7), and these trends hold, in a general way, for the analogous bromine and iodine compounds.

TABLE 19-6   Oxyacids of the Halogens

Oxidation States	Oxyacids of Chlorine	Oxyacids of Bromine	Oxyacids of Iodine
+1	HClO	HBrO	HIO
+3	HClO_2	HBrO_2	
+5	HClO_3	HBrO_3	HIO_3
+7	HClO_4	HBrO_4	HIO_4
			H_5IO_6

TABLE 19-7   Trends in Properties of the Oxyacids of Chlorine and their Sodium Salts

Oxidation State	Acid	Thermal Stability and Acid Strength	Oxidizing Power	Salt	Thermal Stability	Oxidizing Power and Anion Base Strength
+1	HClO			NaClO		
+3	HClO_2	Increase (down)	Increases (up)	NaClO_2	Increases (down)	Increase (up)
+5	HClO_3			NaClO_3		
+7	HClO_4			NaClO_4		

Thermal stability increases greatly →

← Oxidizing power increases greatly

The greater the oxygen content in members of a series of oxygen compounds (oxyacids or their salts) of a halogen, the greater is their thermal stability. The acids decompose at much lower temperatures than do their salts. These rules apply generally to the oxygen compounds of sulfur, nitrogen, and phosphorus, as well as to those of the halogens.

The salts undergo disproportionation upon heating. This, too, is characteristic of the oxysalts of sulfur and phosphorus (but not of nitrogen). Examples include

$$3NaClO \longrightarrow NaClO_3 + 2NaCl$$

Sodium hypochlorite   Sodium chlorate

$$4Na_2SO_3 \longrightarrow 3Na_2SO_4 + Na_2S$$

Sodium sulfite   Sodium sulfate

$$4Na_2HPO_3 \longrightarrow 2Na_3PO_4 + PH_3 + Na_2HPO_4$$

Disodium hydrogen   Sodium phosphate   Disodium hydrogen
phosphite   phosphate

The activity of the oxyacids of the halogens as oxidizing agents is related inversely to the number of oxygen atoms in the molecule; i.e., the activity is greatest for the HXO acids and least for the HXO_4 acids. Note that the oxidizing strength is greatest for the acids of lowest thermal stability.

In a series of oxyacids of any element, the acidic strength, as measured in terms of extent of ionization in aqueous solution, is greater the larger the number of oxygen atoms in the molecule. Phosphoric acid is stronger than phosphorous acid, sulfuric is stronger than sulfurous, perchloric is stronger than chloric, and so on.

The electronic (Lewis) formulas for the oxyacids of chlorine are

$$H\!:\!\ddot{O}\!:\!\ddot{Cl}\!: \qquad H\!:\!\ddot{O}\!:\!\overset{\displaystyle :\ddot{O}:}{\ddot{Cl}}\!: \qquad H\!:\!\ddot{O}\!:\!\overset{\displaystyle :\ddot{O}:}{\underset{\displaystyle :\ddot{O}:}{Cl}} \qquad H\!:\!\ddot{O}\!:\!\overset{\displaystyle :\ddot{O}:}{\underset{\displaystyle :\ddot{O}:}{Cl}}\!:\!\ddot{O}\!:$$

Hypochlorous acid   Chlorous acid   Chloric acid   Perchloric acid

The strengths of these acids are directly proportional to the oxidation states of the chlorine atom. The oxidation state of chlorine increases as the oxygen content of the molecule increases. In all four acids the hydrogen is bonded through an oxygen atom. The hydrogen is attached to the oxygen by a polar covalent bond and the oxygen atoms are attached to the chlorine atom by bonds that are essentially covalent in character. As the positive oxidation state of chlorine increases, the bonding electrons are shifted away from the hydrogen atom, and the tendency for the molecule to lose a proton to water, and thereby form hydronium ions in solution, increases (see Section 14.15). This accounts for the correlation of the strength of the oxyacids as acids with the number of oxygen atoms in the molecule. Conversely, it follows that the tendency of an oxyhalogen anion to accept protons increases as the oxygen content of the anion decreases. As we have seen in Section 14.15, this tendency of a negative ion to accept protons is a measure of its strength as a base.

The oxyacids of chlorine are more acidic than the corresponding oxyacids of bromine; and those of bromine are more acidic than the corresponding ones of iodine. These facts can be accounted for by the decrease in electronegativity with increasing size of the halogen and the resultant decreases in the tendency for the oxyacid to lose a proton to water.

19.15   Hypohalous Acids and Hypohalites

Because of their thermal instabilities, the **hypohalous acids** (HXO) have not been obtained in the pure form and are known only in aqueous solution. The hypohalous acids are all weak, with the acid strength decreasing with increasing size of the halogen atom. Thus, HClO is a stronger acid than HBrO, which in turn is stronger than HIO (Section 19.14). The hypohalite ions (XO^-) readily hydrolyze by accepting protons from water. The solutions of alkali metal hypohalites are basic due to an excess of hydroxide ions.

$$XO^- + H_2O \rightleftharpoons HXO + OH^-$$

$$\begin{array}{ccc} \text{Hypohalite} & & \text{Hypohalous} \\ \text{ion} & & \text{acid} \end{array}$$

Aqueous solutions of the hypohalous acids are prepared by hydrolysis of the free halogens according to the general equation

$$X_2 + H_2O \rightleftharpoons H^+ + X^- + HOX$$

This method of preparation yields a mixture of the hypohalous acid and the hydrohalic acid. The extent to which the reaction proceeds to the right decreases in the series chlorine, bromine, iodine. The equilibrium may be shifted to the right by the addition of a base, such as NaOH, that neutralizes the acids formed. The reaction may then be represented by

$$X_2 + [2Na^+] + 2OH^- \longrightarrow [2Na^+] + X^- + XO^- + H_2O$$

For these products to be formed, the base must be cold and dilute. The hypohalites are unstable at elevated temperatures and disproportionate, forming halides and halates.

$$3XO^- \xrightarrow{\triangle} 2X^- + XO_3^-$$

$$\qquad\qquad\quad \text{Halide} \quad \text{Halate}$$

Sodium hypochlorite is produced commercially by the electrolysis of cold, dilute aqueous sodium chloride solutions under conditions where the sodium hydroxide and chlorine can mix. The hypochlorite ion is shown in Fig. 19-11.

$$2Cl^- + 2H_2O + \text{electrical energy} \longrightarrow 2OH^- + Cl_2 + H_2(g)$$
$$Cl_2 + 2OH^- \longrightarrow Cl^- + ClO^- + H_2O$$

Net reaction: $Cl^- + H_2O + \text{electrical energy} \xrightarrow[\text{dil.}]{\text{Cold,}} ClO^- + H_2(g)$

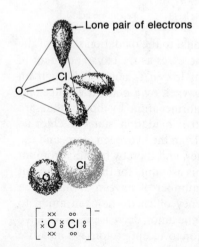

Lone pair of electrons

Figure 19-11 The structure of the hypochlorite ion, ClO⁻.

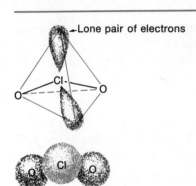

Lone pair of electrons

$$\left[\begin{array}{ccc} \overset{xx}{\underset{xx}{\times}}O\overset{oo}{\underset{oo}{\times}}Cl\overset{xx}{\underset{xx}{\times}}O\overset{xx}{\underset{xx}{\times}} \end{array}\right]^{-}$$

Figure 19-12 The structure of the chlorite ion, ClO_2^-.

19.16   Halous Acids and Halites

None of the **halous acids,** HXO_2, have been isolated in pure form. Of the halites, only the chlorites have been adequately characterized, although the bromite ion was finally produced in small quantities in aqueous solution in 1969.

Chlorous acid, $HClO_2$, has no anhydride; Cl_2O_3 is unknown. However, chlorine dioxide, ClO_2, disproportionates to give two oxyacid anions of chlorine with alkalies, one of which is the anion of chlorous acid.

$$2ClO_2 + 2OH^- \longrightarrow ClO_2^- + ClO_3^- + H_2O$$

Chlorine   Chlorite   Chlorate
dioxide   ion   ion

An aqueous solution of chlorous acid can be prepared by the action of sulfuric acid on a suspension of barium chlorite.

$$Ba(ClO_2)_2 + [2H^+] + SO_4^{2-} \longrightarrow BaSO_4(s) + [2H^+] + 2ClO_2^-$$

The insoluble barium sulfate may be removed from the solution of chlorous acid by filtering. Attempts to isolate pure chlorous acid result in its decomposition into chlorine dioxide, chloric acid, and hydrochloric acid.

The structure of the bent chlorite ion is shown in Fig. 19-12.

19.17   Halic Acids and Halates

Although **chloric acid,** $HClO_3$, and **bromic acid,** $HBrO_3$ decompose when attempts are made to isolate them, **iodic acid,** HIO_3, can be obtained in the form of stable colorless crystals, which melt with decomposition at 110°.

Solutions of chloric acid and bromic acid are readily obtained by the reaction of their barium salts with sulfuric acid.

$$Ba^{2+} + 2XO_3^- + [2H^+] + SO_4^{2-} \longrightarrow BaSO_4(s) + [2H^+] + 2XO_3^-$$

Iodic acid is easily prepared by oxidizing elemental iodine with concentrated nitric acid according to the equation

$$I_2 + 10H^+ + 10NO_3^- \longrightarrow 2H^+ + 2IO_3^- + 10NO_2(g) + 4H_2O$$

Iodic acid

All the halic acids are strong acids and very active oxidizing agents.

The water solubility of the halates decreases with increasing weight of the halogen. Most chlorates are water soluble, bromates are generally much less soluble than the chlorates, and many iodates are insoluble in water. All halates decompose when heated. Potassium chlorate yields potassium chloride and oxygen at high temperatures.

$$2KClO_3 \longrightarrow 2KCl + 3O_2(g)$$

At moderate temperatures potassium perchlorate and potassium chloride are formed when potassium chlorate is heated.

$$4KClO_3 \longrightarrow 3KClO_4 + KCl$$

Thermal decomposition of bromates and iodates results in a number of different

← Lone pair of electrons

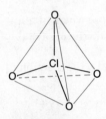

Figure 19-13 The structure of the chlorate ion, ClO_3^-.

Figure 19-14 The structure of the perchlorate ion, ClO_4^-.

products, depending on conditions under which the reactions occur. In some cases, the metal oxides and the free halogens are formed.

The structure of the trigonal pyramidal chlorate ion is shown in Fig. 19-13.

19.18 Perhalic Acids and Perhalates

▶ **1. PERCHLORIC ACID AND ITS SALTS.** **Perchloric acid** ($HClO_4$) may be obtained by treating a perchlorate, such as potassium perchlorate, with sulfuric acid.

$$KClO_4 + H_2SO_4 \longrightarrow KHSO_4 + HClO_4$$

To isolate the perchloric acid the reaction mixture is distilled under reduced pressure. The acid explodes if its temperature is raised above 92°. At reduced pressure, it boils below this temperature with little decomposition and distills without exploding. Aqueous solutions of perchloric acid up to concentrations of about 60% are quite stable thermally. The pure acid is unstable and unsafe to use and handle in concentrations above 60%. It is a powerful oxidizing agent, and serious explosions may occur when concentrated solutions of it are heated with substances that are easily oxidized. However, its reactions as oxidizing agent are very slow when it is cold and dilute.

Perchoric acid is a valuable analytical reagent because it is among the strongest of all acids; it is a good oxidizing agent, and perchlorates are quite soluble in the acid. Perchloric acid is used in the electropolishing of metals and alloys such as nickel, copper, brass, and steel, and in the quantitative determination of potassium, with which it forms slightly soluble potassium perchlorate.

The **perchlorates** of sodium and potassium are produced commercially by the prolonged electrolysis of hot solutions of their chlorides. The final step in the process is one of anodic oxidation of the chlorate ion according to the equation

$$ClO_3^- + H_2O \longrightarrow ClO_4^- + 2H^+ + 2e^-$$

The structure of the tetrahedral perchlorate ion is shown in Fig. 19-14.

▶ **2. PERBROMIC ACID AND ITS SALTS.** All attempts to prepare perbromic acid, $HBrO_4$, and its salts failed until as recently as 1968. That the acid would be so difficult to prepare was unexpected and puzzling, especially since analogous compounds of both chlorine and iodine had been known for some time and methods of preparing oxides of bromine had been discovered. Finally, in 1968, perbromate ion, BrO_4^-, was prepared by the oxidation of bromate ion with xenon difluoride, XeF_2, in aqueous solution. Later, crystalline rubidium perbromate was isolated and found to be stable at room temperature. Shortly after the successful preparation of the perbromate ion, perbromic acid in aqueous solution was prepared by acidifying rubidium perbromate with sulfuric acid. Perbromic acid is a stable compound in dilute solution at room temperature. It is easily decomposed by heat.

▶ **3. PERIODIC ACIDS AND THEIR SALTS.** Several series of periodic acids and periodates are known. Most important of these are **metaperiodic** acid, HIO_4, and **paraperiodic** acid, H_5IO_6, and their salts. The size of the iodine atom is sufficient to

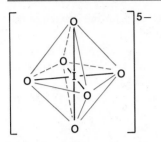

Figure 19-15 The octahedral structure of the paraperiodate ion, IO_6^{5-}.

allow space for six oxygen atoms around it (Fig. 19-15), whereas the smaller chlorine atom can accommodate no more than four oxygen atoms. The periodic acids are strong acids and active oxidizing agents. The oxidizing properties of HIO_4 make it useful in numerous analytical procedures.

Sodium metaperiodate, $NaIO_4$, may be prepared by the oxidation of sodium iodate, $NaIO_3$, with chlorine in a hot alkaline solution. Periodates may also be prepared by the electrolytic oxidation of iodates, a method analogous to the preparation of perchlorates. The free paraperiodic acid, H_5IO_6, may be obtained as a white solid by evaporating a filtered solution resulting from the reaction of barium periodate with sulfuric acid. Salts of paraperiodic acids such as $Na_2H_3IO_6$, $Na_3H_2IO_6$, and Ag_5IO_6 have been prepared.

Questions

1. Arrange the halogens in order of increasing (a) atomic radii, (b) ionic radii, (c) electronegativity, (d) boiling points, (e) oxidizing activity, and (f) depth of color.
2. Describe three properties of fluorine and iodine that show that iodine is less nonmetallic than fluorine.
3. Why is it difficult to prepare elemental fluorine? How is its preparation accomplished?
4. Why must the chlorine and sodium hydroxide resulting from the electrolysis of brine be kept separate in the manufacture of chlorine?
5. Describe the chemistry of the extraction of bromine from sea water.
6. Describe the commercial production of iodine from sodium iodate.
7. Write equations showing the action of chlorine upon lithium, calcium, hydrogen, water, sulfur, and iodine.
8. What product is formed when phosphorus is burned in a limited supply of chlorine? in an excess of chlorine?
9. Show by means of changes in oxidation state that the reaction of chlorine with water is a disproportionation reaction.
10. Illustrate the gradation of properties shown by the family of halogens with regard to three physical properties and two chemical properties.
11. If chlorine, iodine, a chloride, an iodide, and a starch suspension were available to you, how would you demonstrate the relative activity of chlorine and iodine?
12. Would you expect astatine to oxidize iodide ions to form elemental iodine? Why or why not?

13. How can fluorine, which is an extremely powerful oxidizing agent, be contained and stored?
14. Why is iodine monochloride (ICl) more polar than iodine monobromide (IBr)?
15. What products would you expect if BrCl gas were dissolved in water? Write a balanced chemical equation to describe the reaction.
16. Write chemical equations describing:
    (a) The burning of sulfur in a chlorine atmosphere.
    (b) The reaction of bromine with water.
    (c) The dissolution of iodine in potassium iodide solution.
    (d) The hydrolysis of PCl_3.
    (e) The electrolysis of fused (melted) sodium chloride.
    (f) The preparation of hydrogen fluoride by the reaction of calcium fluoride and sulfuric acid.
    (g) The oxidation of sulfide ion by elemental iodine.
    (h) The disproportionation reaction of ClO_2 with a basic solution.
    (i) The anodic oxidation of chlorate ion to perchlorate ion.
    (j) The light-induced decomposition of HClO.
    (k) The reduction of sulfate ion by bromide ion in acid solution.
    (l) The dissolution of HI in water.
17. Write equations for the interaction of hydrochloric acid and (a) MnO_2, (b) $KMnO_4$, and (c) $Na_2Cr_2O_7$.
18. Why can pure HBr and HI not be prepared by the action of concentrated sulfuric acid upon metal bromides and iodides, respectively? Write the equations involved.

19. When a hydrogen halide functions as a reducing agent, which element in the molecule is oxidized? Compare HI, HBr, HCl, and HF with regard to their strengths as reducing agents.

20. Describe the unique properties of hydrogen fluoride and hydrofluoric acid.

21. Describe the hydrogen bond. Between what kind of atoms are hydrogen bonds formed?

22. Why does hydrogen bonding have an effect on such properties as boiling point, vapor pressure, and heat of vaporization?

23. Why is a doubly charged chloride ion not observed?

24. Show by an equation that the hydrogen ion acts as an oxidizing agent toward metals above hydrogen in the activity series.

25. Contrast the chemical and physical properties of ionic and covalent binary metal halides. Give typical examples of each of these classes of halides.

26. Write equations for the reaction of hydrochloric acid with Ca, CaO, $Ca(OH)_2$, and $CaCO_3$.

27. Why is OF_2 called oxygen difluoride rather than fluorine oxide?

28. Two fluorides of sulfur are stable at room temperature, SF_4 and SF_6. How would you prepare these sulfides from sulfur and fluorine?

29. Write the formulas for five compounds in which chlorine is linked by covalent bonds; by ionic bonds.

30. Show by suitable equations that chlorine monoxide and chlorine heptaoxide are acid anhydrides.

31. What is the oxidation state of bromine in each of the oxides listed in Table 19-5?

32. Write the valence electronic structures for the oxyacids of chlorine. Calculate the oxidation state of chlorine in each of these acids.

33. Write the formulas and names for all the oxyacids of chlorine and the corresponding calcium salts.

34. Write equations for the preparation of each of the oxyacids of chlorine and its sodium salt.

35. An aqueous solution of NaClO is basic while an aqueous solution of $NaClO_4$ is neutral. Explain.

36. Why is perchloric acid distilled under reduced pressure?

37. Predict the products in each of the following cases:
    (a) A solution of hypochlorous acid is heated.
    (b) Potassium sulfite is treated with potassium periodate in dilute sulfuric acid solution.
    (c) Oxygen difluoride, OF_2, is brought in contact with powdered aluminum.

38. Which is the stronger acid, $HClO_3$ or $HBrO_3$? Why?

39. Which is the stronger acid, $HClO_2$ or $HClO_3$? Why?

40. Compare the structure of ICl_4^- ion with that of the IO_4^- ion and explain why they differ.

41. List as many products as you can that can be obtained by the electrolysis of aqueous halide solutions. Explain the general procedure involved in the production of each of the substances listed.

Problems

1. A sample of cadmium chloride with a mass of 1.776 g gave 1.089 g of cadmium upon electrolysis. Determine the formula of cadmium chloride.
    Ans. $CdCl_2$

2. Chlorine is composed of two isotopes, ^{35}Cl (atomic mass 34.96885 amu) and ^{37}Cl (atomic mass 36.96590 amu). If a naturally occurring sample of chlorine is 24.229% ^{37}Cl, calculate the average atomic weight of a chlorine atom?   *Ans. 35.453*

3. What mass of PCl_3 can be prepared by the reaction of chlorine with 13.55 g of phosphorus, P_4?
    Ans. 60.08 g

4. What mass of calcium carbonate can be converted to calcium chloride by 0.500 l of 1.50 M hydrochloric acid?   *Ans. 37.5 g*

5. What mass of HI can be prepared from the hydrolysis of 1.00 kg of phosphorus(III) iodide?
    Ans. 932 g

6. What volume (in liters) of 0.1191 M HBr solution is required to react with 1.472 g of ZnO to give $ZnBr_2$?
    Ans. 0.3037 l

7. What volume of HBr(g) at 25°C and 0.977 atm would be required to prepare the 0.3037 l of 0.1191 M HBr solution used in Problem 6?
    Ans. 0.906 l

8. Titanium metal was treated with HF(g) at 250°C for

2 days. A product that contained 45.67% titanium and 54.38% fluorine was obtained. Write the balanced equation for the reaction.

$$Ans.\ 2Ti + 6HF \longrightarrow 2TiF_3 + 3H_2$$

9. Iodine reacts with liquid chlorine at $-40°C$ to give an orange compound containing 54.5% iodine and 45.7% chlorine. Write the balanced equation for the reaction. $Ans.\ I_2 + 3Cl_2 \longrightarrow 2ICl_3$

10. What volume of HCl gas measured at $100°C$ and 1.00 atm can be prepared from 500 g of sodium chloride?   *Ans. 262 ℓ*

11. Show by calculation which hydrogen halide diffuses 0.395 times as fast as HF.   *Ans. HI*

12. How many kilograms of sodium iodate are required to prepare 1.00 kg of iodine by reduction with sodium hydrogen sulfite?   *Ans. 1.56 kg*

13. (a) What mass of sodium dichromate ($Na_2Cr_2O_7$) would be required to react with 50 g of sodium chloride in acid solution? (b) What volume of chlorine is produced by this reaction at standard conditions of temperature and pressure?

Ans. (a) 37 g; (b) 9.6 ℓ

14. The reaction of vanadium(III) oxide, V_2O_3, with chlorine gives a volatile yellow liquid that contains 29.42% vanadium, 61.3% chlorine, and the remainder oxygen. A sample of the liquid with a mass of 0.433 g had a pressure of 390 torr at $19°C$ in a flask with a volume of 115 ml. What is the molecular formula of the compound?   *Ans. $VOCl_3$*

15. What is the molarity of a solution of hydrobromic acid with a density of 1.48 g/cm^3 and which is 47.5% HBr by mass?   *Ans. 8.69 M*

16. Calculate the quantity of sodium chloride that must react completely with sulfuric acid to produce sufficient hydrogen chloride to make 350 ml of 40% hydrochloric acid solution (sp. gr. 1.2).

Ans. 2.7×10^2 g

17. What volume of hydrochloric acid (40.0% by mass; sp. gr. 1.20) would be required to dissolve 250 g of magnesium?   *Ans. 1.56 ℓ*

18. From the data in Appendix J, calculate the bond energies of F_2, Cl_2, Br_2, and I_2.

Ans. 157.98; 243.36; 192.85; 151.24 kJ mol^{-1}

19. From the data in Appendix J, calculate the bond energies of HF, HCl, HBr, and HI.

Ans. 568; 431.95; 366.25; 298.32 kJ mol^{-1}

20. Using free energies of formation from Appendix J, determine which of the hydrogen halides form spontaneously from the reaction of the gaseous elements.   *Ans. HF; HCl; HBr; HI*

21. What is the calculated freezing point of a solution of 1.19 g of NaCl in 100 g of water? ($K_f = 1.86$ for water)   *Ans. $-0.758°$*

22. At equilibrium, at 1375 K, a gas mixture exhibits a partial pressure of HBr of 0.998 atm, of H_2 of 3.82×10^{-3} atm, and of Br_2 of 3.82×10^{-3} atm. What is K_p for the reaction $H_2 + Br_2 \longrightarrow 2HBr$?

Ans. 6.83×10^4

23. What is the molar solubility of TlCl for which the solubility product constant is 1.8×10^{-4}?

Ans. 1.3×10^{-2} M

24. Predict the molecular structures of the following molecules or ions: IF_5; ClO_3^-; $SnBr_4$; $SeCl_3^+$; ICl_4^-.

Ans. square pyramid; trigonal pyramid; tetrahedron; trigonal pyramid; square plane

References

"Chlorine" (Staff), *Chem. and Eng. News,* February 26, 1979, p. 11.

"Carl Wilhelm Scheele and the Discovery of Chlorine," T. I. Williams, *Endeavour,* May 1974, p. 54.

"The Naming of Fluorine," W. H. Waggoner, *J. Chem. Educ.,* **53,** 27 (1976).

"Chlor-Alkali Membrane Cell" (Staff), *Chem. and Eng. News,* March 20, 1978, p. 20.

"Chlorine and Hydrochloric Acid: They Touch Our Lives Daily" (Staff), *J. Chem. Educ.,* **55,** 466 (1978).

"The Strength of Hydrohalogenic Acids," L. Pauling, *J. Chem. Educ.,* **53,** 762 (1976).

"Liquid HF Solvent Facilitates Novel Reactions" (Staff), *Chem. and Eng. News,* February 23, 1970, p. 42.

"Humphry Davy and the Elementary Nature of Chlorine," R. Siegfried, *J. Chem. Educ.,* **36,** 568 (1959).

"Chemical Lasers" (Concerns use of chlorine and iodine), G. C. Pimentel, *Sci. American,* April 1966, p. 32.

"Principles of Halogen Chemistry," R. T. Sanderson, *J. Chem. Educ.,* **41,** 361 (1964).

"Fluorine Chemistry," A. K. Barbour, *Chem. in Britain,* **5,** 250 (1969).

"Chlorine Trifluoride," N. V. Steere, *J. Chem. Educ.,* **44,** A1057 (1967).

"Thermochemistry of Hypochlorite Oxidations," M. J. Bigelow, *J. Chem. Educ.,* **46,** 378 (1969).

"Nucleophilic Reactivities of the Halide Anions," M. S. Puar, *J. Chem. Educ.,* **47,** 473 (1970).

"Orbital Symmetry in Photochemical Transformations," H. Katz, *J. Chem. Educ.,* **48,** 84 (1971).

"Chlorinated Hydrocarbons and the Environment," G. McConnell, D. M. Ferguson, and C. R. Pearson, *Endeavour,* January, 1975, p. 13.

"Chemistry in Oral Health" (Staff), *J. Chem. Educ.,* **55,** 736 (1978).

"Chlorine in Your Water," J. G. Smith, *J. Chem. Educ.,* **52,** 656 (1975).

20

Electrochemistry and Oxidation-Reduction

Electrochemistry is that field of chemistry that deals with chemical changes produced by an electric current and with the production of electricity by chemical reactions. In **electrolytic cells,** electrical energy is used to bring about desired chemical changes. In **voltaic cells,** chemical reactions are used to produce electrical energy. In both types of cells, oxidation occurs at one electrode (the anode) and reduction at the other electrode (the cathode).

Much of our quantitative knowledge concerning the energy changes involved in chemical reactions (see Chapter 18, Chemical Thermodynamics) has come from the study of electrochemistry, because the quantity of electrical energy produced or consumed during electrochemical changes can be measured very accurately. Furthermore, an understanding of the reactions that take place at the electrodes of electrochemical cells throws much light on the process of oxidation and reduction and chemical activity.

Many electrochemical processes are important in science and industry. The use of electrical energy in the commercial production of hydrogen, oxygen, ozone, hydrogen peroxide, chlorine, sodium hydroxide, and the oxygen compounds of the halogens has been noted in previous chapters. Additional technical applications of electrochemistry include production of many other chemicals, electrorefining of metals, electroplating of metals and alloys, the production of metal articles by electrodeposition, and the generation of electrical current by battery cells.

Figure 20-1 An electrolytic cell being used to determine the conductivity of a solution.

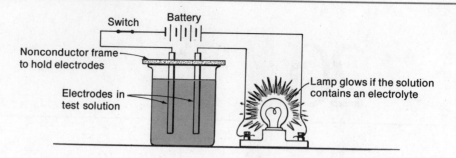

20.1   The Conduction of Electricity

A current of electricity is the movement of electrically charged particles through a conductor. These charged particles may be either electrons or ions (positive or negative). An electric current in a metal conductor is the movement of electrons through the metal without any obvious chemical changes in the metal and with no similar movement of the atoms composing the metal. This type of conduction of electricity is called **metallic conduction.**

Certain compounds, usually fused or in aqueous solution but sometimes as solids, conduct an electric current when placed in an electrolytic cell (Fig. 20-1). These compounds, which conduct an electric current by the movement of ions, are called **electrolytes.** Two electrodes, which are good conductors, dip into the electrolyte. Electrons are forced onto one of the electrodes and withdrawn from the other electrode by a battery or generator. The electrode onto which the electrons are forced becomes negatively charged, and the electrode from which electrons are withdrawn becomes positively charged. The positive ions (cations) in the electrolyte are attracted by the negatively charged cathode and move toward it, and the negative ions (anions) are attracted by the positively charged anode and move toward it. This sort of motion of the ions through an electrolyte is known as **ionic,** or **electrolytic, conduction.** Electrolytic conduction in solutions of strong and weak electrolytes was described more fully in Section 13.11.

Electrolytic Cells

20.2   The Electrolysis of Molten Sodium Chloride

A consideration of the reactions that take place at the electrodes will help explain the way in which the electric current passes between the electrodes and through the electrolyte.

Fused sodium chloride consists of equal numbers of positive sodium ions and negative chloride ions that can move about with considerable freedom. When the cell is connected to a battery, sodium ions are attracted to the cathode, where they combine with electrons on the cathode to form sodium atoms, i.e., metallic sodium. The electrode reaction, referred to as a **half-reaction,** may be represented for the cathode by

$$Na^+ + e^- \longrightarrow Na$$

The sodium decreases in oxidation state from $+1$ to 0. This is a reduction reaction and is sometimes referred to as cathodic reduction. *The cathode is the electrode at which reduction occurs.* The chloride ions of the fused NaCl give up one electron each to the anode and become chlorine atoms, which then combine with each other to form molecules of chlorine gas. The anode half-reaction is

$$2Cl^- \longrightarrow Cl_2 + 2e^-$$

The chlorine is oxidized from -1 to 0, and the reaction is known as anodic oxidation. *The anode is the electrode at which oxidation occurs.*

The net reaction for the electrolytic decomposition of the fused sodium chloride is the sum of the two electrode half-reactions. To balance the equation, the cathode reaction must be multiplied by 2 so that the number of electrons produced by the oxidation is equal to the number of electrons consumed in the reduction.

$$
\begin{array}{ll}
\textit{Cathode:} & 2Na^+ + 2e^- \longrightarrow 2Na \\
\textit{Anode:} & \underline{2Cl^- \longrightarrow Cl_2 + 2e^-} \\
& 2Na^+ + 2Cl^- \longrightarrow 2Na + Cl_2 \\
\end{array}
$$

or

$$2NaCl \longrightarrow 2Na + Cl_2$$

The production of an oxidation-reduction reaction by a direct current (dc) of electricity is known as **electrolysis.**

20.3   The Electrolysis of Aqueous Hydrochloric Acid

Let us consider now the reactions that occur at the electrodes of an electrolytic cell when an aqueous solution of fairly concentrated hydrochloric acid is electrolyzed. The positive hydrogen ions of the hydrochloric acid are attracted to and reduced at the cathode by accepting electrons from it.

$$2H^+ + 2e^- \longrightarrow H_2 \quad \text{(cathodic reduction)}$$

The negative chloride ions lose electrons to the anode and thus are oxidized to chlorine gas.

$$2Cl^- \longrightarrow Cl_2 + 2e^- \quad \text{(anodic oxidation)}$$

The overall cell reaction may be obtained by adding the electrode reactions.

$$
\begin{array}{ll}
\textit{Cathode:} & 2H^+ + 2e^- \longrightarrow H_2 \\
\textit{Anode:} & \underline{2Cl^- \longrightarrow Cl_2 + 2e^-} \\
& 2H^+ + 2Cl^- \longrightarrow H_2 + Cl_2 \\
\end{array}
$$

20.4   The Electrolysis of an Aqueous Sodium Chloride Solution

Suppose we consider the electrode reactions that take place during the electrolysis of an aqueous solution of sodium chloride (Fig. 20-2). If the difference in electrical potential (Section 20.8) between the anode and cathode is sufficiently great and the solution is fairly concentrated, chlorine is formed at the anode, hydrogen is evolved at the cathode, and a test of the solution around the cathode with

Figure 20-2 The electrolysis of an aqueous solution of NaCl. The net reaction produces hydrogen gas and hydroxide ions at the cathode and chlorine gas at the anode, leaving a solution of sodium hydroxide.

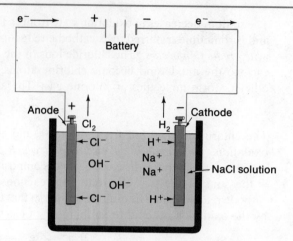

phenolphthalein (the phenolphthalein solution turns red) indicates that hydroxyl ion is also being formed at the cathode. In order to account for these products it is necessary to consider all the possible electrode reactions.

The possible cathodic (reduction) reactions are the reduction of sodium ion or the reduction of water:

$$Na^+ + e^- \longrightarrow Na \tag{1}$$

$$2H_2O + 2e^- \longrightarrow H_2 + 2OH^- \tag{2}$$

The water is more easily reduced than the sodium ion, so Reaction (2) occurs and hydrogen gas forms at the cathode. The hydroxide ions, which are also formed in the vicinity of the cathode, begin to migrate toward the anode.

The possible anodic (oxidation) reactions are either the oxidation of chloride ions or the oxidation of water:

$$2Cl^- \longrightarrow Cl_2 + 2e^- \tag{3}$$

$$2H_2O \longrightarrow O_2 + 4H^+ + 4e^- \tag{4}$$

Chloride ion and water are oxidized with almost equal ease. Hence concentration of the salt solution plays the dominant role in determining the product. In a concentrated sodium chloride solution, Reaction (3) takes place and chlorine is formed at the anode. If the concentration of chloride ion is low, Reaction (4) can take place to some extent also and some oxygen may be formed in addition to the chlorine. Little chlorine is produced when the solution of sodium chloride is very dilute.

The net reaction for the electrolysis of concentrated aqueous sodium chloride can be obtained by adding the electrode reactions.

$$
\begin{array}{ll}
Cathode: & 2H_2O + 2e^- \longrightarrow H_2 + 2OH^- \\
Anode: & \underline{2Cl^- \longrightarrow Cl_2 + 2e^-} \\
& 2H_2O + 2Cl^- \longrightarrow H_2(g) + Cl_2(g) + 2OH^-
\end{array}
$$

Note that hydroxide ions are formed during the electrolysis and that chloride ions are removed from solution. Because the sodium ions remain unchanged, sodium hydroxide accumulates in the cell as electrolysis proceeds.

The electrolysis of aqueous sodium chloride is an important commerical process for the production of hydrogen, chlorine, and sodium hydroxide (Section 19.1).

20.5 The Electrolysis of an Aqueous Sulfuric Acid Solution

The ionization of sulfuric acid in water proceeds stepwise as follows:

$$H_2SO_4 \longrightarrow H^+ + HSO_4^- \quad \text{(essentially complete)}$$
$$HSO_4^- \rightleftharpoons H^+ + SO_4^{2-}$$

In dilute solution the anion furnished by the sulfuric acid is principally the sulfate ion. On electrolysis, the possible choices for the anode reaction are either oxidation of the sulfate ion or oxidation of water.

$$2SO_4^{2-} \longrightarrow S_2O_8^{2-} + 2e^-$$
$$2H_2O \longrightarrow O_2(g) + 4H^+ + 4e^-$$

In practice oxygen and hydrogen ions are formed at the anode because the water molecule is more easily oxidized than the sulfate ion.

The only possible cathode reaction is

$$2H^+ + 2e^- \longrightarrow H_2$$

The net reaction is the sum of the anode and cathode reactions after multiplying the cathode reaction by 2 in order to balance the number of electrons involved in the oxidation and reduction.

$$\begin{aligned} \textit{Anode:} \quad & 2H_2O \longrightarrow O_2 + 4H^+ + 4e^- \\ \textit{Cathode:} \quad & 4H^+ + 4e^- \longrightarrow 2H_2 \\ \hline & 2H_2O(l) \longrightarrow 2H_2(g) + O_2(g) \end{aligned}$$

Hence the electrolysis of an aqueous solution of sulfuric acid is the same as the electroylsis of water (see Section 9.2). Although the hydrogen ion from the sulfuric acid is consumed by cathodic reduction, this same ion is being generated at the anode at the same rate at which it disappears at the cathode. The net result is that water is used up and the sulfuric acid becomes more concentrated as electrolysis proceeds.

20.6 Electrolytic Refining of Metals

In the few examples of electrolysis we have considered thus far, the electrodes used in the electrolytic cells were made of electrically conducting materials, such as graphite or platinum, which are unreactive under the conditions employed. Whenever a metal is made the anode of an electrolytic cell, the anodic oxidation may involve oxidation of the metal composing the electrode. For example, in the electrolysis of a solution of copper(II) sulfate (Fig. 20-3) with a strip of metallic copper as the anode, the following anodic electrode reaction takes place because the oxidation of copper is easier than the oxidation of water:

$$Cu \longrightarrow Cu^{2+} + 2e^-$$

Figure 20-3 The electrolytic refining of copper.

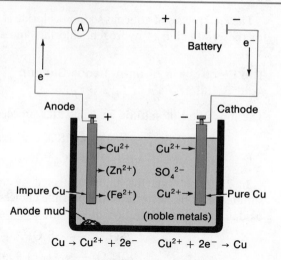

$$Cu \rightarrow Cu^{2+} + 2e^- \qquad Cu^{2+} + 2e^- \rightarrow Cu$$

Thus the electrode itself goes into solution as copper(II) ions. The copper(II) ions formed at the anode migrate to the cathode, where they are more readily reduced than the water.

$$Cu^{2+} + 2e^- \longrightarrow Cu$$

Thus pure copper plates out on the cathode.

This process is the basis of the electrolytic refining of crude copper. When impure copper is the anode, less active impurities in the copper, such as gold and silver, do not dissolve (are not oxidized) and fall to the bottom of the cell, forming a "mud" from which they are readily recovered. The more active metals, such as zinc and iron, which may be present in the crude copper, are oxidized and go into the solution as ions. If the electrical potential between the electrodes is carefully regulated, these metallic ions are not reduced at the cathode; only the copper is reduced and deposited. Deposition of the copper in pure form is made either upon a thin sheet of pure copper, which serves as a cathode, or upon some other metal from which the deposit may be stripped.

20.7   Faraday's Law of Electrolysis

In 1832–1833 Michael Faraday found that *the quantity of substances undergoing chemical change at each electrode during electrolysis is directly proportional to the quantity of electricity that passes through the electrolytic cell* (Fig. 20-4).

The quantity of electricity can be expressed as number of electrons. The quantity of a substance undergoing chemical change at an electrode is related to the number of electrons involved in the change and can be expressed in terms of the moles of substance or in terms of the **equivalent weight** of the substance. The equivalent weight of a substance is the mass of the substance, in grams, that combines with or releases one mole of electrons. A liter of solution that contains one equivalent of a substance is a **one-normal solution** (Section 14.12). The **normality**, N, of a solution is the number of equivalents of solute per liter of solution (equiv solute/ℓ solution).

Figure 20-4 Amounts of various metals discharged at the cathode by 1 faraday (96,487 coulombs) of electricity (one mole of electrons).

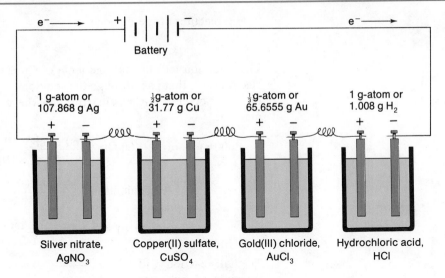

Experimental results show that one electron reduces one silver ion, while two electrons are required to reduce one copper(II) ion.

$$Ag^+ + 1e^- \longrightarrow Ag$$
$$Cu^{2+} + 2e^- \longrightarrow Cu$$

Therefore 6.022×10^{23} electrons reduce 6.022×10^{23} silver ions (one mole) to silver atoms. The mass of silver produced is 107.868 g; thus the equivalent weight of silver is the same as the atomic weight. Because two electrons are required to reduce one Cu^{2+} ion to atomic copper, 6.022×10^{23} electrons would reduce only one-half of a mole of copper. One-half of the atomic weight would be the equivalent weight of this ion.

The term **faraday** refers to the quantity of electricity (electrons) that will reduce one equivalent weight of a substance at the cathode and oxidize one equivalent weight of a substance at the anode, which amounts to the gain or loss of one mole of electrons, or 6.022×10^{23} electrons.

$$1 \text{ faraday} = 6.022 \times 10^{23} \text{ electrons} = \text{one mole of electrons}$$
$$= 96,487 \text{ coulombs (C)}$$

$$1 \text{ coulomb} = \text{quantity of electricity involved when a current}$$
$$\text{of 1 ampere (A) flows past a given point for 1 second}$$
$$= 1 \text{ A sec}$$

Calculations involving passage of an electric current through a cell may be handled like any other stoichiometry problem: Convert the current in amperes to coulombs, then to faradays, and, finally, to moles of electrons. Use the moles of electrons as you would moles of any other reactant.

EXAMPLE   **Let us calculate the mass of copper produced by the cathodic reduction of Cu^{2+} ion during the passage of 1.600 A of current through a solution of copper(II) sulfate for 1.000 hr.**

The reaction at the cathode is

$$Cu^{2+} + 2e^- \longrightarrow Cu$$

This equation tells us that one mole of Cu is produced for each two moles of electrons, e^-, that pass through the solution (1 mol Cu/2 mol e^-). The number of moles of electrons passing through the cell is obtained in the following manner:

$$1.600 \text{ amperes} \times 1.000 \text{ hr} \times \frac{60 \text{ min}}{1 \text{ hr}} \times \frac{60 \text{ sec}}{1 \text{ min}} = 5760 \text{ coulombs}$$

$$5760 \text{ coulombs} \times \frac{1 \text{ faraday}}{96,487 \text{ coulombs}} = 5.970 \times 10^{-2} \text{ faraday}$$

$$5.970 \times 10^{-2} \text{ faraday} \times \frac{1 \text{ mol } e^-}{1 \text{ faraday}} = 5.970 \times 10^{-2} \text{ mol } e^-$$

The half-reaction ($Cu^{2+} + 2e^- \longrightarrow Cu$) is used to relate moles of electrons to the mass of copper produced.

$$5.970 \times 10^{-2} \text{ mol } e^- \times \frac{1 \text{ mol Cu}}{2 \text{ mol } e^-} = 2.985 \times 10^{-2} \text{ mol Cu}$$

$$2.985 \times 10^{-2} \text{ mol Cu} \times \frac{63.55 \text{ g Cu}}{1 \text{ mol Cu}} = 1.897 \text{ g Cu}$$

◆ ◆ ◆ ◆ ◆

Electrode Potentials

In dealing with electricity we need to introduce the concept of **electrical potential.** The electrical potential is the energy change involved in the transfer of one coulomb of electricity from one location to another. The unit of electrical potential is the volt (V), which is one joule per coulomb. If an energy change of one joule accompanies the transfer of one coulomb of electricity, then we say that the potential difference is one volt (1 volt = 1 joule per coulomb). The energy changes associated with the operation of electrolytic and voltaic cells are expressed in volts.

20.8   Single Electrodes and Electrode Potentials

When a metal such as zinc is placed in water, there is a tendency for the atoms to leave the metal and pass into solution as ions. The change may be expressed for zinc by

$$Zn \longrightarrow Zn^{2+} + 2e^-$$

A few zinc ions accumulate in the solution, but the valence electrons of those that do remain behind on the metal. The zinc strip thereby becomes negatively charged, and the solution becomes positive. Thus a difference in electrical potential is established between the metal and the solution. On the other hand, the negative charge that has been generated on the zinc attracts the positive zinc ions in solution and tends to reduce them to the metallic state, according to the equation

$$Zn^{2+} + 2e^- \longrightarrow Zn$$

Also, an increase in the concentration of the zinc ions in the solution, brought about by the addition of a zinc salt, tends to reverse the process. Thus the change is reversible, as indicated by double arrows, and an equilibrium can be established.

$$Zn \rightleftharpoons Zn^{2+} + 2e^-$$

At equilibrium, the difference in electrical potential between the metal and the solution depends on (a) the tendency of the metal to form ions in aqueous solution, (b) the concentration of the metal ions in solution, and (c) the tendency of the metal ions to be reduced to the metal. The more active the metal (the more easily it is oxidized) the greater is its tendency to form ions and the more negative is the potential of the electrode with respect to the solution. A zinc electrode is thus more negative than a hydrogen electrode (see next section). Both the hydrogen electrode and zinc electrode are more negative than a copper electrode, because zinc is more easily oxidized than hydrogen and hydrogen is more easily oxidized than copper.

As mentioned previously, the size of the electrode potential varies with the concentration of the reactants. Thus the potential of the zinc electrode can be varied by changing the concentration of the zinc ions in solution. With an increase in the zinc ion concentration, the equilibrium $Zn \rightleftharpoons Zn^{2+} + 2e^-$ shifts to the left, and the difference in potential between the metal and the solution decreases. A decrease in the concentration of zinc ions shifts the equilibrium to the right, and the potential difference increases. Since the zinc electrode is a solid, its concentration is constant, and the size of the electrode does not influence the electrode potential. To be precise, we should use the term *activity* rather than *concentration* of zinc ion (Section 13.27). However, the approximation that the activity of an ionic species is equal to its concentration will be satisfactory for our purposes.

20.9 Standard Electrode Potentials

No satisfactory method has been devised for measuring the exact difference in electrical potential between a single metal electrode and a solution of its ions. However, the difference between the potentials of two electrodes can be readily measured. We therefore compare the potentials of cells in which one of the electrodes is always the same, and this electrode is used as a standard to which the potentials of all other electrodes are compared. The voltage between this electrode and the solution is arbitrarily taken to be zero. (This is similar to establishing an arbitrary elevation as sea level, which we call zero.) The standard hydrogen electrode (Fig. 20-5) is the electrode chosen for comparison; its potential is defined as zero volt.

Electrodes involving gases can be set up by bubbling the gas around an inert metallic conductor that will conduct electrons but will not itself enter the electrode reaction. For example, the hydrogen electrode can be constructed so that hydrogen gas is bubbled around a strip or wire of platinum covered with very finely

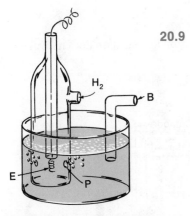

Figure 20-5 Diagram of a simple standard hydrogen electrode. E, platinized platinum helix; P, port for escape of hydrogen bubbles; B, part of salt bridge.

divided platinum and immersed in a solution containing hydrogen ions. There is also a tendency for the hydrogen ion to be reduced to hydrogen by acquiring electrons from the electrode. Thus a difference in potential between the electrode and the solution is established in the same manner as described for zinc.

$$H_2 \rightleftharpoons 2H^+ + 2e^-$$

The platinum acts as an electrode and as a catalyst that quickly brings the reaction to equilibrium.

With gas electrodes, the potential increases with an increase in the partial pressure (activity) of the gas. Therefore the **standard hydrogen electrode** is prepared by bubbling hydrogen gas at a temperature of 25 °C and a pressure of 1 atm around platinized platinum immersed in a solution containing hydrogen ions at a concentration of 1 M. We arbitrarily assign the value zero to the potential for the standard hydrogen electrode.

20.10   Measurement of Electrode Potentials

To measure the electrode potential of zinc, for example, an electrochemical cell is set up consisting of a zinc rod in contact with a 1 M solution of zinc ions as one-half of the cell and a standard hydrogen electrode as the other half (Fig. 20-6).

The two halves of the cell are connected by a **salt bridge** consisting of a saturated solution of potassium chloride in a gel of agar-agar. This bridge permits K^+ and Cl^- ions to migrate through it freely from cell to cell and thus permits a current to flow, but it does not allow the two solutions to mix. The zinc rod in the Zn^{2+} solution and the platinum strip of the hydrogen electrode are connected to a voltmeter by wire. The potential difference, in volts, between the two electrodes is read on the voltmeter.

Before the circuit is closed, i.e., before the zinc and platinum strips are connected by wire, we have the following electrode half-reactions, each at equilibrium.

$$H_2 \rightleftharpoons 2H^+ + 2e^-$$
$$Zn \rightleftharpoons Zn^{2+} + 2e^-$$

Because zinc has a greater tendency to go into solution as ions than does hydrogen,

Figure 20-6 Measurement of the electrode potential of zinc, using the standard hydrogen electrode as reference.

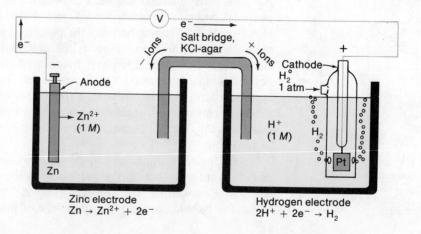

Zinc electrode
$Zn \rightarrow Zn^{2+} + 2e^-$

Hydrogen electrode
$2H^+ + 2e^- \rightarrow H_2$

the zinc strip acquires a higher electron density (from the electrons released as the zinc ions are formed) than does the platinum of the hydrogen electrode. Therefore, when the electrical circuit is closed the greater electron density, or electron pressure, at the zinc electrode causes electrons to flow from the zinc electrode to the hydrogen electrode through the wire connecting the two electrodes. Then, as the electron density increases at the hydrogen electrode, the equilibrium is shifted to the left, causing hydrogen ions to be reduced to free hydrogen. The reaction that occurs at the hydrogen electrode as a current flows is

$$2H^+ + 2e^- \longrightarrow H_2$$

At the same time, the electron density at the zinc electrode decreases as the result of the flow of electrons to the hydrogen electrode, and thus the equilibrium at the zinc electrode is shifted to the right, causing more zinc to be oxidized to zinc ions. The reaction that occurs at the zinc electrode as a current flows is

$$Zn \longrightarrow Zn^{2+} + 2e^-$$

The difference in potential between the two half-cells as measured by a voltmeter is 0.76 V. The potential that causes electrons to move from the zinc electrode to the hydrogen electrode is called the **electromotive force (emf)** and is 0.76 V. The zinc is the negative electrode because its electron density is higher than that of the hydrogen electrode. If we assign the entire electromotive force of the cell to the zinc electrode (the potential of the hydrogen electrode is assumed to be zero), then we may say that the potential of the zinc electrode for the oxidation of zinc metal to zinc ion is 0.76 V above that of the hydrogen electrode (which, by common agreement, is taken to be zero) and is positive ($+0.76$ V). For the reverse reaction, the reduction of zinc ion to metallic zinc,

$$Zn^{2+} + 2e^- \longrightarrow Zn$$

the electrode potential has the same absolute value but opposite sign and therefore is -0.76 V.

   It is customary, in accord with the official recommendation of the International Union of Pure and Applied Chemistry (IUPAC), to express the values of electrode potentials in terms of the reduction process and to refer to them as standard electrode potentials, or standard reduction potentials. The symbol E° is used for standard electrode potentials; that is, potentials at electrodes in which all components are in standard states [ion concentrations (activities) of one molar and gas pressures (activities) of one atmosphere]. Hence the standard electrode potential, E°, for zinc is -0.76 V. The negative sign corresponds to the negative charge assumed by the zinc electrode as its electron density increases. The standard electrode potential for any metal electrode having an electron density greater than the hydrogen electrode is given a negative sign. An electrode that has less electron density than the hydrogen electrode (copper, for example) is positive with reference to the hydrogen electrode and hence is given a positive sign. It should be noted, however, that the assignment of positive or negative signs to the standard electrode potentials is arbitrary, and some scientists follow the opposite convention, defining standard electrode potentials in terms of the oxidation process and hence assigning the opposite signs to standard electrode potentials.

The overall reaction that takes place when the zinc-hydrogen cell is operating is one of oxidation and reduction.

$$\begin{array}{ll} \mathrm{Zn} \longrightarrow \mathrm{Zn^{2+}} + 2e^- & \textit{Anodic oxidation} \\ \underline{2\mathrm{H^+} + 2e^- \longrightarrow \mathrm{H_2}} & \textit{Cathodic reduction} \\ \mathrm{Zn} + 2\mathrm{H^+} \longrightarrow \mathrm{Zn^{2+}} + \mathrm{H_2}(g) & \textit{Cell reaction} \end{array}$$

This same reaction will take place when zinc metal is immersed in a solution of hydrochloric acid. However, in the zinc-hydrogen cell, the reaction takes place without contact of the reactants. The electrons involved in the oxidation-reduction reaction are transferred from the zinc to the hydrogen ions through a wire. This is important from a practical point of view because the current can be used to do work in an electric motor, to light an electric lamp, or to produce some other form of energy.

The standard potential of any electrode, then, **is the potential of the electrode with respect to the hydrogen electrode, measured at 25°C when the concentration of the ions in the solution is 1 M and the pressure of any gas involved is 1 atm.** Compared with the standard hydrogen electrode potential of 0.00 V, the standard electrode potential of sodium is -2.71 V, that of zinc is -0.76 V, and that of silver is $+0.7991$ V. As the standard electrode potential for a series of half-cells becomes more negative, it becomes progressively more difficult to carry out the reduction. Thus $\mathrm{Na^+}$ is more difficult to reduce than $\mathrm{Zn^{2+}}$, which is more difficult to reduce than $\mathrm{Ag^+}$. Also, the more negative is the standard electrode potential, the more positive is $E°$ for the oxidation reaction ($\mathrm{M} \longrightarrow \mathrm{M^{n+}} + ne^-$), and the more easily oxidized is the metal.

Any cell that generates an electric current by an oxidation-reduction reaction is known as a **voltaic cell.** The cell composed of zinc and hydrogen electrodes described previously is a voltaic cell and can be diagrammed

$$\mathrm{Zn} \,|\, \mathrm{Zn^{2+}},\, M = 1 \xrightarrow{\;e^-\;} \| \; \mathrm{H^+},\, M = 1 \,|\, \mathrm{H_2},\, 1 \text{ atm (Pt)}$$

This diagram indicates zinc metal in contact with a 1 M solution of $\mathrm{Zn^{2+}}$. The $\mathrm{Zn^{2+}}$ solution is connected by a salt bridge, represented by $\|$, to a 1 M solution of hydrogen ion in a hydrogen electrode with a gaseous hydrogen pressure of 1 atm. The anode (the electrode at which oxidation occurs) is customarily written on the left in such a diagram.

The purpose of the salt bridge in the zinc-hydrogen cell is to establish an electrical connection between the solutions of the two electrodes. The salt bridge thereby becomes part of the completed circuit. An excess of zinc ions accumulates in the half-cell in which zinc goes into solution. Similarly, an excess of anions accumulates in the half-cell in which hydrogen ions are being reduced. Migration of ions through the salt bridge (negative ions into the zinc half-cell solution and positive ions into the hydrogen half-cell solution) compensates for this situation and maintains neutrality of charge in each solution. Without the salt bridge, no current would flow in the external circuit and the cell reactions would not take place. The migration of ions through the salt bridge results in a net movement of electrons (an electric current) in a direction opposite to the movement of electrons in the external circuit, thereby completing the electrical circuit in the entire cell.

The ions that migrate through the salt bridge to maintain neutrality do not need to be the ions participating in the electrode reactions. In the zinc-hydrogen cell, they are in fact largely chloride ions and potassium ions supplied by the salt bridge itself.

20.11   The Electromotive Series of the Elements

From the standard electrode potentials of different metals whose ion concentrations are $1\,M$, the relative positions of these metals in the electromotive series (Section 9-21) can be determined. It was just shown in the preceding section that zinc has a greater tendency than hydrogen to go into solution as ions and leave electrons on the electrode. When copper metal in contact with a $1\,M$ solution of copper(II) ions is made the other electrode of a cell containing a standard hydrogen electrode, we find that hydrogen has a greater tendency to form positive ions than does the copper. The formation of hydrogen ions from the hydrogen leaves more electrons on the platinum than are left on the copper when the latter forms positive ions. Consequently, electrons flow from the hydrogen electrode to the copper electrode where they reduce copper(II) ions.

$$
\begin{aligned}
H_2 &\rightleftharpoons 2H^+ + 2e^- \qquad \textit{Anodic oxidation} \\
Cu^{2+} + 2e^- &\rightleftharpoons Cu \qquad\qquad\quad \textit{Cathodic reduction} \\
\hline
H_2 + Cu^{2+} &\rightleftharpoons Cu + 2H^+ \quad\;\; \textit{Cell reaction}
\end{aligned}
$$

The emf of the cell

$$
(Pt)H_2,\ 1\ atm\,|\,H^+,\ M = 1 \xrightarrow[\;\|\;]{e^-} Cu^{2+},\ M = 1\,|\,Cu
$$

as measured by a voltmeter is 0.337 V. Since the copper ion is reduced to the free element more easily than the hydrogen ion, the potential of the copper electrode has a value of $+0.337$ V with respect to the hydrogen electrode.

Standard electrode potentials of other metals can be determined in the same manner. When the standard electrode potentials of a number of metals are arranged in order from the most negative to the most positive, we have what is called the **Electromotive Series** or the **Reduction Potential Series** (Table 20-1). Table 20-1 also gives the electrode potentials, often referred to as reduction potentials, of some of the nonmetals as well as the common metals. Additional electrode potential values are given in Table 20-2 and in Appendix I.

20.12   Uses of the Electromotive Series

The electromotive series of the elements correlates many chemical properties; some of the more important ones are given in the following list.

1. The metals with high negative electrode potentials at the top of the series are good reducing agents in the free state. They are the metals most easily oxidized to their ions by the removal of electrons (active metals).
2. The elements with positive electrode potentials at the bottom of the series are good oxidizing agents when in the oxidized form, that is, when the metals are in the form of ions and the nonmetals are in the elemental state.

TABLE 20-1   Electromotive Series

	Electrode Reaction		Standard Electrode (Reduction) Potential, $E°$, volts
	Oxidized Form	Reduced Form	
Potassium	$K^+ + e^-$	$\rightleftharpoons$ K	−2.925
Calcium	$Ca^{2+} + 2e^-$	$\rightleftharpoons$ Ca	−2.87
Sodium	$Na^+ + e^-$	$\rightleftharpoons$ Na	−2.714
Magnesium	$Mg^{2+} + 2e^-$	$\rightleftharpoons$ Mg	−2.37
Aluminum	$Al^{3+} + 3e^-$	$\rightleftharpoons$ Al	−1.66
Manganese	$Mn^{2+} + 2e^-$	$\rightleftharpoons$ Mn	−1.18
Zinc	$Zn^{2+} + 2e^-$	$\rightleftharpoons$ Zn	−0.763
Chromium	$Cr^{3+} + 3e^-$	$\rightleftharpoons$ Cr	−0.74
Iron	$Fe^{2+} + 2e^-$	$\rightleftharpoons$ Fe	−0.440
Cadmium	$Cd^{2+} + 2e^-$	$\rightleftharpoons$ Cd	−0.40
Cobalt	$Co^{2+} + 2e^-$	$\rightleftharpoons$ Co	−0.277
Nickel	$Ni^{2+} + 2e^-$	$\rightleftharpoons$ Ni	−0.250
Tin	$Sn^{2+} + 2e^-$	$\rightleftharpoons$ Sn	−0.136
Lead	$Pb^{2+} + 2e^-$	$\rightleftharpoons$ Pb	−0.126
Hydrogen	$2H^+ + 2e^-$	$\rightleftharpoons$ H_2	0.00
Copper	$Cu^{2+} + 2e^-$	$\rightleftharpoons$ Cu	+0.337
Iodine	$I_2 + 2e^-$	$\rightleftharpoons$ $2I^-$	+0.5355
Silver	$Ag^+ + e^-$	$\rightleftharpoons$ Ag	+0.7991
Mercury	$Hg^{2+} + 2e^-$	$\rightleftharpoons$ Hg	+0.854
Bromine	$Br_2(l) + 2e^-$	$\rightleftharpoons$ $2Br^-$	+1.0652
Platinum	$Pt^{2+} + 2e^-$	$\rightleftharpoons$ Pt	+1.2
Oxygen	$O_2 + 4H^+ + 4e^-$	$\rightleftharpoons$ $2H_2O$	+1.23
Chlorine	$Cl_2 + 2e^-$	$\rightleftharpoons$ $2Cl^-$	+1.3595
Gold	$Au^+ + e^-$	$\rightleftharpoons$ Au	+1.68
Fluorine	$F_2 + 2e^-$	$\rightleftharpoons$ $2F^-$	+2.87

Increasing ease of reduction of the reactant (↑) / Increasing ease of oxidation of the product (↑)

3. The reduced form of any element will reduce the oxidized form of any element below it in the series. For example, metallic zinc will reduce copper(II) ions according to the equation

$$Zn + Cu^{2+} \rightleftharpoons Cu + Zn^{2+}$$

For a voltaic cell made up of a standard zinc electrode and a standard copper electrode (Fig. 20-7), the net cell reaction is obtained by adding the two half-reactions together.

Anode half-reaction:	$Zn \rightleftharpoons Zn^{2+} + 2e^-$
Cathode half-reaction:	$Cu^{2+} + 2e^- \rightleftharpoons Cu$
Net cell reaction:	$Zn + Cu^{2+} \rightleftharpoons Zn^{2+} + Cu$

We can measure the electromotive force (emf) of the cell by means of a voltmeter. Alternatively, the emf of the cell can be calculated by adding, algebraically, the standard potential of the copper electrode to that of the zinc electrode, taking into account the proper sign on each potential, based upon whether the half-reaction is taking place as oxidation or as reduction. The standard electrode potential

Figure 20-7   A zinc-copper voltaic cell.

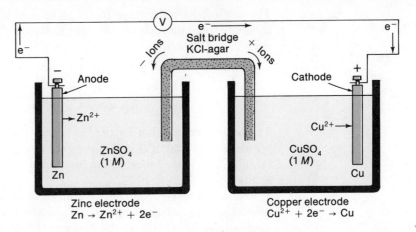

Zinc electrode
$Zn \rightarrow Zn^{2+} + 2e^-$

Copper electrode
$Cu^{2+} + 2e^- \rightarrow Cu$

(reduction) for zinc is -0.76 V. This is the potential for the half-reaction $Zn^{2+} + 2e^- \longrightarrow Zn$. The oxidation of Zn to Zn^{2+}, the reverse half-reaction $Zn \longrightarrow Zn^{2+} + 2e$, has a potential of $+0.76$ V. At the copper electrode, Cu^{2+} is being reduced to Cu; hence the potential is the standard electrode potential (reduction) for copper, $+0.34$ V. The emf for the cell $= 0.76$ V $+ 0.34$ V $= 1.10$ V.

The positive emf value for the zinc-copper cell indicates that the reaction will proceed spontaneously to the right as written and that the cell will deliver an electric current. A negative emf value for a cell would indicate that the reaction will not proceed spontaneously to the right, but would proceed spontaneously to the left.

Suppose we wish to calculate the emf of a cell composed of copper and silver electrodes in 1 M solutions of their respective ions. Because copper is higher in the electromotive series than silver (Cu has a lower $E°$ value than Ag), copper metal will reduce the silver ion; $Cu + 2Ag^+ \rightleftharpoons Cu^{2+} + 2Ag$. To obtain this net cell reaction, we may add the half-reactions as follows:

$$
\begin{array}{ll}
Cu \rightleftharpoons Cu^{2+} + 2e^- & E° = -0.34 \text{ V} \\
2Ag^+ + 2e^- \rightleftharpoons 2Ag & E° = +0.80 \text{ V} \\
\hline
Cu + 2Ag^+ \rightleftharpoons Cu^{2+} + 2Ag & E° = +0.46 \text{ V}
\end{array}
$$

Hence the calculated emf of the cell is 0.46 V. Since the emf of the cell is positive, we can assume that the reaction proceeds spontaneously to the right as written, and the cell delivers current. Note that the potential of the half-reaction $Cu \longrightarrow Cu^{2+} + 2e^-$ is equal to but of opposite sign to that for the half-reaction $Cu^{2+} + 2e^- \longrightarrow Cu$.

Let us next calculate the emf of a cell made up of standard bromine and chlorine electrodes.

$$(Pt)Br_2(l)\,|\,Br^-,\ M = 1 \quad \overset{e^-}{\underset{\|}{\longrightarrow}} \quad Cl^-,\ M = 1\,|\,Cl_2,\ 1 \text{ atm (Pt)}$$

The half-reactions and standard potentials are

$$
\begin{array}{ll}
2Br^- \rightleftharpoons Br_2 + 2e^- & E° = -1.06 \text{ V} \\
Cl_2 + 2e^- \rightleftharpoons 2Cl^- & E° = +1.36 \text{ V} \\
\hline
2Br^- + Cl_2 \rightleftharpoons Br_2 + 2Cl^- & E° = +0.30 \text{ V}
\end{array}
$$

The emf of the cell is thus found to be 0.30 V, meaning that the cell reaction proceeds spontaneously to the right as written. Suppose the half-reactions had been added as follows:

$$
\begin{array}{ll}
2Cl^- \rightleftharpoons Cl_2 + 2e^- & E° = -1.36 \text{ V} \\
\underline{Br_2 + 2e^- \rightleftharpoons 2Br^-} & \underline{E° = +1.06 \text{ V}} \\
2Cl^- + Br_2 \rightleftharpoons Cl_2 + 2Br^- & E° = -0.30 \text{ V}
\end{array}
$$

The negative emf value for the cell means that the reaction proceeds to the left spontaneously, rather than to the right. In other words, chlorine will oxidize the bromide ion to free bromine, but bromine will not oxidize the chloride ion to free chlorine (see Section 19.2).

The standard electrode potentials given in the electromotive series (Table 20-1) refer only to ion concentrations of 1 M, pressures of 1 atm for any gases involved, and temperatures of 25°C. If the concentrations are not 1 M, the potentials are different and members close together in the series may even change places in the table. Furthermore, the values given do not apply to nonaqueous solutions or to fused salts. An illustration of the effect of concentration on potential is provided by the oxidation potentials for the oxidation of hydrogen to hydrogen ion at two different concentrations.

$$
\begin{array}{ll}
2H^+ (1 \text{ M}) + 2e^- \rightleftharpoons H_2 (1 \text{ atm}) & E° + 0.00 \text{ V} \\
2H^+ (10^{-7} \text{ M}) + 2e^- \rightleftharpoons H_2 (1 \text{ atm}) & E° = -0.41 \text{ V}
\end{array}
$$

Similar electromotive series can be constructed for solvents other than water. The electrode potentials would be quite different from those for water, and the members would fall in a different order in the series. The differences occur because of differences in energy of solvation.

20.13   Electrode Potentials for Other Half-Reactions

In addition to redox reactions between elements, there are many other redox reactions that can take place at the electrodes of electrochemical cells. The electrode consists of a relatively inactive substance such as carbon, which undergoes no change itself and merely serves as a carrier of electrons to or from the solution. The standard electrode potentials for the half-reactions of a number of redox reactions are given in Table 20-2 and Appendix I.

The standard electrode potential of the Pt, Sn^{4+}, Sn^{2+} electrode (see Table 20-2) is that of the inert platinum electrode immersed in a solution that is 1 M with respect to both Sn^{4+} and Sn^{2+} ions. Similarly, the standard electrode potential of a Pt, Fe^{3+}, Fe^{2+} electrode pertains to its potential when the Fe^{3+} and Fe^{2+} are both 1 M. The emf of a cell composed of the two electrodes Pt, Sn^{4+}, Sn^{2+} and Pt, Fe^{3+}, Fe^{2+} can be calculated as follows:

$$
\begin{array}{ll}
Sn^{2+} \rightleftharpoons Sn^{4+} + 2e^- & E° = -0.15 \text{ V} \\
\underline{2Fe^{3+} + 2e^- \rightleftharpoons 2Fe^{2+}} & \underline{E° = +0.77 \text{ V}} \\
Sn^{2+} + 2Fe^{3+} \rightleftharpoons Sn^{4+} + 2Fe^{2+} & E° = +0.62 \text{ V}
\end{array}
$$

Thus the emf of the cell is 0.62 V when standard electrodes are used. This means that the reaction will proceed spontaneously to the right as written and will deliver

TABLE 20-2   Standard Electrode Potentials for Some Important Half-Reactions

Electrode	Electrode Reaction		Standard Electrode Potential, volts
	Oxidized Form	Reduced Form	
Fe, Fe(OH)$_2$, OH$^-$	Fe(OH)$_2$ + 2e$^-$ $\rightleftharpoons$ Fe + 2OH$^-$		-0.877
Pb, PbSO$_4$, SO$_4^{2-}$	PbSO$_4$ + 2e$^-$ $\rightleftharpoons$ Pb + SO$_4^{2-}$		-0.356
Pt, Sn^{4+}, Sn^{2+}	Sn^{4+} + 2e$^-$ $\rightleftharpoons$ Sn^{2+}		$+0.15$
Ag, AgCl, Cl$^-$	AgCl + e$^-$ $\rightleftharpoons$ Ag + Cl$^-$		$+0.222$
Hg, Hg$_2$Cl$_2$, Cl$^-$	Hg$_2$Cl$_2$ + 2e$^-$ $\rightleftharpoons$ 2Hg + 2Cl$^-$		$+0.27$
NiO$_2$, Ni(OH)$_2$, OH$^-$	NiO$_2$ + 2H$_2$O + 2e$^-$ $\rightleftharpoons$ Ni(OH)$_2$ + 2OH$^-$		$+0.49$
Pt, Fe^{3+}, Fe^{2+}	Fe^{3+} + e$^-$ $\rightleftharpoons$ Fe^{2+}		$+0.771$
Pt, Cr$_2$O$_7^{2-}$, H$^+$, Cr^{3+}	Cr$_2$O$_7^{2-}$ + 14H$^+$ + 6e$^-$ $\rightleftharpoons$ 2Cr^{3+} + 7H$_2$O		$+1.33$
Pt, MnO$_4^-$, H$^+$, Mn^{2+}	MnO$_4^-$ + 8H$^+$ + 5e$^-$ $\rightleftharpoons$ Mn^{2+} + 4H$_2$O		$+1.51$
PbO$_2$, PbSO$_4$, H$_2$SO$_4$	PbO$_2$ + SO$_4^{2-}$ + 4H$^+$ + 2e$^-$ $\rightleftharpoons$ PbSO$_4$ + 2H$_2$O		$+1.685$

a current. It also means that when solutions containing tin(II) and iron(III) are mixed, the Sn^{2+} will reduce the Fe^{3+}.

20.14   The Nernst Equation

Standard electrode potentials, abbreviated $E°$, are for what is referred to as standard state conditions: 1 M solutions of ions, a pressure of 1 atm for gases, and a temperature of 25°C. However, as pointed out in Section 20.12, the electromotive force value for a half-reaction changes when the concentration changes. An equation referred to as the **Nernst equation** makes it possible to calculate emf values at other concentrations. The Nernst equation for calculating the emf of a half-cell at 25°C is

$$E = E° - \frac{0.05915}{n} \log Q$$

where E = the emf for the reaction at the new concentration

$E°$ = the *standard* electrode potential

n = the number of electrons stated in the half-reaction

Q = the reaction quotient (Section 18.12) for the half-reaction, an expression that takes the same form as the equilibrium constant for the half-reaction without indicating the electrons.

For example, a typical half-reaction and the corresponding Nernst equation is

$$Sn^{4+} + 2e^- \longrightarrow Sn^{2+}$$

$$E = E° - \frac{0.05915}{2} \log \frac{[Sn^{2+}]}{[Sn^{4+}]}$$

or, with the standard electrode potential, $E°$, from Table 20-2 inserted,

$$E = 0.15 - \frac{0.05915}{2} \log \frac{[Sn^{2+}]}{[Sn^{4+}]}$$

where the brackets as usual mean "concentration of" and "log" signifies the common logarithm.

If the concentrations of both Sn^{2+} and Sn^{4+} are 1 M (concentration of the standard state),

$$E = E° - \frac{0.05915}{2} \log \frac{1}{1}$$

Since $\log 1 = 0$,

$$E = E° - \frac{0.05915}{2}(0) \quad \text{or} \quad E = E°$$

This of course shows that the electrode potential, E, is equal to the *standard* electrode potential, $E°$, when the concentrations are the standard state concentration of 1 M. (Note also that $E = E°$ if $[Sn^{2+}] = [Sn^{4+}]$, even if the concentrations are not 1 M.)

EXAMPLE 1   Calculate E for the half-reaction $Sn^{4+} + 2e^- \longrightarrow Sn^{2+}$ at 25°C if the concentration of Sn^{2+} is four times that of Sn^{4+}.

$$\frac{[Sn^{2+}]}{[Sn^{4+}]} = 4$$

Therefore

$$E = 0.15 - \frac{0.05915}{2} \log 4$$

$$= 0.15 - \frac{0.05915}{2}(0.6021)$$

$$= 0.15 - 0.01781$$

$$= 0.13 \text{ V}$$

◆ ◆ ◆ ◆ ◆

In Section 20.12, we noted that whereas the standard electrode potential $(E°)$ for the reduction of $2H^+$ to H_2 is 0.00 V, the emf value (E) at 10^{-7} M is -0.41 V. The following example shows the calculation of the -0.41 V value.

EXAMPLE 2   Calculate the emf value for the reduction from H^+ at 1.0×10^{-7} M to H_2 at 1.0 atm.

$$2H^+ (1.0 \times 10^{-7} M) + 2e^- \rightleftharpoons H_2 (1.0 \text{ atm}) \quad E° = 0.00 \text{ V}$$

$$E = E° - \frac{0.05915}{n} \log \frac{p_{H_2}}{[H^+]^2}$$

$$= 0.00 - \frac{0.05915}{2} \log \frac{1.0}{(1.0 \times 10^{-7})^2}$$

$$= 0.00 - \frac{0.05915}{2} \log \frac{1.0}{1.0 \times 10^{-14}}$$

$$= 0.00 - \frac{0.05915}{2} \log (1.0 \times 10^{14})$$

$$= 0.00 - \frac{0.05915}{2}(14)$$

$$= -0.41 \text{ V}$$

Hence $2H^+ (1.0 \times 10^{-7} M) \rightleftharpoons H_2 (1.0 \text{ atm})$ $E = -0.41$ V

◆ ◆ ◆ ◆ ◆

20.15 Relationship of the Free Energy Change, the Standard Reduction Potential, and the Equilibrium Constant

For a half-cell reaction, the standard free energy change, $\Delta G°$ (see Section 18.10), is related to the standard electrode potential, $E°$, and to the equilibrium constant, K (see Section 18.13), by means of the following two equations:

$$\Delta G° = -nFE° \quad \text{and} \quad \Delta G° = -RT \ln K$$

where n = the number of electrons stated in the half-reaction

F = a constant known as the Faraday constant, with a value of 96.487 kJ V^{-1}, or 96,487 coulombs

$E°$ = the standard electrode potential

R = the gas constant (8.314 J K^{-1})

T = the Kelvin temperature

K = the equilibrium constant

and ln signifies the natural logarithm (to the base e)

By combining the two relationships for $\Delta G°$,

$$-nFE° = -RT \ln K$$

Then

$$E° = \frac{RT}{nF} \ln K$$

For 25°C (298.15 K), after conversion to the common logarithm (to the base 10), and with substitution of the numerical value of R, the expression becomes

$$E° = \frac{0.05915}{n} \log K$$

EXAMPLE 1   **Calculate the standard free energy change at 25°C for the reaction**

$$Cd + Pb^{2+} \rightleftharpoons Cd^{2+} + Pb$$

The half-reactions and calculation of the emf of the cell are

$$
\begin{array}{lll}
& Cd \rightleftharpoons Cd^{2+} + 2e^- & E° = +0.40 \text{ V} \\
\textit{Adding} \quad & Pb^{2+} + 2e^- \rightleftharpoons Pb & E° = -0.13 \text{ V} \\
\hline
& Cd + Pb^{2+} \rightleftharpoons Cd^{2+} + Pb & E° = +0.27 \text{ V}
\end{array}
$$

Hence $E°$ for the cell = +0.27 V.

$$\Delta G° = -nFE° = -2(96.487 \text{ kJ } V^{-1})(0.27 \text{ V}) = -52 \text{ kJ}$$

The negative value for $\Delta G°$ indicates that the reaction proceeds spontaneously to the right (Section 18.9). Note that a positive $E°$ implies a negative $\Delta G°$ so a positive $E°$ also implies a spontaneous forward reaction.

◆ ◆ ◆ ◆ ◆

EXAMPLE 2   **Calculate the equilibrium constant at 25°C for the reaction in the preceding example.**

$$E° = \frac{0.05915}{n} \log K$$

$$0.27 = \frac{0.05915}{2} \log K$$

$$\log K = 0.27 \times \frac{2}{0.05915} = 9.13$$

$$K = \text{antilog } 9.13 = 1.3 \times 10^9$$

Thus at equilibrium the molar concentration of cadmium ion in the solution is more than a billion times that of lead ion!

◆ ◆ ◆ ◆ ◆

EXAMPLE 3   **Calculate the free energy change and the equilibrium constant for the following reaction at 25°C:**

$$Zn + Cu^{2+} \, (0.20 \, M) \rightleftharpoons Zn^{2+} \, (0.50 \, M) + Cu$$

$$\begin{array}{lll} & Zn \rightleftharpoons Zn^{2+} + 2e^- & E° = +0.76 \text{ V} \\ Adding & Cu^{2+} + 2e^- \rightleftharpoons Cu & E° = +0.34 \text{ V} \\ \hline & Zn + Cu^{2+} \rightleftharpoons Zn^{2+} + Cu & E° = +1.10 \text{ V} \end{array}$$

Hence, for the cell reaction *under standard state conditions* (Cu^{2+} and Zn^{2+} at concentrations of 1 M, or more accurately at unit activity), $E° = +1.10$ V.

The Nernst equation enables us to calculate the potential (E) when Zn^{2+} and Cu^{2+} concentrations are 0.50 M and 0.20 M, respectively.

Zinc half-reaction:

$$E = E° - \frac{0.05915}{n} \log \frac{[Zn^{2+}]}{[Zn°]} = 0.76 - \frac{0.05915}{2} \log \frac{0.50}{1}$$

$$= 0.76 - \frac{0.05915}{2}(0.699 - 1) = 0.76 + 0.01 = +0.77 \text{ V}$$

Copper half-reaction:

$$E = E° - \frac{0.05915}{n} \log \frac{[Cu°]}{[Cu^{2+}]} = +0.34 - \frac{0.05915}{2} \log \frac{1}{0.20}$$

$$= +0.34 - \frac{0.05915}{2}(0.699) = +0.34 - 0.02 = +0.32 \text{ V}$$

Thus, for the reaction under the specified conditions, $E = 0.77 + 0.32 = +1.09$ V (as compared to the $E°$ value of $+1.10$ V for standard state conditions).

$$\Delta G = -nFE = -2(96.487)(1.09) = -210 \text{ kJ}$$

The negative value for ΔG indicates that the reaction proceeds spontaneously to the right.

The value of K at a given temperature (25°C in this case) is calculated from

$E°$ and is constant for the various concentrations.

$$E° = \frac{0.05915}{n} \log K$$

$$1.10 = \frac{0.05915}{2} \log K$$

$$\log K = 1.10 \times \frac{2}{0.05915} = 37.19$$

$$K = 1.6 \times 10^{37}$$

◆ ◆ ◆ ◆ ◆

Voltaic Cells

20.16   Primary Voltaic Cells

A voltaic cell is an electrochemical cell in which a chemical reaction is used to produce electrical energy. Once the chemicals in some voltaic cells are consumed, further chemical action is not possible, because the electrodes and electrolytes are of such a nature that the reaction cannot be reversed to restore them to their original states by the application of an external electrical potential. Such cells are called **primary voltaic cells,** or **irreversible cells.** The Daniell cell and the "dry" cell are examples of primary cells.

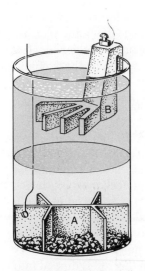

Figure 20-8   The Daniell, or gravity, cell. (A) sheet copper surrounded by copper(II) sulfate solution; (B) zinc plate surrounded by zinc sulfate solution.

▶**1. THE DANIELL CELL.** The construction of a Daniell cell is shown in Fig. 20-8. Metallic copper (A) is placed in a saturated solution of copper(II) sulfate on the bottom of the cell with crystals of copper(II) sulfate to keep the solution saturated. A zinc electrode (B) is suspended near the top of the cell in a dilute solution of zinc sulfate, which floats on the more dense solution of copper(II) sulfate. When the two metals are connected by means of a wire, electrons flow through the wire from the more easily oxidized zinc to the less easily oxidized copper. The zinc is oxidized and goes into solution as zinc ions, while the copper(II) ions are reduced to metallic copper. This is the same reaction that takes place when metallic zinc becomes plated with copper upon being placed in a solution of copper(II) sulfate.

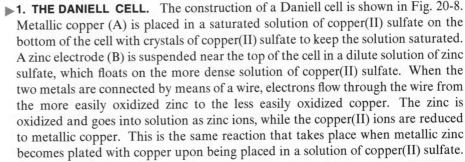

$Zn \rightleftharpoons Zn^{2+} + 2e^-$	*Anodic oxidation*
$Cu^{2+} + 2e^- \rightleftharpoons Cu$	*Cathodic reduction*
$Zn + Cu^{2+} \rightleftharpoons Zn^{2+} + Cu$	*Cell reaction*

If the zinc ion concentration is increased, the anode reaction is shifted to the left, resulting in a lower voltage being produced by the cell.

A cell similar to the Daniell cell in principle, but using cadmium and nickel instead of zinc and copper, is being utilized increasingly in batteries where longer cell life is desirable.

▶**2. THE "DRY" CELL.** The container of the dry cell (Fig. 20-9) is made of zinc, which also serves as one of the electrodes. The container is lined with porous paper, which separates the metal from the materials within the cell. A carbon

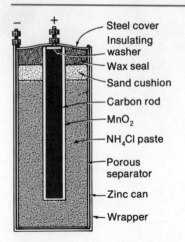

Figure 20-9 Cross section of a "dry" cell.

(graphite) rod in the cell acts as the other electrode. The space around the carbon rod contains a moist mixture of ammonium chloride, manganese dioxide, zinc chloride, and some porous inactive solid such as sawdust. The cell is sealed with some sealing compound to keep the moisture in. When the cell delivers a current the zinc goes into solution as zinc ions and leaves electrons on the zinc container so that it becomes the negative electrode. This electrode, then, is the anode, since this is where oxidation occurs. The oxidation half-reaction is

$$Anode: \quad Zn \longrightarrow Zn^{2+} + 2e^-$$

Hydrogen ions in ammonium ions are reduced at the carbon electrode, and this reaction is thus at the cathode.

$$Cathode: \quad 2NH_4^+ + 2e^- \longrightarrow 2NH_3 + H_2$$

The net cell reaction may be expressed by the equation

$$Zn + 2NH_4^+ \longrightarrow Zn^{2+} + 2NH_3 + H_2$$

The manganese dioxide in the cell oxidizes the hydrogen as it is formed. The hydrogen would otherwise collect on the cathode and stop the action of the cell, a condition called **polarization.**

$$H_2 + MnO_2 + 2H^+ \longrightarrow Mn^{2+} + 2H_2O$$

The ammonia formed at the cathode when the cell is working unites with the zinc ions, forming the complex $[Zn(NH_3)_4]^{2+}$. This reaction holds down the increase in the concentration of zinc ions, thereby keeping the potential of the zinc electrode more nearly constant. It also prevents polarization caused by the accumulation of ammonia gas on the cathode.

▶ **3. CELLS FOR SPECIAL PURPOSES.**   Many cells are designed for batteries that must meet special requirements for one specific purpose. For example, during World War II a torpedo battery was developed that had to fit into a small space and produce a lot of power, but only once over a period of just 30 seconds. A battery cell suitable for a camera lightmeter must produce a small amount of current but must last for a long period. A wristwatch battery cell must produce a very small current without interruption for at least one year. A special cell for a battery used in weather balloons, which was developed to meet the special requirements of small size, light weight, and operation over a period of a few hours, is made by laying a mixture of dry magnesium powder and dry silver chloride on a strip of paper which is then rolled up. When this roll is dipped into water, the reaction begins.

The Rubin-Mallory cell, which is very small, receives widespread use in electrically operated hearing aids. It utilizes a zinc container as the anode and a carbon rod as the cathode. The electrolyte is a paste of moist mercury(II) oxide mixed with a little sodium, or potassium, hydroxide. The cell produces a potential of 1.35 V. The half-reactions and net cell reaction are

$$Anode: \qquad Zn + 2OH^- \longrightarrow Zn(OH)_2 + 2e^-$$
$$Cathode: \qquad HgO + H_2O + 2e^- \longrightarrow Hg + 2OH^-$$
$$\overline{Net\ cell\ reaction: \quad Zn + HgO + H_2O \longrightarrow Zn(OH)_2 + Hg}$$

20.17   Secondary Voltaic Cells

In **secondary,** or **reversible, cells,** the substances consumed in producing electricity may be regenerated in their original forms by causing a direct current of electricity to flow in the reverse direction of the discharge. This reversal recharges the cell. The lead storage battery and the Edison storage battery both consist of several secondary cells connected together.

▶ **1. THE LEAD STORAGE BATTERY.**   The electrodes of the cells in a lead storage battery consist of two sets of lead alloy plates in the form of grids. The openings of one set of grids are filled with lead dioxide and the openings of the other with spongy lead metal. Dilute sulfuric acid serves as the electrolyte. When the battery is delivering a current, the spongy lead is oxidized to lead ions, and these plates become negatively charged.

$$Pb \longrightarrow Pb^{2+} + 2e^-   \text{(oxidation)}$$

The lead ions thus formed combine with sulfate ions of the electrolyte and coat the lead electrode with lead sulfate.

$$Pb^{2+} + SO_4^{2-} \longrightarrow PbSO_4(s)$$

The lead sulfate, being quite insoluble in dilute sulfuric acid, can be regenerated to spongy lead on the electrode during the recharging process.

The net lead electrode reaction for each cell when the battery is delivering a current is

$$Pb + SO_4^{2-} \longrightarrow PbSO_4(s) + 2e^-$$

The negatively charged lead electrode is the anode since it is the electrode at which oxidation occurs.

Electrons flow from the negatively charged lead electrode through the external circuit and enter the lead dioxide electrode. The lead dioxide in the presence of hydrogen ions from the electrolyte is reduced to lead(II) ions.

$$PbO_2 + 4H^+ + 2e^- \longrightarrow Pb^{2+} + 2H_2O   \text{(reduction)}$$

Again lead ions combine with sulfate ions of the electrolyte and the lead dioxide plate becomes coated with lead sulfate.

$$Pb^{2+} + SO_4^{2-} \longrightarrow PbSO_4(s)$$

The net reaction at the positive lead dioxide electrode of each cell when the battery is delivering a current is

$$\overset{+4}{Pb}O_2 + 4H^+ + SO_4^{2-} + 2e^- \longrightarrow \overset{+2}{Pb}SO_4(s) + 2H_2O$$

The positively charged lead dioxide electrode is the cathode since it is the electrode at which reduction occurs.

The lead storage battery can be recharged by passing electrons through each cell in the reverse direction by applying an external potential. The cell is thereby made an electrolytic cell during the recharging process. The reactions are just the reverse of those that occur when the cell is operating as a voltaic cell producing a current. Electrons are forced into the lead electrode, making it the cathode, at which reduction of lead ions from the lead sulfate takes place. Electrons are withdrawn from the lead dioxide electrode, making it the anode, at which oxidation of lead ions from the lead sulfate takes place.

The charge and discharge at the two plates may be summarized by

$$Pb + SO_4{}^{2-} \underset{\text{Charge}}{\overset{\text{Discharge}}{\rightleftharpoons}} PbSO_4(s) + 2e^- \qquad \text{(at the lead plate)}$$

$$PbO_2 + 4H^+ + SO_4{}^{2-} + 2e^- \underset{\text{Charge}}{\overset{\text{Discharge}}{\rightleftharpoons}} PbSO_4(s) + 2H_2O \quad \text{(at the lead dioxide plate)}$$

Thus during the charging process the sulfate ion of the electrolyte is regenerated and lead sulfate is converted back to spongy lead at the lead electrode; hydrogen ions and sulfate ions of the electrolyte are regenerated, and lead sulfate is converted back to lead dioxide at the lead dioxide electrode.

The net cell reaction of the lead storage battery is given by

$$Pb + PbO_2 + 4H^+ + 2SO_4{}^{2-} \underset{\text{Charge}}{\overset{\text{Discharge}}{\rightleftharpoons}} 2PbSO_4(s) + 2H_2O$$

The above equation indicates that the amount of sulfuric acid decreases as the cell discharges. Conversely, charging the cell regenerates the acid. Since sulfuric acid is much heavier than water, the condition of a lead storage battery may be tested conveniently by determining the specific gravity of the electrolyte. When the specific gravity is 1.25 to 1.30, the battery is charged; when the specific gravity falls much below 1.20, the battery needs charging. As can be calculated from the potentials in Table 20-2, a single, fully charged lead storage cell has a potential of $0.36 + 1.69 = 2.05$ V. The potential falls off slowly as the cell is used. A 12-V automobile battery contains six lead storage cells, whereas a 6-V battery contains three.

▶ **2. THE EDISON STORAGE BATTERY.** In the cells of the Edison battery, the negative electrode consists of steel plates packed with finely divided iron and the positive electrode of steel plates with hydrated nickel dioxide. The electrolyte is a 21% potassium hydroxide solution containing some lithium hydroxide. When the cell is delivering a current of electricity, oxidation takes place at the anode according to the equation

$$\textit{Anode:} \quad Fe + 2OH^- \longrightarrow Fe(OH)_2(s) + 2e^-$$

The reduction at the cathode is expressed by

$$\textit{Cathode:} \quad NiO_2 + 2H_2O + 2e^- \longrightarrow Ni(OH)_2(s) + 2OH^-$$

These reactions are reversed while the cells of the battery are being charged. The net reaction is

$$Fe + NiO_2 + 2H_2O \underset{\text{Charge}}{\overset{\text{Discharge}}{\rightleftharpoons}} Fe(OH)_2(s) + Ni(OH)_2(s)$$

It can be seen from this equation that the electrodes are restored to their original

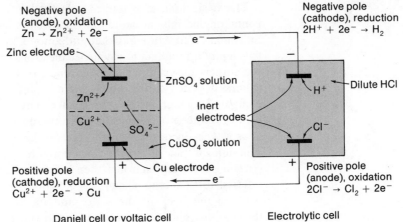

Figure 20-10 Illustration of the relationship between a voltaic cell and an electrolytic cell. The voltaic cell (*left*) provides the electric current to run the electrolytic cell (*right*). Note that the signs for the electrodes are opposite for voltaic and electrolytic cells, resulting in electrodes of like signs being connected. Note also that regardless of whether the cell is voltaic or electrolytic, oxidation occurs at the anode and reduction at the cathode.

Negative pole (anode), oxidation
$Zn \rightarrow Zn^{2+} + 2e^-$

Zinc electrode

e^-

Negative pole (cathode), reduction
$2H^+ + 2e^- \rightarrow H_2$

$ZnSO_4$ solution

Zn^{2+}

Cu^{2+}

SO_4^{2-}

Inert electrodes

Dilute HCl

H^+

Cl^-

$CuSO_4$ solution

Positive pole (cathode), reduction
$Cu^{2+} + 2e^- \rightarrow Cu$

Cu electrode

e^-

Positive pole (anode), oxidation
$2Cl^- \rightarrow Cl_2 + 2e^-$

Daniell cell or voltaic cell

Electrolytic cell

condition on recharging. The electromotive force of each cell in an Edison storage battery, as can be calculated from the potentials in Table 20-2, is about 1.4 V. The cell must be sealed, for otherwise the potassium hydroxide and lithium hydroxide would change to carbonates by action with CO_2 from the air. The cell can be sealed because no gases are produced during reaction. This battery is lighter and more durable than the lead storage battery but more costly.

Because of environmental considerations, considerable interest has arisen recently in connection with the possibility of developing new varieties of secondary cells suitable for powering an electric automobile.

Note that the anode is positive and the cathode negative in an electrolytic cell (see Fig. 20-2, for example), whereas the anode is negative and the cathode positive in a voltaic cell (Fig. 20-6). *For all types of cell, however, the electrode where oxidation occurs is the anode and where reduction occurs is the cathode.*

Figure 20-10 shows diagrammatically the relationships between a voltaic cell and an electrolytic cell. The sign convention is such that when a voltaic cell is supplying the current for an electrolytic cell, the negative electrode of one cell is hooked to the negative electrode of the other cell, and likewise the positive electrodes of the two cells are hooked together.

20.18 The Solar Cell

It has long been known that an enormous amount of energy is given off by the sun. The earth alone receives more energy from sunlight in two days than is stored in all known reserves of fossil fuels. In order to use part of this energy, several devices are used or are under study to convert sunlight into electrical energy. The conventional photoelectric cell, used in electric-eye doors, transforms light into electrical energy (Section 4.1) but delivers as power only about one-half of one per cent of the total light energy it absorbs. A newer type of photoelectric device currently in use is the so-called **solar cell.** It is several times more efficient than the photoelectric cell and is capable of generating electric power from sunlight at the rate of 90 watts per square yard of illuminated surface.

The basic unit of a typical solar cell is a thin wafer of very pure silicon, containing initially no more than one part of impurity per million parts of silicon. Silicon has a structure like that of diamond, in which each silicon atom is bonded covalently through its four valence electrons to four neighboring silicon atoms at the corners of a regular tetrahedron. For use in a solar cell, a tiny amount of arsenic is added to the silicon wafer. Since arsenic has five valence electrons, compared to four for silicon, replacement of silicon atoms with arsenic atoms creates an excess of free electrons within the body of the wafer (commonly referred to as n-type silicon). On the surface of the wafer is placed a thin layer of silicon containing a trace of boron. Since boron has only three valence electrons, "holes" (vacant spots for electrons) then exist at the surface of the wafer (commonly referred to as p-type silicon). A junction (referred to as a p-n junction) exists between the body of the wafer and the thin surface layer. Electrons diffuse through the junction from the wafer to "holes" in the surface layer; at the same time, "holes," or electron vacancies, move to the body of the wafer. This results in a net positive charge within the body of the wafer, which was neutral prior to the diffusion of electrons (negative charges) out of the wafer. A net negative charge, from electrons diffusing in, is correspondingly produced in the previously neutral surface layer. Thus an electrostatic potential develops between the two regions, which builds up until the negatively charged region at the surface repels any further diffusion of electrons. A counter potential, opposite to the diffusion potential, results as the electric field tends to attract the negative electrons back to the positively charged body of the wafer and to attract "holes" to the negatively charged surface of the wafer. When the diffusion of electrons is just balanced by the electric field, equilibrium is established between the two forces and hence also between electrons and "holes." At equilibrium, a net difference in potential exists between the two regions in the wafer.

In a solar cell, one electrical lead is attached to the body of the wafer and one lead to the surface. When the wafer is exposed to sunlight, energy from the sunlight causes electrons to be released from their positions in the lattice near the p-n junction, thereby upsetting the equilibrium between electrons and "holes" and causing additional electrons to move across the junction into the body of the wafer. Thus a sufficient electron "pressure" (electromotive force) is built up to cause electrons to move through the electrical leads, and a current flows through the wire. The device is thus an electric cell, with the positive terminal at the p-contact and the negative terminal at the n-contact (Fig. 20-11).

In actual practice, a series of such wafers makes up a solar cell. The first practical application of the solar cell was in powering eight telephones on a rural telephone line in Georgia in 1955. Such cells now are used to power communication devices in spacecraft, especially those designated to remain in space for lengthy periods of time.

Two new types of solar cells show excellent promise for use in telephone circuits and other applications. One of these cells utilizes a layer of crystalline cadmium sulfide deposited on indium phosphide. The cell has a conversion efficiency of 12.5%, which means that 12.5% of the sunlight that strikes the cell is converted into electrical energy. This is considered very good conversion efficiency, in terms of present technology. The other of the two new cells consists of a layer of cadmium

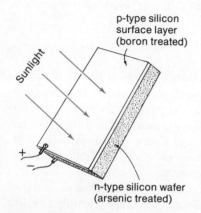

Figure 20-11   A solar cell.

sulfide deposited on copper indium diselenide and has a conversion efficiency of 8%.

Note that a solar cell differs from an electrochemical cell in that it converts the energy of electromagnetic radiation (light) from the sun, rather than the energy of a chemical reaction, into electrical energy.

20.19   Fuel Cells

Fuel cells are voltaic cells in which electrode materials, usually in the form of gases, are supplied continuously to an electrochemical cell and consumed to produce electricity.

A typical fuel cell, of the type currently used extensively in spacecraft, is based upon the reaction of hydrogen (the "fuel") and oxygen (the "oxidizer") to form water. At the anode, hydrogen gas is diffused through a porous electrode, in the surface of which is embedded a catalyst such as finely divided particles of platinum or palladium. At the cathode, oxygen is diffused through a porous electrode impregnated with cobalt oxide, platinum, or silver as catalyst. The two electrodes are separated by an electrolyte such as a concentrated solution of sodium hydroxide or potassium hydroxide (Fig. 20-12).

As hydrogen diffuses through the anode, it is absorbed on the electrode surface in the form of hydrogen atoms, which react with hydroxyl ions of the electrolyte to form water.

$$H_2 \xrightarrow{\text{Catalyst}} 2H$$

$$2H + 2OH^- \longrightarrow 2H_2O + 2e^-$$

Net anode reaction: $\quad H_2 + 2OH^- \longrightarrow 2H_2O + 2e^-$

Figure 20-12 Diagram of a hydrogen-oxygen fuel cell. In actual operation, it would take several such cells to light an ordinary 110-V bulb.

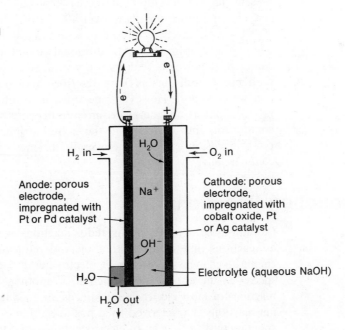

H$_2$ in →

← O$_2$ in

Anode: porous electrode, impregnated with Pt or Pd catalyst

H$_2$O

Na$^+$

OH$^-$

Cathode: porous electrode, impregnated with cobalt oxide, Pt or Ag catalyst

Electrolyte (aqueous NaOH)

H$_2$O

H$_2$O out

The electrons produced at the anode flow through the external circuit to the cathode. The oxygen, diffused through the cathode, is adsorbed on the electrode surface where it is reduced to hydroxyl ions.

$$\textit{Cathode reaction:} \quad O_2 + 2H_2O + 4e^- \longrightarrow 4OH^-$$

The hydroxyl ions complete the cycle by migrating through the electrolyte from the oxygen electrode (cathode) to the hydrogen electrode (anode). The electrical output of the cell results, as in all voltaic cells, from the flow of electrons through the external circuit from anode to cathode.

The overall cell reaction is the combination of hydrogen and oxygen to produce water.

$$
\begin{array}{ll}
2H_2 + 4OH^- \longrightarrow 4H_2O + 4e^- & \textit{Anodic oxidation} \\
\underline{O_2 + 2H_2O + 4e^- \longrightarrow 4OH^-} & \textit{Cathodic reduction} \\
2H_2 + O_2 \longrightarrow 2H_2O & \textit{Cell reaction}
\end{array}
$$

A great deal of research effort is being expended in investigating other fuels such as methane and other hydrocarbons, other electrode systems, and cells that will operate efficiently under extremes of temperature. A fuel such as methane or other hydrocarbons can be made to react in the presence of air to produce hydrogen and carbon dioxide. The hydrogen then is fed into the cell at the anode, and air (to provide oxygen) is introduced into the cell at the cathode, to produce the reactions described earlier.

The efficiency of the fuel cell is potentially greater than that of conventional generating equipment, because of producing electric current directly from the reaction of the fuel and oxidizer without going through the inherently wasteful intermediate conversion of chemical energy to heat, which in conventional power plants can waste 60–80% of the available chemical energy. Fuel cells can, at the present stage of development, increase conversion efficiencies from about 30% in conventional power plants to over 40%. It is estimated that sufficient power for 20,000 people can be produced in a unit that is 18 feet high and covers less than one-half acre of ground. An additional advantage of the fuel cell as a source of commercial power is that its use produces no pollution, because the chemical products of a cell such as those described here are essentially air, carbon dioxide, and water vapor. However, one must take into account power needs for the process, such as the energy required to decompose the fuel and to pump the cell components into the cell and the pollution inherent in the production of such power needs, in assessing the overall efficiency of fuel cells for producing power commercially.

Oxidation-Reduction Reactions

Oxidation-reduction reactions (redox reactions) were defined in Section 8.2, part 4, as reactions that involve oxidation and reduction of the reactants. Since the chemical changes occurring in electrochemical cells involve oxidation of one species at the anode and reduction of another species at the cathode, the net reactions of most electrochemical cells are oxidation-reduction reactions. Those net reactions characterized by a positive value of the cell potential will proceed spontaneously either in an electrochemical cell or when the reactants are mixed together. For example, metallic zinc will transfer electrons directly to a copper(II) ion and reduce it to metallic copper leaving zinc(II) ions when a piece of zinc metal is placed in a solution of copper nitrate, or it will transfer electrons through an external wire when a zinc half-cell and a copper half-cell are combined in an electrochemical cell, as described in Section 20.12.

20.20   Spontaneity, Free Energy Change, and Equilibrium Constants of Redox Reactions

Since a spontaneous oxidation-reduction reaction will occur either in an electrochemical cell or upon direct mixing of the reactants, standard reduction potentials and the Nernst equation can be used to predict the spontaneity of an oxidation-reduction reaction as well as the free energy change of the reaction and the equilibrium constant of the reaction (Section 20.15). The oxidation-reduction reaction, however, must be written as the sum of two half-reactions with known standard reduction potentials.

EXAMPLE   The chloride ion can be oxidized to chlorine in an acid solution by the permanganate ion:

$$2MnO_4^- + 10Cl^- + 16H^+ \longrightarrow 2Mn^{2+} + 5Cl_2 + 8H_2O$$

Show that this reaction is spontaneous and calculate ΔG° and K for the reaction.

In order to solve this problem, we need to write the reaction as the sum of two half-cell reactions. The following half-reactions are found in Appendix I:

$$Cl_2 + 2e^- \longrightarrow 2Cl^- \qquad E^\circ = 1.3595 \text{ V}$$
$$MnO_4^- + 8H^+ + 5e^- \longrightarrow Mn^{2+} + 4H_2O \qquad E^\circ = 1.51 \text{ V}$$

The oxidation-reduction reaction can be written as the following sum:

$$
\begin{array}{ll}
5(2Cl^- \longrightarrow Cl_2 + 2e^-) & E^\circ = -1.3595 \text{ V} \\
2(MnO_4^- + 8H^+ + 5e^- \longrightarrow Mn^{2+} + 4H_2O) & E^\circ = +1.51 \text{ V} \\
\hline
2MnO_4^- + 16H^+ + 10Cl^- \longrightarrow 2Mn^{2+} + 5Cl_2 + 8H_2O & E^\circ = +0.15 \text{ V}
\end{array}
$$

Since the potential for a cell using this net reaction is positive, the reaction is spontaneous.

ΔG° and K can be determined as described in Section 20.15.

$$\Delta G^\circ = -nFE^\circ$$
$$= -10 \times (96.487 \text{ kJ V}^{-1}) \times (0.15 \text{ V})$$
$$= 1.4 \times 10^2 \text{ kJ}$$

$$E^\circ = \frac{RT}{nF} \ln K = \frac{0.05915}{n} \log K$$

$$\log K = E^\circ \times \frac{n}{0.05915} = 0.15 \times \frac{10}{0.05915}$$

$$= 25.359$$

$$K = 2.3 \times 10^{25}$$

◆ ◆ ◆ ◆ ◆

20.21   Balancing Redox Equations by the Half-Reaction Method

Oxidation-reduction reactions can become quite complicated, and sometimes the equations describing them can be difficult to balance by trial and error. The half-reaction method provides a systematic approach for balancing redox equations.

In the half-reaction method of balancing redox equations, the net reaction is broken down into two half-reactions, one representing the oxidation step and the other the reduction step. Each half-reaction is balanced separately, then the two half-reactions are added to give the overall equation. This method was used in the example in the previous section. Additional examples follow.

EXAMPLE 1   **Iron(II) is oxidized to iron(III) by chlorine. Use the half-reaction method to balance the following: $Fe^{2+} + Cl_2 \longrightarrow Fe^{3+} + Cl^-$.**

The oxidation half-reaction is for the oxidation of Fe^{2+} to Fe^{3+}.

$$Fe^{2+} \longrightarrow Fe^{3+}$$

To balance this half-reaction in terms of ion charges and electrons, it is necessary to add one electron to the right side of the equation.

$$Fe^{2+} \longrightarrow Fe^{3+} + e^-$$
$$+2 \quad = \quad +3 \quad -1$$

To balance the reduction half-reaction, which is for the reduction of Cl_2 to Cl^-, it is necessary to add two electrons to the left side of the equation.

$$Cl_2 + 2e^- \longrightarrow 2Cl^-$$
$$-2 \quad = \quad -2$$

In the oxidation half-reaction one electron is lost, while in the reduction half-reaction two electrons are gained. To balance the electrons the oxidation half-reaction must be multiplied by 2.

$$2Fe^{2+} \longrightarrow 2Fe^{3+} + 2e^-$$

On addition of the balanced half-reactions, the two electrons on either side of the equation cancel, and the balanced equation is obtained.

$$2Fe^{2+} \longrightarrow 2Fe^{3+} + 2e^-$$
$$\underline{Cl_2 + 2e^- \longrightarrow 2Cl^-}$$
$$2Fe^{2+} + Cl_2 \longrightarrow 2Fe^{3+} + 2Cl^-$$

◆ ◆ ◆ ◆ ◆

EXAMPLE 2   **The dichromate ion will oxidize iron(II) to iron(III) in acid solution. Balance the equation $Cr_2O_7{}^{2-} + H^+ + Fe^{2+} \longrightarrow Cr^{3+} + Fe^{3+} + H_2O$ by the half-reaction method.**

The oxidation half-reaction is balanced as shown in Example 1.

$$Fe^{2+} \longrightarrow Fe^{3+} + e^-$$

The reduction half-reaction involves the change

$$Cr_2O_7{}^{2-} \longrightarrow 2Cr^{3+}$$

In acid solution, excess oxygen combines with H^+ to give H_2O; thus this half-reaction also involves H^+ and H_2O. By inspection, the 7 atoms of oxygen from the $Cr_2O_7{}^{2-}$ ion require $14 H^+$ ions to produce 7 molecules of water.

$$Cr_2O_7{}^{2-} + 14H^+ \longrightarrow 2Cr^{3+} + 7H_2O$$

To balance this reduction half-reaction in terms of ion charges and electrons, it is necessary to add 6 electrons on the left side of the equation.

$$Cr_2O_7{}^{2-} + 14H^+ + 6e^- \longrightarrow 2Cr^{3+} + 7H_2O$$
$$\phantom{Cr_2O_7{}}{-2} {+14} {-6} = {+6}$$

In order that both half-reactions involve the same number of electrons, the oxidation half-reaction must be multiplied by 6.

$$6Fe^{2+} \longrightarrow 6Fe^{3+} + 6e^-$$

On addition of the balanced half-reactions, the six electrons on either side of the equation will cancel and we have

$$6Fe^{2+} \longrightarrow 6Fe^{3+} + 6e^-$$
$$Cr_2O_7{}^{2-} + 14H^+ + 6e^- \longrightarrow 2Cr^{3+} + 7H_2O$$
$$\overline{Cr_2O_7{}^{2-} + 14H^+ + 6Fe^{2+} \longrightarrow 2Cr^{3+} + 6Fe^{3+} + 7H_2O}$$

◆ ◆ ◆ ◆ ◆

EXAMPLE 3   **Hydrogen peroxide in acidic solution oxidizes Fe^{2+} to Fe^{3+}. Use the half-reaction method to balance the following: $H_2O_2 + H^+ + Fe^{2+} \longrightarrow H_2O + Fe^{3+}$.**

The oxidation half-reaction is

$$Fe^{2+} \longrightarrow Fe^{3+} + e^-$$

The reduction half-reaction involves the reduction of H_2O_2 to H_2O in acid solution.

$$H_2O_2 \longrightarrow ?H_2O$$

By inspection, it is seen that there is sufficient oxygen in one molecule of H_2O_2 to form two molecules of water. This will require two hydrogen ions.

$$H_2O_2 + 2H^+ \longrightarrow 2H_2O$$

Two electrons are required to balance the positive charges on the hydrogen ions.

$$H_2O_2 + 2H^+ + 2e^- \longrightarrow 2H_2O$$

In order that both half-reactions involve the same number of electrons, the oxidation half-reaction must be multiplied by 2.

$$2Fe^{2+} \longrightarrow 2Fe^{3+} + 2e^-$$

Now let us add the two half-reactions.

$$2Fe^{2+} \longrightarrow 2Fe^{3+} + 2e^-$$
$$H_2O_2 + 2H^+ + 2e^- \longrightarrow 2H_2O$$
$$\overline{H_2O_2 + 2H^+ + 2Fe^{2+} \longrightarrow 2H_2O + 2Fe^{3+}}$$

◆ ◆ ◆ ◆ ◆

When using the half-reaction method in basic solution, remember that in *a basic solution* excess oxygen combines with water to give hydroxide ion ($O^{2-} + H_2O \longrightarrow 2OH^-$).

EXAMPLE 4   **Hydrogen peroxide reduces MnO_4^- to MnO_2 in basic solution. Write a balanced equation for the reaction.**

Initially we have the following information.

$$H_2O_2 + MnO_4^- \longrightarrow MnO_2$$

and the knowledge that the reaction is in basic solution. The reduction half-reaction must involve the change

$$MnO_4^- \longrightarrow MnO_2$$

Since the reaction is in basic solution, the excess oxygen combines with water to give hydroxide ion.

$$MnO_4^- + 2H_2O \longrightarrow MnO_2 + 4OH^-$$

Adding the electrons to balance the charge gives the following half-reaction:

$$MnO_4^- + 2H_2O + 3e^- \longrightarrow MnO_2 + 4OH^-$$

The oxidation half-reaction involves oxidation of H_2O_2. Since oxygen is already in an oxidation state of -1 in H_2O_2, it can only be oxidized to O_2, with an oxidation state of zero. The oxidation half-reaction must involve

$$H_2O_2 \longrightarrow O_2$$

In basic solution, excess hydrogen combines with hydroxide ion to give water.

$$H_2O_2 + 2OH^- \longrightarrow O_2 + 2H_2O$$

Addition of electrons gives

$$H_2O_2 + 2OH^- \longrightarrow O_2 + 2H_2O + 2e^-$$

After multiplying to get the same number of electrons in both half-reactions, they are added to give the overall equation

$$2MnO_4^- + 4H_2O + 6e^- \longrightarrow 2MnO_2 + 8OH^-$$
$$\underline{3H_2O_2 + 6OH^- \longrightarrow 3O_2 + 6H_2O + 6e^-}$$
$$2MnO_4^- + 4H_2O + 3H_2O_2 + 6OH^- \longrightarrow 2MnO_2 + 8OH^- + 3O_2 + 6H_2O$$

Cancellation of identical species on the left and right sides of the equation gives

$$2MnO_4^- + 3H_2O_2 \longrightarrow 2MnO_2 + 3O_2 + 2OH^- + 2H_2O$$

◆ ◆ ◆ ◆ ◆

20.22   Some Half-Reactions

When redox equations are to be balanced by the half-reaction method, it is necessary that the proper half-reactions be selected. The following half-reactions should be studied carefully, for they are frequently encountered in the balancing of redox equations. The products listed are the *major products* under the conditions given, but in many cases are not the sole products.

$H_2O_2 + 2H^+ + 2e^- \longrightarrow 2H_2O$	Hydrogen peroxide as an oxidizing agent in acid solution
$H_2O_2 \longrightarrow O_2 + 2H^+ + 2e^-$	Hydrogen peroxide as a reducing agent in acid solution.
$MnO_4^- + 8H^+ + 5e^- \longrightarrow Mn^{2+} + 4H_2O$	Permanganate ion as an oxidizing agent in acid solution.
$MnO_4^- + 2H_2O + 3e^- \longrightarrow MnO_2 + 4OH^-$	Permanganate ion as an oxidizing agent in neutral or basic solution.
$Cr_2O_7^{2-} + 14H^+ + 6e^- \longrightarrow 2Cr^{3+} + 7H_2O$	Dichromate ion as an oxidizing agent in acid solution.
$NO_3^- + 2H^+ + e^- \longrightarrow NO_2 + H_2O$	Concentrated nitric acid as an oxidizing agent toward less active metals, such as Cu, Ag, and Pb, and certain anions (such as Br$^-$).
$NO_3^- + 4H^+ + 3e^- \longrightarrow NO + 2H_2O$	Dilute nitric acid as an oxidizing agent toward less active metals and certain non-metals (such as Br$^-$).
$2NO_3^- + 10H^+ + 8e^- \longrightarrow N_2O + 5H_2O$	Dilute nitric acid as an oxidizing agent toward moderately active metals such as Zn and Fe.
$H_2SO_4 + 2H^+ + 2e^- \longrightarrow SO_2 + 2H_2O$	Concentrated sulfuric acid as an oxidizing agent toward HBr, C, and Cu.
$H_2SO_4 + 6H^+ + 6e^- \longrightarrow S + 4H_2O$	Concentrated sulfuric acid as an oxidizing agent toward H_2S.
$H_2SO_4 + 8H^+ + 8e^- \longrightarrow H_2S + 4H_2O$	Concentrated sulfuric acid as an oxidizing agent toward HI.
$NO_2^- + 2H^+ + e^- \longrightarrow NO + H_2O$	Nitrous acid as an oxidizing agent.
$NO_2^- + H_2O \longrightarrow NO_3^- + 2H^+ + 2e^-$	Nitrous acid as a reducing agent.
$HClO + H^+ + 2e^- \longrightarrow Cl^- + H_2O$	Hypochlorous acid as an oxidizing agent.
$X_2 + 2e^- \longrightarrow 2X^-$	The halogens F_2, Cl_2, Br_2, and I_2 as oxidizing agents.
$O_2 + 4H^+ + 4e^- \longrightarrow 2H_2O$	Oxygen as an oxidizing agent in acid solution.
$2H^+ + 2e^- \longrightarrow H_2$	The hydrogen ion as an oxidizing agent.
$H_2 \longrightarrow 2H^+ + 2e^-$	Molecular hydrogen as a reducing agent in aqueous solution.
$M \longrightarrow M^{n+} + ne^-$	Metals as reducing agents.
$M^{n+} + ne^- \longrightarrow M$	Metal ions as oxidizing agents.
$M^{n+} \longrightarrow M^{n+x} + xe^-$	Oxidation of a metal ion to a higher oxidation number. For example, $Cr^{2+} \longrightarrow Cr^{3+} + e^-$; $Fe^{2+} \longrightarrow Fe^{3+} + e^-$; $Sn^{2+} \longrightarrow Sn^{4+} + 2e^-$.
$M^{n+} + xe^- \longrightarrow M^{n-x}$	Reduction of a metal ion to a lower oxidation number. For example, $Fe^{3+} + e^- \longrightarrow Fe^{2+}$.

20.23   Balancing Redox Equations by the Change in Oxidation Number Method

The fact that oxidation and reduction always occur together and to the same extent makes it possible to use the changes in oxidation number in balancing redox equations. In other words, proportions of the reactants and products in the balanced redox equation must be such that the total increase in units of oxidation number must equal the total decrease in units of oxidation number.

EXAMPLE 1   Use the change in oxidation number method to balance the equation for the reaction of antimony and chlorine to form antimony(III) chloride.

First write the reactants and products of the reaction.

$$Sb + Cl_2 \longrightarrow SbCl_3$$

Indicate the oxidation number of each atom in the reaction (see Sections 5.11 and 8.5). Remember that elements in the free state are always assigned an oxidation number of zero.

$$\overset{0}{Sb} + \overset{0}{Cl_2} \longrightarrow \overset{+3\ -1}{SbCl_3}$$

Designate the changes in oxidation number that occur during the reaction. Use an ascending arrow ($\uparrow$) to denote an increase in oxidation number and a descending arrow ($\downarrow$) to denote a decrease in oxidation number.

$$\overset{0}{Sb} \longrightarrow \overset{+3}{Sb} \uparrow 3 \qquad \text{(total increase of 3 in oxidation number per formula unit, Sb)}$$

$$\overset{0}{Cl_2} \longrightarrow \overset{-1}{2Cl} \downarrow 2 \qquad \text{(total decrease of 2 in oxidation number per formula unit, } Cl_2\text{)}$$

To balance the equation we must select the proper numbers of antimony and chlorine atoms so that the total increase in units of oxidation number will equal the total decrease. If we use two atoms of antimony and three molecules (six atoms) of chlorine, then there will be both an increase and a decrease of six units of oxidation number.

$$\overset{0}{Sb} \longrightarrow \overset{+3}{Sb} \uparrow \qquad 3 \times 2 = 6$$

$$\overset{0}{Cl_2} \longrightarrow \overset{-1}{2Cl} \downarrow \qquad 2 \times 3 = 6$$

Our final equation then becomes

$$2Sb + 3Cl_2 \longrightarrow 2SbCl_3$$

◆ ◆ ◆ ◆ ◆

EXAMPLE 2   The chloride ion is oxidized by the permanganate ion in acid solution to give the manganese(II) ion, molecular chlorine, and water. Balance the equation

$MnO_4^- + Cl^- + H^+ \longrightarrow Mn^{2+} + Cl_2 + H_2O$, using the change in oxidation number method.

Upon assignment of oxidation numbers, it becomes evident that manganese is reduced from $+7$ to $+2$ and chlorine is oxidized from -1 to 0.

$$\overset{+7}{Mn}O_4^- + Cl^- + H^+ \longrightarrow Mn^{2+} + \overset{0}{Cl_2} + H_2O$$

$$2Cl^- \longrightarrow \overset{0}{Cl_2} \uparrow \qquad 2 \times 5 = 10$$

$$\overset{+7}{Mn} \longrightarrow Mn^{2+} \downarrow \qquad 5 \times 2 = 10$$

Ten chloride ions will increase by ten units of oxidation number in going to five chlorine molecules (ten chlorine atoms). Two manganese atoms will decrease by ten units of oxidation number.

$$2MnO_4^- + 10Cl^- + ?H^+ \longrightarrow 2Mn^{2+} + 5Cl_2 + ?H_2O$$

To complete the balancing of this ionic equation, we may now balance the ion charges on both sides of the equation. On the right side of the equation, the only charged particles are the two Mn^{2+} with a total ionic charge of $+4$. The algebraic sum of the charges on the left must also be equal to $+4$. Excluding the H^+ for the moment, the total charge on the left is -12 ($2MnO_4^- + 10Cl^-$). Sixteen H^+ are necessary to give an algebraic sum of $+4$ ($-2 - 10 + 16 = +4$). The 16 hydrogen ions are enough to produce 8 molecules of water.

$$2MnO_4^- + 10Cl^- + 16H^+ \longrightarrow 2Mn^{2+} + 5Cl_2 + 8H_2O$$

The balancing of the equation is checked by noting that there are 8 oxygen atoms on each side of the arrow.

◆ ◆ ◆ ◆ ◆

EXAMPLE 3   **Very active reducing agents such as zinc reduce dilute nitric acid to nitrous oxide (N_2O) or ammonium ion. Balance the equation for the reduction of HNO_3 with zinc, to give NH_4^+.**

Write the reactants and products of the reaction producing NH_4^+, and assign oxidation numbers to the atoms that undergo a change in oxidation number.

$$\overset{0}{Zn} + \overset{+5}{N}O_3^- + H^+ \longrightarrow Zn^{2+} + \overset{-3}{N}H_4^+ + H_2O$$

Balance the equation with regard to the atoms that change in oxidation number.

$$\overset{0}{Zn} \longrightarrow Zn^{2+} \uparrow \qquad 2 \times 4 = 8$$

$$\overset{+5}{N} \longrightarrow \overset{-3}{N} \downarrow \qquad 8 \times 1 = 8$$

$$4Zn + NO_3^- + ?H^+ \longrightarrow 4Zn^{2+} + NH_4^+ + ?H_2O$$

Balance the ion charges to find the number of H^+ required.

$$NO_3^- + ?H^+ \longrightarrow 4Zn^{2+} + NH_4^+$$

$$4Zn + \underset{-1}{NO_3^-} + \underset{+10}{10H^+} \underset{=}{\longrightarrow} \underset{+8}{4Zn^{2+}} + \underset{+1}{NH_4^+} + ?H_2O$$

Of the 10 hydrogen atoms on the left, 4 are found in the NH_4^+ on the right. Thus 6 hydrogen atoms are left to form 3 molecules of water on the right.

$$4Zn + NO_3^- + 10H^+ \longrightarrow 4Zn^{2+} + NH_4^+ + 3H_2O$$

As a check on balancing the equation, we note there are 3 oxygen atoms on each side of the arrow.

◆ ◆ ◆ ◆ ◆

Questions

1. How does conduction of electricity in a metal differ from conduction of electricity in a solution of an electrolyte?
2. How does a voltaic cell differ from an electrolytic cell?
3. Define an anode and a cathode. Under what conditions does an anode bear a positive charge? a negative charge?
4. Complete and balance the following half-reactions. (In each case indicate whether oxidation or reduction occurs.)

   (a) $Sn^{4+} \longrightarrow Sn^{2+}$
   (b) $[Ag(NH_3)_2]^+ \longrightarrow Ag$
   (c) $Hg_2Cl_2 \longrightarrow Hg$
   (d) $O_2 \longrightarrow H_2O$ (in acid)
   (e) $O_2 \longrightarrow OH^-$ (in base)
   (f) $SO_3^{2-} \longrightarrow SO_4^{2-}$ (in acid)
   (g) $MnO_4^- \longrightarrow Mn^{2+}$ (in acid)
   (h) $Cl^- \longrightarrow ClO_3^-$ (in base)

5. Diagram voltaic cells with the following net reactions:

   (a) $Mn + 2Ag^+ \longrightarrow Mn^{2+} + Ag$
   (b) $Sn^{4+} + H_2 \longrightarrow Sn^{2+} + 2H^+$
   (c) $Cr_2O_7^{2-} + 14H^+ + 6I^- \longrightarrow$
   $ 3I_2 + 2Cr^{3+} + 7H_2O$

6. By means of description and chemical equations, explain the electrolytic purification of copper from impurities such as silver, zinc, and gold.

7. Which metals in Table 20-1 could be purified by electrolysis in a manner similar to that used to purify copper?

8. Why are different products obtained when molten $ZnCl_2$ is electrolyzed than when a solution of $ZnCl_2$ is electrolyzed? Why do (1) a solution of $CuCl_2$ and (2) molten $CuCl_2$ give the same products upon electrolysis?

9. When gold is plated electrochemically from a basic solution of $[Au(CN)_4]^-$, O_2 forms at one electrode and Au is deposited at the other. Write the half-reactions occurring at each electrode and the net reaction for the electrochemical cell. (The cyanide ion, CN^-, is not oxidized or reduced under these conditions.)

10. Why is the Nernst equation important in electrochemistry?

11. Describe the construction of a dry cell, and write the half-reactions and the net reaction of such a cell.

12. A lead storage battery is used to electrolyze a solution of hydrogen chloride. Sketch the two cells, label the cathode and anode in each cell, give the sign of each electrode, indicate the direction of flow of electrons through the system, show the movement of ions in the cells, and write the half-reactions occurring at each electrode.

13. What are the energy conversions that take place

during the charging of a lead storage battery with electrons from a coal-fired electric generating station?

14. Why does the potential of a single electrode change as the concentrations of the species involved in the half-reaction change?

15. Which is the better oxidizing agent in each of the following pairs at standard conditions?
    (a) Cr^{3+} or Co^{2+}
    (b) I_2 or Mn^{2+}
    (c) MnO_4^- or $Cr_2O_7^{2-}$ in acid solution
    (d) Pb^{+2} or Sn^{4+}

16. Which is the better reducing agent in each of the following pairs at standard conditions?
    (a) F^- or Cu
    (b) H_2 or I^-
    (c) Sn^{2+} or Fe^{2+}
    (d) Fe^{2+} or Cr

17. State Faraday's Law of Electrolysis. What is a faraday of electricity?

18. Define the phrase "standard electrode potential."

19. What is the origin of the zero value for the standard hydrogen electrode?

20. In this and previous chapters we have noted commercial electrolytic processes for the preparation of several substances, some of which are hydrogen, oxygen, chlorine, and sodium hydroxide. Write individual electrode half-reaction equations describing these commercial processes.

21. Soon after a copper metal rod is placed in a silver nitrate solution, copper ions are observed in the solution and silver metal is observed to have deposited on the rod. Can the copper rod be considered an electrode? If so, is it an anode or a cathode?

22. Show by suitable equations that the electrode reaction for the zinc electrode is a reversible one.

23. The Nernst equation for the net reaction of an electrochemical cell is

$$E = E° - \frac{0.05915}{n} \log Q$$

where Q is the reaction quotient for the reaction and n is the total number of electrons transferred in the net reaction. Show that the Nernst equation for the net reaction

$$2Au + 3Cl_2 \longrightarrow 2Au^{3+} + 6Cl^-$$

is identical to that derived by addition of the Nernst

equations for the two half-reactions.

$$Au \longrightarrow Au^{3+} + 3e^- \quad E° = -1.50 \text{ V}$$
$$Cl_2 + 2e^- \longrightarrow 2Cl^- \quad E° = 1.3595 \text{ V}$$

24. Balance the following redox equations by either the half-reaction method or the change of oxidation number method:
    (a) $IF_5 + Fe \longrightarrow FeF_3 + IF_3$
    (b) $Sn^{2+} + Cu^{2+} \longrightarrow Sn^{4+} + Cu^+$
    (c) $H_2S + Hg_2^{2+} \longrightarrow Hg + S + H^+$
    (d) $CN^- + ClO_2^- \longrightarrow CNO^- + Cl^-$
    (e) $Zn + BrO_4^- + OH^- + H_2O \longrightarrow$
       $[Zn(OH)_4]^{2-} + Br^-$
    (f) $H_2SO_4 + HBr \longrightarrow SO_2 + Br_2 + H_2O$
    (g) $MnO_4^- + S^{2-} + H_2O \longrightarrow$
       $MnO_2 + S + OH^-$
    (h) $NO_3^- + I_2 + H^+ \longrightarrow IO_3^- + NO_2 + H_2O$
    (i) $Cu + H^+ + NO_3^- \longrightarrow Cu^{2+} + NO_2 + H_2O$
    (j) $Zn + H^+ + NO_3^- \longrightarrow Zn^{2+} + N_2O + H_2O$
    (k) $Cu + H^+ + NO_3^- \longrightarrow Cu^{2+} + NO + H_2O$
    (l) $MnO_4^- + H_2S + H^+ \longrightarrow Mn^{2+} + S + H_2O$
    (m) $MnO_4^- + NO_2^- + H_2O \longrightarrow$
       $MnO_2 + NO_3^- + OH^-$
    (n) $MnO_4^{2-} + H_2O \longrightarrow$
       $MnO_4^- + OH^- + MnO_2$
    (o) $Br_2 + SO_2 + H_2O \longrightarrow H^+ + Br^- + SO_4^{2-}$

25. Balance the following redox equations by either the half-reaction method or the change of oxidation number method:
    (a) $Al + [Sn(OH)_4]^{2-} \longrightarrow$
       $[Al(OH)_4]^- + Sn + OH^-$
    (b) $Cl^- + H^+ + NO_3^- \longrightarrow Cl_2 + NO_2 + H_2O$
    (c) $H_2S + H_2O_2 \longrightarrow S + H_2O$
    (d) $MnO_4^- + Se^{2-} + H_2O \longrightarrow$
       $MnO_2 + Se + OH^-$
    (e) $MnO_4^{2-} + Cl_2 \longrightarrow MnO_4^- + Cl^-$
    (f) $OH^- + NO_2 \longrightarrow NO_3^- + NO_2^- + H_2O$
    (g) $HBrO \longrightarrow H^+ + Br^- + O_2$
    (h) $Br_2 + CO_3^{2-} \longrightarrow Br^- + BrO_3^- + CO_2$
    (i) $NH_3 + O_2 \longrightarrow NO + H_2O$
    (j) $C + HNO_3 \longrightarrow NO_2 + H_2O + CO_2$
    (k) $HClO_3 \longrightarrow HClO_4 + ClO_2 + H_2O$
    (l) $ClO_3^- + H_2O + I_2 \longrightarrow IO_3^- + Cl^- + H^+$
    (m) $Cr_2O_7^{2-} + HNO_2 + H^+ \longrightarrow$
       $Cr^{3+} + NO_3^- + H_2O$

26. Complete and balance the following equations. (Remember that when a reaction occurs in acidic solution, H^+ and/or H_2O may be added on either

side of the equation, as necessary, to balance the equation properly; when a reaction occurs in basic solution, OH^- and/or H_2O may be added, as necessary, on either side of the equation. No indication of the acidity of the solution is given if neither H^+ nor OH^- is involved as a reactant or product.)

[S](a) $Zn + NO_3^- \longrightarrow Zn^{2+} + N_2$ (acidic solution)

(b) $Zn + NO_3^- \longrightarrow Zn^{2+} + NH_3$ (basic solution)

[S](c) $CuS + NO_3^- \longrightarrow Cu^{2+} + S + NO$ (acidic solution)

(d) $NH_3 + O_2 \longrightarrow NO_2$ (gas phase)

(e) $H_2SO_4 + HI \longrightarrow I_2 + SO_2$ (acidic solution)

[S](f) $Cl_2 + OH^- \longrightarrow Cl^- + ClO_3^-$ (basic solution)

(g) $H_2O_2 + MnO_4^- \longrightarrow Mn^{2+} + O_2$ (acidic solution)

[S](h) $NO_2 \longrightarrow NO_3^- + NO_2^-$ (basic solution)

(i) $KClO_3 \longrightarrow KCl$ (no solvent)

[S](j) $Fe^{3+} + I^- \longrightarrow Fe^{2+} + I_2$

(k) $P_4 \longrightarrow PH_3 + HPO_3^{2-}$ (basic solution)

[S](l) $P_4 \longrightarrow PH_3 + HPO_3^{2-}$ (acidic solution)

27. Complete and balance the following reactions (see the instructions for Question 26):

(a) $Zn + H^+ \longrightarrow$ (acidic solution)

(b) $MnO_2 + Cl^- \longrightarrow$ (acidic solution)

(c) $Pb^{4+} + Sn^{2+} \longrightarrow$

(d) $Fe^{2+} + H_2O_2 \longrightarrow$ (acidic solution)

(e) $Cl_2 + SO_2 \longrightarrow$ (acidic solution)

(f) $ZnS + O_2 \xrightarrow{\Delta}$ (no solvent)

(g) $ClO^- + Sn^{2+} \longrightarrow$ (acidic solution)

(h) $PbO_2 + SeO_3^{2-} \longrightarrow$ (basic solution)

(i) $Cr_2O_7^{2-} + Br^- \longrightarrow$ (acidic solution)

Problems

Standard reduction potentials for these problems may be found in Tables 20-1 and 20-2 and Appendix I.

[S]1. Calculate the value of the Faraday constant, *F*, from the charge on a single electron, 1.6021×10^{-19} coulombs. *Ans.* 9.648×10^4 *coulombs*

2. How many moles of electrons are involved in the following electrochemical changes?

[S](a) 1.0 mol of Al^{3+} is converted to Al. *Ans. 3.0 mol e⁻*

[S](b) 0.800 mol of I_2 is converted to I^-. *Ans. 1.60 mol e⁻*

(c) 118.7 g of Sn^{2+} is converted to Sn^{4+}. *Ans. 2.000 mol e⁻*

[S](d) 27.6 g of SO_3 is converted to SO_3^{2-}. *Ans. 0.690 mol e⁻*

(e) 0.174 g of MnO_4^- is converted to Mn^{2+}. *Ans.* 7.31×10^{-3} *mol e⁻*

(f) 1.0 ℓ of O_2 at STP is converted to H_2O in acid solution. *Ans. 0.18 mol e⁻*

(g) 100 ml of 0.50 *M* Cu^{2+} is converted to Cu. *Ans. 0.10 mol e⁻*

[S](h) 15.80 ml of 0.1145 *M* MnO_4^- is converted to Mn^{2+}. *Ans.* 9.046×10^{-3} *mol e⁻*

3. How many faradays of electricity are involved in the electrochemical changes described in Problem 2?
Ans. [S] *(a) 3.0;* [S] *(b) 1.60; (c) 2.000; (d) 0.690; (e)* 7.31×10^{-3}*; (f) 0.18; (g) 0.10;* [S] *(h)* 9.046×10^{-3}

4. How many coulombs of electricity are involved in the electrochemical changes described in Problem 2?
Ans. [S] *(a)* 2.9×10^5*;* [S] *(b)* 1.54×10^5*; (c)* 1.930×10^5*; (d)* 6.66×10^4*; (e)* 7.05×10^2*; (f)* 1.7×10^4*; (g)* 9.6×10^3*;* [S] *(h)* 8.728×10^2

5. Calculate the emf for cells with the following net reactions run at standard conditions: (Consult Tables 20-1 and 20-2, or Appendix I for standard reduction potentials.)

[S](a) $Mn + Cd^{2+} \longrightarrow Cd + Mn^{2+}$ *Ans. +0.78 V*

(b) $2Al + 3Co^{2+} \longrightarrow 2Al^{3+} + 3Co$ *Ans. +1.38 V*

[S](c) $2Br^- + I_2 \longrightarrow Br_2 + 2I^-$ *Ans. −0.5297 V*

(d) $Cr_2O_7^{2-} + 3Fe + 14H^+ \longrightarrow$
$$2Cr^{3+} + 3Fe^{2+} + 7H_2O$$
Ans. +1.77 V

(e) $2Mn^{2+} + 8H_2O + 5Fe^{2+} \longrightarrow$
$$2MnO_4^- + 16H^+ + 5Fe$$
Ans. −1.95 V

6. Determine the standard emf for the following cells. (As is customary, the anode is written on the left.)

S (a) $Co, Co^{2+} M = 1 \overset{e^-}{\parallel} Cr^{3+} M = 1, Cr$
Ans. −0.46 V

(b) $Ni, Ni^{2+} M = 1 \overset{e^-}{\parallel} Br_2(l), Br^- M = 1, (Pt)$
Ans. +1.315 V

(c) $Pb, PbSO_4(s), SO_4^{2-} M = \overset{e^-}{\parallel} H^+ M = 1,$
$$H_2(Pt) \text{ 1 atm}$$
Ans. +0.356 V

(d) $(Pt) Mn^{2+} M = 1, MnO_4^- M = 1, H^+ M = 1 \overset{e^-}{\parallel}$
$$Fe^{3+} M = 1, Fe^{2+} M = 1 (Pt)$$
Ans. −0.74 V

7. Write the net reactions for the cells in Problem 6. Which of these reactions will proceed spontaneously.

Ans. (a) $3Co + 2Cr^{3+} \longrightarrow 3Co^{2+} + 2Cr;$
(b) $Ni + Br_2(l) \longrightarrow Ni^{2+} + 2Br^-;$
(c) $Pb + 2H^+ + SO_4^{2-} \longrightarrow PbSO_4 + H_2;$
(d) $Mn^{2+} + 5Fe^{3+} + 4H_2O \longrightarrow$
$$MnO_4^- + 8H^+ + 5Fe^{2+};$$
Reactions (b) and (c) are spontaneous.

8. Calculate the standard free energy change and equilibrium constant for each net reaction in Problem 5.

Ans. S (a) $\Delta G° = −150 kJ, K = 2.4 \times 10^{26};$
(b) $\Delta G° = −799 kJ, K = 9.62 \times 10^{139};$
S (c) $\Delta G° = 102.2 kJ, K = 1.229 \times 10^{-18};$
(d) $\Delta G° = −1020 kJ, K = 3.49 \times 10^{179};$
(e) $\Delta G° = 1880 kJ, K = 2.14 \times 10^{-330}$

9. Calculate the emf for each of the following half-reactions:

S (a) $Sn^{2+} (0.0100 M) + 2e^- \longrightarrow Sn$
Ans. −0.195 V

(b) $Hg \longrightarrow Hg^{2+} (0.2500 M) + 2e^-$
Ans. −0.836 V

S (c) $O_2 (0.0010 atm) + 4H^+ (0.100 M) + 4e^- \longrightarrow$
$$2H_2O(l) \quad \text{Ans. 1.13 V}$$

S (d) $Cr_2O_7^{2-}(0.150 M) + 14H^+(0.100 M) + 6e^- \longrightarrow$
$$2Cr^{3+} (0.000100 M) + 7H_2O(l)$$
Ans. 1.26 V

(e) $Mn^{2+} (0.0125 M) + 4H_2O(l) \longrightarrow$
$$MnO_4^- (0.0125 M) + 8H^+ (0.100 M) + 5e^-$$
Ans. −1.42 V

(f) $Sn^{4+} (0.00010 M) + 2e^- \longrightarrow Sn^{2+} (4.0 M)$
Ans. 0.01 V

10. Chromium metal can be plated electrochemically from an acidic aqueous solution of CrO_3. (a) What is the half-reaction for the process? (b) What mass of chromium, in grams, will be deposited by a current of 2.50 A passing for 20.0 min? (c) How long will it take to deposit 1.0 g of chromium using a current of 10.0 A
Ans. $CrO_3 + 6H^+ \longrightarrow Cr + 3H_2O$; 0.269 g; 1100 sec

S 11. How many grams of zinc will be deposited from a solution of zinc(II) sulfate by 3.40 faradays of electricity? *Ans. 111 g*

12. An experiment is conducted using the apparatus depicted in Fig. 20-4. How many grams of gold could be plated out of solution by the current required to plate out 4.97 g of copper? How many moles of hydrogen would simultaneously be released from the hydrochloric acid solution? How many moles of oxygen would be freed at each anode in the copper sulfate and silver nitrate solutions?
Ans. 10.3 g Au; 0.0782 mole H_2; 0.0391 mole O_2

S 13. How many grams of cobalt will be deposited from a solution of cobalt(II) chloride electrolyzed with a current of 20.0 A for 54.5 min? *Ans. 20.0 g*

14. How many faradays of electricity would be required to reduce 21.0 g of $Na_2[CdCl_4]$ to metallic cadmium? How long would this take (in minutes) with a current of 7.5 A? *Ans. 0.140 faraday; 30 min*

15. A total of 69,500 coulombs of electricity was required to reduce 37.7 g of M^{3+} to the metal. What is M? *Ans. Gd.*

S 16. A current of 10.0 A is applied to 1.0 l of a solution containing 1.0 mol of HCl for 1.0 hr. Calculate the pH of the solution at the end of this time.
Ans. 0.20

17. A current of 10.0 A is applied to 1.0 l of a solution containing 10.0 mol of NaCl for 1.0 hr. Calculate the pH of the solution at the end of this time.
Ans. 13.57

18. A current of 9.0 A flowed for 45 min through water containing a small quantity of sodium hydroxide. How many liters of gas were formed at the anode at 27.0°C and 750 torr pressure?

Ans. 1.57 l

19. A lead storage battery has initially 200 g of lead and 200 g of PbO_2, plus excess H_2SO_4. Theoretically, how long could this cell deliver a current of 10.0 A, without recharging, if it were possible to operate it so that the reaction goes to completion?

Ans. 4.48 hours

20. Should the following compounds be stable in aqueous solution at a 1 M concentration? (*Hint:* Check the possibility of oxidation or reduction of the anion by the cation.)
    (a) $Ba(MnO_4)_2$ *Ans. Yes*
    (b) FeI_3 *Ans. No*
    (c) $Pd[HgBr_4]$ *Ans. Yes*
    (d) $[Co(NH_3)_6](ClO)_2$ *Ans. No*
    (e) $Na_2[Cd(CN)_4]$ *Ans. Yes*

21. In reading the chapter on halogens (Chapter 19), you learned that when chlorine dissolves in water, it disproportionates, producing chloride ion and hypochlorous acid. At what hydrogen ion concentration does the potential (emf) for the disproportionation of chlorine change from a negative value to a positive value, assuming 1.00 atm pressure and 1.00 M concentration for all species except hydrogen ion? (The standard electrode potential for the reduction of chlorine to chloride ion is 1.36 V and for hypochlorous acid to chlorine, 1.63 V.) Could chlorine be produced from hypochlorite and chloride ions in solution, through a reverse of the disproportionation reaction, by acidifying the solution with strong acid? Explain.   *Ans. 2.7×10^{-5} M; yes*

22. Calculate the electromotive force (emf) under standard conditions of a nickel-cadmium cell; of an Edison storage cell; of a single lead storage cell.

Ans. 0.15 V; 1.37 V; 2.041 V

[s]23. What is the cell with the highest emf that could be constructed from the metals, iron, nickel, copper, and silver?   *Ans. $Fe, Fe^{2+} \parallel Ag^+, Ag$*

[s]24. Calculate the free energy change and equilibrium constant for the reaction $2Br^- + F_2 \longrightarrow 2F^- + Br_2$.

Ans. $\Delta G° = -349$ kJ; $K = 1.59 \times 10^{61}$

25. Using the standard reduction potentials for the half-reactions in the hydrogen-oxygen fuel cell, calculate the free energy change and equilibrium constant for the combustion of hydrogen $(2H_2 + O_2 \longrightarrow 2H_2O)$.

Ans. $\Delta G° = -475$ kJ; $K = 1.51 \times 10^{83}$

26. Copper(I) salts disproportionate in water to form copper(II) salts and copper metal: $2Cu^+ \longrightarrow Cu^{2+} + Cu$. What concentration of Cu^+ remains at equilibrium in 1.00 l of a solution prepared from 1.00 mol of Cu_2SO_4?

Ans. 5.48×10^{-4} M

[s]27. The standard reduction potentials for the reactions

$$Ag^+ + e^- \longrightarrow Ag$$

and $$AgCl + e^- \longrightarrow Ag + Cl^-$$

are 0.7991 and 0.222 V, respectively. From these data and the Nernst equation, calculate a value for the solubility product constant (K_{sp}) for AgCl. Compare your answer to the value given in Appendix E.   *Ans. 1.76×10^{-10}*

28. Calculate the standard reduction (electrode) potential for the reaction $H_2O + e^- \longrightarrow \frac{1}{2}H_2 + OH^-$ using the Nernst equation and the fact that the standard reduction potential for the reaction $H^+ + e^- \longrightarrow \frac{1}{2}H_2$ is by definition equal to zero volt.   *Ans. -0.83 V*

29. The standard reduction potentials for the reactions

$$Ag^+ + e^- \longrightarrow Ag$$

and $$[Ag(NH_3)_2]^+ + e^- \longrightarrow Ag + 2NH_3$$

are 0.7991 and 0.373 V, respectively. From these values and the Nernst equation, determine K_f for the $[Ag(NH_3)_2]^+$ ion. Compare your answer to the value given in Appendix F.   *Ans. 1.60×10^7*

30. A standard zinc electrode is combined with a hydrogen electrode with H_2 at 1 atm. If the emf of the cell is 0.46 V, what is the pH of the electrolyte in the hydrogen electrode?   *Ans. 5.07*

31. The standard reduction potential of oxygen in acidic solution is 1.23 V ($O_2 + 4H^+ + 4e^- \longrightarrow 2H_2O$). Calculate the standard reduction potential of oxygen in basic solution and compare your result with the value in Appendix I. (*Hint:* What is $[H^+]$ when $[OH^-] = 1$ M?)   *Ans. 0.40 V*

References

"The Motion of Ions in Solution under the Influence of an Electric Field," C. A. Vincent, *J. Chem. Educ.,* **53,** 490 (1976).

"On the Relationship between Standard Electrode and Ionization Potentials of Metal Ions," H. D. Burrows, *J. Chem. Educ.,* **53,** 365 (1976).

"On the Relationship between Cell Potential and Half-Cell Reactions," D. N. Bailey, O. A. Moe, and J. N. Spenser, *J. Chem. Educ.,* **53,** 77 (1976).

"Electrolyte Theory and SI Units," R. I. Holliday, *J. Chem. Educ.,* **53,** 21 (1976).

"Electrochemical Double Cells," P. A. Rock, *J. Chem. Educ.,* **52,** 787 (1975).

"Raku: A Redox Experiment in Glass," R. S. Cichowski, *J. Chem. Educ.,* **52,** 616 (1975).

"Faraday's Contribution to Electrolytic Solution Theory," Ollin K. Drennan, *J. Chem. Educ.,* **42,** 679 (1965).

"The Chemistry of Michael Faraday, 1791–1867," R. H. Cragg, *Chem. in Britain,* **3,** 482 (1967).

"Superconductivity at High Pressure," N. B. Brandt and N. I. Ginsburg, *Sci. American,* April 1971, p. 83.

"The Electronics Industry—A Sizable Challenge to Chemists," L. W. Dunlap, *Chem. Eng. News,* November 30, 1970, p. 42.

"Ionic Conduction and Diffusion in Solids," J. Kummer and M. E. Milberg, *Chem. Eng. News,* May 12, 1969, p. 90.

"Electromotive Force of Molten Salt Concentration Cells and Association Equilibria in Solution," J. Braunstein, *J. Chem. Educ.,* **44,** 223 (1967).

"Molten Carbonate Electrolytes as Acid-Base Solvent Systems," G. J. Janz, *J. Chem. Educ.,* **44,** 581 (1967).

"Electrochemistry in the Solid State," J. S. McKechnie *et al., J. Chem. Educ.,* **55,** 418 (1978)

"Batteries and Fuel Cells" (Staff), *J. Chem. Educ.,* **55,** 399 (1978).

"Fuel-Cell Power Plants," A. P. Fickett, *Sci. American,* December 1978, p. 70.

"New Materials Boost Battery Research" (Staff), *Chem. and Eng. News,* October 2, 1978, p. 14.

"Advanced Storage Batteries: Progress, But Not Electrifying," A. L. Robinson, *Science,* **192,** 541 (1976).

"The Photovoltaic Generation of Electricity," B. Chalmers, *Sci. American,* October 1976, p. 34.

"Fundamental Principles of Semiconductors," E. F. Gurnee, *J. Chem. Educ.,* **46,** 80 (1969).

"The Electrical Properties of Materials," H. Reiss, *Sci. American,* September 1967, p. 210.

"Chemistry of Semiconductors," S. J. Bass, *Chem. in Britain,* **5,** 100 (1969).

"High-Purity Metals for the Electronics Industry," J. E. Wardill and D. J. Dowling, *Chem. in Britain,* **5,** 226 (1969).

"The Standard Electrode Potential of the Silver-Silver Bromide Electrode," R. L. Venable and D. V. Roach, *J. Chem. Educ.,* **46,** 741 (1969).

"Electrochemical Principles Involved in a Fuel Cell," A. K. Vijh, *J. Chem. Educ.,* **47,** 680 (1970).

"Large Scale Integration in Electronics," F. G. Heath, *Sci. American,* February 1970, p. 22.

"The Magnetic Structure of Superconductors," U. Essmann and H. Träuble, *Sci. American,* March 1971, p. 75.

"Solar Energy: A Feasible Source of Power?" A. L. Hammond, *Science,* **172,** 660 (1971).

"Electrochemical Cells for Space Power," R. M. Lawrence and W. H. Bowman, *J. Chem. Educ.,* **48,** 359 (1971).

"Electrodeless Plating of Plastics," G. A. Krulik, *J. Chem. Educ.,* **55,** 361 (1978).

"Everyday Examples of Oxidation-Reduction Processes" (Staff), *J. Chem. Educ.,* **55,** 332 (1978).

"The Recovery of Silver from Laboratory Wastes," K. J. Bush and H. Diehl, *J. Chem. Educ.,* **56,** 54 (1979).

"DOE Gears up for Electric Car Program" (Staff), *Chem. and Eng. News,* January 2, 1978, p. 25.

21

Sulfur and Its Compounds

The elements sulfur, selenium, tellurium, and polonium follow oxygen in Group VIA of the Periodic Table. These elements, oxygen through polonium, constitute the group known as the **chalcogen family.** Oxygen (discussed in Chapter 9) is the first member of this group and exhibits properties that set it apart from the other elements of the chalcogens, just as fluorine, the first member of the halogen family, is different in many respects from the other elements of its group. Polonium is formed only as a product of radioactive change; it is itself highly radioactive. Some properties are given for the elements of the chalcogen family, excluding polonium, in Table 21-1.

Each atom of each element in this periodic group has six valence electrons. Oxygen, more electronegative than any other element except fluorine, usually exhibits an oxidation state of -2 in its compounds, but the larger and less electronegative elements of the chalcogen family exhibit both negative and positive oxidation states. As the electronegativity decreases with increasing atomic size, the strength of the elements as oxidizing agents decreases, a property reflected very strikingly in the heats of formation of their respective hydrides (Table 21-1).

As would be expected from the change in electronegativity, metallic character within the group increases with increasing atomic size. Oxygen is a typical nonmetal, selenium evidences some metallic character, and polonium exhibits definite metallic characteristics.

Of the elements composing Group VIA, oxygen and sulfur are the most useful commercially. The largest single use for sulfur is in the production of sulfuric acid,

TABLE 21-1   Some Properties of the Chalcogens

Property	Oxygen	Sulfur	Selenium	Tellurium
Atomic number	8	16	34	52
Atomic weight	15.9994	32.064	78.96	127.60
Electronic structure	2, 6	2, 8, 6	2, 8, 18, 6	2, 8, 18, 18, 6
Radius of divalent anion, Å	1.40	1.84	1.98	2.21
Radius of covalent atom, Å	0.66	1.04	1.17	1.37
Physical state	Gas	Solid	Solid	Solid
Color	Colorless	Yellow	Red or gray	Silvery
Melting point, °C	−218.4	112.8, 119.25	217 (gray)	450
Boiling point, °C	−183	444.6	688	1390
Electronegativity	3.5	2.4	2.5	2.0
Enthalpy of formation of hydride, kJ mol^{-1}	−285.83	−20.6	29.7	99.6
Common oxidation states	−2, −1	−2, +2, +4, +6	−2, +4, +6	−2, +4, +6

the most important chemical in industry. A considerable amount of sulfur is also used in vulcanizing rubber and in producing gunpowder, sulfites, thiosulfates, fertilizers, and medicines.

Selenium is used in photoelectric cells, since it has a sufficiently low ionization potential that light can remove an electron from the outer cell. The use of a selenide in a new solar cell has been mentioned in Section 20.18. The addition of a little selenium to ordinary glass offsets the green color such glass usually has due to the presence of iron(II) silicate. If selenium is added in larger amounts, it colors the glass red. Sodium selenide is used when selenium is to be incorporated in glass, because elemental selenium is volatile at the temperatures employed in glassmaking and is very toxic. In the electronics industry, selenium of high purity is used as an efficient rectifier, a device for changing alternating electric current to direct current. Selenium is also used in the production of certain stainless steels and special copper alloys.

Tellurium is used to some extent in coloring glass blue, brown, or red. It is alloyed in small percentages with lead to increase the hardness of the metal for use in such things as battery plates and printing type. Traces of tellurium in stainless steel increase its machinability and in cast iron give a hard, wear-resistant surface.

Sulfur

21.1   Occurrence of Sulfur

Sulfur has been known from very early times because it occurs free in nature as a solid. The principal deposits that have been utilized in the United States are in Texas and Louisiana. There are also extensive deposits in Mexico and in the volcanic regions of Italy and Japan. Minerals containing sulfur in the combined

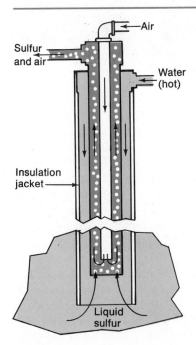

Figure 21-1 Diagram illustrating the Frasch method of mining sulfur.

Figure 21-2 Crystals of orthorhombic sulfur.

Figure 21-3 Crystals of monoclinic sulfur.

form are numerous and widely distributed, and many of the metallic sulfides serve as valuable sources of the element. Sulfides of iron, zinc, lead, and copper and sulfates of sodium, calcium, barium, and magnesium are common and abundant. Hydrogen sulfide is a common component of natural gas and occurs in many volcanic gases. Sulfur is also commonly found in coal in a combined state. Sulfur is a constituent of some proteins and therefore exists in the combined state in animal and vegetable matter.

21.2   Extraction of Sulfur

Sulfur is extracted by the **Frasch process** from enormous underground deposits in Texas and Louisiana (Fig. 21-1). In this process, superheated water (170° and 100 pounds per square inch pressure) is forced down the outermost of three concentric pipes to the deposit of sulfur located several hundred feet below the surface of the earth. When the hot water melts the sulfur, compressed air is forced down the innermost pipe. The liquid sulfur mixed with air forms a foam that is less dense than water, and this mixture readily flows up through the outlet pipe. The emulsified sulfur is conveyed to large settling vats, where it solidifies on cooling. Sulfur produced by this method is remarkably pure, 99.5 to 99.9%, and for most of its uses requires no purification.

Increasing quantities of sulfur as a by-product from the purification of "sour" natural gas and petroleum refinery gases have caused a decline in the Frasch method of obtaining sulfur. This by-product sulfur now exceeds the total production by the Frasch process. With air pollution by sulfur dioxide a major problem, increasing effort is being applied toward eliminating sulfur dioxide from the exhaust gases of power plants. The reaction of sulfur dioxide with salts of magnesium and calcium shows promise for reducing this source of pollution (see Sections 22.3–22.5). Perhaps income from the sale of sulfur or sulfuric acid extracted from the gases can partially offset the costs involved.

21.3   Allotropic Modifications of Sulfur

Sulfur exists in several allotropic forms (Section 9.10). Native sulfur is a yellow solid that forms crystals that belong to the orthorhombic crystal system; it is called **rhombic sulfur** (Fig. 21-2). This form of sulfur is soluble in carbon disulfide, producing a solution from which well-formed crystals separate when it is allowed to evaporate slowly. When heated to 112.8°, crystals of rhombic sulfur melt and form a straw-colored liquid known as **λ-sulfur.** When this liquid cools and crystallizes, long transparent needles of **monoclinic sulfur** are formed (Fig. 21-3). This modification of sulfur, the stable form of sulfur above 96°, melts at 119.25° and is soluble in carbon disulfide. Upon standing at room temperature, it gradually changes to the rhombic form.

Rhombic sulfur, monoclinic sulfur, the straw-colored liquid, and solutions of rhombic sulfur in carbon disulfide all contain S_8 molecules in which the atoms form eight-membered, puckered rings (Fig. 21-4a). Each atom of sulfur is linked to each of its two neighbors in the ring by single electron-pair bonds; each atom thus has a completed octet of electrons.

Figure 21-4 Perspective drawings of (a) an S_8 molecule and (b) a chain of sulfur atoms.

(a)   (b)

The straw-colored liquid form of sulfur is quite mobile, i.e., its viscosity is low, because the S_8 molecules are essentially spherical in shape and offer relatively little resistance to their motion past one another. As the temperature is raised, the S_8 rings of the yellow mobile sulfur rupture, and long chains of sulfur atoms (Fig. 21-4b) are formed. As these chains become entangled with one another and combine end to end to form still longer chains with larger numbers of sulfur atoms, an increase in the viscosity of the liquid occurs. The liquid gradually darkens in color and becomes so viscous that finally (at about 230°) it will not pour from its container. The unpaired electrons at the ends of the chains of sulfur atoms are responsible for the dark red color. This liquid form of the element is known as **μ-sulfur.** When it is cooled rapidly, a rubberlike amorphous mass, insoluble in carbon disulfide, results. This supercooled liquid is known as **plastic sulfur.** On standing for some time at room temperature, plastic sulfur, like all of the other allotropes, changes to the rhombic form.

Sulfur boils at 444.6°C and forms a vapor consisting of S_2, S_6, and S_8 molecules; at about 1000° the vapor density corresponds to the formula S_2.

21.4   Chemical Properties of Sulfur

Elemental sulfur is quite reactive, even at ordinary temperatures, although generally less so than oxygen. Many of the metals will react with solid sulfur on coming in contact with it at room temperature. Mercury has been shown to combine with sulfur at temperatures as low as $-180°$. Only the noble gases and the elements iodine, nitrogen, tellurium, gold, platinum, and palladium do not combine directly with elemental sulfur.

Sulfur, with the electronic configuration $1s^2 2s^2 2p^6 3s^2 3p^4$, exhibits distinctly nonmetallic behavior. It oxidizes metals, giving metal sulfides.

$$16Na + S_8 \longrightarrow 8Na_2S$$
$$4Na + S_8 \longrightarrow 2Na_2S_4$$
$$8Ca + S_8 \longrightarrow 8CaS$$
$$8Fe + S_8 \longrightarrow 8FeS$$
$$4Fe + S_8 \longrightarrow 4FeS_2$$

A great variety of such binary compounds of sulfur with metals can be prepared. The products of these reactions depend upon the ratios of reactants, the temperature of the reaction, and other conditions. The simplest metal sulfides contain the **sulfide ion,** S^{2-}. With additional sulfur, compounds containing **polysulfide ions** (S_2^{2-}, S_4^{2-}, S_5^{2-}—Section 21.10) result.

With nonmetals, sulfur can form compounds such as SCl_2, CS_2, P_4S_{10}, CH_3—SH, and H_2S. In such compounds, each sulfur atom forms two covalent bonds. The oxidation state of sulfur in these compounds is either $+2$ or -2, depending upon whether the nonmetal is more electronegative (Cl) or less electronegative (C, H, P) than sulfur (see Section 5.6). Sulfur also forms compounds containing four, five, or six bonding electron pairs with more electronegative nonmetals. In such compounds (SO_3, H_2SO_4, $SOCl_2$, SF_4, SF_6), sulfur exhibits an oxidation state of $+4$ or $+6$.

Elemental sulfur acts as an oxidizing agent toward hydrogen, forming **hydrogen sulfide** (H_2S) in low yield; toward carbon at elevated temperatures, forming **carbon disulfide** (CS_2) and toward phosphorus when heated, forming a variety of phosphorus sulfides depending on the stoichiometry.

$$8H_2 + S_8 \longrightarrow 8H_2S$$
$$4C + S_8 \longrightarrow 4CS_2$$
$$8P_4 + 10S_8 \longrightarrow 8P_4S_{10}$$
$$8P_4 + 5S_8 \longrightarrow 8P_4S_5$$

From this point on in this book, elemental sulfur generally will be written as S in chemical equations rather than as S_8. Although sulfur is stable as an eight-membered ring at room temperature and should be written as S_8 to be perfectly correct, chemists commonly use the symbol S in order to reduce the complexity of the coefficients in chemical equations.

Sulfur acts as a reducing agent toward those nonmetals that are more electronegative than it is, such as oxygen and the halogens. When sulfur is ignited in the air, it burns with a blue flame forming **sulfur dioxide** and a little **sulfur trioxide.**

$$S(s) + O_2(g) \longrightarrow SO_2(g) \qquad \Delta H° = -296.83 \text{ kJ}$$
$$2SO_2(g) + O_2(g) \longrightarrow 2SO_3(g) \qquad \Delta H° = -197.8 \text{ kJ}$$

As we shall see later in this chapter, the oxidation of sulfur is important in the production of **sulfurous acid,** H_2SO_3, and **sulfuric acid,** H_2SO_4. Sulfur in moist air is slowly oxidized to sulfuric acid.

$$2S + 2H_2O + 3O_2 \longrightarrow 4H^+ + 2SO_4{}^{2-}$$

Sulfur reduces chlorine in a series of reactions, the products of which depend on the temperature and relative quantities of the reactants.

$$2S + Cl_2 \longrightarrow S_2Cl_2 \qquad \text{("sulfur monochloride," a misnomer)}$$
$$S_2Cl_2 + Cl_2 \rightleftharpoons 2SCl_2 \qquad \text{(sulfur dichloride)}$$
$$SCl_2 + Cl_2 \rightleftharpoons SCl_4 \qquad \text{(sulfur tetrachloride)}$$

The valence electronic structure of S_2Cl_2 is

$$: \overset{..}{\underset{..}{Cl}} - \overset{..}{\underset{..}{S}} - \overset{..}{\underset{..}{S}} - \overset{..}{\underset{..}{Cl}} :$$

Only one bromide of sulfur is known; it is the "monobromide," S_2Br_2, a garnet-red liquid. Sulfur forms a series of fluorides that includes the hexafluoride, SF_6, a very stable gaseous compound. In SF_6, the fluorine atoms are arranged around the sulfur atom at the corners of a regular octahedron (Fig. 21-5), and the bonding is highly covalent in character. There are twelve bonding electrons in the valence shell of sulfur in SF_6 rather than the usual eight. This is possible because

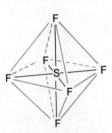

Figure 21-5 The octahedral molecular structure of the SF_6 molecule.

of the availability of unfilled $3d$ orbitals in the valence shell of sulfur; sulfur exhibits d^2sp^3 hybridization in SF_6. In its binary compounds with the halogens, sulfur attains its maximum oxidation state of $+6$ only with fluorine (1) because of the small size and high electronegativity of the fluorine atom compared with the size and electronegativity of the atoms of the other halogens, and (2) because sulfur in its $+6$ oxidation state is a sufficiently strong oxidizing agent to oxidize bromide ions and iodide ions.

Sulfur reduces strong oxidizing agents such as concentrated nitric acid and hot, concentrated sulfuric acid. The equations are

$$S + 6H^+ + 6NO_3^- \longrightarrow 2H^+ + SO_4^{2-} + 2H_2O + 6NO_2(g)$$
$$S + 2H_2SO_4 \longrightarrow 3SO_2(g) + 2H_2O(g)$$

Hydrogen Sulfide

Each member of Group VIA of the Periodic Table forms a hydride. The strengths of these hydrides as acids and their reducing powers increase, whereas their thermal stabilities decrease in the order H_2O, H_2S, H_2Se, H_2Te, H_2Po.

21.5   Preparation of Hydrogen Sulfide

The gas hydrogen sulfide occurs dissolved in the water of sulfur springs and as one of the gases issuing from volcanoes. It is a product of the decay of animal matter in the absence of air. The offensive odor of spoiled eggs is responsible for the name "rotten-egg gas" that is sometimes applied to hydrogen sulfide.

Large quantities of hydrogen sulfide are produced in the refining of petroleum and thereby contribute to the air pollution problem. The production of hydrogen sulfide by the direct union of its elements is unsatisfactory because the reaction is reversible and, as it is usually carried out, not more than 2% of the elements are combined at equilibrium.

$$H_2(g) + S(s) \rightleftharpoons H_2S(g) \qquad \Delta H^\circ = -20.63 \text{ kJ}$$

Hydrogen sulfide can be prepared by treating metal sulfides with a dilute strong acid; for example,

$$FeS + 2H^+ \longrightarrow Fe^{2+} + H_2S$$

When an aqueous solution of hydrogen sulfide is desired, it may be prepared conveniently by the hydrolysis of the organic sulfur-containing compound **thio-acetamide**, CH_3CSNH_2. The equation for the hydrolysis is

$$CH_3CSNH_2 + 2H_2O \longrightarrow CH_3COO^- + NH_4^+ + H_2S(aq)$$

Heating accelerates the hydrolysis considerably.

21.6   Physical Properties of Hydrogen Sulfide

Hydrogen sulfide is a colorless gas with an offensive odor. It is toxic and great care must be exercised in handling it. Hydrogen sulfide is nearly as toxic as hydrogen cyanide (prussic acid), which is used in death chambers in some states for capital punishment. Small amounts of H_2S gas in the air cause headaches, while larger amounts cause paralysis in the nerve centers of the heart and lungs, which results in fainting and death. Hydrogen sulfide is particularly deceptive in that it also paralyzes the olfactory nerves, so that after a short exposure one does not smell it. Many persons have died because of this lack of warning.

Small amounts of H_2S can be produced in the catalytic converters installed in automobiles. At the high operating temperatures of the converter and under fuel-rich conditions, some of the hydrocarbons (binary compounds of carbon and hydrogen) of which the gasoline is composed react with sulfur compounds present in most gasolines to produce the toxic hydrogen sulfide (see Sections 22.4 and 22.5). It is evident that exhaust gases from modern automobiles can be perhaps just as dangerous to life as those in the past. Liquid hydrogen sulfide freezes at $-82.9°C$ and boils at $-61.8°$. The gas is slightly more dense than air and its solubility in water is 0.1 mole per liter at $18°$.

The boiling points of H_2S ($-61.8°$) and H_2Se ($-42°$) are much lower than that of water ($100°$), indicating that the effect of molecular association through hydrogen bonding decreases greatly in the series H_2O, H_2S, H_2Se. Of the three hydrides, only water with its highly electronegative oxygen atom has pronounced hydrogen bonding (see Sections 12.1 and 19.10). Hydrogen sulfide is much less polar than water, and hence liquid hydrogen sulfide is not a good solvent for ionic or polar compounds; it does dissolve many nonpolar compounds.

21.7   Chemical Properties of Hydrogen Sulfide

The sulfur in hydrogen sulfide readily gives up electrons, making the hydrogen sulfide a good reducing agent. In acidic solutions, hydrogen sulfide reduces Fe^{3+} to Fe^{2+}, Br_2 to Br^-, MnO_4^- to Mn^{2+}, $Cr_2O_7^{2-}$ to Cr^{3+}, and HNO_3 to NO_2. When acting as a reducing agent the sulfur of the H_2S is usually oxidized to elemental sulfur, unless a large excess of the oxidizing agent is present. In this case the sulfide may be oxidized to SO_3^{2-} or SO_4^{2-} (or to SO_2 and SO_3 in the absence of water).

Hydrogen sulfide decomposes into hydrogen and sulfur when heated. It burns in air, forming water and sulfur dioxide. The equation for the combustion of hydrogen sulfide is

$$2H_2S(g) + 3O_2(g) \longrightarrow 2H_2O(g) + 2SO_2(g) \qquad \Delta H° = -103.6 \text{ kJ}$$

When hydrogen sulfide is ignited in a limited supply of air or when a burning jet of the gas is impinged upon a cold surface, free sulfur is deposited. In this way, tons of sulfur are recovered at oil refineries each day.

$$2H_2S(g) + O_2(g) \longrightarrow 2H_2O(g) + 2S(s) \qquad \Delta H° = -442.4 \text{ kJ}$$

Hydrogen sulfide is a weak acid. Thus most of the more reactive metals will displace hydrogen from hydrogen sulfide. Lead sulfide is formed by the action of hydrogen sulfide on metallic lead, according to the equation

$$Pb(s) + H_2S(g) \longrightarrow PbS(s) + H_2(g) \qquad \Delta H° = -79.91 \text{ kJ}$$

Hydrogen sulfide will cause metallic silver to tarnish, black silver sulfide being formed. Hence, silver tableware tarnishes when it is used with eggs and other foods containing certain sulfur compounds.

$$4Ag(s) + 2H_2S(g) + O_2(g) \longrightarrow 2Ag_2S(s) + 2H_2O(l) \qquad \Delta H° = -595.8 \text{ kJ}$$

In the presence of moisture, hydrogen sulfide reduces sulfur dioxide to sulfur.

$$2H_2S + SO_2 \longrightarrow 2H_2O + 3S$$

The deposits of sulfur in volcanic regions may be the result of this reaction, because both H_2S and SO_2 are constituents of volcanic gases.

The reaction of H_2S and SO_2 is particularly important as the basis of one method being investigated for the removal of sulfur dioxide, a major air pollutant, from petroleum refinery and power plant exhaust gases. For example, the reaction goes almost to completion and is very rapid in molten sulfur with ethylenediamine ($H_2NCH_2CH_2NH_2$) as a catalyst. Temperatures of 120° to 160° are utilized not only to keep the sulfur molten but also to prevent the undesirable increase in viscosity of sulfur that occurs at higher temperatures (Section 21.3). The recovery of sulfur from stack gases is becoming significant as a source of the element (Section 21.2).

21.8   Hydrogen Sulfide in Aqueous Solution

An aqueous solution of hydrogen sulfide is known as **hydrosulfuric acid.** The acid is weak and diprotic, i.e., it ionizes in two stages. Hydrogen sulfide yields hydrosulfide ions, HS^-, in the first stage and sulfide ions, S^{2-}, in the second.

$$H_2S \rightleftharpoons H^+ + HS^- \qquad K_1 = 1.0 \times 10^{-7}$$
$$HS^- \rightleftharpoons H^+ + S^{2-} \qquad K_2 = 1.3 \times 10^{-13}$$

When a solution of hydrogen sulfide is exposed to the air for a time, sulfur precipitates as the result of oxidation of the sulfide ion.

$$4H^+ + 2S^{2-} + O_2 \longrightarrow 2H_2O + 2S(s)$$

21.9   Sulfides

Because hydrogen sulfide is diprotic, two series of salts are possible. For example, the **normal sulfide,** Na_2S, and the **hydrogen sulfide,** $NaHS$, of sodium are known.

Sodium hydrogen sulfide may be made by passing an excess of hydrogen sulfide into a solution of sodium hydroxide.

$$H_2S + [Na^+] + OH^- \rightleftharpoons [Na^+] + HS^- + H_2O$$

By adding a stoichiometric amount of sodium hydroxide to a solution of sodium hydrogen sulfide, normal sodium sulfide, Na_2S, is formed.

$$[Na^+] + OH^- + [Na^+] + HS^- \rightleftharpoons [2Na^+] + S^{2-} + H_2O$$

Many metal sulfides can be prepared by the direct union of sulfur with the metal (Section 21.4).

Because of the strong tendency of the sulfide ion and the hydrogen sulfide ion to act as proton acceptors, as indicated by the weakness of hydrogen sulfide as an acid, aqueous solutions of soluble sulfides and hydrogen sulfides are basic.

$$S^{2-} + H_2O \rightleftharpoons HS^- + OH^-$$
$$HS^- + H_2O \rightleftharpoons H_2S + OH^-$$

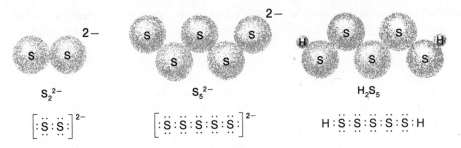

Figure 21-6 Lewis formulas and geometric structures of S_2^{2-}, S_5^{2-}, and H_2S_5.

21.10   Polysulfides

When elemental sulfur is added to a solution of a soluble metal sulfide, the sulfur dissolves by combining with the sulfide ion to form complex polysulfide ions, formulated as S_n^{2-} (n = 2 to 5). **Disulfide ions,** S_2^{2-}, are analogous to peroxide ions in structure (Section 12.15). The sulfur atoms of these complex ions are linked together through shared electron pairs. The electronic and geometric structures of S_2^{2-}, S_5^{2-}, and H_2S_5 are shown in Fig. 21-6.

The polysulfide ions, like the peroxide ion, are oxidizing agents. For example, the disulfide ion oxidizes tin(II) sulfide to the thiostannate(IV) ion.

$$SnS + S_2^{2-} \longrightarrow SnS_3^{2-}$$

When solutions containing polysulfide ions are acidified, free sulfur, in a white, very finely divided form (milk of sulfur), and hydrogen sulfide are produced.

$$S_2^{2-} + 2H^+ \longrightarrow H_2S(g) + S(s)$$

The Oxygen Compounds of Sulfur

The principal oxides of sulfur are sulfur dioxide, SO_2, and sulfur trioxide, SO_3.

21.11   Sulfur Dioxide

▶ **1. PHYSICAL PROPERTIES.** The odor of burning sulfur is that of **sulfur dioxide,** SO_2. It is a colorless gas, 2.26 times as heavy as air, and very soluble in water (80 volumes of gas to 1 volume of water at STP). It readily condenses to the liquid, which boils at $-10°$ and freezes to a white solid that melts at $-75.5°$. Liquid sulfur dioxide can be stored in steel cylinders or shipped in steel tank cars because its vapor pressure is only 5 atm at $32°$, but it must be quite dry, because with moisture, even in trace amounts, enough sulfurous acid is formed to corrode the steel container.

▶ **2. OCCURRENCE AND PREPARATION.** Sulfur dioxide occurs in volcanic gases and in the atmosphere near industrial plants that use coal or oil as energy sources. The oxide forms when sulfur compounds in oil and coal react with oxygen in the air during combustion. Sulfur dioxide is also formed in small quantities during combustion of gasoline, from sulfur compounds present in the gasoline. In the catalytic converters in automobiles the sulfur dioxide is oxidized to sulfur trioxide in the presence of the converters' catalysts at the relatively high temperatures of the converters. The sulfur trioxide then reacts with water to form a small amount of sulfuric acid mist, which can be a problem when emitted in the exhaust.

For commercial purposes, sulfur dioxide is produced by burning free sulfur and by roasting (heating in air) certain sulfide ores, such as ZnS, FeS_2, and Cu_2S. The roasting of metal sulfide ores (to oxidize the sulfur to the dioxide and to form the oxide of the metal) is the first step in the metallurgy of zinc and copper (Section 30.3).

Sulfur dioxide may be prepared conveniently in the laboratory by the action of sulfuric acid upon either sodium sulfite or sodium hydrogen sulfite. Sulfurous acid is first formed, but it quickly decomposes into sulfur dioxide and water.

$$2H^+ + SO_3^{2-} \longrightarrow H_2SO_3 \longrightarrow H_2O + SO_2(g)$$
$$H^+ + HSO_3^- \longrightarrow H_2SO_3 \longrightarrow H_2O + SO_2(g)$$

Many reducing agents react with hot concentrated sulfuric acid with the formation of sulfur dioxide. Three examples are

$$Cu(s) + 2H_2SO_4(l) \longrightarrow CuSO_4(s) + SO_2(g) + 2H_2O(l) \quad \Delta H^\circ = -11.88 \text{ kJ}$$
$$S(s) + 2H_2SO_4(l) \longrightarrow 3SO_2(g) + 2H_2O(l) \quad \Delta H^\circ = +165.83 \text{ kJ}$$
$$2HBr(aq) + H_2SO_4(aq) \longrightarrow Br_2(l) + SO_2(g) + 2H_2O(l) \quad \Delta H^\circ = 49.20 \text{ kJ}$$

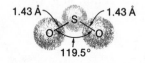

1.43 Å   1.43 Å

119.5°

Figure 21-7 The molecular structure of sulfur dioxide.

▶ **3. THE STRUCTURE OF THE SULFUR DIOXIDE MOLECULE.** Molecules of sulfur dioxide are bent, with two equal sulfur-oxygen bond distances of 1.43 Å and a bond angle of 119.5° (Fig. 21-7). In order to indicate that the two sulfur-oxygen bonds are equivalent, the electronic structure of the sulfur dioxide molecule must be described as a resonance hybrid (Section 5.8) of the following two Lewis structures:

$$ \ddot{O}\!=\!\overset{\displaystyle \ddot{S}}{}\!-\!\ddot{\underset{..}{O}}: \longleftrightarrow :\ddot{\underset{..}{O}}\!-\!\overset{\displaystyle \ddot{S}}{}\!=\!\ddot{O} $$

(a)   (b)

The resonance hybrid electronic structure is not structure (a) part of the time and structure (b) part of the time, as would be the case with an equilibrium mixture of the two forms. Instead, the electronic structure of the molecule is an average of the two Lewis structures. A double-headed single arrow ($\longleftrightarrow$) is utilized, therefore, to distinguish the resonance notation from the double arrow ($\rightleftharpoons$) notation for an equilibrium.

▶ **4. SOLVENT PROPERTIES OF LIQUID SULFUR DIOXIDE.** Liquid sulfur dioxide is a good solvent for a wide variety of substances, including many salts, and it is used as a medium for carrying out certain types of reactions (see Section 14.18). Its electrical conductivity is about twice that of pure water.

▶ **5. REACTIONS OF SULFUR DIOXIDE.** Sulfur dioxide is a weak reducing agent. It is slowly oxidized by oxygen of the air to sulfur trioxide, according to the equation

$$2SO_2(g) + O_2(g) \rightleftharpoons 2SO_3(g) \quad \Delta H^\circ = -197.8 \text{ kJ}$$

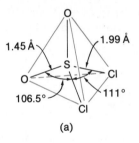

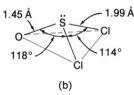

Figure 21-8 The molecular structures (a) of sulfuryl chloride (tetrahedral) and (b) of thionyl chloride (trigonal pyramidal).

This oxidation is much faster in the presence of suitable catalysts, and the reaction is one step in the production of sulfuric acid (Section 21.14). Sulfur dioxide also reduces chlorine when the mixture is exposed to sunlight; sulfuryl chloride, SO_2Cl_2, is the compound formed (Fig. 21-8a).

$$SO_2(g) + Cl_2(g) \longrightarrow SO_2Cl_2(l) \qquad \Delta H^\circ = -97.5 \text{ kJ}$$

Thionyl chloride, $SOCl_2$ (Fig. 21-8b), is formed by the metathetical reaction of sulfur dioxide with phosphorus(V) chloride.

$$SO_2(g) + PCl_5(s) \longrightarrow SOCl_2(l) + POCl_3(l) \qquad \Delta H^\circ = -103 \text{ kJ}$$

Like most nonmetal halides, $SOCl_2$ and SO_2Cl_2 hydrolyze, giving hydrogen chloride and the sulfur dioxide or sulfur trioxide, respectively. The reaction of sulfur dioxide with water will be considered in the next section.

Sulfur dioxide reacts with metal oxides to give compounds containing the sulfite ion, SO_3^{2-}; for example,

$$Na_2O + SO_2 \longrightarrow Na_2SO_3$$

Sulfur dioxide functions as a Lewis acid (Sections 8.1 and 14.17) in these reactions.

21.12   Sulfurous Acid and Sulfites

As you might expect for the oxide of a nonmetal, sulfur dioxide dissolves in water to form a weakly acidic solution of sulfurous acid.

$$H_2O + SO_2 \rightleftharpoons H_2SO_3$$

Sulfurous acid is unstable, and anhydrous H_2SO_3 cannot be isolated. Boiling a solution of sulfurous acid expels the sulfur dioxide. Like other diprotic acids, sulfurous acid ionizes in two steps.

$$H_2SO_3 \rightleftharpoons H^+ + HSO_3^- \qquad K_1 = 1.2 \times 10^{-2}$$
$$HSO_3^- \rightleftharpoons H^+ + SO_3^{2-} \qquad K_2 = 6.2 \times 10^{-8}$$

Since sulfurous acid is a moderately weak acid, the amount of ionization is not very large in either stage, but it is much less in the secondary stage than in the primary.

Both normal and hydrogen salts are formed by sulfurous acid. Sulfurous acid acts as a reducing agent toward strong oxidizing agents. Oxygen of the air oxidizes it slowly to the more stable sulfuric acid.

$$2H_2SO_3 + O_2 \longrightarrow 4H^+ + 2SO_4^{2-}$$

Solutions containing the permanganate ion (purple in color) rapidly turn colorless when sulfurous acid is added, the purple MnO_4^- ion being reduced to the colorless Mn^{2+} ion.

$$2MnO_4^- + 5H_2SO_3 \longrightarrow 2Mn^{2+} + 4H^+ + 5SO_4^{2-} + 3H_2O$$

Solid sodium hydrogen sulfite forms sodium sulfite, sulfur dioxide, and water when heated.

$$2NaHSO_3 \xrightarrow{\Delta} Na_2SO_3 + SO_2 + H_2O$$

When solid sodium sulfite is heated to high temperatures, disproportionation (auto-oxidation-reduction) occurs with the formation of sodium sulfide and sodium sulfate.

$$\overset{+4}{4Na_2SO_3} \longrightarrow \overset{-2}{Na_2S} + \overset{+6}{3Na_2SO_4}$$

This reaction is analogous to the disproportionation reactions that metal hypochlorites and metal chlorates undergo (Sections 19.15 and 19.17).

Solutions of sulfites are very susceptible to air oxidation, as is sulfurous acid, and sulfates are formed. Thus solutions of sulfites always contain sulfates after standing in contact with the air.

21.13   Sulfur Trioxide

▶ 1. PREPARATION. When sulfur burns in air, small amounts of **sulfur trioxide** are formed in addition to sulfur dioxide. When sulfur dioxide and oxygen are heated together, a small amount of the trioxide is formed according to the equation

$$2SO_2(g) + O_2(g) \rightleftharpoons 2SO_3(g) \qquad \Delta H° = -197.8 \text{ kJ}$$

If the temperature of the system is raised to about 400°C the equilibrium is reached more rapidly, but even at this temperature the time required to attain equilibrium is too great for the reaction to be commercially useful. The presence of a catalyst such as finely divided platinum or vanadium(V) oxide greatly decreases the time required for attainment of equilibrium. The higher the temperature, the more the equilibrium shifts to the left. See van't Hoff's law (Section 15.21).

▶ 2. STRUCTURE. Sulfur trioxide in the vapor state is **monomeric,** i.e., its molecules are single SO_3 units, with the sulfur atom at the center and the oxygen atoms at the corners of an equilateral triangle as shown in Fig. 21-9a. Because the sulfur-oxygen bond distance is less than that of a single bond, the electronic structure of sulfur trioxide must be described as a resonance hybrid of the following three Lewis structures:

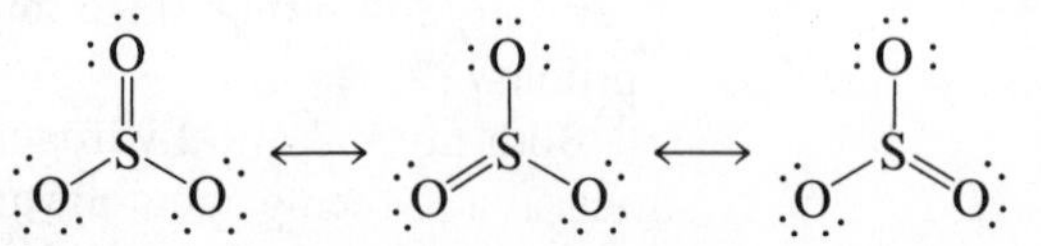

There are three distinct solid forms of sulfur trioxide. One form is icelike in appearance and has as building units trimeric molecular units (Fig. 21-9b). A second form is an asbestoslike solid, in which tetrahedral SO_4 units are joined to each other in long chains that extend the length of the crystal (Fig. 21-9c). In a third form (not shown in Fig. 21-9) SO_3 chains (SO_4 groups joined together through common oxygen atoms at one corner of each tetrahedron) are joined together to give a layerlike arrangement. Upon changing from the gaseous state to the solid state, a sulfur trioxide molecule changes its hybridization from sp^2 (trigonal planar) to sp^3 (tetrahedral) as it functions as a Lewis acid to form a bond to a fourth oxygen atom.

Figure 21-9 Sulfur trioxide structures.

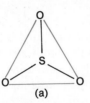

(a)

Structure of the SO_3 vapor molecule (symmetrical planar molecule).

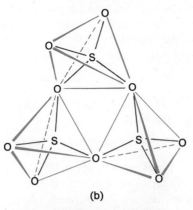

(b)

The orthorhombic "icelike" solid form with formula S_3O_9 (cyclic with three tetrahedral SO_4 groupings joined through three of the nine oxygen atoms).

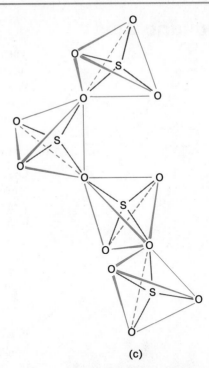

(c)

The "asbestoslike" solid form with tetrahedral SO_4 groupings, each joined to another through a common oxygen atom to form infinite chain molecules.

▶ **3. PROPERTIES.** Liquid sulfur trioxide boils at 43° and freezes at 17° to the icelike solid form. When a trace of moisture is added to liquid sulfur trioxide, it changes to the asbestoslike solid form.

Liquid sulfur trioxide fumes in moist air and dissolves in water in a highly exothermic reaction to form sulfuric acid, H_2SO_4. With a limited amount of water, the molecular acid is formed according to the equation

$$SO_3(l) + H_2O(l) \longrightarrow H_2SO_4(l) \qquad \Delta H° = -87.11 \text{ kJ}$$

Sulfur trioxide dissolves readily in concentrated sulfuric acid and forms **pyrosulfuric acid**, $H_2S_2O_7$, also known as "fuming sulfuric acid," or "oleum."

$$H_2SO_4 + SO_3 \longrightarrow H_2S_2O_7$$

At temperatures of 900° or higher, sulfur trioxide decomposes into the dioxide and oxygen. The trioxide reacts as a Lewis acid with many oxides and hydroxides in Lewis acid-base reactions with the formation of sulfates and hydrogen sulfates, respectively.

$$BaO + SO_3 \longrightarrow BaSO_4$$

Sulfuric Acid

The amount of sulfuric acid used in industry exceeds that of any other manufactured compound. Its importance is so great that the amount of it produced from year to year is a fairly accurate index of industrial prosperity. The 39.6 million tons produced in 1978 represented a 10.5% increase over the 35.8 million tons produced the year earlier.

21.14   Manufacture of Sulfuric Acid

Over 90% of the sulfuric acid prepared industrially is prepared by the **contact process.** This process involves oxidizing sulfur dioxide to sulfur trioxide and combining sulfur trioxide with water. Usually, the sulfur dioxide is obtained by burning nearly pure sulfur in air.

$$S(s) + O_2(g) \longrightarrow SO_2(g) \qquad \Delta H° = -296.83 \text{ kJ}$$

The sulfur dioxide is then oxidized, by means of air in the presence of a suitable catalyst, to the trioxide.

$$2SO_2(g) + O_2(g) \longrightarrow 2SO_3(g) \qquad \Delta H° = -197.8 \text{ kJ}$$

Finely divided platinum was originally used as the "contact" catalyst, but it now has been largely replaced by vanadium(V) oxide (V_2O_5), which is more resistant to poisoning (inactivation by impurities). Although the oxidation of the dioxide to the trioxide is exothermic and thus favored by low temperatures, the reaction is too slow to be commercially feasible at low temperatures. Consequently, the oxidation step is carried out at about 400°. A yield of about 98% is obtained at this temperature.

In spite of the vigorous reaction of sulfur trioxide vapor with water, the sulfur trioxide if mixed with air reacts slowly with water, because the bubbles of O_2 and N_2 (which go through the solution as *big* bubbles) contain particles of liquid sulfur trioxide; the sulfur trioxide particles in the middle of these bubbles do not come in contact with the water. For this reason, the sulfur trioxide vapor is first absorbed in concentrated sulfuric acid, pyrosulfuric acid being formed.

$$H_2SO_4(l) + SO_3(g) \longrightarrow H_2S_2O_7(s) \qquad \Delta H° = -64.02 \text{ kJ}$$

The addition of water to pyrosulfuric acid gives sulfuric acid.

$$H_2S_2O_7(s) + H_2O(l) \longrightarrow 2H_2SO_4(l) \qquad \Delta H° = -66.62 \text{ kJ}$$

Since the sulfur dioxide used in this process must be quite pure (to avoid poisoning of the catalyst), the contact process gives pure acid, which may be highly concentrated.

21.15   Physical Properties of Sulfuric Acid

Pure sulfuric acid (or **hydrogen sulfate**) is a colorless, oily liquid that freezes at 10.5°. It fumes when heated due to decomposition of the acid to water and sulfur trioxide. More sulfur trioxide than water is lost during the heating, until a concentration of 98.33% acid is reached. The acid of this concentration boils at 338° without further change in concentration and is sold as concentrated H_2SO_4.

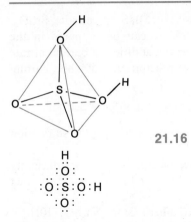

Figure 21-10 The tetrahedral molecular structure and electronic structure of the sulfuric acid molecule.

Concentrated sulfuric acid dissolves in water with the evolution of a large amount of heat. The dilution may be carried out safely by pouring the concentrated acid slowly into water while the solution is stirred in order to distribute the heat of dilution. **Caution.** The addition of water to the concentrated acid may cause dangerous spattering of the acid. Figure 21-10 shows the electronic and molecular structure (tetrahedral) of the sulfuric acid molecule.

21.16   Chemical Properties of Sulfuric Acid

The sulfur atom in the sulfuric acid molecule is surrounded tetrahedrally by two oxygen atoms and two hydroxyl groups. The hydrogen atoms form hydrogen bonds between these tetrahedra, binding them together and giving the liquid a high boiling point.

The large heat of dilution of sulfuric acid is caused by hydrate and hydronium ion formation. The hydrates $H_2SO_4 \cdot H_2O$, $H_2SO_4 \cdot 2H_2O$, and $H_2SO_4 \cdot 4H_2O$ are known. The acid ionizes in two stages.

$$H_2SO_4 \longrightarrow H^+ + HSO_4^- \qquad \text{(100\% ionized)}$$
$$HSO_4^- \rightleftharpoons H^+ + SO_4^{2-} \qquad K = 1.2 \times 10^{-2}$$

In dilute solution sulfuric acid undergoes almost complete primary ionization. The secondary ionization is less complete, but even so HSO_4^- is a strong acid.

The strong affinity of concentrated sulfuric acid for water makes it a good dehydrating agent. Gases that do not react with the acid may be dried by being passed through it. So great is the affinity of concentrated sulfuric acid for water that it will remove hydrogen and oxygen, in the form of water, from many compounds containing these elements. Organic substances containing hydrogen and oxygen in the proportion of 2 to 1, such as cane sugar, $C_{12}H_{22}O_{11}$, and cellulose, $(C_6H_{10}O_5)_x$, are charred by concentrated sulfuric acid.

$$C_{12}H_{22}O_{11} \longrightarrow 12C + 11H_2O$$

Concentrated sulfuric acid is highly destructive to human flesh because of its reactions with organic compounds in the flesh, and is very dangerous on that account.

Sulfuric acid acts as an oxidizing agent, particularly when hot and concentrated. Depending upon its concentration, the temperature, the strength of the reducing agent with which it acts, and other factors, sulfuric acid oxidizes many compounds and, in the process, undergoes reduction to either SO_2, HSO_3^-, SO_3^{2-}, S, H_2S, or S^{2-}. Its oxidizing action towards hydrogen bromide and hydrogen iodide was noted in Section 19.7, and towards metals and sulfur in Section 21.11. The displacement of volatile acids from their salts by means of concentrated sulfuric acid has been mentioned in Section 14.3. Aqueous solutions of sulfuric acid exhibit the characteristic properties of strong acids.

21.17   Sulfates

Being a diprotic acid, sulfuric acid forms both **sulfates,** such as Na_2SO_4, and **hydrogen sulfates,** such as $NaHSO_4$.

The sulfates of barium, strontium, calcium, and lead are only slightly soluble in water. These salts occur in nature as the minerals barite, $BaSO_4$; celestite, $SrSO_4$; gypsum, $CaSO_4 \cdot 2H_2O$; and anglesite, $PbSO_4$. They can be prepared in the laboratory by metathetical reactions. For example, the addition of barium nitrate to a solution containing sodium sulfate causes the precipitation of white barium sulfate.

$$Ba^{2+} + [2NO_3{}^-] + [2Na^+] + SO_4{}^{2-} \longrightarrow BaSO_4(s) + [2Na^+] + [2NO_3{}^-]$$

This reaction is the basis of a qualitative and quantitative test for the sulfate ion and the barium ion.

Although barium and its salts are very toxic, barium sulfate is sufficiently insoluble ($K_{sp} = 1.08 \times 10^{-10}$) that it can be safely put into the alimentary tract for x-ray studies of that portion of the body.

Among the important soluble sulfates are **Glauber's salt,** $Na_2SO_4 \cdot 10H_2O$; **Epsom salt,** $MgSO_4 \cdot 7H_2O$; **blue vitriol,** $CuSO_4 \cdot 5H_2O$; **green vitriol,** $FeSO_4 \cdot 7H_2O$; and **white vitriol,** $ZnSO_4 \cdot 7H_2O$. The hydrogen sulfates, such as $NaHSO_4$, are acids as well as salts. Sodium hydrogen sulfate is the primary ingredient in some household cleansers.

21.18   Uses of Sulfuric Acid

The 1978 production of sulfuric acid in the United States was 39,594,000 tons. For many years it has been, and it still is, the top chemical in amount produced. The major uses are in the production of ammonium sulfate (1,943,000 tons in the United States in 1978) and soluble phosphate fertilizers; in the refining of petroleum to remove impurities from such products as gasoline and kerosene; in the pickling of steel (cleaning its surface of iron rust) before coating it with tin, zinc, or enamel; in the production of dyes, drugs, and disinfectants from coal tar; in the electrometallurgy of certain metals as the electrolyte or in the production of sulfates of metals to be used as electrolytes; in the manufacture of other chemicals, such as hydrochloric and nitric acids, and the sulfates of metals; and in the production of textiles, paints, pigments, plastics, explosives, and lead storage batteries.

Other Acids of Sulfur

In addition to the acids of sulfur that we have thus far discussed, there are several others that may be regarded as being derived from sulfuric acid; that is, compounds in which the oxygen or hydroxyl group of a sulfuric acid molecule has been replaced by some other atom or combination of atoms.

21.19   Thiosulfate Ion and Thiosulfuric Acid

The electronic formula and the structure of the thiosulfate ion are compared to that of the sulfate ion in Fig. 21-11 to show that one of the oxygen atoms of the sulfate is replaced by a sulfur atom. The average oxidation state of sulfur in the thiosulfate ion is $+2$, but it is evident from Fig. 21-11 that the central sulfur atom

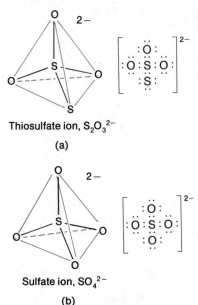

Thiosulfate ion, $S_2O_3^{2-}$

(a)

Sulfate ion, SO_4^{2-}

(b)

Figure 21-11 Tetrahedral molecular structures and electronic structures of (a) the thiosulfate ion and (b) the sulfate ion.

of this ion is in the $+6$ oxidation state and the other sulfur atom (like oxygen in the sulfate ion) is in the -2 state.

It was pointed out in Section 21.12 that sulfites are slowly oxidized to sulfates by oxygen. Sulfur plays a role similar to that of oxygen by transforming sulfites to thiosulfates. For example, when a mixture of sulfur and a solution of sodium sulfite is boiled, **sodium thiosulfate**, $Na_2S_2O_3$, is formed.

$$2Na^+ + SO_3^{2-} + S \longrightarrow 2Na^+ + S_2O_3^{2-}$$

Crystals of the **pentahydrate**, $Na_2S_2O_3 \cdot 5H_2O$, separate when the solvent is evaporated.

When a solution of sodium thiosulfate is acidified, unstable **thiosulfuric acid**, $H_2S_2O_3$, is formed.

$$[2Na^+] + S_2O_3^{2-} + 2H^+ + [2Cl^-] \longrightarrow H_2S_2O_3 + [2Na^+] + [2Cl^-]$$

The acid decomposes immediately into sulfurous acid and sulfur, the sulfur appearing either as a precipitate or in colloidal suspension.

$$H_2S_2O_3 \longrightarrow H_2SO_3 + S(s)$$

Sodium thiosulfate, also known as "hypo," is used in the photographic process as a fixing solution to dissolve from the plate or film any silver halides that have not been reduced to metallic silver by the developer.

$$AgX + 2S_2O_3^{2-} \rightleftharpoons [Ag(S_2O_3)_2]^{3-} + X^- \qquad (X = a\ halogen)$$

The thiosulfate ion is a reducing agent. It is oxidized by iodine to the **tetrathionate ion**, $S_4O_6^{2-}$.

$$2S_2O_3^{2-} + I_2 \longrightarrow S_4O_6^{2-} + 2I^-$$

This reaction is used extensively in analytical chemistry. For example, in the quantitative analyses for chlorine and the copper ion, iodine is often produced by the following reactions and then titrated with a standard thiosulfate solution using a suitable indicator to determine the equivalence point (end point).

To determine Cl_2: $Cl_2 + 2I^- \longrightarrow I_2 + 2Cl^-$

To determine Cu^{2+}: $Cu^{2+} + 3I^- \longrightarrow I_2 + CuI(s)$

21.20   Peroxymonosulfuric Acid and Peroxydisulfuric Acid

Peroxymonosulfuric and peroxydisulfuric acids have the formulas H_2SO_5 and $H_2S_2O_8$, respectively (see Fig. 21-12). Both of these acids are derivatives of sulfuric acid. A hydroxyl group of the sulfuric acid molecule has been replaced by a —OOH group in peroxymonosulfuric acid and by a —OOSO$_3$H group in peroxydisulfuric acid.

Peroxydisulfuric acid is produced commercially by the anodic oxidation of hydrogen sulfate ions in 45–55% sulfuric acid at a low temperature.

$$2HSO_4^- \longrightarrow H_2S_2O_8 + 2e^-$$

Electrolysis of potassium hydrogen sulfate in aqueous solution gives potassium

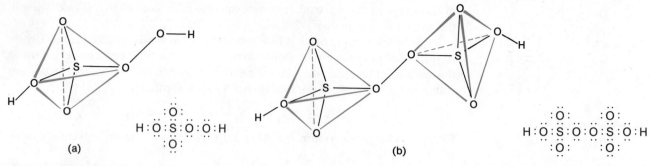

$$H:\overset{\overset{\displaystyle ..}{O}:}{\underset{\underset{\displaystyle ..}{O}:}{\overset{|}{\underset{|}{O:S:O:O:}}}}H$$

(a)

$$H:\overset{\overset{\displaystyle ..}{O}:}{\underset{\underset{\displaystyle ..}{O}:}{\overset{|}{\underset{|}{O:S:O:O:}}}}\overset{\overset{\displaystyle ..}{O}:}{\underset{\underset{\displaystyle ..}{O}:}{\overset{|}{\underset{|}{S:O:}}}}H$$

(b)

Figure 21-12 Structures of (a) peroxymonosulfuric acid and (b) peroxydisulfuric acid. Each acid possesses an oxygen-oxygen (peroxide) linkage; hence the use of the prefix peroxy- in the name.

peroxydisulfate, $K_2S_2O_8$. Treatment of this salt at low temperatures with concentrated sulfuric acid produces peroxymonosulfuric acid, commonly called **Caro's acid,** H_2SO_5. This acid is also produced by the reaction of sulfur trioxide with H_2O_2.

$$H_2O_2 + SO_3 \longrightarrow SO_2(OH)(OOH)$$

Notice that this reaction is analogous to that of sulfur trioxide with water.

$$H_2O + SO_3 \longrightarrow SO_2(OH)_2$$

The peroxysulfuric acids and their salts are useful as strong oxidizing agents and are used to prepare hydrogen peroxide (Section 12.13).

21.21 Chlorosulfonic Acid and Sulfamic Acid

Chlorosulfonic acid, HSO_3Cl, and sulfamic acid, SO_3NH_3, are related structurally to sulfuric acid, as shown in Fig. 21-13. The formula for sulfamic acid, sometimes written HSO_3NH_2, is more properly designated SO_3NH_3.

Chlorosulfonic acid is a colorless liquid that fumes in moist air, reacts vigorously with water, and is used to introduce the sulfonate group, SO_3H, into many organic compounds. Sulfamic acid is a white, crystalline, nonhygroscopic solid. It is one of the few strong monoprotic acids that can be weighed without special drying precautions being required. It is of some importance, therefore, in analytical chemistry. Sulfamates are important ingredients in some weed killers.

Figure 21-13 Structure of a sulfuric acid molecule compared to the structures of two related acids. The normal bond angle in a tetrahedral structure is 109°. The O—S—N angle in sulfamic acid is 108°, indicating a slight distortion of the tetrahedron. Such a distortion arises when the structure is not symmetrical, i.e., the atoms at the corners of the tetrahedron are not all the same.

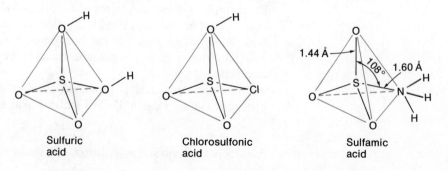

Questions

1. Based on the location of sulfur in the Periodic Table, which common oxidation states would be expected for sulfur? Give two examples of compounds in which sulfur exhibits each of these oxidation states.

2. Describe five chemical properties of sulfur or of compounds of sulfur that characterize it as a nonmetal.

3. What kinds of sulfur compounds are most nearly ionic? Which kinds are covalent? Give two examples of each type.

4. Based on the location of sulfur in the Periodic Table, predict the products of the following reactions and write balanced equations describing the reactions. (There may be more than one correct answer depending upon your choice of stoichiometries.)

   (a) $Li + S \longrightarrow$ (g) $Na_2O + SO_2 \longrightarrow$
   (b) $Ga + S \longrightarrow$ (h) $KClO + Na_2S \longrightarrow$
   (c) $P + S \longrightarrow$ (i) $O_2 + H_2S \longrightarrow$
   (d) $H_2 + S \longrightarrow$ (j) $Cl_2 + H_2S \longrightarrow$
   (e) $F_2 + S \longrightarrow$ (k) $I^- + SO_2 \longrightarrow$
   (f) $F_2 + SO_2 \longrightarrow$

5. Suggest an explanation for the existence of SF_6 but not SCl_6 and the existence of SCl_4 but not SBr_4.

6. (a) Which is a stronger acid, H_2O or H_2S?
   (b) Which is a stronger base, OH^- or SH^-?
   (c) Which is a stronger base, SH^- or S^{2-}? Explain each.

7. Account for the fact that hydrogen sulfide is a gas at room temperature whereas water, which has a smaller molecular weight, is a liquid.

8. Why are solutions of sulfides and hydrogen sulfides alkaline? Write equations.

9. Why do the two sulfur-oxygen bonds in sulfur dioxide have the same length?

10. Describe the hybridization of the sulfur atom in gaseous molecules of SO_2 and SO_3.

11. How does the hybridization of the sulfur atom change when gaseous SO_3 condenses to give the asbestoslike form of solid SO_3?

12. Why is sulfuric acid a stronger acid than sulfurous acid?

13. Which is the stronger acid, $NaHSO_3$ or $NaHSO_4$? Which is the stronger base, Na_2SO_4 or Na_2SO_3?

14. The following compounds can be considered to be derived from sulfurous acid, sulfuric acid, or their salts, by replacement of one or more atoms in these molecules. Suggest structures for these molecules and write their Lewis structures. (a) SOF_2, (b) $Ca(FSO_3)_2$, (c) $(CH_3)_2SO_4$, (d) $Na_2S_2O_8$, (e) $H_2S_2O_7$, (f) $S_2O_6F_2$.

15. The average oxidation state of sulfur is not one of its common oxidation states in Na_2S_2, H_2S_2, K_2S_5, and $Na_2S_2O_3$. Calculate the average oxidation state of sulfur in these compounds. What is the common structural feature in these compounds? In view of your answer to the preceding question, write a Lewis structure for S_2F_{10}. What is the oxidation state of sulfur in S_2F_{10}?

16. Explain, with the aid of a diagram, the Frasch process for extracting sulfur from underground deposits.

17. What are the allotropic forms of solid sulfur and how may they be produced?

18. What is the molecular structure of sulfur in the liquid, solid, and gaseous states?

19. Write chemical equations describing three chemical reactions in which elemental sulfur acts as an oxidizing agent; three in which it acts as a reducing agent.

20. Write equations for the reaction of sulfur with H_2, C, Fe, O_2, and HNO_3.

21. Write equations for the reaction of hydrogen sulfide with Fe^{3+}, MnO_4^-, Br_2, and $Cr_2O_7^{2-}$ in acidic solution.

22. How can sodium sulfite be made in the laboratory?

23. Determine the oxidation state of sulfur in each of the following species: HS^-, $SOCl_2$, SF_6, Na_2S, $S_2O_3^{2-}$, $S_4O_6^{2-}$.

24. Why do solutions of sulfites usually contain sulfate ions?

25. Write chemical equations describing, respectively, the reaction of water and of sulfuric acid with sulfur trioxide. What are the products of these reactions?

26. What single chemical test could be used to distinguish sulfides, polysulfides, sulfites, and sulfates from one another?

27. Interpret the following reaction in terms of an acid-base relationship according to the Lewis theory:

$$CaO + SO_3 \longrightarrow CaSO_4$$

28. Show by equations the dehydrating action of sulfuric acid upon cane sugar and upon cellulose.

29. What are the possible reduction products of sulfuric acid? What is the oxidation state of sulfur in each of these products?

30. Show by writing formulas that the peroxysulfuric acids may be considered as derivatives of hydrogen peroxide; that they may be considered as derivatives of sulfuric acid.

31. Account for the formation of sulfur in a solution of sodium thiosulfate in contact with air (air contains carbon dioxide, the anhydride of carbonic acid).

32. Write the Lewis electronic formula of each of the following: S^{2-}, H_2S_2, SO_2, H_2SO_3, SO_3, Na_2SO_4, H_2SO_5, $H_2S_2O_7$, and $H_2S_2O_8$.

33. Devise the best valence electron structure you can for the SCl molecule. How satisfactory is your structure compared to that of S_2Cl_2? Comment on the relative probabilities for the existence of SCl and S_2Cl_2.

Problems

1. In 1978, 1.943×10^6 tons of ammonium sulfate were produced in the United States. If all of this ammonium sulfate resulted from the reaction of ammonia with sulfuric acid, what per cent of the 3.9594×10^7 tons of sulfuric acid produced in 1978 was used in ammonium sulfate manufacture?
   Ans. 3.642%

2. Sulfur(IV) fluoride, SF_4, is an active fluorinating agent; that is, it reacts with many halides and oxides, converting them to fluorides. The reaction of $NaAsO_3$ with SF_4 gives a compound that contains 10.8% sodium, 35.35% arsenic, and 53.80% fluorine. What is the empirical formula of this compound?
   Ans. $NaAsF_6$

3. The reaction of an excess of sulfur with metallic copper at 400°C gives a copper sulfide, which contains 50.2% sulfur. Write the equation for the reaction.   *Ans. $Cu + 2S \longrightarrow CuS_2$*

4. What volume of hydrogen sulfide at STP can be produced by the reaction of 308 g of Al_2S_3 with an excess of phosphoric acid?   *Ans. 138 l*

5. Air contains 20.99% oxygen by volume. What volume of air in cubic meters at 27.0° and 0.9868 atm is required to oxidize 1.00 metric ton (1000 kg) of sulfur to sulfur dioxide, assuming that all of the oxygen reacts with sulfur?   *Ans. $3.71 \times 10^3 m^3$*

6. A compound of sulfur, oxygen, and fluorine was found to contain approximately 27.0% sulfur and 32.3% fluorine. A 0.106-g sample of this compound occupied a volume of 20.2 ml at STP. What is the molecular formula of the compound? Suggest a Lewis structure for the compound.   *Ans. SO_3F_2*

7. A volume of 22.85 ml of a 0.1023 M standard sodium thiosulfate solution is required to titrate a 25.00-ml sample of a solution containing iodine. What is the iodine concentration in the 25.00-ml sample?   *Ans. 0.04675 M*

8. The reactions involved in the preparation of sulfuric acid are highly exothermic. Using the data in Appendix J, calculate the enthalpy changes for the following reactions:
   (a) $S(s) + O_2(g) \longrightarrow SO_2(g)$
   (b) $2SO_2(g) + O_2(g) \longrightarrow 2SO_3(g)$
   (c) $SO_3(g) + H_2O(l) \longrightarrow H_2SO_4(l)$
   Ans. (a) −296.83 kJ;
   (b) −197.8 kJ;
   (c) −132.5 kJ

9. Determine if the following reactions are spontaneous at standard conditions using either $\Delta G°$ values or cell potentials. The equations are not complete.
   (a) $NiS + Sn \longrightarrow SnS + Ni$
   Ans. Spontaneous
   (b) $S + NO_2^- \longrightarrow S^{2-} + NO_3^-$ (in basic solution)
   Ans. Not spontaneous
   (c) $H_2S + CrO_4^{2-} \longrightarrow S + Cr(OH)_3$ (in basic solution)
   Ans. Not spontaneous
   (d) $Al_2O_3(s) + 3H_2S(g) \longrightarrow Al_2S_3(s) + 3H_2O(l)$
   Ans. Not spontaneous

10. A fused mixture of rubidium fluoride and uranium(IV) fluoride can be oxidized with fluorine to produce a uranium compound in which the uranium is mainly but not entirely in the +5 oxidation state. The product is found to contain 54.43% uranium. A 1.0357-g sample of the product immersed in 100.0 ml of 0.1007 M acidified potassium iodide solution reacted according to the following equation:

$$2I^- + 2UF_6^- \longrightarrow 2UF_4 + I_2 + 4F^-$$

The iodine produced was titrated with 14.80 ml of 0.1494 M sodium thiosulfate solution. What per cent of the original uranium was oxidized to the +5 oxidation state? *Ans. 93.36%*

References

"Sulfur" (Staff), *Chem. and Eng. News,* April 24, 1978, p. 12.

"Sulfuric Acid" (Staff), *Chem. and Eng. News,* March 27, 1978, p. 11.

"Sulfur Pavement Material Begins Road Test" (Staff), *Chem. and Eng. News,* February 26, 1979, p. 30.

"Sulfur Isotope Distribution in Solfatoras, Yellowstone National Park," R. Schoen and R. O. Rye, *Science,* **170,** 1082 (1970).

"Elemental Sulfur: Accumulation in Different Species of Fungi," R. Pezet and V. Pont, *Science,* **196,** 428 (1977).

Manufacture of Sulfuric Acid (American Chemical Society Monograph No. 144), edited by W. W. Duecker and J. R. West, Reinhold Publ. Corp., New York, 1959.

"Chemistry of Solutions in Liquid Sulfur Dioxide," P. J. Elving and J. M. Markowitz, *J. Chem. Educ.,* **37,** 75 (1960).

"The Sulphur Cycle," J. R. Postgate, *Educ. in Chemistry,* **2,** 58 (1965).

"The Chemistry of Tetrasulfur Tetranitride," C. W. Allen, *J. Chem. Educ.,* **44,** 38 (1967).

"Sulfur," C. J. Pratt, *Sci. American,* May 1970, p. 62.

"Modern Sulfuric Acid Technology," T. J. Browder, *Chem. Eng. Progress,* May 1971, p. 45.

"SO$_2$ Emission Control from Acid Plants," W. G. Tucker and J. R. Burleigh, *Chem. Eng. Progress,* May 1971, p. 57.

"An SO$_2$ Removal and Recovery Process," J. J. Humphries, S. B. Zdonik, and E. J. Parsi, *Chem. Eng. Progress,* May 1971, p. 64.

"Reducing SO$_2$ Emission from Stationary Sources," T. H. Chilton, *Chem. Eng. Progress,* May 1971, p. 69.

"The Preservation of Stone," K. Lal Gauri, *Sci. American,* June 1978, p. 126.

"Odor Generation in the Kraft Process," M. A. Karnofski, *J. Chem. Educ.,* **52,** 490 (1975).

"Chalcogenide Glasses: A Decade of Dissension and Progress," A. L. Robinson, *Science,* **197,** 1068 (1977).

22

The Atmosphere,
Air Pollution,
and the Noble Gases

The Atmosphere

22.1 The Composition of the Atmosphere

The **atmosphere,** or air, is the mixture of gaseous substances that surrounds the earth. For the most part, the atmosphere consists of uncombined elements. Nitrogen, oxygen, and the noble gases are present in the atmosphere in almost constant proportion. The percentage of carbon dioxide in the air varies somewhat and that of dust and water vapor is widely variable. There are also traces of hydrogen, ammonia, hydrogen sulfide, oxides of nitrogen, sulfur dioxide, and other gases. The average composition of dry air at sea level is given in Table 22-1.

The per cent composition of dry air does not vary much with location on the earth's surface or with altitude. However, the density of air, as reflected in pressure, varies greatly with altitude. The average pressure at sea level and 45° latitude is 1.0 atm, at 15,000 feet it is about 0.53 atm, at 10 miles it is about 0.05 atm, and at 30 miles is only about 1×10^{-4} atm.

TABLE 22-1 Composition of Dry Air			
Component	Per Cent by Volume	Component	Per Cent by Volume
Nitrogen	78.03	Helium	0.0005
Oxygen	20.99	Krypton	0.0001
Argon	0.94	Ozone	0.00006
Carbon dioxide	0.035–0.04	Hydrogen	0.00005
Neon	0.0012	Xenon	0.000009

22.2   Liquid Air

▶ **1. PREPARATION.**   Before air can be liquefied it must be freed of moisture and carbon dioxide, because these substances change to the solid state when cooled and thus clog the pipes of the liquid-air machine (Fig. 22-1).   Once free of water and

Figure 22-1 Diagram of the liquefier used in the commercial preparation of liquid air.

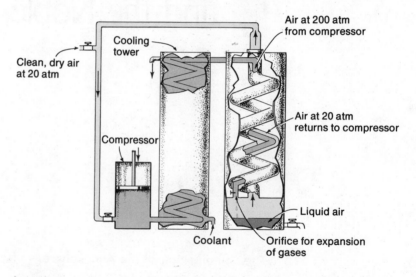

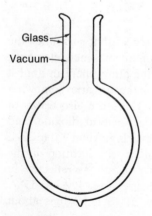

Figure 22-2 Cross section of a Dewar flask.

carbon dioxide, the air is compressed to about 200 atm and cooled to remove the heat produced by compression. The cold, compressed air is next passed into the liquefier in which it escapes through a valve and expands to a pressure of about 20 atm. This expansion is accompanied by the absorption of heat from other compressed air in the small inner coil of the liquefier. As the process continues, each quantity of air that escapes from the valve is colder than that which preceded it, and finally liquefied air begins to escape.

▶ **2. PROPERTIES.**   Liquid air is a mobile liquid with a faint blue color. It evaporates rapidly in an open container with the absorption of a large quantity of heat. To reduce the rate of evaporation, liquid air is stored in Dewar flasks. These flasks (Fig. 22-2) have the space between the inner and outer walls evacuated. Heat is not transmitted through a vacuum and without the heat the liquid air cannot

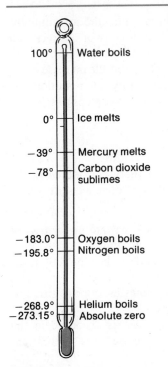

Figure 22-3 Comparative Celsius temperatures for changes of state of several substances.

evaporate. Dewar flasks are frequently silvered on the walls of the evacuated space to reflect heat waves that fall upon them. Thermos bottles are Dewar flasks.

Because liquid air is a mixture, it has no definite boiling point. Oxygen boils at −183.0 and nitrogen at −195.8° (Fig. 22-3). Commercial supplies of nitrogen, oxygen, and argon are obtained by the fractional distillation of liquid air. The low temperature of liquid air is of great practical value in scientific research requiring low temperatures; the study of the effects of very low temperatures on substances is called **cryogenics.** Liquid air is also a source of liquid oxygen used in rocket planes and missiles.

Air Pollution

22.3   The Problem

Air pollution is a subject of much current interest. It is a much worse problem than in years past, but it is not a new problem. William Shakespeare, for example, wrote in *Hamlet* in about 1601:

> This most excellent canopy, the air, look you, this brave o'erhanging firmament, this majestical roof fretted with golden fire, why, it appears no other thing to me than a foul and pestilent congregation of vapors.*

In 1272, King Edward I banned the use of a smoke-producing variety of coal in an effort to clear the smoky air of London, and the British Parliament ordered a man tortured and hanged for illegally burning the banned coal. Later, King Richard III put a high tax on the use of coal. These efforts nevertheless accomplished little, and London remained smoky. In 1952, in London, a combination of smoke and fog was responsible for the deaths of some 4000 people during a four-day period. In Donora, Pennsylvania, in 1948, twenty people died and 5900 were made ill during a bad period of air pollution. These are dramatic instances. Less dramatic, but no less tragic for those involved, were the years 1953, 1963, and 1966, when air pollution in New York City was a serious problem. It is estimated that during each of these years smog caused approximately 700 more deaths than the normal number from other causes. Polluted air may ultimately result in shortened lives for all of us, but it is a special problem for those suffering from respiratory diseases such as emphysema and bronchitis. Lung cancer is found to be twice as prevalent among people living in air-polluted cities as among those living in rural areas with cleaner air.

The problems resulting from air pollution extend well beyond those related to human health. Various types of air pollution can damage or ruin flowers and crops, make paint peel and discolor, damage textiles, discolor dyes, add to the expense of cleaning clothes, poison animals by contaminating feed, corrode metals, damage rubber tires, and increase lighting costs through partially blocking out the sun—to cite only some of the problems. In the United States alone, the estimate is

---

Hamlet, Prince of Denmark, William Shakespeare (1601); Act II, Scene 2.

that well over 200,000,000 tons of pollutants are emitted into the air annually. Strictly on an economic basis, it has been estimated that directly or indirectly air pollution costs the people in the United States over 12 billion dollars per year. Furthermore, over the past several years, we have had an incredible increase in the average per capita use of consumer goods and power. Truly, air pollution is a major problem. It behooves us to learn more about it and to take steps to solve the problem.

22.4   The Causes

The six major air pollutants commonly recognized are those considered in the federal standards for air quality. The six, with a partial list of the effects attributed to each, are*:

1. *Sulfur oxides*—cause acute leaf injury and attack trees; irritate upper part of human respiratory tract; corrode metals and other building materials; ruin hosiery and other textiles; disintegrate book pages and leather; destroy paint pigments; erode statues.
2. *Particulates* (*solids*)—obscure vision; aggravate lung illnesses; deposit grime on buildings and personal belongings; corrode metals.
3. *Carbon monoxide*—causes headaches, dizziness, nausea; is absorbed in the blood, reducing oxygen content, impairing mental processes, and in sufficient quantity causing death.
4. *Nitrogen oxides*—cause leaf damage; stunt plants; irritate eyes and nose membranes; cause brown pungent haze; corrode metals and other building materials; damage rubber.
5. *Hydrocarbons*—may be carcinogenic (cancer-producing); retard plant growth, cause abnormal leaf and bud development.
6. *Photochemical oxidants:* (a) Ozone and resultant chemical products—discolor upper surface of leaves of many crops, trees, and shrubs; damage and fade textiles; reduce physical activity, including athletic performance; speed deterioration of rubber; disturb lung function; irritate eyes, nose, and throat; induce coughing. (b) Peroxyacetyl nitrate (PAN)—discolors lower leaf surface; irritates eyes; disturbs lung function.

Table 22-2 lists the average amount of daily emissions put in the air from all sources and Table 22-3 the annual amount from different sources, according to data from the Department of Health, Education, and Welfare.

The above general categories can be broken down into 119 specific polluting compounds, in itself one indication of the magnitude of the problem facing us. In addition to these compounds, many other materials can be major contaminants in specific locations under special conditions. These include, to list only a few, fluorides, lead compounds, nuclear radioactive emissions, chlorine, beryllium, mercury, and pesticides.

*Air Conservation, American Association for the Advancement of Science, No. 80; The Effects of Air Pollution, Health, Education, and Welfare (HEW)—National Air Pollution Control Administration (NAPCA), No. 1556; Air Pollution Primer, National Tuberculosis and Respiratory Disease Association; Facts and Issues, League of Women Voters, No. 393; Air Pollution Injury to Vegetation, NAPCA, No. AP-71.

TABLE 22-2 Daily Average Emissions in the United States, in Tons

Emission	All Sources	Automobiles Only
Sulfur oxides	90,000	2000–3000
Particulates	75,000	2000–3000
Carbon monoxide	270,000	115,000
Nitrogen oxides	55,000	15,000
Hydrocarbons	85,000	25,000

TABLE 22-3 Annual Emissions in the United States, in Millions of Tons

Source	Sulfur Oxides	Partic- ulates	Carbon Monoxide	Nitrogen Oxides	Hydro- carbons	Total	Per Cent of Total
Transportation	0.8	1.2	63.8	8.1	16.6	90.5	42.3%
Fuel combustion in stationary sources (power generation, industry, space heating)	24.0	8.9	1.9	10.0	0.7	45.5	21.3
Industrial processes	7.3	7.5	9.7	0.2	4.6	29.3	13.7
Solid waste disposal	0.1	1.1	7.8	0.6	1.6	11.2	5.2
Miscellaneous (agricultural burning, forest fires, etc.)	0.6	9.6	16.9	1.7	8.5	37.3	17.5
Total	32.8	28.3	100.1	20.6	32.0	213.8	100.0%

Most of the air pollution problem arises from combustion. Coal is largely carbon; fuel oils, gasoline, and natural gas are largely hydrocarbons (compounds containing only carbon and hydrogen). If combustion were complete, carbon would burn in air to produce only carbon dioxide, and a hydrocarbon would burn to produce only carbon dioxide and water, as illustrated below for carbon (C), methane (CH_4), and ethane (C_2H_6).

$$C + O_2 \longrightarrow CO_2$$
$$CH_4 + 2O_2 \longrightarrow CO_2 + 2H_2O$$
$$2C_2H_6 + 7O_2 \longrightarrow 4CO_2 + 6H_2O$$

However, combustion is usually incomplete as a result of a deficiency of oxygen and the presence of impurities such as elemental sulfur or pyrite (FeS_2) in coal. Carbon monoxide is formed when a deficiency of oxygen is present in relation to the carbon or the hydrocarbon.

$$2C + O_2 \longrightarrow 2CO$$
$$2CH_4 + 3O_2 \longrightarrow 2CO + 4H_2O$$
$$2C_2H_6 + 5O_2 \longrightarrow 4CO + 6H_2O$$

Sulfur oxides are produced when sulfur-containing fuels are burned.

$$S + O_2 \longrightarrow SO_2$$
$$2FeS_2 + 5O_2 \longrightarrow 2FeO + 4SO_2$$

The SO_2 either reacts with water to form sulfurous acid or is oxidized by nitrogen oxides, or other catalysts, to SO_3, which unites with water to form sulfuric acid (see Section 21.14). These acids are among the main substances responsible for ruining nylon hosiery and for corroding metals. Electrical generating plants using sulfur-containing coal as fuel contribute greatly to air pollution. The catalytic converters of automobile exhaust systems produce small quantities of sulfur dioxide, sulfur trioxide, sulfuric acid, and hydrogen sulfide from reactions of the sulfur compounds present in gasoline. Nitrogen oxides are produced with all combustion because of reaction of nitrogen in the air with oxygen, particularly at high temperatures (see Sections 23.15 and 23.22).

$$N_2(g) + O_2(g) \longrightarrow 2NO(g) \qquad \Delta H° = 180.5 \text{ kJ}$$
$$2NO(g) + O_2(g) \longrightarrow 2NO_2(g) \qquad \Delta H° = -114.1 \text{ kJ}$$
$$3NO_2 + H_2O \longrightarrow 2HNO_3 + NO$$

Much of the particulate matter in polluted air is noncombustible or unburned material in fuel (carbon particles, ash, etc.). Particulate matter, such as dust, carbon, and ash, may be present either as large particles that fall out of the air quickly (nevertheless creating problems where they fall) or as small particles ("aerosols") which when suspended in air as part of the colloidal system (Section 13.29) move about in air just as gases do.

Some materials not resulting from combustion that contribute to air pollution are from vaporization of liquids (or sometimes sublimation of solids) to gases, which then disperse in the air. Solid particles, other than those from combustion, are introduced into the air by abrasion or grinding of one material in contact with another. Examples are metal filings introduced into the air when a metal is polished on a grinding wheel, dust put into the air during a dust storm or by automobile wheels on a dusty road, and (what is said to be the most common particulate matter in human lungs) finely divided rubber added to the air by friction of automobile tires on pavements.

Secondary reactions produce additional pollution problems. One example, mentioned earlier, is the oxidation of sulfur dioxide to sulfur trioxide and the subsequent reaction with moisture in the atmosphere to produce sulfuric acid. Another example is the photochemical reaction by which unburned hydrocarbons and nitrogen oxides in automobile exhausts react in the presence of sunlight to form peroxyacetyl nitrate (PAN) and ozone. These compounds are primary constituents of smog, which is increasingly becoming a problem in many of our larger cities.

The problem of smog is intensified by the so-called air "inversion" common in Los Angeles and some other cities. Normally air near the earth's surface becomes warmer than the air above and rises, allowing cooler air to take its place. Sometimes, however, a warmer layer of air develops above, as when cool air flows down a mountain into the valleys and underneath a mass of warm air. When this happens, the cooler air is trapped and held at the earth's surface. Pollutants from auto exhausts, industrial stacks, and other sources are also trapped, sometimes for days at a time.

Tetraethyl lead has been used commonly in gasoline to provide greater antiknock capability. Its use has been curtailed greatly, however, with the introduction of catalytic converters, because lead compounds poison the catalysts used in the converters, quickly rendering them ineffective. When used, tetraethyl lead is converted during the combustion process to lead oxide, which forms harmful deposits in the engine. Therefore, an additional compound is added, ethylene dibromide, which reacts in the engine with the lead oxide to form lead bromide, a volatile substance that is emitted along with other gases to the air through the exhaust. Lead compounds are known to have serious physiological effects on humans.

22.5   Some Solutions to the Air Pollution Problem

Removal of particulate (solid) pollutants from smokestack gases is easily accomplished using electrostatic Cottrell precipitators (see Section 13.33). Processes for removing sulfur oxides, however, are not well developed, though several methods are being studied and tested. But even at the present stage of development of such processes, sulfur oxides recovered from stack gases are now becoming a significant source of sulfur for the production of sulfuric acid (see Section 21.14). The "wet limestone" process for removal of sulfur oxides from stack gases is one that shows promise:

$$SO_2 + CaCO_3 \xrightarrow{H_2O} CaSO_3 + CO_2$$

$$SO_3 + CaCO_3 \xrightarrow{H_2O} CaSO_4 + CO_2$$

The use of low-sulfur coal, natural gas, or fuel oil, instead of high-sulfur coal, would alleviate the problem of sulfur oxide emission, but the quantity of such fuels is severely limited. It is likely that such low-sulfur fuels might most logically be used for small pollution sources such as homes and small industries. These sources individually do not produce a large amount of pollution, but collectively are a big factor because of the large numbers, the discharge of pollutants near ground level, and the economic problem of providing pollution control devices for so many small sources.

A possible solution for the smaller industrial sources, and perhaps large ones as well, is the removal of sulfur from sulfur-containing fuels. Fuel oil can be fairly readily desulfurized, but only 12% of the sulfur oxides presently emitted by combustion sources originates in heavy fuel oil, whereas 65% originates in coal. Coal contains 2.5–3% sulfur on the average. Twenty-five per cent of the coal mined in the United States for use as a fuel is such that its sulfur content can be reduced to 1% by existing processes. For the other 75% of coal mined, it is possible to remove only about 40% of its sulfur content with present technology.

Over 95% of nitrogen oxide emissions results from combustion; 25% of the emissions come from combustion in steam-electric power plants. Some reduction in nitrogen oxide emissions can be made by modifying the combustion process, though with some loss of efficiency, but the technology is not well defined yet. The technology for removal of nitrogen oxides from stack gases and auto exhausts is also still in the laboratory stage.

Carbon monoxide emission can be lessened by finer control of combustion,

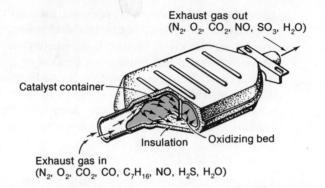

Exhaust gas out
(N_2, O_2, CO_2, NO, SO_3, H_2O)
Catalyst container
Insulation
Oxidizing bed
Exhaust gas in
(N_2, O_2, CO_2, CO, C_7H_{16}, NO, H_2S, H_2O)

Figure 22-4 A catalytic converter used to decrease pollutants in the exhaust of automobiles.

including better tuning of auto engines, but the technology to eliminate the carbon monoxide problem completely, either by more efficient combustion or by removal from stacks and exhausts, is not available at present.

Catalytic converters (Fig. 22-4) decrease carbon monoxide and unburned hydrocarbon pollutants to very low levels. One type of converter contains refractory beads coated with a catalytic material consisting of platinum and palladium metal in an extremely finely divided form, thereby providing a very large total surface area. The pollutants react with oxygen from the air on the surface of the solid catalyst.

$$2CO + O_2 \longrightarrow 2CO_2$$

$$\underset{\text{(Typical hydrocarbon in gasoline)}}{C_7H_{16}} + 11O_2 \longrightarrow 7CO_2 + 8H_2O$$

In addition to decreasing carbon monoxide and hydrocarbon pollutants in the exhaust, however, the converters have been found to catalyze reactions between sulfur compounds in the gasoline and hydrocarbons, which are the principal constituents of gasoline, and between the sulfur compounds and oxygen. As mentioned previously, reaction of hydrocarbons with the sulfur compounds produces the very toxic hydrogen sulfide, but this only occurs with an abnormally rich fuel-air mixture (high ratio of fuel to air). Hence careful tuning of the engine ordinarily eliminates that problem, though sometimes at the expense of less satisfactory performance of the engine. More serious are the catalysis of the reaction of sulfur compounds with oxygen of the air to produce sulfur dioxide and the catalysis of the oxidation of sulfur dioxide to sulfur trioxide. The sulfur trioxide then reacts with water to form small quantities of sulfuric acid mist. Fortunately, technology is available for removing sulfur compounds from gasoline during the refining process, but there is certain to be an increase in the cost of gasoline to the consumer. The extra expense may well become necessary, however, if catalytic converters continue to be used extensively.

As pointed out in the previous section, lead compounds poison the relatively expensive catalyst in the converter after short use. Hence, low-lead or no-lead gasolines must be produced in large quantity. The undesirable emission of toxic lead compounds in the exhaust is also eliminated by the use of unleaded gasolines.

An additional pollutant, nitric oxide, results from the reaction of nitrogen and oxygen from the air during the combustion process. This reaction is promoted both by the high temperature of the converter (see Section 23.15, Part 2) and by the lean fuel-air mixture (high ratio of air to fuel) employed for more complete combustion. Some effort has been directed toward finding a suitable catalyst for the conversion of nitric oxide to harmless substances, such as elemental nitrogen and carbon dioxide. Reduction of the oxide with carbon monoxide is one possibility.

$$2NO + 2CO \longrightarrow N_2 + 2CO_2$$

Some success has been achieved recently in developing an engine that uses a stratified fuel concept to reduce undesirable emissions. The design of the engine is based on the principle that good performance requires a fuel-rich mixture only in the region of the initial spark. Each cylinder is divided into two sections, a small auxiliary combustion chamber of high temperature near the spark source and a larger main combustion chamber of lower temperature comprising the remainder of the cylinder. A fuel intake valve is provided in each chamber. The auxiliary chamber is connected to the main chamber by a small opening called a torch nozzle, through which the initial flame can spread. The spark plug is located in the auxiliary chamber.

To operate the engine, a fuel mixture with a relatively high ratio of fuel to air is introduced into the auxiliary chamber. When the mixture is ignited, even at the high temperature, very little nitric oxide is produced because of the relatively low proportion of air as a source of nitrogen and oxygen. At about the same time, an extremely lean fuel mixture that is compatible with more complete combustion is introduced into the main chamber. The lean mixture is ignited by the flame spreading into it through the torch nozzle, and the resulting explosion drives the piston. The lower temperature in the main chamber offsets the effect of the large volume of air, thereby keeping the production of nitric oxide to a low level. In addition, the more complete combustion of the leaner mixture in the main chamber reduces the amounts of carbon monoxide and unburned hydrocarbon pollutants produced.

Evaporation of gasoline from automobile gasoline tanks during filling and during use contributes surprising amounts of hydrocarbon vapors to the air. It is estimated that in one large U.S. city alone, 16 tons of hydrocarbons are introduced into the air each day by evaporation during the filling of automobile tanks at service stations. By combustion, automobile engines in the same city put over 1700 tons of hydrocarbons and approximately 650 tons of nitrogen oxides into the air each day.

Considerable effort is being expended to study alternatives to the use of the internal combustion engine, as well as modifications of existing internal combustion engines. Alternatives include electric automobiles with rechargeable batteries, steam automobiles (or ones using gases other than steam), and even wind-up automobiles in which a large spring mechanism would be wound periodically by electrical or mechanical means. It should be noted, however, that recharging of batteries or even periodic winding in turn require electrical or other forms of power and so necessitate additional generation facilities, with the subsequent increase in pollution. A change to avoid one type of pollution often introduces another type. The complete solution to the pollution dilemma involves solving exceedingly complex and intertwined problems.

Limited supplies of crude oil, low-sulfur coal, and natural gas have placed the United States, and indeed the whole world, in an energy crisis. There are, however, vast coal deposits still untouched, and a possible partial solution is gasification of coal at the mine to produce clean-burning pipeline gas. But this would introduce several technological problems, including those of cleaning the emissions in the gasification process itself at the mine.

To many people, nuclear energy is the answer to our energy crisis. The use of nuclear energy is now increasing steadily, but with it come problems of radioactive emissions (Chapter 29) and of thermal discharge, which raises the temperature of lakes and streams and of air. Thermal pollution (sometimes called "thermal loading") by both nuclear and conventional power plants is a serious problem. With nuclear power plants there is also the possibility of escape of radioactive materials. Although safety measures have been quite well worked out, radioactive wastes must be handled very carefully to prevent such a loss. Some scientists believe that thermonuclear fusion is the best hope for a clean power source. Thermonuclear fusion has a much less potential hazard from accidental explosion than fission and also a much reduced radioactive emission problem. A fission reaction, however, with its inherent problems, is often used to provide the large initial amounts of energy to initiate the fusion reaction. Nuclear fission and nuclear fusion are discussed in Chapter 29.

The use of solar energy (energy from the sun) is in a very rudimentary state of technology, but solar energy holds high promise as a clean method for solving some of our power needs in the long-term future (see Section 20.18). Windmills may once again become a useful and clean means of generating small quantities of electrical power in areas where sufficient and reasonably consistent wind velocities are commonly present. Meanwhile, we must solve, at least on a short-term basis, the environmental problems of combustion of conventional fossil fuels.

The United States government, on April 30, 1971, announced strict air quality standards, originally specified to go into effect by July 1, 1975, but then in part postponed several times to later dates. These standards establish limits for the amounts of sulfur oxides, particulate matter, carbon monoxide, photochemical oxidants, nitrogen oxides, and hydrocarbons that can safely exist in the air. Full enforcement of the limits would undoubtedly cause considerable changes such as closing some sections of large cities to automobile traffic at certain hours, a greater use of public transportation and car pools, and very significant changes in the fuels used by electric generating plants and other industries.

In summary, air pollution involves the loss of natural resources, degrades the environment, and destroys health. Methods presently known to control pollution cost money, time, and effort. Development of new technology necessary for really adequate control of pollution carries a high price. It appears, nevertheless, that we really have no choice but to pay the necessary price in dollars and inconvenience.

Another equally important part of the pollution problem, that of water pollution, was discussed in Section 12.9.

The Noble Gases

22.6 Discovery of the Noble Gases

The discovery of the noble gases makes an interesting story. Helium was first discovered in the atmosphere that surrounds the sun. In 1868, the French astronomer Pierre-Jules-César Janssen went to India to study a total eclipse of the sun by using a spectroscope. He observed among other bright lines one new yellow line, which caused the English chemist Professor Edward Frankland and the English astronomer Sir Norman Lockyer to conclude that the sun contained an undiscovered element. They called this new element **helium,** from the Greek word *helios,* meaning "the sun." All attempts to find this element on the earth were unsuccessful until 1895, when the Scottish chemist and physicist Sir William Ramsay showed that the gas driven off by heating the uranium mineral cleveite has a spectrum identical with that of helium.

In 1894, the British chemist Lord Rayleigh observed that a liter of nitrogen that he had prepared by removing the oxygen, carbon dioxide, and water vapor from air weighed 1.2572 grams, while a liter of nitrogen prepared from ammonia weighed only 1.2506 grams under the same conditions. This discrepancy caused Rayleigh to suspect the presence of a previously undiscovered element in the atmosphere. By passing nitrogen obtained from the air over red-hot magnesium ($3Mg + N_2 \longrightarrow Mg_3N_2$), he found a small amount of residual gas that he could not cause to combine with any other element. Lord Rayleigh and Sir William found that the residual gas showed spectral lines never before observed. In 1894, they announced the discovery of the first noble gas, which they called **argon,** meaning "the lazy one." It was later found that this residual gas was a mixture of the noble gases helium, neon, argon, krypton, and xenon.

In 1898, Sir William Ramsay and his assistant Professor Morris W. Travers isolated **neon** (meaning "the new element") by the fractional distillation of impure liquid oxygen. Shortly thereafter they showed the less volatile fractions from liquid air to contain two other new elements, **krypton** ("the hidden element") and **xenon** ("the stranger").

In 1900, the Prussian physicist Professor Friedrich E. Dorn discovered that one of the disintegration products of radium is a gas similar in chemical properties to the noble gases. At first this gas was called radium emanation, but later its name was changed to **radon.**

22.7 Production of the Noble Gases

Helium is isolated from certain natural gases, which sometimes contain as much as 2 to 5% of the element by mass. Professor Hamilton P. Cady and D. F. McFarland of the University of Kansas first discovered helium in natural gas in 1905 when asked to analyze a sample of natural gas that would not ignite from a new well in southern Kansas. They found that the gas contained, by mass, 1.84% helium, and so what seemed at first to be a financial fiasco turned out to be the beginning of a profitable business in the sale of helium. To isolate the helium, the condensable components of the gas are liquefied in a liquid-air machine, leaving helium in the gaseous condition. The United States has most of the world's commercial supply of this element in its helium-bearing gas fields.

Argon, neon, krypton, and xenon are produced by the fractional distillation of liquid air. Radon gas is collected from radium salts.

22.8   Physical Properties of the Noble Gases

The principal properties of the noble gases are given in Table 22-4. The boiling points and melting points of the noble gases are extremely low in comparison to those of other substances of comparable atomic or molecular weights. The reason for this is that no strong chemical bonds hold the atoms together in the liquid and solid states. Only weak London forces (Section 11.4) exist between atoms in noble gases. London forces are effective only when molecular motion is very slight, as it is at very low temperatures, so as to permit close approach of neighboring atoms. Liquid helium is the only liquid known that does not solidify on cooling. It remains liquid down to absolute zero (-273.15 C) at ordinary pressures; however, it can be solidified by increasing the pressure.

TABLE 22-4   Properties of the Noble Gases

Property	Helium	Neon	Argon	Krypton	Xenon	Radon
Atomic number	2	10	18	36	54	86
Electrons in outer shell	2	8	8	8	8	8
Atomic weight	4.0	20.18	39.95	83.8	131.3	222
Melting point, °C	-269.7[a]	-248.6	-189.3	-157.2	-111.9	-71
Boiling point, °C	-268.9	-246.1	-186	-153.4	-108.1	-62

[a]At a pressure of 26 atm.

22.9   Chemical Properties of the Noble Gases

The elements of Group 0 were formerly called the inert gases, because until 1962 it was thought that they were chemically inert. When it was discovered that they really are not inert, the term **noble gases** was adopted. Besides being called noble gases, they are sometimes referred to as rare gases. The stability of the noble gas type of electronic structure has been mentioned frequently in the preceding chapters. The two electrons of helium represent a completed shell, while a complement of eight electrons in the outer shell of each of the other noble gases gives them a stable configuration. Many elements form stable ions that have noble gas structures. These include the elements of high electron affinities (the halogens), which immediately precede the noble gases in the Periodic Table, and the elements of low ionization potentials (the alkali and alkaline earth metals), which immediately follow them.

Almost from the time the noble gases were first discovered, chemists have tried to force them into chemical combination by using unusual conditions. Early efforts were successful to only a limited extent. To date, helium has been forced to combine with certain other elements by using the large energies obtainable in electrical discharge tubes. Examples of species thus formed are He_2^+ and such combinations with hydrogen as HeH^+ and HeH_2^+. The molecular orbital structure of He_2^+ was discussed in Chapter 6 (Section 6.4). These combinations are not stable and have only momentary existence.

A strong dipole, such as the water molecule, may polarize a noble gas atom so that it acts as a dipole itself and thereby attracts the original strong dipole. The larger the noble gas atom, the greater the susceptibility toward dipole induction. Thus hydrates of argon, krypton, xenon, and radon have been prepared. These compounds have formulas such as $Ar \cdot xH_2O$, $Kr \cdot xH_2O$, and $Xe \cdot xH_2O$, where x is 1 to 6 but equals 6 only with noble gases of largest radii. These compounds are crystalline solids, which form when water and the noble gases are brought together at low temperatures and high pressures.

The fact that the noble gases have higher ionization potentials than those of any of the other elements in their respective horizontal rows of the Periodic Table (see Section 4.19) has encouraged the widely held concept of their inertness. However, if the ionization potentials of the larger noble gases are compared with the ionization potentials of other elements, it is apparent that these values are of the order of magnitude of those of certain other elements and indeed are even smaller than the ionization potentials for some elements. In June 1962, Neil Bartlett, then in Canada, reported the preparation of a yellow compound of xenon to which he ascribed the formula $Xe[PtF_6]$. This discovery led a group of chemists at the Argonne National Laboratory to think that perhaps xenon might be oxidized by a highly electronegative element such as fluorine, and in September 1962, they reported the preparation of **xenon tetrafluoride,** XeF_4. This was the first report of a stable compound of a noble gas with another single element. The compound was made by the surprisingly simple procedure of mixing xenon gas and fluorine gas at $400°C$ for one hour and cooling to $-78°$. The material has the form of colorless crystals, which melt at about $90°$ and which are stable at room temperature, showing no evidence of reaction or decomposition after prolonged storage in glass vessels. Shortly after reporting the preparation of xenon tetrafluoride, the same group of chemists reported the preparation of XeF_2, $XeOF_4$, and XeO_3. Since that time, quite a number of additional compounds have been prepared. Table 22-5 lists the formulas for some of the noble gas compounds prepared to date.

The preparation of compounds of the noble gases not only opened the door to an exciting new area of research, but it also serves to point out that the ideas of

TABLE 22-5 Formulas for Some Compounds of Three Noble Gases

Xenon			Krypton	Radon
$XePtCl_6$	XeO_3	Na_2XeF_8	KrF_2	RnF_x(where x
$XeCl_2$	XeO_4	K_2XeF_8	$KrF_2 \cdot 2SbF_5$	is not known)
XeF_2	$XeF_2 \cdot 2SbF_5$	Rb_2XeF_8	KrF_4?	
XeF_4	$XeF_2 \cdot 2TaF_5$	Cs_2XeF_8		
XeF_6	$XeF_6 \cdot 2SbF_5$	$(NO)_2^+[XeF_8]^{2-}$		
$XeOF_2$	$XeF_6 \cdot AsF_5$	$Na_4XeO_6 \cdot xH_2O$		
$XeOF_4$	$XeF_6 \cdot BF_3$	(where $x = 6$ and 8)		
XeO_2F_2	$RbXeF_7$	$Ba_2XeO_6 \cdot 1.5H_2O$		
	$CsXeF_7$	$K[XeO_3F]$		

science are subject to change. We should always keep our minds open to new possibilities and not blindly accept preconceived or entrenched ideas.

22.10   Stability of Noble Gas Compounds

The xenon fluorides are formed in exothermic reactions, reflected in negative values for heats of formation (Table 22-6). The energy release is sufficient to represent a substantial bond, as shown by the average bond energies listed in Table 22-6. By way of comparison, the energies of strong ionic and covalent bonds tend to be in the range of 170–500 kJ/mol.

TABLE 22-6 Heats of Formation and Average Bond Energies for Three Noble Gas Compounds		
Compound	Heat of Formation for Gaseous State, kJ/mol	Average Bond Energy, kJ/mol
XeF_2	−108	131
XeF_4	−215	130
XeF_6	−298	126

The oxyfluorides of xenon, in general, are formed by endothermic reactions, have positive heats of formation, and are quite unstable. $XeOF_4$ is an exception, with a heat of formation of −38 kJ/mol; it is the only oxyfluoride of a noble gas that is relatively easy to prepare and characterize.

Fluorides of krypton and radon have positive heats of formation. As would be expected, therefore, they are more difficult to make and less stable than the corresponding xenon fluorides. The heat of formation of KrF_2 is +60.2 kJ/mol.

22.11   Chemical Properties of Noble Gas Compounds

Stable compounds of xenon are formed when xenon is bonded to either fluorine or oxygen, two highly electronegative elements. The compounds of xenon with oxygen are prepared by replacing fluorine atoms in the xenon fluorides with oxygen. Present knowledge of the chemistry of xenon compounds is limited. They are very strong oxidizing agents, which may disproportionate in water, and they react with strong Lewis acids by donating a fluoride ion. Xenon difluoride is quite soluble in water, going into solution as undissociated molecules that soon oxidize the water to produce xenon, hydrogen fluoride, and oxygen.

$$2XeF_2 + 2H_2O \longrightarrow 2Xe + 4HF + O_2$$

The hydrolysis of xenon tetrafluoride is more complicated because it is accompanied by disproportionation of the xenon(IV) and oxidation of water. With a stoichiometric amount of water, $XeOF_4$ is formed, but the oxide, XeO_3, forms in an excess of water.

$$6XeF_4 + 8H_2O \longrightarrow 2XeOF_4 + 4Xe + 16HF + 3O_2$$
$$6XeF_4 + 12H_2O \longrightarrow 2XeO_3 + 4Xe + 24HF + 3O_2$$

Dry, solid XeO_3 is an extremely sensitive explosive that must be handled with care.

In the reaction of XeF_6 with water, the $+6$ oxidation state for the xenon is retained.

$$XeF_6 + 3H_2O \longrightarrow XeO_3 + 6HF$$

Both XeF_6 and XeO_3 disproportionate in basic solution, producing salts of the perxenate ion, XeO_6^{4-}. For example,

$$2XeF_6 + 4Na^+ + 16OH^- \longrightarrow Na_4XeO_6 + Xe + O_2 + 12F^- + 8H_2O$$

$$2XeO_3 + 4Na^+ + 4OH^- \longrightarrow Na_4XeO_6 + Xe + O_2 + 2H_2O$$

Xenon difluoride reacts with acids such as HSO_3F, F_5TeOH, and $HClO_4$, which are resistant to oxidation with evolution of hydrogen fluoride and formation of Xe—O bonds. The product depends upon the stoichiometry of the reaction.

$$XeF_2 + HSO_3F \longrightarrow F—Xe—O—SO_2F + HF$$

$$XeF_2 + F_5TeOH \longrightarrow F—Xe—O—TeF_5 + HF$$

$$XeF_2 + 2F_5TeOH \longrightarrow F_5Te—O—Xe—O—TeF_5 + 2HF$$

Standard reduction potentials indicate that oxidation of Xe is more difficult than oxidation of hydrogen. Hence Xe is below hydrogen in the activity series. The oxidation reactions take place more readily in basic solution than in acidic solution. The following equations show the oxidation half-reactions and the potentials corresponding to the oxidation.

In acid solution:

$$Xe + 3H_2O \rightleftharpoons XeO_3 + 6H^+ + 6e^- \qquad E° = -1.8 \text{ V}$$

$$XeO_3 + 3H_2O \rightleftharpoons H_4XeO_6 + 2H^+ + 2e^- \qquad E° = -3.0 \text{ V}$$

In basic solution:

$$Xe + 7OH^- \rightleftharpoons HXeO_4^- + 3H_2O + 6e^- \qquad E° = -0.9 \text{ V}$$

$$HXeO_4^- + 4OH^- \rightleftharpoons HXeO_6^{3-} + 2H_2O + 2e^- \qquad E° = -0.9 \text{ V}$$

22.12   Structures of Noble Gas Compounds

The structures of XeF_2 (linear), XeF_4 (square planar), and $XeOF_4$ (slightly distorted square base pyramidal) are shown in Chapter 19 (Fig. 19-1) and are like those illustrated for polyhalides (Fig. 19-3).

Oxygen derivatives of XeF_2 also have a linear structure. The structure of XeO_3 (Fig. 22-5a) is interpreted as being a trigonal pyramid with the four atoms at the corners. The XeO_4 molecule has a tetrahedral structure (Fig. 22-5b). The shape of the XeF_6 molecule has been difficult to establish experimentally. Evidence indicates that the molecule has a somewhat distorted octahedral structure.

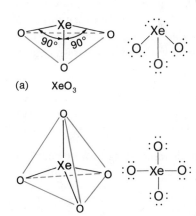

Figure 22-5 The Lewis structures and the molecular structures of XeO_3 (trigonal pyramidal) and XeO_4 (tetrahedral). See Fig. 19-1 for structures of XeF_2, XeF_4, and $XeOF_4$.

22.13   Uses of the Noble Gases

▶ **1. HELIUM.** This gas is used for filling observation balloons and other lighter-than-air craft, particularly by the U.S. Navy. The fact that helium does not burn makes it safer to use than hydrogen. Although it is twice as dense as hydrogen, helium has a lifting power that is 92.6% that of hydrogen. (See Problem 1 at the end of this chaper.)

Because helium at high pressures is not a narcotic while nitrogen is, mixtures of oxygen and helium are used by divers working under high pressures in order to avoid a disoriented mental condition known as nitrogen narcosis, the so-called rapture of the deeps. This disorientation can result from the narcotic effect of nitrogen when air is breathed at high pressures.

Mixtures of helium and oxygen are beneficial in the treatment of certain respiratory diseases such as asthma. The lightness and rapid diffusion of helium decrease the muscular effort involved in breathing.

Helium is used as an inert atmosphere for the melting and welding of easily oxidizable metals and for many chemical processes that are sensitive to air. Liquid helium is of extremely great value in attaining low temperatures for cryogenic research.

▶ **2. NEON.** Neon is used in the familiar neon lamps and signs. When an electric current is passed through a tube containing neon under very low pressure, a brilliant orange-red glow is emitted. The color of the light given off by a neon tube may be changed by mixing argon or mercury vapor with the neon and by utilizing tubes made of glasses of special color. Neon lamps cost less to operate than ordinary electric lamps and their light penetrates fog better.

▶ **3. ARGON.** This gas is used in gas-filled electric light bulbs, where its lower heat conductivity and chemical inertness make it preferable to nitrogen for inhibiting volatilization of the tungsten filament and prolonging the life of the bulb. Fluorescent tubes commonly contain a mixture of argon and mercury vapor. Many Geiger-counter tubes are also filled with argon.

▶ **4. KRYPTON AND XENON.** A krypton-xenon photographic flash tube has been developed for taking high-speed photographic exposures. An electric discharge through the tube gives a very intense light which lasts only 1/50,000 of a second.

▶ **5. RADON.** This gas is collected in hospitals from disintegrating radium supplies and sealed in small tubes; the gas is then used in the radiotherapy of malignant growths.

Questions

1. What evidence can you cite to show that air is a mixture rather than a compound?
2. Name the six principal air pollutants and identify a source of each. Briefly describe one technique for controlling each of these pollutants.
3. Write balanced equations for each of the following processes:
   (a) The formation of sulfur dioxide from zinc sulfide

present in burning coal
   (b) The formation of nitrogen dioxide in an internal combustion engine
   (c) The formation of sulfuric acid mist from sulfur dioxide
   (d) The removal of carbon monoxide from an automobile's exhaust using a catalytic converter
4. Based on the periodic properties of the elements,

write balanced chemical equations for the reactions, if any, between the components of air given in Table 22-1 and hot magnesium.

5. Write the Lewis structures, for XeF_2, XeF_4, XeF_6, F—Xe—O—SO_2F, and XeO_3.

6. Write complete and balanced equations for the following reactions:

(a) $XeO_3 + Ba(OH)_2$ (in water) $\longrightarrow$
(b) $XeF_2 + HClO_4 \longrightarrow$
(c) $XeF_2 + 2HClO_4 \longrightarrow$
(d) $XeF_6 + Na \longrightarrow$

Problems

1. (a) Calculate the density of air, of hydrogen, and of helium at STP in g l^{-1}. (Assume the average molecular weight of air as 29.10.)   *Ans. Air, 1.30 g l^{-1}; hydrogen, 0.0900 g l^{-1}; helium, 0.179 g l^{-1}*
(b) The "lifting power" of a gas with a density less than that of air is determined by the difference between its mass and that of an equal volume of air. Calculate the lifting power of hydrogen, the lifting power of helium [which by calculation in part (a) is shown to be approximately twice as heavy as hydrogen for a given volume], and the percentage lifting power for helium compared to that of hydrogen.
   Ans. Hydrogen, 1.21; helium, 1.12; 92.6%

2. A concentration of 750 parts per million by volume of carbon monoxide is lethal. What mass of carbon monoxide is required to provide a lethal uniform concentration in a room 6.8 meters long, 5.3 meters wide, and 2.2 meters high (about 22 × 17 × 7 feet) if the temperature is 22° and the atmospheric pressure 740 torr?   *Ans. 67 g*

3. A mixture of xenon and fluorine was heated at 400° for 1 hour. A sample of the white solid that formed reacted with hydrogen to give 54.0 ml of xenon at STP and hydrogen fluoride, which was collected in water, giving a solution of hydrofluoric acid. The hydrofluoric acid solution was titrated, and 45.62 ml of 0.2115 M sodium hydroxide was required to reach the end point. Determine the empirical formula for the white solid and write balanced chemical equations for the reactions involving xenon.
   Ans. Xe + 2F_2 $\longrightarrow$ XeF_4
   XeF_4 + 2H_2 $\longrightarrow$ Xe + 4HF

4. A 0.492-gram sample of the XeF_4 described in Problem 3 exerted a pressure of 47.8 torr at 127° in a bulb with a volume of 1238 ml. Determine the molecular formula of xenon tetrafluoride from these data.
   Ans. XeF_4

5. Basic solutions of Na_4XeO_6 are powerful oxidants. What mass of $Mn(NO_3)_2 \cdot 6H_2O$ will react with 125.0 ml of a 0.1717 M basic solution of Na_4XeO_6 if the products include Xe and a solution of sodium permanganate?   *Ans. 9.857 g*

References

"The Earth's Evolving Atmosphere" (Staff), *Chemistry*, February 1979, p. 22.

"Atmospheric Chemistry," M. L. Corrin, *J. Chem. Educ.,* **55,** 210 (1978).

"Medieval Uses of Air," L. White, Jr., *Sci. American,* August 1970, p. 92.

"How Have Sea Water and Air Got Their Present Compositions?" L. G. Sillén, *Chem. in Britain,* **3,** 291 (1967).

"The Global Circulation of Atmospheric Pollutants," R. E. Newell, *Sci. American,* January 1971, p. 32.

"The Carbon Dioxide Question," G. M. Woodwell, *Sci. American,* January 1978, p. 34.

"Air Quality Is Improving in the U.S." (Staff), *Chem. and Eng. News,* February 5, 1979, p. 6.

"Environmental Impact: Controlling the Overall Level," W. E. Westman and R. M. Gifford, *Science,* **181,** 819 (1973).

"Analysis of an Important Air Pollutant: Peroxyacetyl Nitrate," E. R. Stephens and M. A. Price, *J. Chem. Educ.,* **50,** 351 (1973).

"Pollution of the Air," H. J. M. Bowen, *Educ. in Chem.,* **11,** 157 (1974).

"The Automobile and Air Pollution: A Chemical Review of the Problem," T. R. Wildeman, *J. Chem. Educ.,* **51,** 290 (1974).

"Gasohol: Energy Mountain or Molehill," E. V. Anderson, *Chem. and Eng. News,* July 21, 1978, p. 8.

"Alternative Automobile Engines," D. G. Wilson, *Sci. American,* July 1978, p. 39.

"Oil and Gas from Coal," N. P. Cochran, *Sci. American,* May 1976, p. 24.

"Cathedral Chemistry—Conserving the Stained Glass," R. G. Newton, *Chem. in Britain,* **10,** 89 (1974).

"Atmospheric Effects of Pollutants," P. V. Hobbs, H. Harrison, and E. Robinson, *Science,* **183,** 909 (1974).

"Chemical Waste Disposal a Costly Problem," C. Murray, *Chem. and Eng. News,* March 12, 1979, p. 12.

"Coal and the Present Energy Situation," E. F. Osborn, *Science,* **183,** 477 (1974).

"Energy and Life-Style," A. Mazur and E. Ross, *Science,* **186,** 607 (1974).

"Solar and Geothermal Energy: New Competition for the Atom," L. J. Carter, *Science,* **186,** 811 (1974).

"Solar Biomass Energy: An Overview of U.S. Potential," C. C. Burwell, *Science,* **199,** 1041 (1978).

"Energy Policy in the U.S.," D. J. Rose, *Sci. American,* January 1974, p. 20.

"Stratospheric Ozone Depletion and Solar Ultraviolet Radiation on Earth," P. Cutchis, *Science,* **184,** 13 (1974).

"Stratospheric Ozone Destruction by Aircraft-Induced Nitrogen Oxides," F. N. Alyea, D. M. Cunnold, and R. G. Prinn, *Science,* **188,** 177 (1975).

"Long Term Baseline Atmospheric Monitoring," M. A. Goldman, *J. Chem. Educ.,* **52,** 365 (1975).

"Chemistry and Pollution of the Stratosphere," R. J. Donovan, *Educ. in Chem.,* July 1978, p. 110.

"The Great Spray Can Debate," H. Bassow, *J. Chem. Educ.,* **54,** 371 (1977).

"Atmospheric Halocarbons, Hydrocarbons, and Sulfur Hexafluoride; Global Distribution, Sources, and Sinks," H. B. Singh *et al., Science,* **203,** 899 (1979).

"The Fuel Consumption of Automobiles," J. R. Pierce, *Sci. American,* January 1975, p. 34.

"Energy Conservation: Better Living through Thermodynamics," W. D. Metz, *Science,* **188,** 820 (1975).

"The Correct Sizes of the Noble Gas Atoms," J. E. Huheey, *J. Chem. Educ.,* **45,** 791 (1968).

"Discovery of the Noble Gases and Foundations of the Theory of Atomic Structure," J. E. Frey, *J. Chem. Educ.,* **43,** 371 (1966).

"The Chemistry of the Noble Gases," J. Selig, J. G. Malm, and H. H. Claassen, *Sci. American,* May 1964, p. 66.

"Bonding in Xenon Hexafluoride," J. J. Kaufman, *J. Chem. Educ.,* **41,** 183 (1964).

"The Noble Gases and the Periodic Table," J. H. Wolfenden, *J. Chem. Educ.,* **46,** 569 (1969).

"Helium Conservation Program: Casting it to the Winds," W. D. Metz, *Science,* **183,** 59 (1974).

"Argon: Its Market Is Anything but Inert," W. J. Storck, *Chem. and Eng. News,* August 21, 1978, p. 11.

"Noble Gas Chemistry," J. H. Holloway, *Educ. in Chem.,* **10,** 140 (1973).

"A Decade of Xenon Chemistry," G. J. Moody, *J. Chem. Educ.,* **51,** 628 (1974).

"Noble Gases in the Murchison Meteorite: Possible Relics of S-Process Nucleosynthesis," S. E. Palmer and E. Anders, *Science,* **201,** 51 (1978).

"Xenon Becomes a Metal" (Staff), *Chemistry,* February 1979, p. 5.

23

Nitrogen and Its Compounds

The elements of Periodic Group VA make up the **nitrogen family.** Nitrogen is the first member of the family, which also includes phosphorus, arsenic, antimony, and bismuth. Nitrogen and phosphorus are among our most important and useful nonmetals. There is a regular gradation with increasing atomic weight in the nitrogen family, from the characteristics of a true nonmetal (nitrogen) to those of an almost true metal (bismuth). As has been noted for fluorine and oxygen, the lightest members of their families, some of the properties of nitrogen are anomalous, while more expected trends are found for the other members of the family (discussed in later chapters of this book).

The chemistry of nitrogen is sufficiently important and interesting to warrant its study apart from the other members of the family.

23.1 The Chemical Properties of Nitrogen

The electronic structure of a nitrogen atom is $1s^2 2s^2 2p^3$, with the three $2p$ electrons distributed in the p_x, p_y, and p_z orbitals. Since there are no d orbitals in the valence shell of nitrogen, in covalent compounds nitrogen contains a maximum of eight electrons and forms either sp, sp^2, or sp^3 hybrid orbitals. Nitrogen is one of the most electronegative elements; only fluorine and oxygen have higher electro-negativities.

Nitrogen fills its valence shell in one of several different ways. It gains three electrons to form the **nitride ion,** N^{3-}, with active metals such as lithium and magnesium. This ion is a very strong base. Many other metallic nitrides such as vanadium nitride, VN, are known, but they are not ionic. The octet can also be completed by the formation of three single bonds as in NH_3 and H_2NOH, or by multiple-bond formation, as in molecular nitrogen (N_2), in the cyanide ion (CN^-), and in nitrous acid (HONO).

Since nitrogen is a nonmetal, several of its oxides are acidic, and it forms oxyacids such as nitric acid (HNO_3) and nitrous acid (HNO_2). Nitrogen-hydrogen bonds are weakly acidic; for example, anhydrous ammonia (NH_3) reacts with strong bases like N^{3-} and H^- with loss of a proton and formation of the amide ion (NH_2^-). The lone pair of electrons on nitrogen in compounds with three single bonds confers Lewis base character to these molecules.

Compounds of nitrogen in all of the possible oxidation states from -3 to $+5$ are known (Table 23-1). Much of the chemistry of nitrogen involves oxidation-reduction reactions, which convert nitrogen from one oxidation state to another.

The various oxidation states of nitrogen emphasize the formal nature of the concept of oxidation state (Section 5.11). Although oxidation states are useful in balancing equations, predicting empirical formulas, and organizing types of chemical behavior, they must not be taken literally as indicating charge on an atom. For example, NF_3 and NCl_3 have identical Lewis structures and similar chemistry, yet nitrogen has an oxidation state of $+3$ in NF_3 and -3 in NCl_3. This difference results from the arbitrary conventions that fluorine can have only a negative oxidation state and that the more electronegative atom in a binary compound is assigned a negative oxidation state. The distribution of electrons in a molecule is a more fundamental property of a molecule than the oxidation states of the atoms involved. We should always think of oxidation states as a useful but quite arbitrary concept.

The remaining sections of this chapter will illustrate the chemical behavior of nitrogen by examination of elemental nitrogen and several nitrogen compounds.

TABLE 23-1	Oxidation States of Nitrogen in Some Compounds and Ions	
Oxidation State	**Example**
-3	N^{3-}, NCl_3, NH_4^+, NH_2^-, $(CH_3)_3N$
-2	N_2H_4 (H_2N—NH_2), $N_2H_5^+$ (H_2N—NH_3^+)
-1	H_2NOH, CH_3N=NCH_3
0	N_2
$+1$	N_2O
$+2$	NO
$+3$	HNO_2, ClNO, NF_3
$+4$	NO_2, N_2O_4
$+5$	N_2O_5, HNO_3, NF_4^+

Elemental Nitrogen

23.2 History and Occurrence of Elemental Nitrogen

Nitrogen was first recognized as an element by the Scotch botanist Daniel Ruth-erford in 1772. He demonstrated that this gas does not support either life or combustion. Nitrogen composes 78% of the atmosphere by volume and 75% by weight. It has been estimated that there are more than 20 million tons of nitrogen over every square mile of the earth's surface. Natural gas also contains some free nitrogen. The most important mineral sources of nitrogen in combined form are the **saltpeter** (KNO_3) deposits in India and other countries of the Far East and **Chile saltpeter** ($NaNO_3$) deposits in South America. Nitrogen is an essential element of the proteins of all plants and animals. An important source of nitrogen compounds is the coal used in the production of coke and illuminating gas.

23.3 Preparation of Elemental Nitrogen

On an industrial scale, nitrogen is obtained by the fractional distillation of liquid air (Section 22.2). Nitrogen prepared in this way usually contains small amounts of oxygen and the noble gases, particularly argon.

Pure nitrogen is prepared in the laboratory by heating a solution of ammonium nitrite.

$$NH_4NO_2(aq) \longrightarrow 2H_2O(l) + N_2(g) \qquad \Delta H° = -305.1 \text{ kJ}$$

In actual practice, a mixture of sodium nitrite and ammonium chloride is used to furnish the ions of ammonium nitrite because ammonium nitrite is so unstable that it cannot be stored. Nitrogen may also be prepared by the oxidation of ammonia. One such method involves passing ammonia gas over red-hot copper oxide.

$$2NH_3(g) + 3CuO(s) \longrightarrow 3H_2O(l) + N_2(g) + 3Cu(s) \qquad \Delta H° = -161.1 \text{ kJ}$$

A convenient laboratory method of obtaining atmospheric nitrogen (nitrogen plus the noble gases) is to burn phosphorus in air that is confined over water. The P_4O_{10} formed dissolves in the water, and the residual gas is mainly nitrogen.

23.4 Properties of Elemental Nitrogen

Under ordinary conditions nitrogen is a colorless, odorless, and tasteless gas. It boils at $-195.8°$ under 1 atm pressure and freezes at $-210.0°$. Its density under standard conditions is 1.2506 g/ℓ; it is slightly less dense than air, for air contains the heavier molecules of oxygen as well as those of nitrogen. Under standard conditions, 100 ml of water dissolves 2.4 ml of nitrogen.

Nitrogen molecules contain a triple bond, $:N\equiv N:$. The molecular orbital description of the bonding in N_2 was discussed in Section 6.9. The triple bond in a nitrogen molecule is very strong; the nitrogen molecule is the most stable diatomic molecule known. The bond energy of the nitrogen-nitrogen triple bond is 944.7 kJ per mole of triple bonds, which compares with 498 kJ per mole of double bonds in the oxygen molecule and 159 kJ per mole of single bonds in the fluorine molecule (Section 6.14).

Nitrogen molecules are very unreactive. Upon heating with active metals they form nitrides; heating with hydrogen gives low yields of ammonia, and heating with oxygen followed by rapid cooling (quenching) produces nitric oxide, NO. A great deal of effort has been expended in order to improve the yields and rates of conversion of molecular nitrogen into compounds with other elements, primarily into ammonia for use in fertilizers and into nitric acid for industrial use.

23.5   Uses of Elemental Nitrogen

Large quantities of elemental nitrogen are used in the various processes for "fixing" atmospheric nitrogen, i.e., bringing about the combination of atmospheric nitrogen with other elements. These processes are discussed later. Several uses of nitrogen are based on its inactivity. Frequently, nitrogen is used when a chemical process requires an inert atmosphere. Nitrogen was once widely used to fill electric light bulbs to inhibit vaporization of the filament, but now it has been displaced to a large extent by argon, which is even more inert. Canned foods such as coffee and hydrogenated vegetable oils (Crisco and Spry) retain their flavor and color much better if sealed with nitrogen instead of air. Sliced luncheon meats are often packed with nitrogen gas in sealed plastic containers to retard spoilage.

Ammonia

Ammonia, NH_3, is produced in nature when any nitrogen-containing organic material is decomposed in the absence of air. The decomposition may be brought about by heat or by the bacteria that produce decay. The odor of ammonia is prevalent in stables and is frequently detected in sewage where decay is taking place.

23.6   Preparation of Ammonia

▶ **1. LABORATORY METHODS.** Ammonia is usually prepared in the laboratory by heating an ammonium salt with sodium hydroxide or slaked lime. The acid-base reaction with the weakly acidic ammonium ion produces ammonia.

$$NH_4^+(aq) + OH^-(aq) \longrightarrow NH_3(g) + H_2O(l)$$

Ammonia is also formed by the hydrolysis of ionic nitrides. The nitride ion is a much stronger base than the hydroxide ion.

$$Mg_3N_2 + 6H_2O \longrightarrow 3Mg(OH)_2 + 2NH_3(g)$$
$$Li_3N + 3H_2O \longrightarrow 3Li^+ + 3OH^- + NH_3(g)$$

▶ **2. THE HABER PROCESS.** Ammonia is produced in large quantities by the direct union of its elements by the Haber process.

$$N_2(g) + 3H_2(g) \xrightarrow{\text{Catalyst}} 2NH_3(g) \qquad \Delta H^\circ = -92.2 \text{ kJ}$$

TABLE 23-2 The Equilibrium Concentrations of Ammonia (Per Cent by Volume) at Several Temperatures and Pressures for the Reaction $N_2 + 3H_2 \rightleftharpoons 2NH_3$

Pressure, atm	Temperature, °C					
	200°	300°	400°	500°	600°	800°
1	15.3	2.18	0.44	0.129	0.05	0.012
100	80.6	52.1	25.1	10.4	4.47	1.15
200	85.8	62.8	36.3	17.6	8.25	2.24
1000	98.3	92.5	80.0	57.5	31.5	—

The equilibrium concentrations of ammonia at several temperatures and pressures are shown in Table 23-2. Note that the reaction is exothermic and reversible, so that the yield of ammonia becomes less and less as the temperature of the system is raised. However, at low temperatures the reaction is too slow to be of practical value even when a catalyst is present. Because four volumes of the reactants ($1N_2$ and $3H_2$) give two volumes of the product ($2NH_3$), high pressure causes an increase in the yield of ammonia at any given temperature. It would seem, then, that the process should be carried out at the lowest temperature and highest pressure practicable. Of the many catalysts that have been tried, the most efficient is a mixture of iron oxide and potassium aluminate.

In actual practice a pressure of about 200 to 600 atm and a temperature of 450 to 600°C are ordinarily used. Pressures as high as 1000 atm are sometimes employed, but the added cost of equipment designed to withstand the higher pressures may more than offset the advantages of higher yield.

The hydrogen and nitrogen used in the Haber process must be very pure to avoid "poisoning" the catalyst. Nitrogen for the process is obtained from liquid air, and much of the hydrogen is obtained from water gas (Section 9.15). The mixture of hydrogen and nitrogen is compressed and then passed over the catalyst. The ammonia is removed by liquefaction and the residual hydrogen and nitrogen are recycled through the catalyst chamber.

Fritz Haber, a German chemist, received the 1918 Nobel award in chemistry for his success in developing the direct synthesis of ammonia on a commercial scale.

23.7 Physical Properties of Ammonia

Ammonia is a colorless gas with a characteristic, irritating odor. It is a powerful heart stimulant, and people have died from inhaling it. It is lighter than air, one liter of it weighing 0.7710 g under standard conditions. Ammonia gas is readily liquefied by cooling and compressing. The liquid boils at −33.43° and is colorless. The solid is white and crystalline; it melts at −77.76°. The heat of vaporization of liquid ammonia (1374 J/g) is higher than that of any other liquid except water, so ammonia is valuable as a refrigerant. Ammonia is quite soluble in water, alcohol, and ether. One liter of water at 0° dissolves 1185 liters of the gas. This fact makes possible the ammonia fountain (Fig. 23-1), triggered by squirting a few drops of water into the flask filled with ammonia gas. Ammonia can be completely expelled from an aqueous solution of ammonia by boiling.

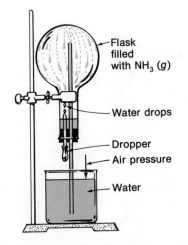

Flask filled with NH_3 (g)

Water drops

Dropper
Air pressure

Water

Figure 23-1 The "ammonia fountain." The fountain results when air pressure forces the water into the space left when the gaseous ammonia dissolves in water introduced from the dropper.

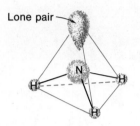

Lone pair

Figure 23-2 Geometric structure of the ammonia molecule.

In the ammonia molecule, the three hydrogen atoms are linked to the nitrogen atom by covalent bonds.

$$\text{H}-\text{N}: \quad (\text{with two additional H bonded above and below})$$

The ammonia molecule is tetrahedral in shape (Fig. 23-2), with the unshared pair of electrons occupying one of the sp^3 hybrid orbitals (Chapter 7). The ammonia molecule is highly polar, a property partly responsible for its physical and chemical behavior.

23.8   Chemical Properties of Ammonia

Of the hydrides of the members of the nitrogen family, ammonia is the most highly associated in the liquid state, the most stable toward decomposition by heat, and the most strongly basic.

The unshared electron pair of the ammonia molecule can readily be shared by an atom or ion that has an incomplete outer shell, with the formation of a coordinate covalent bond. The chemical properties of ammonia may be outlined as follows:

1. Ammonia is a Brönsted base since it readily accepts protons and a Lewis base in that it acts as an electron-pair donor (Chapter 14). When ammonia dissolves in water some of the molecules react to form ammonium and hydroxide ions.

$$\text{H}-\text{N}: + \text{H}-\ddot{\text{O}}: \rightleftharpoons \left[\text{H}-\text{N}-\text{H}\right]^{+} + [:\ddot{\text{O}}-\text{H}]^{-}$$

The reaction is reversible and the equilibrium lies far to the left so that ionization is only slight. **Aqueous ammonia** is therefore only weakly basic.

Ammonia accepts protons from acids and hydronium ions, as shown by the following equations:

$$\text{NH}_3 + \text{HCl} \rightleftharpoons \text{NH}_4^+ + \text{Cl}^-$$

$$\text{NH}_3 + \text{H}_3\text{O}^+ \rightleftharpoons \text{NH}_4^+ + \text{H}_2\text{O}$$

The ammonium ion (NH_4^+) is similar in size to the potassium ion, and ionic compounds of the two ions exhibit many similarities. Salts of the ammonium ion are called **ammonium salts.**

Gaseous ammonia and gaseous hydrogen chloride react and form a cloud of very small crystals of ammonium chloride. This reaction is used in forming smoke screens; it also accounts largely for the white deposit so apparent on windowpanes and glassware in chemical laboratories.

Ammonia forms **ammines** by sharing electrons with certain metallic ions, just as water forms hydrates (see Chapter 31).

$$\text{Cu}^{2+} + 4\text{NH}_3 \longrightarrow [\text{Cu}(\text{NH}_3)_4]^{2+} \quad \text{[tetraamminecopper(II) ion]}$$

$$\text{Ag}^+ + 2\text{NH}_3 \longrightarrow [\text{Ag}(\text{NH}_3)_2]^+ \quad \text{[diamminesilver ion]}$$

2. Ammonia also displays acidic behavior, although it is a weaker acid than water. It will react with very strong bases such as the CH_3^- ion.

$$LiCH_3 + NH_3 \longrightarrow LiNH_2 + CH_4$$

Like other acids ammonia reacts with metals, although it is so weak an acid that high temperatures are often required. Depending upon the stoichiometry, **amides** (salts of NH_2^-), **imides** (salts of NH^{2-}), or **nitrides** (salts of N^{3-}) are formed.

$$2Na + 2NH_3(l) \xrightarrow[\text{catalyst}]{Fe} 2NaNH_2 + H_2$$

$$2Li + NH_3(g) \xrightarrow{\Delta} Li_2NH + H_2$$

$$3Mg + 2NH_3(g) \xrightarrow{\Delta} Mg_3N_2 + 3H_2$$

Compounds that hydrolyze in water often will also undergo **ammonolysis,** the equivalent acid-base reaction with ammonia. The equation for the ammonolysis of mercury(II) chloride is

$$HgCl_2 + 2NH_3 \longrightarrow Hg(NH_2)Cl + NH_4Cl$$

Many ammonolysis reactions are complicated by subsequent reactions of the initial product.

3. The nitrogen atom in ammonia is in its lowest possible oxidation state and thus is not susceptible to reduction. However, the nitrogen atom can be oxidized; hot ammonia and the ammonium ion are active reducing agents.

$$2NH_3(g) + 3CuO(s) \xrightarrow{\Delta} N_2(g) + 3Cu(s) + 3H_2O(g) \qquad \Delta H° = -161.1 \text{ kJ}$$

$$(NH_4)_2Cr_2O_7(s) \xrightarrow{\Delta} N_2(g) + 4H_2O(g) + Cr_2O_3(s) \qquad \Delta H° = -300.4 \text{ kJ}$$

$$NH_4NO_3(s) \xrightarrow{\Delta} N_2O(g) + 2H_2O(g) \qquad \Delta H° = -36.03 \text{ kJ}$$

4. Ammonia is decomposed into hydrogen and nitrogen at red heat; the action of electric sparks has the same effect. The reaction is reversible.

23.9   Liquid Ammonia as a Solvent

Of the various solvents that have been studied, liquid ammonia most closely resembles water. It dissolves many electrolytes readily, and the solutions so formed are good conductors of electricity. The waterlike character of liquid ammonia is due to association of ammonia molecules through hydrogen bonding. However, because the NHN bond of associated ammonia molecules is weaker than the OHO bond of associated water, the properties that result from association are less pronounced with ammonia than with water. Ammonia is a poorer solvent for electrolytes than water because its dielectric constant is lower than that of water. This means that ammonia does not insulate oppositely charged ions in solution as well as does water (Sections 13.12 and 13.13).

The various types of reactions that occur in water also take place in liquid ammonia. Like water, pure ammonia is a very poor conductor of electricity, but it does ionize slightly according to the equation

$$2NH_3 \rightleftharpoons NH_4^+ + NH_2^-$$

Note the similarity of the ionization of ammonia to that of water.

$$2H_2O \rightleftharpoons H_3O^+ + OH^-$$

Reactions of ammonia and reactions in liquid ammonia as a solvent are usefully studied by means of the concept of solvent systems (see Section 14.18).

Liquid ammonia possesses the ability to dissolve the more electropositive metals such as Na, K, Ba, and Ca. The free metals can be recovered by evaporation of the solvent. If the solutions are concentrated they exhibit a bronze color and conduct electricity like a pure metal. When dilute they are bright blue and conduct electricity like aqueous solutions of salts. It appears that when a metal such as sodium dissolves in liquid ammonia it becomes a positive ion by losing its valence electron.

$$Na \rightleftharpoons Na^+ + e^-$$

The ammonia molecules then solvate the ion and electron reversibly.

$$Na^+ + xNH_3 \rightleftharpoons Na(NH_3)_x^+$$
$$e^- + yNH_3 \rightleftharpoons e^-(NH_3)_y$$

The "ammoniated electron" is responsible for the blue color of dilute solutions and for the strong reducing properties shown by solutions of metals in liquid ammonia. Such solutions are abundant sources of free electrons, which may be taken up readily by reducible ions or compounds. In fact, liquid ammonia solutions of sodium are used widely in both organic and inorganic reduction reactions.

If metal-liquid ammonia solutions are allowed to stand, hydrogen is slowly liberated according to the equation (in the case of sodium)

$$2Na + 2NH_3 \longrightarrow 2Na^+ + 2NH_2^- + H_2$$

Sodium amide

However, if metallic iron is added as a catalyst, the reaction takes place quite rapidly.

The amide ion, NH_2^-, in the liquid ammonia system is analogous to the OH^- ion in the water system. The amide ion is a base in liquid ammonia, just as the hydroxide ion is a base in the water system. The ammonium ion, NH_4^+, is an acid in the liquid ammonia system, analogous to the hydronium ion, H_3O^+, in the water system. In aqueous solutions, neutralization involves the formation of water (the solvent).

$$H_3O^+ + [Cl^-] + [Na^+] + OH^- \longrightarrow [Na^+] + [Cl^-] + 2H_2O$$

or
$$H_3O^+ + OH^- \longrightarrow 2H_2O$$

Neutralization reactions in liquid ammonia result in the formation of ammonia (the solvent).

$$NH_4^+ + [Cl^-] + [Na^+] + NH_2^- \longrightarrow [Na^+] + [Cl^-] + 2NH_3$$

or
$$NH_4^+ + NH_2^- \longrightarrow 2NH_3$$

Metathetical (double decomposition) reactions very often proceed in an entirely different direction in liquid ammonia than in water, because solubility relationships are not always the same in the two solvents. For example, silver chloride is soluble in ammonia but not in water, whereas barium chloride is insoluble in liquid ammonia but soluble in water. The reaction between silver nitrate and barium chloride in water is

$$2Ag^+ + [2NO_3^-] + [Ba^{2+}] + 2Cl^- \longrightarrow 2AgCl(s) + [Ba^{2+}] + [2NO_3^-]$$

The reaction in liquid ammonia is

$$2Ag^+ + [2NO_3^-] + [Ba^{2+}] + 2Cl^- \longrightarrow BaCl_2(s) + [2Ag^+] + [2NO_3^-]$$

Ammonia combines with the proton with the formation of a stronger bond than does water, and hence many ammonium salts are quite stable. On the other hand, most of the corresponding hydronium compounds are unstable and exist only in solution, or at low temperature. The fact that hydronium ions give up protons to ammonia molecules shows that the bond between the proton and water is weaker than the bond between the proton and ammonia. The equation for the reaction is

$$NH_3 + H_3O^+ \rightleftharpoons NH_4^+ + H_2O$$

23.10 Uses of Ammonia

About three-fourths of the 16,944,000 tons of ammonia produced in the United States during 1978 was used in fertilizers, either as the compound itself or as ammonium salts such as the sulfate and nitrate. The practice of the direct application of anhydrous ammonia and of solutions of ammonium compounds to the soil is rapidly increasing. Large quantities of ammonia are used in the production of nitric acid, urea, and other nitrogen compounds. Ammonia is the most common refrigerant used in the production of ice and for the maintenance of low temperatures in refrigerating plants. Household ammonia is an aqueous solution of ammonia. It is used to clean windows and to remove the carbonate hardness from hard water and thus save soap in cleansing and washing.

Other Compounds of Nitrogen and Hydrogen

23.11 Hydrazine

When ammonia is oxidized by sodium hypochlorite in the presence of sodium hydroxide and either gelatin or glue, **hydrazine** (N_2H_4) is produced. The reactions involved may be described by the equations

$$NH_3 + OCl^- \longrightarrow NH_2Cl + OH^-$$
$$NH_2Cl + NH_3 + OH^- \longrightarrow N_2H_4 + Cl^- + H_2O$$

A large excess of ammonia is used, and the reactants are thoroughly mixed at low temperatures. Chloramine, NH_2Cl, is an intermediate in the process.

Electronic formulas show that hydrazine may be regarded as the nitrogen analogue of hydrogen peroxide.

$$\begin{array}{ccc}
\overset{\displaystyle H\ \ H}{\underset{\displaystyle ..\ \ ..}{H-\overset{|}{\underset{|}{N}}-\overset{|}{\underset{|}{N}}-H}} & \overset{}{\underset{\displaystyle ..\ \ ..}{H-\overset{..}{O}-\overset{..}{O}-H}} & \overset{\displaystyle H}{H-\overset{|}{\underset{\displaystyle ..}{N}}-H} \\
\text{Hydrazine} & \text{Hydrogen peroxide} & \text{Ammonia}
\end{array}$$

It is also useful to recognize that hydrazine is an ammonia derivative in which one hydrogen of the ammonia molecule is replaced by an NH_2 group. Anhydrous hydrazine is thermally stable but very reactive toward many reagents. It burns in air and reacts vigorously with the halogens. Like ammonia, hydrazine is a base, although it is weaker than ammonia and forms two series of salts.

$$N_2H_4 + H_2O \rightleftharpoons N_2H_5^+ + OH^-$$

$$N_2H_4 + HCl \rightleftharpoons N_2H_5^+ + Cl^-$$

$$N_2H_4 + 2HCl \rightleftharpoons N_2H_6^{2-} + 2Cl^-$$

Anhydrous hydrazine is readily oxidized to free nitrogen and water by hydrogen peroxide.

$$N_2H_4(l) + 2H_2O_2(l) \longrightarrow N_2(g) + 4H_2O(g) \qquad \Delta H^\circ = -642.2 \text{ kJ}$$

Because the products of the reaction between these two liquids are both gaseous at the temperature of the reaction and the reaction is highly exothermic, there is a temendous increase in volume of the products. Hydrazine and hydrogen peroxide have therefore been used as the propellant in rockets.

23.12   Hydroxylamine

Hydroxylamine, NH_2OH, may also be considered to be an ammonia derivative in which an OH group has replaced one hydrogen atom of ammonia.

$$\begin{array}{cc}
\overset{}{\underset{\displaystyle |}{H-\overset{..}{N}-\overset{..}{\underset{\displaystyle ..}{O}}-H}} & \overset{}{\underset{\displaystyle |}{H-\overset{..}{N}-H}} \\
H & H \\
\text{Hydroxylamine} & \text{Ammonia}
\end{array}$$

As a solvent, hydroxylamine exhibits waterlike character. The pure substance is an unstable white solid that undergoes thermal decomposition at about 15°C to ammonia, water, and a mixture of nitrogen and nitrous oxide. The decomposition can be explosive at elevated temperatures. Aqueous solutions of hydroxylamine are more stable. Hydroxylamine is a base with a base strength somewhat less than that of ammonia. Hydroxylamine is an active reducing agent. It is usually prepared and handled as a **hydroxylammonium salt,** such as the chloride, $NH_3OH^+Cl^-$.

23.13   Hydrazoic Acid

When nitrous acid is reduced by hydrazine, **hydrazoic acid,** HN_3, a colorless liquid, is formed.

$$N_2H_4(aq) + HNO_2(aq) \longrightarrow HN_3(aq) + 2H_2O(l)$$

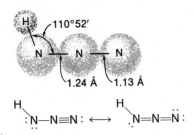

$$:\overset{H}{\underset{\cdot\cdot}{N}}-N\equiv N: \longleftrightarrow \overset{H}{\underset{\cdot\cdot}{N}}=N=\overset{\cdot\cdot}{N}:$$

Figure 23-3 Resonance formulas and the geometric structure of the hydrazoic acid molecule, HN_3.

Hydrazoic acid detonates violently when subjected to shock. It is a weak acid and reacts with both oxidizing and reducing agents. Its salts are called **azides;** like the acid, the azides are unstable. Azides of electropositive metals decompose smoothly to nitrogen and the metal when heated; other metal azides explode. Lead azide, $Pb(N_3)_2$, is used as a detonator for military explosives. Structures of hydrazoic acid are shown in Fig. 23-3. The three nitrogen atoms of the azide ion, N_3^-, lie in a straight line.

Oxyacids of Nitrogen

23.14   Nitric Acid

Nitric acid, HNO_3, was known to the alchemists of the eighth century as *aqua fortis* (meaning "strong water"). It was prepared from KNO_3 and was used in the separation of gold and silver; it will dissolve the silver from alloys rich in silver, leaving the gold. Traces of nitric acid occur in the atmosphere after thunderstorms, and its salts are widely distributed in nature. Chile saltpeter, $NaNO_3$, is found in tremendous deposits (2 by 200 miles, and to a thickness of 5 feet) in the desert region near the boundary of Chile and Peru. Bengal saltpeter (KNO_3) is found in India and other countries of the Far East.

23.15   Preparation of Nitric Acid

▶**1. FROM NITRATES.** Nitric acid can be prepared in the laboratory by heating a mixture of sodium nitrate and concentrated sulfuric acid.

$$NaNO_3 + H_2SO_4 \xrightarrow{\triangle} NaHSO_4 + HNO_3(g)$$

At ordinary temperatures the reaction is reversible, and equilibrium is soon established. Because nitric acid boils at 86°C while sulfuric acid boils at 338° ($NaNO_3$ and $NaHSO_4$ are nonvolatile), the nitric acid is readily removed from the reaction mixture by gentle heating. Consequently, the equilibrium is shifted to the right and the reaction goes to completion.

▶**2. FROM AMMONIA.** The **Ostwald process** for producing nitric acid consists of the oxidation of ammonia to nitric oxide, NO, the oxidation of nitric oxide to nitrogen dioxide, NO_2, and finally the conversion of nitrogen dioxide to nitric acid. Ammonia is mixed with 10 times its volume of air. The mixture is first heated to 600°–700° and then brought into contact with platinum-rhodium alloy gauze (90% Pt, 10% Rh), which acts as a catalyst. The reaction is highly exothermic; the ammonia literally burns with a flame, and the temperature of the system is raised to about 1000°C.

$$4NH_3(g) + 5O_2(g) \longrightarrow 4NO(g) + 6H_2O(g) \qquad \Delta H° = -905.5 \text{ kJ}$$

Additional air is admitted to the reaction chamber to cool the mixture and oxidize the nitric oxide to nitrogen dioxide

$$2NO(g) + O_2(g) \longrightarrow 2NO_2(g) \qquad \Delta H° = -114.1 \text{ kJ}$$

The nitrogen dioxide, excess oxygen, and the unreacted nitrogen from the air are passed through a spray of water in an absorption tower, where nitric acid and nitric oxide are formed as the nitrogen dioxide disproportionates.

$$3NO_2 + H_2O \longrightarrow 2H^+ + 2NO_3^- + NO$$

The nitric oxide is combined with more oxygen and returned to the absorption tower. The nitric acid is drawn off and concentrated. Most of the 8,051,000 tons of our nitric acid produced in 1978 was produced by the Ostwald process.

23.16   Physical Properties of Nitric Acid

Electronic and structural formulas for the nitric acid molecule are shown in Fig. 23-4.

Pure nitric acid is a colorless liquid that boils at 86°C, has a density of 1.50 g/ml, and freezes to a white solid at −42°. It fumes in moist air, forming a cloud of very small droplets of aqueous nitric acid. An aqueous solution containing 68% of the acid has a constant boiling point of 120.5° at 1 atm pressure. This is the "concentrated" acid of commerce; it has a specific gravity of 1.4048 at 20°.

23.17   Chemical Properties of Nitric Acid

Pure nitric acid is unstable; it decomposes in light or when heated to produce a mixture of nitrogen oxides, of which nitrogen dioxide is predominant.

$$4HNO_3 \xrightarrow{\triangle} 4NO_2 + 2H_2O + O_2$$

It is often yellow or brown in color due to the NO_2 and N_2O_4 formed as it decomposes. In aqueous solution nitric acid is stable and exhibits both the properties of a strong acid and the properties of a strong oxiding agent. The following types of reactions illustrate the ability of nitric acid to act as an oxidizing agent.

▶1. **ACTION WITH METALS.** The products formed when nitric acid reacts with a metal depend upon the concentration of the acid, the activity of the metal, and the temperature. A mixture of oxides and other reduction products is usually produced, but the less active metals, such as copper, silver, and lead, reduce dilute nitric acid primarily to nitric oxide. The following three equations show the oxide that is produced in largest quantity in the particular reactions.

The reaction of dilute nitric acid with copper is

$$3Cu + 8H^+ + 2NO_3^- \longrightarrow 3Cu^{2+} + 2NO(g) + 4H_2O$$

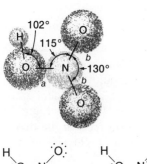

Figure 23-4 Resonance formulas and the geometric structure showing the arrangement of atoms in the HNO_3 molecule, which in the vapor state is planar. N—O distances are (a) 1.41 Å and (b) 1.22 Å.

With concentrated nitric acid, nitrogen dioxide is formed predominantly.

$$Cu + 4H^+ + 2NO_3^- \longrightarrow Cu^{2+} + 2NO_2(g) + 2H_2O$$

The more active metals, such as zinc and iron, reduce dilute nitric acid to nitrous oxide, N_2O. With zinc, we have

$$4Zn + 10H^+ + 2NO_3^- \longrightarrow 4Zn^{2+} + N_2O(g) + 5H_2O$$

When the acid is very dilute, either nitrogen or ammonium ions may be formed, depending on the conditions.

$$5Zn + 12H^+ + 2NO_3^- \longrightarrow 5Zn^{2+} + N_2(g) + 6H_2O$$
$$4Zn + 10H^+ + NO_3^- \longrightarrow 4Zn^{2+} + NH_4^+ + 3H_2O$$

In concentrated nitric acid, zinc, as well as copper, produces nitric oxide and nitrogen dioxide. The nitrate salts of these metals separate as crystals when the resulting solutions are evaporated. The action of nitric acid on a metal rarely produces hydrogen in more than small amounts.

▶ **2. ACTION WITH NONMETALS.** Nonmetallic elements, such as sulfur, carbon, iodine, and phosphorus, are oxidized by concentrated nitric acid to their oxygen acids (or acid anhydrides), with the formation of NO_2.

$$S + 6HNO_3 \longrightarrow H_2SO_4 + 6NO_2(g) + 2H_2O$$
$$C + 4HNO_3 \longrightarrow CO_2 + 4NO_2(g) + 2H_2O$$

▶ **3. ACTION WITH INORGANIC COMPOUNDS.** Many compounds are oxidized by nitric acid. Hydrochloric acid is readily oxidized by concentrated nitric acid to chlorine and chlorine dioxide. A mixture of nitric and hydrochloric acids reacts vigorously with metals. This mixture is particularly useful in dissolving gold and platinum and other metals that lie below hydrogen in the Activity Series. A mixture of three parts of concentrated hydrochloric acid and one part concentrated nitric acid by volume is called **aqua regia** (meaning "royal water"). The action of aqua regia on gold may be represented, in a somewhat simplified form, by the equation

$$Au + 4HCl + 3HNO_3 \longrightarrow HAuCl_4 + 3NO_2(g) + 3H_2O$$

Dilute nitric acid oxidizes hydrogen sulfide to elemental sulfur and water.

$$3H_2S + 2H^+ + 2NO_3^- \longrightarrow 3S + 2NO(g) + 4H_2O$$

Sulfur dioxide is oxidized to sulfuric acid by nitric acid.

$$3SO_2 + 2NO_3^- + 2H_2O \longrightarrow 4H^+ + 3SO_4^{2-} + 2NO(g)$$

It is interesting to note that nitric acid reacts with proteins, such as those in the skin, to give a yellow material by what is called the **xanthoprotein reaction.** You may have noticed in the laboratory that if you inadvertently get nitric acid on your fingers they turn a yellow color. During World War I, the explosive plants in the United States were not well ventilated, and the people who worked in them were exposed to fumes of nitric acid all day. Those persons turned a bright yellow from head to foot and consequently were referred to as "canaries."

23.18   Uses of Nitric Acid

Nitric acid is used extensively in the laboratory and in the chemical industries as a strong acid and as an active oxidizing agent. It is used in the manufacture of explosives, dyes, plastics, and drugs. Salts of nitric acid (nitrates) are valuable as fertilizers. **Black gunpowder** is a mixture of potassium nitrate, sulfur, and charcoal. **Ammonal,** an explosive, is a mixture of ammonium nitrate and aluminum powder.

23.19   Nitrates

The nitrate ion has a trigonal planar structure (see Section 7.2), with each bond between nitrogen and oxygen being a single-bond, double-bond hybrid. The three resonance structures are shown.

The salts of nitric acid may be prepared by treating either metals or their oxides, hydroxides, or carbonates with nitric acid. All normal nitrates are soluble in water, and the oxynitrates, such as those of bismuth and mercury, are soluble in nitric acid. All nitrates decompose when heated. When sodium or potassium nitrate is heated, a nitrite is formed and oxygen is evolved.

$$2NaNO_3 \longrightarrow 2NaNO_2 + O_2$$

A nitrate of a heavy metal produces the oxide of the metal when heated. With copper(II) nitrate, the equation is

$$2Cu(NO_3)_2 \longrightarrow 2CuO + 4NO_2 + O_2$$

Ammonium nitrate produces nitrous oxide, N_2O, when heated (see Section 23.21).

The nitrates, including especially ammonium nitrate, have a high explosive hazard.

23.20   Nitrous Acid and Nitrites

Nitrous acid, HNO_2, is obtained in solution (pale blue in color) when an acid, such as sulfuric, is added to a cold solution of a nitrite.

$$2Na^+ + 2NO_2^- + 2H^+ + SO_4^{2-} \longrightarrow 2HNO_2 + 2Na^+ + SO_4^{2-}$$

Nitrous acid is very unstable and is known only in solution. It changes slowly at room temperature (rapidly when heated) into nitric acid and nitric oxide.

$$3HNO_2 \longrightarrow H^+ + NO_3^- + 2NO(g) + H_2O$$

Nitrous acid is a weak acid. It acts as an active oxidizing agent toward strong reducing agents, but it is oxidized to nitric acid by active oxidizing agents. Figure 23-5 shows the structure of nitrous acid.

Sodium nitrite is the most important salt of nitrous acid. It is usually made by melting the nitrate with lead.

$$NaNO_3 + Pb \longrightarrow NaNO_2 + PbO$$

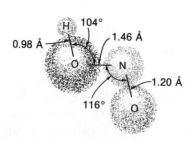

Figure 23-5 Structure of the nitrous acid molecule, HNO_2.

The nitrites are much more stable than the acid (the salts of all oxyacids are more stable than the acids themselves). Like the nitrates, they are soluble in water ($AgNO_2$ is sparingly soluble). Two resonance structures of the nitrite ion are known.

$$:\ddot{O}\diagdown\overset{..}{N}\diagup\overset{-}{\ddot{O}:} \longleftrightarrow :\ddot{O}\diagup\overset{..}{N}\diagdown\overset{-}{\ddot{O}:}$$

The nitrites, like the nitrates, have a high explosive hazard.

Oxides of Nitrogen

23.21   Nitrous Oxide

When ammonium nitrate is heated, **nitrous oxide,** N_2O, is formed.

$$NH_4NO_3(s) \xrightarrow{\triangle} N_2O(g) + 2H_2O(g) \qquad \Delta H° = -36.03 \text{ kJ}$$

This is an oxidation-reduction reaction in which the nitrogen in the ammonium ion is oxidized by the nitrogen in the nitrate ion. **Caution!** If ammonium nitrate is heated too strongly an explosion will occur.

Nitrous oxide is a colorless gas possessing a mild, pleasing odor and a sweet taste. It is used as an anesthetic for minor operations, especially in dentistry, under the name "laughing gas." The structure and resonance forms of nitrous oxide are shown in Fig. 23-6.

Nitrous oxide resembles oxygen in its behavior as an oxidizing agent with combustible substances. It decomposes when heated to form nitrogen and oxygen.

$$2N_2O(g) \xrightarrow{\triangle} 2N_2(g) + O_2(g) \qquad \Delta H° = -164.1 \text{ kJ}$$

Because one-third of the gas liberated is oxygen, nitrous oxide supports combustion better than air. A glowing splint will burst into flame when thrust into a bottle of this gas.

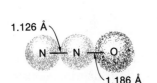

1.126 Å

1.186 Å

$:\ddot{N}{=}N{=}\ddot{O}: \longleftrightarrow :N{\equiv}N{-}\ddot{O}:$

Figure 23-6 Resonance formulas and the geometric structure of the nitrous oxide molecule, N_2O.

23.22   Nitric Oxide and Nitric Oxide Dimer

When copper is treated with dilute nitric acid, **nitric oxide,** NO, is formed as the chief reduction product of the nitric acid.

$$3Cu + 8H^+ + 2NO_3^- \longrightarrow 3Cu^{2+} + 2NO(g) + 4H_2O$$

Prepared in this way the nitric oxide is usually mixed with other oxides of nitrogen. Pure nitric oxide may be obtained by reducing nitric acid with iron(II) sulfate in dilute sulfuric acid solution.

$$3Fe^{2+} + NO_3^- + 4H^+ \longrightarrow 3Fe^{3+} + NO(g) + 2H_2O$$

The method of producing nitric oxide on a commercial scale by the Ostwald process, in which ammonia is oxidized in the presence of platinum at high temperatures, was discussed in Section 23.15. Nitric oxide is formed by direct

1.15 Å

Figure 23-7 Structure of the nitric oxide molecule, NO.

union of the elements nitrogen and oxygen in the air by lightning during thunderstorms.

$$N_2(g) + O_2(g) \rightleftharpoons 2NO(g) \qquad \Delta H° = +180.5 \text{ kJ}$$

It has been estimated that as a result of this electrical phenomenon 40,000,000 tons of nitrogen are fixed annually by the series of reactions described in Section 23.26.

Nitric oxide is a colorless gas that is only slightly soluble in water. It is the most stable thermally of the oxides of nitrogen. It is one of the air pollutants from the exhaust of the internal combustion engine, due to the reaction of nitrogen and oxygen from the air at the region of the spark in the combustion process (see Sections 22.3–22.5). Its structure is illustrated in Fig. 23-7.

Nitric oxide will act both as an oxidizing agent or as a reducing agent.

$$\textit{Oxidizing agent:} \quad P_4 + 6NO \longrightarrow P_4O_6 + 3N_2$$
$$\textit{Reducing agent:} \quad Cl_2 + 2NO \longrightarrow 2ClNO$$

Two resonance forms can be written for nitric oxide although the electronic structure of this molecule also can be described by the molecular orbital model (Section 6.13).

$$:\ddot{N}=\ddot{O}: \longleftrightarrow :\dot{N}=\ddot{O}:$$

Note that the molecule contains an unpaired electron that makes the substance paramagnetic. As is often, though not always, the case with molecules containing an unpaired electron, a combination of one molecule of nitric oxide with another to form a dimer can take place with some of the molecules, thereby pairing the unpaired electrons of each molecule that takes part in the dimerization. The dimer, N_2O_2, hence, is diamagnetic.

23.23   Dinitrogen Trioxide

When the temperature of a mixture of one part of nitric oxide and one part of nitrogen dioxide is lowered to $-21°C$, the gases become a blue liquid consisting of N_2O_3 molecules. Dinitrogen trioxide exists only in liquid and solid forms. When vaporized by heating, it forms the brown equilibrium mixture of NO and NO_2 ($N_2O_3 \rightleftharpoons NO + NO_2$).

Dinitrogen trioxide is the anhydride of nitrous acid. The resonance and geometric structures of N_2O_3 are shown in Fig. 23-8.

23.24   Nitrogen Dioxide and Dinitrogen Tetraoxide

Nitrogen dioxide is prepared in the laboratory by heating the nitrate of a heavy metal:

$$2Pb(NO_3)_2 \xrightarrow{\triangle} 2PbO + 4NO_2 + O_2$$

or by the action of concentrated nitric acid upon copper metal:

$$Cu + 4H^+ + 2NO_3^- \longrightarrow Cu^{2+} + 2NO_2 + 2H_2O$$

Nitrogen dioxide is prepared commercially by exposing nitric oxide to air.

$$2NO + O_2 \longrightarrow 2NO_2$$

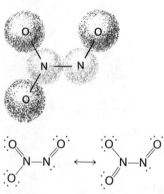

Figure 23-8 Resonance formulas and the geometric structure of the dinitrogen trioxide molecule, N_2O_3.

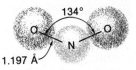

134°

O — N — O

1.197 Å

Nitrogen dioxide, NO₂

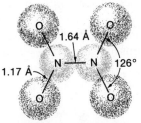

Dinitrogen tetraoxide
(nitrogen dioxide dimer),
N_2O_4

$O_2N—NO_2$

Figure 23-9 Formulas and geometric structures of nitrogen dioxide and dinitrogen tetraoxide molecules.

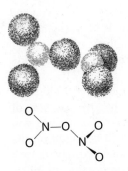

Figure 23-10 Formula and geometric structure of the dinitrogen pentaoxide molecule, N_2O_5.

Depending upon the conditions, nitrogen dioxide disproportionates in one of two ways when it reacts with water. In cold water, a mixture of nitric and nitrous acids is formed.

$$2NO_2 + H_2O \longrightarrow H^+ + NO_3^- + HNO_2$$

At higher temperatures, nitric acid and nitric oxide form (Section 23.15, Part 2).

At 140° nitrogen dioxide has a deep brown color and a density corresponding to the formula NO_2. At low temperatures the color almost entirely disappears, and the density corresponds to the formula N_2O_4. At ordinary temperatures an equilibrium exists between the colored and the colorless compounds.

$$2NO_2(g) \rightleftharpoons N_2O_4(g) \qquad \Delta H° = -57.20 \text{ kJ}; \Delta G° = -4.77 \text{ kJ}$$

Nitrogen dioxide contains an odd number of electrons and therefore an unpaired electron. The unpaired electron accounts for the fact that NO_2 is paramagnetic and highly colored. Two molecules of NO_2 share their unpaired electrons with each other when they combine to form N_2O_4; the tetraoxide is not paramagnetic. The resonance and geometric structures of NO_2 are shown in Fig. 23-9, along with the atomic arrangement for N_2O_4.

23.25   Dinitrogen Pentaoxide

This compound is a white solid formed by the dehydrating action of phosphorus(V) oxide upon pure nitric acid.

$$P_4O_{10} + 4HNO_3 \longrightarrow 4HPO_3 + 2N_2O_5$$

Dinitrogen pentaoxide decomposes above 30°.

$$2N_2O_5 \longrightarrow 4NO_2 + O_2$$

It is the anhydride of nitric acid.

$$N_2O_5 + H_2O \longrightarrow 2HNO_3$$

Figure 23-10 shows the electronic structure and the arrangement of atoms in N_2O_5 in the gaseous state. Crystals are made up of equal numbers of planar NO_2^+ ions (N—O bond length 1.15 Å) and NO_3^- ions (N—O bond length 1.24 Å).

23.26   The Cycle of Nitrogen in Nature

Nitrogen is an essential constituent of all plants and animals. It is present principally in the proteins, which are complex organic materials containing carbon, hydrogen, and oxygen as well as nitrogen (Chapter 26, Biochemistry). Most plants obtain the nitrogen necessary for their growth through their roots as nitrogen compounds, mainly in the form of ammonium and nitrate salts. However, certain legumes, such as clover, alfalfa, peas, and beans, are able to obtain nitrogen indirectly from the air by means of nitrogen-fixing bacteria, which are found in the nodules on their roots. These bacteria convert atmospheric nitrogen into nitrites and nitrates, which are assimilated by the "host" plant in the form of proteins. Generally, the bacteria fix more nitrogen in the form of soluble compounds than is used by the plant itself. When legumes die, much of the combined nitrogen remains in the soil and is available for use by nonleguminous plants.

As mentioned in Section 23.22, tremendous quantities of nitrogen are fixed as nitric oxide by lightning. The nitric oxide is then oxidized to nitrogen dioxide by oxygen of the air, and the reaction of the dioxide with water forms nitrous and nitric acids by the same reactions as in the Ostwald process (Section 23.15). These acids are carried by rain to the soil, where they react with oxides and carbonates of metals to form nitrites and nitrates, respectively. Certain soil bacteria oxidize the nitrites to nitrates, a process called **nitrification.** Certain other kinds of bacteria change ammonia into nitrites. There are also denitrifying bacteria, which decompose nitrates and other nitrogen compounds, thereby returning free nitrogen to the air.

Animals obtain their nitrogenous compounds by feeding on plants and other animals. The decay of both plant and animal matter returns nitrogen to the soil in the form of nitrates, and either ammonia or free nitrogen is produced. Thus, we see that nitrogen passes through a cycle of fundamental importance to all plants and animals.

Questions

1. Why does nitrogen form a maximum of four single covalent bonds?
2. Write balanced equations illustrating the acidic and basic character of the two binary compounds of hydrogen and nitrogen.
3. Write balanced equations illustrating three ways in which elemental nitrogen can be converted into ammonia. (Two of these paths involve more than one reaction.)
4. What is the oxidation state of nitrogen in each of the following: N_2, NH_4^+, $NaNO_2$, N_2H_4, NH_2OH, NO_2, N_2O_4, NH_4NO_3, N_2O, NCl_3, NF_3, Li_3N, HN_3?
5. Write electronic structures for N_2, NH_3, NH_4^+, N_2H_4, HN_3, and NH_2OH.
6. What are the products of the chemical reaction occurring between an ionic nitride and water?
7. Explain the effects of temperature, pressure, and a catalyst upon the direct synthesis of ammonia.
8. Account for the solubility of electrolytes in liquid ammonia.
9. What compounds act as acids in liquid ammonia?

as bases? What is the essential reaction involved in neutralizations in liquid ammonia?
10. What evidence is there that the bond between the proton and an ammonia molecule is stronger than that between the proton and water?
11. Draw a valence electron formula for the azide ion, N_3^-.
12. The salt hydroxylammonium chloride, $NH_2OH \cdot HCl$, can be formed by the reaction of hydrogen chloride and hydroxylamine. Would you expect the proton from the HCl to be found on the oxygen or on the nitrogen of the hydroxylamine?
13. Outline the chemistry of the production of nitric acid from ammonia.
14. List the possible reduction products of nitric acid. What factors determine which product will be formed when nitric acid is reduced?
15. Write equations for the reaction of concentrated HNO_3 with each of the following: Cu, S, C, and SO_2.
16. Write electronic formulas for HNO_3, NO_3^-, HNO_2,

N_2O, NO, NO_2, N_2O_3, N_2O_5, NCl_3, and $NOCl$.

17. Write equations for the preparation of nitrous acid starting with sodium nitrate.

18. What is the anhydride of nitrous acid? of nitric acid?

19. What would be the result of bubbling SO_2 through a strongly acid barium nitrate solution? Write equations for the reactions which occur.

20. Write equations for the preparation of each of the oxides of nitrogen.

21. Using the data in Appendix J, determine the $\Delta H°$ and $\Delta G°$ values for the conversion, $2NO_2(g) \rightleftharpoons N_2O_4(g)$. The values are widely different. Discuss possible reasons for this difference, and discuss the significance of each of the two values as completely as you can. (You may wish to refer to Chapter 18 in answering this question.) Check your answers against the values for $\Delta H°$ and $\Delta G°$ given in Section 23.24.

22. Describe the nitrogen cycle in nature.

23. Which is the stronger oxidizing agent, hydrazine or hydrogen peroxide? Give evidence for your answer.

Problems

1. What volume of nitrogen measured at 75°C and 740 torr is formed by the decomposition of 20.0 g of ammonium nitrite? *Ans. 9.16 l*

2. What mass of ammonia is produced by adding water to the product of the reaction of 93.0 g of lithium with elemental nitrogen? *Ans. 76.1 g*

3. What volume of ammonia measured at 1.20 atm and 25.0°C is required to react with 0.800 l of 0.510 M HCl solution? *Ans. 8.31 l*

4. What is the maximum volume of ammonia that could be collected at 27°C and 750 torr by the treatment of 2.00 g of ammonium chloride with 0.500 l of 0.500 M sodium hydroxide? *Ans. 0.932 l*

5. What mass of ammonia would be required to neutralize 0.350 l of 0.750 M HNO_3 solution?
   Ans. 4.47 g

6. What mass of hydrazine can be obtained from 0.0500 l of ammonia, measured at 740 mm and 245 K, if a 65.0% yield is obtained? *Ans. 0.252 g*

7. What quantity of nitric acid could be prepared by the Ostwald process from 300 cubic feet of ammonia, measured at 4.00 atm and 250°C, if a 93.0% yield is obtained based upon the original quantity of ammonia? *Ans. 46.4 kg*

8. The oxidation of ammonia and of nitric oxide are exothermic processes. Using the data in Appendix J, calculate $\Delta H°$ for these reactions.
   Ans. −905.50 kJ; −114.1 kJ

9. At 25°C and 1.0 atm, a mixture of $N_2O_4(g)$ and $NO_2(g)$ contains 30.0% $NO_2(g)$ by volume. Calculate the partial pressures of the two gases when they are at equilibrium at 25°C with a total pressure of 9.0 atm. *Ans. $P_{N_2O_4} = 8.0$ atm; $P_{NO_2} = 1.0$ atm*

10. An equilibrium mixture at 400°C contained 0.63 mole of hydrogen, 0.45 mole of nitrogen, and 0.24 mole of ammonia per liter. Calculate the equilibrium constant for the system. *Ans. 0.51*

References

"Nitrogen" (Staff), *Chem. and Eng. News,* June 26, 1978, p. 12.

"Ammonia" (Staff), *Chem. and Eng. News,* January 22, 1979, p. 11.

"Urea" (Staff), *Chem. and Eng. News,* January 9, 1978, p. 12.

"The Nitrogen Cycle," C. C. Delwiche, *Sci. American,* September, 1970, p. 136.

"The Physical Chemistry of Ammonia Manufacture and Nitrogenous Fertilizer Production," J. T. Gallagher and F. M. Tayler, *Educ. in Chemistry,* **4,** 30 (1967).

"Ammonia Synthesis," P. L. Bayless, *J. Chem. Educ.,* **53,** 318 (1976).

"Synthesis of Ammonium Cyanate and Urea on Pt, Rh, and Ru Catalysts," R. J. H. Voorhoeve *et al., Science,* **200,** 759 (1978).

"Wohler's Synthesis of Artificial Urea," G. B. Kauffman and S. H. Chooljian, *J. Chem. Educ.,* **56,** 197 (1979).

"Hydrazine," L. P. Lessing, *Sci. American,* July 1953, p. 30.

"High Temperatures: Chemistry," F. Daniels, *Sci. American,* September, 1954, p. 109 (describes production of nitric oxide directly from air).

"Is Ammonia Like Water?" J. B. Gill, *J. Chem. Educ.,* **47,** 619 (1970).

"Chemistry in the Technology of Explosives and Propellants," I. Dunstan, *Chem. in Britain,* **7,** 62 (1971).

"Nitrogen Fixation: Research Efforts Intensify," J. L. Marx, *Science,* **185,** 132 (1974).

"Nitrogen Fixation," D. R. Safrany, *Sci. American,* October, 1974, p. 64.

"Fixation of Nitrogen in the Prebiotic Atmosphere," Y. L. Yung and M. B. McElroy, *Science,* **203,** 1002 (1979).

"The Fixation of Nitrogen," S. P. S. Andrew, *Educ. in Chem.,* July 1978, p. 114.

"Biological Nitrogen Fixation," W. J. Brill, *Sci. American,* March 1977, p. 68.

"Nitrogen Fixation: Prospects for Genetic Manipulation," J. L. Marx, *Science,* **196,** 638 (1977).

"Marine Phosphorite Deposits and the Nitrogen Cycle," D. Z. Piper and L. A. Codispoti, *Science,* **188,** 15 (1975).

24

Phosphorus and Its Compounds

Phosphorus—The Element

Phosphorus and nitrogen are both members of Group VA and have similar valence shell electronic configurations, ns^2np^3. However, aside from the analogous stoichiometries of some of their simpler compounds (NH_3, PH_3 and NF_3, PF_3, for example), marked physical and chemical differences exist between the two elements and their compounds. These differences result from the larger size of the phosphorus atom, its lower electronegativity (2.1) relative to nitrogen (3.1), and the presence of empty d orbitals in the valence shell ($n = 3$) of phosphorus. Like nitrogen, phosphorus is a nonmetal, and its chemistry is essentially that of covalent compounds. However, nitrogen is quite inert except at elevated temperatures, whereas phosphorus (at least, the white modification) is very active even at low temperatures. The union of nitrogen with oxygen is an endothermic reaction, whereas phosphorus burns readily in air with the evolution of considerable heat. Nitric acid and nitrous acid are relatively strong oxidizing agents, but phosphoric acid and phosphorous acid are very weak oxidizing agents. Ammonia, NH_3, is a much weaker reducing agent than phosphine, PH_3. Furthermore, nitrogen forms bonds to only two or three oxygen atoms in its oxyacids and their salts, and thus exhibits a coordination number (Section 11.15) of only 2 or 3. Phosphorus exhibits a coordination number of 4 in its oxyacids and their salts. Nitrogen is limited to a maximum coordination number of 4 (in NH_4^+, for example), while phosphorus exhibits higher coordination numbers (in PCl_5 and PF_6^-, for example).

Phosphorus is essential to both plants and animals. Bones, teeth, and nerve and muscle tissue contain combined phosphorus. The nucleic acids, molecules that provide genetic information to an organism, contain phosphorus. Phosphorus compounds are also important in the metabolism of sugar. Such foods as eggs, beans, peas, and milk furnish phosphorus for our body requirements. Plants obtain phosphorus from soluble phosphates in the soil. Soils that have become deficient in phosphorus may be enriched by the addition of soluble phosphates.

24.1   Preparation of Phosphorus

The principal commercial source of phosphorus is the so-called **phosphate rock.** It is mostly calcium phosphate, $Ca_3(PO_4)_2$, but always contains several other phosphates as well. Phosphorus is produced commercially by heating calcium phosphate with sand and coke in an electric furnace.

$$2Ca_3(PO_4)_2 + 6SiO_2 + 10C \longrightarrow 6CaSiO_3 + 10CO(g) + P_4(g)$$

The phosphorus distills off at the quite high furnace temperature and is condensed to the solid state, or burned to P_4O_{10}, from which other compounds of phosphorus may be manufactured. The calcium silicate, along with any calcium fluoride or ferrophosphorus, melts and collects at the bottom of the furnace, where it is drawn off as a slag. The elemental phosphorus is shipped to industrial plants, where it is to be made into phosphoric acid and phosphates. In this way, the cost of the freight on the oxygen and water used in the manufacture of phosphorus compounds is avoided. The profits of chemical industries often depend on such considerations.

24.2   Physical Properties of Phosphorus

Phosphorus exists in several allotropic modifications, of which only two are of general interest and importance; these are **white phosphorus** and **red phosphorus.**

As prepared by the method described in the previous section, phosphorus is a white, translucent, soft, waxlike solid. When exposed to light, white phosphorus slowly turns yellow, due to the formation of a superficial coating of red phosphorus. For this reason, white phosphorus is sometimes called **yellow phosphorus.** White phosphorus is very soluble in carbon disulfide and less so in ether, chloroform, and other organic solvents. It is very nearly insoluble in water, 0.0033 g dissolving in one liter at 0°. White phosphorus melts at 44.2° and boils at 280°. Either as a solid, in solution, or as the vapor at just above the boiling point, white phosphorus exists as P_4 molecules. The four phosphorus atoms are arranged at the corners of a regular tetrahedron, and each atom is covalently bonded to the three other atoms of the molecule (Fig. 24-1). At very high temperatures phosphorus gas consists of P_2 molecules; these are assumed to have the electronic structure $:P\equiv P:$, which is similar to that of the N_2 molecule.

White phosphorus is very poisonous; a dose of about 0.15 g produces acute pains and convulsions, and may result in death. Necrosis, or death, of the bones of the jaw and nose results from continued breathing of small amounts of the fumes of phosphorus. In very small doses phosphorus stimulates the nervous system.

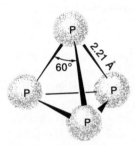

Figure 24-1 The white phosphorus molecule, P_4.

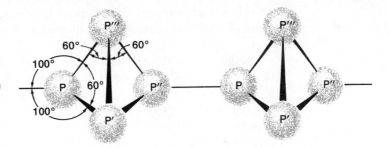

Figure 24-2 Probable structure of red phosphorus. Note that no bond exists between P and P″ in the first tetrahedron or between the two corresponding phosphorus atoms in the other tetrahedron; these bonds have been ruptured in the formation of red phosphorus from white phosphorus. The tetrahedra with ruptured bonds then join to form a chain structure.

Red phosphorus is made by heating the white modification between 230° and 300°, with air excluded.

$$P \text{ (white)} \rightleftharpoons P \text{ (red)} \qquad \Delta H° = -18 \text{ kJ}$$

This change is speeded up when catalyzed by a trace of iodine. Because the change from white to red phosphorus is exothermic, the heat of combustion of red phosphorus is less than that of the white form. Red phosphorus sublimes when heated, forming a vapor identical in molecular structure and properties to that from the white variety. The red modification is insoluble in the solvents that dissolve white phosphorus.

In red phosphorus the atoms are bonded together in an infinite polymer, the length of the polymer chain being limited only by the size of the individual piece under consideration. The exact structure of red phosphorus is not definitely known, but it is thought to be an infinite chain formed from the tetrahedral white phosphorus by the rupture of one P—P bond of each tetrahedron and a subsequent combination of the ruptured tetrahedra (Fig. 24-2). Half the phosphorus atoms (P and P″ in the figure, for example) have two bond angles of 100° and one of 60°. The other half (P′ and P‴, for example) have two bond angles of 60° and one of 100°.

24.3   Chemical Properties of Phosphorus

The chemistry of phosphorus is that of an active nonmetal. The oxidation states most commonly observed with phosphorus are −3 with metals and less electronegative nonmetals and +3 and +5 with more electronegative nonmetals. Phosphorus exhibits oxidation states that are unusual for a Group VA element in compounds such as diphosphorus tetrahydride, P_2H_4, and tetraphosphorus trisulfide, P_4S_3 (Fig. 24-3). These compounds contain phosphorus-phosphorus bonds.

The most important chemical property of phosphorus is its chemical activity toward oxygen. Slow oxidation of white phosphorus at room temperature causes a rise in its temperature, and spontaneous combustion results when the ignition temperature (35–45°) is reached. Because of the readiness with which white phosphorus ignites, it must be stored under water. Burns caused by phosphorus are very painful and slow to heal. *It should be handled only with forceps and put in water immediately after use.* When moist phosphorus is allowed to oxidize slowly in air, a faint light is emitted (phosphorescence) and ozone, phosphoric acid, and phosphorous acid are formed. When phosphorus is burned in either an excess of

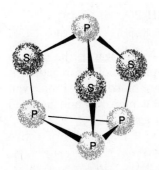

Figure 24-3 The molecular structure of P_4S_3. Note the P—P bonds.

air or in oxygen, covalent phosphorus(V) oxide, P_4O_{10}, is formed. In moist air, the cloud of solid phosphorus(V) oxide forms a fog of minute droplets of phosphoric acid.

Phosphorus combines with the halogens exothermically. When heated, it combines with sulfur and many of the metals. Concentrated nitric acid oxidizes phosphorus to orthophosphoric acid.

Red phosphorus is less active than the white form; it does not ignite in air until it is heated to about 250°. However, the products of the reactions of red phosphorus are the same as those of the white form.

24.4   Uses of Phosphorus

Formerly, white phosphorus was used in the manufacture of matches. Because it is poisonous and the workers who were exposed to the fumes suffered from necrosis of the bones, it has been replaced, in the heads of "strike anywhere" matches, by tetraphosphorus trisulfide, P_4S_3 (Fig. 24-3).

Large quantities of phosphorus are converted into acids and salts to be used in fertilizers, in baking powder, and in the chemical industries. Other uses are in the manufacture of special alloys such as ferrophosphorus and phosphorbronze. Considerable quantities of phosphorus are used in making fireworks, bombs, and rat poisons. Burning phosphorus has been used in the production of smoke screens during warfare.

Compounds of Phosphorus

24.5   Phosphine

Phosphorus forms a series of hydrides, the most important of which is **phosphine,** PH_3, a gaseous compound analogous to ammonia in formula and structure. Unlike ammonia, phosphine cannot be made by the direct union of the elements. It is prepared by heating white phosphorus in a concentrated solution of sodium hydroxide.

$$3Na^+ + 3OH^- + P_4 + 3H_2O \longrightarrow \underset{\text{Sodium hypophosphite}}{3Na^+ + 3H_2PO_2^-} + \underset{\text{Phosphine}}{PH_3(g)}$$

Pure phosphine is not spontaneously flammable in air. However, the diphosphorus tetrahydride, P_2H_4, which is formed as a by-product of the reaction, is spontaneously flammable, and ignition of this substance on contact with air will ignite the phosphine. Consequently, phosphine is best prepared and handled under an atmosphere of nitrogen or some other inert gas.

Phosphine can also be prepared by the hydrolysis of phosphides of active metals. The phosphide ion, P^{3-}, is a very strong base.

$$Ca_3P_2 + 6H_2O \longrightarrow 3Ca(OH)_2 + 2PH_3(g)$$

Phosphine is a colorless, very poisonous gas, which has an odor like that of decaying fish. It is easily decomposed by heat ($4PH_3 \longrightarrow P_4 + 6H_2$). Like ammonia, phosphine unites with the hydrogen halides forming the corresponding **phosphonium compounds,** PH_4Cl, PH_4Br, and PH_4I. Unlike the ammonium halides, however, the phosphonium compounds do not give the phosphonium ion, PH_4^+, in solution. Instead, phosphine escapes and the hydrogen halide remains in solution as the hydrohalic acid. Phosphine is only very slightly soluble in water and is a much weaker base than ammonia.

Phosphides of several of the metals may be prepared by reducing the corresponding phosphates with carbon

$$Ca_3(PO_4)_2 + 8C \longrightarrow Ca_3P_2 + 8CO$$

or by the direct reaction of the metals with phosphorus.

24.6   Halides of Phosphorus

With one exception, phosphorus unites directly with all of the halogens, forming trihalides, PX_3, and pentahalides, PX_5; only the pentaiodide is not known. These halides are much more stable than the corresponding compounds of nitrogen.

Nitrogen is the only member of Group VA that cannot form pentahalides. This can be explained in terms of the availability of orbitals for bond formation. The Group VA atom, X, in the trihalide XF_3 uses four orbitals, three for σ bonds and one for the lone pair as shown by the Lewis structure of PF_3.

$$:P{-}F \qquad F{-}P$$

The pentafluorides, XF_5, require five orbitals on the Group VA atom. Since there are only four orbitals in the valence shell of nitrogen, it cannot form pentahalides. Phosphorus and heavier members of Group VA have empty d orbitals in their valence shells. One of these orbitals can be used with the s orbital and three p orbitals to form dsp^3 hybrid orbitals to bond five halogen atoms.

The trichloride, PCl_3, and the pentachloride, PCl_5, are the most important halides of phosphorus. **Phosphorus trichloride** is prepared by passing dry chlorine over molten phosphorus ($P_4 + 6Cl_2 \longrightarrow 4PCl_3$). The trichloride is a colorless liquid which boils at 76° and freezes at −92°. The PCl_3 molecule is pyramidal, with the phosphorus atom at one corner of the pyramid (Fig. 24-4). As with most other nonmetal halides, phosphorus trichloride is irreversibly hydrolyzed by water, with the formation of phosphorous acid and hydrogen chloride.

$$PCl_3 + 3H_2O \longrightarrow H_3PO_3 + 3HCl(g)$$

Because of hydrolysis, all of the halides of phosphorus fume in moist air, much like concentrated hydrochloric acid.

Phosphorus pentachloride is prepared by oxidizing the trichloride with excess chlorine ($PCl_3 + Cl_2 \longrightarrow PCl_5$). The pentachloride is a straw-colored solid, which sublimes when heated in air and decomposes reversibly into the trichloride and chlorine upon heating ($PCl_5 \rightleftharpoons PCl_3 + Cl_2$).

In both the gaseous and liquid states the PCl_5 molecule has the structure of a triangular bipyramid, with the phosphorus atom in the center and the chlorine atoms at each of the five corners (Fig. 24-5).

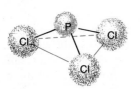

Figure 24-4 The molecular structure of PCl_3.

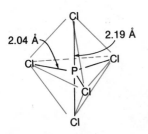

Figure 24-5 The molecular structure of PCl_5 in the gaseous and liquid states.

Figure 24-6 The structures of the ions PCl_4^+ and PCl_6^- found in PCl_5 in the solid state.

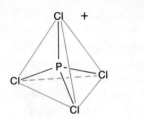

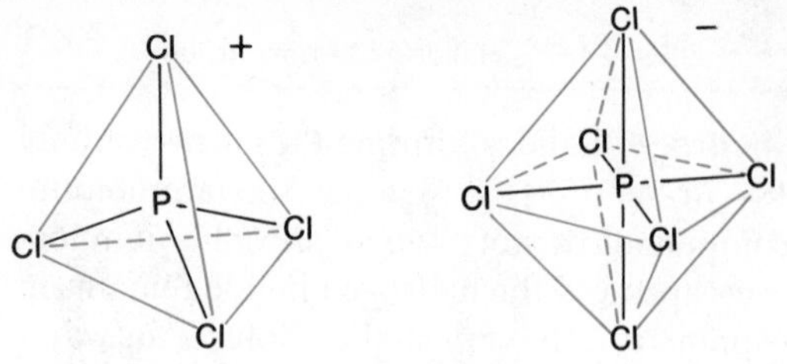

The pentahalides of phosphorus are Lewis bases. They pick up a halide ion to give the anion PX_6^-. X-ray studies show solid phosphorus pentachloride to be an ionic compound, $[PCl_4]^+[PCl_6]^-$, in which the cation has a tetrahedral structure and the anion an octahedral one (Fig. 24-6). The structure is the same when PCl_5 is in solution in solvents of high dielectric constant.

24.7 Phosphorus(V) Oxyhalides

Figure 24-7 The molecular structure of POX_3.

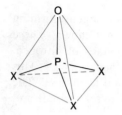

Compounds of the type POX_3 are called **phosphorus(V) oxyhalides,** or **phosphoryl halides.** Their molecules have tetrahedral structures (Fig. 24-7). The chloride and bromide are readily obtained by the action of the pentahalide upon phosphorus(V) oxide.

$$P_4O_{10} + 6PX_5 \longrightarrow 10POX_3$$

Partial hydrolysis of a phosphorus pentahalide also produces the corresponding oxyhalide.

$$PX_5 + H_2O \longrightarrow POX_3 + 2HX(g)$$

The iodide POI_3 apparently does not exist. The POX_3 compounds, especially $POCl_3$, are used in replacing OH groups in organic compounds with halogens. They hydrolyze to give orthophosphoric acid and HX.

$$POX_3 + 3H_2O \longrightarrow H_3PO_4 + 3HX(g)$$

24.8 Oxides of Phosphorus

Phosphorus forms two significant oxides, P_4O_6 and P_4O_{10}. The oxidation states of phosphorus in these compounds are +3 and +5, respectively. The structures of P_4O_6 and P_4O_{10} are shown in Fig. 24-8. Note that the phosphorus atom in these

Figure 24-8 The molecular structures of P_4O_6 and P_4O_{10}. Phosphorus atoms are shown in color.

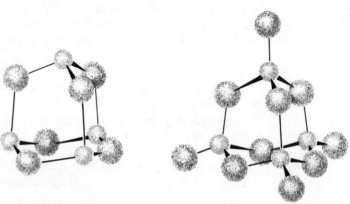

compounds is hybridized to form sp^3 hybrid orbitals. In the oxides of nitrogen(III) and nitrogen(V), the nitrogen exhibits sp^2 hybridization.

The combustion of phosphorus in air generally produces a mixture of P_4O_6 and P_4O_{10}. However, it is virtually impossible to get much of the P_4O_6 by burning phosphorus, even in a limited supply of air.

▶ **1. PHOSPHORUS(III) OXIDE, P_4O_6.** This compound is often written as its empirical formula, P_2O_3, and called phosphorus trioxide, but vapor density measurements have shown that the molecular formula is actually P_4O_6. The name tetraphosphorus hexaoxide clearly indicates the true formula. Phosphorus(III) oxide is a white crystalline solid that melts at 23.8°C and boils at 175.3°. Phosphorus(III) oxide has a garliclike odor, and its vapor is very poisonous. It oxidizes slowly in air and inflames when heated to 70°, forming P_4O_{10}. Phosphorus(III) oxide dissolves slowly in cold water to form phosphorous acid, H_3PO_3.

▶ **2. PHOSPHORUS(V) OXIDE, P_4O_{10}.** This oxide of phosphorus also is frequently identified by its empirical formula, P_2O_5, and called phosphorus pentaoxide. In fact, however, it has the molecular formula P_4O_{10}. The name tetraphosphorus decaoxide is best for clearly indicating the true formula. It is a white flocculent powder that melts at 420°. Its enthalpy of formation is very high (-2984 kJ), and for this reason it is quite stable and is a very poor oxidizing agent. With a limited amount of water it forms metaphosphoric acid, $(HPO_3)_x$.

$$xP_4O_{10} + 2xH_2O \longrightarrow 4(HPO_3)_x$$

When more water is added, the metaphosphoric acid slowly changes to orthophosphoric acid, H_3PO_4. The net reaction is given by the equation

$$P_4O_{10}(s) + 6H_2O(l) \longrightarrow 4H_3PO_4(l) \qquad \Delta H° = -368.6 \text{ kJ}$$

When a piece of P_4O_{10} is dropped into water, it reacts with a hissing sound, and much heat is liberated. Because of its great affinity for water, phosphorus(V) oxide is used extensively for drying gases and removing water from many compounds.

Acids of Phosphorus and Their Salts

The important oxygen acids of phosphorus are orthophosphoric acid, H_3PO_4; diphosphoric (pyrophosphoric) acid, $H_4P_2O_7$; triphosphoric acid, $H_5P_3O_{10}$; metaphosphoric acid, $(HPO_3)_x$; phosphorous acid, H_3PO_3; and hypophosphorous acid, H_3PO_2.

24.9   Orthophosphoric Acid

The various phosphoric acids all have the same anhydride, P_4O_{10}. Each acid represents a different degree of hydration of this oxide but not a different oxidation state of phosphorus, which is $+5$ in every case. These acids are stable toward most reducing agents.

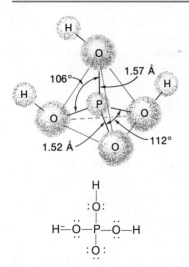

Figure 24-9 The electronic formula and the molecular structure of the orthophosphoric acid molecule.

One commercial method of preparing **orthophosphoric acid** (commonly called **phosphoric acid**) is the treatment of calcium rock (calcium phosphate) with concentrated sulfuric acid.

$$Ca_3(PO_4)_2 + 3H_2SO_4 \longrightarrow 2H_3PO_4 + 3CaSO_4(s)$$

The products are diluted with water and the calcium sulfate removed by filtration. This method gives a dilute acid that is impure with calcium dihydrogen phosphate, $Ca(H_2PO_4)_2$, and certain fluorine compounds that are formed from the fluorapatite, $Ca_{10}F_2(PO_4)_6$, usually associated with the phosphate rock.

Pure orthophosphoric acid is manufactured by oxidizing phosphorus from the electric furnace (Section 24.1) to P_4O_{10} and dissolving the product in water.

The pure acid forms colorless, deliquescent crystals that melt at 42.4°. A common commercial form of the acid is an 82% aqueous solution of H_3PO_4 and is known as **syrupy phosphoric acid.** Orthophosphoric acid is tetrahedral (Fig. 24-9); the phosphorus atom in orthophosphoric acid has a coordination number of 4. Large quantities of crude (impure) orthophosphoric acid are used in the manufacture of fertilizers. The acid is also used in medicine as an astringent, as an antipyretic, and as a stimulant.

24.10   Salts of Orthophosphoric Acid

Because orthophosphoric acid is a triprotic acid it forms three series of salts, corresponding to the three stages of ionization.

$$H_3PO_4 \rightleftharpoons H^+ + H_2PO_4^- \quad \text{(primary ionization, } K_1 = 7.5 \times 10^{-3})$$
$$H_2PO_4^- \rightleftharpoons H^+ + HPO_4^{2-} \quad \text{(secondary ionization, } K_2 = 6.2 \times 10^{-8})$$
$$HPO_4^{2-} \rightleftharpoons H^+ + PO_4^{3-} \quad \text{(tertiary ionization, } K_3 = 3.6 \times 10^{-13})$$

The formulas for the sodium salts of orthophosphoric acid and their names are NaH_2PO_4, sodium dihydrogen phosphate; Na_2HPO_4, disodium hydrogen phosphate; and Na_3PO_4, trisodium phosphate.

Sodium dihydrogen phosphate forms aqueous solutions that are weakly acidic because $H_2PO_4^-$ is a weak acid.

$$H_2PO_4^- + H_2O \rightleftharpoons HPO_4^{2-} + H_3O^+$$

Disodium hydrogen phosphate solutions are alkaline because the monohydrogen phosphate ion is stronger as a base than it is as an acid.

$$HPO_4^{2-} + H_2O \rightleftharpoons H_2PO_4^- + OH^-$$

Trisodium phosphate solutions are strongly alkaline as a result of hydrolysis.

$$PO_4^{3-} + H_2O \rightleftharpoons HPO_4^{2-} + OH^-$$

The hydrogen and ammonium phosphate salts are decomposed by heating, with volatile products such as water and ammonia being given off. The phosphorus-containing products resulting from the reactions have P—O—P bonds (see Sections 24.11–24.13). Several examples are

$$x\text{NaH}_2\text{PO}_4 \xrightarrow{\triangle} (\text{NaPO}_3)_x + x\text{H}_2\text{O}(g)$$

$$2\text{Na}_2\text{HPO}_4 \xrightarrow{\triangle} \text{Na}_4\text{P}_2\text{O}_7 + \text{H}_2\text{O}(g)$$

$$x\text{NaNH}_4\text{HPO}_4 \xrightarrow{\triangle} (\text{NaPO}_3)_x + x\text{NH}_3(g) + x\text{H}_2\text{O}(g)$$

$$2\text{MgNH}_4\text{PO}_4 \xrightarrow{\triangle} \text{Mg}_2\text{P}_2\text{O}_7 + 2\text{NH}_3(g) + \text{H}_2\text{O}(g)$$

Sodium dihydrogen phosphate is used as a boiler cleansing compound (to prevent the formation of boiler scale) and in some baking powders; disodium hydrogen phosphate as a boiler cleansing compound and in the weighting of silk; and trisodium phosphate as a water softener and boiler cleansing compound. **Calcium dihydrogen phosphate,** $\text{Ca}(\text{H}_2\text{PO}_4)_2$, is used as a fertilizer and as a constituent of baking powders. **Calcium monohydrogen phosphate,** CaHPO_4, is added to animal food for its mineral properties and is used as a polishing agent in toothpastes.

Phosphorus, in the form of soluble orthophosphates, is essential to plant growth. Tricalcium phosphate is too insoluble to furnish an adequate supply of phosphorus to plants. Calcium dihydrogen phosphate, however, is soluble in water and therefore suitable as a fertilizer. This compound is prepared commercially by treating tricalcium phosphate with sulfuric acid.

$$\text{Ca}_3(\text{PO}_4)_2 + 2\text{H}_2\text{SO}_4 + 4\text{H}_2\text{O} \longrightarrow \text{Ca}(\text{H}_2\text{PO}_4)_2 + 2(\text{CaSO}_4 \cdot 2\text{H}_2\text{O})$$

The mixture of calcium dihydrogen phosphate and gypsum, $\text{CaSO}_4 \cdot 2\text{H}_2\text{O}$, is sold as **superphosphate of lime.** A mixture containing a higher percentage of phosphorus, **triple superphosphate of lime,** is manufactured by treating pulverized tricalcium phosphate with orthophosphoric acid.

$$\text{Ca}_3(\text{PO}_4)_2 + 4\text{H}_3\text{PO}_4 \longrightarrow 3\text{Ca}(\text{H}_2\text{PO}_4)_2$$

24.11   Diphosphoric Acid

Diphosphoric acid, $\text{H}_4\text{P}_2\text{O}_7$, may be prepared by heating the ortho acid to 250°C. The equation for the reaction is

$$2\text{H}_3\text{PO}_4 \xrightarrow{\triangle} \text{H}_4\text{P}_2\text{O}_7 + \text{H}_2\text{O}$$

This reaction involves the elimination of a molecule of water from two molecules of orthophosphoric acid, as shown by the equation

The structure of the diphosphoric acid molecule is shown in Fig. 24-10. The substance is a white crystalline solid that melts at 61°. When it is dissolved in

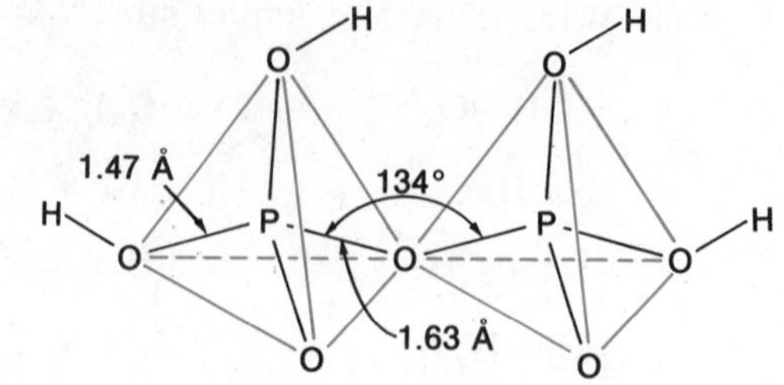

Figure 24-10 The molecular structure of the diphosphoric acid molecule, $H_4P_2O_7$, consists of two tetrahedra joined at a corner.

water, it gradually hydrolyzes to the ortho form of the acid. Among the salts of diphosphoric acid that have been prepared are tetrasodium diphosphate, $Na_4P_2O_7$, trisodium hydrogen diphosphate, $Na_3HP_2O_7$, and disodium dihydrogen diphosphate, $Na_2H_2P_2O_7$. As you might expect, the acid strengths of the hydrogen diphosphate anions ($H_3P_2O_7^-$, $H_2P_2O_7^{2-}$, $HP_2O_7^{3-}$) decrease and the base strengths increase with increasing negative charge.

24.12 Triphosphoric Acid

Triphosphoric acid (Fig. 24-11), $H_5P_3O_{10}$, is formed by the elimination of two molecules of water from three molecules of orthophosphoric acid.

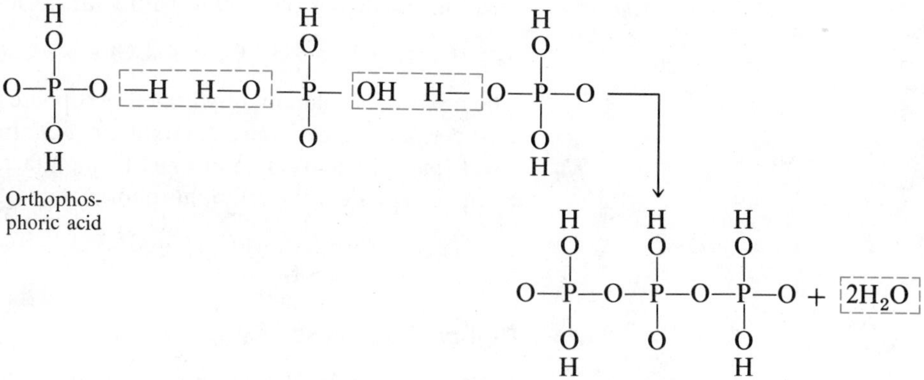

Orthophosphoric acid

Triphosphoric acid

The sodium salt of triphosphoric acid, $Na_5P_3O_{10}$, is used as a water softener. It may be prepared by fusing equimolar quantities of sodium di- and metaphosphates.

$$x\text{Na}_4\text{P}_2\text{O}_7 + (\text{NaPO}_3)_x \longrightarrow x\text{Na}_5\text{P}_3\text{O}_{10}$$

Figure 24-11 The molecular structure of triphosphoric acid, $H_5P_3O_{10}$.

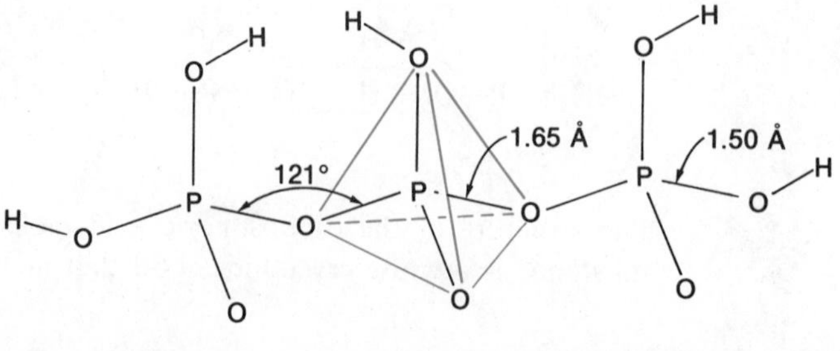

Polyphosphate linkages of the type existing in pyrophosphates and triphosphates are of great biochemical importance. The energy required for the contraction of muscles results from hydrolysis of "energy-rich" P—O—P bonds of a complex organic polyphosphate, existing in muscle tissue, known as **adenosine triphosphate.** The enzyme-catalyzed hydrolysis is slow but involves the release of 29 kJ of energy per mole of adenosine triphosphate. The equation for the reaction is given below (R represents the complex organic portion of the molecule):

$$
\underset{\text{Adenosine triphosphate}}{RO-\overset{\overset{\displaystyle O}{\|}}{\underset{\underset{\displaystyle OH}{|}}{P}}-O-\overset{\overset{\displaystyle O}{\|}}{\underset{\underset{\displaystyle OH}{|}}{P}}-O-\overset{\overset{\displaystyle O}{\|}}{\underset{\underset{\displaystyle OH}{|}}{P}}-OH} + H_2O \longrightarrow \underset{\text{Adenosine diphosphate}}{RO-\overset{\overset{\displaystyle O}{\|}}{\underset{\underset{\displaystyle OH}{|}}{P}}-O-\overset{\overset{\displaystyle O}{\|}}{\underset{\underset{\displaystyle OH}{|}}{P}}-OH} + \underset{\text{Phosphoric acid}}{HO-\overset{\overset{\displaystyle O}{\|}}{\underset{\underset{\displaystyle OH}{|}}{P}}-OH}
$$

This subject will be discussed in greater detail in Chapter 26.

24.13   Metaphosphoric Acid and Its Salts

The elimination of one mole of water per mole of orthophosphoric acid gives metaphosphoric acid, $(HPO_3)_x$.

$$xH_3PO_4 \longrightarrow (HPO_3)_x + xH_2O(g)$$

Metaphosphoric acid is usually manufactured by heating orthophosphoric acid above 400°C. It solidifies from the liquid as a glassy product called **glacial phosphoric acid.**

In metaphosphoric acid oxygen bridges are formed between adjacent phosphorus atoms to build up rings and chains, in which there are four oxygen atoms linked to every phosphorus atom. By combining with each other, the HPO_3 molecules polymerize into multiple units, which may be represented by the formula $(HPO_3)_x$. Metaphosphoric acid dissolves readily in water, in which it slowly changes into orthophosphoric acid.

Sodium metaphosphate is made by heating sodium dihydrogen orthophosphate.

$$xNaH_2PO_4 \xrightarrow{\Delta} (NaPO_3)_x + xH_2O(g)$$

If the product is heated to about 700° and then cooled rapidly, a water soluble, **glassy polymetaphosphate** is formed. This is a long-chain polymer which is best formulated as $(NaPO_3)_x$, although it has been called "sodium hexametaphosphate" because it was once thought to have the composition $(NaPO_3)_6$. The metaphosphates form soluble complexes with the calcium ion and reduce the concentration of the calcium ion in water solutions to such a low value that it cannot be precipitated by soaps. For this reason the sodium polymetaphosphates and other forms of phosphate have been used extensively as water softeners. However, their use has recently been seriously questioned because they enhance the growth of algae in bodies of water (see Chapter 12).

24.14   Phosphorous Acid and Its Salts

Phosphorous acid, H_3PO_3, can be prepared by the action of water upon P_4O_6, PCl_3, PBr_3, or PI_3. Pure phosphorous acid is most readily obtained by hydrolyzing

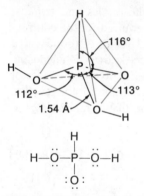

Figure 24-12 Phosphorous acid, H_3PO_3. Note that one hydrogen atom is bonded directly to the phosphorus atom. The other two hydrogen atoms are bonded to oxygen atoms. Only those hydrogen atoms bonded through an oxygen atom are acidic.

phosphorus trichloride.

$$PCl_3 + 3H_2O \longrightarrow H_3PO_3 + 3HCl(g)$$

The resulting solution is heated to expel the hydrogen chloride and to evaporate the water to a point where white crystals of phosphorous acid appear upon cooling. The crystals are deliquescent, very soluble in water, and have an odor like that of garlic. The solid crystals melt at 70.1° and decompose at about 200° by an auto-oxidation-reduction reaction into phosphine and orthophosphoric acid.

$$4H_3PO_3 \longrightarrow PH_3 + 3H_3PO_4$$

The electronic formula and tetrahedral structure of phosphorous acid is shown in Fig. 24-12.

Phosphorous acid and its salts are active reducing agents because they are readily oxidized to phosphoric acid and phosphates, respectively. Phosphorous acid reduces the silver ion to free silver, mercury(II) salts to the mercury(I) stage, and sulfurous acid to sulfur.

Phosphorous acid forms only two series of salts, such as **sodium dihydrogen phosphite**, NaH_2PO_3, and **disodium hydrogen phosphite**, Na_2HPO_3. The third atom of hydrogen cannot be replaced by a cation. The nonacidic character of the third hydrogen atom is due to its direct linkage to the phosphorus atom rather than to an oxygen atom.

A statement is in order here regarding a point of nomenclature that is often a source of confusion. You may have noticed the difference in spelling, in the preceding sections, of the name of the element (phosphor*us*) and the name of phosphor*ous* acid, H_3PO_3. In the latter case, the word *phosphorous* is made up of the stem of the element's name plus the *-ous* ending, which indicates a lower oxidation number for phosphorus (+3) than that in phosphoric acid, H_3PO_4 (+5). The name phosphor*ous* acid is therefore analogous to the names chlor*ous* acid, nitr*ous* acid, and sulfur*ous* acid. You may find it helpful to review the nomenclature discussion in Section 5.14.

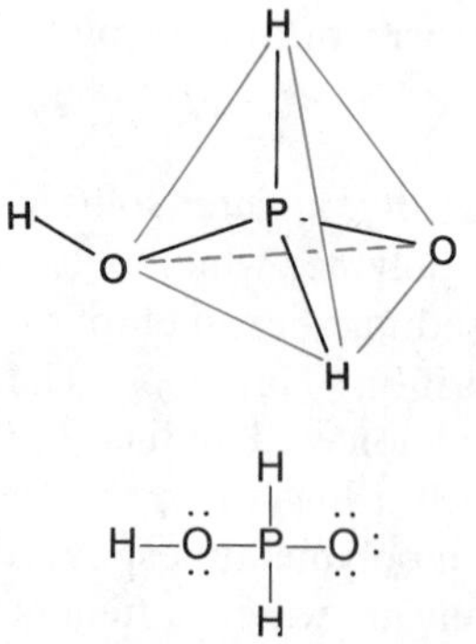

24.15   Hypophosphorous Acid and Its Salts

The solution remaining from the preparation of phosphine from white phosphorus and sodium hydroxide contains sodium hypophosphite, NaH_2PO_2 (Section 24.5). The corresponding barium salt may be obtained by replacing sodium hydroxide with barium hydroxide in the preparation of phosphine. When barium hypophosphite is treated with sulfuric acid, barium sulfate precipitates and **hypophosphorous acid**, H_3PO_2, is formed in solution.

$$Ba^{2+} + 2H_2PO_2^- + 2H^+ + SO_4^{2-} \longrightarrow BaSO_4(s) + 2H_3PO_2$$

The acid is weak and monoprotic, forming only one series of salts. The two nonacidic hydrogen atoms are linked directly to the phosphorus atom (see Fig. 24-13). Hypophosphorous acid and its salts are strong reducing agents inasmuch as phosphorus is in the unusually low oxidation state of +1.

Figure 24-13 Hypophosphorous acid, H_3PO_2. Note that two of the three hydrogen atoms are attached directly to the phosphorus atom. Hence only one hydrogen atom is acidic; i.e., the one bonded through an oxygen atom.

Questions

1. Compare and contrast the chemical properties of elemental phosphorus and nitrogen; of phosphoric acid and nitric acid; and of phosphine and ammonia.
2. Determine the oxidation state of phosphorus in each of the following: PH_3, P_2H_4, H_3PO_2, H_3PO_3, H_3PO_4, $H_4P_2O_7$, PH_4I, P_4, PCl_3.
3. Write the Lewis structures for each of the compounds given in Question 2.
4. Contrast the properties of white and red phosphorus.
5. Write equations showing the stepwise ionization of phosphoric acid.
6. Write equations for the preparation and hydrolysis of sodium phosphide. Compare the hydrolysis of sodium phosphide to that of sodium nitride.
7. How may the phosphorus in insoluble tricalcium phosphate be made available for plant nutrition?
8. Write equations showing the hydrolysis of PO_4^{3-}, HPO_4^{2-}, and $H_2PO_4^-$.
9. Relate $H_5P_3O_{10}$, $H_4P_2O_7$, H_3PO_4, and HPO_3 to P_4O_{10} in terms of extent of hydration of the latter compound.
10. Explain why H_3PO_4 is a stronger acid than H_3PO_3.
11. Write equations to show the action of heat upon NaH_2PO_4, $NaNH_4HPO_4$, and $MgNH_4PO_4$.
12. Explain the action of sodium polymetaphosphates as water softeners. Why is the use of such materials in detergents being questioned (see Chapter 12)?
13. Why does phosphorous acid form only two series of salts although its molecule contains three hydrogen atoms?
14. Write equations for the preparation of hypophosphorous acid, starting with white phosphorus.
15. Show that the decomposition of phosphorous acid by heat involves an auto-oxidation-reduction reaction.
16. Explain the difference in spelling in the name of the element, phosphorus, and the name of H_3PO_3, phosphorous acid.
17. Name each of the following compounds: K_2HPO_4, $FePO_4$, NaH_2PO_2, $(NaPO_3)_x$, $Na_4P_2O_7$, $Ca(H_2PO_4)_2$, and Na_2HPO_3.
18. Write equations for each of the following preparations: (a) P_4 from $Ca_3(PO_4)_2$; (b) P_4O_{10} from P_4; (c) H_3PO_4 from P_4O_{10}; (d) Na_2HPO_4 from H_3PO_4, and (e) $Na_4P_2O_7$ from Na_2HPO_4.
19. Draw valence electron structures for diphosphine and P_4.
20. Compare the structures of PCl_4^+, PCl_5, PCl_6^-, and $POCl_3$.
21. Complete and balance the following equations (xs indicates that the reactant is present in excess):
    (a) $P_4 + xs\ S \longrightarrow$
    (b) $P_4 + Li \longrightarrow$
    (c) $NaH_2PO_4 + NH_3 \longrightarrow$
    (d) $PF_5 + KF \longrightarrow$
    (e) $P_4O_6 + CaO \longrightarrow$
    (f) $P_4O_{10} + K_2O \longrightarrow$
    (g) $xs\ K_3PO_4 + HCl \longrightarrow$
    (h) $Na_3P + xs\ H_2O \longrightarrow$
    (i) $Na_4P_2O_7 + xs\ H_2O \longrightarrow$
    (j) $PCl_3 + CH_3OH \longrightarrow$
    (Treat CH_3OH as a derivative of water in which one H is replaced by the CH_3 group.)

Problems

1. How much $Ca_3(PO_4)_2$ would be needed to prepare 1.0 ton of phosphorus in the electric furnace process, if a yield of 94% is obtained?   *Ans. 5.3 tons*
2. A detergent is advertised as containing "only 10% phosphate, expressed as phosphorus pentaoxide." Is the name misleading? If so, explain why. Calculate what this percentage would be if expressed as sodium tripolyphosphate ($Na_5P_3O_{10}$), a common form for detergent phosphate.   *Ans. 26%*
3. How many pounds of phosphate ion would be added to the water system of the United States in one year if there is one load of clothes washed every week for every person in the country? Assume a population of 200 million people and that 4 ounces of detergent

that is 20% sodium orthophosphate (Na_3PO_4) is used per washing.   *Ans. 300,000,000 pounds*

4. How much $POCl_3$ can be produced from 50.0 g of PCl_5 and the appropriate amount of P_4O_{10}?

Ans. 61.4 g

5. (a) When phosphine burns to give liquid orthophosphoric acid, an unusually large amount of energy is involved. Calculate $\Delta H°$ for the reaction using the data in Appendix J.   *Ans. −1272 kJ*

   (b) Calculate the value of $\Delta H°$ for production of *solid* orthophosphoric acid by this reaction.

Ans. −1284 kJ

   (c) Calculate $\Delta G°$ for the reaction giving solid orthophosphoric acid.   *Ans. −1132 kJ*

   (d) Compare the values of $\Delta H°$ and $\Delta G°$, and on the basis of the comparison draw as specific conclusions as possible about the reaction.

6. At 422 K, the partial pressures of gaseous PCl_5, PCl_3, and Cl_2 in a closed system were 0.453 atm, 6.08 × 10^{-2} atm, and 6.08 × 10^{-2} atm, respectively. What is K for the dissociation reaction [$PCl_5(g) \rightleftharpoons PCl_3(g) + Cl_2(g)$]?   *Ans. 8.16 × 10^{-3}*

7. The solubility product constant of $Ca_3(PO_4)_2$ is 1 × 10^{-25}. What is the concentration of $Ca_3(PO_4)_2$ in a saturated solution of $Ca_3(PO_4)_2$?   *Ans. 4 × 10^{-6} M*

8. What volume of 0.100 M NaOH will be required to neutralize the solution produced by dissolving 1.00 g of PCl_3 in 200 ml of water? Note that when H_3PO_3 is titrated under these conditions, only one proton of the phosphorous acid molecule reacts.   *Ans. 291 ml*

References

"Phosphorus" (Staff), *Chem. and Eng. News,* April 24, 1978, p. 11.

"Phosphoric Acid" (Staff), *Chem. and Eng. News,* March 12, 1978, p. 12.

"Phosphate Cycles," J. Emsley, *Chem. in Britain,* **13,** 459 (1977).

"TVA Displays Energy-Saving Phosphoric Acid Process," (Staff), *Chem. and Eng. News,* November 13, 1978, p. 32.

"Mineral Cycles," E. S. Deevey, Jr. *Sci. American,* September 1970, p. 148.

"Chemical Fertilizers," C. J. Pratt, *Sci. American,* June 1965, p. 62.

"Ionic and Molecular Halides of the Phosphorus Family," R. R. Holmes, *J. Chem. Educ.,* **40,** 125 (1963).

"The Structures and Reactions of Phosphorus Sulfides," A. H. Cowley, *J. Chem. Educ.,* **41,** 530 (1964).

"Some Aspects of *d*-Orbital Participation in Phosphorus and Silicon Chemistry," J. E. Bissey, *J. Chem. Educ.,* **44,** 95 (1967).

"Historical Aspects of the Tetrahedron in Chemistry," D. F. Larder, *J. Chem. Educ.,* **44,** 661 (1967).

"The Chemistry of Orthophosphoric Acid and its Sodium Salts," H. A. Neidig, T. G. Teates, and R. T. Yingling, *J. Chem. Educ.,* **45,** 57 (1968).

"Development of Diphosphonates as Significant Health Care Products," M. D. Francis and R. L. Centner, *J. Chem. Educ.,* **55,** 760 (1978).

"Phosphate Replacements: Problems with the Washday Miracle," A. L. Hammond, *Science,* **172,** 361 (1971).

"Inorganic Polymers," H. R. Allcock, *Sci. American,* March 1974, p. 66.

"Conodont Pearls?" B. F. Glenister, G. Klapper, and K. M. Chauff, *Science,* **193,** 571 (1976).

"Flame Retardants," H. J. Sanders, *Chem. and Eng. News,* April 24, 1978, p. 22.

"Chemical Warfare: One of the Dilemmas of the Arms Race," W. Lepkowski, *Chem. and Eng. News,* January 2, 1978, p. 16.

25

Carbon and Its Compounds

Carbon—The Element

Carbon is the first member of Group IVA of the Periodic Table; the other members of this group are silicon, germanium, tin, and lead. Carbon is predominantly nonmetallic in character, silicon and germanium are metalloids with both nonmetallic and metallic character, and tin and lead are metallic. Each of the elements in Group IVA has four valence electrons with a valence shell electronic configuration of ns^2np^2. Each element exhibits a maximum oxidation state of $+4$; these elements, especially the heavier ones, also can show an oxidation state of $+2$ when combining with other elements.

With few exceptions, carbon is covalent in its compounds. Generally, all four of its valence electrons are used in bonding, with either sp, sp^2, or sp^3 hybridization at the carbon atom. Thus carbon can have a coordination number of 2, 3, or 4 with a linear, trigonal planar, or tetrahedral geometry, respectively.

Carbon is nineteenth among the elements in abundance; it constitutes only about 0.027% of the earth's crust. It is found in the free state as diamond and graphite; it occurs combined in natural gas, petroleum, coal, plants and animals, limestone, dolomite, coral, and chalk. Carbon is found in the air as carbon dioxide and in natural waters as carbon dioxide, carbonic acid, and carbonates.

25.1 Diamond

Carbon exists in two allotropic forms, **diamond** and **graphite.** Studies have shown that amorphous carbon has the same structure as graphite. Charcoal, coke, and carbon black are also microcrystalline, or amorphous, forms of carbon.

653

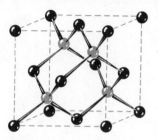

Figure 25-1 The crystal structure of diamond. Each sphere represents a carbon atom, with the darker spheres indicating the atoms at the corners and the centers of the faces of the cubic unit cell and the colored spheres, the atoms in tetrahedral sites. The covalent bonds are wedge-shaped to help show the perspective.

Diamonds are found in South Africa, the Congo, the Gold Coast, Brazil, India, Australia, and in the United States in Arkansas. Most of the world's supply comes from Africa. The accepted theory is that diamonds are formed when very pure carbon is subjected to a high temperature and very great pressure.

Many attempts have been made during the past 100 years to synthesize diamonds, but apparently none were successful until 1954. In that year, H. Tracy Hall and his co-workers successfully produced the first authentic artificial diamonds. They found that when graphite, in molten FeS as a solvent, was subjected to high temperature and pressure under carefully controlled conditions, small crystals of diamond were formed. This method of making diamonds has become increasingly important because manufactured diamonds are superior to the natural stones for certain industrial uses.

Diamond is very brittle, and (with the possible exception of boron carbide, B_4C) it is the hardest substance known. Diamond is inert to all chemicals at ordinary temperatures. However, diamonds burn when heated in air or oxygen. When heated to 1000° in the absence of air, diamond changes to graphite.

A diamond crystal belongs to the cubic system. Each atom is sp^3 hybridized and covalently bonded to four others located at the corners of a regular tetrahedron (Fig. 25-1). These covalent bonds bind all of the atoms in the diamond crystal into a single giant molecule. Because the carbon-carbon bonds are very strong and extend throughout the crystal in its three dimensions, the crystal is very hard and has a high melting point (probably the highest of all elements). The diamond is a nonconductor of electricity, because there are no mobile electrons in the crystal; its electrons are all used in bond formation.

25.2   Graphite

Natural graphite is mined in Mexico, Madagascar, Ceylon, and Canada. The United States leads in the production of synthetic graphite. To synthesize graphite, a mixture of amorphous carbon and a little sand and iron oxide (as catalysts) is heated in an electric furnace for 24 to 30 hours at a temperature of about 3500°. The carbon vapor that is formed condenses as graphite. Graphite, which is also known as **plumbago,** or **black lead,** is distinctly different in crystalline form and physical properties from diamond. It is a soft, grayish black solid that crystallizes in hexagonal plates that have a metallic luster and conduct electricity. The word *graphite* comes from the Greek meaning "to write," because the substance is used for this purpose. Graphite melts at 3527°, and it is inert to most chemical reagents. The internal energy of graphite is less than that of diamond. This means that graphite is the more stable of the two crystalline forms, in accord with the principle that substances seek the lowest energy form possible (see Section 18.9). More energy is released in the combustion of diamond, therefore, than in the combustion of graphite.

$$C \text{ (diamond)} + O_2(g) \longrightarrow CO_2(g) \qquad \Delta H° = -395.40 \text{ kJ}$$
$$C \text{ (graphite)} + O_2(g) \longrightarrow CO_2(g) \qquad \Delta H° = -393.51 \text{ kJ}$$
$$C \text{ (diamond)} \longrightarrow C \text{ (graphite)} \qquad \Delta H° = -1.89 \text{ kJ}$$

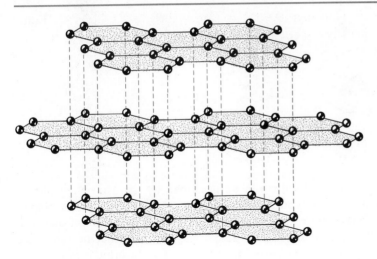

Figure 25-2   The structure of graphite.

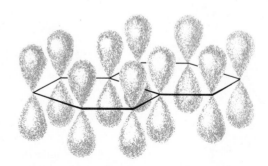

Figure 25-3   The orientation of unhybridized p orbitals on the carbon atoms in graphite. Each p orbital is perpendicular to the plane of carbon atoms.

The comparative softness and electrical conductivity of graphite can be related to its structure (Fig. 25-2). The crystal consists of layers of atoms, each atom of which has three carbon atoms as its near neighbors; each atom thus exhibits sp^2 hybridization. The hybrid orbitals are used by each carbon atom to form three σ bonds, one to each of its three nearest neighbors in the layer. The unhybridized p orbital on each carbon atom projects above and below the layer of atoms (Fig. 25-3). Each unhybridized p orbital contains one electron which is used in π bonding to adjacent carbon atoms in the same layer. The π bonds are not fixed but resonate to give some double-bond character to each bond between carbon atoms within a layer. Two of the many resonance formulas necessary to describe the electronic structure of a graphite layer are shown in Fig. 25-4. The combination of σ and π bonds bind the atoms tightly together within the layers. However, the forces holding the layers to each other are weak, and the layers can be separated easily. The flaky, soft character of graphite results from this weak binding. The loosely held electrons of the mobile double bonds within the layers of carbon atoms are responsible for the electrical conductivity and the color of graphite.

Figure 25-4   Two of the resonance formulas necessary to describe the resonance hybrid electronic structure of graphite.

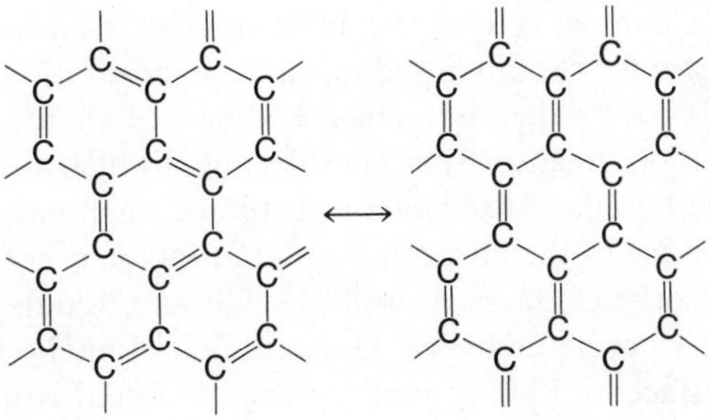

Figure 25-5 A portion of one delocalized molecular orbital in graphite which contains π electrons.

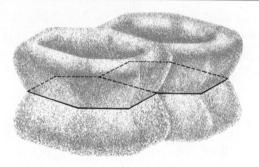

The π bonding in graphite also can be described by the Molecular Orbital Theory (Chapter 6). The unhybridized p orbitals on each carbon atom overlap to give molecular orbitals that extend over the entire layer. A portion of one such molecular orbital is illustrated in Fig. 25-5. The electrons that occupy these orbitals are free to move over the entire layer inasmuch as the orbitals extend over the entire layer. Since the electrons can move, electrical conductivity is one of the characteristics of graphite.

25.3   Properties and Uses of Carbon

All forms of carbon are almost chemically inert toward most reagents at ordinary temperatures. However, graphite is slowly oxidized by a mixture of nitric acid and sodium chlorate. The activity of carbon increases rapidly with rising temperatures, and at elevated temperatures it is very active. At high temperatures carbon unites with oxygen, forming either carbon monoxide, CO, or carbon dioxide, CO_2, depending on the amount of oxygen present. Carbon with sulfur forms carbon disulfide, CS_2; with certain metals, the carbides, such as iron carbide, Fe_3C; and with fluorine, carbon tetrafluoride, CF_4. Hot carbon combines directly with hydrogen in the presence of a suitable catalyst, forming such products as methane, CH_4, and acetylene, C_2H_2, in very low yields.

Because of its resistance to heat and chemical action, its high vaporization temperature, and its ability to conduct electricity, graphite is used in making electrodes and crucibles. It is also used in the manufacture of paints, commutator brushes, and "lead" pencils. Suspensions of colloidal graphite in water or in oil provide excellent lubricants for many purposes. Diamond is used in cutting, grinding, and polishing operations because of its hardness.

A third form of carbon, wood charcoal, is produced by the destructive distillation of wood, i.e., the heating of wood in the absence of air.

Charcoal is used to a limited extent as a fuel. It is also used as a decolorizing agent for certain liquids such as sugar solutions, alcohol, and petroleum products. This use depends upon the ability of charcoal (and other amorphous forms of carbon) to adsorb certain substances which are responsible for the discoloration of the liquids. **Adsorption** is a surface phenomenon in which forces existing on the surface of the adsorbing agent attract and hold molecules of the substance being adsorbed. Gases, as well as solids and liquids, are adsorbed on the surface of the very porous charcoal. One milliliter of finely powdered charcoal may have a total surface of 1000 square meters. Charcoal is used in removing odoriferous gases from petroleum products and in water purification to remove chloroform and other organic molecules that may be present in small amounts in drinking water.

Inorganic Compounds of Carbon

25.4 Carbon Monoxide

Carbon monoxide, CO, is produced when carbon is burned with insufficient oxygen to fully oxidize it to carbon dioxide, CO_2.

$$2C(s) + O_2(g) \xrightarrow{\Delta} 2CO(g)$$

A mixture of carbon monoxide and nitrogen in a ratio of about 1 to 2 by volume results when a limited supply of air is passed through hot coke, the product of the destructive distillation of coal. This product, called **producer gas,** is used as an inexpensive and economical gaseous fuel for certain industrial operations.

When steam is passed through a bed of red-hot coke, a mixture of hydrogen and carbon monoxide, called **water gas,** is formed.

$$C(s) + H_2O(g) \longrightarrow H_2(g) + CO(g) \qquad \Delta H° = +131.3 \text{ kJ}$$

Because the reaction is endothermic, the coke soon becomes too cool to reduce the steam. Hence oxygen and steam are passed together through the coke to keep it heated to red heat. Water gas is used extensively as an industrial fuel and as a source of hydrogen.

Carbon monoxide is prepared in the laboratory by heating crystals of oxalic acid, $H_2C_2O_4$, with concentrated sulfuric acid; the latter compound dehydrates the oxalic acid.

$$H_2C_2O_4 \xrightarrow{\text{Concd } H_2SO_4} H_2O + CO_2 + CO$$

The carbon monoxide in the resulting mixture may be removed from the mixture of gases by passing the mixture through solid sodium hydroxide, which absorbs the carbon dioxide.

Carbon monoxide is a colorless, odorless, and tasteless gas that is only slightly soluble in water. Liquid carbon monoxide boils at $-192°$ and freezes at $-205°$.

The valence electron formula for carbon monoxide is $:C\equiv O:$, in which the arrangement of electrons gives a completed shell to each atom. The molecular orbital model for carbon monoxide is mentioned in Section 6.13.

Carbon monoxide readily burns in oxygen, forming the dioxide.

$$2CO(g) + O_2(g) \longrightarrow 2CO_2(g) \qquad \Delta H° = -565.97 \text{ kJ}$$

This reaction makes carbon monoxide valuable as a gaseous fuel. At high temperatures carbon monoxide will reduce many metallic oxides and is important, therefore, as a reducing agent in metallurgical processes. Two examples are

$$CuO + CO \longrightarrow Cu + CO_2$$
$$FeO + CO \longrightarrow Fe + CO_2$$

Carbon monoxide combines with chlorine in the presence of sunlight and charcoal (as a catalyst) forming **carbonyl chloride,** $COCl_2$, which is also called **phosgene.**

Carbon monoxide is a very dangerous poison. It is especially dangerous because it is odorless and tasteless and therefore gives no warning of its presence. It combines with the hemoglobin of the blood and forms a compound that is too stable to be broken down by the body processes. In this way carbon monoxide destroys the ability of the blood to carry oxygen.

25.5   Carbon Dioxide

Figure 25-6 The electronic and molecular structures of CO_2.

Carbon dioxide (Fig. 25-6) is produced when any form of carbon is burned in an excess of oxygen ($C + O_2 \longrightarrow CO_2$). The same is true of almost all compounds of carbon. Some examples are

$$CH_4 + 2O_2 \longrightarrow CO_2 + 2H_2O$$
$$C_2H_5OH + 3O_2 \longrightarrow 2CO_2 + 3H_2O$$

Many carbonates liberate carbon dioxide when they are heated. An example of commercial importance is the reaction that takes place when quicklime, CaO, is produced by heating limestone.

$$CaCO_3 \longrightarrow CaO + CO_2(g)$$

Large quantities of carbon dioxide are obtained for commercial purposes as a by-product of the fermentation of sugar (glucose) during the preparation of alcohol and alcoholic beverages. The net reaction is given by

$$\underset{\text{Glucose}}{C_6H_{12}O_6} \xrightarrow{\text{Yeast}} \underset{\text{Ethanol}}{2C_2H_5OH} + 2CO_2(g)$$

In the laboratory, carbon dioxide is produced by the action of acids on carbonates.

$$CaCO_3 + 2H^+ \longrightarrow Ca^{2+} + H_2O + CO_2(g)$$

Carbon dioxide is a colorless and odorless gas that is 1.5 times as heavy as air. It is a component of all "carbonated" beverages. With water it forms carbonic acid, which has a mildly acid taste. One liter of water at 20° dissolves 0.9 liter of carbon dioxide. The gas is easily liquefied by compression because its critical temperature is relatively high (31.1°). When the liquid is allowed to evaporate, the vapor freezes to a snowlike solid at −56.2°. The solid vaporizes without melting (sublimes) because its vapor pressure is 1 atm at −78.5°. This property makes solid carbon dioxide valuable as a refrigerant that is always free from the liquid; for this reason it is called **Dry Ice.**

The valence electron formula for carbon dioxide is $:\overset{..}{O}=C=\overset{..}{O}:$, in which the double bonds are strong, and the molecule is quite stable toward heat. At 2000°C the dissociation into carbon monoxide and oxygen is only 1.8% ($2CO_2 \rightleftharpoons 2CO + O_2$). When the heating is carried out in the presence of carbon, carbon monoxide is formed in good yield ($CO_2 + C \longrightarrow 2CO$). Burning magnesium in a carbon dioxide atmosphere reduces carbon dioxide to carbon.

$$CO_2 + 2Mg \longrightarrow 2MgO + C$$

Large quantities of carbon dioxide are used in the manufacture of **washing soda,** $Na_2CO_3 \cdot 10H_2O$, and **baking soda,** $NaHCO_3$, by the Solvay process (discussed later); in the production of **white lead,** $Pb(OH)_2 \cdot 2PbCO_3$, a paint pigment; and in the preparation of carbonated soft drinks.

Carbon dioxide is a valuable fire extinguisher because it does not support the combustion of ordinary substances, it is easily generated, and it is cheap. Air containing as little as 2.5% carbon dioxide will extinguish a flame.

Carbon dioxide is not toxic; it can be harmful to life, however, because a large concentration in air can cause suffocation, i.e., lack of oxygen. Breathing air that contains a higher than usual percentage of carbon dioxide stimulates the respiratory centers and causes rapid breathing.

The atmosphere contains about 0.04% by volume of carbon dioxide and thus serves as a huge reservoir of this compound. Green plants absorb carbon dioxide from the air or water, and under the influence of sunlight and chlorophyll (as a catalyst) the carbon dioxide and water become sugar and free oxygen. The reactions that take place during this process, called **photosynthesis,** are complex, but the process may be summarized by

$$6CO_2 + 6H_2O \longrightarrow C_6H_{12}O_6 + 6O_2$$

The sugar produced during photosynthesis is glucose, $C_6H_{12}O_6$; it is the sugar that green plants use to synthesize substances such as other sugars, starches, cellulose, and fats.

Carbon dioxide is a product of respiration and is returned to the air by the green plants and by animals. The organisms (mostly nongreen plants) that produce decay of both plant and animal matter and those that cause fermentation of sugars also assist in the production of carbon dioxide, as does the combustion of carbon-containing fuels. The air from volcanoes and from certain other geological formations are other sources of carbon dioxide in the atmosphere. Because of the solubility of carbon dioxide in water, oceans and lakes are great reservoirs of this compound. The ocean reservoir is to a great extent responsible for the fact that the carbon dioxide content of the air is almost constant. Nevertheless, the carbon dioxide content of the atmosphere has increased detectably in the last few years because we burn so much fuel to generate power.

25.6 Carbonic Acid and Carbonates

Carbon dioxide is the anhydride of **carbonic acid,** H_2CO_3, which forms slowly when carbon dioxide dissolves in water. Carbonic acid is a diprotic acid; the following ionization constants are reported for aqueous carbonic acid:

$$H_2CO_3(aq) \rightleftharpoons H^+(aq) + HCO_3^-(aq) \qquad K_{i_1} = \frac{[H^+][HCO_3^-]}{[H_2CO_3]} = 4.3 \times 10^{-7}$$

$$HCO_3^-(aq) \rightleftharpoons H^+(aq) + CO_3^{2-}(aq) \qquad K_{i_2} = \frac{[H^+][CO_3^{2-}]}{[HCO_3^-]} = 7 \times 10^{-11}$$

The first value is not strictly correct. It assumes that all of the dissolved carbon dioxide has reacted with water and is present as carbonic acid. In actual fact, most of the carbon dioxide in a solution of CO_2 in water is present as CO_2 molecules rather than H_2CO_3 molecules. The actual ionization constant for carbonic acid, taking into account the true concentration of H_2CO_3 molecules in a carbon dioxide–water solution, is about 2×10^{-4}. In calculations of equilibria involving carbonic acid, it is convenient to assume that all of the carbon dioxide in solution is present as carbonic acid, and thus the value of 4.3×10^{-7} is commonly used for K_{i_1} in such calculations. This assumption leads to correct values for the concentration of hydronium ion and hydrogen carbonate ion in solution.

The rate at which carbon dioxide reacts with water is slow enough to be observed by eye. If an excess of a saturated solution of carbon dioxide in water and an excess of a dilute solution of acetic acid are added to separate beakers, each containing a dilute solution of sodium hydroxide and phenolphthalein indicator, the sodium hydroxide is neutralized by the acetic acid (indicated by the fading or the disappearance of color) as rapidly as the two solutions can be mixed. The neutralization by the aqueous carbon dioxide solution takes several seconds, as evidenced by the time it takes the color of the indicator to fade.

When a solution of sodium hydroxide is saturated with carbon dioxide, **sodium hydrogen carbonate** is formed.

$$Na^+ + OH^- + CO_2 \longrightarrow Na^+ + HCO_3^-$$

The compound $NaHCO_3$ is also called **sodium bicarbonate,** or **baking soda.** The hydrogen carbonate anion is stronger as a base than it is as an acid, so solutions of salts of hydrogen carbonate are weakly alkaline.

$$HCO_3^- + H_2O \rightleftharpoons H_2CO_3 + OH^-$$

When an equivalent amount of sodium hydroxide is added to a solution of sodium hydrogen carbonate and crystals are allowed to form, the hydrate $Na_2CO_3 \cdot 10H_2O$ is obtained. The hydrate, washing soda, is commonly used as a water softener in the home. If this substance is heated gently, the anhydrous salt called **soda ash,** Na_2CO_3, is formed. Solutions of sodium carbonate are strongly basic due to extensive hydrolysis of the carbonate ion.

$$CO_3^{2-} + H_2O \rightleftharpoons HCO_3^- + OH^-$$

25.7   Carbon Disulfide

At the temperature of the electric furnace, carbon is oxidized by sulfur to **carbon disulfide** (Fig. 25-7).

$$C + 2S \longrightarrow CS_2$$

Air must be excluded because the volatile carbon disulfide is highly flammable and burns according to the equation

$$CS_2 + 3O_2 \longrightarrow CO_2 + 2SO_2$$

Carbon disulfide is also produced commercially by burning methane in sulfur vapor at about 700° in the presence of either silica gel or activated alumina.

Pure carbon disulfide is a colorless liquid, which boils at 46.3°. The commercial product is yellow and has a disagreeable odor. The liquid is heavy (specific gravity

Figure 25-7 The electronic and molecular structures of CS_2.

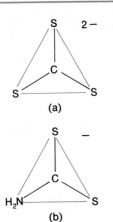

Figure 25-8 The structures of (a) the thiocarbonate ion and (b) the dithiocarbamate ion.

1.27 to 1.29), has a high index of refraction, and is immiscible with water. The vapor is heavier than air, very poisonous, and highly flammable.

Carbon disulfide, like carbon dioxide, is a Lewis acid. Carbon disulfide reacts with the sulfide ion, giving the thiocarbonate ion CS_3^{2-} (Fig. 25-8a) and with the amide ion, giving the dithiocarbamate ion, $H_2NCS_2^-$ (Fig. 25-8b).

$$S^{2-} + CS_2 \longrightarrow CS_3^{2-}$$
$$\cdot NH_2^- + CS_2 \longrightarrow H_2NCS_2^-$$

When sulfur replaces part or all of the oxygen in a molecule or ion, the prefix *thio-* is used. Thus CO_3^{2-} is the carbonate ion, whereas CS_3^{2-} is the thiocarbonate ion.

Large quantities of carbon disulfide are used in making rayon by the viscose process and in the manufacture of cellophane.

25.8 Carbon Tetrachloride

Carbon tetrachloride, CCl_4, is manufactured by the reaction of chlorine with methane, CH_4.

$$CH_4 + 4Cl_2 \xrightarrow{\triangle} CCl_4 + 4HCl$$

Carbon tetrachloride (bp 76.7°) is a colorless, pleasant-smelling liquid, with a specific gravity of 1.58, and a freezing point of −23°. It is an excellent solvent for fats, oils, and greases; therefore it is used in the dry cleaning of various fabrics. Because its vapor is about five times as heavy as air and it does not burn, it makes a good fire extinguisher. The fact that it is a nonconductor of electricity makes carbon tetrachloride particularly useful in fighting fires involving electrical equipment where water might cause short circuits. It is a poison, and the vapor should not be inhaled.

25.9 Calcium Carbide

Calcium carbide, CaC_2, is an important commercial product made by heating calcium oxide with coke in an electric furnace.

$$CaO + 3C \longrightarrow CaC_2 + CO$$

The product, which is liquid at the temperature of the furnace, is drawn off and solidified by cooling. Calcium carbide is an ionic compound with the valence electron formula $Ca^{2+}[:C{\equiv}C:]^{2-}$. It hydrolyzes in water forming acetylene (ethyne), $HC{\equiv}CH$ (Section 25.14).

$$CaC_2 + 2H_2O \longrightarrow Ca(OH)_2 + C_2H_2(g)$$

Acetylene is an important compound that is used in the synthesis of many organic materials such as ethyl alcohol, synthetic fabrics, synthetic rubber, and plastics. It is used also as an illuminant and as a fuel for cutting and welding metals.

25.10 Cyanides

The reaction of calcium carbide with nitrogen at about 1100° gives **calcium cyanamide,** $CaCN_2$.

$$CaC_2 + N_2 \xrightarrow{1100°} CaCN_2 + C$$

Figure 25-9 The electronic and molecular structures of the cyanamide ion. Note the similarity to CO_2 (Fig. 25-6) and CS_2 (Fig. 25-7).

The linear cyanamide ion, NCN^{2-}, is isoelectronic and isostructural with carbon dioxide and carbon disulfide (Compare Fig. 25-9 with Fig. 25-6 and Fig. 25-7).

The fusion of calcium cyanamide with carbon and sodium carbonate produces **sodium cyanide,** NaCN.

$$CaCN_2 + C + Na_2CO_3 \longrightarrow CaCO_3 + 2NaCN$$

The sodium cyanide is separated from the insoluble calcium carbonate by dissolving it in water. Sodium cyanide can also be prepared by the reaction of sodium amide with carbon at 500° to 600°.

$$NaNH_2 + C \xrightarrow{500°-600°} NaCN + H_2$$

The cyanide ion is a basic ion. It combines with acid to form **hydrogen cyanide,** HCN.

$$H^+(aq) + CN^-(aq) \longrightarrow HCN(g)$$

Hydrogen cyanide is a gas with an odor like that of bitter almonds. The liquid boils at 25.7° and freezes at $-13.2°$. When hydrogen cyanide is dissolved in water, hydrocyanic acid is formed. This acid is so weak ($K = 4 \times 10^{-10}$) that it will not turn blue litmus red. Its salts, such as sodium cyanide, form solutions that are alkaline by hydrolysis.

Hydrogen cyanide is very poisonous. A dose of about 0.05 g is fatal to human beings; moreover, its toxic action is very rapid. It paralyzes the central nervous system and inhibits all tissue respiration by combining with the iron in the enzymes that promote respiration.

Both cyanamides and cyanides find extensive utilization in the plastics and polymer industries. Calcium cyanamide is used to prepare melamine, a raw material for the production of plastics such as Melmac. Hydrogen cyanide is used in the preparation of acrylonitrile polymers used in synthetic fibers such as Dynel and Orlon.

Organic Compounds of Carbon

The term "organic" appears to have been used for the first time about 1777 and was applied to those materials occurring in or derived from living organisms. Accordingly, such substances as starch, alcohol, and urea were classified as organic, for starch is produced by living plants, alcohol is a product of fermentation caused by microorganisms, and urea is contained in urine. In 1828, however, the German chemist Friedrich Wöhler synthesized urea from materials obtained from inanimate sources, and the original meaning of the term "organic" no longer applied. **Organic compounds,** in the modern sense, are compounds of carbon that either contain carbon-carbon bonds and/or carbon-hydrogen bonds. Many thousands of carbon compounds that are not found in or derived from living organisms have been produced by chemists, and well over a million organic compounds are already known.

The existence of so many organic compounds is due primarily to the ability of carbon atoms to combine with other carbon atoms, forming chains of different lengths and rings of different sizes. The elements, other than carbon, most frequently found in organic compounds are hydrogen, oxygen, nitrogen, sulfur, the halogens, phosphorus, and some of the metals.

The simplest organic compounds are those containing only carbon and hydrogen. Such compounds are known as **hydrocarbons.** Many hydrocarbons are found in nature, where they were derived from plant or animal life. Many other hydrocarbons have been prepared in the laboratory. Several types of hydrocarbons have been characterized. The various types are distinguished by the type of bonding and the hybridization (Sections 7.3–7.7) of the carbon atoms in each.

25.11   Saturated Hydrocarbons

Saturated hydrocarbons contain only single covalent bonds. Consequently, all of the carbon atoms in a saturated hydrocarbon display sp^3 hybridization and are bonded to four other carbon or hydrogen atoms.

A saturated hydrocarbon with the general molecular formula C_nH_{2n+2}, where n is an integer, is called an **alkane,** or **paraffin** (from the Latin for "having little affinity," or being not very reactive). A few of the alkanes and some of their properties are listed in Table 25-1. With the exception of methane, each of the alkanes listed in the table contains a chain of carbon atoms. Since all of the carbon atoms are tetrahedrally hybridized (sp^3) and bonded either to other carbon atoms or to hydrogen atoms by single bonds, all of the C—C bond distances are about 1.54 Å, all of the C—H distances are about 1.09 Å, and the bond angles are close to 109.5° (the tetrahedral angle). Thus the chains of carbon atoms have a staggered, or zigzag, configuration. The Lewis structures and the molecular structures of methane, ethane, and pentane are illustrated in Fig. 25-10. In the molecular structures, dashed-line bonds point behind the page, and the wedge-shaped bonds point in front of the page. Each carbon atom can rotate about its carbon-carbon

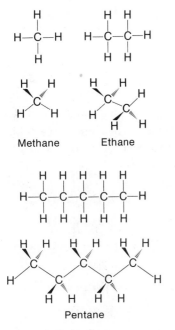

Methane   Ethane

Pentane

Figure 25-10 The electronic and molecular structures of methane, ethane, and pentane. Dashed-line bonds point behind the page; wedge-shaped bonds in front of the page.

TABLE 25-1   Some Alkanes or Paraffins[a]

	Molecular Formula	Melting Point, °C	Boiling Point, °C	Usual Form	Number of Possible Isomers
Methane	CH_4	−182.5	−161.5	Gas	1
Ethane	C_2H_6	−183.2	−88.6	Gas	1
Propane	C_3H_8	−187.7	−42.1	Gas	1
Butane	C_4H_{10}	−138.3	−0.5	Gas	2
Pentane	C_5H_{12}	−129.7	36.1	Liquid	3
Hexane	C_6H_{14}	−95.3	68.7	Liquid	5
Heptane	C_7H_{16}	−90.6	98.4	Liquid	9
Octane	C_8H_{18}	−56.8	125.7	Liquid	18
Nonane	C_9H_{20}	−53.6	150.8	Liquid	35
Decane	$C_{10}H_{22}$	−29.7	174.0	Liquid	75
Undecane	$C_{11}H_{24}$	−25.6	195.8	Liquid	159
Dodecane	$C_{12}H_{26}$	−9.6	216.3	Liquid	355
Tridecane	$C_{13}H_{28}$	−5.4	235.4	Liquid	802
Tetradecane	$C_{14}H_{30}$	5.9	253.5	Liquid	1,858
Octadecane	$C_{18}H_{38}$	28.2	316.1	Solid	60,523

[a] Physical properties for C_4H_{10} and the heavier molecules are those of the *normal* isomer, *n*-butane, *n*-pentane, etc.

single bonds; hence the chain of carbon atoms need not lie in a plane. It is possible for a long chain to twist itself into many different shapes. For clarity, the carbon atoms are usually written in straight lines in the Lewis structures, but it must be remembered that Lewis structures show only the distribution of electrons and which atoms are bonded together, not the geometry about each atom.

There are two hydrocarbons having the formula C_4H_{10}; these are known by the common names normal butane and isobutane. The two butanes are **structural isomers.** They have the same molecular formula, and hence the same composition, but different physical and chemical properties because they differ in the arrangement of the atoms in their molecules. Normal butane, or *n*-butane, is a "straight chain" molecule, and isobutane is a "branched chain" molecule.

n-Butane   Isobutane

The number of possible isomers increases with increasing molecular weight. The hydrocarbons of large molecular weight have great numbers of isomers (see Table 25-1). There are three structural isomers of the hydrocarbon C_5H_{12} with the common names *n*-pentane, isopentane, and neopentane.

n-Pentane, or   Isopentane, or   Neopentane, or
pentane   2-methylbutane   2,2-dimethylpropane

$CH_3CH_2CH_2CH_2CH_3$ $CH_3CH_2CH(CH_3)CH_3$ $CH_3C(CH_3)_2CH_3$

Note that the following three structures, however, all represent the same molecule, *n*-pentane, and hence do not constitute separate isomers; they are identical.

Close examination of Fig. 25-10 will show that all of the hydrogen atoms in the —CH$_3$ groups at the end of a hydrocarbon chain are equivalent, since the carbon atom can rotate freely around the C—C bond.

The members of a second series of saturated hydrocarbons with the molecular formula C$_n$H$_{2n}$, where n is an integer, are called **cycloalkanes.** As the name implies, molecules of these substances are cyclic, possessing ring structures. The smallest member of the series is cyclopropane; next are cyclobutane, cyclopentane, etc.

Cyclopropane 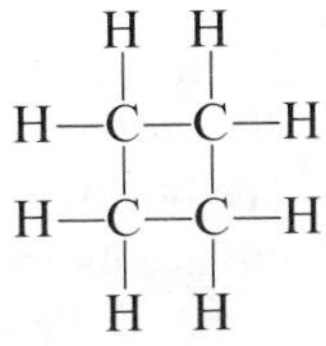   Cyclobutane 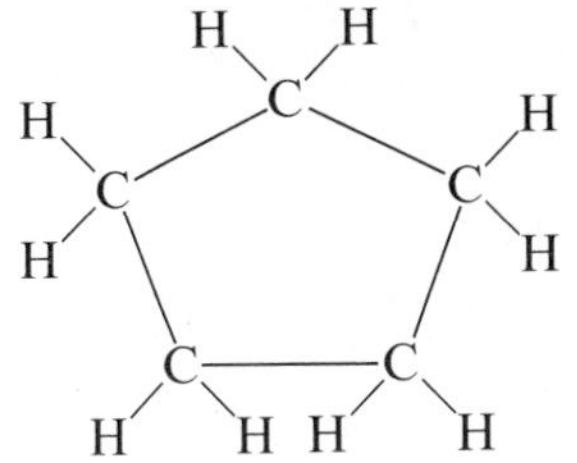   Cyclopentane

Note that cyclobutane is not an isomer of either n-butane or isobutane, because the molecular formula of cyclobutane is different.

The reactions of alkanes all involve breaking C—H or C—C single bonds. The following reaction is a typical one for the alkanes and is referred to as a substitution reaction.

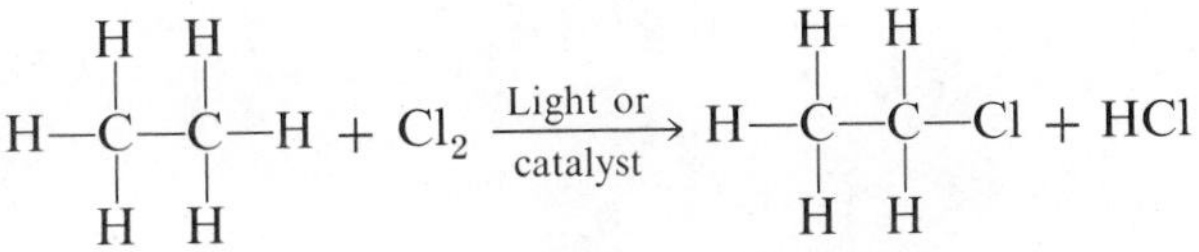

Ethane   Ethyl chloride

This reaction is a chain reaction that proceeds in several steps (Section 15.14). An important point concerning this particular reaction is that it transforms an alkane molecule into one that has a more reactive "functional group" on it—a halogen in this specific case. The functional group makes it possible for the molecule to take part in many different kinds of reactions.

Another very important reaction of alkanes is based upon their ability to burn in air in a highly exothermic oxidation-reduction reaction. A typical combustion reaction is

$$2C_2H_6(g) + 7O_2(g) \longrightarrow 4CO_2(g) + 6H_2O(g) \qquad \Delta H° = -2856 \text{ kJ}$$

Methane is the principal component of cooking gas. Gasoline is a mixture of straight and branched chain alkanes containing 5–9 carbon atoms plus various additives. Kerosene, diesel oil, and fuel oil are primarily mixtures of alkanes with higher molecular weights.

25.12   The Nomenclature of Saturated Hydrocarbons

The hydrocarbons, their isomers, and derivatives of hydrocarbons (compounds in which one or more hydrogen atoms have been replaced by other atoms or groups of atoms) can be named by a system of nomenclature devised by the International

Union of Pure and Applied Chemistry, IUPAC. The IUPAC nomenclature for saturated hydrocarbons is based on two rules.

1. The longest *continuous* chain of more than four carbon atoms in a saturated hydrocarbon is indicated by one of the following prefixes: five carbons, *penta;* six carbons, *hexa;* seven carbons, *hepta;* eight carbons, *octa;* nine carbons, *nona;* and ten carbons, *deca.* The suffix *ane* is added at the end (an *a* is dropped if two *a*'s follow in succession). A two-carbon chain is indicated as ethane, a three-carbon chain as propane, and a four-carbon chain as butane. Thus the names of the hydrocarbons listed in Table 25-1 indicate that these compounds, which are the normal isomers when more than one form exists, contain a single chain of carbon atoms with the number of carbon atoms in the chain equal to the number of carbon atoms in the molecule.

2. The position and the names of branches from the longest chain, or of atoms that replace hydrogen atoms on the chain (substituents), are added as prefixes to the name of the chain. The position of attachment is identified by the number of the carbon atom to which a substituent is attached; the number of the carbon atom is found by counting from the end of the chain nearest the substituent.

$$\overset{3}{C}H_3\overset{2}{C}H\overset{1}{C}H_3$$
$$|$$
$$Cl$$

2-Chloropropane

$$\overset{3}{C}H_3\overset{2}{C}H\overset{1}{C}H_3$$
$$|$$
$$CH_3$$

2-Methylpropane

$$\overset{6}{C}H_3\overset{5}{C}H_2\overset{4}{C}H\overset{3}{C}H_2\overset{2}{C}H\overset{1}{C}H_3$$
$$|\qquad\quad|$$
$$F\qquad\quad F$$

2,4-Difluorohexane

$$\overset{5}{C}H_3\overset{4}{C}H_2\overset{3}{C}Br_2\overset{2}{C}H_2\overset{1}{C}H_2Cl$$

3,3-Dibromo-1-chloropentane

Notice that *o* replaces *ide* at the end of the name of an electronegative substituent and that the number of substituents of the same type is indicated by the prefixes *di* (2), *tri* (3), *tetra* (4), etc. (for example, "difluoro" for two fluoride substituents). Also note that the longest continuous chain of carbon atoms in *n*-butane is four, but in isobutane the longest *continuous* chain of carbon atoms is only three even though there are a total of four carbon atoms in the molecule. Hence, by IUPAC nomenclature, *n*-butane is named butane, but isobutane becomes 2-methyl-propane.

A substituent that contains one less hydrogen than an alkane is called an **alkyl group.** The name of an alkyl group is obtained by dropping the *ane* of the alkane name and adding *yl.* For example, methane becomes *methyl;* ethane becomes *ethyl.*

Methane   (A methyl group)   Ethane   (An ethyl group)

The open bonds in the methyl and ethyl groups indicate that these groups are bonded to another atom.

Removal of any one of the four hydrogen atoms from a methane molecule forms a methyl group with the same geometry. These four hydrogen atoms are equivalent. Likewise, removal of any one of the six equivalent hydrogen atoms in ethane gives an ethyl group. However, there are two different "kinds" of hydrogen atoms in both propane and 2-methylpropane, in terms of the kinds of immediately adjacent atoms or groups of atoms.

Propane   2-Methylpropane

The six equivalent hydrogen atoms of one type in propane and the nine equivalent hydrogen atoms of one type in 2-methylpropane (all shown in black) are each bonded to a **primary carbon atom,** a carbon atom that is bonded to only one other carbon atom. The two colored hydrogen atoms in propane are of the second type. They differ from the six hydrogen atoms of the first type in terms of the nature of the carbon atoms to which they are bonded, since they are bonded to a **secondary carbon atom,** a carbon atom that is bonded to two other carbon atoms. The single colored hydrogen atom in 2-methylpropane similarly differs from the other nine hydrogen atoms in that molecule; it is bonded to a **tertiary carbon atom,** a carbon atom that is bonded to three other carbon atoms. Two different alkyl groups can arise in each of these molecules, depending upon which hydrogen atom is removed. The names and structures of these and several other alkyl groups are listed in Table 25-2. Note that alkyl groups cannot exist as stable independent entities. They are always combined with some other group.

TABLE 25-2 Some Alkyl Groups

Alkyl Group	Structure
Methyl	CH_3-
Ethyl	CH_3CH_2-
n-propyl	$CH_3CH_2CH_2-$
Isopropyl	$CH_3\overset{\shortmid}{C}HCH_3$
n-butyl	$CH_3CH_2CH_2CH_2-$
sec-butyl (where sec stands for secondary)	$CH_3CH_2\overset{\shortmid}{C}HCH_3$
Isobutyl	$CH_3\overset{\shortmid}{C}HCH_2-$ $\ \ \ \ \ CH_3$
t-butyl (where t stands for tertiary)	$CH_3\overset{\shortmid}{C}CH_3$ $\ \ \ \ \ CH_3$

The location of an alkyl group on a hydrocarbon chain is indicated in the same way as any other substituent. If more than one substituent is present, they are listed alphabetically or in order of increasing complexity.

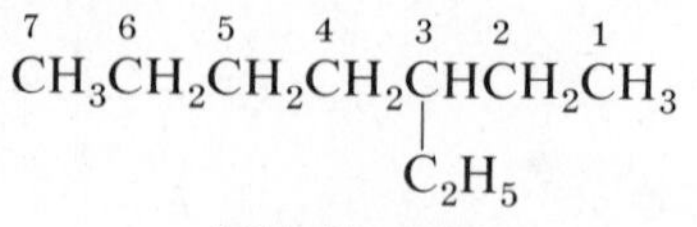

$$\overset{7}{C}H_3\overset{6}{C}H_2\overset{5}{C}H_2\overset{4}{C}H_2\overset{3}{C}HCH_2\overset{1}{C}H_3$$
$$\underset{C_2H_5}{|}$$

3-Ethylheptane

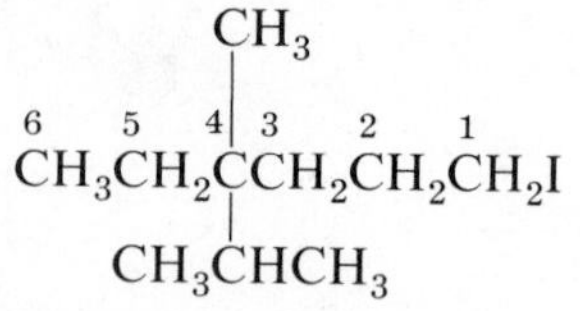

$$\overset{CH_3}{|}$$
$$\overset{6}{C}H_3\overset{5}{C}H_2\overset{4}{C}CH_2\overset{2}{C}H_2\overset{1}{C}H_2I$$
$$\underset{CH_3CHCH_3}{|}$$

4-Methyl-4-isopropyl-1-iodohexane

Note that in the IUPAC nomenclature system the longest continuous chain of carbon atoms is numbered in the direction that produces the lowest number or sum of the numbers used to indicate the substituted atoms or groups. For example, the six carbon atoms of the chain in the 4-methyl-4-isopropyl-1-iodo-hexane (above) could be numbered from left to right, resulting in the name 3-methyl-3-isopropyl-6-iodohexane; however, this would result in a sum of 12 for the numbers of the substituents, whereas the first name has a sum of 9 and is hence the preferred one.

Cycloalkanes are named in the same way as alkanes. The carbons are counted with a substituted carbon as the first carbon in the ring.

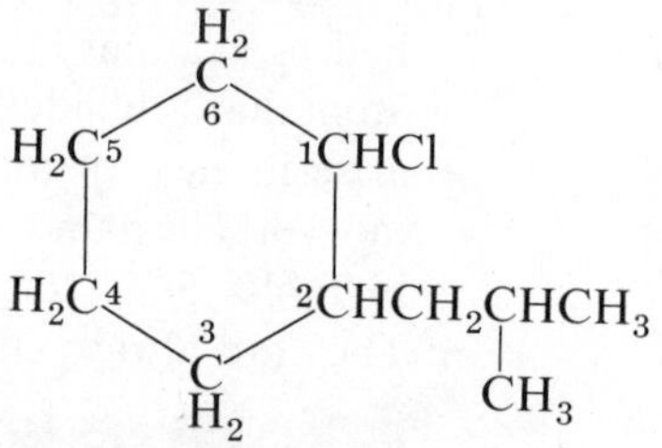

1-Chloro-2-isobutylcyclohexane

25.13   Alkenes

Hydrocarbon molecules that contain a double bond are members of a second homologous series, referred to as **alkenes.** As was pointed out in Chapter 7, the two carbon atoms linked by a double bond are bound together by two kinds of bonds, one σ C—C bond and one π C—C bond (see Section 7.6).

The alkenes have the general empirical formula C_nH_{2n}. **Ethene,** C_2H_4, commonly called ethylene, is the simplest alkene. Each carbon atom in ethene has a trigonal planar structure, as illustrated in Fig. 7-18, Section 7.6. The second member of the series is **propene** (propylene) and then come the butene isomers. The series builds up in a manner analogous to the alkane series.

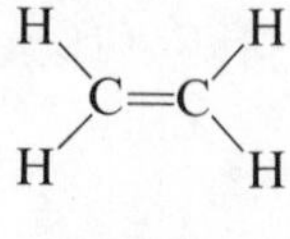

Ethene
(Ethylene)

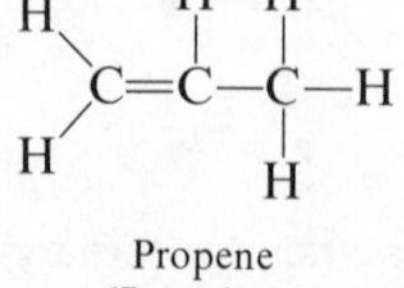

Propene
(Propylene)

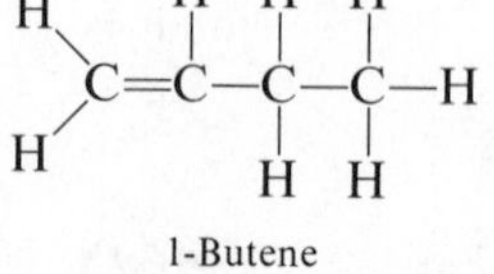

1-Butene

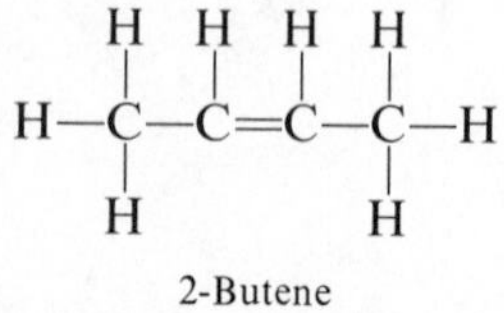

2-Butene

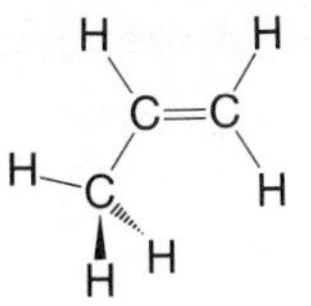

Figure 25-11 The molecular structure of propene.

The IUPAC names of the alkenes are derived from the name of the alkane with the same number of carbon atoms. The presence of a double bond in the chain is indicated by replacing the suffix *ane* by the suffix *ene*. The location of the double bond in the chain is indicated by a number that locates the position in the chain of the first carbon atom in the double bond.

The carbon atoms with single bonds in alkenes have sp^3 hybridization, and those involved in the double bond have sp^2 hybridization. Thus the geometry around the carbon atoms at the double bonds is trigonal planar. The molecular structure of propene is illustrated in Fig. 25-11.

Although the carbon atoms in a single bond are free to rotate around the bond, carbon atoms do not rotate around a double bond except under rare special conditions because a double bond is quite rigid. This makes it possible to separate two isomers of 2-butene, one with both methyl groups on the same side of the double bond and one with the two methyl groups on opposite sides of the double bond. Writing the formulas for the butene molecules in a slightly different fashion makes it possible to show the three butene isomers more clearly.

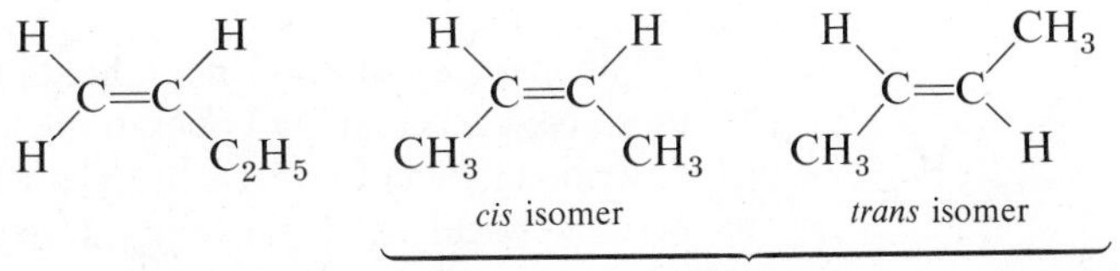

1-Butene   *cis* isomer   *trans* isomer

2-Butene

The 2-butene isomer that has the two CH_3 groups next to each other is referred to as the *cis* isomer; the one with the two CH_3 groups opposite each other is called the *trans* isomer. In these **geometric isomers,** the same types of atoms are attached to each other, but the geometries of the two molecules differ. The different geometries give different properties that make separation of the isomers possible.

Alkenes are much more reactive than alkanes. The π bond, being a relatively weaker bond, is disrupted much more easily than a σ bond. Thus the reactions characteristic of alkenes are those in which the π bond is broken.

In the presence of a suitable catalyst (Pt, Pd, or Ni, for example), alkenes add hydrogen to form the corresponding alkanes.

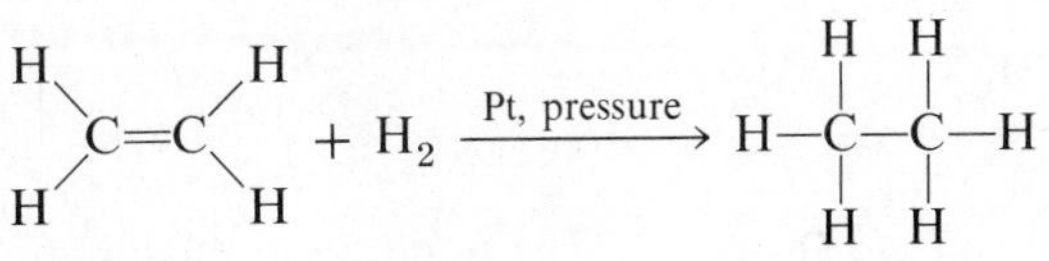

Ethene   Ethane

Chlorine breaks the double bond and adds to the two carbons adjacent to the double bond in the alkenes, instead of replacing a hydrogen as in the alkanes.

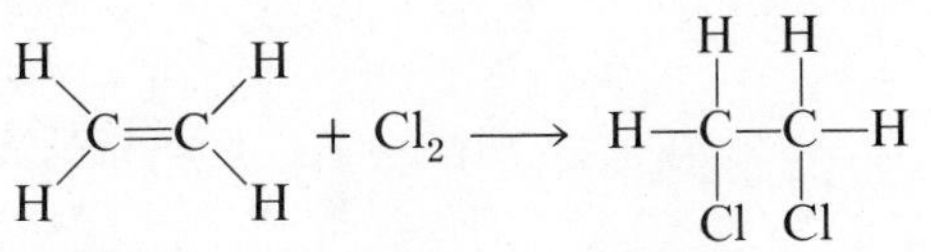

Ethene   1,2-Dichloroethane

Many other reagents react with alkenes by breaking the double bond. An example is the acid-catalyzed addition of water to an alkene.

$$\underset{\text{Ethene}}{\begin{array}{c}\text{H}\\\diagdown\\\text{H}\end{array}\text{C}=\text{C}\begin{array}{c}\text{H}\\\diagup\\\text{H}\end{array}} + \text{H}_2\text{O} \xrightarrow[\text{Catalyst}]{\text{Acid}} \underset{\substack{\text{Ethanol}\\\text{(Ethyl alcohol)}}}{\text{H}-\overset{\text{H}}{\underset{\text{H}}{\text{C}}}-\overset{\text{H}}{\underset{\text{OH}}{\text{C}}}-\text{H}}$$

A very important property of the alkenes is their ability to add to themselves. This reaction is a very significant one in the polymer industry.

$$n\left(\begin{array}{c}\text{H}\\\diagdown\\\text{H}\end{array}\text{C}=\text{C}\begin{array}{c}\text{H}\\\diagup\\\text{H}\end{array}\right) \xrightarrow{\text{Catalyst}} \underset{\text{Polyethylene}}{\left(\begin{array}{cc}\text{H} & \text{H}\\| & |\\\text{C}-\text{C}\\| & |\\\text{H} & \text{H}\end{array}\right)_n} \quad \text{(where } n \text{ is large)}$$

The majority of ethylene (ethene) used commercially is produced from alkanes by so-called "cracking" operations. The preparation of ethylene from ethane is described in Section 15.14. Ethylene is the basic raw material used in the polymer, petrochemical, and plastics industries. The production of 14,066,500 tons of ethylene in 1978 made it the sixth highest of all chemicals produced in the United States for that year.

25.14   Alkynes

Hydrocarbon molecules that contain a triple bond are called **alkynes;** they make up another series of unsaturated hydrocarbons. As was discussed in Chapter 7, two carbon atoms joined by a triple bond are bound together by one σ C—C bond and two π C—C bonds (see Section 7.7). The alkynes have the general empirical formula C_nH_{2n-2}.

The simplest and most important member of the alkyne homologous series is **ethyne,** C_2H_2, commonly called acetylene. The Lewis structure for ethyne is

$$\text{H}-\text{C}\equiv\text{C}-\text{H}$$

Ethyne
(Acetylene)

The bonding in the ethyne molecule, which has a linear structure, is illustrated in Fig. 7-20, Section 7.7.

The nomenclature of alkynes according to the IUPAC system is similar to that used for alkenes except that the suffix *yne* is used to indicate the presence of a triple bond in the chain.

$$\overset{4}{\text{CH}_3}\overset{3}{\text{CH}_2}\overset{2}{\text{C}}\equiv\overset{1}{\text{CH}} \qquad \overset{6}{\text{CH}_3}\overset{5}{\text{CH}_2}\overset{4}{\text{C}}\equiv\overset{3}{\text{C}}\overset{2}{\text{CH}}\overset{1}{\text{CH}_3}$$
$$\underset{\text{1-Butyne}}{} \qquad\qquad \underset{\substack{|\\ \text{CH}_3\\ \text{2-Methyl-3-hexyne}}}{}$$

Chemically, the alkynes are similar to alkenes except that with two π bonds they react even more readily, adding twice as much reagent in addition reactions. The reaction of acetylene with bromine is a typical example.

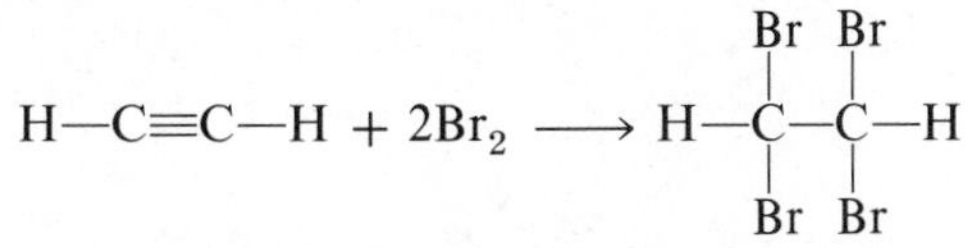

$$H-C\equiv C-H + 2Br_2 \longrightarrow H-\underset{\underset{Br}{|}}{\overset{\overset{Br}{|}}{C}}-\underset{\underset{Br}{|}}{\overset{\overset{Br}{|}}{C}}-H$$

Tetrabromoethane

Acetylene and all the other alkynes burn very easily.

$$2H-C\equiv C-H(g) + 5O_2(g) \longrightarrow 4CO_2(g) + 2H_2O(g) \qquad \Delta H° = -2511.2 \text{ kJ}$$

The triple bond is a high-energy bond, and this energy is released when the compound is transformed into carbon dioxide and water. Thus the flame produced by burning acetylene is very hot; it is used in welding and cutting metal. When acetylene burns, some of it breaks down a second way according to the equation $C_2H_2 \longrightarrow 2C + H_2$. At the temperature of the flame, the particles of carbon become heated to incandescence; i.e., they give off a brilliant luminous white light. At one time, acetylene was used to light homes and in lamps for bicycles and automobiles.

25.15 Aromatic Hydrocarbons

Benzene, C_6H_6, is the simplest member of a large family of hydrocarbons known as **aromatic hydrocarbons.** The benzene molecule consists of a hexagonal ring of sp^2 hybridized carbon atoms. The unhybridized p orbital on each carbon atom is perpendicular to the ring. The electronic structure of benzene can be described as a resonance hybrid of two Lewis structures.

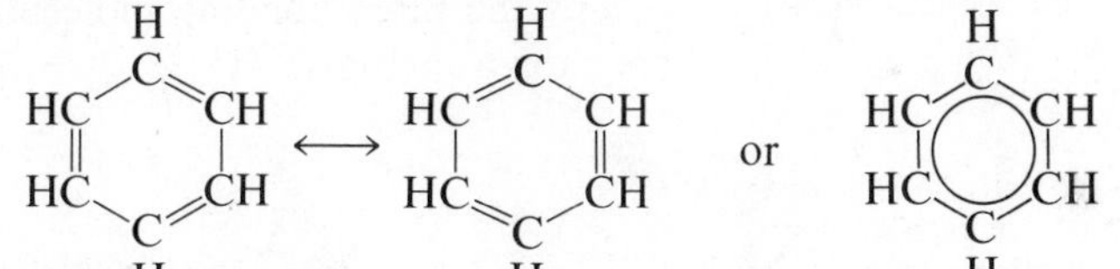

As shown in Fig. 25-12 and Fig. 25-13, the electrons in the π bonds in benzene are delocalized around the ring inasmuch as the electronic structure is the average of the resonance forms. As a result of this delocalization, benzene does not have the character of an alkene. Each bond between two carbon atoms is actually neither a single nor a double bond; each bond is equivalent to each of the others and is intermediate in character **(hybrid)** between a single and a double bond. It is said to have a bond order of $1\frac{1}{2}$. Since all six carbon-carbon bonds in a benzene ring are equivalent, the structure of benzene is often written with a circle (right-hand formula in Lewis structures shown above) to emphasize the fact that the electrons are delocalized.

The π bonding in benzene can also be described by the molecular orbital model. Three of the valence electrons on each carbon atom and the valence electron on

Figure 25-12   The π electronic structure of a benzene molecule. Hydrogen atoms have been omitted for clarity. The six sp^2 hybridized carbon atoms lie in a plane. The six remaining unhybridized p orbitals, shown in color in (a) and (c), extend above and below the plane and are perpendicular to it. The dashed lines between lobes of p orbitals indicate side-by-side overlap. The overlap can occur in either direction between adjacent lobes as shown in (a) and (c). The resulting π orbitals are shown in (b) and (d), with the orbital portion of each (in color) lying above and below the hexagonal plane. The resonance structures shown in (a) and (c) and again in (b) and (d) are exactly equivalent. The true structure is intermediate in character (hybrid) between the two (see Fig. 25-13).

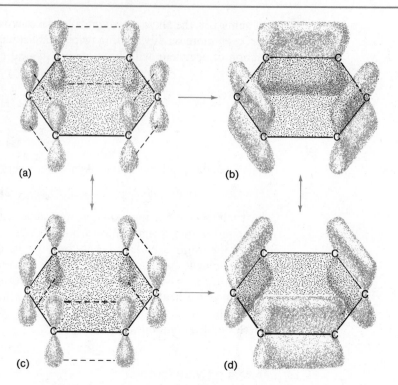

(a)   (b)

(c)   (d)

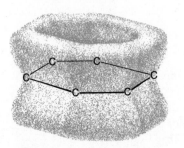

Figure 25-13   The distribution of π electrons in a benzene molecule.

Figure 25-14   The overlap of sp^2 hybrid orbitals of the carbon atoms and s orbitals of the hydrogen atoms that form σ bonds in a benzene ring.

each hydrogen atom are used to form the framework of σ bonds in the benzene molecule (Fig. 25-14). This leaves each unhybridized p orbital with one electron, giving a total of six electrons in the six unhybridized p orbitals. The six p orbitals can be combined to give three bonding π-type molecular orbitals and three π-type antibonding molecular orbitals. The six electrons (as three pairs of electrons) occupy the three bonding molecular orbitals. These bonding molecular orbitals are shown in Fig. 25-15.

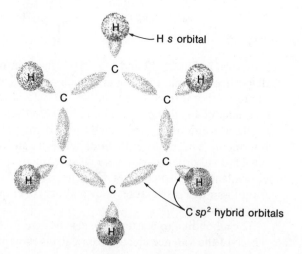

H s orbital

C sp^2 hybrid orbitals

Although benzene is usually written as one of its resonance forms in most chemical formulas, it must be remembered that all of the carbon-carbon bonds in benzene are equivalent.

There are many derivatives of benzene. The hydrogen atoms can be replaced by many different kinds of groups. The following are typical examples.

Toluene   Styrene   Bromobenzene

If the hydrogen atom on each of two different carbon atoms of benzene is replaced by a CH_3 group, the product is xylene. Thus, xylene also may be thought of as a derivative of benzene. Xylene has three isomers since substitutions to introduce the two CH_3 groups may be made in three different relative positions on the ring.

Orthoxylene   Metaxylene   Paraxylene

Any disubstituted benzene can take one of these three structurally different (isomeric) forms. The relative positions are indicated by the prefixes *ortho, meta,* and *para*.

Some of the important aromatic hydrocarbons and their derivatives contain more than one ring (polynuclear aromatic hydrocarbons). Examples are naphthalene, $C_{10}H_8$, familiarly known in moth balls; anthracene, $C_{14}H_{10}$; and benzpyrene, $C_{20}H_{12}$.

Naphthalene   Anthracene   Benzpyrene

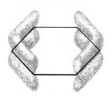

Figure 25-15 The three filled π molecular orbitals in benzene.

The aromatic hydrocarbons are widely used in the manufacture of dyes, synthetic drugs, explosives, plastics, and many other substances. For example, the important explosive trinitrotoluene (TNT), $C_6H_2(CH_3)(NO_2)_3$, is synthesized from toluene by replacing three hydrogen atoms with nitro groups, $-NO_2$.

Trinitrotoluene

25.16   Isomerism

Thus far we have seen two kinds of isomerism, structural isomerism and geometrical isomerism, both of which involve a set of molecules with the same molecular formula. Structural isomers (Section 25.11) are molecules with different arrangements of bonds. Examples of structural isomers are *n*-butane and 2-methylpropane, or the three isomers of xylene (Section 25.15). Geometric isomers have the same atoms bonded in the same order, but some parts of the two molecules are arranged differently in space relative to each other. In *cis-trans* isomers of organic compounds, for example, groups are arranged either on the same side (*cis*) or on different sides (*trans*) at either end of a bond that cannot rotate (Section 25.13).

A third type of isomerism is detected by using what is known as polarized light and is called **optical isomerism.** Optical isomers are detected by their effect on polarized light.

Ordinary light is composed of rays that vibrate in all planes parallel to the direction of travel of the rays. When ordinary light is passed through a polarizer such as found in Polaroid sunglasses or a Nicol prism, only light that vibrates in one plane is transmitted; such light is called **plane-polarized light.** The direction of the plane in which the light vibrates can be detected using another polarizer. When a beam of plane-polarized light passes through a solution containing molecules of a single form of an optical isomer, the plane of vibration of the light is rotated. Molecules that rotate the plane of plane-polarized light are said to be **optically active.** The amount of rotation and thus the optical activity varies from compound to compound. A compound that forms optical isomers exists as at least two different kinds of molecules (isomers). One of these isomers rotates plane-polarized light clockwise; such an isomer is called **dextrorotatory,** or a + isomer (*d*-isomer). The other isomer rotates plane-polarized light counterclockwise by an amount exactly equal to the clockwise rotation of the + isomer. The second isomer is called **levorotatory,** or a − isomer (*l*-isomer). A mixture of equal amounts of a dextrorotatory isomer and of a levorotatory isomer of the same substance shows no rotation of plane-polarized light because the effects of the two isomers cancel. In most properties such as melting point, boiling point, and solubility, optical isomers exhibit identical properties. However, + and − isomers

Figure 25-16 Mirror images of a glove, the molecule CHClBrI, and the molecule alanine. Although these objects have the same composition and are the same size and shape, their mirror images cannot be superimposed.

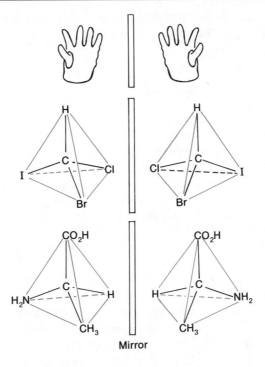

Mirror

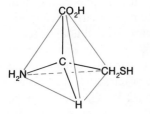

Figure 25-17 The structure of L-cystine.

may exhibit different degrees of reactivity with other optically active isomers.

Optical isomers of the same compound have the same relationship to each other as a right hand and a left hand. Both have the same shape and the same order of attachment of the various parts to each other; they look much alike. However, they differ in that they are mirror images of each other. Just as your right and left hands cannot be exactly superimposed, the two optical isomers of a molecule cannot be superimposed, no matter how they are rotated. Figure 25-16 illustrates the mirror relationship for a pair of gloves, for the molecule CHClBrI, and for the amino acid alanine. Note that simple molecules that contain sp^3 hybridized carbon atoms form optical isomers only when the carbon atom is bonded to four different groups; such a carbon atom within a molecule is referred to as an **asymmetric** carbon atom. The $CH(CH_3)_2Br$ molecule does not form optical isomers because no carbon atom within it is bonded to four different groups; interchanging the hydrogen, methyl, and bromine groups on the central carbon atom always produces a molecule that is superimposable on the first and hence identical to it.

Practically all molecules found in nature that can form optical isomers are found as only one kind of optically active isomer. For example, the amino acids found in proteins (Chapter 26) are all of one type designated L. The structure of L-cysteine, one of the amino acids, is shown in Fig. 25-17. A few D-amino acids are found in antibiotics, but these acids cannot be incorporated into proteins in living systems. A striking example of the effect of optical activity may be found in the oils that give the scent of spearmint and caraway seeds. Both oils are composed of carvone

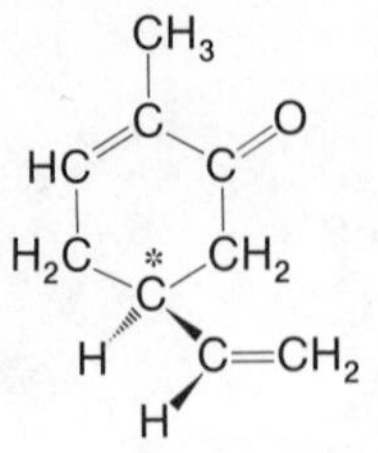

Figure 25-18 The molecule carvone. The asymmetric carbon atom (the one with four different kinds of groups attached) is indicated by the asterisk.

(Fig. 25-18). However, caraway seed oil contains D-carvone, and spearmint oil contains the other optical isomer, L-carvone.

25.17   Petroleum

Petroleum is composed chiefly of saturated (paraffin) hydrocarbons, but may also contain unsaturated hydrocarbons, aromatic hydrocarbons and their derivatives. The main step in the refining of petroleum is its separation by distillation into a number of fractions (fractional distillation, Section 13.5), each of which is a complex mixture of hydrocarbons and has properties that make it commercially valuable. The composition of each fraction depends upon the temperature range over which it is collected. Some of the more important products obtained from the refining of petroleum, classified according to number of carbon atoms, are listed in Table 25-3. Before the distillation products of petroleum are ready for use, further purification is usually necessary.

Ordinary gasoline is principally a mixture of *n*-hexane, *n*-heptane, *n*-octane, and their isomers.

Gasolines are rated upon an arbitrary scale in which isooctane (2,2,4-trimethyl-pentane), once thought to be the ideal fuel for the gasoline engine, is given the rating "100 octane." Gasolines with octane numbers less than 100 are less efficient than isooctane and those with octane numbers greater than 100 are more efficient. Gasolines of low octane number are more prone to cause engine knock than gasoline with higher octane number. Gasoline has in the past usually been made by blending **aliphatic** (open chain) and aromatic hydrocarbons and adding tetra-ethyl lead. However, manufacturers now are producing increasing quantities of gasoline without adding the tetraethyl lead because of possible toxic effects of lead compounds in the atmosphere and the poisoning effect of tetraethyl lead on the catalysts in the catalytic converters (Sections 22.4 and 22.5). Gasoline can be produced without tetraethyl lead by blending in more highly branched hydrocarbon isomers. Studies are being made of possible nonlead additives that will permit engines to run without knocking, two such additives being methanol and ethanol. At the same time, engines are being designed to run efficiently on lower octane gasolines, and to run on sources of power other than gasoline, such as electricity.

TABLE 25-3 Some Hydrocarbon Products from Petroleum			
	Approximate Composition	Boiling Point Range, °C	Uses
Petroleum ether	C_4 to C_{10}	35–80°	Solvent
Gasoline	C_4 to C_{13}	40–225°	Motor fuel
Kerosene	C_{10} to C_{16}	175–300°	Fuel, lighting
Lubricating oils	C_{20} up	350° up	Lubrication
Paraffin	C_{23} to C_{29}	50–60° (mp)	Candles, waxed paper
Asphalt		Viscous liquids	Paving, roofing
Coke		Solid	Fuel

Derivatives of the Hydrocarbons

In previous sections of this chapter we have seen examples of such derivatives of hydrocarbons as chloroethane, ethanol, and xylene. Derivatives of the hydrocarbons are formed, as noted previously, by replacing one or more hydrogen atoms of the hydrocarbon by other atoms or groups of atoms referred to as functional groups.

An alcohol is closely related to a hydrocarbon in that it contains an OH group in place of a hydrogen atom. Thus, methanol (common name, methyl alcohol), CH_3OH (Fig. 25-19a) is a derivative of methane; and ethanol (common name, ethyl alcohol), C_2H_5OH (Fig. 25-19b), is a derivative of ethane.

Figure 25-19 (a) The molecular structures of methane and methanol. (b) The molecular structures of ethane and ethanol.

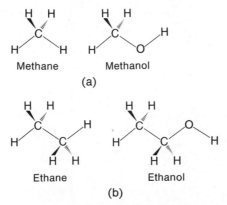

Methane   Methanol

(a)

Ethane   Ethanol

(b)

Table 25-4 lists several of the more important types of compounds derived from the hydrocarbons, the corresponding functional groups, and the general formula. In the general formula, R is the portion of the molecule other than the functional group; R might represent an alkyl group (Section 25.12) or a phenyl group (C_6H_5—), for example.

The following sections describe the properties of some specific compounds of hydrocarbon derivatives. The functional groups that are present should be noted carefully in all the examples discussed. The reactions of these hydrocarbon derivatives are summarized in Sections 25.24–25.28.

25.18   Alcohols, R—OH

Although all alcohols have one or more OH groups, they differ from bases such as sodium hydroxide and potassium hydroxide in that they do not furnish hydroxide ions in water, nor do they have the other usual properties of bases.

Methanol (methyl alcohol) is produced commercially from either carbon monoxide or carbon dioxide.

$$CO + 2H_2 \xrightarrow{\text{Ag or Cu}} CH_3OH$$

$$CO_2 + 3H_2 \longrightarrow CH_3OH + H_2O$$

Methanol is a colorless liquid, boiling at 65°. In odor and taste it is similar to ethyl alcohol. Methanol is very poisonous, however; intoxication, blindness, and death

TABLE 25-4   Several Types of Derivatives of the Hydrocarbons

Derivative Type	Functional Group	Type Formula	Typical Examples
Alcohols	—OH	R—OH	CH_3CH_2OH Ethanol (a primary alcohol) OH CH_3CHCH_3 2-Propanol (a secondary alcohol) OH CH_3CCH_3 CH_3 2-Methyl-2-propanol (a tertiary alcohol)
Ethers	—O—	R—O—R	$CH_3—O—C_2H_5$ $C_2H_5—O—C_2H_5$ Methyl ethyl ether Diethyl ether
Amines	$\overset{\|}{—N—}$	$R—N\overset{H}{\underset{H}{}}$	CH_3NH_2 Methyl amine (a primary amine)
		$R—\overset{R}{N}—H$	$(C_2H_5)_2NH$ Diethyl amine (a secondary amine)
		$R—\overset{R}{N}—R$	$(C_3H_7)_3N$ Tri-*n*-propyl amine (a tertiary amine)
Thiols	—SH	R—SH	C_2H_5SH Ethanethiol
Aldehydes	$\overset{O}{—CH}$	$R—\overset{O}{CH}$	$H\overset{O}{CH}$ $CH_3\overset{O}{CH}$ Formaldehyde Acetaldehyde
Ketones	$—\overset{O}{C}—$	$R—\overset{O}{C}—R$	$CH_3—\overset{O}{C}—CH_3$ $CH_3—\overset{O}{C}—C_6H_5$ Dimethyl ketone Methyl phenyl ketone
Acids	$—\overset{O}{C}OH$	$R\overset{O}{C}OH$	$CH_3\overset{O}{C}OH$ $C_6H_5\overset{O}{C}OH$ Acetic acid Benzoic acid

TABLE 25-4   (Continued)

Derivative Type	Functional Group	Type Formula	Typical Examples	
Esters	$\overset{\displaystyle O}{-\overset{\|}{C}OR}$	$R\overset{\displaystyle O}{\overset{\|}{C}}-OR$	$CH_3\overset{\displaystyle O}{\overset{\|}{C}}OC_2H_5$ Ethyl acetate	$C_3H_7\overset{\displaystyle O}{\overset{\|}{C}}OC_2H_5$ Ethyl butyrate
Salts	$\overset{\displaystyle O}{-\overset{\|}{C}O^-M^+}$	$R\overset{\displaystyle O}{\overset{\|}{C}}O^-M^+$	$CH_3\overset{\displaystyle O}{\overset{\|}{C}}O^-Na^+$ Sodium acetate	$C_2H_5\overset{\displaystyle O}{\overset{\|}{C}}O^-K^+$ Potassium pro-pionate
Amides	$\overset{\displaystyle O}{-\overset{\|}{C}-NH_2}$	$R\overset{\displaystyle O}{\overset{\|}{C}}NH_2$	$CH_3\overset{\displaystyle O}{\overset{\|}{C}}NH_2$ Acetamide	$C_6H_5\overset{\displaystyle O}{\overset{\|}{C}}NH_2$ Benzamide

may result when its vapors are breathed in quantities or when the liquid is taken internally.

Methanol is used in the manufacture of formaldehyde and other organic products; as a solvent for resins, gums, and shellac; and as a denaturant for ethyl alcohol (to make the ethyl alcohol unfit for human consumption).

Ethanol (ethyl alcohol), C_2H_5OH, is the most important of the alcohols. It is also known as **grain alcohol** or simply as **alcohol.** It has long been prepared from starch, cellulose, and sugars of certain plants by the process of fermentation.

$$C_6H_{12}O_6 \xrightarrow[\text{yeast}]{\text{Enzymes in}} 2C_2H_5OH + 2CO_2(g)$$

$$\text{Glucose} \qquad\qquad\qquad \text{Ethanol}$$

Large quantities of ethanol are produced synthetically from both ethylene and acetylene. The synthesis from ethylene is summarized by the equation

$$\underset{\text{H}\quad\text{H}}{H-C=C-H} + HOH \underset{}{\overset{H^+}{\rightleftharpoons}} \underset{\text{H}\quad\text{OH}}{\overset{\text{H}\quad\text{H}}{H-C-C-H}}$$

Ethanol is a colorless liquid with a characteristic and somewhat pleasant odor. It is miscible with water in all proportions. The boiling point of the pure alcohol is 78.37°, but it forms a constant-boiling mixture with water that contains 95.57% alcohol by weight and boils at 78.15°. Ethanol is the least toxic of all the alcohols and is present in all alcoholic beverages. It is used as a solvent in the preparation of tinctures, essences, extracts, and varnishes. It is used in the preparation of iodoform, ether, medicinals, dyes, perfumes, collodion, and solvents for the lacquer industry. It is used to some extent as a motor fuel and is currently in use as an additive to gasolines **(gasohol).**

Alcohols containing two or more hydroxyl groups can be made. 1,2-Ethanediol (ethylene glycol), $C_2H_4(OH)_2$, and 1,2,3-propanetriol (glycerol), $C_3H_5(OH)_3$, are important examples.

Ethylene glycol   Glycerol

Ethylene glycol is used as a solvent and as an antifreeze in automobile radiators. Glycerol (also called glycerin) is produced, along with soap, when either fats or oils are heated with alkali.

An alcohol is named by replacing the final e in the name of the hydrocarbon from which it is derived by the suffix ol and indicating the carbon atom to which the —OH group is bonded by a number preceding the name. An alcohol in which the —OH group is bonded to a primary carbon atom (Section 25.12) is called a **primary alcohol.** A **secondary alcohol** has the —OH group bonded to a secondary carbon atom, while a **tertiary alcohol** has the —OH group bonded to a tertiary carbon atom.

25.19 Ethers, R—O—R

The ethers are compounds obtained from alcohols, by the elimination of a molecule of water from two molecules of the alcohol. For example, when ethanol is treated with a limited amount of sulfuric acid and heated to 140°, **diethyl ether** (ordinary ether) is formed in a series of reactions which result also in the formation of water.

Diethyl ether

In the general formula for ethers, R—O—R, the hydrocarbon radicals (R) may be the same or different. Diethyl ether is the most important compound of this class. It is a colorless volatile liquid (boiling point 35°), and highly flammable as a vapor. It has been used since 1846 as an anesthetic. Diethyl ether and other ethers are valuable solvents for gums, fats, waxes, and resins.

25.20 Aldehydes, $R{-}\overset{\text{O}}{\underset{\|}{C}}{-}H$

The alcohols represent the first stage of oxidation of hydrocarbons. Further oxidation leads to the production of compounds containing the group, —CHO. These compounds are known as **aldehydes.** Aldehydes are generally prepared by the oxidation of primary alcohols. When a mixture of methanol and air is passed through a heated tube containing either silver or a mixture of iron powder and molybdenum oxide, an aldehyde called formaldehyde is formed.

$$2H-\overset{\overset{\displaystyle H}{|}}{\underset{\underset{\displaystyle H}{|}}{C}}-OH + O_2 \xrightarrow{\text{Catalyst}} 2H-\overset{\overset{\displaystyle H}{\diagup}}{\underset{\underset{\displaystyle O}{\diagdown}}{C}} + 2H_2O$$

Methanol   Formaldehyde

Formaldehyde, HCHO, is a colorless gas with a pungent and irritating odor; it is soluble in water in all proportions. Formaldehyde is sold in an aqueous solution which contains about 37–37.5% of formaldehyde by weight and is known as **formalin.** The solution also contains 7% of methyl alcohol, which is added to inhibit the reaction of formaldehyde molecules with each other to form an insoluble polymer (compound of high molecular weight). The formaldehyde polymer is *Bakelite,* an important material that makes formaldehyde an industrially important compound. Formaldehyde causes coagulation of proteins, making it useful as a preservative of anatomical specimens and in embalming fluids. It is also useful as a disinfectant and as a reducing agent in the production of silvered mirrors.

Acetaldehyde, CH_3CHO, is a liquid that boils at 20.2°. It is colorless, water-soluble, and has an odor like that of freshly cut green apples. It is used in the manufacture of aniline dyes, synthetic rubber, and other organic materials.

The systematic names of aldehydes are formed by replacing the *e* ending of the parent alkane by *al*. The chain is numbered with the —CHO group as 1. Thus $CH_3C(CH_3)HCH_2CHO$ is 3-methylpropanal. Acetaldehyde is the common name for ethanal.

25.21   Ketones, R—$\overset{\overset{\displaystyle O}{\|}}{C}$—R

Ketones resemble aldehydes in certain respects, since both contain the **carbonyl group,** $>C=O$. In fact, a ketone may be regarded as an aldehyde in which the hydrogen in the aldehyde group is replaced by a hydrocarbon radical. In the general structural formula for ketones (see section heading) the R groups may be the same or different. Ketones are often prepared by the oxidation of secondary alcohols.

Dimethyl ketone, CH_3COCH_3, commonly called **acetone,** is the simplest and most important ketone. It is made commercially by the fermentation of corn or molasses, or by the oxidation of petroleum gases. It is also one of the products of the destructive distillation of wood. Acetone is a colorless liquid, boiling at 56.5° and possessing a characteristic pungent odor and a sweet taste. Among the many uses of acetone are as a solvent for cellulose acetate, cellulose nitrate, acetylene, plastics, and varnishes; as a remover of paints, varnishes, and fingernail polish; and as a solvent in the manufacture of drugs, chemicals, smokeless powder, and the high explosive cordite.

25.22   Carboxylic Acids, R—$\overset{\overset{\text{O}}{\|}}{\text{C}}$—OH

The **carboxyl group,** —CO_2H, is characteristic of **carboxylic acids.** In general, the carboxylic acids are weak acids, yet they readily form metallic salts. Carboxylic acids may be prepared by the oxidation of primary alcohols or of aldehydes.

$$RCH_2OH \xrightarrow{\text{Oxid}} R-\overset{\overset{\text{H}}{|}}{\underset{\underset{\text{O}}{\|}}{C}} \xrightarrow{\text{Oxid}} R-\overset{\overset{\text{OH}}{|}}{\underset{\underset{\text{O}}{\|}}{C}}$$

Primary alcohol   Aldehyde   Carboxylic Acid

The simplest carboxylic acid is **formic acid,** HCO_2H, which was first isolated in 1670 by the distillation of red ants, and its name was derived from the Latin word *formicus* for ant. It is partially responsible for the irritation of ant bites and bee stings.

Acetic acid, CH_3CO_2H, constitutes 3 to 6% of vinegar. Cider vinegar is produced by allowing cider (apple juice) to ferment, causing the sugar present to change to ethyl alcohol. Then certain bacteria present in the juice produce an enzyme (see Chapter 26) that catalyzes the air oxidation of the alcohol to acetic acid. Pure, anhydrous acetic acid is a liquid boiling at 118.1°. It freezes at 16.6°, forming a solid resembling ice in appearance; for this reason the pure acid is usually called **glacial acetic acid.** Acetic acid has a penetrating odor and a sour taste, and produces painful burns on the skin. It is an excellent solvent for many organic compounds and some inorganic compounds. In addition to its use as a solvent, acetic acid is essential in the production of cellulose acetate and has wide application in the textile and rubber industries.

Oxalic acid is a dicarboxylic acid represented by the formula $(CO_2H)_2$, or $H_2C_2O_4$. Its molecule consists of two carboxyl groups (dicarboxylic) bonded together.

$$\underset{\underset{\text{O}}{\|}}{\overset{\overset{\text{HO}}{|}}{C}}-\underset{\underset{\text{O}}{\|}}{\overset{\overset{\text{OH}}{|}}{C}}$$

Oxalic acid

Oxalic acid is a colorless crystalline solid, which is found as a metal hydrogen salt in sorrel, rhubarb, and other plants to which it imparts a sour taste. In concentrated form it is poisonous.

Benzoic acid, $C_6H_5CO_2H$, is a colorless crystalline solid with the cyclic structural formula shown. It is a monocarboxylic acid that occurs in cranberries and coal

Benzoic acid

tar. The sodium salt of benzoic acid is used in the preservation of certain foods, such as tomato ketchup and fruit juices.

A type of organic acid commonly called a **fatty acid,** since this type of acid is commonly obtained from fats, results when a carboxyl group replaces one of the hydrogen atoms in a high molecular weight hydrocarbon. Examples are palmitic acid ($C_{15}H_{31}CO_2H$), stearic acid ($C_{17}H_{35}CO_2H$), and oleic acid ($C_{17}H_{33}CO_2H$).

25.23   Esters, R—$\overset{\overset{\text{O}}{\|}}{\text{C}}$—OR

Esters are the products of reaction of acids with alcohols. For example, the ester **ethyl acetate,** $CH_3CO_2C_2H_5$, is formed when acetic acid reacts with ethanol.

$$CH_3-\overset{\overset{\text{O}}{\|}}{\underset{\underset{\text{OH}}{}}{C}} \quad + HOC_2H_5 \longrightarrow CH_3-\overset{\overset{\text{O}}{\|}}{\underset{\underset{\text{O}-C_2H_5}{}}{C}} \quad + H_2O$$

The distinctive and attractive odors and flavors of many flowers and ripe fruits are due to the presence of one or more esters. Among the most important of the natural esters are fats (such as lard, tallow, and butter) and oils (such as linseed, cottonseed, and olive). Fats and oils are esters of the trihydroxyl alcohol glycerol, $C_3H_5(OH)_3$, utilizing such high molecular weight acids as palmitic acid, stearic acid, and oleic acid.

The Reactivity of Organic Compounds

Organic compounds generally react by breaking covalent bonds and forming other covalent bonds. It is possible to break a covalent bond between a carbon atom and another atom, X, in a molecule like R_3C—X in one of three ways depending upon the nature of the other atom.

1. The bonding electron pair can remain on the carbon atom giving a negatively charged R_3C^- ion, a **carbanion,** and a positively charged X^+ ion. Carbanions are generally strong Lewis bases.
2. The bonding electron pair in the C—X bond can remain on the X atom to give a positively charged R_3C^+ ion, a **carbonium ion,** or **carbocation,** and a negatively charged X^- ion. Carbonium ions are generally strong Lewis acids.
3. The C—X bond can break so that the carbon atom and the X atom each receives one of the bonding electrons. The resulting species, $R_3C\cdot$ and $\cdot X$, each with an unpaired electron are highly reactive species called **free radicals.**

The formation of a carbanion, a carbonium ion, or a free radical in an organic reaction depends upon the difference in electronegativity of C and X, the other reactants in the reaction, and the conditions under which the reaction is run. In the following sections we shall examine a few examples of the ways in which bonds break in different types of reactions of organic compounds.

25.24   Reactions Involving Nucleophilic Attack at Carbon

A carbon atom that is bonded to a more electronegative element such as nitrogen, oxygen, sulfur, or a halogen is located at the positive end of a polar bond (Section 5.9). Consequently, the carbon atom bears a fractional positive charge. When reactive species such as an ammonia molecule, a hydroxide ion, or a carbanion (Lewis bases that are rich in electrons) attack a compound containing a polar bond they approach the positive end of the bond and form a bond to the carbon atom. Such an attack on a partially positively charged atom is called **nucleophilic attack.** Nucleophilic steps are common steps in the mechanisms of the reactions of organic compounds. The reactions of the compounds given below all involve nucleophilic steps.

1. *Nucleophilic Reactions on Alkyl Halides.* Alkyl halides react with a variety of Lewis bases to form alcohols, thiols, amines, or ethers. The general reaction may be written

$$Y^- + \ -\overset{|}{\underset{|}{C}}{}^{\delta+}-X^{\delta-} \longrightarrow Y-\overset{|}{\underset{|}{C}}- \ + X^-$$

where Y^- is the **nucleophile,** the species that attacks the partially positive carbon atom, and X is a halogen atom. Specific examples include the following reactions:

$$OH^- \ + CH_3CH_2Br \longrightarrow CH_3CH_2OH + Br^-$$

Hydroxide ion   Bromoethane   Ethanol

$$CH_3O^- \ + CH_3CH_2CH_2I \longrightarrow CH_3CH_2CH_2OCH_3 + I^-$$

Methoxide ion   Iodopropane   Methyl propyl ether

$$HS^- \ + \ CH_3Br \longrightarrow CH_3SH \ + Br^-$$

Hydrogen sulfide ion   Bromomethane   Methanethiol

$$NH_3 \ + CH_3CHClCH_3 \longrightarrow \left[CH_3\overset{\overset{NH_3}{|}}{C}HCH_3 \right]^+ + Cl^-$$

Ammonia   2-Chloropropane   Isopropylammonium ion

Isopropylamine can be prepared from isopropylammonium chloride by reaction with base.

$$[C_3H_7NH_3{}^+]Cl^- + OH^- \longrightarrow C_3H_7NH_2 + Cl^- + H_2O$$

The reaction is analogous to the preparation of ammonia from ammonium chloride (Section 23.6).

2. *Nucleophilic Reactions on Alcohols.* Alcohols in which the —OH group is bonded to a secondary carbon atom (secondary alcohols) or in which the —OH group is bonded to a tertiary carbon atom (tertiary alcohols) react with hydrochloric or hydrobromic acid to give the corresponding alkyl halides. These reactions involve formation of a carbonium ion followed by nucleophilic attack on the carbonium ion.

$$\underset{\substack{\text{2-Methyl-2-butanol}}}{\overset{\overset{\displaystyle OH}{|}}{CH_3CH_2\underset{\underset{\displaystyle CH_3}{|}}{C}CH_3}} + HBr \longrightarrow \overset{\overset{\displaystyle \overset{+}{O}H_2}{|}}{CH_3CH_2\underset{\underset{\displaystyle CH_3}{|}}{C}CH_3} + Br^-$$

$$\overset{\overset{\displaystyle \overset{+}{O}H_2}{|}}{CH_3CH_2\underset{\underset{\displaystyle CH_3}{|}}{C}CH_3} \longrightarrow \underset{\substack{\text{Carbonium ion}}}{CH_3CH_2\overset{+}{\underset{\underset{\displaystyle CH_2}{|}}{C}}CH_3} + H_2O$$

$$Br^- + CH_3CH_2\overset{+}{\underset{\underset{\displaystyle CH_3}{|}}{C}}CH_3 \longrightarrow \underset{\substack{\text{2-Bromo-2-methylbutane}}}{CH_3CH_2\underset{\underset{\displaystyle CH_3}{|}}{\overset{\overset{\displaystyle Br}{|}}{C}}CH_3}$$

3. *Nucleophilic Reactions on Aldehydes and Ketones.* The reactions of aldehydes and ketones can give a wide variety of products depending upon the nucleophile that attacks the carbon in the **carbonyl group,** the C=O unit in the molecule. All of these reactions, however, involve the same process, addition of a Lewis base to the carbon of the carbonyl group by nucleophilic attack.

$$Y:^- + \underset{\underset{\displaystyle R'}{|}}{\overset{\overset{\displaystyle R}{|}}{C}}=O \longrightarrow Y-\underset{\underset{\displaystyle R'}{|}}{\overset{\overset{\displaystyle R}{|}}{C}}-O^-$$

The final product of the reaction depends upon how the intermediate anion in the process reacts in subsequent steps. For example, if Y^- is a carbanion, addition of water or acid in a subsequent step will give an alcohol.

$$CH_3^- + \underset{\substack{\text{Dimethyl ketone}\\\text{(Acetone)}}}{\underset{\underset{\displaystyle CH_3}{|}}{\overset{\overset{\displaystyle CH_3}{|}}{C}}=O} \longrightarrow CH_3-\underset{\underset{\displaystyle CH_3}{|}}{\overset{\overset{\displaystyle CH_3}{|}}{C}}-O^-$$

$$CH_3-\underset{\underset{\displaystyle CH_3}{|}}{\overset{\overset{\displaystyle CH_3}{|}}{C}}-O^- + H^+ \longrightarrow \underset{\substack{\text{2-Methyl-2-propanol}\\\text{(t-Butanol)}}}{CH_3-\underset{\underset{\displaystyle CH_3}{|}}{\overset{\overset{\displaystyle CH_3}{|}}{C}}-OH}$$

$$CH_3CH_2^- + \underset{\substack{\text{Ethanal}\\\text{(Acetaldehyde)}}}{\overset{\overset{\displaystyle O}{\|}}{CH_3CH}} \longrightarrow CH_3CH_2-\underset{\underset{\displaystyle H}{|}}{\overset{\overset{\displaystyle O^-}{|}}{C}}-CH_3$$

$$CH_3CH_2-\underset{\underset{\displaystyle H}{|}}{\overset{\overset{\displaystyle O^-}{|}}{C}}-CH_3 + H_2O \longrightarrow \underset{\substack{\text{2-Butanol}\\\text{(Isobutanol)}}}{CH_3CH_2-\underset{\underset{\displaystyle H}{|}}{\overset{\overset{\displaystyle OH}{|}}{C}}-CH_3} + OH^-$$

The reactions of acetylide $(RC{\equiv}C^-)$ and cyanide (CN^-) ions with aldehydes proceed in a similar fashion. With $HC{\equiv}C^-$

$$HC{\equiv}C^- + \underset{\substack{\text{Propanal}}}{\overset{\overset{\displaystyle O}{\|}}{CH_3CH_2CH}} \longrightarrow CH_3CH_2\underset{\underset{\displaystyle H}{|}}{\overset{\overset{\displaystyle O^-}{|}}{C}}-C{\equiv}CH$$

$$CH_3CH_2-\underset{\underset{\displaystyle H}{|}}{\overset{\overset{\displaystyle O^-}{|}}{C}}-C{\equiv}CH + H^+ \longrightarrow \underset{\substack{\text{3-Hydroxy-1-pentyne}}}{CH_3CH_2-\underset{\underset{\displaystyle H}{|}}{\overset{\overset{\displaystyle OH}{|}}{C}}-C{\equiv}CH}$$

4. *Nucleophilic Reactions on Carboxylic Acids.* In the presence of protons (which may be provided by the acid itself), a carboxylic acid reacts with an alcohol to give an ester. The mechanism of the reaction involves the initial protonation of the carbonyl oxygen of the acid

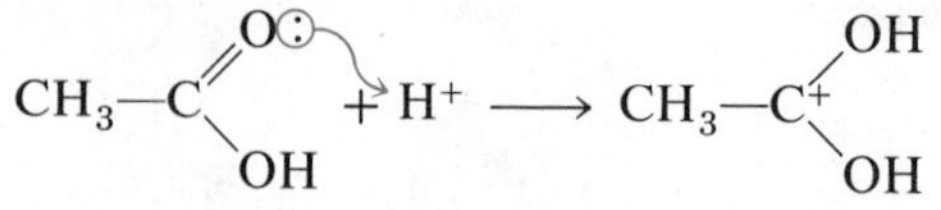

$$CH_3-C\overset{\displaystyle \ddot{O}\colon}{\underset{\displaystyle OH}{}} + H^+ \longrightarrow CH_3-\overset{\displaystyle OH}{\underset{\displaystyle OH}{\overset{+}{C}}}$$

Acetic acid

This protonation is followed by nucleophilic attack of the alcohol on the positive carbon atom

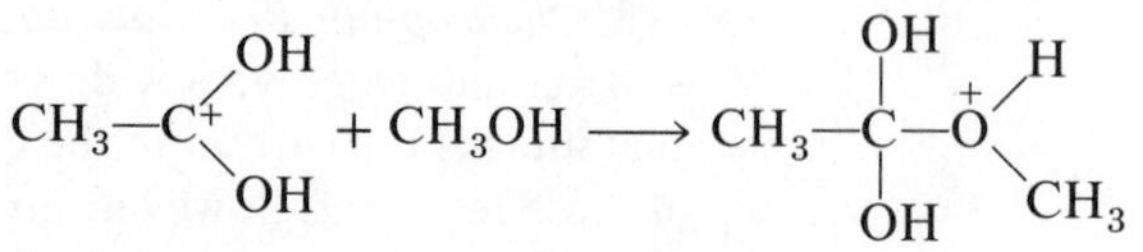

$$CH_3-\overset{\displaystyle OH}{\underset{\displaystyle OH}{\overset{+}{C}}} + CH_3OH \longrightarrow CH_3-\overset{\displaystyle OH}{\underset{\displaystyle OH}{C}}-\overset{+}{O}\overset{\displaystyle H}{\underset{\displaystyle CH_3}{}}$$

The resulting positive ion loses water spontaneously to give the ester.

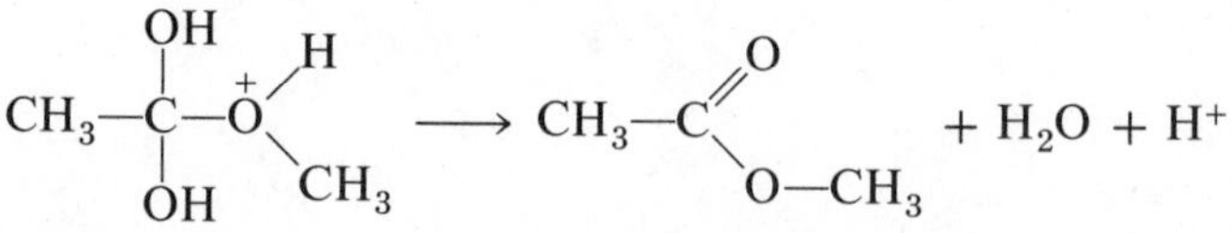

$$CH_3-\overset{\displaystyle OH}{\underset{\displaystyle OH}{C}}-\overset{+}{O}\overset{\displaystyle H}{\underset{\displaystyle CH_3}{}} \longrightarrow CH_3-C\overset{\displaystyle O}{\underset{\displaystyle O-CH_3}{}} + H_2O + H^+$$

Methyl acetate

5. *Nucleophilic Reactions on Carbon Dioxide.* Although they are not generally identified as such, the Lewis acid-base reactions of oxide ion and hydroxide ion with carbon dioxide are nucleophilic reactions which give the carbonate or hydrogen carbonate ions, respectively. The analogous reaction of a carbanion with carbon dioxide gives a salt of a carboxylic acid.

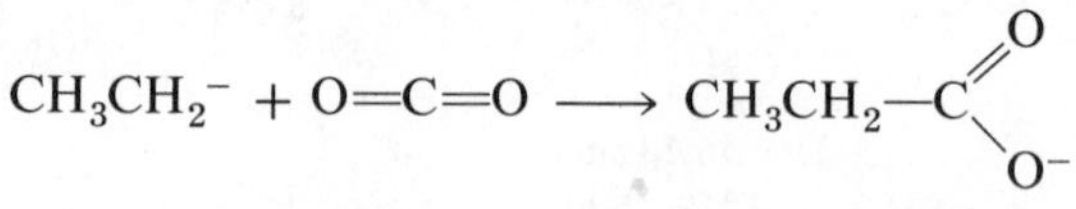

$$CH_3CH_2^- + O=C=O \longrightarrow CH_3CH_2-C\overset{\displaystyle O}{\underset{\displaystyle O^-}{}}$$

Propionate ion

Since carboxylic acids are weak acids, addition of an acid will convert the carboxylate anion to the carboxylic acid.

$$CH_3CH_2CO_2^- + H^+ \longrightarrow CH_3CH_2CO_2H$$

Propionate ion   Propionic acid

25.25   Reactions Involving Electrophilic Attack at Carbon

Carbon atoms that are involved in π bonds with other carbon atoms are rich in electrons and are subject to attack by Lewis acids. Such an attack of a Lewis acid on an electron-rich carbon atom is called an **electrophilic attack.**

The addition of a hydrogen halide to the multiple bond of an alkene or an alkyne is an electrophilic reaction involving an attack by H^+.

$$H^+ + CH_3-CH{=}CH_2 \longrightarrow CH_3-\overset{+}{C}H-CH_3$$

Propene   (Secondary carbonium ion)

$$CH_3-\overset{+}{C}H-CH_3 + Br^- \longrightarrow CH_3CHBrCH_3$$

2-Bromopropane

This reaction might have given another product with the bromine attached to the end carbon of the propane chain. However, in electrophilic addition reactions, the more electronegative group always ends up on the carbon atom of the multiple bond initially having the fewest hydrogen atoms. This rule was first formulated by the Russian chemist V. W. Markovnikoff and bears his name. Addition of two molecules of a hydrogen halide to an alkyne also obeys Markovnikoff's Rule.

$$CH_3CH_2C{\equiv}CH + 2HCl \longrightarrow CH_3CH_2\overset{\displaystyle Cl}{\underset{\displaystyle Cl}{\overset{|}{\underset{|}{C}}}}CH_3$$

1-Butyne   2,2-Dichlorobutane

25.26 Free Radical Reactions

A reaction that proceeds by breaking of a chemical bond to give free radicals is a **free radical reaction.** The mechanism of formation of ethene from ethane ($C_2H_6 \longrightarrow C_2H_4 + H_2$) was discussed in Section 15.14. This reaction involves the formation of the free radicals $\cdot H$, $\cdot CH_3$, and $\cdot CH_2CH_3$. The substitution reactions of alkanes (Section 25.11) are also free radical reactions. The reaction of ethane with chlorine involves the following steps:

$$Cl_2 \xrightarrow[\text{Heat}]{\text{Light or}} 2Cl\cdot$$

$$Cl\cdot + CH_3CH_3 \longrightarrow HCl + \cdot CH_2CH_3$$

$$\cdot CH_2CH_3 + Cl_2 \longrightarrow \cdot Cl + CH_2ClCH_3$$

The oxidation of hydrocarbons also involves a variety of complex free radical steps.

25.27 Acid-Base Reactions

Many organic compounds can be viewed as derivatives of simple inorganic molecules such as NH_3, H_2O, H_2S, and H_2CO_3. Like their inorganic "parents," these organic derivatives undergo acid-base reactions involving transfer of protons. Like water (HOH), alcohols (ROH) are weak acids and weak bases, although they are weaker proton acceptors than water. The reaction of an alcohol as an acid gives an **alkoxide ion, RO^-.** Since alcohols are weak acids, alkoxides are strong bases. An alcohol can be converted to an ionic alkoxide by reaction with an active metal.

$$2CH_3CH_2OH + 2Na \longrightarrow 2CH_3CH_2O^-Na^+ + H_2$$

Ethanol   Sodium ethoxide

Alcohols behave as bases toward *very* strong acids.

$$CH_3OH + H_2SO_4 \longrightarrow CH_3OH_2^+ + HSO_4^-$$

Thiols (RSH) may be regarded as derivatives of hydrogen sulfide (HSH). Thiols are stronger acids than alcohols or water; and a **thiolate anion** (RS$^-$) is a weaker base than RO$^-$ or HO$^-$.

The acidity and basicity of primary amines (RNH$_2$) and secondary amines (R$_2$NH) parallels that of ammonia (NH$_3$). An N—H bond is much less acidic than an O—H bond; thus the anions RNH$^-$ and R$_2$N$^-$ are extremely strong bases. These anions can be prepared by the reaction of the parent amines with lithium or sodium.

$$2CH_3NH_2 + 2Na \longrightarrow 2CH_3NH^-Na^+ + H_2$$

Methyl amine   Sodium
                                methylamide

$$2(CH_3CH_2)_2NH + 2Li \longrightarrow 2(CH_3CH_2)_2N^-Li^+ + H_2$$

Diethylamine   Lithium
                                diethylamide

The important property of amines is their basicity, which is due to the unshared pair of electrons on the nitrogen atom; amines are stronger bases than water or alcohols. With protons, amines give ammonium ions.

$$(CH_3)_3N\text{:} + H^+ \longrightarrow (CH_3)_3NH^+$$

Trimethylamine   Trimethylammonium ion

Carboxylic acids may be regarded as derivatives of carbonic acid in which an OH group is replaced by an organic group.

Carbonic acid   A carboxylic acid

Like carbonic acid, carboxylic acids are weak acids that will give salts with strong bases.

$$CH_3C + NaOH \longrightarrow CH_3CO_2^-Na^+ + H_2O$$

Acetic acid   Sodium acetate

A hydrogen bonded to a carbon atom involved in a triple bond is also acidic. Such alkynes react with active metals, giving ionic salts.

$$2CH_3C\equiv CH + 2Li \longrightarrow 2CH_3C\equiv C^-Li^+ + H_2$$

25.28   Oxidation-Reduction Reactions

The addition of oxygen to an organic molecule by direct reaction with O$_2$ is difficult to control and usually leads to a variety of products. Oxygen is usually introduced into a molecule to give an aldehyde, a ketone, or a carboxylic acid by oxidation of an alcohol with an oxidizing agent such as chromium(VI) oxide, CrO_3. Oxidation of a primary alcohol first gives an aldehyde, then a carboxylic acid.

$$CH_3CH_2CH_2OH \xrightarrow{[O]} CH_3CH_2\overset{\displaystyle O}{\overset{\|}{C}}H \xrightarrow{[O]} CH_3CH_2\overset{\displaystyle O}{\overset{\|}{C}}OH$$

1-Propanol   Propanal   Propanoic acid

The [O] in this equation represents an oxidizing agent such as CrO_3. Oxidation of a secondary alcohol gives a ketone

$$CH_3CH_2\overset{\displaystyle OH}{\overset{|}{C}}HCH_3 \xrightarrow{[O]} CH_3CH_2\overset{\displaystyle O}{\overset{\|}{C}}CH_3$$

2-Butanol   Methyl ethyl ketone

A number of organic compounds are produced by reduction. Active metals reduce alkyl halides, giving salts containing carbanions.

$$CH_3CH_2Cl + 2Na \longrightarrow CH_3CH_2{}^-Na^+ + NaCl$$

Chloroethane   Ethyl sodium

Amines can be prepared by the reduction of a nitrile ($R-C\equiv N$) using gaseous hydrogen catalyzed by metallic nickel.

$$CH_3CH_2C\equiv N + 2H_2 \xrightarrow{Ni} CH_3CH_2CH_2NH_2$$

Ethyl nitrile   Propylamine

Polymers

Polymers and plastics, compounds of very high molecular weight, are built up of a large number of simple molecules, or monomers, which have been caused to react with each other. Among the polymers that occur in nature are rubber, cellulose, starch, and proteins (see Chapter 26). Familiar synthetic polymers include synthetic rubber, nylon, rayon, polyethylene, and Dacron. Polymers are used in clothing, fibers, insulation, and construction materials. A recent development has been the use of polymers in automobile construction in order to save weight, and in turn, gasoline. In some 1979 model automobiles, for example, a savings of 100 pounds was realized by substituting polyurethane fascia for steel in bumpers. It is estimated that as much as 600 pounds of polymers may be used in each automobile by 1985. However, it should be noted that the polymers used to reduce the weight, thereby saving fuel in operating the automobile, are themselves made from petroleum. Thus part of the advantage is offset.

25.29   Rubber

Natural rubber is obtained mainly from the sap, called latex, of the rubber tree. Rubber consists of very long molecules, which are polymers formed by the union of isoprene units, C_5H_8.

$$CH_2\!=\!\underset{\underset{CH_3}{|}}{C}\!-\!CH\!=\!CH_2 + CH_2\!=\!\underset{\underset{CH_3}{|}}{C}\!-\!CH\!=\!CH_2$$

Isoprene   Isoprene

$$\cdots\!-\!CH_2\!-\!\underset{\underset{CH_3}{|}}{C}\!=\!CH\!-\!CH_2\!\vdots\!CH_2\!-\!\underset{\underset{CH_3}{|}}{C}\!=\!CH\!-\!CH_2\!-\!\cdots$$

The number of isoprene units in the rubber molecule is about 2000, giving it a molecular weight of approximately 136,000.

Rubber has the undesirable property of becoming sticky when warmed, but the stickiness is eliminated by the process of **vulcanization.** Rubber is vulcanized by heating it with sulfur to about 140°. During the process, sulfur atoms add at the double bonds in the linear polymer and form bridges that bind one rubber molecule to another. In this way, a linear polymer is converted into a three-dimensional polymer. During vulcanization, fillers are also added to increase the wearing qualities of the rubber and to yield colored products. Among the substances used as fillers are carbon black, zinc oxide, antimony(V) sulfide, barium sulfate, and titanium dioxide. Most rubber also contains oil and is said to be "oil extended."

Synthetic rubbers are ordinarily not identical with natural rubber, although they resemble it and often are superior to it in several properties. For example, neoprene is a synthetic elastomer with rubberlike properties.

$$n\,CH_2\!=\!CH\!-\!\underset{\underset{Cl}{|}}{C}\!=\!CH_2 \xrightarrow{\text{Polymerization}} \left[-CH_2\!-\!CH\!=\!\underset{\underset{Cl}{|}}{C}\!-\!CH_2\!-\right]_n$$

Chloroprene   Neoprene

The repeating unit, chloroprene, is similar to isoprene except for a chlorine atom instead of a methyl group, $-CH_3$. Neoprene is more elastic than natural rubber, resists abrasion well, and is less affected by oil and gasoline. It is used for making gasoline and oil hose, automobile and refrigerator parts, and electrical insulation. There are many other synthetic elastomers of this type.

25.30   Synthetic Fibers

A synthetic fiber familiar to everyone is *nylon,* which is a **condensation polymer.** Two kinds of molecules take part in the condensation reaction that produces nylon, each of which contains two indentical groups. The identical groups are $-NH_2$ and $-CO_2H$, respectively. One of the reactants is hexamethylenediamine and the other is adipic acid, which is a dicarboxylic acid. During condensation, molecules are formed of the type $R\!-\!NH\!-\!CO\!-\!R$, and water is eliminated. The part shown in color is an amide linkage.

$$NH_2\!-\!(CH_2)_6\!-\!\underset{\underset{H}{}}{\overset{\overset{H}{|}}{N}}\,\boxed{H + HO}\,OC\!-\!(CH_2)_4\!-\!CO_2H \longrightarrow$$

Hexamethylenediamine   Adipic acid

$$NH_2\!-\!(CH_2)_6\!-\!NH\!-\!CO\!-\!(CH_2)_4\!-\!CO_2H + H_2O$$

Because the molecule resulting from the condensation has the $—NH_2$ group at one end and the $—CO_2H$ group at the other, the condensation process can be repeated many times to form a linear polymer of great length.

$$(n + 1) \ \ H_2N(CH_2)_6NH_2 \ + (n + 1) \ HO_2C(CH_2)_4CO_2H \longrightarrow$$

Hexamethylenediamine   Adipic acid

$$H_2N(CH_2)_6NH \left[\overset{O}{\overset{\|}{C}}(CH_2)_4 \overset{O}{\overset{\|}{C}} NH(CH_2)_6NH \right]_n \overset{O}{\overset{\|}{C}}(CH_2)_4CO_2H + 2n \ H_2O$$

Nylon

Nylon may be made into fine threads by melting and extruding through a spinneret. It is used in making hosiery and other clothing, bristles for toothbrushes, surgical sutures, strings for tennis rackets, fishing line leaders, and many other products.

Dacron is made by a similar condensation process from ethylene glycol and terephthalic acid to form an ester linkage. Dacron is one of the family of polyesters.

Ethylene glycol   Terephthalic acid

Dacron

25.31 Polyethylene

The use of polyethylene in plastic bottles, bags for fruits and vegetables, and many other items has become commonplace. Polyethylene is a flexible, tough polymer that has high water resistance and excellent insulating properties. **Polyethylene** results from the polymerization of ethylene (Section 25.13).

If the ethylene molecules contain substituents, the polymer will form with the substituents attached.

$$n \ CH_2{=}CHR \xrightarrow{\text{Catalysts}} \left[CH_2{-}\underset{R}{\overset{}{CH}} \right]_n$$

Polypropylene ($R = CH_3$) is the polymer of propylene, $CH_3CH{=}CH_2$. **Polyvinyl-chloride, PVC** ($R = Cl$), is the polymer of vinyl chloride, $CH_2{=}CHCl$; **Teflon** results from polymerization of tetrafluoroethylene, $CF_2{=}CF_2$; and **polystyrene** is the polymer of styrene, $C_6H_5CH{=}CH_2$ ($R = C_6H_5$).

Questions

1. Describe the crystal structure of graphite and diamond and relate the physical properties of each to their structures.
2. What experimental evidence shows that diamond contains more internal energy than graphite? How is this fact related to the manufacture of synthetic diamonds?
3. Compare and contrast the bonding in graphite, benzene, ethylene, and acetylene.
4. Write the equation describing the chemical reaction that occurs when carbon dioxide is bubbled through sodium hydroxide solution.
5. Explain the toxicity of carbon monoxide.
6. Compare the heat of combustion of carbon monoxide to that of carbon. Explain.
7. Write the electronic structures for carbon monoxide and for phosgene.
8. Write balanced equations for the complete combustion of CH_4, CH_3OH, CH_2O_2, and C_2H_2.
9. How are "producer gas" and "water gas" manufactured?
10. Write equations for the production of carbon disulfide, carbon tetrachloride, calcium carbide, acetylene, cyanogen, and hydrogen cyanide.
11. On the basis of orbital hybridization in the carbon atoms, explain why cyclohexane exists in the form of a "puckered" (bent) ring and benzene in the form of a planar ring.
12. Write the general formulas for and the names of the hydrocarbon derivatives studied in this chapter.
13. What is the difference between the electronic structures of saturated and unsaturated hydrocarbons?
14. Why is CH_3CH_2OH soluble in water while its isomer CH_3OCH_3 is not?
15. Identify the functional groups present in the molecule carvone, Fig. 25-18.
16. Write the Lewis structures, indicating resonance structures where appropriate, for the following molecules or ions: ethane, propyne, methanol, formaldehyde, methanethiol, acetate ion, benzene.
17. Indicate the types of hybrid orbitals used by carbon and the geometry about each carbon atom in the following molecules: $H_2C=O$, C_2H_4, C_2H_6, $CH_3C\equiv CH$, CH_3CONH_2.
18. Draw three-dimensional structures, using wedge-shaped and dashed-line bonds where desirable, for the following molecules: C_3H_4, C_3H_8, CH_3CO_2H, n-C_4H_{10}, $C_6H_5CH_3$, CH_3OCH_3, $C_2H_5NH_2$.
19. Write Lewis structures for all of the isomers of C_4H_9Br. Do any of these isomers exist as optical isomers?
20. Write Lewis structures for all of the isomers of the alkene C_5H_{10}.
21. Write Lewis structures for all of the isomers of the alkyne C_5H_8.
22. Write the Lewis structures and the names of all isomers of the following alkyl groups: C_3H_7—, C_4H_9—.
23. Indicate which of the following molecules can form optical isomers:
   (a) 2-butanethiol
   (b) dichlorodifluoromethane
   (c) 1-cyanoethanol [$CH_3CH(CN)OH$]
   (d) trans-2-pentene
   (e) 3-ethyloctane
   (f) 4-ethyloctane
24. Draw three-dimensional structures, using wedge-shaped and dashed-line bonds, for the optical isomers of phenylalanine, $H_2NCH(R)CO_2H$, where R is $C_6H_5CH_2$.
25. Write Lewis structures for the following molecules: propane, 3-ethyl-1-heptyne, trans-4-chloro-2-octene, 2,3,4,-trimethylpentene, methylcyclohexane, ortho-dichlorobenzene, methyl ethyl ether, propanethiol, isopropyl alcohol.
26. Give complete names for the following compounds:

   (a) $CH_3CH_2CH_2CH=CHCH_3$

   (b) $CH_3CH_2CHBrCHICH_3$

   (c) $CH_3CH_2CHCH_3$
   $\quad\quad\quad\quad |$
   $\quad\quad CH_3CHCH_3$

   (d) $CH_3CH_2CHClCHCH_2CH_2CH_3$
   $\quad\quad\quad\quad\quad\quad\quad |$
   $\quad\quad\quad\quad\quad\quad CH=CH_2$

   (e) $(CH_3)_3CH$

   (f)
   $$H_2C \begin{array}{c} CH_2-CH_2 \\ \\ CH_2-CH_2 \end{array} CHCH_2CH(CH_3)CH_3$$

(g)

$$CH_3 \diagdown \diagup H$$
$$C=C$$
$$H \diagup \diagdown C_2H_5$$

(h) $(CH_3)_2CHCH_2C\equiv CH$

27. Does the following molecule exist as *cis-trans* isomers? optical isomers?

$$CH_2$$
$$H_2C \diagup \diagdown CHCl$$
$$H_2C \diagdown \diagup CHBr$$
$$CH_2$$

28. The molecular formula C_5H_{11} with an odd number of hydrogen atoms corresponds to a free radical. Write structures of all possible isomers for this formula.

29. Write the Lewis structures for all carbonium ions with the formula $C_4H_9^+$.

30. Write complete and balanced equations for the following reactions:
    (a) $CH_3I + OH^- \longrightarrow$
    (b) $CH_3CH_2I + NH_3 \longrightarrow$
    (c) Water is added to the product of the reaction of

    $$NaC_2H_5 \text{ with } CH_3CH_2\overset{\displaystyle O}{\overset{\displaystyle \|}{C}}CH_2CH_3.$$

(d) $CH_3CH_2CH_2CO_2H + (CH_3)_2CHCH_2OH \longrightarrow$
(e) $(CH_3)_2CH^-Na^+ + CO_2 \longrightarrow$
(f) Water is added to the product of the reaction of CN^- with $CH_3C(O)CH_3$.

31. Write complete and balanced equations for the following reactions:
    (a) $CH_3CH_2CH=CH_2 + HBr \longrightarrow$
    (b) $(CH_3)_2C=CH_2 + HCl \longrightarrow$
    (c) $CH_3C\equiv CH + 2HI \longrightarrow$

32. Write complete and balanced equations for the following reactions.
    (a) $(CH_3)_2CHCH_2OH + Na \longrightarrow$
    (b) $CH_3CH_2CO_2H + (CH_3)_3N \longrightarrow$
    (c) $(CH_3)_3CCH_2SH + NaOH \longrightarrow$
    (d) $(CH_3)_2NH + Li \longrightarrow$
    (e) $CH_3Li + H_2O \longrightarrow$
    (f) $CH_3OLi + CS_2 \longrightarrow$

33. Which alcohol can be oxidized to give the acid $(CH_3)_2CHCH_2CO_2H$? which aldehyde?

34. What are the products formed by the stepwise oxidation of methanol? of ethanol?

35. Does the reaction that leads to the formation of Dacron involve an electrophilic attack or a nucleophilic attack on carbon?

Problems

1. What volume of water gas at 600 K and 750 torr would result from the action of 100 g of water on carbon? *Ans. 554 ℓ*

2. How much $CaCO_3$ would be formed upon the addition of 10.0 ℓ of CO_2, measured at 27° and 770 torr, to an excess of a solution of calcium hydroxide? *Ans. 41.2 g*

3. How much heat energy would be released by the combustion of one gram of C_2H_6? C_2H_2? *Ans. 47.49 kJ g^{-1}; 48.22 kJ g^{-1}*

4. (a) What volume of $C_2H_2(g)$(STP) would result from the hydrolysis of 10.0 g of CaC_2? *Ans. 3.49 ℓ*
   (b) How many grams of magnesium would have to

react with HCl to provide enough hydrogen to reduce the C_2H_2 of part (a) to ethane? *Ans. 7.58 g*

5. (a) If all of the 24.65 billion pounds of ethylene produced in the United States during 1977 were produced by the pyrolysis of ethane $(C_2H_6 \longrightarrow C_2H_4 + H_2)$, what mass of hydrogen, in pounds, would be produced? *Ans. 1.771 billion pounds*
   (b) Is this enough hydrogen to prepare the 33.89 billion pounds of ammonia $(3H_2 + N_2 \longrightarrow 2NH_3)$ manufactured during 1978? *Ans. No; 6.017 billion pounds of H_2 was required*

6. (a) From the reactions

$$C_2H_4(g) + 2H_2(g) \longrightarrow 2CH_4(g)$$
$$C(g) + 2H_2(g) \longrightarrow CH_4(g)$$

and by considering the bonds that must be broken and formed to get from reactants to products, show that the C=C bond energy, D(C=C), is given by

$$D(C{=}C) = \Delta H^\circ_{f_{C(g)}} + \Delta H^\circ_{f_{CH_4(g)}} - \Delta H^\circ_{f_{C_2H_4(g)}}$$

Calculate, using data in Appendix J, a value for D(C=C).   *Ans. 590 kJ*

(b) By considerations similar to those in (a) show that the reaction

$$C_2H_4(g) + H_2(g) \longrightarrow C_2H_6(g)$$

leads to the conclusion that

$$D(C{=}C) = D(C{-}C) + 2D(C{-}H) \\ - D(H{-}H) + \Delta H^\circ_{f_{C_2H_6(g)}} - \Delta H^\circ_{f_{C_2H_4(g)}}$$

and calculate, using data in Table 18-2 and Appendix J, a value for D(C=C).   *Ans. 602 kJ*

7. Assuming a value of (n) equal to six in the formula for nylon, calculate how many pounds of hexamethylenediamine would be needed to produce 1.00 ton of nylon. (See Appendix C for conversion factors.)   *Ans. 1.02×10^3 lb*

8. How much acetic acid, in grams, is in exactly one quart of vinegar if the vinegar contains 3.00% acetic acid by volume? The density of acetic acid is 1.049 g/ml. (See Appendix C for conversion factors.)   *Ans. 29.8 g*

References

"The Shapes of Organic Molecules," J. B. Lambert, *Sci. American,* January 1970, p. 58.

"Criteria for Optical Activity in Organic Molecules," D. F. Mowery, Jr., *J. Chem. Educ.,* **46,** 269 (1969).

"Methanol" (Staff), *Chem. and Eng. News,* January 22, 1979, p. 13.

"Key Polymers: Polybutadiene" (Staff), *Chem. and Eng. News,* March 5, 1979, p. 11.

"Carbon Dioxide" (Staff), *Chem. and Eng. News,* June 26, 1978, p. 13.

"Formaldehyde" (Staff), *Chem. and Eng. News,* January 22, 1979, p. 14.

"Cyclohexane" (Staff), *Chem. and Eng. News,* March 26, 1979, p. 13.

"Ethylene Glycol" (Staff), *Chem. and Eng. News,* March 26, 1979, p. 12.

"Dimethyl Terephthalate/Terephthalic Acid" (Staff), *Chem. and Eng. News,* March 26, 1979, p. 11.

"Organic Chemicals: Angels or Goblins?" L. N. Ferguson, *J. Chem. Educ.,* **55,** 553 (1978).

"Future Sources of Organic Raw Materials," I. S. Shapiro, *Science,* **202,** 287 (1978).

"Ethylene by Naphtha Cracking," P. Wiseman, *J. Chem. Educ.,* **54,** 154 (1977).

"The Chemistry of Oil Recovery from Bituminous Sands," M. B. Hocking, *J. Chem. Educ.,* **54,** 725 (1977).

"Chemicals from Wood," P. E. Childs, *Educ. in Chem.,* May 1978, p. 79.

"Guayule, The Rubber Shrub," E. Campos-Lopez *et al., Chemtech,* **9,** 50 (1979).

"Donor-Acceptor Interactions in Organic Chemistry," S. G. Sunderwith, *J. Chem. Educ.,* **47,** 728 (1970).

"Control Elements in Organic Synthesis," S. Turner, *Chem. in Britain,* **7,** 191 (1971).

"Kinetics and Mechanisms in Organic Chemistry," R. Baker, *Chem. in Britain,* **4,** 250 (1968).

"Organometallic Chemistry," M. F. Lappert, *Chem. in Britain,* **5,** 342 (1969).

"The Nature of Polymeric Materials," H. F. Mark, *Sci. American,* September 1967, p. 149.

"Plastics, Utilizing the Properties of String-like Molecules," (Staff), *J. Chem. Educ.,* **56,** 42 (1979).

"The Carbon Cycle," B. Bolin, *Sci. American,* September 1970, p. 124.

"The Stereochemical Theory of Odor," J. E. Amoore, J. W. Johnston, Jr., and M. Rubin, *Sci. American,* February 1964, p. 42.

"Flower Pigments," S. Clevenger, *Sci. American,* June 1964, p. 85.

"Molecular Isomers in Vision," R. Hubbard and A. Kropf, *Sci. American,* June 1967, p. 64.

"The Synthesis of Diamond," H. T. Hall, *J. Chem. Educ.,* **38,** 484 (1961).

"Hydrocarbon Reactions on Transition Metal Catalysts," J. J. Rooney, *Chem. in Britain,* **2,** 242 (1966).

"Kekulé and Benzene," C. A. Russell, *Chem. in Britain,* **1,** 141 (1965).

"Free Radicals and Aging," W. A. Pryor, *Chem. & Eng. News,* June 7, 1971; p. 34.

"Artificial Organs," H. J. Sanders, *Chem. and Eng. News:* Part

I, April 6, 1971, p. 32. Part II, April 12, 1971, p. 68.

"Catalysis," V. Haensel and R. L. Burwell, Jr., *Sci. American,* December 1971, p. 46.

"The Nature of Aromatic Molecules," R. Breslow, *Sci. American,* August 1972, p. 32.

"The Gasification of Coal," H. Perry, *Sci. American,* March 1974, p. 19.

"Asymmetric Synthesis," J. W. Scott and D. Valentine, Jr., *Science,* **184,** 943 (1974).

"Superhard Materials," F. P. Bundy, *Sci. American,* August 1974, p. 62.

"Spatial Configuration of Macromolecular Chains," P. J. Flory, *Science,* **188,** 1268 (1975).

"Cyanate and Sickle-Cell Disease," A. Cerami and C. M. Peterson, *Sci. American,* April 1975, p. 45.

"Discovery of the Anesthetic Properties of Chloroform," C. G. Moseley, *J. Chem. Educ.,* **55,** 581 (1978).

"Emil Fischer's Discovery of Phenylhydrazine," G. B. Kauffman and R. P. Cinla, *J. Chem. Educ.,* **54,** 295 (1977).

"Formation of Nitrosamines in Food and in the Digestive System," J. S. Wishnok, *J. Chem. Educ.,* **54,** 440 (1977).

"Pittacal—The First Synthetic Dyestuff," G. B. Kauffman, *J. Chem. Educ.,* **54,** 753 (1977).

"1-Butanethiol and the Striped Skunk," K. K. Andersen and D. T. Bernstein, *J. Chem. Educ.,* **55,** 159 (1978).

"The Discovery of Saccharin," J. H. Wotiz, *J. Chem. Educ.,* **55,** 161 (1978).

"Sweet Organic Chemistry," R. W. Bragg, *et al., J. Chem. Educ.,* **55,** 281 (1978).

"Isoniazid—Destroyer of the White Plague," G. B. Kauffman, *J. Chem. Educ.,* **55,** 448 (1978).

"A Pill for Birth Control," D. Kolb, *J. Chem. Educ.,* **55,** 591 (1978).

26

Biochemistry

Biochemistry is the study of the chemical composition and structure of living organisms and of the chemical reactions that take place within these organisms. Considering the diverse and difficult functions that a cell must carry out, it is not surprising to find that the molecules contained in living cells are among the most complicated substances known.

The biomolecules are often divided into four classes—proteins, carbohydrates, lipids, and nucleic acids—on the basis of similarities in either structure or function. Proteins, carbohydrates, and nucleic acids are found in the form of macromolecules with molecular weights of 10^4 to 10^6, or larger. Molecules of this size and complexity are formed by the polymerization of monomers of relatively low molecular weight. Molecules of the fourth major class, the lipids, are considerably smaller than those of the three other classes, although they can associate to form clusters of molecules of enormous size.

26.1 The Cell

All cells are basically similar. They contain many of the same structures, they have the same or similar systems of enzymes, and they contain the same kind of genetic material. From the standpoint of mass, the principal constituent of living matter is water. Most plants and animals contain 60 to 90% water by mass. The principal function of water in living organisms is that of a solvent, in which both organic and inorganic substances can be dissolved and therefore transported either through the cell or from cell to cell in the organism.

697

Figure 26-1 Simplified drawing of a eukaryotic cell.

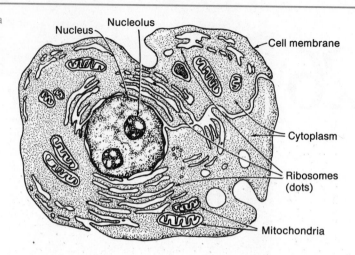

An understanding of the role played by water in biological systems requires that a distinction be made between substances that are **hydrophilic** (literally water-loving), and **hydrophobic** (water-hating). Water is a polar solvent that readily dissolves either polar or ionic substances. In general, as a substance becomes less polar, it becomes more hydrophobic. The nonpolar hydrocarbons are excellent examples of hydrophobic compounds, and they are almost totally immiscible with water. Some biomolecules contain polar or ionic substituents that impart hydrophilic character, and they are therefore water soluble. Those biomolecules that do not contain polar or ionic substituents are hydrophobic and are effectively insoluble in water.

All cells, regardless of their origin, are surrounded by a **cell membrane** whose primary function is to insulate the cell from the fluid that surrounds the cell (Fig. 26-1). There is considerable evidence to suggest that the inner and outer surfaces of the cell membrane are different, providing one of the means whereby the cell establishes and maintains a difference between the concentration of various substances within the cell and in the extracellular fluid. An erythrocyte, or red-blood cell, for example, may contain 10 to 100 times more potassium ion, and 10 to 100 times less sodium ion, than the fluid surrounding the cell. Bacteria and plant cells, but not animal cells, contain a **cell wall,** which surrounds the cell membrane. The cell wall is composed primarily of carbohydrates, such as cellulose, and is quite porous. The cell wall provides some of the rigidity of plant tissue, and it appears to connect cells, thereby facilitating the exchange of material between cells.

Cells are commonly divided into two classes on the basis of the presence or absence of a discrete cell **nucleus. Eukaryotic cells** such as that represented in Fig. 26-1 contain a nucleus, **prokaryotic cells** do not. The nucleus, when present, contains the chromosomes, which are a mixture of two substances, deoxyribonucleic acid (DNA) and protein. The nucleus serves as the storage center for the genetic information that the cell requires for replication. The nucleus also directs the synthesis of the various proteins that the cell needs for biological activity. The **nucleolus** is a small, dense body within the nucleus; it is rich in ribonucleic acid

(RNA), containing as much as 20% of the total RNA of the cell, and serves as the site for the synthesis of ribosomal RNA.

The **cytoplasm** is the name given to the entire contents of the cell, with the exception of the cell nucleus. It can in turn be divided into the **cytosol,** which represents the water-soluble contents of the cell, and a variety of cellular organelles such as the mitochondria, ribosomes, and chloroplasts. All cells contain **ribosomes,** which serve as the site of the directed polymerization (genetic coding) of amino acids to form enzymes and other proteins. The **mitochondria,** which are present in eukaryotic cells, can be considered to be the "power plant" of the cell. The mitochondria serve as the site for the degradation of a variety of biomolecules, including the oxidation of carbohydrates, amino acids, and fatty acids to carbon dioxide and water. A portion of the energy released during this oxidation is stored as chemical energy through the synthesis of certain chemical compounds known as the nucleoside triphosphates, for example, ATP and GTP. **Chloroplasts** are present in all higher plants and algae that are capable of photosynthesis. The chloroplast represents the site where light-induced synthesis of carbohydrates from carbon dioxide and water takes place.

26.2   Amino Acids

When proteins are heated with aqueous acid or base, or are treated with certain hydrolytic enzymes, they undergo a hydrolysis reaction (Section 12.3, part 5), producing compounds called **amino acids.** Each member of this class of compounds contains both an amine group, $-NH_2$, and a carboxylic acid function, $-CO_2H$ (Section 25.22). At least 150 naturally occurring amino acids are known; the most important, however, are the 20 genetically coded amino acids used in the synthesis of proteins. The genetically coded amino acids are all **α-amino acids,** which means that the amino group is attached to the carbon atom adjacent, or α, to the carbon of the carboxylic acid group. All but one of the genetically coded amino acids are primary amines in which the nitrogen is bound to only a single carbon atom. Proline is the only genetically coded amino acid that contains a secondary amine, a nitrogen that is bound to two carbon atoms. The primary α-amino acids can be represented by the following structure:

$$\overset{\displaystyle R}{\underset{\displaystyle |}{H_3N^+ - CH - CO_2^-}}$$

Since the structures of these amino acids differ only in the nature of the R substituent, they are usually grouped on the basis of similarities in the structure of the R side chain. The names and structures of the genetically coded amino acids are given in Fig. 26-2.

It is tempting to write the formulas of the amino acids such as alanine (abbreviated as ALA) in the form

$$\overset{\displaystyle CH_3}{\underset{\displaystyle |}{H_2N - CH - CO_2H}}$$

However, it is important to recognize that the basic H_2N- end of this molecule

Figure 26-2 The 20 genetically coded amino acids at $pH = 7$ (pp. 700–702).

Name	Structure	Isoelectric Point (pI)
Nonpolar R Groups		
Glycine (GLY)	$H_3N^+ - \overset{\overset{\displaystyle H}{|}}{C}H - CO_2^-$	5.97
Alanine (ALA)	$H_3N^+ - \overset{\overset{\displaystyle CH_3}{|}}{C}H - CO_2^-$	6.00
Valine (VAL)	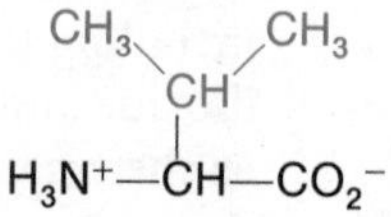 $H_3N^+ - \overset{\overset{\displaystyle CH}{\overset{\diagup \hspace{4pt} \diagdown}{CH_3 \hspace{10pt} CH_3}}}{C}H - CO_2^-$	5.96
Leucine (LEU)	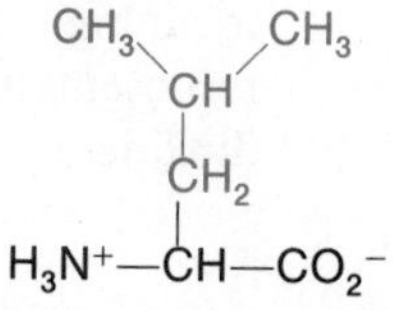 $H_3N^+ - CH - CO_2^-$ with side chain $CH_2 - CH(CH_3)_2$	5.98
Isoleucine (ILE)	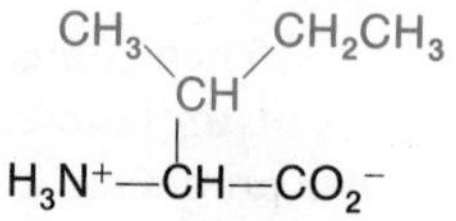 $H_3N^+ - CH - CO_2^-$ with side chain $CH(CH_3)(CH_2CH_3)$	6.02
Proline (PRO)	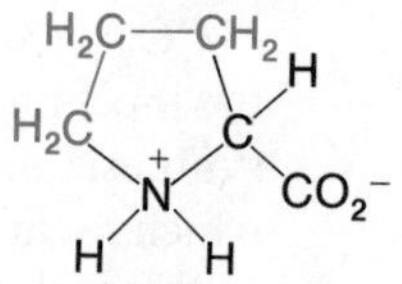 pyrrolidine ring structure	6.30
Aromatic R Groups		
Phenylalanine (PHE)	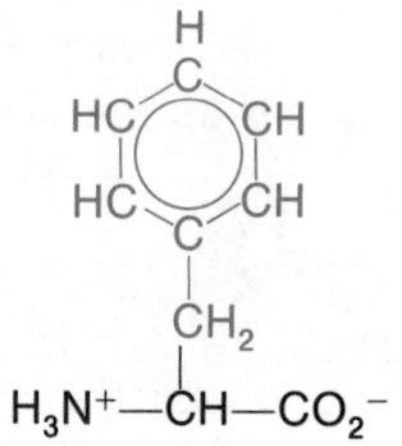 $H_3N^+ - CH - CO_2^-$ with side chain CH_2—phenyl	5.48
Tyrosine (TYR)	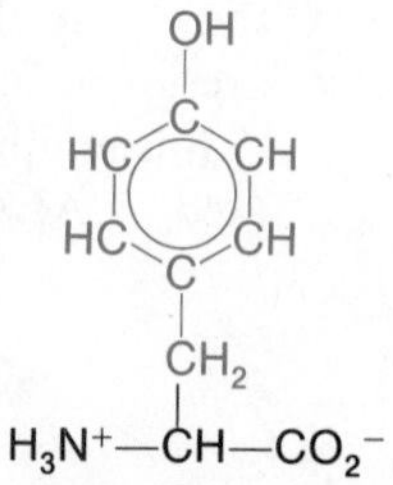 $H_3N^+ - CH - CO_2^-$ with side chain CH_2—(4-hydroxyphenyl)	5.66

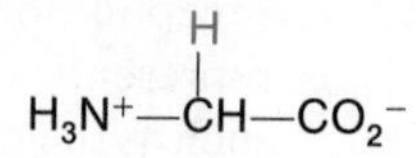

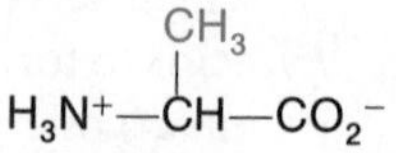

Name	Structure	Isoelectric Point (pI)
Aromatic R Groups (Cont.)		
Tryptophan (TRP)	$H_3N^+\!-\!CH\!-\!CO_2^-$	5.89
Carboxylic Acid or Amide R Groups		
Aspartic Acid (ASP)	CO_2^- CH_2 $H_3N^+\!-\!CH\!-\!CO_2^-$	2.77
Asparagine (ASN)	$\overset{\displaystyle O}{\underset{}{C}}\!-\!NH_2$ CH_2 $H_3N^+\!-\!CH\!-\!CO_2^-$	5.41
Glutamic Acid (GLU)	CO_2^- CH_2 CH_2 $H_3N^+\!-\!CH\!-\!CO_2^-$	3.22
Glutamine (GLN)	$\overset{\displaystyle O}{\underset{}{C}}\!-\!NH_2$ CH_2 CH_2 $H_3N^+\!-\!CH\!-\!CO_2^-$	5.65
Basic R Groups		
Lysine (LYS)	NH_3^+ CH_2 CH_2 CH_2 CH_2 $H_3N^+\!-\!CH\!-\!CO_2^-$	9.74

Name	Structure	Isoelectric Point (pI)
Basic R Groups (Cont.)		
Arginine (ARG)	NH_2 $C=NH_2^+$ NH CH_2 CH_2 CH_2 $H_3N^+—CH—CO_2^-$	10.76
Histidine (HIS)	(imidazole ring) CH_2 $H_3N^+—CH—CO_2^-$	7.59
Alcoholic R Groups		
Serine (SER)	CH_2OH $H_3N^+—CH—CO_2^-$	5.68
Threonine (THR)	$CH_3\quad OH$ CH $H_3N^+—CH—CO_2^-$	5.64
Sulfur-Containing R Groups		
Cysteine (CYS)	CH_2SH $H_3N^+—CH—CO_2^-$	5.07
Methionine (MET)	CH_3 S CH_2 CH_2 $H_3N^+—CH—CO_2^-$	5.74

acts as a proton acceptor, while the —CO_2H end functions as a proton donor, to form a species known as a **zwitterion** (from German, meaning *hybrid ion*). Such zwitterions possess the property of containing ionic charges within the molecule, which is electrically neutral overall.

$$H_2\ddot{N}-\underset{\underset{CH_3}{|}}{CH}-CO_2\text{(H)} \longrightarrow H_3N^+-\underset{\underset{CH_3}{|}}{CH}-CO_2^-$$

Zwitterion

The zwitterion is the dominant form for amino acids in aqueous solution at or near neutral pH ($pH = 7$). In the presence of an acid, however, the amino acid becomes positively charged as the carboxylate ion is protonated. Conversely, upon addition of a base, the amino acid becomes negatively charged as the amino group is deprotonated.

$$H_3N^+-\underset{\underset{CH_3}{|}}{CH}-CO_2H \underset{OH^-}{\overset{H^+}{\rightleftharpoons}} H_3N^+-\underset{\underset{CH_3}{|}}{CH}-CO_2^- \underset{H^+}{\overset{OH^-}{\rightleftharpoons}} H_2N-\underset{\underset{CH_3}{|}}{CH}-CO_2^-$$

Acidic
solution   Neutral
solution   Basic
solution

When an electric current is passed through an acidic aqueous solution of an amino acid (low pH), the positively charged ion migrates toward the negative electrode. At high pH (basic solution), the negatively charged amino acid migrates toward the positive electrode. At some intermediate pH the amino acid has no net charge, and it therefore does not move in the presence of an electric field. This intermediate pH is known as the *p*I, or **isoelectric point** of the amino acid. The *p*I of simple amino acids is slightly less than 7 since the α-NH_3^+ ion is slightly more acidic than the α-RCO_2^- ion is basic. Amino acids whose R groups contain acidic substituents, such as aspartic acid and glutamic acid, have unusually low values for *p*I. The predominant form of glutamic acid, for example, in aqueous solutions at neutral pH is negatively charged.

$$H_3N^+-\underset{\underset{CH_2CH_2CO_2^-}{|}}{CH}-CO_2H$$

Thus the pH of this solution must be significantly less than 7 before the isoelectric point is reached. Conversely, amino acids whose R groups contain basic substituents have unusually high values for *p*I, since the amino acid is still positively charged at neutral pH. For example, at neutral pH, lysine exists in the form

$$H_3N^+-\underset{\underset{CH_2CH_2CH_2CH_2NH_3^+}{|}}{CH}-CO_2^-$$

The variation in the charge on amino acids with pH provides a method for separating amino acids via ion exchange (Section 12.12). A mixture of amino acids at low pH is poured onto a column that contains a resin with strongly acidic sulfonic acid substituents. The positively charged amino acid molecule displaces protons from the sulfonic acid substituents, and then binds to the resultant negative sites.

$$\langle Resin\rangle-SO_3H + H_3N^+\overset{\overset{\displaystyle R}{|}}{C}HCO_2H(aq) \longrightarrow \langle Resin\rangle-SO_3^-\left\{H_3N^+\overset{\overset{\displaystyle R}{|}}{C}HCO_2H\right\} + H^+(aq)$$

The column is then washed with buffered solutions of gradually increasing pH. In general, the amino acids with low values of pI will wash through the column first, since they are converted to their zwitterion form at relatively low pH, while the amino acids with high values of pI come off last.

With the exception of glycine, the α-amino acids are optically active (Section 25.16). A pair of optical isomers, which differ primarily in their interaction with plane-polarized light, can be imagined for each optically active substance. As pointed out previously, the dextrorotatory (d, or $+$) isomer rotates plane-polarized light to the right, whereas the levorotatory (l, or $-$) isomer rotates light to the left. Unfortunately, there is no simple relationship between the molecular structure of an isomer and its classification as either d or l. Optical isomers whose molecular structures are related to the structure of the ($+$) isomer of glyceraldehyde, HOCH$_2$CH(OH)CHO, are labeled D isomers, while their mirror images, which are related structurally to the ($-$) isomer of glyceraldehyde, are labeled L isomers. Every amino acid isolated from a naturally occurring protein is found to have the same molecular configuration. Such amino acids are invariably found to be L amino acids. Looking down the H—C bond, we see that the —CO$_2^-$, R, and —NH$_3^+$ substituents are arranged in a clockwise fashion. In D amino acids these substituents would be arranged in a counterclockwise manner (Fig. 26-3). Though occasionally found in nature, D amino acids are not genetically coded into proteins.

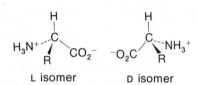

L isomer   D isomer

Figure 26-3 The molecular structures of the D and L isomers of an amino acid.

26.3   Peptides and Proteins

The structures of peptides and proteins were first determined by the German chemist Emil Fischer between 1900 and 1910. **Peptides** are polymers formed from two or more amino acids linked by covalent amide bonds (Table 25-4) that have been given the special name **peptide bonds.** The nature of the peptide bond (in color) can be seen in the following structure for a dipeptide.

$$H_3N^+-\overset{\overset{\displaystyle R}{|}}{C}H-\overset{\overset{\displaystyle O}{\|}}{C}-NH-\overset{\overset{\displaystyle R}{|}}{C}H-CO_2^-$$

The name peptide is usually restricted to polymers that contain only a small number of amino acids, and such polymers are often called **oligopeptides** (from the Greek *oligo,* meaning *few*). Polymers containing between approximately 10 and 40 amino acids are then called **polypeptides.** Relatively large polypeptides, containing 40 to 10,000 or more amino acid residues, are called **proteins.**

The peptide bond results from combining the α-carboxyl end of one amino acid with the α-amine end of another. Starting with two different amino acids, for example, glycine (GLY) and alanine (ALA), two different dipeptides can be formed.

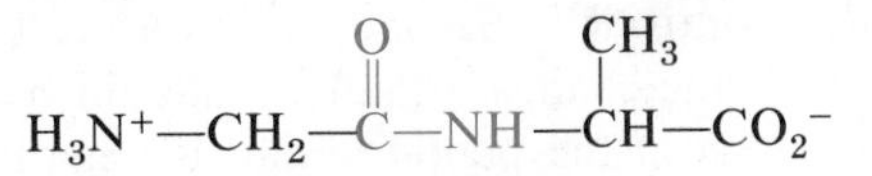

$$H_3N^+-CH_2-\overset{\overset{\displaystyle O}{\|}}{C}-NH-\overset{\overset{\displaystyle CH_3}{|}}{CH}-CO_2^-$$

Glycylalanine
(GLY-ALA)

$$H_3N^+-\overset{\overset{\displaystyle CH_3}{|}}{CH}-\overset{\overset{\displaystyle O}{\|}}{C}-NH-CH_2-CO_2^-$$

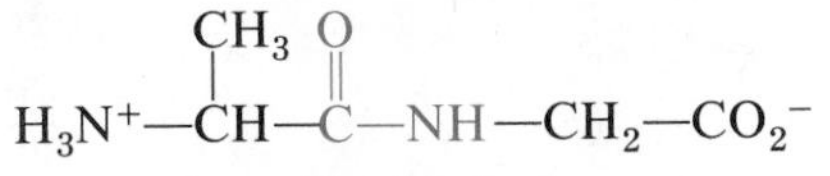

Alanylglycine
(ALA-GLY)

The glycine residue in the dipeptide GLY-ALA has a free amine, and this residue is therefore called the **N-terminal** amino acid. In the dipeptide ALA-GLY, however, the glycine residue is the **C-terminal** amino acid. Peptides and proteins are both written and named starting with the N-terminal amino acid. The tetrapeptide

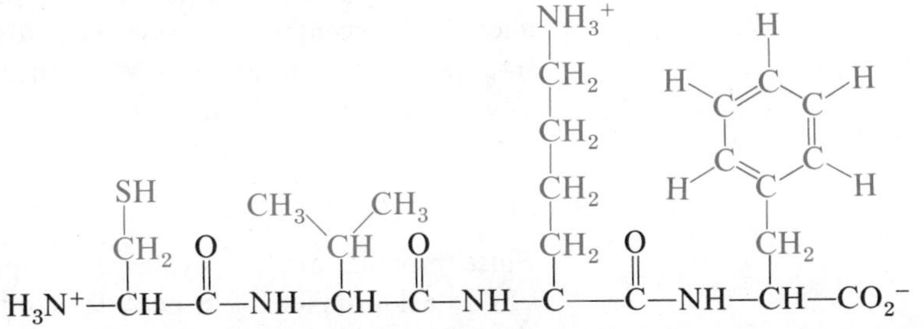

is therefore called cysteinylvalyllysylphenylalanine, or CYS-VAL-LYS-PHE. The side chains of the four amino acids in this tetrapeptide are shown in color.

There is no reason why peptide linkages cannot also form at the side chains of certain amino acids. In fact, a limited number of oligopeptides are known in which such branched peptide linkages are formed; for example, the tripeptide glutathione contains glutamic acid bonded through the carboxylic acid group in the side chain.

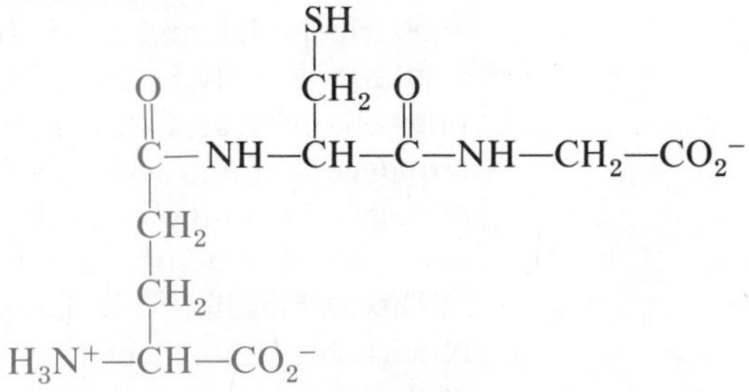

Glutathione

Proteins, however, seldom form peptide bonds at amino acid side chains.

It would be very useful to be able to synthesize proteins at will, since a variety of diseases result from the inability of the body to synthesize one or more proteins, for example, hemophilia, sickle-cell anemia, and related hereditary diseases. However, even the ability to synthesize oligopeptides and small proteins would be beneficial, since a number of these substances are either important antibiotics or hormones. The hormone insulin, for example, which is required for the control of glucose metabolism, and which is used in the treatment of diabetes, is a relatively small protein containing only 51 amino acid residues. Unfortunately, however, the synthesis of peptides or proteins is a difficult task. There are three major obstacles to overcome. First, the order in which the amino acids are incorporated must be rigorously controlled. Second, reaction must be prevented from occurring at the amino acid side chains. Finally, it should be recognized that the free energy for the hydrolysis of the peptide bond is negative,

$$H_3N^+—\overset{R}{\underset{|}{CH}}—\overset{O}{\underset{||}{C}}—NH—\overset{R}{\underset{|}{CH}}—CO_2^- + H_2O \longrightarrow H_3N^+—\overset{R}{\underset{|}{CH}}—CO_2^- + H_3N^+—\overset{R}{\underset{|}{CH}}—CO_2^-$$
$$\Delta G^\circ = -17 \text{ kJ/mol}$$

and therefore the formation of a peptide bond is an uphill battle. Because the chemical synthesis of large amounts of peptides or proteins has not been very successful, recent efforts have been devoted to using the protein biosynthesis apparatus of bacterial cells to synthesize mammalian peptides or proteins via genetic engineering techniques.

26.4   The Structures of Proteins

Since proteins are simply large polypeptides containing a linear sequence of amino acids linked by α-peptide bonds, and there are only a limited number of amino acids with very similar structures that are used in the synthesis of proteins, we might expect that the structures of proteins would be remarkably similar. It is therefore somewhat surprising to note an extraordinary diversity in protein structures. This diversity in the structure of proteins can be understood by recognizing that the sequence of amino acid side chains on the polypeptide backbone controls the way in which this chain folds upon itself in three dimensions. There are only 20 amino acids that are incorporated during the synthesis of a protein. However, this means that there are 20^2 or 400 possible dipeptides, 20^3 or 8000 tripeptides, and 20^4 or 160,000 tetrapeptides. For a protein even as small as insulin, with only 51 amino acid residues, there are 2.3×10^{66} possible combinations of amino acid sequences. Since the folding of a protein is controlled by the sequence of amino acid side chains, and there is an almost infinite variability in the sequence of amino acids in proteins, an essentially infinite number of three-dimensional structures are available for proteins.

The extraordinary variations possible in the structures of proteins allow for remarkable differences in their type and in their function. Among the types of proteins and their functions are the following:

1. *Enzymes.* Enzymes, which catalyze biochemical reactions, invariably contain a protein component.
2. *Hormones.* A number of oligopeptides and small proteins, such as insulin, bind to specific sites in or on the cell to stimulate biological activity.
3. *Structural proteins.* A wide variety of proteins are involved in the formation of various kinds of connective tissue including skin, hair, horn, and hoof.
4. *Food-storage proteins.* The caseins found in mammalian milk, the albumins in eggs, and a variety of proteins in seeds have as their primary function the storage of food energy.

5. *Transport proteins*. Proteins are involved in the transport of both organic and inorganic nutrients across cell membranes. The metal-containing proteins, or metalloproteins, such as the cytochromes, are involved in electron transport.

6. *Antibodies*. The antibodies produced by the immune response system are proteins.

7. *Muscle tissue*. Muscles are composed of fibrous bundles of the proteins actin and myosin, which are involved in the contraction and relaxation of muscle tissue.

The most fundamental, or **primary,** structure of a protein is simply the sequence of amino acids, listed as if printed on ticker tape, starting with the N-terminal amino acid and proceeding toward the C-terminal residue. This polypeptide chain many then fold upon itself to produce a **secondary** structure, in which the number of hydrogen bonds between peptide linkages is maximized. The peptide bond is best described as the resonance hybrid of the following Lewis structures:

There are three important consequences of this resonance. First, the partial double-bond character restricts rotation around the carbon-nitrogen bond in the same way that rotation around the carbon-carbon double bond in an alkene is restricted (Section 25.13). Second, the double-bond character requires that the peptide bond, and the atoms immediately adjacent to this bond, lie in the same plane. This in turn limits the number of ways in which the protein chain can fold upon itself. Third, the increased polarity of the peptide bond increases the basicity of the carbonyl oxygen, and the acidity of the amide hydrogen, thereby increasing the strength of the hydrogen bonds between peptides.

$$\diagdown N{-}H^{\delta+}{\cdots}^{\delta-}O{=}C\diagdown$$

There are only a limited number of secondary structures that a polypeptide chain can adopt that maximize the number of such hydrogen bonds between peptide linkages. Two of the most common forms of secondary structure are the **α-helix** (Fig. 26-4) and the **β-** or **pleated sheet** (Fig. 26-5). In the α-helix, the polypeptide backbone forms a helical coil with approximately 3.6 amino acids per 360° turn. The α-helix is the dominant secondary structure in fibrous proteins such as collagen or the α-keratins of hair and wool. The β-sheeted structure results from hydrogen bonds between adjacent strands of the polypeptide backbone. Extended arrays of the β secondary structure are found in the β-keratins such as silk.

Proteins often contain regions of both helical and pleated-sheet structures that are folded into a complex conformation known as the **tertiary** structure

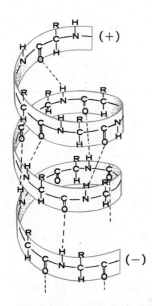

Figure 26-4 The α-helical structure of a peptide chain, one form of the secondary structure of protein molecules. The hydrogen bonds that generate this structure are indicated by dashed lines.

Figure 26-5 The β- or pleated
sheet form of the secondary
structure of protein molecules.
[*From H. D. Springall, The
Structural Chemistry of Proteins*
(Butterworth Scientific Publishers,
London, 1954.)]

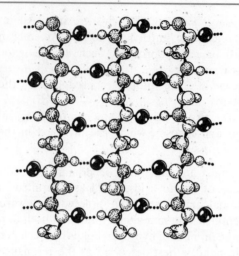

Figure 26-6 The tertiary structure
of a globular protein molecule. The
distance between the marks on the
vertical scale at the right represents
1 nm.

of the protein (Fig. 26-6). This tertiary structure results from interactions
between amino acid side chains. Factors of importance leading to such
tertiary interactions are:

1. *Disulfide linkages.* When the folding of a polypeptide chain brings two
   cysteine side chains into close proximity, mild oxidation of the two —SH
   side-chain substituents can lead to the formation of a covalent disulfide
   bond (S—S) that "freezes" this conformation into place.

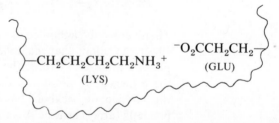

2. *Ionic bonding.* In a similar fashion we can envision ionic bonds between
   adjacent side chains of opposite charge.

$$-CH_2CH_2CH_2CH_2NH_3{}^+ \qquad {}^-O_2CCH_2CH_2-$$
$$\text{(LYS)} \qquad\qquad\qquad \text{(GLU)}$$

The attraction between charged ions in aqueous solution is relatively
small, because of the high dielectric constant of water. However, when the
folding of the polypeptide chain brings two such charged substituents
together in a region of the molecule where water is not abundant, the ionic
attraction between these substituents can help to stabilize the protein
structure.

3. *Hydrogen bonding.* In addition to the hydrogen bonds between peptide linkages that generate the secondary structure of the protein, hydrogen bonds can also form between pairs of amino acid side chains. Whereas individual hydrogen bonds are relatively unimportant, the large number of such interactions within a protein makes this factor important.

4. *Hydrophobic interactions.* A number of amino acid side chains are nonpolar (ALA, VAL, LEU, ILE, MET, PHE, TYR, and TRP), and these hydrophobic side chains tend to avoid the polar water molecules of the solvent. These side chains can associate to form hydrophobic pockets or cavities in the interior of globular proteins that stabilize the protein structure by the exclusion of water. These hydrophobic regions reflect the observation that hydrocarbon molecules are more soluble in other hydrocarbons than in water.

5. *Proline residues.* The incorporation of a proline (PRO) residue into a polypeptide chain disturbs the direction in which the chain propagates by placing a right angle bend into the primary structure. Thus PRO residues are never found within regions of either α-helix or β-sheet conformations. They are often, but not always, found at points where the α and β forms of the secondary structure interchange, or at points in a protein where changes in the direction of folding take place.

Many of the larger proteins contain more than one polypeptide chain. The **quaternary** structure of a protein describes how these independent polypeptide chains are related to each other. Whereas hydrophobic side chains tend to point toward the center of globular proteins to form hydrophobic pockets from which water may be excluded, the charged or hydrophilic side chains of the amino acids ARG, LYS, ASP, and GLU tend to occupy the exterior of the protein, and thus maximize their contact with the polar solvent. This leads to a distribution of ionic charges across the surface of the polypeptide chain, which in turn can lead to weak ionic forms of attraction between individual polypeptide units to produce the quaternary structure of the protein. It is also possible that hydrophobic interactions between side-chain substituents on individual polypeptide chains can lead to the quaternary structure. There is no evidence, however, for covalent bonds between the individual polypeptide chains. The best known example of a protein containing more than one polypeptide chain held together by quaternary forces is the oxygen-carrier protein hemoglobin (Section 31.5). Hemoglobin consists of four separate polypeptide chains, two α subunits and two β subunits.

Factors that **denature** a protein disturb the natural conformation of the protein, without necessarily cleaving the primary structure, and may lead to the loss of biological activity. Among the factors that may lead to denaturation are

1. *Temperature.* As the temperature increases, so does motion within the protein. Internal motion tends to disrupt secondary and tertiary structure within the protein, thereby inducing denaturation.

2. *Hydrogen ion concentration.* Extremes of pH alter the charges on amino acid side chains, thereby interfering with both the tertiary and quaternary structures of the protein.

3. *Redox agents.* Oxidizing or reducing agents can alter both the number

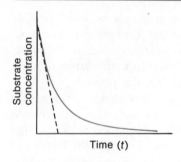

Figure 26-7 A plot of the dependence of the concentration of substrate on time for an enzyme-catalyzed reaction. The velocity of reaction at any moment in time is equal to the slope of a line tangent to this curve at that time. The velocity of reaction at $t = 0$ is known as the initial velocity of reaction, v_0, and is given by the slope of the curve at $t = 0$ (broken line).

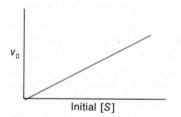

Figure 26-8 The predicted dependence of the initial velocity of reaction (v_0) on the initial substrate concentration for a first-order reaction.

Figure 26-9 The observed dependence of the initial rate of reaction on the initial substrate concentration for many enzyme-catalyzed reactions. The maximum value of the initial rate of reaction at a given enzyme concentration is known as v_{max}. As the enzyme concentration increases, so does v_{max}.

and location of disulfide linkages, or alter the nature of side-chain substituents, thereby altering the conformation of the protein.

4. *Detergents.* Detergents act to dissolve hydrophobic substances such as grease and oil in aqueous solution (Section 13.31). They therefore disrupt the hydrophobic pockets in the tertiary structure of a protein, thereby denaturing the protein.

26.5   Enzymes and Enzyme Kinetics

Enzymes are proteins that catalyze chemical reactions in living systems. To qualify as catalysts (Section 15.15), enzymes must increase the rate of a reaction without changing the magnitude of the equilibrium constant, and without being consumed in the course of the reaction.

The molecule upon which the enzyme acts is called the **substrate** (S). The simplest enzyme-catalyzed reactions appear to convert a single molecule of substrate into a single molecule of product (P). Regardless of the mechanism of this reaction, the rate, or velocity (v), of this reaction should decrease with time as substrate is transformed into product (Fig. 26-7). Extrapolating the dependence of substrate concentration on time (t) to $t = 0$ yields the initial rate of reaction v_0. If this reaction is first-order in substrate concentration (Section 15.6), $v = k[S]$ and the initial rate of reaction should increase with increasing initial concentration of substrate (Fig. 26-8). Unfortunately, plots of the initial rate of reaction versus the initial concentration of substrate for enzyme-catalyzed reactions are far from linear (Fig. 26-9). At high concentrations of substrate the initial rate of reaction is independent of the initial substrate concentration. There appears to be a maximum initial rate of reaction, v_{max}, which cannot be exceeded, regardless of the initial substrate concentration. The value of v_{max}, however, does increase as the concentration of the enzyme is increased.

In 1913, L. Michaelis and M. L. Menten proposed a mechanism for enzyme-catalyzed reactions that not only explained the appearance of a maximum initial rate of reaction, but also why v_{max} increased with increasing enzyme concentration. They proposed the formation of an intermediate, known as the **enzyme-substrate complex, ES,** in which the substrate is bound to the enzyme. The equation for enzyme-catalyzed reactions then becomes

$$E + S \rightleftharpoons ES \rightleftharpoons E + P$$

As the substrate concentration is increased, the equilibrium shifts toward the enzyme-substrate complex. Since there are only a limited number of substrate-

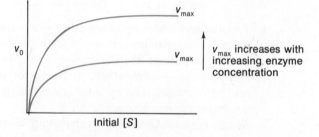

binding sites on the enzyme, when all of the sites are occupied by substrate molecules, the rate of reaction cannot increase further, and the rate of the reaction becomes zero order (Section 15.6) in substrate concentration. As the concentration of the enzyme is increased, the number of substrate-binding sites increases, and therefore v_{max} increases as well. In the Michaelis-Menten model the initial velocity of reaction at a given enzyme concentration is given by the following equation:

$$v_0 = \frac{v_{max}[S]}{K_m + [S]}$$

where K_m is a constant that reflects the affinity of the enzyme for its substrate.

26.6   The Control of Enzyme Activity

Enzymes are remarkably efficient catalysts. The enzyme carbonic anhydrase, for example, which catalyzes the reaction,

$$CO_2 + H_2O \rightleftharpoons H^+ + HCO_3^-$$

is such a good catalyst that a single enzyme molecule can transform over 36,000,000 molecules of substrate per minute into product. This corresponds to a 10 to 100 millionfold increase in the rate of this reaction in the presence of the enzyme. One explanation for the efficiency of enzyme catalysis suggests that the binding of the substrate to the enzyme can stretch or strain the substrate until the conformation of the substrate more closely resembles the transition state for the reaction (Section 15.8), thereby reducing the activation energy for this reaction. Enzymes are not only efficient catalysts; they can also be remarkably selective catalysts. The ability of some enzymes to be specific for a very few substances (substrates) in a variety of substances with related structures suggests that the enzyme and substrate fit together in much the same manner that a lock is opened by only a very limited number of keys.

The Michaelis-Menten model suggests that enzyme-catalyzed reactions proceed through at least two steps. First, the substrate binds to the enzyme to form an enzyme-substrate complex. Second, the enzyme acts upon the substrate to convert the substrate into product. Many substances may bind to an enzyme, but only a limited number of substances can undergo reaction once bound. The enzyme succinate dehydrogenase, for example, not only binds its normal substrate, succinate, but also binds other dicarboxylates such as malonate, which cannot undergo reaction once bound. By occupying a portion of the substrate binding sites on the enzyme, malonate slows down the rate of reaction, in effect inhibiting the activity of the enzyme. Since the substrate, succinate, and the inhibitor, in this case malonate, compete for the same binding sites on the enzyme, this form of enzyme inhibition is called **competitive inhibition.** Enzymes can also be inhibited by substances that have no resemblance to the substrate, for example, the heavy metal ions Ag^+, Pb^{2+}, and Hg^{2+}. These inhibitors can bind in either the presence or the absence of substrate-binding, and this form of enzyme inhibition is called **noncompetitive inhibition.**

In order for a cell to function at its maximum efficiency, it is essential that the rate of chemical reactions be adjusted to the needs of the cell. It is not enough for the cell to possess enzymes that catalyze the conversion of a given substrate into a particular product. It is also essential for the cell to possess control mechanisms that can turn on enzyme activity when the product of an enzyme-catalyzed reaction is needed, and switch off enzyme activity when the product is present in excess. There are four ways in which the cell can control the rate of enzyme-catalyzed reactions. First, it can influence reaction rate by changing the concentration of the substrate. In the absence of substrate, the enzyme is effectively switched off. Second, the cell can produce natural inhibitors of enzyme activity, which operate by either competitive or noncompetitive inhibition mechanisms. Third, the cell can take advantage of the existence of **allosteric** enzymes, which contain two or more binding sites. Allosteric enzymes bind both a substrate and a small molecule known as the **effector.** Binding of the effector is thought to alter the structure of the enzyme, thereby either switching on or switching off enzyme activity. Binding of the effector usually does not influence the binding of the substrate; it either facilitates or prevents conversion of the substrate into a product. There are several forms of control of allosteric enzymes. In so-called **feedback inhibition,** one or more of the enzymes involved in the synthesis of an end-product are inhibited by binding the end-product as an effector. Alternatively, when the synthesis of a particular product should occur only when the substrate is present in excess, the enzyme may exhibit **substrate activation.** More than one substrate must bind to the enzyme before the enzyme is active. Finally, the cell can control the rate of enzyme-catalyzed reactions by controlling the population of enzyme molecules. This can be done by controlling either the rate with which the enzyme is synthesized, or the rate with which the enzyme is degraded or destroyed.

26.7   Carbohydrates

The class of compounds known as **carbohydrates** was originally limited to compounds with the empirical formula CH_2O, and which therefore appeared quite literally to represent hydrates of carbon. This definition has since been broadened to include all polyhydroxyaldehydes and ketones. Among the compounds that fall into the class of carbohydrates are the starches, cellulose, glycogen, and the great variety of sugars. The carbohydrates serve as an important source of energy for all organisms, and form the supporting tissue of plants and some animals. There are three important classes of carbohydrates: (a) the monosaccharides, (b) the disaccharides, and (c) the polysaccharides.

26.8   Monosaccharides

The simplest carbohydrates contain only a single aldehyde or ketone functional group (Sections 25.20 and 25.21), and are known as **monosaccharides.** They are colorless, crystalline, and frequently possess a sweet taste. The monosaccharides are classified as either **aldose** or **ketose** on the basis of whether they contain an aldehyde or a ketone functional group, respectively. With the exception of the simplest ketose, the three-carbon monosaccharide known as dihydroxyacetone, all of the monosaccharides are optically active. Although both D and L isomers are possible, the dominant form for carbohydrates isolated from natural sources is the

D configuration. Structures of the D and L forms of the simplest aldose, glyceraldehyde, are

D-Glyceraldehyde   L-Glyceraldehyde

The formulas of monosaccharides are often written using a convention first proposed by Emil Fischer, in which the carbon skeleton is written vertically, with the aldehyde or ketone function toward the top, and with the second to the last —OH group pointed toward the right for the D isomer, and toward the left for the L isomer. Fischer projections for the two isomers of glyceraldehyde are

D-Glyceraldehyde   L-Glyceraldehyde

For purposes of reference, the Fischer projections for the heaviest members of the D families of aldose and ketose sugars are given in Tables 26-1 and 26-2.

TABLE 26-1 Fischer Projections for the Heaviest Members of the D-aldose Family of Monosaccharides

D-allose   D-altrose   D-glucose   D-mannose   D-gulose   D-idose   D-galactose   D-talose

TABLE 26-2 Fischer Projections for the Heaviest Members of the D-ketose Family of Monosaccharides

D-psicose   D-fructose   D-sorbose   D-tagatose

Monosaccharides can undergo a reversible intramolecular reaction to form cyclic compounds. Glucose and galactose form six-membered **pyranose** rings preferentially (Fig. 26-10), whereas fructose and ribose preferentially form five-membered **furanose** rings (Fig. 26-11). In each case, two isomers known as the α

Figure 26-10 Glucose and galactose undergo reversible reactions, leading to the formation of six-membered glucopyranose or galactopyranose rings. Two isomers are formed, differing only in the orientation at the C(1) carbon atom.

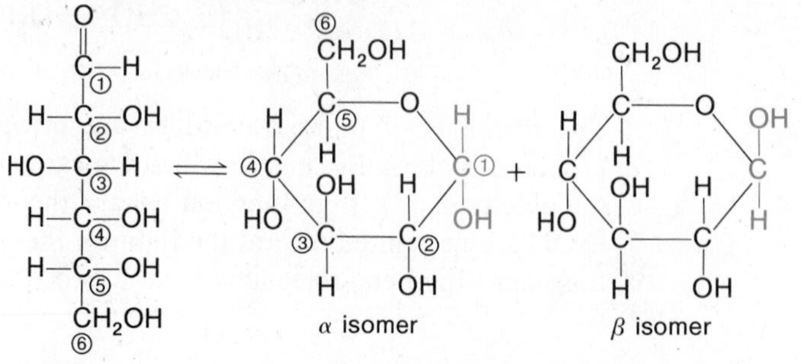

D-Glucose   D-Glucopyranose

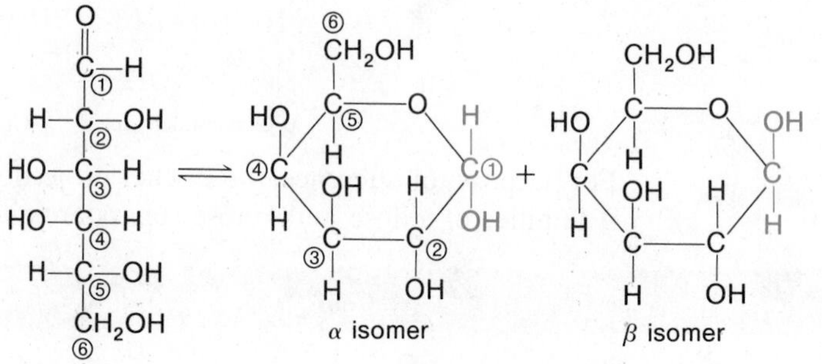

D-Galactose   D-Galactopyranose

Figure 26-11 Fructose and ribose undergo reversible reactions, leading to the formation of α and β isomers of five-membered fructofuranose and ribofuranose rings.

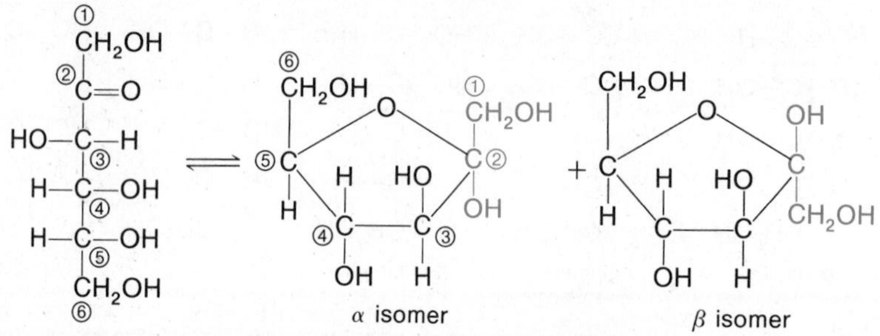

D-Fructose   D-Fructofuranose

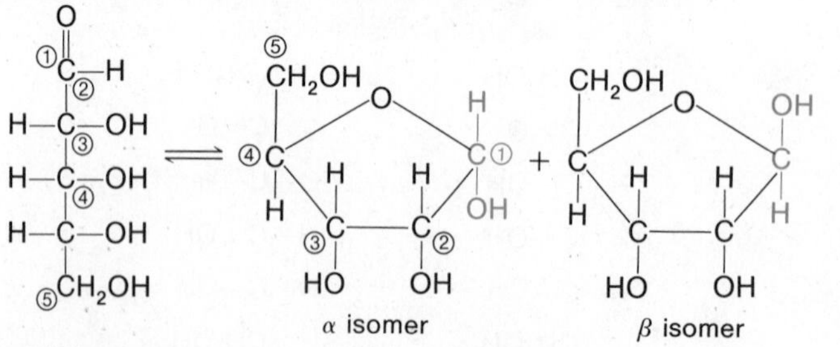

D-Ribose   D-Ribofuranose

and β forms are produced which differ only in the orientation at the first, or C(1), carbon atom.

Glucose (Table 26-1), $C_6H_{12}O_6$, also known as **dextrose,** is the single most abundant organic compound in nature. It is the sole component of the polysaccharides cellulose, starch, and glycogen. In medical work, glucose is often called "blood sugar," since it is by far the most abundant carbohydrate in the bloodstream. Human blood normally contains about one gram of glucose per liter of blood. Individuals suffering from diabetes are unable to assimilate glucose, and this sugar is eliminated through the kidneys. One hundred milliliters of a diabetic's urine may contain as much as 8 to 10 grams of glucose, and its presence is one symptom of the disease. **Fructose** (Table 26-2) also known as **levulose** and **fruit sugar,** has the same molecular formula as glucose, but contains a ketone rather than an aldehyde function. Fructose occurs naturally in both fruits and honey, and it is found combined with glucose in the disaccharide known as sucrose, or cane sugar. Fructose is nearly twice as sweet per gram as sucrose, and, in fact, is the sweetest of all sugars.

26.9   Disaccharides and Polysaccharides

In addition to the reversible intramolecular reactions that lead to pyranose and furanose rings, monosaccharides can also undergo an intermolecular reaction that leads to polymeric chains of saccharide residues. However, the polymerization reaction can be reversed only in the presence of acid or a suitable enzyme catalyst. Therefore polysaccharides are relatively stable to hydrolysis.

The most important oligosaccharide is the disaccharide **sucrose,** or **cane sugar** (Fig. 26-12). Hydrolysis of sucrose ($C_{12}H_{22}O_{11}$) yields a 50:50 mixture of the monosaccharides fructose and glucose (both $C_6H_{12}O_6$), a mixture often called **invert sugar,** which is a major component of honey. **Maltose,** or **malt sugar** (Fig. 26-13), is a disaccharide of α-D-glucopyranose rings that does not occur in large amounts in nature, but can be produced by the partial hydrolysis of starch. Maltose is an important constituent of corn syrup. **Lactose,** or **milk sugar,** is a

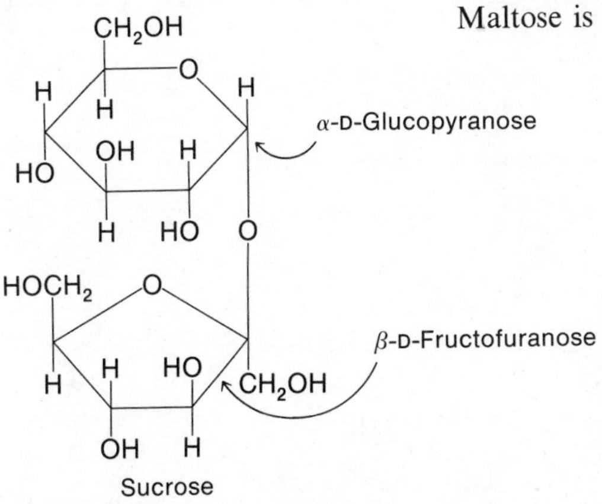

Figure 26-12 The structure of the disaccharide sucrose formed from reaction of α-D-glucopyranose with β-D-fructofuranose. (Carbon atoms in the rings are not shown.)

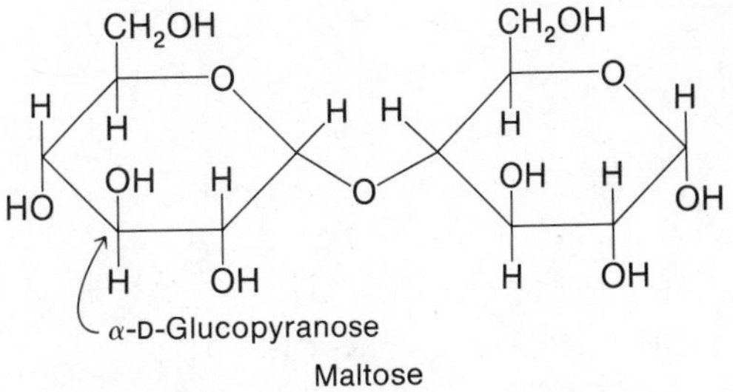

Figure 26-13 The structure of the disaccharide maltose formed by the dimerization of α-D-glucopyranose. (Carbon atoms in the rings are not shown.)

Figure 26-14 The structure of the disaccharide lactose formed by reaction of β-D-galactopyranose with α-D-glucopyranose. (Carbon atoms in the rings are not shown.)

Figure 26-14 The structure of the disaccharide lactose formed by reaction of β-D-galactopyranose with α-D-glucopyranose. (Carbon atoms in the rings are not shown.)

α-D-Glucopyranose

β-D-Galactopyranose

Lactose

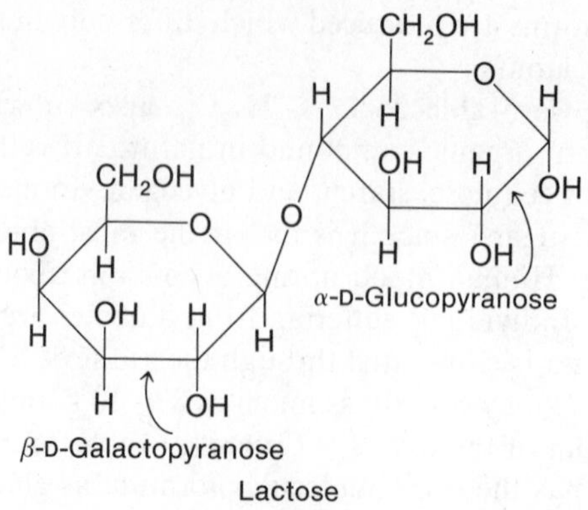

disaccharide that differs from maltose in the replacement of an α-D-glucose ring by a β-D-galactose residue (Fig. 26-14). It is interesting to note that many individuals of African or Asiatic origin develop a sensitivity to milk related to the absence of enzymes that can aid in the digestion of the β-linkage in lactose.

There are three major classes of polysaccharides that are formed by the polymerization of D-glucopyranose rings: starch, glycogen, and cellulose. **Starch** serves primarily as a long-term storage center for food energy. It is accumulated by plants in seeds, tubers, and fruits. It is often the main food supply for the young plant until it has developed a leaf system and can manufacture its own food through photosynthesis. Starch is composed of a mixture of two polysaccharides: amylose and amylopectin. **Amylose** is a linear polysaccharide of α-D-glucopyranose residues that twists into a helical coil. **Amylopectin** is a branched polysaccharide in which polymerization occasionally occurs at the C(6) carbon atom of an α-D-glucopyranose ring as well as at the C(1) and C(4) carbons (Fig. 26-15).

Glycogen is a branched polysaccharide containing α-D-glucopyranose residues, like amylopectin, but with approximately twice as many branches. Glycogen

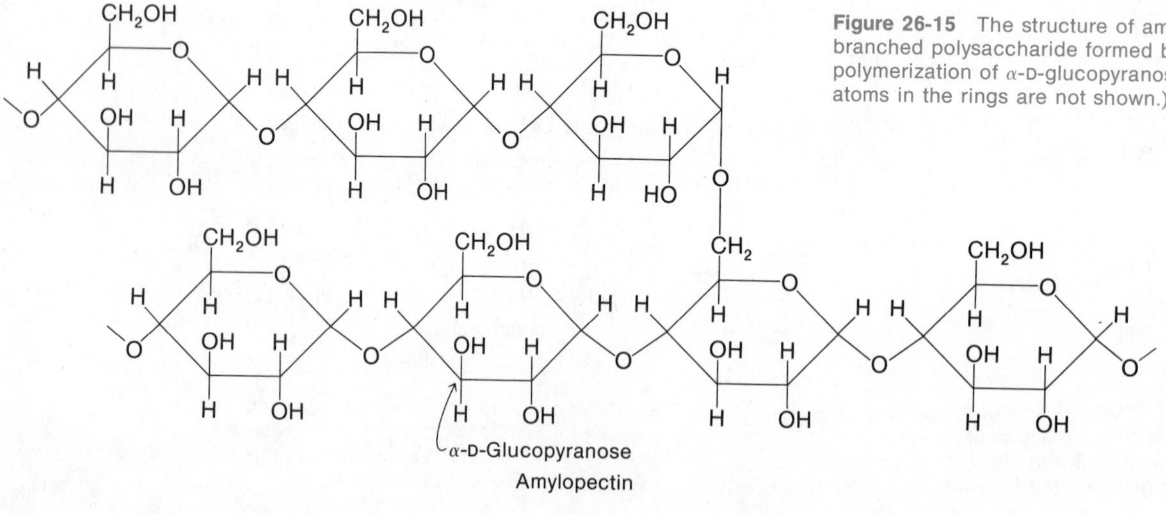

Figure 26-15 The structure of amylopectin, a branched polysaccharide formed by the polymerization of α-D-glucopyranose. (Carbon atoms in the rings are not shown.)

α-D-Glucopyranose

Amylopectin

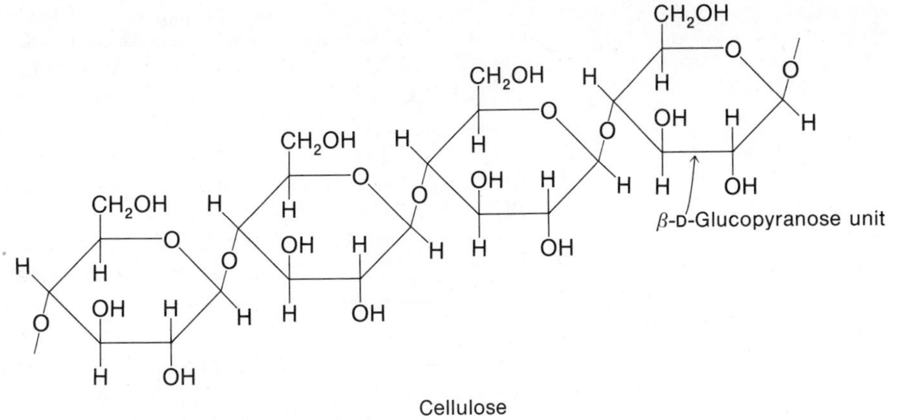

Figure 26-16 The structure of cellulose, a linear polysaccharide formed by the polymerization of β-D-glucopyranose. (Carbon atoms in the rings are not shown.)

Cellulose

serves the same function in animals that starch serves in plants and is found primarily in the liver and muscle tissue.

Cellulose is a linear polysaccharide of β-D-glucopyranose residues (Fig. 26-16) and is found predominantly in plants. Cellulose is the principal component of the rigid cell wall of plant cells, and therefore serves to provide structure in plant cells rather than to store food energy. It is fascinating that only the absence of an enzyme capable of cleaving the β-linkage in cellulose stops humans from digesting either the paper of this text or cotton, the latter being at least 90% cellulose by weight. Most other animals also cannot digest cellulose, due to the lack of suitable enzymes to catalyze its hydrolysis. Termites, however, and the ruminant animals such as cattle, sheep, and goats, are able to use cellulose as food, due to the presence of bacteria in their digestive tracts that contain enzymes that can cleave the β-linkages in cellulose. Termites are commonly destroyed by killing these bacteria. The termites then literally starve to death.

26.10   The Complex Lipids

The **lipids** are substances that can be extracted from living systems with nonpolar solvents such as chloroform, ether, or benzene. The class of lipids is the most diverse we have seen so far, containing compounds of widely differing structures. The only structural similarity common to the lipids is the presence of a hydrophobic hydrocarbon backbone, which may or may not contain a highly charged hydrophilic head. Lipids find roles in both the structure of membranes and the long-term storage of food energy. The word "lipid" is derived from the Greek word *lipos* (meaning fat), although not all lipids are fats.

Lipids are often divided into two classes, the complex and the simple, on the basis of the presence or absence of a fatty acid residue. The **complex lipids** are derivatives of long-chain carboxylic acids, or **fatty acids,** such as stearic acid.

$$CH_3CH_2CH_2CH_2CH_2CH_2CH_2CH_2CH_2CH_2CH_2CH_2CH_2CH_2CH_2CH_2CH_2 CO_2H$$

Stearic acid

The family of fatty acids can be subdivided into saturated and unsaturated derivatives. Among the *saturated* fatty acids we find

Formula	Common Name	IUPAC Name
$CH_3(CH_2)_2CO_2H$	Butyric acid	Butanoic acid
$CH_3(CH_2)_4CO_2H$	Caproic acid	Hexanoic acid
$CH_3(CH_2)_6CO_2H$	Caprylic acid	Octanoic acid
$CH_3(CH_2)_8CO_2H$	Capric acid	Decanoic acid
$CH_3(CH_2)_{10}CO_2H$	Lauric acid	Dodecanoic acid
$CH_3(CH_2)_{12}CO_2H$	Myristic acid	Tetradecanoic acid
$CH_3(CH_2)_{14}CO_2H$	Palmitic acid	Hexadecanoic acid
$CH_3(CH_2)_{16}CO_2H$	Stearic acid	Octadecanoic acid

The low-molecular-weight carboxylic acids are characterized by distinctive odors. Butyric acid, for example, is the essence of rancid butter, and one of the components of the odor of rotting meat, while the next three homologs take their names from the Greek word for goat, *capro*. The naturally occurring *unsaturated* fatty acids contain one or more carbon-carbon double bonds with a *cis* geometry (Section 25.13).

Formula	Common Name	IUPAC Name
$CH_3(CH_2)_7CH{=}CH(CH_2)_7CO_2H$	Oleic acid	9-Octadecenoic acid
$CH_3(CH_2)_4CH{=}CHCH_2CH{=}CH(CH_2)_7CO_2H$	Linoleic acid	9,12-Octadecadienoic acid
$CH_3CH_2CH{=}CHCH_2CH{=}CHCH_2CH{=}CH(CH_2)_7CO_2H$	Linolenic acid	9,12,15-Octadecatrienoic acid

Although fatty acids are available to a cell from a wide variety of sources, the concentration of the free fatty acid in the cell is negligible. Most fatty acids occur in plant and animal cells in the form of triesters (Section 25.23) of the alcohol glycerol ($HOCH_2CHCH_2OH$).
                              |
                             OH

$$CH_3(CH_2)_{16}CO\overset{\displaystyle O}{\overset{\|}{C}}\begin{array}{l} CH_2O\overset{\displaystyle O}{\overset{\|}{C}}(CH_2)_{16}CH_3 \\ CH \\ CH_2O\underset{\displaystyle O}{\underset{\|}{C}}(CH_2)_{16}CH_3 \end{array}$$

Tristearylglycerol

Such lipids are commonly known as **triglycerides,** or as neutral fats and oils, although the proper name is **triacylglycerols.** The primary function of these triglycerides is the long-term storage of food energy. Triglycerides formed from saturated fatty acids, such as tristearylglycerol, are typically solids and are known as **fats,** or sometimes **animal fats.** Triglycerides formed from predominantly unsaturated fatty acids are more frequently liquids and are known as **oils;** examples include cottonseed oil, linseed oil, and palm oil.

One of the characteristic reactions of fats and oils is **saponification,** which is the hydrolysis of the triester in the presence of base to form glycerol and a salt of the fatty acid.

$$CH_3(CH_2)_{16}\overset{\displaystyle O}{\overset{\displaystyle \|}{C}}OCH \begin{array}{l} CH_2O\overset{\displaystyle O}{\overset{\displaystyle \|}{C}}(CH_2)_{16}CH_3 \\[2mm] \\[2mm] CH_2O\underset{\displaystyle \underset{\displaystyle O}{\|}}{C}(CH_2)_{16}CH_3 \end{array} + 3NaOH(aq) \longrightarrow \begin{array}{l} CH_2OH \\[2mm] CHOH \\[2mm] CH_2OH \end{array} + 3CH_2(CH_2)_{16}CO_2^-Na^+(aq)$$

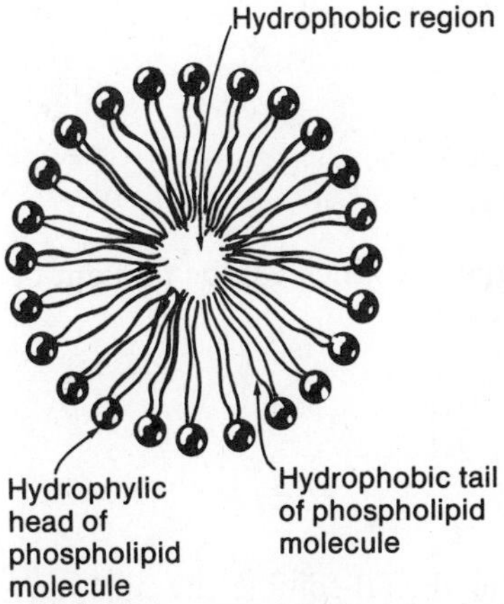

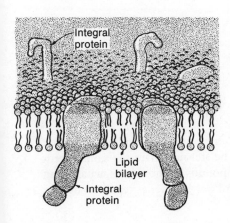

Hydrophobic region

Hydrophylic head of phospholipid molecule

Hydrophobic tail of phospholipid molecule

Figure 26-17 Amphipathic substances, such as the phospholipids, which possess a charged hydrophilic head and a nonpolar hydrophobic tail, can associate to form micelles in which the hydrophilic heads are oriented towards the surrounding water molecules, and the hydrophobic tails share a region of the solution where water molecules are partially excluded.

Integral protein

Lipid bilayer

Integral protein

Figure 26-18 The phospholipids can associate to form bilayers that serve as the basis for cell membranes. The nonpolar tails associate to form hydrophobic pockets, and the hydrophilic heads point toward the polar solvent, water. Membranes may also contain other lipids, such as cholesterol, and proteins that are involved in the transport of molecules across the cell membrane. (*From "The Assembly of Cell Membranes," by Harvey F. Lodish and James E. Rothman.* Copyright © 1979 by Scientific American, Inc. All rights reserved.)

The salts of the fatty acids thus produced are soaps (Section 13.31).

A second major class of complex lipids are the **phosphoglycerides,** or **glycerophosphatides,** which are major components of cell membranes. These lipids can be thought of as derivatives of phosphoric acid and glycerol, with the following structure:

$$\begin{array}{l} R-\overset{\displaystyle O}{\overset{\displaystyle \|}{C}}-O-CH_2 \\[3mm] R-\underset{\displaystyle \underset{\displaystyle O}{\|}}{C}-O-CH \qquad \overset{\displaystyle O}{\overset{\displaystyle \|}{}} \\[3mm] \qquad\qquad\quad CH_2-O-\overset{\displaystyle \|}{P}-OH \\[3mm] \qquad\qquad\qquad\qquad OH \end{array}$$

Among the most important of the phosphoglycerides are the **phosphatidyl cholines,** or **lecithins,** of which the following is a representative example:

$$\begin{array}{l} CH_3(CH_2)_{16}\overset{\displaystyle O}{\overset{\displaystyle \|}{C}}OCH_2 \\[3mm] CH_3(CH_2)_{16}\underset{\displaystyle}{C}OCH \quad O \\[3mm] \qquad\qquad\quad O \quad CH_2OPOCH_2CH_2N(CH_3)_3{}^+ \\[3mm] \qquad\qquad\qquad\qquad O^- \end{array}$$

Like soaps, phosphatidyl cholines are substances that contain a highly charged hydrophilic head and a nonpolar hydrophobic tail. These phospholipids can spontaneously organize to form **micelles** in which the ionic heads face the surrounding water solvent and the nonpolar tails form a hydrophobic pocket (Fig. 26-17). Alternatively, they can form bilayers which serve as the basis for membranes that separate the cell from its surroundings or segment the interior of the cell (Fig. 26-18).

26.11   The Simple Lipids

Lipids that do not contain a fatty acid are given the rather misleading name **simple lipids.** The simple lipids include substances such as the **fatty alcohols** (cetyl alcohol and myricyl alcohol, for example), which are long-chain alcohols.

$$CH_3(CH_2)_{14}CH_2OH \qquad CH_3(CH_2)_{29}CH_2OH$$

Cetyl alcohol   Myricyl alcohol

With fatty acids, these fatty alcohols form esters (Section 25.23) known as **waxes.** One of the principal components of beeswax, for example, is myricyl palmitate, the ester of myricyl alcohol and palmitic acid.

$$CH_3(CH_2)_{14}\overset{\displaystyle O}{\overset{\|}{C}}OCH_2(CH_2)_{29}CH_3$$

Myricyl palmitate

The family of simple lipids also includes the **terpenes,** which can be thought of as polymers of the five-carbon unit **isoprene.**

$$CH_2{=}\overset{\displaystyle CH_3}{\overset{|}{C}}{-}CH{=}CH_2$$

Isoprene

These terpenes include pigments such as β-carotene, the fat-soluble vitamins such as vitamin A, and forms of natural rubber (Section 25.29). Stick structures of the carbon skeleton of β-carotene and vitamin A are shown below. Only bonds between carbon atoms are shown in these structures. A CH_3, CH_2, or CH group is located at the end of each bond, as needed. The location in the molecules of units derived from isoprene molecules is shown in color.

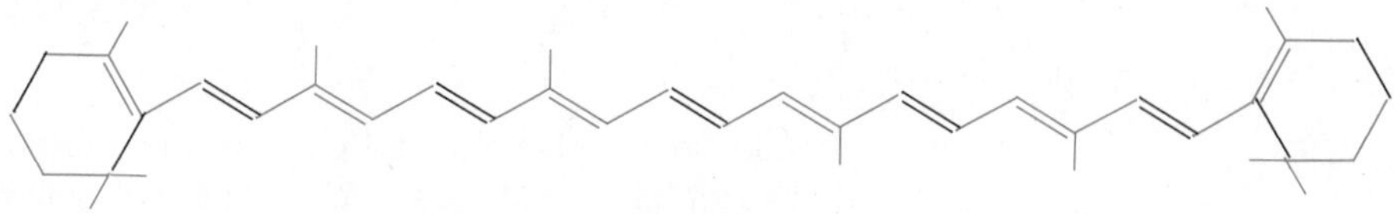

β-Carotene

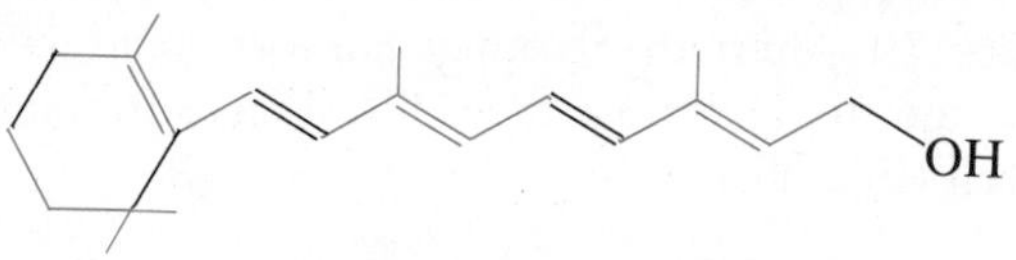

Vitamin A

A remarkably diverse group of lipids, known as the *steroids,* can be synthesized from terpenes by a complex set of reactions. The steroids share the same fused-ring structure. Among the steroids are such compounds as cholesterol, which is an important component of cell membranes, and the sex hormones testosterone and estradiol-17 (Fig. 26-19).

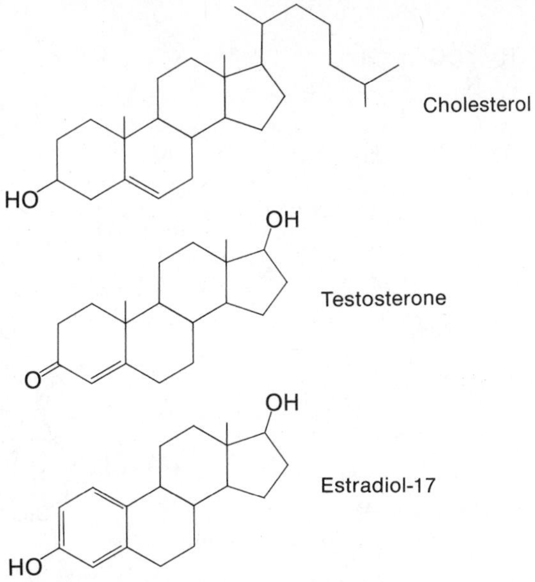

Figure 26-19 The steroids represent a class of compounds each with the same essential fused-ring structure. The steroids function as important components of cell membranes, such as cholesterol; as precursors in the synthesis of vitamins, such as vitamin D_2; as hormones, for example, the male and female sex hormones testosterone and estradiol-17; and more recently as ingredients of oral contraceptives.

Cholesterol

Testosterone

Estradiol-17

26.12   The Nucleic Acids

Our understanding of the chemical nature of the nucleic acids forms the basis for an understanding of biochemical genetics. The **nucleic acids** are macromolecules of very high molecular weight whose sole function is the storage and transfer of genetic information. There are two types of nucleic acids: **ribonucleic acid (RNA)**, which is located primarily in the cytoplasm of the cell; and **deoxyribonucleic acid (DNA)**, which is concentrated in the nucleus of the cell.

The ribonucleic acids (RNA) are built upon the framework of a β-D-ribofuranose ring (Section 26.8), whereas the deoxyribonucleic acids (DNA) contain a modified furanose ring in which the —OH group on the second, or 2′, carbon of a ribose molecule is replaced by a hydrogen atom.

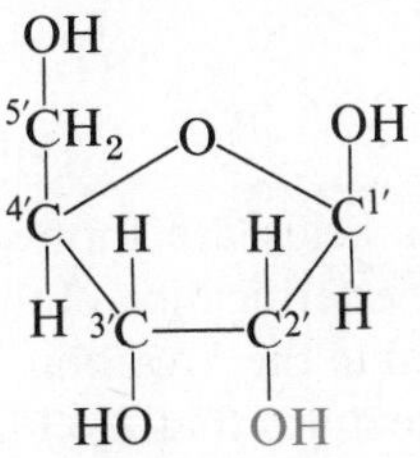

β-D-Ribofuranose 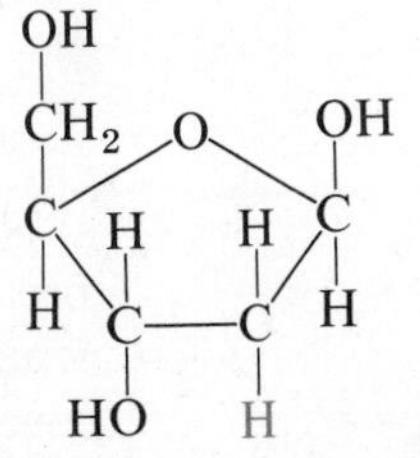   2-Deoxy-β-D-ribofuranose

Nucleosides are formed by combining one of these monosaccharides with a nitrogen-containing base. The five most common nitrogenous bases fall into two classes: the pyrimidines [Fig. 26-20(a)] and the purines [Fig. 26-20(b)]. **Nucleotides** are phosphate esters of nucleosides, with esterification occurring predominantly at the 5′ carbon. For example, the base **uracil,** which is found only in RNA, not DNA, can be joined to a ribose ring at the 1′ carbon to form the nucleoside

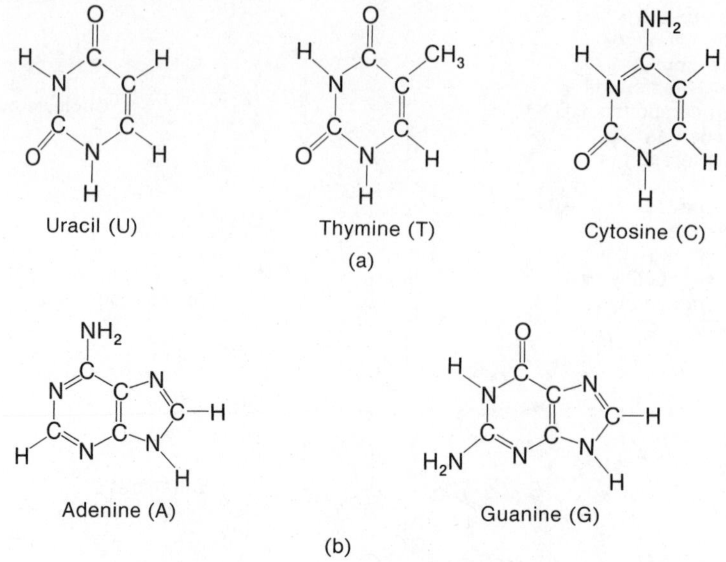

Figure 26-20 The five most common nitrogen-containing bases found in nucleic acids are divided into two families: (a) the pyrimidines uracil (U), thymine (T), and cytosine (C); and (b) the purines adenine (A) and guanine (G).

Uracil (U)   Thymine (T)   Cytosine (C)

(a)

Adenine (A)   Guanine (G)

(b)

uridine. This nucleoside may then combine with the phosphate ion, PO_4^{3-}, to form the nucleotide **uridine monophosphate.**

Uracil   Uridine   Uridine monophosphate

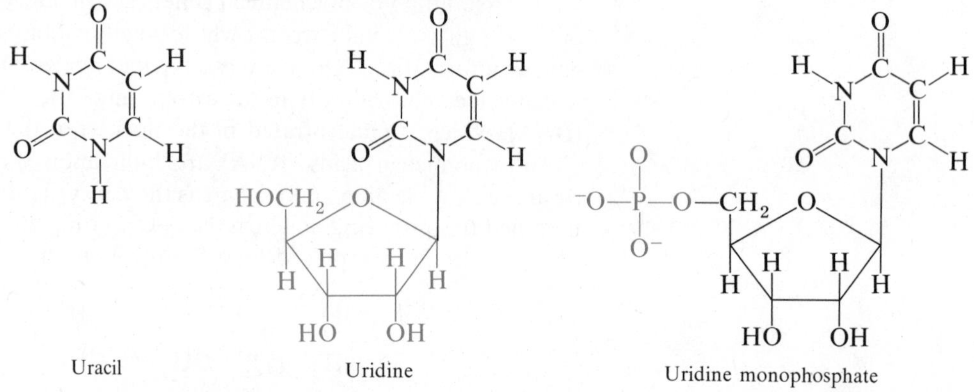

Nucleic acids are formed by polymerization of nucleotides. Polymerization occurs by esterification of the 3′ alcohol of one nucleotide with the phosphate that is attached to the 5′ carbon of another. A segment of a DNA chain containing the four nucleotides that are found in deoxyribonucleic acids is shown in Fig. 26-21. The sequence of nucleotides in Fig. 26-21, reading from the 5′ end of this chain toward the 3′ end, can be symbolized as T-A-C-G. This tetranucleotide is drawn as the molecule would exist at or about neutral pH. The phosphodiester linkages between adjacent nucleosides in nucleic acids are strong acids. The phosphate groups would therefore be expected to carry a significant negative charge in aqueous solution at or near neutral pH. The nucleic acids are therefore normally found associated with either highly charged positive proteins known as **histones,** or with cations such as the Mg^{2+} ion.

Figure 26-21 The structure of a tetranucleotide segment of a single strand of deoxyribonucleic acid (DNA). Reading from the 5' to the 3' end, this tetranucleotide can be symbolized as T-A-C-G. (Carbon atoms in the rings are not shown.)

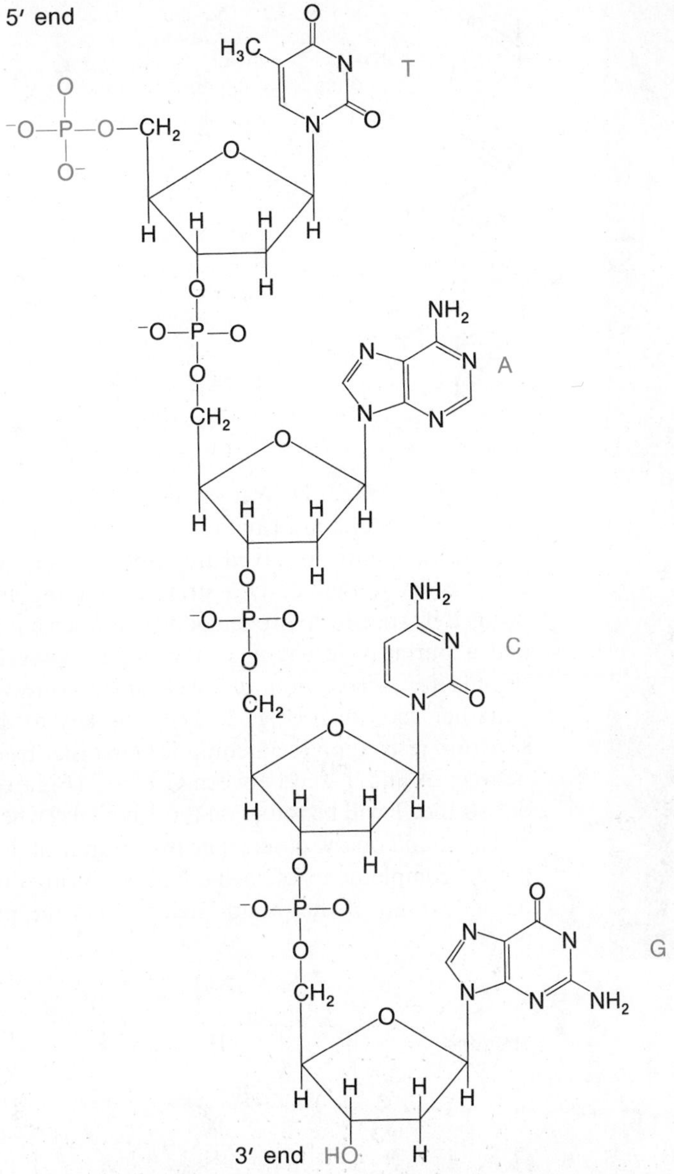

There are several important structural differences between the two nucleic acids DNA and RNA. As noted previously, each RNA nucleotide contains an —OH group on the 2' carbon of the furanose ring whereas the —OH group is absent in DNA nucleotides. Furthermore, the two forms of nucleic acid differ in the identity of the nitrogenous bases used to form nucleotides. DNA contains exclusively four nucleotides based upon two purines, adenine (A) and guanine (G), and two pyrimidines, cytosine (C) and thymine (T). In RNA the pyrimidine uracil (U) is used instead of thymine (T). There are also a number of other nucleotides that can be found in RNA. Evidence suggests, however, that these additional nucleotides are formed by modification of A, G, C, or U after the RNA polymer is produced.

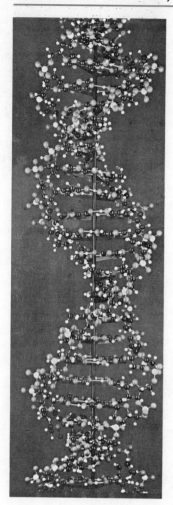

Figure 26-22 The double helix structure of deoxyribonucleic acids (DNA). Schematic drawing (*right*) and model of the double-stranded α-helix (*left*).

In April 1953, J. D. Watson and F. H. C. Crick proposed a mechanism for the storage of genetic information based upon the structure of DNA. For this work they subsequently received the Nobel Prize. Watson and Crick suggested that DNA is composed of two strands, running in opposite directions, which are bridged by specific hydrogen bonds between a pyrimidine side chain on one strand and a purine side chain on the other. They further suggested that these two strands are twisted into an α-helical structure with approximately 10 nucleotide pairs per 360° turn (Fig. 26-22). The key to the storage of information in this structure rests upon the assumption that the strongest hydrogen bonds are formed between A and T and between G and C (Figure 26-23). Wherever A is found on one strand, T will be found on the other, and wherever G is found on one strand, C will be found on the other. The two strands of the DNA double helix are therefore perfect complements of each other. Knowing the sequence of nucleotide residues on one strand of the double helix allows the prediction of the sequence on the other.

Figure 26-23 The hydrogen bonds in A-T and G-C base pairs. (Carbon atoms in the ring are not shown.)

Figure 26-24 The replication of the DNA double helix. Both strands are copied simultaneously, and each strand is copied in small segments that are then joined together. The overall process is referred to as semiconservative since each copy retains one strand from the parent molecule.

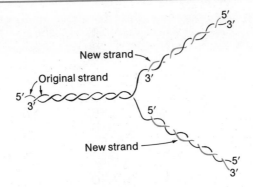

26.13   The Genetic Code

Modern genetics, and much of modern biochemistry, rests upon the dogmatic belief that the double helix of DNA serves as the center for the storage of the information necessary for controlling the sequence of incorporation of amino acids during protein biosynthesis. To transmit this information to the daughter cells during cell division, the DNA molecule must be capable of faultless **replication,** or reproduction, so that each of the daughter cells gets a perfect copy of the information present in the parent cell. For this information to be useful to the cell, the information coded within the structure of the DNA double helix must be capable of being **transcribed,** or read. Finally, a mechanism must exist which can **translate** this code into a sequence of amino acids during the assembly of a protein.

The successful replication of the DNA double helix rests upon the observation that the presence of the nitrogenous base A on one strand of the double helix implies the presence of T on the complementary strand, whereas the presence of G on one strand implies the presence of C on the complementary strand. The first step in the replication of DNA during cell division involves the binding of certain proteins that separate a portion of the double helix into two single strands of DNA. Each strand is then copied by the formation of short segments, which are finally joined to form the complements of the strand being copied. Replication proceeds until eventually two identical copies of the original DNA molecule are present. Replication is referred to as a semiconservative process since each copy retains one strand from the parent DNA molecule (Fig. 26-24).

DNA functions exclusively as the center for storage of the information that tells a cell how to synthesize any of several thousand proteins. RNA functions in the processing of this information. This processing occurs by a two-step mechanism. The first step is known as transcription. An RNA complement of one strand of the DNA double helix is synthesized under the control of certain enzymes. Since this segment of RNA carries the message contained in the DNA molecule, it is known as messenger RNA, or m-RNA. Just as the two strands of the DNA double helix run in opposite directions, the m-RNA segment synthesized during transcription is an antiparallel copy of the original DNA strand (Fig. 26-25), with one major difference. As noted previously, RNA contains the nucleotide U wherever DNA would contain T. Thus the base pairs used during transcription are A-U and G-C.

The second step in the processing of the information contained within the DNA double helix is translation. During translation, the m-RNA fragment serves

Figure 26-25 The transcription of a sequence of messenger RNA on the template of one strand of a DNA double helix. Note that the messenger RNA chain is the complement of the DNA strand being copied.

TABLE 26-3   The Amino Acids Coded by Specific Triplets of Nucleotide Residues on Messenger RNA

First Position (5′ End)	Second Position				Third Position (3′ End)
	U	C	A	G	
U	PHE	SER	TYR	CYS	U
	PHE	SER	TYR	CYS	C
	LEU	SER	TERM[a]	TERM[a]	A
	LEU	SER	TERM[a]	TRP	G
C	LEU	PRO	HIS	ARG	U
	LEU	PRO	HIS	ARG	C
	LEU	PRO	GLN	ARG	A
	LEU	PRO	GLN	ARG	G
A	ILE	THR	ASN	SER	U
	ILE	THR	ASN	SER	C
	ILE	THR	LYS	ARG	A
	MET	THR	LYS	ARG	G
G	VAL	ALA	ASP	GLY	U
	VAL	ALA	ASP	GLY	C
	VAL	ALA	GLU	GLY	A
	VAL	ALA	GLU	GLY	G

[a]The three codons UAA, UGA, and UAG code for the termination of the polypeptide chain.

as the template upon which the protein is synthesized. The synthesis of the protein takes place on the ribosome (Section 26.1), which contains both proteins and a second form of RNA known as ribosomal RNA. The first step in translation therefore involves binding of the m-RNA fragment to a ribosome. The amino acids are transported to the ribosome by a third form of RNA known as transfer RNA, or t-RNA. Individual t-RNA molecules, each carrying an amino acid, enter the ribosome and bind to the m-RNA chain at specific positions through hydrogen bonds analogous to the base-pairing used to form the DNA double helix. The amino acids are then joined to form a polypeptide chain that grows from the N-terminal amino acid toward the C-terminal amino acid.

The sequence of nucleotides in the DNA double helix is the basis for coding the sequence of amino acids in proteins. However, since there are only four nucleotides in DNA, and 20 genetically coded amino acids to be coded, it is obvious that a one-to-one relationship cannot exist between amino acids and nucleotide bases. There are, however, $4 \times 4 \times 4$, or 64, possible combinations of three nucleotides, which is more than enough. Each three-letter sequence, or **codon,** is specific for a single amino acid. Since there are 64 possible codons, and only 20 amino acids to be coded, this genetic code is somewhat redundant. Among the codons are three that signal chain termination. When a termination codon is encountered during protein biosynthesis the ribosome releases both the finished polypeptide chain and the m-RNA fragment.

The identity of the amino acid coded by any three-nucleotide sequence on a m-RNA fragment can be determined from the genetic code in Table 26-3.

EXAMPLE   **What is the sequence of amino acids that would be coded for by the following sequence of nucleotides on a strand of DNA, reading from the 5′ to the 3′ end: A-T-C-G-C-T-A-C-G-A-A-T?**

The m-RNA fragment that forms on this DNA template is the complement of this sequence and runs in the opposite direction. Thus the m-RNA chain, reading from the 3′ to the 5′ end, would be U-A-G-C-G-A-U-G-C-U-U-A. Reading this message in the other direction, from the 5′ to the 3′ end, yields the triplets A-U-U, C-G-U, A-G-C, and G-A-U, which codes for the sequence ILE-ARG-SER-ASP, reading from the N- to the C-terminal amino acid.

◆ ◆ ◆ ◆ ◆

All of the information necessary for the synthesis of two to three thousand different proteins found in the simplest cells is contained in the DNA molecules of the **chromosomes** of the cell. It is believed that the information from only a small portion of one of the strands of the DNA molecule is used in producing each protein. The portion of the DNA molecule that codes for a single polypeptide chain is known as a **gene.** Each gene is specific for the synthesis of a single kind of protein, although multiple copies of a gene may be found on the chromosome.

26.14   Metabolism

The term **metabolism** is used to describe the chemical reactions that take place in living organisms. Metabolism therefore includes the reactions the cell uses to produce energy, to capture nutrients for the synthesis of biomolecules, to degrade macromolecules to produce the basic building blocks for future construction, and to shuffle intermediates between the more abundant and the less abundant. Metabolism is divided into pathways that result in either the synthesis or degradation of a compound. **Catabolic** pathways lead to the degradation of a metabolic intermediate, whereas **anabolic** pathways result in the synthesis of a compound.

The energy required for the growth and maintenance of cellular structures is provided by metabolic reactions of two general types: photosynthesis and respiration, or fermentation. **Photosynthesis** is the endothermic process whereby light energy is converted into chemical energy in the chloroplasts of green plants and algae. The most important set of photosynthetic reactions eventually lead to the reduction of carbon dioxide and water to produce oxygen and carbohydrates such as glucose.

$$6CO_2 + 6H_2O + \text{light energy} \longrightarrow C_6H_{12}O_6 + 6O_2 \qquad \Delta G° = 2870 \text{ kJ/mol}$$

Respiration, or **fermentation,** reactions provide energy through the metabolic degradation or catabolism of organic compounds. The reactions involved in respiration can be divided into two categories: (a) those that can occur in the absence of molecular oxygen **(anaerobic reactions),** and (b) those that cannot occur in the absence of oxygen **(aerobic reactions).** Anaerobic metabolic pathways typically do not lead to a net oxidation or reduction of a metabolic intermediate, and liberate significantly smaller amounts of energy than the oxidation-reduction reactions that accompany aerobic metabolic pathways.

The sequence of reactions involved in the anaerobic degradation of the most abundant carbohydrate, glucose, is known collectively as **glycolysis,** or the glycolytic pathway. Glycolysis ultimately degrades a single molecule of glucose into two pyruvate ions, with the release of more than 200 kJ of energy per mole of glucose consumed.

Glucose   Pyruvate ion

Anaerobic organisms, such as the yeast cells used in the production of alcoholic beverages by the fermentation of sugars, convert the pyruvate produced in glycolysis to either acetaldehyde or ethanol. Aerobic organisms ultimately convert pyruvate into carbon dioxide and water through a sequence of reactions known as the **citric acid cycle,** or the **Krebs cycle.** The net reaction in aerobic organisms is thus the opposite of the photosynthetic reaction, and liberates approximately 2870 kJ/mol of glucose consumed.

$$C_6H_{12}O_6 + 6O_2 \longrightarrow 6CO_2 + 6H_2O \qquad \Delta G° = -2870 \text{ kJ/mol}$$

A significant portion of the energy released during the degradation of glucose is lost in the form of heat. The remainder, however, is stored as chemical energy through the synthesis of nucleoside triphosphates such as **adenosine triphosphate, ATP.**

ATP, and the related nucleoside triphosphates GTP, CTP, and UTP, can react with water to produce adenosine diphosphate, ADP (or GDP, CDP and UDP) and the phosphate ion, with the release of approximately 31 kJ/mol of ATP hydrolyzed. The nucleoside diphosphates, such as ADP, can undergo further hydrolysis to form nucleoside monophosphates, such as AMP, with the release of an additional 30 kJ/mol. The nucleoside triphosphates, and to some extent the diphosphates as well, serve as the cell's short-term energy-storage mechanism. ATP, for example, is synthesized from AMP and ADP during the degradation of carbohydrates, proteins, and lipids. When the cell needs energy to either drive a chemical reaction, such as the synthesis of proteins from amino acids, or for a mechanical process, such as the contraction of a muscle, ATP or its equivalent is consumed.

Glycolysis and the citric acid cycle are important means of producing biological energy in the form of ATP. This is by no means the sole function of these cycles, however. There are several points in these cycles where intermediates are produced that can be used in the synthesis of amino acids, or where intermediates from the degradation of amino acids and proteins can be inserted for eventual conversion to carbohydrates. The metabolism of carbohydrates and proteins are therefore intimately related. The metabolism of carbohydrates and the complex lipids are similarly connected. The complete degradation of either carbohydrates or lipids leads eventually to the same products, carbon dioxide and water. However, since lipids contain carbon in a lower oxidation state than carbohydrates, more energy is released during the oxidation of lipids. As much as $2\frac{1}{2}$ times as much energy is released per gram. In times of plenty, that is, when biological energy is abundant, the cell first captures this energy through the synthesis of ATP from ADP and/or AMP. Once this short-term storage capacity is exceeded, the cell may capture additional energy through the synthesis of carbohydrates. These carbohydrates may then be linked to form polysaccharides such as glycogen or starch, which serve as a more efficient means of storing energy for times of shortage to come. In times of extreme plenty the cell stores excess energy in the form of fatty acids and complex lipids, or fats. In each case the cell searches for a more and more efficient means of storing excess energy during times of plenty for the times of shortage that are likely to come. It is an unfortunate fact, in modern times, that for some humans these times of shortage never come, whereas for others, times of plenty are all too infrequent.

Questions

1. List the four principal chemical constituents of living cells. Identify the specific functions of each of these constituents.
2. Define the terms "hydrophilic" and "hydrophobic." What factor or factors determine whether a compound is hydrophilic or hydrophobic? Give an example of each class of compound.
3. Into what simpler components may a protein be hydrolyzed?
4. Define the following terms: amino acid, peptide, and protein.
5. At a pH of 6.0, which amino acid in a mixture of arginine, alanine, and aspartic acid would migrate toward a negatively charged electrode, and which toward a positively charged electrode?
6. Write symbolic formulas for all of the possible tri-

peptides that could be formed from the amino acids GLY, ALA and VAL.
7. Write the structure of the tripeptide CYS-LYS-GLU at neutral pH.
8. Distinguish between the primary, secondary, tertiary, and quaternary structures of a protein.
9. Define what is meant by the term "denaturation." What factor or factors can lead to the denaturation of a protein?
10. Define the following terms: enzyme, substrate, and product.
11. Describe what happens to the rate of an enzyme-catalyzed reaction with time from the instant that the enzyme and substrate are mixed until the substrate is consumed.
12. Describe what happens to the rate of an enzyme-

catalyzed reaction as the concentration of the enzyme is increased.

13. What is the literal meaning of the term "carbohydrate," and how did this name arise?

14. Write the name and formula for one example each of a monosaccharide, a disaccharide, and a polysaccharide.

15. In what sense can sucrose be considered an anhydride?

16. Compare and contrast the structures of cellulose and starch.

17. Define saponification and write an equation for the saponification of tripalmitylglycerol.

18. Write the general formula for a neutral fat. How do fats and oils differ in composition?

19. Explain why the triacylglycerols do not form micelles.

20. Describe the common feature or features of complex and simple lipids, and provide an example of each class of compound. In what way do complex and simple lipids differ?

21. What is the connection between nucleic acids, amino acids and proteins?

22. What is the difference between the nucleotides adenosine monophosphate and deoxyadenosine monophosphate?

23. If a single strand of DNA contains the nucleotide sequence A-G-G-C-T-C-A-G-C-T-A-G, reading from the 5' to the 3' carbon, what would be the sequence of nucleotides on the complementary strand of the DNA double helix, reading in the same direction?

24. What sequence of amino acids would be specified by a single strand of DNA containing the nucleotide sequence A-G-G-C-T-C-A-G-C-T-A-G, reading from the 5' to the 3' carbon?

25. Distinguish between aerobic and anaerobic processes.

26. The synthesis of organic molecules in nature from carbon dioxide and water is an endothermic process. What is the source of the required energy?

27. Discuss the manner in which the energy requirements of a cell are satisfied.

28. The degradation both of fatty acids, such as palmitic acid, and of carbohydrates, such as glucose, leads to the formation of carbon dioxide and water. However, the degradation of palmitic acid produces $2\frac{1}{2}$ times as much energy as glucose, per gram of substance consumed. Explain this observation in terms of the average oxidation states of the carbon atoms in fatty acids, carbohydrates, and carbon dioxide.

29. What evidence do we have that a twenty-first genetically coded amino acid will not be found?

30. How would you explain the observation that proteins can contain additional amino acids that are structurally related to the 20 genetically coded amino acids?

Problems

1. If the degradation of one mole of glucose and one mole of palmitic acid yields 38 and 129 moles of ATP, respectively, calculate the moles of ATP produced per gram of lipid or carbohydrate consumed.

    Ans. *0.21 mol ATP/g glucose;*
    0.504 mol ATP/g palmitic acid

2. How many grams of invert sugar can be produced by the hydrolysis of 1.00 g of sucrose, $C_{12}H_{22}O_{11}$?

    Ans. *1.05 g of invert sugar*

3. The molecular weight of an unknown triglyceride can be estimated by determining the mass of KOH required to saponify a known mass of lipid. What is the molecular weight of a triglyceride if 3.12 mg of the triglyceride requires 0.590 mg of KOH for saponification?   Ans. *890 g/mole (tripalmitylglycerol)*

4. Fungal lactase, a blue protein found in wood-rotting fungi, contains approximately 0.397% copper by mass. If the molecular weight of fungal lactase is approximately 64,000, how many copper atoms are there in each protein molecule?

    Ans. *4 Cu atoms/protein molecule*

References

"Structure of Protein Molecules," L. Pauling, R. B. Corey, and R. Hayward, *Sci. American,* July 1954, p. 51.

"Proteins," P. Doty, *Sci. American,* September 1957, p. 173.

"The Nucleotide Sequences of a Nucleic Acid," R. W. Holley, *Sci. American,* February 1966, p. 30.

"The Genetic Code," F. H. C. Cricke, *Sci. American,* October 1966, p. 55.

"The Three-Dimensional Structure of an Enzyme Molecule," D. C. Phillips, *Sci. American,* November 1966, p. 78.

"The Synthesis of DNA," A. Kornberg, *Sci. American,* October 1968, p. 64.

"The Oxygen Cycle," P. Cloud and A. Gibor, *Sci. American,* September 1970, p. 110.

"RNA-Directed DNA Synthesis," H. M. Temin, *Sci. American,* January 1972, p. 24.

"The Structure of Cell Membranes," C. F. Fox, *Sci. American,* February 1972, p. 31.

"The Sources of Muscular Energy," R. Margaria, *Sci. American,* March 1972, p. 84.

"The Structure and History of an Ancient Protein (Cytochrome-c)," R. E. Dickerson, *Sci. American,* April 1972, p. 58.

"The Chemical Elements of Life," E. Frieden, *Sci. American,* July 1972, p. 52.

"The Visualization of Genes in Action," O. L. Miller, *Sci. American,* March 1973, p. 34.

"The Isolation of Genes," D. D. Brown, *Sci. American,* August 1973, p. 20.

"Protein Shape and Biological Control," D. E. Koshland, Jr., *Sci. American,* October 1973, p. 52.

"The Cell Cycle," D. Mazia, *Sci. American,* January 1974, p. 54.

"The Cooperative Action of Muscle Proteins," J. M. Murray and A. Weber, *Sci. American,* February 1974, p. 58.

"Nutrition and the Brain," J. D. Fernstrom and R. J. Wurtman, *Sci. American,* February 1974, p. 84.

"A Dynamic Model of Cell Membranes," R. A. Capaldi, *Sci. American,* March 1974, p. 27.

"Glycoproteins," N. Sharon, *Sci. American,* May 1974, p. 78.

"Hybrid Cells and Human Genes," F. H. Ruddle and R. A. Kucherlapati, *Sci. American,* July 1974, p. 36.

"A Family of Protein-Cutting Proteins," R. M. Stroud, *Sci. American,* July 1974, p. 74.

"How Actinomycin Binds to DNA," H. M. Sobell, *Sci. American,* August 1974, p. 82.

"The Absorption of Light in Photosynthesis," Govindjee and R. Govindjee, *Sci. American,* December 1974, p. 68.

"A Mechanism of Disease Resistance in Plants," G. A. Strobel, *Sci. American,* January 1975, p. 81.

"Cyanate and Sickle-Cell Disease," A. Cerami and C. M. Peterson, *Sci. American,* April 1975, p. 44.

"The Walls of Growing Plant Cells," P. Albersheim, *Sci. American,* April 1975, p. 81.

"The Molecular Biology of Poliovirus," D. H. Spector and D. Baltimore, *Sci. American,* May 1975, p. 25.

"The Manipulation of Genes," S. N. Cohen, *Sci. American,* July 1975, p. 24.

"The Metabolism of Alcohol," C. S. Lieber, *Sci. American,* March 1976, p. 25.

"Neutron-Scattering Studies of the Ribosome," D. M. Engelman and P. B. Moore, *Sci. American,* October 1976, p. 44.

"Opiate Receptors and Internal Opiates," S. H. Snyder, *Sci. American,* March 1977, p. 44.

"Biological Nitrogen Fixation," W. J. Brill, *Sci. American,* March 1977, p. 68.

"The Recombinant DNA Debate," C. Grobstein, *Sci. American,* July 1977, p. 22.

"The Nucleotide Sequence of a Viral DNA," J. C. Fiddes, *Sci. American,* December 1977, p. 55.

"The Three-Dimensional Structure of Transfer RNA," A. Rich and S. H. Kim, *Sci. American,* January 1978, p. 52.

"How Cells Make ATP," P. C. Hinkle and R. E. McCarty, *Sci. American,* March 1978, p. 104.

"Chemical Evolution and the Origin of Life," R. E. Dickerson, *Sci. American,* September 1978, p. 70.

"The Assembly of a Virus," P. J. G. Butler and A. Klug, *Sci. American,* November 1978, p. 62.

"Hemoglobin Structure and Respiratory Transport," M. F. Perutz, *Sci. American,* December 1978, p. 92.

"Hemoglobin: Model Systems Shed Light on Oxygen Binding," T. H. Maugh, II, *Science,* **187,** 154 (1975).

"Structural Domains of t-RNA Molecules," G. Quigley and A. Rich, *Science,* **194,** 796 (1976).

"The Hydrophobic Effect and the Organization of Living Matter," C. Tanford, *Science,* **200,** 1012 (1978).

"Electrostatic Effects in Proteins," M. F. Perutz, *Science,* **201,** 1187 (1978).

"From Peptide Synthesis to Protein Synthesis," M. Bodansky and A. A. Bodansky, *Am. Scientist,* **55,** 2 (1967).

"Photosynthesis as a Resource for Energy and Material," M. Calvin, *Am. Scientist,* **64,** 270 (1976).

"Three-Dimensional Structures and Chemical Mechanisms of Enzymes," W. N. Lipscomb, *Chem. Soc. Rev.,* **1,** 319 (1972).

"Prostaglandins: Tomorrow's Drugs," W. E. Horton, *Chem. Soc. Rev.,* **4,** 589 (1975).

"Chemical Carcinogens and their Significance for Chemists," C. E. Searle, *Chem. in Britain,* **6,** 5 (1970).

"Thermodynamics of the Hydrolysis of Adenosine Triphosphate," R. A. Alberty, *J. Chem. Educ.,* **46,** 713 (1969).

"Recollections of Personalities Involved in the Early History of American Biochemistry," W. C. Rose, *J. Chem. Educ.,* **46,** 759 (1969).

"Solid Phase Synthesis," D. C. Neckers, *J. Chem. Educ.,* **52,** 695 (1975).

"Non-Covalent Interactions, Key to Biological Flexibility and Specificity," E. Frieden, *J. Chem. Educ.,* **52,** 754 (1975).

27

Silicon and
Its Compounds

In the preceding two chapters we have examined the chemical behavior of carbon, the first member of Group IVA, and its compounds. This chapter deals with the second member of the group, silicon. Whereas carbon, with its ability to form compounds containing carbon-carbon bonds, plays the dominant structural role in the animal and vegetable worlds, silicon, which readily forms compounds containing Si—O—Si bonds, is of prime importance in the mineral world. Indeed, the name silicon is derived from the Latin word for flint, *silex*.

27.1 The Occurrence of Silicon

The earth's crust is composed almost entirely of minerals in which silicon atoms are connected by oxygen atoms in complex structures involving chains, layers, and three-dimensional frameworks. The minerals constitute the bulk of most common rocks (except limestone and dolomite), and of soils, clays, and sands, which are the products of the weathering of rocks. Inorganic building materials such as granite, bricks, cement, mortar, ceramics, and glasses are composed of silicon compounds. Silicon composes nearly one-fourth of the mass of the earth's crust, being second only to oxygen in abundance. Its binary compound with oxygen is called **silica,** SiO_2. Familiar forms of impure silica are sand and sandstone. Silica also occurs as quartz, amethyst, agate, and flint.

Silica is acidic in character and, at high temperatures, it combines with basic

TABLE 27-1   Some Important Minerals
Containing Silicon

Quartz	SiO_2
Asbestos	$H_4Mg_3Si_2O_9$
Natrolite (a zeolite)	$Na_2(Al_2Si_3O_{10}) \cdot 2H_2O$
Garnet	$Ca_3Al_2(SiO_4)_3$
Mica	$K_2Al_2(AlSi_3O_{10})(OH)_2$
Talc	$Mg_3(Si_4O_{10})(OH)_2$
Kaolin (a clay)	$Al_2Si_2O_5(OH)_4$
Feldspar	$KAlSi_3O_8$
Beryl	$Be_3Al_2Si_6O_{18}$

metal oxides forming silicates, in an acid-base reaction.

$$CaO + SiO_2 \longrightarrow \underset{\text{Calcium silicate}}{CaSiO_3}$$

Most silicate rocks are built up of the common metal cations and complex silicate anions. A great variety of silicates exists in nature. A few of the more important silicates are given in Table 27-1. Most of these silicates have important specific uses, but they are such stable compounds that it is not economically feasible (except for beryl) to use them as sources of the metals they contain.

27.2   Preparation of Silicon

Elemental silicon was first prepared in an impure, amorphous form by Berzelius in 1823 by heating silicon tetrafluoride with potassium. It can also be obtained by the action of strong reducing agents at high temperatures upon silicon dioxide. With carbon and magnesium as the reducing agents, the equations are

$$SiO_2(s) + 2C(s) \longrightarrow Si(s) + 2CO(g)$$
$$SiO_2(s) + 2Mg(s) \longrightarrow Si(s) + 2MgO(s)$$

Extremely pure silicon, such as that required for the fabrication of semiconductor electronic devices, is prepared by the decomposition of silicon tetrahalides or silane, SiH_4, at high temperatures.

27.3   Properties and Uses of Silicon

Silicon crystallizes with a diamondlike structure in which each silicon atom is covalently bonded to four neighboring silicon atoms at the corners of a regular tetrahedron. Thus a single crystal of silicon is a three-dimensional giant molecule. Silicon is hard enough to scratch glass, melts at 1410°, and is a brittle, gray-black, metallic-appearing solid.

Silicon is inactive at low temperatures and resists attack by air, water, and acids. However, it is sufficiently active that it does not occur free on the earth's surface. It reacts with strong oxidizing agents and strong bases. It dissolves in hot sodium hydroxide or potassium hydroxide solutions, forming silicates and hydrogen.

$$Si + 2OH^- + H_2O \longrightarrow SiO_3^{2-} + 2H_2(g)$$

Silicon is attacked by the halogens at high temperatures, with the formation of volatile tetrahalides, such as SiF_4, and it oxidizes in air at elevated temperatures to give the dioxide, SiO_2.

Elemental silicon is used as a deoxidizer in the production of steel, copper, and bronze, and in the manufacture of certain acid-resistant alloys such as Duriron and Tantiron. Highly purified silicon, containing no more than one part impurity per million parts silicon, is used in semiconductor electronic devices such as transistors, microcomputers, and solar cells (Section 20.18). See Fig. 27-1.

Figure 27-1 Single crystals of silicon. Such crystals are produced for use in solar cells, semiconductors, and other devices. (*Courtesy Dow Corning Corporation*)

27.4   The Chemical Behavior of Silicon

Silicon is a metalloid; it exhibits both nonmetallic and metallic behavior. Although both carbon and silicon have a valence electron shell containing two s and two p electrons, little of the chemistry of silicon can be inferred from that of carbon except for the stoichiometry of some simple halides and hydrides. Silicon forms compounds in which it is sp^3 or d^2sp^3 hybridized with four or six single covalent bonds, respectively. Silicon does not form double or triple bonds. Silicon forms a series of hydrides and a series of halides, both with the formulas Si_nX_{2n+2} (where X is hydrogen or a halide) containing Si—Si single bonds. However, these compounds are much more reactive, much less stable, and much more limited in number than the corresponding carbon compounds inasmuch as the Si—Si bond (D_{Si-Si} 175 kJ/mol) is much weaker than the C—C bond (D_{C-C} 350 kJ/mol). Si—O, Si—C, and Si—F bonds are particularly stable. Exposure of most silicon compounds to water or oxygen results in their decomposition to compounds containing Si—O bonds, unless the compounds are stabilized by the presence of Si—O, Si—C, or Si—F bonds.

The valence shell of silicon contains d orbitals, and the valence shell can be expanded beyond the octet. Thus silicon compounds of the general formula SiX_4, where X is a highly electronegative group, can act as Lewis acids giving six-coordinate silicon (six bonds) with d^2sp^3 hybridization. For example, silicon tetrafluoride (SiF_4) reacts with sodium fluoride to give $Na_2[SiF_6]$, which contains the octahedral $[SiF_6]^{2-}$ ion.

The presence of d orbitals in the valence shell of silicon also contributes to the reactivity of silicon compounds. For example, when $SiCl_4$ is exposed to water, it reacts rapidly in a stepwise fashion:

$$\underset{\substack{\text{Cl}\\ \text{|}\\ \text{Cl}}}{\text{Cl}-\text{Si}-\text{OH}} + \text{H}_2\text{O} \longrightarrow \begin{array}{c} \overset{\text{H}}{\underset{\text{Cl}}{\text{Cl}-\text{Si}}} \overset{\text{O}-\text{H}}{\underset{\text{OH}}{}} \end{array} \longrightarrow \underset{\substack{\text{Cl}}}{\overset{\text{HO}}{\text{Cl}-\text{Si}-\text{OH}}} + \text{HCl}$$

This process is repeated until all four chlorine atoms are replaced by hydroxide groups, $Si(OH)_4$ being the final product of the hydrolysis. The overall reaction is

$$SiCl_4 + 4H_2O \longrightarrow Si(OH)_4 + 4HCl$$

Silicic acid, $Si(OH)_4$ or H_4SiO_4, is unstable and gradually dehydrates to SiO_2.

$$Si(OH)_4 \longrightarrow SiO_2 + 2H_2O$$

Carbon cannot expand its valence shell beyond the octet. Thus carbon tetrachloride cannot form five-coordinate intermediates of the type suggested for silicon tetrachloride. Hence tetrahedral carbon compounds are much less reactive than silicon compounds.

27.5   Silicon Hydrides

Silicon forms a series of hydrides that includes SiH_4, Si_2H_6, Si_3H_8, Si_4H_{10}, and Si_6H_{14}. The chemical behavior of the hydrides of silicon is decidedly different from that of the hydrocarbons of similar formulas. For example, the silicon hydrides are spontaneously combustible in air whereas the hydrocarbons are not. Only silicon hydrides corresponding to the alkanes are known; none are known that are analogous to the alkenes or alkynes with double or triple bonds. (Only carbon, nitrogen, and oxygen form multiple bonds readily.)

Acids react with magnesium silicide to form **silane,** SiH_4, analogous in formula to methane, CH_4.

$$Mg_2Si + 4H^+ \longrightarrow 2Mg^{2+} + SiH_4(g)$$

Silane is a colorless gas, thermally stable at ordinary temperatures, but spontaneously combustible in air. Silicon dioxide and water are its combustion products.

$$SiH_4(g) + 2O_2(g) \longrightarrow SiO_2(s) + 2H_2O(g) \qquad \Delta H° = -1429 \text{ kJ}$$

Stepwise halogenation of silane is possible, though the reactions are difficult to control. A better method of obtaining partially halogenated silanes involves the use of hydrogen halides in the presence of the corresponding aluminum halide as a catalyst. With HBr, using $AlBr_3$ as the catalyst, the reactions are given by the equations

$$SiH_4 + HBr \longrightarrow SiH_3Br + H_2$$
$$SiH_3Br + HBr \longrightarrow SiH_2Br_2 + H_2$$
$$SiH_2Br_2 + HBr \longrightarrow SiHBr_3 + H_2$$
$$SiHBr_3 + HBr \longrightarrow SiBr_4 + H_2$$

Silane is extremely sensitive to alkalies, reacting easily to give silicates and hydrogen.

$$SiH_4 + 2OH^- + H_2O \longrightarrow SiO_3^{2-} + 4H_2(g)$$

In contrast, methane is unreactive to hydroxides.

Perhaps the most important reaction of compounds containing a Si—H bond, at least from a commercial standpoint, is the reaction with alkenes (Section 25.13).

$$CH_3CH{=}CH_2 + H_2SiCl_2 \longrightarrow CH_3CH_2CH_2SiHCl_2$$

$$CH_3CH{=}CH_2 + CH_3CH_2CH_2SiHCl_2 \longrightarrow (CH_3CH_2CH_2)_2SiCl_2$$

These reactions are used in the preparation of silicones (Section 27.11).

27.6   Silicon Carbide

When a mixture of sand and a large excess of coke is heated in an electric furnace, **silicon carbide (carborundum),** SiC, is produced according to the equation

$$SiO_2(s) + 3C(s) \longrightarrow SiC(s) + 2CO(g) \qquad \Delta H^\circ = 624.7 \text{ kJ}$$

The product comes from the furnace in the form of blue-black iridescent crystals, nearly as hard as diamonds and very stable at high temperatures. The crystals are crushed, the particles are screened to uniform size, mixed with a binder of clay or sodium silicate, molded into various shapes such as grinding wheels, and fired. Silicon carbide is used as an abrasive for cutting, grinding, and polishing.

Silicon carbide exists in many different crystalline forms; yet in each of these forms carbon and silicon atoms have alternate positions, and each atom is surrounded tetrahedrally by four others. One crystalline form has the diamond structure (Section 25.1). In order to rupture a crystal of silicon carbide, a number of very strong covalent bonds must be broken. The high decomposition temperature (above 2200°), the extreme hardness, the brittleness, and the chemical inactivity of silicon carbide are in accord with a diamondlike structure.

27.7   Silicon Halides

All the tetrahalides of silicon, SiX_4, have been synthesized, and several mixed halides of the type $SiCl_2F_2$ have also been prepared. **Silicon tetrachloride** can be prepared by either direct chlorination at elevated temperatures or by heating silicon dioxide with chlorine and carbon. The equations are

$$Si(s) + 2Cl_2(g) \longrightarrow SiCl_4(g)$$

$$SiO_2(s) + 2C(s) + 2Cl_2(g) \longrightarrow SiCl_4(g) + 2CO(g)$$

Silicon tetrachloride is a covalent tetrahedral molecule containing four covalent Si—Cl bonds. It is a low-boiling (57°), colorless liquid that fumes strongly in moist air to produce a dense smoke of finely divided silica as the Si—Cl bonds are replaced by Si—O bonds.

$$SiCl_4(g) + 2H_2O(g) \longrightarrow SiO_2(s) + 4HCl(g)$$

The Si—Cl bonds in silicon tetrachloride can be replaced by Si—C bonds in a stepwise fashion by reaction with a stoichiometric amount of a carbanion such as the methide anion (CH_3^-) in $Na^+CH_3^-$ or with a partially negatively charged alkyl group in a polar covalent bond as found in $\overset{\delta-}{Cl}—\overset{\delta 2+}{Mg}—\overset{\delta-}{C_2H_5}$.

$$SiCl_4 + Na^+CH_3^- \longrightarrow Cl_3SiCH_3 + NaCl$$

$$SiCl_4 + 2ClMgC_2H_5 \longrightarrow Cl_2Si(CH_2CH_3)_2 + 2MgCl_2$$

Elemental silicon ignites spontaneously in an atmosphere of fluorine, forming **silicon tetrafluoride**, SiF_4, which is a gas. This compound is also readily prepared by the action of hydrofluoric acid upon silica or a silicate.

$$SiO_2(s) + 4HF(g) \longrightarrow SiF_4(g) + 2H_2O(l) \qquad \Delta H° = -191.2 \text{ kJ}$$

$$CaSiO_3(s) + 6HF(g) \longrightarrow SiF_4(g) + CaF_2 + 3H_2O$$

Silicon tetrafluoride hydrolyzes in water and produces **fluosilicic acid** as well as **orthosilicic acid.**

$$3SiF_4 + 4H_2O \longrightarrow \underset{\substack{\text{Orthosilicic} \\ \text{acid}}}{H_4SiO_4(s)} + 4H^+ + \underset{\substack{\text{Fluosilicic} \\ \text{acid}}}{2SiF_6^{2-}}$$

Fluosilicic acid is a stronger acid than sulfuric acid. It is stable only in solution, however, and upon evaporation decomposes according to the equation

$$H_2SiF_6(aq) \longrightarrow 2HF(g) + SiF_4(g)$$

27.8   Silicon Dioxide (Silica)

The usual crystalline form of silicon dioxide is **quartz**—a hard, brittle, clear, colorless solid. It is used in many ways—for architectural decorations, semiprecious jewels, optical instruments, and frequency control in radio transmitters. The contrast in structure and physical properties between silica and its carbon analog, carbon dioxide, is interesting. The unit of structure in solid carbon dioxide (Dry Ice) is the single CO_2 molecule with very weak intermolecular forces holding the building units at the points of the crystal lattice. The low melting point and volatility of Dry Ice reflect these weak crystal forces between its building units. Each of the two oxygen atoms is attached to the central carbon atom by double covalent bonds. In contrast, silicon does not form double bonds; hence quartz crystallizes as a hexagonal closest packed array of oxygen atoms, with silicon atoms occupying one-fourth of the tetrahedral holes. Since silicon does not form double bonds, a silicon atom in quartz is linked to four oxygen atoms by single bonds directed toward the corners of a regular tetrahedron, and the SiO_4 tetrahedra are bonded together by sharing the oxygen atom at the corners of two tetrahedra. This structure extends outward in a three-dimensional, continuous, giant silicon-oxygen network to give a macromolecule of silicon dioxide, the quartz crystal. The overall ratio of silicon to oxygen atoms is 1 to 2, and the simplest formula for the compound is SiO_2, but the formula SiO_2 does not represent a single molecule as does the formula CO_2.

At 1600°, quartz melts to give a viscous liquid with a random internal structure. When the liquid is cooled it does not crystallize readily, but usually undercools and forms a glass, called **silica glass.** The SiO_4 tetrahedra in this glass have the random arrangement characteristic of undercooled liquids, and the glass has some very interesting and useful properties. Silica glass is highly transparent to both visible and ultraviolet light and so finds use in the manufacture of mercury vapor lamps, which give radiation rich in the ultraviolet region of the spectrum. It is also used in the production of certain optical instruments that operate in the ultraviolet region. The coefficient of expansion of silica glass is very low, so it is not easily fractured by sudden changes in temperature. It is also insoluble in water and inert toward acids except hydrofluoric.

$$SiO_2(s) + 4HF(aq) \longrightarrow SiF_4(g) + 2H_2O$$

This reaction is used in the quantitative separation of silica from other oxides, both products of the reaction being volatile. Hot alkali hydroxides and fused alkali carbonates convert silica into soluble silicates (SiO_4^{4-} and SiO_3^{2-}). Typical examples are

$$SiO_2 + 4OH^- \longrightarrow SiO_4^{4-} + 2H_2O$$
$$SiO_2 + Na_2CO_3 \longrightarrow Na_2SiO_3 + CO_2$$

The latter reaction is employed in the conversion of silicate rocks to soluble forms for analysis and is part of the glass-making process (Section 27.12).

27.9   Silicic Acids

Orthosilicic acid, H_4SiO_4, is an extremely weak acid. It cannot be formed from its acid anhydride, SiO_2, by hydration, due to the fact that silica is very nearly insoluble in water. However, the addition of a strong mineral acid to a solution of an alkali metal silicate results in the formation of orthosilicic acid.

$$SiO_4^{4-} + 4H_3O^+ \longrightarrow H_4SiO_4(s) + 4H_2O$$

The orthosilicic acid first appears as a colloidal dispersion and then very shortly forms a colloidal gelatinous mass. In this form orthosilicic acid adsorbs certain ions to the extent that they cannot be removed by washing with water. It is thought that this is the mechanism by which colloidal silicic acids present in soils retain ions of soluble salts essential to plant growth.

The loss of water from orthosilicic acid by heating results in silica being formed as the final product of dehydration. Although there is no evidence to support the mechanism, it is thought that the dehydration is stepwise, with the formation of **metasilicic acid** (H_2SiO_3), **disilicic acid** ($H_6Si_2O_7$), **trisilicic acid** ($H_4Si_3O_8$), and perhaps other polysilicic acids as intermediates. Salts of these and other polysilicic acids are found among the naturally occurring silicates.

Silica gel is obtained when gelatinous orthosilicic acid is dehydrated until it contains only a small percentage of moisture. This gel has an open, porous structure with a large surface area per unit of mass. It has a great tendency to adsorb gases and to catalyze certain chemical reactions involving substances in the gaseous state.

27.10   Natural Silicates

As a group, the silicates are characterized by large variations in the silicon-oxygen ratios that occur from one silicate to another. In the molecules of all of the silicates, however, sp^3 hybridized silicon atoms are to be found at the centers of oxygen tetrahedra, and thus the tetravalency of silicon is maintained. The variation in the silicon-oxygen ratio is due to the fact that the silicon-oxygen tetrahedra may exist as discrete and independent building units, or they may share atoms at corners, edges, and more rarely faces, in a variety of ways. The silicon-oxygen ratio varies with the *extent* of sharing of oxygen atoms by silicon atoms in the linking together of the tetrahedra.

It is convenient to classify the silicates whose structures are known into a few groups, based upon the manner of linking of the silicon-oxygen tetrahedra.

Figure 27-2 The tetrahedral structure of the $SiO_4{}^{4-}$ ion. (Only the oxygen atoms are shown; the silicon atom, not shown, is at the center.)

▶ **1. INDIVIDUAL SiO_4 TETRAHEDRA EXISTING AS INDEPENDENT GROUPS IN THE CRYSTAL LATTICE.**   Examples are **olivine**, Mg_2SiO_4, and **zircon**, $ZrSiO_4$. The positively charged metallic ions (Mg^{2+}, Zr^{4+}) serve to bind together the negative $SiO_4{}^{4-}$ ions, which have the tetrahedral structure shown in Fig. 27-2. Note that only the oxygen atoms are shown. A silicon atom is in the center of each tetrahedron and is not shown.

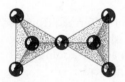

Figure 27-3 The structure of the $Si_2O_7{}^{6-}$ ion. (A silicon atom, not shown, is at the center of each tetrahedron.)

▶ **2. TWO SiO_4 TETRAHEDRA SHARING ONE OXYGEN CORNER AND THUS FORMING Si_2O_7 GROUPS WHICH ACT AS DISCRETE BUILDING UNITS.**   Examples are **hardystonite**, $Ca_2ZnSi_2O_7$, and **hemimorphite**, $Zn_4(OH)_2Si_2O_7 \cdot H_2O$. The cations are to be found between the negative $Si_2O_7{}^{6-}$ ions (Fig. 27-3), binding them together.

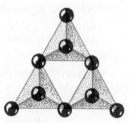

Figure 27-4 The structure of the $Si_3O_9{}^{6-}$ ion. (A silicon atom, not shown, is at the center of each tetrahedron.)

▶ **3. THREE TETRAHEDRA SHARING CORNERS WITH EACH OTHER AND FORMING CLOSED RINGS.**   An example is **benitoite**, $BaTiSi_3O_9$. The $Si_3O_9{}^{6-}$ ions (Fig. 27-4) are held together by the positive metallic ions. The ring ions are arranged in sheets with their planes parallel.

Beryl (emerald) has the formula $Be_3Al_2Si_6O_{18}$; its structure involves six SiO_4 tetrahedra sharing corners to form a closed ring.

▶ **4. SINGLE SILICON-OXYGEN ENDLESS CHAINS FORMED OF SiO_4 TETRAHEDRA, EACH TETRAHEDRON SHARING TWO OXYGEN ATOMS.**   This structure gives an empirical composition of $SiO_3{}^{2-}$ although there are no $SiO_3{}^{2-}$ ions present as independent groups. An example is **diopside**, $CaMg(SiO_3)_2$. The chains (Fig.

27-5) extend the full length of the crystal, and the parallel chains are held together by the attraction of the positively charged metal ions lying between them.

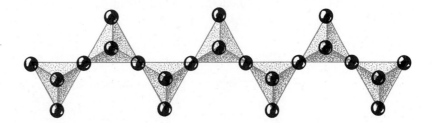

Figure 27-5 A portion of an SiO_3^{2-} chain. (A silicon atom, not shown, is at the center of each tetrahedron.)

▶ **5. DOUBLE SILICON-OXYGEN CHAINS.** In the single- and double-chain crystals the metallic ions link the parallel chains together. The fact that these ionic linkages are not as strong as the silicon-oxygen bonds within the chains accounts for the fibrous nature of **asbestos,** a chain-type silicate. An example is **tremoline,** $Ca_2Mg_5(Si_4O_{11})_2(OH)_2$. These silicates always contain some hydroxyl groups, which are attached to the metal atoms and never to the silicon atoms. The double chains may be represented as shown in Fig. 27-6.

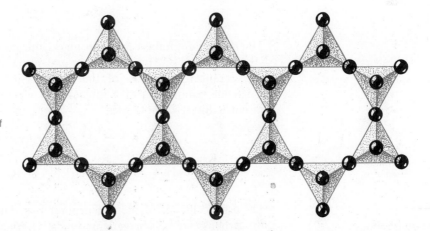

Figure 27-6 A portion of a double silicon-oxygen chain. (A silicon atom, not shown, is at the center of each tetrahedron.)

▶ **6. SILICON-OXYGEN SHEETS FORMED BY THE EXTENSIONS OF DOUBLE CHAINS.** The metal ions form ionic bonds between the sheets. These ionic bonds are weaker than the silicon-oxygen bonds within the sheets. Thus silicate minerals with this structure tend to cleave into thin layers, a property characteristic of the micas. Examples are **talc,** $Mg_3Si_4O_{10}(OH)_2$ and **muscovite,** $KAl_2(AlSi_3O_{10})(OH,F)_2$. Note that aluminum atoms can substitute for some of the silicon atoms, as in muscovite; when this occurs, a larger cationic charge is required to produce a neutral entity.

▶ **7. THREE-DIMENSIONAL SILICON-OXYGEN NETWORKS IN WHICH A PORTION OF THE TETRAVALENT SILICON IS REPLACED BY TRIVALENT ALUMINUM.** The negative charge that results is neutralized by a distribution of positive ions throughout the network. Examples are **feldspar,** $K(AlSi_3O_8)$, and the **zeolites,** such as $Na_2(Al_2Si_3O_{10}) \cdot 2H_2O$.

27.11   Silicones

A modern development in the field of silicon chemistry has been the production of polymeric organosilicon compounds containing Si—O—Si linkages. These compounds are known as **silicones,** and have the general formula $(R_2SiO)_x$. They may be linear, cyclic, or cross-linked polymers of the types shown.

Linear silicone   Cyclic silicone

Cross-linked silicone

The R in the formulas represents an organic group, such as methyl (CH_3), ethyl (C_2H_5), or phenyl (C_6H_5). The linear and cyclic silicones are produced by hydrolyzing organochlorosilanes of the type R_2SiCl_2, followed by polymerization through the elimination of a molecule of water from two hydroxyl groups of adjacent $R_2Si(OH)_2$ molecules.

$$R_2SiCl_2 + 2H_2O \longrightarrow R_2Si(OH)_2 + 2HCl$$

The organosilicon polymers incorporate, to some extent, the properties of both hydrocarbons and oxysilicon compounds. Organosilicones are remarkably stable toward heat and chemical reagents and are not wetted by water. Depending upon the extent of polymerization and molecular complexity, the silicones may take the form of oils, greases, rubberlike substances, and resins. They are used as lubricants, as hydraulic fluids, for electrical insulators, and as moisture-proofing agents.

A particularly valuable property of silicone oils is their very low coefficient of viscosity. Thus the viscosity of these oils changes so little as the temperature changes that they can be employed as lubricants where there are extreme variations in temperature.

Such materials as paper, wool, glass, silk, and porcelain can be coated with a water-repellent film by simply exposing them for a second or two to the vapor of trimethylchlorosilane, $(CH_3)_3SiCl$. As the OH groups on the surface react, the surface becomes coated with a thin film of $(CH_3)_3Si$—O— groups and repels water because it is covered with a monolayer similar to that of a hydrocarbon film.

$$(Surface)—OH + Cl—Si(CH_3)_3 \longrightarrow (Surface)—O—Si(CH_3)_3 + HCl$$

27.12   Glass

The common glass used for windowpanes, bottles, dishes, and the like is a mixture of sodium and calcium silicates with an excess of silica. It is made by heating together sand, sodium carbonate (or sodium sulfate), and calcium carbonate.

$$Na_2CO_3 + SiO_2 \longrightarrow Na_2SiO_3 + CO_2(g)$$
$$Na_2SO_4 + SiO_2 \longrightarrow Na_2SiO_3 + SO_3(g)$$
$$CaCO_3 + SiO_2 \longrightarrow CaSiO_3 + CO_2(g)$$

After the bubbles of gas have been expelled, a clear viscous melt results. This material is poured into molds or stamped with dies to produce pressed glassware. Articles such as bottles, flasks, and beakers are formed by taking a lump of the molten glass on a hollow tube, inserting it into a mold and blowing with compressed air until the outline of the mold is assumed. The conventional method of making high-quality plate glass is to draw the glass from the furnace in broad strips, roll it to the desired thickness, grind it flat, and polish it on both sides. A new technique for making plate glass (float glass) involves pouring the molten glass on a layer of very pure molten tin. The tin surface is perfectly smooth, so when the glass hardens the glass surface floating on the molten tin is also perfectly smooth and does not need to be ground and polished.

Glassware is "annealed" by heating it for a time just below the softening temperature and then cooling it slowly. Annealing avoids internal strains and thereby reduces the danger of breakage from shock or temperature change.

Glass is a complex mixture of silicates and is classified as an undercooled liquid. It is transparent, brittle, and entirely lacking in the ordered internal structure characteristic of crystals. When heated, it does not melt sharply. Instead it gradually softens until it reaches the liquid state.

When a colored glass is desired, an appropriate substance is added during the manufacture of the glass. Table 27-2 lists a few of the substances used in coloring glass.

TABLE 27-2   Substances Used in the Production of Colored Glass

Substance	Color Produced	Substance	Color Produced
Iron(II) compounds	Green	Cobalt(II) oxide	Blue
Iron(III) compounds	Yellow	Manganese dioxide	Violet
Uranium compounds	Yellow, green, fluorescence	Calcium fluoride	Milky
		Tin(IV) oxide	Opaque
Colloidal selenium	Ruby	Copper(I) oxide	Red, green, blue
Colloidal gold	Red, purple, blue		

If sodium is replaced by potassium in a glass melt, a higher-melting, harder, and less soluble glass is obtained. If part of the calcium is replaced by lead a glass of high density and high refractive index is formed. This variety of glass is called **flint glass** and is used in making lenses and cut-glass articles. **Pyrex glass,** very suitable for articles subject to sudden changes in temperature and resistant to chemical action, is a borosilicate glass in which some of the silicon atoms are replaced by boron atoms.

One form of **safety glass,** used in the manufacture of automobile windshields, consists of a thin layer of plastic held between two pieces of thin plate glass. Adhesion of the glass to the flexible plastic decreases the danger from flying glass and jagged edges when the glass is broken.

Enamels on iron kitchen utensils, sinks, and bathtubs and glazes on pottery are made of easily fusible glass containing opacifiers such as titanium dioxide and tin(IV) oxide.

27.13   Cement

Portland cement is essentially powdered calcium aluminosilicate which sets to a hard mass when treated with water. It is made by pulverizing a mixture of limestone ($CaCO_3$) and clay (an aluminosilicate), and roasting the powder in a rotary kiln heated by gas or powdered coal to a temperature of about 1500°. This treatment yields sintered lumps called "clinker" about the size of small marbles. The clinker is then ground with a little gypsum, $CaSO_4 \cdot 2H_2O$, to a very fine powder.

Roads, walks, foundations, and floors are constructed of **concrete,** a substance made by adding water to a mixture of Portland cement, sand, and stone or gravel. Experimental mixtures incorporating such substances as rubber or plastic and powdered glass have been tried in efforts, at least partially successful, to improve the structural properties of concrete.

The reactions taking place as the Portland cement mixture "sets" (starts the hardening process to form concrete) are complex and not completely understood. It is known that during the process calcium aluminate hydrolyzes, forming calcium hydroxide and aluminum hydroxide. These compounds then react with the calcium silicates present, forming calcium aluminosilicate in the form of interlocking crystals. Portland cement sets rapidly (within 24 hours), and then "hardens" slowly, years being required for completion of the reactions.

Questions

1. What is the electronic configuration of a silicon atom? What orbitals are present in the valence shell of silicon? Include the unfilled orbitals in the valence shell.

2. Describe the hybridization and the bonding of a silicon atom in elemental silicon.

3. Describe the molecular structure and the hybridization of silicon in each of the following molecules or

ions: $SiCl_4$, SiO_4^{4-}, $(CH_3)_2SiH_2$, $[SiF_6]^{2-}$, Si_3H_8.

4. Write Lewis structures for the species given in Question 3.

5. Write two equations for reactions in which silicon exhibits metallic behavior and two equations for reactions in which silicon exhibits nonmetallic behavior.

6. By heating a mixture of sand and coke in an electric furnace, either silicon or silicon carbide may be obtained. What determines which will be formed?

7. In terms of molecular and crystal structure, account for the low melting point of Dry Ice (solid CO_2) and the high melting point of SiO_2.

8. Write balanced equations describing at least two different reactions in which solid silica exhibits acidic character.

9. How does the following reaction show the acidic character of SiO_2? Which acid-base system applies?

$$CaO + SiO_2 \xrightarrow{\Delta} CaSiO_3$$

10. Describe the conversion of silicate rocks to soluble forms for the purpose of analysis.

11. Which would you expect to be a better Lewis base: the carbide ion, C^{4-}, or the silicide ion, Si^{4-}?

12. Account for the existence of the great variety of silicates in terms of the manner of linking of SiO_4 tetrahedra.

13. Give equations for the reactions involved in the production of common window glass.

14. How does the incorporation of each of the following substances affect the properties of glass? potassium oxide, cobalt(II) oxide, lead(II) oxide, boric acid, iron(II) oxide, colloidal gold

15. How does the internal structure of silica glass differ from that of quartz?

16. In what ways does the chemistry of silicates resemble that of phosphates?

17. How does the structure of acetone $(CH_3)_2CO$ differ from the structure of the silicone $[(CH_3)_2SiO]_3$?

18. Compare and contrast the chemistry of carbon and that of silicon.

19. Write equations for the reaction of silicon with the following (heated if needed for reaction to occur): F_2, $NaOH$, O_2, N_2.

20. Write equations to contrast the hydrolysis of SiF_4, $SiCl_4$, and CCl_4.

21. Write equations for the combustion of SiH_4 and Si_3H_8.

22. Carbon forms carbonic acid, H_2CO_3, whereas silicon forms silicic acid, H_4SiO_4. Explain. (*Hint:* Draw the Lewis formulas for the two compounds.)

Problems

1. Silicon reacts with sulfur at elevated temperatures. If 0.0923 g of silicon reacts with sulfur to give 0.3030 g of silicon sulfide, determine the empirical formula of silicon sulfide.   *Ans.* SiS_2

2. A hydride of silicon prepared by the reaction of Mg_2Si with acid exerted a pressure of 306 torr at 26°C in a bulb with a volume of 57.0 ml. If the mass of the hydride was 0.0861 g, what is its molecular weight? What is the molecular formula of the hydride?   *Ans.* 92.0; Si_3H_8

3. (a) From the table of bond energies given in Section 18.6, calculate the approximate enthalpy change for the following reaction:

$$SiH_4(g) + 4HF(g) \longrightarrow SiF_4(g) + 4H_2(g)$$

Ans. $-452\ kJ$

(b) Using bond energy data calculate the approximate enthalpy change for the corresponding reaction with methane.

$$CH_4(g) + 4HF(g) \longrightarrow CF_4(g) + 4H_2(g)$$

Ans. $+435\ kJ$

(c) Why do the two enthalpy changes differ so greatly?

(d) Which reaction is most likely to proceed spontaneously?

References

"Silicon," E. Abel, *Educ. in Chem.,* March 1978, p. 48.

"Some Aspects of *d*-Orbital Participation in Phosphorus and Silicone Chemistry," J. E. Bissey, *J. Chem. Educ.,* **44,** 95 (1967).

"*d*-Orbitals in Main Group Elements," T. B. Brill, *J. Chem. Educ.,* **50,** 392 (1973).

"The Solidification of Cement," D. D. Double and A. Hellawell, *Sci. American,* July 1977, p. 82.

"Asbestiform Chain Silicates: New Minerals and Structural Groups," D. R. Veblen, P. R. Buseck, and C. W. Burnham, *Science,* **198,** 359 (1977).

"The Chemistry of Concrete," S. Brunauer and L. E. Copeland, *Sci. American,* April 1964, p. 80.

"One Hundred Years of Organosilicon Chemistry," R. Müller (transl. by E. G. Rochow), *J. Chem. Educ ,* **42,** 41 (1965).

"The Nature of Ceramics," J. J. Gilman, *Sci. American,* September 1967, p. 112.

"Microelectronics," B. N. Noyce, *Sci. American,* September, 1977, p. 62.

"The Small Electronic Calculator," E. W. McWhorter, *Sci. American,* March 1976, p. 88.

"Amorphous Silicon: A New Direction for Semiconductors," A. L. Robinson, *Science,* **197,** 851 (1977).

"Artificial Organs," H. J. Sanders, *Chem. and Eng. News:* Part I, April 6, 1971, p. 32. Part II, April 12, 1971, p. 68.

"A New Approach to Inorganic Polymers," N. H. Ray, *Endeavour,* January, 1975, p. 9.

"Chemistry of Fossilization," L. Huestis, *J. Chem. Educ.,* **53,** 270 (1976).

"The Chemistry of Silicon Difluoride," D. L. Perry and J. L. Margrave, *J. Chem. Educ.,* **53,** 696 (1976).

"Ancient Glass," R. H. Brill, *Sci. American,* November 1963, p. 120.

"The Nature of Glasses," R. J. Charles, *Sci. American,* September 1967, p. 126.

"Chemistry in Glass Technology," R. W. Douglas, *Chem. in Britain,* **5,** 349 (1969).

"Glass Formation and Crystal Structure," J. F. G. Hicks, *J. Chem. Educ.,* **51,** 28 (1974).

"Desert Varnish: The Importance of Clay Minerals," R. M. Potter and G. R. Rossman, *Science,* **196,** 1447 (1977).

28

Boron and
Its Compounds

Although boron and silicon are members of different groups of the Periodic Table (the periodic groups IIIA and IVA, respectively), in many respects they resemble each other rather closely in chemical behavior. In addition to the similarities between elements in the same family, similarities also occur in a number of cases between elements in pairs, one element of which is in both the next family and the next group from the other element—a diagonal relationship in the Periodic Table. For example, lithium is similar in chemical properties to magnesium; beryllium is similar to aluminum; and, as we have just noted, boron is similar to silicon. The explanation of this phenomenon lies in the fact that in each case the ratio of the atomic radius to the nuclear charge for each of the two elements is nearly the same, which means that the outer-shell electrons are attracted to the nucleus with about the same force.

A comparison of some of the properties of boron with those of aluminum (the second member of Group IIIA) and with those of silicon (the second member of Group IVA) illustrates this behavior. Boron resembles aluminum in forming compounds in which these elements are trivalent—for example, B_2O_3 and Al_2O_3, $B(OH)_3$ and $Al(OH)_3$, and BCl_3 and $AlCl_3$. However, BCl_3 is more like $SiCl_4$ than $AlCl_3$ in physical and chemical properties. Boron and silicon chlorides have melting and boiling points much lower than those of aluminum chloride. Aluminum chloride hydrolyzes reversibly,

$$AlCl_3 + 3H_2O \rightleftharpoons Al(OH)_3(s) + 3H^+ + 3Cl^-$$

whereas boron trichloride and silicon tetrachloride hydrolyze almost completely and irreversibly.

$$BCl_3 + 3H_2O \longrightarrow B(OH)_3(s) + 3H^+ + 3Cl^-$$

$$SiCl_4 + 4H_2O \longrightarrow Si(OH)_4(s) + 4H^+ + 4Cl^-$$

Boron and silicon are primarily nonmetallic in their chemical properties, but both show some metallic character in the elemental state. For this reason, boron and silicon are included in the elements referred to as **semimetallic elements,** or as **metalloids** (Section 8.3), which are elements intermediate in nature between metals and nonmetals. Other elements commonly classified as metalloids are germanium, arsenic, antimony, tellurium, polonium, and astatine.

28.1 Occurrence and Preparation of Boron

The boron-containing compound "borax" is referred to in early Latin works on chemistry. The element boron was first prepared in an impure form in 1808 by Gay-Lussac and Thénard in France and by Davy in England, by reducing boric acid with potassium.

Boron constitutes less than 0.001% of the earth's crust. It does not occur in the free state in nature, but is found in compounds with oxygen. Boron is widely distributed as **boric acid,** $B(OH)_3$ (or H_3BO_3), in volcanic regions; and it occurs as borates, such as **borax,** $Na_2B_4O_7 \cdot 10H_2O$, **kernite,** $Na_2B_4O_7 \cdot 4H_2O$, and **colemanite,** $Ca_2B_6O_{11} \cdot 5H_2O$, in dry lake regions including the desert areas of southern California.

Impure boron is usually prepared by reducing the oxide with powdered magnesium.

$$B_2O_3 + 3Mg \longrightarrow 2B + 3MgO$$

The magnesium oxide is removed by dissolving it in hydrochloric acid. Prepared in this way, boron is an impure brown amorphous powder. Pure boron may be obtained by passing a mixture of the trichloride and hydrogen either through an electric arc or over a hot tungsten filament.

$$2BCl_3(g) + 3H_2(g) \xrightarrow{1500°} 2B(s) + 6HCl(g) \qquad \Delta H° = 253.7 \text{ kJ}$$

28.2 Properties of Elemental Boron

Figure 28-1 The icosahedron, an important geometric figure in the structure of boron and several of its compounds.

Crystalline boron is transparent and nearly as hard as diamond. Elemental boron crystals have a rather unusual structure, being made up of tightly packed small icosahedra. An icosahedron is a symmetrical geometric figure with 20 faces, each of which is an equilateral triangle (Fig. 28-1). The faces meet at 12 corners. In the boron structure, a boron atom occupies each of the 12 corners. In the most common form of boron (α-rhombohedral), the icosahedra are packed together in a manner similar to cubic closest packing of spheres. The length of the bonds between adjacent boron atoms in each icosahedron is 1.76 Å. However, between the icosahedra themselves, two kinds of bonds of different lengths (1.71 Å and 2.03 Å) are present (Fig. 28-2).

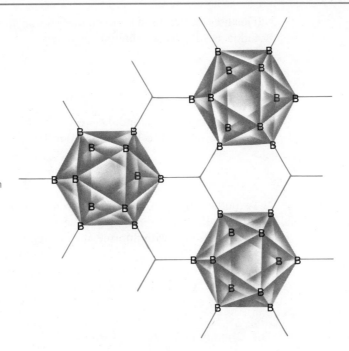

Figure 28-2 The structure of α-rhombohedral boron. Each icosahedron contains twelve boron atoms, six of which are each bonded to a boron atom in another icosahedron by two-center bonds (1.71 Å) and six of which are each bonded to two boron atoms, each in a separate icosahedron, by three-center bonds (2.03 Å). Only the three-center bonds between icosahedra are shown here.

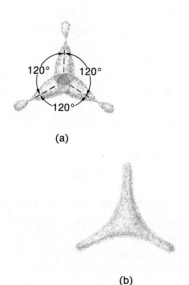

Figure 28-3 The three-center B—B—B bond involving the overlap of three sp^3 atomic orbitals, which share one pair of electrons. (a) The overlap diagram for the atomic orbitals. (b) The overall outline of the bonding molecular orbital.

Half of the twelve boron atoms of an icosahedron are each joined by regular B—B two-center bonds (1.71 Å in length) to a boron atom of another icosahedron (these bonds are not shown in Fig. 28-2). In Chapters 5 and 6, we discussed the *two-electron two-center* bond, in which two atoms are bonded together by one pair of electrons in a bond that results from the overlap of two atomic orbitals (see Section 6.1). The other half of the twelve boron atoms of each icosahedron are *each* bonded to *two* boron atoms, one in each of two other icosahedra, by a *two-electron three-center bond* (2.03 Å in length), as indicated in Fig. 28-2.

In a two-electron three-center bond, *three* atoms are bonded together by one pair of electrons in a molecular orbital that results from the overlap of three atomic orbitals. The three-center bond is especially likely for atoms that do not have sufficient electrons to satisfy ordinary two-center bond requirements. Of all the known elements, boron makes the greatest use of the three-center bond. It exhibits this type of bond not only in the elemental boron structure but also in several boron hydrides (Section 28.4). As was pointed out in Chapter 6, in the regular two-center bond, one bonding molecular orbital and one antibonding molecular orbital result from the overlap of two atomic orbitals (Section 6.1). In the three-center bond, one bonding molecular orbital and two antibonding molecular orbitals result from the overlap of *three* atomic orbitals. The two electrons occupy the bonding molecular orbital. The formation of the bonding orbital is shown in Fig. 28-3. The atomic orbitals involved for boron are largely *p* in character but are slightly less symmetric due to some hybridization with the *s* orbital to give hybrid *sp* orbitals.

The three-center bond between three boron atoms (sometimes referred to by the notation B—B—B) is usually designated

and the three-center bond between two boron atoms and a hydrogen atom (sometimes referred to by the notation B—H—B) is usually designated

Two other forms of elemental boron also are icosahedral in nature but have somewhat different bonding between the individual icosahedra.

Boron combines with fluorine at room temperature and with chlorine, bromine, oxygen, and sulfur at elevated temperatures. It does not react with iodine directly. Boron reacts with carbon at the temperature of the electric arc and forms boron carbide, B_4C, which ranks next to diamond in hardness and is used in cutting tools in industry.

28.3   The Chemical Behavior of Boron

The valence shell electronic configuration of a boron atom is $2s^2 2p^1$; thus boron forms compounds such as BF_3 and B_2O_3 (Fig. 11-7) with three single covalent bonds to boron. The geometry about boron in such compounds is trigonal planar, and the boron atom is sp^2 hybridized. The p orbital, which is not hybridized, is empty. Consequently, boron compounds with three covalent bonds are strong Lewis acids. The addition of a fourth group in a Lewis acid-base reaction gives a four-coordinate boron atom with sp^3 hybridization with a tetrahedral geometry.

Boron is covalent in its compounds. Its first ionization potential is rather high (8.3 eV), and the next two are much higher. Thus the total ionization energy needed to produce B^{3+} ions is very large, and the formation of these ions plays no part in boron chemistry.

Boron has a large affinity for fluorine and oxygen. With the exception of BF_3, boron compounds hydrolyze or oxidize readily to give boric acid, $B(OH)_3$, or boric oxide, B_2O_3, respectively.

Boron acts as a reducing agent when heated with water, sulfur dioxide, nitric oxide, carbon dioxide, or many other oxides. It is used as a deoxidizing agent in the purification of fused copper before it is cast. It also reduces both concentrated nitric and sulfuric acid, in which reactions the boron is oxidized to boric acid. In most of its chemical properties, boron acts as a nonmetal; its halides are volatile and irreversibly hydrolyzed. However, boron is intermediate in character between nonmetals and metals, as evidenced in part by $B(OH)_3$, which is a weak acid.

28.4   Boron Hydrides

Boron forms a series of volatile hydrides that are quite different from the hydrides of carbon and silicon. No hydride having the molecular formula BH_3 is stable at

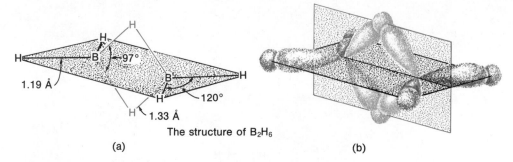

The structure of B_2H_6

(a)   (b)

Figure 28-4 The structure of the diborane molecule, B_2H_6. (a) The spatial arrangement of atoms showing the bridging hydrogens (in color). (b) The spatial arrangement of the bonding orbitals.

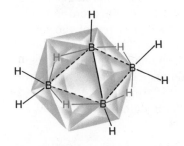

Figure 28-5 The B_4H_{10} molecule. Dashed lines do not indicate bonds, but are given to indicate the portion of the icosahedron occupied by the molecule.

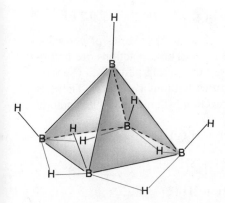

Figure 28-6 The B_5H_9 molecule.

room temperature although there is evidence for the existence of BH_3 (borane) in the gas phase at elevated temperatures. Lewis acid-base adducts, such as H_3BCO, are known. The simplest and most important of the boron hydrides is diborane, B_2H_6, the dimer of BH_3. However, an electronic structure cannot be written for B_2H_6 that would be in keeping with our theory of regular covalent bonds as outlined in Chapter 5 and with the properties of the compound. Fourteen valence electrons (seven electron pairs) would be required for a covalent structure of the type

$$
\begin{array}{ccc}
\text{H} & \text{H} \\
| & | \\
\text{H}-\text{B}-\text{B}-\text{H} \\
| & | \\
\text{H} & \text{H}
\end{array}
$$

Actually, only twelve electrons are available for bond formation in B_2H_6. Compounds of this sort are sometimes referred to as being "electron deficient." The proposed structure most in keeping with the properties of diborane is one in which two hydrogen bridges bond the two halves of the molecule together (Fig. 28-4). Each boron atom is connected to two hydrogen atoms, all in the same flat plane, by regular two-center single covalent bonds. These four bonds use eight of the twelve available electrons. The two bridge hydrogens are above and below the plane, respectively. Each of the two bridge hydrogens is connected to the two boron atoms by a two-electron three-center B—H—B bond. The three atoms in each bond are bonded by a molecular orbital that contains one pair of electrons, analogous to the B—B—B three-center bond in elemental boron. The two three-center bonds thereby together utilize the remaining four electrons.

Higher hydrides than B_2H_6 are also possible, and many have been made. The B_4H_{10} molecule is pictured in Fig. 28-5. Six of the ten hydrogen atoms are bonded by regular single covalent bonds; the other four by B—H—B two-electron three-center bonds. Note that the boron atoms are arranged at the corners of triangular planes, reminiscent of the icosahedral structure.

The B_5H_9 structure, shown in Fig. 28-6, is a square pyramid with boron atoms at each of the five corners. Five hydrogen atoms are bonded, one to each boron, by a regular single covalent bond. The other four hydrogen atoms are bonded through four two-electron three-center B—H—B bonds.

Figure 28-7 The B_6H_{10} molecule. Dashed lines do not indicate bonds, but are given to indicate the portion of the icosahedron occupied by the molecule.

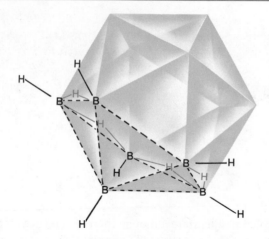

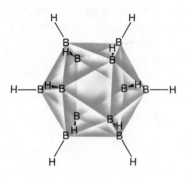

Figure 28-8 The $B_{12}H_{12}^{2-}$ ion.

In the B_6H_{10} structure, diagrammed in Fig. 28-7, six hydrogen atoms are bound by regular covalent bonds and four by three-center bonds. It is interesting to note that the B_6H_{10} structure can be recognized as half of an icosahedral structure.

Several, but not all, of the known higher boron hydrides also exhibit a partial or complete icosahedral structure, with the structure of the icosahedron becoming more complete for higher members of the series. In $B_{10}H_{14}$ ten of the twelve corners of the icosahedron are occupied by boron atoms. The structure has ten two-center B—H bonds and four three-center B—H—B bonds. The $B_{12}H_{12}^{2-}$ ion, in which all B—H bonds are two-center bonds, exhibits a complete icosahedron (Fig. 28-8).

Diborane is readily prepared by the action of lithium aluminum hydride upon boron trifluoride in ether solution.

$$4BF_3 + 3LiAlH_4 \longrightarrow 2B_2H_6(g) + 3LiF + 3AlF_3$$

Above 300° diborane rapidly decomposes into boron and hydrogen. It ignites spontaneously in moist air, the products of combustion being boric oxide and water. Diborane reacts violently with chlorine to form the trichloride and hydrogen chloride. It is hydrolyzed by water to boric acid and hydrogen.

$$B_2H_6(g) + 6H_2O(l) \longrightarrow 2H_3BO_3(s) + 6H_2(g) \qquad \Delta H° = -509.2 \text{ kJ}$$

The higher boron hydrides are prepared from diborane by a series of complex, reversible reactions at temperatures of 100° to 250°. These reactions are analogous to the cracking and reforming reactions used in refining petroleum (Section 25.17) although they occur at lower temperatures without catalysts. There is evidence that BH_3 is an intermediate in these reactions. Some examples include

$$2B_2H_6(g) \underset{}{\overset{100°}{\rightleftharpoons}} B_4H_{10}(g) + H_2(g)$$

$$5B_2H_6(g) \underset{}{\overset{120°}{\rightleftharpoons}} B_{10}H_{14}(g) + 8H_2(g)$$

$$5B_4H_{10}(g) \underset{}{\overset{200°}{\rightleftharpoons}} 4B_5H_9(g) + 7H_2(g)$$

When an excess of sodium hydride acts upon boron trifluoride, the complex

sodium borohydride, $NaBH_4$, is formed. The four hydrogen atoms are covalently bonded to the boron atom in the tetrahedral complex anion $[BH_4]^-$. The formula is shown below. Sodium borohydride and other similar complex hydrides, such as **lithium aluminum hydride,** $LiAlH_4$, are very useful reducing agents, particularly toward organic compounds.

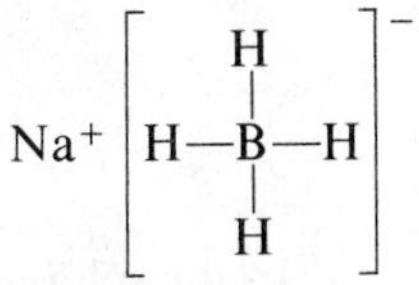

$$Na^+ \left[\begin{array}{c} H \\ | \\ H-B-H \\ | \\ H \end{array} \right]^-$$

Sodium borohydride

28.5   The Boron Halides

As mentioned in Section 28.2, the trifluoride, trichloride, and tribromide of boron can be prepared by direct union of the elements. These are trigonal planar molecules in which the boron is sp^2 hybridized. The fluoride is a colorless gas that hydrolyzes in water to form boric acid and hydrofluoric acid.

$$BF_3(g) + 3H_2O(l) \longrightarrow H_3BO_3(s) + 3HF(g) \qquad \Delta H^\circ = 86.6 \text{ kJ}$$

The hydrofluoric acid thus formed combines with excess boron trifluoride and gives **fluoboric acid,** HBF_4.

$$BF_3 + HF \longrightarrow H^+ + BF_4^-$$

In this latter reaction the BF_3 molecule acts as a Lewis acid (electron-pair acceptor) and accepts a pair of electrons from a fluoride ion, as shown by the equation

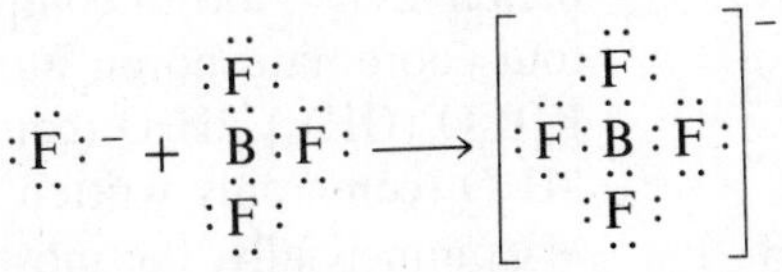

$$:\!\overset{..}{\underset{..}{F}}\!:^- \; + \; \overset{..}{\underset{..}{B}}\!:\!\overset{..}{\underset{..}{F}}\!: \; \longrightarrow \; \left[:\!\overset{..}{\underset{..}{F}}\!:\!\overset{..}{\underset{..}{B}}\!:\!\overset{..}{\underset{..}{F}}\!: \right]^-$$

Fluoboric acid is a strong acid that has not been isolated in the free condition. In dilute aqueous solutions at room temperature, it undergoes slow hydrolysis according to the equation

$$BF_4^- + 3H_2O \rightleftharpoons H_3BO_3 + 3H^+ + 4F^-$$

Crystal structure studies show the fluoborate ion, BF_4^-, to be tetrahedral.

The trichloride of boron is a colorless mobile liquid, the bromide is a viscous liquid, and the iodide is a white crystalline solid.

28.6   Oxide and Oxyacids of Boron

Boron burns at 700° in oxygen, forming **boric oxide,** B_2O_3. Boric oxide finds use in the production of chemical-resistant glass and certain optical glasses. The oxide dissolves in hot water to form **boric acid,** H_3BO_3.

$$B_2O_3 + 3H_2O \longrightarrow 2H_3BO_3$$

The boron atom in H_3BO_3 is at the center of an equilateral triangle with oxygen atoms at the corners.

$$
\begin{array}{c}
H \\
O \\
120° \nearrow \nwarrow 120° \\
B \\
O \swarrow \searrow O \\
H \quad 120° \quad H
\end{array}
$$

Orthoboric acid

In the solid acid these triangular units are held together by hydrogen bonding. When boric acid is heated to 100°C, a molecule of water is split out between a pair of adjacent OH groups to form **metaboric acid,** HBO_2. With further heating at 140° to 160°, additional B—O—B linkages form, connecting the BO_3 groups together with shared oxygen atoms, to form **tetraboric acid,** $H_2B_4O_7$. At still higher temperatures, boric oxide is formed.

$$H_3BO_3 \longrightarrow HBO_2 + H_2O$$

$$4HBO_2 \longrightarrow H_2B_4O_7 + H_2O$$

$$H_2B_4O_7 \longrightarrow 2B_2O_3 + H_2O$$

28.7   Borates

Borates are salts of the oxyacids of boron. Borates result from the reactions of bases with the oxyacids of boron or from the fusion of boric acid or boron oxide with metal oxides or hydroxides. Borate anions range from the simple trigonal planar BO_3^{3-} ion to complex species containing chains and rings of three- and four-coordinate boron atoms. The structures of the anions found in CaB_2O_4, $K[B_5O_6(OH)_4] \cdot 2H_2O$ (commonly written $KB_5O_8 \cdot 4H_2O$), and $Na_2[B_4O_5(OH)_4] \cdot 8H_2O$ (commonly written $Na_2B_4O_7 \cdot 10H_2O$) are shown in Fig. 28-9.

Commercially, the most important borate is **borax,** or **sodium tetraborate 10-hydrate,** $Na_2B_4O_7 \cdot 10H_2O$. Most of the supply of borax comes directly from dry lakes, such as Searles Lake in California, or is prepared from **kernite,** $Na_2B_4O_7 \cdot 4H_2O$. Aqueous solutions of borax are basic, since it is a salt of a strong base and weak acid. The hydrolysis of the tetraborate ion may be represented by

$$B_4O_7^{2-} + 7H_2O \rightleftharpoons 4H_3BO_3 + 2OH^-$$

Borax is used to soften water and to make washing compounds, uses that depend upon the alkaline character of its solutions and the insolubility of the borates of calcium and magnesium.

When heated, borax fuses to form a glass that has the property of dissolving fused metal oxides. An example is given by the equation

$$Na_2B_4O_7 + CuO \longrightarrow 2NaBO_2 + Cu(BO_2)_2$$

The use of borax as a flux to remove oxides from metal surfaces in soldering and welding depends upon this property. Different metals give borax glasses of different colors, a fact used in detecting certain metals. Cobalt borax glass is blue, for example.

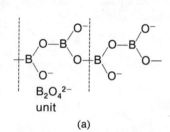

Figure 28-9 The borate anions found in (a) CaB_2O_4, (b) $KB_5O_8 \cdot 4H_2O$, and (c) $Na_2B_4O_7 \cdot 10H_2O$. The anion in CaB_2O_4 is an infinite chain.

28.8   Boron-Nitrogen Compounds

A number of compounds containing boron-nitrogen linkages have been prepared and studied. Two of particular interest are **boron nitride**, BN, and **triborinetria-mine**, $B_3N_3H_6$ (also called **borazole** and **borazine**).

Boron nitride is the final product of the thermal decomposition of many boron-nitrogen compounds, such as $B(NH_2)_3$ and $BF_3 \cdot NH_3$. It can also be prepared by heating boron with nitrogen or ammonia, or by heating borax with ammonium chloride. Crystalline boron nitride is a white solid that sublimes somewhat below 3000°; it melts at this temperature under pressure. It is inert to most reagents but can be decomposed by fusion with alkalies. The crystalline structure of boron nitride is analogous to that of graphite (Section 25.2 and Fig. 25-2). In fact, this form of boron nitride has been called **inorganic graphite.** The layers of atoms in boron nitride and graphite are made up of analogous hexagonal rings as shown.

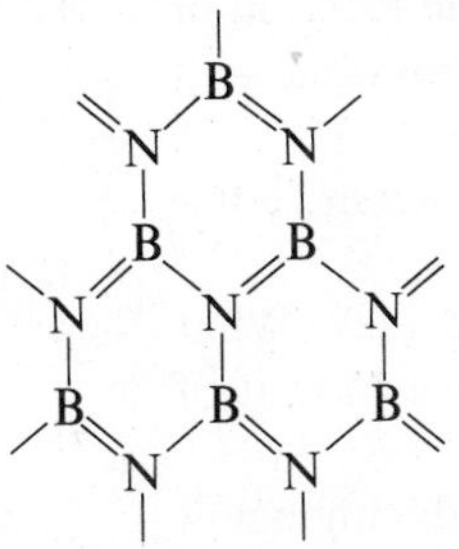

Boron nitride
layer

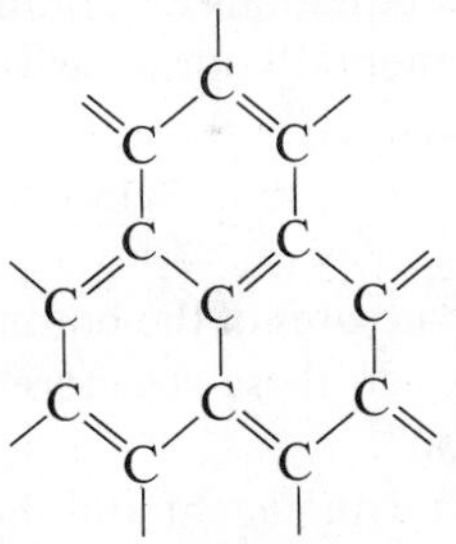

Graphite
layer

A second form of boron nitride, Borazon, has been synthesized. It has the diamond cubic structure (Section 25.1 and Fig. 25–1) and is second to diamond in hardness on a relative hardness scale. It is now produced commercially.

Borazine, $B_3N_3H_6$, is formed when ammoniates of the boron hydrides, such as $B_2H_6 \cdot 2NH_3$, are heated at 180° to 200°; hydrogen is the other product of the reaction. Borazine has been termed "inorganic benzene," because its structure and physical properties are similar to those of benzene. The structures of these two compounds are given.

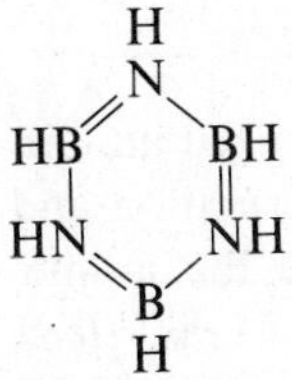

Borazine

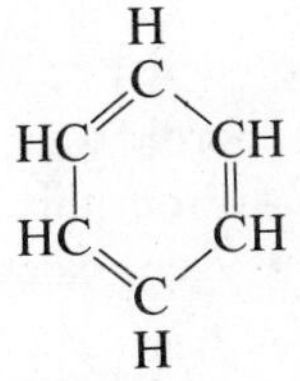

Benzene

Because a boron atom and a nitrogen atom bonded together have the same number of electrons as two carbon atoms and the sizes of these two groups are about the same, it is not surprising that the forms of boron nitride are similar to graphite and diamond, nor that borazine resembles benzene. Groups of this type are called **isosteres.** The concept of **isosterism** has been of particular value in accounting for similarities among the properties of compounds with apparently unrelated structures.

Questions

1. What is the electron configuration of a boron atom?
2. Why is boron limited to a maximum coordination number of 4 in its compounds?
3. Describe the molecular structure and the hybridization of boron in each of the following molecules or ions: BCl_3, BH_4^-, B_2H_6, $B(OH)_3$, $B(OH)_4^-$, H_3BCO.
4. Write Lewis structures for the molecules and ions given in Question 3. (Omit B_2H_6.)
5. Write two equations for reactions in which boron exhibits metallic behavior and two equations for reactions in which boron exhibits nonmetallic behavior.
6. Write balanced equations describing the reaction of boron with elemental fluorine, carbon, oxygen, and nitrogen, respectively.
7. Why is B_2H_6 said to be an "electron deficient" compound?
8. Describe the structures of the boron hydrides. What relation, if any, do these structures have to that of elemental boron?
9. Identify the oxidizing agent and the reducing agent in the chemical reaction for the preparation of diborane from boron trifluoride and lithium aluminum hydride.
10. The experimentally determined atomic ratio of hydrogen to boron in a gas is 3 to 1. How could it be proved that the gas actually consists of B_2H_6 molecules?
11. Explain what is meant by the "irreversibility" of the hydrolysis of the halides of boron.
12. Write equations to show the formation of fluoboric acid from boron trifluoride.
13. Show by equations the relationship of the boric acids to boric oxide.
14. Why does an aqueous solution of borax turn red litmus blue?
15. Explain the chemistry of the use of borax as a flux in soldering and welding.
16. Why is boron nitride sometimes referred to as "inorganic graphite"?
17. Why should the physical properties of borazine be similar to those of benzene?
18. Complete and balance the following equations. (In some instances, two or more reactions may be correct, depending upon the stoichiometry.)
    (a) $B + Cl_2 \longrightarrow$
    (b) $B + S_8 \longrightarrow$
    (c) $B + SO_2 \longrightarrow$
    (d) $B_2H_6 + NaH \longrightarrow$
    (e) $BCl_3 + C_2H_5OH \longrightarrow$
    (f) $HBO_2 + H_2O \longrightarrow$
    (g) $B_2O_3 + CuO \longrightarrow$
    (h) $BF_3 + NH_3 \longrightarrow$

Problems

1. A 0.7849-g sample of boron was oxidized in air, giving 2.5274 g of B_2O_3. From this information and the atomic weight of oxygen, calculate the atomic weight of boron.   *Ans. 10.81*
2. A hydride of boron contains 88.5% boron and 11.5% hydrogen by mass. The density of the hydride at 125°C and 450 torr is 2.21 g/ℓ. A 5.0-ml gaseous sample of this hydride at 125°C and 450 torr decomposed to the elements, giving 35 ml of H_2 under the same conditions. Determine the molecular formula of the hydride and show that all data are consistent with that formula.   *Ans. $B_{10}H_{14}$*

3. From the data given in Appendix J, determine the standard enthalpy change and standard free energy change for each of the following reactions:
   (a) $BF_3(g) + 3H_2O(l) \longrightarrow B(OH)_3(s) + 3HF(g)$
       Ans. $\Delta H° = 87\,kJ;\ \Delta G° = 44\,kJ$
   (b) $BCl_3(g) + 3H_2O(l) \longrightarrow B(OH)_3(s) + 3HCl(g)$
       Ans. $\Delta H° = -109.9\,kJ;\ \Delta G° = -154.7\,kJ$
   (c) $B_2H_6(g) + 6H_2O(l) \longrightarrow 2B(OH)_3(s) + 6H_2(g)$
       Ans. $\Delta H° = -510\,kJ;\ \Delta G° = -601.5\,kJ$

References

"The Boranes and Their Relatives," W. N. Lipscomb, *Science,* **196,** 1047 (1977).

"NMR of Boron Compounds," G. R. Eaton, *J. Chem. Educ.,* **46,** 547 (1969).

"Boron Chemistry in Industrial Perspective," R. Thompson, *Chem. in Britain,* **7,** 140 (1971).

"The Hydrides of Boron," A. G. Massey, *Educ. in Chemistry,* **11,** 20 (1974).

"Hydride Reductions: A 40-Year Revolution in Organic Chemistry," H. C. Brown, *Chem. and Eng. News,* March 5, 1979, p. 24.

"Cage Compounds of Boron," M. S. Gaunt and A. G. Massey, *Educ. in Chemistry,* **11,** 118 (1974).

"Inorganic Polymers," H. R. Allcock, *Sci. American,* March 1974, p. 66.

"The Chemical Elements of Life," E. Frieden, *Sci. American,* July 1972, p. 52.

"Superhard Materials," F. P. Bundy, *Sci. American,* August 1974, p. 62.

"Organoboranes and Organoborate Anions: New Classes of Electrophiles and Nucleophiles in Organic Synthesis," E. Negishi, *J. Chem. Educ.,* **52,** 159 (1975).

"Boron Compounds May Find Therapeutic Uses," R. L. Rawls, *Chem. and Eng. News,* Aug. 21, 1978, p. 21.

29

Nuclear Chemistry

The chemical changes (transformations of one form of matter into another) of the types we have studied thus far have involved changes only in the electronic structures of atomic systems without alteration of their nuclear structures. Another type of transformation of matter involving changes in atomic nuclei is the basis for a branch of science that is on the borderline between physics and chemistry: **nuclear chemistry.** It had its beginning with the discovery of radioactivity and has become increasingly important during the past 40 years.

The Stability of Nuclei

29.1 The Nucleus

As we have seen in Sections 4.6–4.8, the nucleus of an atom is composed of protons and, with the exception of ^1_1H, neutrons. The number of protons in the nucleus determines what chemists call the atomic number (Z) of the element, and the sum of the number of protons and number of neutrons determines the mass number (A) of the element. Atoms with the same atomic number but with different mass numbers are isotopes of the same element. When talking about a single nuclear species, the term **nuclide** is often used.

Protons and neutrons, collectively called **nucleons,** are packed together tightly in a nucleus, leaving little free space. Thus the radius of a nucleus, which can be regarded as approximately spherical, is proportional to the number of nucleons that it contains. It has been determined experimentally that the nuclear radius (R) is approximately proportional to the cube root of the mass number of the nucleus:

$$R = \sqrt[3]{A} \times 1.3 \times 10^{-13} \text{ cm}$$

Thus the size of a nucleus is quite small when compared to that of the entire atom (10^{-8} cm).

The nuclear density can be determined from the volume of a nucleus and the masses of the nucleons it contains. On the average, nuclei are extremely dense, 1.8×10^{14} g/cm^3. This density is very, very large compared to the densities of familiar materials. Water, for example, has a density of 1 g/cm^3. Osmium, the densest element known, has a density of 22.6 g/cm^3. If the density of the earth were equal to the average density of a nucleus, the earth would have a radius of only about 200 meters (650 feet).

To hold positively charged protons together in the very small volume of a nucleus requires a very strong force inasmuch as, at very short distances, the positive charges repel each other strongly. The **nuclear force** is the force of attraction between nucleons that holds the nucleus together. This force acts between protons, between neutrons, and between protons and neutrons. The nuclear force is very different from and much stronger than the electrostatic force that holds negatively charged electrons around a positively charged nucleus. In fact, the nuclear force is the strongest force discovered, about 30 to 40 times stronger than electrostatic repulsions between protons in a nucleus. Although the exact nature of the nuclear force is not known, it is known that it is only a short-range force, effective only up to distances of about 10^{-13} cm.

29.2 Nuclear Binding Energy

Although the nature of the nuclear force is unknown, the magnitude of the energy changes associated with the action of this force of attraction can be determined. As an example, let us consider the nucleus of the helium atom, ^4_2He, which consists of two protons and two neutrons. If a helium atom were to be formed by the combination of two protons and two neutrons without change of mass, the mass of the helium atom (including the mass of the two electrons outside the nucleus) should be 4.0330 (see Section 4.4 for masses of the proton, neutron, and electron); however, it is only 4.0026. This difference represents a loss in mass of 0.0304 amu. The union of protons and neutrons to form a nucleus involves the conversion of some of the nuclear mass to **nuclear binding energy.** The loss in mass accompanying the formation of a heavier atom from hydrogen atoms and neutrons is due to the fact that such reactions involve a change of mass into energy; they are thus strongly exothermic.

Einstein, in his theory of relativity, related energy and mass by means of the equation

$$E = mc^2$$

in which E represents energy in joules, m stands for mass in kilograms, and c is the speed of light in meters per second. The fact that the speed of light is very high, 3.00×10^8 m sec^{-1}, and that this term is squared in Einstein's equation, makes it evident that a tremendous quantity of energy results from the destruction of a small quantity of matter.

EXAMPLE   **Calculate the binding energy for the nuclide $_2^4$He in units of joules per mole and electronvolts per nuclide.**

The difference in mass between that of a $_2^4$He nucleus and that of two protons plus two neutrons is 0.0304 amu. In order to use the Einstein equation to convert this mass loss into the equivalent energy in joules, the mass loss must be expressed in kilograms (1 amu $= 1.66 \times 10^{-27}$ kg, Section 2.4).

$$0.0304 \text{ amu} \times \frac{1.66 \times 10^{-27} \text{ kg}}{1 \text{ amu}} = 5.05 \times 10^{-29} \text{ kg}$$

The nuclear binding energy in *one* nucleus is found from the Einstein equation ($E = mc^2$).

$$E = 5.05 \times 10^{-29} \text{ kg} \times (3.00 \times 10^8 \text{ m sec}^{-1})^2 = 4.54 \times 10^{-12} \text{ kg m}^2 \text{ sec}^{-2} = 4.54 \times 10^{-12} \text{ J}$$

Recall that units of kg m^2 sec^{-2} are equivalent to J (Section 1.21). The binding energy in one mole of nuclei is

$$4.54 \times 10^{-12} \text{ J/nucleus} \times 6.022 \times 10^{23} \text{ nuclei/mol} = 2.73 \times 10^{12} \text{ J mol}^{-1} = 2.73 \times 10^9 \text{ kJ mol}^{-1}$$

Binding energies are usually expressed in millions of electron volts (MeV) per nuclide instead of joules. The conversion factor is 1 MeV $= 1.602189 \times 10^{-13}$ J, giving

$$4.54 \times 10^{-12} \text{ J} \times \frac{1 \text{ MeV}}{1.602189 \times 10^{-13} \text{ J}} = 28.3 \text{ MeV per atom}$$

◆ ◆ ◆ ◆ ◆

The changes in mass in all ordinary chemical changes are negligibly small because these changes only involve the formation or breakage of chemical bonds. On the other hand, the breakdown of a nucleus into its component protons and neutrons requires very large amounts of energy and, conversely, the formation of a stable nucleus by bringing together the proper number of protons and neutrons involves the release of very large amounts of energy. If the nuclear reaction of two moles of neutrons with two moles of hydrogen atoms to give one mole of helium atoms could be made to occur, 2.73×10^9 kJ of energy would be released (see calculation in the example). As a comparison, the chemical reaction of one mole of methane with two moles of oxygen to give carbon dioxide and liquid water produces only 8.9×10^2 kJ, about three million times less energy.

The binding energy per nucleon (i.e., the total binding energy for the nucleus divided by the sum of the numbers of protons and neutrons present in the nucleus) is greatest for the nuclei of elements of mass numbers between 40 and 100, and decreases with mass numbers less than 40 or greater than 100 (Fig. 29-1). The most stable nuclei are those with the largest binding energy per nucleon, those in the vicinity of iron, cobalt, and nickel.

Figure 29-1 Binding energy curve for the elements. Note the branching in the curve for the lighter elements. Elements with even atomic numbers are on the upper branch; those with odd atomic numbers are on the lower branch.

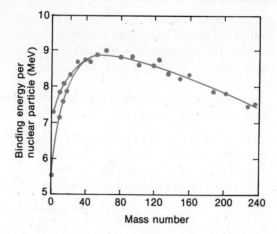

Returning to the example worked out previously in this section, we see that the binding energy of helium is 28.3 MeV per atom or 28.3/4 = 7.08 MeV per nucleon.

29.3   Nuclear Stability

A nucleus is stable if it cannot be transformed into another configuration without the addition of energy from the outside. A plot of the number of protons versus the number of neutrons for stable nuclei (the curve shown in Fig. 29-2) is referred to as a **stability curve.** The straight line in Fig. 29-2 represents equal numbers of protons and neutrons and is given for comparison. The curve indicates that the lighter, stable nuclei, in general, have equal numbers of protons and neutrons. For example, nitrogen-14 has seven protons and seven neutrons. Heavier stable nuclei, however, have somewhat larger numbers of neutrons than protons, as the curve

Figure 29-2 A plot of number of neutrons versus number of protons for stable, naturally occurring nuclei. Broken portion of the curve represents elements with atomic number above 83. All isotopes of these heavier elements are unstable.

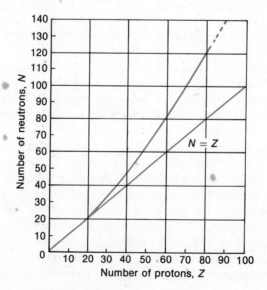

shows. For example, iron-56 has 26 protons and 30 neutrons. Lead-207 has 82 protons and 125 neutrons.

The isotopes that fall to the left or right of the stability curve have unstable nuclei and are referred to as being **radioactive.** Note that all isotopes with atomic number higher than 83 are radioactive; this is indicated in Fig. 29-2 by the dashed portion of the stability curve. Unstable (radioactive) nuclei tend to change spontaneously to stable configurations that would fall on or near the stability curve. The specific nature of these nuclear changes toward more stable nuclear configurations will be discussed in subsequent sections.

29.4   The Half-Life

The number of atoms of a specific element that undergo a radioactive change per unit of time is a constant fraction of the total number of atoms present. The time required for one-half of the atoms in a sample to disintegrate is called its **half-life.** The half-lives of the various radioactive nuclides vary widely. For example, the half-life of $^{142}_{58}Ce$ is 5×10^{15} years, that of $^{226}_{88}Ra$ is 1590 years, that of $^{222}_{86}Rn$ is 3.82 days, and that of $^{216}_{84}Po$ is 0.16 seconds. Only one-half of a sample of radium-226 will remain unchanged after 1590 years, and at the end of another 1590 years the sample will be reduced to one-fourth its initial mass, and so on. The half-lives of a number of nuclides are listed in Appendix L.

Radioactive disintegrations are first-order reactions. It can be shown that the rate constant k (see Section 15.5) for a radioactive disintegration reaction (or any first-order chemical reaction) can be expressed in terms of the half-life, $t_{1/2}$, by the following equation, the same as that which applies to all first-order reactions (Section 15.7).

$$k = \frac{0.693}{t_{1/2}}$$

As would be expected, the relationship between the initial concentration of a radioactive isotope and the concentration remaining after a given period of time is expressed by the same logarithmic relationship that applies to all other first-order reactions (Section 15.7).

$$\log \frac{c_o}{c_t} = \frac{kt}{2.303}$$

where c_o is the initial concentration and c_t is the concentration remaining at time t.

EXAMPLE 1   **Calculate the rate constant for the radioactive disintegration of $^{60}_{27}Co$, an isotope used in cancer therapy. It disintegrates with a half-life of 5.2 years to produce $^{60}_{28}Ni$.**

$$k = \frac{0.693}{t_{1/2}}$$

$$= \frac{0.693}{5.2 \text{ years}} = 0.13 \text{ year}^{-1}$$

◆ ◆ ◆ ◆ ◆

EXAMPLE 2   **Calculate the fraction and the per cent of a sample of the $^{60}_{27}$Co isotope that will remain after 15 years.**

$$\log \frac{c_o}{c_t} = \frac{kt}{2.303}$$

$$= \frac{(0.13 \text{ year}^{-1})(15 \text{ years})}{2.303} = 0.847$$

$$\frac{c_o}{c_t} = \text{antilog } 0.847 = 7.031$$

The fraction remaining will be the concentration at time t divided by the initial concentration, or c_t/c_o.

$$\frac{c_t}{c_o} = \frac{1}{7.031} = 0.14$$

Hence, 0.14 of the $^{60}_{27}$Co originally present, or 14%, would still remain after 15 years.

◆ ◆ ◆ ◆ ◆

EXAMPLE 3   **How long would it take for a sample of $^{60}_{27}$Co to disintegrate to the extent that only 2.0% of the original concentration remains?**

$$\frac{c_t}{c_o} = 0.020$$

Therefore,

$$\frac{c_o}{c_t} = \frac{1}{0.020} = 50$$

$$\log 50 = \frac{(0.13 \text{ year}^{-1})t}{2.303}$$

$$t = \frac{2.303 \log 50}{0.13 \text{ year}^{-1}} = 30 \text{ years}$$

◆ ◆ ◆ ◆ ◆

EXAMPLE 4   **The half-life of $^{216}_{84}$Po is 0.16 sec. How long would it take to reduce its concentration to the negligible value of $1.0 \times 10^{-5}\%$ of its original concentration (the fraction 1.0×10^{-7} times its original concentration)?**

$$k = \frac{0.693}{0.16 \text{ sec}} = 4.33 \text{ sec}^{-1}$$

$$\frac{c_t}{c_o} = 1.0 \times 10^{-7}$$

Hence,
$$\frac{c_o}{c_t} = 1.0 \times 10^7$$

$$\log(1.0 \times 10^7) = \frac{(4.33 \ \text{sec}^{-1})t}{2.303}$$

$$t = \frac{2.303 \log(1.0 \times 10^7)}{4.33 \ \text{sec}^{-1}} = 3.7 \ \text{sec}$$

◆ ◆ ◆ ◆ ◆

These examples indicate part of the problem of disposal of radioactive isotopes. In 3.7 sec, any likely quantity of $^{216}_{84}\text{Po}$ would be reduced to a negligible quantity and would itself no longer be a problem, but the burst of radiation during the first second or so would be extremely intense and very hazardous. Furthermore, as indicated in Table 29-1, $^{216}_{84}\text{Po}$ is a part of a series, disintegrating into $^{212}_{82}\text{Pb}$, which is itself radioactive with a half-life of 10.6 sec, producing $^{212}_{83}\text{Bi}$, which has a half-life of 60.5 min, and so on until finally the stable isotope $^{208}_{82}\text{Pb}$ is reached. Hence, in coping with radiation effects and isotope disposal, all steps in the disintegration of the material must be taken into account. In contrast to the time for disintegration of $^{216}_{84}\text{Po}$, it would take 30 years for 98% of a sample of $^{60}_{27}\text{Co}$ to disintegrate, leaving 2% still present. A similar calculation for $^{226}_{88}\text{Ra}$, an isotope with a half-life of 1590 years used commonly in cancer treatment, shows that, even after 200 years, 91.6% of the original isotope would still be present. While the radiation from $^{226}_{88}\text{Ra}$ is less intense than that of an equivalent quantity of an isotope of short half-life, the radiation is nevertheless extremely dangerous, and the level of the radiation would be diminished less than 10% even after two centuries of emission. Thus it is obvious that we cannot simply leave $^{226}_{88}\text{Ra}$ or other radioactive isotopes lying around on the assumption that the radiation danger will disappear within a short period.

29.5   The Age of the Earth

One of the most interesting applications of radioactivity has been its use in determining the age of the earth. An estimate of the lower limit of the earth's age can be made by determining the age of various rocks and minerals, assuming that the earth must be at least as old as the rocks and minerals in its crust. Several methods involving radioactivity have been used for this purpose. One method involves the disintegration of naturally occurring $^{87}_{37}\text{Rb}$ to $^{87}_{38}\text{Sr}$ with a half-life of 47 billion years. For example, 1 g of $^{87}_{37}\text{Rb}$ would produce 0.5000 g of $^{87}_{38}\text{Sr}$ and leave 0.5000 g of $^{87}_{37}\text{Rb}$ after decaying for 47 billion years. Comparison of the amounts of $^{87}_{37}\text{Rb}$ and $^{87}_{38}\text{Sr}$ in the Amitsog gneiss, a rock formation found in southwestern Greenland, has shown this formation to have an age of 3.75 billion years. This is the oldest known rock on earth.

Nuclear Reactions

The following sections describe reactions of nuclei that result in changes in the atomic number, mass number, and energy states of nuclei. Such reactions are called **nuclear reactions.** Nuclear reactions result (1) from the spontaneous decay of naturally occurring or artificially produced radioactive nuclei, (2) from the fission of unstable heavy nuclei, (3) from the fusion of light nuclei, and (4) from the bombardment of nuclei with other nuclei or with other fast-moving particles.

29.6 Equations for Nuclear Reactions

An equation that describes a nuclear reaction identifies the nuclides involved in the reaction as well as the mass numbers and atomic numbers of these nuclides. In addition, such an equation also identifies the particles involved in the reaction. These particles include (1) **alpha particles** ($_2^4He$, or α), which consist of two neutrons and two protons and are identical to helium nuclei; (2) **beta particles** ($_{-1}^0e$, or β), which are high-speed electrons; (3) **positrons** ($_{+1}^0e$, or β^+), which are particles with the same mass as an electron but with one unit of *positive* charge; (4) **protons** ($_1^1H$, or **p**), which are nuclei of hydrogen atoms; and (5) **neutrons** ($_0^1n$, or **n**), which are particles with a mass approximately equal to that of a proton but which bear no charge. Nuclear reactions also often involve **gamma rays** (γ). Gamma rays are electromagnetic radiation, somewhat like x rays in character but of higher energies and shorter wavelengths.

Examples of equations for different types of nuclear reactions of historical interest follow.

1. The radioactive decay of naturally occurring polonium, the first radioactive element discovered (in 1898) by the Polish scientist Marie Curie and her husband Pierre, proceeds by the reaction

$$_{84}^{212}Po \longrightarrow \ _{82}^{208}Pb + _2^4He \tag{1}$$

2. Technetium, an artificial radioactive element that does not occur naturally on the earth and was first prepared in 1937, decays by the reaction

$$_{43}^{98}Tc \longrightarrow \ _{44}^{98}Ru + _{-1}^0e \tag{2}$$

3. One of the many different fission reactions of uranium in the first nuclear reactor (1942) is represented by the reaction

$$_{92}^{235}U + _0^1n \longrightarrow \ _{35}^{87}Br + _{57}^{146}La + 3\,_0^1n \tag{3}$$

4. One of the fusion reactions present during the detonation of a hydrogen bomb is represented by the reaction

$$_1^3H + _1^2H \longrightarrow \ _2^4He + _0^1n \tag{4}$$

5. The first element prepared by artificial means (1919) was prepared by bombarding nitrogen atoms with alpha particles

$$_7^{14}N + _2^4He \longrightarrow \ _8^{17}O + _1^1H \tag{5}$$

A nuclear reaction is not correctly written unless it is balanced; that is, both the total nuclear mass and the total charge balance. The total number of protons and neutrons in the reactants must be equal to the total number of protons and neutrons in the products. Additionally, the total charge of the reactants must be equal to the total charge of the products. For example, in Equation (5) we see that the reactants contain 9 protons and 9 neutrons (7 protons and 7 neutrons in $^{14}_{7}N$, and 2 protons and 2 neutrons in $^{4}_{2}He$); and the products contain 9 protons and 9 neutrons (8 protons and 9 neutrons in $^{17}_{8}O$ and one proton in $^{1}_{1}H$). The charges also balance, with a total of 9 protons in the reactants (the $^{14}_{7}N$ nucleus and the $^{4}_{2}He$ nucleus) and a total of 9 protons in the products (the $^{17}_{8}O$ nucleus and the $^{1}_{1}H$ nucleus).

If the atomic numbers and the mass numbers of all but one of the particles in a nuclear reaction are known, the particle can be identified by balancing the reaction.

EXAMPLE   **The reaction of an alpha particle with magnesium-25 ($^{25}_{12}Mg$) produces a proton ($^{1}_{1}H$) and a nuclide of another element. Identify the new nuclide produced.**

The nuclear reaction may be written

$$^{25}_{12}Mg + ^{4}_{2}He \longrightarrow ^{A}_{Z}X + ^{1}_{1}H$$

where A is the mass number and Z the atomic number of the new nuclide, X. Since the sum of the masses of the reactants must equal the sum of the masses of the products, $25 + 4 = A + 1$, or $A = 28$. Similarly, the charges must balance so $12 + 2 = Z + 1$, and $Z = 13$. The element of atomic number 13 is aluminum. Thus the product is $^{28}_{13}Al$.

Aluminum-28 is unstable, with a half-life of 2.3 min, and decays by β emission to $^{28}_{14}Si$.

◆ ◆ ◆ ◆ ◆

29.7   Radioactive Decay

The spontaneous change of an unstable nuclide into another nuclide is called **radioactive decay.** The unstable nuclide is often called the **parent nuclide;** the nuclide that results from the decay, the **daughter nuclide.** The daughter nuclide may be stable or it may undergo subsequent decay. The different types of radioactive decay can be classified by the particles or electromagnetic radiation involved in the equation that describes the process.

▶ **1. α DECAY, OR α EMISSION.** α decay, the loss of an α particle by a radioactive nuclide, occurs primarily in heavy nuclei ($A \geq 200$, $Z > 83$). Loss of an α particle gives a daughter nuclide with a mass number 4 units less and an atomic number 2 units less than those of the parent nuclide. Consequently, the daughter nuclide has a larger neutron-to-proton ratio than the parent nuclide, and, if the parent nuclide lay below the stability curve (Fig. 29-2), the daughter nuclide would lie closer to the curve.

The lightest naturally occurring nuclide in which α decay has been observed is $^{144}_{60}\text{Nd}$. The decay of neodymium-144 is slow; $t_{1/2} = 2.4 \times 10^{15}$ years. At least one nuclide of every element beginning with lithium undergoes α decay. All such nuclides with atomic numbers less than 60 are artificially produced, however, rather than occurring naturally.

▶ **2. β DECAY, OR β EMISSION.** A nuclide with a large neutron-to-proton ratio (one that lies above the nuclear stability curve, Fig. 29-2) can reduce this ratio by **electron emission,** or **β decay.** The electron is created as a neutron in the nucleus decays into a proton and an electron.

$$\,^{1}_{0}\text{n} \longrightarrow \,^{1}_{1}\text{H} + \,^{0}_{-1}\text{e}$$

The emission of an electron does not change the mass number of a nuclide but does increase the number of protons and decrease the number of neutrons in the nuclide. Consequently, the neutron-to-proton ratio is decreased, with the result that the daughter nuclide lies closer to the stability curve than does the parent nuclide.

▶ **3. β^+ (POSITRON) DECAY, OR β^+ EMISSION.** Certain artificially produced nuclides in which the neutron-to-proton ratio is low (nuclides that lie below the nuclear stability curve, Fig. 29-2) are converted to daughter nuclides by **β^+ decay,** the emission of a positron. During the course of β^+ decay, a proton is converted into a neutron with the emission of a **positron,** a particle that is identical to an electron except that its charge is positive rather than negative.

$$\,^{1}_{1}\text{H} \longrightarrow \,^{1}_{0}\text{n} + \,^{0}_{+1}\text{e}$$

The result is an increase in the neutron-to-proton ratio, which brings the daughter nuclide up closer than the parent nuclide to the nuclear stability curve.

When a positron and an electron interact, they annihilate each other. All of their mass is converted into energy—two 0.511-MeV γ rays are produced.

$$\,^{0}_{-1}\text{e} + \,^{0}_{+1}\text{e} \longrightarrow 2\gamma \quad (0.511 \text{ MeV})$$

▶ **4. ORBITAL ELECTRON CAPTURE.** By orbital electron capture (EC), a proton is converted to a neutron when one of the electrons in an atom is captured by the nucleus.

$$\,^{1}_{1}\text{H} + \,^{0}_{-1}\text{e} \longrightarrow \,^{1}_{0}\text{n}$$

This, as with β^+ decay, occurs when the neutron-to-proton ratio is low (nuclide lies below the nuclear stability curve, Fig. 29-2). Electron capture has the same effect on the nucleus as positron decay; the atomic number is decreased by one unit as a proton is converted into a neutron. This increases the neutron-to-proton ratio and produces a daughter nuclide that lies closer to the nuclear stability curve than the parent nuclide.

▶ **5. γ-RAY EMISSION.** A parent nuclide does not always decay to the ground state of the daughter nuclide by loss of an α, β, or β^+ particle; it may decay to an excited

Figure 29-3 The decay of $^{233}_{92}$U to $^{229}_{90}$Th. The decay can occur by three paths. Two paths give excited states of the $^{229}_{90}$Th nucleus which decay to the ground state by emission of γ rays.

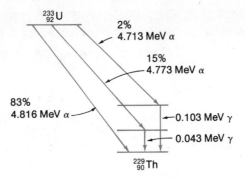

nuclear state of the daughter nuclide. The daughter nuclide in its excited state then decays to its ground state with the emission of a γ ray, a quantum of high-energy electromagnetic radiation. Figure 29-3 illustrates the relationships for the decay of uranium-233. Note that there is no change of mass number or atomic number during emission of a γ ray. γ-ray emission is observed only when a nuclear reaction gives a daughter nuclide that is not in its ground state.

29.8 Radioactive Series

All isotopes of the elements with atomic numbers larger than that of bismuth (83) are radioactive. A few elements of lower atomic number, such as potassium and rubidium, have naturally occurring isotopes that also are radioactive.

The naturally occurring radioisotopes of the heavier elements belong to chains of successive disintegrations, or "decays," and all the species in one chain constitute a radioactive family, or series. Three of these series include most of the naturally radioactive elements of the Periodic Table. They are the **uranium series,** the **actinium series,** and the **thorium series.** Each series is characterized by a parent (first member) of long half-life and a series of decay processes that ultimately lead to a stable end product; i.e., an isotope on the stability curve of Fig. 29-2. In all three natural series, the end-products are isotopes of lead: $^{206}_{82}$Pb in the uranium series, $^{207}_{82}$Pb in the actinium series, and $^{208}_{82}$Pb in the thorium series.

Successive transformations in the natural disintegration series take place in a manner described by the so-called **displacement laws** originally formulated by Rutherford, Soddy, and Fajans: (1) When an atom emits an alpha particle, the product is an isotope of an element two places to the left of the parent element in the Periodic Table. (2) When a beta particle is emitted, the product is an isotope of an element one place to the right of the parent in the Periodic Table.

The steps in the thorium series are given in Table 29-1 as an illustration of one natural radioactive decay series.

A fourth radioactive series was discovered during World War II. This series is called the **neptunium series,** after its member of longest half-life, and was discovered through the production of its members by artificial means (Section 29.9). The end product of the series is an isotope of bismuth, $^{209}_{83}$Bi. Both the parent and end product of this series have been detected in uranium ores in recent years.

The four radioactive series are referred to as the 4n (or thorium), 4n + 1 (or

TABLE 29-1   The Thorium Series[a]

$$^{232}_{90}\text{Th} \xrightarrow[1.39 \times 10^{10} \text{ y}]{\alpha} {}^{228}_{88}\text{Ra} \xrightarrow[6.7 \text{ y}]{\beta} {}^{228}_{89}\text{Ac} \xrightarrow[6.13 \text{ h}]{\beta} {}^{228}_{90}\text{Th}$$

$$\alpha \Big| 1.90 \text{ y}$$

$$^{212}_{82}\text{Pb} \xleftarrow[0.16 \text{ sec}]{\alpha} {}^{216}_{84}\text{Po} \xleftarrow[54.5 \text{ sec}]{\alpha} {}^{220}_{86}\text{Rn} \xleftarrow[3.64 \text{ d}]{\alpha} {}^{224}_{88}\text{Ra}$$

$$\beta \Big| 10.6 \text{ h}$$

$$^{212}_{83}\text{Bi} \xrightarrow[60.5 \text{ min}]{\beta} {}^{212}_{84}\text{Po}$$

$$\alpha \Big| 60.5 \text{ min} \qquad \alpha \Big| 3 \times 10^{-7} \text{ sec}$$

$$^{208}_{81}\text{Tl} \xrightarrow[3.1 \text{ min}]{\beta} {}^{208}_{82}\text{Pb}$$

[a] Type of emission and half-life time is shown for each step (y means years; h, hours; min, minutes; and sec, seconds).

neptunium), $4n + 2$ (or uranium), and the $4n + 3$ (or actinium) series. These numerical designations indicate whether the mass numbers of the members of the series are exactly divisible by 4, or by 4 with a remainder of 1, 2, or 3.

29.9   Synthesis of Nuclides

Atoms with stable nuclei can be converted to other atoms **(transmutation)** by bombarding their nuclei with other nuclei or with high-speed particles. The first person to produce a nucleus artificially in the laboratory was Lord Rutherford, in 1919. Using high-speed α particles emanating from a naturally radioactive isotope of radium as projectiles, he bombarded nitrogen atoms and observed the nuclear reaction

$$^{14}_{7}\text{N} + {}^{4}_{2}\text{He} \longrightarrow {}^{17}_{8}\text{O} + {}^{1}_{1}\text{H} \tag{1}$$

Thus two new nuclei are formed, an $^{17}_{8}\text{O}$ nucleus and a proton. The $^{17}_{8}\text{O}$ nucleus is stable, so that this nuclear reaction does not lead to further radioactive changes.

Many elements undergo artificially induced nuclear reactions with the formation of *unstable* nuclei, which then undergo radioactive disintegration. The disintegration in these cases is referred to as **artificial radioactivity.** For example, the first member of the neptunium series, $^{239}_{93}\text{Np}$, is prepared by the reaction

$$^{238}_{92}\text{U} + {}^{1}_{0}\text{n} \longrightarrow {}^{239}_{93}\text{Np} + {}^{0}_{-1}\text{e} \tag{2}$$

Neptunium-239 is radioactive and decays by α decay:

$$^{239}_{93}\text{Np} \longrightarrow {}^{235}_{91}\text{Pa} + {}^{4}_{2}\text{He}$$

An abbreviated notation for transmutation reactions is sometimes used. In this notation the bombarding particle and the product particle are written in parentheses between the symbols for the reactant and product nuclides. Thus equations

(1) and (2) can be written $^{14}_{7}N$ (α, p) $^{17}_{8}O$ and $^{238}_{92}U$ (n, β) $^{239}_{93}Np$, respectively, where α, p, n, and β are symbols for an α particle, a proton, a neutron, and a β particle; $^{14}_{7}N$ and $^{238}_{92}U$ are symbols for the reactants; and $^{17}_{8}O$ and $^{239}_{93}Np$ are symbols for the products, respectively.

Many transmutation reactions require the absorption of energy, as compared with radioactive decay, which gives up energy.

For example, for the reaction $^{14}_{7}N$ (α, p) $^{17}_{8}O$:

$$^{14}_{7}N + {}^{4}_{2}He + energy \longrightarrow {}^{17}_{8}O + {}^{1}_{1}H$$

$^{14}_{7}N$ = 14.00307 amu	$^{17}_{8}O$ = 16.99914 amu
$^{4}_{2}He$ = 4.00260 amu	$^{1}_{1}H$ = 1.00782 amu
Total mass: 18.00567 amu	Total mass: 18.00696

This reaction requires that an additional mass of 0.00129 amu be formed by the conversion of energy to mass during the course of the reaction of one $^{14}_{7}N$ nucleus with one α particle. The energy required can be determined from the Einstein equation $E = mc^2$, Section 29.2.

$$E = mc^2 = \left(0.00129 \; amu \times \frac{1.66 \times 10^{-27} \; kg)}{1 \; amu}\right)(3.00 \times 10^8 \; m \; sec^{-1})^2$$

$$= 1.93 \times 10^{-13} \; J$$

$$1.93 \times 10^{-13} \; J \times \frac{1 \; MeV}{1.602 \times 10^{-13} \; J} = 1.20 \; MeV$$

Thus 1.20 MeV, or 1.93×10^{-13} J, must be provided for the collision of one $^{14}_{7}N$ nucleus with an α particle to give a reaction. This amounts to 1.16×10^8 kJ/mol of reactant.

The energy required for a transmutation reaction is usually provided as kinetic energy (pages 3, 19) by one of the reactants. The energy required for the reaction $^{14}_{7}N$ (α, p) $^{17}_{8}O$ comes from the kinetic energy of the α particle; an α particle moving with a velocity of 7.62×10^6 m sec^{-1} has a kinetic energy of 1.20 MeV. The decay of $^{226}_{88}Ra$ provides α particles with kinetic energies in excess of 1.20 MeV; thus such particles can be used in the reaction (as was done by Lord Rutherford in 1919).

In addition to α particles and neutrons, particles such as β particles, protons, and other light nuclei can be used to produce transmutation reactions. Generally, the charged particles are accelerated to the necessary kinetic energies to produce reactions by machines (**accelerators**) that use magnetic and electric fields to accelerate the particles. Such machines include the cyclotron (Figs. 29-4 and 29-5), the synchrocyclotron, and the linear accelerator. Both cyclotrons and synchrocyclotrons accelerate particles in a spiral path. In linear accelerators, particles pass down a linear tube through a series of charged cylinders so that they are continuously accelerated. In all accelerators, the particles move in a vacuum so as to avoid collisions with gas molecules.

Since neutrons are not charged, they cannot be accelerated, by such means as just described, to the velocities needed for nuclear transmutations. Neutrons used in transmutation reactions are therefore obtained from radioactive decay and other nuclear reactions occurring in a nuclear reactor (Section 29.12).

Figure 29-4 A beam of deuterons, $_1^2H$ (bright area), from a cyclotron at the University of California. Nuclear reactions take place when deuterons and other atomic particles strike the nuclei of atoms. (*Courtesy of the Rockefeller Foundation*)

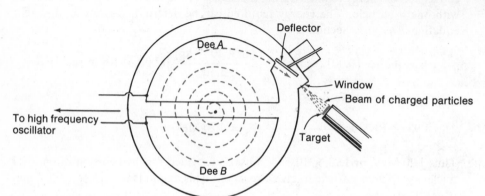

Figure 29-5 Diagram of the vacuum chamber of a cyclotron. By making the dees, *A* and *B*, alternately positively and negatively charged and correctly adjusting the strength of the magnetic field, accelerating charged particles are caused to follow an ever-widening spiral path until they are finally deflected and pass through the thin metal window to the target.

Not only have many previously known elements been transmuted into other known elements, but a considerable number of new elements have been synthesized in the laboratory. These include, among others, element 43, technetium (Tc); element 85, astatine (At); element 87, francium (Fr); and element 61, promethium (Pm).

Prior to 1940 the heaviest known element was uranium, atomic number 92. In 1940, McMillan and Abelson were able to make element 93, neptunium (Np), by bombarding uranium with high-velocity neutrons. The nuclear reaction is given by

$$_{92}^{238}U + _0^1n \longrightarrow _{92}^{239}U$$

$$_{92}^{239}U \xrightarrow{\text{23 min}} _{93}^{239}Np + _{-1}^0e$$

Neptunium-239 is also radioactive, with a half-life period of 2.3 days, and converts to plutonium (Pu), atomic number 94.

$$_{93}^{239}Np \longrightarrow _{94}^{239}Pu + _{-1}^0e$$

Elements 95 through 106 have likewise been prepared artificially. The elements 93 through 106 are known as the **transuranium elements.** Elements 93 through 103, along with those of atomic numbers 89, 90, 91, and 92, make up the **actinide series.** Equations describing the preparation of some isotopes of the transuranium elements are given in Table 29-2.

The name *rutherfordium,* with the symbol Rf, has been suggested for element 104 by American scientists who have synthesized the element. Soviet scientists, who also have worked on the synthesis of the element, suggest the name *kurchatovium.* It has been proposed that element 105 be named *hahnium,* with the symbol Ha, to honor Otto Hahn, the late German scientist who won the Nobel prize for the discovery of nuclear fission.

Both a Soviet and an American group have announced the synthesis of element 106, by quite different methods. The two groups have agreed not to propose any name for the element until a decision is reached on which group will get credit for being the first to synthesize it.

The Soviet group used relatively heavy ions to bombard smaller nuclei than those used by the American group. The Soviet scientists bombarded a lead isotope with chromium-54 ions, producing an isotope reported to be either $^{257}_{106}X$ or $^{258}_{106}X$ and hence containing either 151 or 152 neutrons; it decays by spontaneous fission with the very short half-life of four to ten milliseconds.

The American group bombarded a target of californium-249 with oxygen-18 ions, resulting in emission of four neutrons and production of an isotope of

TABLE 29-2   Preparation of the Transuranium Elements

Name	Symbol	Atomic Number	Reaction
Neptunium	Np	93	$^{238}_{92}U + ^{1}_{0}n \longrightarrow ^{239}_{93}Np + ^{0}_{-1}e$
Plutonium	Pu	94	$^{238}_{92}U + ^{2}_{1}H \longrightarrow ^{238}_{93}Np + 2\,^{1}_{0}n$
			$^{238}_{93}Np \longrightarrow ^{238}_{94}Pu + ^{0}_{-1}e$
Americium	Am	95	$^{239}_{94}Pu + ^{1}_{0}n \longrightarrow ^{240}_{95}Am + ^{0}_{-1}e$
Curium	Cm	96	$^{239}_{94}Pu + ^{4}_{2}He \longrightarrow ^{242}_{96}Cm + ^{1}_{0}n$
Berkelium	Bk	97	$^{241}_{95}Am + ^{4}_{2}He \longrightarrow ^{243}_{97}Bk + 2\,^{1}_{0}n$
Californium	Cf	98	$^{242}_{96}Cm + ^{4}_{2}He \longrightarrow ^{245}_{98}Cf + ^{1}_{0}n$
Einsteinium	Es	99	$^{238}_{92}U + 15\,^{1}_{0}n \longrightarrow ^{253}_{99}Es + 7\,^{0}_{-1}e$
Fermium	Fm	100	$^{239}_{94}Pu + 15\,^{1}_{0}n \longrightarrow ^{254}_{100}Fm + 6\,^{0}_{-1}e$
Mendelevium	Md	101	$^{253}_{99}Es + ^{4}_{2}He \longrightarrow ^{256}_{101}Md + ^{1}_{0}n$
Nobelium	No	102	$^{246}_{96}Cm + ^{12}_{6}C \longrightarrow ^{254}_{102}No + 4\,^{1}_{0}n$
Lawrencium	Lr	103	$^{250}_{98}Cf + ^{11}_{5}B \longrightarrow ^{257}_{103}Lr + 4\,^{1}_{0}n$
Rutherfordium	Rf	104	$^{249}_{98}Cf + ^{12}_{6}C \longrightarrow ^{257}_{104}Rf + 4\,^{1}_{0}n$
Hahnium	Ha	105	$^{249}_{98}Cf + ^{15}_{7}N \longrightarrow ^{260}_{105}Ha + 4\,^{1}_{0}n$
—	—	106	$^{206}_{82}Pb + ^{54}_{24}Cr \longrightarrow ^{257}_{106}X + 3\,^{1}_{0}n$
			$^{249}_{98}Cf + ^{18}_{8}O \longrightarrow ^{263}_{106}X + 4\,^{1}_{0}n$

element 106 having a mass number of 263 and containing 157 neutrons; it decays by α emission with a half-life of 0.9 sec, followed by two additional α-emission steps.

$$^{249}_{98}\text{Cf} + ^{18}_{8}\text{O} \longrightarrow ^{263}_{106}\text{X} + 4\,^{1}_{0}\text{n}$$
$$\longrightarrow ^{259}_{104}(\text{Rf}) + ^{4}_{2}\alpha$$
$$\longrightarrow ^{255}_{102}\text{No} + ^{4}_{2}\alpha$$
$$\longrightarrow ^{251}_{100}\text{Fm} + ^{4}_{2}\alpha$$

Because elements 104 and 105 are more stable than expected and by other reasoning based on experiment and extrapolation of known facts, it is believed that element 114 can be synthesized. This element, chemically similar to lead, could be so stable that tiny amounts may be found to occur naturally on earth—or, perhaps, to be present in regions of outer space where new elements seem to be forming. It is interesting to note, however, that isotopes containing more than 157 neutrons have proved to be exceedingly unstable, an experimental observation for which there is as yet no explanation.

Glenn T. Seaborg has suggested an extension of the Periodic Table (see Table 29-3) to include new elements whose synthesis is considered possible. (It will probably be helpful to review the material in Chapter 4 on filling energy shells and subshells by electrons.) With element 104 we enter the unexplored region of the periodic system with the first member of what Seaborg has chosen to call the "trans-actinide" elements, that is, all elements beyond the inner transition or actinide series. It becomes possible to locate the position of elements 104 through 121 and, in the style of Mendeleev, to predict their chemical properties by comparing them with their analogs in the Periodic Table. Element 104 should be an analog of hafnium; 105, an analog of tantalum; and so forth until element 118, a noble gas analogous to radon, is reached.

The most striking feature of Seaborg's extension of the Periodic Table is the addition of another inner transition series of elements starting with atomic number 121 and extending through atomic number 153. He calls this grouping the "superactinide" series. This series, except for element 121, is postulated as including those elements characterized by the progressive filling of the $5g$ electron subshell (see Table 4-3). Element 121 can be considered as receiving the first $7d$ electron (the $n = 7$ major shell being the second from the outside), analogous to scandium, yttrium, lanthanum, and actinium. Perhaps the first of eighteen $5g$ electrons would enter at element 122, for which the $n = 5$ major shell is the fourth from the outside; the eighteenth $5g$ electron would enter at element 139. This would presumably be followed by the filling in of fourteen $6f$ electrons (third major shell from the outside) for elements 140 through 153, making this latter series an inner transition series analogous to those series for elements 58 through 71 and 90 through 103. Following the superactinide inner transition series, elements 154 through 168 would be analogous to elements 104 through 118, with entrance of the remainder of the ten $7d$ electrons in the second major shell from the outside and the six $8p$ electrons in the outermost major shell. Element 168 would be another noble gas—or should we expect a "noble liquid"?

TABLE 29-3   Predicted Locations of New Elements in the Periodic Table

																	s
																1 H	2 He

s-block, d-block, p-block:

1 H																	
3 Li	4 Be						*d*					5 B	6 C	7 N	8 O	9 F	10 Ne
11 Na	12 Mg											13 Al	14 Si	15 P	16 S	17 Cl	18 Ar
19 K	20 Ca	21 Sc	22 Ti	23 V	24 Cr	25 Mn	26 Fe	27 Co	28 Ni	29 Cu	30 Zn	31 Ga	32 Ge	33 As	34 Se	35 Br	36 Kr
37 Rb	38 Sr	39 Y	40 Zr	41 Nb	42 Mo	43 Tc	44 Ru	45 Rh	46 Pd	47 Ag	48 Cd	49 In	50 Sn	51 Sb	52 Te	53 I	54 Xe
55 Cs	56 Ba	[57–71] *	72 Hf	73 Ta	74 W	75 Re	76 Os	77 Ir	78 Pt	79 Au	80 Hg	81 Tl	82 Pb	83 Bi	84 Po	85 At	86 Rn
87 Fr	88 Ra	[89–103] †	104	105	106	107	108	109	110	111	112	113	114	115	116	117	118
119	120	[121–153] ‡	154	155	156	157	158	159	160	161	162	163	164	165	166	167	168

f

* LANTHANIDE SERIES	57 La	58 Ce	59 Pr	60 Nd	61 Pm	62 Sm	63 Eu	64 Gd	65 Tb	66 Dy	67 Ho	68 Er	69 Tm	70 Yb	71 Lu
† ACTINIDE SERIES	89 Ac	90 Th	91 Pa	92 U	93 Np	94 Pu	95 Am	96 Cm	97 Bk	98 Cf	99 Es	100 Fm	101 Md	102 No	103 Lr

g   *f*

‡ SUPER ACTINIDES	121	122	123		139	140				153

29.10   Nuclear Fission

The greater stability of the nuclei of elements with mass numbers of intermediate values suggests the possibility of spontaneous splitting of the less stable nuclei of the heavy elements into fragments of approximately half size, and that such nuclear fissions would be accompanied by the release of large quantities of energy. Two German scientists, Hahn and Strassman, reported in 1939 that when they bombarded uranium with slow-moving neutrons, the uranium-235 atoms split into smaller fragments, consisting of elements about in the middle of the Periodic

Table, and several neutrons. The process is referred to as **fission.** Among the fission products identified were barium, krypton, lanthanum, and cerium, the nuclei of all of which are more stable than the nucleus of uranium.

$$^{235}_{92}U + ^{1}_{0}n \longrightarrow \quad \text{fission fragments} \quad + \text{neutrons} + \text{energy}$$
$$\text{(Isotopes of Ba, Kr, etc.)}$$

The sum of the atomic numbers of the fission products is 92, the atomic number of the original uranium nucleus.

In this type of disintegration a loss of mass of about 0.2 amu occurs, which corresponds to a fantastic quantity of energy that is released in the reaction. The fission of a pound of uranium-235 produces about 2.5 million times as much energy as the burning of a pound of coal.

The fission products of a uranium-235 nucleus include, on the average, 2.5 neutrons as well as the fission fragments. These neutrons may cause the fission of neighboring uranium-235 atoms, which in turn provide more neutrons for setting up a **chain reaction.** Nuclear fission can become self-substaining when the number of neutrons produced by fission is equal to or exceeds the number of neutrons absorbed by fissioning nuclei plus the number lost to the surroundings. The amount of a fissionable material that will support a self-sustaining chain reaction is called a **critical mass.** The critical mass of a fissionable material depends upon the shape of the sample of material as well as the type of material.

An atomic bomb consists of several pounds of fissionable material, $^{235}_{92}U$ or $^{239}_{94}Pu$, and an explosive device for compressing the material quickly into a small volume. With small pieces of fissionable material the proportion of neutrons that escape at the relatively large surface area is great and a chain reaction does not take place. When the small pieces of fissionable material are brought together quickly to form a larger body with a mass larger than the critical mass, the relative number of escaping neutrons decreases, and a chain reaction and an explosion result, with the fission of nearly all the nuclei within a very short period of time. The explosion of an atomic bomb can release more energy than that resulting from the explosion of a million tons of TNT (trinitrotoluene).

Chain reactions of fissionable materials can be controlled in a nuclear reactor (Section 29.12). The energy produced in a reactor can be used to meet some of the energy requirements of modern society.

29.11   Nuclear Fusion

In Section 29.10 we discovered that the nuclear binding energy of heavy atoms may be increased by fission of these atoms into fragments of lower mass numbers, and that such nuclear fissions are accompanied by the liberation of extremely large amounts of energy. The process of converting very light nuclei into heavier nuclei is likewise accompanied by the conversion of mass into large amounts of energy. Such reactions are known as **nuclear fusion** and are the basis of an intensive research effort to develop a practical thermonuclear reactor (Section 29.14). The principal source of energy of the sun is the fusion of four hydrogen nuclei into one helium nucleus. Four hydrogen nuclei have a greater mass (0.7%) than the helium nucleus, so the fusion converts the extra matter into energy.

It has been found that a deuteron, ^2_1H, and a triton, ^3_1H, which are the positive ions of the heavy isotopes of hydrogen, will undergo fusion at extremely high temperatures (**thermonuclear fusion**) to form a helium nucleus and a neutron.

$$^2_1H + {}^3_1H \longrightarrow {}^4_2He + {}^1_0n$$

This change is accompanied by a conversion of a portion of the mass into energy and is the nuclear reaction of the hydrogen bomb. In the hydrogen bomb, a fission type of atomic bomb (uranium or plutonium) is exploded inside a charge of deuterium and tritium to provide the temperature of many millions of degrees required for the fusion of the deuterium and tritium.

If the fusion of heavy isotopes of hydrogen can be controlled, hydrogen from the water of the oceans will provide an inexhaustible supply of energy for future generations.

Nuclear Energy

29.12   Nuclear Power Reactors

Any nuclear reactor that produces power by the fission of uranium or plutonium by bombardment with slow neutrons must have at least five components. (See Fig. 29-6.)

Figure 29-6 A light-water nuclear reactor. This reactor uses pressurized liquid water at 280°C and 150 atm as both a coolant and a moderator.

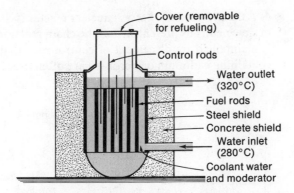

▶ **1. A NUCLEAR FUEL.** A fissionable isotope (commonly $^{235}_{92}U$, $^{233}_{92}U$, or $^{239}_{93}Pu$) must be present in sufficient quantity to provide a self-sustaining chain reaction. Most reactors used in the United States use pellets of U_3O_8 in which the amount of uranium-235 has been increased from its natural level to about 3%. The U_3O_8 pellets are contained in a tube (a fuel rod) of a protective material, usually a zirconium alloy. A typical nuclear power reactor in the United States uses about 40,000 kg of enriched U_3O_8, contained in several hundred fuel rods, as its core.

As it occurs in nature, uranium consists of a mixture of several isotopes; the most abundant of these is uranium-238. About one atom in every 140 atoms of uranium is the uranium-235 isotope. Because a higher concentration of uranium-235 is needed, it is necessary to separate uranium-235. This proved to be a major task, because there is relatively very little difference in the masses of these isotopes. Electromagnetic methods, based upon the principles of the mass spectrograph, were found to be effective, but the yields were small. The most successful of the several methods tried makes use of the separation of $^{235}_{92}UF_6$ from $^{238}_{92}UF_6$ by fractional diffusion of large volumes of gaseous UF_6 at low pressure through porous diffusion barriers of very large areas. This method is based upon the fact that the lighter $^{235}_{92}UF_6$ molecules diffuse through a porous barrier faster than the heavier molecules of $^{238}_{92}UF_6$. The enriched UF_6 is converted to U_3O_8.

▶ **2. A MODERATOR.** Neutrons produced by nuclear reactions must be slowed down before they will be absorbed by nuclides in a reactor's fuel rod and produce additional nuclear reactions. Neutrons lose their kinetic energy by collision with the nuclei of a moderator without reacting with, or being absorbed by, the moderator. The moderator must have a very low neutron absorption and should contain light nuclei so that the speed of the neutrons will be reduced in a minimum number of collisions. Materials commonly used as moderators include heavy water (D_2O, Section 12.6), graphite, carbon dioxide, and light (ordinary) water. Most reactors in operation in the United States use light water as the moderator.

▶ **3. A COOLANT.** The coolant carries the heat from the fission reaction to an external boiler and turbine system, where it is transformed into some other form of energy (generally electricity, Fig. 29-7). The coolant must be a gas or a liquid so it can be pumped into and out of the reactor core. The coolant may also serve as the moderator if it has the right properties.

▶ **4. A CONTROL SYSTEM.** A nuclear reactor is controlled by adjusting the number of slow neutrons present so that the chain reaction proceeds at a safe rate. Control is maintained by placement of control rods that absorb neutrons. Control rods containing cadmium or boron-10 are often used. Boron-10, for example, absorbs

Figure 29-7 Power-generating plant employing a nuclear power reactor. In a coal-fired power plant the steam would be generated in a boiler. Note that the coolant is contained in a closed system and does not come in contact with outside cooling water. The reactor shielding has been omitted for clarity.

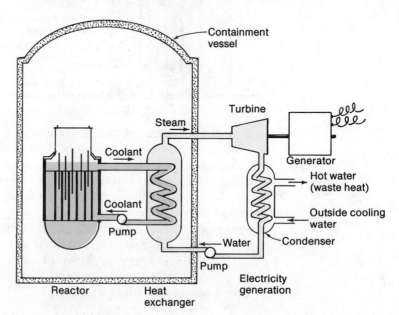

neutrons by the reaction $^{10}_{5}B\,(n, \alpha)\,^{7}_{3}Li$. Enough control rods are available that the chain reaction can be completely stopped by inserting them into the core between the fuel rods.

▶ **5. A SHIELD AND CONTAINMENT SYSTEM.** A nuclear reactor produces neutrons and other particles due to radioactive decay of the products resulting from fusion. In addition, high temperatures are present, and high pressures result from the circulation of water or other coolants through the hot regions of the reactor. A reactor must be constructed to endure these high temperatures and pressures and to protect operating personnel from the radiation produced. A reactor container often consists of three parts: (1) the reactor vessel, a steel shell ranging from 3 to 20 cm in thickness which absorbs much of the radiation produced by the reactor; (2) a main shield of 1 to 3 m of high-density concrete; and (3) a personnel shield of lighter materials to absorb γ rays and x rays. In addition, reactors are often covered with a steel or concrete dome designed to contain radioactive materials that might be released by a serious reactor accident.

The energy produced by a reactor fueled with enriched uranium results from the fission of $^{235}_{92}U$ (Section 29.10) as well as from the fission of $^{239}_{94}Pu$. Plutonium-239 forms from $^{238}_{92}U$ present in the fuel (Section 29.9). In any nuclear reactor, only about 0.1% of the mass of the fuel is converted into energy. The other 99.9% remains in the fuel rods as fission products and unused fuel. All of the fission products absorb neutrons, and, after a period of several months to a year, depending upon the reactor, the fission products must be removed by changing the fuel rods. Otherwise, the concentration of these fission products will increase to the point where the reactor will no longer operate due to the absorption of neutrons by the fission products.

Spent fuel rods contain a variety of products consisting of unstable nuclei ranging in atomic number from 25 to 60, some transuranium elements including $^{239}_{94}Pu$ and $^{241}_{95}Am$, and unreacted $^{235}_{92}U$ and $^{238}_{92}U$. The unstable nuclei and the transuranium elements give the spent fuel a dangerously high level of radioactivity. The long-lived isotopes $^{90}_{38}Sr$ and $^{137}_{55}Cs$ and the shorter lived isotope $^{131}_{53}I$ are particularly dangerous in that they can be incorporated into the body if the radioactive material is dispersed in the environment. Consequently, it is absolutely essential that this material not be allowed to be released into the biosphere. It takes about 400 years for the level of radioactivity from $^{90}_{38}Sr$ and $^{137}_{55}Cs$ to decrease to a reasonably safe level. Other nuclides such as $^{241}_{95}Am$ and $^{239}_{94}Pu$ have much longer half-lives and require thousands of years to decay to a safe level. The ultimate fate of the nuclear reactor as a significant source of energy in the United States probably rests upon whether or not a scientifically and politically satisfactory technique for processing and storage of the components of spent fuel rods can be developed.

29.13   Breeder Reactors

A **breeder reactor** is a nuclear reactor that produces more fissionable material than it consumes. Because the supply of naturally occurring uranium is limited (some estimates suggest that the known reserves will last for only another 50 years of full-scale use), the conversion of nonfissionable material to nuclear fuel in a breeder reactor may provide a long-term energy supply. In contrast to the regular reactor, the breeder reactor is constructed with a blanket of "fertile" material such as $^{238}_{92}U$ or $^{232}_{90}Th$, surrounding a core of concentrated fissionable material ($^{235}_{92}U$, $^{239}_{94}Pu$, or $^{233}_{92}U$). Extra neutrons arising from the fission of the core material are captured by the blanket of fertile material to form ("breed") more fissionable atoms. Hence the breeder reactor is a source of energy and at the same time produces more fuel than it consumes.

Breeder reactors are based upon one of two sets of nuclear reactions. One begins with $^{238}_{92}U$ and the other with $^{232}_{90}Th$. Both nuclides are the most abundant isotopes of their respective elements. $^{238}_{92}U$ produces the fissionable nuclide $^{239}_{94}Pu$, according to the reactions described in Section 29.9. $^{232}_{90}Th$ produces the fissionable nuclide $^{233}_{92}U$ by the following series of reactions:

$$^{232}_{90}Th + {}^{1}_{0}n \longrightarrow {}^{233}_{90}Th$$

$$^{233}_{90}Th \xrightarrow{\text{23 min}} {}^{233}_{91}Pa + {}^{0}_{-1}e$$

$$^{233}_{91}Pa \xrightarrow{\text{27 days}} {}^{233}_{92}U + {}^{0}_{-1}e$$

Experimental breeder reactors are in operation in the United States, Great Britain, Russia, and France.

Breeder reactors present additional challenges to those outlined in Section 29.12. In addition to the problems of storage of the spent fuel of a breeder reactor, there are problems associated with processing the blanket of fertile material (by remote control) in order to separate the radioactive fuel from other radioactive by-products, as well as questions of safeguarding the large amounts of plutonium that would be produced. Plutonium-239 decays with a half-life of 24,000 years by emission of an α particle; thus, it will remain in the biosphere for a very long time if it is dispersed. Plutonium is one of the most toxic substances known, estimates of a fatal dose being as low as one microgram (10^{-6} g). Moreover, plutonium can be used to make bombs (Section 29.10), and the existence of quantities of the pure material could be attractive to terrorist groups and to some countries not yet capable of producing their own atomic weapons.

29.14 Fusion Reactors

A fusion reactor is a nuclear reactor in which fusion reactions of light nuclei (Section 29.11) are controlled. At the time of this writing, there are no self-sustaining fusion reactors operating in the world, although small-scale fusion reactions have been run in such devices for very brief periods.

In order to produce a fusion reaction, very high temperatures of about 10^8 K are required. At these temperatures, all molecules dissociate into atoms, and the atoms ionize, forming a new state of matter called a **plasma.** Since no solid materials are stable at 10^8 K, a plasma cannot be contained by mechanical devices. Two techniques to contain a plasma at the necessary density and temperature for a fusion reaction to occur are currently under study. These techniques involve containment by a magnetic field and the use of focused laser beams.

Once a plasma of hydrogen isotopes is generated and contained, it will undergo fusion reactions when the temperature exceeds about 10^8 K. One such reaction is

$$^2_1H + {}^3_1H \longrightarrow {}^4_2He + {}^1_0n$$

which proceeds with a mass loss of 0.0188 amu, corresponding to the release of 1.69×10^9 kJ per mole of ^4_2He formed. Deuterium (^2_1H) is available from heavy water. The tritium (^3_1H) can be prepared by reaction of the neutrons from the fusion reaction with lithium.

$$^1_0n + {}^6_3Li \longrightarrow {}^3_1H + {}^4_2He$$

Questions

1. Describe the three types of radiation emitted from nuclei of naturally radioactive elements.
2. What is the change in the nucleus that gives rise to a β particle?
3. The loss of an α particle by a nucleus causes what change in the atomic number and the mass of the nucleus? What is the change in the atomic number and mass when a β particle is emitted?
4. Define and illustrate the term "half-life."
5. How do nuclear reactions differ from ordinary chemical changes?
6. Many nuclides of atomic number greater than 83 decay by processes such as β emission. Rationalize the observation that the radioactive emissions resulting from these unstable nuclides normally include α emission also.
7. How may charged particles be accelerated artificially for use in promoting nuclear reactions?
8. Identify the various particles that may be produced in a nuclear reaction.
9. Complete the following equations:
   [s](a) $^{27}_{13}Al + {}^2_1H \longrightarrow \quad + {}^4_2He$
   [s](b) $^7_3Li + \quad \longrightarrow 2\,{}^4_2He$
   [s](c) $^9_4Be + {}^4_2He \longrightarrow {}^{12}_6C +$
   (d) $^{23}_{11}Na + {}^2_1H \longrightarrow {}^{24}_{11}Na +$
10. Fill in the atomic number of the initial nucleus and write out the complete nuclear symbol for the product of the following nuclear reactions:
    [s](a) ^{65}Cu (n, 2n)
    [s](b) ^{54}Fe (α, 2p)
    [s](c) ^{33}S (n, p)
    (d) ^{33}S (p, n)
    (e) ^{106}Rh (n, p)

(f) ^{27}Al (α, n)
[s](g) ^{14}N (p, γ)
11. Complete the following notations by filling in the missing parts:
    [s](a) ^2H (d, n)
    [s](b) $\quad$ (α, n) ^{30}P
    (c) $\quad$ (d, n) ^{238}Np
    (d) ^{10}B (α, $\quad$) ^{13}C
    [s](e) ^{232}Th ($\quad$, n) ^{235}U
    (f) ^2H (^{12}C, $\quad$) ^{10}B
    (g) ^{24}Mg (α, n)
    (h) ^{238}U ($^{12}_6C$, 4n)
12. Use the abbreviated notation as in the question above to describe
    (a) the production of ^{17}O from ^{14}N by α bombardment
    (b) the production of ^{14}C from ^{14}N by neutron bombardment
    (c) the neutron induced fission of $^{235}_{92}U$
    (d) the production of ^{233}Th from ^{232}Th by neutron bombardment
    (e) the production of ^{239}U from ^{238}U by deuteron bombardment
13. Below are listed several unstable isotopes. Predict by what mode(s) spontaneous radioactive decay might proceed in each case.
    (a) ^{34}P (n/p ratio too large)
    (b) ^{156}Eu (n/p ratio too large)
    (c) ^{235}Pa (too much mass, n/p ratio too large)
    (d) ^3_1H
    (e) ^{18}F
    (f) ^{129}Ba
    (g) ^{237}Pu

14. Give two reasons why the atomic weights of the naturally occurring elements are not whole numbers.
15. Explain in terms of Fig. 29-2 how unstable heavy nuclides (atomic number greater than 83) may decompose to form nuclides of greater stability (a) if they are on the dashed portion of the stability curve, and (b) if they are to the left of the dashed portion of the stability curve.
16. Verify 18 as the maximum number of *g* electrons theoretically possible in a given major shell, based upon permissible values for the four quantum numbers (see Chapter 4).
17. Write out, using standard notation ($1s^2$, $2s^2$, etc.), logical predicted electron structures of elements of atomic numbers 104, 105, 106, 114, 118, 122, 130, 139, 140, 146, 153, 159, 166, 168, 170. In each case, to what known elements would the predicted element be analogous in chemical properties?
18. Distinguish between nuclear fission and nuclear fusion. Why are both processes exothermic?
19. How are atomic bombs and hydrogen bombs detonated?
20. Describe the construction and operation of a uranium nuclear reactor.
21. Describe how the potential energy of uranium is converted into electrical energy in a nuclear power plant.
22. What is meant by the term "breeder reactor"?
23. List advantages and disadvantages of nuclear energy as a source of domestic power as compared to coal, fuel oil, natural gas, and water.
24. Discuss and compare the problems of radioactive wastes for radioactive substances of short half-life and those of long half-life.

Problems

[S] 1. The isotopic mass of $^{27}_{13}Al$ is 26.98154. (a) Calculate its binding energy per atom in MeV. (b) Calculate its binding energy per nucleon. (See Appendix D.)
Ans. (a) 224.96 MeV; (b) 8.332 MeV

2. A $^{7}_{4}Be$ atom (mass 7.0169 amu) decays to a $^{7}_{3}Li$ atom (mass 7.0160 amu) by electron capture. How much energy (in MeV) is produced by this reaction?
Ans. 0.8 MeV

[S] 3. The mass of a deuteron ($^{2}_{1}H$) is 2.01355 amu; that of an α particle, 4.00150 amu. How much energy per mole of $^{4}_{2}He$ produced is released by the reaction

$$^{2}_{1}H + ^{2}_{1}H \longrightarrow ^{4}_{2}He$$

Ans. 2.301×10^9 kJ

[S] 4. What percentage of $^{212}_{82}Pb$ remains of a 1.00-g sample, 1.0 min after it is formed (half-life of 10.6 sec)? 10 min after it is formed? *Ans. 2.0%; 9.1×10^{-16}%*

5. The isotope of ^{208}Tl undergoes β decay with a half-life of 3.1 min.
[S] (a) What isotope is the product of the decay?
[S] (b) Is ^{208}Tl more stable or less stable than an isotope with a half-life of 54.5 sec?
[S] (c) How long will it take for 99.0% of a sample of pure ^{208}Tl to decay? *Ans. 20.6 min*
[S] (d) What percentage of a sample of pure ^{208}Tl will remain undecayed after an hour?
Ans. 1.5×10^{-4}%

6. Calculate the time required for 99.999% of each of the following radioactive isotopes to decay:
[S] (a) $^{226}_{88}Ra$ (half-life, 1590 years)
Ans. 26,400 years
(b) $^{214}_{84}Po$ (half-life, 1.6×10^{-4} second)
Ans. 0.0027 second
(c) $^{232}_{90}Th$ (half-life, 1.39×10^{10} years)
Ans. 2.31×10^{11} years

[S] 7. The isotope $^{90}_{38}Sr$ is an extremely hazardous isotope in the fallout from a nuclear fission explosion. A 0.500-g sample diminishes to 0.393 g in 10.0 years. Calculate the half-life. *Ans. 28.8 years*

8. One gram of $^{226}_{88}Ra$ (atomic weight 226) produces 0.0001 ml of the gas $^{222}_{86}Rn$ (atomic weight 222) at STP in 24 hours. What is the half-life of ^{226}Ra in years? *Ans. 2×10^3 years*

9. Calculate the density of each of the following nuclei and compare the values to the average nuclear density value given in Section 29.1: $^{7}_{3}Li$; $^{12}_{6}C$; $^{59}_{28}Ni$; $^{108}_{47}Ag$; $^{192}_{77}Ir$; $^{238}_{92}U$; $^{260}_{103}Lr$.
Ans. 1.8×10^{14} g/cm³ for each nucleus

References

"Some Recollections of Early Nuclear Age Chemistry," G. T. Seaborg, *J. Chem. Educ.*, **45**, 278 (1968).

"Discovery of Fission," O. Hahn, *Sci. American,* February 1958, p. 76.

"The Size and Shape of Atomic Nuclei," M. Baranger and R. A. Sorensen, *Sci. American,* August 1969, p. 59.

"The Structure of the Proton and the Neutron," H. W. Kendall and W. Panofsky, *Sci. American,* June 1971, p. 60.

"Concepts of Nuclear Structure," A. Bohr, *Science,* **172**, 17 (1971).

"Structure of the Proton," R. F. Feynman, *Science, 183,* 601 (1974).

"Extranuclear Effects on Nuclear Decay Rates," P. K. Hopke, *J. Chem. Educ.,* **51**, 517 (1974).

"Nucleophilic Substitution—A Radioactive Study," B. A. Shaw and M. E. Shaw, *Educ. in Chemistry,* **11**, 77 (1974).

"Tin(IV) Iodide as a Radioactive Tracer," P. W. Wiggans, *Educ. in Chemistry,* **11**, 194 (1974).

"Carbon-13 as a Label in Biosynthetic Studies," U. Séquin and A. I. Scott, *Science,* **186**, 101 (1974).

"The Age of the Elements," D. N. Schramm, *Sci. American,* January 1974, p. 69.

"Deuterium in the Universe," J. M. Pasachoff and W. A. Fowler, *Sci. American,* May 1974, p. 108.

"Electron-Positron Collisions," A. M. Litke and R. Wilson, *Sci. American,* October 1973, p. 104.

"Exotic Atoms," C. E. Wiegand, *Sci. American,* November 1972, p. 102.

"Dual-Resonance Models of Elementary Particles," J. H. Schwarz, *Sci. American,* February 1975, p. 61.

"Proton Interactions at High Energies," U. Amaldi, *Sci. American,* November 1973, p. 36.

"Radiocarbon Dating," W. F. Libby, *Chem. in Britain,* **5**, 548 (1969).

"Elemental Evolution and Isotopic Composition," J. Rydberg and G. R. Choppin, *J. Chem. Educ.,* **54**, 742 (1977).

"Assembly of Greek Marble Inscriptions by Isotopic Methods," N. Herz and D. B. Wenner, *Science,* **199**, 1071 (1978).

"The Batavia Accelerator," R. R. Wilson, *Sci. American,* February 1974, p. 72.

"Collective-Effect Accelerators," D. Keefe, *Sci. American,* April 1972, p. 22.

"The First Isolations of the Transuranium Elements—A Historical Survey," J. C. Wallmann, *J. Chem. Educ.,* **36**, 340 (1959).

"The Synthetic Elements," G. T. Seaborg and J. L. Bloom, *Sci. American,* April 1969, p. 56.

"Element 106: Soviet and American Claims in Muted Conflict," T. H. Maugh II, *Science,* **186**, 42 (1974).

"Superheavy Elements, A Crossroads," G. T. Seaborg,

W. Loveland, and D. J. Morrissey, *Science, 203,* 711 (1979).

"Energy from Uranium" (Staff), *J. Chem. Educ.,* **56,** 119 (1979).

"Energy for the Long Run: Fission or Fusion," G. L. Kulcinski *et al., Amer. Scientist,* **67,** 78 (1979).

"Fission Power, An Evolutionary Strategy," H. A. Feiveson, F. von Hippel, and R. H. Williams, *Science, 203,* 330 (1979).

"Natural Uranium Heavy-Water Reactors," H. C. McIntyre, *Sci. American,* October, 1975, p. 17.

"Superphénix: A Full-Scale Breeder Reactor," G. A. Vendryes, *Sci. American,* March 1977, p. 26.

"Fusion Power Research Gains Momentum" (Staff), *Chem. and Eng. News,* August 21, 1978, p. 7.

"Fusion Power with Particle Beams," G. Yonas, *Sci. American,* November 1978, p. 50.

"The Tokamak: Model T Fusion Reaction," D. Steiner and J. F. Clarke, *Science,* **199,** 1395 (1978).

"Engineering Limitations of Fusion Power Plants," W. E. Parkins, *Science,* **199,** 1403 (1978).

"Plutonium: (I) Questions of Health in a New Industry" and "—(II) Watching and Waiting for Adverse Effects," R. Gillette, *Science,* **185,** 1027 and 1140 (1974).

"Plutonium: Biomedical Research," W. J. Bair and R. C. Thompson, *Science, 183,* 715 (1974).

"Recycling Plutonium: The National Research Council Proposes a Second Look," R. Gillette, *Science,* **188,** 818 (1975).

"Fast Breeder Reactors," G. T. Seaborg and J. L. Bloom, *Sci. American,* November 1970, p. 13.

"The Energy Resources of the Earth," M. K. Hubbert, *Sci. American,* September, 1971, p. 60.

"Fusion Power," D. A. Dingee, *Chem. and Eng. News,* April 2, 1979, p. 32.

"Fusion Power by Laser Implosion," J. L. Emmett, J. Nuckolls, and L. Wood, *Sci. American,* June 1974, p. 24.

"Energy Policy in the U.S.," D. J. Rose, *Sci. American,* January 1974, p. 20.

"Nuclear Accident—Twenty Years Late" (Staff), *Chemistry,* February 1979, p. 4.

"New Problems for Nuclear Waste Storage" (Staff), *Chem. and Eng. News,* June 12, 1978, p. 28.

"The Disposal of Radioactive Wastes from Fission Reactors," B. L. Cohen, *Sci. American,* June 1977, p. 21.

"The Reprocessing of Nuclear Fuels," W. P. Bebbington, *Sci. American,* December 1976, p. 30.

"New Mexicans Debate Nuclear Waste Disposal," W. Lepkowski, *Chem. and Eng. News,* January 1, 1979, p. 20.

"Natural Nuclear Reactors; The Oklo Phenomenon," R. West, *J. Chem. Educ.,* **53,** 336 (1976).

"A Natural Fission Reactor," G. A. Cowan, *Sci. American,* July 1976, p. 36.

30

The Metallic Elements

Many times in the preceding chapters we have had occasion to classify the elements as either metallic or nonmetallic. (See Section 8.3.) Thus far, however, our study has been confined chiefly to the nonmetals and their compounds. Before taking up the study of the metallic elements and their compounds, a general consideration of the metals as a class of elements will be of value in correlating the study of the individual metals. Several of the chapters that follow describe the chemistry of the metals.

30.1 Occurrence of the Metals

The metals above hydrogen in the activity series (Section 9.21) are rarely found in the **free,** or **native, state.** The less active metals, which are below hydrogen in the activity series, often occur in the free state. Among the latter are copper, silver, gold, and platinum. Some metals, such as copper and silver, are found both free and combined.

It is not surprising that the metallic compounds occurring in the earth's crust are low in water solubility, while the more soluble compounds are found in sea water and in large salt beds which have been formed by the evaporation of inland seas. Water is such an excellent solvent that rains soaking into the earth dissolve many of the salts, carrying them slowly to the nearest stream, then to the nearest river, and finally down to the sea. The ocean is continually becoming richer in salts and other minerals, although human beings have not been observing it long enough to notice any real change. It is interesting that the age of the ocean as calculated from its salt content is between 1 and 7 billion years. This is of the same order of magnitude as that (3.8 billion years) calculated for the age of the earth by methods involving radioactive elements (Section 29.5).

The materials in which metals or their compounds occur in the earth and from which the metals may be extracted economically are known as **ores.** Ores usually contain large percentages of rocky material called **gangue.** The important classes of ores, based on the nonmetallic element or acid anion with which the metal is combined (if any), are listed below.

1. *Native ores:* Gold, silver, platinum, copper, mercury, arsenic, antimony, and bismuth.
2. *Oxide ores:* Iron, aluminum, manganese, and tin.
3. *Sulfide ores:* Zinc, cadmium, mercury, copper, lead, nickel, cobalt, silver, arsenic, and antimony.
4. *Carbonate ores:* Iron, lead, zinc, and copper. These ores are less important as sources of the metals than are the oxides and sulfides of the same metals. The carbonates of calcium, barium, strontium, and magnesium are important as sources for the preparation of various compounds of these metals.
5. *Halides:* The chlorides of sodium and potassium are important sources of these metals and their compounds. The halides of magnesium, calcium, and silver are also important.
6. *Sulfates:* Calcium, strontium, barium, and lead.
7. *Silicates:* Most silicates are unsuitable as ores because of difficulty in extracting the metals from them. However, the silicates of beryllium, zinc, and nickel are important.

30.2   The Ocean as a Source of Minerals

The oceans, which cover a large fraction of the earth's surface, have been termed "the world's greatest mine." A ton of sea water contains about 55 lb of sodium chloride, 2.54 lb of magnesium, 1.75 lb of sulfur, 0.8 lb of calcium, 0.75 lb of potassium, 0.125 lb of bromine, and lesser quantities of such elements as strontium, boron, fluorine, iodine, iron, copper, lead, zinc, uranium, silver, gold, and even radium. The amount of gold in the oceans is estimated to be 8.5 million tons, worth over 110,000 billion dollars at the October 1979 price of $400 an ounce. Our mineral deposits in the earth's crust are being depleted at an ever increasing rate, so it is necessary to look to the oceans for future supplies of the metals.

The first mineral to be "mined" from sea water was undoubtedly salt. If sea water is trapped at high tide and the sun evaporates the water, "solar" salt with all the other solids that were in the water is obtained. This salt can be purified by fractional crystallization. Solar salt is made in large quantities on our west coast and on the coasts of Lebanon, Israel, and other Near East countries. Many chemical industries use salt as their raw material, especially the alkali and chlorine industries. Scandinavian sea salt has served as a starting point for the manufacture of soda ash (sodium carbonate) for many years.

Processes, described in the magnesium and bromine sections, are now in use by which magnesium and bromine are extracted from sea water. Extraction methods for other elements such as gold, copper, silver, potassium (for fertilizer), and especially uranium, from sea water, are as yet too expensive to be used extensively. Indirect methods of extracting these metals may be the final answer. For example, living plants have the curious ability to remove certain materials from soil or water and to store them in their tissues. The horsetail (*Equisetum*) removes silica and gold from the soil, the locoweed takes up barium, and certain seaweeds extract large amounts of potassium and iodine from sea water. When the weeds are burned, these elements are left in the ashes, from which they can be removed. The ocean is a potential source of uranium, which may well be our chief energy fuel when our deposits of coal and petroleum are exhausted. It is also the source of "manganese nodules," which occur in great quantity on the ocean floor. Much research is going into devising economically practical methods for their recovery, because they contain, in addition to manganese, other valuable metals such as nickel and copper.

30.3   Extractive Metallurgy

Extractive metallurgy pertains to the processes involved in the production of metals from their ores. Most such metallurgical processes include three principal steps: (1) preliminary treatment, (2) smelting, and (3) refining.

▶ **1. PRELIMINARY TREATMENT.**   Generally, ores must be subjected to preliminary processing to render them suitable for chemical extraction of the metals. The first step is usually that of pulverizing the ore by crushing and grinding. This is followed by **concentration** of the metal-bearing portions by removing most of the gangue. Sometimes this is accomplished by washing away the lighter gangue particles. Slightly inclined shaking tables are used to separate the heavier ore particles from the lighter rocky material. Metal-bearing particles which are affected by a magnetic field are often separated from nonmagnetic impurities by passing the finely ground ore through a magnetic field. Certain ores are affected by an electrostatic field, and in such cases electrostatic separation may be employed. The "flotation" process is a concentration method particularly applicable to sulfide ores of lead, zinc, and copper, but it is used also for carbonates and silicates. In this process, the pulverized ore is mixed with water to which a carefully selected oil has been added. A froth is produced by blowing air into the mixture. The metal-bearing particles have little or no attraction for water (such substances are referred to as **hydrophobic**). The particles do have an attraction for the oil and are therefore preferentially coated by the oil and adhere to the air bubbles in the froth which floats on the surface. The particles of gangue (sand, rock, clay, etc.) are wetted by water (such substances are **hydrophilic**) more readily than by the oil and sink to the bottom of the flotation vat. The froth containing the metal-bearing particles is removed, and the ore is recovered in a highly concentrated form.

The preliminary treatment of ores often includes chemical changes that convert the metallic compounds into substances that may be more readily reduced to give the metals. Ores containing moisture or water chemically combined in hydrates or hydroxides are heated to expel water. Ores containing metal carbonates are heated to decompose the carbonates and drive off carbon dioxide. This treatment of heating to drive off a volatile substance is known as **calcination.** The following equations illustrate typical changes that may occur during calcination.

$$2M(OH)_3 \longrightarrow M_2O_3 + 3H_2O \qquad (M = metal)$$
$$MCO_3 \longrightarrow MO + CO_2$$

Most sulfide ores are heated in air to change the sulfides to oxides, and to expel sulfur as sulfur dioxide. This treatment is known as **roasting.** The following equation illustrates the reaction.

$$2MS + 3O_2 \longrightarrow 2MO + 2SO_2$$

▶ **2. SMELTING.** The next step in the metallurgical process involves the extraction of the metal in the molten state, a process called **smelting.** The basic chemical reaction involved in smelting is one of reduction of the metallic compound. Most ores still contain some gangue at this stage of the process; this material is removed by addition of a **flux** (from the Latin *fluere,* meaning "to flow"). The flux is a chemical that will react with the gangue to form a substance of low melting point—a **slag.** When the gangue is silica or a silicate, a basic flux such as lime or limestone is used. At the high temperature of the furnace in which the reduction is carried out, the lime reacts with the silica to produce fused calcium silicate. The molten slag is easily separated from the fused metal because of the difference in their densities and because they are insoluble in each other.

The methods commonly employed in the smelting of various ores are:

a. *The reduction of oxide ores by carbon.*

$$ZnO + C \longrightarrow Zn + CO$$
$$SnO_2 + 2C \longrightarrow Sn + 2CO$$

When carbon is for some reason unsatisfactory, reducing agents such as aluminum, hydrogen, or iron are used.

$$Cr_2O_3 + 2Al \longrightarrow 2Cr + Al_2O_3$$
$$WO_3 + 3H_2 \longrightarrow W + 3H_2O$$
$$Sb_2S_3 + 3Fe \longrightarrow 2Sb + 3FeS$$

b. *The electrolytic reduction of halides of such active metals as sodium, potassium, magnesium, and calcium.* Aluminum is produced by the electrolysis of aluminum oxide dissolved in fused cryolite, Na_3AlF_6.

c. *Direct heating of some metals that occur in the native state.* The metals are smelted by heating the ore until the metals are melted. The fused metals are then drained away from the gangue or the gangue is removed as a slag.

▶ **3. REFINING OF METALS.** Smelting usually results in the production of metals that contain greater or lesser amounts of impurities such as other metals (or nonmetals), slag, and dissolved gases. The removal of impurities is usually necessary in the preparation of a metal for at least some of its uses. The low-boiling metals, such as zinc and mercury, may be purified from less volatile impurities by distillation. Other low-melting metals, such as tin, when fused on an inclined table may flow away from higher melting impurities. Electrolysis is the most widely used method of refining metals. Electrolytic refining is accomplished by making the impure metal the anode and a piece of the pure metal the cathode of an electrolytic cell (see Section 20.6). Among the metals that are purified by this process are copper, gold, lead, zinc, and cadmium.

A process known as **zone refining** provides ultrahigh-purity materials essential to semiconductor and transistor devices, such as those described earlier for the solar cell (Section 20.18). When a solid bar of metal is melted at one section (zone) and then the melted zone is made to progress longitudinally through the bar, the zone carries certain impurities with it. The procedure can be repeated with increased purification each time. The process depends upon the fact that the impurities are more soluble in the molten zone than in the solid metal. Zone refining is applicable to a wide range of substances and impurities.

▶ **4. HYDROMETALLURGY.**   Certain metals can be extracted from their ores by causing the metal to go into aqueous solution, as the result of chemical changes, and then precipitating the metal in the free state using a suitable reducing agent. The following equations illustrate the method.

$$4Ag(s) + 8CN^- + O_2 + 2H_2O \longrightarrow 4[Ag(CN)_2]^- + 4OH^-$$
or
$$AgCl + 2CN^- \longrightarrow [Ag(CN)_2]^- + Cl^-$$
Then,
$$2[Ag(CN)_2]^- + Zn \longrightarrow 2Ag(s) + [Zn(CN)_4]^{2-}$$

30.4   Chemical Bonding in Metals

We have noted that the atoms of the nonmetals have a sufficient number of valence electrons to combine with each other by sharing electron pairs. For example, two chlorine atoms share an electron pair in the Cl_2 molecule, two nitrogen atoms share three electron pairs in the N_2 molecule, and each carbon atom in the diamond crystal shares electron pairs with each of four neighboring carbon atoms. However, many metal atoms have less than four valence electrons (many only one or two), and each atom is surrounded by eight or twelve nearest neighbors in the crystal. With only one valence electron it would be impossible for an atom of sodium, for example, to be linked by electron-pair bonds to eight neighboring atoms. Thus a theory of the bonds in metals must explain how a relatively small number of electrons from a given atom can be shared by eight or more other atoms.

The modern theory of bonding in metals assumes that in solid or liquid metals the valence electrons are free to pass from one atom to another in such a way that they may be shared by several atoms (or ions, in the sense that each atom has given up its own specific valence electrons). These shared electrons serve to hold the atoms (or ions) together and constitute what is known as a **metallic bond,** which may be considered as a modified covalent bond with more or less mobile electrons being shared by more than two atoms. This arrangement has sometimes been rather picturesquely referred to as being made up of "metallic ions buried in a sea of electrons."

Such characteristic properties of metals as high electrical and thermal conductivity, luster, and high reflectivity are thought to be due to these relatively free, mobile valence electrons. These considerations are in accord with the fact that metallic character exists only when the material is in the massive solid or liquid state. Gaseous mercury, zinc, and sodium do not have metallic luster and the other characteristic metallic properties; their gaseous molecules are monatomic, and metallic bonding is, of course, absent.

The physical properties that permit us to distinguish metals from nonmetals may be readily accounted for by the **band theory** of metals. To aid us in understanding metallic bonding from the standpoint of the band concept it will be helpful to consider the following points:

a. Metals crystallize in structures of high coordination number, usually 8 or 12 (Section 11.15), yet have only a small number of valence electrons (usually 1, 2, or 3) to use in bonding. It is evident that no assignment of valence electrons to metallic bonding can account for such high coordination numbers, unless the valence electrons are "delocalized" (do not belong to any particular pairs of atoms).

b. Atomic orbitals of the metal atoms combine to yield molecular orbitals that extend over an entire metal fragment. As the number of metal atoms (and thus the number of atomic orbitals) in a fragment of a metal is increased, the number of molecular orbitals is increased (see Section 6.1 and Fig. 30-1).

c. As the number of molecular orbitals, or **energy levels** as they are sometimes called, increases, the difference in energy between these molecular orbitals becomes smaller and smaller until there is very little difference in energy between adjacent molecular orbitals. Effectively, a continuous *band* of molecular orbitals, or energy levels, results that extends over the entire crystal. There is one energy level in the band for each atomic orbital that participates in forming the energy levels. Each of the energy levels in the band can contain two electrons. However, only the electrons in the higher energy portion of a band are sufficiently free, or mobile, to be of significance for such properties as electrical and thermal conductivity, luster, and reflectivity.

To gain a better understanding of the band theory, consider the case of the lightest metal, lithium, whose electronic structure is $1s^2, 2s^1$. A very small crystal of lithium contains about 10^{18} atoms. For such a large number of atoms the energy difference between successive energy levels is so small that the levels are essentially continuous. These energy levels taken together constitute a band. In a lithium crystal, one band arises from the $1s$ atomic orbitals and is fully occupied with electrons. However, the second band, arising from the $2s$ orbitals, is only half-filled with electrons (Fig. 30-1). Between the two bands is a range of energy in which no energy levels, or molecular orbitals, are located. No electrons can be found with energies in this range inasmuch as there are no energy levels for electrons to occupy with these energies. Such energy gaps are called **forbidden zones.**

Metals are characterized by having incompletely filled bands—for example, lithium with its half-filled band arising from the $2s$ orbitals. Because there are so many electronic energy levels within a given band and because some bands are only partly filled, any electron within an unfilled band can absorb energy (for example, from an applied electric field) and be energetically promoted to other levels within the band.

In beryllium ($1s^2, 2s^2$) an isolated $2s$ band would be filled, and we might expect that this element should be nonmetallic. According to band theory a substance

Figure 30-1 Atomic orbitals in a Li atom; molecular orbitals in Li_2, Li_3, and Li_4; and bands in a lithium crystal (Li_n).

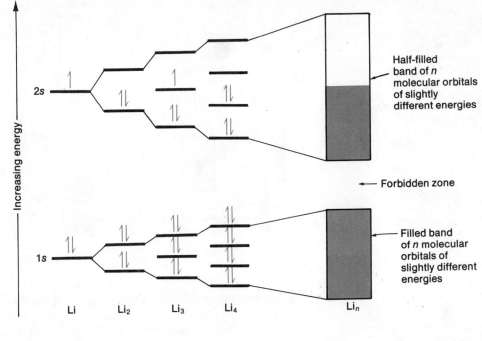

Half-filled band of n molecular orbitals of slightly different energies

← Forbidden zone

Filled band of n molecular orbitals of slightly different energies

Figure 30-2 Valence atomic orbitals in a Be atom; molecular orbitals in Be_2, Be_3, and Be_4; and overlapping bands in a crystal of beryllium (Be_n).

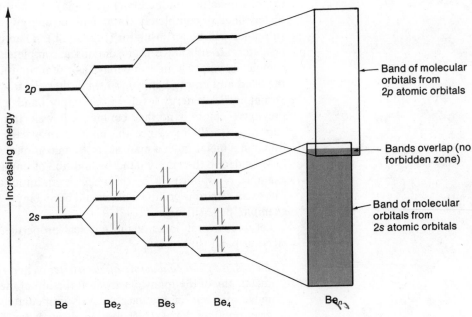

← Band of molecular orbitals from $2p$ atomic orbitals

← Bands overlap (no forbidden zone)

← Band of molecular orbitals from $2s$ atomic orbitals

containing only completely filled and empty bands is an **insulator.** Beryllium does, however, have appreciable metallic character. In beryllium, the $2s$ and $2p$ bands overlap, there is no forbidden zone, and electrons from the top (higher energy) part of the $2s$ band "spill over" into the bottom (lower energy) part of the $2p$ band (Fig. 30-2) resulting in incompletely filled bands and imparting metallic character.

Figure 30-3 (a) Partially filled or overlapping bands are found in metals. (b) Completely filled or completely empty bands separated by a large gap are found in insulators. (c) A completely filled band and a completely empty band separated by a small gap are found in semiconductors at 0 K. Upon warming to room temperature, a few electrons are thermally promoted from the previously filled band to the previously empty band, as shown.

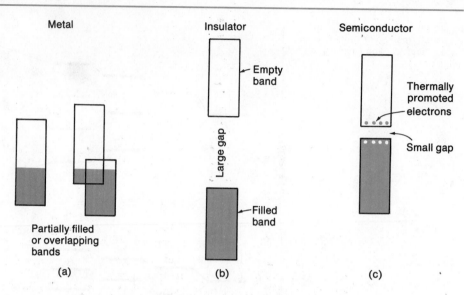

The spacing of the bands and their filling determines whether a substance is a conductor, nonconductor (insulator), or semiconductor (a poor conductor). A substance such as lithium or beryllium, which contains partially filled bands, exhibits metallic conduction (Fig. 30-3a). If the bands in a substance are completely filled or completely empty and the energy gap between bands is large, the substance will be an insulator (Fig. 30-3b). Diamond is an example of such an insulator. A substance that contains a completely filled band separated by a completely empty band can behave as a semiconductor if the energy gap between the filled and empty bands is so small that electrons from the filled band can be "promoted" to energy levels in the empty band by thermal energy (heat). The previously empty band then contains a few electrons that can conduct an electric current (Fig. 30-3c). Since the number of electrons is much smaller than that found in a metal, such a material is a semiconductor. Alternatively, electrons can be provided to the empty band by addition of impurities with extra electrons (for example, phosphorus $1s^2 2s^2 2p^6 3s^2 3p^3$, as an impurity in silicon, $1s^2 2s^2 2p^6 3s^2 3p^2$). The extra electrons occupy what would be an empty band in the pure material (for example, pure silicon).

Let us interpret some of the physical properties of metals in the light of the electron band theory.

a. *Metals are good conductors of electricity.* When an electric field is applied to a metal, any of the many electrons at the top of the filled portion of a band may move into one of the vacant levels immediately above it in the band. The electrons thus have a net motion through the lattice in the direction of the applied field, producing conductivity.

b. *Metals are good conductors of heat.* The mobile electrons of the bands can also absorb heat energy. Transport of this energy by these electrons accounts for the high thermal conductivity of metals.

c. *Metals are good reflectors of radiant energy.* The absorption and subsequent emission of photons of radiant energy by the mobile electrons of the bands accounts for the high reflectivity of metals.

d. *Metals are ductile and malleable.* When a metallic crystal is subjected to mechanical stress, bonds are readily broken and new ones formed by the promotion of electrons from lower to higher electronic levels in the conduction bands and subsequent return of these electrons to their original levels. Thus the crystal lattice is not altered by mechanical stress but remains essentially the same, because the bonding electrons are delocalized (do not belong to any particular atoms) and can form other bonds between any adjacent atoms.

30.5   Classification of the Metals

It is convenient to classify the metals in terms of electronic structure as representative and transition metals (see Section 4.18). Gaseous atoms of the representative metals have all their valence electrons in one shell, whereas those of the transition metals have valence electrons in more than one shell. Those representative metals with one or two valence electrons lose them in forming ionic bonds (Section 5.2) and usually exhibit only one oxidation state: e.g., sodium, $+1$; calcium, $+2$; and magnesium, $+2$. Some of the representative metals that have three or more valence electrons have two oxidation states: e.g., lead, $+2$ and $+4$; tin, $+2$ and $+4$; bismuth, $+3$ and $+5$. The periodic variation of oxidation states of main group metals is discussed in Section 8.5.

The transition metals include the elements of subgroups IIIB through IB in each of the long periods of the Periodic Table. The first series of transition metals includes Sc, Ti, V, Cr, Mn, Fe, Co, Ni, and Cu. Because of the carry-over in properties, elements in subgroup IIB (Zn, Cd, and Hg) have many characteristics analogous to those of the transition elements, and for this reason they are sometimes classed with the transition elements. However, in terms of our definition of transition and representative metals, the metals of subgroup IIB are classified as representative metals. The electron distribution and common oxidation states of the transition metals of the fourth period of the Periodic Table are given in Table 30-1.

TABLE 30-1   The Transition Metals of the Fourth Period

Atomic Number	Metal	Electronic Structure	Some Oxidation States
21	Scandium	$1s^2 2s^2 2p^6 3s^2 3p^6 3d^1\, 4s^2$	$+3$
22	Titanium	$3d^2\, 4s^2$	$+2, +3, +4$
23	Vanadium	$3d^3\, 4s^2$	$+2, +3, +4, +5$
24	Chromium	$3d^5\, 4s^1$	$+2, +3, +6$
25	Manganese	$3d^5\, 4s^2$	$+1, +2, +3, +4, +6, +7$
26	Iron	$3d^6\, 4s^2$	$+2, +3, +6$
27	Cobalt	$3d^7\, 4s^2$	$+2, +3, +4$
28	Nickel	$3d^8\, 4s^2$	$+2, +3, +4$
29	Copper	$3d^{10} 4s^1$	$+1, +2, +3$

TABLE 30-2 Some Physical Properties of the Transition Metals of the Fourth Period

Property	Sc	Ti	V	Cr	Mn	Fe	Co	Ni	Cu
Atomic number	21	22	23	24	25	26	27	28	29
Density, g/cm³	2.5	4.5	5.96	7.19	7.21	7.87	8.90	8.90	8.91
Melting point, °C	1200	1725	1710	1890	1260	1535	1490	1452	1083
Atomic radius, Å	1.60	1.46	1.31	1.25	1.29	1.26	1.26	1.24	1.28
Ionization potential, eV	6.56	6.83	6.74	6.76	7.432	7.896	7.86	7.633	7.723

The transition metals are noted for their variability in oxidation state; this is attributed to the presence of valence electrons in more than one shell. Thus manganese has two electrons in its outside shell and five electrons in the next underlying $3d$ subshell and exhibits oxidation states of $+1$, $+2$, $+3$, $+4$, $+5$, $+6$, and $+7$. An atom of iron may lose the two electrons in its outermost shell and form an iron(II) ion, Fe^{2+}, or it may lose an additional electron from its underlying $3d$ subshell and form an iron(III) ion, Fe^{3+}. When it shares electrons with oxygen in K_2FeO_4, iron exhibits an oxidation state of $+6$.

The transition metals are further characterized by the fact that well into the series going from left to right across the Periodic Table, the properties of succeeding metals do not differ greatly from preceding ones (Table 30-2). This is attributed to the fact that, generally, succeeding elements in the series differ in electronic structure by one electron in the next to the outer valence shell rather than the outer valence shell (with two exceptions). Differences in electronic structure in the outer shell have more significant effects on properties than do differences within inner shells.

In contrast to the properties of the transition metals, those of succeeding representative metals in a period differ extensively (Table 30-3). The electronic structures of succeeding elements in this series differ by one electron in the outer shell.

The transition metal ions containing incomplete underlying electron shells are usually colored, both in solid salts and in solution. The color exhibited changes as

TABLE 30-3 Some Physical Properties of Three Representative Metals in the Third Period

Property	Na	Mg	Al
Atomic number	11	12	13
Density, g/cm³	0.97	1.74	2.71
Melting point, °C	97.6	651	658.7
Atomic radius, Å	1.86	1.60	1.43
Ionization potential, eV	5.138	7.644	5.984

the oxidation state of the metal changes, as indicated by the following examples: Cr^{2+} (blue), Cr^{3+} (violet); Mn^{2+} (pink), Mn^{3+} (violet); and Fe^{2+} (green), Fe^{3+} (yellow). The color is also modified by the nature of the nonmetallic element or acid radical with which the metal is combined. For example, CuO is black; $Cu(OH)_2$, pale blue; $CuCl_2$, yellow; $CuBr_2$, black; CuS, black; $CuSO_4$, white; and $CuSO_4 \cdot 5H_2O$, blue. When the electron shell underlying the outermost shell is filled, as in the case of Cu^+ and Zn^{2+}, the substance containing the ion is usually colorless. This is also usually true of compounds containing the representative metals, i.e., those with valence electrons only in the outer shells. The transition metals have two other general properties that should be mentioned—they have marked catalytic properties, and they form more stable coordination compounds (see Chapter 31) than do the representative metals.

30.6   Compounds of the Metals

The metals that have relatively large atomic radii and only one or two valence electrons show a marked tendency to react with nonmetals by electron transfer, with the formation of ionic compounds. For example, $NaCl$, KBr, CaF_2, MgO, and BaS are ionic compounds. However, those metals with relatively small atomic radii and having three or more valence electrons tend to share electrons with nonmetals and form covalently bonded molecules. For example, anhydrous aluminum chloride, anhydrous tin(IV) chloride, and titanium(IV) chloride are covalent compounds.

The most metallic metals are distinguished from the nonmetals by the basic nature of their hydroxyl compounds (Section 8.3). The hydroxides of the highly electropositive metals such as sodium and potassium are ionic in the solid state and yield, therefore, large quantities of hydroxide ions in solution; they thus are strong bases. The bonding in the less soluble metal hydroxides, such as magnesium hydroxide and calcium hydroxide, is largely ionic, and they are fairly strong bases. The hydroxides of metals such as aluminum, iron, and tin are largely covalent; they are sparingly soluble in water and extremely weak as hydroxide bases.

The hydroxides of some metals such as zinc, tin, and aluminum are **amphoteric;** i.e., they can act as either acids or bases and hence are soluble in either strongly basic or strongly acidic solutions.

$$[Al(H_2O)_3(OH)_3] + OH^- \rightleftharpoons [Al(H_2O)_2(OH)_4]^- + H_2O$$

$$[Al(H_2O)_3(OH)_3] + H_3O^+ \rightleftharpoons [Al(H_2O)_4(OH)_2]^+ + H_2O$$

The hydroxyl compounds of metals that have several oxidation states, such as manganese, become less basic and more acidic as the oxidation state is increased (see Table 30-4). The larger charge, as the oxidation number becomes higher, makes the central metal ion attract electrons more readily, thereby strengthening the metal-oxygen bond and weakening the O—H bond. Hence, the compound releases hydrogen ions more readily.

Note that the behavior of manganese in the +2 oxidation state is typically metallic, as indicated by the basic character of $Mn(OH)_2$. In permanganic acid, $HMnO_4$, manganese has an oxidation state of +7 and behaves as a nonmetal; it is similar to the chlorine in perchloric acid, $HClO_4$.

An important difference between the metals and the nonmetals is demonstrated

TABLE 30-4   Acid-Base Character of Hydroxyl Compounds of Manganese

Hydroxide	Oxidation State	Character
$Mn(OH)_2$	Mn(II)	Moderately basic
$Mn(OH)_3$	Mn(III)	Weakly basic
H_2MnO_3, $(HO)_2MnO$	Mn(IV)	Weakly acidic
H_2MnO_4, $(HO)_2MnO_2$	Mn(VI)	Definitely acidic
$HMnO_4$, $(HO)MnO_3$	Mn(VII)	Strongly acidic

in the extent and reversibility of the hydrolysis of their chlorides. The chlorides of the more nonmetallic elements usually hydrolyze extensively—in many cases, irreversibly.

$$PCl_3 + 3H_2O \longrightarrow H_3PO_3 + 3HCl \qquad \text{(irreversible)}$$
$$AlCl_3 + 3H_2O \rightleftharpoons Al(OH)_3 + 3HCl \qquad \text{(reversible)}$$

The chlorides of very active metals (Na, K) are not hydrolyzed appreciably at ordinary temperatures.

The metal hydroxides, except those of the alkali and alkaline earth metals, yield oxides upon heating. The heating of metallic carbonates and nitrates, except those of the alkali metals, produces oxides. Metals react with the halogens and sulfur to form halides and sulfides, respectively. Mercury and the metals preceding it in the activity series (Section 9.21) react directly with oxygen. Sodium and the metals above it in the activity series displace hydrogen from cold water while the metals above hydrogen liberate hydrogen from acids and form salts. The metals below hydrogen in the series ordinarily will not liberate hydrogen from acids, but most of them react with oxidizing acids, forming salts or oxides. Salts of metals may also be produced by the reaction of oxides, hydroxides, carbonates, or other salts with acids or other salts (see Chapter 14).

30.7 Metal-to-Metal Bonds

In some compounds, bonds occur between one metal and another. For example, the ions Hg_2^{2+}, $Re_2Br_8^{2-}$, and $Mo_2Cl_8^{4-}$ are known to contain metal-to-metal bonds (Fig. 30-4). Several divalent metal acetates also are known to contain metal-to-metal bonds. One typical example is chromium(II) acetate monohydrate, $[Cr(CH_3CO_2)_2 \cdot H_2O]_2$, which occurs as a dimer (Fig. 30-5). Each chromium atom in this molecule is in an octahedral environment.

Bonds between metal atoms in molecules and in simple ions are covalent bonds and can be described by simple molecular orbital theory. Band theory does not apply to these molecules.

30.8 Alloys

An **alloy** is usually composed of two or more metals, but it also can be made up of a metallic element and one of the less nonmetallic elements such as carbon, silicon, nitrogen, or phosphorus. Such systems are considered alloys only when they have metallic properties.

(a)

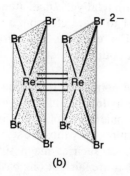

(b)

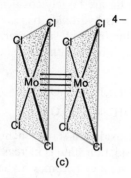

(c)

Figure 30-4 The structures of (a) Hg_2^{2+}, (b) $Re_2Br_8^{2-}$, and (c) $Mo_2Cl_8^{4-}$. In both (b) and (c) each metal atom is surrounded by a square plane of halide ions with another metal atom occupying the apex of a square pyramid. The two planes are parallel to each other and are perpendicular to the Re—Re or Mo—Mo bonds.

Figure 30-5 The structure of the chromium(II) acetate monohydrate dimer, which has a metal-to-metal bond. Each chromium atom is in an octahedral environment. The octahedron is shown in color for one of the two chromium atoms.

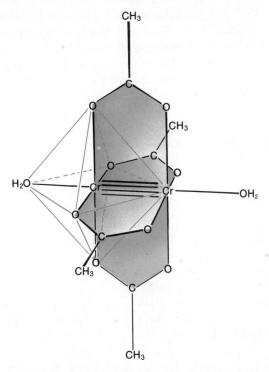

Alloys of two metals are usually prepared by fusing the metals together and allowing the melt to cool, but they are sometimes produced by simultaneous electrodeposition of two or more metals at a cathode. For example, many iron hardware fittings such as doorknobs are plated with brass (Cu 60–80%, Zn 40–20%) by electrodeposition from a bath containing copper and zinc in the form of soluble cyanide complexes. In general, alloys are solids, but they may be liquids, as in the case of certain amalgams, e.g., alloys of mercury. The study of alloys by x-ray methods has shown that they may be grouped into three classes, based upon their structures.

1. *Simple mixtures,* in which the component metals are mutually insoluble and the solid alloy is composed of an intimate mixture of crystals of each metal. For example, tin and lead (in plumber's solder) are insoluble in each other in the solid state.
2. *Solid solutions,* in which the atoms of one of the component metals take up positions in the crystal lattice of the other. The solute atoms may randomly replace some of the atoms at the lattice points of the solvent crystal to form **substitutional solid solutions.** For example, chromium dissolves in nickel to form a solid solution in which the chromium atoms replace nickel atoms in the face-centered cubic nickel lattice. The solubility may be limited (zinc and copper; chromium and nickel) or practically infinite (nickel and copper). If the solubility limit is exceeded, crystals of an alloy mixed with crystals of one of the metals are formed, as in class (1). The very small atoms of the elements H, C, B, and N may occupy the holes in the lattice of a metal, forming **interstitial solid solutions.** The solid solution of carbon in gamma-iron (austenite) is an example

of this type; the iron atoms are on the face-centered cubic lattice points, and the carbon atoms occupy the interstitial positions.

3. *Intermetallic compounds,* in which atoms of the components of the alloy appear in stoichiometric proportions that seem to be related to the ratio of the total number of electrons in the outer shell to the total number of atoms: Cu_5Zn_8 $(\frac{21}{13})$, Ag_3Al $(\frac{6}{4} = \frac{3}{2} = \frac{21}{14})$, and Cu_3Sn $(\frac{7}{4} = \frac{21}{12})$. The ratios for many intermetallic compounds (though not all) are either $\frac{21}{12}$, $\frac{21}{13}$, or $\frac{21}{14}$. Intermetallic compounds with the same ratio tend to be alike in crystal structure. In general, the formulas of such intermetallic compounds are not those that might be predicted on the basis of valence rules.

30.9   Properties of Alloys

Alloys that are solid solutions are generally harder than the pure solvent metal and are apt to be less ductile; for example brass, which is a solution of zinc in copper, is much harder than pure copper and less ductile. As one metal is dissolved in another, the electrical conductivity of the solvent metal is usually lowered very sharply. Thus copper must be quite free of impurities if it is to be used for transmission of electricity. The thermal conductivity of a solid solution alloy is less than that of the solvent metal. The melting point of a metal is usually lowered when another metal is dissolved in it. The practical application of a great many alloys depends upon their low melting, or freezing, points. For example, fuse metal, type metal, solder, and the alloys for sprinkler heads are low-melting alloy systems. Alloys of iron, cobalt, and nickel with each other are strongly magnetic. The color of a solid solution alloy is impossible to predict. Copper turns gray with the addition of 23% of nickel or 50% of zinc. Certain gold-iron solid solutions are blue; copper-antimony is green; and gold-silver-cadmium is green. The compositions of some alloys are given in Table 30-5.

TABLE 30-5 Composition of Some Common Alloys

Trade Name	Composition, Per Cent by Weight
White cast iron	97 Fe, 3 C
Stainless steel	82.5 Fe, 16.5 Cr, 0.65 C, 0.35 Mn
Bronze	70–95 Cu, 1–25 Zn, 1–18 Sn
Type metal	70 Pb, 18 Sb, 10 Sn, 2 Cu
Plumber's solder	67 Pb, 33 Sn
Battery plate	94 Pb, 6 Sb
Dentist amalgam	50 Hg, 35 Ag, 13 Sn, 1.5 Cu, 0.5 Zn
Sterling silver	92.5 Ag, 7.5 Cu
18 carat yellow gold	75 Au, 12.5 Ag, 12.5 Cu
18 carat white gold	75 Au, 3.5 Cu, 16.5 Ni, 5 Zn
14 carat gold	58 Au, 4–28 Ag, 14–28 Cu
Yellow brass	67 Cu, 33 Zn
Red brass	90 Cu, 10 Zn

Questions

1. Is the distinction beween metals and nonmetals well defined? Explain.

2. Compare metals with nonmetals with regard to (a) number of valence electrons, (b) ionization potential, (c) electron affinity, (d) electronegativity, and (e) acidic or basic nature of hydroxides.

3. What physical properties distinguish metallic elements from nonmetallic elements?

4. What chemical properties distinguish metallic elements from nonmetallic elements?

5. What is accomplished by roasting an ore before smelting?

6. Explain the concentration of sulfide ores by flotation.

7. Which metals are likely to form ionic compounds and which ones are apt to form covalent compounds?

8. Compare the packing structure of the atoms in face-centered cubic close-packed, hexagonal close-packed, and body-centered cubic crystal lattices.

9. In what ways are metallic and covalent bonds similar and how do they differ?

10. Contrast the metallic bond with the electron-deficient bond discussed in Chapter 28 for certain boron compounds. Explain how metallic bonding holds the atoms together in a crystal lattice, and discuss how various characteristic metallic properties can be explained by means of the metallic bond.

11. Contrast the hydrolysis of the chlorides of metals and nonmetals.

12. How does the basicity of the hydroxyl compounds of manganese vary with the oxidation state of the metal? Why is this so?

13. Compare the thermal stability of the hydroxides of the alkali and alkaline earth metals with that of the other metal hydroxides.

14. Why is it that most compounds of the transition metals are colored?

15. What is the basis for the classification of metals as representative and transition?

16. Compare the electron configurations of calcium and zinc, both as atoms and as dipositive (2+) ions.

17. Show that cobalt(III) and iron(II) are isoelectronic.

18. Vanadium, chromium, and manganese are elements with atomic numbers 23, 24, and 25, respectively, but have 3, 6, and 5 unpaired electrons, respectively, in the gaseous atoms. Explain the irregular progression in the number of unpaired electrons.

19. Is it correct to state that the electronic configuration of a beryllium atom in a crystal of beryllium is $1s^2 2s^2$?

20. List five physical properties of metals that may be modified by alloying.

21. How may alloys be produced? What are alloys containing mercury called?

22. In Section 30.8, mention was made that many (but not all) intermetallic compounds have a ratio of total number of electrons in the outer shell to total number of atoms equal to either $\frac{21}{12}$, $\frac{21}{13}$, or $\frac{21}{14}$. Determine the ratio for each of the following known intermetallic compounds: $CuZn$, Ag_5Zn_8, Cu_3Al, $AgCd_3$, $Na_{31}Pb_8$ (note that lead has four electrons in its outer shell), Cu_9Al_4, Cu_5Sn, $CuBe_3$, Cu_5Sn, Cu_3Ge, Ag_5Al_3, $LiAg$, and $Li_{10}Pb_3$. (Two in the group do not fall into any of the three classifications.)

23. Write balanced chemical reactions for the following processes, which involve metals or compounds of metals:

    (a) Magnetite, Fe_3O_4, is reduced to iron by carbon monoxide.

    (b) Copper sulfide, CuS, is roasted to produce copper oxide, and the oxide is reduced to copper with carbon.

    (c) Zinc carbonate is calcined.

    (d) Magnesium is prepared by electrolysis of molten magnesium chloride.

    (e) Silver is separated from its ore by oxidation with air in the presence of a solution of cyanide ion.

    (f) Titanium is oxidized to its highest oxidation state by heating in air.

    (g) Titanium dioxide (a white pigment) is prepared in two steps: hydrolysis of $TiCl_4$ followed by calcination of the product.

    (h) Lime, CaO, is prepared from limestone, $CaCO_3$.

    (i) Aluminum oxide in bauxite is separated from gangue by dissolving the oxide in a sodium hydroxide solution. Aluminum hydroxide is then precipitated by careful addition of hydrochloric acid.

Problems

1. A 4.53-g sample of lead gives 5.00 g of a lead oxide when heated in air. What is the formula of the lead oxide?   *Ans. Pb_3O_4*

2. A 1.10-g sample of chromium gives 1.61 g of chromium oxide when heated in air. What is the oxidation state of chromium in this oxide?   *Ans. $+3$*

3. What mass of hydrogen is formed when 11.46 g of the active metal scandium reacts with excess hydrochloric acid? Write the chemical equation for the reaction.   *Ans. 0.7708 g*

4. What mass of nickel will be deposited from a solution of nickel(II) chloride by 51,400 coulombs of electricity?   *Ans. 15.6 g*

5. Calculate the emf of the following cell, which uses copper and silver electrodes.

$$Cu \,|\, Cu^{2+}, \; M = 1 \,\|\, Ag^+, \; M = 1 \,|\, Ag$$

Write the equation for the cell reaction.
   Ans. 0.462 V

6. When 49.7 g of manganese is converted to Mn_3O_4 at standard state conditions, 418.4 kJ of heat is released to the surroundings. What is the standard enthalpy of formation, $\Delta H^\circ_{f_{298}}$, of $Mn_3O_4(s)$?
   Ans. 1390 kJ mol^{-1}

7. Consider the following reactions as possible means of preparing copper (I) oxide:
   (a) $Cu(s) + CuO(s) \longrightarrow Cu_2O(s)$
   (b) $2CuO(s) \longrightarrow Cu_2O(s) + \frac{1}{2}O_2(g)$
   Using the data in Appendix J, calculate ΔH°_{298} and ΔG°_{298} for these reactions. Would either of these reactions be feasible for preparing $Cu_2O(s)$ at 298°? at higher temperatures?   *Ans. (a) $\Delta H^\circ_{298} = -12$ kJ; $\Delta G^\circ_{298} = -16$ kJ; (b) $\Delta H^\circ_{298} = +145$ kJ; $\Delta G^\circ_{298} = +114$ kJ*

References

"Composition and Evolution of the (Earth's) Mantle and Core," D. L. Anderson, C. Sammis, and T. Jordan, *Science,* **171,** 1103 (1971).

"The Origins of Metal Deposits in the Oceanic Lithosphere," E. Bonatti, *Sci. American,* February 1978, p. 54.

"Superplastic Metals," H. W. Hayden, R. C. Gibson, and J. H. Brophy, *Sci. American,* March 1969, p. 28.

"The Nature of Metals," A. H. Cottrell, *Sci. American,* September 1967, p. 90.

"Electrical Conduction in Metals," P. B. Allen and W. H. Butler, *Phys. Today,* December 1978, p. 44.

"An Analogy for Elementary Band Theory Concept in Solids," P. F. Weller, *J. Chem. Educ.,* **44,** 391 (1967).

"Chemistry of Alloy Development," J. N. Pratt and S. G. Glover, *Chem. in Britain,* **5,** 207 (1969).

"Extraction Metallurgy," J. H. E. Jeffes, *Chem. in Britain,* **5,** 189 (1969).

"Production and Use of High-Purity Metals," J. C. Chaston, *Chem. in Britain,* **5,** 224 (1969).

"High-Gradient Magnetic Separation," H. Kolm, J. Oberteuffer, and D. Kelland, *Sci. American,* November 1975, p. 47.

"Magnetic Separation Finding New Uses" (Staff), *Chem. and Eng. News,* July 10, 1978, p. 26.

"Hydrometallurgy Attracting New Interests" (Staff), *Chem. and Eng. News,* April 3, 1978, p. 29.

"Chemical Aspects of Dislocations in Solids," J. M. Thomas, *Chem. in Britain,* **6,** 60 (1970).

"The Forming of Sheet Metal," S. S. Hecker and A. K. Ghosh, *Sci. American,* November 1976, p. 100.

"Permanent Magnets," J. J. Becker, *Sci. American,* December 1970, p. 92.

"Zone Refining," W. C. Pfann, *Sci. American,* December 1967, p. 62.

"Chemical Reactions and the Composition of Sea Water," K. E. Chave, *J. Chem. Educ.,* **48,** 148 (1971).

"Superconductors for Power Transmission," D. P. Snowden, *Sci. American,* April 1972, p. 84.

"Conduction Electrons in Metals," M. Y. Azbel, M. I. Kaganov, and I. M. Lifshitz, *Sci. American,* January 1973, p. 88.

"Metal-Oxide-Semiconductor Technology," W. C. Hittinger, *Sci. American,* August 1973, p. 48.

"Electron Tunnelling and Superconductivity," I. Giaever, *Science,* **183,** 1253 (1974).

"How Productive Is the Sea?" I. Morris, *Chem. in Britain,* **10,** 198 (1974).

"Manganese Nodules (I): Mineral Resources on the Deep Sea

Bed" and "(II): Prospects for Deep Sea Mining," A. L. Hammond, *Science,* **183,** 502 and 644 (1974).

"Molecular Metal Clusters," E. L. Muetterties, *Science,* **196,** 839 (1977).

"The Chemistry of Liquid Metals," C. C. Addison, *Chem. in Britain,* **10,** 332 (1974).

"The Deformation of Metals at High Temperatures," H. J. McQueen and W. J. McG. Tegart, *Sci. American,* April 1975, p. 116.

"Hydrocarbon Reactions on Transition Metal Catalysts," J. J. Rooney, *Chem. in Britain* **2,** 242 (1966).

"Lighter Flint Chemistry," F. C. Hentz, Jr. and G. G. Long, *J. Chem. Educ.,* **53,** 651 (1976).

31

Coordination Compounds

The hemoglobin in your blood, the blue dye in your ballpoint pen ink and in your blue jeans, chlorophyll, vitamin B-12, and the catalyst used in the manufacture of polyethylene all contain **coordination compounds,** or **complexes**—compounds in which negative ions and/or neutral molecules are attached to metal ions. Simple examples of complexes include $[Ag(NH_3)_2]^+$, $[Cu(NH_3)_4]^{2+}$, $[Fe(CN)_6]^{4-}$, $[Fe(CN)_6]^{3-}$, $[Co(NH_3)_6]^{3+}$, $[Pt(NH_3)_2Cl_2]$, and species such as $[Al(H_2O)_6]^{3+}$ and $[Zn(OH)_4]^{2-}$ mentioned in previous chapters.

Formation of a complex requires the combination of two kinds of species: (1) an ion or molecule that has at least one pair of electrons available for coordinate covalent bonding, and (2) a metal ion or atom that has a sufficient attraction for electrons to form a coordinate covalent bond with the attaching group. Ions of the transition metals, inner transition metals, and a few metals near these series in the Periodic Table are especially prone to react in this way.

Enthalpy values are given for several coordination compounds in Appendix J, and, for the ones for which reliable values have been established, free energy and entropy values are also given. From these values it is evident that, although different coordination compounds vary greatly in stability, many are exceedingly stable.

Nomenclature, Structures, and Properties of Coordination Compounds

31.1   Definitions of Terms

It is important to understand the meaning of several terms commonly used in discussing coordination compounds.

The metal ion or atom in a complex is frequently called the **central metal ion** or **atom.** The groups attached to the metal ion are referred to as **ligands** The ligands may be either ions or neutral molecules. Within the ligand, the atom attached directly to the metal through a coordinate covalent bond (Section 5.10) is called the **donor atom.**

The **coordination sphere,** which is usually enclosed in brackets in the formula, includes the central metal ion plus the ligands attached to it.

The **coordination number** of the central metal ion is the number of donor atoms that are bonded to the central atom. The coordination number for the silver ion in $[Ag(NH_3)_2]^+$ is 2; that for the copper ion in $[Cu(NH_3)_4]^{2+}$ is 4; and that for the iron(III) ion in $[Fe(CN)_6]^{3-}$ is 6. In each of these examples, the coordination number is also equal to the number of ligands attached to the metal ion, but such is not always the case. Some ligands, such as ethylenediamine,

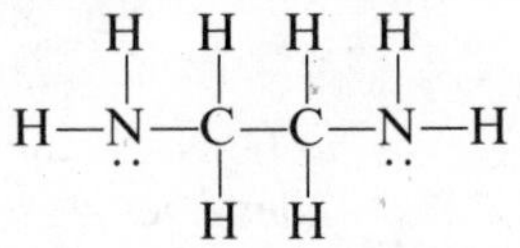

$$H-\underset{\underset{\textstyle\cdot\cdot}{H}}{\overset{\textstyle H}{N}}-\underset{\underset{\textstyle H}{|}}{\overset{\textstyle H}{C}}-\underset{\underset{\textstyle H}{|}}{\overset{\textstyle H}{C}}-\underset{\underset{\textstyle\cdot\cdot}{H}}{\overset{\textstyle H}{N}}-H$$

contain two donor atoms. Thus the coordination number for cobalt in $[Co(H_2NCH_2CH_2NH_2)_3]^{3+}$ is 6. Although the coordination sphere of this complex contains only three ligands, *six donor atoms* are bonded to the cobalt. The most common coordination numbers are 2, 4, and 6, but in some complex molecules or ions coordination numbers of 3, 5, 7, or 8 occur, and sometimes (though much less commonly) 9, 10, 11, and 12.

When a ligand attaches itself to a central metal ion by the use of two or more donor atoms, it is referred to as a **polydentate ligand,** or a **chelating group.** The resulting complex is referred to as a **metal chelate;** examples are

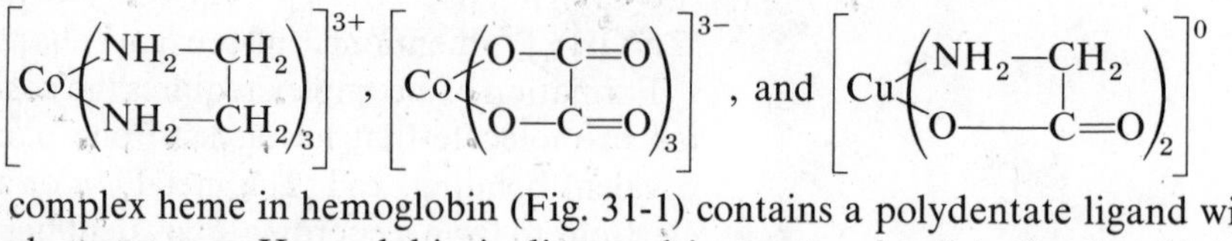

$$\left[Co\left(\!\!\begin{array}{c}NH_2-CH_2\\NH_2-CH_2\end{array}\!\!\right)_3\right]^{3+},\ \left[Co\left(\!\!\begin{array}{c}O-C=O\\O-C=O\end{array}\!\!\right)_3\right]^{3-},\ \text{and}\ \left[Cu\left(\!\!\begin{array}{c}NH_2-CH_2\\O\!-\!\!-C=O\end{array}\!\!\right)_2\right]^{0}$$

The complex heme in hemoglobin (Fig. 31-1) contains a polydentate ligand with four donor atoms. Hemoglobin is discussed in greater detail in Section 31.5.

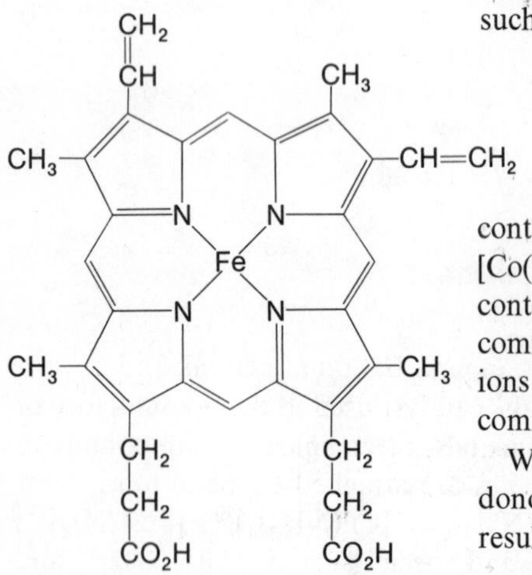

Figure 31-1 Heme, the square planar complex of iron found in hemoglobin.

31.2   The Naming of Complex Compounds

The nomenclature of complexes is patterned after a system originally suggested by Alfred Werner, a Swiss chemist and Nobel laureate, whose outstanding work in the latter part of the nineteenth century and early part of the twentieth century laid the foundation for a clearer understanding of these compounds. The following rules for naming complexes are used.

1. If the compound is ionic, the cation is named first and then the anion, in accord with usual nomenclature rules.

2. In naming the complex, whether a cation, an anion, or a neutral molecule, the ligands are named first and then the central metal.

3. The preferred order of indicating the ligands within the coordination sphere is to name the ligands alphabetically. A former, but still much used order, is to name negative ligands alphabetically first, neutral ligands alphabetically next, and positive ligands alphabetically last. Negative ligands (anions) have names formed by adding *o* to the stem name of the group. The names used for some anionic ligands are fluoro (F^-), chloro (Cl^-), bromo (Br^-), iodo (I^-), cyano (CN^-), nitro (NO_2^-), nitrito (ONO^-), nitrato (NO_3^-), hydroxo (OH^-), oxo (O^{2-}), amido (NH_2^-), oxalato ($C_2O_4^{2-}$), and carbonato (CO_3^{2-}). For most neutral ligands the name of the molecule is used. Four exceptions are aqua (H_2O), ammine (NH_3), carbonyl (CO), and nitrosyl (NO).

4. If more than one ligand of a given type is present within a complex, the number of that type of ligand is given by the prefixes *di* (for two), *tri* (for three), *tetra* (for four), *penta* (for five), and *hexa* (for six). Sometimes the prefixes *bis* (for two) *tris* (for three) and *tetrakis* (for four) are used when the name of the ligand contains numbers or already includes *di*, *tri*, etc.

5. When the complex is either a cation or a neutral molecule, the name of the central metal remains intact, followed by a Roman numeral designation in parentheses to indicate the oxidation state of the metal. When the complex is an anion, the ending *-ate* is added to the stem for the name of the central metal (or sometimes to the stem of the Latin name for the metal), followed by the Roman numeral designation of the oxidation state of the metal.

Examples in which the complex is a cation:

$[Co(NH_3)_6]Cl_3$	Hexaamminecobalt(III) chloride
$[Pt(NH_3)_4Cl_2]^{2+}$	Tetraamminedichloroplatinum(IV) ion
$[Ag(NH_3)_2]^+$	Diamminesilver(I) ion
$[Cr(H_2O)_4Cl_2]Cl$	Tetraaquadichlorochromium(III) chloride
$[Co(H_2NCH_2CH_2NH_2)_3]_2(SO_4)_3$	Tris(ethylenediamine)cobalt(III) sulfate

Examples in which the complex is neutral:

$[Pt(NH_3)_2Cl_4]$	Diamminetetrachloroplatinum(IV)
$[Co(NH_3)_3(NO_2)_3]$	Triamminetrinitrocobalt(III)
$[Ni(H_2NCH_2CH_2NH_2)_2Cl_2]$	Bis(ethylenediamine)dichloronickel(II)

Examples in which the complex is an anion:

$K_3[Co(NO_2)_6]$	Potassium hexanitrocobaltate(III)
$[PtCl_6]^{2-}$	Hexachloroplatinate(IV) ion
$Na_2[SnCl_6]$	Sodium hexachlorostannate(IV)

31.3   The Structures of Complexes

In Chapter 7, we studied the structures of a considerable number of simple compounds and ions. We noted that many spatial arrangements of the atoms are possible, some of these being linear, trigonal planar, tetrahedral, square planar, square pyramidal, trigonal bipyramidal, and octahedral. We are now ready to

Figure 31-2 Octahedral structures of the $[Co(H_2O)_6]^{2+}$, the $[Co(NH_3)_6]^{3+}$, and the $[PtCl_6]^{2-}$ ions.

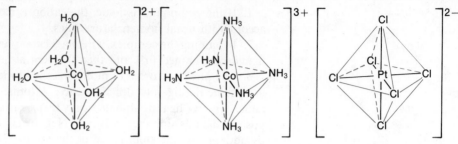

Figure 31-3 Steps in drawing an octahedron.

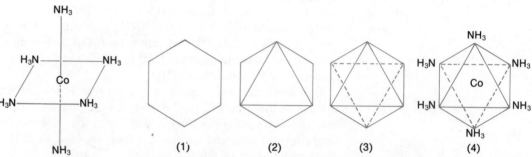

(a) An abbreviated drawing of an octahedron.

(b) To construct a complete octahedron quickly and simply: (1) Draw a regular hexagon. (2) Draw a triangle inside the hexagon. (3) Draw an upside-down triangle with dashed lines inside the hexagon and back of the triangle. (4) Add symbols for central metal and ligands.

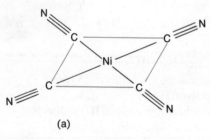

(a)

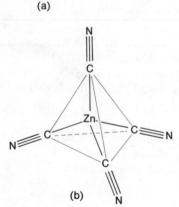

(b)

Figure 31-4 (a) The square planar configuration of the $[Ni(CN)_4]^{2-}$ ion, and (b) the tetrahedral configuration of the $[Zn(CN)_4]^{2-}$ ion.

extend our study to the structures of coordination molecules and ions. It may be helpful for you to reread Chapter 7 to refresh your memory of the principles and structures discussed there. You will find that many of the concepts discussed in Chapter 7 for simple compounds apply also to coordination compounds.

In 1893, Alfred Werner presented a concept of coordination compounds that laid the foundation for the further study of these substances. He suggested, as one part of his theory, that when ions or polar molecules are coordinated to a metal ion they are arranged in a definite geometrical pattern about the metal ion. With this concept he was able to account for the properties of hydrates, ammoniates, and various so-called double salts of the types $CoCl_2 \cdot 6H_2O$, $CoCl_3 \cdot 6NH_3$, and $PtCl_4 \cdot 2KCl$, respectively. Werner assigned the formulas $[Co(H_2O)_6]Cl_2$, $[Co(NH_3)_6]Cl_3$, and $K_2[PtCl_6]$ to these compounds and pointed out that the properties of the complexes $[Co(H_2O)_6]^{2+}$, $[Co(NH_3)_6]^{3+}$, and $[PtCl_6]^{2-}$ could be explained by postulating that the six coordinated groups are arranged about the central ion at the corners of a regular octahedron (Fig. 31-2).

Many chemists use an abbreviated drawing of an octahedron (Fig. 31-3a) when describing the geometry about a metal ion. However, a more realistic octahedron can be drawn by following the steps outlined in Fig. 31-3b.

Complexes in which the metal shows a coordination number of 4 exist in one of two different geometric arrangements, the square planar or the tetrahedral configurations. Examples of 4-coordinate complex ions with the planar configuration are $[Ni(CN)_4]^{2-}$ [Fig. 31-4(a)] and $[Cu(NH_3)_4]^{2+}$, and with the tetrahedral configuration are $[Zn(CN)_4]^{2-}$ (Fig. 31-4b) and $[Zn(NH_3)_4]^{2+}$.

Many other geometries are also possible. Table 31-1 shows several of the known types.

TABLE 31-1   Geometric Shapes and Corresponding Hybridization of Orbitals for Several Complexes

Geometric Shape	Diagram of Shape	Coordination Number	Hybridization	Examples
Linear		2	sp	$[Ag(NH_3)_2]^+$, $Cu(CN)_2^-$
Trigonal planar		3	sp^2	$[HgCl_3]^-$
Tetrahedral		4	sp^3 (or sd^3)	$[Zn(CN)_4]^{2-}$, $[FeCl_4]^-$, $[CoBr_4]^{2-}$
Square planar		4	dsp^2	$[Ni(CN)_4]^{2-}$, $[Cu(NH_3)_4]^{2+}$
Square pyramidal		5	d^2sp^2 (or d^4s)	$[VOCl_4]^{2-}$, $[Ni(Br)_3\{(C_2H_5)_3P\}_2]$
Trigonal bipyramidal		5	dsp^3 (or d^3sp)	$[Fe(CO)_5]$, $[Mn(CO)_4NO]$
Octahedral		6	d^2sp^3	$[Co(NH_3)_6]^{3+}$, $[PtCl_6]^{2-}$, $[MoF_6]^-$

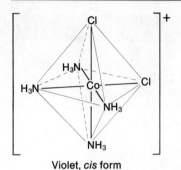

Violet, *cis* form

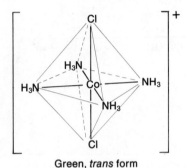

Green, *trans* form

Figure 31-5 The *cis* and *trans* isomers of $[Co(NH_3)_4Cl_2]^+$.

31.4   Isomerism in Complexes

Certain complexes, such as $[Co(NH_3)_4Cl_2]^+$—the tetraamminedichlorocobalt(III) ion—have more than one form. These different forms of the substance, possessing the same formula, are referred to as **isomers** (see Section 25.16). The $[Co(NH_3)_4Cl_2]^+$ ion has two isomers, one of which is violet and the other green. The violet form has been shown by crystal structure analysis, using x-ray methods and other experimental techniques, to have the *cis* configuration (the chloride ions on adjacent corners of the octahedron), and the green form to have the *trans* configuration (the chloride ions on opposite corners), as shown in Fig. 31-5. Isomers such as these, which differ only in the way that the atoms are oriented in space relative to each other, are called **geometric isomers,** or **stereoisomers.**

A complex such as $[Cr(NH_3)_2(H_2O)_2(Br)_2]^+$ has a variety of different geometric isomers. Each coordinating group can be *trans* to one like it:

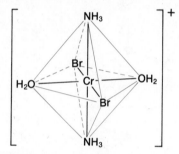

The ammonia molecules can be *trans* to each other but the water molecules and bromine atoms *cis* to groups like themselves:

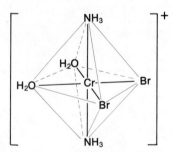

Similarly, the water molecules or the bromine atoms can be *trans* to each other with the other groups *cis* to groups like themselves:

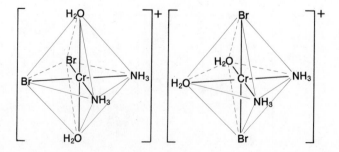

Finally, each group can be *cis* to one like itself. In this latter case, it can be shown that two different arrangements are possible:

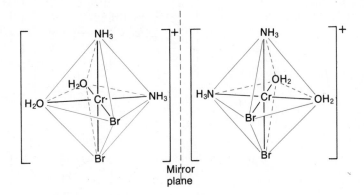

These two geometric isomers have the same arrangement of ligands in the coordination sphere (*cis* H_2O, *cis* Br, *cis* NH_3), but they are mirror images of each other, and are not superimposable. Therefore they are not geometrically identical. Such isomers are called **optical isomers,** a term applied to isomers that are mirror images of each other but not identical. Diagrams for mirror images of the other geometrical isomers of $[Cr(NH_3)_2(H_2O)_2(Br)_2]^+$ can be drawn, but it can be shown in each case that the additional form can be superimposed upon the one for which it is a mirror image and is therefore identical with it. For example, the following mirror image forms are actually identical to each other, inasmuch as either may be turned 90° on an axis through the two corners occupied by the ammonia molecules to superimpose it upon the other.

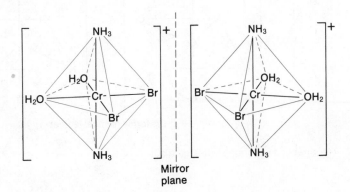

The $[Cr(NH_3)_2(H_2O)_2(Br)_2]^+$ complex thus has a total of six geometrical isomers, two of which are a pair of optical isomers.

The tris(ethylenediamine)cobalt(III) ion, $[Co(H_2NCH_2CH_2NH_2)_3]^{3+}$, has two optical isomers, as shown in Fig. 31-6 (see also abbreviated form, where $N \frown N = H_2NCH_2CH_2NH_2$).

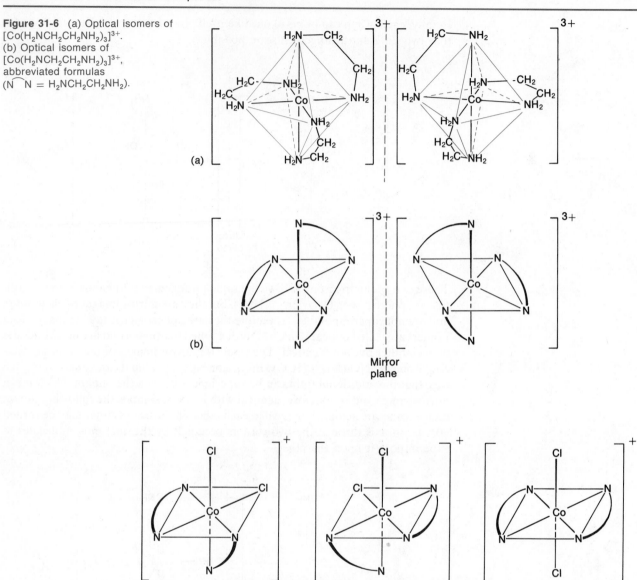

Figure 31-6 (a) Optical isomers of [Co(H$_2$NCH$_2$CH$_2$NH$_2$)$_3$]$^{3+}$. (b) Optical isomers of [Co(H$_2$NCH$_2$CH$_2$NH$_2$)$_3$]$^{3+}$, abbreviated formulas (N⌒N = H$_2$NCH$_2$CH$_2$NH$_2$).

Mirror plane

Cis forms (optical isomers)   *Trans* form

Figure 31-7 The three isomeric forms of [Co(en)$_2$Cl$_2$]$^+$.

The [Co(en)$_2$Cl$_2$]$^+$ ion, where en refers to ethylenediamine, has two *cis* isomers, which are optical isomers of each other, and one *trans* isomer. The *trans* form is symmetrical and has no possible optical isomerism. Its mirror image is superimposable on it and is therefore identical to the original *trans* form. The three stereoisomers are shown in Fig. 31-7.

See Section 25.16 for a discussion of optical isomers in organic compounds.

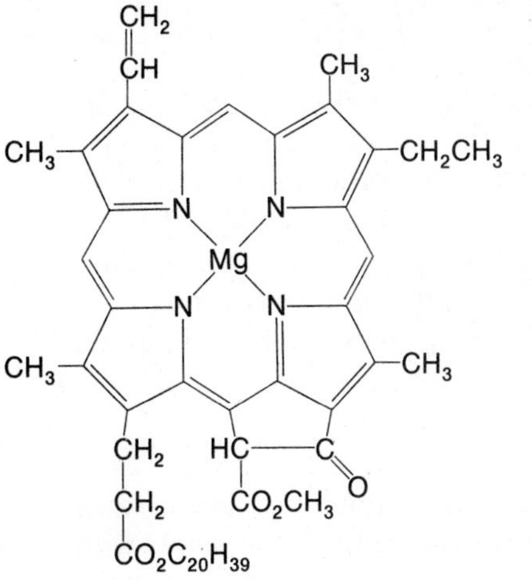

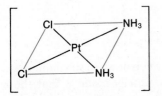

Figure 31-8 Chlorophyll, a square planar magnesium complex.

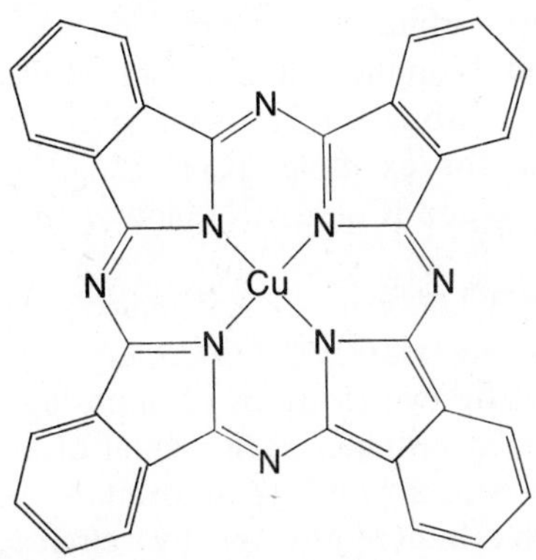

Figure 31-9 Copper phthalocyanine blue, a square planar copper complex.

Figure 31-10 The square planar structure of the anticancer agent Platinol, cis-[Pt(NH_3)_2Cl_2].

31.5   Uses of Complex Compounds

Many of the uses of complexes are based on their color, their solubility, or the change in chemical behavior of metal ions and ligands when they form complexes.

Chlorophyll (Fig. 31-8), the green pigment in plants, is a complex that contains magnesium. Plants appear green because chlorophyll absorbs yellow light; the reflected light consequently appears green (Section 31.10). The energy resulting from the absorption of this light is used in the process of photosynthesis (Section 25.5). The square planar copper(II) complex phthalocyanine blue (Fig. 31-9), is one of many complexes used as pigments or dyes. This complex is used in blue ballpoint pen ink, in blue jeans, and in certain blue paints.

The structure of heme (Fig. 31-1), the iron-containing complex in hemoglobin, is very similar to that of chlorophyll. In hemoglobin, the red heme complex is bonded to a large protein molecule (globin) by coordination of the protein to a position above the plane of the heme molecule. Oxygen is transported by hemoglobin in the blood when the oxygen molecules are bound to the coordination site opposite the binding site of the globin molecule.

Complexing agents are often used in water softening because they tie up such ions as Ca^{2+}, Mg^{2+}, and Fe^{2+}, which are responsible for the hardness in water (see Section 12.12). Complexing agents that tie up metal ions are also used as drugs. British Anti-Lewisite, $HSCH_2CH(SH)CH_2OH$, a drug developed during World War I as an antidote for the arsenic-based war gas Lewisite, is now used to treat poisoning by heavy metals such as arsenic, mercury, thallium, and chromium. The drug, called BAL for short, is a ligand and functions by making a soluble chelate of these metals; this chelate is eliminated by the kidneys. Another polydentate ligand, enterobactin, which is isolated from certain bacteria, is used to form complexes of iron and thereby to control the severe iron buildup found in patients suffering from blood diseases such as Cooley's anemia. With this disease, the patient's own blood is unable to transport oxygen adequately. Such patients must have regular blood transfusions to survive, but as the new blood breaks down, the usual metabolic processes that remove excess iron are overloaded. This excess iron can build up to fatal levels in the heart, kidneys, and liver. Enterobactin forms a soluble complex with the excess iron, and this complex can be eliminated by the body.

A new anticancer drug, Platinol, was approved for general use by the Food and Drug Administration in 1979. Platinol is cis-[Pt(NH_3)_2Cl_2], a square planar platinum(II) complex (Fig. 31-10). This substance, whose biological activity was first observed by scientists at Michigan State University in 1969, is believed to cross-link DNA strands (Section 26.12) and thereby interfere with cellular mitosis.

Some complexing groups, when coordinated to certain metals, make the metal more easily assimilated by plants; in other cases, they tie up the metal so that it cannot be effectively utilized by the plant. Hence complexes can be an important aid to the farmer in effective soil treatment.

In the electroplating industry, it has been found that many metals plate out in a smoother, more uniform plate which adheres better and has better appearance if the metal is plated from a bath that contains the metal in the form of a complex ion. Thus complexes such as $[Ag(CN)_2]^-$ and $[Au(CN)_2]^-$ are used extensively in the electroplating industry.

Bonding in Coordination Compounds

Any theory of the bonding in coordination compounds must explain four important properties: their stabilities, their structures, their colors, and their magnetic properties. The first modern attempt to explain the properties of complex compounds was made by applying the concepts of hybridization (Sections 7.3–7.9) and valence bond theory. More recent models of bonding in such molecules use crystal field theory, a model that examines the electrostatic attractions and repulsions between the central metal ion and the ligands, or use molecular orbital theory.

31.6   Valence Bond Theory

The valence bond theory treats metal-ligand bonds as coordinate covalent bonds (Section 5.10). Electron pairs from the ligands are considered as being shared with the metal atom or ion. According to this theory, each of these electron pairs occupies both an atomic orbital on a ligand and one of several equivalent hybrid orbitals on the metal.

In an octahedral complex such as $[Co(NH_3)_6]^{3+}$, a total of six electron pairs from the six ligands may be considered as entering hybrid orbitals of the metal ion. As discussed in Section 7.9, six equivalent hybrid orbitals on the metal ion can result from the hybridization of two d, one s, and three p orbitals; this is spoken of as d^2sp^3 hybridization. All octahedral complexes, for example, $[Co(NH_3)_6]^{3+}$, $[SnCl_6]^{2-}$, $[Co(H_2O)_6]^{2+}$, $[Co(CN)_6]^{3-}$, $[Fe(CN)_6]^{3-}$, and $[Co(en)_3]^{3+}$, show d^2sp^3 hybridization.

Let us look more closely at the octahedral $[Co(NH_3)_6]^{3+}$ ion. An isolated cobalt atom in its ground state has the electron configuration $1s^22s^22p^63s^23p^63d^74s^2$ (see Section 4.16). Each atomic orbital can accommodate two electrons of opposing spin. As we learned earlier (Section 4.15), the electrons enter each orbital of a given type singly before any pairing of electrons occurs within those orbitals—Hund's Rule. The orbitals for the cobalt atom, each orbital represented by a circle, can be diagrammed as follows:

Co atom

$1s$ $2s$ $2p$ $3s$ $3p$ $3d$ $4s$ $4p$

When the cobalt(III) ion, Co^{3+}, is formed, the atom loses three electrons, resulting in a structure possessing four unpaired electrons.

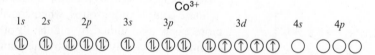

Co^{3+}

$1s$ $2s$ $2p$ $3s$ $3p$ $3d$ $4s$ $4p$

Six pairs of electrons are available for sharing with the cobalt when six ammonia molecules bond to a cobalt(III) ion. The postulation usually made is that the unpaired electrons that are already present in the $3d$ orbitals of the metal ion pair up as the electron pairs in the ligands approach, and the two empty $3d$ orbitals that result hybridize with the $4s$ and the three $4p$ orbitals to form six d^2sp^3 hybrid orbitals which are directed toward the corners of an octahedron. These hybrid orbitals accept the incoming electrons from the ammonia (indicated below in color). It should be noted, however, that once the structure is formed it is impossible to distinguish the source of the electrons, since (as far as we know) all electrons are identical regardless of their origin. The distribution of electrons on cobalt in the $[Co(NH_3)_6]^{3+}$ ion is represented as follows:

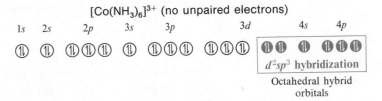

Octahedral complexes of cobalt(II), $[Co(CN)_6]^{4-}$ for example, contain one more electron than those of cobalt(III). To permit d^2sp^3 bonding, this extra electron

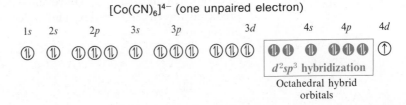

must be promoted to a $4d$ orbital (or perhaps $5s$), where it is loosely held. This electron is readily removed, as indicated by the ease with which most Co(II) complexes are oxidized to Co(III) complexes.

For 4-coordinate complexes with a tetrahedral structure, the four bonds of the central atom arise from the hybridization of one s and three p orbitals; this is referred to as sp^3 hybridization. This is analogous to filling the carbon orbitals in methane or carbon tetrachloride to produce a tetrahedral sp^3 configuration (see Section 7.3), except that in the present instance coordinate covalent bonds are being formed with both electrons of each shared pair coming from the ligand.

The distribution of electrons in the tetrahedral complex $[Zn(CN)_4]^{2-}$ is represented diagrammatically as follows:

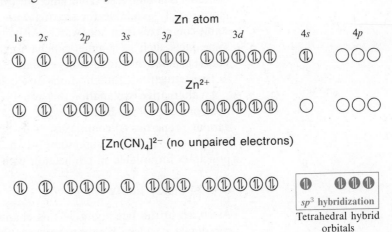

In the case of 4-coordinate structures with a square planar configuration, the four bonds of the central atom arise from dsp^2 hybridization, as illustrated for $[Ni(CN)_4]^{2-}$.

$[Ni(CN)_4]^{2-}$ (no unpaired electrons)

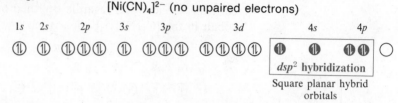

Square planar hybrid orbitals

Table 31-1 shows other types of hybridization in coordination compounds.

Some properties of complexes cannot be explained satisfactorily assuming hybridization and coordinate covalent bonding. For example, measurement of magnetic moments (Section 31.9) for a group of iron(II) complexes indicate that $[Fe(CN)_6]^{4-}$ has no unpaired electrons, whereas $[Fe(H_2O)_6]^{2+}$ has four unpaired electrons. On the assumption of covalent bonding, no unpaired electrons would be predicted for either complex, and indeed $[Fe(CN)_6]^{4-}$ does not contradict the assumption.

$[Fe(CN)_6]^{4-}$

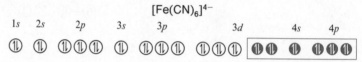

However, a contradiction does arise in the case of the $[Fe(H_2O)_6]^{2+}$ ion with its four unpaired electrons. If $[Fe(H_2O)_6]^{2+}$ is assumed to be ionically bonded by the electrostatic attraction of the Fe^{2+} ion for the negative end of the polar water molecule, the four unpaired electrons can be justified. Just as in the simple iron(II) ion, there are four unpaired electrons in the d orbitals since the ligand electrons do not enter the orbitals of the Fe^{2+} ion through sharing.

Fe^{2+} and $[Fe(H_2O)_6]^{2+}$

A differentiation between $[Fe(CN)_6]^{4-}$ and $[Fe(H_2O)_6]^{2+}$, in terms of type of bonding, is not unreasonable, since the cyanide ion usually makes an electron pair more readily available for sharing with the central metal ion and forms more stable complexes than does water. Although the concept of ionic bonding explains some facts such as magnetic behavior more satisfactorily than does covalent bonding, it does not adequately explain certain other properties of complexes such as their structural configurations and colors.

An alternative explanation that provides a useful modification of the concept of ionic or electrostatic bonding has proved to be of much value in explaining the various properties of complexes. This theory is referred to as the Crystal Field Theory, so named because some of the ideas used in the theory derive from principles formulated in connection with crystals.

31.7　Crystal Field Theory

As stated at the beginning of this chapter, a coordination compound involves coordinate covalent bonding between the metal and the ligands. Amazingly, however, it is possible to understand, interpret, and predict the properties of many coordination compounds, especially those of transition metals, on the assumption that there is no covalent bonding between the metal and the ligands, but only simple electrostatic interactions. This ionic model of the bonding in complexes is called **Crystal Field Theory.**

According to the postulates of Crystal Field Theory, the basic reason for the formation of a complex ion or molecule is the electrostatic attraction of a positively charged metal ion for negative ions or for the negative ends (the electron pairs) of dipolar molecules. The colors of complexes and their magnetic properties result from the electrostatic interactions of the electron pairs on the ligands with electrons in the *d* orbitals of the metal ion.

In Section 4.12, the atomic orbitals were pictured for *s*, *p*, and *d* orbitals. It will be recalled that the orbital of an *s* electron is spherical, whereas that of a *p* electron is a dumbbell shape, and that the three *p* orbitals for a given major energy level are oriented at right angles to each other along the *x*, *y*, and *z* axes (see Fig. 4-12).

The *d* orbitals, which occur in sets of five, each consist of lobe-shaped regions, and are arranged in space as shown in Fig. 4-13 and reproduced within an octahedral structure in Fig. 31-11.

Figure 31-11 Diagrams showing the directional characteristics of the five *d* orbitals. *L* indicates a ligand at each corner of the octahedron.

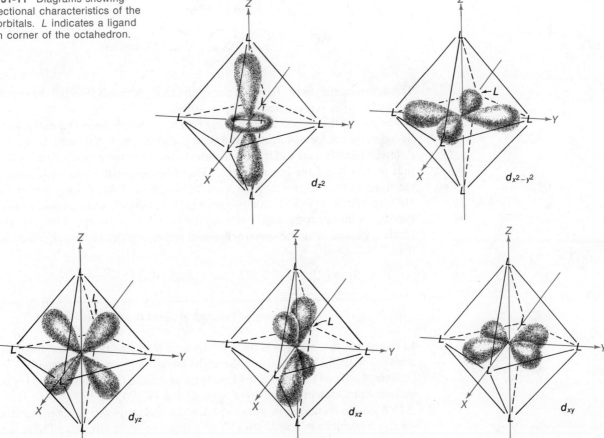

As shown in Fig. 31-11, the lobes in two of the five orbitals, the d_{z^2} and $d_{x^2-y^2}$ orbitals, point toward corners of the octahedron around the metal. These two orbitals are referred to as the e_g **orbitals** (this symbol refers to the symmetry of the orbitals, but we will use it simply as the name of these two orbitals in an octahedral complex). The other three orbitals, the d_{xy}, d_{xz}, and d_{yz} orbitals, whose lobes point in between the corners of the octahedron, are called the t_{2g} **orbitals.** (Again the symbol refers to the symmetry of the orbitals.) As six ligands approach the metal ion along the axes of the octahedron, the electron pairs on the ligands repel the electrons in the d orbitals of the metal ion at the center of the octahedron. However, the repulsions between electrons in the e_g orbitals (the d_{z^2} and $d_{x^2-y^2}$ orbitals) and the electron pairs on the ligands are greater than the repulsions between electrons in the t_{2g} orbitals (the d_{xy}, d_{xz}, and d_{yz} orbitals) and the electron pairs on the ligands; this is because the lobes of the e_g orbitals point directly at the electron pairs on the ligands whereas the lobes of the t_{2g} orbitals point between the electron pairs on the ligands. Thus electrons in e_g orbitals of a metal ion in an octahedral complex have higher potential energies than electrons in the t_{2g} orbitals of the metal ion. The difference in energy may be represented as

$$\underset{d_{z^2}}{\text{O}} \quad \underset{d_{x^2-y^2}}{\text{O}} \qquad e_g \text{ orbitals}$$

$$\underset{d_{xy}}{\text{O}} \quad \underset{d_{xz}}{\text{O}} \quad \underset{d_{yz}}{\text{O}} \qquad t_{2g} \text{ orbitals}$$

The difference in energy between the e_g and the t_{2g} orbitals is called the **crystal field splitting** and is represented by Δ, or **10Dq.**

The size of the crystal field splitting (10Dq) depends upon the nature of the six ligands that are located around the central metal ion. It has been discovered that different ligands have different concentrations of electrons in the pairs of electrons that point toward the central atom. Those ligands that project a low density of electrons toward the metal produce a small crystal field splitting (small value of 10Dq) while those ligands that project a high density of electrons toward the metal produce a large crystal field splitting (large value of 10Dq). This experimental result is expressed in the **spectrochemical series,** a short version of which is given here.

$$\text{I}^- < \text{Br}^- < \text{Cl}^- < \text{F}^- < \text{H}_2\text{O} < \text{C}_2\text{O}_4{}^{2-} < \text{NH}_3 < \text{en} < \text{NO}_2{}^- < \text{CN}^-$$

$$\longrightarrow$$

Increasing field strength of ligand electron pairs

In this series, the ligands on the left project a low density of electrons toward the metal (low field), and those on the right have a large concentration of electrons (high field). Thus the crystal field splitting produced by an iodide ion (I^-) is much smaller than that produced by a cyanide ion (CN^-).

In a simple metal ion in the gas phase, the electrons will be distributed among the five $3d$ orbitals in accord with Hund's Rule, since the orbitals all have the same energy.

Ion	3d Orbitals	Ion	3d Orbitals
Ti^{3+}	↑ ○ ○ ○ ○	Fe^{2+}	↑↓ ↑ ↑ ↑ ↑
V^{3+}	↑ ↑ ○ ○ ○	Co^{3+}	↑↓ ↑ ↑ ↑ ↑
Cr^{3+}	↑ ↑ ↑ ○ ○	Co^{2+}	↑↓ ↑↓ ↑ ↑ ↑
Cr^{2+}	↑ ↑ ↑ ↑ ○	Ni^{2+}	↑↓ ↑↓ ↑↓ ↑ ↑
Mn^{2+}	↑ ↑ ↑ ↑ ↑	Cu^{2+}	↑↓ ↑↓ ↑↓ ↑↓ ↑
Fe^{3+}	↑ ↑ ↑ ↑ ↑	Zn^{2+}	↑↓ ↑↓ ↑↓ ↑↓ ↑↓

However, if six ligands, each with a free electron pair, approach a metal ion along the axes of an octahedron with those electron pairs pointing toward the metal ion, the energies of the d orbitals are no longer the same, and two opposing forces are set up. One force tends to keep the electrons of the metal ion distributed with unpaired spins within all of the d orbitals according to Hund's Rule (Section 4.15). The other force tends to reduce the average energy of the d electrons by placing as many d electrons as possible in the lower energy t_{2g} orbitals. The d electrons will end up in the lowest possible energy situation. If it requires less energy for the electrons to be excited to the upper e_g orbitals than to pair in the lower t_{2g} orbitals, then the d electrons will remain unpaired. If it requires more energy to put d electrons in the upper e_g orbitals than to pair them in the lower t_{2g} orbitals, then they will pair in the lower orbitals.

In $[Fe(CN)_6]^{4-}$, the strong field of the six cyanide ions produces a large crystal field splitting. Under these conditions less energy is required for the electrons to pair than to be excited to the e_g orbitals. Thus the six $3d$ electrons of the Fe^{2+} ion pair in the three t_{2g} orbitals. The result is in agreement with the experimentally measured magnetic moment (no unpaired electrons).

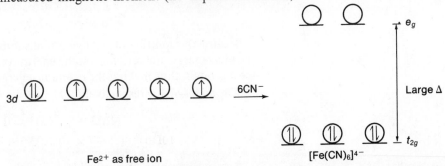

Fe²⁺ as free ion   [Fe(CN)₆]⁴⁻

In $Fe(H_2O)_6^{2+}$, on the other hand, the weak field of the water molecules produces only a small crystal field splitting. Since it does not require much energy to excite electrons to the e_g orbitals, Hund's Rule predominates and the electrons remain distributed through all five $3d$ orbitals.

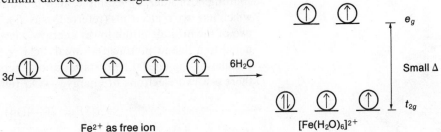

Fe²⁺ as free ion   [Fe(H₂O)₆]²⁺

Thus four unpaired electrons would be present, in agreement with the magnetic moment measurement, in Fe^{2+} and in $[Fe(H_2O)_6]^{2+}$. A similar line of reasoning can be followed to show why in complexes for the $+3$ oxidation state of iron the $[Fe(CN)_6]^{3-}$ has only one unpaired electron, whereas $[Fe(H_2O)_6]^{3+}$ and $[FeF_6]^{3-}$ possess five unpaired electrons each.

Actually, as pointed out in Section 5.9, there is no sharp dividing line between covalent and ionic bonding. The Molecular Orbital Theory, discussed in Section 31.11, seeks to introduce a covalent component into the ionic viewpoint.

31.8   Crystal Field Stabilization Energy

As we have said, the difference in energy between the e_g and the t_{2g} levels in an octahedral complex is commonly designated either Δ or 10Dq.

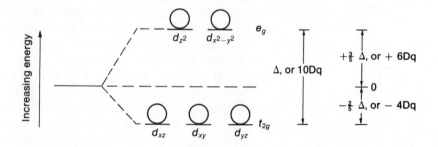

Because the splitting of the five orbitals into two energy levels does not change the total energy, the zero line on the energy axis is the weighted average energy of the d orbitals, thus making the e_g and t_{2g} energies, $+\frac{3}{5}\Delta$ and $-\frac{2}{5}\Delta$, respectively (or $+6Dq$ and $-4Dq$, respectively). Hence

$$(\tfrac{3}{5}\Delta)2 + (-\tfrac{2}{5}\Delta)3 = 0$$

or $$(\tfrac{3}{5} \times 10Dq)2 + (-\tfrac{2}{5} \times 10Dq)3 = (+6Dq)2 + (-4Dq)3 = 0$$

For each complex, it is possible to calculate a value, referred to as the **crystal field stabilization energy** (CFSE), which is related to the stability of the complex. Actually the crystal field stabilization energy is a measure of how much more stable the complex is than an identical hypothetical complex with no crystal field splitting. Consider, for example, an octahedral complex of the vanadium(III) ion, which has two d electrons (referred to as d^2). The two unpaired electrons occupy two of the orbitals of the lower energy t_{2g} level, whether the vanadium is coordinated to a ligand producing a weak field or to one producing a strong field. In calculating the crystal field stabilization energy, we take account of the fact that there are two electrons of energy $-4Dq$ relative to the zero energy line.

$$\text{CFSE} = 2(-4Dq) = -8Dq$$

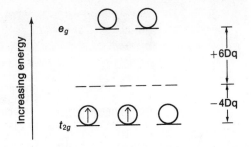

Now suppose that the metal ion is the manganese(II) ion, which has five d electrons (d^5). In the free ion, these electrons will be distributed evenly thoughout the five d orbitals in accord with Hund's Rule. If the ion is coordinated to ligands that produce a weak field, the five electrons will remain distributed one in each of the five orbitals; i.e., three unpaired electrons will be at the t_{2g} level and two unpaired electrons will be at the e_g level.

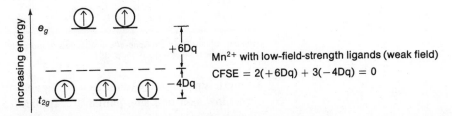

Mn^{2+} with low-field-strength ligands (weak field)

CFSE $= 2(+6\text{Dq}) + 3(-4\text{Dq}) = 0$

The two electrons in the e_g level will be at the higher energy level, corresponding to $+6\text{Dq}$, and three at the lower energy level, corresponding to -4Dq. Hence, the crystal field stabilization energy is zero in this weak-field case.

$$\text{CFSE} = 2(+6\text{Dq}) + 3(-4\text{Dq}) = 0$$

If, however, the Mn^{2+} ion is coordinated to ligands of high field strength, all five electrons will be forced into the lower energy nonbonding t_{2g} level and none will be at the higher energy e_g level.

Mn^{2+} with high-field-strength ligands (strong field)

CFSE $= 5(-4\text{Dq}) = -20\text{Dq}$

The crystal field stabilization energy for Mn^{2+} in this strong-field case is -20Dq.

$$\text{CFSE} = 5(-4\text{Dq}) = -20\text{Dq}$$

In general, the more negative the value of CFSE, the more stable will be the complex. As we would expect, a complex of Mn^{2+} with ligands of high field

TABLE 31-2   Crystal Field Stabilization Energy [CFSE] of Octahedral Complexes[a]

Configuration	Weak Field (High Spin)				Strong Field (Low Spin)				Examples
	t_{2g}	e_g	Number of Unpaired Electrons	CFSE (−Dq)	t_{2g}	e_g	Number of Unpaired Electrons	CFSE (−Dq)	
d^0	0	0	0	0	0	0	0	0	Ca^{2+}, Sc^{3+}
d^1	1	0	1	4	1	0	1	4	Ti^{3+}
d^2	2	0	2	8	2	0	2	8	V^{3+}
d^3	3	0	3	12	3	0	3	12	Cr^{3+}, V^{2+}
d^4	3	1	4	6	4	0	2	16	Cr^{2+}, Mn^{3+}
d^5	3	2	5	0	5	0	1	20	Mn^{2+}, Fe^{3+}
d^6	4	2	4	4	6	0	0	24	Fe^{2+}, Co^{3+}
d^7	5	2	3	8	6	1	1	18	Co^{2+}
d^8	6	2	2	12	6	2	2	12	Ni^{2+}
d^9	6	3	1	6	6	3	1	6	Cu^{2+}
d^{10}	6	4	0	0	6	4	0	0	Cu^+, Zn^{2+}

[a] Used by permission from *Concepts and Models of Inorganic Chemistry*, B. E. Douglas and D. H. McDaniel. Ginn, Blaisdell, New York, 1965, p. 351.

strength is more stable than one with ligands of low field strength. Sometimes the terms **low spin** and **spin-paired** are applied to complexes of a given metal with ligands of high field strength; conversely, the terms **high spin** and **spin-free** are sometimes used for the complexes with ligands of low field strength.

In the weak-field case for Mn^{2+}, the splitting of the two energy levels (10Dq) is much smaller than for the strong-field case. One can say that the possible gain in energy that would result if the fourth and fifth electrons were in the lower energy level, t_{2g}, is not sufficient to provide the energy necessary to overcome the repulsion of the electrons for each other and to pair them up. Hence the fourth and fifth electrons go into the higher energy level, e_g, in preference to pairing.

Table 31-2 shows the crystal field stabilization energy for a variety of octahedral complexes of weak and strong fields. The calculated CFSE values in the table can be experimentally verified by spectroscopic methods. The values for CFSE in the table are given in −Dq units; hence, −6Dq is listed as 6, −20Dq is listed as 20, etc. On this basis, a larger negative number indicates that more energy is lost in formation of the complex and that, in general, the complex is more stable for a given ligand. We must bear in mind, however, that each ligand has its own specific 10Dq value, and this must be taken into account in evaluating the stabilities of complexes with different ligands. Weak-field ligands cause less splitting of the energy levels and have smaller 10Dq values, in general, than strong-field ligands.

The largest value for CFSE in the table for strong-field complexes is for the d^6 complexes, since the lower energy t_{2g} orbitals are completely occupied and the higher energy e_g orbitals are empty. The largest CFSE values for weak-field complexes are for the d^3 and d^8 cases. In d^3 complexes the lower energy t_{2g} orbitals are each singly occupied, and the higher energy e_g orbitals are empty. The d^8 complexes represent the lowest number of electrons for which the lower energy

t_{2g} orbitals are completely filled; however, since two electrons are present in the higher energy e_g orbitals, the highest CFSE value for weak-field complexes (12) is lower than that for strong-field complexes (24). Note also that the CFSE is zero, indicating very low complex stability, at two places in the table for strong-field complexes, i.e., in the cases when no d electrons are present and when 10 d electrons are present. In the latter case, all t_{2g} and e_g orbitals are filled, so that the stability gained in the complexing process by the electrons going to the lower energy orbitals is exactly counterbalanced by the stability lost by electrons going to the higher energy orbitals. For weak-field complexes, the d^0 and d^{10} cases have CFSE values of zero, and in addition the d^5 complexes, in which all t_{2g} and e_g orbitals are singly occupied, have a CFSE value of zero.

31.9   Magnetic Moments of Molecules and Ions

Molecules such as O_2 (Section 6.10) or NO (Section 6.13) and ions such as $[Co(CN)_6]^{4-}$ (Section 31.6) and $[Fe(H_2O)_6]^{2+}$ (Section 31.7) which contain unpaired electrons are referred to as **paramagnetic.** Most transition metal ions have unpaired electrons and hence are paramagnetic. Paramagnetic materials tend to move into a magnetic field, such as that between the poles of a magnet. Molecules such as N_2 (Section 6.9) and ethane (Section 7.3) and ions such as Na^+, $[Co(NH_3)_6]^{3+}$ (Section 31.6), and $[Fe(CN)_6]^{4-}$ (Section 31.7), which contain orbitals that are either completely filled or empty (and thus contain no unpaired electrons), are referred to as **diamagnetic.** Diamagnetic materials have a slight tendency to move out of a magnetic field.

An electron in an atom spins about its own axis. Because the electron is electrically charged, the spin about its axis gives it the properties of a small magnet, with north and south poles. Two electrons in the same orbital spin in opposite directions. This means that their magnetic moments will cancel each other because their north and south poles are opposed. On the other hand, when an electron in an atom is unpaired the magnetic moment due to its spin confers paramagnetism upon the entire atom or ion, and in turn upon the specimen containing such atoms or ions.

The size of the magnetic moment of a system containing unpaired electrons is related directly to the number of unpaired electrons; i.e., the greater the number of such electrons, the larger the magnetic moment. Therefore the observed magnetic moment is used to determine the number of unpaired electrons present.

31.10   Color of Transition Metal Complexes

In Chapter 4 we found that atoms can absorb light of the proper frequency and excite their electrons to higher energy levels. The same thing can happen in coordination compounds. Electrons can be excited from the lower t_{2g} to the upper e_g orbitals, provided that the latter are not all filled with paired electrons.

The human eye perceives a mixture of all the colors, in the proportion present in sunlight, as white light. But the eye has a further property of importance: It utilizes complementary colors in color vision. The eye perceives a mixture of two complementary colors, in the proper proportions, as white light. Likewise, when a color is removed from white light the eye sees its complement. For example, as

TABLE 31-3 Complementary Colors

Wavelength, Å	Spectral Color	Complementary Color[a]
4100	Violet	Lemon-yellow
4300	Indigo	Yellow
4800	Blue	Orange
5000	Blue-green	Red
5300	Green	Purple
5600	Lemon yellow	Violet
5800	Yellow	Indigo
6100	Orange	Blue
6800	Red	Blue-green

[a]The complementary color is seen when the spectral color is removed from white light.

shown in Table 31-3, if red light is removed from white light, the eye sees the color blue-green; if violet is removed from white light, the eye sees lemon yellow; if green light is removed, the eye sees purple.

Let us consider the $[Fe(CN)_6]^{4-}$ complex (Section 31-7). The electrons in the lower t_{2g} orbitals can absorb energy and be excited to the upper level. The energy necessary corresponds to photons of violet light. If white light impinges on $[Fe(CN)_6]^{4-}$, violet light is absorbed (to accomplish the excitation), and the eye sees the complement, lemon yellow. $K_4[Fe(CN)_6]$ is lemon yellow in color.

In contrast, if white light strikes $[Fe(H_2O)_6]^{2+}$, red light of a longer wavelength, and lower energy, is absorbed. The eye sees its complement, blue-green. $[Fe(H_2O)_6]^{2+}$, in $[Fe(H_2O)_6]SO_4$ for example, is therefore blue-green in color.

As shown by Table 31-2, a coordination compound of the Cu^+ ion has a d^{10} configuration, and all the upper e_g orbitals are filled. In order to excite an electron to a higher level such as the $4s$ orbital, very high energy photons are needed. This energy corresponds to very short wavelengths in the ultraviolet. Since no visible light is absorbed, the eye can perceive no change, and the compound appears white or colorless. For example, a solution containing $[Cu(CN)_2]^-$ is colorless. On the other hand, Cu^{2+} complexes have a vacancy in the e_g orbitals, and electrons can be excited to this level. The wavelength (energy) absorbed corresponds to the visible part of the spectrum, and Cu^{2+} complexes are almost always colored: blue, blue-green, violet, or yellow.

As we have noted earlier, high-field ligands cause a large split in the energies of the d orbitals of the metal (large 10Dq). Transition metal coordination compounds with these ligands should be yellow, orange, or red since higher energy violet or blue light is absorbed. On the other hand, coordination compounds of transition metals with low-field ligands should be blue-green, blue, or indigo-violet since they absorb lower energy yellow, orange, or red light.

31.11   Molecular Orbital Theory

We have seen that a bonding theory based primarily on covalent bonding is especially useful in considering the structures of metal complexes, and that a bonding theory based primarily on ionic bonding is particularly useful in explaining magnetic properties and the energies involved in the electron transitions that give rise to the spectra of complexes. However, neither of these theories is adequate to explain certain other properties of complexes. Since most bonds have both ionic and covalent character, an improvement in the theory of bonding in complexes can logically be made by introducing some contributions of covalent bonding into the Crystal Field Theory (ionic) by utilizing molecular orbital concepts. In Chapter 6 we discussed the Molecular Orbital Theory and in later chapters applied the theory to a variety of substances. It is an especially useful concept in understanding the nature of the bonding in coordination compounds.

The molecular orbital energy diagrams for complexes are the same in principle as those shown in Chapter 6 for diatomic species. The ones for coordination compounds are, of course, more complicated, since several atoms are involved instead of just two, and a larger number of electrons must be considered. The molecular orbital energy diagrams for $[CoF_6]^{3-}$, possessing weak-field ligands, and for $[Co(NH_3)_6]^{3+}$, possessing relatively strong field ligands, are shown in Fig. 31-12. Only the six d electrons for the Co^{3+} and the free pair of electrons for each of the six ligands (a total of 18 electrons) are indicated.

The arrangement and symbols used in Fig. 31-12 are, in general, those described in Chapter 6. In addition, the terms (t_{1u}, a_{1g}, e_g, t_{2g}, etc.) alongside the circles (orbitals) are commonly used to label the various energy levels. In some cases, the orbital description ($d_{x^2-y^2}$, d_{z^2}, d_{xy}, etc.) corresponding to the directional characteristics of the orbitals, as indicated in Fig. 31-11, are also given. It can be noted that the antibonding and nonbonding energy levels use the e_g and t_{2g} designations, respectively, in the molecular orbital energy level diagram. The energy difference

Figure 31-12 Molecular orbital energy diagrams for $[CoF_6]^{3-}$ and $[Co(NH_3)_6]^{3+}$.

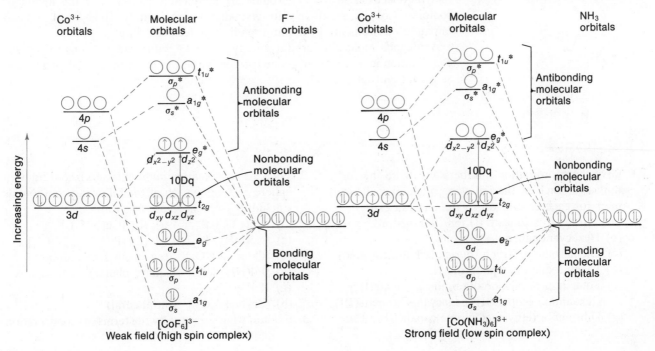

[CoF₆]³⁻
Weak field (high spin complex)

[Co(NH₃)₆]³⁺
Strong field (low spin complex)

between them is the Δ, or 10Dq, value referred to in Sections 31.7 and 31.8.

An inspection of Fig. 31-12 shows that the ligand atomic orbitals are lower in energy than are the corresponding metal atomic orbitals. Hence, as discussed in Section 6.13, this is indicative of some ionic character; that is, some electronic charge has been transferred from the metal to the ligands.

Two of the d orbitals (the e_g orbitals: $d_{x^2-y^2}$ and d_{z^2}), are directed, as indicated in Section 31.7, toward the corners of the octahedron where the ligands are located. In addition, the one $4s$ atomic orbital and the three $4p$ orbitals of the metal are also so oriented with respect to the octahedron corners. Hence metal atomic orbital overlap with ligand orbitals occurs, which results in the formation of six anti-bonding and six bonding molecular orbitals. The other three d orbitals (the t_{2g} orbitals: d_{xy}, d_{xz}, and d_{yz}) do not point toward the ligand orbitals and hence are not involved in σ bonding; these molecular orbitals are called nonbonding orbitals. In complexes, therefore, a key observation is that the t_{2g} *d orbitals are unaffected by σ bonding, whereas each of the e_g d orbitals combines with ligand orbitals to give an antibonding molecular orbital and a bonding molecular orbital.*

In both $[CoF_6]^{3-}$ and $[Co(NH_3)_6]^{3+}$, the bonding molecular orbitals are closer to the ligand atomic orbitals with respect to energy content than to the metal atomic orbitals and hence most closely resemble the ligand orbitals. On the other hand, the antibonding molecular orbitals are closer in energy to the metal atomic orbitals and hence resemble them more closely. In general, a molecular orbital formed from overlap with two atomic orbitals most closely resembles (and is said to receive a larger contribution from) the atomic orbital to which it is closer in energy.

Note in Fig. 31-12 that the 10Dq value is larger for the low-spin complex, $[Co(NH_3)_6]^{3+}$. Measurements show it to be approximately 290 kJ/mol. Note also that the electrons in the t_{2g} orbitals are completely paired. For the high-spin complex, $[CoF_6]^{3-}$, the 10Dq value is smaller (approximately 150 kJ/mol), and two unpaired electrons are in the antibonding molecular orbitals, e_g^*.

Analogous molecular orbital energy level diagrams can be constructed for coordination ions that have geometric configurations other than octahedral, such as planar and tetrahedral.

Questions

1. Give the coordination numbers and write the formulas for each of the following, including all isomers for each case in which isomers can exist:
   (a) Tetrahydroxozincate(II) ion (tetrahedral)
   (b) Hexacyanopalladate(IV) ion
   (c) Dichloroaurate(I) ion (*aurum* is Latin for gold)
   (d) Dichlorodiammineplatinum(II)
   (e) Potassium tetrachlorodiamminechromate(III)
   (f) Hexaamminecobalt(III) hexacyanochromate(III)
   (g) Dibromobis(ethylenediamine)cobalt(III) nitrate

2. Give the coordination number for each coordination sphere and name each of the following compounds (including isomers where applicable):
   (a) $[Co(CO_3)_3]^{3-}$ (b) $[Cu(NH_3)_4]^{2+}$
   (c) $[Co(NH_3)_4Br_2]_2SO_4$ (d) $[Pt(NH_3)_4][PtCl_4]$
   (e) $[Cr(en)_3](NO_3)_3$ (en = ethylenediamine)
   (f) $[Pd(NH_3)_2Br_2]$ (square planar)
   (g) $K_3[Fe(CN)_6]$
   (h) $[Zn(NH_3)_2Cl_2]$ (tetrahedral)

3. Explain what is meant by the terms (a) coordination

sphere, (b) ligand, (c) donor atom, (d) coordination number, (e) coordination compound, and (f) chelating group.

4. Sketch the structures of the following complexes. Indicate any possible *cis, trans,* and optical isomers.
   (a) $[Pt(H_2O)_2Br_2]$ (square planar)
   (b) $[Pt(NH_3)(py)(Cl)(Br)]$ (square planar, py = pyridine, C_5H_5N)
   (c) $[Zn(NH_3)_2Cl_2]$ (tetrahedral)
   (d) $[Zn(NH_3)(py)(Cl)(Br)]$ (tetrahedral)
   (e) $[Ni(H_2O)_4Cl_2]$ (f) $[Co(C_2O_4)_2Cl_2]^{3-}$

5. Show orbital diagrams, indicate the type of hybridization you would expect, and draw a diagram showing the geometrical structure for each of the following:
   (a) $[Cu(NH_3)_4]^{2+}$ (b) $[Zn(NH_3)_4]^{2+}$
   (c) $[Cd(CN)_4]^{2-}$ (d) $[Co(NH_3)_6]^{3+}$
   (e) $[Fe(CN)_6]^{4-}$

6. Show by means of orbital diagrams the hybridization for each of the examples given in Table 31-1.

7. Draw diagrams for any *cis, trans,* and optical isomers that could exist for each of the following (en is ethylenediamine):
   (a) $[Co(en)_2(NO_2)Cl]^+$ (b) $[Co(en)_2Cl_2]^+$
   (c) $[Cr(NH_3)_2(H_2O)_2Br_2]^+$ (d) $[Pt(NH_3)_2Cl_4]$
   (e) $[Cr(en)_3]^{3+}$ (f) $[Pt(NH_3)_2Cl_2]$

8. In this and other chapters we have discussed substances whose structures encompass a considerable variety of geometric shapes. (a) Name each shape; (b) draw a diagram of each; (c) give a specific example of each; (d) indicate the relation of atomic or molecular orbitals to each of the geometric shapes; (e) discuss the relationship of the geometric shape to the number of groups attached to the central element. (*Hint:* You will find it useful to make use of coordination compounds in your answer, but do not confine your answer to coordination compounds nor to the material in this chapter alone.)

9. Determine the number of unpaired electrons you would expect for $[Fe(CN)_6]^{3-}$ and for $[Fe(H_2O)_6]^{3+}$. Explain in terms of the Crystal Field Theory.

10. (a) Verify, by calculation, the crystal field stabilization energies listed in Table 31-2; (b) discuss the relationship of crystal field stabilization energies to the stabilities of complexes, and the limitations to the concept.

11. How many unpaired electrons will be present in $[CoF_6]^{3-}$ (high spin), $[Co(en)_3]^{3+}$ (low spin), $[Mn(CN)_6]^{3-}$ (low spin), $[Mn(CN)_6]^{4-}$ (low spin), $[MnCl_6]^{4-}$ (high spin), and $[RhCl_6]^{3-}$ (low spin)?

12. Is it possible for a complex of a metal of the first transition series to have 6 or 7 unpaired electrons? Explain.

13. The crystal field splitting in a tetrahedral complex is

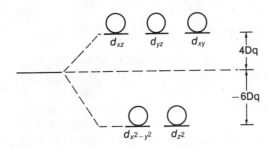

   (a) Show that the crystal field stabilization energy for a tetrahedral complex such as VCl_4 with a d^1 configuration is $-6Dq$.
   (b) All known tetrahedral complexes are high spin. For what configurations will the value of the crystal field stabilization energy be greatest?
   (c) Calculate the CFSE (in $-Dq$) for the following tetrahedral complexes: $[NiCl_4]^{2-}$, $[FeCl_4]^-$, $[CoBr_4]^{2-}$, $[Zn(OH)_4]^{2-}$, $[MnCl_4]^{2-}$.

14. What is the significance of the energy values, relative to each other, of ligand atomic orbitals, metal atomic orbitals, and the corresponding molecular orbitals for $[CoF_6]^{3-}$ and $[Co(NH_3)_6]^{3+}$?

15. For complexes of the same metal ion with no change in oxidation state, the stability increases as the CFSE of the complexes increase. Which complex in each of the following pairs of complexes is the more stable? $[Fe(H_2O)_6]^{2+}$ or $[Fe(CN)_6]^{4-}$; $[Co(NH_3)_6]^{3+}$ or $[CoF_6]^{3-}$; $[Mn(H_2O)_6]^{2+}$ or $[MnCl_6]^{4-}$; $[Co(NH_3)_6]^{3+}$ or $[Co(en)_3]^{3+}$

16. Explain how the diphosphate ion, $P_2O_7^{4-}$ (Section 24.11), can function as a water softener by complexing Fe^{2+}.

17. Determine the crystal field splitting of the *d* orbitals in a 2-coordinate complex as the two ligands approach the metal along the *z* axis.

18. What is the crystal field splitting of the nickel ion in NiO? in the Fe^{3+} ion in $FeCl_3$? (See Section 11.16.)

19. The association constant for $[Fe(CN)_6]^{3-}$ is 1×10^{44}; that for $[Fe(CN)_6]^{4-}$ is 1×10^{37}. Is this consistent with the difference in CFSE between the two complexes? CN^- is a strong-field ligand in both complexes. What other feature of the crystal field model can explain the difference in association constants?

20. Assume that you have a complex of a transitional metal ion with a d^6 configuration. Could you tell whether the complex is octahedral or tetrahedral if a measurement of the magnetic moment establishes that it has no unpaired electrons? (See Question 13 for the crystal field splitting in a tetrahedral complex.)

Problems

1. Calculate the concentration of free copper ion present in equilibrium with 1.0×10^{-3} M $[Cu(NH_3)_4]^{2+}$ and 1.0×10^{-1} M NH_3.   *Ans. 8.5×10^{-12} M*

2. Trimethylphosphine, $:P(CH_3)_3$, can act as a ligand by donating the lone pair of electrons on the phosphorus atom. If trimethylphosphine is added to a solution of nickel(II) chloride in acetone, a blue compound with a molecular weight of approximately 270 and containing 21.5% Ni, 26.0% Cl, and 52.5% $P(CH_3)_3$ can be isolated. This blue compound does not have any isomeric forms. What is the structure and molecular formula of the blue compound?

Ans. Tetrahedral; $NiCl_2[P(CH_3)_3]_2$

3. The standard reduction potential for the reaction $[Co(H_2O)_6]^{3+} + e^- \longrightarrow [Co(H_2O)_6]^{2+}$ is about 1.8 V. The reduction potential for the reaction $[Co(NH_3)_6]^{3+} + e^- \longrightarrow [Co(NH_3)_6]^{2+}$ is $+0.1$ V. Show by calculation of the cell potentials which of the complex ions, $[Co(H_2O)_6]^{2+}$ or $[Co(NH_3)_6]^{2+}$, can be oxidized to the corresponding cobalt(III) complex by oxygen.   *Ans. $[Co(NH_3)_6]^{2+}$*

4. Using the radius ratio rule and the ionic radii given inside the back cover, predict whether complexes of Ni^{2+}, Mn^{2+}, and Sc^{3+} with Cl^- will be tetrahedral or octahedral. Write the formulas for these ions.

Ans. $[NiCl_4]^{2-}$, tetrahedral; $[MnCl_6]^{4-}$, octahedral; $[ScCl_6]^{3-}$, octahedral

References

"An Ingenious Impudence: Alfred Werner's Coordination Theory," G. B. Kauffman, *J. Chem. Educ.*, **53**, 445 (1976).

"D. I. Mendeleev's Conceptions Concerning the Structure of Complex Compounds," Yu. I. Solov'ev, *J. Chem. Educ.*, **55**, 494 (1978).

"The Stereochemistry of Complex Inorganic Compounds," D. H. Busch, *J. Chem. Educ.*, **44**, 77 (1964).

"Symmetry," C. A. Coulson, *Chem. in Britain*, **4**, 113 (1968).

"Seven Coordinate Complexes of First-Row Transition Metals," R. K. Boggess and W. D. Wiegele, *J. Chem. Educ.*, **55**, 156 (1978).

"Identification of Geometrical Isomers of the Cobalt(III) Iminodiacetate System," G. A. Lawrance and C. J. Rix, *J. Chem. Educ.*, **56**, 211 (1979).

"Rearrangements in Five- and Six-Coordinate Systems," J. I. Musher, *J. Chem. Educ.*, **51**, 94 (1974).

"Tricatecholamides, A New Class of Iron Sequestering Agents," (Staff), *Chem. and Eng. News*, February 19, 1979, p. 34.

"Caged Metal Ions," A. M. Sargeson, *Chem. in Britain*, January 1979, p. 23.

"Therapeutic Chelating Agents," M. M. Jones and T. H. Pratt, *J. Chem. Educ.*, **53**, 342 (1976).

"Chelation in Medicine," J. Schubert, *Sci. American*, May 1966, p. 40.

"Optically Active Coordination Compounds," R. D. Gillard, *Chem. in Britain*, **3**, 205 (1967).

"Effect of Complexing Agents on Oxidation Potentials," J. Helsen, *J. Chem. Educ.*, **45**, 518 (1968).

"Molecular Models of Metal Chelates to Illustrate Enzymatic Reactions," H. S. Hendrickson and P. A. Srere, *J. Chem. Educ.*, **45**, 539 (1968).

"A Simple Approach to Crystal Field Theory," R. C. Johnson, *J. Chem. Educ.*, **42**, 147 (1965).

"Ligand Field Theory," F. A. Cotton, *J. Chem. Educ.*, **41**, 467 (1964).

"Crystal Field Potentials," S. F. A. Kettle, *J. Chem. Educ.*, **46**, 339 (1969).

"A Detailed, Simple Crystal Field Consideration of the Normal Spinel Structure of Co_3O_4," L. Suchow, *J. Chem. Educ.,* **53,** 560 (1976).

"Molecular Orbital Theory for Transition Metal Complexes," H. B. Gray, *J. Chem. Educ.,* **41,** 2 (1964).

"Simplified Molecular Orbital Approach to Inorganic Stereochemistry," R. M. Gavin, *J. Chem. Educ.,* **46,** 413 (1969).

"Magnetic Moment Measurement of a Coordination Complex," T. G. Dunne, *J. Chem. Educ.,* **44,** 142 (1967).

"An Inexpensive, Convenient Demonstration of Magnetic Susceptibility," S. S. Eaton and G. R. Eaton, *J. Chem. Educ.,* **56,** 170 (1979).

"Mechanisms of Ligand Substitution Reactions of Cobalt(III) Complexes," R. G. Pearson, *J. Chem. Educ.,* **55,** 720 (1978).

"Mechanisms of Oxidation-Reduction Reactions," H. Taube, *J. Chem. Educ.,* **45,** 452 (1968).

"Metal Ion Control of Chemical Reactions," D. H. Busch, *Science,* **171,** 241 (1971).

"Out-of-Plane Metallophorphyrin Complexes," G. A. Taylor and M. Tsutsui, *J. Chem. Educ.,* **52,** 715 (1975).

"Liquid Lasers" (describes chelate lasers which utilize rare earth chelates), A. Lempicki and H. Samelson, *Sci. American,* June 1967, p. 81.

"Shapes of the Future," V. Klee, *Amer. Scientist,* **59,** 84 (1971).

"Hydrido Transition-Metal Cluster Complexes," H. D. Kaesz, *Chem. in Britain,* **9,** 344 (1973).

"Molecular Oxygen Adducts of Transition Metal Complexes: Structure and Mechanism," L. Klevan, J. Peone, Jr., and S. K. Madan, *J. Chem. Educ.,* **50,** 670 (1973).

"Hemoglobin: Isolation and Chemical Properties," S. F. Russo and R. B. Sorstokke, *J. Chem. Educ.,* **50,** 347 (1973).

"Homogeneous Models Throw Light on Industrial Catalysts," D. G. H. Ballard, *Chem. in Britain,* **10,** 20 (1974).

"Binuclear Cobalt(III) Complexes—A Survey of Reactions," A. G. Sykes, *Chem. in Britain,* **10,** 170 (1974).

"Chemotherapy: Renewed Interest in Platinum Compounds," J. L. Marx, *Science,* **192,** 774 (1976).

"Platinum Anti-Tumor Compounds; An Unexpected Discovery," C. F. Ledger and J. Webb, *J. Chem. Educ.,* **53,** 174 (1976).

"The Role of Metal Ions in Proteins and Other Biological Molecules," E. W. Ainscough and A. M. Brodie, *J. Chem. Educ.,* **53,** 156 (1976).

"Biochemical Roles of Some Essential Metal Ions" (Staff), *J. Chem. Educ.,* **54,** 761 (1977).

"Some Aspects of the Bioinorganic Chemistry of Molybdenum," K. B. Swedo and J. H. Enemark, *J. Chem. Educ.,* **56,** 70 (1979).

"Assigning Oxidation States to Some Metal Dioxygen Complexes of Biological Interest," D. A. Summerville, R. D. Jones, B. M. Hoffman, and F. Basolo, *J. Chem. Educ.,* **56,** 157 (1979).

"Metal Ions in Enzymes Using Ammonia or Amides," N. E. Dixon *et al., Science,* **191,** 1144 (1976).

"Metals as Regulators of Heme Metabolism," M. D. Maines and A. Kappas, *Science,* **198,** 1215 (1977).

32

Spectroscopy and Chromatography

In earlier chapters we discussed molecular structures and considered the properties of a significant number of compounds. It is appropriate now to look into the methods by which chemists characterize each compound that they prepare—that is, how they establish its composition, purity, and structure. The characterization of a compound is of course especially important if the compound has never been made before, but it is extremely important even after the synthesis of a compound previously known and prepared. A compound, whether known previously or not, is seldom of use unless at least its composition and purity are known, and for many purposes it is highly desirable, and sometimes essential, that its molecular structure be known as well.

Before World War II, a compound was usually characterized by the method of synthesis, by chemical analysis to determine the per cent composition, and (if a previously known compound) by a comparison of properties, such as melting point, color, and density, with those previously established for a pure sample of the same compound. Finally, an attempt was made to convert a sample of the compound into one or more **derivatives,** which could result only if the original compound was actually what it was thought to be. Characterization of the derivatives was, of course, essential to the method. Such procedures are still used, but in addition, during the last 30 years, several instrumental methods for characterizing compounds have been developed or improved to high levels of usefulness. Chief among these are the various **spectroscopic methods,** most of which involve the effects of electromagnetic radiation upon matter as detected and recorded by a spectrometer. In the following discussion, we will consider a variety of kinds of spectroscopy, often referred to also as **spectrometry:** (1) ultraviolet and visible spectroscopy, (2) infrared spectroscopy, (3) nuclear magnetic resonance spectroscopy, (4) electron paramagnetic spectroscopy, and (5) mass spectrometry. We will not consider here other important kinds of spectroscopy, such as Raman spectroscopy, Mössbauer spectroscopy, and photoelectron spectroscopy. The major emphasis will be placed on ultraviolet and visible, infrared, and nuclear magnetic resonance spectroscopy.

Another very useful method for characterizing compounds and also for purifying compounds and separating mixtures of compounds is chromatography. **Chromatography** is a method of separating the components of a mixture by arranging a system wherein the different components of the mixture migrate through the system at different rates. We shall consider both **gas phase chromatography** and **liquid phase chromatography** later in this chapter. As we will see, adsorption plays an important role in chromatographic procedures.

It should be mentioned that another technique, diffraction of x rays by crystals, though not a topic of this chapter, is also used for compound characterization (Section 11.22). X-ray diffraction by crystals provides one of the most direct sources of structural information, although the technique is limited to substances that form suitable crystals. X-ray diffraction involves the collection and interpretation of a large amount of data for each structural determination. Hence recent development of automated data collection systems and high-speed computers has changed the method from a time-consuming and laborious process to one that is widely used and exceedingly valuable for the characterization of chemical substances.

Spectroscopy

32.1   Introduction

Most spectroscopic methods involve the absorption of energy in the form of electromagnetic radiation from a particular region of the spectrum (Section 4.10). When a molecule absorbs radiation it gains an amount of energy that corresponds to some form of motion within the molecule. These motions within the molecule, which are referred to as **transitions,** can involve movement of electrons from one energy level to another (electronic transitions—see Sections 4.10 and 31.7), vibrations and rotations of molecules, spins of electrons, or spins of nuclei. The various portions of the spectrum, in order of decreasing energy, are shown in Fig. 32-1.

How much energy the molecule gains depends upon the frequency of the radiation that it absorbs. As pointed out in Chapter 4, radiation can be characterized either by its wavelength, λ, or by its frequency, ν (Section 4.10). The relationship between the two and the energy relationship, ΔE, pertaining to both, are

$$\nu_{sec^{-1}} = c_{m/sec}/\lambda_m$$

$$\Delta E_{joules/molecule} = h\nu = hc/\lambda$$

where c is the velocity of light (2.998×10^8 m/sec), and h is Planck's constant (6.626×10^{-34} J sec). The higher the energy of the radiation, therefore, the higher the frequency and the shorter the wavelength.

By means of a spectrophotometer, we can determine the wavelengths at which a

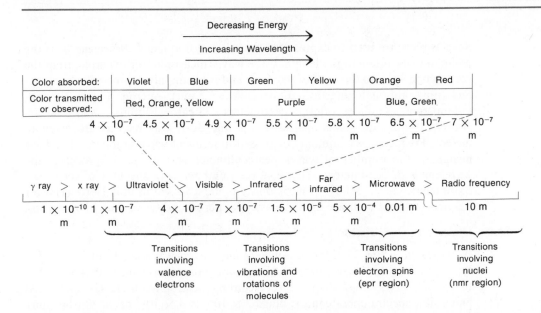

Color absorbed:	Violet	Blue	Green	Yellow	Orange	Red
Color transmitted or observed:	Red, Orange, Yellow		Purple		Blue, Green	

Decreasing Energy →

Increasing Wavelength →

4×10^{-7} m 4.5×10^{-7} m 4.9×10^{-7} m 5.5×10^{-7} m 5.8×10^{-7} m 6.5×10^{-7} m 7×10^{-7} m

γ ray > x ray > Ultraviolet > Visible > Infrared > Far infrared > Microwave > Radio frequency

1×10^{-10} m 1×10^{-7} m 4×10^{-7} m 7×10^{-7} m 1.5×10^{-5} m 5×10^{-4} m 0.01 m 10 m

Transitions involving valence electrons

Transitions involving vibrations and rotations of molecules

Transitions involving electron spins (epr region)

Transitions involving nuclei (nmr region)

Figure 32-1 The various portions of the electromagnetic spectrum in order of decreasing energy and increasing wavelength.

molecule absorbs energy *due to the presence of a transition in the molecule corresponding to that energy.* In addition, we can determine how much energy (in the form of electromagnetic radiant energy) is absorbed at that wavelength. The wavelength at which the molecule absorbs the maximum amount of energy is given the symbol λ_{max}; the quantity of radiant energy absorbed at λ_{max} is called the **molar extinction coefficient,** or **molar absorptivity,** and is given the symbol ε_{max}.

$$\varepsilon_{max} = \frac{\log (I_0/I)}{cl}$$

where I_0 = intensity of incident radiation (i.e., radiant energy put into the system)

I = intensity of transmitted, or emitted, radiation (i.e., the portion of incident radiation that is not absorbed by the sample and hence gets through the sample)

c = concentration of the sample solution in moles per liter

l = length of the light path for the absorption cell in centimeters (i.e., the distance the radiation travels through the sample solution)

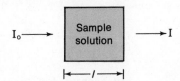

$I_0 \longrightarrow$ Sample solution $\longrightarrow I$

|← l →|

The concentration factor, c, in the expression for the molar extinction coefficient makes possible quantitative analysis by spectroscopic methods, if the other quantities in the expression are known. Log (I_0/I) is referred to as the **absorbance.** The ratio I/I_0 (note that this is the inverse of the ratio in the log term for absorbance) is the ratio of transmitted radiation to incident radiation; hence, $(I/I_0) \times 100$ is the **per cent transmission,** sometimes referred to as the **per cent transmittance,** or the **per cent transmittancy.**

32.2   Units

Several units are used to express wavelength and frequency. **Wavelength,** as the name implies, refers to the length of the wave (measured, for example, from the crest of one wave to the crest of the next) and is commonly expressed in nanometers (nm), although angstrom units ($1 \text{ Å} = 0.1$ nm), microns ($1 \mu = 10^3$ nm), millimicrons ($1 \text{ m}\mu = 1$ nm), centimeters ($1 \text{ cm} = 10^7$ nm), and meters ($1 \text{ m} = 10^9$ nm) are sometimes used. In Fig. 32-1, wavelengths are given in meters. **Frequency** is commonly expressed in terms of wave numbers. The **wave number** is the number of waves per centimeter and is the reciprocal of the wavelength in centimeters. Hence, wave numbers have the unit of reciprocal centimeters, cm^{-1}. Many spectroscopists prefer to use the unit Hertz (Hz) for frequency. Hertz units are the same as cycles per second (cps); thus 1 Hz = 1 cps.

32.3   Ultraviolet and Visible Spectra

As indicated in Fig. 32-1, the ultraviolet range covers wavelengths of 1×10^{-7} m to 4×10^{-7} m (or 100–400 nm), and the visible range wavelengths of 4×10^{-7} m to 7×10^{-7} m (or 400–700 nm). Most recording spectrophotometers for ultraviolet and visible spectra operate in the range 2×10^{-7} to 8×10^{-7} m (or 200–800 nm). Such relatively high energies are sufficient to move an electron from its ground state orbital to an orbital of higher energy. The electron can be one that is initially either in a σ orbital or in a π orbital. Often the electron transitions are from a π orbital to the antibonding π^* orbital ($\pi \longrightarrow \pi^*$) or from a nonbonding orbital (Section 31.11) to a π^* orbital ($n \longrightarrow \pi^*$).

Ultraviolet and visible light have limited use in characterizing organic compounds but are highly useful for compounds of metals in which the metal has an unfilled d subshell. Hence ultraviolet and visible light are especially useful in characterizing compounds of the transition metals and the inner transition metals.

In organic compounds, electronic transitions of electrons in either single bonds or nonconjugated double bonds usually occur at energies that correspond to light with a wavelength below 200 nm, which is below the operating range of ordinary spectrometers; but systems with conjugated double and single bonds have $\pi \longrightarrow \pi^*$ electronic transitions at lower energies, that correspond to light with a wavelength above 200 nm and so can be analyzed with ordinary instruments. Conjugated systems are ones in which double bonds and single bonds occur alternately along the chain, as in isoprene (Section 25.29).

$$
\begin{array}{ccccc}
 & \text{H} & \text{CH}_3 & \text{H} & \text{H} \\
 & | & | & | & | \\
\text{H}-&\text{C}=&\text{C}-&\text{C}=&\text{C}-\text{H}
\end{array}
$$

Isoprene

In general, the longer the conjugated system, the lower the energy of the $\pi \longrightarrow \pi^*$ transition and the longer the wavelength at which the system absorbs light. For example,

$$
\begin{array}{ccc}
\text{H}-\text{C}=\text{C}-\text{H} & \text{H}-\text{C}=\text{C}-\text{C}=\text{C}-\text{H} & \text{H}-\text{C}=\text{C}-\text{C}=\text{C}-\text{C}=\text{C}-\text{H}
\end{array}
$$

Ethylene   1,3-Butadiene   1,3,5-Hexatriene
$\lambda = 162$ nm $\lambda = 215$ nm $\lambda = 265$ nm

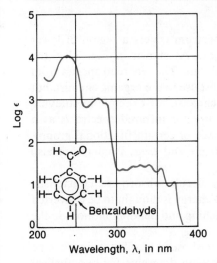

Figure 32-2   The ultraviolet spectrum of benzaldehyde. (*From Theory and Practice in the Organic Laboratory, 2nd ed. J. A. Landgrebe, D. C. Heath, 1977.*)

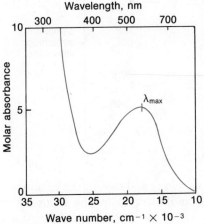

Figure 32-3   The visible spectrum of $[Ti(H_2O)_6]^{3+}$.

Examples of other conjugated systems are aromatic rings, such as in benzene, (which have conjugated systems through resonance) and ketones with the $>C=O$ group conjugated to carbon-carbon double bonds.

Hence the main use of ultraviolet and visible light for characterizing organic compounds involves establishing whether or not the structure includes a conjugated system. The ultraviolet spectrum for benzaldehyde, which has a conjugated structure, is shown in Fig. 32-2.

In the use of visible light for characterization of octahedral coordination compounds of the transition metals, the splitting of the d orbitals into the three lower-energy nonbonding orbitals (t_{2g}) and the higher-energy antibonding orbitals (e_g), as described in Section 31.7, occurs at the relatively high energies of the visible absorption range. The recorded spectra that result from use of visible light energies provide a method for obtaining the 10Dq (or Δ) values for metal coordination compounds (Section 31.8). Hence such spectra provide considerable information concerning the bonding strengths of various ligands with different metals and the distributions of the valence electrons within the complex. From the valence electron distribution, information relative to the molecular structures of the metal complexes can be deduced.

One of the simplest cases of visible-light energy absorption is that of a metal complex ion in which the metal has a d^1 configuration and an octahedral structure. An example is the Ti(III) ion in $[Ti(H_2O)_6]^{3+}$, in which the single d electron occupies a t_{2g} orbital. Absorption of light in the visible range raises the electron to the e_g orbital. The absorption band occurs in the visible light portion of the electromagnetic spectrum of $[Ti(H_2O)_6]^{3+}$ (Fig. 32-3) and accounts for the violet color of the ion.

The band, or peak, in the spectrum of the $[Ti(H_2O)_6]^{3+}$ ion is broad, a characteristic common to visible spectra. The peak maximum occurs at 20,000 cm^{-1}, corresponding in this simplest case to the Δ, or 10Dq, value for the splitting of the orbital levels. In more complicated cases with large numbers of d electrons, the Δ, or 10Dq, value can likewise be obtained from the visible spectrum but by a somewhat more involved procedure. The visible and near-ultraviolet spectra of the octahedral complexes $[Co(NH_3)_6]^{3+}$ and $[Co(en)_3]^{3+}$ are shown in Fig. 32-4.

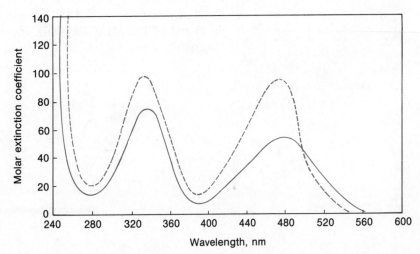

Figure 32-4   The visible and near-ultraviolet spectra for $[Co(NH_3)_6]^{3+}$ (*solid line*) and $[Co(en)_3]^{3+}$ (*broken line*). Each complex has an octahedral configuration (*en* stands for ethylenediamine).

32.4   Infrared Spectra

The infrared portion of the electromagnetic spectrum covers a region of lower energy than the visible and ultraviolet portions. As shown in Fig. 32-1, the infrared region extends from 7×10^{-7} to 5×10^{-4} m. The recorded spectra from infrared energy absorption are of perhaps the most use to the organic chemist, but they are also highly valuable to the inorganic chemist, for example in studying coordination compounds. The most common use for infrared spectra is as a "fingerprint method" for establishing the presence of certain functional groups, but other uses such as differentiating between *cis* and *trans* isomers are also important.

Many kinds of functional groups in organic molecules absorb at characteristic frequencies in the infrared region. The absorbed energy is thought to be converted into vibrations that take the form of several stretching and bending motions in the bonds within the molecule.

The stretching motions, which can be likened to those in a coiled spring, can be either unsymmetrical or symmetrical. The following diagrams for hypothetical atoms *A*, *B*, and *C* illustrate the two types of stretching vibrations:

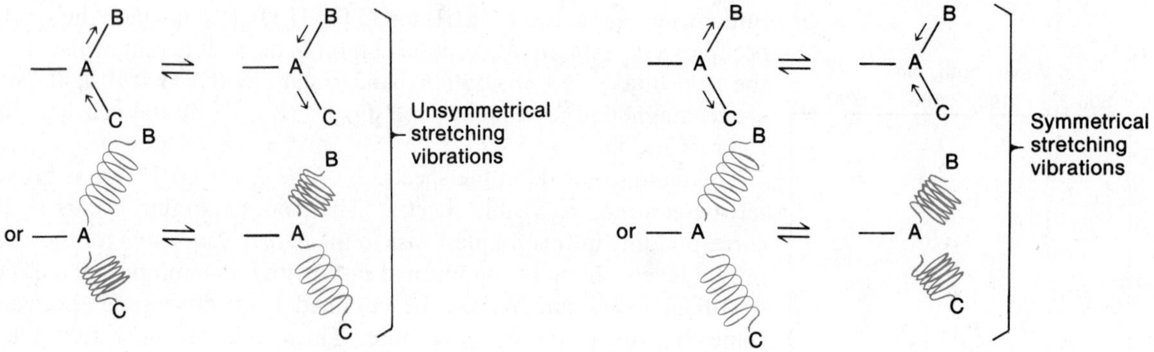

Bending vibrations can be bending in the plane (*scissoring* and *rocking*) or bending in or out of the plane (*wagging* and *twisting*), as illustrated by the following diagrams:

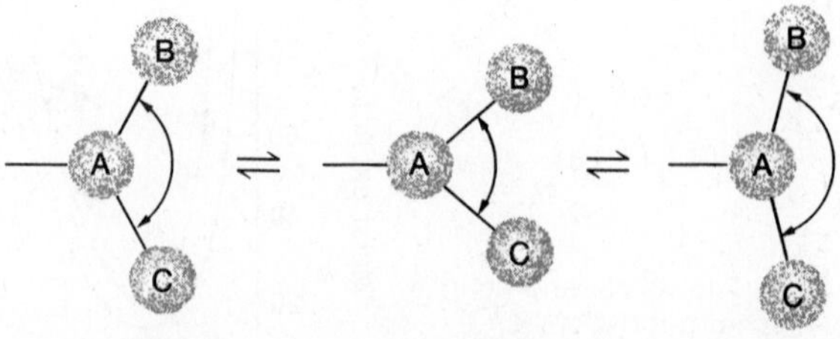

Scissoring bending vibrations in the same plane

(The *B—A—C* bonds alternately open up and close partially, in a motion similar to that of a pair of scissors.)

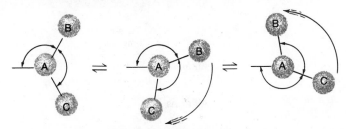

Rocking bending vibrations in the same plane

(*B* and *C* move in the same direction, alternately down and then up so that the *B—A—C* group rocks down and up within the plane.)

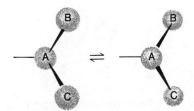

Wagging bending vibrations in and out of the plane

(Alternately both *B* and *C* bend forward and back together from the plane.)

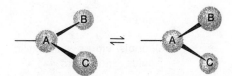

Twisting bending vibrations in and out of the plane

(Alternately, *B* bends back from the plane while *C* bends forward, then *B* bends forward while *C* bends back.)

The spectrometer detects and records the absorbed energies involved in these bond motions. Various functional groups have characteristic absorption values. Hence, by studying the absorption peaks (or bands) in the spectrum of a compound, various functional groups can be shown to be present.

It is important to note that vibrational frequencies are observable in infrared spectra, only if the vibrational stretching or bending motion results in a change in the dipole moment of the molecule. The larger the change in the dipole moment, the larger will be the peak observed for the stretching or bending motion.

Figure 32-5 and Table 32-1 show the frequencies of infrared radiation absorbed by some common bonds and the corresponding stretching, bending, and rocking vibrations.

TABLE 32-1 Characteristic Infrared Absorption Peaks for Some Common Bonds

Type of Compound	Bond	Range of Frequency for Absorption Maximum, cm^{-1}
Alkanes	—C—H	2850–2960, 1350–1470
Alkenes	=C—H	3010–3090, 675–1000
	C=C	1600–1680
Alkynes	≡C—H	3270–3300
	C≡C	2100–2260
Aromatic rings		1450–1600
		3000–3100, 675–870
Alcohols and phenols		
Not hydrogen bonded	—O—H	3500–3700
Hydrogen bonded	—O—H--O	3100–3700
Thio alcohols	—S—H	2500
Amines	—N—H	3300–3500
Alcohols, ethers, carboxylic acids, and esters	—C—OR	1080–1300
Aldehydes, ketones, carboxylic acids, and esters	—C=O	1680–1750
Aldehydes	—C=O	1695–1740
Ketones	—C=O	1700–1730
Carboxylic acids	—C=O	1680–1730
Esters	—C=O	1720–1750
Amines	—C—NR$_2$	1180–1360
Nitriles	—C≡N	2210–2260
Nitro compounds		1515–1560, 1345–1385

Figure 32-5 Regions by wavelength (bottom horizontal axis) and wave number (top horizontal axis) for some infrared stretching and bending vibrations. (*From* Theory and Practice in the Organic Laboratory, *2nd ed. J. A. Landgrebe, D. C. Heath, 1977.*)

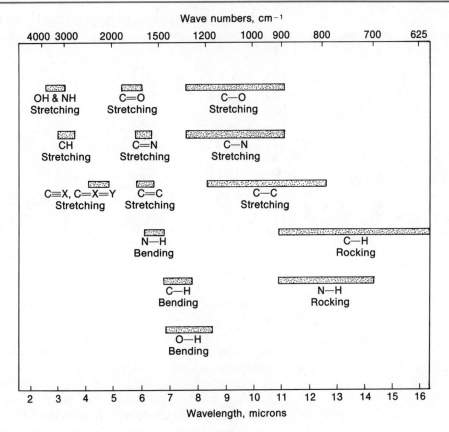

Three examples of infrared spectra are shown in Figs. 32-6, 32-7, and 32-8. Note that infrared spectrometers ordinarily plot the spectra upside down compared to ultraviolet or visible spectra. The vertical axis is in terms of per cent transmittance in Figs. 32-6, 32-7, 32-8, a common method of plotting infrared spectra. Alternatively, infrared spectra are plotted in terms of absorbance. The horizontal axis is plotted in terms of both wave numbers (in cm^{-1}, across the top) and wavelength (in microns, across the bottom) in Figs. 32-6, 32-7, and 32-8, and this is quite customary. It should be noted that infrared spectra are often quite complex with many peaks, or bands. Usually, no effort is made to identify all of the bands. Identification is made of a few selected bands that identify key functional groups or types of bonds.

Notice in the spectrum of di-*n*-hexyl ether (C_6H_{13}—O—C_6H_{13}), Fig. 32-6, that a strong absorption appears at about 2900 cm^{-1}, corresponding to alkane C—H stretching in the C_6H_{13} groups,

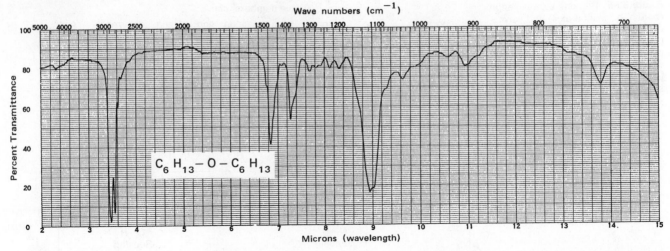

Figure 32-6 The infrared spectrum of di-*n*-hexyl ether.

Bands arising from alkane C—H bending occur at about 1460 cm^{-1} and 1380 cm^{-1}. At 1110 cm^{-1} there is a strong absorption for the C—O ether stretching vibration. The infrared spectrum for C_6H_{13}—O—C_6H_{13} is a relatively simple one.

The spectrum for 4-isopropenyl-1-methylenecyclohexane (Fig. 32-7) is somewhat more complex. Note that the molecule contains two olefinic double bonds. The spectrum shows a sharp olefinic =C—H stretching band, therefore, at about 3050 cm^{-1} and a broader peak at 890 cm^{-1}. It also shows a sharp olefinic C=C stretching band at 1640 cm^{-1}. The bands in the region 2850–2960 cm^{-1} can be ascribed to alkane C—H stretching in the CH_3 group. It should be pointed out that although there is a band at 1450 cm^{-1}, it cannot be ascribed to a benzene ring, inasmuch as the ring is not the benzene ring but rather the saturated cyclohexane ring. The band at 1450 cm^{-1} is instead due, at least in part, to alkane (methyl) C—H bending. These facts emphasize the need for care in making assignments for identification of unknown compounds.

Figure 32-7 The infrared spectrum of 4-isopropenyl-1-methylene-cyclohexane. (Note that the ring is the saturated cyclohexane ring—not the benzene ring.)

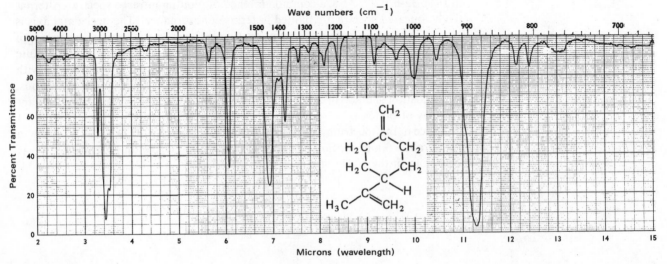

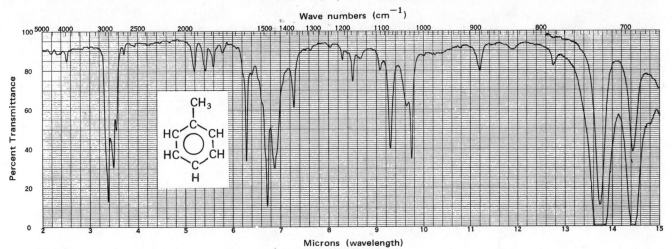

Figure 32-8 The infrared spectrum of toluene (toluene does contain the benzene ring, in contrast to the compound of Fig. 32-7).

The spectrum in Fig. 32-8 is for toluene, which does have a benzene ring. The benzene ring is responsible in part for the band at 1460 cm^{-1}, but alkane (methyl) C—H bending vibrations also contribute to the 1460 cm^{-1} band. In addition, strong alkane (methyl) C—H stretching bands occur in the region 2850–2960 cm^{-1}. Ar—H (hydrogen attached to an aromatic ring) stretching vibrations probably account for the band at 3010 cm^{-1} and perhaps also for the bands in the 690–730 cm^{-1} region.

Infrared spectra have been run for many hundreds, even thousands, of organic compounds. Infrared spectroscopy is one of the principal methods used by chemists for characterization of organic compounds.

Although the largest use of infrared spectroscopy is in the study of organic compounds, the method is also highly useful for a variety of other compounds, including metal coordination compounds. Many of the ligands in coordination compounds are organic in nature and hence give rise to infrared bands such as we have described. Moreover, a number of ligands other than organic ligands also give rise to infrared bands, —NO_2 groups being one example. Also, when a metal attaches to the carbonyl groups of a compound such as 2,4-pentanedione to form a metal chelate of the type

$$
\begin{array}{c}
\text{H} \\
\text{C} \\
CH_3-C \quad\quad C-CH_3 \\
\text{O} \quad\quad \text{O} \\
\text{Cu} \\
\text{O} \quad\quad \text{O} \\
CH_3-C \quad\quad C-CH_3 \\
\text{C} \\
\text{H}
\end{array}
$$

the carbonyl (C=O) absorption bands are shifted in frequency as compared to their absorption frequencies in the uncoordinated molecule. This provides a qualitative measure of the strength of the bonds between the metal and the donor oxygen atoms; the greater the shift, the greater is the stability of the compound.

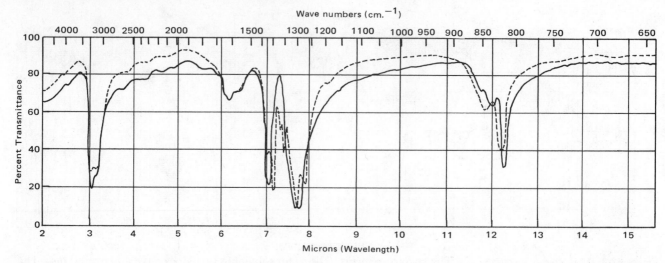

Figure 32-9 The infrared spectra of *cis*-[Co(NH$_3$)$_4$(NO$_2$)$_2$]$^+$ (*broken line*) and *trans*-[Co(NH$_3$)$_4$(NO$_2$)$_2$]$^+$ (*solid line*). [*Reprinted with permission from the* Journal of the American Chemical Society, *Vol. 76 (1954), p. 5346: J. P. Faust and J. V. Quagliano. Copyright by the American Chemical Society.*]

Another interesting application of infrared spectra to metal coordination compounds is in distinguishing between *cis* and *trans* isomers of a given complex. The infrared spectra of *cis* and *trans* isomers usually differ although many similarities also exist. Very often, the spectrum of the more unsymmetric *cis* isomer is more complex, with a larger number of bands, than that of the *trans*. This is true, for example, for the *cis* and *trans* dinitrotetraamminecobalt(III) ions, [Co(NH$_3$)$_4$(NO$_2$)$_2$]$^+$, for which the spectra are shown in Fig. 32-9.

It is interesting to note that even the attachment of different atoms of a ligand to the central metal ion in a metal complex can cause a change in the infrared spectrum. For example, Fig. 32-10 shows the infrared spectra of *nitro*pentaamminecobalt(III) chloride, [Co(NH$_3$)$_5$NO$_2$]Cl$_2$, (in which the nitrite ion is attached to the cobalt ion through the nitrogen atom) and *nitrito*pentaamminecobalt(III) chloride, [Co(NH$_3$)$_5$ONO]Cl$_2$ (in which the nitrite ion is attached through one of the oxygen atoms). Iomers of this type are known as **linkage isomers.**

Figure 32-10 The infrared spectra of [Co(NH$_3$)$_5$NO$_2$]Cl$_2$, in which the NO$_2^-$ ion is attached to cobalt through the nitrogen atom (*solid line*), and [Co(NH$_3$)$_5$ONO]Cl$_2$, in which the NO$_2^-$ ion is attached to cobalt through one of the oxygen atoms (*broken line*). [*Reprinted with permission from the* Journal of American Chemical Society, *Vol. 78 (1956), p. 887: R. B. Penland, F. J. Lane, and J. V. Quagliano. Copyright by the American Chemical Society.*]

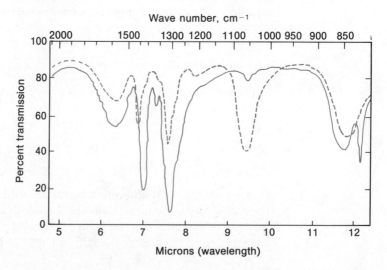

32.5   Nuclear Magnetic Resonance Spectra

We have learned that electrons spin about their own axes (Section 4.12), some clockwise and some counterclockwise, with spin quantum numbers of $+\frac{1}{2}$ and $-\frac{1}{2}$. Since the spin creates a magnetic moment along the axis of spin, the electrons act as tiny bar magnets.

Nuclei of certain types also spin on their own axes, and these nuclei, like electrons, act as very small bar magnets. The hydrogen nucleus (a proton) has spin quantum numbers, like the electron, of $+\frac{1}{2}$ or $-\frac{1}{2}$, commonly referred to simply as spin number $\frac{1}{2}$. If the proton is placed in a magnetic field, it will align itself either *with* the field (lower energy condition) or *against* the field (higher energy condition). The spin quantum number $+\frac{1}{2}$ corresponds to alignment with the field and $-\frac{1}{2}$ to alignment against the field. Energy is required, and thus must be absorbed, to change the spin of the nucleus from alignment with the applied magnetic field to the higher energy, less stable, condition of alignment against the field. This energy absorption results in a signal on a nuclear magnetic resonance spectrometer and forms the basis for the **nuclear magnetic resonance** (nmr) method.

The proton is the most commonly investigated nucleus in nuclear magnetic resonance. The nmr method when applied to the proton is sometimes referred to as **proton magnetic resonance** (pmr). Certain other nuclei, such as ^{13}C, ^{19}F, and ^{31}P, each of which also has spin quantum numbers of $+\frac{1}{2}$ and $-\frac{1}{2}$, behave similarly and are sometimes used in nmr studies. However, some other nuclei have spin numbers other than $\frac{1}{2}$, such as 0, 1, $\frac{3}{2}$, and 2. The spin number of a particular nucleus depends, in a rather complex way, on the total number of protons and neutrons it contains. Nuclei with a spin number of zero, such as ^{12}C and ^{16}O, do not behave as bar magnets and an nmr signal does not result. This actually is fortunate because these very common nuclei thus do not interfere with the nmr signals from protons or other suitable nuclei that do cause a signal. All nuclei with spin numbers other than zero act as bar magnets, but, except for $+\frac{1}{2}$ and $-\frac{1}{2}$, only in rather special cases do useful spectra result from their signals. Examples of such nuclei that are sometimes used are ^{2}H (deuterium) and ^{14}N, each with spin number 1, and ^{11}B, with spin number $\frac{3}{2}$. In the remainder of our discussion of the nuclear magnetic resonance method we will consider only the proton.

The amount of energy necessary to change the alignment of the proton depends upon the strength of the applied magnetic field. Typically, the magnetic field in nmr instruments is arranged so that the instruments operate in either the 60 megahertz frequency range (60 million cycles per second) or the 100 megahertz range (100 million cycles per second). This field energy is in the radiofrequency range; hence, it is energy of lower frequency and longer wavelength than that for infrared radiation.

To obtain an nmr spectrum, the sample is dissolved in an appropriate solvent, such as carbon tetrachloride, and placed in a container between the poles of a magnet where it is usually spun at high speed. The sample is irradiated with sufficient radiofrequency energy to promote the flip in alignment of the nuclei. When protons go into the higher energy state, absorption of energy takes place, producing an nmr signal that is detected and recorded by the instrument as a peak. To accomplish the process, either a constant magnetic field and a gradually increasing frequency of radiation can be used or, as is more common, the frequency can be held constant and the strength of the magnetic field gradually increased.

One would expect, offhand, that all protons would absorb at the same magnetic field strength. However, a proton is "shielded" from the full force of the applied magnetic field by the electrons in the bond between the proton and another atom. In such a case, a larger *applied* field strength must be used to create the same *effective* field strength as would be required for absorption to take place if shielding did not occur. At a constant radiofrequency, all protons absorb energy at the same *effective* magnetic field strength, but at different *applied* magnetic field strengths depending upon the environment in which the proton is located. It is the *applied* field strength that is measured and plotted by the instrument against the absorption. The spectrum that results, therefore, has absorption peaks whose relative positions are dependent upon differences in environments of the various protons present in the sample. Hence the nmr spectrum can provide considerable information about the molecular structure of the sample.

An nmr spectrum, in fact, provides even more information than the foregoing discussion suggests. Not only are the recorded positions of the nmr signals useful but also the *number* of signals, the *intensities* of the signals, and a *splitting* of signals that occurs such that a particular signal is split into several peaks. The number of signals indicates how many different kinds of proton environments are present, the intensities of the signals indicate how many protons are in each kind of environment, and the splitting of signals provides additional information about the environment of a proton with respect to other protons. We will next consider each of these factors in somewhat more detail.

▶ **1. POSITION OF SIGNALS.** The positions of the nmr signals tell us what "kinds" of protons, in terms of their electronic environments, are present. For example, protons bonded to carbon atoms in an aromatic ring, those bonded to carbon atoms in alkanes, and those attached to halogens or other atoms or groups of atoms all have different electronic environments and produce nmr signals at different energies. Electrons moving about the proton itself shield the proton from the full applied magnetic field by generating a field that opposes the applied field. If the effective field, felt by the proton, is thus less than the applied field, the proton is said to be **shielded.** For a shielded proton the energy required for the applied field to reach the effective field necessary for absorption is thus larger, and the signal is said to be **shifted upfield,** as compared to that required for an unaffected proton. Electrons moving about adjacent protons can also generate a magnetic field that either opposes or reinforces the applied field. If opposition of the field takes place, the proton is again shielded; if reinforcement of the field occurs, the effective field felt by the proton will be larger than the applied field and the proton is said to be **deshielded.** In the latter case, the applied field required for absorption is less than for an unaffected proton, and the signal is **shifted downfield.** These shifts upfield and downfield in nmr spectra are referred to as **chemical shifts.**

The reference point for chemical shifts is usually a signal from a reference compound added to the system as an internal standard. The most common reference compound is tetramethylsilane [TMS, $(CH_3)_4Si$], which has only one kind of proton environment and hence only one signal. Considerable shielding of protons occurs in TMS, resulting in the one signal being farther upfield than the nmr signals for most other compounds. Hence nmr signals for protons of most compounds are downfield from the TMS signal. On one common scale of measurement of chemical shifts, the delta (δ) scale, the TMS signal is arbitrarily set as zero and shifts are measured from that point on a scale of units known as ppm (parts per million). Most chemical shifts are less than 10 ppm. Sometimes, a tau (τ) scale is used, in which the TMS signal is set as 10 ppm. Thus $\tau = 10 - \delta$.

Characteristic portions of the structure of a molecule cause particular shifts. For example, a chlorine atom attached to a carbon atom causes a downfield shift for a proton attached to the same carbon atom; two chlorine atoms cause a larger downfield shift. A chlorine atom attached to an adjacent carbon atom also causes a downfield shift but the shift is smaller than when the chlorine atom is attached to the same carbon atom. The proton in the —COOH group of an organic acid, RCOOH, has an nmr peak downfield from that in the —CHO group of an aldehyde, RCHO, with the same R group. The —CHO shift is in turn downfield from that of a proton attached to an aromatic ring. Other structural variations result in characteristic differences in the nmr spectrum.

▶2. **NUMBER OF SIGNALS.** Within a molecule, protons with the same environment are referred to as **equivalent protons.** Equivalent protons absorb radiofrequency energy at the same applied magnetic field strength. Protons with different environments absorb at different applied magnetic field strengths. Hence, by counting the nmr signals the number of "kinds" of protons in the molecule, in terms of their environments, can be determined.

For example, ethanol has three different kinds of protons in terms of environment.

$$\begin{array}{ccc} & H & H \\ & | & | \\ H- & C- & C-OH \\ & | & | \\ & H & H \end{array}$$

The three protons in the methyl group, CH_3—, are all equivalent, with an environment made up of the —CH_2—OH grouping; the two protons in the methylene group, —CH_2—, likewise are equivalent but are different in environment from those in the methyl group; and finally, the one proton attached to oxygen in the —OH group has a different environment from either of the other two types. Figure 32-11 shows the three peaks, corresponding to the three kinds of protons, in a low-resolution nmr spectrum of ethanol. The three peaks in a high-resolution spectrum are shown in Fig. 32-12.

The least shielded proton is on the electronegative oxygen atom and its signal is thus farthest downfield. The two methylene protons are somewhat more shielded than the —OH proton and their signal is upfield from that of the —OH proton. The three methyl protons are still more shielded, and hence their signal is still farther upfield.

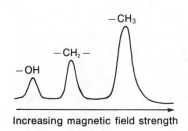

Increasing magnetic field strength

Figure 32-11 The three peaks in a low-resolution nmr spectrum of ethanol.

▶3. **INTENSITY OF SIGNALS.** In Fig. 32-11 it is apparent that, judging roughly the sizes of the signal peaks by their heights, the sizes are approximately in proportion

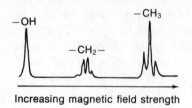

Figure 32-12 The three peaks in the high-resolution nmr spectrum of ethanol.

to the number of protons corresponding to each peak. This is reasonable, because the greater the number of protons to be lined up against the magnetic field the larger would be the amount of energy absorbed, and hence a larger absorption peak would result. A more accurate judgment than one based on height, for the size of a peak, is based on the area under the peak. It happens that, under normal operating conditions, the area under an nmr peak is directly proportional to the number of protons responsible for the peak. Hence the areas under the three peaks for ethanol, in either the low-resolution or the high-resolution spectrum, are in the ratio 1:2:3 for the protons in the hydroxyl, methylene, and methyl groups, respectively.

The nmr spectrometer normally includes an integrator that draws a stepped curve superimposed over the spectrum, in which the heights of the steps are proportional to the areas under the peaks. Figure 32-14 shows the stepped integration curve superimposed on the spectrum for ethyl malonic acid.

▶ **4. SPLITTING OF SIGNALS AND SPIN-SPIN COUPLING.** A comparison of the drawing of the low-resolution spectrum of ethanol in Fig. 32-11 with that of the high-resolution spectrum in Fig. 32-12 shows that the signal for the —OH proton is recorded as one peak in both, but the signal for the $>CH_2$ protons is split into four peaks in the high-resolution spectrum and the signal for the —CH_3 protons is split into three peaks.

The splitting of a signal into a group of peaks arises because of neighboring protons in the molecule. It is sufficient here to note that the magnetic field felt by a particular proton is either slightly increased or slightly decreased by the spin of a neighboring proton—increased if the neighboring proton at that moment is aligned with the applied magnetic field, and decreased if the neighboring proton is aligned against the applied field. The absorption by the absorbing proton in half the molecules at a given moment is thus shifted slightly downfield and for the other half slightly upfield for each neighboring proton, resulting in a splitting of the signal. The net result is that **a set of equivalent neighboring protons will split each proton signal into a number of peaks equal to one more than the number of equivalent neighboring protons.**

In the high-resolution spectrum of ethanol the signal for the —CH_3 protons is split, because of the two equivalent protons on the neighboring methylene group, into $2 + 1 = 3$ peaks. The signal for the methylene protons, however, is split, because of the three neighboring methyl protons, into $3 + 1 = 4$ peaks. In a very pure sample of ethanol without acidic or basic impurities, the —OH signal is split into $2 + 1 = 3$ peaks, because of the two neighboring equivalent methylene protons (also the methylene proton signal is affected somewhat by the neighboring —OH proton). However, even a small amount of acidic or basic impurity catalyzes a rapid proton exchange between the hydroxyl groups of many ethanol molecules, which makes only the average signal (one overall peak) detectable for —OH in ethanol.

It might be mentioned here that even the relative heights of the individual peaks for a given signal are significant and useful in the determination of structure, but we will not go further into those details here.

Figure 32-13 shows the nmr spectrum for ethyl iodide. Note that, as in ethanol,

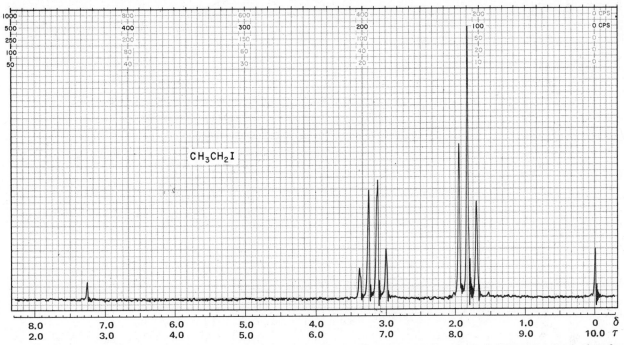

CH_3CH_2I

Figure 32-13 The high-resolution nmr spectrum for ethyl iodide, with TMS (tetramethylsilane) as internal standard. The TMS peak is the one at $\delta = 0$ ($\tau = 10$). The peak at $\delta = 7.27$ ($\tau = 2.73$) is due to an impurity.

(1) the signal for the methylene protons is downfield (3.2δ, 6.8τ) from that for the methyl protons (1.8δ, 8.2τ), (2) the overall signal area for the methylene protons is less than that for the methyl protons (two-thirds the area actually), in proportion to the numbers of equivalent protons for each signal, and (3) the methylene signal is split into four peaks and the methyl signal into three peaks. The peak at 7.27δ (2.73τ) is due to an impurity.

Figure 32-14 is the nmr spectrum of the dicarboxylic acid ethyl malonic acid. The —COOH proton signals are off the scale in Figure 32-14 and are not shown. The signal for the —CH proton appears farthest downfield (3.6δ, 6.4τ) and is split into three peaks (a triplet) because of the two protons in the neighboring —CH$_2$ (methylene) group. The methylene proton signal, which is next upfield (2.2δ, 7.8τ), is more complex in that it is influenced by the three protons on the neighboring methyl group *and also* by the one proton on the neighboring —CH group. The signal is split into four peaks (a quartet) by the three protons of the methyl group, and each of the four components of the quartet is split into two peaks (a doublet) by the one proton of the —CH group. Some of the resulting eight peaks overlap to give the apparent five peaks (a quintet). Finally, the methyl proton signal is still farther upfield (1.1δ, 8.9τ) and is split into three peaks (a triplet) because of the two protons on the neighboring methylene group. The areas under the overall signals are in the ratios of $1:2:3$, reflecting the numbers of equivalent protons of each type. In this spectrum, the stepped integration curve is shown, with the step for the —CH protons designated by (*a*), that for the —CH$_2$ protons designated by (*b*), and that for the —CH$_3$ protons designated by (*c*). Note that step (*a*) has a height of 6 divisions on the recording paper, step (*b*) a height of 12 divisions, and step (*c*) a height of 18 divisions, reflecting the ratios $1:2:3$ for the numbers of equivalent protons of each type.

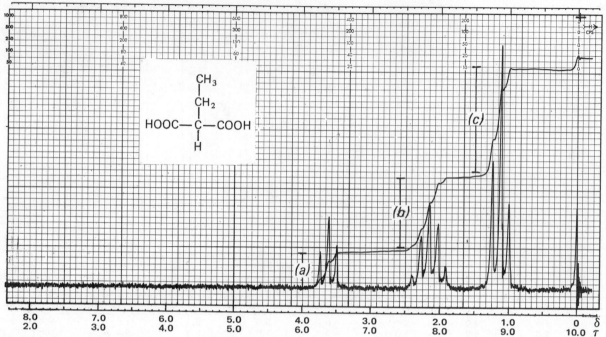

Figure 32-14 The high-resolution nmr spectrum for ethyl malonic acid. The stepped integration curve is superimposed on the spectrum. The TMS internal standard peak is at $\delta = 0$ ($\tau = 10$).

The many effects that protons and electrons within various portions of a molecule have upon the nmr spectrum make possible the determination of a very large amount of useful information concerning the structure of the molecule. Since protons are present in many different kinds of inorganic and organic molecules, and also since a number of other nuclei can be used as the basis for nmr spectra, nuclear magnetic spectroscopy has come to be one of the most useful tools the chemist has for the characterization of compounds.

32.6   Electron Paramagnetic Resonance

Electron paramagnetic resonance (epr), sometimes referred to as **electron spin resonance** (esr), is analogous to nuclear magnetic resonance except that epr involves electrons rather than nuclei. When a compound contains an unpaired electron, the unpaired electron spins about its own axis and creates a magnetic moment (Section 31.9). Just as in the nmr method with nuclei, the magnetic moment of the electron can be lined up *with* an external magnetic field or *against* the field, and energy is required to change the spin of the electron from alignment with the field to the higher energy condition of alignment against the field. In the epr method, as in the nmr method, the energy is provided by electromagnetic radiation of the proper frequency. The absorption of the radiation gives rise to an epr, or esr, spectrum.

Although the epr and nmr methods are quite analogous, they differ in the amount of energy required. A much greater amount of energy is required to reverse the spin of an electron than to reverse the spin of a hydrogen nucleus (proton), because of the much larger magnetic moment created by the electron. Hence, epr absorption occurs at a much higher frequency, or lower wavelength (in the microwave region), than does nmr absorption (see Fig. 32-1).

As with nmr signals, splitting and coupling also occur with epr signals, the spin-spin coupling occurring between electrons and neighboring hydrogen nuclei in much the same way and for essentially the same reasons as between two hydrogen nuclei for nmr signals. Electron paramagnetic resonance is useful in the detection of organic free radicals (organic structures with an odd, or unpaired electron) and in the study of their structures, especially with respect to where the lone electron is located within the system. The method also can be used in the study of some metal complexes and other kinds of molecules possessing unpaired electrons.

32.7   Mass Spectra

A **mass spectrum** can be produced by bombarding a sample of a compound in the gas phase with a beam of high-energy electrons; hence, unlike the spectral methods we have studied thus far in this chapter, mass spectrometry does not utilize absorption of electromagnetic radiation.

In the mass spectrometer (Fig. 32-15), bombardment of the sample in the gas phase by high-energy electrons knocks one electron from a sample molecule,

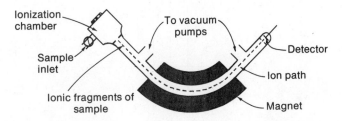

Figure 32-15 A diagram of a mass spectrometer. Ions leave the ionization chamber and move into the magnetic field. Light ions experience a large deflection and collide with the upper wall of the tube. Heavy ions have greater momentum, experience less deflection, and collide with the lower wall. Only those ions with a specific value of m/e pass through the field into the detector.

thereby producing a singly charged positive ion referred to as the **parent ion.** The parent ion in turn fragments to form a variety of other fast-moving positive ion fragments, mainly singly charged, which are stable only in the gas phase under the high-vacuum conditions used in a mass spectrometer. These ions are directed in the mass spectrometer through a magnetic field, which deflects their paths to an extent dependent upon their mass-to-charge ratios (m/e), the ions with lowest mass-to-charge ratio being deflected the most. By varying the magnetic field, ions of any mass-to-charge ratio can be made to strike the detector. Since the ions are principally singly charged ($e = 1$), the separation is primarily on the basis of mass. Taking into account the magnetic field and other characteristics of the instrument, the m/e ratio can be determined, and hence the mass, of each fragment.

A mass spectrum is determined by (1) varying the magnetic field so that ions with progressively higher masses strike the detector, and (2) plotting the relative intensities (proportional to the numbers of ions of particular masses), as indicated by the detector, versus their m/e ratios. A low-resolution mass spectrum of methanol, CH_3OH, is shown in Fig. 32-16. This spectrum shows a parent ion peak at $m/e = 32$, a very strong peak at $m/e = 31$ resulting from the ion CH_3O^+, a strong peak at $m/e = 29$ resulting from the ion COH^+, and a peak at $m/e = 15$ resulting from the ion CH_3^+.

Figure 32-16 The mass spectrum of methanol.

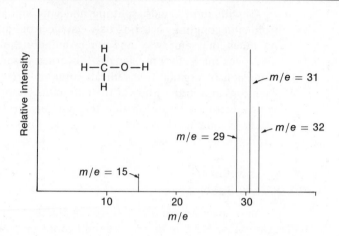

Figure 32-17 The mass spectrum of mercury, showing seven peaks due to the seven naturally occurring isotopes of mercury.

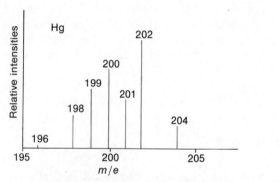

The m/e ratio of the parent ion is the molecular weight of the molecule. Therefore the mass spectrometer provides one of the best methods for obtaining very accurate molecular weights. More than that, however, the pattern of fragmentation and the masses of the parent ion and of each fragment ion provide an excellent "fingerprint method" for identifying a compound; only in a very small number of cases, such as with optical isomers, do different compounds fragment in exactly the same manner. In addition, a study of the fragmentation pattern yields considerable information about the structure of the compound and about the relative strengths of its bonds.

The mass spectrometer was first used to verify the existence of isotopes (Sections 2.3 and 4.8). For example, the seven naturally occurring isotopes of mercury show up very clearly in a mass spectrum (Fig. 32-17).

Chromatography

We mentioned earlier that chromatography involves the separation of a mixture of compounds by means of a system in which the components of the mixture migrate through the system at different rates. The method can be applied whether the sample is in the liquid phase or the gas phase. The method is based principally upon selective adsorption; in addition, other factors such as solubility, complexing, and hydrogen bonding play important roles in the process.

32.8   Gas Phase Chromatography

Gas phase, or vapor phase, chromatography utilizes a volatile mixture. Normally the gaseous mixture, referred to as the **moving phase,** is passed through a heated column containing a **stationary phase** consisting of an inert support material impregnated with a high-boiling liquid. The support material is typically either finely divided silicon dioxide, polymers of tetrafluoroethylene, glass beads, or porous polymeric beads of polyfluorolefins. An example of a commonly used high-boiling liquid is silicone oil.

Separation of the gaseous mixture depends upon the fact that the various components will be distributed in different proportions between the moving vapor phase and the liquid part of the stationary phase. The component that has the lowest solubility in the stationary phase (or the least attraction otherwise for the stationary phase) will go through the column most rapidly, other components following in inverse order of solubility (or other attractive forces). Hence, separation is effected. As each component is eluted from the column, a detector produces an electrical signal that is in proportion to the amount of component being eluted. The electrical signal for the concentrations of the eluted components is plotted against time on a record referred to as a **chromatogram.**

32.9   Liquid Phase Chromatography

Liquid phase chromatography is especially useful for separating mixtures of components that are not volatile or heat-sensitive and hence are not suitable for gas phase chromatography. The method can be applied either with a column **(column chromatography)** or with thin layers on a flat surface **(thin layer chromatography).**

Liquid phase chromatography can be accomplished either by **partition chromatography,** in which the stationary phase is an inert support impregnated with a liquid, or by **adsorptive chromatography,** in which the stationary inert support is a solid adsorbent material on which the components will be adsorbed but to different extents. In either partition or adsorptive chromatography the moving phase is a liquid.

A still different form of liquid phase chromatography is **ion exchange chromatography,** in which the stationary phase is a polymer containing acidic or basic groups located along its backbone. The acidic or basic groups react with ions present in the moving phase. Usually, a column is employed. The system can be set up so as to produce a quantitative exchange of ions, as in water softening by ion exchange (see Section 12.12), or so as to produce a difference in the rate at which various ionic substances in the moving phase proceed through the column. The method can be adapted to either cation or anion exchange. Acidic sites on the polymer are often carboxylic or sulfonic acids; basic sites are often substituted ammonium groups or amines.

The chromatographic process can be used to separate components of complex mixtures, such as 15 to 20 amino acids obtained from a protein, to separate components of very small samples, and to separate components in samples of very low concentrations.

A refinement and improvement in liquid chromatography is accomplished by operating the column at high pressures, making possible more efficient columns. The method, referred to as **high performance liquid chromatography** (HPLC), is receiving a considerable amount of research effort currently and is proving to be of immense value in certain types of difficult mixture separations.

Combinations of Methods for Characterization of Compounds

The power of instrumental methods for characterization of compounds is often greatly increased by combining two or more methods together. One of several examples is the combination of gas chromatography with mass spectrometry (**gas chromatographic mass spectrometry,** GCMS). The rapidly increasing availability of small "on-line" computers suitable for use with an individual piece of equipment and the rapid development of a wide variety of automated data collection systems have in recent years made such combinations much more feasible than in the past.

Questions

1. Explain what is meant by the *characterization* of a compound.
2. List five forms of spectroscopy and distinguish between them.
3. What is the meaning of each of the following terms? molar extinction coefficient, absorbance, per cent transmittance, molar absorptivity, wavelength, wave number
4. State which spectroscopic methods cover, respectively, the various portions of the spectrum from 100 nanometers wavelength to above 1×10^{10} nanometers.
5. (a) Relate the following units to each other: millimicrons, centimeters, nanometers, reciprocal centimeters, microns.
   (b) Which of the above units are used for wave number? Which for wavelength?
   (c) How many microns are in a nanometer? How many centimeters in a nanometer? How many nanometers in an angstrom unit? How many microns in an angstrom unit?
6. How can ultraviolet spectra be used in the characterization of organic compounds?
7. How can visible spectra be used in the characterization of metal coordination compounds?
8. What is meant by the terms *stretching vibrations* and *bending vibrations,* and what is their significance to (a) spectroscopy and (b) characterization of compounds?
9. How may certain functional groups in a molecule be identified by spectroscopy?
10. Of what use is infrared spectroscopy in the characterization of metal coordination compounds?
11. Explain why nuclear magnetic resonance is helpful in determining the molecular structures of some types of compounds. What are some kinds of compounds that cannot be studied by nmr?
12. Distinguish between gas phase chromatography and

liquid phase chromatography.

13. Identify as many bands (peaks) as possible in the following infrared spectrum for 2-methyl-1-penten-3-ol, $C_6H_{12}O$, considering the molecular

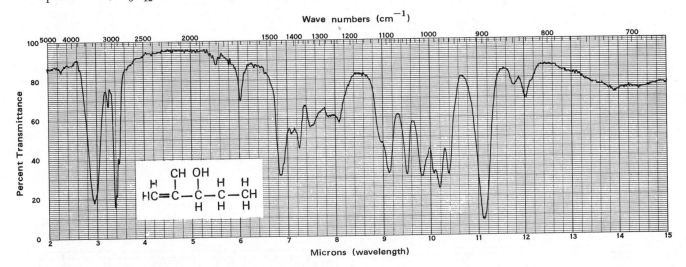

formula given in the box on the spectrum and using the data provided in Fig. 32-5 and Table 32-1.

14. Identify as many bands (peaks) as possible in the following infrared spectrum for isobutyl cinnamate, $C_{13}H_{16}O_2$, considering the molecular formula given in the box on the spectrum and using the data provided in Fig. 32-5 and Table 32-1.

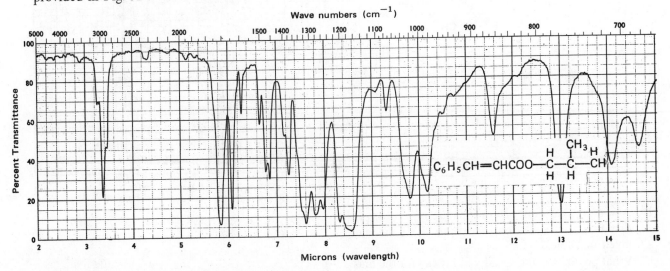

15. In the following nmr spectrum for triethylamine, $(C_2H_5)_3N$, two signals, each split into several peaks,

are present; in addition there is the small tetra-methylsilane peak at 0δ, or 10τ.

(a) Explain why two signals, other than the one for TMS, would be expected.

(b) Determine what portion of the molecule is responsible for each signal and explain the basis for your decision.

(c) State how many peaks in theory should result from the splitting of each signal at high resolution, explain why, and determine if the spectrum agrees with the theory?

(d) Determine the pertinent heights in the stepped integration curve, by counting divisions, to determine the relative total signal areas; then check whether this agrees with the theory based upon numbers of equivalent protons of each environmental type.

(e) Explain the purpose of the tetramethylsilane peak.

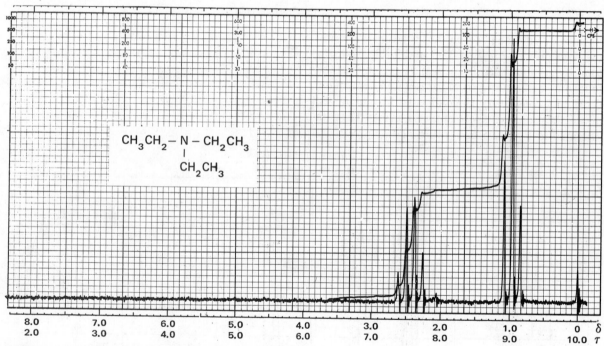

16. Assuming that only parent ions are observed, predict the mass spectrum of dry air from the data given in Table 22-1.

17. What possible peaks would you predict for the mass spectrum of methane, CH_4, and for that of ethane, C_2H_6?

References

"Instrumentation," J. H. Krieger, M. Waldrop, and W. Worthy, *Chem. and Eng. News,* March 13, 1978, p. 32.

"Analytical Chemistry Using Lasers to Detect Less and Less," A. L. Robinson, *Science,* **199,** 1191 (1978).

"The Laser Revolution," W. J. Cromie, *Chemistry,* February 1979, p. 13.

"Chemical Analysis by Infrared," B. Crawford, Jr., *Sci. American,* October 1953, p. 42.

"Infrared Determination of Stereochemistry in Metal Complexes," M. Y. Darensbourg and D. J. Darensbourg, *J. Chem. Educ.,* **47,** 33 (1970).

"Electronic Spectra of Some Transition Metal Complexes: Derivation of Dq and B," A. B. P. Lever, *J. Chem. Educ.,* **45,** 711 (1968).

"Spectroscopy and Structure Elucidation," C. W. Jefford, R. McCreadie, P. Muller, and J. Pfyffer, *J. Chem. Educ.,* **50,** 181 (1973).

"Spectroscopic Determination of Thermodynamic Quantities," M. Berger, J. A. Bell, and C. Steel, *J. Chem. Educ.,* **52,** 191 (1975).

"A Combined Infrared and Kinetic Study of Linkage Isomers: An Inorganic Experiment," W. H. Hohman, *J. Chem. Educ.,* **51,** 553 (1974).

"Infrared Determination of Stereochemistry in Metal Complexes: The Determination of Symmetry Coordinates," D. J. Darensbourg and M. Y. Darensbourg, *J. Chem. Educ.,* **51,** 787 (1974).

"Two Spectrophotometric Experiments with Alpha-Chymotrypsin," E. R. Kantrowitz and G. Eisele, *J. Chem. Educ.,* **52,** 410 (1975).

"Isomer Analysis by Spectral Methods," G. A. Poulton, *J. Chem. Educ.,* **52,** 397 (1975).

"^{13}C NMR Spectra of Styrene Derivatives," J. W. Blunt and D. A. R. Happer, *J. Chem. Educ.,* **56,** 56 (1979).

"Advances Boost Use of NMR of Solids," J. L. Fox, *Chem. and Eng. News,* October 16, 1978, p. 22.

"Biological NMR Spectroscopy," S. J. Opella, *Science,* **198,** 158 (1977).

"Whole-Cell Metabolism Studied with NMR," J. L. Fox, *Chem. and Eng. News,* February 26, 1979, p. 26.

"NMR of Boron Compounds," G. R. Eaton, *J. Chem. Educ.,* **46,** 547 (1969).

"Troublesome Concepts in NMR Spectrometry," R. M. Silverstein and R. G. Silberman, *J. Chem. Educ.,* **50,** 484 (1973).

"Test for Chemical Shift and Magnetic Equivalence in NMR," A. Ault, *J. Chem. Educ.,* **51,** 729 (1974).

"Calculation of Complex NMR Spectra in the Undergraduate Laboratory," R. J. Seyse, H. L. Pearce, and T. L. Rose, *J. Chem. Educ.,* **52,** 194 (1975).

"Structural Assignment of a $C_{10}H_{12}O_3$ Ester by Mass Spectroscopy," A. P. Marchand and D. Jackson, *J. Chem. Educ.,* **53,** 390 (1976).

"Mass Spectrometry of Biochemical Materials," A. H. Jackson, *Endeavour,* **1,** 75 (1977).

"Mass Spectra of Organic Compounds," D. H. Williams, *Chem. in Britain* **4,** 5 (1968).

"Modern Aspects of Mass Spectrometry," J. F. J. Todd, *Educ. in Chemistry,* **10,** 89 (1973).

"Venus Lower Atmosphere Composition, Analysis by Gas Chromatography," V. I. Oyama *et al., Science,* **203,** 802 (1979).

"Gas Chromatographic Analysis of Gasoline," R. F. Cassidy, Jr. and C. Schuerch, *J. Chem. Educ.,* **53,** 51 (1976).

"An Early Application of Paper Chromatography," F. Kurzer, *J. Chem. Educ.,* **55,** 321 (1978).

"Thin-layer Chromatography," C. L. Stong, *Sci. American,* February 1976, p. 128.

"Chromatography on Chalk," A. Wollrab, *J. Chem. Educ.,* **52,** 809 (1975).

"The First Complete Nucleotide Sequencing of an Organism's DNA," M. Smith, *Amer. Scientist,* **67,** 57 (1979).

"Gas Chromatography," A. Keller, *Sci. American,* October 1961, p. 58.

"Identifying Alcohols by Thin Layer Chromatography," M. Crawford and A. K. Keenan, *Educ. in Chemistry,* **10,** 98 (1973).

"On Thermodynamics of Solution by Gas-Liquid Chromatography," E. F. Meyer, *J. Chem. Educ.,* **50,** 191 (1973).

"Automatic Analysis of Blood Cells," M. Ingram and K. Preston, Jr., *Sci. American,* November 1970, p. 72.

"Chemistry and the Spinning Electron" (Staff), *New Scientist,* **60,** 128 (1973).

"Qualitative Emission Spectroscopy in Freshman Chemistry," L. F. Druding and R. A. Lalancette, *J. Chem. Educ.,* **51,** 527 (1974).

"Lead in Blood—The Analyst's Problem," A. A. Cernik, *Chem. in Britain,* **10,** 58 (1974).

"Recent Developments in the Analysis of Toxic Elements," D. J. Lisk, *Science,* **184,** 1137 (1974).

"Electron Spin Resonance in Transition-Metal Chemistry," J. B. Raynor, *Chem. in Britain,* **10,** 254 (1974).

"Laser Spectroscopy: Probing Biomolecular Functions," J. L. Marx, *Science,* **188,** 1002 (1975).

33

The Alkali Metals-
Group IA, and
the Ammonium Ion

33.1 Periodic Relationships of the Alkali Metals

The metallic elements lithium, sodium, potassium, rubidium, cesium, and francium constitute Group IA of the Periodic Table and are known as the **alkali metals.** The heaviest of them, francium, occurs in nature in very small quantities as a short-lived radioactive isotope. The alkali metals and the metals of Group IIA (the alkaline earth metals) are often called the **light metals** because of their relatively low densities. The alkali metals have the largest atomic radii (Table 33-1) of the elements in their respective periods of the Periodic Table. There is a single electron in the outermost shell of each of the alkali metals. Since this

TABLE 33-1 Alkali Metals

	Atomic Number	Atomic Weight	Atomic Radius, Å	Ionic Radius, Å	Density at 20°	Melting Point, °C	Boiling Point, °C
Lithium	3	6.941	1.52	0.60	0.534	180	1326
Sodium	11	22.98977	1.86	0.95	0.97	98	889
Potassium	19	39.0983	2.31	1.33	0.86	63.4	757
Rubidium	37	85.4678	2.44	1.48	1.53	38.8	679
Cesium	55	132.9054	2.62	1.69	1.87	28.7	690
Francium	87	223	2.7	—	—	—	—

electron is far removed from the nucleus of the atom, it is easily lost, and the alkali metals readily form stable positive ions. The difficulty with which their ions are reduced to the metallic state accounts for the fact that none of the alkali metals were isolated until Sir Humphry Davy, in 1807, produced sodium and potassium by the electrolytic reduction of their respective hydroxides. As expected, the ease with which the valence electron is lost increases with increasing atomic radius. Of those members of the group that have been extensively studied, cesium is the most reactive; however, francium, with atomic radius even larger than that of cesium, should be found to be more reactive than cesium.

The alkali metals all react vigorously with water, liberating hydrogen and giving solutions of the corresponding strong bases. They also react directly with oxygen, sulfur, nitrogen, hydrogen, and the halogens, giving ionic compounds. Most of the salts of the alkali metals are readily soluble in water; however, Li_2CO_3, Li_3PO_4, and LiF are relatively insoluble. In this respect, lithium resembles magnesium, the second member of Group IIA.

Sodium and Potassium

33.2   Occurrence of Sodium and Potassium

Sodium and potassium are not found free in nature because of their high ease of oxidation. Their compounds, however, occur in abundance and are widely distributed in nature. They appear in many complex silicate rocks such as feldspars, of which the very abundant **albite,** $NaAlSi_3O_8$, and **orthoclase,** $KAlSi_3O_8$, are typical. As these minerals are disintegrated by weathering, the sodium and potassium are converted to soluble compounds, most of which are leached out of the soil by water and are eventually carried to the sea. In passing through the soil, some of the potassium compounds are adsorbed by the colloidal particles of the soil and held there until used by plants. Sodium compounds are much less retained by the soil than those of potassium, and little sodium is needed by plants. Changes in the earth's crust during past geological ages have, at various times and places, caused sections of the sea to become isolated and gradually evaporate. Such geological changes have resulted in the formation of large deposits of sodium and potassium salts.

Sodium and potassium occur also in various parts of the world in chloride, sulfate, and borate minerals.

33.3   Preparation of Sodium and Potassium

For many years metallic sodium was obtained by reducing sodium carbonate with carbon.

$$Na_2CO_3 + 2C \longrightarrow 2Na + 3CO$$

With the development of the dynamo for the production of cheap electricity, Davy's original electrolytic method of obtaining sodium from fused sodium hydroxide became the industrial process for producing the metal. Most of our sodium is now prepared by the electrolysis of fused sodium chloride, admixed with either sodium carbonate or calcium chloride. The energy required to melt sodium chloride and the high vapor pressure of metallic sodium at this temperature prohibit the use of sodium chloride alone; the melting point of sodium chloride is 801°, while that of the mixture is approximately 600°.

Sodium is shipped in the form of 12-pound bricks in air-tight, steel barrels or in tank cars which are loaded by pumping in the fused metal. To unload the cars, the metal is melted and the liquid forced out by nitrogen under pressure.

Potassium is produced on a relatively small scale, since this element offers few technical advantages over the less costly sodium. Potassium may be made by the electrolytic reduction of either molten potassium chloride or potassium hydroxide but most of the production in the United States is by the reaction of sodium with fused potassium chloride ($Na + KCl \longrightarrow K + NaCl$). The sodium is fed into the bottom of a column of melted potassium chloride, and potassium metal escapes at the top. It should be noted that for fused salts the regular electromotive series does not apply exactly. A variety of chemical factors determine the course of the reaction. Although potassium is above sodium in the electromotive series, this particular reaction apparently proceeds because metallic potassium is more volatile than is sodium, thereby causing a shift in the equilibrium to the right.

33.4   Properties of Sodium and Potassium

Both sodium and potassium are soft enough to be cut by a knife; they are silvery white in color and are excellent conductors of heat and electricity. Both metals exhibit the reactivity that is characteristic of very active metals. They tarnish immediately in moist air, and react vigorously with water and acids, forming hydroxides or salts of the acids, respectively, and liberating hydrogen. The heat of reaction between these metals and water may cause the evolved hydrogen to ignite. One should never touch sodium or potassium with the hands, because the heat of reaction with the moisture on the skin may cause the metal to ignite. These metals are stored under kerosene or in sealed containers.

Sodium dissolves in mercury, forming an amalgam. This alloy is an active reducing agent that is more suitable for many uses than pure sodium because the mercury, being an inactive metal, retards the action of the sodium. Sodium forms alloys with many other metals such as potassium, tin, and antimony, but not with iron. Potassium and sodium, if mixed in correct proportion, give an alloy that is liquid at room temperature.

In general, potassium is more reactive than sodium; otherwise the chemical properties of the two metals are quite similar. Potassium and sodium differ primarily in regard to the solubilities of their salts, the potassium salts being in general less soluble than those of sodium; there are exceptions to this generalization, however.

33.5   Uses of Sodium and Potassium

Sodium is used as a reducing agent in the production of other metals, such as titanium and zirconium, from their chlorides or oxides. Its properties as a reducing agent make it useful in the manufacture of certain dyes, drugs, and perfumes. It is used in the preparation of sodium-lead alloys, which in turn are used in the production of tetraethyl lead from ethyl chloride. It is used in sodium lights for lighting highways because its yellow light penetrates fog well. The synthetic rubber industry consumes a considerable amount of sodium. The largest uses, however, are in the manufacture of compounds such as sodium peroxide and sodium cyanide, which cannot be made directly from sodium chloride.

Potassium has no major uses for which sodium cannot be substituted. Sodium-potassium alloys, however, are growing in importance. Their low density, low viscosity, wide liquid range, and high thermal conductivity combine to make them particularly useful as industrial heat-transfer media.

33.6 Sodium Chloride

Sodium chloride is one of our most abundant minerals. Seawater contains 2.7% sodium chloride, and the waters of the Dead Sea and the Great Salt Lake contain 23%. There are vast deposits of rock salt in the Stassfurt region of Germany, and in the United States large deposits are found in New York, Michigan, West Virginia, California, and a very extensive bed (about 400 to 500 feet thick) underlies parts of Oklahoma, Texas, and Kansas. Most of the salt consumed comes from beds that lie below the surface of the earth, rather than from seawater. The salt is removed from these underground beds by mining or by forcing water down into the deposits to form saturated brines, which are then pumped to the surface. Natural salt contains other soluble salts that make it unsuitable for many uses. Salt is purified by **recrystallization**—dissolving it in water, concentrating the solution by evaporation (often under reduced pressure), and allowing the crystals to form again. Some of the impurities are more soluble than the salt, and for this reason remain in solution when the salt crystallizes. Some of the impurities that are less soluble than salt do not crystallize out because they are present in small amounts. Other of the less soluble impurities, such as calcium sulfate, do crystallize out with the salt, but are then removed by taking advantage of the decreasing solubility of calcium sulfate in sodium chloride solution as the temperature is raised above 80°C. Thus, if the brine is filtered hot, the resulting solution will not contain enough calcium sulfate to be saturated when cooled. Hence calcium sulfate cannot precipitate with the sodium chloride, and the sodium chloride crystallizes relatively free of the impurity.

Pure sodium chloride, when the relative humidity is about 75%, or sodium chloride contaminated with chlorides of magnesium and calcium, absorbs water causing the salt to "cake." When in a salt shaker, the salt cakes and clogs the shaker. Basic magnesium carbonate or calcium aluminosilicate is usually added to serve as a drier or "anticaking" agent.

Sodium chloride is the usual source of almost all other compounds containing sodium or chlorine. Its most extensive use is in the maufacture of chlorine, hydrochloric acid, sodium hydroxide, and sodium carbonate. Sodium chloride is an essential constituent of the foods of animals. It not only makes food more palatable, but is the source of chlorine from which hydrochloric acid, a constituent of gastric juice, is produced. Sodium chloride is also a constituent of the blood and is essential for the life processes of the human body, although care must be taken not to use too much salt, especially in diets of those with potential heart problems or high blood pressure.

33.7   Hydrides of Sodium and Potassium

The hydrides NaH and KH are prepared by the direct union of the elements at slightly elevated temperatures and are decomposed at higher temperatures. They are saltlike compounds, with a crystal structure like that of sodium chloride (Section 11.18). When the fused hydrides are electrolyzed, the metal is formed at the cathode and hydrogen is evolved at the anode, so the hydrides must be composed of metal ions and hydride ions, H^-. The hydrides are readily decomposed by water with the evolution of hydrogen, the hydride ions combining with hydrogen ions of the water ($H^+ + H^- \longrightarrow H_2$), and leaving hydroxide ions in solution.

$$NaH + H_2O \longrightarrow Na^+ + OH^- + H_2(g)$$

Sodium hydride is used widely in organic chemistry as a reducing agent and in the descaling of steel before it is plated or enameled.

33.8   Oxides and Peroxides of Sodium and Potassium

Oxides of most metals are produced by heating either the hydroxides, nitrates, or carbonates of the corresponding metals. The oxides of the more active alkali metals cannot be prepared in this way, however, because their hydroxides, nitrates, and carbonates are stable toward heat, decomposing only at extremely high temperatures.

Sodium forms both the **oxide,** Na_2O, and the **peroxide,** Na_2O_2. Sodium oxide is prepared in commercial quantities by heating sodium in a limited supply of air under carefully controlled conditions. The product is an extremely alkaline white powder. It is a very effective drying agent and is also used as a strong caustic.

The peroxide is the more important of the two oxides of sodium. It is manufactured by the reaction of sodium with an excess of air at 300° to 400°. Sodium peroxide is a yellowish white powder, used extensively as an oxidizing agent and as a bleaching agent. Solutions of the compound are alkaline by hydrolysis, as shown by the equations

$$O_2^{2-} + H_2O \rightleftharpoons HO_2^- + OH^-$$
$$HO_2^- + H_2O \rightleftharpoons H_2O_2 + OH^-$$

Potassium oxide, K_2O, may be obtained by heating potassium nitrate with elementary potassium in the absence of air. The **peroxide,** K_2O_2, is formed by burning the metal in the calculated quantity of air. When exposed to an excess of air, potassium oxide combines readily with oxygen to give **potassium superoxide,** KO_2, in which the superoxide ion has a charge of -1. The superoxide of potassium is most readily formed by burning the metal in an excess of oxygen. The KO_2 is used in gas masks for mine safety work. Moisture from the air in contact with the superoxide reacts with the superoxide to give O_2 for the wearer of the mask to breathe.

$$4KO_2 + 2H_2O \longrightarrow 4KOH + 3O_2$$

When the breath is exhaled through the mask, the CO_2 is absorbed by the KOH.

$$KOH + CO_2 \longrightarrow KHCO_3$$

The gas-mask user breathes no air directly from outside the mask.

33.9   Hydroxides of Sodium and Potassium

Sodium hydroxide is frequently called **caustic soda** in commerce, and its solutions are sometimes referred to as **lye**, or **soda lye**. It is the eighth highest chemical in terms of production in the United States (10,710,000 tons in 1978). Its production commercially is by two processes. The older method involves the action of slaked lime, $Ca(OH)_2$, on soda ash, Na_2CO_3, and is called the **lime-soda process.**

$$[2Na^+] + CO_3^{2-} + Ca^{2+} + [2OH^-] \longrightarrow CaCO_3(s) + [2Na^+] + [2OH^-]$$

The insoluble calcium carbonate is filtered out, and the filtrate is evaporated to yield solid sodium hydroxide. The residue is fused to remove the last traces of water and is then molded into sticks or pellets.

A second and more important industrial process for the production of sodium hydroxide is the electrolysis of aqueous sodium chloride (see Sections 19.1 and 20.4).

$$[2Na^+] + 2Cl^- + 2H_2O \longrightarrow [2Na^+] + 2OH^- + H_2(g) + Cl_2(g)$$

Hydrogen is formed at the cathode and chlorine is evolved at the anode. Sodium ions and hydroxide ions accumulate in the solution. The cell is so designed that the chlorine is not permitted to react with the hydroxide ions to form hypochlorite.

Electrolytic sodium hydroxide is impure with sodium chloride, most of which can be removed by a process called **fractional crystallization** because sodium chloride is less soluble than sodium hydroxide in water.

Pure sodium hydroxide can be obtained electrolytically from aqueous sodium chloride by using a mercury cathode cell. In this cell sodium, rather than hydrogen, is liberated at the cathode; the sodium, as it is reduced, forms an amalgam with the mercury. The amalgam is decomposed by water in a separate compartment to produce hydrogen and very pure sodium hydroxide, containing no chlorides.

$$2Na^+ + 2Cl^- + 2xHg \xrightarrow{\text{Electrolysis}} 2Na(Hg)_x + Cl_2(g)$$
$$2Na(Hg)_x + 2H_2O \longrightarrow 2Na^+ + 2OH^- + 2xHg + H_2(g)$$

Sodium hydroxide is an ionic compound that melts and boils without decomposition. When fused or in aqueous solution, sodium hydroxide attacks silicate minerals and glass. The solid hydroxide and its solutions readily absorb carbon dioxide from the air, with the formation of sodium carbonate and water.

$$2OH^- + CO_2 \longrightarrow H_2O + CO_3^{2-}$$

Sodium hydroxide dissolves extensively in water, with the evolution of a great deal of heat, and yields strongly basic solutions. It is used extensively in the production of chemicals, rayon, lye, cleaners, textiles, soap, paper and pulp, and in petroleum refining.

Potassium hydroxide is manufactured in a manner similar to that described for the production of sodium hydroxide. The method involving the treatment of aqueous potassium carbonate with slaked lime has largely given way to the electrolytic method. Potassium hydroxide is very soluble in water, and being strongly deliquescent is used as a dehydrating agent. It forms strongly alkaline solutions with water. Because potassium hydroxide is more expensive than sodium hydroxide, which almost always serves equally well, its use is limited.

33.10   Carbonates of Sodium and Potassium

Sodium carbonate is, after sodium hydroxide, perhaps the most important manufactured compound of sodium. The total amount used in the United States in 1978 was 7,581,000 tons, making it eleventh among all industrially produced chemicals. It has the formula Na_2CO_3 and it is commonly called **soda ash,** or simply **soda.**

Eighty per cent of the sodium carbonate produced in the United States is prepared from the mineral **trona,** $Na_2CO_3 \cdot NaHCO_3 \cdot 2H_2O$. Following recrystallization from water, trona is roasted to produce Na_2CO_3.

$$2(Na_2CO_3 \cdot NaHCO_3 \cdot 2H_2O) \xrightarrow{\Delta} 3Na_2CO_3 + 5H_2O + CO_2$$

Sodium carbonate is also manufactured by the **Solvay process.** This process is based upon the reaction of ammonium hydrogen carbonate with a saturated solution of sodium chloride. Sodium hydrogen carbonate precipitates, since it is only slightly soluble in the reaction medium.

$$Na^+ + [Cl^-] + [NH_4^+] + HCO_3^- \longrightarrow NaHCO_3(s) + [NH_4^+] + [Cl^-]$$

The basic raw materials used in the process are limestone and common salt. The carbon dioxide is generated by heating limestone.

$$CaCO_3 \longrightarrow CaO + CO_2 \tag{1}$$

The ammonia is obtained by treating ammonium chloride, formed as a by-product of the process, with calcium hydroxide.

$$2NH_4^+ + [2Cl^-] + [Ca^{2+}] + 2OH^- \longrightarrow 2NH_3(g) + 2H_2O + [Ca^{2+}] + [2Cl^-] \tag{2}$$

Calcium hydroxide results from slaking the lime produced in Reaction (1).

$$CaO + H_2O \longrightarrow Ca^{2+} + 2OH^- \tag{3}$$

Aqueous ammonia reacts with an excess of carbon dioxide, yielding ammonium hydrogen carbonate.

$$NH_3 + CO_2 + H_2O \longrightarrow NH_4^+ + HCO_3^- \tag{4}$$

Sodium hydrogen carbonate is the first desired product of the Solvay process; and, after being freed of ammonium chloride by recrystallization, it is made commercially available. The major portion of the sodium hydrogen carbonate, however, is converted to sodium carbonate by heating.

$$2NaHCO_3 \longrightarrow Na_2CO_3 + H_2O + CO_2 \tag{5}$$

The carbon dioxide from this reaction is used to produce more sodium hydrogen carbonate. The only by-product of the entire process not reused in the process is calcium chloride.

Sodium carbonate is produced to a limited extent from sodium hydroxide, a by-product of the chlorine industry. We saw in Section 20.4 that the electrolysis of aqueous sodium chloride produces chlorine, hydrogen, and sodium hydroxide. By saturating the sodium hydroxide solution with carbon dioxide, sodium hydrogen carbonate is produced.

$$[Na^+] + OH^- + CO_2 \longrightarrow [Na^+] + HCO_3^-$$

The use of less carbon dioxide gives rise to the normal carbonate.

$$[2Na^+] + 2OH^- + CO_2 \longrightarrow [2Na^+] + CO_3^{2-} + H_2O$$

Sodium carbonate is used extensively in the manufacture of glass, other chemicals, soap, paper and pulp, cleansers, and water softeners, and in petroleum refining.

When a solution of sodium carbonate is evaporated below 35.2°, the **deca-hydrate,** $Na_2CO_3 \cdot 10H_2O$, known as **washing soda,** crystallizes out; above this temperature the **monohydrate,** $Na_2CO_3 \cdot H_2O$, separates. Heating either hydrate produces the anhydrous compound. Solutions of sodium carbonate are basic, due to hydrolysis of the carbonate ion, as shown by the equation

$$CO_3^{2-} + H_2O \longrightarrow HCO_3^- + OH^-$$

For this reason, sodium carbonate is generally used in commercial processes requiring an alkali that is not as strong as sodium hdyroxide.

Sodium hydrogen carbonate, $NaHCO_3$ is commonly known as **bicarbonate of soda,** or **baking soda.** Its solutions are weakly basic due to hydrolysis of the hydrogen carbonate.

$$HCO_3^- + H_2O \rightleftharpoons H_2CO_3 + OH^-$$

Acids act upon sodium hydrogen carbonate with the formation of carbon dioxide.

$$HCO_3^- + H_3O^+ \longrightarrow 2H_2O + CO_2(g)$$

This reaction is involved in the leavening, or rising, process in baking. Baking powders contain baking soda mixed with an acidic substance, such as either potassium hydrogen tartrate [$KHC_4H_4O_6$ (cream of tartar)], calcium dihydrogen phosphate [$Ca(H_2PO_4)_2$], or sodium aluminum sulfate [$NaAl(SO_4)_2 \cdot 12H_2O$]. Starch or flour is also added to keep the powder dry. When the mixture is dry no reaction takes place, but as soon as water is added carbon dioxide is given off. The acidic substance and water produce hydronium ions, which in turn react with the baking soda to form carbon dioxide.

$$HC_4H_4O_6^- + H_2O \rightleftharpoons H_3O^+ + C_4H_4O_6^{2-}$$
$$H_2PO_4^- + H_2O \rightleftharpoons H_3O^+ + HPO_4^{2-}$$
$$[Al(H_2O)_x]^{3+} + H_2O \rightleftharpoons H_3O^+ + [Al(OH)(H_2O)_{x-1}]^{2+}$$
$$HCO_3^- + H_3O^+ \longrightarrow CO_2(g) + 2H_2O$$

The carbon dioxide when trapped in bread dough will expand if warmed and produce spaces that give the bread a desirable lightness.

The lactic acid of sour milk and buttermilk or the acetic acid in vinegar will also furnish hydronium ions and thus serve the same purpose as the acidic constituent of baking powder. Thus it is possible to make bread rise by using either sour milk or buttermilk with ordinary sodium bicarbonate (baking soda).

Potassium carbonate cannot be prepared by the Solvay process because of the high solubility of potassium hydrogen carbonate in solutions of aqueous ammonia. Instead, it is obtained either by treating electrolytic potassium hydroxide with carbon dioxide, or from potassium chloride. The latter method involves heating potassium chloride under pressure with magnesium carbonate, water, and carbon dioxide, whereby the slightly soluble double salt $KHCO_3 \cdot Mg(HCO_3)_2 \cdot 4H_2O$ is formed.

$$2K^+ + [2Cl^-] + 3MgCO_3 + 3CO_2 + 11H_2O \longrightarrow 2\{KHCO_3 \cdot Mg(HCO_3)_2 \cdot 4H_2O\} + Mg^{2+} + [2Cl^-]$$

The complex salt is removed from the solution by filtration and heated to $120°$, which causes the following reaction to take place:

$$2\{KHCO_3 \cdot Mg(HCO_3)_2 \cdot 4H_2O\} \longrightarrow K_2CO_3 + 2MgCO_3 + 11H_2O + 3CO_2$$

Finally, the potassium carbonate is leached from the solid mass, leaving a residue of insoluble magnesium carbonate, which is used again in the first step.

Potassium carbonate forms three hydrates, containing one, two, and three molecules of water of crystallization, respectively. Potassium carbonate is usually sold in the form of an anhydrous powder. It is deliquescent and very soluble in water, forming a strongly alkaline solution. It is used in the manufacture of soft soap, glass, pottery, and various potassium compounds.

When a solution of potassium carbonate is saturated with carbon dioxide and concentrated by evaporation, crystals of **potassium hydrogen carbonate,** $KHCO_3$, are deposited. Solutions of this salt are weakly basic. When heated, the hydrogen carbonate is converted to the normal carbonate.

33.11   Nitrates of Sodium and Potassium

Sodium nitrate, or **Chile saltpeter,** is a very important compound that was discussed in connection with nitrogen and nitric acid (Chapter 23).

Potassium nitrate was one of the principal reagents used by the alchemists. It is formed in nature by the decay of organic matter and is prepared commercially from sodium nitrate and potassium chloride.

$$Na^+ + [NO_3^-] + [K^+] + Cl^- \longrightarrow NaCl(s) + [K^+] + [NO_3^-]$$

The sodium nitrate and potassium chloride are dissolved in hot water, and the solution is evaporated by boiling. Of the four compounds possible in a solution containing Na^+, NO_3^-, K^+, and Cl^-, sodium chloride is the least soluble in hot water, and hence it is the first to crystallize out when the concentration of the solution is increased by evaporation. On the other hand, potassium nitrate is very soluble in hot water, making it possible to separate the two salts by filtration. Sodium chloride has about the same solubility in cold as in hot water, so very little more of it crystallizes out when the filtrate is cooled. However, potassium nitrate is only slightly soluble in cold water, and thus it crystallizes out as the filtrate is cooled.

Potassium nitrate is a valuable fertilizer, because it furnishes both potassium and nitrogen in forms that are readily utilized by growing plants. It is also used to some extent in preserving ham and corned beef, to which it imparts a red color, but studies are being conducted on the safety of using potassium nitrate as an additive to meat.

33.12   Sulfates of Sodium and Potassium

Extensive deposits of **sodium sulfate** occur in Canada, North Dakota, and the southwestern section of the United States, but most of the sodium sulfate used commercially is obtained as a by-product in the manufacture of hydrochloric acid from salt and sulfuric acid. The sodium sulfate thus formed is commonly called **salt cake** and is used in the manufacture of glass, paper, rayon, coal-tar dyes, and soap.

When sodium sulfate is crystallized from solution at temperatures below $32.28°C$, the decahydrate, $Na_2SO_4 \cdot 10H_2O$, is formed. This hydrate is called **Glauber's salt,** in honor of the alchemist Glauber, who used it as a medicine in the seventeenth century. The anhydrous salt, Na_2SO_4, crystallizes from solutions at temperatures above the transition point of $32.28°C$.

Sodium hydrogen sulfate is prepared by heating either sodium chloride or sodium nitrate to a moderate temperature with sulfuric acid.

$$NaCl + H_2SO_4 \longrightarrow NaHSO_4 + HCl(g)$$

$$NaNO_3 + H_2SO_4 \longrightarrow NaHSO_4 + HNO_3(g)$$

Potassium also forms two sulfates: the **normal salt,** K_2SO_4, and the **hydrogen salt,** $KHSO_4$. The normal sulfate is obtained from natural salt deposits, and it is used as a fertilizer and in the preparation of alums. **Potassium hydrogen sulfate** is made by heating the normal sulfate with the proper quantity of sulfuric acid. Heat converts potassium hydrogen sulfate to potassium pyrosulfate.

$$2KHSO_4 \longrightarrow K_2S_2O_7 + H_2O$$

When the pyrosulfate is strongly heated, it gives up sulfur trioxide.

$$K_2S_2O_7 \longrightarrow K_2SO_4 + SO_3$$

The Ammonium Ion and Its Salts

33.13   The Ammonium Ion

Salts of the ammonium ion in general exhibit physical properties like salts of the potassium ion, the two ions being of nearly the same size and having the same charge. The resemblance of ammonium and potassium salts is especially noticeable with respect to the formation of slightly soluble salts. There are two notable exceptions to the similarity between the salts of the ammonium and potassium ions: (1) The ammonium ion undergoes a slight hydrolysis whereas the potassium ion does not,

$$NH_4^+ + H_2O \rightleftharpoons NH_3 + H_3O^+$$

and (2) ammonium salts decompose when heated whereas the potassium salts melt.

$$NH_4Cl(s) \longrightarrow NH_3(g) + HCl(g) \qquad \Delta H^\circ = 176.0 \text{ kJ}$$

$$NH_4NO_3(s) \longrightarrow N_2O(g) + 2H_2O(g) \qquad \Delta H^\circ = -36.0 \text{ kJ}$$

$$NH_4NO_2(s) \longrightarrow N_2(g) + 2H_2O(g) \qquad \Delta H^\circ = -227.2 \text{ kJ}$$

$$(NH_4)_2Cr_2O_7(s) \longrightarrow N_2(g) + 4H_2O(g) + Cr_2O_3(s) \qquad \Delta H^\circ = -300 \text{ kJ}$$

The ammonium group, NH_4, thus far has not been isolated as a neutral species; it is always found as a positively charged ion, NH_4^+, in combination with a negative ion. When attempts have been made to isolate the neutral NH_4 unit, ammonia and hydrogen have always resulted.

33.14   Ammonium Chloride

Ammonium chloride, NH_4Cl, is known in commerce as **sal ammoniac.** It is produced by the reaction of ammonia with hydrochloric acid and is purified by sublimation. Ammonium chloride decomposes at 350° into ammonia and hydrogen chloride.

$$NH_4Cl(s) \Longrightarrow NH_3(g) + HCl(g) \qquad \Delta H^\circ = 176.0 \text{ kJ}$$

For this reason it is used as a flux in soldering, because the hydrogen chloride that is formed when the salt is heated reacts with the films of metal oxides, converting them into chlorides that are either fusible or volatile, thus cleansing the metal surfaces to be joined by the solder. Ammonia and hydrogen chloride gases in the air of chemical laboratories combine to form the familiar white deposits of ammonium chloride so commonly seen on laboratory glassware and windowpanes.

Much of the ammonium chloride produced is used in the manufacture of dry cells (Section 20.16). It is also used in medicine, in dyeing, in calico printing, as a laboratory reagent, and as a fertilizer.

33.15   Ammonium Nitrate

Ammonia reacts with nitric acid to form **ammonium nitrate,** NH_4NO_3, a white crystalline salt. Its production in the United States in 1978 was 7,207,000 tons ranking it twelfth among all chemicals. When heated to 166° ammonium nitrate fuses and decomposes smoothly, with the formation of nitrous oxide and water. When it detonates, which sometimes happens simply on being heated, it decomposes with explosive violence. The enormous explosion that destroyed a large part of Texas City and took 576 lives in April 1947 was due to the decomposition of ammonium nitrate that was being loaded onto a ship in the harbor. Ammonium nitrate has been used for many years as an ingredient of explosives for warfare and mining. Its principal use, however, is as a fertilizer.

33.16   Ammonium Carbonates

Ammonium hydrogen carbonate, NH_4HCO_3, is prepared by evaporating a solution made by treating an aqueous solution of ammonia with an excess of carbon dioxide. When heated, the white crystalline salt decomposes rapidly.

$$NH_4HCO_3 \Longrightarrow NH_3 + H_2O + CO_2$$

Even at ordinary temperatures, ammonium hydrogen carbonate has a faint odor of ammonia. By treating a solution of the hydrogen carbonate with an excess of ammonia, the **normal carbonate** results.

$$HCO_3^- + NH_3 \rightleftharpoons NH_4^+ + CO_3^{2-}$$

When exposed to the air, $(NH_4)_2CO_3$ gives off ammonia more readily than does NH_4HCO_3 and thus is useful in the form of "smelling salts" since ammonia is an effective heart stimulant.

33.17   Other Ammonium Salts

Large quantities of **ammonium sulfate**, $(NH_4)_2SO_4$, are produced when ammonia is absorbed in sulfuric acid. Its chief use is as a fertilizer to supply nitrogen to the soil.

When an aqueous solution of ammonia is saturated with hydrogen sulfide, **ammonium hydrogen sulfide**, NH_4HS, is formed. If the resulting solution is then treated with an equivalent amount of the aqueous ammonia, the **normal sulfide**, $(NH_4)_2S$, is formed.

Questions

1. Why do the alkali metals not occur in nature?
2. Correlate the positive oxidation states of the alkali metals with their atomic structures.
3. Why should metallic sodium never be handled with the fingers?
4. Why is sodium carbonate or calcium chloride added to the electrolye in the electrolysis of fused sodium chlorides?
5. By analogy with its reaction with water, suggest a chemical formula for the product of reaction between sodium metal and ethyl alcohol.
6. Why must the chlorine and sodium hydroxide be kept separate in the manufacture of sodium hydroxide by the electrolysis of aqueous sodium chloride?
7. Outline the chemistry of the Solvay process for the production of sodium carbonate.
8. Why cannot potassium hydrogen carbonate be manufactured by the Solvay process?
9. Compare the solubilities of potassium and ammonium salts.
10. Give two examples of slightly soluble sodium salts.
11. What is the principal reaction involved when baking powder acts as a leavening agent?
12. What evidence is available to show that hydrogen is present as the negative hydride ion in sodium hydride?
13. Cite evidence that the hydride and peroxide ions are strongly basic.
14. Why is a consideration of the chemistry of the ammonium ion and its salts introduced along with a discussion of the alkali metals and their ions?
15. When potassium ion is reduced, potassium metal results. What is (are) the product(s) when ammonium ion is reduced?
16. Write equations for the thermal decomposition of the following salts: NH_4Cl, NH_4NO_3, NH_4NO_2, and $(NH_4)_2Cr_2O_7$.
17. Suggest in terms of oxidation states of nitrogen why ammonium nitrate might explode.
18. Write chemical equations describing the conversion of ammonium ion to nitride ion; of nitride ion to ammonium ion.

Problems

1. "Normal saline solution," a solution of sodium chloride in water used for washing blood cells, contains 0.85 g of NaCl in 0.100 ℓ of solution. What is the sodium ion concentration in "normal saline solution"?   *Ans. 0.15 M*
2. What volume of hydrogen measured at 25°C and 735 torr would be produced by the passage of a current of 0.674 amperes through an aqueous solution of sodium chloride for a period of 2.00 hours?   *Ans. 0.636 ℓ*
3. How much anhydrous sodium carbonate contains the same number of moles of sodium as 100 grams of $Na_2CO_3 \cdot 10H_2O$?   *Ans. 37.0 g*
4. A 25.00-ml sample of KOH solution is exactly neutralized with 35.27 ml of 0.1062 *M* HCl. What is the concentration of the KOH solution?   *Ans. 0.1498 M*

References

"Sodium Hydroxide" (Staff), *Chem. and Eng. News*, February 6, 1979, p. 10.

"Caustic Soda" (Staff), *Chem. and Eng. News*, February 26, 1979, p. 12.

"Soda Ash" (Staff), *Chem. and Eng. News*, February 26, 1979, p. 14.

"Hydrolysis of Sodium Carbonate," F. S. Nakayama, *J. Chem. Educ.*, **47**, 67 (1970).

"Industrial Emergence of Sodium and Aluminum," M. Schofield, *Chem. in Britain*, **4**, 98 (1968).

"The Social Influence of Salt," M. R. Bloch, *Sci. American*, July 1963, p. 89.

"Alkali Metal—Water Reactions," M. M. Markowitz, *J. Chem. Educ.*, **40**, 633 (1963).

"Alkali Metal Nitrides," G. L. Moody and J. D. R. Thomas, *J. Chem. Educ.*, **43**, 205 (1966).

"X-Ray Crystallography Experiment: Powder Patterns for Alkali Halides," F. P. Boer and T. H. Jordan, *J. Chem. Educ.*, **42**, 76 (1965).

"Chemical Bonds—X-Rays and the Alkali Metal Chlorides," N. Booth, *Educ. in Chemistry*, **1**, 66 (1964).

"Lithium," H. Gilman and J. J. Eisch, *Sci. American*, January 1963, p. 89.

"The Lithium Story," J. Webb, *J. Chem. Educ.*, **53**, 291 (1976).

"Lithium: Will Short Supply Constrain Energy Technology?" A. L. Hammond, *Science*, **191**, 1037 (1976).

"Lithium: Effects on Subjective Functioning," D. R. Jasinski *et al.*, *Science*, **195**, 583 (1977).

"A Contribution to the History of Francium," J. A. Cranston, *Chem. in Britain*, **4**, 66 (1968).

"Lithium and Mental Health," M. T. Doig III, M. G. Heyl, and D. F. Martin, *J. Chem. Educ.*, **50**, 343 (1973).

"Life's Essential Elements," D. R. Williams, *Educ. in Chemistry*, **10**, 56 (1973).

"How Productive Is the Sea?" I. Morris, *Chem. in Britain*, **10**, 198 (1974).

"Salt Solution" (Ocean as a source of drugs, food, salt, and water), K. Jacques, *Chem. in Britain*, **11**, 12 (1975).

"Odor Generation in the Kraft Process," M. A. Karnofski, *J. Chem. Educ.*, **52**, 490 (1975).

"Anions of the Alkali Metals," J. L. Dye, *Sci. American*, July 1977, p. 92.

"The Absorption Spectra of Alkali Metal Vapors," R. A. Ashby, *J. Chem. Educ.*, **55**, 500 (1978).

34

The Alkaline Earth Metals-Group IIA

34.1 Periodic Relationships of the Alkaline Earth Metals

Each of the elements of Periodic Group IIA (Be, Mg, Ca, Sr, Ba, and Ra) has two electrons in its outer shell. Each exhibits a single oxidation state, +2. They are all very reactive metals, the reactivity increasing with increasing atomic number. The members of the B subgroup (Zn, Cd, and Hg), on the other hand, are heavier than the corresponding A subgroup elements of the same series, and the reactivities of the B subgroup elements decrease with increasing atomic number. Beryllium, the first member of Group IIA, resembles aluminum, the second member of Group IIIA of the Periodic Table, in its chemical behavior. Beryllium and magnesium are somewhat different in chemical properties from the other elements of Group IIA. For example, the hydroxides of beryllium and magnesium are nearly insoluble in water and are easily decomposed by heat into water and metallic oxides. The hydroxide of beryllium, the smallest and most electronegative element of the group, is amphoteric, whereas the hydroxides of barium, strontium, and calcium are strong bases and their hydrated ions do not hydrolyze appreciably. Beryllium ions form stable complex ions, but with a few exceptions the ions of other Group IIA elements do not. In contrast to the salts of the alkali metals (Chapter 33), many of the common salts of the alkaline earth metals are insoluble in water. Most of the salts that the alkaline earth metals form with weak or moderately strong acids are sparingly soluble in water—e.g., the sulfates, phosphates, oxalates, and carbonates. Many are quite soluble, however, especially the cyanides, sulfides,

TABLE 34-1 Important Physical Properties of the Alkaline Earth Metals

	Atomic Number	Atomic Weight	Atomic Radius, Å	Ionic Radius, Å	Density at 20°	Melting Point, °C	Boiling Point, °C
Beryllium	4	9.01218	1.11	0.31	1.86	1283	1500
Magnesium	12	24.305	1.60	0.65	1.74	650	1120
Calcium	20	40.08	1.97	0.99	1.54	850	1490
Strontium	38	87.62	2.15	1.13	2.60	770	1384
Barium	56	137.33	2.17	1.35	3.5	704	1638
Radium	88	226.0254	2.20	1.52	about 5	700	1500

and acetates. The solubility of the hydroxides of these metals increases with increasing atomic number, whereas the solubility of the carbonates and sulfates decreases in this order.

Some of the important physical properties of the alkaline earth metals are given in Table 34-1.

Magnesium

34.2   Occurrence of Magnesium

Magnesium never occurs free in nature because of its reactivity, but, in combination, it is abundant and widely distributed. The chloride and sulfate of magnesium are readily soluble in water and, consequently, are found in ground waters, to which they give noncarbonate hardness (Section 12.12). Seawater contains both the chloride and the sulfate, and compounds of magnesium concentrate in the mother liquor of seawater and underground brines from which sodium chloride has been crystallized. Typical magnesium minerals are **carnallite** ($KCl \cdot MgCl_2 \cdot 6H_2O$), **magnesite** ($MgCO_3$), **dolomite** ($MgCO_3 \cdot CaCO_3$), **asbestos** ($H_4Mg_3Si_2O_9$), **meershaum** ($Mg_2Si_3O_8 \cdot 2H_2O$), **talc** or **soapstone** [$Mg_3Si_4O_{10}(OH)_2$], and **brucite** [$Mg(OH)_2$].

34.3   Preparation of Magnesium

Magnesium metal is obtained from several different sources and prepared by several different methods.

▶ **1. FROM UNDERGROUND BRINES.** Magnesium chloride is obtained from the underground brines of Michigan, which contain about 3% magnesium chloride, 9% calcium chloride, 14% sodium chloride, and 0.1% bromine as bromide ion. The bromine is extracted first (Section 19.1), and the brine is then treated with a suspension of magnesium hydroxide to precipitate the hydroxides of iron and certain other metals, after which the filtrate is evaporated to crystallize out the sodium chloride. The magnesium and calcium chlorides are separated by fractional crystallization. Crystalline magnesium chloride hexahydrate, $MgCl_2 \cdot 6H_2O$, is then completely dehydrated in an atmosphere of hydrogen chloride. The hydrogen chloride prevents the formation of basic magnesium salts by hydrolysis

(Section 34.6). Electrolysis of the fused chloride produces magnesium metal and chlorine. Sodium chloride is usually added to lower the melting point of the electrolyte and increase the electrical conductivity. The product has a purity of 99.9%.

▶ **2. FROM SEAWATER.** Seawater now serves as an important and inexhaustible source of magnesium. Nearly 6 million tons of magnesium, as the chloride and sulfate, are contained in each cubic mile of seawater. The raw materials for the process are seawater, oyster shells ($CaCO_3$) from shallow waters along the coast, salt from salt domes, fresh water, and natural gas. Nearly 800 tons of seawater must be processed to obtain one ton of magnesium metal. Oyster shells are calcined to produce lime, which is slaked by adding water. In some operations, the same seawater that was used for the recovery of bromine (Section 19.1) is treated with slaked lime to precipitate magnesium hydroxide. After filtration the magnesium hydroxide may be treated with hydrochloric acid to produce the chloride. The magnesium chloride is crystallized from solution by evaporation. The crystallized salt is then partially dehydrated by heating. The resulting magnesium chloride, which has a composition corresponding to $MgCl_2 \cdot 1\frac{1}{2}H_2O$, is electrolyzed to produce magnesium and chlorine. The by-product chlorine is used in the burning of natural gas to produce the hydrogen chloride used in the process. In some seawater operations, magnesium carbonate and magnesium oxide are produced instead of magnesium metal.

34.4 Properties and Uses of Magnesium

Magnesium is a silvery white metal that is malleable and ductile at high temperatures. It is the lightest of the widely used structural metals. Although it is very active, magnesium is not readily attacked by air or by water, due to the formation of a protective basic carbonate film on its surface. Magnesium is soluble in acids, including carbonic acid, evolving hydrogen.

$$Mg + H_2CO_3 \longrightarrow MgCO_3(s) + H_2(g)$$
$$MgCO_3 + H_2CO_3 \longrightarrow Mg^{2+} + 2HCO_3^-$$

It is also attacked by the alkali metal hydrogen carbonates and by various salts, which give an acid reaction by hydrolysis.

Magnesium decomposes boiling water very slowly, but it rapidly reduces the hydrogen in steam, forming magnesium oxide and hydrogen. The affinity of magnesium for oxygen is so great that it will react when heated in an atmosphere of carbon dioxide, reducing the carbon to elemental carbon.

$$2Mg + CO_2 \longrightarrow 2MgO + C$$

The great reducing power of hot magnesium is utilized in separating many metals and nonmetals, such as silicon and boron, from their oxides.

The metal will unite with most nonmetals. The brilliant white light emitted from burning magnesium makes it useful in flashlight powders, military flares, and incendiary bombs. When magnesium burns in air, both the oxide, MgO, and the nitride, Mg_3N_2, are formed.

Most of the magnesium produced commercially is used in making lightweight alloys, the most important of which are those of aluminum and zinc. **Magnalium** (1–15% Mg, 0–1.75% Cu, and the remainder Al) is lighter, harder, stronger, and more easily machined than pure aluminum. Other important magnesium alloys include **duralumin** (0.5% Mg, 0.5% Mn, 3.5–5.5% Cu, remainder Al), and **Dowmetal** (8.5% Al, 0.15% Mn, 2.0% Cu, 1.0% Cd, 0.5% Zn, 87.85% Mg).

34.5   Magnesium Oxide and Hydroxide

The **oxide** of magnesium, MgO, is sometimes called **magnesia.** It is formed commercially by heating **magnesite,** $MgCO_3$, to 600–800°C, which drives off the carbon dioxide from most of the magnesium carbonate.

$$MgCO_3 \xrightarrow{\triangle} MgO + CO_2(g)$$

The product is a light, fluffy powder that still contains a small percentage of magnesium carbonate. It reacts slowly with water, forming magnesium hydroxide, and it is soluble in aqueous carbon dioxide, forming magnesium hydrogen carbonate, a constituent of hard water. On the other hand, when magnesite is heated to 1400°C or above and ignited, the product contains no magnesium carbonate and is a powder that is much more dense than the light form of the oxide. This pure form of magnesium oxide melts at 2800°C; it is used in making fire brick and crucibles, as a lining in furnaces, and in heat insulation.

Since magnesium oxide does not slake readily, the hydroxide is best prepared by treating a solution of a magnesium salt with an alkali hydroxide. The hydroxide is slightly soluble in water, but readily soluble in solutions of ammonium salts because of the acidity of the NH_4^+.

$$Mg(OH)_2 + 2NH_4^+ \longrightarrow Mg^{2+} + 2NH_3 + 2H_2O$$

A suspension of magnesium hydroxide in water is called **milk of magnesia** and is used as a medicine to correct hyperacidity of the stomach.

34.6   Other Magnesium Compounds

The **normal carbonate,** $MgCO_3$, is found in nature as the mineral magnesite. A basic carbonate of approximately the composition $3MgCO_3 \cdot Mg(OH)_2 \cdot 3H_2O$ is known commercially as **magnesia alba,** and is used as a dental abrasive, as a medicine, as a cosmetic, and as a silver polish. **Magnesium hydrogen carbonate,** $Mg(HCO_3)_2$, is a constituent of many hard waters.

Magnesium chloride crystallizes from aqueous solutions as the hydrated salt, $MgCl_2 \cdot 6H_2O$. The hexahydrate is deliquescent, becoming moist in damp air. When heated, the salt undergoes hydrolysis and forms the oxide, hydrogen chloride, and water.

$$MgCl_2 \cdot 6H_2O \longrightarrow MgO + 2HCl(g) + 5H_2O$$

The anhydrous salt may be obtained by heating the hydrate in a current of hydrogen chloride, by burning magnesium in chlorine, or by carefully heating the double salt $NH_4Cl \cdot MgCl_2 \cdot 6H_2O$, in which case the water of crystallization is driven off first and then the ammonium chloride is volatilized off and recondensed as a very pure white powder.

Magnesium sulfate is found as the minerals **kieserite,** $MgSO_4 \cdot H_2O$, and **epsomite,** $MgSO_4 \cdot 7H_2O$. The heptahydrate in pure form is familiar as **Epsom salts.** It is used in medicine as a purgative, particularly in veterinary practice, in weighting cotton and silk, in powders for polishing, and in heat insulation.

Magnesium ammonium phosphate, $MgNH_4PO_4$, is a slightly soluble crystalline salt, which is formed whenever a soluble phosphate is added to a solution containing magnesium, ammonium, and hydroxide ions.

Anhydrous magnesium perchlorate, $Mg(ClO_4)_2$, is a highly efficient drying agent called **Anhydrone.** It rapidly absorbs up to 35% of its weight of water. The anhydrous salt is easily regenerated by heating the hydrate.

Several silicates of magnesium are of commercial importance. The mineral known as **talc,** or **soapstone,** is a hydrated magnesium silicate that feels greasy to the touch. It can be sawed and turned on a lathe, so is useful in the fabrication of tables, sinks, switchboards, and window sills. **Asbestos** is a calcium-magnesium silicate with a fibrous structure (Section 27.10), from which incombustible fabrics of considerable strength and durability can be made. Such materials have been widely used in making automobile brake linings, paper, drop curtains for theaters, cardboard, flooring, roofing, and covering for heating pipes and boilers, but their use has been curtailed due to the tendency for asbestos dust to induce lung cancers when inhaled over a period of time.

Calcium, Strontium, and Barium

34.7   Occurrence of Calcium, Strontium, and Barium

Because of their high chemical activity, these metals never occur free in nature. Their most abundant native compounds are the carbonates and the sulfates.

▶ **1. CALCIUM.** Calcium is the fifth most abundant element in the earth's crust and is widely distributed in nature. The common rock known as **limestone** is an impure calcium carbonate. Other natural forms of calcium carbonate include **calcite, aragonite, chalk, marl, marble, marine shells,** and **pearls. Dolomitic limestone** (dolomite) is $CaCO_3 \cdot MgCO_3$. Calcium sulfate occurs as **anhydrite,** $CaSO_4$, and **gypsum,** $CaSO_4 \cdot 2H_2O$. Other common minerals include **fluorite,** or **fluorspar,** which is CaF_2, and **apatite,** which may be formulated as $Ca_{10}(PO_4)_6(OH, Cl, F)_2$. **Hydroxyapatite** is the chief constituent of the bones and teeth of animals.

▶ **2. STRONTIUM.** The minerals containing strontium are much rarer than those of calcium. The chief ones are **celestite,** $SrSO_4$, and **strontianite,** $SrCO_3$.

▶ **3. BARIUM.** Barium occurs more abundantly than strontium, but much less so than calcium. Its chief minerals are **barite,** or **heavy spar,** $BaSO_4$, and **witherite,** $BaCO_3$.

34.8   Preparation of Calcium, Strontium, and Barium

The most important methods for producing these metals are those in which their molten salts (generally the chlorides) are electrolyzed. Because of the high chemical activity of these metals, their salts are not readily reduced by other methods. However, in the King process for producing barium a mixture of barium oxide and barium peroxide is reduced with aluminum in a vacuum furnace. A temperature of 950° to 1100°C is required for the reaction to take place, and at the very low pressure (10^{-3} to 10^{-4} mm) that is maintained in the furnace the metallic barium distills off and is collected on a cold surface.

34.9   Properties and Uses of Calcium, Strontium, and Barium

These metals are all silvery white, crystalline, malleable, and ductile. Calcium is harder than lead, strontium is about as hard as lead, and barium is quite soft. Calcium, strontium, and barium are all active metals, their activity increasing in the order named. For example, calcium does not react with oxygen unless it is heated, while strontium oxidizes rapidly when exposed to air, and barium is spontaneously flammable in moist air. When heated, these metals combine with hydrogen, forming hydrides; with sulfur, forming sulfides; with halogens, forming halides; with nitrogen, forming nitrides; and with phosphorus, forming phosphides. Each of the metals displaces hydrogen from water, and the corresponding hydroxide is formed.

Calcium is used as a dehydrating agent for certain organic solvents, as a reducing agent in the production of certain metals, as a scavenger (to remove gases in fused metals) in metallurgy, as a hardening agent for lead used for covering cables and making storage battery grids and bearings, in steelmaking when alloyed with silicon, and for many other purposes.

Elemental strontium is not abundant and has no commercial uses. Barium is used as a degassing agent in the manufacture of vaccum tubes, and alloys of barium and nickel are used in vacuum tubes and spark plugs because of their high thermionic electron emission. It is interesting that Mg^{2+} and Ca^{2+} are not poisonous, whereas Be^{2+} and Ba^{2+} are very much so.

34.10   Oxides of Calcium, Strontium, and Barium

When the carbonates of these metals are heated sufficiently, they undergo a decomposition that yields their respective oxides and carbon dioxide ($MCO_3 \longrightarrow MO + CO_2$). The temperature required for these decompositions increases with increasing basicity of the oxide. The temperatures required to attain equilibrium pressures of carbon dioxide of 1 atm are about 900°C for $CaCO_3$, 1250°C for $SrCO_3$, and 1450°C for $BaCO_3$.

Pure calcium oxide is a white substance that emits an intense light, called **limelight,** when heated to a high temperature. It reacts vigorously and exothermally with water (Section 34.11) and therefore is often employed as a drying agent—for example, in the preparation of anhydrous alcohol and in the drying of ammonia.

In the production of barium oxide, the carbonate is heated with finely divided carbon. The purpose of the carbon is to reduce the carbon dioxide that is produced, and thus to shift the equilibrium $BaCO_3 \rightleftharpoons BaO + CO_2$ to the right. This makes possible the conversion of the carbonate to the oxide at a lower temperature than the 1450°C otherwise necessary. Most of the barium oxide produced is used in the preparation of barium peroxide.

Calcium, strontium, and barium form peroxides of the general formula MO_2. Only barium peroxide is easily formed and important. It is produced when barium oxide is heated in air or oxygen to 500–600°C.

$$2BaO + O_2 \rightleftharpoons 2BaO_2$$

When barium peroxide is heated to 700–800°C the reaction is reversed, and oxygen is liberated. The equilibrium can also be shifted by varying the partial pressure of the oxygen. Barium peroxide is used for bleaching various materials of both plant and animal origin.

34.11   Hydroxides of Calcium, Strontium, and Barium

When the oxides of calcium, strontium, and barium react with water the corresponding hydroxides are formed, and large amounts of heat are evolved.

$$CaO(s) + H_2O(l) \longrightarrow Ca(OH)_2(s) \qquad \Delta H° = -65.3 \text{ kJ}$$
$$SrO(s) + H_2O(l) \longrightarrow Sr(OH)_2(s) \qquad \Delta H° = -83.3 \text{ kJ}$$
$$BaO(s) + H_2O(l) \longrightarrow Ba(OH)_2(s) \qquad \Delta H° = -103 \text{ kJ}$$

The heat of hydration and the reactivity of the oxides toward water both increase with increasing cation size. The solubility of the hydroxides in water increases in the same direction: the solubilities in moles per liter at 20°C are $Ca(OH)_2$, 0.025; $Sr(OH)_2 \cdot 8H_2O$, 0.043; and $Ba(OH)_2 \cdot 8H_2O$, 0.23.

▶ **1. CALCIUM HYDROXIDE.**   Calcium oxide is known by various names, such as **lime, quicklime,** and **unslaked lime,** while the product formed from its reaction with water, $Ca(OH)_2$, is known commercially as **slaked,** or **hydrated, lime.** Calcium hydroxide is a dry white powder that forms a pasty mass with an excess of water. Because of the high heat of reaction of quicklime with water, it should always be stored where it cannot come in contact with water, since the slaking reaction (reaction of the oxide with water to form the hydroxide) generates enough heat to ignite paper or wood. Dry hydrated lime rather than quicklime is usually sold, because it can be shipped and stored in paper bags without risk of fire.

Calcium hydroxide is classed as a strong base, and, because of its activity and cheapness, it is used more extensively in commercial processes than any other base. During 1978, 19,389,000 tons of lime was produced in the United States, making $Ca(OH)_2$ second only to sulfuric acid (Section 21.14) in production. Calcium hydroxide is sparingly soluble in water, only 1.7 grams dissolving in a liter of water at room temperature. A saturated solution of calcium hydroxide is often called **limewater.** A suspension of calcium hydroxide in water is known as **milk of lime.** Even though calcium hydroxide has a limited solubility, milk of lime furnishes a ready supply of hydroxide ions, because the solid continues to dissolve as hydroxide ions are removed from solution during chemical reaction; i.e., the equilibrium shifts to the right.

$$Ca(OH)_2 \rightleftharpoons Ca^{2+} + 2OH^-$$

Solid   Dissolved

Mortar, widely used in the construction industry, is prepared by mixing slaked lime with sand and water. One part of slaked lime is used with about three or four parts of sand, and enough water is added to give the mixture the consistency of paste. It dries and becomes hard (sets) on exposure to the air by adsorbing carbon dioxide and forming crystalline calcium carbonate.

$$Ca(OH)_2 + CO_2 \longrightarrow CaCO_3 + H_2O$$

The calcium carbonate cements together the particles of sand and unchanged calcium hydroxide. Many years are required for the complete conversion of the calcium hydroxide to carbonate, especially in heavy wall construction.

Ordinary **lime plaster** for coating walls and ceilings is similar to mortar, with some binding material such as hair or fiber to help hold it in place. Cement was considered in Section 27.13.

The manufacture of at least 150 important industrial chemicals requires the use of lime. In fact, only five other raw materials are used more frequently than lime (or limestone, from which lime is made); these are salt, coal, sulfur, air, and water.

▶ **2. STRONTIUM HYDROXIDE. Strontium hydroxide** is a rather active base, which is made by treating strontium carbonate with superheated steam.

$$SrCO_3 + H_2O \longrightarrow Sr(OH)_2 + CO_2$$

▶ **3. BARIUM HYDROXIDE. Barium hydroxide** is made by slaking barium oxide and by treatment of the carbonate with superheated steam. Barium hydroxide is the most soluble of the alkaline earth hydroxides, and it crystallizes from solution in the form of the hydrate $Ba(OH)_2 \cdot 8H_2O$.

34.12   Calcium Carbonate

Calcium carbonate occurs in two distinctly different crystal forms, which are known as **calcite** and **aragonite.** The first is common while the second is comparatively rare.

Calcite is the "low-temperature" form of calcium carbonate and results when precipitation occurs below 30°C. Above this temperature calcium carbonate crystallizes in rhombic prisms of aragonite, the high-temperature form. Calcite is known in many varieties, one of which is **onyx.** This crystal form possesses the property of **birefringence,** or **double refraction;** i.e., when a beam of light enters the crystal it is broken up into two beams (Fig. 34-1).

Limestone is by far the most abundant source of calcium carbonate, being second in abundance only to silicate rocks in the earth's crust. Limestone is used

Figure 34-1 Calcite has the unusual property of birefringence, or double refraction. Notice that calcite produces two images of the word CALCITE; the piece of glass (*below*) produces only one image. (*Courtesy of Bausch and Lomb Optical Co.*)

Figure 34-2 Stalactites and stalagmites, formations of calcium carbonate, in Luray Caverns in Virginia. (*Courtesy Caverns of Luray.*)

more extensively than any other building stone in the United States; more than 80% of it is supplied by southern Indiana. Coral reefs are composed largely of calcium carbonate of marine-animal origin. A pearl is built up of concentric layers of calcium carbonate deposited upon a foreign particle, such as a small grain of sand, which has entered the shell of an oyster.

Calcium carbonate is only slightly soluble in pure water. However, it dissolves readily in water that contains dissolved carbon dioxide because of its conversion to the more soluble **calcium hydrogen carbonate,** $Ca(HCO_3)_2$.

$$CaCO_3 + H_2O + CO_2 \rightleftharpoons Ca^{2+} + 2HCO_3^-$$

This reaction is involved in the formation of limestone caves. When water containing carbon dioxide comes in contact with limestone rocks, it dissolves the rocks over which it flows. If water containing $Ca(HCO_3)_2$ finds its way into a cave where the $Ca(HCO_3)_2$ may liberate carbon dioxide, calcium carbonate will again be deposited, according to the equation

$$Ca^{2+} + 2HCO_3^- \longrightarrow CaCO_3(s) + H_2O + CO_2$$

This reaction is the basis for the formation of **stalactites,** which hang from the ceiling of limestone caves, and of **stalagmites,** which grow upward from the floor of the cave where the water has dripped (Fig. 34-2).

Calcium hydrogen carbonate is readily converted to the normal carbonate when its solutions are heated. This reaction is involved in the formation of scale in kettles and boilers, and is the basis for one method of removing carbonate hardness from water (Section 12.12).

34.13 Chlorides of Calcium, Strontium, and Barium

▶ **1. CALCIUM CHLORIDE.** This salt is obtained commercially as a by-product of the Solvay process for the manufacture of sodium carbonate (see Section 33.10). Calcium chloride may be produced in the laboratory by treating the oxide, hydroxide, or carbonate of calcium with hydrochloric acid. The salt crystallizes from water as the hexahydrate, $CaCl_2 \cdot 6H_2O$, which is converted upon heating to the monohydrate, $CaCl_2 \cdot H_2O$. Complete dehydration occurs upon heating to a higher temperature, but some hydrolysis always takes place during the dehydration.

$$CaCl_2 + H_2O \longrightarrow CaO + 2HCl(g)$$

The anhydrous salt and the monohydrate are used extensively as drying agents for gases and liquids. Calcium chloride forms compounds with ammonia, $CaCl_2 \cdot 8NH_3$, and with alcohol, $CaCl_2 \cdot 4C_2H_5OH$, and so cannot be used for drying these substances. Calcium chloride hexahydrate is very soluble in water and with ice makes an excellent freezing mixture, giving temperatures as low as $-55°C$. Solutions of calcium chloride are quite commonly used as the cooling brine in refrigeration plants. Because it is a very deliquescent salt, $CaCl_2 \cdot 6H_2O$ is used to keep dust down on highways and in coal mines. Calcium chloride is frequently used to melt snow and ice on roads and walks.

▶ **2. STRONTIUM CHLORIDE HEXAHYDRATE.** This salt, $SrCl_2 \cdot 6H_2O$, is made by treating the carbonate with hydrochloric acid and evaporating the resulting solution. It is used, as are other salts of strontium, in the production of a red color in fireworks, railway fusees, and military flares.

▶ **3. BARIUM CHLORIDE.** This salt is prepared by the high-temperature reduction of barium sulfate with carbon to form barium sulfide, which is then treated with hydrochloric acid.

$$BaSO_4 + 4C \longrightarrow BaS(s) + 4CO(g)$$
$$BaS(s) + 2H^+(aq) + [2Cl^-(aq)] \longrightarrow Ba^{2+}(aq) + [2Cl^-(aq)] + H_2S(g)$$

Upon evaporation of the solution of the chloride, the dihydrate, $BaCl_2 \cdot 2H_2O$, separates. When a solution of barium ion is required, this salt is usually employed. Like all other soluble barium salts, the chloride is very poisonous.

34.14 Sulfates of Calcium, Strontium, and Barium

The solubilities of the sulfates of Ca, Sr, and Ba in water decrease with increasing atomic weight of the metal; in moles per liter at 20°C they are $CaSO_4 \cdot 2H_2O$, 4.9×10^{-3}; $SrSO_4$, 5.3×10^{-4}; and $BaSO_4$, 1×10^{-5}.

▶ **1. CALCIUM SULFATE. Anhydrous calcium sulfate** occurs in nature as the mineral **anhydrite,** $CaSO_4$, and as the dihydrate, **gypsum,** $CaSO_4 \cdot 2H_2O$. The latter is found in enormous deposits. Even though it is low in solubility, calcium sulfate is largely responsible for the noncarbonate hardness of ground waters (Section 12.12).

When gypsum is heated it loses water, forming the hemihydrate, $(CaSO_4)_2 \cdot H_2O$.

$$2(CaSO_4 \cdot 2H_2O) \longrightarrow (CaSO_4)_2 \cdot H_2O + 3H_2O$$

The hemihydrate is known as **plaster of paris.** When it is ground to a fine powder and mixed with water, it sets by forming small interlocking crystals of gypsum. The setting results in an increase in volume and hence the plaster fits tightly any mold into which it is poured. Plaster of paris is used extensively as a component of plaster for the interiors of buildings. It is also used in making statuary, stucco, wallboard, and casts of various kinds.

Gypsum is also used in making Portland cement (Section 27.13), blackboard crayon (erroneously called "chalk"), plate glass, terra cotta, pottery, and orthopedic and dental plasters. Anhydrous calcium sulfate is sold under the name Drierite as a drying agent for gases and organic liquids.

▶ **2. STRONTIUM SULFATE.** This compound occurs in nature as the mineral **celestite,** which is sometimes found in the form of beautiful crystals in caves, where its faint tinge of blue suggested the name of the mineral.

▶ **3. BARIUM SULFATE.** The mineral **barite,** $BaSO_4$, is usually the source of barium when producing other barium compounds. Since the sulfate is insoluble in acids, it is first converted to the sulfide by a high-temperature reduction with carbon.

The sulfide can then be dissolved in acids such as hydrochloric or nitric, to produce the desired salts.

A mixture of $BaSO_4$ and ZnS is known as **lithopone** and is used as a white paint pigment. It is made by the reaction of aqueous solutions of barium sulfide and zinc sulfate.

$$Ba^{2+} + S^{2-} + Zn^{2+} + SO_4^{2-} \longrightarrow BaSO_4(s) + ZnS(s)$$

Barium sulfate is used in taking x-ray photographs of the intestinal tract because it is opaque to x rays. Even though Ba^{2+} is poisonous, $BaSO_4$ is so slightly soluble that it is nonpoisonous.

Questions

1. Account for the considerable difference in size between the alkaline earth metal atoms and their dipositive ions (see table, inside back cover of text).
2. How do the metals of Group IIA differ from those of Group IA in atomic structure and general properties?
3. List several specific differences in properties between elements in the IIA and IIB families. Discuss the role of electron structure, with respect to these differences in properties.
4. The total of the first and second ionization potentials for beryllium atoms is larger than the corresponding sum for helium. Why does beryllium exhibit extensive chemistry and helium very little?
5. Outline the extraction of magnesium from seawater.
6. Magnesium is an active metal; it is burned in the form of ribbons and filaments to provide flashes of brilliant light. Why is it possible to use magnesium in construction and even for fabrication of cooking grills?
7. Why cannot a magnesium fire be extinguished by either water or carbon dioxide? Suggest a method of extinguishing such a fire.
8. Write a chemical equation describing the dehydrating activity of magnesium perchlorate.
9. Explain the dissolution of magnesium hydroxide in solution of ammonium salts.
10. Why cannot $MgCl_2 \cdot 6H_2O$ be dehydrated by heating in air? How may it be dehydrated?
11. How is metallic calcium produced commercially?
12. Identify each of the following: quicklime, slaked lime, milk of lime, and soda lime.
13. Give the formula for each of the following naturally occurring calcium compounds: limestone, chalk, gypsum, fluorite, and hydroxyapatite.
14. Explain the formation of limestone caves, stalactites, and boiler scale.
15. Why cannot anhydrous calcium chloride be used for drying alcohol or ammonia?
16. Write a chemical equation describing the dehydrating action of calcium metal.
17. Give the chemistry of the preparation and setting of plaster of paris.
18. What is the essential reaction involved in the setting of mortar?
19. Crushed limestone is used in the treatment of soils that are acid. Write equations to explain the reactions involved.
20. On the basis of atomic structure, explain why barium is more reactive than calcium.
21. Give equations for the reaction of barium with each of the following: oxygen, carbon dioxide, hydrogen, nitrogen, sulfur, and water.
22. The barium ion is poisonous. Why, then, can barium sulfate be taken internally with safety in making x-ray photographs of the intestinal tract?
23. Why is carbon added to barium carbonate when the latter is calcined to make barium oxide?
24. Explain how calcium chloride hexahydrate can be used to melt snow and ice on roads and walks and yet when mixed with ice can provide an excellent "freezing mixture," much better than ice alone, for refrigeration plants and making ice cream. Are these two uses compatible?

Problems

1. How much water could be heated from 25°C to 100°C with the heat given off in hydrating 1.0 kg of BaO?   *Ans. 2.1 kg*

2. When $MgNH_4PO_4$ is heated to 1000°C it is converted to $Mg_2P_2O_7$. A 1.203-g sample containing magnesium yielded 0.5275 g of $Mg_2P_2O_7$ after precipitation of $MgNH_4PO_4$ and heating. What per cent magnesium was present in the original sample?

Ans. 9.577%

3. Calculate the per cent magnesium in asbestos.

Ans. 26.31%

4. Calculate the solubility of $SrSO_4$ in grams per 100 ml of water.   *Ans. 9.7×10^{-3} g/100 ml*

5. Calculate the pH of a saturated solution of strontium hydroxide.   *Ans. 12.93*

References

"Lime" (Staff), *Chem. and Eng. News,* April 24, 1978, p. 10.

"Beryllium and Berylliosis," J. Schubert, *Sci. American,* August 1958, p. 27.

"An Historical Note on Beryllium," H. Gilman, *J. Chem. Educ.,* **46,** 276 (1969).

"Calcium and Life," L. V. Heilbrunn, *Sci. American,* June 1951, p. 60.

"How an Eggshell Is Made," T. G. Taylor, *Sci. American,* March 1970, p. 89.

"How Is Muscle Turned on and off?" G. Hoyle, *Sci. American,* April 1970, p. 85.

"Calcitonin," H. Rasmussen and M. M. Pechet, *Sci. American,* October 1970, p. 42.

"Calcium Carbonate Equilibria in Lakes," S. D. Morton and G. F. Lee, *J. Chem. Educ.,* **45,** 511 (1968).

"Calcium Carbonate Equilibria in the Oceans—Ion Pair Formation," S. D. Morton and G. F. Lee, *J. Chem. Educ.,* **45,** 513 (1968).

"Texture and Composition of Bone," D. McConnell and D. W. Foreman, Jr., *Science,* **172,** 972 (1971).

"Life's Essential Elements," D. R. Williams, *Educ. in Chemistry,* **10,** 56 (1973).

"Photosynthesis and Plant Productivity," I. Zelitch, *Chem. and Eng. News,* February 5, 1979, p. 28.

"On the History of Portland Cement," C. Hall, *J. Chem. Educ.,* **53,** 222 (1976).

"The Solidification of Cement," D. D. Double and A. Hellawell, *Sci. American,* July 1977, p. 82.

"Fertilization of Sea Urchins Needs Magnesium Ions in Seawater," K. Sane and H. Mohri, *Science,* **192,** 1339 (1976).

"Calcium Oxalate: Occurrence and Effect in Soils," W. C. Graustein, K. Cromack, and P. Sollins, *Science,* **198,** 1252 (1977).

"Intracellular Calcium," E. Sugaya and M. Onozuka, *Science,* **202,** 1195 (1978).

35

The Coinage Metals—
Group IB

35.1 Periodic Relationships of the Coinage Metals

Copper, silver, and gold are known as **the coinage metals** and have been used from very early times for the manufacture of ornamental objects and coins. They make up Group IB of the Periodic Table and may be classified as transition elements in that they have valence electrons in two shells. Each has one electron in its outermost shell and eighteen in the underlying shell. Because both the alkali metals of Group IA and the coinage metals of Group IB each has one electron in its outermost shell, the members of these two families might be expected to behave similarly. The elements of both families are good conductors of electricity and form series of univalent compounds with analogous formulas (Na_2O, Ag_2O; NaCl, AgCl; and Na_2SO_4, Ag_2SO_4). But beyond this, similarities between the two families are almost entirely lacking. For example, the alkali metals are highly reactive while the coinage metals are quite unreactive. This results, in part, from the fact that the atoms of the IB family are smaller than those of the IA family and hold their valence electrons more closely to the nucleus. Copper, silver, and gold, unlike the alkali metals, can use one or two electrons from the underlying shell in bond formation. Thus copper, silver, and gold each exhibit $+1$, $+2$, and $+3$ oxidation states. However, the most common oxidation state for copper is $+2$; for silver, $+1$; and for gold, $+3$. Many of the ions and compounds of the coinage metals are colored, whereas the ions of the alkali metals are colorless and their only colored compounds are those having a color associated with the anion (e.g., $K_2Cr_2O_7$ and $KMnO_4$). The hydroxides of the alkali metals are soluble and

TABLE 35-1   Some Physical Properties of the Coinage Metals

	Atomic Number	Atomic Weight	Atomic Radius, Å	Ionic (M^+) Radius, Å	Density at 20°	Melting Point, °C	Boiling Point, °C
Copper	29	63.546	1.28	0.96	8.92	1083	2582
Silver	47	107.868	1.44	1.26	10.50	960.8	2193
Gold	79	196.9665	1.44	1.37	19.3	1063	2660

strongly basic whereas those of copper and gold are insoluble and weakly basic. (Silver hydroxide is slightly soluble and strongly basic.) Most compounds of the alkali metals are soluble in water, while the majority of those of silver and gold are insoluble. Finally, the alkali metals show little tendency toward the formation of complex ions, while many stable complexes, such as $[Cu(CN)_2]^-$, $[Ag(CN)_2]^-$, $[Au(CN)_2]^-$, $[Cu(NH_3)_2]^+$, $[Cu(NH_3)_4]^{2+}$, and $[Ag(NH_3)_2]^+$, are known for the coinage metals.

Among the coinage metals, trends in properties are not very regular nor readily interpreted. The chemical activity, elasticity, tensile strength, and specific heat decrease in the order copper, silver, gold; the density increases in this order. All three metals in their $+1$ oxidation states form insoluble chlorides. Compounds of univalent copper are not common and are not as important as those of divalent copper. Gold is considered a less common metal, and its compounds are relatively unimportant. Some physical properties of the coinage metals are given in Table 35-1.

35.2   History of the Coinage Metals

These metals were the first to be used by primitive races, because they are found in the native state, they tarnish slowly, and their appearance is pleasing to the eye. Ornaments of gold have been found in Egyptian tombs constructed in the prehistoric Stone Age. Gold and silver coins were used as a medium of exchange in India and Egypt long before the Christian era.

It is believed that copper was the first metal to be fashioned into utensils, instruments, and weapons; its use for such purposes began sometime during the Stone Age. The practice of alloying copper with tin and the use of the resulting bronze in place of stone introduced the Bronze Age.

The chemical symbol for gold (Au) is derived from the Latin name *aurum*, that for silver (Ag) from *argentum*, and that for copper (Cu) from *cuprum*.

Copper

35.3   Occurrence of Copper

Copper occurs both in the native and combined forms. The most important deposit of native copper in the world is that in the Michigan peninsula near Houghton. The largest single mass of native copper yet found weighs 420 tons, and it is now on display in the Smithsonian Institution in Washington. The most important copper ores are the sulfides, such as **chalcocite,** Cu_2S, and **chalcopyrite,** $CuFeS_2$. Other ores are **cuprite,** Cu_2O; **melaconite,** CuO; and **malachite,** $Cu_2(OH)_2CO_3$.

Copper is found in trace amounts in some plants. It is also found in the brightly colored feathers of certain birds, and in the blood of certain marine animals such as lobsters, oysters, and cuttlefish, where it serves the same oxygen-carrying function as iron in the blood of higher animals (Section 31.5).

35.4   Metallurgy of Copper

In the metallurgy of copper, native copper ores are pulverized, the gangue is washed away, and the copper melted and poured into molds to cool. Oxide and carbonate ores are often leached with sulfuric acid to produce copper(II) sulfate solutions, from which copper metal may be obtained by electrodeposition. High-grade oxide or carbonate ores are reduced by heating them with coke mixed with a suitable flux.

Sulfide ores are usually low grade, containing less than 10% copper. These ores are first concentrated by the flotation process (Section 30.3). The concentrate (or a high-grade sulfide ore if concentration is unnecessary) is then roasted in a furnace, at a temperature below the fusion point of the ore, to drive off the moisture and remove part of the sulfur as sulfur dioxide. The remaining mixture, which is called **calcine** and consists of Cu_2S, FeS, FeO, and SiO_2, is smelted by mixing it with limestone (to serve as a flux) and heating the mixture above its melting point. The reactions taking place during the formation of the slag are

$$CaCO_3 + SiO_2 \longrightarrow CaSiO_3 + CO_2$$
$$FeO + SiO_2 \longrightarrow FeSiO_3$$

The Cu_2S, impure with FeS, that remains after smelting is called **matte.** Reduction of the matte is accomplished in a **converter** by blowing air through the molten material. The air first oxidizes the iron(II) sulfide to iron(II) oxide and sulfur dioxide. Sand is added to form a slag of iron(II) silicate with the iron(II) oxide. After the iron has been removed, the air blast converts part of the Cu_2S to Cu_2O. As soon as copper(I) oxide is formed, it is reduced by the remaining copper(I) sulfide to metallic copper. The equations are

$$2Cu_2S + 3O_2 \longrightarrow 2Cu_2O + 2SO_2$$
$$2Cu_2O + Cu_2S \longrightarrow 6Cu + SO_2$$

The last traces of copper(II) oxide produced by the air blast are removed by reduction with H_2 and CO produced catalytically from methane or natural gas. The copper obtained in this way has a characteristic appearance due to the air blisters that it contains, and is called **blister copper.**

The impure copper is cast into large plates, which are used as anodes in the electrolytic purification of the metal (Section 20.6). Thin sheets of pure copper serve as the cathodes; copper(II) sulfate, acidified with sulfuric acid, serves as the electrolyte. The impure copper passes into solution from the anodes and pure copper plates out on the cathodes as electrolysis proceeds. Gold and silver in the anodes do not oxidize but fall to the bottom of the electrolytic cell as anode mud, along with bits of slag and Cu_2O. Silver, gold, and the platinum group metals are recovered from the anode mud as valuable by-products. The metals more active than copper, such as zinc and iron, are oxidized at the anode and their cations pass into solution, where they remain. The copper deposited on the cathode has an average purity of 99.955%. The value of the precious metals recovered from the anode mud is often sufficient to pay the cost of the electrolytic refining.

35.5   Properties of Copper

Copper is a reddish yellow metal. It is ductile and malleable, so that it is readily fashioned into wire, tubing, and sheets. Copper is the best electrical conductor of the cheaper metals, but when used for this purpose it must be quite pure, since small amounts of impurities reduce its conductivity greatly.

Copper is relatively inactive chemically. In moist air it first turns brown, due to the formation of a very thin, adherent film of oxide or sulfide. Prolonged weathering of copper causes it to become coated with a green film of the basic carbonate, $Cu_2(OH)_2CO_3$, which is similar to the mineral malachite. This compound, or the basic sulfate, is responsible for the green color of copper roofs, gutters, or downspouts that have been weathered for a considerable time. When heated in air, copper oxidizes and forms copper(II) oxide along with some copper(I) oxide. Oxidizing acids, and nonoxidizing acids in the presence of air, convert copper to the corresponding copper(II) salts. Aqueous ammonia in the presence of air will dissolve copper and give a blue solution containing $[Cu(NH_3)_4]^{2+}$. Alkali metal hydroxides do not act upon the metal. Sulfur vapor reacts with hot copper to give both Cu_2S and CuS. Hot copper burns in chlorine to give $CuCl$.

35.6   Uses of Copper

The extent and wide range of the uses of copper cause it to be regarded as second in importance to iron. The chief use of copper is in the production of electrical wiring.

The fact that copper may be electrolytically deposited in thin sheets that are smooth and tough makes it useful in **electrotyping.** An impression of the type is made in wax, and this is rubbed with graphite to make its surface an electrical conductor. The wax impression is used as the cathode in an electrolytic cell containing copper sulfate as the electrolyte and an anode of copper. Copper is plated out on the graphite film until the deposit becomes the thickness of a sheet of paper. The copper sheet is then removed from the wax and strengthened by covering its back surface with lead. Plates prepared in this way were at one time widely used in printing books. Modern printing methods use a photographic process of producing the printing plates.

Copper is also used in the production of a great many alloys. Among the more important ones are **brass** (Cu, 60–82%; Zn, 18–40%), **bronze** (Cu, 70–95%; Zn, 1–25%; Sn, 1–18%), **aluminum bronze** (Cu, 90–98%; Al, 2–10%), and **German silver** (Cu, 50–60%; Zn, 20%; Ni, 20–25%). Other alloys containing copper are bell metal, gun metal, silver coin, sterling silver, gold coin, jewelry gold, and jewelry silver.

35.7   Oxidation States of Copper

Copper forms two principal series of compounds, which are based upon the oxidation states of $+1$ and $+2$, respectively. When the one electron in the outermost shell of the copper atom is involved in bond formation, the oxidation state of $+1$ is exhibited. When an additional electron from the underlying $3d$ subshell is removed, the copper(II) ion results. When two electrons from the underlying shell are used in bonding, copper exhibits an oxidation state of $+3$; copper (III) compounds are rare and relatively unimportant, however. Copper(I) forms a complete series of binary compounds such as the halides and the oxide, but the ternary oxygen salts are few in number and readily decomposed by water. In contrast, copper(II) forms a complete series of oxygen salts, while some of its binary compounds are unstable and break down spontaneously.

35.8   Copper(I) Compounds

Solid salts of copper(I) show a variety of colors, but generally the solutions are colorless.

▶1. COPPER(I) OXIDE. Cu_2O may be prepared as a reddish brown precipitate by boiling copper(I) chloride with sodium hydroxide.

$$2CuCl + 2OH^- \longrightarrow Cu_2O(s) + 2Cl^- + H_2O$$

When basic solutions of copper(II) salts are heated with reducing agents, copper(I) oxide precipitates.

$$2Cu^{2+} + 2OH^- + 2e^- \text{ (reducing agent)} \longrightarrow Cu_2O(s) + H_2O$$

This reaction is the basis for the test for the presence of reducing sugars in the urine in the diagnosis of diabetes. The reagent most often used for this purpose is Benedict's solution, which is a solution of copper(II) sulfate, sodium carbonate, and sodium citrate. The addition of a reducing sugar, such as glucose, causes the precipitation of the reddish brown copper(I) oxide.

▶2. COPPER(I) HYDROXIDE. This compound is formed as a yellow precipitate when a cold solution of copper(I) chloride in hydrochloric acid is treated with sodium hydroxide. The hydroxide is unstable and decomposes into the oxide and water upon heating.

▶3. COPPER(I) CHLORIDE. When copper metal is heated with a solution of copper(II) chloride acidified with hydrochloric acid, an oxidation-reduction reaction takes place with the precipitation of copper(I) chloride.

$$Cu^{2+} + 2Cl^- + Cu \longrightarrow 2CuCl(s)$$

This white crystalline compound is readily soluble in concentrated hydrochloric acid, forming the complex ion $[CuCl_2]^-$.

$$CuCl + Cl^- \rightleftharpoons [CuCl_2]^-$$

The two chlorides of the $[CuCl_2]^-$ ion are bonded to the copper by covalent bonds. Dilution with water brings about reprecipitation of the copper(I) chloride.

Among other stable and insoluble copper(I) binary salts are Cu_2S (black), CuCN (white), CuBr (white), and CuI (white).

35.9   Copper(II) Compounds

The more common compounds of copper are those in which the metal is bivalent. In the solid state the copper(II) compounds may be blue, black, green, yellow, or white. However, dilute solutions of all soluble copper(II) compounds have the blue color of the hydrated copper(II) ion $[Cu(H_2O)_4]^{2+}$.

▶ **1. COPPER(II) OXIDE.** Copper(II) oxide, CuO, is a black, insoluble compound that may be obtained by heating the carbonate or nitrate, or by heating finely divided copper in oxygen.

$$CuCO_3 \longrightarrow CuO + CO_2$$
$$2Cu(NO_3)_2(s) \longrightarrow 2CuO(s) + 4NO_2(g) + O_2(g) \qquad \Delta H° = 420.5 \text{ kJ}$$
$$2Cu(s) + O_2(g) \longrightarrow 2CuO(s) \qquad\qquad\qquad \Delta H° = -157 \text{ kJ}$$

Because of the ease with which the copper can be reduced either to oxidation state +1 or to the metal, hot copper(II) oxide is a good oxidizing agent. As such, it is used in the analysis of organic compounds to determine the percentage of carbon that they contain. Copper(II) oxide oxidizes the carbon to carbon dioxide, which is absorbed by sodium hydroxide, and the gain in weight of the sodium hydroxide is taken as a measure of the quantity of carbon dioxide produced and absorbed. The copper oxide is continuously regenerated by passing a stream of oxygen through the tube in which the organic matter and the copper oxide are reacting.

▶ **2. COPPER(II) HYDROXIDE.** When hydroxide ions are added to cold solutions of copper(II) salts, a bluish green gelatinous precipitate of the hydroxide, $Cu(OH)_2$, is formed. If hot solutions are employed, CuO is obtained. Copper(II) hydroxide is somewhat amphoteric and dissolves slightly in solutions of concentrated alkalies, the cuprate ion, $[Cu(OH)_4]^{2-}$, being formed. The hydroxide is also soluble in aqueous ammonia, giving the **tetraamminecopper(II) hydroxide,** $[Cu(NH_3)_4](OH)_2$, which is a strong base with a deep blue color.

$$Cu(OH)_2(s) + 4NH_3 \longrightarrow [Cu(NH_3)_4]^{2+} + 2OH^-$$

▶ **3. COPPER(II) SULFATE.** This sulfate is the most important of the copper salts. The anhydrous salt is colorless, but when it is crystallized from aqueous solution the blue **pentahydrate,** $[Cu(H_2O)_4]SO_4 \cdot H_2O$, known as **blue vitriol,** is formed. As indicated by the formula, four of the water molecules are coordinated to the copper(II) ion and the fifth is attached to both the sulfate and two of the coordinated water molecules by hyrogen bonds. The structure of the pentahydrate may be represented as shown.

The dehydration of this salt proceeds in steps, yielding successively $CuSO_4 \cdot 3H_2O$ (with loss of the two nonhydrogen-bonded waters), $CuSO_4 \cdot H_2O$ [with loss of the other two water molecules coordinated to the Cu(II) ion], and finally $CuSO_4$ (with loss of the water molecule attached to the sulfate ion).

Copper(II) sulfate is produced commercially by oxidizing the sulfide either directly to the sulfate, or to the oxide which is then converted to the sulfate by reaction with sulfuric acid. In the laboratory the anhydrous sulfate may be prepared by oxidizing copper with hot concentrated sulfuric acid.

$$Cu(s) + 2H_2SO_4(l) \longrightarrow CuSO_4(s) + SO_2(g) + 2H_2O(l) \quad \Delta H^\circ = -11.9 \text{ kJ}$$

Aqueous solutions of copper(II) sulfate are slightly acidic by hydrolysis.

$$[Cu(H_2O)_4]^{2+} + H_2O \rightleftharpoons [Cu(H_2O)_3OH]^+ + H_3O^+$$

or simply, $\qquad Cu^{2+} + H_2O \rightleftharpoons CuOH^+ + H^+$

Copper(II) sulfate is used in the electrolytic refining of copper, in electroplating, in the Daniell cell, in the manufacture of pigments, as a mordant in the textile industry, and in the prevention of algae in reservoirs and swimming pools. Because anhydrous copper(II) sulfate is insoluble in alcohol and ether and readily takes up water, turning blue, this salt is used in detecting water in liquids such as alcohol and ether; it is also used to remove water from these liquids.

▶ **4. COPPER(II) HALIDES.** Anhydrous **copper(II) chloride**, $CuCl_2$, may be prepared as a yellow crystalline salt by direct union of the elements. The blue-green hydrated salt $Cu(H_2O)_2Cl_2$ may be obtained by treating either the carbonate or the hydroxide of copper(II) with hydrochloric acid and evaporating the solution. Concentrated solutions of copper(II) chloride are green, because they contain both the hydrated copper(II) ion $[Cu(H_2O)_4]^{2+}$, which is blue, and the tetrachlorocuprate(II) ion $[CuCl_4]^{2-}$, which is yellow. (The combination of blue and yellow looks green to the eye.) Upon dilution, water molecules replace the chloro groups of the $[CuCl_4]^{2-}$ ion, and the solution becomes blue.

Copper(II) bromide, $CuBr_2$, is a black solid that may be obtained by direct union of the elements or by the action of hydrobromic acid upon the oxide or carbonate.

It is interesting that **copper(II) iodide**, CuI_2, does not exist. When a solution containing the copper(II) ion is treated with an excess of iodide, an oxidation-reduction reaction occurs, with precipitation of **copper(I) iodide** and the formation of elementary iodine.

$$2Cu^{2+} + 4I^- \longrightarrow 2CuI(s) + I_2$$

This reaction is used in the quantitative determination of copper by titration of the liberated iodine with a standard solution of sodium thiosulfate. Once the amount of elemental iodine is known, the amount of copper can be calculated.

▶**5. COPPER(II) SULFIDE.** CuS, black, is precipitated by passing hydrogen sulfide into acid, alkaline, or neutral solutions of copper(II) salts. The sulfide readily dissolves in warm dilute nitric acid with the formation of elemental sulfur and nitric oxide.

$$3CuS + 8H^+ + 2NO_3^- \rightleftharpoons 3Cu^{2+} + 3S(s) + 2NO(g) + 4H_2O$$

Copper(II) sulfide dissolves only to a slight extent in solutions of alkali metal sulfides.

35.10 Complex Copper Salts

Copper possesses the property of forming both complex cations and anions that are quite stable. Copper(II) ions coordinate with four neutral ammonia molecules to form the deep blue **tetraamminecopper(II) ion.**

$$[Cu(H_2O)_4]^{2+} + 4NH_3 \rightleftharpoons [Cu(NH_3)_4]^{2+} + 4H_2O$$

The fact that the ammonia molecules displace the coordinated water molecules indicates the greater stability of the ammine complex. In fact, the ammine complex is so stable that most slightly soluble copper(II) compounds are dissolved by aqueous ammonia. The copper ion is located at the center of a square formed by the four attached groups, whether they are water, $[Cu(H_2O)_4]^{2+}$; or ammonia, $[Cu(NH_3)_4]^{2+}$.

Monovalent copper forms linear complexes with ammonia, $[Cu(NH_3)_2]^+$, with the halides, $[CuX_2]^-$, and with cyanide ions, $[Cu(CN)_2]^-$. In contrast to the copper(II) complexes, most copper(I) complexes are colorless.

Silver

35.11 Occurrence and Metallurgy of Silver

Silver is sometimes found in large nuggets but more frequently in veins and related deposits. It is often found alloyed with gold, copper, or mercury. It occurs combined as the chloride, AgCl **(horn silver),** and as the sulfide, Ag_2S, which is usually mixed with the sulfides of lead, copper, nickel, arsenic, or antimony. Most of the silver of commerce is a by-product of the mining of other metals such as lead and copper.

When lead is obtained from lead sulfide ore, it usually contains some silver. The silver is extracted from the fused lead by the use of zinc, in which silver is about 3000 times more soluble than it is in lead. To accomplish this, the lead is fused and thoroughly mixed with a small quantity of zinc. Lead and zinc are immiscible. Most of the silver leaves the lead and dissolves in the zinc. When mixing is stopped, the zinc rises to the surface of the lead and solidifies. The zinc-silver alloy is removed from the lead and the more volatile zinc is separated from the silver by distillation. This extraction procedure is known as the **Parkes process.**

The extraction of silver from its ores is dependent upon the formation of the complex **dicyanoargentate ion,** $[Ag(CN)_2]^-$. Silver metal and all of its compounds are readily dissolved by alkali cyanides in the presence of air. Representative equations for the hydrometallurgy of silver are

$$4Ag + 8CN^- + O_2 + 2H_2O \longrightarrow 4[Ag(CN)_2]^- + 4OH^-$$

$$2Ag_2S + 8CN^- + O_2 + 2H_2O \longrightarrow 4[Ag(CN)_2]^- + 2S + 4OH^-$$

$$AgCl + 2CN^- \longrightarrow [Ag(CN)_2]^- + Cl^-$$

The silver is precipitated from the cyanide solution upon the addition of either zinc or aluminum.

$$2[Ag(CN)_2]^- + Zn \longrightarrow 2Ag + [Zn(CN)_4]^{2-}$$

Much silver is also obtained from the anode mud formed during the electrolytic refining of copper (Section 35.4).

35.12   Properties of Silver

Silver is a white, lustrous metal and its polished surface is an excellent reflector of light. It is the best conductor of heat and electricity that we have other than gold, which is better but too expensive to use as a conductor. Silver is noted for its ductility and malleability.

Silver is not attacked by oxygen of the air under ordinary conditions but tarnishes quickly in the presence of hydrogen sulfide or upon contact with sulfur-containing food materials such as eggs and mustard. Its reaction with hydrogen sulfide in the presence of air is

$$4Ag(s) + 2H_2S(g) + O_2(g) \longrightarrow 2Ag_2S(s) + 2H_2O(l) \qquad \Delta H^\circ = -595.59 \text{ kJ}$$

Silver tarnish is a thin film of silver sulfide. The halogens react with silver forming the halides, and the metal is soluble in such oxidizing acids as nitric and hot sulfuric.

$$3Ag + 4H^+ + NO_3^- \longrightarrow 3Ag^+ + NO(g) + 2H_2O$$

Because silver lies below hydrogen in the activity series, it is not soluble in nonoxidizing acids such as hydrochloric. Most silver salts are sparingly soluble in water, but the nitrate is quite soluble. The common oxidation state exhibited by silver is $+1$, although the oxides AgO and Ag_2O_3 and some complex compounds containing Ag(II) and Ag(III) are known.

35.13   Uses of Silver

A considerable amount of silver is used for the making of coins, silverware, and ornaments. For most of its uses silver is too soft to wear well, and it is therefore hardened by alloying it with other metals, particularly copper. **Sterling silver** contains 7.5% copper, and jewelry silver contains 20% copper. Large amounts of silver are used in the preparation of dental alloys, photographic films, and mirrors.

35.14   Plating with Silver

The electroplating industry uses a large percentage of the metal produced. The object to be plated with silver is made the cathode in an electrolytic cell containing a solution of sodium dicyanoargentate, $Na[Ag(CN)_2]$, as the electrolyte. The anode is a bar of pure silver, which dissolves to replace the silver ions removed from solution as plating at the cathode proceeds. The electrode reactions are

Cathodic reduction: $\quad [Ag(CN)_2]^- + e^- \longrightarrow Ag + 2CN^-$

Anodic oxidation: $\qquad\quad Ag + 2CN^- \longrightarrow [Ag(CN)_2]^- + e^-$

The film of deposited silver has a flat white appearance, but it can be made to assume a brilliant luster by burnishing.

The characteristic luster can be restored to tarnished silverware by the following application of the activity series. The tarnished silver object is immersed in a solution of baking soda and salt containing a teaspoonful of each to a quart of water held in an aluminum vessel. (A piece of aluminum metal foil in a glass vessel will serve equally well.) The aluminum and silver must be in contact. After the tarnish is removed, the silver object should be thoroughly washed. The chemistry involved may be explained in this way: The silver sulfide of the film of tarnish dissolves somewhat in the solution, forming silver and sulfide ions. The more active aluminum displaces the silver from solution as the metal, causing it to plate out on the silver object, according to the equation

$$3Ag^+ + Al \longrightarrow 3Ag + Al^{3+}$$

The tarnish is removed with little loss of silver; however, the "antiquing" (actually silver sulfide tarnish), which often gives a desirable appearance to the crevices of the design, will be removed also. Silver polish can be used to remove tarnish but with loss of the silver in the tarnish.

Silver mirrors are formed by depositing a thin layer of silver on glass. This is accomplished by reducing an ammoniacal solution of silver nitrate with some mild reducing agent, such as glucose or formaldehyde. The equation for the reaction may be written as

$$[Ag(NH_3)_2]^+ + e^- \text{ (from reducing agent)} \longrightarrow Ag + 2NH_3$$

The film of silver deposited on the glass is washed, dried, and varnished.

35.15   Compounds of Silver

▶ **1. SILVER OXIDE. Silver oxide** is formed when silver is exposed to ozone, or when finely divided silver is heated in oxygen under pressure. The alkalies act upon silver nitrate to give a dark brown amorphous precipitate of silver oxide. Although the oxide is but slightly soluble in water, it dissolves enough to give a distinctly alkaline solution. The equilibrium is given by

$$Ag_2O(s) + H_2O \rightleftharpoons 2Ag^+ + 2OH^-$$

Silver oxide is a convenient reagent, both in inorganic and organic chemistry, for preparing soluble hydroxides from the corresponding halides, because the silver halide formed at the same time may be removed conveniently by filtration. Cesium hydroxide, for example, may be prepared according to the reaction

$$2Cs^+ + 2Cl^- + Ag_2O + H_2O \longrightarrow 2Cs^+ + 2OH^- + 2AgCl(s)$$

This reaction proceeds to the right because AgCl is much less soluble than Ag_2O. Silver oxide dissolves readily in aqueous ammonia to form the strong base, **diamminesilver hydroxide.**

$$Ag_2O + 4NH_3 + H_2O \longrightarrow 2[Ag(NH_3)_2]^+ + 2OH^-$$

When silver oxide is heated at atmospheric pressure in air, it readily gives up its oxygen according to the equation

$$2Ag_2O(s) \longrightarrow 4Ag(s) + O_2(g) \qquad \Delta H^\circ = 62.09 \text{ kJ}$$

▶**2. THE SILVER HALIDES.**   It is of interest that **silver fluoride,** AgF, is very soluble in water, whereas the chloride, bromide, and iodide are relatively insoluble—the higher the atomic weight of the halogen, the lower the solubility of the corresponding silver halide. The insoluble silver halides are formed as curdy precipitates when halide ions are added to solutions of silver salts. The chloride is white, silver bromide is pale yellow, and silver iodide is yellow. Upon exposure to light the halides of silver turn violet at first and finally black; they are decomposed by light into their elements.

$$2AgX + light \longrightarrow 2Ag + X_2 \qquad (X = halogen)$$

Silver chloride is readily soluble in an excess of dilute aqueous ammonia to form the diamminesilver complex.

$$AgCl + 2NH_3 \longrightarrow [Ag(NH_3)_2]^+ + Cl^-$$

▶**3. NITRATE.**   **Silver nitrate,** $AgNO_3$, is the only simple silver salt that is found to be usefully soluble. It is obtained by dissolving silver in nitric acid and evaporating the solution. Organic materials, such as the skin, readily reduce silver nitrate to give free silver, which forms a black stain. Much silver nitrate is used in the production of photographic materials, as a laboratory reagent, and in the manufacture of other silver compounds.

35.16   The Photographic Process (Black and White)

▶**1. THE PHOTOGRAPHIC FILM.**   Photographic films are thin sheets of cellulose acetate coated with a colloidal suspension of small silver halide crystals in gelatin. Pure solutions of silver nitrate, potassium bromide, and potassium iodide are mixed in the presence of the gelatin in the preparation of the photographic emulsion ($Ag^+ + X^- \longrightarrow AgX$). The size of the silver halide particles and the relative amounts of bromide and iodide used determine the sensitivity, or speed, of the film.

▶ **2. EXPOSURE.** The film is exposed by focusing an image upon the light-sensitive photographic emulsion on the film. The silver halide crystals become "activated" when exposed to light, so that they are more readily reduced than before. The chemical change occurring during exposure is not well understood, but the silver halide crystals receiving the most light (from the light areas of the object) are more readily reduced, during development of the film, than those receiving little light (from the dark areas of the object). Exposure causes no visible change in the appearance of the film.

▶ **3. DEVELOPING.** After exposure the film is developed by placing it in an alkaline solution of an organic reducing agent such as either hydroquinone or pyrogallol. The reducing agent acts upon the grains of silver halide with a speed proportional to the intensity of the illumination during exposure and reduces them to metallic silver.

$$AgX(s) + e^- \text{ (reducing agent)} \longrightarrow Ag(s) + X^-$$

▶ **4. FIXING.** After developing, the film is fixed by treating it with a solution of sodium thiosulfate (hypo) to dissolve the unreduced silver halide.

$$AgX(s) + 2S_2O_3^{2-} \rightleftharpoons [Ag(S_2O_3)_2]^{3-} + X^-$$

The metallic silver remaining on the film forms the visible image and is called a **negative,** since the light portions of the original image are now dark and the dark portions of the original are now light.

▶ **5. PRINTING.** The process of printing is essentially the same as that of making a negative, but due to the fact that the sensitive printing paper is illuminated through the negative, the image is once again reversed, and now corresponds to the original image as regards light and dark areas on the print.

The print may be **toned** by replacing part of the silver of the image by gold or platinum. To do this, the print is treated with solutions of $Na[AuCl_4]$ or $K_2[PtCl_6]$. The more active silver displaces these noble metals from their salts to give a thin deposit of gold (red tone) or platinum (dark gray).

Gold

35.17   Occurrence of Gold

Gold is found chiefly as the metal, occasionally in nuggets scattered through gravel but more frequently as small particles in veins of quartz or in sands formed by the disintegration of gold-bearing rock. Native gold always contains silver and some platinum metals (Ru, Rh, Pd, Os, Ir, and Pt). In the combined state, gold is found in a few minerals such as the telluride, $AuTe_2$, and as the double telluride, $AuAgTe_4$, which is called **silvanite.**

35.18   Metallurgy of Gold

For centuries gold was obtained by processes known as **panning, sluicing,** or **placer mining,** in which the gold-bearing sand and gravel was washed with water to separate the lighter particles from the heavier gold particles. Nowadays, powerful streams of water are thrown against the deposits, which are washed into the sluices (long troughs). This process is called **hydraulic mining.**

In the amalgamation process for the extraction of gold, finely powdered ore is washed over plates of copper coated with mercury, in which about half of the gold dissolves. At intervals, the amalgam is scraped off, the mercury is removed by distillation, and the gold residue refined.

The cyanide process for gold is similar to that used for obtaining silver from its ores (Section 35.11); in fact, the two metals are usually extracted together. Oxygen of the air is essential to the reaction. The gold is recovered from the solution by displacement, using zinc.

$$4Au + 8CN^- + 2H_2O + O_2 \longrightarrow 4[Au(CN)_2]^- + 4OH^-$$

$$Zn + 2[Au(CN)_2]^- \longrightarrow 2Au + [Zn(CN)_4]^{2-}$$

Gold obtained by any of the above processes is impure with silver; and other metals, such as lead, copper, and zinc may be present. The refining, or "parting," of the impure gold may be accomplished by electrolysis or by dissolving the impurities with sulfuric or nitric acid.

35.19   Properties of Gold

Gold is a yellow, soft metal, and is the most malleable and ductile of all metals. Gold foil can be made by hammering the metal into sheets so thin that 300,000 of them would be required to make a pile one inch thick; one gram of gold can be drawn into a wire more than a mile and a half in length. Pure gold is too soft to be used for jewelry and coinage; for such purposes it is always alloyed with copper, silver, or some other metal. The purity of gold is expressed in **carats,** a designation that indicates the number of parts by weight of gold in 24 parts of alloy. Thus, 24-carat gold is the pure metal, while a 10-carat alloy is $\frac{10}{24}$ gold by weight. Red or yellow gold alloys contain copper, whereas white gold alloys are made using palladium, nickel, or zinc.

Gold is a very inactive metal. It neither combines directly with oxygen nor corrodes in the atmosphere. The metal is not affected by any single common acid nor by alkalies. It is acted upon, however, by selenic acid, H_2SeO_4, and it dissolves readily in aqua regia.

$$Au + 6H^+ + 3NO_3^- + 4Cl^- \longrightarrow [AuCl_4]^- + 3NO_2(g) + 3H_2O$$

It is also dissolved by a solution of chlorine, and by cyanide in the presence of air.

Gold is a transition metal with one valence electron in its outermost shell. It forms principally compounds in which it is univalent or trivalent. When gold dissolves in aqua regia and the solution is evaporated, yellow cystals of **chlorauric acid,** $HAuCl_4 \cdot 4H_2O$, are formed. When this complex compound is heated, hydrogen chloride is evolved, and red crystalline **gold(III) chloride,** $AuCl_3$, remains. When this compound is heated to 175° it is changed to **gold(I) chloride,** $AuCl$, and at higher temperatures the metal is obtained. The thermal instability of these compounds is typical, for all compounds of gold are decomposed by heat. Gold(I) chloride undergoes an auto-oxidation-reduction reaction in water, forming gold(III) chloride and the metal. Gold forms two oxides, Au_2O and Au_2O_3, and the corresponding hydroxides, $AuOH$, which is a weak base, and $Au(OH)_3$. The trihydroxide is a weak acid capable of reacting with strong bases to form **aurates,** such as $NaAuO_2$. Potassium cyanide acts upon gold(I) and gold(III) compounds giving the complex soluble salts $Na[Au(CN)_2]$ and $Na[Au(CN)_4]$, which are important in the extraction of gold from its ores and in gold plating operations.

A sensitive test for gold consists of the treatment of a dilute solution of gold(III) chloride with tin(II) chloride. Colloidal gold with a deep purple color, known as the **purple of Cassius,** is formed by the reducing action of the divalent tin.

Questions

1. Describe the electrolytic refining process for metallic copper.
2. Copper plate makes up the surface of the Statue of Liberty in New York harbor. Why does it present a green appearance?
3. In terms of electronic structure, explain the existence of copper in three oxidation states.
4. Why must copper be quite free of impurities when used as an electrical conductor?
5. Explain the use of copper(II) sulfate in urine analysis.
6. Compare the stability of the copper(I) halides to that of copper(II) halides.
7. A white precipitate forms when copper metal is added to a solution of copper(II) chloride and hydrochloric acid. The white precipitate dissolves in excess concentrated hydrochloric acid. Dilution with water results again in a white precipitate. Write equations for the reactions that are involved.
8. Why are most slightly soluble copper salts readily dissolved by aqueous ammonia?
9. Write an equation to explain the acidic nature of copper(II) sulfate solutions.
10. Sketch the structure of $Cu(H_2O)_4SO_4 \cdot H_2O$.
11. A solution made by dissolving anhydrous copper sulfate in pure water has a pH that is less than 7.0. Explain.
12. Would you expect CuS to dissolve in a 1 M ammonia solution?
13. What properties of silver have made it valuable as a coinage metal down through the ages?

14. Explain the tarnishing of silver in the presence of materials containing sulfur.
15. Compare the solubilities of the silver halides in water and in aqueous ammonia.
16. Describe the Parkes process for extracting silver from lead.
17. Why does silver dissolve in nitric acid but not in hydrochloric acid?
18. Write equations for the electrode reactions when $Na[Ag(CN)_2]$ is used as the electrolyte in silver plating.
19. Outline the chemistry of the photographic process.
20. Dilute sodium cyanide solution is slowly dripped into a slowly stirred silver nitrate solution. A white precipitate forms temporarily but dissolves as the sodium cyanide addition continues. Use chemical equations to explain these observations.
21. Account for the variable oxidation state of gold in terms of the electronic structure of its atoms.
22. Would you expect salts of the gold(I) ion, Au^+, to be colored? Explain.
23. Write balanced equations for the following changes, which occur during the recovery of gold from an ore:

$$Au \longrightarrow [Au(CN)_2]^- \longrightarrow Au$$

24. Write a chemical equation describing the preparation of "purple of Cassius."
25. Show several ways in which coordination compounds play an important part in the chemistry of copper, silver, and gold.

Problems

1. How many grams of $CuCl_2$ contain the same mass of copper as 100 g of CuCl? *Ans. 136 g*

2. A solid 1.008-g sample of a silver-copper alloy is dissolved and treated with excess iodide ion. The liberated I_2 is titrated with a $0.1052\ M\ S_2O_3^{2-}$ solution. If 14.92 ml of this solution is required, what is the % Cu in the alloy? *Ans. 9.894%*

3. How many gold atoms are there in 5.0 g of 20 carat gold? *Ans. 1.3×10^{22} atoms*

4. The formation constant of $[Ag(CN)_2]^-$ is 1.0×10^{20}. What concentration of Ag^+ will be in equilibrium if 1.0 g of Ag is oxidized and put into 1.0 ℓ of solution with $1.0 \times 10^{-1}\ M\ CN^-$? *Ans. $1.4 \times 10^{-20}\ M$*

5. A solid 1.4820-g sample of a pure alkali metal chloride is dissolved in water and treated with excess silver nitrate. The resulting precipitate is filtered, dried, and found to weigh 2.849 g. What per cent chloride is in the original chloride compound? What is the identity of the salt? *Ans. 47.55%; KCl*

6. The formation constant of $[Cu(NH_3)_4]^{2+}$ is 1.2×10^{12}. What concentration of Cu^{2+} will be in equilibrium if 1.0 g of Cu is oxidized and put into 1.0 ℓ of solution with $0.25\ M\ NH_3$? *Ans. $1.1 \times 10^{-11}\ M$*

References

"Ailing Copper Industry Seeks New Technology" (Staff), *Chem. and Eng. News*, March 13, 1978, p. 30.

"The Biochemistry of Copper," E. Frieden, *Sci. American*, May 1968, p. 102.

"Hemocyanin: The Copper Blood," N. M. Senozan, *J. Chem. Educ.*, **53**, 684 (1976).

"Cryoscopic Measurement of the Coordination Number of Copper(II)," H. C. de Lorenzo, J. A. C. Frugoni, and V. Lopez, *J. Chem. Educ.*, **46**, 113 (1969).

"Developments in Copper Smelting," E. G. West, *Chem. in Britain*, **5**, 199 (1969).

"The Gold Content of the Sea," G. L. Putnam, *J. Chem. Educ.*, **30**, 576 (1953).

"Photographic Development," T. H. James, *Sci. American*, November 1952, p. 30.

Discovery of the Elements, 7th ed., M. E. Weeks and H. M. Leicester, published by the Journal of Chemical Education, Easton, Pa., 1968, pp. 3–29.

"A Potentiometric Copper Assay in Normal and Copper-Poisoned Humans," I. A. Matheson and D. R. Williams, *J. Chem. Educ.*, **50**, 345 (1973).

"Catalytic Determination of Trace Copper Using Kinetic Analysis," J. T. Bartis and J. R. Wiesenfield, *J. Chem. Educ.*, **53**, 666 (1976).

"Recent Developments in Copper Production," J. V. Huxley, *Educ. in Chemistry*, **10**, 94 (1973).

"Hallmarking Gold and Silver," J. S. Forbes, *Chem. in Britain*, **7**, 98 (1971).

"Resource Utilization—Copper from Low Grade Ores," M. J. Cahalan, *Chem. in Britain*, **9**, 392 (1973).

"Inside Color Photography," J. R. Thirtle, *Chemtech*, **9**, 25 (1979).

36

The Metals of Group IIB

36.1 Periodic Relationships of Zinc, Cadmium, and Mercury

The three elements of Periodic Group IIB are zinc, cadmium, and mercury. Each of these elements has two electrons in its outer shell and eighteen in the underlying shell. They are **representative** metals and the $+2$ oxidation state prevails, although the $+1$ state of mercury (Hg_2^{2+}) is important. The melting points, boiling points, and heats of vaporization for the Group IIB members are lower than for any other group of metals except the alkali metals (Group IA). Of the three elements, zinc is the most reactive and mercury the least. The ions of all three elements are small in relation to their charges, so they show strong tendencies toward the formation of complex ions like the transition metals, which they immediately follow in the Periodic Table. The most frequently encountered complex ions of this family are the halo-, cyano-, and ammine complexes. Coordination numbers of both 4 and 6 are quite common in the complexes of these elements. The 4-coordinate complexes are tetrahedral (sp^3 hybridization). The 6-coordinate complexes have octahedral structures, but the bonding must be of the sp^3d^2 rather than the d^2sp^3 type, for the inner d orbitals of the simple ions are completely filled with electrons. Hybridization of the sp^3d^2 type is represented diagrammatically for $[Zn(NH_3)_6]^{2+}$ as

$$[Zn(NH_3)_6]^{2+} \quad 3d^2 \; 3d^2 \; 3d^2 \; 3d^2 \; 3d^2 \quad \boxed{4s^2 \; 4p^2 \; 4p^2 \; 4p^2 \; 4d^2 \; 4d^2}$$

$$\boxed{sp^3d^2 \text{ hybridization}}$$

Octahedral hybridization

TABLE 36-1 Some Physical Properties of the Metals of Group IIB

	Atomic Number	Atomic Weight	Atomic Radius, Å	Ionic (M^{2+}) Radius, Å	Density at 20°	Melting Point, °C	Boiling Point, °C
Zinc	30	65.38	1.33	0.74	7.14	419.5	907
Cadmium	48	112.41	1.49	0.97	8.65	320.9	767
Mercury	80	200.59	1.55	1.10	13.546	−38.87	356.57

Some of the physical properties of zinc, cadmium, and mercury are given in Table 36-1.

Zinc

36.2   Occurrence and Metallurgy of Zinc

Sphalerite, or **zinc blende**, ZnS, is the principal ore of zinc. Less important ores include **zincite**, ZnO; **smithsonite**, $ZnCO_3$; **franklinite**, a mixture of oxides of zinc, iron, and manganese; and **willemite**, Zn_2SiO_4.

In the metallurgy of zinc, sulfide ores are first concentrated by flotation and then roasted to convert the sulfide to the oxide. From carbonate ores the oxide is obtained by simple heating.

$$2ZnS(s) + 3O_2(g) \longrightarrow 2ZnO(s) + 2SO_2(g) \qquad \Delta H° = -878.3 \text{ kJ}$$
$$ZnCO_3(s) \longrightarrow ZnO(s) + CO_2(g) \qquad \Delta H° = 71.0 \text{ kJ}$$

The zinc oxide formed by roasting the sulfide or carbonate ores is reduced by heating it with coal in a fire-clay retort. As rapidly as the zinc is produced it distills out and is condensed. The zinc so produced is impure with cadmium, iron, lead, and arsenic. It may be purified by careful redistillation.

An electrolytic process is also employed to produce zinc. The ore is roasted in special furnaces under such conditions that most of the sulfide is converted to the sulfate and the remainder changed to the oxide. The roasted ore is leached with dilute sulfuric acid to dissolve the zinc sulfate and zinc oxide, as well as the sulfates and oxides of other metals. Silver sulfate and lead sulfate remain in the residue, for they are insoluble. The solution of zinc sulfate is treated with powdered zinc to reduce the less active metals such as copper, arsenic, and antimony to the elemental condition. Any iron(II) present is oxidized by passing air through the solution. Manganese and iron are precipitated as hydroxides by adding the calculated amount of lime. After the solution is purified in this way, it is electrolyzed, zinc being deposited on aluminum cathodes. Sulfuric acid is regenerated as electrolysis proceeds and is used in the leaching of more roasted ore. At intervals, the cathodes are removed and the zinc scraped off, melted, and cast into ingots. Electrolytic zinc has a purity of 99.95%.

36.3   Properties and Uses of Zinc

Zinc is a silvery metal that quickly tarnishes to a blue-gray appearance. This color is due to an adherent coating of a basic carbonate, $Zn_2(OH)_2CO_3$, which protects the underlying metal from further corrosion. The metal is hard and brittle at ordinary temperatures but ductile and malleable at 100–150°C. When molten zinc is poured into cold water it solidifies in irregular masses called **granulated,** or **mossy, zinc.** Zinc is a fairly active metal and will reduce steam at high temperatures.

$$Zn(s) + H_2O(g) \longrightarrow ZnO(s) + H_2(g) \qquad \Delta H° = -106.4 \, kJ$$

Zinc dissolves in strong bases liberating hydrogen and forming zincate ions.

$$Zn + 2OH^- + 2H_2O \longrightarrow [Zn(OH)_4]^{2-} + H_2(g)$$

A considerable amount of zinc is used in the manufacture of dry cells (Section 20.16) and in the production of alloys such as brass and bronze (Section 30.8). About half of the zinc metal produced is used to protect iron and other metals from corrosion by air and water. Its protective action depends upon the property of forming a basic carbonate film on its surface. Furthermore, it protects less active metals such as iron, because it forms an electrochemical cell with the iron, in which zinc serves as the anode and iron as the cathode (see Section 41.9). The zinc coating on iron may be applied in several ways, and the product is called **galvanized iron.**

36.4 Compounds of Zinc

▶ **1. ZINC OXIDE AND ZINC HYDROXIDE.** The **oxide,** ZnO, is produced when zinc vapor is burned and when zinc ores are roasted in an excess of air. The oxide is yellow when hot but pure white when cold. It is used as a paint pigment, called **zinc white,** or **Chinese white;** unlike white lead it does not blacken with hydrogen sulfide, because zinc sulfide is white. It is also used in the manufacture of automobile tires and other rubber goods, and in the preparation of medical ointments.

Zinc hydroxide, $Zn(OH)_2$, is formed as a white, gelatinous precipitate when a soluble hydroxide is added to a solution of a zinc salt.

$$Zn^{2+} + 2OH^- \longrightarrow Zn(OH)_2(s)$$

The hydroxide is amphoteric and dissolves in an excess of alkali as well as in acids.

$$Zn(OH)_2 + 2OH^- \longrightarrow [Zn(OH)_4]^{2-} \quad \text{(zincate ion)}$$
$$Zn(OH)_2 + 2H^+ \longrightarrow Zn^{2+} + 2H_2O$$

Zinc hydroxide is soluble in aqueous ammonia, forming **tetraamminezinc hydroxide.**

$$Zn(OH)_2 + 4NH_3 \longrightarrow [Zn(NH_3)_4]^{2+} + 2OH^-$$

▶ **2. ZINC CHLORIDE.** When zinc and chlorine are brought together **anhydrous zinc chloride,** $ZnCl_2$, is formed as a white deliquescent solid. It is also made by dissolving either the metal, the oxide, or the carbonate in hydrochloric acid, evaporating the solution to dryness, and then fusing the residue to remove the last traces of moisture. The product, which melts at 262°C, is cast into sticks. The anhydrous salt is used as a caustic in surgery and as a dehydrating agent in certain organic reactions. Aqueous solutions of zinc chloride are acidic due to hydrolysis. Because of its acidic nature a solution of zinc chloride (along with ammonium chloride) is used to dissolve the oxides on metal surfaces before soldering. Concentrated solutions of zinc chloride dissolve cellulose, forming a gelatinous mass, which may be molded into various shapes. Fiberboard is made of this material. Zinc oxide dissolves in concentrated solutions of zinc chloride, producing a basic zinc chloride, **zinc oxychloride,** Zn_2OCl_2, which sets to a hard mass and is used as a cement.

▶ **3. ZINC SULFATE.** **Zinc sulfate** is produced in large quantities by roasting zinc blende at low red heat.

$$ZnS(s) + 2O_2(g) \longrightarrow ZnSO_4(s) \qquad \Delta H° = -777.0 \text{ kJ}$$

The product is extracted with water and recovered by recrystallization to form crystals of the **heptahydrate,** $ZnSO_4 \cdot 7H_2O$, which is called **white vitriol.** It is used principally in the production of the white paint pigment **lithopone,** which is a mixture of barium sulfate and zinc sulfide, formed by the reaction

$$Ba^{2+} + S^{2-} + Zn^{2+} + SO_4{}^{2-} \longrightarrow BaSO_4(s) + ZnS(s)$$

▶ **4. ZINC SULFIDE.** **Zinc sulfide,** ZnS, is found in nature as zinc blende and is formed in the laboratory when ammonium sulfide is added to a solution of a zinc salt. The sulfide is insoluble in acetic acid but dissolves readily in stronger acids such as hydrochloric or sulfuric.

$$ZnS(s) + 2H^+ \longrightarrow Zn^{2+} + H_2S(g)$$

Zinc sulfide is used as a white paint pigment, either alone or mixed with zinc oxide.

Cadmium

36.5   Occurrence, Metallurgy, and Uses of Cadmium

Cadmium is found in the rare mineral **greenockite,** CdS, and in small amounts (less than 1%) in several zinc ores. Most of our supply of cadmium comes from zinc smelters and from the sludge obtained from the electrolytic refining of zinc.

In the smelting of cadmium-containing zinc ores, the two metals are reduced together. Because cadmium is more volatile (bp 767°C) than zinc (bp 907°C) they may be separated by fractional distillation. Separation of the two metals is also possible by selective electrolytic deposition. Cadmium, being less active than zinc, is deposited at a lower voltage.

Cadmium is a silvery, crystalline metal resembling zinc. It is only slightly tarnished by air or water to form the oxide at ordinary temperatures. A large part of the cadmium produced is used in electroplating metals, such as iron and steel, to protect them from corrosion. The electrolytic bath contains tetracyanocadmate ions, $[Cd(CN)_4]^{2-}$, made by mixing cadmium cyanide and sodium cyanide. The equation for the cathodic reduction is

$$[Cd(CN)_4]^{2-} + 2e^- \longrightarrow Cd + 4CN^-$$

Cadmium-plated metals are more resistant to corrosion, more easily soldered, and more attractive in appearance than galvanized (zinc-coated) metals.

Cadmium is used in making a number of alloys. Certain ones are easily fusible; these include **Wood's metal** (12.5% Cd), melting at 65.5°C, and **Lipowitz alloy** (10% cadmium), melting at 70°C. Some antifriction bearing metals contain cadmium. These alloys melt at a higher temperature and have a lower friction coefficient than **Babbitt metal** (Section 39.8). Amalgamated cadmium, along with cadmium sulfate, is used in the Weston standard cell for measuring electrical potentials. Rods of cadmium are used in nuclear reactors (Section 29.12) to absorb neutrons and thus control the chain reaction.

36.6   Properties of Compounds of Cadmium

Cadmium dissolves slowly in either hot, moderately dilute hydrochloric or sulfuric acid, with the evolution of hydrogen. It is also dissolved by nitric acid, the oxides of nitrogen being evolved.

Cadmium oxide, CdO, is formed as a brown solid when cadmium burns in air. Alkali metal hydroxides react with cadmium salts to give the white hydroxide $Cd(OH)_2$, which is soluble in aqueous ammonia because of the formation of the soluble tetraammine complex, $[Cd(NH_3)_4]^{2+}$. Solutions of cadmium chloride are poor conductors of electricity; the compound dissolves as covalent molecules that ionize only slightly. As would be expected from the electronic configuration of the cadmium ion, the four-coordinate complexes of cadmium, unlike those of copper, are tetrahedral (see Section 31.3).

Cadmium chloride is converted almost completely into $[CdCl_4]^{2-}$ by high concentrations of the chloride ion. **Cadmium sulfide,** CdS, is formed as a bright yellow precipitate by the action of hydrogen sulfide upon solutions of cadmium salts. It is used as a paint pigment called **cadmium yellow.** The carbonate, phosphate, cyanide, and ferrocyanide are all insoluble in water. All cadmium compounds are soluble in an excess of sodium iodide, due to the formation of the soluble complex, $[CdI_4]^{2-}$.

Mercury

36.7   History of Mercury

Mercury was one of the few metals known to the ancients. The Greek philosopher Aristotle refers to the metal as "liquid silver," or "quick silver," because of its appearance and liquid state. The alchemists named the element after *Mercury,* the swift messenger of the gods of Roman mythology. One of the early uses of mercury was for the extraction of gold from its ores, inasmuch as gold is quite soluble in mercury. The alchemists used mercury in their attempts to transmute base metals into gold, assuming that the mercury had unusual powers due to its liquid state and the property of readily forming amalgams.

36.8   Occurrence and Metallurgy of Mercury

Elemental mercury is sometimes found in rocks and as an amalgam of silver and gold. However, the most important commercial source of the metal is the dark red sulfide, HgS, known as **cinnabar.** When cinnabar is roasted, metallic mercury distills from the furnace and is condensed to the liquid state.

$$HgS(s) + O_2(g) \longrightarrow Hg(l) + SO_2(g) \qquad \Delta H° = -238 \, kJ$$

Ordinarily, the roasting of a mineral in air results in the oxidation of the metal. However, the oxides of mercury are thermally unstable and decompose at the temperature of the furnace. Mercury may be purified by filtering it through chamois or a gold seal (the mercury "wets" the gold seal and hence passes through it; the impurities do not). It may also be purified by washing with nitric acid to oxidize the metallic impurities, or by distilling it in an atmosphere of oxygen to remove the more active metals.

36.9 Properties and Uses of Mercury

Mercury is a silvery white metal and is the only metal that is liquid at room temperature. Due to its low freezing point ($-38.87°C$), its high boiling point ($356.6°C$), its uniform coefficient of expansion, and the fact that it does not wet glass, mercury is an excellent thermometric substance. Because of its chemical inactivity, mobility, high density, and electrical conductivity, it is used extensively in barometers, vacuum pumps, and liquid seals, and for electrical contacts. With the exception of iron and platinum, all metals readily dissolve in, or are wet by mercury, to form amalgams. Sodium amalgam is used as a reducing agent, because it is less active than the alkali metal alone. Amalgams of tin, silver, and gold are used in dentistry.

Mercury does not change chemically in air at ordinary temperatures but is slowly oxidized when heated, forming **mercury(II) oxide,** HgO, which decomposes at still higher temperatures. Being below hydrogen in the activity series, mercury is not attacked by hydrochloric acid but does dissolve in oxidizing acids such as cold nitric or hot concentrated sulfuric acid. With excess acid, mercury(II) salts are formed.

$$3Hg + 8H^+ + 2NO_3^- \longrightarrow 3Hg^{2+} + 2NO(g) + 4H_2O$$
$$Hg + 2H_2SO_4 \longrightarrow Hg^{2+} + SO_4^{2-} + SO_2(g) + 2H_2O$$

When the mercury is in excess these oxidizing acids convert the metal to the corresponding mercury(I), Hg_2^{2+}, salts rather than to the mercury(II), Hg^{2+}, salts.

The halogens attack mercury, forming the halides, and sulfur combines with it to produce mercury(II) sulfide. Mercury is displaced from solutions of its ions by all metals exept silver, gold, and the platinum metals (Ru, Rh, Pd, Os, Ir, and Pt). A copper wire soon becomes amalgamated when immersed in a solution of a mercury compound.

Mercury forms two series of compounds; in these series the metal shows oxidation states of $+1$ and $+2$. The mercury atom uses both its valence electrons in bonding in all of its compounds. For example, in mercury(II) chloride, $HgCl_2$, the chlorine atoms are covalently bonded to the mercury atom in a linear molecule, as shown by the electronic formula

$$:\overset{..}{\underset{..}{Cl}}:Hg:\overset{..}{\underset{..}{Cl}}:$$

With the strongly electronegative sulfate and nitrate groups, mercury(II) forms ionic compounds in which dipositive mercury(II) ions, Hg^{2+}, are present. In the series of compounds in which mercury exhibits a $+1$ oxidation state, one valence electron of each of the mercury atoms serves to form an electron-pair bond between the two mercury atoms. The Hg_2Cl_2 molecule has a linear covalent structure.

$$:\overset{..}{\underset{..}{Cl}}:Hg:Hg:\overset{..}{\underset{..}{Cl}}:$$

The equal sharing of the electron pair between the two mercury atoms does not contribute to the oxidation state of either atom, but the unequal sharing of an electron pair by each mercury atom with an electronegative chlorine atom gives each mercury atom an oxidation state of $+1$. When the second electron of each mercury atom is lost completely by electron transfer, as in the formation of $Hg_2(NO_3)_2$, the dimercury(I) ion, $(Hg:Hg)^{2+}$, is formed.

The Hg_2^{2+} ion is the most common ionic species with a **metal-metal bond** (Section 30.7). Evidence from various sources points to the dimeric nature of this ion. X-ray studies of crystalline mercury(I) chloride show that the mercury atoms occur in pairs. Magnetic studies of solutions containing univalent mercury show that it is not paramagnetic, meaning that there are no unpaired electrons as there would be if such solutions were to contain Hg^+ ions with an odd number of electrons per ion. On the contrary, $Hg:Hg^{2+}$ contains an even number of electrons so that all the electrons present are assumed to exist in pairs. Thus it is correct to write the formulas of mercury(I) compounds as Hg_2X_2, instead of HgX, where X is an acid radical.

36.10   Mercury(I) Compounds

A number of compounds containing univalent mercury are known. One of these, **mercury(I) chloride,** Hg_2Cl_2 (commonly called mercurous chloride), is a white, insoluble crystalline compound which is manufactured by heating a mixture of mercury(II) sulfate, mercury, and sodium chloride.

$$HgSO_4 + Hg + 2NaCl \longrightarrow Hg_2Cl_2 + Na_2SO_4$$

Mercury(I) chloride is volatile and sublimes from the reaction mixture. It is used in the field of medicine under the name **calomel** as a cathartic and diuretic (i.e., it stimulates the organs of secretion).

When exposed to light, mercury(I) chloride slowly decomposes ($Hg_2Cl_2 \longrightarrow HgCl_2 + Hg$); hence, it is usually stored in amber-colored bottles.

When mercury dissolves in cold, dilute nitric acid (with the mercury in excess), **mercury(I) nitrate,** $Hg_2(NO_3)_2$, is formed. This compound is important because it is the only readily soluble salt of univalent mercury. This salt hydrolyzes in aqueous solution to form the basic nitrate, $Hg_2(OH)NO_3$. The addition of nitric acid reverses the hydrolysis.

$$Hg_2(NO_3)_2 + H_2O \rightleftharpoons Hg_2(OH)NO_3(s) + H^+ + NO_3^-$$

On standing in contact with air, solutions of mercury(I) nitrate readily oxidize to mercury(II) nitrate. The accumulation of Hg^{2+} can be prevented by keeping a small amount of metallic mercury in the bottle in which the solution is stored ($Hg^{2+} + Hg \longrightarrow Hg_2^{2+}$).

Mercury(I) sulfide, Hg_2S, is unstable and immediately decomposes into mercury and mercury(II) sulfide when it is formed by passing hydrogen sulfide into a solution containing mercury(I) ions.

$$Hg_2^{2+} + H_2S \longrightarrow Hg_2S(s) + 2H^+$$
$$\hookrightarrow Hg(l) + HgS(s)$$

Aqueous ammonia reacts with mercury(I) chloride in an oxidation-reduction reaction to form metallic mercury, in the form of a finely divided black powder, and a complex white salt, **mercury(II) amido chloride,** containing bivalent mercury.

$$Hg_2Cl_2 + 2NH_3 \longrightarrow Hg + HgNH_2Cl + NH_4^+ + Cl^-$$

Alkali metal hydroxides precipitate the mercury(I) ion as black Hg_2O, which is unstable and decomposes into Hg and HgO.

36.11   Mercury(II) Compounds

When a strong base is added to a solution of a mercury(II) compound the oxide, HgO, precipitates. When precipitated from cold solutions **mercury(II) oxide** is yellow, but precipitated from hot solutions it is red. This difference in color has been attributed to a difference in the state of subdivision of the substance, inasmuch as the crystal structure is the same in the two cases. **Mercury(II) hydroxide** is unstable, losing the elements of a molecule of water to form the oxide.

Mercury(II) chloride, $HgCl_2$, may be formed by heating the metal with an excess of chlorine, by dissolving mercury(II) oxide in hydrochloric acid, or by the action of aqua regia upon mercury. It is prepared commercially by heating mercury(II) sulfate with sodium chloride; the $HgCl_2$ sublimes from the reaction mixture.

$$2NaCl + HgSO_4 \longrightarrow Na_2SO_4 + HgCl_2$$

The compound $HgCl_2$ is also called **bichloride of mercury** and **corrosive sublimate.** In dilute solutions it is used as an antiseptic. It is moderately soluble in water but only slightly ionized, as indicated by the low electrical conductivity of its solutions. Its solubility can be increased by adding an excess of chloride ions, which causes the formation of the complex **tetrachloromercurate(II) ion** according to the equation

$$HgCl_2 + 2Cl^- \rightleftharpoons [HgCl_4]^{2-}$$

Mercury(II) chloride hydrolyzes in water somewhat and ammonolyzes in aqueous ammonia. The reactions are similar:

$$Cl-Hg-Cl + 2H_2O \longrightarrow Cl-Hg-OH + H_3O^+ + Cl^-$$
$$Cl-Hg-Cl + 2NH_3 \longrightarrow Cl-Hg-NH_2 + NH_4^+ + Cl^-$$

Iodide ions precipitate mercury(II) ions from solution as HgI_2, which is red in color. It is soluble in a solution containing an excess of iodide ions, tetra-iodomercurate ions being formed, according to

$$HgI_2 + 2I^- \rightleftharpoons [HgI_4]^{2-}$$

Hydrogen sulfide precipitates black **mercury(II) sulfide,** HgS, from solutions of mercury(II) salts, even in strongly acidic solutions. The precipitate formed by the interaction of $HgCl_2$ in solution and H_2S is first white, then yellow, then red, and finally black. The white compound has the formula $HgCl_2 \cdot 3HgS$. When heated, HgS becomes bright red; the red sulfide is isomeric with the black and has, in the past, been used as pigment under the name of **vermilion.**

Mercury(II) sulfide is dissolved by aqueous solutions of sodium sulfide, Na_2S, in the presence of an excess of hydroxyl ions and forms the **thiomercurate ion.**

$$HgS + S^{2-} \rightleftharpoons [HgS_2]^{2-}$$

Mercury(II) fulminate, $Hg(ONC)_2$, is an explosive compound formed by the action of nitric acid on mercury in the presence of ethyl alcohol. It is used in making detonators and percussion caps because it explodes when struck. Mercury fulminate probably has the electron structure shown.

$$:C{\equiv}N{-}\ddot{O}{-}Hg{-}\ddot{O}{-}N{\equiv}C:$$

Mercury fulminate

36.12 Physiological Action of Mercury

Metallic mercury is an accumulative poison; the breathing of air containing mercury vapor or contact of the liquid metal with the skin should be avoided. All soluble mercury compounds are poisonous, but in small doses they may be medicinal. The fatal dose of mercury(II) chloride is 0.2 to 0.4 gram. The mercury(II) ion combines with the protein tissue of the kidney and destroys the ability of this organ to remove waste products from the blood. Antidotes for mercury poisoning are egg white and milk; their proteins precipitate the mercury in the stomach. Mercury poisoning is treated with chelates like British anti-Lewisite (Section 31.5).

Questions

1. How is zinc metal separated from the impurities cadmium, iron, lead, and arsenic during the refining process?
2. Why does pure zinc react only slowly with dilute hydrochloric or sulfuric acids?
3. What is galvanized iron, how is it produced, and what properties does it have that make it useful?
4. A dilute solution is dripped into a slowly stirred solution of sodium zincate, $Na_2[Zn(OH)_4]$. A white gelatinous precipitate is formed which, upon analysis, proves to be a hydroxide. Use a chemical equation to explain these observations.
5. What is lithopone and how is it produced? In what way is lithopone superior to white lead as paint pigment?

6. Why is cadmium chloride considered to be a weak electrolyte?
7. What property makes cadmium effective as a protective coating for other metals?
8. Predict the favored direction of the reaction shown by the equation

$$[Cd(NH_3)_4]^{2+} + Cu^{2+} \rightleftharpoons Cd^{2+} + [Cu(NH_3)_4]^{2+}$$

9. The roasting of an ore of a metal usually results in the conversion of the metal to the oxide. Why does the roasting of cinnabar produce metallic mercury rather than an oxide of mercury?
10. Why is mercury not attacked by hydrochloric acid even though it dissolves readily in nitric acid or hot sulfuric acid?

11. Why are mercury(II) halides spoken of as weak electrolytes?

12. What properties make mercury valuable as a thermometric substance?

13. Write electronic formulas for $HgCl_2$ and Hg_2Cl_2.

14. Hydrogen sulfide gas is bubbled through a solution marked mercury(I) nitrate. A precipitate is observed to form. Of what substance(s) is the precipitate composed?

15. How many moles of ionic species would you predict to be present in a solution marked 1.0 M mercury(I) nitrate? How would you demonstrate the accuracy of your prediction?

16. Write balanced chemical equations for the following reactions:

(a) Zinc is burned in air.

(b) An aqueous solution of ammonia is added dropwise to a solution of zinc chloride until the mixture becomes clear again.

(c) Zinc is heated with sulfur.

(d) Mossy zinc is added to an aqueous solution of hydrobromic acid.

(e) Mossy zinc is added to a solution of an excess of mercury(II) nitrate.

(f) An excess of mossy zinc is added to a solution of mercury(II) nitrate.

(g) Metallic cadmium is added to an aqueous solution of hydrogen chloride.

(h) Cadmium chloride is added to a concentrated solution of potassium cyanide.

(i) A solution of cadmium nitrate is mixed with a solution of sodium sulfide.

(j) Mercury(II) oxide is added to a solution of nitric acid.

(k) A solution of mercury(II) nitrate is stirred with liquid mercury.

(l) A solution of mercury(I) nitrate is added to a solution of potassium sulfide.

Problems

1. How many pounds of cadmium can be obtained from ten tons of ore that is 5.0% greenockite?

   Ans. 7.8 × 10² lb

2. How many grams of CdS will precipitate upon treatment with an excess of sulfide ion in a solution obtained by dissolving 2.5 g of Wood's metal?

   Ans. 0.40 g

3. What is the minimum number of grams of fish that one would have to consume to obtain a fatal dose of mercury, if the fish contains 5.0 parts per million mercury by weight? (Assume all the mercury from the fish ends up as mercury(II) chloride in the body and that a fatal dose is 0.20 g of $HgCl_2$.) How many pounds of fish would this be? *Ans. 30 kg; 66 lb*

4. How many pounds of mercury can be obtained from ten tons of ore that is 7.4% cinnabar?

   Ans. 1.3 × 10³ lb

References

"Zinc Extraction Metallurgy in the UK," A. W. Richards, *Chem. in Britain,* **5,** 203 (1969).

"Alfred E. Stock and the Insidious 'Quecksilbervergiftund' (Poisoning)," E. K. Mellon, *J. Chem. Educ.,* **54,** 211 (1977).

"The Chemistry of Methylmercury Toxicology," D. L. Rabenstein, *J. Chem. Educ.,* **55,** 292 (1978).

"The Determination of Zinc in Hair," R. K. Pomeroy, N. Drikitis, and Y. Koga, *J. Chem. Educ.,* **52,** 544 (1975).

"Elevated Plasma Zinc: A Heritable Anomaly," J. C. Smith, Jr., J. A. Zeller, E. D. Brown, and S. C. Ong, *Science,* **193,** 497 (1976).

"Mercury in the Environment: Natural and Human Factors,"

A. L. Hammond, *Science,* **171,** 788 (1971).

"Mercury Stirs More Pollution Concern" (Staff), *Chem. and Eng. News,* June 22, 1970, p. 36.

"Mercury in Swordfish" (Staff), *Chem. and Eng. News,* May 17, 1971, p. 11.

"Mercury in the Environment," L. J. Goldwater, *Sci. American,* May 1971, p. 15.

"Mercury in British Fish," G. Grimstone, *Chem. in Britain,* **8,** 244 (1972).

"Hair Element Content in Learning Disabled Children," R. O. Phil and M. Parkes, *Science,* **198,** 205 (1977).

37

The Metals of
Groups IIIA and IIIB

37.1 Periodic Relationships of the Metals of Group III

Group IIIA of the Periodic Table consists of boron, aluminum, gallium, indium, and thallium. Boron is a nonmetallic element that has already been considered (Chapter 28). Aluminum is a common and abundant metal, whereas gallium, indium, and thallium are all scarce and almost without useful application. Consequently, for Group IIIA only aluminum and some of its compounds are discussed in detail in this chapter. Some physical properties of all the elements in Group IIIA are given in Table 37-1. The Group IIIB elements (scandium, yttrium, the lanthanide elements, and the actinide elements) are considered in the latter part of this chapter. All the elements of these two groups are generally trivalent. Oxidation states other than $+3$ are exhibited also by some of these elements. The inner transition elements (lanthanide elements and actinide elements) are similar in properties because the atoms of these elements have the same number of electrons in their outer two shells; differences in electronic structure occur in the third from the outermost shell. Some physical properties of the metals of Group IIIB are given in Table 37-2 (see Sections 37.10–37.12).

TABLE 37-1   Important Physical Properties of Group IIIA Metals

	Atomic Number	Atomic Weight	Atomic Radius, Å	Ionic (M^{3+}) Radius, Å	Density at 20°	Melting Point, °C	Boiling Point, °C
Aluminum	13	26.98154	1.43	0.50	2.70	660	2327
Gallium	31	69.72	1.22	0.62	5.91	29.78	1983
Indium	49	114.82	1.62	0.81	7.36	157	2000
Thallium	81	204.37	1.71	0.95	11.85	303.6	1457

Aluminum

37.2   Occurrence and Metallurgy of Aluminum

Aluminum is the most abundant metal and the third most abundant element in the earth's crust. It is too active chemically to occur free in nature and is usually found combined with oxygen. The most important ore of aluminum is the hydrated oxide, $Al_2O_3 \cdot 2H_2O$, which is called **bauxite.** The mineral **cryolite,** Na_3AlF_6, occurs in nature and is used in the metallurgy of aluminum, though it is not an ore of aluminum. Very large and widely distributed quantities of aluminum are found in complex alumino silicate minerals, such as **clays** and **feldspars.** Anhydrous aluminum oxide occurs as **corundum, ruby,** and **sapphire. Emery** is a mixture of Al_2O_3 and Fe_3O_4.

Aluminum is produced by an electrolytic process invented in 1886 by Charles M. Hall, who began work on the problem while he was a student at Oberlin College. It was invented independently a month or two later, by Paul L.. T. Héroult in France.

The first step in the production of metallic aluminum from bauxite involves the purification of the mineral. The bauxite is dried and pulverized and then digested with sodium hydroxide under steam pressure. The aluminum oxide is thereby converted to soluble sodium aluminate, while the iron oxide and silicates, both of which are present as impurities, remain undissolved.

$$Al_2O_3 \cdot 2H_2O + 2OH^- + H_2O \longrightarrow 2Al(OH)_4^- \quad \text{(aluminate ion)}$$

After the insoluble iron oxide and silicates are filtered from the solution, the aluminum hydroxide is precipitated by hydrolysis of the aluminate. The precipitated aluminum hydroxide is ignited to the oxide, which is then dissolved in fused cryolite, Na_3AlF_6. This solution is electrolyzed, whereupon aluminum metal is liberated at the cathode, and oxygen, carbon monoxide, and carbon dioxide are formed at the carbon anode. A mixture of fluorides sometimes replaces cryolite as the electrolyte. A typical mixture is $2AlF_3 \cdot 6NaF \cdot 3CaF_2$.

Aluminum prepared by electrolysis is about 99% pure, the impurities being small amounts of copper, iron, silicon, and aluminum oxide. If aluminum of exceptionally high purity (99.9%) is needed, it may be obtained by the **Hoopes electrolytic process** (Fig. 37-1), which employs a fused bath consisting of three layers. The bottom layer is a fused alloy of copper and the impure aluminum, the

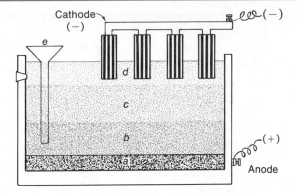

Figure 37-1 Electrolytic cell for the purification of aluminum. *a* is bottom portion of cell in contact with the anode layer; *b* is fused impure aluminum and copper alloy, which acts as the anode and in which the anodic oxidation takes place; *c* is an electrolyte of fused fluorides; *d* is the layer of pure molten aluminum at the cathodes; *e* is a funnel arrangement by which impure aluminum may be added to the bottom of the cell.

top layer is pure molten aluminum, and the middle layer (the electrolyte) consists of a fused mixture of the fluorides of barium, aluminum, and sodium, and nearly enough aluminum oxide to saturate it. The densities of the layers are such that their separation is maintained during electrolysis. The bottom layer serves as the anode and the top layer as the cathode. As electrolysis proceeds, the aluminum in the bottom layer passes into solution in the electrolyte as Al^{3+}, leaving the copper, iron, and silicon behind in the anode, for they are not oxidized under these conditions. The aluminum ion is reduced at the cathode. During electrolysis, purified aluminum is drawn off from the upper layer, and the impure molten metal is added to the lower layer through a carbon-lined funnel.

37.3   Properties and Uses of Aluminum

When freshly cut, aluminum has a silvery appearance, but it soon becomes superficially oxidized and assumes a dull white luster. The metal is very light and possesses high tensile strength. Weight for weight it is twice as good a conductor of electricity as copper. Although aluminum is a relatively reactive metal, the tenacious coating of the oxide that forms on it prevents further atmospheric corrosion. When heated in the air, finely divided aluminum burns with a brilliant light. Because of the formation of a protective coat of oxide, aluminum does not decompose water. It is readily dissolved by hydrochloric acid and sulfuric acid, but is rendered passive by nitric acid. Aluminum dissolves in concentrated alkalies with the formation of aluminates and hydrogen.

$$2Al + 2OH^- + 6H_2O \longrightarrow 2Al(OH)_4^- + 3H_2(g)$$

This behavior is in accord with the amphoteric character of aluminum hydroxide. When heated, aluminum combines directly with the halogens, nitrogen, carbon, and sulfur.

The fact that aluminum is an excellent conductor of heat, together with its light weight and resistance to corrosion, accounts for its use in the manufacture of cooking utensils. The most important uses of aluminum are in the airplane and other transportation industries which depend upon the lightness, toughness, and high tensile strength of the metal. Aluminum is also used in the manufacture of electrical transmission wire, as a paint pigment, and, in the form of foil, as a wrapping material.

Aluminum is one of the best reflectors of heat and light, including the wavelengths in the ultraviolet. For this reason it is used as an insulating material, and as a mirror in reflecting telescopes. The 200-inch mirror in the telescope at Mount Palomar, California, is coated with a thin film of aluminum.

About half of the aluminum produced in this country is converted to alloys for special uses. **Duralumin** is an alloy containing aluminum, copper, manganese, and magnesium. It is light and nearly as strong as steel, so it is useful in the construction of aircraft. **Aluminum bronzes** contain copper, and occasionally some silicon, manganese, iron, nickel, and zinc. These lightweight alloys have high tensile strength and great resistance to corrosion, so they are used extensively in the manufacture of crankcases and connecting rods for gasoline engines. **Alnico** is a magnetic alloy containing 50% iron, 20% aluminum, 20% nickel, and 10% cobalt; it will lift more than 4000 times its own weight of iron. It is made by pressing together the constituent metals in powder form and heating the mixture just below the melting point. This is an example of **powder metallurgy**—a branch of science that is growing rapidly in importance.

When powdered aluminum and iron(III) oxide are mixed and ignited by means of a magnesium fuse a vigorous and highly exothermic reaction occurs.

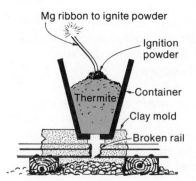

Figure 37-2 Thermite welding. Molten metal, mostly iron, flows out of the container at the bottom and welds a broken rail without having to remove the rail from the roadbed.

$$2Al(s) + Fe_2O_3(s) \longrightarrow 2Fe(s) + Al_2O_3(s) \qquad \Delta H^\circ = -851.4 \text{ kJ}$$

This is an example of the **thermite,** or **Goldschmidt, reaction.** The temperature of the reaction mixture rises to about 3000°C, so the iron and aluminum oxide become liquid. The process (Fig. 37-2) is frequently used in welding large pieces of iron or steel. Thermite bombs were used for incendiary purposes during World War II, because of the high temperature generated by the thermite reaction and because the reaction is not readily quenched by water. The thermite reaction is used in the reduction of metallic oxides (such as MoO_3 and WO_3) which are not readily reduced by carbon, and others (such as MnO_2 and Cr_2O_3) which do not give pure metal when reduced by carbon. The reaction is also of particular value when carbon-free alloys, such as ferrotitanium, are desired.

37.4   Aluminum Oxide

The oxide Al_2O_3 occurs in nature in pure form as the mineral **corundum,** a very hard substance used as an abrasive for grinding and polishing. **Emery** is aluminum oxide mixed with Fe_3O_4; it is also used as an abrasive. Several precious stones are composed of aluminum oxide and certain impurities that impart color: ruby (red, by chromium compounds); sapphire (blue, by compounds of cobalt, chromium, and titanium); oriental amethyst (violet, by manganese compounds); and oriental topaz (yellow, by iron). Artificial rubies and sapphires are now manufactured by melting aluminum oxide (mp 2050°C) with small amounts of oxides to produce the desired color, and cooling the melt so as to produce large crystals. These gems are indistinguishable from natural stones, except for microscopic, rounded air bubbles in the synthetic gems and flattened bubbles in the natural stones. The synthetic gems are used not only as jewelry but as bearings ("jewels") in watches and other instruments, and as dies through which wires are drawn. Very finely divided aluminum oxide, called **activated alumina,** is used as a dehydrating agent and as a catalyst.

Various aluminum articles, such as drinking tumblers, are given a wear resistant coating of oxide by anodic oxidation in a bath of chromic, sulfuric, or oxalic acid. This surface readily absorbs dyes and pigments that give a pleasing decorative effect to the article.

37.5   Aluminum Hydroxide

When an alkali is added to a solution of an aluminum salt, a white, gelatinous precipitate is formed.

$$Al^{3+} + 3OH^- \longrightarrow Al(OH)_3(s)$$

or
$$[Al(H_2O)_6]^{3+} + 3OH^- \longrightarrow [Al(OH)_3(H_2O)_3](s) + 3H_2O$$

Aqueous solutions of sulfides or carbonates also precipitate aluminum hydroxide, because such solutions contain hydroxide ions in considerable concentrations due to hydrolysis (see Chapter 16).

$$S^{2-} + H_2O \rightleftharpoons HS^- + OH^-$$

$$CO_3{}^{2-} + H_2O \rightleftharpoons HCO_3{}^- + OH^-$$

Aluminum hydroxide is amphoteric and is dissolved by both acids and bases.

$$[Al(OH)_4]^- \underset{}{\overset{OH^-}{\rightleftharpoons}} Al(OH)_3(s) \underset{}{\overset{3H^+}{\rightleftharpoons}} Al^{3+} + 3H_2O$$

or
$$[Al(H_2O)_2(OH)_4]^- \underset{}{\overset{OH^-}{\rightleftharpoons}} Al(H_2O)_3(OH)_3(s) \underset{}{\overset{3H^+}{\rightleftharpoons}} [Al(H_2O)_6]^{3+}$$

Aluminum oxide, obtained by heating the hydroxide above 850°C, is insoluble in acids and alkalies. Products obtained below 600°C are soluble in these reagents.

Aluminum hydroxide is widely used to "fix" dyes to fabrics. Cloth that is soaked in a hot solution of aluminum acetate becomes impregnated with the aluminum hydroxide formed by hydrolysis of the salt. The aluminum hydroxide absorbs the dye and holds it fast to the cloth. When used in this way, the aluminum hydroxide is called a **mordant,** and the colored precipitate is called a **lake.**

Aluminum hydroxide is also used in the purification of water because its gelatinous character enables it to carry down with it any suspended material in the water, including most of the bacteria. The aluminum hydroxide for water purification is ordinarily produced by the reaction of aluminum sulfate with lime.

$$2Al^{3+} + 3SO_4{}^{2-} + 3Ca^{2+} + 6OH^- \longrightarrow 2Al(OH)_3(s) + 3CaSO_4(s)$$

37.6   Aluminum Chloride

Anhydrous aluminum chloride is a white crystalline solid that sublimes at 180°C and is soluble in organic solvents. Measurements of its density show that in the vapor state it is dimeric, and the formula should be written Al_2Cl_6. In these molecules each aluminum atom is tetrahedrally bonded to four chlorine atoms; two of these chlorine atoms are each bonded to two aluminum atoms.

Aluminum chloride

During formation of the dimer, one chlorine atom of each monomeric covalent molecule donates a pair of electrons to be shared with the aluminum atom of the other molecule, and thus two "chlorine bridges" are formed; each aluminum atom then posseses an octet of electrons. X-ray studies show that there are no discrete Al_2Cl_6 molecules in solid aluminum chloride. The formula for the compound in the solid state is usually written $AlCl_3$, although the structure is actually a polymeric layer lattice that does not possess individual $AlCl_3$ molecules.

Aluminum chloride can be prepared either by direct chlorination of aluminum or by heating bauxite with carbon and chlorine.

$$2Al(s) + 3Cl_2(g) \longrightarrow 2AlCl_3(s) \qquad \Delta H° = -1408 \text{ kJ}$$
$$Al_2O_3(s) + 3C(s) + 3Cl_2(g) \longrightarrow 2AlCl_3(s) + 3CO(g) \qquad \Delta H° = -64.0 \text{ kJ}$$

If aluminum or aluminum hydroxide is treated with hydrochloric acid and the solution is evaporated, crystals of the hexahydrate, $AlCl_3 \cdot 6H_2O$, are formed. When this salt is heated the oxide is produced.

$$2(AlCl_3 \cdot 6H_2O)(s) \longrightarrow Al_2O_3(s) + 6HCl(g) + 9H_2O(g) \qquad \Delta H° = 977.4 \text{ kJ}$$

37.7   Aluminum Sulfate

The sulfate $Al_2(SO_4)_3 \cdot 18H_2O$ is prepared by treating bauxite or clay with sulfuric acid.

$$Al_2Si_2O_5(OH)_4 + 6H^+ + [3SO_4{}^{2-}] \longrightarrow 2H_2SiO_3(s) + 2Al^{3+} + [3SO_4{}^{2-}] + 3H_2O$$

The metasilicic acid (H_2SiO_3) is removed by filtration. The sulfate is the cheapest soluble salt of aluminum, so it is used as a source of aluminum hydroxide for the purification of water, the waterproofing of fabrics, the sizing of paper, and the dyeing of fabrics.

37.8   Alums

When solutions of potassium sulfate and aluminum sulfate are mixed and concentrated by evaporation, crystals of a salt called **potassium alum** are formed. The salt may best be formulated as $KAl(SO_4)_2 \cdot 12H_2O$. It is the commonest of a class of double salts known as **alums** and is frequently referred to simply as "alum." The univalent ion in an alum may be an alkali metal or silver ion, or the ammonium ion. The trivalent ion may be aluminum, chromium, iron, manganese, or some other. The selenate ion, $SeO_4{}^{2-}$, may replace the sulfate ion. Thus, we have **ammonium alum**, $NH_4Al(SO_4)_2 \cdot 12H_2O$; **sodium iron(III) alum,**

$NaFe(SO_4)_2 \cdot 12H_2O$, and **potassium chrome alum,** $KCr(SO_4)_2 \cdot 12H_2O$. The alums are all isomorphous; i.e., they have the same crystal structure.

37.9   Aluminum Silicates

As we discovered in Section 27.10, many of the most important silicate rocks contain aluminum. Clay and sand are formed as disintegration products by the weathering of these rocks. Weathering involves the thawing and freezing of water in the rocks, and the chemical action of water and carbon dioxide upon them. For example, the chemical disintegration of feldspar may be represented by

$$2KAlSi_3O_8 + 2H_2O + CO_2 \longrightarrow 2K^+ + CO_3^{2-} + Al_2Si_2O_5(OH)_4 + 4SiO_2$$

| Feldspar | Potassium carbonate | Clay | Sand |

The soluble potassium carbonate formed is largely removed by water, and the residue of sand and clay remains as soil. One kind of pure clay has the formula shown above; it is white and is called **kaolin.** Ordinary clay is colored by compounds of iron and other metals. **Porcelain** and **china** are made of pure kaolin, while impure clays are used in the manufacture of earthenware products such as **tile** and **brick.** The buff or red color of such products is due to the presence of iron(III) silicate.

The Metals of Group IIIB

The elements of Group IIIB include scandium, yttrium, the lanthanide elements (rare earths), and the elements of the actinide series. Because scandium and yttrium resemble the rare earths so closely, they are often included in that series.

Some physical properties of scandium, yttrium, lanthanum, and actinium are given in Table 37-2.

TABLE 37-2   Some Physical Properties of the Metals of Group IIIB

	Atomic Number	Atomic Weight	Atomic Radius, Å	Ionic (M^{3+}) Radius, Å	Density at 20°	Boiling Point, °C
Scandium	21	44.9559	1.60	0.81	3.02	1400
Yttrium	39	88.9059	1.80	0.93	4.47	1500
Lanthanum	57	138.9055	1.88	1.15	6.15	880
Actinium	89	227.0278	2.0	1.11	—	—

37.10   The Rare Earth Elements (the Lanthanide Series)

The **rare earths** include elements of atomic numbers 57 to 71, inclusive. These elements are very similar in physical and chemical properties. This unusual similarity is attributed to the fact that their electron configurations differ principally in the number of electrons in the $4f$ and $5d$ subshells rather than in their outer, $6s$, subshell (see Table 37-3). Although lanthanum (at. no. 57) is a transition

TABLE 37-3 Electron Configurations and Ionic Radii of the Rare Earth Elements

Atomic Number	Element	Electron Configuration	Ionic (M^{3+}) Radius, Å
57	La	$1s^2 2s^2 2p^6 3s^2 3p^6 3d^{10} 4s^2 4p^6 4d^{10} 4f^0\ 5s^2 5p^6 5d^1 6s^2$	1.15
58	Ce	$4f^2\ 5s^2 5p^6 5d^0 6s^2$	1.034
59	Pr	$4f^3\ 5s^2 5p^6 5d^0 6s^2$	1.013
60	Nd	$4f^4\ 5s^2 5p^6 5d^0 6s^2$	0.995
61	Pm	$4f^5\ 5s^2 5p^6 5d^0 6s^2$	0.979
62	Sm	$4f^6\ 5s^2 5p^6 5d^0 6s^2$	0.964
63	Eu	$4f^7\ 5s^2 5p^6 5d^0 6s^2$	0.950
64	Gd	$4f^7\ 5s^2 5p^6 5d^1 6s^2$	0.938
65	Tb	$4f^9\ 5s^2 5p^6 5d^0 6s^2$	0.923
66	Dy	$4f^{10} 5s^2 5p^6 5d^0 6s^2$	0.908
67	Ho	$4f^{11} 5s^2 5p^6 5d^0 6s^2$	0.894
68	Er	$4f^{12} 5s^2 5p^6 5d^0 6s^2$	0.881
69	Tm	$4f^{13} 5s^2 5p^6 5d^0 6s^2$	0.869
70	Yb	$4f^{14} 5s^2 5p^6 5d^0 6s^2$	0.858
71	Lu	$4f^{14} 5s^2 5p^6 5d^1 6s^2$	0.848

element (see Section 4.17), it is often included as a rare earth along with elements 58 to 71, which are inner transition elements, because of the similarity in properties.

The rare earths are found in many minerals, and some of the rare earths are actually more common than several of the most familiar elements. Many of these rare earth minerals are complex silicates or phosphates and usually contain several of the rare earth elements. **Monazite sand** is the most important source of these minerals; it contains the various elements of the group in the form of phosphates, except promethium (at. no. 61), which does not occur in nature at all.

The rare earth elements resemble each other so closely that their separation has been an exceedingly laborious task. Formerly, the only effective method of separation was that of fractional crystallization of certain of their salts, involving many hundreds of crystallizations. Now, ion-exchange liquid chromatography techniques (see Section 32.9) provide a much more rapid and effective separation. This technique employs the selective adsorption of the ions of the various rare earths on a chromatographic column containing an ion-exchange resin (Sections 12.12 and 32.9), followed by fractional elution of the adsorbed material by a suitable solvent.

The rare earth elements are metallic in character, and all form oxides of the general formula M_2O_3. Their usual oxidation state is +3. but some of the elements have other oxidation states as well, such as +2 or +4.

An alloy of iron with a mixture of the rare earth metals, known commercially as **misch metal,** is used in making the "flints" in cigarette lighters. A mixture of the fluorides of these metals forms the core of the carbon arcs used in motion picture projectors. Recently, the rare earths have been found to have great catalytic power in some important industrial reactions such as petroleum cracking and reforming.

Relatively large quantities of some of the rare earth elements are used in color television tubes to improve the color rendition (especially reds).

37.11   The Lanthanide Contraction

The rare earth elements decrease in size with increasing atomic number, as is shown by the atomic radii and the ionic radii (+3 ion) listed in Table 37-3. This size decrease, referred to as the **Lanthanide Contraction,** results in an expected regular gradation in properties through the series, including a decrease in the basicity of the hydroxides, $M(OH)_3$, with increasing atomic number of the elements (Section 14.16). An interesting additional result of the Lanthanide Contraction is that zirconium and hafnium are the same size (atomic radius, 1.57 Å), hafnium being smaller than would normally be expected because of the contraction in size through the lanthanide elements immediately preceding it. Zirconium and hafnium, since they are not only in the same family but also of the same size, are similar in chemical properties; they occur together in nature (Sections 38.13 and 38.14). Similarly, the niobium and tantalum pair and the molybdenum and tungsten pair are more similar in size than would normally be expected, with a resulting enhanced similarity in properties for each pair. The Lanthanide Contraction also accounts for the fact that yttrium has nearly the same radius as gadolinium, which is about midway through the lanthanide series; with very similar charge and size characteristics as compared to the lanthanide elements, yttrium has very similar properties and occurs with the rare earth elements in nature.

37.12   The Actinide Series

The series of elements beginning with actinium, element 89, and extending through lawrencium, element 103, is called the **actinide series.** All the elements of this series are radioactive. They are analogous to the rare earths in that the electron configurations of succeeding elements in the series differ by one electron, in this case in either the $5f$ or $6d$ subshells. The properties of the actinide elements are similar to those of the lanthanide rare earths. The actinide elements have a characteristic oxidation state of +3 but show a much wider variation in oxidation states than do the elements of the rare earth series. Thorium resembles cerium in that it shows an oxidation state of +4.

Elements 93 through 103 (and three elements beyond the actinides, through 106), are prepared synthetically by nuclear reactions (Section 29.9). However, since they were first synthesized, it has been discovered that plutonium and neptunium occur in nature in small amounts. The most important elements in the actinide series are thorium, uranium, and plutonium, the importance of the latter two being related to their use in atomic energy (Sections 29.10–29.14).

Thorium occurs as the dioxide in monazite sand, in some cases to the extent of 10%, along wih the rare earths. Metallic thorium is obtained by reducing the tetrachloride, $ThCl_4$, with sodium; the thorium tetrachloride is prepared by the action of chlorine on a heated mixture of the dioxide and carbon. Thorium is very similar to the elements of Group IVB, which are discussed in Chapter 38. It is tetravalent in its compounds, its hydroxide, $Th(OH)_4$, is highly basic, and its salts are hydrolyzed to a limited extent in water. Tungsten filaments for electric lamps are thoriated to increase their efficiency and retard their disintegration.

Uranium, the first radioactive element to be discovered, is found in the minerals **pitchblende,** U_3O_8, and **carnotite,** $K_2(UO_2)_2(VO_4)_2 \cdot 8H_2O$. The principal source areas of uranium minerals are Colorado, Utah, the Congo, and the Great Bear Lake region of Canada. The metal may be obtained by reduction of the oxide with either carbon, calcium, or aluminum. It is a white, lustrous metal of high density (18.9 g/cm³), and it melts at $1132°$. The metal is moderately active and forms compounds in which it shows oxidation states of $+2$, $+3$, $+4$, $+5$, and $+6$. The "uranium" salts of commerce contain the dipositive uranyl ion, UO_2^{2+}. These include uranyl nitrate, $UO_2(NO_3)_2 \cdot 6H_2O$, and the acetate, $UO_2(C_2H_3O_2)_2 \cdot 2H_2O$. There are several series of uranates, the most important of which are the diuranates, such as sodium diuranate, $Na_2U_2O_7 \cdot 6H_2O$. Uranium compounds are used for producing yellow glazes on ceramic ware, as mordants, and in the dye industry. The chief interest in uranium at present is in its release of nuclear energy and its converstion to elements that will release nuclear energy (see Sections 29.10–29.14).

The actinide elements exhibit a contraction in size with increasing atomic number analogous to the Lanthanide Contraction.

Questions

1. Describe the production of metallic aluminum by electrolytic reduction.
2. What is the action of sodium hydroxide upon aluminum?
3. Illustrate the amphoteric nature of aluminum hydroxide by suitable equations.
4. How is advantage taken of the amphoteric nature of aluminum ion in the processing of bauxite?
5. Why can aluminum, which is an active metal, be used so successfully as a structural metal?
6. What is the composition of alums as a class of double salts?
7. Explain the function of aluminum hydroxide in water purification.
8. Why is it impossible to prepare anhydrous aluminum chloride by heating the hexahydrate to drive off the water?
9. What is the Goldschmidt, or thermite, process?
10. From the electron structure of thallium, what would

you predict concerning the solubility of thallium(I) chloride? the basicity of thallium(I) hydroxide?
11. Why are the rare earths associated in their natural occurrence?
12. Point out the analogy in structure between the elements of the rare earths and those of the actinide series.
13. Complete and balance the following equations:
    (a) $Al + H^+ \rightleftharpoons Al^{3+} + H_2$
    (b) $NO_2^- + Al + OH^- + H_2O \rightleftharpoons$
    $NH_3 + [Al(OH)_4]^-$
    (c) $Cr_2O_7^{2-} + U^{4+} + H^+ \rightleftharpoons Cr^{3+} + UO_2^{2+}$
    (d) $Al + OH^- \rightleftharpoons [Al(OH)_4]^- + H_2$
14. What are the highest stable oxidation states you would predict for uranium and protactinium respectively?
15. Write the electronic structures that you would expect elements 107, 108, and 109 (if and when they are synthesized) to possess.

16. Write balanced chemical equations for the following reactions:

    (a) Metallic gallium is burned in air.
    (b) Gaseous hydrogen fluoride is bubbled through a suspension of bauxite in molten sodium fluoride.
    (c) Metallic aluminum is added to a solution of sulfuric acid, ammonia is added, and the solution is evaporated to give crystals.
    (d) Aluminum is heated in an atmosphere of chlorine.
    (e) Aluminum sulfide is added to water.
    (f) Aluminum hydroxide is added to a solution of nitric acid.
    (g) Metallic scandium is heated in an atmosphere of oxygen.
    (h) Metallic gadolinium is added to a solution of hydrogen fluoride.

Problem

1. How much uranium is it possible to obtain from 1.0 ton of ore that is 2.0% carnotite?   *Ans. 19 lb*

References

"Studying the Chemical Properties of Metallic Aluminum," N. Feifer, *J. Chem. Educ.*, **45**, 648 (1968).

"The Story of Hall and Aluminum," H. N. Holmes, *J. Chem. Educ.*, **7**, 232 (1930).

"Some Developments in Aluminum Chemistry," K. Wade, *Chem. in Britain*, **4**, 503 (1968).

"Industrial Emergence of Sodium and Aluminum," M. Schofield, *Chem. in Britain*, **4**, 98 (1968).

"The Structure of Solid Aluminum Chloride," M. J. Bigelow, *J. Chem. Educ.*, **46**, 495 (1969).

"Carbonate Inhibits the Crystallization of Aluminum Hydroxide in Bauxite," G. Bardossy and J. L. White, *Science*, **203**, 355 (1979).

"Carnotite: What's in a Name?" L. Badash, *Chem. in Britain*, **2**, 240 (1966).

"A Solid-State Source of Microwaves," R. Bowers, *Sci. American*, August 1966, p. 22.

"Flame-Grown Gem Stones Enjoy Broadened Use in Optics and Fashion Jewelry," *Chem. Eng.*, December 25, 1961. p. 26.

"Gallium," G. H. Wagner and W. H. Gitzen, *J. Chem. Educ.*, **29**, 162 (1952).

"The Rare Earths," F. H. Spedding, *Sci. American*, November 1951, p. 26.

"Carl Auer von Welsbach" (History of the rare earths), F. Lieber, *J. Chem. Educ.*, **35**, 230 (1958).

The Chemistry of Uranium, J. Katz and E. Rabinowitch, McGraw-Hill Book Company, Inc., New York, 1951. (Reprinted, paperbound, by Dover Publications, New York.)

"Uranium Extraction from Sea Water," R. B. Fischer, *J. Chem. Educ.*, **46**, 430 (1969).

"Periodicity and the Lanthanides and Actinides," T. Moeller, *J. Chem. Educ.*, **47**, 417 (1970).

"Coordination of Trivalent Lanthanide Ions," D. G. Karraker, *J. Chem. Educ.*, **47**, 424 (1970).

"Uranium Enrichment: Laser Methods Nearing Full-Scale Test," W. D. Metz, *Science*, **185**, 602 (1974).

38

The Metals of Groups IVA and IVB

Group IVA

38.1 Periodic Relationships of Germanium, Tin, and Lead

The first two elements of Periodic Group IVA, carbon and silicon, have been discussed in Chapters 25 and 27, respectively. The remaining three members of this family are germanium, tin, and lead.

Carbon and silicon are primarily nonmetallic in character, while germanium, tin, and lead, as would be expected, become increasingly metallic in properties with increasing atomic weight. In ionic compounds carbon and silicon are always found in the anion. On the other hand, germanium, tin, and lead form divalent cations, Ge^{2+}, Sn^{2+}, and Pb^{2+}. The fact that the hydroxides of these ions are amphoteric is a reflection of some nonmetallic character. All members of this periodic group form covalent compounds or anions in which the +4 oxidation state is exhibited; for example, CCl_4, $SiCl_4$, $GeCl_4$, $SnCl_4$, and $PbCl_4$ are low-boiling covalent liquids; Na_2CO_3, Na_2SiO_3, Na_2GeO_3, Na_2SnO_3, and Na_2PbO_3 are ionic compounds.

Germanium and the members of Periodic Group IVB (Ti, Zr, and Hf) are among the less familiar elements. The members of the B group usually show an oxidation state of +4 in their compounds; they are transition metals with two electrons in their outer shells and ten electrons in the next to the outer shells.

Some physical properties of the Group IVA metals are given in Table 38-1.

TABLE 38-1   Important Physical Properties of the Metals of Group IVA

	Atomic Number	Atomic Weight	Atomic Radius, Å	Ionic (M^{4+}) Radius, Å	Density at 20°	Melting Point, °C	Boiling Point, °C
Germanium	32	72.59	1.22	0.53	5.36	960	—
Tin	50	118.69	1.4	0.71	7.31	231.9	2337
Lead	82	207.2	1.75	0.84	11.34	327.4	1750

Tin

38.2   History and Occurrence of Tin

Tin has been found in early Egyptian tombs, evidence that the metal was known during the very early periods of history. The metal was obtained from deposits of cassiterite in England by the Romans and Phoenicians.

The most important and abundant ore of tin is called **cassiterite,** or **tinstone,** SnO_2. No important deposits have been found in the United States; yet more than half of the world's production of tin is used in this country. Some tin is reclaimed from the bright scrap left from the manufacture of tin cans. (Tin cans are actually made of steel, plated with tin.) Most detinning of tin plate is based upon the removal of the tin by alkali metal hydroxide solutions, with the formation of stannites and hydrogen.

38.3   Metallurgy of Tin

Tin ore is crushed and washed with water to separate any lighter rocky material from the heavier ore. Roasting of the ore removes arsenic and sulfur as volatile oxides; oxides of other metals are extracted by means of hydrochloric acid. The purified ore is reduced by carbon.

$$SnO_2(s) + 2C(s) \longrightarrow Sn(s) + 2CO(g) \qquad \Delta H° = 360 \, kJ$$

The molten tin, which collects on the bottom of the furnace, is drawn off and cast into blocks (block tin). The crude tin is remelted and permitted to flow away from the higher-melting impurities, which are chiefly compounds of iron and arsenic. Further purification of the tin is accomplished by electrolysis, using impure tin anodes, pure tin cathodes, and a bath of fluosilicic acid (H_2SiF_6) and sulfuric acid.

38.4   Properties of Tin

Tin exists in three solid allotropic forms (see Section 9.10). They are **gray tin** (cubic crystals), **malleable tin** (tetragonal crystals), and **brittle tin** (rhombic crystals). The malleable form is silvery white with a bluish tinge. When a rod of it is bent, the crystals slip over one another, producing a sound described as "tin cry." When malleable tin, or "white tin," is heated it changes to the brittle form. White tin changes slowly at low temperatures (below 13.2°C) into gray tin, a powdery form of the element. Consequently, articles made of tin are likely to disintegrate in cold weather, particularly if kept cold for a long time. The change progresses slowly from a spot of origin, with the gray tin that is formed catalyzing its own formation. An effect is thus produced which, in a way, is similar to the spread of an infection in a plant or animal body. For this reason, it is called **tin disease,** or **tin pest.**

The metal dissolves slowly in hydrochloric acid when the acid is cold and dilute but rapidly when hot and concentrated, $SnCl_2$ and H_2 being produced. Very dilute cold nitric acid converts tin to tin(II) nitrate. Concentrated nitric acid rapidly converts tin into insoluble, hydrated metastannic acid, H_2SnO_3. Alkali hydroxides dissolve tin, forming stannites and hydrogen.

$$Sn + 2OH^- + 2H_2O \longrightarrow [Sn(OH)_4]^{2-} + H_2(g)$$

The reactions with concentrated nitric acid and sodium hydroxide show that tin is not entirely metallic in character. Dry chlorine oxidizes tin to the tetrachloride, $SnCl_4$, and oxygen converts it to tin(IV) oxide, SnO_2. Tin burns with a white flame.

38.5   Uses of Tin

The principal use of tin is in the production of **tin plate,** i.e., sheet iron coated with tin. Most tin is now plated electrolytically. Iron is more active than tin and will corrode more rapidly when in contact with tin than otherwise. This is because iron and tin in contact with each other, and both in contact with moist air, comprise an electrolytic cell that promotes corrosion (Section 41.9). As a result, when a scratch through the tin exposes the iron, corrosion sets in rapidly. Tin is also used in making alloys (Section 30.8) such as **bronze** (Cu, Zn, and Sn), **solder** (Sn and Pb), and **type metal** (Sn, Pb, Sb, and Cu). A plate of nickel and tin, in a ratio of about 1:2, is sometimes used as a substitute for chromium plate.

38.6   Compounds of Tin

▶ **1. OXIDES OF TIN.** Depending upon the method of preparation, **tin(II) oxide** (stannous oxide), SnO, occurs as either a black or a green powder. It may be prepared by treating a hot solution of a tin(II) compound with an alkali carbonate or by heating tin(II) oxalate in the absence of air. The equations are

$$Sn^{2+} + CO_3^{2-} \longrightarrow SnO(g) + CO_2(g)$$
$$SnC_2O_4(s) \longrightarrow SnO(s) + CO_2(g) + CO(g)$$

When tin is burned in air **tin(IV) oxide** (stannic oxide), SnO_2, is formed. The product is white when cold but yellow when hot.

▶ **2. TIN(II) HYDROXIDE.** Soluble bases precipitate tin(II) ions from solution as **tin(II) hydroxide,** $Sn(OH)_2$, which is white and gelantinous in character. An excess of alkali causes the hydroxide to dissolve, with the formation of the soluble **stannite ion.**

$$Sn(OH)_2(s) + 2OH^- \longrightarrow Sn(OH)_4^{2-}$$

The stannite ion is an active reducing agent.

▶3. TIN(II) CHLORIDE (STANNOUS CHLORIDE).   Tin(II) chloride can be obtained as the dihydrate ($SnCl_2 \cdot 2H_2O$) by evaporating a solution formed by the reaction of tin or tin(II) oxide with hydrochloric acid. The salt is hydrolyzed by water, forming the basic hydroxychloride, Sn(OH)Cl.

$$SnCl_2 + H_2O \rightleftharpoons Sn(OH)Cl(s) + H^+ + Cl^-$$

Tin(II) chloride finds wide use as a reducing agent, because of the ease with which tin is oxidized to the tin(IV) condition. It follows that aqueous solutions of tin(II) must be protected against air oxidation.

▶4. TIN(IV) CHLORIDE (STANNIC CHLORIDE).   Tin(IV) chloride is formed when an excess of chlorine reacts with metallic tin. It is a colorless liquid, soluble in organic solvents such as carbon tetrachloride, and it is a nonconductor of electricity. These properties are typical of covalent compounds. When dissolved in water the tetrachloride is highly hydrolyzed (Section 38.7), whereas in hydrochloric acid **hexachlorostannic acid** is produced.

$$SnCl_4 + 2HCl \longrightarrow H_2SnCl_6$$

▶5. SULFIDES OF TIN.   Dark brown **tin(II) sulfide,** SnS, is precipitated when hydrogen sulfide is passed into a solution of a tin(II) salt. Tin(II) sulfide is not dissolved by alkali metal sulfides, whereas tin(IV) sulfide is dissolved. Tin(II) sulfide does dissolve in alkaline polysulfides.

Tin(IV) sulfide is prepared as a yellow precipitate by the interaction of hydrogen sulfide with tin(IV) ions in a moderately acid solution. It dissolves in alkali metal sulfides with the formation of **thiostannate ions,** SnS_3^{2-}. The addition of acid reprecipitates the sulfide.

$$SnS_3^{2-} + 2H^+ \longrightarrow SnS_2(s) + H_2S(g)$$

Concentrated hydrochloric acid dissolves tin(IV) sulfide according to

$$SnS_2(s) + 4H^+ + 6Cl^- \longrightarrow SnCl_6^{2-} + 2H_2S(g)$$

38.7   Hydrolysis of Covalent Halides

Except for the tetrahalides of carbon, the tetrahalides of the Group IVA elements hydrolyze when dissolved in water, for the most part as indicated by the general equation

$$MX_4 + 2H_2O \longrightarrow MO_2(s) \text{ (or hydrate thereof)} + 4HX$$

The mechanism by which the tetrahalides of silicon, germanium, tin, and lead hydrolyze may be as follows, using silicon tetrachloride as an example. The silicon atom carries a partial net positive charge, because it is less electronegative than chlorine in the $SiCl_4$ molecule. During hydrolysis the positively charged silicon atom may attract hydroxide ions of water, forming a pentacovalent intermediate, according to the equation

$$\begin{array}{c} \text{Cl} \\ | \\ \text{Cl}-\text{Si}-\text{Cl} \\ | \\ \text{Cl} \end{array} + \; :\ddot{\text{O}}\text{H}^- \; \longrightarrow \; \left[\begin{array}{c} \text{Cl}\;\ddot{:}\ddot{\text{O}}\text{H} \\ | \\ \text{Cl}-\text{Si}-\text{Cl} \\ | \\ \text{Cl} \end{array}\right]^-$$

One of the highly negative chlorines is then easily lost from the intermediate as a chloride ion.

$$\left[\begin{array}{c} \text{Cl}\;\ddot{:}\ddot{\text{O}}\text{H} \\ | \\ \text{Cl}-\text{Si}-\text{Cl} \\ | \\ \text{Cl} \end{array}\right]^- \; \longrightarrow \; \begin{array}{c} \text{Cl} \\ | \\ \text{Cl}-\text{Si}-\text{OH} \\ | \\ \text{Cl} \end{array} + \text{Cl}^-$$

This process is repeated until all four chlorine atoms are replaced by hydroxide groups, $Si(OH)_4$ being the final hydrolytic product. The conversion of the hydroxide ions of water to coordinated hydroxo groups leaves an excess of hydrogen ions in solution, and hydrochloric acid is formed. Resistance of the carbon tetrahalides to hydrolysis is probably due to the fact that carbn is **coordinately saturated** (all its binding orbitals are filled with electrons) in these compounds, and it cannot form intermediates of the type suggested for silicon tetrachloride. However, carbon tetrachloride, when hot, does react with water to produce phosgene $(COCl_2)$, which is extremely poisonous. Hence carbon tetrachloride fire extinguishers must be used with caution, particularly if water is also being used to fight a fire. Other halides, such as sulfur hexafluoride, SF_6, do not hydrolyze because the central element is already coordinately saturated. On the other hand, tungsten hexachloride, WCl_6, hydrolyzes, because the maximum coordination number for tungsten is 8 and the central element is therefore not coordinately saturated.

The silicon halides undergo complete hydrolysis, but with halides of more ionic character, hydrolysis is often less complete and may be suppressed by the addition of acids. Thus aqueous solutions of tin tetrachloride can be prepared in the presence of hydrochloric acid.

38.8   Stannic Acids and Stannates

When an alkali metal hydroxide is added to a solution of a tin(IV) compound, a white precipitate forms. We would expect this compound to have the formula $Sn(OH)_4$ and to be acidic in character. However, neither the acid nor the salts derived from the acid have been found. On the other hand, alkali metal stannates are known that appear to be derived from an **orthostannic acid** which may be formulated as $H_2[Sn(OH)_6]$. The sodium salt of the acid, for example, may be written as $Na_2[Sn(OH)_6]$. Orthostannic acid readily loses water, yielding a **metastannic acid,** H_2SnO_3, and finally the anhydride, SnO_2. When this anhydride is fused with sodium hydroxide, **sodium metastannate,** Na_2SnO_3, is formed. The metastannic acid mentioned above is sometimes called **alpha-stannic acid** to distinguish it from **beta-stannic acid,** which is formed when hot concentrated nitric acid reacts with tin. Beta-stannic acid has the same composition as the alpha form, but it is insoluble in acids and only very slightly soluble in strongly alkaline solutions.

Lead

38.9   History, Occurrence, and Metallurgy of Lead

Because lead is easily extracted from its ores, it was known to the early Egyptians and Babylonians. It is mentioned in the Bible in Job and in Numbers. Lead pipes were commonly used by the Romans for conveying water, and in the Middle Ages lead was used as a roofing material. The stained glass windows of the great cathedrals of this age were set in lead.

The principal lead ore is the sulfide, PbS, commonly called **galena.** Other common ores of lead are the carbonate, **cerrusite** ($PbCO_3$), and the sulfate, **anglesite** ($PbSO_4$), both of which appear to have been formed by the weathering of sulfide ores.

Lead ores are first concentrated by a series of selective flotation processes to remove the gangue materials and a large part of the zinc sulfide that is usually associated with lead ores. The concentrated ore is then roasted in air to convert most of the sulfide to the oxide.

$$2PbS(s) + 3O_2(g) \longrightarrow 2PbO(s) + 2SO_2(g) \qquad \Delta H^\circ = -710.4 \text{ kJ}$$

The roasted product is reduced in a blast furnace with coke and scrap iron.

$$PbO(s) + C(s) \longrightarrow Pb(s) + CO(g) \qquad \Delta H^\circ = 106.8 \text{ kJ}$$
$$PbO(s) + CO(g) \longrightarrow Pb(s) + CO_2(g) \qquad \Delta H^\circ = -65.7 \text{ kJ}$$
$$PbS(s) + Fe(s) \longrightarrow Pb(s) + FeS(s) \qquad \Delta H^\circ = 0.4 \text{ kJ}$$

The lead obtained from the blast furnace contains copper, antimony, arsenic, bismuth, gold, and silver. The crude lead is melted and stirred to bring about the oxidation of antimony, arsenic, and bismuth. The oxides of these metals rise to the surface, and the molten lead is drained off for further refining. Gold and silver may be extracted from the lead by the Parkes process (Section 35.11), or by the electrolytic **Betts process.** In the Betts process thin sheets of pure lead are made the cathodes and plates of impure lead the anodes. The electrolyte is a solution containing lead hexafluosilicate, $PbSiF_6$, and hexafluosilicic acid, H_2SiF_6.

An alternative method of metallurgy for lead involves the electrolysis of lead sulfide dissolved in molten lead chloride. Lead is liberated at the cathode and sulfur at the anode.

38.10   Properties and Uses of Lead

Lead is a soft metal having little tensile strength, and it is the heaviest of the common metals except gold and mercury. Lead has a metallic luster when freshly cut but quickly acquires a dull gray color when exposed to moist air. In air that contains moisture and carbon dioxide, lead becomes oxidized on the surface, forming a protective layer that is both compact and adherent; this film is probably the basic carbonate.

Lead lies above hydrogen in the activity series and dissolves slowly in dilute nonoxidizing acids. Concentrated nitric acid attacks it readily. It is not dissolved by pure water in the absence of air, but in the presence of air it reacts with water to form the hydroxide.

$$2Pb + 2H_2O + O_2 \longrightarrow 2Pb(OH)_2$$

When heated in a stream of air, lead burns. It also reacts with sulfur, fluorine, and chlorine.

The uses of lead depend mainly upon the ease with which it is worked, its low melting point, its great density, and its resistance to corrosion.

38.11   Compounds of Lead

Lead forms two well-defined series of compounds in which its oxidation states are +2 and +4. An atom of lead (like those of its congeners carbon, silicon, germanium, and tin) has four valence electrons; two of these are s electrons and the other two are p electrons. All four of the valence electrons are seldom, if ever, completely removed from the atom but are very often shared with electronegative elements. This latter fact accounts for the +4 oxidation state, which is typical of the Group IVA elements. Lead (and also tin) forms many compounds in which its two s valence electrons do not participate in the bonding but remain associated with the core of the atom as a stable electron pair. When this happens the element assumes the +2 oxidation state. Because of the high oxidation potential of tetravalent lead ($Pb^{2+} \longrightarrow PbO_2$, −1.5 volts), the +2 oxidation state of this element is often considered to be the characteristic oxidation state.

▶ **1. OXIDES OF LEAD.** When lead is heated in air the yellow, powdery **monoxide,** PbO, is obtained.

Lead dioxide, PbO_2, is a chocolate-brown powder formed by oxidizing lead(II) compounds in alkaline solution. With sodium hypochlorite as the oxidizing agent, the equation for the oxidation of the plumbite ion is

$$Pb(OH)_3^- + ClO^- \longrightarrow PbO_2(s) + Cl^- + OH^- + H_2O$$

Lead dioxide is the principal constituent of the cathode of the charged lead storage battery (Section 20.17). Since lead +4 tends to revert to the stable +2 state by gaining two electrons, lead dioxide is a powerful oxidizing agent.

The oxide Pb_3O_4, called **red lead** or **trilead tetraoxide,** is prepared by carefully heating the monoxide in air at temperatures between 400° and 500°C. When red lead is treated with nitric acid, two-thirds of the lead dissolves as lead nitrate (oxidation state +2), and the remaining third remains as lead dioxide (oxidation state +4). The equation for the reaction is

$$Pb_3O_4 + 4H^+ + [4NO_3^-] \longrightarrow PbO_2(s) + 2Pb^{2+} + [4NO_3^-] + 2H_2O$$

▶ **2. LEAD HYDROXIDE.** When alkali metal hydroxides are added to solutions of lead(II) compounds, white **lead hydroxide,** $Pb(OH)_2$, precipitates. An excess of alkali causes the hydroxide to dissolve according to the equation

$$Pb(OH)_2(s) + OH^- \longrightarrow Pb(OH)_3^-$$

which shows the amphoteric character of the compound. Its reaction with hydrogen ions is given by the equation

$$Pb(OH)_2(s) + 2H^+ \longrightarrow Pb^{2+} + 2H_2O$$

▶ **3. LEAD CHLORIDE.** Lead chloride, $PbCl_2$, may be formed by the direct union of the elements, by the action of hydrochloric acid upon lead monoxide, or by precipitation from solutions containing lead(II) ions and chloride ions. It is soluble in hot water and in solutions of high chloride concentration, **tetrachloroplumbate(II) ions** being formed in the latter case.

$$PbCl_2(s) + 2Cl^- \rightleftharpoons PbCl_4^{2-}$$

▶ **4. LEAD NITRATE.** When either metallic lead or lead monoxide, PbO, is dissolved in nitric acid, **lead nitrate,** $Pb(NO_3)_2$ is formed. This salt is readily soluble in water, but unless the solution is slightly acid with nitric acid, hydrolysis occurs and basic nitrates are precipitated.

$$Pb^{2+} + NO_3^- + H_2O \rightleftharpoons Pb(OH)NO_3(s) + H^+$$

Lead nitrate is unstable at moderately high temperatures and decomposes in the same manner as the nitrates of other heavy metals.

$$2Pb(NO_3)_2 \longrightarrow 2PbO(s) + 4NO_2 + O_2$$

▶ **5. CARBONATES OF LEAD.** The normal carbonate of lead, $PbCO_3$, may be prepared by the action of sodium hydrogen carbonate upon lead chloride. The basic carbonate, $Pb_3(OH)_2(CO_3)_2$, is formed when alkali metal carbonates are added to solutions containing the lead ion. This compound is important commercially as the paint pigment **white lead,** though because of the danger of lead poisoning its use is restricted (especially for objects used by children). It is prepared commercially by the action of air, carbon dioxide, and acetic acid vapor upon lead metal. The essential reactions for the process may be represented by

$$2Pb + O_2 + 2HOAc \longrightarrow 2Pb(OH)OAc \quad \text{(HOAc represents acetic acid)}$$
$$6Pb(OH)OAc + 2CO_2 \longrightarrow Pb_3(OH)_2(CO_3)_2 + 3Pb(OAc)_2 + 2H_2O$$

The $Pb(OAc)_2$ that is formed in the second reaction is eventually used up in the process, with the regeneration of acetic acid.

Although the covering power of white lead paints is excellent, they have the disadvantage of turning dark in the presence of hydrogen sulfide due to the formation of black lead sulfide.

▶ **6. OTHER LEAD COMPOUNDS.** Lead sulfate, $PbSO_4$, is formed by ionic combination. It is insoluble in water but readily dissolves in solutions containing an excess of alkali metal or acetate ions. **Lead acetate,** $Pb(OAc)_2$, is one of the few soluble compounds of lead; it is a weak electrolyte, indicating that it dissolves mainly as a covalent compound rather than as an ionic one. It is extremely toxic. The **chromate,** $PbCrO_4$, is insoluble in water but dissolves readily in acid and alkali metal hydroxides. Hydrogen sulfide precipitates black **lead sulfide,** PbS, which is insoluble in dilute acids and alkali metal sulfides.

The organometallic **lead tetraethyl,** $Pb(C_2H_5)_4$, is a covalent compound that is liquid at ordinary temperatures. It has been used extensively in the production of "antiknock" gasoline but with the present concern over air pollution and the fact that it poisons the catalysts in the catalytic converters its use in gasoline is diminishing rapidly (Section 22.4). It is prepared by the reaction of ethyl chloride, C_2H_5Cl, with a sodium-lead alloy, $Na_{31}Pb_8$, which is an intermetallic compound (Section 30.8).

Group IVB

The elements of Group IVB (titanium, zirconium, and hafnium) are transition metals, with valence electrons in their two outermost electron shells. They most commonly exhibit the oxidation state of $+4$, two electrons of the incomplete inner subshell and the two electrons of the outer shell being used in bonding. These elements, however, also show the oxidation states of $+2$, and $+3$. Zirconium and hafnium are even more similar in chemical properties than would be expected from their being in the same family. This remarkable similarity results from their similarity in size, which arises because of the Lanthanide Contraction (Section 37.11). Some physical properties of titanium, zirconium, and hafnium are given in Table 38-2.

TABLE 38-2   Some Physical Properties of the Metals of Group IVB

	Atomic Number	Atomic Weight	Atomic Radius, Å	Ionic (M^{4+}) Radius, Å	Density at 20°	Melting Point, °C	Boiling Point, °C
Titanium	22	47.90	1.46	0.64	4.49	1812	3260
Zirconium	40	91.22	1.57	0.87	6.4	1852	4375
Hafnium	72	178.49	1.57	0.84	11.4	2227	about 5200

38.12   Titanium

Although titanium is commonly classed as a "less common" element, it is actually one of the more abundant elements. It is tenth among the elements in abundance in the earth's crust, thus ranking in abundance above such useful metals as nickel, copper, zinc, lead, tin, and mercury.

The most important ores of titanium from the commercial standpoint are **ilmenite,** $FeTiO_3$, and **rutile,** TiO_2. The dioxide is an excellent white pigment for paints and is used in the preparation of white rubber and white leather. In production of chemicals in the United States titanium dioxide ranks 49th, with 718 thousand tons produced in 1978.

Titanium metal is very difficult to obtain in the pure state because it has a high melting point (1812°), it combines readily with such nonmetals as oxygen, nitrogen, hydrogen, and carbon, and it readily forms alloys with the common metals. The free metal has been most successfully prepared by reducing titanium tetrahalides with active metals such as sodium, potassium, magnesium, and calcium. Titanium metal is very strong, light (specific gravity is 4.49), high melting, and resistant to corrosion. These properties make the metal and its special alloys valuable in the production of jet motors and high speed aircraft. An alloy, ferrotitanium, is used in making special steels of great strength and toughness, with the titanium acting as a "scavenger" to remove nitrogen and other undesirable impurities.

Titanium forms three oxides, TiO, Ti_2O_3, and TiO_2, and their corresponding hydroxides. The dioxide may be prepared from ilmenite by the reaction

$$2FeTiO_3 + 4HCl + Cl_2 \longrightarrow 2FeCl_3 + 2TiO_2 + 2H_2O$$

The temperature is held high enough to volatilize the iron(III) chloride that is formed.

Titanium(IV) chloride, $TiCl_4$, is a liquid boiling at 136.4°. It is produced by passing chlorine over a heated mixture of carbon and titanium dioxide.

$$TiO_2(s) + C(s) + 2Cl_2(g) \longrightarrow TiCl_4(l) + CO_2(g) \qquad \Delta H° = -252.9 \, kJ$$

The tetrachloride hydrolyzes quickly in moist air, producing a dense white smoke consisting of finely divided TiO_2. Titanium tetrachloride is used in producing smoke screens and in skywriting.

The **trichloride,** $TiCl_3$, and the **sulfate,** $Ti_2(SO_4)_3$, are powerful reducing agents, and in this capacity are used in the chemical laboratory. Fused alkalies react with titanium dioxide, forming titanates such as **sodium titanate,** Na_2TiO_3.

38.13   Zirconium

Zirconium is not as abundant as titanium but it is far more abundant than such familiar metals as lead, copper, nickel, zinc, mercury, and tin. Its chief ores are **zircon,** $ZrSiO_4$, and **baddeleyite,** ZrO_2. The metal is obtained by heating K_2ZrF_6 with sodium, potassium, or aluminum. Ferrozirconium has been used with some success in the steel industry in the production of a tough steel.

Zirconium dioxide, ZrO_2, which is called **zirconia,** is the most important compound of the element because of its excellent refractory qualities. These include a high melting point (about 2700°), a low coefficient of expansion, and a high resistance to corrosion.

Although zirconia is attacked by scarcely any acid except hydrofluoric, it reacts with fused alkalies, forming **zirconates** such as Na_2ZrO_3. Insoluble zirconates are formed when the dioxide is used in the production of enamels and opaque glass.

Zirconium forms the **tetrachloride,** $ZrCl_4$, which is not as readily hydrolyzed as titanium tetrachloride; hydolysis results in the formation of derivatives such as the **oxychloride,** $ZrOCl_2$, instead of the oxide as is the case with $TiCl_4$.

Zircon, the naturally occurring silicate, $ZrSiO_4$, is found in a variety of colors, and because of its beauty and hardness, the mineral is used as a semiprecious stone in jewelry.

38.14   Hafnium

This element was discovered in 1923 by Coster and Hevesy in zircon from Norway by means of spectroscopic analysis. It is found in nearly all zirconium minerals, most of which contain about 5% hafnium. The fact that hafnium has chemical properties very similar to those of zirconium makes the separation of the two elements difficult (see Section 37.11).

Questions

1. How and why is tin sometimes plated on the surface of iron?
2. What is meant by *tin pest,* which is also known as *tin disease?*
3. What is the action of metallic tin with HCl, with HNO_3, and with NaOH?
4. Write the equations for the reactions involved when an excess of aqueous NaOH is slowly added to a solution of tin(II) chloride.
5. Write the equation for the thermal decomposition of tin(II) oxalate.
6. Write balanced chemical equations describing:
   (a) The preparation of sodium metastannate
   (b) The purification of tin by electrolysis
   (c) The action of dry chlorine on tin
   (d) The formation of thiostannate ion
   (e) The dissolution of tin by alkali solution
7. Why cannot $SnCl_4$ be classified as a salt?
8. Why must aqueous solutions of $SnCl_2$ be protected from the air?
9. Describe the Betts process for the refining of lead.
10. Compare the nature of the bonds in $PbCl_2$ to those in $PbCl_4$. Would you expect the existence of Pb^{4+} ions? Explain.
11. Show by suitable equations that lead(II) hydroxide is amphoteric.
12. What is the composition of the tarnish on lead?
13. Why should water to be used for human consumption not be conveyed in lead pipes?
14. Account for the solubility of $PbCl_2$ in solutions of high chloride concentration.
15. When elements in the first transition metal series form dipositive ions, they usually give up their *s* electrons. Does this apply to lead when it forms the +2 ion?
16. How is the paint pigment *white lead* prepared commercially?
17. Mercury(I) solutions can be protected from air oxidation to mercury(II) by placing metallic mercury in contact with the solutions. Can tin(II) solutions be similarly stabilized against oxidation to tin(IV) by keeping metallic tin in contact with the solutions? (The standard reduction potential for tin(II) to tin(IV) is +0.15 V.)
18. What chemistry is involved in skywriting using titanium tetrachloride?
19. Why is titanium difficult to obtain in the pure state?
20. What are the hydrolysis products of $TiCl_4$?
21. What properties make zirconia effective as a refractory?

Problems

1. One hundred milliliters of a saturated solution of lead(II) iodide contains 0.41 g of solute at 100°C. Calculate the solubility product at this temperature.
   Ans. 2.8×10^{-6}
2. A 1.497-g sample of type metal is dissolved in nitric acid, whereupon metastannic acid precipitates. This is dehydrated by heating to stannic oxide which is found to weigh 0.4909 g. What per cent tin was the original type-metal sample?   *Ans. 25.83%*

References

"Organotin Chemistry," A. G. Davies, *Chem. in Britain,* **4,** 403 (1968).

"Resolution and Stereochemistry of Asymmetric Silicon, Germanium, Tin, and Lead Compounds," R. Belloli, *J. Chem. Educ.,* **46,** 640 (1969).

"Hydrolysis of Group IV Chlorides," C. H. Yoder, *J. Chem. Educ.,* **46,** 382 (1969).

"Electronegativities and Group IV A Chemistry," D. A. Payne, Jr. and F. H. Fink, *J. Chem. Educ.,* **43,** 654 (1966).

"Applied Research in the Development of Anticaries Dentifrices," W. E. Cooley, *J. Chem. Educ.,* **47,** 177 (1970).

"Tin(IV) Iodide as a Radioactive Tracer," P. W. Wiggans, *Educ. in Chemistry,* **11,** 194 (1974).

"Biologically Active Organotin Polymers," Z. M. O. Rzaev, *Chemtech,* **9,** 58 (1979).

"Titanium(III) Citrate as a Nontoxic Oxidation-Reduction

Buffering System for the Culture of Obligate Anaerobes," A. J. B. Zehnder and K. Wuhrmann, *Science,* **194,** 1165 (1976).

"Chemicals in the Manufacture of Paint," W. C. Weber, *J. Chem. Educ.,* **37,** 323 (1960).

"Lead Poisoning," J. J. Chisolm, Jr., *Sci. American,* February 1971, p. 15.

"Central Nervous System Disfunction Due to Lead Exposure," J. A. Valciukas *et al., Science,* **201,** 465 (1978).

"Lead Pollution—A Growing Hazard to Public Health," D. Bryce-Smith, *Chem. in Britain,* **7,** 54 (1971).

"Lead in the Environment," A. L. Mill, *Chem. in Britain,* **7,** 160 (1971).

"Unleaded Gasoline" (Staff), *Chem. and Eng. News,* March 8, 1971, p. 14.

"A Simple Titrimetric Determination of Lead in Gasoline," S. L. Watt, T. M. Martina, M. A. Chamberlin, and Patty H.

Laswick, *J. Chem. Educ.,* **54,** 262 (1977).

"Our Daily Lead," T. J. Chow, *Chem. in Britain,* **9,** 258 (1973).

"Lead in Blood—the Analyst's Problem," A. A. Cernik, *Chem. in Britain,* 10, **58** (1974).

"Lead in Petrol," F. D. Porter, *Chem. in Britain,* **10,** 61 (1974).

"Lead in Food: Are Today's Regulations Sufficient?" D. Bryce-Smith and H. A. Waldron, *Chem. in Britain,* **10,** 202 (1974).

"Titanium Dioxide." Staff, *Chem. and Eng. News,* May 29, 1978, p. 9.

"Zirconium Chemistry in Industry," W. B. Blumenthal, *J. Chem. Educ.,* **39,** 604 (1962).

"Homogeneous Models Throw Light on Industrial Catalysts" (Information on zirconium), D. G. H. Ballard, *Chem. in Britain,* **10,** 20 (1974).

"Hair Element Content in Learning Disabled Children," R. O. Phil and M. Parkes, *Science,* **198,** 205 (1977).

39

The Metals of Group VA

39.1 Periodic Relationships

Periodic Group VA includes the elements nitrogen, phosphorus, arsenic, antimony, and bismuth. These elements furnish an excellent illustration of the gradation in properties characteristic of groups of the periodic system. Nitrogen, the lightest member of the group, is a typical nonmetal in that it accepts electrons from active metals and forms the nitride ion N^{3-}, as in magnesium nitride, Mg_3N_2, and lithium nitride, Li_3N. Bismuth is the heaviest member of the group, and it is almost entirely metallic in character. It gives up three of its five valence electrons to active nonmetals to form the tripositive ion, Bi^{3+}. The nonmetallic character of nitrogen and the metallic character of bismuth, each of which has five valence electrons, can be attributed, for the most part, to differences in the distance between the valence shell and the nucleus for the two atoms. Because the distance between the valence shell and the nucleus is greater for bismuth, the attraction of the positive nucleus for the valence electrons is much smaller in an atom of bismuth than in an atom of nitrogen.

The properties of phosphorus, arsenic, and antimony are intermediate between those of nitrogen and bismuth, with the nonmetallic character becoming less pronounced with increasing atomic weight. Atoms of each member of this family have five valence electrons and commonly exhibit oxidation states of $+3$ and $+5$.

Some of the more important physical properties of arsenic, antimony, and bismuth are given in Table 39-1.

TABLE 39-1   Important Physical Properties of Group VA Metals

	Atomic Number	Atomic Weight	Atomic Radius, Å	Ionic (M^{5+}) Radius, Å	Density at 20°	Melting Point, °C	Boiling Point, °C
Arsenic	33	74.9216	1.21	0.47	5.73	817 (36 atm)	610 (sublimes)
Antimony	51	121.75	1.41	0.62	6.68	630.5	1440
Bismuth	83	208.9804	1.46	0.74	9.78	271.0	1420.0

Arsenic

39.2   Occurrence and Preparation of Arsenic

Arsenic is found both as the free element and combined. Its principal native compounds are **arsenopyrite,** FeAsS; **realgar,** As_2S_2; and **orpiment,** As_2S_3. Arsenic is widely distributed in trace amounts in the sulfide ores of many metals; consequently, the metals and the sulfuric acid obtained from these sulfides frequently contain arsenic as an impurity. A large part of the arsenic used in the United States is a by-product from the gases of copper furnaces, from which arsenic(III) oxide is collected by Cottrell precipitators (Section 13.33).

Pure arsenic is obtained by subliming native arsenic. When arsenopyrite is heated it decomposes according to the equation

$$4FeAsS(s) \longrightarrow 4FeS(s) + As_4(g) \qquad \Delta H° = -88.7 \text{ kJ}$$

When arsenic ores are roasted, As_4O_6 is formed. Reduction of the oxide by heating with carbon produces the element.

$$As_4O_6(s) + 6C(s) \longrightarrow As_4(g) + 6CO(g) \qquad \Delta H° = 795 \text{ kJ}$$

39.3   Properties and Uses of Arsenic

Elemental arsenic exists in several allotropic modifications. Ordinary **gray arsenic** is monatomic (As). It is metallic in appearance and sublimes at 615°C, forming tetraatomic molecules, As_4, which are tetrahedral in structure like P_4 (Section 24.2). When arsenic vapor is cooled rapidly an unstable yellow crystalline allotrope is produced, which also consists of As_4 molecules and is soluble in carbon disulfide. Arsenic vapor is yellow in color, has the odor of garlic, and is very poisonous. Arsenic is a member of an intermediate class of elements that possess, to some degree, the properties of both metals and nonmetals and hence is a metalloid (Section 8.3). Arsenic ignites when heated, producing white clouds of As_4O_6. It is slowly attacked by hot hydrochloric acid in the presence of air, forming arsenic(III) chloride, $AsCl_3$. Hot nitric acid readily oxidizes arsenic to arsenic acid, H_3AsO_4. Arsenic combines directly with sulfur, the halogens, and many metals. The trihalides (AsX_3) are either liquids or low-melting solids with properties suggestive of covalent rather than ionic bonding, as observed in the absence of saltlike character.

Certain alloys contain arsenic as a hardening agent. Examples are bronze and lead shot.

39.4   Compounds of Arsenic

▶ **1. ARSINE.** Arsine, AsH_3, a compound that has a formula analogous to that of ammonia (NH_3), is formed by reducing arsenic compounds with zinc in acid solution or by the action of a metallic arsenide, such as zinc arsenide, with acids.

$$As_4O_6 + 12Zn + 24H^+ \longrightarrow 4AsH_3(g) + 12Zn^{2+} + 6H_2O$$
$$Zn_3As_2 + 6H^+ \longrightarrow 3Zn^{2+} + 2AsH_3(g)$$

Arsine does not react with or dissolve in water or acids to form compounds corresponding to ammonium compounds. It is unstable and decomposes into its elements when heated ($4AsH_3 \longrightarrow As_4 + 6H_2$).

▶ **2. ARSENIC(III) OXIDE AND ARSENOUS ACID.** The substance commonly referred to as "arsenic," or "white arsenic," is actually the oxide As_4O_6. It is the product of the combustion of the element or its compounds. It has a sweet taste and is a violent poison. Arsenic(III) oxide is slowly dissolved by water with the formation of **arsenous acid,** H_3AsO_3, an acid of unknown structure.

$$As_4O_6 + 6H_2O \rightleftharpoons 4H_3AsO_3 \quad [\text{or } As(OH)_3]$$

Arsenous acid is amphoteric as shown by the equations

$$As(OH)_3 \rightleftharpoons As(OH)_2^+ + OH^- \rightleftharpoons AsOH^{2+} + OH^- \rightleftharpoons As^{3+} + OH^-$$
$$H_3AsO_3 \rightleftharpoons H^+ + H_2AsO_3^-$$

Hence it acts as a base in the presence of strong acids and as an acid in the presence of strong bases.

$$As(OH)_3 + 3H^+ + 3Cl^- \rightleftharpoons AsCl_3 + 3H_2O$$
$$H_3AsO_3 + [Na^+] + OH^- \rightleftharpoons [Na^+] + H_2AsO_3^- + H_2O$$

Arsenous acid is known only in solution; when attempts are made to isolate it as a solid, it loses water and forms the oxide. All arsenites except those of the alkali metals are insoluble in water.

▶ **3. ARSENIC(V) OXIDE AND ARSENIC ACID.** **Arsenic(V) oxide,** which has the empirical formula As_2O_5 (the structure is not known), can be produced by heating **arsenic acid,** H_3AsO_4.

$$2H_3AsO_4 \longrightarrow As_2O_5 + 3H_2O$$

Arsenic acid is formed when arsenic(III) oxide is oxidized by concentrated nitric acid.

$$As_4O_6 + 8HNO_3 + 2H_2O \longrightarrow 4H_3AsO_4 + 8NO_2(g)$$

By careful heating, H_3AsO_4 may be converted stepwise into pyroarsenic acid, $H_4As_2O_7$, **meta-arsenic acid,** $HAsO_3$, and finally the oxide, As_2O_5.

Arsenic acid is a much stronger acid than arsenous acid, as would be expected from the higher oxidation state (see Section 14.16), and it does not react as a base with strong acids as does arsenous acid. Arsenic acid is rather easily reduced, so that it can be used as an oxidizing agent. The arsenates of calcium and lead are used as insecticides.

▶4. **SULFIDES OF ARSENIC.** When hydrogen sulfide is passed into a hydrochloric acid solution of arsenic(III) chloride, the corresponding yellow sulfide is precipitated.

$$2AsCl_3 + 3H_2S \longrightarrow As_2S_3(s) + 6H^+ + 6Cl^-$$

Arsenic(III) sulfide dissolves readily in sodium sulfide, forming the thioarsenite, Na_3AsS_3.

$$As_2S_3 + 3S^{2-} \longrightarrow 2AsS_3^{3-}$$

Acids reprecipitate the sulfide from solutions of thioarsenites.

$$2AsS_3^{3-} + 6H^+ \longrightarrow As_2S_3(s) + 3H_2S(g)$$

Arsenic(V) sulfide, As_2S_5, is a yellow compound obtained very slowly as a precipitate when hydrogen sulfide is passed into a solution of arsenic acid in hydrochloric acid.

$$2H_3AsO_4 + 5H_2S \longrightarrow As_2S_5(s) + 8H_2O$$

The arsenic acid oxidizes hydrogen sulfide in cold, weakly acidic solutions.

$$2H_3AsO_4 + 5H_2S \longrightarrow As_2S_3(s) + 2S(s) + 8H_2O$$

Iodide ions are a catalyst in the precipitation of arsenic(III) sulfide from the arsenate ion.

$$2I^- + AsO_4^{3-} + 8H^+ \longrightarrow As^{3+} + 4H_2O + I_2$$

$$I_2 + H_2S \longrightarrow 2H^+ + 2I^- + S(s)$$
$$2As^{3+} + 3H_2S \longrightarrow As_2S_3(s) + 6H^+$$

Arsenic(V) sulfide dissolves in sodium sulfide with the formation of the **thioarsenate,** Na_3AsS_4. The addition of an acid reprecipitates the sulfide, As_2S_5.

39.5   Uses of Arsenic Compounds

The most important uses of arsenic compounds depend upon their poisonous character. They are used as weed killers, cattle and sheep dips, and insecticides. The compounds most widely used for these purposes are **Paris green** $[Cu_3(AsO_3)_2 \cdot Cu(C_2H_3O_2)_2]$, lead arsenate, calcium arsenate, sodium arsenite, and arsenous acid.

Antimony

39.6   History of Antimony

Antimony has been used, both as a metal and in compounds, for at least 5000 years. The Chaldeans used the metal in making vases and ornaments. The sulfide, Sb_2S_3, was used by the Egyptians as a pigment and for painting eyebrows at least as early as 3000 B.C. Compounds of antimony were also used in medicine. Basil Valentine, a Benedictine monk of the fifteenth century, collected all that the alchemists knew about this element in his treatise, *The Triumphal Chariot of Antimony*.

39.7   Occurrence and Metallurgy of Antimony

The principal ore of antimony is **stibnite**, Sb_2S_3, most of which is supplied to the world by China, Mexico, Argentina, Bolivia, and Chile. The sulfide is separated from earthy material of the ore by melting the Sb_2S_3; it is then reduced by heating with iron.

$$Sb_2S_3(s) + 3Fe(s) \longrightarrow 2Sb(s) + 3FeS(s) \qquad \Delta H° = -125 \text{ kJ}$$

The ore may also be roasted to the oxide, from which antimony may be obtained by reduction with carbon.

$$Sb_4O_6(s) + 6C(s) \longrightarrow 4Sb(s) + 6CO(g) \qquad \Delta H° = 777.4 \text{ kJ}$$

39.8   Properties and Uses of Antimony

Antimony is a lustrous silver-white metal, brittle, and readily pulverized. It tarnishes only slightly in dry air but does oxidize slowly in moist air. The metal is readily attacked by the halogens, phosphorus, and sulfur. It is attacked, but not dissolved, by hot nitric acid, forming Sb_4O_6. It is slowly dissolved by hot concentrated sulfuric acid, forming $Sb_2(SO_4)_3$ and evolving SO_2. The fact that the metal and its oxides do not react with nitric acid to give the nitrate indicates a lack of marked base-forming properties. However, the metallic nature of the element is more pronounced than that of arsenic, which lies immediately above it in Group VA.

Lead hardened by the addition of 10 to 20% of antimony is suitable for shrapnel, bullets, and bearings. Because of its resistance to corrosion by acids, it is used in making storage battery plates. Babbitt metal is an antifriction alloy of antimony with tin and copper, which is used in machine bearings. Since antimony, like bismuth, expands on freezing, it is used as a constituent of type metal, to which it confers this property, thus giving a sharp and distinct imprint of the type.

39.9   Compounds of Antimony

▶ **1. STIBINE.** The compound **stibine,** SbH_3, is prepared by reducing other antimony compounds with zinc in acid solution. Stibine is poisonous and unstable toward heat. It produces a mirror on a solid surface placed in the vapor when heated in the absence of air, as does arsine. The two compounds may be distinguished by the fact that the antimony deposit is insoluble in sodium hypochlorite, whereas the arsenic deposit is dissolved by this reagent.

▶ **2. OXIDES AND ACIDS OF ANTIMONY.**   Antimony(III) oxide, Sb_4O_6, and **antimony(V) oxide**, Sb_2O_5, resemble the oxides of arsenic in their preparation and properties. Both antimony oxides form acids with water, and antimony(III) oxide exhibits some basic properties in forming salts with acids. The antimony in **antimonic acid** has a coordination number of 6, the formula being $HSb(OH)_6$. **Sodium antimonate**, $NaSb(OH)_6$, is one of the very few relatively insoluble salts of sodium. Sb_2O_4 is also known; it is formed when either Sb_4O_6 or Sb_2O_5 is heated with a free access to air.

▶ **3. ANTIMONY HALIDES.**   Antimony forms all of the **trihalides** and **pentahalides** except the pentabromide and the pentaiodide. The trihalides and pentahalides that can be made are prepared either by direct union of the elements or by the action of the hydrohalic acids on the oxides of antimony. Hydrolysis of antimony(III) chloride forms the oxychloride, which is white and insoluble.

$$SbCl_3 + H_2O \rightleftharpoons SbOCl(s) + 2H^+ + 2Cl^-$$

An excess of hydrochloric acid changes the oxychloride to **tetrachloroantimonous acid**, $HSbCl_4$.

▶ **4. SULFIDES OF ANTIMONY.**   Antimony(III) sulfide, Sb_2S_3, is interesting in that it occurs in two strikingly different modifications: a black form, found in nature as the mineral stibnite and also formed when antimony and sulfur are heated together; and a brilliant orange-red form, obtained by the reaction of hydrogen sulfide with slightly acidified solutions of compounds of trivalent antimony. The orange-red modification changes to the black one on being heated or on standing in contact with a dilute solution of an acid.

Antimony(III) sulfide dissolves in sodium sulfide solutions, forming **thioantimonite ions.**

$$Sb_2S_3(s) + 3S^{2-} \longrightarrow 2SbS_3^{3-}$$

Acidification of the solution causes the sulfide to reprecipitate.

$$2SbS_3^{3-} + 6H^+ \longrightarrow Sb_2S_3(s) + 3H_2S(g)$$

Excess hydrochloric acid dissolves antimony(III) sulfide with the formation of the **tetrachloroantimonate(III) ion**, $SbCl_4^-$, and hydrogen sulfide.

$$Sb_2S_3(s) + 6H^+ + 8Cl^- \longrightarrow 2SbCl_4^- + 3H_2S(g)$$

Antimony(V) sulfide, Sb_2S_5, is orange in color and is similar to Sb_2S_3 in many of its reactions. Treatment of antimony(V) sulfide with an excess of hydrochloric acid causes reduction of the antimony to the trivalent state.

$$Sb_2S_5(s) + 6H^+ + 8Cl^- \longrightarrow 2SbCl_4^- + 3H_2S(g) + 2S(s)$$

▶ **5. POTASSIUM ANTIMONYL TARTRATE.** When antimony(III) oxide is heated with a solution of potassium hydrogen tartrate, $KHC_4H_4O_6$, **potassium antimonyl tartrate,** $KSbOC_4H_4O_6$, known as "tartar emetic," is formed. Like antimony oxychloride, this compound contains the univalent **antimonyl group** (SbO).

Bismuth

39.10   Occurrence and Metallurgy of Bismuth

Bismuth is most often found in the free state in nature. It sometimes occurs in combination, as **bismuth ocher,** Bi_2O_3, and **bismuth glance,** Bi_2S_3. Bismuth is produced in the United States as a by-product of the refining of other metals, particularly lead. Ores containing the free metal are treated by heating them in inclined iron pipes, whereupon the metal melts and flows away from the gangue. The oxide and sulfide ores are roasted and then heated with charcoal. As the bismuth is set free, it melts and collects beneath the less dense material.

39.11   Properties and Uses of Bismuth

Bismuth is a lustrous, hard, and brittle metal with a reddish tint. The metal burns when heated in air, giving the oxide, Bi_2O_3, but oxidizes only superficially in moist air at ordinary temperatures to form a coating of oxide, which protects it against further oxidation. Bismuth reduces the hydrogen in steam and combines directly with the halogens and sulfur. Oxidizing acids, such as hot sulfuric and nitric, dissolve it forming the respective salts. It is dissolved slowly by hydrochloric acid in the presence of air, forming the chloride, $BiCl_3$. An unstable hydride of bismuth, BiH_3, exists; it is called **bismuthine.**

Like antimony, melted bismuth expands upon solidifying, a most unusual property. Because of this, it is used in the formation of alloys to prevent them from shrinking upon solidification. Alloys of bismuth, tin, and lead have low melting points, which makes them useful for electrical fuses, safety plugs for boilers, and automatic sprinkler systems. Some of these alloys melt even in hot water. For example, **Rose's metal** (Bi, 50%; Pb, 25%; Sn, 25%) melts at 94°; and **Wood's metal** (Bi, 50%; Pb, 25%; Sn, 12.5%; Cd, 12.5%) melts at 65.5°.

39.12   Compounds of Bismuth

▶ **1. OXIDES AND HYDROXIDES.** When bismuth is burned in air or when the nitrate is strongly heated, yellow **bismuth(III) oxide,** Bi_2O_3, is formed. This oxide is basic, in contrast to the more acidic character of the lower oxide of the smaller elements in the family. Because of its basic nature, Bi_2O_3 dissolves in acids to form salts, and it does not exhibit acidic properties such as the ability to react with bases. Alkali metal hydroxides or aqueous ammonia precipitate the white **hydroxide,** $Bi(OH)_3$, from solutions of trivalent bismuth salts. When a suspension of bismuth hydroxide is boiled it loses the elements of a molecule of water, forming **bismuth oxyhydroxide,** BiO(OH).

Bismuth(V) oxide, Bi_2O_5, is produced by the action of very strong oxidizing agents upon the trioxide. As expected on the basis of the higher oxidation state of the bismuth, this oxide is more acidic than Bi_2O_3 and shows its acidic character by dissolving in concentrated sodium hydroxide to form **sodium bismuthate**, $NaBiO_3$. This salt reacts with nitric acid to give **bismuthic acid**, $HBiO_3$, which is a very strong oxidizing agent.

▶**2. BISMUTH(III) SULFIDE.** When hydrogen sulfide is passed into a solution of a bismuth salt, brown **bismuth(III) sulfide**, Bi_2S_3, is precipitated. The Bi_2S_3 does not dissolve in concentrated sulfide solutions to form thiosalts. This property is utilized in separating bismuth from arsenic and antimony, which do form soluble thiosalts.

▶**3. SALTS OF BISMUTH.** Bismuth(III) oxide dissolves in acids, forming the corresponding salts such as the chloride, $BiCl_3 \cdot 2H_2O$; the sulfate, $Bi_2(SO_4)_3$; and the nitrate, $Bi(NO_3)_3 \cdot 5H_2O$. Bismuth salts hydrolyze readily when water is added, forming hydroxysalts.

$$BiCl_3 + 2H_2O \rightleftharpoons Bi(OH)_2Cl(s) + 2H^+ + 2Cl^-$$

$$Bi(NO_3)_3 + 2H_2O \rightleftharpoons Bi(OH)_2NO_3(s) + 2H^+ + 2NO_3^-$$

The dihydroxychloride loses the elements of a molecule of water to form **bismuth oxychloride.**

$$Bi(OH)_2Cl \longrightarrow BiOCl(s) + H_2O$$

The dihydroxynitrate, when repeatedly washed with water, is converted to **bismuth hydroxide.**

$$Bi(OH)_2NO_3 + H_2O \longrightarrow Bi(OH)_3(s) + H^+ + NO_3^-$$

When dried, bismuth dihydroxynitrate forms the **oxynitrate**, $BiONO_3$.

39.13   Uses of Bismuth Compounds

The oxycarbonate, $(BiO)_2CO_3$, and the oxynitrate (under the names **bismuth subcarbonate** and **bismuth subnitrate**) are used in medicine for the treatment of stomach disorders such as gastritis and ulcers and of skin diseases such as eczema.

Questions

1. Arsenic is classed as a metalloid. Explain.
2. Compare the metallic character of arsenic with that of antimony.
3. Can the term "amphoteric" be applied propertly to the sulfides of antimony and arsenic? Explain why or why not.
4. Compare and contrast arsine with ammonia.
5. Account for the oxidation states $+3$ and $+5$ of arsenic in terms of electronic structure.
6. Draw structural representations of the pyroarsenic acid and meta-arsenic acid molecules.
7. Which is the more acidic, As_4O_6 or As_2O_5?

8. Write balanced equations to show the action of Na_2S upon As_2S_3 and As_2S_5.
9. What are the composition and use of Paris green?
10. Write balanced chemical equations describing the action of hot nitric acid on elemental arsenic; the action of hot concentrated sulfuric acid on elemental antimony; the oxidation of manganous ion to permanganate ion by the action of a solution made by acidifying sodium bismuthate with nitric acid.
11. Why can it be said that iodide ions serve as a catalyst in the precipitation of arsenic(III) sulfide from arsenate ion?
12. Why is antimony used as a constituent of type metal?
13. Why does not antimony dissolve in hot nitric acid?
14. What is the coordination number of antimony in the antimonate ion?
15. What are the properties of metallic bismuth that make it commercially useful?
16. Compare the basicity and acidity of Bi_2O_3 and Bi_2O_5.

17. Write equations to show the hydrolysis of $BiCl_3$.
18. Write balanced chemical equations for the following reactions:
    (a) Arsenic is burned in air.
    (b) Arsenic is added to a hot solution of hydrochloric acid.
    (c) As_2S_3 is roasted in air.
    (d) Arsenic(III) chloride is added to a solution of sodium sulfide.
    (e) Stibnite, Sb_2S_3, is heated in air.
    (f) Stibnite, Sb_2S_3, is added to an excess of hot sulfuric acid.
    (g) Bismuth is heated in air.
    (h) Bismuth is added to hot concentrated sulfuric acid.
    (i) BiH_3 is burned in air.
    (j) Bi_2O_3 is dissolved in a concentrated solution of lithium hydroxide.
    (k) Bi_2S_3 is roasted in air.

References

"Synthesis, Properties, and Hydrolysis of Antimony Trichloride," F. C. Hentz, Jr. and G. G.. Long, *J. Chem. Educ.*, **52**, 189 (1975).
"Factors Involved in the Stereochemistry of AX_6E Systems of the Heavy Main Group Elements," K. J. Wynne, *J. Chem. Educ.*, **50**, 328 (1973).

"*d*-Orbitals in Main Group Elements," T. B. Brill, *J. Chem. Educ.*, **50**, 392 (1973).
"It Isn't Easy Being King" (Arsenic poisoning), T. H. Maugh, *Science,* **203**, 637 (1979).

40

The Metals of
Groups VIB and VIIB

Group VIB—Chromium, Molybdenum, and Tungsten

40.1 Periodic Relationships of Group VIB Metals

The elements that constitute Group VIB of the Periodic Table are chromium, molybdenum, and tungsten. These metals differ distinctly from the nonmetals of Periodic Group VIA, oxygen and sulfur, which we have already considered (Chapters 9 and 21). Chromium, molybdenum, and tungsten are each used in the production of commercially important alloy steels. These metals are typical transition metals; they show a variety of oxidation states, form highly colored compounds, and enter into the formation of stable complex ions. Like those of the other transition metals their lowest oxides are basic, the intermediate ones are amphoteric, and the highest ones are primarily acidic. The members of Groups VIA and VIB show similarities in the higher oxidation states; thus, the sulfates (e.g. K_2SO_4) and the chromates (K_2CrO_4) are analogous in formula and in physical and chemical properties.

In Table 40-1 are listed some of the physical properties of all of the metals of Group VIB. However, only chromium will be discussed in detail in the following sections.

TABLE 40-1   Some of the Physical Properties of Group VIB Metals

	Atomic Number	Atomic Weight	Atomic Radius, Å	Ionic (M^{6+}) Radius, Å	Density at 20°	Melting Point, °C	Boiling Point, °C
Chromium	24	51.996	1.25	0.52	7.1	1900	2642
Molybdenum	42	95.94	1.36	0.62	9.0	2610	4825
Tungsten	74	183.85	1.37	0.68	19.3	3380	5900

40.2   Occurrence, Metallurgy, and Properties of Chromium

Chromium does not occur free in nature. Its most important ore is **chromite,** $FeCr_2O_4$, sometimes called **chrome iron ore.** Practically all of the chromium used in the United States is imported. Because chromium is essential to the production of special alloy steels and the supply is often short, it is classed as a strategic material by the United States Government.

Reduction of chromite by carbon in an electric furnace yields an alloy of iron and chromium, called **ferrochrome,** which is used in making chromium steels.

$$FeCr_2O_4 + 4C \longrightarrow Fe + 2Cr + 4CO$$

Pure chromium is prepared by a thermite reaction, reducing chromium(III) oxide with aluminum.

$$Cr_2O_3(s) + 2Al(s) \longrightarrow Al_2O_3(s) + 2Cr(s) \qquad \Delta H° = -536.0 \text{ kJ}$$

Chromium can also be produced by the electrolytic reduction of its compounds from aqueous solution, a process used in chromium plating.

Chromium is a very hard, silvery white, crystalline metal. It assumes a passive state by becoming coated with a thin layer of oxide, which protects it against further corrosive attack. This property, along with its metallic luster, accounts for its extensive use in the plating of iron and copper objects such as plumbing fixtures and automobile trim. Passivity is produced by the action of concentrated nitric acid, chromic acid, or exposure of the metal to air. When not in the passive state chromium dissolves readily in dilute acids with the evolution of hydrogen. Chromium is a transition metal that shows several oxidation states, the principal ones being +2, +3, and +6. The compounds of chromium are highly colored, each color exhibited being dependent to some extent upon a particular oxidation state and on the structure of the compound.

40.3   Compounds of Chromium

▶ **1. CHROMIUM(II) COMPOUNDS. Chromium(II) chloride,** $CrCl_2$, can be prepared by dissolving chromium in hydrochloric acid or by reducing a solution of chromium(III) chloride by zinc in the presence of an acid. The anhydous salt is colorless, but its solutions have the bright blue color of the hydrated ion, $[Cr(H_2O)_6]^{2+}$. This ion is rapidly oxidized in air to the chromium(III) ion, $[Cr(H_2O)_6]^{3+}$; for this reason solutions of chromium(II) salts are frequently employed to remove the last traces of oxygen from gases. The **sulfate**

$CrSO_4 \cdot 7H_2O$ is prepared by dissolving the metal in sulfuric acid. Chromium(II) **acetate**, $Cr(C_2H_3O_3)_2$, forms as a red precipitate when a saturated solution of sodium acetate is added to a solution of chromium(II) chloride. Chromium(II) acetate is one of few exceptions to the rule that acetates are soluble. **Chromium(II) hydroxide** is basic, and chromium(II) salts are only slightly hydrolyzed.

▶ **2. CHROMIUM(III) COMPOUNDS.** Green **chromium(III) oxide,** Cr_2O_3, is formed either by heating the metal in air, by igniting ammonium dichromate, or by reducing a dichromate with sulfur or another similar reducing agent.

$$(NH_4)_2Cr_2O_7(s) \longrightarrow N_2(g) + 4H_2O(g) + Cr_2O_3(s) \qquad \Delta H° = -300 \text{ kJ}$$
$$Na_2Cr_2O_7 + S \longrightarrow Na_2SO_4 + Cr_2O_3$$

This oxide finds use as a pigment called **chrome green.** The bluish green, gelatinous **chromium(III) hydroxide** is precipitated when either an alkali, a soluble sulfide, or a carbonate is added to a solution of a chromium(III) compound. Chromium(III) hydroxide is amphoteric, dissolving in both acids and bases.

$$Cr(OH)_3 + 3H^+ \longrightarrow Cr^{3+} + 3H_2O$$
$$Cr(OH)_3 + OH^- \longrightarrow [Cr(OH)_4]^- \quad \text{(chromite ion)}$$

Chromium(III) chloride 6-hydrate, $CrCl_3 \cdot 6H_2O$, and the sulfate octadecahydrate, $Cr_2(SO_4)_3 \cdot 18H_2O$, are the best known salts of trivalent chromium. Dilute solutions of the chloride are violet, the hexaaquochromium(III) ion, $[Cr(H_2O)_6]^{3+}$ being responsible for the color. In more concentrated solutions of the chloride the green $[Cr(H_2O)_4Cl_2]^+$ is formed. The hydrates of chromium(III) sulfate also exist in violet and green modifications. Trivalent chromium forms alums of the type $KCr(SO_4)_2 \cdot 12H_2O$, solutions of which are bluish violet when cold but green when hot. **Chrome alum** is used in the tanning of leather, the printing of calico, the waterproofing of fabrics, and as a mordant. It is the most important of the soluble chromium salts.

▶ **3. CHROMIUM(VI) COMPOUNDS.** In the +6 oxidation state chromium is usually combined with oxygen in the form of the oxide, **chromium trioxide,** CrO_3, or the oxyanions chromate, CrO_4^{2-}, and dichromate, $Cr_2O_7^{2-}$. **Potassium chromate,** K_2CrO_4, is produced commercially by the atmospheric oxidation of chromite ore admixed and heated with potassium carbonate.

$$4FeCr_2O_4 + 8K_2CO_3 + 7O_2 \longrightarrow 2Fe_2O_3 + 8K_2CrO_4 + 8CO_2$$

Lead chromate, $PbCrO_4$, and **barium chromate,** $BaCrO_4$, are both insoluble in water and are used as yellow pigments.

When an acid is added to a solution containing chromate ions, the solution changes from yellow to orange-red due to the formation of the **dichromate ion,** $Cr_2O_7^{2-}$.

$$2CrO_4^{2-} + 2H^+ \rightleftharpoons Cr_2O_7^{2-} + H_2O$$

The addition of a base or dilution with water reverses the reaction.

$$Cr_2O_7^{2-} + 2OH^- \rightleftharpoons 2CrO_4^{2-} + H_2O$$

The conversion of chromate to dichromate involves the formation of an oxygen linkage between two chromium atoms. The electronic structure of the dichromate ion is

$$
\left[
\begin{array}{c}
\ddot{\text{O}} \quad\quad \ddot{\text{O}} \\
\ddot{\text{O}}\!:\!\text{Cr}\!:\!\ddot{\text{O}}\!:\!\text{Cr}\!:\!\ddot{\text{O}} \\
\ddot{\text{O}} \quad\quad \ddot{\text{O}}
\end{array}
\right]^{2-}
$$

A mixture of sodium dichromate and sulfuric acid is used as a **cleaning solution** for glassware in the laboratory. The cleansing action is due to the strong oxidizing power of the dichromate and dehydrating action of the concentrated sulfuric acid. The dichromates are used as oxidizing agents in many reactions. The dichromate is readily reduced to the chromium(III) ion by reducing agents such as sulfurous acid.

$$Cr_2O_7^{2-} + 3H_2SO_3 + 2H^+ \longrightarrow 2Cr^{3+} + 3SO_4^{2-} + 4H_2O$$

As the reduction progresses, a change from the orange color of the dichromate to the green color of the chromium(III) ion occurs.

The trioxide CrO_3 is produced as scarlet needle-shaped crystals by the action of concentrated sulfuric acid upon a concentrated solution of potassium dichromate.

$$Cr_2O_7^{2-} + 2H^+ \longrightarrow 2CrO_3(s) + H_2O$$

The anhydride CrO_3 reacts with water in different proportions forming **chromic acid,** H_2CrO_4 or dichromic acid, $H_2Cr_2O_7$.

$$2CrO_3 + H_2O \longrightarrow 2H^+ + Cr_2O_7^{2-}$$
$$CrO_3 + H_2O \longrightarrow 2H^+ + CrO_4^{2-}$$

These solutions are used as active oxidizing agents. When heated the anhydride decomposes, liberating oxygen and forming the green chromium(III) oxide.

$$4CrO_3(s) \longrightarrow 2Cr_2O_3(s) + 3O_2(g) \qquad \Delta H^\circ = 497.1\,kJ$$

40.4   Uses of Chromium

A large portion of the chromium produced goes into steel alloys, which are very hard and strong. **Stainless steel,** which usually contains chromium and some nickel, is used in the manufacture of cutlery because of its corrosion resistance. Nonferrous chromium alloys include **nichrome** and **chromel** (Ni and Cr), which are used in various heating devices because of their electrical resistance property. Chromium is widely used as a protective and decorative coating for other metals, such as plumbing fixtures and automobile trim.

The compounds of chromium have many uses. These include uses as paint pigments and mordants and in the tanning of leather. Certain refractories used as lining for high-temperature furnaces are made by mixing pulverized chromite ore with clay or magnesia (MgO).

Group VIIB—Manganese, Technetium, and Rhenium

40.5   Periodic Relationships of Group VIIB Metals

Manganese, technetium, and rhenium make up Periodic Group VIIB. Although rhenium was not discovered until 1925, its chemistry has now been widely studied. No stable isotope of technetium is known; however, the chemistry of the element has been studied by means of radioactive isotopes prepared synthetically. Each of the three elements in Group VIIB has valence electrons in two shells, the outermost one of which contains two electrons. Contrasted with these elements are the halogens of Periodic Group VIIA, which have seven electrons in their outer and only valence shell. The halogens are active nonmetals whereas the Group VIIB elements are metals. However, in the higher oxidation states, there are striking similarities between the elements of these two groups. Thus, Mn_2O_7 and Cl_2O_7 are both volatile, explosively unstable liquids; $KMnO_4$ and $KClO_4$ are both strong oxidizing agents that form isomorphous crystals of nearly the same solubility. In the following sections, only manganese and its compounds will be discussed.

40.6   Occurrence and Metallurgy of Manganese

The most important ore of manganese is **pyrolusite**, MnO_2. Other manganese ores include **braunite**, Mn_2O_3, **manganite**, $MnO(OH)$ or $Mn_2O_3 \cdot H_2O$, **hausmannite**, Mn_3O_4, **franklinite**, $(Fe, Mn, Zn)O$, and **psilomelane**, MnO_2 (BaO, K_2O, H_2O, etc.).

Nearly pure manganese may be produced by the high-temperature reduction of the dioxide by aluminum.

$$3MnO_2(s) + 4Al(s) \longrightarrow 3Mn(s) + 2Al_2O_3(s) \qquad \Delta H^\circ = -1791 \text{ kJ}$$

Since alloys of manganese and iron are extensively used in the production of steel, such alloys are usually produced instead of the pure manganese. The alloys are prepared by reducing the mixed oxides of manganese and iron with coke in a blast furnace. The alloys high in manganese are called **ferromanganese**, and those low in manganese are called **spiegeleisen** (German), meaning *mirror iron.*

40.7   Properties of Manganese

Manganese is a gray-white metal with a slightly reddish tinge. It is brittle and has the general appearance of cast iron. It is readily oxidized by moist air and decomposes water slowly, forming manganese(II) hydroxide and hydrogen. It dissolves readily in dilute acids, forming manganese(II) salts. Manganese forms five oxides and five corresponding series of salts (Table 40.2). These compounds are considered in the following sections.

TABLE 40-2 Classes of Manganese Compounds

Oxidation State	Oxide	Hydroxide	Character	Derivative	Name	Color
+2	MnO	$Mn(OH)_2$	Moderately basic	$MnCl_2$	Manganese(II) chloride	Pink
+3	Mn_2O_3	$Mn(OH)_3$	Weakly basic	$MnCl_3$	Manganese(III) chloride	Violet
+4	MnO_2	H_2MnO_3	Weakly acidic	$CaMnO_3$	Calcium manganite	Brown
+6	MnO_3	H_2MnO_4	Moderately acidic	K_2MnO_4	Potassium manganate	Green
+7	Mn_2O_7	$HMnO_4$	Strongly acidic	$KMnO_4$	Potassium permanganate	Purple

40.8   Compounds of Divalent Manganese

Although manganese forms compounds in which it exhibits oxidation states of $+2$, $+3$, $+4$, $+6$, and $+7$, the only stable cation is the divalent manganese ion, Mn^{2+}. The common soluble manganese(II) salts are the chloride, sulfate, and nitrate. Each imparts a faint pink color to its solutions.

Alkali metal hydroxides precipitate pale pink **manganese(II) hydroxide,** $Mn(OH)_2$, which is oxidized on exposure to air to the dark brown **manganese(III) oxyhydroxide,** MnO(OH). Manganese(II) hydroxide is only partially precipitated by aqueous ammonia and is dissolved by solution of the ammonium salts of strong acids, according to the equation

$$Mn(OH)_2 + 2NH_4^+ \rightleftharpoons Mn^{2+} + 2NH_3 + 2H_2O$$

Manganese(II) hydroxide is entirely basic in character, as indicated by the fact that it dissolves in acids but not in bases such as sodium hydroxide. **Manganese(II) oxide,** MnO, may be produced by heating the corresponding hydroxide in the absence of air.

From manganese(II) solutions, alkali metal sulfides precipitate the pink **sulfide,** MnS, which is readily soluble in dilute acids.

40.9   Compounds of Trivalent Manganese

Dimanganese trioxide, Mn_2O_3, and MnO(OH) occur naturally, but the manganese(III) ion is unstable in aqueous solution and is readily reduced to the manganese(II) ion. **Manganese(III) chloride** is formed in solution by the action of hydrochloric acid upon manganese dioxide at low temperature, but when the solution is warmed the manganese(III) chloride decomposes with the formation of manganese(II) chloride and chlorine.

$$MnO_2 + 4H^+ + 4Cl^- \longrightarrow MnCl_4 + 2H_2O$$
$$2MnCl_4 \longrightarrow 2MnCl_3 + Cl_2(g)$$
$$2MnCl_3 \longrightarrow 2Mn^{2+} + 4Cl^- + Cl_2(g)$$

40.10   Compounds of Tetravalent Manganese

The most important compound of tetravalent manganese is the dioxide, MnO_2. This oxide is amphoteric but is relatively inert toward acids and alkalies. Cold, concentrated hydrochloric acid acts upon the dioxide, giving a green solution of the unstable **manganese(IV) chloride.** The sulfate, $Mn(SO_4)_2$, is also unstable, but the complex salt K_2MnF_6 is not readily decomposed. When the dioxide is fused with calcium oxide, **calcium manganite,** $CaMnO_3$, is formed. This is a salt of the hypothetical **manganous acid,** H_2MnO_3.

When manganese dioxide is heated to 535°C it is transformed to **trimanganese tetraoxide,** Mn_3O_4, and oxygen, according to

$$3MnO_2(s) \longrightarrow Mn_3O_4(s) + O_2(g) \qquad \Delta H° = 172\,kJ$$

A note of historical interest is that Scheele, independently of Priestley and Lavoisier, discovered oxygen by the action of concentrated sulfuric acid on manganese dioxide.

$$MnO_2 + 2H_2SO_4 \longrightarrow Mn(SO_4)_2 + 2H_2O$$
$$2Mn(SO_4)_2 + 2H_2O \longrightarrow 2MnSO_4 + 2H_2SO_4 + O_2$$

40.11   Manganates, Hexavalent Manganese

When an oxide of manganese is fused with an alkali metal hydroxide or carbonate in the presence of air or some other oxidizing agent (such as potassium chlorate or potassium nitrate), a manganate is formed.

$$2MnO_2 + 4KOH + O_2 \longrightarrow 2K_2MnO_4 + 2H_2O$$

Manganates are green in color and stable only in alkaline solution. The addition of water to solutions of manganates may bring about disproportionation, with the precipitation of manganese dioxide and the formation of the purple permanganate ion, MnO_4^-.

$$3MnO_4^{2-} + 2H_2O \longrightarrow 2MnO_4^- + 4OH^- + MnO_2(s)$$

Free **manganic acid,** H_2MnO_4, is too unstable to be prepared and isolated. Solutions containing the manganate ion become active oxidizing agents when acidified. The half-reaction is

$$MnO_4^{2-} + 8H^+ + 4e^- \longrightarrow Mn^{2+} + 4H_2O$$

40.12   Permanganates, Heptavalent Manganese

An important compound of manganese is **potassium permanganate,** $KMnO_4$. It is prepared commercially by oxidizing potassium manganate in alkaline solution with chlorine.

$$2MnO_4^{2-} + Cl_2 \longrightarrow 2MnO_4^- + 2Cl^-$$

The resultant purple solution deposits crystals when sufficiently concentrated. Potassium permanganate is a valuable laboratory reagent because it acts as a strong oxidizing agent. A solution of the free **permanganic acid,** $HMnO_4$, may be prepared by the reaction of dilute sulfuric acid and barium permanganate.

$$Ba^{2+} + 2MnO_4^- + [2H^+] + SO_4^{2-} \longrightarrow BaSO_4(s) + [2H^+] + 2MnO_4^-$$

Mn_2O_7, a dark brown, highly explosive liquid, is the **anhydride of permanganic acid.**

40.13   Uses of Manganese and Its Compounds

Manganese forms a number of important alloys. Two of these were mentioned in Section 40.6; they are **spiegeleisen,** which is a bright and lustrous iron alloy containing 5–20% manganese, and **ferromanganese,** which contains 70–80% manganese. Both of these alloys are used in making very hard steels for the manufacture of rails, safes, and heavy machinery. About 12.5 pounds of manganese metal are used in the production of every ton of steel to remove oxygen, nitrogen, and sulfur. An alloy called **manganin** (Cu, 84%; Mn, 12%; Ni, 4%) is used in instruments for making electrical measurements, because the electrical resistance of the alloy does not change significantly with changes in temperature.

Manganese dioxide is used in glassmaking to correct the green color produced by iron(II) compounds; to color glass and enamels black; to act as an oxidizing agent, or "dryer," in black paints; and as a depolarizer in dry cells. Potassium permanganate is used as a disinfectant, a deodorant, and a germicide.

Questions

1. In what way are the Group VIB metals typical of transition metals in general?
2. Write a balanced chemical equation describing the air oxidation of aqueous chromium(II) to chromium(III).
3. Write the equation for the decomposition of ammonium dichromate by ignition.
4. Account for the extensive use of chromium in the plating of iron and copper objects, such as the trim on automobiles.
5. Reduction of chromite ore by carbon leads to an alloy of iron and chromium. How would you separate the two components of this alloy into separate solutions of their respective cations, if this were necessary?
6. What is the action of hydrogen ions upon chromate ions? This is called a condensation reaction. Explain.
7. Represent the electronic structure of chromic acid.
8. Write an equation to illustrate the action of the dichromate ion as an oxidizing agent.
9. What is the composition and chemical action of the "cleaning solution" of the laboratory?
10. How are the alloys known as ferromanganese produced?
11. Outline the chemistry of the action of manganese dioxide upon hydrochloric acid.
12. Explain the dissolution of manganese(II) hydroxide in aqueous ammonium chloride in terms of an acid-base reaction.
13. Mention two uses of manganese dioxide.
14. Write the names and formulas for compounds in which manganese exhibits each of the following oxidation states: +2, +3, +4, +6, and +7.
15. How is potassium permanganate prepared commercially?
16. Would you expect a calcium manganite solution to have a pH greater or less than 7.0? Justify your answer.
17. How may the insolubility of barium sulfate be used to good advantage in the preparation of free permanganic acid?

References

"The Oxidation States of Molybdenum," J. G. Stark, *J. Chem. Educ.,* **46,** 505 (1969).

"Some Perspectives on Heteropoly Ion Chemistry," P. G. Rasmussen, *J. Chem. Educ.,* **44,** 277 (1967).

"Paints and Plastic Coatings," W. M. Morgans, *Educ. in Chemistry,* **10,** 12 (1973).

"Manganese Nodules (I): Mineral Resources on the Deep Seabed," A. L. Hammond, *Science,* **183,** 502 (1974). "(II):Prospects for Deep Sea Mining," A. L. Hammond, *Science,* **183,** 644 (1974).

"Manganee Nodules: Controversy upon Controversy," D. S. Cronan, *Endeavour,* **2,** 80 (1978).

"Some Aspects of the Bioinorganic Chemistry of Molybdenum," K. B. Swedo and J. H. Enemark, *J. Chem. Educ.,* **56,** 70 (1979).

"Manganese(II) Complexes Mimic Function of Heme," D. A. O'Sullivan, *Chem. and Eng. News,* December 4, 1978, p. 24.

41

The Metals of Group VIII

Group VIII		
Fe	Co	Ni
Ru	Rh	Pd
Os	Ir	Pt

We noted in Section 30.5 that successive members of each of the transition series show remarkable resemblances to each other. This is particularly true in the case of the three triads: iron, cobalt, nickel; ruthenium, rhodium, palladium; and osmium, iridium, platinum. Because of these "horizontal" similarities Mendeleev placed these elements in one group of the Periodic Table—Group VIII. The number of the group might imply a maximum positive oxidation state of 8 for these metals, and, in fact, osmium and ruthenium do exist in this oxidation state in certain of their compounds. However, it is the exception rather than the rule, and should not be considered typical of the group. In the following sections, iron, cobalt and nickel will be discussed in detail.

41.1 Periodic Relationships of Group VIII Metals

Iron, cobalt, and nickel make up the first horizontal triad of Group VIII. As with the metals in the other horizontal series of Group VIII, the tendency of iron, cobalt, and nickel to lose electrons decreases as the nuclear charge increases. Thus, iron exhibits oxidation states of $+2$, $+3$, and $+6$; and cobalt, $+2$, $+3$, and $+4$; while with nickel, oxidation states higher than $+2$ do exist (NiO_2 in the Edison cell, described in Section 20.17, for example) but are rare. This variation in oxidation state is associated with incompletely filled $3d$ subshells. Other properties shown to a marked degree by iron, cobalt, and nickel are magnetism and the tendencies to

TABLE 41-1 Important Physical Properties of Iron, Cobalt, and Nickel

	Atomic Number	Atomic Weight	Atomic Radius, Å	Ionic (M^{2+}) Radius, Å	Density at 20°	Melting Point, °C	Boiling Point, °C
Iron	26	55.847	1.26	0.75	7.86	1535	2800
Cobalt	27	58.9332	1.26	0.72	8.71	1493	3100
Nickel	28	58.71	1.24	0.70	8.9	1455	2800

form colored compounds and complex ions. These properties are also typical of other metals with a partially filled inner electron subshell.

Some of the more important physical properties of iron, cobalt, and nickel are given in Table 41-1.

Iron

41.2 History and Importance of Iron

Iron was known at least as early as 4000 B.C. and probably was used to some extent during prehistoric times. Because free iron is not commonly found in nature, that first used in prehistoric times may well have been of meteoric origin. The early discovery and application of iron to the manufacture of tools and weapons resulted from the wide distribution of iron ores and the ease with which the iron compounds in the ores can be reduced by carbon. Charcoal was the form of carbon used for a long time in the reduction process, but the production and use of iron on a large scale did not begin until about 1620, when coal was introduced as the reducing agent.

Iron is the most widely used of all the metals. This is because its ores are abundant and widely distributed, the metal is easily and cheaply produced from the ores, and its properties can be varied over a wide range by the addition of other substances and by different methods of treatment, such as tempering and annealing. More iron is used than all the other metals combined (14 times as much).

41.3 Occurrence of Iron

Iron ranks second in abundance among the metals (aluminum is first) and fourth among all elements in the earth's crust. The central core of the earth is largely iron, as has been indicated by studies of the earth's density and the rate of transmission of earthquake shock. Metallic meteors are usually about 90% iron, and the remainder is principally nickel. Practically all rocks, minerals, and soils, as well as plants, contain some iron. Iron is present in the hemoglobin of the blood, which acts as a carrier of oxygen (Section 31.5).

The important iron-bearing ores are **hematite,** Fe_2O_3, and **magnetite,** Fe_3O_4 (Fig. 41-1). The presence of Fe_2O_3 accounts largely for the red coloration of many rocks and soils. Hematite often occurs as the reddish brown hydrate $2Fe_2O_3 \cdot 3H_2O$, which is called **limonite.** Magnetite is a black, crystalline mineral,

Figure 41-1 The magnetic properties of a piece of magnetite, Fe_3O_4, or $FeO \cdot Fe_2O_3$. (*Courtesy of Ward's Natural Science Establishment, Inc.*)

which is strongly magnetic, sometimes possessing polarity, in which case it is called **lodestone.** Black crystalline magnetite interlocked with crystals of silica in the form of hard rock is called **taconite. Siderite** is a white or brownish carbonate, $FeCO_3$. **Pyrite,** FeS_2, occurs as pale yellow crystals with a metallic luster; they are easily mistaken for gold, hence the name "fool's gold." Pyrite is quite abundant and useful as a source of sulfur for the manufacture of sulfuric acid; but because roasting does not remove all of the sulfur, pyrite is usually considered unsatisfactory as a source of iron, although low-grade pig iron can be made from it.

With the rather rapid depletion of domestic high-grade iron ore (over 50% iron) deposits, the steel industry has turned to the conquest of the taconite ores (25–30% iron). Taconite is the extremely hard rock that makes up some 95% by volume of the famous Mesabi Range's vast iron-bearing deposits in upper Minnesota. Once regarded as useless for steel production, the taconite ores are now used extensively as a result of the development of satisfactory techniques for mining and concentrating them.

41.4 Metallurgy of Iron

The first step in the metallurgy of iron is usually that of roasting the ore to remove water, decompose carbonates, and oxidize sulfides. The oxides of iron are then reduced with coke in a blast furnance 80–100 feet high and 25 feet in diameter. Near the bottom of the furnace are nozzles, or "tuyères," through which preheated air is blown into the furnace under pressure. The charge of roasted ore, coke, and flux (limestone or sand) is introduced into the top of the furnace through cone-shaped valves arranged so that the charge is deposited uniformly around the circumference of the furnace near the walls. As the charge melts, or is burned in the lower regions of the furnace, the stock column gradually settles, thus leaving room for additional charges at the top, and making the operation of the blast furnace a continuous process. The entire stock in the furnace weighs several hundred tons.

As soon as the preheated air (500°C) under pressure enters the furnace, coke in the region of the tuyères is oxidized to carbon dioxide with the liberation of much heat, which raises the temperature to about 1500°C. As the carbon dioxide passes upward through the overlying layer of white-hot coke, it is reduced to carbon monoxide ($CO_2 + C \longrightarrow 2CO$). The carbon monoxide thus produced serves as the reducing agent in the upper regions of the furnace. The individual reduction reactions are indicated in Fig. 41-2 and can be summarized as a single reversible reaction as follows:

$$Fe_2O_3(s) + 3CO(g) \rightleftharpoons 2Fe(s) + 3CO_2(g) \qquad \Delta H° = -25 \text{ kJ}$$

The presence of an excess of carbon monoxide keeps the equilibrium shifted to the right. The iron ore is completely reduced before the formation of the slag begins in the middle region of the furnace; otherwise, part of the iron would be lost due to the reaction, $FeO + SiO_2 \longrightarrow FeSiO_3$.

Just below the middle of the furnace, the temperature is high enough to melt both the iron and the slag that have been produced. The molten iron and slag collect in two layers at the bottom of the furnace, the less dense slag floating on the

Figure 41-2 Reactions taking place in the blast furnace.

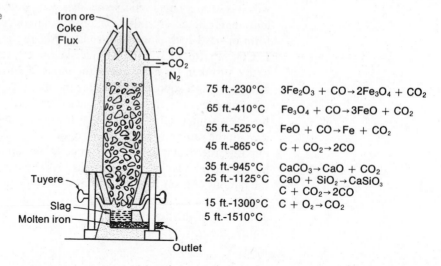

Iron ore
Coke
Flux

CO
CO$_2$
N$_2$

75 ft.-230°C	$3Fe_2O_3 + CO \rightarrow 2Fe_3O_4 + CO_2$
65 ft.-410°C	$Fe_3O_4 + CO \rightarrow 3FeO + CO_2$
55 ft.-525°C	$FeO + CO \rightarrow Fe + CO_2$
45 ft.-865°C	$C + CO_2 \rightarrow 2CO$
35 ft.-945°C	$CaCO_3 \rightarrow CaO + CO_2$
25 ft.-1125°C	$CaO + SiO_2 \rightarrow CaSiO_3$
	$C + CO_2 \rightarrow 2CO$
15 ft.-1300°C	$C + O_2 \rightarrow CO_2$
5 ft.-1510°C	

Tuyere

Slag
Molten iron

Outlet

iron and protecting it against oxidation. Every hour or two, the liquid slag is withdrawn from the furnace. Blast furnace slag is often used in the manufacture of Portland cement. Four or five times a day, the molten iron is withdrawn through a tap hole into a large ladle lined with fire brick that holds several hundred tons of the hot metal. The ladle containing the hot metal is then transferred to the pig casting machine or to the steelmaking plant, which may be several miles away. The daily production of a blast furnace is about 1200 tons of iron.

The hot exhaust gases, which issue from the top of the blast furnace and contain some unoxidized carbon monoxide, are mixed with air and burned to preheat the air used in the operation of the blast furnace.

41.5   Pig Iron and Cast Iron

The metal obtained from the blast furnace is called **pig iron.** It contains as impurities a small percentage of carbon (3–4%), silicon, phosphorus, and manganese, and smaller amounts of sulfur. Iron reacts with carbon at the temperatures of the blast furnace to form **cementite,** Fe$_3$C.

$$3Fe(s) + C(s) \rightleftharpoons Fe_3C(s) \qquad \Delta H° = 25 \text{ kJ}$$

Pig iron is brittle and is usually converted to cast iron or steel. When pig iron is remelted and recooled it is called **cast iron.** When pig iron is cooled rapidly the carbon remains chemically combined with the iron, in the form of the carbide, Fe$_3$C. The product has a light color, because Fe$_3$C is white, and is therefore called **white cast iron.** It is brittle but hard and resistant to wear. When pig iron is allowed to cool slowly, the cementite equilibrium shifts so that some of the carbon separates from the iron as black graphite. This gives the product a gray color so it is called **gray cast iron.** In contrast to white cast iron, the gray variety is soft and tough. It is possible to control the cooling of cast iron in such a way that the surface cools rapidly, while the body of the casting cools slowly. This treatment results in a casting that has the toughnss of gray cast iron and the wearing qualities of white cast iron.

Cast iron is especially adapted to making molded products (castings), because it expands upon solidification and consequently fills all details of the mold completely.

41.6 The Manufacture of Steel

The term **steel** is used to denote many widely different alloys of iron. It is made from pig iron by removing the impurities and later adding substances such as manganese, chromium, nickel, tungsten, molybdenum, and vanadium to produce alloys with properties that make them suitable for specific uses. Most steels also contain small but definite percentages of carbon (0.04–2.5%). Thus a large part of the carbon contained in pig iron must be removed in the manufacture of steel.

Although there are many grades of steel, steels can be classified into three main categories: (a) **carbon steels,** which are made primarily of iron and carbon; (b) **stainless steels,** low-carbon steels containing over 12% chromium; and (c) **alloy steels,** specialty steels that contain large amounts of alloying elements to impart special properties for specific uses. Of the 141,000,000 tons of steel produced in the United States in a recent single year, 124,600,000 tons (88.4%) were carbon steels, 14,800,000 tons (10.5%) were stainless steels, and 1,600,000 tons (1.1%) were alloy steels. This amount, produced in the United States, constituted 22.5% of the world's production of 625,000,000 tons of steel.

Two processes are used chiefly in the production of steel; they are the **open-hearth process** and the **basic oxygen process,** the latter accounting for most of the total production of steel in the United States. The open-hearth process was used for 90% of the steel produced in the United States immediately prior to the introduction of the basic oxygen process in this country in 1954. Steels for special purposes are made by the electric furnace method, which accounts for 14% of total steel production in this country.

The Bessemer process, formerly quite a significant factor in steel production, accounted for only about 5% of total production in 1950 and almost none at the present time.

▶**1. OPEN-HEARTH PROCESS.** A diagram of the open-hearth furnace is given in Fig. 41-3. The charge, which typically is about 200 tons, consists of pig iron, scrap iron, and iron ore. It is placed on a shallow hearth and heated from above by a gaseous fuel containing an excess of air. The mixture of fuel and air required for the oxidation of such impurities as carbon and sulfur in the charge is preheated by passage through a checkerwork of hot brick. The bricks are heated by the hot gases produced during the oxidation of the impurities in the charge. The directions of the entering gases and the gaseous products are reversed at intervals so that each chamber of bricks is alternately heated and used to preheat the incoming fuel and air mixture. The iron ore and the rust on the scrap iron, which are parts of the charge, aid in the oxidation of the impurities.

When the charge contains elements whose oxides are acidic (e.g., silicon, phosphorus, and sulfur) the lining of the hearth is made of basic materials, principally magnesium and calcium oxides. The acidic oxides react with the basic lining, forming a slag. Most of the open-hearth steel is made in basic-lined furnaces. If the charge contains basic oxides an acid lining composed chiefly of sand or siliceous materials is employed.

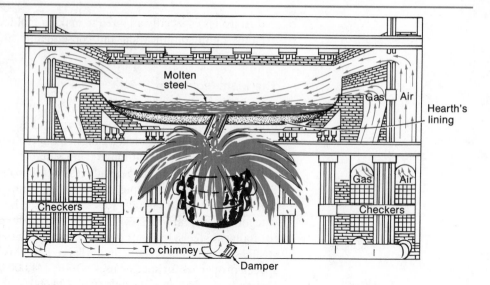

About eight hours are required to oxidize most of the carbon in the charge of an open-hearth furnace. At intervals, samples of the metal are removed for testing, the oxidation being continued until the desired composition is obtained. Before the steel is poured, the required amounts of manganese, chromium, and other alloying substances are added.

Increasing use is being made of pure oxygen in open-hearth furnaces. Introduced through a water-cooled lance during the furnace operation, it speeds up the oxidation of the carbon.

Open-hearth steel is high grade and uniform in quality. It is used in making products that are subject to frequent vibration or sudden strains, such as girders, heavy rails, guns, and armor plate.

▶**2. BASIC OXYGEN PROCESS.** This process utilizes a cylindrical furnace with a basic lining. A typical furnace is charged with 80 tons of scrap iron, 200 tons of molten pig iron, and 18 tons of limestone (to form slag), and operates under computer control. A jet of high-purity (99.5%) oxygen at 140–180 pounds pressure per square inch is directed vertically through a water-cooled lance, 50 feet long and 10 inches in diameter, to a point 4 to 8 feet below the surface of the white-hot molten charge. The reaction produces a vigorous boiling action that reaches all parts of the charge. In the central reaction zone, temperatures reach a level close to the boiling point of iron. The entire steel-making cycle is completed in one hour or less—much faster than the average heat time for open-hearth furnaces. Electrostatic precipitators (see Section 13.33) clean the gases resulting from the furnace reactions, making the furnaces virtually smokeless. The steel produced is of extremely high and uniform quality. The basic oxygen process, because of its speed and efficiency and the high quality of its product, is rapidly increasing in importance relative to the open-hearth process, rising from 6% of the total steel production in 1962 to well over half the total production in 1977.

▶3. OTHER PROCESSES. Various types of electric furnaces are used in the United States to obtain high-grade steel. Three distinct advantages of the electric furnace are (a) the temperature and composition of the metal can be controlled with high precision, (b) high temperatures are easily attainable, making possible the production of special steels with high melting points, and (c) the furnace can accept an all-scrap charge (whereas basic oxygen furnaces are limited to a maximum of about 35% scrap, since excess scrap would have too much cooling effect on the charge). However, the cost of power limits the use of the electric furnace. **Electric steel** is used as tool steel in the making of such articles as files, drills, knives, razor blades, watch springs, and axes.

41.7   Carbon Steels

Carbon is the most important alloying element used in the manufacture of steel. It may be present in three forms—in combination with iron as cementite, Fe_3C; as relatively large crystals of graphite; and as very small crystals of graphite (Sections 25.2 and 41.5).

Steel containing cementite in solid solution in iron contains up to about 1.5% carbon and is made by quenching the hot metal in water or oil. This steel is hard, brittle, and light-colored. If the metal is cooled slowly the carbon is deposited largely as separate crystals of cementite or as particles of graphite; the product is then softer and more pliable than before and has a much higher tensile strength than steel that has been cooled rapidly.

Steels of low carbon content (up to 0.2%) are quite ductile and are known as **mild steels.** These are used in the manufacture of such products as sheet iron, wire, and pipe. **Medium steels** contain 0.2 to 0.6% carbon and are used in rails, structural steel, and boiler plate. Both mild and medium steels can be forged and welded. **High-carbon steels** contain from 0.6 to 1.5% carbon. They are hard and brittle and are used in surgical instruments, razor blades, springs, and cutlery.

Steel that has been heated to redness and allowed to cool slowly is said to be **annealed.** Steel is tempered by rapid cooling in water or oil, followed by controlled reheating to obtain the desired proportion of cementite and graphite. **Case-hardened steel** is made by heating low-carbon steel in closed containers with powdered carbon, followed by quenching in oil. The carbon reacts with the iron on the surface, forming cementite. Such a steel is very hard on its surface and tough in the interior. It is used in making axles and other articles that must resist wear and, at the same time, be tough and flexible so as not to break when subjected to sudden strains and shocks. Steel can also be case-hardened using nitrogen compounds. A very hard iron nitride is formed, and the process is therefore called **nitriding.**

41.8   Alloy Steels

Certain materials, called **scavengers,** are added to iron in the manufacture of steel to remove impurities, especially oxygen and nitrogen, and thus improve the quality of the product. The most important scavengers are aluminum, ferrosilicon, ferromanganese, and ferrotitanium. They react with dissolved oxygen and nitro-

TABLE 41-2   Composition and Properties of Some Alloy Steels

Name	Composition	Characteristic Properties	Uses
Manganese	10–18% Mn	Hard, tough, resistant to wear	Railroad rails, safes, armor plate, rock crushing machinery
Silicon	1–5% Si	Hard, strong, highly magnetic	Magnets
Duriron	12–15% Si	Resistant to corrosion, acids	Pipes, kettles, condensers, etc.
Invar	36% Ni	Low coefficient of expansion	Meter scales, measuring tapes, pendulum rods
Chrome-vanadium	1–10% Cr, 0.15% V	Strong, resistant to stains	Axles
Stainless steel	14–18% Cr, 7–9% Ni	Resistant to corrosion	Cutlery, instruments
Permalloy	78% Ni	High magnetic susceptibility	Ocean cables
High-speed steels	14–20% W or 6–12% Mo	Retain temper at high temperatures	High-speed cutting tools
Nickel	2–4% Ni	Hard and elastic, resistant to corrosion	Drive shafts, gears, cables

gen forming oxides and nitrides, respectively. The oxides and nitrides are removed in the slag.

For every ton of steel produced about 25–30 pounds of nonferrous metals are added or used as coatings. By the appropriate choice of the number and percentages of these elements, steels of widely varying properties can be manufactured. Some important alloy steels and their features are given in Table 41-2.

41.9   Corrosion

Many metals, particularly iron, undergo corrosion when exposed to air and water. Losses caused by corrosion of metals cost the people of the United States billions of dollars annually.

It has been shown that iron will not rust in dry air, nor in water that is free from dissolved oxygen. It follows then that both air and water are involved in the corrosion process. The presence of an electrolyte in the water accelerates corrosion, particularly when the solution is acidic. Strained metals corrode more rapidly than unstrained ones. Heated portions of a metal corrode more rapidly than unheated ones. Finally, iron in contact with a less active metal such as tin, lead, or copper corrodes more rapidly than when either alone or in contact with a more active metal such as zinc.

An electrochemical theory helps to explain why the presence of a less active metal than iron accelerates the corrosion of iron and the presence of a more active metal slows down the corrosion. It appears that tiny electrochemical cells are set up when corrosion takes place. When iron is in contact with water containing an electrolyte, the half-reaction given below tends to occur.

Anodic oxidation: $Fe \longrightarrow Fe^{2+} + 2e^-$

Sometimes, one part of a piece of iron is more active than the rest and tends to go into solution more easily. This means that the more active part has a higher potential than the rest of the iron piece, and this portion tends to dissolve in the electrolyte. On the less active parts of the iron (lower electrode potential), hydrogen tends to form according to the half-reaction

Cathodic reduction: $2H^+ + 2e^- \longrightarrow H_2$

The accumulation of hydrogen on the surface of the less active part of the iron tends to polarize the electrode and stop the cell action. However, oxygen dissolved in any water moisture present gradually removes the hydrogen in the same manner that MnO_2 depolarizes the dry cell (Section 20.16), and the electrochemical reaction proceeds; i.e., corrosion occurs.

$$2H_2 + O_2 \longrightarrow 2H_2O$$

Iron(II) ions combine with hydroxide ions of the electrolytic solutions, forming $Fe(OH)_2$. Iron(II) hydroxide is then readily oxidized by air to $Fe_2O_3 \cdot xH_2O$, which is **iron rust.**

Many methods and devices have been employed to prevent or retard corrosion. Some of these methods are as follows: Iron can be protected against corrosion by coating it with (a) an organic material such as paint, lacquer, grease, or asphalt; (b) another metal such as zinc, copper, nickel, chromium, tin, cadmium, or lead; (c) a ceramic enamel, like that used on sinks, bathtubs, stoves, refrigerators, and washers; or (d) an adherent oxide, which may be formed by exposing iron to superheated steam, thereby giving it an adherent coating of magnetic oxide, Fe_3O_4.

A piece of iron may also be treated with a compound that provides a suitable coordinating group, such as phosphate, which ties up the surface iron in a complex such that it cannot react to form $Fe_2O_3 \cdot xH_2O$ (Chapter 31). Alloys of iron with certain other elements are often corrosion resistant. Typical examples are **stainless steel** (Fe, Cr, and Ni) and **duriron** (Fe and Si).

Another method of preventing the corrosion of iron or steel involves an application of electrochemistry called **cathodic protection.** For example, corrosion of iron or steel water tanks can be retarded by suspending several stainless steel anodes in the tank; the tank serves as a cathode. A small current is passed continuously through the system, with the natural salts in the water making it conducting. A slight cathodic evolution of a protective coat of hydrogen takes place on the wall of the tank. Cathodic protection against corrosion is also used on iron and steel that is in contact with soil, such as underground pipe lines. Other applications of cathodic protection do not involve the use of a current from an external source. In these cases, the iron becomes cathodic when connected by a wire to a more active metal such as zinc, aluminum, or magnesium. The difference in activity of the two metals causes a current to flow between them, producing corrosion on the more active metal and furnishing cathodic protection to the iron.

41.10 Properties of Iron

Pure iron is silvery white, capable of taking a high polish, ductile, relatively soft, and high in tensile strength. These properties are greatly modified by the amount and nature of the impurities present, especially carbon. Pure iron is attracted by a magnet but does not retain magnetism. Iron dissolves in hydrochloric acid and dilute sulfuric acid, forming hydrogen and iron(II) ions, Fe^{2+}. Hot concentrated sulfuric acid forms Fe^{3+}, and SO_2 is evolved. Cold concentrated nitric acid induces **passivity** in iron (see also Section 40.2). When in the passive condition iron does not react with dilute acids that otherwise would readily dissolve it. The passivity is easily destroyed by a scratch or by shock. Hot dilute nitric acid oxidizes iron to Fe^{3+} with the evolution of nitric oxide. When heated to redness iron reduces steam to produce hydrogen and iron(II, III) oxide, Fe_3O_4, which is the same oxide that is produced when iron burns in oxygen. When exposed to moist air at ordinary temperatures, iron becomes oxidized, with a loose coating of partially hydrated iron(III) oxide, $Fe_2O_3 \cdot xH_2O$ (Section 41.9).

Iron forms two principal series of compounds in which the oxidation states are +2 and +3; compounds in which iron is +6 are rare and less important. The +2 and +3 series are called iron(II), or ferrous, and iron(III), or ferric, respectively. The oxides and hydroxides of iron(II) and (III) are basic in character with little or no acid-forming property.

41.11 Oxides and Hydroxides of Iron

Black **iron(II) oxide,** FeO, may be obtained by the thermal decomposition of iron(II) oxalate, FeC_2O_4.

$$FeC_2O_4 \longrightarrow FeO + CO + CO_2$$

Iron(II) hydroxide, $Fe(OH)_2$, is white when pure; it precipitates upon the addition of alkali metal hydroxides to solutions of iron(II) salts. Ammonium salts and many organic acids greatly increase the solubility of iron(II) hydroxide. In the air the white hydroxide quickly turns green and then reddish brown, due to oxidation to **iron(III) hydroxide.**

$$4Fe(OH)_2(s) + O_2(g) + 2H_2O(g) \longrightarrow 4Fe(OH)_3(s) \qquad \Delta H° = -532.2 \text{ kJ}$$

Iron(III) oxide, Fe_2O_3, occurs in nature as the mineral **hematite.** It may also be prepared by igniting $Fe(OH)_3$ or by roasting pyrite, FeS_2, in the air. It is a red powder, which is used as a paint pigment under the names **rouge** and **Venetian red.** When treated with either alkali metal hydroxides or aqueous ammonia, solutions of Fe^{3+} yield $Fe(OH)_3$, which is reddish brown and insoluble in an excess of reagent.

Iron(II, III) oxide, Fe_3O_4, is a mixed oxide, $FeO \cdot Fe_2O_3$, and is called **magnetic oxide of iron,** or **loadestone** (Fig. 41-1). It is a valuable ore of iron.

41.12 Iron(II) Salts

Iron(II) chloride 4-hydrate, $FeCl_2 \cdot 4H_2O$, is produced by the action of dilute hydrochloric acid upon iron. It is pale green in color. Sulfuric acid reacts with iron or iron(II) oxide forming **green vitriol,** or **copperas,** which crystallizes as the **7-hydrate,** $FeSO_4 \cdot 7H_2O$.

Iron(II) ammonium sulfate 6-hydrate, $(NH_4)_2SO_4 \cdot FeSO_4 \cdot 6H_2O$, **Mohr's salt,** is a double salt that crystallizes from an equimolar solution of ammonium sulfate and iron(II) sulfate. It is readily obtained in a pure state and is used extensively in laboratory work as a reducing agent, the iron being readily oxidized to the $+3$ state.

Iron(II) carbonate, $FeCO_3$, occurs in nature as the mineral **siderite.** It may be produced, also, as a white precipitate by the reaction of alkali metal carbonates with solutions of iron(II) salts in the absence of air. The carbonate dissolves in carbonic acid and forms the **hydrogen carbonate** $Fe(HCO_3)_2$, a common constituent of hard waters.

Iron(II) sulfide, FeS, is precipitated when an alkali metal sulfide is added to a solution of an iron(II) salt. It is prepared commercially by heating a mixture of iron and sulfur. Acids act upon it, forming hydrogen sulfide.

41.13   Iron(III) Compounds

An important iron(III) compound is the chloride $FeCl_3$, which may be obtained by heating iron with chlorine. It is appreciably covalent, as indicaed by its volatility and its solubility in nonpolar solvents. When the anhydrous chloride is dissolved in water much heat is liberated, due to hydration of the ions. From water iron(III) chloride crystallizes as the brownish yellow hydrate, $[Fe(H_2O)_6]Cl_3$. This compound, like all soluble iron(III) salts, gives an acid reaction by hydrolysis. The first step in the hydrolysis can be written

$$[Fe(H_2O)_6]^{3+} + H_2O \longrightarrow [Fe(H_2O)_5(OH)]^{2+} + H_3O^+$$

or
$$Fe^{3+} + H_2O \longrightarrow Fe(OH)^{2+} + H^+$$

Several additional steps also occur. Evidence for this is the appearance of the reddish brown gelatinous $[Fe(H_2O)_3(OH)_3]$, or simply hydrated $Fe(OH)_3$.

Hydrated iron(III) chloride is used to coagulate blood and thus stop bleeding, and in the treatment of anemia.

When a solution of an iron(III) salt is added to a solution containing thiocyanate ions a blood-red color appears, while an iron(II) salt yields only a colorless solution. Use is made of these facts in the detection of iron(III) ions in the presence of iron(II) ions. It appears that the species responsible for the red color is the complex cation $[Fe(H_2O)_5NCS]^{2+}$.

Other iron(III) salts include the sulfate, $Fe_2(SO_4)_3$, and the iron **alums** formed from this salt, such as $NH_4Fe(SO_4)_2 \cdot 12H_2O$.

41.14   Inks Containing Iron

Some inks contain **iron(II) tannate,** a soluble and nearly colorless compound made by mixing tannic acid and iron(II) sulfate. Iron(II) tannate is readily oxidized by air to the black insoluble **iron(III) tannate.** A blue dye is added to the ink to make it visible before oxidation takes place. Ink stains due to iron(III) tannate may be removed by soaking the stained material in a solution of ammonium oxalate for several hours. This treatment reduces the iron(III) tannate to the soluble iron(II) tannate, which can be washed away with water.

41.15   Complex Cyanides of Iron

When an excess of potassium cyanide is added to a solution of an iron(II) salt **potassium hexacyanoferrate(II)**, $K_4Fe(CN)_6$ (also called potassium ferrocyanide), is formed in solution.

$$Fe^{2+} + 6CN^- \longrightarrow [Fe(CN)_6]^{4-}$$

The complex hexacyanoferrate(II) ion is so stable that it gives none of the common qualitative tests for either the iron(II) or the cyanide ions. When the hexacyanoferrate(II) ion is oxidized by chlorine the **hexacyanoferrate(III) ion,** $[Fe(CN)_6]^{3-}$ (also called ferricyanide), is produced.

$$2[Fe(CN)_6]^{4-} + Cl_2 \longrightarrow 2[Fe(CN)_6]^{3-} + 2Cl^-$$

Solutions of potassium hexacyanoferrate(III) do not give the usual tests for either the iron(III) ions or the cyanide ions.

The addition of iron(II) ions to hexacyanoferrate(III) solutions yields a blue precipitate known as **Turnbull's blue.** Iron(III) ions react with hexacyanoferrate(II) ions to form **Prussian blue.** These precipitates are evidently the same and have the approximate composition $KFe[Fe(CN)_6] \cdot H_2O$; both have a deep blue color. Three independent experimental methods have shown the iron outside the coordination sphere to have the oxidation state of $+3$, and that inside the coordination sphere to have $+2$. When solutions of iron(III) ions and hexacyanoferrate(III) are mixed no precipitate forms, only a brown solution. Iron(II) ions and hexacyanoferrate(II) ions produce a white precipitate, $K_2FeFe(CN)_6$. These reactions serve as sensitive tests to distinguish between Fe^{2+} and Fe^{3+} in solution.

41.16   Blueprints

Blueprint paper is coated with a mixture of iron(III) ammonium citrate and potassium hexacyanoferrate(III); this gives the paper the bronze-green color of iron(III) hexacyanoferrate(III). When a drawing in black ink on tracing cloth is placed over blueprint paper and the paper exposed to light, the iron(III) ions are reduced by the citrate ions to iron(II) wherever light transmission occurs. The iron(II) ion immediately reacts with the hexacyanoferrate(III) ion to form an insoluble blue precipitate. After exposure, the paper is washed with water to remove the unchanged iron(III) hexacyanoferrate(III), leaving the body of the paper blue with white lines produced by the tracing.

Cobalt

41.17   History of Cobalt

The name of this metal is derived from the German word *Kobold*, which means "goblin." The reason for this was that the early metallurgists believed that goblins carefully guarded the cobalt ores and prevented human beings from liberating the metal. Georg Brandt finally succeeded in isolating the metal in the year 1742. Even today, the metallurgy of both cobalt and nickel is difficult and complicated.

41.18 Occurrence and Metallurgy of Cobalt

The common cobalt minerals are **cobaltite**, CoAsS, and **smaltite**, $CoAs_2$, with the richest deposits of these minerals being found in Ontario, Canada. However, nearly all of the world's production of cobalt is obtained as a by-product of the metallurgy of nickel, copper, iron, silver, and other metals. The metallurgy of cobalt involves its separation from nickel, copper, and iron; conversion to Co_3O_4; and the reduction of the oxide with aluminum or hydrogen.

$$3Co_3O_4(s) + 8Al(s) \longrightarrow 9Co(s) + 4Al_2O_3(s) \qquad \Delta H° = -4029 \text{ kJ}$$

Purification of the metal is effected by electrolytic deposition on a rotating, stainless steel cathode.

41.19 Properties and Uses of Cobalt

Cobalt is similar to iron in appearance, except that it has a faint tinge of pink. Like iron and nickel, it is magnetic. It is slowly soluble in warm dilute hydrochloric or sulfuric acid and more rapidly soluble in dilute nitric acid. Like iron and nickel, cobalt is rendered passive by contact with concentrated nitric acid. It is not oxidized on exposure to the air, but at red heat it reduces hydrogen in steam with the evolution of hydrogen. The halogens, except fluorine, convert it to cobalt(II) halides. When cobalt is heated with fluorine, **cobalt(III) fluoride**, CoF_3, is formed.

Cobalt is alloyed with iron and small percentages of other metals in making high-speed cutting tools and surgical instruments. Permanent magnets are made from the alloys **Alnico** (Al, Ni, Co, Fe), **Hiperco** (Co, Fe, Cr), and **Vicalloy** (Co, Fe, V).

Finely divided cobalt metal is used as a catalyst in the hydrogenation of carbon monoxide and carbon dioxide, with the formation of hdyrocarbons, and for the oxidation of ammonia.

41.20 Compounds of Cobalt

▶**1. HYDROXIDES AND OXIDES OF COBALT.** **Cobalt(II) hydroxide**, $Co(OH)_2$, is formed as a blue flocculent precipitate when an alkali metal hydroxide is added to a solution of a cobalt(II) salt. The blue color of the precipitate changes to violet and then to pink, probably as a result of hydration. The hydroxide is readily soluble in aqueous ammonia to form **hexaamminecobalt(II) hydroxide**, $[Co(NH_3)_6](OH)_2$. Solutions of the latter compound are oxidized by oxygen in air to the various cobalt(III) compounds; the oxidation is accompanied by a darkening of the solution.

When cobalt(II) hydroxide is heated in the absence of air **cobalt(II) oxide**, CoO, results. This oxide is a black substance, but when dissolved in fused glass it gives the glass a blue color; such glass is called "cobalt glass," and it contains **cobalt(II) silicate**. Ignition of the hydroxide or oxide in air gives rise to **cobalt(II, III) oxide**, Co_3O_4. Cobalt(III) oxide may be produced by gently heating cobalt(II) nitrate.

$$4Co(NO_3)_2 \longrightarrow 2Co_2O_3 + 8NO_2 + O_2$$

Note that three elements in this reaction change oxidation state. (You will find that balancing this equation by oxidation-reduction methods will provide a challenging review.)

▶ **2. COBALT(II) CHLORIDE.** Cobalt(II) oxide and hydroxide readily dissolve in hydrochloric acid. Concentration of the solution results in the crystallization of **cobalt(II) chloride 6-hydrate,** $CoCl_2 \cdot 6H_2O$, which is red in color. The hydrated cobalt(II) ion, $[Co(H_2O)_6]^{2+}$, exhibits a pink color in solution. When partially dehydrated, cobalt(II) chloride changes to a deep blue color. This is believed to result from a change in the coordination number of the cobalt(II) ion from six to four.

$$[Co(H_2O)_6]Cl_2 \longrightarrow [Co(H_2O)_4]Cl_2 + 2H_2O$$
$$\text{Pink} \qquad\qquad\qquad \text{Blue}$$

The same change in color is effected by dissolving $CoCl_2 \cdot 6H_2O$ in alcohol. Writing made on paper with a dilute solution of the hydrated salt is almost invisible, but when the paper is warmed, dehydration of the salt occurs and the writing becomes blue. It fades again as hydration takes place from moisture of the air. This is the chemistry of one kind of "invisible" ink.

▶ **3. COBALT(II) SULFIDE.** Black **cobalt(II) sulfide,** CoS, is completely precipitated only from basic solutions. However, once precipitated this sulfide is but slightly soluble in hydrochloric acid. Aqua regia readily dissolves it.

▶ **4. COMPLEX COBALT COMPOUNDS.** In addition to the simple salts, oxides, and hydroxides, cobalt forms a large number of complex compounds. Cobalt(II) simple salts are more stable than are those of cobalt(III); complex cobalt(III) compounds, in contrast, are much more stable than the corresponding cobalt(II) complexes. Thus the majority of the more stable complex salts of cobalt contain the metal in the +3 oxidation state. Some of the more important cobalt(III) complexes are **hexaamminecobalt(III) chloride,** $[Co(NH_3)_6]Cl_3$, **potassium hexacyanocobaltate(III),** $K_3[Co(CN)_6]$, and **sodium hexanitrocobaltate(III),** $Na_3[Co(NO_2)_6]$. In these complex compounds the six coordinated groups, such as NH_3, CN^-, and NO_2^-, occupy the corners of a regular octahedron with the cobalt at the center (see Section 31.3).

Nickel

41.21   History and Occurrence of Nickel

Because nickel ores are difficult to reduce, early attempts to produce the metal from its ores were unsuccessful. It was thought by the seventeenth century metallurgists that nickel ores were copper ores, because of the similarity in appearance; but when the ores failed to yield copper, they were called *Kupfernickel* (German). *Kupfer* means copper, and *nickel* refers to a devil, which was thought to prevent the extraction of copper from the ores. The metal was first obtained pure by Baron Axel Fredric Cronstedt, a Swedish chemist, in 1751.

Nickel occurs, along with iron, as an alloy in meteorites. Metallic nickel and iron probably constitute most of the core of the earth. The most important ores of nickel are **pentlandite,** (Ni, Cu, Fe)$_9$S$_8$, and **garnierite,** which is a hydrated magnesium-nickel silicate of variable composition.

41.22   Metallurgy of Nickel

Pentlandite is roasted and then reduced with carbon. This process results in the production of an alloy containing nickel, iron, and copper. The alloy is known as **Monel metal** and, because of its resistance to corrosion, has many important industrial uses. The process used most extensively in the production of pure nickel in recent years involves the separation of the sulfides of nickel, copper, and iron by selective flotation. After separation the nickel sulfide is converted to the oxide by roasting, and the oxide is reduced by carbon. The metal obtained in this way is approximately 96% pure. It is purified electrolytically to 99.98% purity. Gold, silver, and especially platinum are recovered from the anode mud.

41.23   Properties and Uses of Nickel

Nickel is a silvery white metal that is hard, malleable, and ductile. Like iron and cobalt, it is highly magnetic. It is not oxidized by air under ordinary conditions, and it is resistant to the action of alkalies. Dilute acids slowly dissolve nickel, hydrogen being evolved. Nickel is made passive by treatment with concentrated nitric acid; the nickel then will no longer displace hydrogen from dilute acids.

Because of its hardness, resistance to corrosion, and high reflectivity when polished, nickel is widely used in the plating of iron, steel, and copper. It is also a constituent of many important alloys, such as Monel metal (Ni, Cu, and a little Fe), and **Permalloy** (Ni and Fe), used in instruments for electrical transmission and reproduction of sound. **German silver** is a nickel-zinc-copper alloy. **Nichrome** and **chromel** are alloys containing nickel, iron, and chromium; they are resistant to oxidation at high temperatures and show high electrical resistance, so are used in electrical heating units such as electric stoves, pressing irons, and toasters. **Alnico** contains aluminum, nickel, iron, and cobalt; it is highly magnetic, being able to lift 4000 times its own weight of iron. **Platinite** and **invar** are nickel alloys that have the same coefficient of expansion as glass and are therefore used for "seal-in" wires through glass, such as those in electric light bulbs. Finely divided nickel is used as a catalyst in the hydrogenation of oils.

41.24   Compounds of Nickel

In combination, nickel exhibits an oxidation state of +2 almost exclusively. One notable exception is tetravalent nickel in the dioxide NiO_2, a black hydrous substance formed by the oxidation of nickel(II) salts in alkaline solution. This oxide forms one of the electrodes in the Edison storage cell (Section 20.17).

Nickel(II) oxide, NiO, is prepared by heating either the hydroxide, the carbonate, or the nitrate. When alkali metal hydroxides are added to solutions of nickel(II) salts, pale green **nickel(II) hydroxide,** $Ni(OH)_2$, precipitates. Both the oxide and the hydroxide are dissolved by aqueous ammonia with the formation of the deep blue **hexaamminenickel(II) hydroxide,** $[Ni(NH_3)_6](OH)_2$.

$$Ni(OH)_2(s) + 6NH_3 \longrightarrow [Ni(NH_3)_6]^{2+} + 2OH^-$$

The soluble salts of nickel are prepared from either the oxide, hydroxide, or carbonate by treatment with the proper acid. The important soluble salts of the metal are the acetate, chloride, nitrate, and sulfate, and also the hexaammine-nickel(II) sulfate. Ammoniacal solutions of the latter compound are used in nickel-plating baths. The hydrated nickel(II) ion, $[Ni(H_2O)_6]^{2+}$, imparts a pale green color to the solutions and crystallized salts of the nickel(II) ion.

Nickel(II) sulfide is produced as a black precipitate by the action of ammoniacal sulfide solutions upon nickel(II) salts.

Nickel carbonyl, $Ni(CO)_4$, is formed as a colorless volatile liquid (boiling at 43°C) when carbon monoxide is led over finely divided nickel. The carbonyl readily decomposes at higher temperatures and deposits pure nickel metal. The **Mond** process for separating nickel from other metals is based upon the formation and decomposition of nickel carbonyl. When carbon monoxide is passed over a mixture of nickel and other metals, nickel carbonyl is formed and carried along with the excess carbon monoxide, leaving the other metals behind. When heated to 200°C the nickel carbonyl decomposes, and the metal deposits as a fine dust.

Questions

1. Write the names and formulas of the principal ores of iron?
2. List the equations for the reactions that take place in the reduction of iron(III) oxide in the blast furnace.
3. What use is made of the hot exhaust gases from the blast furnace?
4. Compare the open-hearth, basic oxygen, and electric processes for the production of steel in terms of (a) speed, (b) control of composition and temperature, (c) quality of product, and (d) other factors you consider to be important.
5. The basic oxygen process for making steel has gained favor very rapidly within the last few years. Discuss the reasons why this is so.
6. Give the composition and distinguishing properties of pig iron, cast iron, and steel.
7. Identify each of the following: hematite, magnetic oxide of iron, rouge, Mohr's salt, and Prussian blue.
8. What is the composition of iron rust?
9. Decribe the chemistry of an invisible ink.
10. What compound of iron is a common constituent of hard water?
11. What is the gas produced when iron(II) sulfide is treated with a nonoxidizing acid?
12. Give the general equation for the hydrolysis of iron(III) salts. Write equations showing in steps the hydrolysis of iron(III) chloride to produce a precipitate of hydrated $Fe(OH)_3$ in solution, $[Fe(H_2O)_3(OH)_3]$.
13. Why does cobalt precede nickel in the Periodic Table when cobalt has a higher atomic weight than does nickel?
14. What is meant by the term "passivity"?
15. How is nickel rendered passive?
16. What is the composition of Alnico? What use is made of this alloy?
17. Write the equation for the dissolution of nickel(II) hydroxide in aqueous ammonia.
18. What is the physical nature of nickel carbonyl?
19. Sketch the structure of $[Co(CN)_6]^{3-}$.
20. Balance the following equation by oxidation-reduction methods (you will probably find this to be a challenge):

$$Co(NO_3)_2 \longrightarrow Co_2O_3 + NO_2 + O_2$$

21. On the basis of the reactions of elemental fluorine, chlorine, and oxygen with metallic cobalt, predict which is the strongest oxidizing agent.

Problems

1. Iron(II) can be titrated to iron(III) by dichromate ion, which is reduced to chromic ion in acid solution. A 2.500-g sample of iron ore is dissolved and the iron converted to the iron(II) form. Exactly 19.17 ml of 0.0100 M $Na_2Cr_2O_7$ is required in the titration. What per cent iron was the ore sample?

   Ans. 2.569%

2. How many cubic feet of air is required per ton of Fe_2O_3 to convert the oxide into iron in a blast fur- nace? Assume air is 19% oxygen.

   Ans. 3.5×10^4 ft^3

3. What is the potential of the following electrochemical cell:

   $$Cd\,|\,Cd^{2+},\ M = 0.10\,\|\,Ni^{2+},\ M = 0.50\,|\,Ni?$$

   Ans. 0.17 volt

4. What is the per cent cobalt in sodium hexanitro- cobaltate(III)?   *Ans. 14.59%*

References

"Metallic Corrosion," S. C. Britton, *Chem. in Britain,* **5,** 228 (1969).

"Kinetics of Corrosion of Metals," F. Habashi, *J. Chem. Educ.,* **42,** 318 (1965).

"The Composition of the Earth's Interior," T. Takahashi and W. A. Bassett, *Sci. American,* June 1965, p. 100.

"Nature's Geological Paint Pots and Pigments," R. C. Brasted, *J. Chem. Educ.,* **48,** 323 (1971).

"Turnbull's Blue and Prussian Blue: KFe(III) [Fe(II)(CN)$_6$]," L. D. Hansen, W. M. Litchman, and G. H. Daub, *J. Chem. Educ.,* **46,** 46 (1969).

"The Role of Chelation in Iron Metabolism," P. Saltman, *J. Chem. Educ.,* **42,** 683 (1965).

"The Beneficiation of Iron Ores," M. M. Fine, *Sci. American,* January 1968, p. 28.

"Crystallography of the Hexagonal Ferrites," J. A. Kohn, D. W. Eckart, and C. F. Cook, Jr., *Science,* **172,** 519 (1971).

"Strong and Ductile Steels," E. H. Parker and V. F. Zackay, *Sci. American,* November 1968, p. 36.

"Oxygen in Steelmaking," J. K. Stone, *Sci. American,* April 1968, p. 24.

"Carbon Determination in Modern Steelmaking," R. V. Wil- liams, *Chem. in Britain,* **5,** 213 (1969).

"Basic Oxygen Steelmaking," H. E. McGannon, *J. Chem. Educ.,* **46,** 293 (1969).

"How the Iron Age Began," R. Maddin, J. P. Murphy, and T. S. Wheeler, *Sci. American,* October 1977, p. 122.

"The Direct Reduction of Iron Ore," J. R. Miller, *Sci. Ameri- can,* July 1976, p. 68.

"Steelmaking and its Future," R. S. Barnes, *Endeavour,* **2,** 2 (1978).

"Iron Ore: Energy, Labor, and Capital Changes with Tech- nology," P. J. Kakela, *Science,* **202,** 1151 (1978).

"Hydrometallurgy Attracting New Interests" (Staff), *Chem. and Eng. News,* April 3, 1978, p. 29.

"Magnetic Separation Finding New Uses" (Staff), *Chem. and Eng. News,* July 10, 1978, p. 26.

"Complex Iron Smelting and Prehistoric Culture in Tanza- nia," P. Schmidt and D. H. Avery, *Science,* **201,** 1085 (1978).

"From Cabul to Cobalt—A Historical View of the Mischie- vous Metal," F. R. Morral, *J. Chem. Educ.,* **34,** 185 (1957).

"Hemoglobin: Isolation and Chemical Properties," S. F. Russo and R. B. Sorstokke, *J. Chem. Educ.,* **50,** 347 (1973).

"Bees Have Magnetic Remanence," J. L. Gould, J. L. Kirschvink, and K. S. Deffeyes, *Science,* **201,** 1027 (1978).

"Life's Essential Elements," D. R. Williams, *Educ. in Chemis- try,* **10,** 56 (1973).

"Pollution and Public Health: Taconite Case Poses Major Test," L. J. Carter, *Science,* **186,** 31 (1974).

"Iron and Susceptibility to Infectious Diseases," E. D. Wein- berg, *Science,* **184,** 952 (1974).

"The Production of Cobalt," J. V. Huxley, *Educ. in Chemistry,* **11,** 203 (1974).

"A Detailed, Simple Crystal Field Consideration of the Nor- mal Spinel Structure of Co_3O_4," L. Suchow, *J. Chem. Educ.,* **53,** 560 (1976).

"The Production of Nickel," J. V. Huxley, *Educ. in Chem.,* March 1978, p. 62.

Appendices

Appendix A

Chemical Arithmetic

In the study of general chemistry, elementary mathematics is frequently used. Of particular importance and wide application in the calculations that you make in chemistry are exponential arithmetic, significant figures, and logarithms.

A.1 Exponential Arithmetic

In chemistry we use the exponential method of expressing very large and very small numbers. These numbers are expressed as a product of two numbers. The first number of the product is called the *digit term*. This term is usually a number not less than 1 and not greater than 10. The second number of the product is called the *exponential term* and is written as 10 with an exponent. Some examples of the exponential method of expressing numbers are given below.

$$1000 = 1 \times 10^3$$
$$100 = 1 \times 10^2$$
$$10 = 1 \times 10^1$$
$$1 = 1 \times 10^0$$
$$0.1 = 1 \times 10^{-1}$$
$$0.01 = 1 \times 10^{-2}$$
$$0.001 = 1 \times 10^{-3}$$
$$2386 = 2.386 \times 1000 = 2.386 \times 10^3$$
$$0.123 = 1.23 \times 0.1 = 1.23 \times 10^{-1}$$

The power (exponent) of 10 is equal to the number of places the decimal is shifted to give the digit number. The exponential method is particularly useful as a shorthand for big numbers. For example, $1,230,000,000 = 1.23 \times 10^9$; and $0.000\ 000\ 000\ 36 = 3.6 \times 10^{-10}$.

▶ **1. ADDITION OF EXPONENTIALS.** Convert all the numbers to the same power of 10 and add the digit terms of the number.

EXAMPLE   Add 5×10^{-5} and 3×10^{-3}.

$3 \times 10^{-3} = 300 \times 10^{-5}$
$(5 \times 10^{-5}) + (300 \times 10^{-5}) = 305 \times 10^{-5} = 3.05 \times 10^{-3}$

◆ ◆ ◆ ◆ ◆

▶ **2. SUBTRACTION OF EXPONENTIALS.** Convert all the numbers to the same power of 10 and take the difference of the digit terms.

EXAMPLE   Subtract 4×10^{-7} from 5×10^{-6}.

$4 \times 10^{-7} = 0.4 \times 10^{-6}$
$(5 \times 10^{-6}) - (0.4 \times 10^{-6}) = 4.6 \times 10^{-6}$

◆ ◆ ◆ ◆ ◆

▶ **3. MULTIPLICATION OF EXPONENTIALS.** Multiply the digit terms in the usual way and add algebraically the exponents of the exponential terms.

EXAMPLE   Multiply 4.2×10^{-8} by 2×10^3.

4.2×10^{-8}
$\underline{2 \times 10^3}$
8.4×10^{-5}

◆ ◆ ◆ ◆ ◆

▶ **4. DIVISION OF EXPONENTIALS.** Divide the digit term of the numerator by the digit term of the denominator and subtract algebraically the exponents of the exponential terms.

EXAMPLE   Divide 3.6×10^{-5} by 6×10^{-4}.

$$\frac{3.6 \times 10^{-5}}{6 \times 10^{-4}} = 0.6 \times 10^{-1} = 6 \times 10^{-2}$$

◆ ◆ ◆ ◆ ◆

▶ **5. SQUARING OF EXPONENTIALS.** Square the digit term in the usual way and multiply the exponent of the exponential term by 2.

EXAMPLE   Square the number 4×10^{-6}.

$$(4 \times 10^{-6})^2 = 16 \times 10^{-12} = 1.6 \times 10^{-11}$$

◆ ◆ ◆ ◆ ◆

▶ **6. CUBING OF EXPONENTIALS.** Cube the digit term in the usual way and multiply the exponent of the exponential term by 3.

EXAMPLE   Cube the number 2×10^3.

$$(2 \times 10^3)^3 = 2 \times 2 \times 2 \times 10^9 = 8 \times 10^9$$

◆ ◆ ◆ ◆ ◆

▶ **7. EXTRACTION OF SQUARE ROOTS OF EXPONENTIALS.** Decrease or increase the exponential term so that the power of 10 is evenly divisible by 2. Extract the square root of the digit term and divide the exponential term by 2.

EXAMPLE   Extract the square root of 1.6×10^{-7}.

$$1.6 \times 10^{-7} = 16 \times 10^{-8}$$
$$\sqrt{16 \times 10^{-8}} = \sqrt{16} \times \sqrt{10^{-8}} = 4 \times 10^{-4}$$

◆ ◆ ◆ ◆ ◆

A.2   Significant Figures

A bee keeper reports that he has 525,341 bees. The last three figures of the number are obviously inaccurate, for during the time the keeper was counting the bees, some of them would have died and others would have hatched; this would have made the exact number of bees quite difficult to determine. It would have been more accurate if he had reported the number 525,000. In other words, the last three figures are not significant, except to set the position of the decimal point. Their exact values have no meaning.

In reporting any information in terms of numbers, only as many significant figures should be used as are warranted by the accuracy of the measurement. The accuracy of measurements is dependent upon the sensitivity of the measuring instruments used. For example, if the mass of an object has been reported as 2.13 g, it is assumed that the last figure (3) has been estimated and that the mass lies between 2.125 g and 2.135 g. The quantity 2.13 g represents three significant figures. The mass of this same object as determined by a more sensitive balance may have been reported as 2.134 g. In this case we would assume the correct mass to be between 2.1335 g and 2.1345 g, and the quantity 2.134 g represents 4 significant figures. Note that the last figure is estimated and is also considered as a significant figure.

A zero in a number may or may not be significant, depending upon the manner in which it is used. When one or more zeros are used in locating a decimal point, they are not significant. For example, the numbers 0.063, 0.0063, and 0.00063 have each two significant figures. When zeros appear between digits in a number they are significant. For example, 1.008 g has four significant figures. Likewise, the zero in 12.50 is significant. However, the quantity 1370 cm has four significant figures provided the accuracy of the measurement includes the zero as a significant digit; if the digit 7 is estimated, then the number has only three significant figures.

The importance of significant figures lies in their application to fundamental computation. When adding or subtracting, the last digit that is retained in the sum or difference should correspond to the first doubtful decimal place (as indicated by underscoring).

EXAMPLE   **Add 4.383 g and 0.0023 g.**

$$4.383 \text{ g}$$
$$\underline{0.0023}$$
$$4.385 \text{ g}$$

◆ ◆ ◆ ◆ ◆

When multiplying or dividing, the product or quotient should contain no more digits than the least number of significant figures in the numbers involved in the computation.

EXAMPLE   **Multiply 0.6238 by 6.6.**

$$0.6238 \times 6.6 = 4.1$$

◆ ◆ ◆ ◆ ◆

In rounding off numbers, increase the last digit retained by one if the following digit is 6 or more. Do not change the last digit retained if the following digit is 4 or less. If the last digit retained is followed by 5, increase the last digit retained by 1 if it is odd or leave the last digit retained unchanged if it is even.

A.3   The Use of Logarithms and Exponential Numbers

The common logarithm of a number is the power to which the number 10 must be raised to equal that number. For example, the logarithm of 100 is 2 because the number 10 must be raised to the second power to be equal to 100. Additional examples follow.

Number	Number Expressed Exponentially	Logarithm
10,000	10^4	4
1,000	10^3	3
10	10^1	1
1	10^0	0
0.1	10^{-1}	-1
0.01	10^{-2}	-2
0.001	10^{-3}	-3
0.0001	10^{-4}	-4

What is the logarithm of 60? Because 60 lies between 10 and 100, which have logarithms of 1 and 2, respectively, the logarithm of 60 must lie between 1 and 2. The logarithm of 60 is 1.7782; i.e., $60 = 10^{1.7782}$.

Every logarithm is made up of two parts, called the *characteristic* and the *mantissa*. The characteristic is that part of the logarithm that lies to the left of the decimal point; thus the characteristic of the logarithm of 60 is 1. The mantissa is that part of the logarithm that lies to the right of the decimal point; thus the mantissa of the logarithm of 60 is .7782. The characteristic of the logarithm of a number greater than 1 is one less than the number of digits to the left of the decimal point in the number.

Number	Characteristic	Number	Characteristic
60	1	2.340	0
600	2	23.40	1
6,000	3	234.0	2
52,840	4	2,340.0	3

The mantissa of the logarithm of a number is found in the logarithm table (see Appendix B), and its value is independent of the position of the decimal point. Thus, 2.340, 23.40, 234.0, and 2340.0 all have the same mantissa. The logarithm of 2.340 is 0.3692, that of 23.40 is 1.3692, that of 234.0 is 2.3692, and that of 2340.0 is 3.3692.

The meaning of the mantissa and characteristic can be better understood from a consideration of their relationship to exponential numbers. For example, 2340 may be written 2.34×10^3. The logarithm of $(2.34 \times 10^3) =$ the logarithm of 2.34 + the logarithm of 10^3. The logarithm of 2.34 is 0.3692 and the logarithm of 10^3 is 3. Thus, the logarithm of $2340 = 3 + 0.3692$, or 3.3692.

The logarithm of a number less than 1 has a negative value, and a convenient method of obtaining the logarithm of such a number is given below. For example, we may obtain the logarithm of 0.00234 as follows: When expressed exponentially, $0.00234 = 2.34 \times 10^{-3}$. The logarithm of $2.34 \times 10^{-3} =$ the logarithm of 2.34 + the logarithm of 10^{-3}. The logarithm of 2.34 is 0.3692 and the logarithm of 10^{-3} is -3. Thus the logarithm of $0.00234 = 0.3692 + (-3) = 0.3692 - 3 = -2.6208$. The abbreviated form for the expression $(0.3692 - 3)$ is $\bar{3}.3692$. Note that only the characteristic has a negative value in the logarithm $\bar{3}.3692$, and that the mantissa is positive. The logarithm $\bar{3}.3692$ may also be written as $7.3692 - 10$.

To multiply two numbers we add the logarithms of the numbers. For example, suppose we multiply 412 by 353.

$$
\begin{aligned}
\text{Logarithm of 412} &= 2.6149 \\
\text{Logarithm of 353} &= 2.5478 \\
\hline
\text{Logarithm of product} &= 5.1627
\end{aligned}
$$

The number that corresponds to the logarithm 5.1627 is 145,400 or 1.454×10^5. Thus 1.45×10^5 is the product of 412 and 353.

To divide two numbers we subtract the logarithms of the numbers. Suppose we divide 412 by 353.

$$\begin{aligned}
\text{Logarithm of } 412 &= 2.6149 \\
\text{Logarithm of } 353 &= 2.5478 \\
\hline
\text{Logarithm of quotient} &= 0.0671
\end{aligned}$$

The number that corresponds to the logarithm 0.0671 is 1.17. Thus, 412 divided by 353 is 1.17.

Suppose we wish to multiply 5432 by 0.3124. Add the logarithm of 0.3124 to that of 5432.

$$\begin{aligned}
\text{Logarithm of } 5432 &= 3.7350 \\
\text{Logarithm of } 0.3124 &= \overline{1}.4948 \\
\hline
\text{Logarithm of the product} &= 3.2298
\end{aligned}$$

The number that corresponds to the logarithm 3.2298 is 1697 or 1.697×10^3.

Let us divide 5432 by 0.3124. Subtract the logarithm of 0.3124 from that of 5432.

$$\begin{aligned}
\text{Logarithm of } 5432 &= 3.7350 \\
\text{Logarithm of } 0.3124 &= \overline{1}.4948 \\
\hline
\text{Logarithm of the quotient} &= 4.2402
\end{aligned}$$

The number that corresponds to the logarithm 4.2402 is 17,390 or 1.739×10^4.

The extraction of roots of numbers by means of logarithms is a simple procedure. For example, suppose we extract the cube root of 7235. The logarithm of $\sqrt[3]{7235}$ or $(7235)^{1/3}$ is equal to $\frac{1}{3}$ of the logarithm of 7235.

$$\begin{aligned}
\text{Logarithm of } 7235 &= 3.8594 \\
\tfrac{1}{3} \text{ of } 3.8594 &= 1.2865
\end{aligned}$$

The number that corresponds to the logarithm 1.2865 is 19.34. Thus 19.34 is the cube root of 7235.

A.4   The Solution of Quadratic Equations

Any quadratic equation can be expressed in the following form:

$$aX^2 + bX + c = 0$$

In order to solve a quadratic equation, the following formula is used.

$$X = \frac{-b \pm \sqrt{b^2 - 4ac}}{2a}$$

EXAMPLE   Solve the equation $3X^2 + 13X - 10 = 0$.

Substituting the values $a = 3$, $b = 13$, and $c = -10$ in the formula, we obtain

$$X = \frac{-13 \pm \sqrt{(13)^2 - 4 \times 3 \times (-10)}}{2 \times 3}$$

$$X = \frac{-13 \pm \sqrt{169 + 120}}{6} = \frac{-13 \pm \sqrt{289}}{6} = \frac{-13 \pm 17}{6}$$

The two roots are therefore

$$X = \frac{-13 + 17}{6} = 0.67 \quad \text{and} \quad X = \frac{-13 - 17}{6} = -5$$

• • • • •

Equations constructed upon physical data always have real roots, and of these real roots only those having positive values are usually of any significance.

Problems

In all problems, observe the principle of significant figures.

1. Carry out the following calculations:
   (a) $(4.8 \times 10^{-2}) \times (1.92 \times 10^{-4})$
   (b) $(6.1 \times 10^{-3}) \times (5.81 \times 10^5)$
   (c) $(4.8 \times 10^{-4}) \div (2.4 \times 10^{-8})$
   (d) $(9.6 \times 10^{-4}) \div (2.00 \times 10^3)$

2. Perform the following operations:
   (a) $\sqrt{90 \times 10^{-3}}$
   (b) $\sqrt[3]{2.7 \times 10^{-2}}$
   (c) $(1 \times 10^6)^4$
   (d) $(4.1 \times 10^{-2})^2$
   (e) $(64 \times 10^5)^{1/2}$
   (f) $(5.2 \times 10^{-2})^{1/2}$

3. Add the following numbers: 2.863, 42.580, 0.02316, and 41.33.

4. Multiply 7.850 by 3.2; divide 4.6 by 2.075.

5. Carry out the following calculation:
   $(5 \times 1.008) + (2 \times 12.011) + (126.9045)$

6. Find the logarithms of the following numbers:
   (a) 51.5
   (b) 0.0515
   (c) 512
   (d) 51.2
   (e) 4.80×10^3
   (f) 4.80×10^{-3}
   (g) 1000
   (h) 34,200

7. Write the characteristic and mantissa of each of the following logarithms. Check the values given to verify that they are correct.

   (a) $\log 8 = 0.9031$
   (b) $\log 800 = 2.9031$
   (c) $\log 0.08 = \overline{2}.9031$
   (d) $\log 0.8 = \overline{1}.9031$
   (e) $\log (8 \times 10^{-2}) = -1.0969$

8. Make the following calculations using logarithms:
   (a) 35.4×8.07
   (b) $79.2 \div 0.082$
   (c) $\sqrt[5]{4.270}$
   (d) $\frac{47.2 \times 8.50}{12.1 \times 306}$
   (e) $\frac{1.50 \times 0.082 \times 293}{740/760}$
   (f) $(9.54 \times 1.20 \times 10^2)^{1/2}$

9. Find the value of x in the following:
   (a) $x = \dfrac{-5 \pm \sqrt{5^2 - 4(0.2)(1.7)}}{2(2)}$
   (b) $x = \dfrac{11 \pm \sqrt{(-11)^2 - 4(3.5 \times 10^2)(-8.0 \times 10^{-3})}}{2(3.5 \times 10^2)}$

10. Write the quadratic equations whose solutions are shown in Problem 9.

11. Solve for x if $k = 1.8 \times 10^{-5}$ and $c = 3.8 \times 10^{-3}$:
    $$x^2 + kx - kc = 0$$

12. Solve for x if $k = 5.0 \times 10^{-4}$ and $c = 2.7 \times 10^{-2}$:
    $$k = \frac{x^2}{c - x}$$

Appendix B

Four-Place Table of Logarithms

No.	0	1	2	3	4	5	6	7	8	9	1	2	3	4	5	6	7	8	9
10	0000	0043	0086	0128	0170	0212	0253	0294	0334	0374	4	8	12	17	21	25	29	33	37
11	0414	0453	0492	0531	0569	0607	0645	0682	0719	0755	4	8	11	15	19	23	26	30	34
12	0792	0828	0864	0899	0934	0969	1004	1038	1072	1106	3	7	10	14	17	21	24	28	31
13	1139	1173	1206	1239	1271	1303	1335	1367	1399	1430	3	6	10	13	16	19	23	26	29
14	1461	1492	1523	1553	1584	1614	1644	1673	1703	1732	3	6	9	12	15	18	21	24	27
15	1761	1790	1818	1847	1875	1903	1931	1959	1987	2014	3	6	8	11	14	17	20	22	25
16	2041	2068	2095	2122	2148	2175	2201	2227	2253	2279	3	5	8	11	13	16	18	21	24
17	2304	2330	2355	2380	2405	2430	2455	2480	2504	2529	2	5	7	10	12	15	17	20	22
18	2553	2577	2601	2625	2648	2672	2695	2718	2742	2765	2	5	7	9	12	14	16	19	21
19	2788	2810	2833	2856	2878	2900	2923	2945	2967	2989	2	4	7	9	11	13	16	18	20
20	3010	3032	3054	3075	3096	3118	3139	3160	3181	3201	2	4	6	8	11	13	15	17	19
21	3222	3243	3263	3284	3304	3324	3345	3365	3385	3404	2	4	6	8	10	12	14	16	18
22	3424	3444	3464	3483	3502	3522	3541	3560	3579	3598	2	4	6	8	10	12	14	15	17
23	3617	3636	3655	3674	3692	3711	3729	3747	3766	3784	2	4	6	7	9	11	13	15	17
24	3802	3820	3838	3856	3874	3892	3909	3927	3945	3962	2	4	5	7	9	11	12	14	16
25	3979	3997	4014	4031	4048	4065	4082	4099	4116	4133	2	3	5	7	9	10	12	14	15
26	4150	4166	4183	4200	4216	4232	4249	4265	4281	4298	2	3	5	7	8	10	11	13	15
27	4314	4330	4346	4362	4378	4393	4409	4425	4440	4456	2	3	5	6	8	9	11	13	14
28	4472	4487	4502	4518	4533	4548	4564	4579	4594	4609	2	3	5	6	8	9	11	12	14
29	4624	4639	4654	4669	4683	4698	4713	4728	4742	4757	1	3	4	6	7	9	10	12	13
30	4771	4786	4800	4814	4829	4843	4857	4871	4886	4900	1	3	4	6	7	9	10	11	13
31	4914	4928	4942	4955	4969	4983	4997	5011	5024	5038	1	3	4	6	7	8	10	11	12
32	5051	5065	5079	5092	5105	5119	5132	5145	5159	5172	1	3	4	5	7	8	9	11	12
33	5185	5198	5211	5224	5237	5250	5263	5276	5289	5302	1	3	4	5	6	8	9	10	12
34	5313	5328	5340	5353	5366	5378	5391	5403	5416	5428	1	3	4	5	6	8	9	10	11
35	5441	5453	5465	5478	5490	5502	5514	5527	5539	5551	1	2	4	5	6	7	9	10	11
36	5563	5575	5587	5599	5611	5623	5635	5647	5658	5670	1	2	4	5	6	7	8	10	11
37	5682	5694	5705	5717	5729	5740	5752	5763	5775	5786	1	2	3	5	6	7	8	9	10
38	5798	5809	5821	5832	5843	5855	5866	5877	5888	5899	1	2	3	5	6	7	8	9	10
39	5911	5922	5933	5944	5955	5966	5977	5988	5999	6010	1	2	3	4	5	7	8	9	10
40	6021	6031	6042	6053	6064	6075	6085	6096	6107	6117	1	2	3	4	5	6	8	9	10
41	6128	6138	6149	6160	6170	6180	6191	6201	6212	6222	1	2	3	4	5	6	7	8	9
42	6232	6243	6253	6263	6274	6284	6294	6304	6314	6325	1	2	3	4	5	6	7	8	9
43	6335	6345	6355	6365	6375	6385	6395	6405	6415	6425	1	2	3	4	5	6	7	8	9
44	6435	6444	6454	6464	6474	6484	6493	6503	6513	6522	1	2	3	4	5	6	7	8	9
45	6532	6542	6551	6561	6571	6580	6590	6599	6609	6618	1	2	3	4	5	6	7	8	9
46	6628	6637	6646	6656	6665	6675	6684	6693	6702	6712	1	2	3	4	5	6	7	7	8
47	6721	6730	6739	6749	6758	6767	6776	6785	6794	6803	1	2	3	4	5	5	6	7	8
48	6812	6821	6830	6839	6848	6857	6866	6875	6884	6893	1	2	3	4	4	5	6	7	8
49	6902	6911	6920	6928	6937	6946	6955	6964	6972	6981	1	2	3	4	4	5	6	7	8
50	6990	6998	7007	7016	7024	7033	7042	7050	7059	7067	1	2	3	3	4	5	6	7	8
51	7076	7084	7093	7101	7110	7118	7126	7135	7143	7152	1	2	3	3	4	5	6	7	8
52	7160	7168	7177	7185	7193	7202	7210	7218	7226	7235	1	2	2	3	4	5	6	7	7
53	7243	7251	7259	7267	7275	7284	7292	7300	7308	7316	1	2	2	3	4	5	6	6	7
54	7324	7332	7340	7348	7356	7364	7372	7380	7388	7396	1	2	2	3	4	5	6	6	7
	0	1	2	3	4	5	6	7	8	9	1	2	3	4	5	6	7	8	9

Four-Place Table of Logarithms (continued)

No.	0	1	2	3	4	5	6	7	8	9	1	2	3	4	5	6	7	8	9
55	7404	7412	7419	7427	7435	7443	7451	7459	7466	7474	1	2	2	3	4	5	5	6	7
56	7482	7490	7497	7505	7513	7520	7528	7536	7543	7551	1	2	2	3	4	5	5	6	7
57	7559	7566	7574	7582	7589	7597	7604	7612	7619	7627	1	2	2	3	4	5	5	6	7
58	7634	7642	7649	7657	7664	7672	7679	7686	7694	7701	1	1	2	3	4	4	5	6	7
59	7709	7716	7723	7731	7738	7745	7752	7760	7767	7774	1	1	2	3	4	4	5	6	7
60	7782	7789	7796	7803	7810	7818	7825	7832	7839	7846	1	1	2	3	4	4	5	6	6
61	7853	7860	7868	7875	7882	7889	7896	7903	7910	7917	1	1	2	3	4	4	5	6	6
62	7924	7931	7938	7945	7952	7959	7966	7973	7980	7987	1	1	2	3	3	4	5	6	6
63	7992	8000	8007	8014	8021	8028	8035	8041	8048	8055	1	1	2	3	3	4	5	5	6
64	8062	8069	8075	8082	8089	8096	8102	8109	8116	8122	1	1	2	3	3	4	5	5	6
65	8129	8136	8142	8149	8156	8162	8169	8176	8182	8189	1	1	2	3	3	4	5	5	6
66	8195	8202	8209	8215	8222	8228	8235	8241	8248	8254	1	1	2	3	3	4	5	5	6
67	8261	8267	8274	8280	8287	8293	8299	8306	8312	8319	1	1	2	3	3	4	5	5	6
68	8325	8331	8338	8344	8351	8357	8363	8370	8376	8382	1	1	2	3	3	4	4	5	6
69	8388	8395	8401	8407	8414	8420	8426	8432	8439	8445	1	1	2	2	3	4	4	5	6
70	8451	8457	8463	8470	8476	8482	8488	8494	8500	8506	1	1	2	2	3	4	4	5	6
71	8513	8519	8525	8531	8537	8543	8549	8555	8561	8567	1	1	2	2	3	4	4	5	5
72	8573	8579	8585	8591	8597	8603	8609	8615	8621	8627	1	1	2	2	3	4	4	5	5
73	8633	8639	8645	8651	8657	8663	8669	8675	8681	8686	1	1	2	2	3	4	4	5	5
74	8692	8698	8704	8710	8716	8722	8727	8733	8739	8745	1	1	2	2	3	4	4	5	5
75	8751	8756	8762	8768	8774	8779	8785	8791	8797	8802	1	1	2	2	3	3	4	5	5
76	8808	8814	8820	8825	8831	8837	8842	8848	8854	8859	1	1	2	2	3	3	4	5	5
77	8865	8871	8876	8882	8887	8893	8899	8904	8910	8915	1	1	2	2	3	3	4	4	5
78	8921	8927	8932	8938	8943	8949	8954	8960	8965	8971	1	1	2	2	3	3	4	4	5
79	8976	8982	8987	8993	8998	9004	9009	9015	9020	9025	1	1	2	2	3	3	4	4	5
80	9031	9036	9042	9047	9053	9058	9063	9069	9074	9079	1	1	2	2	3	3	4	4	5
81	9085	9090	9096	9101	9106	9112	9117	9122	9128	9133	1	1	2	2	3	3	4	4	5
82	9138	9143	9149	9154	9159	9165	9170	9175	9180	9186	1	1	2	2	3	3	4	4	5
83	9191	9196	9201	9206	9212	9217	9222	9227	9232	9238	1	1	2	2	3	3	4	4	5
84	9243	9248	9253	9258	9263	9269	9274	9279	9284	9289	1	1	2	2	3	3	4	4	5
85	9294	9299	9304	9309	9315	9320	9325	9330	9335	9340	1	1	2	2	3	3	4	4	5
86	9345	9350	9355	9360	9365	9370	9375	9380	9385	9390	1	1	2	2	3	3	4	4	5
87	9395	9400	9405	9410	9415	9420	9425	9430	9435	9440	0	1	1	2	2	3	3	4	4
88	9445	9450	9455	9460	9465	9469	9474	9479	9484	9489	0	1	1	2	2	3	3	4	4
89	9494	9499	9504	9509	9513	9518	9523	9528	9533	9538	0	1	1	2	2	3	3	4	4
90	9542	9547	9552	9557	9562	9566	9571	9576	9581	9586	0	1	1	2	2	3	3	4	4
91	9590	9595	9600	9605	9609	9614	9619	9624	9628	9633	0	1	1	2	2	3	3	4	4
92	9638	9643	9647	9652	9657	9661	9666	9671	9675	9680	0	1	1	2	2	3	3	4	4
93	9685	9689	9694	9699	9703	9708	9713	9717	9722	9727	0	1	1	2	2	3	3	4	4
94	9731	9736	9741	9745	9750	9754	9759	9763	9768	9773	0	1	1	2	2	3	3	4	4
95	9777	9782	9786	9791	9795	9800	9805	9809	9814	9818	0	1	1	2	2	3	3	4	4
96	9823	9827	9832	9836	9841	9845	9850	9854	9859	9863	0	1	1	2	2	3	3	4	4
97	9868	9872	9877	9881	9886	9890	9894	9899	9903	9908	0	1	1	2	2	3	3	4	4
98	9912	9917	9921	9926	9930	9934	9939	9943	9948	9952	0	1	1	2	2	3	3	4	4
99	9956	9961	9965	9969	9974	9978	9983	9987	9991	9996	0	1	1	2	2	3	3	3	4
	0	1	2	3	4	5	6	7	8	9	1	2	3	4	5	6	7	8	9

Appendix C

Units and Conversion Factors

Base Units of International System of Units (SI)

Physical Property	Name of Unit	Symbol
Length	Meter	m
Mass	Kilogram	kg
Time	Second	s
Electric current	Ampere	A
Thermodynamic temperature	Kelvin	K
Luminous intensity	Candela	cd
Quantity of substance	Mole	mol

Units of Length

Meter (m) = 39.37 inches (in) = 1.094 yards
Centimeter (cm) = 0.01 m
Millimeter (mm) = 0.001 m
Kilometer (km) = 1000 m
Angstrom unit (Å) = 10^{-8} cm = 10^{-10} m

Yard = 0.9144 m
Inch = 2.54 cm

Mile (U.S.) = 1.609 km

Units of Volume

Liter (l) = 0.001 m^3 = 1000 cm^3
Milliliter (ml) = 0.001 liter = 1 cm^3

Liquid quart (U.S.) = 0.9463 liter
Cubic foot (U.S.) = 28.316 liters

Units of Weight

Gram (g) = 0.001 kg

Milligram (mg) = 0.001 g
Kilogram (kg) = 1000 g
Ton (metric) = 1000 kg = 2204.62 lb

Ounce (oz) (avoirdupois) = 28.35 g
Pound (lb) (avoirdupois) = 0.45359237 kg
Ton (short) = 2000 lb = 907.185 kg
Ton (long) = 2240 lb = 1.016 metric tons

Units of Energy

4.184 joule (J) = 1 thermochemical calorie (cal) = 4.184×10^7 erg
Erg = 10^{-7} J
Electron volt (eV) = 1.602189×10^{-12} erg = 23.061 kcal/mol
Liter atmosphere = 24.217 cal = 101.32 J

Unit of Force

Newton (N) = 1 kg m s^{-2} (force which when applied for 1 sec will give to a 1-kilogram mass a speed of 1 meter per second)

Units of Pressure

Torr = 1 mm Hg
Atmosphere (atm) = 760 mm Hg = 760 torr = 101,325 N m^{-2} = 101,325 Pa
Pascal = kg m^{-1} s^{-2} = N m^{-2}

Appendix D

General Physical Constants

Avogadro's Number, N_A	6.022045×10^{23} mol^{-1}
Electron charge, e	$1.6021892 \times 10^{-19}$ coulomb
Electron rest mass, m_e	9.109534×10^{-31} kg
Proton rest mass, m_p	$1.6726485 \times 10^{-27}$ kg
Neutron rest mass, m_n	$1.6749543 \times 10^{-27}$ kg
Charge-to-mass ratio for electron, e/m_e	1.7588047×10^{11} coulomb kg^{-1}
Faraday constant, F	9.648670×10^4 coulomb/g-equivalent weight
Planck constant, h	6.626176×10^{-34} J sec
Boltzmann constant, k	1.380662×10^{-23} J K^{-1}
Gas constant, R	8.20562×10^{-2} ℓ atm mol^{-1} K^{-1}
	$= 8.31441$ J mol^{-1} K^{-1}
Speed of light (in vacuum), c	2.99792458×10^8 m sec^{-1}
Atomic mass unit ($= \frac{1}{12}$ the mass of an atom of the ^{12}C nuclide), amu	$1.6605655 \times 10^{-27}$ kg
Rydberg constant, R_∞	1.097373177×10^7 m^{-1}

Appendix E

Solubility Product Constants

Substance	K_{sp} at 25°	Substance	K_{sp} at 25°
Aluminum		Cobalt	
Al(OH)$_3$	1.9×10^{-33}	Co(OH)$_2$	2×10^{-16}
Barium		CoS(α)	5.9×10^{-21}
BaCO$_3$	8.1×10^{-9}	CoS(β)	8.7×10^{-23}
BaC$_2$O$_4\cdot$2H$_2$O	1.1×10^{-7}	CoCO$_3$	1.0×10^{-12}
BaSO$_4$	1.08×10^{-10}	Co(OH)$_3$	2.5×10^{-43}
BaCrO$_4$	2×10^{-10}	Copper	
BaF$_2$	1.7×10^{-6}	CuCl	1.85×10^{-7}
Ba(OH)$_2\cdot$8H$_2$O	5.0×10^{-3}	CuBr	5.3×10^{-9}
Ba$_3$(PO$_4$)$_2$	1.3×10^{-29}	CuI	5.1×10^{-12}
Ba$_3$(AsO$_4$)$_2$	1.1×10^{-13}	CuCNS	4×10^{-14}
Bismuth		Cu$_2$S	1.6×10^{-48}
BiO(OH)	1×10^{-12}	Cu(OH)$_2$	5.6×10^{-20}
BiOCl	7×10^{-9}	CuS	8.7×10^{-36}
Bi$_2$S$_3$	1.6×10^{-72}	CuCO$_3$	1.37×10^{-10}
Cadmium		Iron	
Cd(OH)$_2$	1.2×10^{-14}	Fe(OH)$_2$	7.9×10^{-15}
CdS	3.6×10^{-29}	FeCO$_3$	2.11×10^{-11}
CdCO$_3$	2.5×10^{-14}	FeS	1×10^{-19}
Calcium		Fe(OH)$_3$	1.1×10^{-36}
Ca(OH)$_2$	7.9×10^{-6}	Lead	
CaCO$_3$	4.8×10^{-9}	Pb(OH)$_2$	2.8×10^{-16}
CaSO$_4\cdot$2H$_2$O	2.4×10^{-5}	PbF$_2$	3.7×10^{-8}
CaC$_2$O$_4\cdot$H$_2$O	2.27×10^{-9}	PbCl$_2$	1.7×10^{-5}
Ca$_3$(PO$_4$)$_2$	1×10^{-25}	PbBr$_2$	6.3×10^{-6}
CaHPO$_4$	5×10^{-6}	PbI$_2$	8.7×10^{-9}
CaF$_2$	3.9×10^{-11}	PbCO$_3$	1.5×10^{-13}
Chromium		PbS	8.4×10^{-28}
Cr(OH)$_3$	6.7×10^{-31}	PbCrO$_4$	1.8×10^{-14}

Appendix E
(continued)

Solubility Product Constants (continued)

Substance	K_{sp} at 25°	Substance	K_{sp} at 25°
$PbSO_4$	1.8×10^{-8}	Silver	
$Pb_3(PO_4)_2$	3×10^{-44}	$\frac{1}{2}Ag_2O(Ag^+ + OH^-)$	2×10^{-8}
Magnesium		$AgCl$	1.8×10^{-10}
$Mg(OH)_2$	1.5×10^{-11}	$AgBr$	3.3×10^{-13}
$MgCO_3 \cdot 3H_2O$	$ca.\ 1 \times 10^{-5}$	AgI	1.5×10^{-16}
$MgNH_4PO_4$	2.5×10^{-13}	$AgCN$	1.2×10^{-16}
MgF_2	6.4×10^{-9}	$AgCNS$	1.0×10^{-12}
MgC_2O_4	8.6×10^{-5}	Ag_2S	1.0×10^{-51}
Manganese		Ag_2CO_3	8.2×10^{-12}
$Mn(OH)_2$	4.5×10^{-14}	Ag_2CrO_4	9×10^{-12}
$MnCO_3$	8.8×10^{-11}	$Ag_4Fe(CN)_6$	1.55×10^{-41}
MnS	5.6×10^{-16}	Ag_2SO_4	1.18×10^{-5}
Mercury		Ag_3PO_4	1.8×10^{-18}
$Hg_2O \cdot H_2O$	1.6×10^{-23}	Strontium	
Hg_2Cl_2	1.1×10^{-18}	$Sr(OH)_2 \cdot 8H_2O$	3.2×10^{-4}
Hg_2Br_2	1.26×10^{-22}	$SrCO_3$	9.42×10^{-10}
Hg_2I_2	4.5×10^{-29}	$SrCrO_4$	3.6×10^{-5}
Hg_2CO_3	9×10^{-17}	$SrSO_4$	2.8×10^{-7}
Hg_2SO_4	6.2×10^{-7}	$SrC_2O_4 \cdot H_2O$	5.61×10^{-8}
Hg_2S	1×10^{-45}	Thallium	
Hg_2CrO_4	2×10^{-9}	$TlCl$	1.9×10^{-4}
HgS	3×10^{-53}	$TlCNS$	5.8×10^{-4}
Nickel		Tl_2S	1.2×10^{-24}
$Ni(OH)_2$	1.6×10^{-14}	$Tl(OH)_3$	1.5×10^{-44}
$NiCO_3$	1.36×10^{-7}	Tin	
$NiS(\alpha)$	3×10^{-21}	$Sn(OH)_2$	5×10^{-26}
$NiS(\beta)$	1×10^{-26}	SnS	8×10^{-29}
$NiS(\gamma)$	2×10^{-28}	$Sn(OH)_4$	$ca.\ 1 \times 10^{-56}$
Potassium		Zinc	
$KClO_4$	1.07×10^{-2}	$ZnCO_3$	6×10^{-11}
K_2PtCl_6	1.1×10^{-5}	$Zn(OH)_2$	4.5×10^{-17}
$KHC_4H_4O_6$	3×10^{-4}	ZnS	1.1×10^{-21}

Appendix F

Association (Formation) Constants for Complex Ions

Equilibrium	K_f
$Al^{3+} + 6F^- \rightleftharpoons [AlF_6]^{3-}$	5×10^{23}
$Cd^{2+} + 4NH_3 \rightleftharpoons [Cd(NH_3)_4]^{2+}$	4.0×10^6
$Cd^{2+} + 4CN^- \rightleftharpoons [Cd(CN)_4]^{2-}$	1.3×10^{17}
$Co^{2+} + 6NH_3 \rightleftharpoons [Co(NH_3)_6]^{2+}$	8.3×10^4
$Co^{3+} + 6NH_3 \rightleftharpoons [Co(NH_3)_6]^{3+}$	4.5×10^{33}
$Cu^+ + 2CN^- \rightleftharpoons [Cu(CN)_2]^-$	1×10^{16}
$Cu^{2+} + 4NH_3 \rightleftharpoons [Cu(NH_3)_4]^{2+}$	1.2×10^{12}
$Fe^{2+} + 6CN^- \rightleftharpoons [Fe(CN)_6]^{4-}$	1×10^{37}
$Fe^{3+} + 6CN^- \rightleftharpoons [Fe(CN)_6]^{3-}$	1×10^{44}
$Fe^{3+} + 6SCN^- \rightleftharpoons [Fe(NCS)_6]^{3-}$	3.2×10^3
$Hg^{2+} + 4Cl^- \rightleftharpoons [HgCl_4]^{2-}$	1.2×10^{15}

$$Ni^{2+} + 6NH_3 \rightleftharpoons [Ni(NH_3)_6]^{2+} \qquad 1.8 \times 10^8$$
$$Ag^+ + 2Cl^- \rightleftharpoons [AgCl_2]^- \qquad 2.5 \times 10^5$$
$$Ag^+ + 2CN^- \rightleftharpoons [Ag(CN)_2]^- \qquad 1 \times 10^{20}$$
$$Ag^+ + 2NH_3 \rightleftharpoons [Ag(NH_3)_2]^+ \qquad 1.6 \times 10^7$$
$$Zn^{2+} + 4CN^- \rightleftharpoons [Zn(CN)_4]^{2-} \qquad 1 \times 10^{19}$$
$$Zn^{2+} + 4OH^- \rightleftharpoons [Zn(OH)_4]^{2-} \qquad 2.9 \times 10^{15}$$

Appendix G

Ionization Constants of Weak Acids

Acid	Formula	K_i at 25°
Acetic	HOAc (CH_3COOH)	1.8×10^{-5}
Arsenic	H_3AsO_4	4.8×10^{-3}
	$H_2AsO_4^-$	1×10^{-7}
	$HAsO_4^{2-}$	1×10^{-13}
Arsenous	H_3AsO_3	5.8×10^{-10}
Boric	H_3BO_3	5.8×10^{-10}
Carbonic	H_2CO_3	4.3×10^{-7}
	HCO_3^-	7×10^{-11}
Cyanic	HCNO	3.46×10^{-4}
Formic	HCOOH	1.8×10^{-4}
Hydrazoic	HN_3	1×10^{-4}
Hydrocyanic	HCN	4×10^{-10}
Hydrofluoric	HF	7.2×10^{-4}
Hydrogen peroxide	H_2O_2	2.4×10^{-12}
Hydrogen selenide	H_2Se	1.7×10^{-4}
	HSe^-	1×10^{-10}
Hydrogen sulfate ion	HSO_4^-	1.2×10^{-2}
Hydrogen sulfide	H_2S	1.0×10^{-7}
	HS^-	1.3×10^{-13}
Hydrogen telluride	H_2Te	2.3×10^{-3}
	HTe^-	1×10^{-5}
Hypobromous	HBrO	2×10^{-9}
Hypochlorous	HClO	3.5×10^{-8}
Nitrous	HNO_2	4.5×10^{-4}
Oxalic	$H_2C_2O_4$	5.9×10^{-2}
	$HC_2O_4^-$	6.4×10^{-5}
Phosphoric	H_3PO_4	7.5×10^{-3}
	$H_2PO_4^-$	6.2×10^{-8}
	HPO_4^{2-}	3.6×10^{-13}
Phosphorous	H_3PO_3	1.6×10^{-2}
	$H_2PO_3^-$	7×10^{-7}
Sulfurous	H_2SO_3	1.2×10^{-2}
	HSO_3^-	6.2×10^{-8}

Appendix H

Ionization Constants of Weak Bases

Base	Ionization Equation	K_i at 25°
Ammonia	$NH_3 + H_2O \rightleftharpoons NH_4^+ + OH^-$	1.8×10^{-5}
Dimethylamine	$(CH_3)_2NH + H_2O \rightleftharpoons (CH_3)_2NH_2^+ + OH^-$	7.4×10^{-4}
Methylamine	$CH_3NH_2 + H_2O \rightleftharpoons CH_3NH_3^+ + OH^-$	4.4×10^{-4}
Phenylamine (aniline)	$C_6H_5NH_2 + H_2O \rightleftharpoons C_6H_5NH_3^+ + OH^-$	4.6×10^{-10}
Trimethylamine	$(CH_3)_3N + H_2O \rightleftharpoons (CH_3)_3NH^+ + OH^-$	7.4×10^{-5}

Appendix I

Standard Electrode (Reduction) Potentials

Half-Reactions	$E°$, volts
$Li^+ + e^- \longrightarrow Li$	-3.09
$K^+ + e^- \longrightarrow K$	-2.925
$Rb^+ + e^- \longrightarrow Rb$	-2.925
$Ra^{2+} + 2e^- \longrightarrow Ra$	-2.92
$Ba^{2+} + 2e^- \longrightarrow Ba$	-2.90
$Sr^{2+} + 2e^- \longrightarrow Sr$	-2.89
$Ca^{2+} + 2e^- \longrightarrow Ca$	-2.87
$Na^+ + e^- \longrightarrow Na$	-2.714
$La^{3+} + 3e^- \longrightarrow La$	-2.52
$Ce^{3+} + 3e^- \longrightarrow Ce$	-2.48
$Nd^{3+} + 3e^- \longrightarrow Nd$	-2.44
$Sm^{3+} + 3e^- \longrightarrow Sm$	-2.41
$Gd^{3+} + 3e^- \longrightarrow Gd$	-2.40
$Mg^{2+} + 2e^- \longrightarrow Mg$	-2.37
$Y^{3+} + 3e^- \longrightarrow Y$	-2.37
$Am^{3+} + 3e^- \longrightarrow Am$	-2.32
$Lu^{3+} + 3e^- \longrightarrow Lu$	-2.25
$\frac{1}{2}H_2 + e^- \longrightarrow H^-$	-2.25
$Sc^{3+} + 3e^- \longrightarrow Sc$	-2.08
$[AlF_6]^{3-} + 3e^- \longrightarrow Al + 6F^-$	-2.07
$Pu^{3+} + 3e^- \longrightarrow Pu$	-2.07
$Th^{4+} + 4e^- \longrightarrow Th$	-1.90
$Np^{3+} + 3e^- \longrightarrow Np$	-1.86
$Be^{2+} + 2e^- \longrightarrow Be$	-1.85
$U^{3+} + 3e^- \longrightarrow U$	-1.80
$Hf^{4+} + 4e^- \longrightarrow Hf$	-1.70
$SiO_3^{2-} + 3H_2O + 4e^- \longrightarrow Si + 6OH^-$	-1.70
$Al^{3+} + 3e^- \longrightarrow Al$	-1.66
$Ti^{2+} + 2e^- \longrightarrow Ti$	-1.63
$Zr^{4+} + 4e^- \longrightarrow Zr$	-1.53
$ZnS + 2e^- \longrightarrow Zn + S^{2-}$	-1.44
$Cr(OH)_3 + 3e^- \longrightarrow Cr + 3OH^-$	-1.3
$[Zn(CN)_4]^{2-} + 2e^- \longrightarrow Zn + 4CN^-$	-1.26
$Zn(OH)_2 + 2e^- \longrightarrow Zn + 2OH^-$	-1.245
$[Zn(OH)_4]^{2-} + 2e^- \longrightarrow Zn + 4OH^-$	-1.216
$CdS + 2e^- \longrightarrow Cd + S^{2-}$	-1.21
$[Cr(OH)_4]^- + 3e^- \longrightarrow Cr + 4OH^-$	-1.2
$[SiF_6]^{2-} + 4e^- \longrightarrow Si + 6F^-$	-1.2
$V^{2+} + 2e^- \longrightarrow V$	ca. -1.18
$Mn^{2+} + 2e^- \longrightarrow Mn$	-1.18
$[Cd(CN)_4]^{2-} + 2e^- \longrightarrow Cd + 4CN^-$	-1.03
$[Zn(NH_3)_4]^{2+} + 2e^- \longrightarrow Zn + 4NH_3$	-1.03
$FeS + 2e^- \longrightarrow Fe + S^{2-}$	-1.01
$PbS + 2e^- \longrightarrow Pb + S^{2-}$	-0.95
$SnS + 2e^- \longrightarrow Sn + S^{2-}$	-0.94
$Cr^{2+} + 2e^- \longrightarrow Cr$	-0.91
$Fe(OH)_2 + 2e^- \longrightarrow Fe + 2OH^-$	-0.877
$SiO_2 + 4H^+ + 4e^- \longrightarrow Si + 2H_2O$	-0.86
$NiS + 2e^- \longrightarrow Ni + S^{2-}$	-0.83
$2H_2O + 2e^- \longrightarrow H_2 + 2OH^-$	-0.828
$Zn^{2+} + 2e^- \longrightarrow Zn$	-0.763
$Cr^{3+} + 3e^- \longrightarrow Cr$	-0.74
$HgS + 2e^- \longrightarrow Hg + S^{2-}$	-0.72

Half-Reactions	$E°$, volts
$[Cd(NH_3)_4]^{2+} + 2e^- \longrightarrow Cd + 4NH_3$	-0.597
$Ga^{3+} + 3e^- \longrightarrow Ga$	-0.53
$S + 2e^- \longrightarrow S^{2-}$	-0.48
$[Ni(NH_3)_6]^{2+} + 2e^- \longrightarrow Ni + 6NH_3$	-0.47
$Fe^{2+} + 2e^- \longrightarrow Fe$	-0.440
$[Cu(CN)_2]^- + e^- \longrightarrow Cu + 2CN^-$	-0.43
$Cr^{3+} + e^- \longrightarrow Cr^{2+}$	-0.41
$Cd^{2+} + 2e^- \longrightarrow Cd$	-0.403
$Se + 2H^+ + 2e^- \longrightarrow H_2Se$	-0.40
$[Hg(CN)_4]^{2-} + 2e^- \longrightarrow Hg + 4CN^-$	-0.37
$ClO_4^- + H_2O + 2e^- \longrightarrow ClO_3^- + 2OH^-$	-0.36
$PbSO_4 + 2e^- \longrightarrow Pb + SO_4^{2-}$	-0.356
$In^{3+} + 3e^- \longrightarrow In$	-0.342
$[Ag(CN)_2]^- + e^- \longrightarrow Ag + 2CN^-$	-0.31
$Co^{2+} + 2e^- \longrightarrow Co$	-0.277
$[SnF_6]^{2-} + 4e^- \longrightarrow Sn + 6F^-$	-0.25
$Ni^{2+} + 2e^- \longrightarrow Ni$	-0.250
$Sn^{2+} + 2e^- \longrightarrow Sn$	-0.136
$CrO_4^{2-} + 4H_2O + 3e^- \longrightarrow Cr(OH)_3 + 5OH^-$	-0.13
$Pb^{2+} + 2e^- \longrightarrow Pb$	-0.126
$MnO_2 + 2H_2O + 2e^- \longrightarrow Mn(OH)_2 + 2OH^-$	-0.05
$[HgI_4]^{2-} + 2e^- \longrightarrow Hg + 4I^-$	-0.04
$2H^+ + 2e^- \longrightarrow H_2$	0.00
$NO_3^- + H_2O + 2e^- \longrightarrow NO_2^- + 2OH^-$	$+0.01$
$[Ag(S_2O_3)_2]^{3-} + e^- \longrightarrow Ag^+ + 2S_2O_3^{2-}$	$+0.01$
$[Co(NH_3)_6]^{3+} + e^- \longrightarrow [Co(NH_3)_6]^{2+}$	$+0.1$
$S + 2H^+ + 2e^- \longrightarrow H_2S$	$+0.141$
$Sn^{4+} + 2e^- \longrightarrow Sn^{2+}$	$+0.15$
$Cu^{2+} + e^- \longrightarrow Cu^+$	$+0.153$
$Co(OH)_3 + e^- \longrightarrow Co(OH)_2 + OH^-$	$+0.17$
$[HgBr_4]^{2-} + 2e^- \longrightarrow Hg + 4Br^-$	$+0.21$
$AgCl + e^- \longrightarrow Ag + Cl^-$	$+0.222$
$Hg_2Cl_2 + 2e^- \longrightarrow 2Hg + 2Cl^-$	$+0.27$
$ClO_3^- + H_2O + 2e^- \longrightarrow ClO_2^- + 2OH^-$	$+0.33$
$Cu^{2+} + 2e^- \longrightarrow Cu$	$+0.337$
$[Fe(CN)_6]^{3-} + e^- \longrightarrow [Fe(CN)_6]^{4-}$	$+0.36$
$[Ag(NH_3)_2]^+ + e^- \longrightarrow Ag + 2NH_3$	$+0.373$
$O_2 + 2H_2O + 4e^- \longrightarrow 4OH^-$	$+0.401$
$[RhCl_6]^{3-} + 3e^- \longrightarrow Rh + 6Cl^-$	$+0.44$
$Ag_2CrO_4 + 2e^- \longrightarrow 2Ag + CrO_4^{2-}$	$+0.446$
$NiO_2 + 2H_2O + 2e^- \longrightarrow Ni(OH)_2 + 2OH^-$	$+0.49$
$Cu^+ + e^- \longrightarrow Cu$	$+0.521$
$TeO_2 + 4H^+ + 4e^- \longrightarrow Te + 2H_2O$	$+0.529$
$I_2 + 2e^- \longrightarrow 2I^-$	$+0.5355$
$[PtBr_4]^{2-} + 2e^- \longrightarrow Pt + 4Br^-$	$+0.58$
$MnO_4^- + 2H_2O + 3e^- \longrightarrow MnO_2 + 4OH^-$	$+0.588$
$[PdCl_4]^{2-} + 2e^- \longrightarrow Pd + 4Cl^-$	$+0.62$
$ClO_2^- + H_2O + 2e^- \longrightarrow ClO^- + 2OH^-$	$+0.66$
$[PtCl_6]^{2-} + 2e^- \longrightarrow [PtCl_4]^{2-} + 2Cl^-$	$+0.68$
$O_2 + 2H^+ + 2e^- \longrightarrow H_2O_2$	$+0.682$
$[PtCl_4]^{2-} + 2e^- \longrightarrow Pt + 4Cl^-$	$+0.73$
$Fe^{3+} + e^- \longrightarrow Fe^{2+}$	$+0.771$
$Hg_2^{2+} + 2e^- \longrightarrow 2Hg$	$+0.789$
$Ag^+ + e^- \longrightarrow Ag$	$+0.7991$
$Hg^{2+} + 2e^- \longrightarrow Hg$	$+0.854$
$HO_2^- + H_2O + 2e^- \longrightarrow 3OH^-$	$+0.88$

Appendix I
(continued)

Standard Electrode (Reduction) Potentials (continued)

Half-Reactions	$E°$, volts
$ClO^- + H_2O + 2e^- \longrightarrow Cl^- + 2OH^-$	+0.89
$2Hg^{2+} + 2e^- \longrightarrow Hg_2^{2+}$	+0.920
$NO_3^- + 3H^+ + 2e^- \longrightarrow HNO_2 + H_2O$	+0.94
$NO_3^- + 4H^+ + 3e^- \longrightarrow NO + H_2O$	+0.96
$Pd^{2+} + 2e^- \longrightarrow Pd$	+0.987
$Br_2(l) + 2e^- \longrightarrow 2Br^-$	+1.0652
$ClO_4^- + 2H^+ + 2e^- \longrightarrow ClO_3^- + H_2O$	+1.19
$Pt^{2+} + 2e^- \longrightarrow Pt$	ca. +1.2
$ClO_3^- + 3H^+ + 2e^- \longrightarrow HClO_2 + H_2O$	+1.21
$O_2 + 4H^+ + 4e^- \longrightarrow 2H_2O$	+1.23
$MnO_2 + 4H^+ + 2e^- \longrightarrow Mn^{2+} + 2H_2O$	+1.23
$Cr_2O_7^{2-} + 14H^+ + 6e^- \longrightarrow 2Cr^{3+} + 7H_2O$	+1.33
$Cl_2 + 2e^- \longrightarrow 2Cl^-$	+1.3595
$HClO + H^+ + 2e^- \longrightarrow Cl^- + H_2O$	+1.49
$Au^{3+} + 3e^- \longrightarrow Au$	+1.50
$MnO_4^- + 8H^+ + 5e^- \longrightarrow Mn^{2+} + 4H_2O$	+1.51
$Ce^{4+} + e^- \longrightarrow Ce^{3+}$	+1.61
$HClO + H^+ + e^- \longrightarrow \frac{1}{2}Cl_2 + H_2O$	+1.63
$HClO_2 + 2H^+ + 2e^- \longrightarrow HClO + H_2O$	+1.64
$Au^+ + e^- \longrightarrow Au$	ca. +1.68
$NiO_2 + 4H^+ + 2e^- \longrightarrow Ni^{2+} + 2H_2O$	+1.68
$PbO_2 + SO_4^{2-} + 4H^+ + 2e^- \longrightarrow PbSO_4 + 2H_2O$	+1.685
$H_2O_2 + 2H^+ + 2e^- \longrightarrow 2H_2O$	+1.77
$Co^{3+} + e^- \longrightarrow Co^{2+}$	+1.82
$F_2 + 2e^- \longrightarrow 2F^-$	+2.87

Appendix J

Standard Molar Enthalpies of Formation, Standard Molar Free Energies of Formation, and Absolute Standard Entropies [298.15 K (25°C), 1 atm]

Substance	$\Delta H^°_{f298.15}$, kJ/mol	$\Delta G^°_{f298.15}$, kJ/mol	$S^°_{298.15}$, J/K mol
Aluminum			
Al(s)	0	0	28.3
Al(g)	326	286	164.4
$Al_2O_3(s)$	−1676	−1582	50.92
$AlF_3(s)$	−1504	−1425	66.44
$AlCl_3(s)$	−704.2	−628.9	110.7
$AlCl_3 \cdot 6H_2O(s)$	−2692	—	—
$Al_2S_3(s)$	−724	−492.4	–
$Al_2(SO_4)_3(s)$	−3440.8	−3100.1	239
Antimony			
Sb(s)	0	0	45.69
Sb(g)	262	222	180.2
$Sb_4O_6(s)$	−1441	−1268	221
$SbCl_3(g)$	−314	−301	337.7
$SbCl_5(g)$	−394.3	−334.3	401.8
$Sb_2S_3(s)$	−175	−174	182
$SbCl_3(s)$	−382.2	−323.7	184
SbOCl(s)	−374	—	—
Arsenic			
As(s)	0	0	35
As(g)	303	261	174.1

Substance	$\Delta H^\circ_{f_{298.15}}$, kJ/mol	$\Delta G^\circ_{f_{298.15}}$, kJ/mol	$S^\circ_{298.15}$, J/K mol
$As_4(g)$	144	92.5	314
$As_4O_6(s)$	−1313.9	−1152.5	214
$As_2O_5(s)$	−924.87	−782.4	105
$AsCl_3(g)$	−258.6	−245.9	327.1
$As_2S_3(s)$	−169	−169	164
$AsH_3(g)$	66.44	68.91	222.7
$H_3AsO_4(s)$	−906.3	—	—
Barium			
$Ba(s)$	0	0	66.9
$Ba(g)$	175.6	144.8	170.3
$BaO(s)$	−558.1	−528.4	70.3
$BaCl_2(s)$	−860.06	−810.9	126
$BaSO_4(s)$	−1465	−1353	132
Beryllium			
$Be(s)$	0	0	9.54
$Be(g)$	320.6	282.8	136.17
$BeO(s)$	−610.9	−581.6	14.1
Bismuth			
$Bi(s)$	0	0	56.74
$Bi(g)$	207	168	186.90
$Bi_2O_3(s)$	−573.88	−493.7	151
$BiCl_3(s)$	−379	−315	177
$Bi_2S_3(s)$	−143	−141	200
Boron			
$B(s)$	0	0	5.86
$B(g)$	562.7	518.8	153.3
$B_2O_3(s)$	−1272.8	−1193.7	53.97
$B_2H_6(g)$	36	86.6	232.0
$B(OH)_3(s)$	−1094.3	−969.01	88.83
$BF_3(g)$	−1137.3	−1120.3	254.0
$BCl_3(g)$	−403.8	−388.7	290.0
$B_3N_3H_6(l)$	−541.0	−392.8	200
$HBO_2(s)$	−794.25	−723.4	40
Bromine			
$Br_2(l)$	0	0	152.23
$Br_2(g)$	30.91	3.142	245.35
$Br(g)$	111.88	82.429	174.91
$BrF_3(g)$	−255.6	−229.5	292.4
$HBr(g)$	−36.4	−53.43	198.59
Cadmium			
$Cd(s)$	0	0	51.76
$Cd(g)$	112.0	77.45	167.64
$CdO(s)$	−258	−228	54.8
$CdCl_2(s)$	−391.5	−344.0	115.3
$CdSO_4(s)$	−933.28	−822.78	123.04
$CdS(s)$	−162	−156	64.9
Calcium			
$Ca(s)$	0	0	41.6
$Ca(g)$	192.6	158.9	154.78
$CaO(s)$	−635.5	−604.2	40
$Ca(OH)_2(s)$	−986.59	−896.76	76.1
$CaSO_4(s)$	−1432.7	−1320.3	107
$CaSO_4 \cdot 2H_2O(s)$	−2021.1	−1795.7	194.0
$CaCO_3(s)$ (calcite)	−1206.9	−1128.8	92.9
$CaSO_3 \cdot 2H_2O(s)$	−1762	−1565	184

Appendix J
(continued)

Standard Molar Enthalpies of Formation, Standard Molar Free Energies of Formation, and Absolute Standard Entropies [298.15 K (25°C), 1 atm]
(continued)

Substance	$\Delta H^\circ_{f\,298.15}$, kJ/mol	$\Delta G^\circ_{f\,298.15}$, kJ/mol	$S^\circ_{298.15}$, J/K mol
Carbon			
C(s) (graphite)	0	0	5.740
C(s) (diamond)	1.897	2.900	2.38
C(g)	716.681	671.289	157.987
CO(g)	−110.52	−137.15	197.56
CO_2(g)	−393.51	−394.36	213.6
CH_4(g)	−74.81	−50.75	186.15
CH_3OH(l)	−238.7	−166.4	127
CH_3OH(g)	−200.7	−162.0	239.7
CCl_4(l)	−135.4	−65.27	216.4
CCl_4(g)	−102.9	−60.63	309.7
$CHCl_3$(l)	−134.5	−73.72	202
$CHCl_3$(g)	−103.1	−70.37	295.6
CS_2(l)	89.70	65.27	151.3
CS_2(g)	117.4	67.15	237.7
C_2H_2(g)	226.7	209.2	200.8
C_2H_4(g)	52.26	68.12	219.5
C_2H_6(g)	−84.68	−32.9	229.5
CH_3COOH(l)	−484.5	−390	160
CH_3COOH(g)	−432.25	−374	282
C_2H_5OH(l)	−277.7	−174.9	161
C_2H_5OH(g)	−235.1	−168.6	282.6
C_3H_8(g)	−103.85	−23.49	269.9
C_6H_6(g)	82.927	129.66	269.2
C_6H_6(l)	49.028	124.50	172.8
CH_2Cl_2(l)	−121.5	−67.32	178
CH_2Cl_2(g)	−92.47	−65.90	270.1
CH_3Cl(g)	−80.83	−57.40	234.5
C_2H_5Cl(l)	−136.5	−59.41	190.8
C_2H_5Cl(g)	−112.2	−60.46	275.9
C_2N_2(g)	308.9	297.4	241.8
HCN(l)	108.9	124.9	112.8
HCN(g)	135	124.7	201.7
Chlorine			
Cl_2(g)	0	0	222.96
Cl(g)	121.68	105.70	165.09
ClF(g)	−54.48	−55.94	217.8
ClF_3(g)	−163	−123	281.5
Cl_2O(g)	80.3	97.9	266.1
Cl_2O_7(l)	238	—	—
Cl_2O_7(g)	272	—	—
HCl(g)	−92.307	−95.299	186.80
$HClO_4$(l)	−40.6	—	—
Chromium			
Cr(s)	0	0	23.8
Cr(g)	397	352	174.4
Cr_2O_3(s)	−1140	−1058	81.2
CrO_3(s)	−589.5	—	—
$(NH_4)_2Cr_2O_7$(s)	−1807	—	—
Cobalt			
Co(s)	0	0	30.0

Substance	$\Delta H^\circ_{f\,298.15}$, kJ/mol	$\Delta G^\circ_{f\,298.15}$, kJ/mol	$S^\circ_{298.15}$, J/K mol
CoO(s)	−237.9	−214.2	52.97
Co$_3$O$_4$(s)	−891.2	−774.0	103
Co(NO$_3$)$_2$(s)	−420.5	—	—
Copper			
Cu(s)	0	0	33.15
Cu(g)	338.3	298.5	166.3
CuO(s)	−157	−130	42.63
Cu$_2$O(s)	−169	−146	93.14
CuS(s)	−53.1	−53.6	66.5
Cu$_2$S(s)	−79.5	−86.2	121
CuSO$_4$(s)	−771.36	−661.9	109
Cu(NO$_3$)$_2$(s)	−303	—	—
Fluorine			
F$_2$(g)	0	0	202.7
F(g)	78.99	61.92	158.64
F$_2$O(g)	−22	−4.6	247.3
HF(g)	−271	−273	173.67
Hydrogen			
H$_2$(g)	0	0	130.57
H(g)	217.97	203.26	114.60
H$_2$O(l)	−285.83	−237.18	69.91
H$_2$O(g)	−241.82	−228.59	188.71
H$_2$O$_2$(l)	−187.8	−120.4	110
H$_2$O$_2$(g)	−136.3	−105.6	233
HF(g)	−271	−273	173.67
HCl(g)	−92.307	−95.299	186.80
HBr(g)	−36.4	−53.43	198.59
HI(g)	26.5	1.7	206.48
H$_2$S(g)	−20.6	−33.6	205.7
H$_2$Se(g)	30	16	218.9
Iodine			
I$_2$(s)	0	0	116.14
I$_2$(g)	62.438	19.36	260.6
I(g)	106.84	70.283	180.68
IF(g)	95.65	−118.5	236.1
ICl(g)	17.8	−5.44	247.44
IBr(g)	40.8	3.7	258.66
IF$_7$(g)	−943.9	−818.4	346
HI(g)	26.5	1.7	206.48
Iron			
Fe(s)	0	0	27.3
Fe(g)	416	371	180.38
Fe$_2$O$_3$(s)	−824.2	−742.2	87.40
Fe$_3$O$_4$(s)	−1118	−1015	146
Fe(CO)$_5$(l)	−774.0	−705.4	338
Fe(CO)$_5$(g)	−733.9	−697.26	445.2
FeSeO$_3$(s)	−1200	—	—
FeO(s)	−272	—	—
FeAsS(s)	−42	−50	120
Fe(OH)$_2$(s)	−569.0	−486.6	88
Fe(OH)$_3$(s)	−823.0	−696.6	107
FeS(s)	−100	−100	60.29
Fe$_3$C(s)	25	20	105
Lead			
Pb(s)	0	0	64.81
Pb(g)	195	162	175.26

Appendix J
(continued)

Standard Molar Enthalpies of Formation, Standard Molar Free Energies of Formation, and Absolute Standard Entropies [298.15 K (25°C), 1 atm] (continued)

Substance	$\Delta H^\circ_{f_{298.15}}$, kJ/mol	$\Delta G^\circ_{f_{298.15}}$, kJ/mol	$S^\circ_{298.15}$, J/K mol
PbO(s) (yellow)	−217.3	−187.9	68.70
PbO(s) (red)	−219.0	−188.9	66.5
Pb(OH)$_2$(s)	−515.9	—	—
PbS(s)	−100	−98.7	91.2
Pb(NO$_3$)$_2$(s)	−451.9	—	—
PbO$_2$(s)	−277	−217.4	68.6
PbCl$_2$(s)	−359.4	−314.1	136
Lithium			
Li(s)	0	0	28.0
Li(g)	155.1	122.1	138.67
LiH(s)	−90.42	−69.96	25.
Li(OH)(s)	−487.23	−443.9	50.2
LiF(s)	−612.1	−584.1	35.9
Li$_2$CO$_3$(s)	−1215.6	−1132.4	90.4
Manganese			
Mn(s)	0	0	32.0
Mn(g)	281	238	173.6
MnO(s)	−385.2	−362.9	59.71
MnO$_2$(s)	−520.03	−465.18	53.05
Mn$_2$O$_3$(s)	−959.0	−881.2	110
Mn$_3$O$_4$(s)	−1388	−1283	156
Mercury			
Hg(l)	0	0	76.02
Hg(g)	61.317	31.85	174.8
HgO(s) (red)	−90.83	−58.555	70.29
HgO(s) (yellow)	−90.46	−57.296	71.1
HgCl$_2$(s)	−224	−179	146
Hg$_2$Cl$_2$(s)	−265.2	−210.78	192
HgS(s) (red)	−58.16	−50.6	82.4
HgS(s) (black)	−53.6	−47.7	88.3
HgSO$_4$(s)	−707.5	—	—
Nitrogen			
N$_2$(g)	0	0	191.5
N(g)	472.704	455.579	153.19
NO(g)	90.25	86.57	210.65
NO$_2$(g)	33.2	51.30	239.9
N$_2$O(g)	82.05	104.2	219.7
N$_2$O$_3$(g)	83.72	139.4	312.2
N$_2$O$_4$(g)	9.16	97.82	304.2
N$_2$O$_5$(g)	11	115	356
NH$_3$(g)	−46.11	−16.5	192.3
N$_2$H$_4$(l)	50.63	149.2	121.2
N$_2$H$_4$(g)	95.4	159.3	238.4
NH$_4$NO$_3$(s)	−365.6	−184.0	151.1
NH$_4$Cl(s)	−314.4	−201.5	94.6
NH$_4$Br(s)	−270.8	−175	113
NH$_4$I(s)	−201.4	−113	117
NH$_4$NO$_2$(s)	−256	—	—
HNO$_3$(l)	−174.1	−80.79	155.6
HNO$_3$(g)	−135.1	−74.77	266.2

Substance	$\Delta H^\circ_{f_{298.15}}$, kJ/mol	$\Delta G^\circ_{f_{298.15}}$, kJ/mol	$S^\circ_{298.15}$, J/K mol
Oxygen			
$O_2(g)$	0	0	205.03
$O(g)$	249.17	231.75	160.95
$O_3(g)$	143	163	238.8
Phosphorus			
$P(s)$	0	0	41.1
$P(g)$	58.91	24.5	280.0
$P_4(g)$	314.6	278.3	163.08
$PH_3(g)$	5.4	13	210.1
$PCl_3(g)$	−287	−268	311.7
$PCl_5(g)$	−375	−305	364.5
$P_4O_6(s)$	−1640	−	−
$P_4O_{10}(s)$	−2984	−2698	228.9
$HPO_3(s)$	−948.5	−	−
$H_3PO_2(s)$	−604.6	−	−
$H_3PO_3(s)$	−964.4	−	−
$H_3PO_4(s)$	−1279	−1119	110.5
$H_3PO_4(l)$	−1267	−	−
$H_4P_2O_7(s)$	−2241	−	−
$POCl_3(l)$	−597.1	−520.9	222.5
$POCl_3(g)$	−558.48	−512.96	325.3
Potassium			
$K(s)$	0	0	63.6
$K(g)$	90.00	61.17	160.23
$KF(s)$	−562.58	−533.12	66.57
$KCl(s)$	−435.868	−408.32	82.68
Silicon			
$Si(s)$	0	0	18.8
$Si(g)$	455.6	411	167.9
$SiO_2(s)$	−910.94	−856.67	41.84
$SiH_4(g)$	34	56.9	204.5
$H_2SiO_3(s)$	−1189	−1092	130
$H_4SiO_4(s)$	−1481	−1333	190
$SiF_4(g)$	−1614.9	−1572.7	282.4
$SiCl_4(l)$	−687.0	−619.90	240
$SiCl_4(g)$	−657.01	−617.01	330.6
$SiC(s)$	−65.3	−62.8	16.6
Silver			
$Ag(s)$	0	0	42.55
$Ag(g)$	284.6	245.7	172.89
$Ag_2O(s)$	−31.0	−11.2	121
$AgCl(s)$	−127.1	−109.8	96.2
$Ag_2S(s)$	−32.6	−40.7	144.0
Sodium			
$Na(s)$	0	0	51.0
$Na(g)$	108.7	78.11	153.62
$Na_2O(s)$	−415.9	−377	72.8
$NaCl(s)$	−411.00	−384.03	72.38
Sulfur			
$S(s)$ (rhombic)	0	0	31.8
$S(g)$	278.80	238.27	167.75
$SO_2(g)$	−296.83	−300.19	248.1
$SO_3(g)$	−395.7	−371.1	256.6
$H_2S(g)$	−20.6	−33.6	205.7
$H_2SO_4(l)$	−813.989	690.101	156.90

Appendix J
(continued)

Standard Molar Enthalpies of Formation, Standard Molar Free Energies of Formation, and Absolute Standard Entropies [298.15 K (25°C), 1 atm] (continued)

Substance	$\Delta H^\circ_{f_{298.15}}$, kJ/mol	$\Delta G^\circ_{f_{298.15}}$, kJ/mol	$S^\circ_{298.15}$, J/K mol
$H_2S_2O_7(s)$	−1274	—	—
$SF_4(g)$	−774.9	−731.4	291.9
$SF_6(g)$	−1210	−1105	291.7
$SCl_2(l)$	−50	—	—
$SCl_2(g)$	−20	—	—
$S_2Cl_2(l)$	−59.4	—	—
$S_2Cl_2(g)$	−18	−32	331.4
$SOCl_2(l)$	−246	—	—
$SOCl_2(g)$	−213	−198	309.7
$SO_2Cl_2(l)$	−394	—	—
$SO_2Cl_2(g)$	−364	−320	311.8
Tin			
$Sn(s)$	0	0	51.55
$Sn(g)$	302	267	168.38
$SnO(s)$	−286	−257	56.5
$SnO_2(s)$	−580.7	−519.7	52.3
$SnCl_4(l)$	−511.2	−440.2	259
$SnCl_4(g)$	−471.5	−432.2	366
Titanium			
$Ti(s)$	0	0	30.6
$Ti(g)$	469.9	425.1	180.19
$TiO_2(s)$	−944.7	−889.5	50.33
$TiCl_4(l)$	−804.2	−737.2	252.3
$TiCl_4(g)$	−763.2	−726.8	354.8
Tungsten			
$W(s)$	0	0	32.6
$W(g)$	849.4	807.1	173.84
$WO_3(s)$	−842.87	−764.08	75.90
Zinc			
$Zn(s)$	0	0	41.6
$Zn(g)$	130.73	95.178	160.87
$ZnO(s)$	−348.3	−318.3	43.64
$ZnCl_2(s)$	−415.1	−369.43	111.5
$ZnS(s)$	−206.0	−201.3	57.7
$ZnSO_4(s)$	−982.8	−874.5	120
$ZnCO_3(s)$	−812.78	−731.57	82.4

Complexes

$[Co(NH_3)_4(NO_2)_2]NO_3$, *cis*	−898.7	—	—
$[Co(NH_3)_4(NO_2)_2]NO_3$, *trans*	−896.2	—	—
$NH_4[Co(NH_3)_2(NO_2)_4]$	−837.6	—	—
$[Co(NH_3)_6][Co(NH_3)_2(NO_2)_4]_3$	−2733	—	—
$[Co(NH_3)_4Cl_2]Cl$, *cis*	−997.0	—	—
$[Co(NH_3)_4Cl_2]Cl$, *trans*	−999.6	—	—
$[Co(en)_2(NO_2)_2]NO_3$, *cis*	−689.5	—	—
$[Co(en)_2Cl_2]Cl$, *cis*	−681.1	—	—
$[Co(en)_2Cl_2]Cl$, *trans*	−677.4	—	—
$[Co(en)_3](ClO_4)_3$	−762.7	—	—
$[Co(en)_3]Br_2$	−595.8	—	—
$[Co(en)_3]I_2$	−475.3	—	—

Substance	$\Delta H^\circ_{f_{298.15}}$, kJ/mol	$\Delta G^\circ_{f_{298.15}}$, kJ/mol	$S^\circ_{298.15}$, J/K mol
Complexes (continued)			
$[Co(en)_3]I_3$	−519.2	—	—
$[Co(NH_3)_6](ClO_4)_3$	−1035	−227	636
$[Co(NH_3)_5NO_2](NO_3)_2$	−1089	−418.4	350
$[Co(NH_3)_6](NO_3)_3$	−1282	−530.5	469
$[Co(NH_3)_5Cl]Cl_2$	−1017	−582.8	366
$[Pt(NH_3)_4]Cl_2$	−728.0	—	—
$[Ni(NH_3)_6]Cl_2$	−994.1	—	—
$[Ni(NH_3)_6]Br_2$	−923.8	—	—
$[Ni(NH_3)_6]I_2$	−808.3	—	—

Appendix K

Composition of Commercial Acids and Bases

Acid or Base	Specific Gravity	Percentage by Mass	Molarity	Normality
Hydrochloric	1.19	38	12.4	12.4
Nitric	1.42	70	15.8	15.8
Sulfuric	1.84	95	17.8	35.6
Acetic	1.05	99	17.3	17.3
Aqueous ammonia	0.90	28	14.8	14.8

Appendix L

Half-Life Times for Several Radioactive Isotopes

(Symbol in parentheses indicates type of emission; E.C. = K-electron capture, S.F. = spontaneous fission; y = years, d = days, h = hours, m = minutes, s = seconds.)

Isotope	Half-life	Emission	Isotope	Half-life	Emission
$^{14}_{6}C$	5770 y	(β^-)	$^{226}_{88}Ra$	1590 y	(α)
$^{13}_{7}N$	10.0 m	(β^+)	$^{228}_{88}Ra$	6.7 y	(β^-)
$^{24}_{11}Na$	15.0 h	(β^-)	$^{228}_{89}Ac$	6.13 h	(β^-)
$^{32}_{15}P$	14.3 d	(β^-)	$^{228}_{90}Th$	1.90 y	(α)
$^{40}_{19}K$	1.3×10^9 y	(β^- or E.C.)	$^{232}_{90}Th$	1.39×10^{10} y	(α, β^-, or S.F.)
$^{60}_{27}Co$	5.2 y	(β^-)	$^{233}_{90}Th$	23 m	(β^-)
$^{87}_{37}Rb$	4.7×10^{10} y	(β^-)	$^{234}_{90}Th$	24.1 d	(β^-)
$^{90}_{38}Sr$	28 y	(β^-)	$^{223}_{91}Pa$	27 d	(β^-)
$^{115}_{49}In$	6×10^{14} y	(β^-)	$^{233}_{92}U$	1.62×10^5 y	(α)
$^{131}_{53}I$	8.05 d	(β^-)	$^{234}_{92}U$	2.4×10^5 y	(α or S.F.)
$^{142}_{58}Ce$	5×10^{15} y	(α)	$^{235}_{92}U$	7.3×10^8 y	(α or S.F.)
$^{198}_{79}Au$	64.8 h	(β^-)	$^{238}_{92}U$	4.5×10^9 y	(α or S.F.)
$^{208}_{81}Tl$	3.1 m	(β^-)	$^{239}_{92}U$	23 m	(β^-)
$^{210}_{82}Pb$	21 y	(β^-)	$^{239}_{93}Np$	2.3 d	(β^-)
$^{212}_{82}Pb$	10.6 h	(β^-)	$^{239}_{94}Pu$	24,360 y	(α or S.F.)
$^{214}_{82}Pb$	26.8 m	(β^-)	$^{240}_{94}Pu$	6.58×10^3 y	(α or S.F.)
$^{206}_{83}Bi$	6.3 d	(β^+ or E.C.)	$^{241}_{94}Pu$	13 y	(α or β^-)
$^{210}_{83}Bi$	5.0 d	(β^-)	$^{241}_{95}Am$	458 y	(α)
$^{212}_{83}Bi$	60.5 m	(α or β^-)	$^{242}_{96}Cm$	163 d	(α or S.F.)
$^{207}_{84}Po$	5.7 h	(α, β^+, or E.C.)	$^{243}_{97}Bk$	4.5 h	(α or E.C.)
$^{210}_{84}Po$	138.4 d	(α)	$^{245}_{98}Cf$	350 d	(α or E.C.)
$^{212}_{84}Po$	3×10^{-7} s	(α)	$^{253}_{99}Es$	20.0 d	(α or S.F.)

Appendix L
(continued)

Half-Life Times for Several Radioactive Isotopes (continued)

$^{216}_{84}Po$	0.16 s	(α)		$^{254}_{100}Fm$	3.24 h	(S.F.)
$^{218}_{84}Po$	3.0 m	$(\alpha$ or $\beta^-)$		$^{255}_{100}Fm$	22 h	(α)
$^{215}_{85}At$	10^{-4} s	(α)		$^{256}_{101}Md$	1.5 h	(E.C.)
$^{218}_{85}At$	1.3 s	(α)		$^{254}_{102}No$	3 s	(α)
$^{220}_{86}Rn$	54.5 s	(α)		$^{257}_{103}Lr$	8 s	(α)
$^{222}_{86}Rn$	3.82 d	(α)		$^{263}_{106}(106)$	0.9 s	(α)
$^{224}_{88}Ra$	3.64 d	(α)				

Index

PERIODS

IA

METALS

TRANSITION METALS

	IA	IIA	IIIB	IVB	VB	VIB	VIIB		VII...

Period 1
- ● 0.3 H
- +1 ?●

Period 2
- ○ Li 1.52   +1 0.60 ○
- ○ Be 1.11   +2 0.31 ○

Period 3
- ○ Na 1.86   +1 0.95 ○
- ○ Mg 1.60   +2 0.65 ○

Period 4
- ○ K 2.31   +1 1.33 ○
- ○ Ca 1.97   +2 0.99 ○
- ○ Sc 1.60   +3 0.81 ○
- ○ Ti 1.46   +4 0.64 ○
- ○ V 1.31   +5 0.4 ○
- ○ Cr 1.25   +6 0.52 ○
- ○ Mn 1.29   +2 0.80 ○
- ○ Fe 1.26   +2 0.75 ○
- ○ 1.2...   0.7...

Period 5
- ○ Rb 2.44   +1 1.48 ○
- ○ Sr 2.15   +2 1.13 ○
- ○ Y 1.80   +3 0.93 ○
- ○ Zr 1.57   +4 0.87 ○
- ○ Nb 1.43   +5 0.69 ○
- ○ Mo 1.36   +6 0.62 ○
- ○ Tc 1.3
- ○ Ru 1.33
- ○ 1....

Period 6
- ○ Cs 2.62   +1 1.69 ○
- ○ Ba 2.17   +2 1.35 ○
- ○ La 1.88   +3 1.15 ○
- ○ Hf 1.57   +4 0.84 ○
- ○ Ta 1.43   +5 0.68 ○
- ○ W 1.37   +6 0.68 ○
- ○ Re 1.37
- ○ Os 1.34
- ○ 1....

Period 7
- ○ Fr 2.7
- ○ Ra 2.20   +2 1.52 ○
- ○ Ac 2.0   +3 1.11 ○